The Laboratory Primate

The Handbook of Experimental Animals

List of Editorial Advisory Board

The Laboratory Primate

Edited by

Sonia Wolfe-Coote

ELSEVIER
ACADEMIC
PRESS

Amsterdam • Boston • Heidelberg • London • New York • Oxford
Paris • San Diego • San Francisco • Singapore • Sydney • Tokyo

Elsevier Academic Press
84 Theobald's Road, London WCIX 8RR, UK
http://books.elsevier.com

Elsevier Academic Press
525 B Street, Suite 1900 San Diego, California 92101-4495, USA
http://books.elsevier.com

ISBN 0-12-080261-9

British Library Cataloguing Publication Data
A catalogue record for this book is available from the British Library

Typeset by Cepha Imaging Pvt. Limited, Bangalore, India
Printed and bound in Italy by Printer Trento

05 06 07 08 09 10 9 8 7 6 5 4 3 2 1

Contents

Part 4 Current Uses in Biomedical Research

Preface

*A*s with each of the volumes in The Laboratory Animal Series, *The Laboratory Primate* is a reference book for anyone who is considering the non-human primate as a research model. The volume provides as much information as possible, or details of where to access it, on everything from psychological well-being and environmental enrichment to procedures and examples of research using the primate model.

Part 1 includes chapters which define the model in such areas as anatomy, reproduction, behaviour, pathology and the non-human primate relationship to the human. Physiological aspects of subject matter are included in each chapter, where appropriate. Part 2 deals with aspects of primate management and medical care, always stressing the importance of the psychological and physical well-being of the non-human primate who is acknowledged and handled as a sentient being throughout the volume. Part 3 details some techniques and procedures that can be used in research in a primate model and Part 4 then provides examples of current uses of nonhuman primates in biomedical research.

With the large number of Primate Units in the United States of America, it is inevitable that most of the chapters in this volume are by authors from the States. An effort was made, however, to make the volume as international as possible and it does, indeed, include contributions by authors from Australia, Japan, Germany, Switzerland, Holland, South Africa, Kenya and the United Kingdom.

Preparing and editing this volume has been both a privilege and an opportunity to learn from the masters in primatology and the various other specialist fields included in the book. Added to this was the bonus of interacting with so many diverse personalities around the globe, all of whom were not only prepared, but also willing, to share their time and expertise with other primatologists, biomedical researchers and newcomers to this field. The human is indeed a very interesting and accommodating primate. My grateful thanks go to each of these chapter authors for their contribution, and also to the publishing team and editors in chief for their patience, encouragement and support.

Finally, I would not even have attempted to edit this volume without two important people. Absolutely invaluable to the design and production of the book has been the knowledge, support, advice, encouragement and humor of the Head of the South African Medical Research Council's Primate Unit, Dr Jürgen Seier; and without the ever gentle reminders and patient assistance of my Personal Assistant, Ms Jeanette Wyeth, in chasing authors, maintaining records and helping to put together the final compilation, the volume would never have made it into print. My grateful thanks to both of them.

Sonia Wolfe-Coote, PhD
September 2004

Part 1

Definition of the Primate Model

Contents

Chapter 1

The Taxonomy of Primates in the Laboratory Context

Colin Groves
School of Archaeology and Anthropology, Australian National University, Canberra, ACT 0200, Australia

Taxonomy: Organizing nature

Taxonomy means classifying organisms. It is nowadays commonly used as a synonym for systematics, though strictly speaking systematics is a much broader sphere of interest – interrelationships, and biodiversity. At the basis of taxonomy lies that much-debated concept, the species.

Because there is so much misunderstanding about what a species is, it is necessary to give some space to discussion of the concept. The importance of what we mean by the word "species" goes way beyond taxonomy as such: it affects such diverse fields as genetics, biogeography, population biology, ecology, ethology, and biodiversity; in an era in which threats to the natural world and its biodiversity are accelerating, it affects conservation strategies (Rojas, 1992). In the present context, it is of crucial importance for understanding laboratory primates and their husbandry.

What are species? The biological species concept

Disagreement as to what precisely constitutes a species is to be expected, given that the concept serves so many functions (Vane-Wright, 1992). We may be interested in classification as such, or in the evolutionary implications of species; in the theory of species, or in simply how to recognize them; or in their reproductive, physiological, or husbandry status.

Most non-specialists probably have some vague idea that species are defined by not interbreeding with each other; usually, that hybrids between different species are sterile, or that they are incapable of hybridizing at all. Such an impression ultimately derives from the definition by Mayr (1940), whereby species are "groups of actually or potentially interbreeding natural populations which are reproductively isolated from other such groups" (the Biological Species Concept). Mayr never

The Laboratory Primate

ISBN 0-1208-0261-9

actually said that species *can't* breed with each other, indeed he denied that that this was in any way a necessary part of reproductive isolation; he merely said that, under natural conditions, they *don't*.

Reproductive isolation, in some form, stands at the basis of what a species is. Having said this, it must be admitted that it is no longer possible to follow Mayr's concept as definitive. In a recent book (Groves, 2001, see especially Chapter 3) I sketched the main reasons why this is so:

- It offers no guidance for the allocation of allopatric populations.
- Many distinct species actually do breed with each other under natural conditions, but manage to remain distinct.
- The interrelationships of organisms under natural conditions are often (usually?) unknown.
- Many species do not reproduce sexually anyway.

Allopatry

To say that two populations are allopatric means that their geographic distributions do not overlap – they are entirely separate. This means that they do not have the chance to breed with each other, even if they wanted to. There is, for example, no way of testing whether *Macaca fuscata* (of Japan), *M.cyclopis* (of Taiwan) and *M.mulatta* (the Rhesus Macaque, of the East Asian mainland) are actually different species or not; they are classified as distinct species in all major checklists, but there is no objective way of testing this classification under the Biological Species Concept.

Indeed, this is the usual situation: populations that differ, in some respect, from one another and are, by relevant criteria, closely related are usually allopatric. To take demonstrable reproductive isolation, the requisite criterion under the Biological Species Concept, as the *sine qua non* of species status would be to leave the majority of living organisms unclassifiable except by some arbitrary fiat.

Natural interbreeding

The two common species of North American deer (*Odocoileus virginanus*, the Whitetail, and *O.hemionus*, the Blacktail) are found together over a wide geographic area, and are always readily distinguishable; yet molecular studies have found evidence that there has been hybridization. For example, in Pecos Country, west Texas, four out of the nine whitetails examined had mitochondrial DNA characteristic of the blacktails with which they share their range (Carr and Hughes, 1993). Evidently in the not-too-distant past blacktail females joined whitetail breeding herds and, while the whitetail phenotype was strongly selected for, the blacktail mtDNA has remained in the population, fossil documentation of the hybridization event.

In Primates, also, there are examples of hybridization in the wild. A good example of the first case, *Cercopithecus ascanius* (Redtail monkey) and *C.mitis* (Blue monkey) in Uganda, has been described in detail by Struhsaker *et al.* (1988). The two monkeys, which are widely sympatric, meaning that they live in the same areas over a wide range, interbreed at quite noticeable levels, yet remain separate and clearly distinguishable and no one has ever proposed to regard them as anything but distinct species. This case is not unlike that of the North American deer, mentioned above.

These are two examples – one non-Primate, one Primate – of pairs of distinct species which manage to remain distinct over wide areas even though there is gene-flow between them. Much more common (or, better, more readily documented) are cases where pairs of species occupy ranges that are largely separate but meet along their margins (parapatric), and interbreed where they do so. Interbreeding varies from occasional to full hybrid zones, and such cases have, unlike the hybridization-in-sympatry cases, been regarded as evidence that reproductive isolation does not exist, so the two species should be merged into one. But there is no difference, in principle, from the hybridization-in-sympatry cases.

The classic study of a hybrid zone is that of two mice, *Mus musculus* and *Mus domesticus*, across the Jutland peninsula, Denmark (see summary in Wilson *et al.*, 1985). The hybrid zone, as measured by morphology and protein alleles, is very narrow; yet the mtDNA of the southern species, *M.domesticus*, introgresses well across the boundary, and across the seaway (the Skagerrak) into Sweden. This suggests both that hybridization has been occurring, and that *M.musculus* has been expanding its range, and the hybrid zone has been moving south since before the sea broke through separating Denmark and Sweden in the early Holocene. There has been no selection against hybridization during this long period.

In a well-studied Primate example, two baboons, *Papio hamadryas* (Hamadryas baboon) and *P.anubis* (Olive baboon), are parapatric and hybridize where their ranges meet in Ethiopia, the hybrid zone being not more than a few kilometres wide. The two taxa are adapted to more arid and more mesic environments, respectively, and the hybrid zone travels up and down the Awash River according to whether there has been a run of dry seasons or a run of wet seasons, but remains more or less the same width. This case is therefore not unlike that of the two mice in Denmark. Unlike the *Cercopithecus* example, the two baboon taxa have been shuffled back and forth between subspecies and species (compare Jolly, 1993 and Groves, 2001). Yet what is the difference, really?

What are species? The phylogenetic species concept

Most attempts to modify the definition of a species have been modifications of the Mayr concept, and relied on reproductive status (see Groves, 2001, Chapter 3). Even without the practical problems summarized above, such definitions seem inherently flawed because they appeal to the *process* of how species come to be, or are maintained, when surely they should be recognized by the *pattern* of what they actually are. It was put succinctly by Cracraft (1983): "Evolution produces taxonomic entities, defined in terms of their evolutionary differentiation from other such forms. These entities should be called species... A species is the smallest diagnosable cluster of individual organisms within which there is a parental pattern of ancestry and descent". This is the Phylogenetic Species Concept.

"Diagnosable" means 100% different in one or more heritable characters. It implies that there are fixed genetic differences, though it does not require that they be demonstrable here and now in the form of DNA sequences (given advances in knowledge, presumably they will be in the fullness of time). It is as nearly objective as the evidence permits. The only query that can arise is whether a "parental pattern of ancestry and descent" exists, and this is as close to inference as the concept approaches.

In this concept, we cease to use the species as a hypothesis of relationship: each diagnosable entity is recognized as a species, and hypotheses of relationships are reserved for some other level, whether a formal taxonomic rank or an informal grouping such as "species-group".

There are three common misunderstandings about the Phylogenetic Species Concept. First, that the diagnostic character states of a species must be evolutionarily derived (evolutionary novelties). A moment's reflection shows that their evolutionary status introduces an element of process, whereas the objective aspect is the mere existence of diagnosable difference.

The second common misunderstanding is that species are now being defined by *degree* of difference. They are not. They are being defined by the *status* of the difference, whether the candidates for species status can be diagnosed or not. There is, for example, no standard genetic distance above which species status is involved and below which it is not.

The third misunderstanding is that species must still be in some way reproductively isolated. The interbreeding criterion obviously dies hard. The baboon example and others show that diagnosability exists irrespective of the persistence of the ability to hybridize. Their separateness is genetic rather than necessarily reproductive.

Groves (2001) noted that some Primate taxonomists have already begun employing the phylogenetic species concept, particularly those working on South American monkeys, and made proposals as to what a full classification of the order might look like. These proposals should be regarded as a first step and are not in any way intended as definitive.

What are subspecies?

Subspecies are geographic segments of species that differ from one another as a whole, but not absolutely. The two criteria are:

- They are geographic populations (or groups of populations), not morphs within a single population.
- They do not differ absolutely; the subspecies allocation of an isolated individual is probabilistic only.

What proportion of one population of a species must be distinguishable from others, in the same species, before they can be considered subspecies? This is rather arbitrary, although Mayr's (1963) 75% rule – that 75% of individuals must be distinct from all those in other populations – is widely adhered to.

Cracraft (1983) argues against the recognition of subspecies: if a population is absolutely distinct, so that all individuals can be objectively allocated, then it is a full species; if not, then it is not an objective entity in any sense. Other adherents of the phylogenetic species concept are less dismissive. Groves (2001) takes the position that if two populations (or groups of populations) are distinguishable most of the time then it is valuable to dignify them with subspecific names.

There is a pleasing symmetry about this. Compared with traditional (subjective) arrangements, a system under which the phylogenetic species is employed and subspecies are considered useful categories (if less objective), the number of taxa does not change but the former subspecies, that are 100% different, are raised to the rank of species. As species, not subspecies, are the units with which conservationists, biogeographers, field biologists and the like – including those concerned with captive husbandry – generally work, this seems entirely appropriate.

Nomenclature

A species has two names (binomial). The first is that of the genus (see below) and the second denotes the species itself. *Macaca mulatta*, *Macaca cyclopis* and *Macaca fuscata* refer to three different species of the genus *Macaca*.

If we need to recognize subspecies, a third word is simply added onto the end of the binomial. *Macaca fuscata yakui* is the subspecies from Yakushima, a small island of southern Japan. It is claimed that most, but not all, of these individuals are distinguishable from those on the main Japanese islands, which must now be designated *Macaca fuscata fuscata*.

The subspecies whose subspecific name repeats the name of the species is called the nominotypical subspecies, and its distribution (by definition) includes the region whence the species was first described (called the type locality).

How to classify species

Once these entities have been delineated, they must be arranged into groups. These groups are called genera (singular, genus). A species may also be so isolated from other species that it occupies a genus by itself.

Genera, in turn, are grouped into families and families into orders. Most families include a large number of genera, so are divided into subfamilies; these may in turn be divided into tribes, if required. The order Primates is divided into suborders, these into infraorders, and these in turn into superfamilies.

A genus is a Latin or Latinized noun in the nominative singular. It forms the first word in the binomial species name (or trinomial subspecies name). Families and orders (and their various sub- and superordinate categories) are plural nouns. Superfamily names end in –oidea, families in –idae, subfamilies in –inae and tribes in –ini.

Genera, families and orders are quite different sorts of categories from species. They are artificial where species are objective and they are groupings where species are entities. It has been recognized, since the work of Hennig (1966) that the function of these "higher categories" is to cluster species according to their degree of relatedness, i.e. into monophyletic groups (monophyletic = descended from a unique common ancestor; a monophyletic group is a group, for example, of species, which include not only but all the descendants of a common ancestor).

Formerly, humans were placed, alone, in the family Hominidae, while "great apes" (chimpanzee, gorilla and orangutan) were placed in a separate family, Pongidae. It is clear, from molecular and other studies, that orangutan, not human, was the first to separate from this group, so the family Pongidae is not monophyletic. Hence, no primatologist with any taxonomic understanding would today recognize the family Pongidae, at least in its traditional sense. Instead, "great apes" and humans are together placed in the family Hominidae.

Formerly, the order Primates was divided into two suborders: Prosimii (incorporating lemurs, lorises and tarsiers) and Anthropoidea (incorporating monkeys and apes, including humans). It is evident today that tarsiers share a common ancestor with "Anthropoidea" which they do not share with lemurs and lorises, so the suborder Prosimii is not monophyletic. Hence, no primatologist with any taxonomic understanding would today recognize the suborder Prosimii. Instead, tarsiers and "Anthropoidea" are placed together in one suborder, Haplorrhini, while lemurs and lorises are placed in a separate suborder, Strepsirrhini.

The question that has recently been asked is where does a genus, or a family, begin and end? It is surprising that such a crucial question has so rarely been posed. The most logical answer, espoused by Goodman *et al.* (1988), is that a genus or a family should have a certain time depth. Their proposal was adapted and modified by Groves (2001), who suggested that a genus is a group of species whose last common ancestor lived around the Miocene-Pliocene boundary, and a family is a group of genera whose last common ancestor lived around the Oligocene-Miocene boundary. Goodman *et al.* (1988) also proposed time depths for tribes and for the infra-, sub- and super-ranks. Groves (2001), however, considered that, as these are ranks inserted only for convenience, as when, for example a family contains a large number of genera which are better handled by grouping them in some way, it is unnecessary to designate time depths for them.

Time depths are sometimes ascertained from fossil evidence, but more usually from molecular clocks. The major families of Primates are unaffected by adopting a time/rank criterion, but there are implications for some genera.

At the end of the chapter an appendix presents a controversial outline classification of Primates to genus.

A commentary on genera commonly housed in laboratories

Microcebus

The nocturnal mouse lemurs (genus *Microcebus*) are the smallest living Primates and are sometimes kept in laboratories, where they must be carefully maintained on an insect and fruit diet. They are studied for their unusual metabolism (Perret and Aujard, 2001) and for their reproductive physiology (Aslam *et al.*, 2002), particularly the environmental cues that control their reproductive seasonality. They commonly, but not invariably, experience lowered metabolic rates during the dry season and, in preparation for this, they store up fat in the tail.

Knowledge of mouse-lemur taxonomy has grown steadily over the past ten years. In 1990, two species were known, one in the dry forests of western and southern Madagascar and one in the rainforests of the east. By 2000, it had been shown that there are seven species in the western forests alone (Rasoloarison *et al.*, 2000) and further biodiversity can be predicted for the eastern forests.

Other genera of Malagasy lemurs are probably housed in a few laboratories but, in general, this cannot be recommended as their husbandry is not problem-free.

Loris and *Nycticebus*

Slender and Slow lorises are the only Asian strepsirrhines. Like mouse lemurs, they are nocturnal. Slender lorises, of which there are two species (*L.tardigradus* in the wet zone of Sri Lanka and *L.lydekkerianus* in the dry zone of Sri Lanka and in southern India), feed almost entirely on insects, especially ants (Nekaris and Rasmussen, 2003), plus some vegetable matter. Slow lorises, of which there are three species (Groves, 2001) are more vegetarian, especially frugivorous, but also require some animal prey. Lorises, despite some being dubbed "slow", can move with surprising speed when pressed; both types can bite hard, and the bite of Slow lorises is toxic (Alterman, 1995).

Galago and *Otolemur*

The four species of Galagidae (bushbabies) that are sometimes kept in laboratories are:

- *Galago senegalensis* (Senegal Bushbaby). They are actually found all over western, northeastern and southeastern Africa, are grey with yellow limbs, and dark eye-rings with a white stripe between them. They are agile and make long hops. Gestation is 142 days and they usually have single births.
- *Galago moholi* (Moholi Bushbaby), from southeastern and southern Africa, are more buffy with larger ears and more prominent face pattern. They are agile and make long hops. Gestation is 125 days and they nearly always give birth to twins.
- *Otolemur crassicaudatus* (Brown Greater Galago), from southeastern Africa. They are very large, bushy-tailed, big-eared and brown with a pale face. They do not hop. Gestation is 135 days and they usually bear twins or triplets.
- *Otolemur garnettii* (Northern Greater Galago), from eastern Africa, is similar to *O.crassicaudatus* but is rather smaller, shorter-eared, more greyish-toned and has a face that is not pale. It sometimes hops. Gestation is 130 days; usually producing single births.

Saimiri

Squirrel monkeys are agouti-coloured ("agouti" means that their hairs are banded, usually with black and yellowish), with characteristic white faces with a black muzzle, hence their German name *Totenkopfaffen* meaning "death's-head monkeys". They are easily kept in laboratories but they need space and to be kept in social groups. Their diet is fruit and some leaves, and they catch insects, often on the wing. A supply of live insects in the laboratory can keep them occupied for quite a while, and so forms a source of behavioural enrichment for them.

There are five species of squirrel monkey, as follows:

1. Gothic type, characterized by a white face-mask, forming a high arch above each eye, and a bushy tail tuft.
   a. *Saimiri sciureus* (Common squirrel monkey). This is the widespread species of the South American rainforests, mainly north of the Amazon, but

extending south of the lower course of the Amazon. Body colour is greyish or greenish to reddish agouti, crown greenish or greyish agouti, often mixed with black, and ears are tufted. Hands, feet and forearms are orange or yellowish or merely tinged with this tone.

b. *Saimiri ustus* (Bare-eared squirrel monkey). This species is from south of the Amazon, west of the R. Xingu. It is larger than the previous species and is distinguished by its untufted ears and the grey colour of the thighs which contrasts with the body tone.

c. *Saimiri oerstedti* (Central American squirrel monkey). This comes from isolated areas in Costa Rica and Panama. The body colour is reddish or orange-red; the crown is generally black (but agouti in some males) and the ears are tufted.

2. Roman type, characterized by white colour restricted to a narrow line above the brows and no bushy tuft.

d. *Saimiri boliviensis* (Black-capped squirrel monkey). These come from Peru, Bolivia and far western Brazil. The body colour is greenish-agouti but with black tones of varying intensity, the crown is black or just black-bordered, in some males, and the tail is often blackish on the dorsal surface.

e. *Saimiri vanzolinii* (Black squirrel monkey). These are restricted to the region where the R. Japurá meets the upper Amazon and are distinguished by their black dorsal colour.

Probably only two of these species (*S.sciureus* and *S.boliviensis*) are likely to be held in laboratories, but it is very important to distinguish them clearly because their behaviours are very different. Boinski and Cropp (1999) have summarized the differences and emphasized their significance for laboratory husbandry and other purposes. In *S.boliviensis*, females are more aggressive, form social coalitions and are dominant to males, whereas in *S.sciureus* the males are dominant. The male in *S.boliviensis* is more peripheral in the social group and, in the wild, males emigrate from their social groups. In *S.sciureus*, it is the females that emigrate and the males defend the young keenly and are well integrated into the group. Males of *S.boliviensis* can be kept in all-male groups, but those of *S.sciureus* cannot. Bisexual groups of *S.sciureus* range in number from about 15 to 50, whereas those of *S.boliviensis* are typically over 50 in number.

S.sciureus are more fertile, breeding every year, and infants are weaned after six months. *S.boliviensis* usually breed every second year and the infants are weaned after about 19 months. In *S.sciureus* the breeding season lasts less than a week, while in *S.boliviensis*, it extends over two months. In both species the males put on fat around the shoulders in the breeding season ("fatted males") and compete vigorously for matings.

Finally, there is a bizarre and unexpected difference in the males' threat display. The male *S.boliviensis* utters a whining threat to other males but, in *S.sciureus*, the dominant male spreads his legs and subjects a subordinate male to a penile erection. This can be elicited by presenting the male with a mirror, and he will perform the display to his mirror image.

The behaviour of *S.oerstedti* is different again, as Boinski and Cropp (1999) record, but this species is IUCN-Endangered and on CITES Appendix I, so cannot be held in laboratories.

Aotus

Night monkeys (Owl monkeys, Douroucoulis) are of special biomedical interest because some of them have proved susceptible to *falciparum* malaria. Because susceptibility has been found to depend on species, correct taxonomic determination is vital, but this is often extremely difficult. In addition, Defler *et al.* (2001) showed that the names of the various species/subspecies from Colombia and Panama have been wrongly applied and they also argue that what have commonly been regarded as subspecies are actually distinct species. The following arrangement according to Groves (2001), was modified according to Defler *et al.* (2001):

1. Grey-necked group, in which the sides of the neck are greyish, like the body.

a. *Aotus zonalis*. Defler *et al.* (2001) showed that this short-haired night monkey, which lacks any interscapular whorl or crest, is the one from lowland Panama. It was formerly, but wrongly, called *Aotus lemurinus lemurinus*, and is probably a distinct species, although generally similar in external appearance to the real *A.lemurinus*. Karyotype: 2n = 55–56. It has low susceptibility to malaria.

b. *Aotus lemurinus*. These come from the Andes of Colombia, and perhaps Panama. The darkest species, it has long shaggy pelage and lacks any interscapular whorl or crest. Underparts are yellowish to pale orange and this colour extends down the inner aspects of limbs to knee and elbow but does not extend forward to the throat. Hands and feet are dark. Individuals may be more greyish or more reddish and the development of a dark dorsal stripe is variable. 2n = 58. They have low susceptibility to malaria.

The species is not the same as the one described by Hershkovitz (1983) and by Groves (2001) as *Aotus lemurinus lemurinus*, but the one that has become known as *Aotus hershkovitzi*.

c. *Aotus griseimembra*. This "species" is confusing and may be a totally spurious association of specimens, from the lowlands of eastern Colombia, which resemble *A.zonalis* but lack the dark hands and feet. It includes specimens of uncertain origin with 2n = 52–54 karyotype, known to be susceptible to *falciparum* malaria (see Defler *et al.,* 2001). Much more information is required on what phenotype occurs where and has what genotype.

d. *Aotus brumbacki*. Known only from the highlands of Meta, Colombia, at 467–1543 metres. It is distinguished by its interscapular crest; its longitudinal gular gland with hairs parted on either side of it; its yellowish, not white, spots above the eyes and its pale orange underparts extending to the posterior part of the throat and by 2n = 50. Highly susceptible to *falciparum* malaria.

e. *Aotus vociferans*. It occurs from southern Colombia into northern parts of Brazil and Peru. It is distinguished by its interscapular whorl and circular throat gland, with hairs radiating from its brownish colour, black hands and feet and thicker crown stripes than other species of the grey-neck group, orange-white underparts extending to wrists and ankles, black tail that is reddish under its proximal half and 2n = 46. Highly susceptible to *falciparum* malaria.

f. *Aotus trivirgatus*. This is found from eastern Colombia east to Guyana. It is distinguished from all other members of the grey-neck group by its strongly contrasting orange dorsal stripe. Hands and feet are dark, face pattern very inconspicuous and underparts are orange. Malaria susceptibility and karyotype are not known.

2. Red-necked group, in which the red of the underside extends not only well forward on the throat, but also up onto the sides of the neck.

g. *Aotus miconax*. These occur in a small region in northwestern Peru. They are light brownish or reddish grey, bushy-tailed, and have an inconspicuous face pattern and pale orange underparts. Malaria susceptibility and their karyotype are not known.

h. *Aotus nancymaae*. These are from a small area along the Peru-Brazil border. They are very like *A.miconax* but greyer, with a dark median dorsal zone and 2n = 54. This species, which is commonly held in laboratories (because it is the species that occurs in the vicinity of the important centre of Iquitos), was even described from laboratory specimens and is resistant to *falciparum* malaria.

i. *Aotus nigriceps*. These come from Brazil south of the Amazon and west of the R. Tapajós, into Peru. These are iron-grey with a brownish wash on the mid dorsal region, and a very conspicuous facial pattern of broad black stripes and white areas. Underparts are whitish and orange, this zone extending up the sides of the neck. 2n = 51 (male), 52 (female).

j. *Aotus azarae*. These occur south of the Amazon, between the Tapajós and Tocantins Rivers, into Bolivia and Paraguay. They are distinguished by an interscapular whorl. They are very like *Aotus nigriceps* but may be more buffy or olive with black digits and deep red underparts but this zone does not extend far up the sides of the neck or the inner aspects of limbs, beyond the elbows and knees. 2n = 49 (male), 50 (female). There are three geographic forms which may even be distinct species: the Paraguayan *A.a.azarae* with its shaggy pelage; *A.a.boliviensis* from Bolivia, more olive with contrasting grey limbs; and *A.a.infulatus* from southern Brazil, with more white on the face.

It will be seen that there is a great deal still to be learned about night monkeys, particularly with regard to matching up karyotypes with place of origin. The differential susceptibility of different species to malaria is perhaps predictable from the altitude at which each lives. They are all thought to be basically nocturnal, and live in mated pairs. Their inquisitiveness, docility and tameability make them very attractive charges for laboratory handlers.

Callithrix

In the 1960s, four species of true marmosets were recognized: *Callithrix jacchus* (Common marmoset) from the Atlantic seaboard of Brazil, *C.argentata* and *C. humeralifer* (Silvery and Tassel-eared marmosets) from the southern Amazonian forests, and *Cebuella pygmaea* (Pygmy marmoset) from the upper Amazon tributaries. The latest count is 21 with 6 in the Atlantic forests, 14 Amazonian, plus the same pygmy species, now called *Callithrix pygmaea*.

Molecular data show that the Atlantic and Amazonian marmosets are not closely related and the pygmy marmoset is more closely related to the

Amazonian group and the three species-groups were separated around five million years ago (Goodman *et al.*, 1998), which is just too recently for generic separation. Consequently, it is best to place them all in one genus, *Callithrix*, with three species-groups or subgenera (*Callithrix* [Atlantic], *Mico* [Amazonian] and *Cebuella* [Pygmy]). The old, "traditional" classification, which placed the pygmy marmoset in the genus *Cebuella,* and the rest combined under *Callithrix*, is unacceptable.

Probably the only species likely to be held in laboratories is the Common Marmoset, *Callithrix jacchus*. This has a pelage mottled in black, grey and yellow, a black-and-white striped tail, a white patch on the forehead and long white tufts arranged in an arc above and in front of the ears. It has a wide range in easternmost Brazil, in dry coastal and inland forest. It lives in mated pairs which bear twins and these are carried about at first by the father and later by the older siblings, being transferred to the mother essentially only for suckling ("helper system"). The sexual maturation of the older offspring is suppressed if they are not allowed to disperse.

An important part of the diet for all marmosets is tree exudates (gum, resin), and the lower incisor and canine teeth are long and narrow, arranged in a semicircle, and used for digging into bark to allow exudates to run out. This notching activity is almost constant, so marmosets must be supplied with sturdy wooden perches and supports in captivity. They scent-mark their home range assiduously and when placed in a new cage a marmoset will spend the initial period marking all over it, substituting the previous owners' scent with its own.

Saguinus

Tamarins are related to true marmosets but lack the adaptations for bark notching. Seventeen species are currently recognized but this is very likely an underestimate. The most familiar species in laboratories is *Saguinus oedipus*, known as the Cottontop or Pinché, which has an almost bare (sparsely haired) black face, an agouti grey-brown body, red rump and thighs, white limbs and underside, and a long "Iroquois" hairstyle. In the wild it is confined to a tiny area in Colombia, from the Atlantic coast to the lower Magdalena and Cauca rivers. It spontaneously develops colon cancer, which has rendered it of great biomedical interest, and from about 1960 to 1975 some 30–40,000 were imported to the USA. It is now regarded as endangered by IUCN, and there are probably more in captivity than there are in the wild.

Other species of tamarin, mainly of the *Saguinus fuscicollis* group (the Saddleback Tamarins, of which there may be one or several species, all very small in size, 350g compared to about 500g in the cottontop), may also be kept in laboratories.

Chlorocebus

These are the monkeys that are often lumped together as vervets, and formerly included in the genus *Cercopithecus*. Unlike the latter, however, they live mainly in open country or savannah-woodland and not in dense forest. The distribution of the genus is throughout the non-forested areas of sub-Saharan Africa. There are several species which may well differ physiologically and in disease susceptibility:

Chlorocebus sabaeus, the African Green Monkey, a West African species, found from Senegal east probably to Ghana. It is a grizzled golden greenish colour with off-white underparts, a yellow tail-tip, and yellow cheek whiskers directed upward from a whorl in front of the ears and over the temples. The scrotum is very pale blue.

Chlorocebus aethiops, the Grivet Monkey. This is found in eastern Sudan and western Ethiopia. It is grizzled olive with the crown yellow, grey limbs, white underparts, and a white tuft at the base of the tail; there is a white brow-band, which is continuous with the very long, white cheek whiskers, and there is a sparse white moustache. The scrotum is sky blue.

Chlorocebus djamdjamensis is a rare species from the Bale Mountains, Ethiopia.

Chlorocebus tantalus, the Tantalus Monkey, is found from Ghana to Sudan, Uganda and northwestern Kenya. It looks very like the Grivet but the white, sinuous brow-band is separated from the cheek whiskers, which are stiff, yellowish and black-tipped, by a black line from eye to temples. The scrotum is sky blue like the Grivet's and is surrounded by long orange hairs.

Chlorocebus pygerythrus, the true Vervet Monkey, is found widespread from eastern Ethiopia south to the southern tip of Africa. This differs from the Grivet and the Tantalus by the limbs not being grey, by the dark hands, feet and tail tip, the short bright red hair in the perineal region, and the short white cheek whiskers which join the brow-band to form a continuous face-ring which grades into the greenish speckled neck and crown. The scrotum is turquoise blue. The Ethiopian and Kenyan subspecies, *C.p.hilgerti*, tends to be pale brownish yellow; the Tanzanian *C.p.rufoviridis* averages more fawn and the underparts are often redder, the cheek whiskers longer; and the southern

African *C.p.pygerythrus* is more grey or olive. There are also some small-sized subspecies on coastal offshore islands.

Chlorocebus cynosuros, the Malbrouck, is from Angola, western Zambia, and the savannah country of southern D.R.Congo. This is like the Vervet but paler, and above all has a pale, blotched face (all other species have a black face), and pale palms and soles; the cheek whiskers are long, and directed upward and backward. The scrotum is lapis blue.

Long regarded as a minor biomedical source, this genus has leapt into prominence because of the discovery of the SIV "African green monkey". There is a need to examine populations of other species for SIVs, but this will be conveniently done by trap-release investigations in the wild, rather than in the laboratory.

Papio

Baboons have long been favoured research subjects. Studies on reproductive biology, in one species of baboon (Birrell *et al.*, 1996), have been facilitated by housing them in their natural social groups, minimizing undue stress and thus enabling apparently normal processes of pregnancy, including hormonal levels, to be continuously monitored, as briefly described by Horam *et al.* (1992). Baboons are widely used for these and other research areas in the laboratory. The potential drawback is their large size, requiring large cages, preferably with outdoor runs, if they are to be kept under humane conditions. They are strikingly intelligent compared to Platyrrhines, and even to Vervets, and, for full behavioural enrichment, they require social interaction and intellectual stimulation. This includes having their food scattered, so that they have to forage for it, rather than simply collecting it from a tray. There are five species of baboon:

Papio hamadryas, the Hamadryas, Mantled or Sacred Baboon. This comes from arid environments around the Red Sea in northeast Africa and Arabia. The male is grey with a huge mane and white cheek whiskers, red face and rump skin. The female is browner and maneless with a black face.

Papio papio, the Guinea Baboon, is from far western Africa. It is reddish, with a large mane in the male.

Papio anubis, the Anubis or Olive Baboon, is from Mali east to Ethiopia and Kenya. It is much larger than Hamadryas and Guinea Baboons and is olive brown, with a mane in adult male.

Papio cynocephalus, the Yellow Baboon, is from Tanzania south to the Zambezi. It is yellowish with white underparts and white cheeks and no mane.

Papio ursinus, the Chacma Baboon, comes from southern Africa (south of the Zambezi). It is as large as the Anubis or larger and is black in the far south, becoming fawn to the north with no mane.

The Anubis, Yellow and Chacma Baboons ("savannah baboons") live in multimale, multifemale troops with dominance hierarchies. The Hamadryas lives in harems, the surplus males associating in bachelor groups, and a number of harems and bachelor groups come together to form large bands. The basic behavioural difference is that hamadryas males herd females, and this has striking consequences for the social organization as well as the temperament of both sexes. Guinea Baboons are poorly studied but may be more like hamadryas.

This may not exhaust the biological differences between baboon taxa. It has been reported that the Cape of Good Hope baboons, which are *P.ursinus*, mate in a mount series, with ejaculation apparently occurring only at the end of the series. In contrast, those in Nairobi National Park (*P.anubis*), typically ejaculate after a single mount (Hall and DeVore, 1965) but these 40-year-old observations need to be confirmed and extended.

Cercocebus

Mangabeys are now regarded as belonging to two distinct genera: the arboreal group, *Lophocebus*, are related to *Papio*, while the semi-terrestrial mangabeys separated from the mandrills only four million years ago and are therefore regarded, by Goodman *et al.* (1998), as congeneric. I shall adopt this classification here. The prior generic name is *Cercocebus*, of which *Mandrillus* is therefore a subgenus.

The only mangabey that is widely kept in the laboratory is the Sooty Mangabey, *Cercocebus atys*, which is found in the far west of Africa, from Senegal east to the Nzo-Sassandra river system in the Ivory Coast. Its endemic SIV is thought likely to be the source of human HIV2.

Macaca

Macaques are, without a doubt, the most widespread laboratory Primates. The species-groups are well-separated, with strong morphological and behavioural differences between them. There is every indication of a considerable time depth (see below).

1. African macaques. There is just one species, the Barbary Macaque (*M.sylvanus*) of the Atlas region, which is now endangered.

2. Asian macaques.

- *M.nemestrina* group: These are short-tailed, long-faced macaques. In the non-Sulawesi species, the crown hairs radiate from a central whorl and there are cheek whiskers. The female experiences sexual skin swellings at mid cycle, like baboons and mangabeys.

  M.silenus, the Lion-tailed Macaque. This is a highly endangered species from the Western Ghats of India. It is black, with long grey cheek whiskers.

  M.nemestrina, the Southern Pigtail, is from Peninsular Malaysia, Sumatra, Bangka and Borneo. It is agouti brown, much darker in the median dorsal region, with blackish crown hair and short cheek whiskers.

  M.leonina, the Northern Pigtail, is from mainland Southeast Asia. It is uniform agouti golden brown with brown crown, fairly long cheek whiskers and a red streak at the lateral corner of each eye.

  Mentawai Island: Macaques are represented by two species, both critically endangered, from the Mentawai Is. west of Sumatra.

  Sulawesi Macaques: These are six or seven species from Sulawesi, most of them now endangered. The Crested Macaque or "Black Ape", *M.nigra*, was formerly kept in a few laboratories to investigate the occurrence of spontaneous diabetes. The only non-endangered species, the Tonkean Macaque, *M.tonkeana*, is kept in a few laboratories.

- *M.fascicularis*: the Long-tailed or Crab-eating Macaque is from Southeast Asia which includes Thailand, southern Vietnam and central Burma, to western and southern Indonesia and the Philippines. Their tail is longer than head and body which is brown with grey or black tones with crown hair directed backward and outward, sometimes with a small crest and there is a light spot at the inner corner of the eyelid. There are no sexual swellings.
- *M.arctoides*: the Stumptail or Bear Macaque is from mainland Southeast Asia. This is an odd species with a short tail, shaggy dark brown pelage, a red face, which becomes dark brown, often blotchily, in sunlight and it becomes bald at maturity. There are no sexual swellings.
- *M.mulatta* group: These are short-tailed, rather short-faced macaques with crown hair directed backwards and a pink or red face. There are no periodic sexual swellings.

  M.mulatta, the Rhesus Monkey, comes from the northern half of the Indian subcontinent, northern Burma, northern Indochina, and much of China. It has a short curly tail and is brown with a reddish tone on its hindparts, including hindlegs.

  M.cyclopis, the Formosan Rock Macaque, comes from Taiwan. It is darker than the Rhesus without its reddish hindparts and has bushy cheek whiskers and a longer tail.

  M.fuscata, the Japanese Macaque, is from Honshu, Shikoku, Kyushu, Yaku, and many offshore islands of central Japan. It is yellowish brown with no reddish tone on its hind parts, and has a very short, furry tail.
- *M.sinica* group: tails of this group vary in length and they usually have a whorl on their crown and a pink or brown face. There is periodic reddening of the genital area at mid cycle, but no sexual swelling.

  M.sinica, the Toque Macaque, is from Sri Lanka. Their tail is longer than the head and body and there is a prominent whorl on the crown, with long hairs radiating from it and reaching forward to brows in a "toupée".

  M.radiata, the Bonnet Macaque, is from southern India (south of the range of Rhesus Monkey; the approximate dividing line is the Tapti and Krishna Rivers). Their tail is usually, but not always, as long as the head and body and the crown hair is short in front, leaving a very short-haired forehead.

  M.assamensis, the Assam Macaque, is from central Nepal east to southern China and north Vietnam. It is much larger and shorter-tailed than others of the group, often lacking the "toupée", but with prominent cheek whiskers.

  M.thibetana, the Milne-Edwards's Macaque, is from central China. It is the largest macaque, very short-tailed, dark brown with a bushy pale beard and cheek whiskers.

Morales and Melnick (1998), noting that the fossil record reports a split between African and Asian macaques (i.e. between *M.sylvanus* and the rest) of at least 5.5 million years, dated the other major separations as follows:

Sulawesi macaques at 4.5 Ma, hence the *nemestrina* group somewhat before this:

M.fascicularis from the *mulatta* + *sinica* groups at 3.5 Ma.

The *mulatta* and *sinica* groups at 2.5 Ma.

M.fuscata from *M.mulatta* at 0.5 Ma—after the beginning of diversification within *M.mulatta* itself (both *M.fuscata* and *M.cyclopis* are closer, in mtDNA, to the Chinese than to the Indian *M.mulatta*).

The really astonishing molecular finding is that *M.arctoides* has the mtDNA of *M.fascicularis* but the Y-chromosome DNA of the *M.sinica* group (Tosi *et al.*, 2000). The favoured hypothesis is that the species is a stabilized hybrid between *M.fascicularis* and proto – *M.assamensis/thibetana* – so far the only plausibly hypothesized case of a species of hybrid origin among Primates.

Homo

Goodman *et al.* (1998) calculated, on the basis of their molecular data, that the ancestors of humans and chimpanzees separated only six million years ago. As this is less than the time for generic separation, they proposed to combine them into one genus: there would thus be a single genus (*Homo*) with two subgenera (*Homo* and *Pan*). Groves (2001), adopting a looser time-frame for generic separation, retained *Pan* as a distinct genus. The proposal to sink *Pan* into *Homo* was, however, endorsed by Watson *et al.* (2001). Chen and Li (2001), on the basis of 53 non-transcribed DNA segments, calculated the human–chimpanzee separation at 4.6 to 6.2 Ma, while Wildman *et al.* (2003) put it at between 5 and 6 Ma. On the time-depth criterion, there seems no longer any reason to keep the two genera separate.

The subgenus *Pan* has two species: *Homo (Pan) troglodytes* and *H.(P.) paniscus*. Both are currently kept in laboratories: *H.paniscus* (Bonobo or Pygmy Chimpanzee) only for language research, and *H.troglodytes* (Common Chimpanzee) which is unfortunately no longer "common" but mostly now in retirement from research, except for a few instances for continuing research in infectious disease such as HIV and hepatitis.

The question of whether it is ethical to use chimpanzees in discomforting, disabling or potentially terminal research, seems more sharply focused if they are now to be regarded as a species of the human genus. In fact, the evidence for self-awareness, and other human-like cognitive and emotional qualities in chimpanzees, gorillas and orangutans has been available for some years (see, for example, Russon *et al.*, 1996) and research in these is not even in question. All that has changed is the taxonomy. Indeed, it is interesting to reflect how taxonomy has moved from:

1. a grade-based, anthropocentric scheme whereby humans belong alone in one family, Hominidae, while chimpanzees, gorillas and orangutans are relegated to the family Pongidae, to
2. a clade-based scheme in which all are placed in Hominidae, with orangutans put in their own subfamily, Ponginae, and the others combined in a subfamily Homininae with three tribes Gorillini, Panini, Hominini, to
3. the time-depth-based scheme in which not only are the three tribes abolished, along with a family division between "great apes" and "lesser apes" (gibbons), but two of the genera are now even combined.

Whereas all Primates have special husbandry needs, for housing, socialisation, and behavioural enrichment, it is certainly true that the needs of apes, whether human or non-human, are beyond those of other Primates. Put simply, there is more potential for an ape, such as a chimpanzee or human, to experience distress, boredom, discomfort or pain than other Primates. Whether or not we consider human rights, such as described in Cavallieri and Singer (1993) and enacted, in modified form, in law in New Zealand, to be appropriate for other apes, humane concerns dictate that, at the very least, we approach their treatment with an extra degree of care.

Correspondence

Any correspondence should be directed to Colin Groves, School of Archaeology and Anthropology, Australian National University, Canberra, ACT 0200. Tel: (+612) 6125 4590. *Colin.Groves@anu.edu.au*

References

Alterman, L. (1995) In Alterman, I., Doyle, G.A. and Izard, M.K. (eds), *Creatures of the Dark: The Nocturnal Prosimians*, 413–424. New York: Plenum Press.

Aslam, H., Schneiders, A., Perret, M., Weinbauer, G.F. and Hodges, J.K. (2002). *Reproduction* 123, 323–332.

Birrell A., Hennessy, A., Gillin, A.G., Horvath, J.S. and Tiller, D.J. (1996). *J. Med. Primatol.* 25, 287–293.

Boinski, S. and Cropp, S.J. (1999). *Int. J. Primatol.* 20, 237–256.

Carr, S.M. and Hughes, G.A. (1993). *J. Mamm.* 74, 331–342.

Cavallieri, P. and Singer, P. (1993). *The Great Ape Project: Equality Beyond Humanity.*

Chen, F-C. and Li, W-H. (2001). *Am. J. Hum.Genet.* 68, 444–456.

Cracraft, J. (1983). *Curr. Ornithol.* 1, 159–187.

Defler, T.R., Bueno, M.L. and Hernández-Camacho, J.I. (2001). *Neotrop. Primates* 9, 37–52.

Goodman, M., Porter, C.A., Czelusniak, J., Page, S.L., Schneider, H., Show, J., Gunnell, G. and Groves, C.P. (1998). *Mol. Phyl. Evol.* 9(3), 585–598.

Groves, C. (2001). *Primate Taxonomy* viii, Washington, DC: Smithsonian Institution Press.

Hall, K.R.L. and DeVore, I. (1965). In I. DeVore (ed.), *Primate Behavior,* pp 53–110. New York: Holt, Rinehart & Winston.

Hennig, W. (1966). *Phylogenetic Systematics.* Urbana: University of Illinois Press.

Hershkovitz, P. (1983). Two new species of night monkeys, genus *Aotus* (Cebidae, Platyrrhini): a preliminary report on *Aotus* taxonomy. *Amer. J. Primatol.* 4, 209–43.

Horam, C.J., Harewood, W.J., Phippard, A.F. and Horvath, J.S. (1992). *Aust. Primatol.* 7, 1:11.

Jolly, C.J. (1993). In Kimbel, W.H. and Martin, L.B. (eds), *Species, Species Concepts, and Primate Evolution,* pp 67–107. New York: Plenum Press.

Mayr, E. (1940). *Amer. Nat.* 74, 249–278.

Mayr, E. (1963). *Animal Species and Evolution.* Harvard: Belknap Press.

Morales, J.C. and Melnick, D.J. (1998). *J. Hum. Evol.* 34, 1–28.

Nekaris, K.A.I. and Rasmussen, D.T. (2003). *Int. J. Primatol.* 24, 33–46.

Perret, M. and Aujard, F. (2001). *Amer. J. Physiol.* 281, R1925–R1933.

Rasoloarison, R., Goodman, S.M. and Ganzhorn, J.U. (2000). *Int. J. Primatol.* 21, 963–1019.

Russon, A.E., Bard, K.A. and Parker, S.T. (1996). In Russon, A.E., Bard, K.A. and Parker, S.T. (eds), *Reaching into Thought: The Minds of the Great Apes.* Cambridge: Cambridge University Press.

Struhsaker, T.T., Butynski, T.M. and Lwanga, J.S. (1988). In Gautier-Hion, A., Bourlière, F., Gautier, J-P. and Kingdon, J. (eds), *A Primate Radiation: Evolutionary Biology of the African Guenons,* pp 477–497. New York: Cambridge University Press.

Tosi, A.J., Morales, J.C. and Melnick, D.J. (2000). *Mol. Phyl. Evol.*, 17, 133–144.

Vane-Wright, R.I. (1992). Species concepts. In *Global Biodiversity 1992: a Report Compiled by the World Conservation Monitoring Centre,* pp 13–16. B. Groombridge (ed.), London: Chapman & Hall.

Watson, E.E., Easteal, S. and Penny, D. (2001). In Tobias, P.V., Raath, M.A., Moggi-Cecchi, J. and Doyle, G.A. (eds), *Humanity from African Naissance to Coming Millenia,* pp. 307–318. Italy: Firenze University Press.

Wildman, D.E., Uddin, M., Liu, G-z., Grossman, L.I. and Goodman, M. (2003). *PNAS* 100, 7181–7188.

Wilson, A.C., Cann, R.L., Carr, S.M., George, M., Gyllensten, U.B., Helm-Bychowski, K.M., Higuchi, R.G., Palumbi, S.R. and Prager, E.M. (1985). *Biol. J. Linn. Soc.* 26(4), 375–400.

Appendix

An outline classification of living Primates:

SUBORDER Strepsirrhini
- INFRAORDER Lemuriformes
  - FAMILY Cheirogaleidae
    - GENERA: *Microcebus, Mirza, Cheirogaleus, Allocebus, Phaner*
  - FAMILY Lemuridae
    - GENERA: *Lemur, Hapalemur, Prolemur, Eulemur, Varecia*
  - FAMILY Lepilemuridae
    - GENUS *Lepilemur*
  - FAMILY Indriidae
    - GENERA: *Indri, Propithecus, Avahi*
- INFRAORDER Chiromyiformes
  - FAMILY Daubentoniidae
    - GENUS *Daubentonia*
- INFRAORDER Lorisiformes
  - FAMILY Lorisidae
    - GENERA: *Loris, Nycticebus, Perodicticus, Arctocebus, Propotto*
  - FAMILY Galagidae
    - GENERA: *Galago, Euoticus, Otolemur, Galagoides*

SUBORDER Haplorrhini
- INFRAORDER Tarsiiformes
  - FAMILY Tarsiidae
    - GENERA: *Tarsius, Cephalopachus*, unnamed third genus
- INFRAORDER Simiiformes
  - Platyrrhini[1]
    - FAMILY Cebidae
      - SUBFAMILY Cebinae
        - GENERA: *Cebus, Saimiri*
      - SUBFAMILY Aotinae
        - GENUS *Aotus*
      - SUBFAMILY Callitrichinae
        - GENERA: *Callithrix, Callimico, Leontopithecus, Saguinus*
    - FAMILY Pitheciidae
      - GENERA: *Pithecia, Chiropotes*
    - FAMILY Atelidae
      - SUBFAMILY Alouattinae
        - GENUS *Alouatta*
      - SUBFAMILY Atelinae
        - GENERA: *Ateles, Brachyteles, Lagothrix, Oreonax*
  - Catarrhini[1]
    - FAMILY Cercopithecidae
      - SUBFAMILY Cercopithecinae
        - TRIBE Cercopithecini
          - GENERA: *Cercopithecus, Allochrocebus, Erythrocebus, Chlorocebus, Miopithecus, Allenopithecus*
        - TRIBE Papionini
          - GENERA: *Papio, Theropithecus, Lophocebus, Cercocebus, Macaca*
      - SUBFAMILY Colobinae
        - GENERA: *Colobus, Procolobus, Piliocolobus, Presbytis, Semnopithecus, Trachypithecus, Pygathrix, Rhinopithecus, Nasalis, Simias*
    - FAMILY Hominidae
      - SUBFAMILY Hylobatinae
        - GENERA: *Hylobates, Hoolock, Symphalangus, Nomascus*
      - SUBFAMILY Homininae
        - TRIBE Pongini
          - GENUS *Pongo*
        - TRIBE Hominini
          - GENERA: *Homo, Gorilla*

[1]Platyrrhini and Catarrhini are left unranked. This is a perfectly acceptable procedure when too many divisions are needed for the number of ranks ordained.

CHAPTER 2

Similarities of Non-human Primates to Humans: Genetic Variations and Phenotypic Associations Common to Rhesus Monkeys and Humans

Gregory M. Miller and Bertha K. Madras
The New England Primate Research Center,
Harvard Medical School, Southborough,
MA 01772-9021, USA

Introduction

Our understanding of the causes and pathogenesis of neuropsychiatric disorders is at an early phase. Based on family, twin, adoption studies and genetic linkage analysis, it is generally accepted that genetics is a significant contributor to the manifestation of the majority of spontaneously occurring or drug-induced neuropsychiatric disorders. Until recently, genetic components were not integrated into animal models designed to develop medications or to clarify the pathophysiology of these disorders. Impediments to this approach are clear: the genetic basis of definitive human traits in affected populations is largely unknown, even though the list of genetic variants associated with neuropsychiatric disorders is mounting (Comings *et al.*, 2000a; 2000b). Equally taxing is the need to digress from convenient genetically identical strains of rats, widely used in the majority of models, and attempts to create genetic strains reflective of human polymorphisms associated with disease states. Spontaneously occurring gene variants, common to humans and animal populations, offer an appealing, albeit elusive option. An animal species with a spontaneously occurring genetic variant that contributes to some identifiable phenotype, also found in a population of humans with a specific brain disorder, would present a unique opportunity and a naturalistic model to explore the role of genetic variants on physiology and behavior. An important caveat is that multiple variants, at diverse gene loci, interact with each other and with non-genetic factors to produce susceptibility

The Laboratory Primate
ISBN 0-1208-0261-9

to brain disorders. Furthermore, the search for relevant gene variants in animal populations may be daunting if the incidence of certain diseases is very low in the population.

Gene manipulation in mice (null mutations, gene insertions, transgenics) presents important and effective approaches to unraveling the contribution of specific genes to definable phenotypes (reviewed in van der Weyden *et al.*, 2002). Because of the evolutionary, social and behavioral distance between mice and humans, there may not be analogous gene variants between humans and mice at loci relevant to neuropsychiatric disorders. Primates offer an intriguing choice for this quest. In our experience, the coding regions of genes that encode key therapeutic targets (receptors and transporters) in brain, share more than 95% homology with humans and contrast with the 75–92% homology common to humans and rodents (Figure 2.1). This chapter outlines an early phase in exploring functional genetic polymorphisms shared by rhesus monkeys and humans. Described herein are polymorphisms that appear to be associated with similar phenotypic characteristics in both species and are strongly implicated in neuropsychiatric disorders.

Polymorphisms

Polymorphisms are natural differences occurring in DNA sequences that are distributed among the individuals of a species. These individual differences in the DNA sequence occur throughout the entire genome, within the protein coding and non-coding regions of genes, in introns and within the vast stretches of DNA that separate individual genes. Polymorphisms within the coding region can alter the amino acid sequence of the encoded proteins, resulting in structural changes (amino acids) that may or may not affect protein function. DNA variations in non-coding regions do not alter the structure of proteins but may result in functional changes by altering parameters of protein expression. Polymorphic DNA sequences range in form from large stretches of repeated DNA sequences to smaller di- or tri-nucleotide repeats or single nucleotide polymorphisms (SNPs). SNPs have become an important focus in biomedical research. The human genome project has revealed over three million SNPs and the functional aspects of these are likely to be directly involved in, or serve as markers for, a wide range of diseases, traits and physiological characteristics.

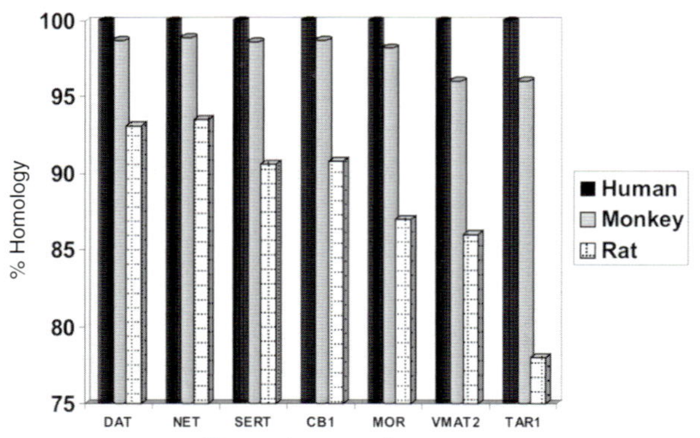

Figure 2.1 Coding sequence percent homology of human, rhesus monkey and rat proteins of relevance to neuropsychiatric disorders. Sequences for the dopamine transporter (DAT), norepinephrine transporter (NET), serotonin transporter (SERT), cannabinoid CB-1 receptor (CB1), mu-opioid receptor (MOR), vesicular monoamine transporter-2 (VMAT2), and trace amine receptor 1 (TAR1) are derived from GenBANK for human and rat. The rhesus monkey sequences were determined in this laboratory and the majority are currently listed in GenBANK.

Rationale for specific SNP studies in monkey

We currently focus on three genes that code for three brain membrane proteins: the mu-opioid receptor, the dopamine transporter and the serotonin transporter. Each encodes proteins implicated in neuropsychiatric disorders and is a primary target of psychotherapeutic drugs and drugs of abuse. The mu-opioid receptor plays a fundamental role in a variety of physiological effects, including analgesia, hormone release, gastric motility and anxiety. It is the principal mediator of opiate analgesics in brain and is implicated as the immediate site of action of heroin. The dopamine transporter is a key regulator of extracellular dopamine levels in the brain, thereby playing a pivotal role in regulating processes triggered by dopamine, including movement, cognition, and reward. The dopamine transporter in brain is a target of both anti-hyperactivity and select antidepressant medications, as well as the psychostimulant drugs of abuse, cocaine and amphetamine. As regulator of the extracellular brain serotonin concentrations, the serotonin transporter is implicated in influencing mood, sleep, and other affective states in the brain. It is the immediate site of action of the majority of antidepressant drugs. The therapeutic role of the serotonin transporter is also balanced by its capacity to transport the hallucinogenic agent MDMA (ecstasy) into serotonin neurons. The genes encoding these clinically relevant proteins in rhesus monkeys are described in detail because each is instructive and representative of genetic variation occurring in different parts of the gene: in a coding region (mu-opiate receptor), in the 3′-untranslated region (dopamine transporter), and the 5′-regulatory (promoter) region (serotonin transporter), respectively. All three genes contain polymorphisms that have been associated with distinct phenotypic parameters. Polymorphisms are described in rhesus monkeys for each of these genes that differ in exact DNA sequence from analogous human polymorphisms. They are, nevertheless, strikingly parallel to those found in humans with regard to type, location in the gene and functional consequence. Taken together, these studies provide intriguing leads to investigate the usefulness of rhesus monkeys as models for deciphering genotype/phenotype relationships relevant to human disorders. Equally significant, this approach may provide novel insights into the relevance of genetic differences between individuals and the resultant effects of these differences on disease susceptibility, treatment and prognosis.

Mu-opioid receptor

Non-human primates are widely used as research models for studying neuroanatomical and behavioral parameters of human drug addiction. Investigation of genotype/phenotype associations in non-human primates may lead to better predictive, diagnostic and clinical assessments with regard to drug addiction and pain management. Accordingly, we asked some fundamental questions specific to our current level of understanding of mu-opioid receptors: 1. Do humans and rhesus monkeys share a structural and functional similarity in the mu-opioid receptor gene? 2. Are there polymorphisms in rhesus monkeys that are analogous to those found in humans, associated with differences in drug binding profiles and other phenotypes? 3. Can rhesus monkeys provide a "naturalistic" model for deciphering genetic associations with behavioral and physiological parameters reported in humans?

Human mu-opioid receptor gene

The human mu-opioid receptor gene contains numerous SNPs, one of which, A118G, alters the structure of the N-terminal extracellular arm of the encoded receptor protein (Bond *et al.,* 1998). This SNP results in enhanced β-endorphin affinity for the receptor and has been implicated in modulating hypothalamic-pituitary-adrenal axis activation (Wand *et al.,* 2002). To explore whether genetic variations of the mu-opioid receptor gene exist in rhesus monkey, we cloned the rhesus monkey mu-opioid receptor coding region (Miller *et al.,* 2004). The finding of a promising ~98% homology to the human coding region was followed by the discovery of a C77G SNP that altered an amino acid in the same region (N-terminal arm) of the A118G SNP in the human mu-opioid receptor (Figure 2.2).

Rhesus monkey mu-opioid receptor gene

Using these data as a lead, we investigated whether the expressed receptors bound an agonist or antagonist differently. The affinity of β-endorphin was 3.5-fold higher for membranes derived from HEK-293 cells transfected with a G77-containing clone versus a C77-containing clone, whereas the affinities of naloxone and buprenorphine did not differ between the two cell lines. Two-site analysis revealed that the 3.5-fold difference in affinity for β-endorphin expanded to 100-fold if the high affinity component was compared with its affinity for the C77-derived receptor. Intriguingly, site-directed mutagenesis, to mimic a G118 allele of the human mu-opioid receptor gene, also resulted in a 3.5-fold increase in β-endorphin affinity for the human mu-opioid receptor but no differences in the affinities of other opioid agonists (Bond *et al.,* 1998). As detailed analysis was not reported, parallel comparisons between our data and the A118G mutant data of the human mu-opioid receptor was not feasible. Nevertheless, the physiological relevance of this difference in β-endorphin affinity warrants further investigation in both human and rhesus monkey defined haplotypes. The higher affinity for β-endorphin by the G77-containing allele may result in altered mu-opioid receptor function (signal transduction, receptor trafficking, regulation, recycling, neurotransmitter and/or hormone release). The amino acid substitution in the monkey receptors was accompanied by other amino acid changes that could also contribute to modifying β-endorphin binding affinity. Accordingly, both the rhesus monkey and the human mu-opioid receptor genes are likely to be highly polymorphic, with numerous haplotypes that may impart distinct phenotypic determinants (Hoehe *et al.,* 2000; Miller *et al.,* 2004).

The allelic frequencies of the C77G SNP in rhesus monkeys indicated that, of 32 animals, 94% had at least one C77-containing allele and only two animals were

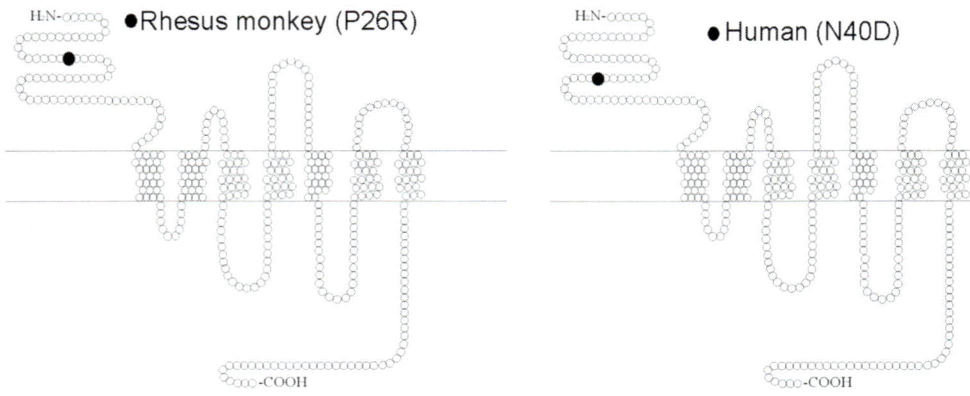

Figure 2.2 Schematic representation of the human and rhesus monkey mu-opioid receptor protein. The seven transmembrane structure of the rhesus monkey (left) and human (right) mu-opioid receptors is depicted with open circles representing amino acids. The similar location of amino acid changes in the mu-opioid receptor proteins resulting from single nucleotide polymorphisms in the rhesus monkey and human mu-opioid receptor genes are shown in filled circles. The schematic illustrates the seven alpha helices that are embedded in the plasma membrane, an intracellular carboxy terminus (COOH), and an extracellular amino terminus (H_2N). The location of the proline-to-arginine (P26R) amino acid change that results from the C77G single nucleotide polymorphism in the rhesus monkey mu-opioid receptor gene, and the asparagine-to-aspartate (N40D) amino acid change that results from the A118G single nucleotide polymorphism in the human mu-opioid receptor gene are in the amino terminal arm of the proteins.

homozygous for G77-containing alleles. Noteworthy is the finding that analogous screening for G118 alleles in the human also demonstrated that homozygosity for the rarer allele is equally uncommon (Bond *et al.*, 1998; Grosch *et al.*, 2001).

Mu-opioid receptor gene: physiological, behavioral association

In humans, the A118G polymorphism in the mu-opioid receptor gene is associated with enhanced HPA axis responses to opioid receptor blockade by naloxone (Wand *et al.*, 2002). More broadly, it has been suggested that persons harboring a G118-containing allele may have abnormal HPA axis responses to stress (Kreek, 1996; Bond *et al.*, 1998; Wand *et al.*, 2002). Persons with enhanced stress responsivity are more prone to addictive disorders as well as to insulin resistance, immunosuppression, osteoporosis and hippocampal injury. If the mu-opioid receptor SNP in rhesus monkeys is relevant to the human SNP, then the physiological consequences should be parallel in both species. Accordingly, we compared the incidence of G77- and C77-containing alleles with plasma cortisol levels in 21 rhesus monkeys. Plasma cortisol levels were measured on two separate occasions, two years apart. Consistently, rhesus monkeys with a G77-containing allele had significantly lower plasma cortisol levels. Upon challenge with ACTH following dexamethasone suppression, rhesus monkeys with a G77-containing allele had significantly lower plasma cortisol increases (Miller *et al.*, 2004). These findings implicate the mu-opioid receptor genotype as a relevant factor in the duration and efficacy of the hormonal cascades occurring in response to stress. Moreover, the mu-opioid receptor polymorphisms in rhesus monkeys and humans appear to be functionally similar.

In rodents, mu-opioid receptor activity is associated with aggression and locomotor activity (Gwynn and Domino, 1984; Benton, 1985; Becker *et al.*, 1997). In rhesus monkeys, we found a statistically significant association between the C77G SNP and the early communicative aspect of aggression (termed aggressive threat which includes behaviors such as staring and an open-mouthed, teeth-baring, ear-flapping, facial display), but not with the physical manifestations of aggression (cage shaking, environment- and self-directed aggression). Animals with one G77-containing allele scored twice the average of animals with two C77-containing alleles. Animals with two C77-containing alleles clustered with low aggression index scores, whereas animals with one G77-containing allele varied widely, perhaps due to a greater diversity of representative haplotypes. This trend intensified with the two animals that harbored two G77-containing alleles. Although not statistically significant, we also observed a trend towards lower total locomotor activity in animals harboring G77-containing alleles (Miller *et al.*, 2004).

Are mu-opioid receptor polymorphisms linked to cortisol levels and aggression in humans? Intriguingly, the association between plasma cortisol levels and aggression are parallel in rhesus monkey and humans (Kalin *et al.*, 1998; Kalin, 1999; McBurnett *et al.*, 2000; Pajer *et al.*, 2001; Miller *et al.*, 2004). An inverse relationship between plasma cortisol levels and aggressive behavior that we discovered in male rhesus monkeys is mirrored by reports in human subjects. Decreased cortisol levels have been associated with antisocial behavior in girls, and early onset aggression in boys (McBurnett *et al.*, 2000; Pajer *et al.*, 2001). Taken together, these data suggest that variations in the mu-opioid receptor gene might contribute to and provide a common link between certain forms of aggression and HPA axis function.

For both C77G and A118G SNPs, the degree to which other SNPs occur in tandem remains to be elucidated. The number of distinct haplotypes of the human mu-opioid receptor gene are unknown, if considering combinations of alleles at A118G and other SNPs. Mu-opioid receptor function may be subtly or profoundly influenced by particular haplotypes in an individual. Altered receptor structure may modify interaction with other receptors that heterodimerize with mu-opioid receptors (e.g. other opioid receptors and of particular emerging interest, cytokine receptors), or receptor function with consequent effects on a range of biochemical sequelae triggered by this receptor. Nevertheless, the significant associations of the C77G SNP with plasma cortisol levels and aggression, reported herein, are independent of the pharmacology of the clones studied.

Taken together, mu-opioid polymorphisms in rhesus monkeys and humans demonstrate similarities in the consequences of SNPs on receptor affinity, cortisol levels and aggression. Although the specific nucleotide affected by the SNPs differs, the conserved location within the gene, the functional effects of the SNP at the level of receptor binding and the similar association of the SNPs with effects on the HPA axis have striking parallels. These data support the use of non-human primates to investigate the physiological and pathological significance of mu-opioid receptor polymorphisms and other functional polymorphisms of relevance to humans.

Dopamine transporter

The brain dopamine transporter (DAT) is a member of a superfamily of Na^+/Cl^- dependent neurotransmitter transporters. By actively sequestering extracellular dopamine to intracellular compartments, the DAT plays a key role in adjusting dopamine availability and consequent dopamine-mediated behaviors. As dopamine is implicated in neuropsychiatric, neurodegenerative disorders and substance abuse, dysregulation of dopamine levels by the DAT may contribute to the etiology of, or susceptibility to, dopamine-related disorders. DAT protein levels vary in normal subjects, particularly as a function of age, and deviate from the normal range in pathological states. DAT is markedly reduced in Parkinson's disease and Lesch-Nyhan syndrome and elevated levels are observed in attention deficit hyperactivity disorder (ADHD) (Dougherty *et al.*, 1999, Dresel *et al.*, 2000, Cheon *et al.*, 2003, Krause *et al.*, 2003, Madras *et al.*, 2002) and Tourette's Syndrome (Malison *et al.*, 1995, Cheon *et al.*, 2004). Chronic use of stimulant drugs such as cocaine leads to increases in DAT levels during withdrawal, whereas amphetamine results in DAT depletion, possibly a consequence of neurotoxicity or amphetamine-induced DAT internalization.

Human dopamine transporter gene

The DAT gene coding sequence is derived from 15 exons distributed across a >64 kb gene in humans. Whereas the human DAT coding region is of fixed length, the 3′-untranslated region (3′-UTR) varies in length due to a polymorphic variable number tandem repeat (VNTR) region (Vandenbergh *et al.*, 1992). This VNTR consists of 3 to >11 copies of a 40-base repeat unit. Numerous reports have attempted to associate the presence or absence of particular DAT alleles, as defined by the size (number of repeats) of the VNTR, with the occurrence of dopamine-related disorders, including Parkinson's disease, schizophrenia, delusional disorder, smoking cessation, polysubstance abuse and alcoholism. The most consistent finding among this literature has been an association of a ten copy allele with attention deficit hyperactivity disorder (ADHD). Although this lead has not clarified the pathophysiology of ADHD, a significant focus of ADHD research has converged on the DAT. The DAT is one principal target of anti-hyperactivity medications in brain and may be elevated in brains of adults with ADHD. High transporter levels can arise from a number of processes, including dysfunctional regulation of DAT protein expression by the transporter gene.

Emergent from these findings is whether the number of repeat sequences, in the 3′-UTR of the DAT gene, influences DAT protein levels in the brain.

The relationship between DAT genotype and phenotype was explored in studies that measured DAT levels in living human brain striatum with single photon emission computed tomography (SPECT) and genotyped DAT alleles, in the same subjects, by the number of tandem repeats in the VNTR region (Jacobsen *et al.*, 2000; Heinz *et al.*, 2000; Martinez *et al.*, 2001). Opposite findings were reported as subjects with the nine-repeats either had lower or higher DAT levels compared with subjects with ten-repeats. We speculated that this discrepancy may arise from the existence of allele diversity independent of the length of the DAT 3′-UTR (Miller *et al.*, 2001; Miller and Madras, 2002).

Rhesus monkey dopamine transporter gene

Although a species of spontaneously hypertensive rat displays hyperactivity and is considered a model for ADHD, rats and mice do not contain analogous repeat sequences in the 3′-UTR of the DAT gene. Accordingly, rodents are inappropriate for investigating the contribution of DAT alleles, of a particular length, to hyperactivity. In view of the evolutionary proximity of rhesus monkeys to humans, we hypothesized that a repeat sequence in the DAT gene may be present in the rhesus monkey genome, and investigated the following: 1. Does the rhesus monkey DAT gene contain a tandem repeat sequence? 2. If so, is it associated with levels of activity in rhesus monkeys? 3. Are other polymorphisms present in the monkey and are these associated with activity levels? 4. Can these polymorphisms in monkey or human influence levels of protein expression? We therefore sought to determine whether a tandem repeat sequence was present in the 3′-untranslated region of the DAT gene in rhesus monkeys, whether the number of repeat units varied between animals, and whether there was an association between the 3′-UTR of the DAT gene with hyperactivity in monkeys.

Dopamine transporter gene: functional, behavioral association

Similar to human, but unlike other species previously studied, we found a fixed number tandem repeat (FNTR) sequence in the 3′-UTR of the monkey DAT gene (Figure 2.3). In the absence of an established animal model of ADHD, we compared, in rhesus monkeys, the five most active with the five most sedate animals from a behaviorally characterized cohort of 22 subjects (Miller *et al.*, 2001). In contrast to the human gene sequence, the FNTR (39 bases/repeat and 12 repeats)

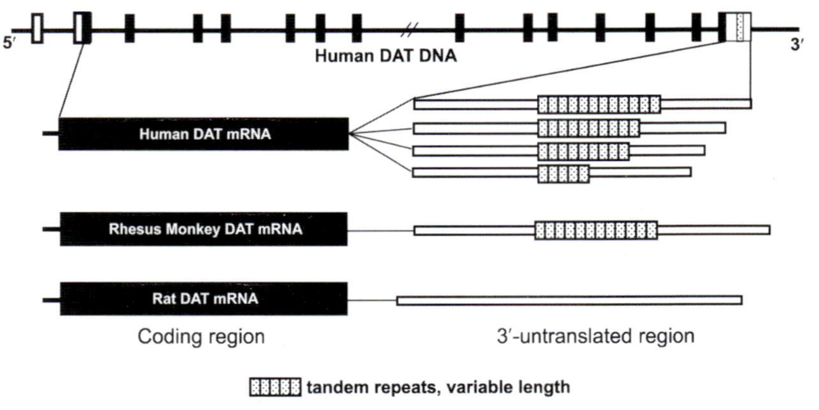

Figure 2.3 Schematic diagram depicting the human dopamine transporter (DAT) gene, implicated in ADHD, and a comparison of the human, rhesus monkey and rat DAT mRNAs. In the human DAT DNA, black boxes depict exons that make up the coding region, empty boxes depict non-coding exons, and the stippled box locates the position of a polymorphic variable number tandem repeat (VNTR) region within the portion of the gene that codes for the 3′-untranslated region. Human DAT mRNAs vary in length depending on how many repeated 40-base-pair sequences are present and each box represents one repeat sequence (e.g. 12 boxes = 12 repeats). Similar to human, the rhesus monkey DAT gene contains a series of tandem repeats in the 3′-untranslated region but, thus far, only a fixed number of tandem repeats (FNTR = 12 repeats) have been identified in >24 rhesus monkeys. The rat DAT gene does not contain analogous repeat sequences in the 3′-UTR.

is present in both very active and sedate animals as well as other monkeys. Accordingly, this FNTR is unbefitting an association of DAT transcript *length* with hyperactivity. However, sequence analysis revealed potential SNPs, one of which affects a *Bst1107I* restriction site. We screened the entire cohort, confirmed that all the rhesus monkeys had repeat regions of the same length, and demonstrated that digestion with *Bst1107I* was sufficient to distinguish two distinct FNTR alleles. *Bst1107I* genotype was suggestive but not predictive of hyperactive behavior (Miller *et al.*, 2001). Based on these data, we speculated that SNPs may exist in human DAT VNTR alleles. To support this hypothesis, we cloned a portion of a novel ten-repeat allele of the human DAT gene and discovered a *DraI* restriction site-sensitive SNP.

If particular alleles of the DAT gene differentially contribute to altered levels of DAT protein, it is important to consider both types of polymorphisms as potential modulators. Based on these considerations, and using a reporter assay, we investigated whether both the number of repeat sequences and the particular SNPs, in the 3′-UTR of the human and rhesus monkey DAT genes, could modify levels of gene expression (Miller and Madras, 2002). In the human sequence, the number of tandem repeat sequences in the VNTR region of the DAT 3′-UTR was an important contributor to levels of reporter gene expression. Vectors containing the nine-repeat sequence yielded higher levels of reporter gene expression than vectors containing a ten-repeat sequence. SNPs also modified reporter gene expression as the human DAT 3′-UTR segment, containing an enzyme-sensitive or insensitive nucleotide (as determined by the allele), yielded significant differences in the levels of reporter gene expression Interestingly, the effect of the SNP was dependent upon the promoter in the vector (Figure 2.4). This observation further illustrates the concept that the functional consequence of a particular SNP is context-dependent and helps to explain the commonality of discrepancies between SNP association studies in the literature and the need to define haplotypes. Although the frequency of the SNP in the human population is unknown, this SNP in the rhesus monkey, DAT 3′-UTR, occurs in about 68% of fixed-length allele PCR products sensitive to *Bst1107I* digestion. Thus, although the specific sequence of the DNA repeat region, the number of repeats and the specific SNP differed between the rhesus monkey and the human DAT 3′-UTR, the polymorphic structure, the location within the gene and, in particular, the functional effects of the polymorphisms, at the level of regulation of gene expression, had striking parallels.

These studies resulted in four major findings. First, a tandem repeat region, previously identified in human but not in rodent, was present in the 3′-UTR of the rhesus monkey DAT gene. Second, the length of the repeat region, which varies in human subjects, was of fixed length in the monkeys. Third, the sequence of the repeat region in the monkey DAT gene varied between animals and both the human and monkey DAT gene contain SNPs in this region. Finally, we related genetic variations in the DAT gene to differences in gene expression and levels of spontaneous activity in monkeys. Most relevant, these data led us to hypothesize that, between individuals, SNPs create a diversity of DAT alleles that extend beyond the *length* of the VNTR region, implying that sequence-defined haplotypes may differentially contribute to dopamine-related disorders.

Is it possible to relate DAT gene polymorphisms with levels of DAT protein in brain? Imaging agents, that label the DAT non-invasively, have enabled quantification of DAT density in living brain. As described previously, contradictory data were reported when comparing individuals harboring nine- and/or ten-repeat length alleles (Jacobsen *et al.*, 2000; Heinz *et al.*, 2000; Martinez *et al.*, 2001). As ADHD has a small but significant association with DAT ten-repeat length alleles, several groups investigated whether DAT levels in ADHD brains deviate from the normal range. In three of four SPECT studies of adults with ADHD, elevated levels of DAT protein were detected but genotyping was not performed in this cohort (Dougherty *et al.*, 1999; Krause *et al.*, 2003; Dresel *et al.*, 2000; van Dyck *et al.*, 2002). As ADHD is most likely polygenic, and the association of the ten-repeat length allele accounts for <4% of the variance in hyperactive-impulsive symptoms and about 1% of the variance in inattentive symptoms in ADHD (Waldman *et al.*, 1998), it is unlikely that a robust association between DAT (length) genotype and DAT density would emerge. However, a detailed analysis of the frequency of SNPs, throughout human ten-repeat length-containing alleles, is needed to further investigate DAT gene polymorphisms and ADHD, as well as other dopamine-related disorders. Accordingly, the relationship between the sequence of the 3′-UTR, DAT gene expression and behavior is yet to be resolved. Other SNPs may also influence levels of gene expression. The relevance of these observations to DAT gene regulation *in vivo* warrants a careful determination of the location and frequency of SNPs in the human DAT 3′-UTR as well as the 5′-UTR and other non-coding regions. The recent discovery of SNPs, in the 5′-flanking region of the human DAT gene, raises the possibility that functional SNPs may exist in the

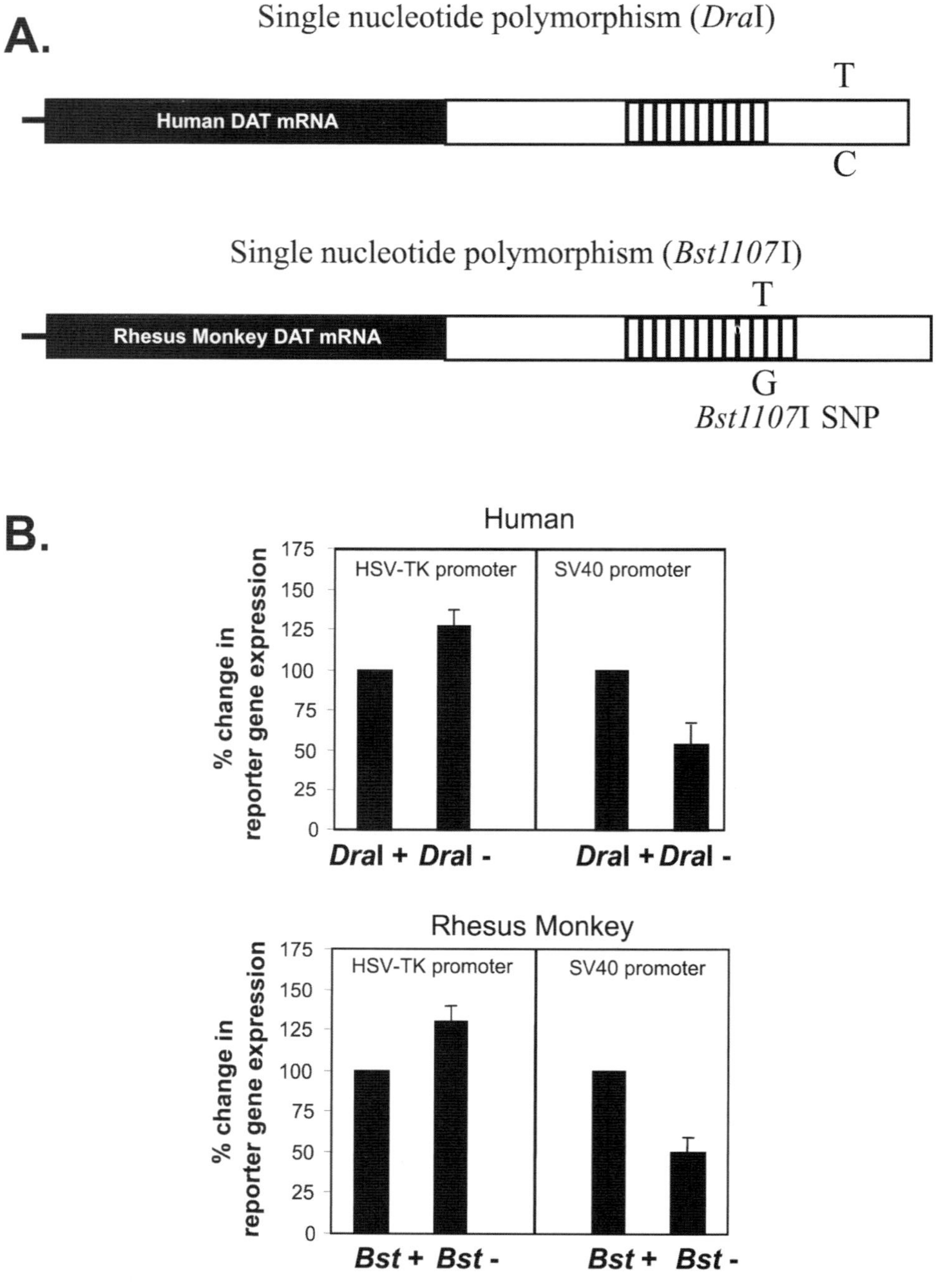

Figure 2.4 Effects of a single nucleotide polymorphism in the 3-untranslated region of the dopamine transporter gene on expression of reporter gene levels. A (top): Schematic comparison of the location of a single nucleotide polymorphism in a region of the DAT gene implicated in ADHD. Both the human DAT mRNA and rhesus monkey DAT mRNA have a tandem repeat region and single nucleotide polymorphisms. The coding region is shown in black and the 3′-untranslated region is shown in white. The location of the tandem repeat regions (open boxes) and the single nucleotide polymorphisms (T/C, human; T/G, monkey) that alter restriction endonuclease-sensitive sequences for *Dra*I (human) and *Bst1107*I (rhesus monkey) are also shown. B (bottom): The effects of polymorphisms of human and rhesus monkey 3′-untranslated regions on luciferase reporter gene expression. Two (ten-repeat-containing) human DAT 3′-untranslated regions (*Dra*I-sensitive and *Dra*I-insensitive, top) are compared to two rhesus monkey 3′-untranslated regions (*Bst1107*I-sensitive vs. *Bst1107*I-insensitive, bottom) are shown. Although the polymorphism in human and rhesus monkey differed in sequence, the magnitude and direction of the promoter-dependent reporter protein levels were parallel. Data is adapted from Miller and Madras, 2002.

promoter or other 5′ regulatory elements in the DAT gene (Rubie *et al.*, 2001). Our data may imply that a DAT 3′-UTR of a particular sequence might function differently, depending on the sequence of the DAT promoter or other regions of the DAT gene, in any given haplotype. The findings support the need to investigate the interaction between the native DAT promoter and 3′-UTR of relevant polymorphisms directly, or in an appropriate cell line that closely mimics dopamine neurons.

Similar to the studies described previously, on mu-opioid polymorphisms in rhesus monkeys and humans, these studies on the DAT also demonstrate similarities between the two species at the level of genetic polymorphisms, their function and phenotypic associations. Although the specific DNA sequences may differ, the conserved location within the gene, the functional effects of the polymorphisms at the level of gene expression, and the similar association of the polymorphisms with behavior, have striking parallels.

Serotonin transporter

The serotonin transporter (SERT) regulates the magnitude and duration of 5-HT signaling by terminating neurotransmission via cellular transport. SERT is the site of action of many centrally active drugs. Of the multiple chemical classes of antidepressants, serotonin-selective re-uptake inhibitors (SSRIs) are the most widely used drugs in the treatment of depression. Variations in the SERT gene may contribute to specific behavioral and neuropsychiatric phenotypes, and may underlie individual responses to drugs.

Human serotonin transporter

Lesch *et al.* (1996) reported that a short-length variant, in the promoter region of the SERT gene, reduces the transcriptional efficiency of the SERT gene promoter, resulting in decreased SERT expression and serotonin transport. The short allele accounted for 3 to 4% of total variation and 7 to 9% of inherited variance for anxiety-related personality traits. This finding launched many efforts to uncover a relationship between long and short alleles of the SERT gene and SERT-related neuropsychiatric disorders (Figure 2.5). Long and short variants of the promoter region of the SERT gene are readily discernible with PCR amplification of genomic DNA samples, followed by size fractionation of the PCR products on agarose gels. With this level of analysis, allelic variations in the promoter region of the SERT gene have been associated with anxiety, depression, aggression-related personality traits and affective disorders in some, but not all, human studies (reviewed in Veenstra-VanderWeele *et al.*, 2000).

Rhesus monkey serotonin transporter gene

The close evolutionary proximity of monkeys and humans is reflected by the observation that rhesus monkeys have a similar length polymorphism in the promoter region of the SERT gene to humans

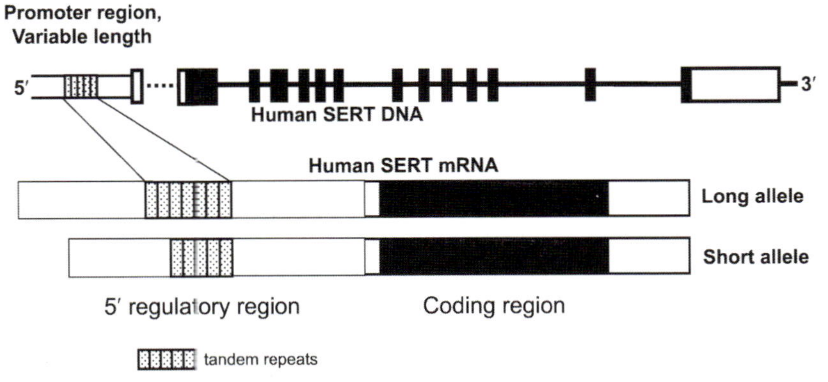

Figure 2.5 Schematic representation of the human serotonin transporter (SERT) DNA, that encodes a protein which is a primary target of antidepressants in brain. mRNAs derived from *long* and *short* alleles illustrate the position of a region of the SERT gene reported to regulate SERT protein expression. Black boxes depict exons that make up the coding region, empty boxes depict non-coding exons, and the stippled box indicates a polymorphic promoter region of variable length, that gives rise to mRNAs with different 5′ regulatory regions and different size. Long and short alleles are also present in rhesus monkeys.

(Trefilov *et al.*, 1999, 2000). Thus, rhesus monkeys provide an excellent model for deciphering genotype/phenotype relationships between SERT gene polymorphisms and behavioral, pharmacological, environmental and endocrine parameters. This is illustrated in a recent study by Bennett *et al.* (2002), in which rhesus monkeys, with deleterious early rearing experiences, were differentiated, by genotype, in cerebrospinal fluid concentrations of the 5-HT metabolite, 5-hydroxyindoleacetic acid, while monkeys reared normally were not. The vast majority of association studies between monoamine transporter gene polymorphisms and monoamine-related phenotypes have focused on segregating genotypes according to length of polymorphisms, disregarding SNPs. With regard to the SERT polymorphism, Nakamura *et al.* (2000) examined the human SERT polymorphism in detail and identified ten novel sequence variants, concluding that the alleles reported as short and long can be divided into four and six kinds of allelic variant, respectively. We would speculate that SNPs impart a differential function in a haplotype-dependent manner. Again, discordance between studies that report associations, or lack of association, between polymorphic length variants and a phenotypic parameter may be explained by this variance in genetic structure. Accordingly, sequence-defined alleles are mandated for a consistent assessment of phenotypic associations across laboratories.

Conclusion

Although relatively little of the rhesus monkey genome has been sequenced, it is probable that the close genetic proximity to human will underlie some degree of conservation of specific polymorphisms shared by humans and rhesus monkeys. An example is the tandem repeat region in the 3′-UTR of the DAT gene, present in both humans and rhesus monkeys, but absent in rodents. The studies, described above, demonstrate that polymorphisms of length or of single nucleotides in genes may contribute to the dynamic processes that regulate protein expression in the brain. Consequently, association studies that group together multiple haplotypes, solely on the basis of the number of repeat sequences (*length*) within the 5′ or 3′-UTR, may obliterate the relevance of other sequence variations which may be involved in gene regulation. An emerging concept is that polymorphisms, in the non-coding regions of transporter and receptor genes, should be investigated as contributors to the etiology of neuropsychiatric disorders, neurodegenerative diseases and susceptibility to drug addiction. For the mu-opioid receptor gene and others, association studies that correlate variation at a single SNP with defined phenotypic variation, should consider that a single SNP will define two heterogeneous groups of haplotypes that differ at the two alleles of the SNP and at alleles at other SNPs. Although it is likely that many CNS disorders are polygenic in nature, detailed analysis of each gene implicated will reveal a comprehensive view of their etiology and expand the range of diagnostic and therapeutic targets for these disorders.

The genetic similarity and common polymorphisms of humans and rhesus monkeys may be exploitable for developing effective and "naturalistic" models to explore, in non-human primates, associations between genotype and phenotype that are of relevance to humans. There will undoubtedly be, as has been found in the mu-opioid receptor gene, specific SNPs or other polymorphisms not shared by humans and rhesus monkeys. The functional consequences of a spontaneously occurring, but non-identical, SNP may be conserved between the species. For specific genes, rhesus monkeys may share a physiogenetic similarity to humans that can transcend identical genomic and proteomic structure. More generally, genotype/phenotype relationships, in closely related species, may not depend solely on a particular specific polymorphism that is shared by the two species, but rather on parallel effects of different polymorphisms on function. Non-human primates are promising for developing models to investigate the physiological and pathological consequences of genetic polymorphisms of relevance to humans.

Correspondence

Any correspondence should be directed to Gregory Miller, The New England Primate Research Center, Harvard Medical School, Southborough, MA 01772-9021, USA. Email: gmiller@hms.harvard.edu

References

Bannon, M.J., Poosch, M.S., Xia, Y., Goebel, D.J., Cassin, B. and Kapatos, G. (1992). *Proc. Natl. Acad. Sci. USA*. 89, 7095–7099.

Becker, A., Schroder, H., Brosz, M., Grecksch, G. and Schneider-Stock, R. (1997). *Pharmacol. Biochem. Behav.* 58, 763–766.

Bennett, A.J., Lesch, K.P., Heils, A., Long, J.C., Lorenz, J.G., Shoaf, S.E., Champoux, M., Suomi, S.J., Linnoila, M.V. and Higley, J.D. (2002). *Mol. Psychiatry* 7(1), 118–122.

Benton, D. (1985). *Pharmacol. Biochem. Behav.* 23, 871–876.

Bond, C., LaForge, K.S., Tian, M., Melia, D., Zhang, S. and Borg, L. (1998). *Proc. Natl. Acad. Sci. USA.* 95, 9608–9613.

Cheon, K.A., Ryu, Y.H., Kim, Y.K., Namkoong, K., Kim, C.H. and Lee, J.D. (2003). *Eur. J. Nucl. Med. Mol. Imaging* 30, 306–311.

Cheon, K.A., Ryu, Y.H., Namkoong, K., Kim, C.H., Kim, J.J. and Lee, J.D. (2004). *Psychiatry Res.* 130, 85–95.

Comings, D.E., Gade-Andavolu, R., Gonzalez, N., Wu, S., Muhleman, D., Blake, H., Chiu, F., Wang, E., Farwell, K., Darakjy, S., Baker, R., Dietz, G., Saucier, G. and MacMurray, J.P. (2000a) *Clin. Genet.* 58(1), 31–40.

Comings, D.E., Gade-Andavolu, R., Gonzalez, N., Wu, S., Muhleman, D., Blake, H., Mann, M.B., Dietz, G., Saucier, G. and MacMurray, J.P. (2000b). *Clin. Genet.* 58(5), 375–385.

Dougherty, D.D., Bonab, A.A., Spencer, T.J., Rauch, S.L., Madras, B.K. and Fischman, A.J. (1999). *Lancet* 354, 2132–2133.

Dresel, S., Krause, J., Krause, K.H., LaFougere, C., Brinkbaumer, K., Kung, H.F., Hahn, K. and Tatsch, K. (2000). *Eur. J. Nucl. Med.* 27(10), 1518–1524.

Frost, J.J., Rosier, A.J., Reich, S.G., Smith, J.S., Ehlers, M.D. and Snyder, S.H. (1993). *Ann. Neurol.* 34, 423–431.

Grosch, S., Niederberger, E., Lotsch, J., Skarke, C. and Geisslinger, G. (2001). *Br. J. Clin. Pharmacol.* 52, 711–714.

Gwynn, G.J. and Domino, E.F. (1984). *J. Pharmacol. Exp. Ther.* 231, 306–311.

Heinz, A., Goldman, D., Jones, D.W., Palmour, R., Hommer, D., Gorey, J.G., Lee, K.S., Linnoila, M. and Weinberger, D.R. (2000). *Neuropsychopharmacology* 22(2), 133–139.

Hoehe, M.R., Kopke, K., Wendel, B., Rohde, K., Flachmeier, C., Kidd, K.K., Berrettini, W.H. and Church, G.M. (2000). *Hum. Mol. Genet.* 9(19), 2895–2908.

Jacobsen, L.K., Staley, J.K., Zoghbi, S.S., Seibyl, J.P., Kosten, T.R., Innis, R.B. and Gelernter, J. (2000). *Am. J. Psychiatry* 157(10), 1700–1703.

Kalin, N.H., Larson, C., Shelton, S.E. and Davidson, R.J. (1998). *Behav. Neurosci.* 112, 286–292.

Kalin, N.H. (1999). *J. Clin. Psychiatry* 60, Suppl 15, 29-32.

Krause, K.H., Dresel, S.H., Krause, J., la Fougere, C. and Ackenheil, M. (2003). *Neurosci. Biobehav. Rev.* 27, 605–613.

Kreek, M.J. (1996). *Neurochem. Res.* 21, 1469–1488.

Lesch, K.P., Bengel, D., Heils, A., Sabol, S.Z., Greenberg, B.D., Petri, S., Benjamin, J., Muller, C.R., Hamer, D.H. and Murphy, D.L. (1996). *Science* 274(5292), 1527–1531.

Madras, B.K., Miller, G.M. and Fischman, A.J. (2002). *Behav. Brain. Res.* 130, 57–63.

Malison, R.T., McDougle, C.J., van Dyck, C.H., Scahill, L., Baldwin, R.M., Seibyl, P., Price, L.H., Leckman, J.F. and Innis, R.B. (1995). *Am. J. Psychiatry* 152, 1359–1361.

Martinez, D., Gelernter, J., Abi-Dargham, A., van Dyck, C.H., Kegeles, L., Innis, R.B. and Laruelle, M. (2001). *Neuropsychopharmacology* 24(5), 553–560.

McBurnett, K., Lahey, B.B., Rathouz, P.J. and Loeber, R. (2000). *Arch. Gen. Psychiatry* 57, 38–43.

Miller, G.M., Bendor, J., Tiefenbacher, S., Yang, D., Novak, M. and Madras, B.K. (2004). *Mol. Psychiatry* 9, 99–108.

Miller, G.M., de la Garza, R. 2nd, Novak, M.A. and Madras, B.K. (2001). *Mol. Psychiatry* 6, 50–58.

Miller, G.M. and Madras, B.K. (2002). *Mol. Psychiatry* 7, 44–55.

Nakamura, M., Ueno, S., Sano, A. and Tanabe, H. (2000). *Mol. Psychiatry* 5(1), 32–38.

Pajer, K., Gardner, W., Rubin, R.T., Perel, J. and Neal, S. (2001). *Arch. Gen. Psychiatry* 58, 297–302.

Rubie, C., Schmidt, F., Knapp, M., Sprandel, J., Wiegand, C., Meyer, J., Jungkunz, G., Riederer, P. and Stober, G. (2001). *Neurosci. Lett.* 297(2), 125–128.

Trefilov, A., Berard, J., Krawczak, M. and Schmidtke, J. (2000). *Behav. Genet.* 30(4), 295–301.

Trefilov, A., Krawczak, M., Berard, J. and Schmidtke, J. (1999). *Electrophoresis* 20(8), 1771–1777.

Vandenbergh, D.J., Persico, A.M., Hawkins, A.L., Griffin, C.A., Li, X. and Jabs, E.W. (1992). *Genomics* 14, 1104–1106.

van der Weyden, L., Adams, D.J. and Bradley, A. (2002). *Physiol. Genomics* 11(3), 133–164.

van Dyck, C.H., Quinlan, D.M., Cretella, L.M., Staley, J.K., Malison, R.T., Baldwin, R.M., Seibyl, J.P. and Innis, R.B. (2002). *Am. J. Psychiatry* 159(2), 309–312.

Veenstra-VanderWeele, J., Anderson, G.M. and Cook, E.H. Jr. (2000). *Eur. J. Pharmacol.* 27, 410(2-3), 165–181.

Waldman, I.D., Rowe, D.C., Abramowitz, A., Kozel, S.T., Mohr, J.H. and Sherman, S.L. (1998). *Am. J. Hum. Genet.* 63, 1767–1776.

Wand, G.S., McCaul, M., Yang, X., Reynolds, J., Gotjen, D. and Lee, S. (2002). *Neuropsychopharmacology* 26, 106–114.

CHAPTER 3

General Anatomy

Laurie R. Godfrey
Department of Anthropology, University of Massachusetts, Amherst, MA 01003, USA

Introduction: Primates as a clade

The Order Primates is a clade – an inclusive group of descendants of a single ancestor whose shared characteristics derive at least in part from that ancestry. Any single species displays a unique combination of "primitive" or "plesiomorphic" (inherited from a remote ancestor) traits and "derived" ("apomorphic") or shared derived ("synapomorphic") traits (inherited from a nearer ancestor, and shared only with relatives descended from that ancestor). Much research has been directed at deciphering traits that are synapomorphic for the Order Primates (see Chapter 1 by Colin Groves, this volume), as these might be considered "defining" primate characteristics. It is important to remember, however, that species are included or excluded from the Order Primates, not on the basis of the presence or absence of any single "primate" characteristic, but on the basis of *inferred close relationship* to other primates. Whereas the latter is based on an analysis of the distribution of molecular and gross morphological traits in mammals, not all similarities are evidence of close relationship. Trait similarities signal close relationship only when they are not the result of either convergent evolution or inheritance from a remote ancestor.

There is a tremendous range of morphological variation within the Order Primates. This stems from the great antiquity of primate origins and numerous subsequent adaptive radiations, each with a unique set of shared derived traits. Although some traits, listed below, such as raised papillary ridges or "fingerprints," are manifested in nonprimates such as tree shrews and other mammals, and must reflect more remote common ancestry or independent derivation: a peculiar set of morphological characteristics may qualify as primate synapomorphies: (1) the housing of the middle and inner ear is petrosal in origin; (2) the orbits (or eye sockets) are convergent and frontated (closely approximated and directed forward rather than laterally) and the eyeballs have lateral bony protection in the form of a postorbital bar (which is expanded in tarsiers, platyrrhines, and catarrhines into a lateral wall that separates the temporal region of the skull from the eye, a condition called postorbital closure); (3) the grasping hands and feet bear flat nails instead of claws, and sensitive tactile pads with a serial arrangement of papillary ridges and grooves (dermatoglyphs, or fingerprints); (4) primate hands and feet are relatively large, and the limbs are long in comparison to a relatively shortened trunk and face, associated with regression of the olfactory apparatus and dental muzzle; (5) the foot has a particular orientation at the point of thrust, its principal fulcrum at the distal end of the mid-tarsal bones instead of at the

The Laboratory Primate
Copyright 2005 Elsevier
ISBN 0-1208-0261-9

distal ends of the metatarsals which is is called tarsi-fulcrumation; (6) the foot has a widely divergent big toe (or hallux). These characters relate to vision, feeding, locomotion, and manipulation or prehension. There is a large literature on their adaptive significance; interested readers might compare the "arboreal theory of primate evolution" of Clark (1963); Cartmill's (1972) "theory of nocturnal visual predation" on solitary flying insects; and Sussman and Raven's (1978) "theory of terminal branch omnivory and primate/angiosperm coevolution." Martin (1990) provides a useful discussion of probable primate synapomorphies.

Of at least equal, if not greater, importance are primate characteristics that relate to primate life histories and social interactions. Primates tend to have relatively large brains with some unique sulcal patterns, low fecundities, slow reproductive turnover, long gestation periods, small litters of precocial, or relatively well-developed, newborns; long maturation periods, long life spans, low instantaneous adult mortality, and other characteristics of growth and development (as, for example, generally slow dental development and prolonged epiphyseal fusion) that are related, at least in part, to their having "slow" life histories (Martin, 2003). Due to their slow life histories, primates tend to form large, semi-closed (or cohesive) social groups comprising multiple generations (Stearns *et al.*, 2003). Rarely do social groups comprising three generations exist outside the Order Primates. Primates are thus social animals and even the so-called "solitary" primate species maintain social networks throughout the year and outside the context of mating (Muller and Thalmann, 2000). Most of the "solitary" are nocturnal (active at night). The exception is the orangutan (*Pongo pygmaeus*), whose dietary preferences (ripe fruit) and requirements (large quantities) apparently constrain foraging party size. Young orangutans, however, spend many years with their mothers, before dispersing, and adult orangutans, in the wild, maintain social contacts through long calls and appear to have large social networks (van Schaik and van Hoof, 1996). It is not surprising that unrelated adult orangutans can form strong social bonds with multiple individuals in captivity. Even nocturnal, "solitary" strepsirrhines can be surprisingly gregarious (Nekaris, 2003). Nocturnal primate species have diverse patterns of social organization, with varying degrees of communication and associations among adult males and females. They may disperse only to feed. A recent study of gray mouse lemurs (*Microcebus murinus*) revealed sleeping clusters of as many as 16 adult and subadult individuals, including multiple males and females in this "solitary" species (Rasoazanabary, submitted).

Most of the nocturnal primate species are strepsirrhines. All bushbabies and lorises, as well as some lemurs (all cheirogaleids, sportive lemurs, and woolly lemurs) are nocturnal. Other lemurs, including brown, and mongoose lemurs, are cathemeral (active day and night), or diurnal (active only during the day, including ringtails, ruffed lemurs, indris or babakotos, and sifakas). Among haplorhines, only tarsiers (*Tarsius*) and owl monkeys (*Aotus*) are nocturnal. All other haplorhines are diurnal although most are well adapted to moderately low light conditions, as may occur in closed canopies.

Strepsirrhines tend to be hypometabolic, expending less energy than one might expect on the basis of their body size, and some, such as *Lepilemur*, the sportive lemur, are energy minimizers par excellence. Some small-bodied, nocturnal strepsirrhines (*Cheirogaleus*) hibernate during the dry season, when resources are limited. Such hibernation can last over six months! Others (*Microcebus*) exhibit differing degrees of torpor by sex because females "hibernate" for several months but males do not. Both sexes can nevertheless experience marked lowering of internal body temperature in the early morning and, even when not hibernating, they enter what has been called "daily" torpor. Most strepsirrhine species, however, neither hibernate nor experience daily torpor and the same is, of course, true of haplorhines.

Several excellent general reviews of primate anatomy are available. These include Ankel-Simons (2000), Fleagle (1999), and Martin (1990).

The musculoskeletal system

Nonhuman primates vary in body size from under 30 g, as in some mouse lemurs, to about 150–200 kg as in normal, non-obese, male gorillas who reach 200 kg. Most people think of strepsirrhines as small-bodied, and haplorhines as large-bodied, primates. In fact, both of these suborders have small and large representatives (Table 3.1). Mouse lemurs and dwarf bushbabies (among strepsirrhines) have adult body masses of 100 g or lower while haplorhine pygmy tarsiers and pygmy marmosets are the same size. Numerous cercopithecoid and hominoid haplorhine species may exceed 10 kg in body mass such as mandrills, gelada baboons, savanna baboons, some leaf monkeys and even *Brachyteles*, the largest platyrrhine. The largest extant

TABLE 3.1: Features of the musculoskeletal and locomotor systems of commonly used laboratory primates

Genus	Adult female body mass (g)	Body mass dimorphism	Intermembral index	Mode of locomotion
Microcebus murinus	63	0.94	72	Arboreal quadruped; employs rapid scurrying & leaping.
Loris tardigradus	255	0.99	90	Quadrupedal climber.
Nycticebus coucang	626	1.08	88	Quadrupedal slow climber.
Galago senegalensis	195	1.14	52	Vertical clinger & leaper.
Otolemur crassicaudatus	1,110	1.07	70	Quadrupedal climber, occasional leaper.
Saimiri sciureus	662	1.18	80	Arboreal quadruped.
Cebus capuchinus	2,540	1.45	81	Arboreal quadruped.
Aotus trivirgatus	724	1.10	74	Arboreal quadruped.
Callithrix jacchus	287	0.98	76	Arboreal quadruped; some clinging & leaping.
Saguinus oedipus	404	1.03	74	Arboreal quadruped; some clinging & leaping.
Chlorocebus aethiops	2,980	1.43	83	Quadruped.
Papio anubis	11,700	1.89	97	Mainly terrestrial quadruped.
Cercocebus torquatus	5,500	1.72	83	Quadruped.
Macaca mulatta	5,370	1.25	93	Mainly terrestrial quadruped.
M. fascicularis	3,574	1.49	93	Mainly arboreal quadruped.
Homo (Pan) troglodytes	30,000	1.27	106	Knuckle-walker, climber, brachiator.
Homo (Homo) sapiens	42,200	1.14	72	Terrestrial biped.

Notes and sources:

Adult female masses were taken from Smith and Jungers (1997).

Body mass dimorphism is calculated here as average adult male mass divided by average adult female mass. Body masses used in the calculation of the values presented here were taken from Smith and Jungers (1997). When, for any species, these authors reported the results for multiple studies, data from the study with the largest samples were used. The dimorphism value provided for *Microcebus* reflects mass differences around during the breeding season, when male and female body masses are approximately equal.

Intermembral indices are taken from Fleagle (1999).

Rowe (1996) and Fleagle (1999) provide descriptions of and sources for primate locomotor behavior.

lemurs weigh little more than 7 kg but giant lemurs, living in Madagascar during the past couple of millennia, rival the size of gorillas and orangutans (Jungers *et al.*, 2002).

Body mass dimorphism also varies considerably in primates. In most "sexually dimorphic" primate species, males are larger than females, sometimes almost twice the mass (see Table 3.1). The highest levels of body size dimorphism are manifested in the catarrhines (baboons, some macaques, gorillas and orangutans). Relatively low or no sexual dimorphism is common in strepsirrhines and, when it exists, females may be larger than males. A few primate species have what has been called "fluctuating" body mass dimorphism – that is, the degree to which the sexes differ (and indeed the direction of that difference) changes seasonally. In the case of mouse lemurs, *Microcebus murinus*, females are much heavier than males just before they (but not the males) enter a prolonged period of seasonal torpor, or hibernation. Female body mass drops precipitously during hibernation, while male body mass remains roughly stable.

Body mass dimorphism also varies seasonally in squirrel monkeys, *Saimiri*, but, in this case, it is the males that seasonally gain (and then lose) the most weight. During the breeding season, "fatted" males have enlarged torsos, high testosterone levels, high testicular mass and increased spermatogenesis.

The degree of prognathism (forward protrusion) of the dental and olfactory muzzle varies in primates, as do the relative sizes of the brain and eye sockets (see below). Strepsirrhines tend to have smaller brains, longer snouts and relatively larger turbinates (thin, scroll-like bones lined with sensory membranes) in the nasal cavity than do monkeys, apes and humans. Nocturnal primates tend to have relatively large eye sockets, but the relatively largest eye sockets exist in nocturnal haplorhine species, *Tarsius* and *Aotus* (the tarsiers and the night monkeys) that lack certain features of the soft anatomy of the eyes of nocturnal strepshirrhines (see below). The lengthening of the snout of some cercopithecoid monkeys, such as baboons, is related not to an enlargement of the turbinals, but to an elongation of the tooth row and enlargement of the canines, especially in males. It is the relative size of the neocortex of the brain that is particularly enlarged in haplorhines (especially apes and humans; see below).

There are differences in the positions of the holes, or foramina, on the base of the skull for the passage of nerves, arteries, and veins. In strepsirrhines, the "foramen magnum," which is the big hole for the passage of the spinal cord where the skull meets the vertebral column, is directed posteriorly. In tarsiers, it tends to be positioned more centrally under the neurocranium, and the same is the case for most anthropoids. A central positioning of the foramen magnum is not necessarily associated with bipedalism, as is sometimes claimed. The quadrupedal squirrel monkey, *Saimiri*, has a very centrally located foramen magnum. However, strepsirrhine "vertical clingers and leapers" that propel their bodies and land on their hind feet when leaping, and often hop or walk bipedally when on the ground, do not.

Primates have clavicles, and unfused radii and ulnae which allow for pronation and supination, to varying degrees and maximally in hominoids. Most have unfused tibiae and fibulae and five digits on the hand and foot. There is fusion of the distal tibia and fibula in the tarsier (an adaptation for leaping), and reduction of the thumb (pollex) in spider monkeys (*Ateles*), woolly spider monkeys (*Brachyteles*) and African colobines. The second digit of the hand and foot bears a grooming claw, instead of a nail, in all strepsirrhines, and this digit is reduced to varying degrees in lorises. All primates have dermatoglyphs, but some "prehensile tailed" platyrrhine species also have sensitive tactile pads with cutaneous ridges on the ventral surfaces of the tail. It is only these primate species that are capable of hanging by their tails. The tails of these species (*Ateles*, *Brachyteles*, *Lagothrix* and *Alouatta*) are very muscular and especially those of atelines which have well-developed flexor muscles. They have a large opening for the spinal cord at the caudal (tail) end of the sacrum and a well-developed system of arteries and veins supplying the ventral tip of the tail (Ankel-Simons, 2000). Capuchin monkeys (*Cebus*) converge, to some extent, in tail morphology and function with the more specialized spider monkeys (*Ateles*) and their closer relatives (Lemelin, 1995).

Body proportions are extremely variable in primates. One of the more useful metrics for describing variation in skeletal form is the *intermembral index* which does not, however, take into consideration the relative lengths of the hand and foot. This index is calculated as the *length of the (humerus + radius) × 100 divided by the length of the (femur + tibia)*. Primates vary in their values for the Intermembral Index from around 50 or 60 when the long bones of the hindlimb are twice the length of those of the forelimb, as in *Tarsius, Galago, Lepilemur* and *Avahi,* to around 140 or 150 when the long bones of the hindlimb are two-thirds the length of those of the forelimb, as in *Pongo* and *Hylobates syndactylus*. Closely related species can be as variable, in this regard, as the entire primate order. Thus, for example, hominoid intermembral indices range from about 70 in *Homo sapiens* to about 150 in *siamangs*. Values for a single clade of lemurs (the indriid-palaeopropithecid clade, including some recently extinct species) range from around 60 in *Propithecus* and *Avahi*) to around 144 in *Palaeopropithecus*. Arm swinging, or some kind of forelimb suspension, occurs in primates with relatively long forelimbs while those with relatively long hindlimbs tend to employ bipedal postures in hopping, walking or leaping. Species with forelimbs and hindlimbs of roughly equal length usually move on all four limbs. Having long forelimbs, however, does not preclude bipedalism. Gibbons, for example, move bipedally on the ground and on massive horizontal branches, but normally engage in ricochetal arm-swinging, a type of forelimb suspensory locomotion. Conversely, having long hindlimbs does not preclude suspension. Sifakas, for example, regularly use suspension in feeding and occasionally use arm-swinging in locomotion, but they are "vertical clingers and leapers," and hop or walk bipedally when on the ground.

Of obvious importance are the hands and feet, which vary in ray proportions, as well as carpal and tarsal morphology, and structure can differ even among species with superficially similar modes of locomotion. Bushbabies, tarsiers, and indriids are all leapers but only bushbabies and tarsiers have elongated tarsals (rear foot bones). Indriid feet are well adapted for grasping and suspension; the hallux (or big toe) is large and divergent, and the digits are elongated. Two broad categories of hand form exist in the Order Primates: strepsirrhines possess *ectaxonic hands* with relatively long fourth rays, while *mesaxonic hands,* with relatively long third rays characterize haplorhines (Lemelin and Schmitt, 1998). Because the hands and feet directly contact substrates (or superstrates), their morphology might be expected to reflect substrate preferences. Behavioral correlates of morphological variation can be found, although they are often more subtle than might be expected on the basis of simple dichotomous classifications of morphology (see Lemelin and Schmitt, 1998 on the ectaxony/mesaxony dichotomy). Some unusual manual adaptations have more to do with feeding than with locomotion and the slender, clawed third digit of the hand of the aye-aye (*Daubentonia*) is a case in point. Others relate to the movements of a single digit and may reflect a variety of functions. Variation in the morphology of the carpo-metacarpal joint of the thumb results in different capacities for thumb "opposability" (Ankel-Simons, 2000), with rotation of the thumb through circumduction at that joint ("true opposability") occurring only in catarrhines. "Knuckle-walking," in a form of quadrupedal locomotion which is characteristic of gorillas and chimpanzees when moving on the ground, and in which the upper body is supported by the dorsal surfaces of the middle phalanges, also has osteological correlates.

Other unusual skeletal adaptations exist, with associated variation in the soft anatomy. All cercopithecoids and some other catarrhines have ischial callosities or sitting pads covering their expanded ischial tuberosities. These structures facilitate sleeping while sitting on branches or other precarious supports. Male gibbons, and especially male howler monkeys, have enlarged hyoid bones that enable them to produce loud vocalizations.

There is remarkable uniformity of the muscles of primates, particularly when compared to other mammalian groups. Primates exhibit minor variation in the presence or absence of particular muscles. For example, the dorsoepitrochlearis muscle (basically, the fourth head of the "triceps" muscle of the upper arm, originating on the latissimus dorsi) is typically absent in humans but present in nonhuman primates, including apes. A superficial muscle of the forearm and hand, palmaris longus, is present in most primate species, but sometimes missing in hominoids, including humans, and is often missing in gorillas. More significant variation exists across primate species in the relative sizes and configuration of muscles. For example, in conjunction with their different styles of leaping, indriids have hip extensors that are larger than their ankle plantarflexors, whereas the reverse is true of bushbabies and tarsiers (Demes *et al.*, 1998). Humans differ from nonhuman primates in having a fully differentiated long flexor of the thumb in contrast to a long pollical flexor which exists as part of the deep flexor stratum (but not as a separate muscle) in most nonhuman primates. Swindler and Wood (1973) provide an excellent comparison of the musculoskeletal systems of humans, chimpanzees, and baboons, and there are detailed descriptions of macaque anatomy (see Hartman and Straus, 1971).

Primates are sometimes described as "hindlimb dominated" or "hindlimb driven," which is considered preadaptive to human bipedalism, in comparison to "forelimb driven" nonprimates (Kimura *et al.*, 1979). Kimura and colleagues argued that primates rely, to a greater degree than nonprimates, on their hindlimbs for both propulsion and support. Demes *et al.* (1994) demonstrated, however, that primates are not unique among mammals in their use of the limbs in propulsion. At most one can say that, due to structural differences in body mass distribution between primates and nonprimates, the former do tend to place larger *vertical* (but not propulsive) forces on their hindlimbs.

Primates are also believed to use relatively larger hindlimb excursions than similarly-sized nonprimates. Experimental evidence for such differences exists, although primates and nonprimates are not as distinct in this regard as some have maintained (Larson *et al.*, 2001). Differences in the long bone geometry of primates and nonprimates of similar size have been noted and related to differences in the "complexity" of arboreal vs. terrestrial habitats (see Kimura, 1995). Primate long bones have greater cross-sectional strength than those of carnivorans and rodents of similar body mass. However, primates also have relatively longer, more gracile, limb bones than similarly-sized nonprimates, and the interaction of bone length and cross-sectional geometry suggests no consistent differences in their resistance to bending forces (Polk *et al.*, 2000).

The dentition

Primates, like other mammals, have two sets of teeth: a primary dentition (comprising all "milk" or deciduous teeth plus the permanent molars) and a replacement (or secondary) dentition. The teeth are *heterodont,* their form varying in association with varying functions such as cutting, puncturing and grinding. Different tooth types are named, as follows, from the front to the back of the mouth: incisors (I), canines (C), premolars (P) and molars (M). Primates also exhibit diphyodonty, i.e., a single set of replacement teeth such that the *deciduous* (or milk) incisors, canines, and premolars have a single set of permanent replacements. None of the so-called "permanent," or adult, teeth have replacements. The molars, which erupt behind the deciduous teeth, have no deciduous precursors. The molars thus belong to both the primary and the permanent dentitions and are neither shed nor replaced.

Stages of dental development can be useful in aging individuals. Each species has a characteristic developmental pattern, with variation, and the teeth form and erupt in a particular sequence and with a particular schedule. Dental crowns always form, from the cusps to the cervix or base of the crown, before root extension begins. Crown formation times vary considerably among primates, as do eruption sequences and schedules. Prior to replacement of the milk dentition by the permanent teeth, the roots of the deciduous teeth are resorbed and the remaining crowns of the deciduous teeth are pushed out by their replacements. Generally, occluding pairs of teeth (uppers and lowers) erupt more or less simultaneously. Dental eruption occurs as the roots grow. Radiographs of unerupted teeth will thus reveal partially or fully formed crowns without roots, or with roots in various stages of development. In many, but not all, primate species, the anterior molars are the first of the permanent teeth to erupt. They erupt before any of the deciduous teeth are replaced, often in association with weaning. Thus, the age at first molar eruption is, in many species, a good life history marker, distinguishing species with "fast" life histories, or rapid reproductive turnover, from those with "slow" life histories. Typically, also, the incisors erupt next as in *Saimiri, Callithrix,* macaques, baboons and apes. The third molars erupt last in humans and many anthropoid primates but, once again, this is not universal for primates. In species with strong dental dimorphism, there may be marked "bimaturism," with males maturing considerably later than females. This will affect dental eruption ages, particularly for the permanent canines.

Teeth preserve an indelible record of their own development, which can be "read" in their enamel microstructure. Disruption of enamel matrix formation causes enamel defects, or "hypoplasias," to occur. Once the crowns are fully formed, they cannot change shape except through breakage and wear. The teeth of different species vary in relative enamel thickness and in enamel prism structure and this variation may affect the functional lifetime of the teeth. In some species, it is normal, even in the wild, for the crowns of cheek teeth to wear down to the roots during the lifetime of the individual. As the enamel wears and dentine is exposed, the primary dentine is replaced with secondary dentine, which helps to prolong the functional lifetime of the teeth.

There is some variation in the "dental formulas" of primates (i.e., the number of each tooth type in each quadrant – upper and lower, right and left). Dental formulas are recorded separately for the milk and permanent dentitions, from the front to the back of the mouth. A dental formula of 2-1-3-2/2-1-3-2 signals the presence of two maxillary incisors, one maxillary canine, three maxillary premolars, and two maxillary molars on each side (right and left), and the same for the mandibular dentition. Such a dental formula, for example, is typical of marmosets and tamarins. Extant primates have a total of 18 to 36 permanent teeth. In most primate species, all of the deciduous teeth have replacements, but this is not universal. For example, sifakas (genus *Propithecus*) have deciduous mandibular teeth (the canine and one premolar) that are vestigial and unreplaced, and sportive lemurs (genus *Lepilemur*) have unreplaced deciduous upper incisors.

Many clades of primates show little variation in dental formulas (Table 3.2). All extant catarrhines, including humans, have two incisors, one canine, two premolars, and three molars in each quadrant. No primate has more than two incisors in each quadrant (reduced from the primitive eutherian mammal formula of three), and none has more than three premolars (reduced from the primitive eutherian mammal formula of four). No primate species has fewer than one lower incisor.

Swindler (2002) provides detailed descriptions of the morphology of the teeth including the number and form of the cusps, valleys, and crests of individual primate species, as well as their metrics (mesiodistal and buccolingual diameters – i.e., lengths and widths). Despite the marked variation he describes, the cheek teeth of primates can be said to be relatively unspecialized

TABLE 3.2: Features of the dentition of commonly used laboratory primates

Taxon	Dental formula (incisors, canines, premolars, and molars in upper and lower quadrants)	Age at M1 eruption in years (maxilla/mandible)	Maxillary and mandibular canine dimorphism
Cheirogaleus medius	2-1-3-3/2-1-3-3	0.07/0.07	–
Lemur catta	2-1-3-3/2-1-3-3	0.33/0.34	1.05/-
Varecia variegata	2-1-3-3/2-1-3-3	0.50/0.48	0.96/-
Propithecus verreauxi	2-1-2-3/2-0-2-3	0.26/0.22	0.96/-
Galago senegalensis	2-1-3-3/2-1-3-3	-/0.10	–
Saimiri sciureus	2-1-3-3/2-1-3-3	0.40/0.37	1.41/1.55
Cebus apella	2-1-3-3/2-1-3-3	1.28/1.15	1.41/1.42
Aotus trivirgatus	2-1-3-3/2-1-3-3	0.40/0.36	1.08/1.06
Callithrix jacchus	2-1-3-2/2-1-3-2	0.31/0.31	1.03/1.07
Saguinus fuscicollis	2-1-3-2/2-1-3-2	0.39/0.38	1.03/0.98
Chlorocebus aethiops	2-1-2-3/2-1-2-3	0.88/0.83	1.80/1.63
*Papio anubis**	2-1-2-3/2-1-2-3	1.63/1.63	2.22/1.81
Cercocebus torquatus	2-1-2-3/2-1-2-3	–	2.46/2.00
*Macaca mulatta**	2-1-2-3/2-1-2-3	1.44/1.32	2.07/1.67
*M. fascicularis**	2-1-2-3/2-1-2-3	1.75/1.50	2.25/1.60
*Homo (Pan) troglodytes**	2-1-2-3/2-1-2-3	3.27/3.19	1.43/1.26
*Homo (Homo) sapiens**	2-1-2-3/2-1-2-3	6.35/6.15	1.08/1.10

Notes and sources:

Data for ages at first molar eruption from Smith *et al.* (1994). *Eruption ages are means for females only in asterisked taxa. Canine dimorphism values (measured as male permanent canine crown height divided by female permanent canine crown height) are taken from Plavcan and van Schaik (1997) for anthropoids and Godfrey *et al.* (2002) for lemurs.

when compared to those of nonprimates. Nevertheless, a large literature attempts to relate variation in tooth crown morphology and relative size to variation in diet and other oral functions. Fruit-eaters (frugivores) tend to have relatively larger incisors and molars, with low, rounded "cusps" or bumps, whereas leaf-eaters (folivores) tend to have relatively smaller incisors and molars with sharp, pointy cusps and elongated shearing crests. Insectivorous primates also have pointy-cusped molars but they are much smaller in body size than are typical folivores. One of the more useful metrics for the analysis of dietary function in teeth is Kay's (1975) shearing quotient ("SQ"), which measures the relative crestiness of teeth. This index clearly distinguishes specialized folivores from frugivores.

It is the anterior dentition that tends to vary the most in conjunction with functions unrelated to food acquisition. In lemurs and lorises, the mandibular incisors and canines (when present) are elongated, tilted forward and laterally flattened to form a "tooth comb," used in grooming as well as feeding. The anterior-most mandibular premolar is "caniniform" (shaped like a canine) and it assumes the canine's function.

Variation in mating systems, male and female agonism and other aspects of social behavior, also impact the relative size and form of the anterior teeth (Table 3.2). While the deciduous canines of males and females differ little, even in highly sexually dimorphic species, primate species vary significantly in the degree to which the permanent upper and lower canines (the weapon teeth) of males and females differ in size and shape. Canine dimorphism, which is usually measured as the ratio of male to female canine crown height, correlates well with body mass dimorphism. It is high in some species, especially the larger, semiterrestrial catarrhines, such as *Papio*, the baboon, or *Mandrillus*, the mandrill, and low or nonexistent in others, such as *Hylobates*, the gibbon; *Aotus*, the owl or night monkey; and all lemurs and lorises. In some species with low canine dimorphism, such as gibbons, both males and females have long, dagger-like canines. In others, such as *Indri*, the babakoto, the canines are short in both sexes. Much has been written about the significance of varying degrees of canine dimorphism in primate species (see, for example, Plavcan and van Schaik, 1997; see review by Plavcan, 2001). The degree of canine dimorphism is particularly informative when considered in conjunction with relative canine size, such as whether the canines of males and females are large or small when benchmarked against some other measure of dental size (Plavcan *et al.,* 1995).

Mean ages for when the tips of the cusps emerge beyond the gum line, i.e. gingival eruption, for the anteriormost upper and lower permanent molars of commonly used laboratory primates, are provided in Table 3.2, along with the dental formulas and mean canine dimorphism for these species. Smith *et al.* (1994) provide dental eruption timing for these and other primate species. They give eruption schedules for the deciduous as well as the permanent teeth.

The digestive system

The generally unspecialized anatomy of the digestive tract of primates suggests that the basal primates ate both plants and animals. Most primate species are mixed feeders that consume large amounts of fruit or other reproductive parts of plants, supplemented by leaves, seeds, insects or other animal matter, for protein. Frugivory, coupled with low-to-moderate supplementary food consumption, does not require unusual gut specializations. Fruits are rich in sugar that is easily digested and the starches of other reproductive parts of plants, such as tubers, are easily converted into sugar during digestion.

Some primate species consume foods that necessitate special processing, however. There is a phylogenetic component to such dietary specializations in that they tend to characterize subclades. Colobines are tree-foliage consumers that consume varying amounts of seeds (Kirkpatrick, 1999). Baboons may include some grass in their diets, and the gelada baboon (*Theropithecus gelada*) specializes on grass consumption (Jolly, 1970). All of the members of the genus *Hapalemur* consume bamboo and other grasses. African and Asian apes tend to prefer fruit. Gorillas are often called "folivores" but they are actually herbivores that prefer fruit when available, and they supplement fruit with terrestrial and/or aquatic herbs as needed (Doran and McNeilage, 1998). Some nonhuman primate species have special adaptations of the masticatory (or chewing) apparatus and cranial architecture for processing fruits with hard pericarps (outer shells) or hard seeds. They tend to have thick molar enamel with a prism structure that helps to prevent enamel crack propagation. Pitheciins (e.g., *Cacajao*, *Pithecia*, *Chiropotes*), some capuchins (*Cebus apella*), and aye-ayes (*Daubentonia*) are hard-object specialists. Some apes (*Pongo*, the orangutan) will also process hard foods.

The mucous membrane covering the hard palate has palatal cross-ridges (rugae) that function in the holding of food items in the mouth. These vary in number and structure across primates being quite variable and reduced in number in humans, who possess no more than four ridges. Cross-ridges are far more numerous in other taxa with up to ten or 11 in tarsiers and typically well-developed in strepsirrhines (Ankel-Simons, 2000).

Differences in taste sensitivities are grounded in differences in the number and type of taste receptors on the surface, sides, and back of the tongue (Ankel-Simons, 2000). Large salivary glands are located to the sides and under the tongue. There is variation across primates in tongue form. For example, red-bellied lemurs (*Eulemur rubriventer*) have a brush-like tip on their tongue, that other lemurs lack. It is used in licking nectar from flowers, without damaging the reproductive parts of the flower; such that red-bellied lemurs are known to be effective pollinators (Overdorff, 1992). Most notably, a sublingua (or "undertongue") occurs in lemurs and lorises. This is a keratinized (or cartilaginous) structure with a serrated edge and thickened medial axis and is, effectively, a second tongue. It is used to clean food and hair debris from between the teeth of the tooth comb. Most other primates have only a median sublingual fold of the mucous membrane, although a small, weakly serrated sublingua exists in marmosets and tarsiers.

Other feeding specializations include the cheek pouches of cercopithecine monkeys (baboons, macaques, vervet monkeys, etc.). These so-called "cheek pouched monkeys" have muscular pockets in the outer membrane of their oral cavities. Using their cheek pouches for temporary storage, these monkeys can gather food quickly in open, unprotected areas, and retreat to safety before processing it. Food-filled pouches appear as bulges in the neck.

Lambert (1998) provides an excellent summary of variations in the digestive system of primates. There are family-specific differences in the form of many digestive organs. The liver, for example, has multiple lobes in strepsirrhines and typically fewer in haplorhines. The shape of the spleen varies by family. There are four main components of the gastrointestinal tract: (1) the stomach, (2) the small intestine, (3) the cecum, and (4) the colon (Chivers and Hladik, 1980; Martin, 1990). The stomach, which plays a key role in digestion, is divided, in humans and other primates, into a fundus, a body, and a pylorus. A sphincter muscle closes the pyloric portion of the stomach, separating it from the small intestine. The small intestine, which functions in digestion and resorption, is also divided into the duodenum, jejunum, and ilium, the latter of which connects to the large intestine. At the point where the small intestine ends and the large intestine begins is a blind pouch, or cecum which terminates, in humans, in a "vermiform" or worm-shaped appendix. The large intestine (or colon) has excretory as well as digestive functions.

Digestive specializations are reflected in the relative proportions of the four major components of the gastrointestinal tract. Leaves, stems, bark, and gums have long-chain carbohydrates that require bacterial decomposition or fermentation. Such fermentation can take place in different parts of the digestive tract, and folivores have digestive tracts dominated by fermentation chambers in the stomach, cecum, or large intestine (colon). Colobines are *foregut fermenters* with enlarged, "sacculated" (compartmentalized) stomachs. The colobine stomach is divided into a large, distended and sacculated "presaccus" and "saccus"; following these is a long gastric tube and a short pylorus (see Hill, 1958).

Strepsirrhine tree-foliage consumers (*Lepilemur*, *Avahi*, and *Indri*) are *cecal or colon fermenters* with enlarged hindguts. The colon is long and coiled into an "ansa coli." These strepsirrhines are notoriously difficult to maintain in captivity. *Propithecus* has similar digestive adaptations, but it consumes more fruit and seeds, and more easily survives in captivity. Howler monkeys (*Alouatta*), like indriids and unlike colobines, have hindgut fermentation but their stomachs are also enlarged and complex in comparison to those of other New World monkeys (Chivers and Hladik, 1980).

Species that depend on animal and other energy-rich foods, particularly capuchin monkeys and humans, among the primates possess gastrointestinal tracts that converge with each other and with those of faunivorous mammals outside the Order Primates. Faunivores tend to have a simple, globular stomach, a greatly twisted small intestine, a short cecum, and a smooth-walled colon. Their gastrointestinal tracts are dominated by the small intestine and the cecum is particularly reduced relative to body size. The colon is also relatively short. Animal matter is quite easily digested and a relatively short, simple gut is adequate for the task.

The brain

Primates are "intelligent" mammals with relatively large brains. Brain size *per se* is a poor measure of intelligence, however, because the relationship of brain size to cognitive capacities is complex, and because many other variables influence brain size. Several indices of "braininess" or encephalization have been devised, and each has its champions. One such "Index of Cranial Capacity" (or "Encephalization Quotient") assesses the relation of the "observed" cranial capacity to an "expected" cranial capacity derived from a brain/body regression analysis for basal insectivores (Martin, 1990). Index values reveal how much larger (or smaller) a species' brain is than what might be "expected" of a like-sized member of the reference population (in this case, basal insectivores). Comparative research confirms that, in general, strepsirrhines and other primates do have larger brains than like-sized basal insectivores (but not than some species of tree shrews). However, within the Order Primates, this index of cranial capacity is remarkably variable (Table 3.3). Values for strepsirrhines range from under 3 to over 6; those for platyrrhines range from just over 4 to under 12; and those for catarrhines, with the exception of humans, have a similar range (though with none over 10). This array bears little apparent relation to social problem-solving or cognitive skills, as understood by specialists studying comparative primate cognition. Despite their impressive problem-solving skills, gorillas have an Index of Cranial Capacity (5.2) lower than that of the aye-aye, *Daubentonia* (6.3), and less than half that of the brown capuchin monkey, *Cebus apella* (11.7).

Of course, the values of indices of cranial capacity devised through regression analysis will depend on the reference sample used to derive "expected" cranial capacities. This means that they are not properties of the species being measured, *per se*. The utility of encephalization quotients can also be questioned because, regardless of reference population, these indices fail to distinguish among different brain parts with different functions. A simple Neocortex Ratio (calculated as neocortex volume divided by the volume of the rest of the brain tissue) may better reflect the capacity for solving problems than any regression-based index of cranial capacity. The neocortex of the brain is relatively larger in haplorhines than in strepsirrhines,

TABLE 3.3: Features of the nervous system of commonly used laboratory primates

Taxon	Mean brain size of adult females (unless otherwise indicated) in cubic centimeters	Index of cranial capacity	Neocortex ratio
Microcebus murinus	1.8 (pooled sexes)	3.4	–
Loris tardigradus	6.4 (pooled sexes)	4.4	–
Nycticebus coucang	9.9	3.1	–
Galago senegalensis	3.9	2.8	–
Otolemur garnettii	9.7 (pooled sexes)	–	–
Otolemur crassicaudatus	11.3	2.3	–
Lemur catta	23.4	–	1.18
Daubentonia madagascariensis	45.2	6.3	–
Saimiri sciureus	23.2	7.0	–
Cebus apella	63.1	11.7	2.25
Aotus trivirgatus	16.1	4.8	–
Callithrix jacchus	7.7	6.0	–
Saguinus fuscicollis	8.2	4.3	–
Chlorocebus aethiops	59.2	–	2.17
Papio anubis	158.9	7.3	2.76
Papio cynocephalus	145.5	–	2.68
Macaca mulatta	81.3	8.2	2.60
Macaca fascicularis	62.5	–	–
Macaca arctoides	105.0	–	2.43
Homo (Pan) troglodytes	371.1	8.2	3.22
Homo sapiens	1,228	23.0	4.10

Notes and sources:

Brain sizes are taken primarily from Godfrey *et al.* (2001) and Martin (1990).

Values for the Index of Cranial Capacity are from Martin (1990). The value for *Saguinus* is based on a pooled species sample.

Neocortex ratios are from Kudo and Dunbar (2001).

and it is relatively largest in hominoids among haplorhines (Table 3.3). Kudo and Dunbar (2001) argue that, across primates, the neocortex ratio is correlated with the size of the social network (and thus reflects problem solving *within a social context*). Most haplorhines are more gregarious than strepsirrhines and have larger social networks. However, this hypothesis is difficult to test, because social network size is difficult to quantify in a meaningful manner across species.

Other factors may correlate with brain size in primates, including the quality of the diet and engagement in problem solving unrelated to social interactions, such as "extractive" foraging. Such factors may help to explain the relatively large brains of the "solitary" *Daubentonia madagascariensis*, among strepsirrhines; the tool-using *Cebus apella* among platyrrhines; and, indeed, modern humans among hominoids. All three have high-quality, omnivorous diets. Aiello and Wheeler (1995) discuss human evolutionary brain expansion and its possible relation to dietary change and their "expensive tissue hypothesis" speaks to this issue. (Note also our discussion of the similarities of the digestive tracts of capuchin monkeys and humans, above.)

Reproduction and life history variation

Primate males have a pendulous penis and testes that descend into a scrotum. There is marked variation, across primate species, in the external morphology of the penis and in its internal structures, such as the penis bone or baculum. Well developed in some species, the baculum is missing in an odd assortment of others, including tarsiers, some platyrrhines (*Ateles* and *Brachyteles*), and humans. Most nonhuman primate males have a baculum, although it is very small in many species. Baculum size bears no relationship to body or penis size across primates.

Many aspects of penile anatomy vary in primates. There is color variation as, for example, the bright red penis in male proboscis monkeys (*Nasalis*). There is variation in the length of the external penis (or pars libera) among primates. Among hominoids, for example, gorillas have an exceptionally short pars libera (relative to female size) and chimpanzees have an exceptionally long pars libera (Dahl, 1994). There is also variation in the development and number of penile spines. Dixson (1991) showed that keratinized spines on the glans penis of common marmosets (*Callithrix jacchus*) overlie dermal tactile receptors, and thus increase male stimulation, shortening intromission duration. Penile spines are also thought to increase copulatory stimulation of females and reduce the duration of female sexual receptivity (Stockley, 2002). There is some evidence for a negative correlation between male penile spinosity and the duration of female sexual receptivity among primates. There is no such negative correlation with baculum length (Stockley, 2002).

Primate males have early descent of the paired testes into a pouch, or scrotum, usually located behind the penis. The testes may descend just before or shortly following birth, but, in some species, re-ascend into the abdomen or inguinal canal until puberty. The timing of the final descent of the testes into the scrotum varies in primates: prior to age one year in callitrichids, eighteen months in sifakas, about three years in macaques and four years in baboons. Also variable across species is the relative size of the testes, a phenomenon that has been studied in relation to different forms of mate competition among males. The males of some species, such as *Brachyteles*, the muriqui, display little or no pre-copulatory intrasexual mate competition, but they have enormous testes and are presumed to maximize their reproductive success through "sperm" rather than pre-mating "contest" competition.

The vagina, located below the uterus, opens onto the perineum, where it is surrounded on either side by lips (or labia), and anteriorly by the clitoris. In some platyrrhines the clitoris is large and pendulous (and easily mistaken for a penis) while in other primates, it is small and hooded. Female spider monkeys (*Ateles*), woolly spider monkeys (*Brachyteles*), woolly monkeys (*Lagothrix*), and capuchin monkeys (*Cebus*) have a long, pendulous clitoris. Females may have a baculum as, for example, some strepsirrhine and platyrrhine species, but this is rare across primates.

Strepsirrhines and haplorhines differ in the form of the uterus, the development of certain fetal structures, the placental type and the structure of the maternal-fetal separating membrane. Within these suborders there is less variation in these features (Tables 3.4 and 3.5). When exceptions occur, they are likely to entail tarsiers that are, in some aspects of their reproductive biology, more like lemurs and lorises and, in others, more like platyrrhines and catarrhines and, in yet others, unique among primates. Strepsirrhines and tarsiers have a *bicornate* uterus (two distinct ducts or horns), whereas haplorhines, with the exception of tarsiers, have a *uterus simplex* with a single duct. In strepsirrhines,

TABLE 3.4: Major reproductive differences between strepsirrhine and haplorhine primates

Trait	Strepsirrhines	Haplorhines
Uterus	Bicornate	Bicornate in tarsiers; Simplex in all platyrrhines and catarrhines
Allantois	Large	Small in tarsiers; absent in anthropoids
Maternal-fetal separating membrane	Epitheliochorial, adeciduate	Hemochorial, deciduate
Placenta type	Diffuse	Discoidal or bidiscoidal
Menstruation	Absent	Present to varying degrees, slight or sporadic in tarsiers and platyrrhines, heavier in catarrhines, especially humans
Sexual skin	Absent	Manifested in some catarrhine species
Reproductive seasonality	Reproductive seasonality is extreme in some species, but not universal	Reproductive seasonality variable across species

TABLE 3.5: Reproductive and life history parameters of commonly used laboratory primates					
Taxon	**Uterus (Bicornate or Simplex)/Placenta (Epitheliochorial or Hemochorial)**	**Gestation length (days)**	**Modal litter size/average mass of each neonate (g)**	**Age at weaning (days)**	**Age at first female reproduction (years)**
Microcebus murinus	B/E	60	2/4.6	40	1
Lemur catta	B/E	135	1/85	179	2
Loris tardigradus	B/E	166	1/10	170	1.5
Nycticebus coucang	B/E	193	1/50.8	180	2.1
Galago senegalensis	B/E	142	1/19	98	1.4
Otolemur crassicaudatus	B/E	135	2/43.2	135	2.2
Saimiri sciureus	S/H	170	1/106.4	168	2.5
Aotus trivirgatus	S/H	133	1/94	75	2.4
Callithrix jacchus	S/H	148	2/27	60	1.4
Saguinus oedipus	S/H	168	2/42.1	50	1.9
Chlorocebus aethiops	S/H	163	1/335.9	201	5
Papio anubis	S/H	180	1/915	584	4.5
Macaca mulatta	S/H	165	1/466.3	192	3
Macaca fascicularis	S/H	160	1/326.1	330	3.9
Homo (Pan) troglodytes	S/H	235	1/1,750	1,680	13
Homo sapiens	S/H	267	1/2,900	730	14

Notes and sources:
Data on neonatal mass.
Data on gestation length, neonatal mass, litter size, weaning age, and age at first reproduction, see Smith and Leigh (1998), and various chapters (plus the appendix) of Kappeler and Pereira (2003).

there is a large embryonic *allantois* which is a bag-like structure that protrudes from the embryonic rectum and functions like a bladder, a small one in tarsiers, and none in platyrrhines and catarrhines.

The placenta serves as the membrane through which materials such as nutrients, oxygen, carbon dioxide, etc., pass from mother to fetus or vice versa. Species have a characteristic number of layers of tissue separating the maternal and fetal vascular systems. In species with *epitheliochorial placentas* (among primates, all strepsirrhines), three fetal tissue layers, collectively known as the chorion, and three layers of the maternal endometrium, which lines the uterine wall, combine to form a six-layered maternal-fetal separating membrane. At birth, the maternal layers remain intact and only the fetal tissue is shed. Because the epithelium of the uterine wall, which is the layer of maternal tissue in closest contact with the fetal tissue and also called the decidua, is not shed, such placentas are called "*adeciduate.*"

In species with *hemochorial placentas* (among primates, all haplorhines), the three layers of the maternal endometrium are resorbed, and the remaining three-layered fetal membrane is bathed directly in maternal blood. At birth, the lining of the uterine wall ruptures, and maternal and fetal tissue is shed, with bleeding, as the "afterbirth." Such placentas are "*deciduate.*"

Tarsiers are said to have hemochorial, deciduate placentas. Like anthropoids, they have three instead of six layers of tissue separating maternal and fetal vascular systems. However, in early pregnancy, the tarsier placenta more closely resembles those of lemurs and lorises, with six layers; it is only as the placenta matures that the number of layers is reduced to three (Ankel-Simons, 2000).

Placentas also vary in form among primates and may be diffuse, discoidal, or bidiscoidal. Species with epitheliochorial placentas have diffuse chorionic protrusions over the whole placenta while those with chorionic protrusions in disc-like regions of the placental surface have either discoidal (one disc) or bidiscoidal (two discs) placentas. While strepsirrhines have diffuse placentas, tarsiers' placentas are discoidal, and platyrrhines and catarrhines bidiscoidal, with the exception of hominoids, which have discoidal placentas. The placentas of marmosets are unique in exhibiting chorionic fusion of the placentas of fraternal twins, with blood chimerism.

Adult females undergo hormonally-regulated changes in their reproductive tissues in association with ovulatory, or "sexual," cycles (generally lasting from 20 to 35 days). These cycles (also called "menstrual" and/or "estrous" cycles depending on whether and how menstruation and/or estrus are manifested) vary in length and in periodicity among primate species. They have three phases: (1) the follicular phase, when one or several ovarian follicles mature and begin to secrete estrogen; (2) the midcycle or ovulatory phase, when estrogen levels increase and then peak, and the now-mature follicle ruptures and releases its ovum or egg; and (3) the luteal phase, when the ruptured follicle, now a "corpus luteum," secretes both progesterone and estrogen in preparation for possible implantation of a fertilized egg. Estrus occurs in the midcycle phase, and menstruation at the end of the sexual cycle. "Estrus" is usually defined as a state of heightened female sexual motivation. It is facilitated by the high estrogen levels at ovulation and inhibited as a result of rising concentrations of progesterone during the luteal phase. The visual, pheromonal and behavioral changes, often associated with estrus, have the effect of making females very attractive to males. In some species, such as macaques, females in estrus have a strong odour.

"Menstruation," the shedding of the endometrial lining of the uterine wall, will only occur if there has been no implantation of a fertilized egg into the uterine wall. Menstruation is manifested only in haplorhines and, even then, bleeding occurs only sporadically in some taxa such as tarsiers, platyrrhines. Humans are said to have a menstrual (and not estrus) cycle because menstrual bleeding can be heavy, and because sexual motivation or attractivity is not tied to ovulation as in many other primates. Estrus is effectively "concealed" in humans. Estrus and menstruation are not mutually exclusive among primates, however, and both occur in many species, including our closest relatives orangutans, gorillas, and chimpanzees. Estrus occurs in strepsirrhines, but no strepsirrhine menstruates.

Females of some primate species, including baboons, macaques, mangabeys, and chimpanzees, have a distinct sex skin that varies in size and coloration in association with the hormonal fluctuations of the sexual cycle. Around the time of menstruation, the sex skin is flat. Around the time of ovulation, it is maximally tumescent or swollen, smooth and shiny, and it can be blue-gray to deep red in color, depending on the species. Its extent varies and it may cover the back of legs as well as the perineum in macaques, such as the rhesus monkey. Areas on the chest can also swell. In other species, only the vulvae show conspicuous cyclical changes in swelling (see Dahl and Nadler, 1992, on gibbons, *Hylobates lar*). The degree of swelling also varies. Baboons (*Papio*) and chimpanzees (*Homo*, subgenus *Pan*) have very large swellings of the sex skin at ovulation. Because the sex skin advertises the female's reproductive condition, it can be used to estimate the timing of ovulation.

There may be a relationship between the degree of swelling of the female sexual skin and the length of the erect penis (Dixson and Mundy, 1994). In chimpanzees, where females display exaggerated sexual swellings, and the penis is long and filiform when erect, the depth of the vagina increases as the sexual skin swells. Dixson and Mundy (1994) suggest that the long penis of chimpanzees, and other species that have a similar penile morphology, may have evolved to penetrate copulatory plugs deposited by other males during prior copulations.

The periodicity of reproductive receptivity also varies among primates. In many lemur species, females are sexually receptive for less than one day annually (there is no sexual cycling at other times of year), and many species, including some macaques, such as the Japanese macaque, *Macaca fuscata* living in highly seasonal habitats, exhibit strong reproductive seasonality. Indeed, it is common among baboons and macaques to have summers "off" with sexual cycling, in non-pregnant females, ceasing or becoming irregular in the early summer and resuming in the early fall. Among lemurs (with some exceptions, such as the aye-aye), females of each species enter estrus, once annually, almost simultaneously, resulting in marked synchrony of birth and weaning. In such species, infants effectively grow in annual cohorts, and late-born infants may be at a distinct developmental disadvantage.

Strepsirrhine neonates are about one-third the mass of neonates born to like-sized haplorhine mothers. The size of neonates depends little on differences in the lengths of the gestation period (which vary in primates from around two to nine months) but rather on differences in rates of prenatal growth. Strepsirrhines and haplorhines do not differ consistently in the length of the gestation period, even when benchmarked against body mass. Tarsiers have long gestation periods and give birth to larger neonates than do strepsirrhines of similar body size; in fact, spectral tarsiers have one of the highest infant/maternal mass ratios of any primate, with single neonates sometimes weighing over 30% of adult weight (Gursky, 2000). Table 3.5 summarizes data on gestation length, neonatal mass, modal number of offspring and weaning age, for commonly used laboratory primates. Most primates, particularly haplorhines, but also many strepsirrhines, give birth to a single young. The larger-bodied strepsirrhines tend to give birth to single young, even though they may possess multiple pairs of mammae (e.g., *Daubentonia*). Primates that regularly give birth to multiple offspring include the cheirogaleids, bushbabies and *Varecia*, among strepsirrhines, and the Callitrichidae, owl monkeys and titi monkeys among haplorhines.

The young of some strepsirrhine species cling to their mothers and the young of other species stay in nests or cling to branches while their mothers forage. Some species transport their young in their mouths, like cats. In contrast, whereas some haplorhines build sleeping nests, the young are not left in nests while the mother forages.

The senses

Due to strong orbital convergence, primates tend to have larger fields of binocular overlap than nonprimates. Each retina receives information from both the binocular field of visual overlap and the monocular field associated with each eye, but only the information from the binocular field can contribute to stereoscopic (or depth) perception. Furthermore, in primates, the ipsilateral nerve fibers (from the eye on the same side) and the contralateral nerve fibers (from the eye on the opposite side) passing to the optic tectum, a structure in the midbrain, are approximately balanced. In nonprimates, contralateral fibers predominate. The balance of ipsilateral and contralateral fibers in primates, coupled with strong orbital convergence, allows for stereoscopic vision as single areas of the brain are able to integrate information coming from different visual fields (Martin, 1990).

The eye develops as a photosensitive lobe of the brain. During development, the lobe becomes indented and forms a two-layered cup, or retina, connected to the brain through the optic nerve that passes through the optic canal at the back of the eye socket. The innermost layer of the retinal cup gives rise both to pigmented rods and cones, which are photosensitive or light-detecting cells, arranged perpendicular to the retinal surface and to neurons that convey visual information to the brain. The outermost layer of the cup becomes heavily pigmented and absorbs light (thus helping to prevent images from blurring). There are two types of photoreceptor cells: rods which are cylindrical rodlike cells that are particularly sensitive to differences between light and dark and thus critical to night vision, and cones which are conical cells that are sensitive to differences in color and thus are critical to day vision). Color perception depends on the number and types of cones present. "Monochromatic" primates such as bushbabies (*Galago*) lack color vision entirely. Most strepsirrhines have limited or no color vision (but see Jacobs *et al.*, 2002, on sifakas).

Primate species vary in the internal structure of the eye. An area of heightened visual *acuity* occurs at the center (the thinnest part) of the retina of haplorhine primates and is called the fovea centralis. The eyes of strepsirrhines lack a central fovea. However, most, including some strictly diurnal species such as *Lemur catta*, *Indri indri* and *Propithecus* spp, have instead, a tapetum lucidum. This is a layer of tissue, containing guanine, in the choroid region of the eye between the lens and the retina, that acts as a reflective membrane and is responsible for the eyeshine characteristic of many nocturnal mammals. The tapetum lucidum enhances visual *sensitivity* under low light conditions. In many nocturnal primate species, visual sensitivity is also enhanced, at the expense of visual acuity, by high "retinal summation" which is a high ratio of photoreceptor cells to optic canal diameter. Species with high retinal summation send visual information to the brain as summed impulses from neighboring photoreceptor cells.

No haplorhine has a tapetum lucidum. Haplorhines, however, do have a central fovea and most have low retinal summation. Haplorhines also have postorbital closure in which the bony socket for the eyeball is closed posteriorly, and greater frontation and convergence of the orbits than do strepsirrhines. Haplorhines vary in cone types and pigmentation. In catarrhines (including humans), there are three types of cones that are capable of distinguishing red, green and blue, respectively and, working together, they allow the perception of many colors. Species with all three types of cones are said to have "trichromatic" color vision. Most platyrrhines have "dichromatic" color vision and thus, excellent perception of blue and green but not red light. Trichromatic color vision does occur, however, in some platyrrhine species such as howler monkeys (*Alouatta*), but rarely in others such as capuchin monkeys, squirrel monkeys and tamarins, and then only in some female individuals (Jacobs and Deegan, 2001; 2003).

Trait	Strepsirrhines	Haplorhines
Tapetum lucidum	Generally present	Absent even in nocturnal forms
Central fovea	Absent	Present
Rhinarium	Moist, shiny, glandular area (or rhinarium) surrounds nostrils, and attaches to the gums of the upper jaw via a median fold, or philtrum. A relatively immobile upper lip is tethered to the underlying gum between the central incisors	No moist, naked rhinarium; mobile upper lip.
Vomeronasal organ	Present	Present in tarsiers and platyrrhines

TABLE 3.6: Major sensory differences between strepsirrhine and haplorhine primates

Nocturnal haplorhines (*Tarsius*, the tarsier, and *Aotus*, the owl monkey) have eyes that are typical of neither strepsirrhines nor diurnal haplorhines. Tarsiers have enormous, protruding eyeballs, only half of which is contained within its bony eye socket, and high retinal summation. Unlike nocturnal strepsirrhines, tarsiers have no tapetum lucidum, and they do have a central fovea (Ross, 1996). Owl monkeys also lack a tapetum lucidum but possess a collagenous tapetum fibrosum, that functions somewhat similarly, and a central fovea. They have the largest eyes (relative to skull size) of any monkey. Unlike other platyrrhines, owl monkeys have only one cone photopigment and thus resemble bushbabies in lacking color vision (Ogden, 1994; Deegan and Jacobs, 1996).

Olfaction is extremely important to all strepsirrhines. The moist, naked rhinarium, that surrounds the nostrils of strepsirrhines, contains nerve receptors for smell as well as touch. Many strepsirrhines also have specialized scent glands for marking physical objects such as branches. Male ringtailed lemurs use their antebrachial (forearm) glands in ritualized "stink fights" during the mating season. A vomeronasal (or Jacobson's) organ is present in strepsirrhines, tarsiers and platyrrhines, but not catarrhines. In nonprimates, this organ, which opens at the front of the roof of the mouth, plays a key role in olfaction. Mouse lemurs (*Microcebus murinus*) appear to have a functional vomeronasal organ (Aujard, 1997). Removal of the vomeronasal organ affects the frequency of sexual behavior and intermale aggression in this species. The role, if any, of the vomeronasal organ in platyrrhines remains controversial.

The morphology of the outer, middle, and inner ear varies across primates. Outer ear morphology is sometimes a reliable species indicator, particularly for nocturnal species that disperse while foraging and that use calls to maintain social contacts or establish territories. A critical factor may be mate recognition, as calls are essential to mate recognition in non gregarious species. An important function of the outer ear is to collect and funnel sound waves (see, for example, Masters, 1991; Masters and Bragg, 2000, and Anderson *et al.*, 2000, on bushbabies). There are also characteristic differences across primates in the bony morphology of the middle and inner ear. This not only includes variation in the position of the tympanic membrane (eardrum) vis-à-vis the external auditory meatus, but also how the membrane is attached to the temporal squamosum, and so on. Differences in the structure of the inner ear result in differences in the ability to hear sounds of different frequencies. Common marmosets (*Callithrix jacchus*) and bushbabies (*Galago*) are able to hear high-frequency sounds that are inaudible to humans (Ankel-Simons, 2000), but humans have excellent discrimination of low-frequency sounds.

Correspondence

Any correspondence should be directed to Laurie R. Godfrey, Department of Anthropology, Machmer Hall, 240 Hicks Way, University of Massachusetts, Amherst, MA 01003, USA. Fax: 413-545-9494. Phone: 413-545-2064. Email: lgodfrey@anthro.umass.edu

References

Aiello, L.C. and Wheeler, P.W. (1995). *Current Anthropology* 36, 199–221.

Anderson, M.J., Ambrose, L., Bearder, S.K., Dixson, A.F. and Pullen, S. (2000). *International Journal of Primatology* 21, 637–655.

Aujard, F. (1997). *Physiology and Behavior* 62, 1003–1008.

Ankel-Simons, F. (2000). *Primate Anatomy: An Introduction.* Second Edition. Academic Press, San Diego.

Cartmill, M. (1972). In Tuttle, R. (ed.) *The Functional and Evolutionary Biology of Primates,* pp. 97–122. Aldine, Chicago.

Chivers, D.J. and Hladik, C.M. (1980). *Journal of Morphology* 166, 337–386.

Clark, W.E. LeGros (1963). *The Antecedents of Man.* Harper & Row, New York.

Dahl, J.F. (1994). *Journal of Mammalogy* 75, 1–9.

Dahl, J.F. and Nadler, R.D. (1992). *American Journal of Physical Anthropology* 89, 101–108.

Deegan, J.F. and Jacobs, G.H. (1996). *American Journal of Primatology* 40, 55–66.

Demes, B., Larson, S.G., Stern, J.T. Jr., Jungers, W.L., Biknevicius, A.R. and Schmitt, D. (1994). *Journal of Human Evolution* 26, 353–374.

Demes, B., Fleagle, J.G. and Lemelin, P. (1998). *Journal of Human Evolution* 34, 385–399.

Dixson, A.F. (1991). *Physiology and Behavior* 49, 557–562.

Dixson, A.F. and Mundy, N.I. (1994). *Archives of Sexual Behavior* 23, 267–280.

Doran, D.M. and McNeilage, A. (1998). *Evolutionary Anthropology* 6, 120–131.

Fleagle, J.G. (1999) "Primate Adaptation and Evolution." Second Edition. Academic Press, San Diego.

Godfrey, L.R., Samonds, K.E., Jungers, W.L. and Sutherland, M.R. (2001). *American Journal of Physical Anthropology* 114, 192–214.

Godfrey, L.R., Petto, A.J. and Sutherland, M.R. (2002). In Plavcan, J.M., Kay, R.F., Jungers, W.L. and van Schaik, C.P. (eds) *Reconstructing Behavior in the Primate Fossil Record,* pp113–157. Kluwer Academic/Plenum Publishers, New York.

Gursky, S. (2000). *Folia Primatologica.* 71, 39–54.

Hartman, C.G. and Straus, W.L. Jr. (1971). *The Anatomy of the Rhesus Monkey (Macaca mulatta).* Hafner, New York.

Hill, W.C.O. (1958). *Primatologia* 3, 139–207.

Jacobs, G.H. and Deegan, J.F. (2001). *Proceedings of the Royal Society of London Series B.* 268, 695–702.

Jacobs, G.H. and Deegan, J.F. (2003). *Vision Research* 43, 227–236.

Jacobs, G.H., Deegan, J.F., Tan, Y. and Li, W.H. (2002). *Vision Research* 42, 11–18.

Jolly, C.J. (1970). *Man (N.S.)* 5, 5–26.

Jungers, W.L., Godfrey, L.R., Simons, E.L., Wunderlich, R.E., Richmond, B.G. and Chatrath, P.S. (2002). In Plavcan, J.M., Kay, R.F., Jungers, W.L. and van Schaik, C.P. (eds) *Reconstructing Behavior in the Primate Fossil Record,* pp. 371–411. Kluwer Academic/Plenum Publishers, New York.

Kappeler, P.M. and Pereira, M.E. (eds). (2003). *Primate Life Histories and Socioecology.* University of Chicago Press, Chicago.

Kay, R.F. (1975). *American Journal of Physical Anthropology* 43, 195–216.

Kimura, T. (1995). *Z. Morph. Anthropol.* 80, 265–280.

Kimura, T., Okada, M. and Ishida, H. (1979). In Morbeck, M.E., Preuschoft, H. and Gomberg, N. (eds) *Environment, Behavior and Morphology: Dynamic Interactions in Primates,* pp. 297–311. G. Fischer, New York.

Kirkpatrick, R.C. (1999). In Dolhinow, P. and Fuentes, A. (eds) *The Nonhuman Primates,* pp. 93–108. Mayfield Publishing Co., Mountain View, CA.

Kudo, H. and Dunbar, R.I.M. (2001). *Animal Behaviour* 62, 711–722.

Lambert, J.E. (1998). *Evolutionary Anthropology* 7, 8–20.

Larson, S.G., Schmitt, D., Lemelin, P. and Hamrick, M. (2001). *Journal of Zoology* 255, 353–365.

Lemelin, P. (1995). *Journal of Morphology* 224, 351–368.

Lemelin, P. and Schmitt, D. (1998). *American Journal of Physical Anthropology* 105, 185–197.

Martin, R.D. (1990). *Primate Origins and Evolution,* Princeton University Press, Princeton, NJ.

Martin, R.D. (2003). In Kappeler P.M. and Pereira, M.E. (eds) *Primate Life Histories and Socioecology,* pp. xi–xx. The University of Chicago Press, Chicago.

Masters, J.C. (1991). *Primates* 32, 153–167.

Masters, J.C. and Bragg, N.P. (2000). *International Journal of Primatology* 21, 793–813.

Muller, A.E. and Thalmann, U. (2000). *Biological Reviews of the Cambridge Philosophical Society* 75, 405–435.

Nekaris, K.A.I. (2003). *American Journal of Physical Anthropology* 121, 86–96.

Ogden, T.E. (1994). In Baer, J.F., Weller, R.E. and Kokoma, I. (eds) *Aotus: The Owl Monkey,* pp. 264–286. Academic Press, San Diego.

Overdorff, D.J. (1992). *American Journal of Primatology* 28, 191–203.

Plavcan, J.M. (2001). *Yearbook of Physical Anthropology* 44, 25–53.

Plavcan, J.M. and van Schaik, C.P. (1997). *Journal of Human Evolution* 32, 345–374.

Polk, J.D., Demes, B., Jungers, W.L., Biknevicius, A.R., Heinrich, R.E. and Runestad, J.A. (2000). *Journal of Human Evolution* 39, 297–325.

Rasoazanabary, E. (Submitted). *International Journal of Primatology.*

Ross, C. (1996). *American Journal of Primatology* 40, 205–230.

Rowe, N. (1996). *The Pictorial Guide to the Living Primates.* Pogonias Press, East Hampton, NY.

Smith, B.H., Crummett, T.L. and Brandt, K.L. (1994). *Yearbook of Physical Anthropology* 37, 177–231.

Smith, R.J. and Jungers W.L. (1997). Body mass in comparative primatology. *Journal of Human Evolution* 32, 523–559.

Smith, R.J. and Leigh, S.R. (1998). *Journal of Human Evolution* 34, 173–201.

Stearns, S.C., Pereira, M.E. and Kappeler, P.M. (2003). Primate life histories and future research. In Kappeler, P.M. and Pereira, M.E. *Primate Life Histories and Socioecology*, pp. 301–312. The University of Chicago Press, Chicago.
Stockley, P. (2002). *Evolutionary Ecology* 16, 123–137.
Sussman, R.W. and Raven, P.H. (1978). *Science* 200, 731–736.
Swindler, D.R. (2002). *Primate Dentition: An Introduction to the Teeth of Non-human Primates.* Cambridge University Press, Cambridge.
Swindler, D.R. and Wood, C.D. (1973). *An Atlas of Primate Gross Anatomy: Baboon, Chimpanzee, and Man.* University of Washington Press, Seattle.
van Schaik, C.P. and van Hoof, J.A.R.A.M. (1996). In McGrew, W.C., Marchant, L.R. and Nishida, T (eds) *Great Ape Societies*, pp. 3–15. Cambridge University Press, Cambridge.

CHAPTER 4

Pathology of Noninfectious Diseases of the Laboratory Primate

Anne D. Lewis and Lois M. A. Colgin
Division of Animal Resources, Oregon National Primate Research Center, Oregon Health and Science University, Beaverton, OR 97006, USA

Introduction

In this chapter, selected entities are described with emphasis on clinical signs, morphologic features and pathogenic mechanisms. It is not intended as a comprehensive account of all aspects of noninfectious pathology of spontaneous disease in nonhuman primates. Discussion is limited primarily to species most often maintained in a laboratory setting. The chapter is arranged by different body systems and, where appropriate, inflammatory, degenerative, metabolic and neoplastic and non-neoplastic proliferative conditions are addressed. Multisystemic and nutritional diseases are covered at the end of the chapter.

Respiratory system

Noninfectious diseases of the respiratory tract are exceedingly rare in nonhuman primates. Pulmonary pneumatocoeles in macaques are unusual complications of infection with lung mites (*Pneumonyssus simicola*) (Gillett *et al.*, 1984). Neoplasms of the respiratory system are generally rare; however, clusters of nasal and oropharyngeal tumors have been reported in the common marmoset (*Callithrix jacchus*) (Baskerville *et al.*, 1984; Betton, 1984; McIntosh *et al.*, 1985). The majority of these tumors (9/13) were squamous cell carcinomas; the remainder were poorly differentiated carcinomas. All were locally aggressive. Half of these

The Laboratory Primate
Copyright 2005 Elsevier
ISBN 0-1208-0261-9

tumors had metastasized to the lung; one undifferentiated carcinoma metastasized to the liver. Sporadic neoplasms reported in the nasal cavity include a carcinosarcoma in a bonnet macaque (*Macaca radiata*), and a nasal adenocarcinoma in an olive baboon (*Papio cyanocephalus*) (Guhad *et al.*, 1997). In the lung, a squamous cell carcinoma in a cynomolgus macaque (*Macaca fascicularis*) (Kaspareit *et al.*, 2001) and a clear cell carcinoma in a pigtailed macaque (*Macaca nemestrina*) (Tsai and Giddens, Jr., 1985) have been reported.

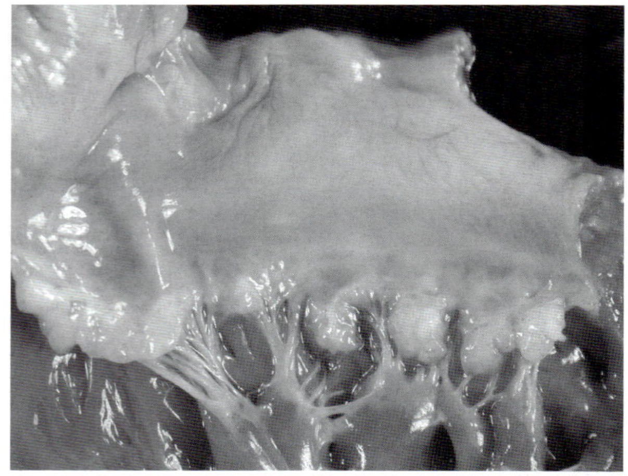

Figure 4.1 Mitral valve, endocardiosis. The free margins of the valve leaflets are nodularly thickened and rolled.

Cardiovascular system

Although there are few reports of cardiac disease in most species of nonhuman primate, a notable exception is the owl monkey (*Aotus sp.*), in which spontaneous cardiomyopathy has been described with relatively high frequency. Cardiomyopathy was reported in 77% (30/39) of animals examined from a single facility over a 41-month period (Gozalo *et al.*, 1992a). Approximately two-thirds had hypertrophic cardiomyopathy; one-third had the dilated form. In another facility, approximately 40% of animals dying of all causes had gross evidence of cardiomyopathy (Weller, 1994). Clinical signs included dyspnea, ascites, and subcutaneous edema. In many, premonitory signs were lacking and "sudden death" was the presenting complaint. Hypertrophic hearts had severe concentric thickening of the left ventricular free wall and interventricular septum. Histologically, there was myocyte hypertrophy, necrosis, and fibrosis. Although the pathogenesis of cardiomyopathy in owl monkeys is unknown, Weller (1994) reports a positive correlation of systemic hypertension in affected monkeys.

Valvular endocardiosis in the atrioventricular valves is a relatively common incidental finding in aged rhesus macaques (personal observation). The free margins of the affected valve leaflets are thickened and rolled (Figure 4.1). Histologically, the atrioventricular valves exhibit fibromyxomatous degeneration as seen in other domestic species. A related lesion, defined clinically as mitral valve prolapse, was observed in a breeding colony of rhesus macaques (Swindle *et al.*, 1985). In contrast to the occurrence of valvular endocardiosis in aged rhesus macaques, valvular prolapse was diagnosed in both subadult and young adult macaques. Auscultation revealed mid-to-late systolic heart murmurs and clicks in affected animals. Valvular prolapse was demonstrated echocardiographically. Fibromyxomatous degeneration of the mitral valve leaflets was observed in those animals that were examined microscopically. An autosomal dominant pattern of inheritance, similar to that seen in man, was proposed in this group of macaques (Swindle *et al.*, 1985).

Aneurysms of the aorta and carotid arteries have been reported in a number of species; the majority of cases have been in New World species, particularly owl and squirrel monkeys (*Saimiri sciureus*) (Baer *et al.*, 1992; Bunton, 1993; Chapman, Jr. *et al.*, 1973; Kessler *et al.*, 1979; Strickland and Bond, 1983). The incidence of aneurysms in one study of squirrel monkeys was 1.5% (11/730) (Strickland and Bond, 1983). In a study of owl monkeys, the incidence was 8.6% (22/257) (Baer *et al.*, 1992). The majority of aneurysms were dissecting; fusiform and saccular types were much less common. Atherosclerosis is a predisposing factor for aneurysm formation in humans. In both studies, many but not all of the affected animals had experimentally induced atherosclerosis.

Atherosclerosis occurs sporadically in many species of nonhuman primate. It is generally clinically inapparent. Grossly, there are minimally raised yellow streaks on the intimal surface of the aorta (Figure 4.2) that correspond histologically with fibrofatty plaques (Clarkson *et al.*, 1976b; Clarkson *et al.*, 1976a; Clarkson, 1998). The ability to induce atherosclerosis by dietary intervention (typically with the feeding of high levels of cholesterol) is variable among species. Cynomolgus macaques and vervet monkeys are considered susceptible and are the current models of choice (Clarkson, 1998; Fincham *et al.*, 1996; Fincham *et al.*, 1998); baboons are relatively resistant (Klumpp *et al.*, 1993).

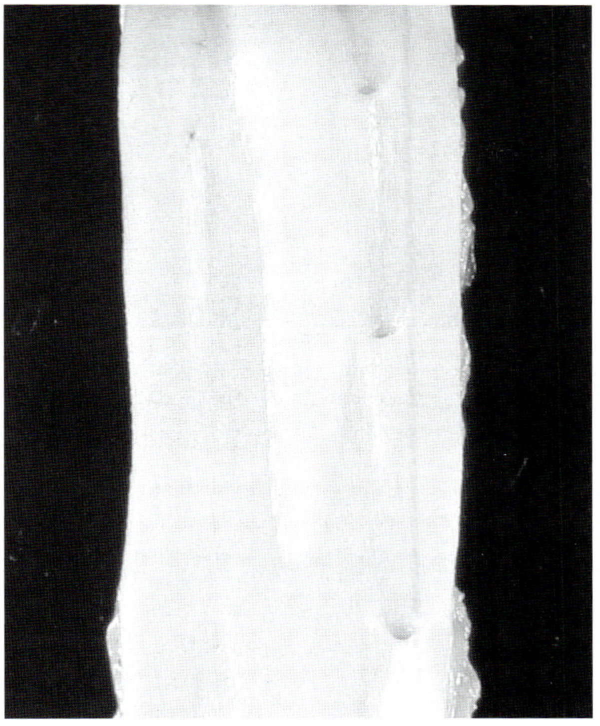

Figure 4.2 Aorta, atherosclerosis, rhesus macaque. A raised fibrofatty plaque is present on the intimal surface of the aorta.

Endocrine system

Pancreas

Islet amyloidosis

Pancreatic islet amyloidosis has been reported in several species of Old World nonhuman primate including the rhesus macaque (de Koning *et al.,* 1993; Palotay and Howard, Jr., 1982), cynomolgus macaque (Davis *et al.,* 1994; O'Brien *et al.,* 1996; Sheldon and Gleiser, 1971), Sulawesi macaque (*Macaca nigra*) (Howard, Jr., 1978; Palotay and Howard, Jr., 1982), baboon (Hubbard *et al.,* 2002), a Formosan rock macaque (*Macaca cyclopis*), a drill baboon (*Mandrillus leucophaeus*) (Palotay and Howard, Jr., 1982) and an orangutan (*Pongo pygmaeus*) (Davis *et al.,* 1994). The precursor protein of islet amyloid is islet associated polypeptide (IAPP), a 37 amino acid polypeptide co-secreted with insulin by beta cells of the endocrine pancreas (reviewed in O'Brien *et al.,* 1993). Unique amino acid residues in the human, macaque and feline IAPP result in the tendency to form amyloid fibrils in tissue (Ohagi *et al.,* 1991; Westermark *et al.,* 1990). Islet amyloid is strongly associated with the development of diabetes mellitus (DM) in these species. The actual role of islet amyloid in the pathogenesis of DM is still unclear (reviewed in Jaikaran and Clark, 2001).

The association of islet amyloid and diabetes is well characterized in rhesus, cynomolgus and Sulawesi macaques; two patterns emerge. The rhesus and cynomolgus macaques develop disease resembling Type 2 or non-insulin-dependent diabetes mellitus (NIDDM) in humans (de Koning *et al.,* 1993; O'Brien *et al.,* 1996). In these species, disease is associated with increasing age and obesity. Initial hyperinsulinemia and glucose intolerance is followed by impaired islet cell function and hyperglycemia. In contrast, *M. nigra* develop diabetes in the absence of obesity. In that species, there is a direct correlation between the degree of islet amyloid deposition and loss of islet secretory cells (Howard, Jr., 1978; Howard, Jr., 1986; Howard, Jr. and Van Bueren, 1986). Clinical signs seen with islet amyloid are related to those of diabetes mellitus. These include weight loss, polyphagia, polydipsia, and polyuria. Laboratory findings include hyperglycemia and glucosuria.

Pancreatic amyloid may be visible grossly as small white to gray dots scattered throughout the pancreas (Figure 4.3), but is generally inconspicuous. Histologically, islet amyloid appears as homogeneous eosinophilic deposits in the islets of Langerhans (Figure 4.4). Initial deposition is centered around islet capillaries. Generally, all islets are affected. With increasing amounts of amyloid deposition, there is often concomitant loss of islet cells. Islet amyloid is congophilic and exhibits apple green birefringence under polarized light.

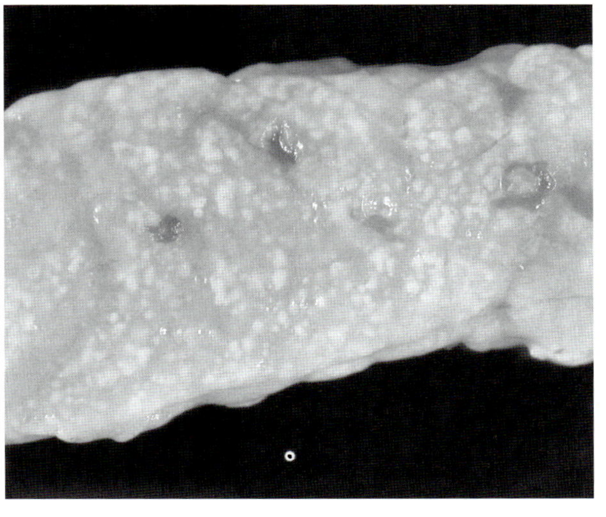

Figure 4.3 Pancreas, islet amyloidosis, rhesus macaque. Multiple white nodular opacities correspond microscopically to accumulations of amyloid within islets.

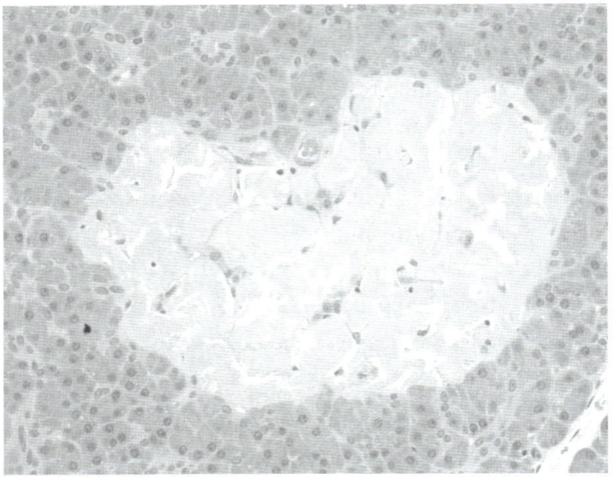

Figure 4.4 Pancreas, islet amyloidosis, rhesus macaque. The islets are expanded and replaced by amyloid. H&E, original magnification 200×.

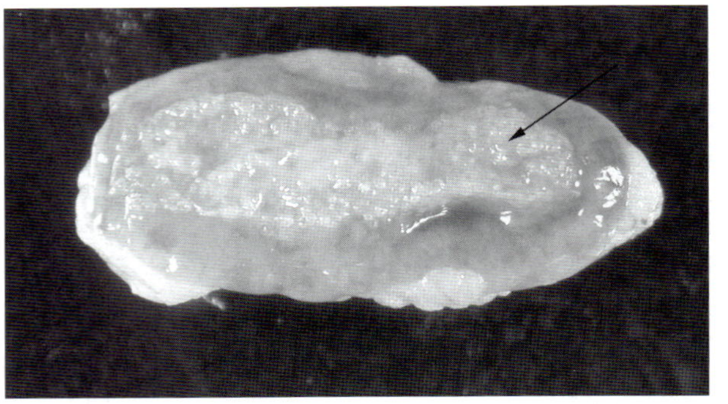

Figure 4.5 Adrenal gland, mineralization, rhesus macaque. Gritty white material is deposited at the corticomedullary junction.

Islet cell hyperplasia

Islet cell hyperplasia is a sporadic lesion of the endocrine pancreas reported in New World monkeys, including five callitrichids (Brunnert *et al.,* 1990a; Juan-Salles *et al.,* 2002), a spider monkey (Brunnert *et al.,*1990a) and two squirrel monkeys (King *et al.,* 1996). Islet cell hyperplasia has been linked clinically in some cases with disorders of glucose regulation, principally hyperglycemia (Fanton *et al.,* 1986; Juan-Salles *et al.,* 2002). Histologically, the condition is characterized by the benign proliferation of pancreatic islet cells.

Adrenal gland

Clinically important primary disorders of the adrenal gland are generally rare. Adrenocortical necrosis and hemorrhage may be seen as a sequela to enterocolitis and gram-negative sepsis, particularly with severe cases of shigellosis. Mineralization of the adrenal is an exceedingly common incidental finding in macaques and baboons (Kast *et al.,* 1994a; Majeed and Gopinath, 1980; Skelton-Stroud and Ishmael, 1985). Inconspicuous to large deposits of mineral are found at the corticomedullary junction (Figure 4.5). The location coincides with the region of the provisional zone in the fetal adrenal. It is thought that mineralization is a consequence of its loss in postnatal life (Kast *et al.,* 1994b). Accessory nodules of adrenocortical tissue in close proximity to the adrenal capsule are also common incidental findings in these species (Skelton-Stroud and Ishmael, 1985).

Alimentary tract

Oral cavity

Squamous cell carcinomas of the oral cavity are reported with some frequency in squirrel monkeys (Montesdeoca *et al.,* 1992; Morris, 1994; Sasseville *et al.,* 1993). The tumors are characterized by rapid growth that deforms facial features. They are locally aggressive with extensive tissue destruction, and frequent metastasis to regional lymph nodes. Distant metastases have not been reported.

Tumors of dental origin have been sporadically reported in several species. Ameloblastic odontomas are seen most commonly and have been reported in a cynomolgus, a Japanese and a rhesus macaque, as well as a baboon, and a cebus monkey (Baskin and Hubbard, 1980; Benjamin and Lang, 1969; Davis *et al.,* 1988; Splitter *et al.,* 1972b; Yanai *et al.,* 1995a).

Stomach

Acute gastric dilatation occurs in numerous species of nonhuman primate in captivity. It occurs with relative frequency as a sporadic disease in macaques (Pond *et al.,* 1982; Smith *et al.,* 1969; Soave, 1978). Affected animals are generally subadult to adult monkeys in good condition. The clinical course is frequently peracute. Clinical signs, when observed, include hunched posture, facial grimacing indicative of abdominal pain, and abdominal distension. Untreated, the disease is invariably fatal. Gross findings at necropsy are marked abdominal distension and gastric dilatation without volvulus (Figure 4.6). Frequently there are terminal hemorrhages. Occasionally, there is postmortem gastric rupture. Death is attributed to impaired venous return

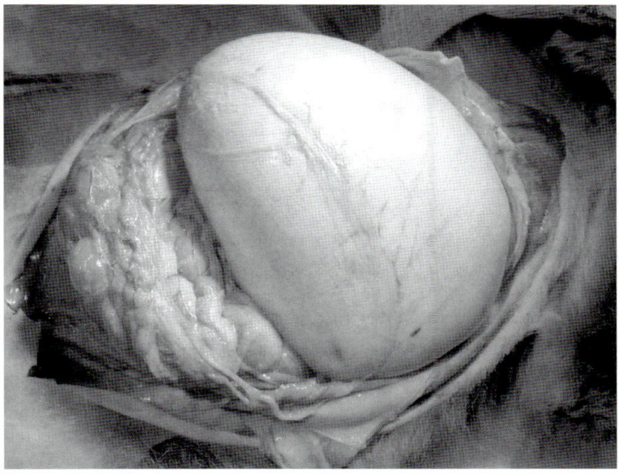

Figure 4.6 Stomach, acute gastric dilatation, rhesus macaque. The markedly distended stomach is visible upon opening the abdominal cavity.

and cardiopulmonary compromise. Historical factors associated with the development of gastric dilatation include overfeeding and/or feed restriction, and prior anesthesia (Pond *et al.*, 1982).

Gastric infarction has been reported in cynomolgus macaques (Fikes *et al.*, 1996). Five cases were observed in a 20-month period. In each, there was gastric necrosis involving the fundus and pylorus. Microscopically, there were thrombi within the gastric venous vasculature. A common finding in each of the cases was the presence of predisposing factors for disseminated intravascular coagulation.

Intestine

Diarrhea is a pervasive clinical problem in the husbandry of many nonhuman primate species. Both infectious and noninfectious etiologies are contributory. As a rule, the small intestine is less commonly affected than the large intestine in most disorders of the intestinal tract of nonhuman primates. Of notable exception is generalized amyloidosis, which frequently manifests as a protein losing enteropathy and most severely affects the proximal small intestine. Systemic amyloidosis is discussed in greater detail in this chapter under systemic diseases.

Inflammatory diseases

Chronic colitis of juvenile rhesus macaques

Adler *et al.* (1993) described a syndrome of chronic colitis in juvenile rhesus macaques raised in nurseries. A similar disease is seen in group-housed, maternally raised animals (personal observation). The disease occurs primarily in young animals between one and three years of age. Clinical signs include persistent or recurrent liquid diarrhea, weight loss and poor growth. Adler *et al.* (1993) reported an 8% to 10% incidence in nursery reared infants. The disease is poorly responsive to therapy over the long term and repeated bouts of diarrhea and dehydration are seen. Initial episodes of diarrhea may be associated with positive fecal cultures for *Campylobacter* or *Shigella* and the presence of pathogenic protozoa. Subsequent episodes are characterized by normal enteric flora and the variable presence of protozoa.

Grossly, colons may be thickened or flaccid and dilated. There are often copious amounts of liquid colonic content. The mucosa may be multifocally eroded or ulcerated. In other animals, the predominant change is thickened and rugose mucosal surfaces (Figure 4.7). Typically, the cecum and proximal colon are most severely affected. Colonic lymph nodes are enlarged. Histologically, there is a diffuse lymphoplasmacytic infiltrate in the lamina propria, numerous crypt abscesses, and mucosal hyperplasia.

The pathogenesis is thought to be complex, involving repeated enteric infections, malnutrition associated with enteric disease, compromised mucosal defenses, environmental stresses and possible hypersensitivity to dietary antigens (Adler *et al.*, 1993). Recently a novel *Helicobacter* species has been isolated which is implicated in the pathogenesis (Fox *et al.*, 2001).

Ulcerative cicatrizing colitis

A syndrome of ulcerative colitis with marked fibroplasia is seen occasionally in macaques. Affected animals

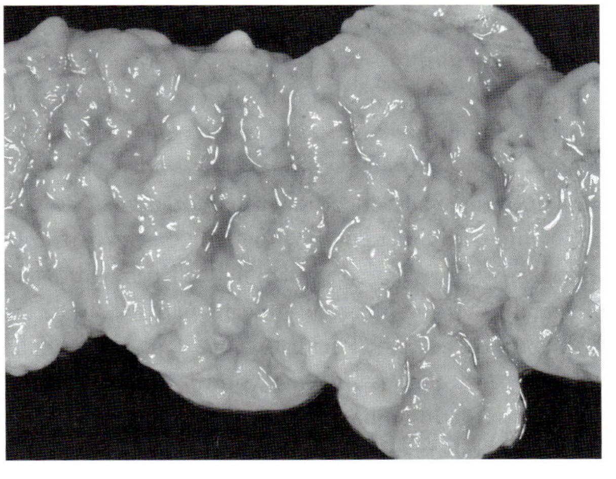

Figure 4.7 Colon, chronic colitis, juvenile rhesus macaque. The mucosal surface of the colon is thickened and edematous.

exhibit weight loss and chronic diarrhea. Signs of colonic obstruction, such as scant feces, vomiting, and abdominal distention, may develop in some animals. Grossly, the lesions appear as linear to focally extensive regions of deep mucosal ulceration in the cecum and proximal colon. Chronic ulcers are surrounded by marked deposition of fibrous connective tissue resulting in stricture formation. Strictures can result in partial to complete obstruction and resulting distention of the anterior intestinal tract occurs. The lesion must be distinguished microscopically from intestinal adenocarcinomas that frequently arise in similar anatomic locations. Histologically, there is deep ulceration of the mucosa and underlying submucosa. Lymphoid aggregates are prominent in the adjacent mucosa, submucosa, and subserosal tissues. The lesion is frequently associated with the secondary development of generalized amyloidosis (Blanchard *et al.,* 1986).

Colitis of callitrichids

Colitis occurs commonly in several species of callitrichids including the cotton top tamarin (*Saguinis oedipus*), saddle back tamarin (*Saguinus fuscicollis*), and common marmoset (*Callithrix jacchus*) (Chalifoux and Bronson, 1981; Chalmers *et al.,* 1983; Clapp *et al.,* 1985b). It has been most extensively studied in cotton top tamarins in which it is associated with intestinal adenocarcinoma and is proposed as a model of colitis and adenocarcinoma in humans. In callitrichids, colitis affects both sexes equally. It appears as early as four months of age and incidence increases with age (Johnson *et al.,* 1996b). Clinical signs include diarrhea and weight loss. The disease course is characterized by acute exacerbations followed by periods of remission. Response to sulfasalazine treatment is reported (Johnson *et al.,* 1996b; Madara *et al.,* 1985). The gross appearance is variable, ranging from no grossly visible changes to multiple areas of hemorrhage and ulceration of the mucosa. The entire large intestine from the cecum to rectum is affected. The initial histologic changes are small numbers of neutrophils within the lamina propria that cross the mucosal epithelium and create crypt abscesses. Crypt abscesses rupture to create microulcerations that ultimately lead to confluent ulcerations of the mucosa. Chronic changes include atrophy and loss of crypts, decreased numbers of goblet cells and chronic inflammatory cell infiltrates in the lamina propria.

The etiology of colitis in callitrichids is unknown. An infectious etiology has not been identified. It appears to be a disease of captivity; free ranging tamarins were found to have no or only mild colitis when evaluated by endoscopic biopsy (Wood *et al.,* 1998). Various factors have been implicated. Cotton top tamarins have markedly reduced amounts of mucin component IV, a mucin fraction decreased in humans with ulcerative colitis (Podolsky *et al.,* 1985). Unidentified environmental factors were implicated in a prospective study of newborn tamarins. Possible factors include unidentified transmissible agents, social stress, caging, ambient temperature and humidity (Johnson *et al.,* 1996b). Diet does not appear to play a primary role in the development of colitis but may influence chronic changes (Johnson *et al.,* 1996b). Stress is thought to play an important role in the pathogenesis (Wood *et al.,* 2000).

Colitis of other species

Syndromes of chronic colitis similar to that described in the rhesus macaque have also been described in cynomolgus macaques and baboons (Rubio and Hubbard, 2001; Rubio and Hubbard, 2002). In these species, chronic intractable diarrhea develops in young animals from infancy to young adulthood. Gross findings include a reddened and edematous colonic mucosa with multiple erosions and ulcerations. In most animals, the entire length of the colon is affected. In a smaller percentage, only the proximal colon is involved. In the majority of animals, chronic inflammatory infiltrates of lymphocytes and plasma cells are found throughout the lamina propria of the mucosa. Chronic ulcerative colitis is seen in a smaller number of animals. In these cases, there is disruption of normal mucosal architecture with shortened and irregular crypts. In addition to chronic inflammatory infiltrates, there are also acute superficial mucosal neutrophilic infiltrates and crypt abscesses.

Neoplastic diseases

Intestinal adenocarcinoma of cotton top tamarins

There is a relatively high incidence of colonic adenocarcinoma in cotton top tamarins in captivity. The disease was initially reported in two colonies (Chalifoux and Bronson, 1981; Lushbaugh *et al.,* 1978) but has since been reported in multiple colonies in the United States and Europe (Brack, 1998; DePaoli and McClure, 1982; Kirkwood *et al.,* 1986; Richter *et al.,* 1980). Incidence rates are estimated at between 3% and 8% of animals at risk (Brack and Weber, 1995). There is no sex predisposition (Chalifoux *et al.,* 1993). Age of occurrence is generally between three and eight years (Chalifoux *et al.*, 1985; Clapp *et al.*, 1985b). The majority of cases exhibit metastasis to the local lymph nodes (Chalifoux and Bronson, 1981; Clapp *et al.*, 1985a). Metastasis to lungs, pancreas and adrenals is also frequently reported (Brack, 1998; Chalifoux and Bronson, 1981; Clapp *et al.*, 1985a).

Clinical signs are weight loss, poor body condition, and chronic diarrhea that is not responsive to therapy. Discrete abdominal masses and thickened bowels may be palpable percutaneously. Intestinal obstruction may occur.

The gross appearance of tumors may range from inapparent mucosal plaques to thickened fibrotic masses that may be found anywhere from the ileocecal junction to the rectum. Metastases result in firm, enlarged colonic lymph nodes. Histologically, tumors are composed of neoplastic epithelial cells arranged in sheets and cords, and forming poorly organized glands. Neoplastic cells generally arise from crypt bases and invade laterally and deeply into the submucosa and muscularis, variably inciting a dense fibroblastic (desmoplastic) response. The cells often produce mucin and are uniformly periodic acid-Schiff (PAS) positive. Multiple tumors and foci of carcinoma restricted to the mucosal layer (carcinoma in situ) are frequent (Lushbaugh *et al.*, 1985).

The development of intestinal carcinomas in cotton top tamarins with colitis has been of particular interest to investigators as a model of some forms of colonic cancer in humans. In cotton top tamarins, there is a clear association of adenocarcinoma with colitis (Chalifoux and Bronson, 1981; Clapp *et al.*, 1985b; Johnson *et al.*, 1996b; Lushbaugh *et al.*, 1985). An increased risk of cancer is also seen in human ulcerative colitis (Johnson *et al.*, 1996b). Other proposed factors include a genetic, possibly familial predisposition (Clapp *et al.*, 1982; DuFrain, 1985). A dietary influence was demonstrated in a prospective study in which a semipurified diet was associated with a reduced incidence of cancer compared to age matched and sibling controls fed a standard diet (Johnson *et al.*, 1996b).

Intestinal adenocarcinoma of rhesus macaques

Intestinal adenocarcinoma is the most common life threatening malignant neoplasm of rhesus macaques. Several recent studies document the increased prevalence of this disease in aged rhesus macaques (Rodriguez *et al.*, 2002; Uno *et al.*, 1998; Valverde *et al.*, 2000). The majority of tumors occur at the ileocecal junction, cecum and proximal colon. Intestinal adenocarcinomas from the duodenum, jejunum and ileum have been reported in the rhesus (DePaoli and McClure, 1982; Rodriguez *et al.*, 2002; Valverde *et al.*, 2000). In published studies, the incidence of tumors arising from the ileocecal junction and cecum ranged from 44% to 69% and from the colon, 33% to 50% (Rodriguez *et al.*, 2002; Uno *et al.*, 1998; Valverde *et al.*, 2000). The majority of published cases occurred in animals 13 years or older (DePaoli and McClure, 1982; Johnson *et al.*, 1996a; Kerrick and Brownstein, 2000; Lembo *et al.*, 1997; O'Sullivan and Carlson, 2001; Rodriguez *et al.*, 2002; Uno *et al.*, 1998; Valverde *et al.*, 2000). In one study, the incidence increased with advancing age from 3.2% in animals 13–19 years of age to 20.7% to animals 30–37 years of age (Uno *et al.*, 1998). Metastasis to local lymph nodes is relatively uncommon, occurring in approximately 10% of cases. Distant metastases are rare and have been reported most often in the liver (DePaoli and McClure, 1982; Kerrick and Brownstein, 2000; Rodriguez *et al.*, 2002; Valverde *et al.*, 2000).

Clinical findings include marked weight loss, partial to complete anorexia, scant or no feces, diarrhea and vomiting. A palpable abdominal mass is occasionally noted on physical examination. Clinicopathologic findings are nonspecific and include hypoproteinemia, hypoalbuminemia, and anemia. Evaluations for fecal occult blood are frequently positive.

Grossly, the masses appear as circumferential or nodular thickenings of the wall of the intestine. Examination of the luminal surface may reveal a sessile to polypoid mass with variable degrees of ulceration (Figures 4.8, 4.9). The lumen is frequently constricted.

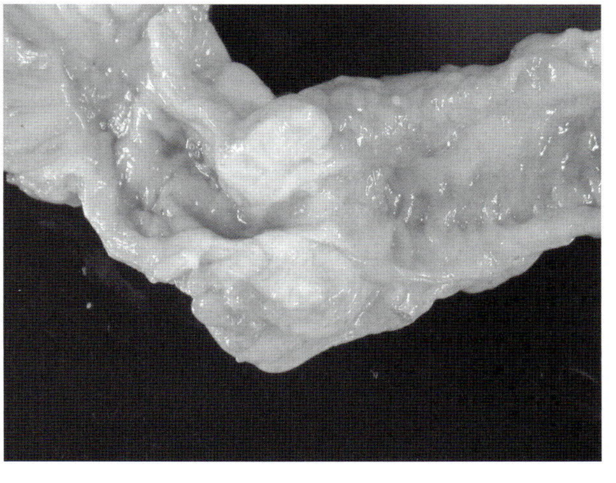

Figure 4.8 Ileocecal junction, adenocarcinoma, rhesus macaque. An annular, constrictive neoplasm arising in the cecum.

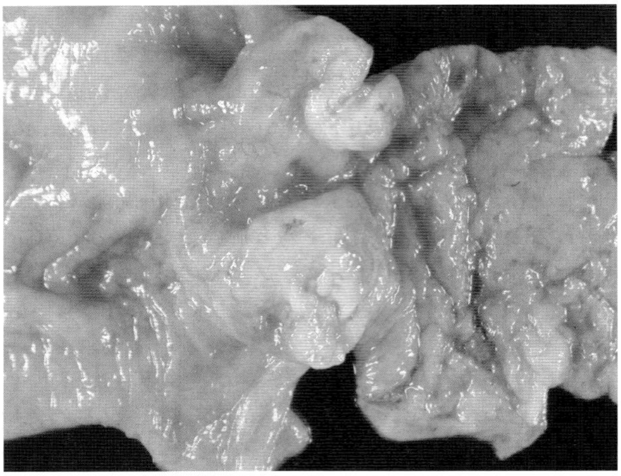

Figure 4.9 Colon, adenocarcinoma, rhesus macaque. A circumferential, stenosing neoplasm of the colon. The mucosa overlying and surrounding the neoplasm is irregular, thickened and ulcerated.

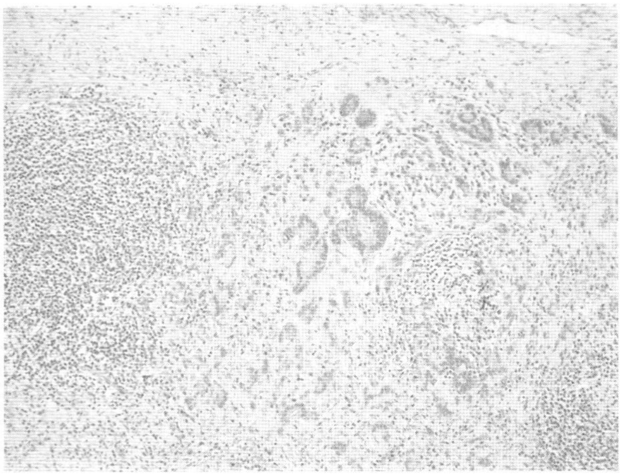

Figure 4.11 Mesenteric lymph node, metastatic adenocarcinoma, rhesus macaque. Neoplastic cells forming tubules invade the subcapsular sinus of the regional lymph node. H&E, original magnification 100×.

The gross appearance of annular constrictions corresponds to the classically described "napkin-ring" lesion.

The typical microscopic appearance of tumors is of neoplastic cuboidal to columnar cells forming poorly organized tubules and acini (Figures 4.10, 4.11). There is often a desmoplastic response. Tumors arise from the mucosa and generally extend into the muscularis and often to the subserosa. Although there is mucin production by neoplastic cells in most of these tumors, in some there are large lakes of mucin in dilated cystic structures prompting classification as mucinous adenocarcinomas (Figure 4.12).

Because these tumors are locally aggressive but slow to metastasize they are often amenable to surgical intervention. The mean survival rate for surgical resection in one study was 483 postoperative days (Valverde *et al.*, 2000).

Increasing age appears to be a primary risk factor for the development of intestinal adenocarcinoma (Rodriguez *et al.*, 2002; Suzuki *et al.*, 2000; Uno *et al.*, 1998; Valverde *et al.*, 2000). The increased use of rhesus macaques for aging studies may contribute to the increased observations of these tumors. There does not appear to be any sex predilection. More cases are reported in females; however, this probably reflects that most colonies of aged rhesus frequently have larger numbers of females because the animals are derived from breeding programs. Earlier case reports included animals that had been exposed to varying types and doses of radiation but it is not clear if there is a causal

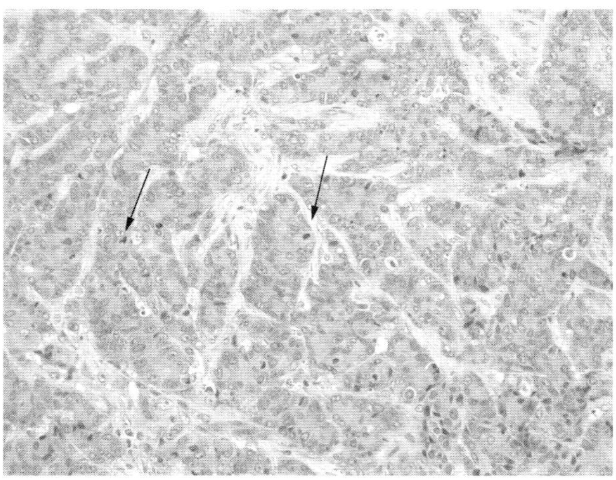

Figure 4.10 Cecum, adenocarcinoma, rhesus macaque. Neoplastic cell forming tubules in the submucosa. Note the numerous mitotic figures (arrows). H&E, original magnification 200×.

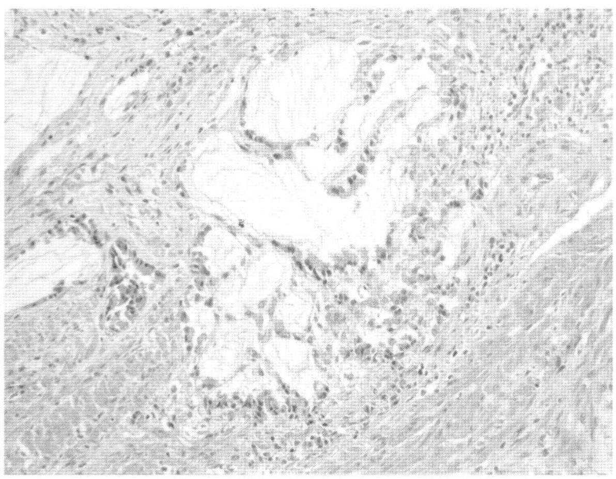

Figure 4.12 Cecum, mucinous adenocarcinoma, rhesus macaque. Lakes of mucin are present within lumina of tubules formed by neoplastic epithelial cells. H&E, original magnification 200×.

relationship (DePaoli and McClure, 1982; Fanton *et al.*, 1984; Lembo *et al.*, 1997). The most frequent sites of tumors in rhesus correspond to those regions most frequently affected by chronic inflammatory colitis and it is likely that inflammation may contribute to the development of adenocarcinomas in this location.

Intestinal adenocarcinoma in other species

Although the majority of cases of adenocarcinoma in macaques have been reported in rhesus, Valverde *et al.* (2000) include two aged cynomolgus macaques. Rubio and Hubbard (Rubio and Hubbard, 1998) report ten cases of cecal adenocarcinoma in baboons with ulcerative typhlocolitis.

Urinary system

Kidney

An IgM mesangial nephropathy has been studied in callitrichids (Brack, 1988; Brack, 1995; Brack and Rothe, 1981). Multiple species of marmosets and tamarins are affected (Brack, 1988). In one study, the disease was reported to occur in 91% of callitrichids over six months of age and thought to have contributed to mortality in 20% (Brack, 1988). The incidence was equal between sexes (Brack, 1988). Glomerular changes characterizing this disease are mesangial proliferation starting at the mesangial hilus, intraglomerular adhesions and thickening of Bowman's capsule. Advanced lesions include global glomerulosclerosis. Immunohistochemistry consistently revealed IgM deposits within the mesangium (Brack, 1988). Glomerular changes are generally accompanied by a subacute to chronic tubulointerstitial nephritis. The disease is progressive and mesangial IgM deposits precede the development of nephropathy in young animals (Brack, 1995). An immune-mediated pathogenesis is proposed (Brack, 1988).

A morphologically similar disease is reported in owl monkeys (Gozalo *et al.*, 1992a; King, Jr. *et al.*, 1976). Glomerular changes include mesangial proliferation, increased mesangial matrix, thickening of the glomerular basement membrane and glomerulosclerosis (Hunt *et al.*, 1976; King, Jr. *et al.*, 1976). In one study, glomerular lesions were observed in approximately 60% of the animals examined (Chalifoux *et al.*, 1981). Inflammatory interstitial infiltrates are highly correlated with the presence of glomerular lesions.

Renal oxalosis has been reported in wild caught macaques (Skelton-Stroud and Glaister, 1994; Yanai *et al.*, 1995b). Oxalate crystals were found primarily in the proximal convoluted tubules. They were generally described as incidental findings in apparently healthy animals and were not associated with renal disease. A dietary etiology was proposed.

Acute tubular necrosis is a common finding in macaques with a variety of underlying conditions. Grossly, kidneys are mildly enlarged and pale tan. The cortical surface bulges on the cut surface (Figure 4.13). Microscopically, renal tubules are lined by flattened cuboidal epithelium with pyknotic nuclei and there is patchy to diffuse loss of epithelial cells. Hyaline and granular casts are common. The condition is most often seen in animals with diarrhea and attendant dehydration and those with severe muscle trauma. Although early studies implicate nephrotoxic antibiotic usage for the treatment of diarrhea (Giddens, Jr. *et al.*, 1979), the condition occurs routinely in the absence of these drugs. It is likely to be the result of hypovolemic shock and renal ischemia to which macaques are particularly susceptible.

Reproductive system

Female genital system

Reported disorders and diseases related to the female nonhuman primate reproductive system include, but are not limited to trauma, sexually transmitted disease,

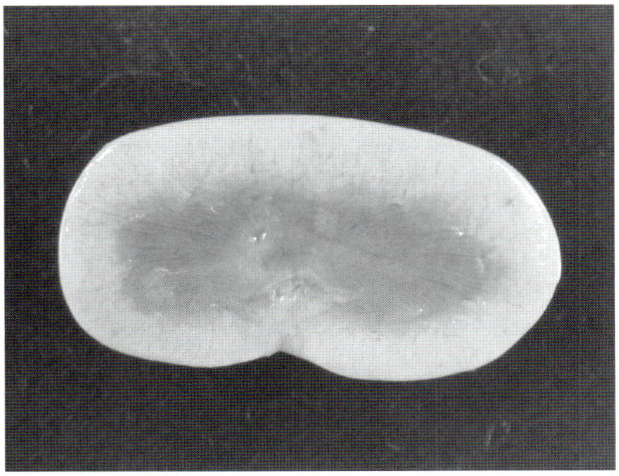

Figure 4.13 Kidney, acute tubular necrosis, juvenile rhesus macaque. The renal cortex is pale and swollen.

menstrual problems, infertility, abortion, fetal death, stillbirth, obstetrical complications, endometriosis, and neoplasia (DiGiacomo, 1977; DiGiacomo and McCann, 1970; Ford *et al.,* 1998; Hertig *et al.,* 1983; Johnson *et al.,* 1978; Mueller-Heubach and Battelli, 1981; Rouse *et al.,* 1981; Ruch, 1959b; Swindle *et al.,* 1981). Selected noninfectious reproductive findings in nonhuman primates are described.

Endometriosis

Endometriosis develops in menstruating female primates and is one of the more common gynecologic disorders in Old World nonhuman primates. It has been reported in rhesus macaques (Bertens *et al.,* 1982; Fanton *et al.,* 1986; MacKenzie and Casey, 1975; McCann and Myers, 1970), cynomolgus macaques (Ami *et al.,* 1993) and baboons (D'Hooghe *et al.,* 1991; Dick, Jr. *et al.,* 2003). Other reported laboratory species include pigtailed macaques (*Macaca nemestrina*) (DiGiacomo *et al.,* 1977; McClure, 1979) and the African green monkey (*Chlorocebus*) (McClure, 1979).

Endometriosis is characterized by the presence of ectopic endometrial tissue. Ectopic growth limited to the myometrium is termed adenomyosis or internal endometriosis. Endometriotic tissue usually contains both epithelial and stromal components and undergoes regular cyclical changes under the influence of estrogen and progesterone similar to normal endometrium. Clinical signs are often vague and not specific but cyclic depression and anorexia, weight loss, menorrhagia, irregular menstrual cycles, abdominal distension, constipation, absence of feces, decreased fertility, and/or anemia are suggestive of endometriosis. Masses, uterine distortion or other pelvic abnormalities palpated via the rectum or abdomen on physical examination may support a preliminary diagnosis of endometriosis. Mildly affected animals often do not present with clinical signs.

Grossly, endometriotic tissue is most frequently found over the visceral and peritoneal surfaces of the pelvis. The serosa of the uterus, urinary bladder, omentum, distal colon and the uterine ligaments are often affected. Ovarian involvement in macaques is common. It is reported as rare in baboons by one group of authors (D'Hooghe *et al.,* 1991; D'Hooghe, 1997); however, in another study, ovarian involvement was reported in 16 of 43 baboons with endometriosis (Dick, Jr. *et al.,* 2003). Endometriotic foci have also been observed throughout the abdominal cavity, the thoracic and abdominal surfaces of the diaphragm and the intercostal musculature. Ectopic endometriotic tissue may appear as firm, white to tan, punctate, puckered

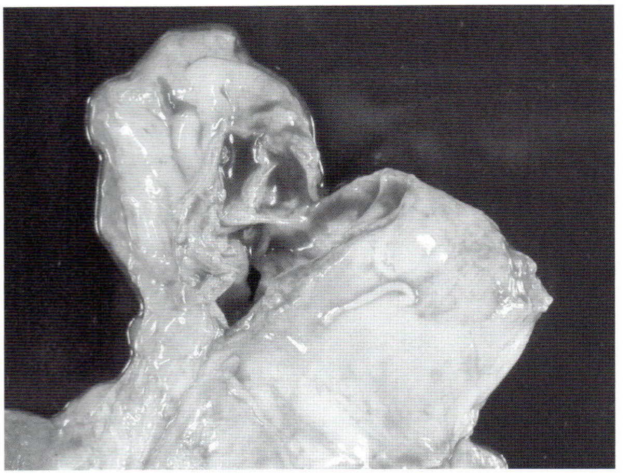

Figure 4.14 Female reproductive tract, endometriosis, rhesus macaque. An endometriotic cyst lies between the uterus and oviduct. Opened cyst reveals turbid brown fluid ("chocolate cyst").

foci or soft red-brown raised masses adhered to serosal surfaces. Variably sized cysts containing brown fluid that are frequently described as "chocolate cysts" are often present (Figure 4.14). The cysts have a thick wall with a shaggy, irregular brown to yellow-brown lining. Fibrous adhesions may compromise gastrointestinal and genitourinary function. Bowel obstruction, perforation and peritonitis, infertility, and hydroureter may ensue. Hemoperitoneum is another complication.

Histologically, endometriosis is characterized by ectopic endometrial glands and stroma (Figure 4.15) and typically involves the serosal surfaces of the pelvic viscera. Penetration of surrounding tissue may be observed. Glandular lumina often contain blood and

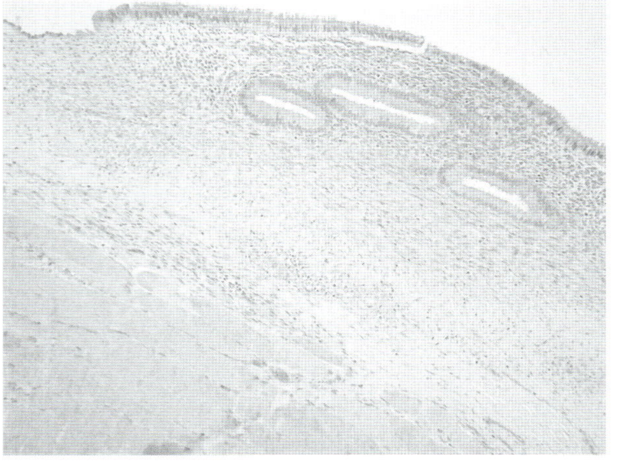

Figure 4.15 Diaphragm, endometriosis, rhesus macaque. Ectopic endometrial glands and stroma are implanted on the abdominal surface of the diaphragm. H&E, original magnification 100×.

cellular debris. Fibrosis, variable numbers of lymphocytes and plasma cells, and accumulations of hemosiderin-laden macrophages are associated with endometriotic foci.

Various risk factors have been identified in the development of endometriosis in nonhuman primates. Hysterotomies were associated with an increased prevalence of endometriosis in four studies with rhesus macaques, (Bertens *et al.,* 1982; Coe *et al.,* 1998; DiGiacomo, 1977; Hadfield *et al.,* 1997; McCann and Myers, 1970) and two with baboons (D'Hooghe *et al.,* 1991; Dick, Jr. *et al.,* 2003). However, in one study involving cynomolgus macaques, no significant difference was observed in the incidence of endometriosis between macaques with or without a history of a hysterotomy (Ami *et al.,* 1993). Laparoscopy did not appear to be a significant risk factor in a study with rhesus macaques (Hadfield *et al.,* 1997); however, the converse was speculated in a study with baboons (D'Hooghe *et al.,* 1992). Other risk factors include age, increased time since last pregnancy, parity (DiGiacomo, 1977; McCann and Myers, 1970), exposure to nonionizing radiation (Fanton and Golden, 1991; Splitter *et al.,* 1972a) treatment with ovarian steroids (Bertens *et al.,* 1982; Hadfield *et al.,* 1997; Lindberg and Busch, 1984; Ruch, 1959a) and chronic exposure to 2,3,7,8-tetrachlorodibenzo-*p*-dioxin (Rier *et al.,* 1993). The results of one study suggested that the prevalence of endometriosis in baboons may increase with the time spent in captivity (D'Hooghe *et al.,* 1996a). The authors speculated that more menstrual cycles uninterrupted by pregnancy occurred because females had limited exposure to males. This allowed more opportunity for the reflux of menstrual products and possible seeding of the peritoneum. One author noted few problems with endometriosis in a colony of rhesus macaques that were regularly bred. A genetically related population of rhesus macaques utilized in reproduction research and thus exposed to multiple risk factors such as surgical and hormonal procedures exhibited a high prevalence of endometriosis (Coe *et al.,* 1998).

The etiology and pathogenesis of endometriosis are unknown but several theories exist as to the origin of the endometrial tissue and the factors that affect subsequent implantation and proliferation. Retrograde menstruation (tubal reflux) is the most widely accepted theory as to the origins of endometrial implants; however, no single theory can explain all cases. Other theories include coelomic metaplasia, transformation of embryonic rests of Müllerian origin and lymphatic and vascular metastasis. Retrograde menstruation in baboons was documented in one study and its prevalence was higher in baboons with spontaneous endometriosis than in animals with a normal pelvis (D'Hooghe *et al.,* 1996b). Retrograde menstruation is a common physiologic phenomenon in women (Liu and Hitchcock, Blumenkrantz *et al.,* Halme *et al.,* cited in D'Hooghe *et al.,* 1996b); however, the frequency of endometriosis in women is much lower suggesting that additional factors are involved in the development of the disorder. Lesion development following tubal reflux presumably involves impaired resorption of endometrial fragments, endometrial cell survival and attachment, invasion of the mesothelium, inflammatory cell recruitment, macrophage activation, angiogenesis, proliferation of endometriotic implants and scarring (Giudice *et al.,* 1998; Seli *et al.,* 2003; van der Linden, 1997; Vinatier *et al.,* 2001).

Neoplasms and other proliferative lesions

Proliferative lesions of the female genital tract are not uncommon in nonhuman primates (Beniashvili, 1989; Lowenstine, 1986; McClure, 1973; Moore *et al.,* 2003; O'Gara and Adamson, 1972; Ruch, 1959c; Seibold and Wolf, 1973; Squire *et al.,* 1978).

Uterine tumors, generally benign, occur more often than ovarian tumors in Old World species, particularly macaques; conversely, gonadal tumors are more common than uterine tumors in New World primates. Endometrial polyps have been reported in nine rhesus macaques (DiGiacomo and McCann, 1970) and one baboon, (cited in O'Gara and Adamson, 1972). These are solitary to multiple, soft sessile tissue masses that project from the endometrial mucosa (Figure 4.16) and are composed of hyperplastic endometrium. They may be clinically inapparent or may cause abnormal uterine bleeding. Their significance with regards to infertility is unknown. Leiomyomas have been reported in rhesus macaques, (DiGiacomo and McCann, 1970; McClure, 1973; Moody *et al.,* 1996; Seibold and Wolf, 1973), a squirrel monkey, sooty mangabey, an African green monkey (Lowenstine, 1986), and a marmoset (Beniashvili, 1989). Leiomyomas are usually multiple, circumscribed, fleshy to firm masses that may distort the normal outline of the uterus. Microscopically, they are composed of interlacing bundles of smooth muscle cells with varying amounts of collagen. Clinical signs are frequently absent; however, abnormal bleeding may occur and, with large tumors, urinary incontinence, constipation, varicose leg veins, abortion, and dystocia may be observed. Other benign uterine neoplasms that have been reported include hemangiomas in rhesus macaques (DiGiacomo and McCann, 1970; Lowenstine, 1986) and cited in Squire *et al.* (1978),

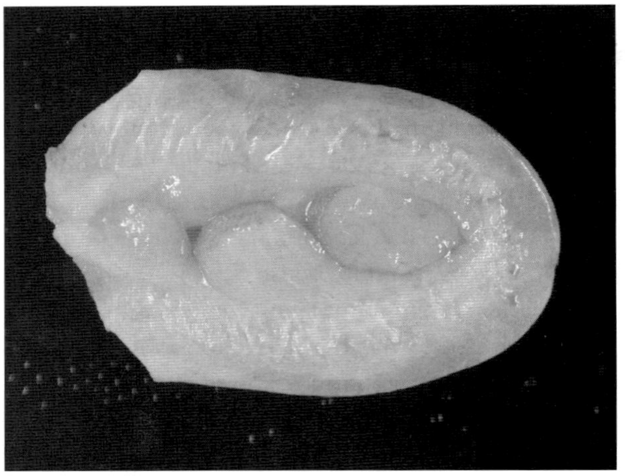

Figure 4.16 Uterus, endometrial polyps, rhesus macaque. Proliferative endometrial masses project into the uterine lumen.

fibromyomata in a squirrel monkey, fibroma in a Japanese macaque, and an endometrial cystadenoma in a baboon (cited in O'Gara and Adamson, 1972). Few malignant uterine neoplasms appear in the literature. They include a leiomyosarcoma in a rhesus macaque (Jackson, 1996) an epithelioid leiomyosarcoma in a pig-tailed macaque (Birkebak *et al.,* 1996), an endometrial adenocarcinoma in a rhesus macaque (Strozier *et al.,* 1972), a sarcoma in a rhesus macaque (Plesker *et al.,* 2002), and a carcinoma in a rhesus macaque (DiGiacomo and McCann, 1970; McClure, 1980).

Cervical polyps have been reported in rhesus macaques (DiGiacomo, 1977; DiGiacomo and McCann, 1970). Malignant neoplasms arising in the cervix include squamous cell carcinoma in a macaque (Lowenstine, 1986), carcinoma in macaques (Lowenstine, 1986; O'Gara and Adamson, 1972) and an African green monkey (Simkins and Bullock, 1995), and metastatic carcinoma in a rhesus macaque (Hisaw and Hisaw, 1958).

Abundant luteal and interstitial tissue is present in the ovaries of pubescent cebids (*Ateles*) and may account for the so-called "luteomas" described in woolly monkeys (Lowenstine, 1986; Lowenstine, 2003). Parovarian cysts are among reported tumor-like lesions and have been noted in rhesus macaques (DiGiacomo, 1977; DiGiacomo and McCann, 1970) and a squirrel monkey (Brown and Kupper, 1972). They may be of Müllerian, mesonephric, or mesothelial origin.

Ovarian tumors are uncommon in nonhuman primates. Benign and malignant tumors have been reported and include surface epithelial-stromal and serous tumors, sex cord-stromal tumors, germ cell tumors, stromal tumors and malignant and, not otherwise specified, neoplasms. Granulosa cell tumors and teratomas are the more commonly reported neoplasms (Moore *et al.,* 2003). Age is a predisposing factor. In one study with baboons, eight of 12 neoplasms occurred in animals over 17 years of age; whereas germ cell tumors, especially teratomas, were seen in younger animals (Moore *et al.,* 2003). Clinical manifestations referable to ovarian neoplasms may be absent (Chalifoux, 1993; Lichtenwalner, 1997; Moore, 1998; Chalifoux, 1993) and fertility is often not affected. Abdominal distension may be noted and a mass revealed on abdominal palpation.

Granulosa cell tumors have been reported in a spider monkey, a squirrel monkey, rhesus macaques, and four baboons (Lowenstine, 1986; Moore *et al.,* 2003). A malignant granulosa cell tumor with metastasis to the uterus and lung was described in a stumptailed macaque (*Macaca arctoides*) (Chalifoux, 1993a). These neoplasms vary in size from microscopic to enormous multinodular masses that expand the abdomen (Chalifoux, 1993a; Moore *et al.,* 2003). On a cut surface, the tumor may be uniformly solid or multilocular with both solid and cystic foci and areas of necrosis and hemorrhage. Microscopically, an admixture of patterns may be observed. A microfollicular pattern characterized by the presence of Call-Exner bodies, which are small fluid-filled spaces surrounded by neoplastic granulosa cells, may be present. A macrofollicular pattern resembling graafian follicles may be present with granulosa cells clustering around large blood or fluid-filled spaces. Neoplastic granulosa cells may also be arranged in sheets and cords. These tumors may produce estrogen with a consequent endometrial hyperplasia. Luteinized granulosa cells may produce progesterone and, under this influence, a decidual reaction of the stroma or secretory glandular activity may be seen.

Ovarian teratomas have been reported in an African green monkey, rhesus macaques (Chalifoux, 1993b; Crews *et al.,* 1967; Eydelloth and Swindle, 1983; Martin, Jr. *et al.,* 1970; Rohovsky *et al.,* 1977; Scott *et al.,* 1975) and baboons (Moore *et al.,* 1998; Moore *et al.,* 2003). A teratoma and a choriocarcinoma in the contralateral ovary were diagnosed in a cynomolgus macaque (Toyosawa *et al.,* 2000). Teratomas are variably sized, encapsulated, solid to cystic neoplasms. The sectioned surface may exude mucinous, serous or hemorrhagic fluid. Hair, bone, cartilage or teeth may be present. Microscopically it is composed of differentiated tissue representing at least two and often all three germ layers (ectoderm, mesoderm and endoderm). In humans, tumors containing mature tissue are considered benign whereas, those containing immature tissues are designated immature teratomas and may become malignant (Scully *et al.,* 1998). Mature teratomas are the only ones that have been described in nonhuman primates (cited in Moore *et al.*, 2003). Other diagnosed ovarian tumors include a cavernous hemangioma and papillary cystadenoma (Martin, Jr. *et al.*, 1970), and a cystadenofibroma and papilloma in a rhesus macaque (Amin, 1974). Carcinoma has been diagnosed in two cynomolgus macaques, one rhesus macaque (Lowenstine, 1986), one bonnet macaque (Bunton and Lollini, 1983) and one baboon (Moore *et al.*, 2003).

Placenta

Spontaneous placental disorders associated with adverse consequences to the dam or fetus have been reported in nonhuman primates. Vaginal hemorrhage may be seen during the second or third trimester in cases of placenta previa where the placenta implants near or over the internal cervical os (Lunn, 1980; Myers, 1972). The attenuation of the cervix that occurs in the latter part of pregnancy may result in premature separation of the placenta with attendant fetal anoxia. The dam usually dies from hemorrhage during or after delivery. Grossly, the placenta appears cone-shaped with associated blood clots (Myers, 1972). Abruptio placentae, or detachment of the placenta from the uterine wall, may be partial or complete. If the affected area is small, the remaining placenta may be sufficient to oxygenate the fetal blood. Death occurs quickly if a large portion is separated acutely. Clinical findings include depression, pale mucous membranes, abdominal pain, possibly vaginal hemorrhage and fetal distress. Grossly, blood clots are present between the placenta and the uterine wall and there is hemorrhage infiltrating the wall of the uterus. Retained placentae are not an uncommon occurrence in nonhuman primates. Clinical findings may include depression and a flaccid uterus upon abdominal palpation. Sepsis may occur. Retention may be related to dysfunctional uterine contractions, placenta accreta or uterine inversion. Placenta accreta refers to abnormal placental adherence to the uterine wall. Microscopically, there is an absence of the decidual portion of the placenta with attachment and, occasionally, penetration of the villi directly into the myometrium.

Lesions described in a study of 21 placentae from normal third trimester rhesus macaques included subchorionic, perivillous and perilobular fibrin accumulation, solitary infarcts, retroplacental hematoma, intervillous thrombosis, and calcification (Bunton, 1986). Microscopic findings included villous fibrinoid necrosis, acute inflammation in perilobular and basal plate regions, excessive cytotrophoblasts, syncytial knot formation and intimal hyperplasia of umbilical arteries or larger branches within the chorion. All were considered incidental findings but may become significant, with regard to fetal survival, if present to a large degree.

Male genital system

The literature contains few reports of noninfectious diseases or neoplasia involving male gonads or accessory sex glands. Squirrel monkeys and rhesus macaques, and a number of other nonhuman primate species, can demonstrate a physiologic seasonal variation in the reproductive cycle. Testicular size may increase during the mating season and microscopic examination reveals active spermatogenesis (Hunt *et al.*, 1993). Conversely, aspermatogenesis marks the regressive phase and should be distinguished from arrested spermatogenesis due to pathologic events such as cryptorchidism, hypopituitarism, estrogen-secreting neoplasms, malnutrition, obstruction of the ductus deferens or certain genetic disorders.

Testicular tumors, reported in nonhuman primates, include an interstitial cell tumor in a squirrel monkey, seminoma in an owl monkey and a rhesus macaque, a Sertoli cell tumor in an owl monkey and a cotton top tamarin and testicular adenocarcinoma in a rhesus macaque (cited from Gozalo *et al.*, 1992b; Lowenstine, 1986; Squire *et al.*, 1978). Clinically, testicular enlargement may be observed. The Sertoli cell tumor in the cotton top tamarin presented as testicular enlargement without clinical signs of feminization.

There have been sporadic reports of glandular (Adams and Bond, 1979; Baskerville *et al.*, 1992; Lewis *et al.*, 1981; Roberts, 1972; Steiner *et al.*, 1999) and fibromuscular hyperplasia of the prostate (Baskerville *et al.*, 1992; Lewis *et al.*, 1981). Prostatic basal cell hyperplasia has been reported once in the literature in a young adult cynomolgus macaque (Wakui *et al.*, 1989) and has been identified in two aged rhesus macaques (personal observation). Prostatic adenoma in two baboons (Lapin, 1982) and prostatic carcinoma in rhesus macaques (Engle and Stout, 1940; Hubbard *et al.*, 1985; O'Gara and Adamson, 1972) and squirrel monkeys (Lewis *et al.*, 1981; McClure *et al.*, 1999) have been reported.

Nervous system

Spontaneous neurologic disease in nonhuman primates, due to noninfectious causes, is sporadic. Many of the findings described in this section are incidental.

Aging and degenerative changes

Reported gross aging changes of the brain include narrowing of the gyri and widening of the sulci in aged macaques (cited in Cork and Walker, 1993). These are not consistent findings and were not observed in two studies involving rhesus macaques (Uno *et al.*, 1996) and baboons (Schultz *et al.*, 2000a). Increased pigmentation of the globus pallidus is another common gross finding that has been reported in the brains of rhesus macaques and baboons (Bronson and Schoene, 1980; Cork and Walker, 1993; Gearing *et al.*, 1994; Gliatto and Bronson, 1993b). Some age-associated histological or immunohistochemical lesions include spheroids, senile plaques and amyloid deposition, cerebral amyloid angiopathy, lipofuscin accumulation, myelin sheath alterations and tau pathology.

Spheroids

Spheroids have been described in the squirrel monkey (Cork and Walker, 1993), various macaque species (Bronson and Schoene, 1980; Cork and Walker, 1993; Gliatto and Bronson, 1993b; Schultz *et al.*, 2001; Willwohl *et al.*, 2002) and baboons (Bronson and Schoene, 1980; Willwohl *et al.*, 2002). They occur most frequently in the nucleus gracilis, globus pallidus, pars reticularis of the substantia nigra, medial vestibular nuclei and the ventral nucleus of the thalamus (Gliatto and Bronson, 1993b; Schultz *et al.*, 2001) and have been reported in the pars compacta of the substantia nigra in the rhesus macaque (Siddiqi and Peters, 1999). These are irregularly round to oval bodies that measure up to 50μ in diameter and contain eosinophilic granules or globules that are often PAS positive. An earlier study did not identify iron within the spheroids (Bronson and Schoene, 1980); however, a later study demonstrated that the spheroids observed contained iron and were immunoreactive for heat shock proteins (Schultz *et al.*, 2001; Willwohl *et al.*, 2002). These proteins orchestrate protective responses to stress at the cellular level and are present at low levels under physiologic conditions. Increased expression is observed under pathologic conditions such as anoxia, fever, oxidative stress or heavy metal (Santoro, 2000). The stress that precipitates this accumulation of heat shock proteins is unknown. There is speculation that iron-mediated oxidative stress may play a role (Schultz *et al.*, 2001; Willwohl *et al.*, 2002).

Senile plaques

Senile plaques commonly occur in the brains of aged rhesus, cynomolgus, and squirrel monkeys (Cork and Walker, 1993; Nakamura *et al.*, 1998; Sani *et al.*, 2003; Walker *et al.*, 1988). They can be detected in rhesus macaques beginning at 20 to 25 years of age (Cork and Walker, 1993; Struble *et al.*, 1985; Uno *et al.*, 1996) and cynomolgus macaques after the age of 20 years (Nakamura *et al.*, 1998). They are not easily seen with the typical hematoxylin and eosin stain. Special histochemical and immunohistochemical stains reveal that the constituents of the plaques generally include abnormal neuronal processes (neurites) and amyloid deposits in varying proportions. The main component of this amyloid is amyloid-β peptide, which is cleaved from the larger amyloid precursor protein, a membrane-associated glycoprotein found in serum, brain and cerebrospinal fluid (cited in Uno *et al.*, 1996). Plaques that occur in squirrel monkeys tend to be smaller than those in rhesus macaques and the neuritic component is not as pronounced (Walker, 1997). Two studies describe deposits of β-amyloid in the brains of marmosets, seven years and older (Geula *et al.*, 2002) and baboons (Schultz *et al.*, 2000a). The anatomic distribution of senile plaques is variable but they are located primarily in the gray matter of the frontal and temporal cortices and amygdala of rhesus and cynomolgus macaques (Cork and Walker, 1993; Martin *et al.*, 1994; Nakamura *et al.*, 1998; Struble *et al.*, 1985; Uno *et al.*, 1996). The distribution in squirrel monkeys involves predominantly the cerebral cortex, amygdala, parts of the septum verum and claustrum (Walker *et al.*, 1990). Amyloid plaques may co-exist with amyloid angiopathy (Gearing *et al.*, 1994; Geula *et al.*, 2002; Hartig *et al.*, 1997; Nakamura *et al.*, 1998; Schultz *et al.*, 2000a; Uno *et al.*, 1996). The mechanisms of plaque formation have not been fully elucidated. Neurites may reflect neuronal response to injury and its subsequent degeneration. Abnormalities of neuronal processes precede the amyloid deposits and exceed the frequency of them in rhesus macaques (Cork *et al.*, 1990). The pathologic significance of plaques in nonhuman primates has not been determined. Rhesus macaques do display a diminished cognitive status with advancing age (cited in Walker *et al.*, 1988); however, one study reports no correlation between cognitive deficits and the density of amyloid deposits (Sloane *et al.*, 1997). Plaques of somewhat similar appearance, distribution and chemical composition have been described in aged humans and patients with Alzheimer's disease (Sani *et al.*, 2003). Neurofibrillary tangles comprising paired helical filaments is a feature of Alzheimer's disease but has not been reported in the neurites of nonhuman primates (Gearing *et al.,* 1994; Geula *et al.,* 2002; Poduri *et al.,* 1994; Selkoe *et al.,* 1987; Uno *et al.,* 1996).

Cerebrovascular amyloidosis

Cerebrovascular amyloidosis or cerebral amyloid angiopathy has been reported in rhesus and cynomolgus macaques, chimpanzees, squirrel monkeys, baboons and marmosets (Cork and Walker, 1993; Gearing *et al.,* 1994; Geula *et al.,* 2002; Nakamura *et al.,* 1998; Sani *et al.,* 2003; Schultz *et al.,* 2000b; Schultz *et al.,* 2000a; Walker *et al.,* 1990). Advancing age is a risk factor. Cerebral amyloid angiopathy usually develops in squirrel monkeys by approximately 15 years of age (Walker, 1997). Affected areas of the brain include the neocortex, amygdala, neostriatum, hippocampus and septum verum (Walker, 1997). The rostral areas have a higher density of β-amyloid than the more caudal regions. In rhesus and cynomolgus macaques, cerebrovascular amyloidosis occurs in the mid to late twenties (Cork and Walker, 1993; Nakamura *et al.,* 1998). Microscopically, homogeneous, eosinophilic deposits of amyloid expand the walls of large arterioles of the neocortex and leptomeninges with some involvement of capillaries and venules in rhesus, cynomolgus and squirrel monkeys (Cork and Walker, 1993; Nakamura *et al.,* 1998; Sani *et al.,* 2003); (Walker *et al.,* 1990). Mechanisms of cerebrovascular amyloidosis and the source of β-amyloid have not been fully elucidated. Potential sources include vascular smooth muscle cells, cerebrospinal fluid, serum, and cells within the brain (Walker, 1997). Cerebrovascular amyloidosis occurs in humans and dogs and there is an increased risk of cerebral hemorrhage with this condition in those species (Walker, 1997).

Lipofuscin deposition

The increasing accumulation of lipofuscin pigment within neurons and glial cells is another age-related change (Cork and Walker, 1993; Peters, 2002; Siddiqi and Peters, 1999). Lipofuscin or "wear and tear" pigment is a brown-yellow complex of lipid and protein. It is a product of lipid peroxidation involving subcellular membranes and its presence is indicative of past free radical injury.

Miscellaneous

Storage diseases

Lysosomal storage diseases affecting only nervous tissue are rare in nonhuman primates. Globoid cell leukodystrophy caused by deficient galactocerebrosidase activity has been documented in a family of rhesus macaques (Baskin *et al.,* 1998). The three affected infants presented with tremors, hypertonia, incoordination, poor weight gain and died within the first four months of life. Grossly, one had enlarged peripheral nerves. Microscopically, there was rarefaction of the white matter of the cerebrum, cerebellum and spinal cord. High numbers of periodic acid-Schiff positive macrophages and multinucleated globoid cells were present. Luxol fast blue stains revealed a marked decrease in myelin.

Cerebral venous thrombosis

Cerebral venous thrombosis has been infrequently reported in rhesus macaques (Cork and Adams, 1993; Sheffield *et al.,* 1981). Clinical signs are variable and may be inapparent. Animals may present with mild to moderate neurological and cognitive deficits. Signs referable to an acute onset cerebral vascular accident may be seen. Abnormal gaits have been reported (Cork and Adams, 1993) and seizures have been observed (Sheffield *et al.,* 1981). Grossly, the white matter of the cerebrum contains multiple hemorrhagic foci (Figure 4.17). Chronic lesions appear as tan to brown foci. Lesions predominate in the white matter but can extend to the gray matter. The cerebellum is only rarely affected and the spinal cord is not affected (Cork and Adams, 1993). Acute and chronic lesions generally coexist and this temporal feature, as well as the lesion distribution, characterizes this entity. Microscopically, cerebral veins are often ectatic, congested and/or contain fibrin thrombi. There is rarefaction of the surrounding

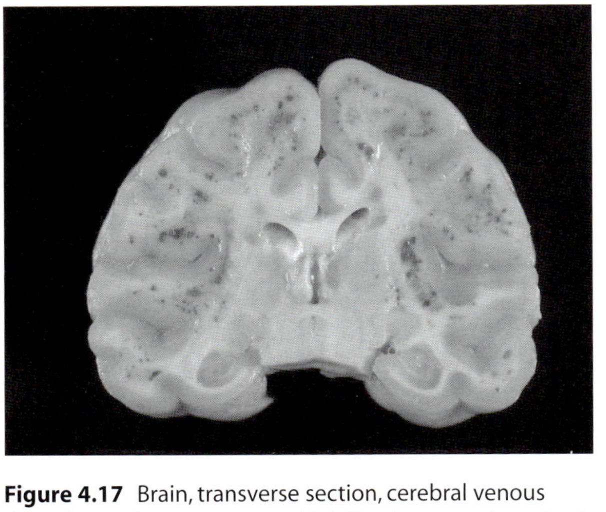

Figure 4.17 Brain, transverse section, cerebral venous thrombosis, rhesus macaque. Multifocal venous thrombosis and hemorrhage are concentrated in the white matter bordering the gray matter.

white matter that stains poorly for myelin (Cork and Adams, 1993), a gemistocytic astrogliosis, hemorrhage and accumulations of hemosiderin-laden macrophages. In more chronic lesions, veins may contain organized, fibrotic and occlusive thrombi that are often recanalized and occasionally mineralized. The etiology is unknown. Some macaques have a history of clinical disease, which could lead to hemoconcentration or a hypercoagulable state such as diarrhea (Cork and Adams, 1993; Sheffield *et al.*, 1981). Presumably, these conditions alter cerebral hemodynamics resulting in thrombosis. Others lack any predisposing history. Risk factors in humans include regional infections, oral contraceptives, pregnancy, genetic prothrombotic conditions, mechanical factors (trauma, neurosurgery, lumbar puncture), some chemotherapeutic regimens and systemic disease (vasculitis, neoplasia, hematologic disorders, congestive heart failure, nephrotic syndrome) (Carvalho *et al.*, 2001; Stam, 2003).

Cerebral hemorrhage

Cerebral hemorrhage and malacia were reported in two cynomolgus macaques with hypernatremia resulting from accidental water deprivation (Harber *et al.*, 1996). Microscopic findings included areas of necrosis and hemorrhage with an acute inflammatory infiltrate and venous thrombi, both fibrinous and organizing. Unlike cerebral venous thrombosis, the gray matter rather than the white matter was primarily involved. These lesions are similar to those described in humans with hypernatremia (cited in Harber *et al.*, 1996). The vascular changes are postulated to occur because of the close vascular attachments of the brain to the cranium. Hypernatremia causes cellular dehydration with overall shrinkage of the brain and consequent tearing of vessels resulting in hemorrhage and thrombosis.

Neoplasia

Nervous system neoplasms are rare and some that appear in the literature include astrocytomas in rhesus macaques, a cynomolgus macaque and baboon (Herring *et al.*, 1990; Lowenstine, 1986; McClure, 1975; Yanai *et al.*, 1992), neurohypophyseal astrocytoma in a rhesus macaque (HogenEsch *et al.*, 1992), meningioma in a baboon (Lowenstine, 1986), olfactory nerve esthesioblastoma in a cynomolgus macaque (Beniashvili, 1989), medulloblastoma in a baboon (Berthe *et al.*, 1980), choroid plexus lipoma in a baboon (Fiori *et al.*, 1994), and an oligodendroglioma in an owl monkey (cited in Weller, 1994).

Integumentary system

Skin disease due to infectious and noninfectious causes is relatively common in nonhuman primates (Hubbard, 2001).

Physicochemical injury

Physical injury to skin may occur with exposure to cold, supplemental heat, sunlight or trauma. Cold injury may involve the digits, ear tips or distal tail. Lesions result primarily from vascular damage and ensuing tissue ischemia. Grossly, alopecia, scaling and pigmentary changes may be seen. Severely damaged sites may slough due to ischemic necrosis.

Burns secondary to thermal injury may be classified as partial or full-thickness according to the depth of injury. Erythema is seen initially with possible progression to erosion, ulceration, and necrosis depending on the level of trauma. Microscopically, there is coagulative necrosis and edema. Serocellular crusts overlie ulcerated areas. In full-thickness burns, the adnexa, blood vessels and even the subcutis may be involved. Vascular thrombosis may be present. Severely affected skin lesions heal by extensive scar tissue formation with loss of adnexa.

Solar radiation may result in cutaneous injury to hairless or sparsely haired, nonpigmented, exposed skin. Erythema is present grossly. The sequelae for more severe lesions are alopecia, scarring and pigmentary changes.

Self-inflicted injury or trauma, due to intraspecific aggression, may be observed with extensive damage to the underlying soft tissues. Secondary bacterial infections are common. Severe trauma with muscle necrosis and hemorrhage may result in shock, sepsis and acute renal tubular necrosis.

An irritant contact dermatitis may result from chemical exposure. The distribution of lesions typically involves surface contact points such as the lateral aspects of the limbs, pelvis, ischial callosities and shoulders. Gross lesions include erythema, edema, and papulovesicular eruptions (Yager and Scott, 1993). Sloughing of damaged skin may occur in severe cases. The sequelae for more severe lesions are similar to those described for sun exposure.

Calcinosis circumscripta has been reported in rhesus macaques (Line *et al.*, 1984; Marini *et al.*, 1999). Firm, raised nodules containing white, pasty material were described within the subcutis of the feet. Microscopically, coalescing foci of calcareous lakes were surrounded by a mixed inflammatory infiltrate and mature fibrous connective tissue. The pathogenesis is not resolved but the lesion may represent dystrophic calcification at sites of possible previous trauma.

Aging changes

The more common dermatologic changes reported in a group of aged rhesus macaques were increased amounts of erythematous skin and focal scaling (Huneke *et al.,* 1996). Other lesions included wrinkling, increased laxity of the skin, foot calluses, thinning hair, proliferative growths, and exudative and erosive lesions. Microscopic examination revealed increased epidermal thickness, telangiectasis, superficial subacute dermatitis, and exophytic proliferation of hyperkeratotic stratified squamous epithelium on fibrous stalks. Flat, round, red foci measuring up to 3 mm in diameter are often present on the face and chest of aged macaques and correspond microscopically to foci of cutaneous telangiectasis.

Miscellaneous

Disseminated intravascular coagulation may present as petechial cutaneous hemorrhages and is caused by a variety of factors including hyperthermia, sepsis, shock, tissue necrosis, vascular stasis, and acidosis. Histologically, fibrin thrombi may be present within blood vessels. A prolonged interval between death and the postmortem examination may preclude visualization of thrombi due to ongoing fibrinolysis.

Some nonhuman primates such as macaques, baboons and chimpanzees have sex skin that can swell dramatically during the reproductive cycle. Sex skin is prominent in the perineal region and around the tail base of females but may be seen on the dorsal aspect of the thighs, on the dorsal and ventral midline and the face, even in males. Microscopically, the skin is edematous and highly vascularized during this tumescent phase.

Neoplasia

Reports of spontaneous cutaneous neoplasia in Old World primates are relatively common in the literature. Benign and malignant neoplasms of the skin and subcutis are reported with similar frequency (Brack and Martin, 1984; Lowenstine, 1986; O'Gara and Adamson, 1972); however, those of mammary glands are predominantly malignant. Epithelial tumors are documented more frequently than mesenchymal tumors.

Skin tumors are infrequent in New World monkeys and one published survey cites three benign tumors; hemangioma in a spider monkey and brown capuchin, and a papilloma on the eyelid of a spider monkey (Lowenstine, 1986). Malignant tumors include subcutaneous adenocarcinoma and leiomyosarcoma in squirrel monkeys, a paratrichial sweat gland adenocarcinoma in a marmoset, a papillary carcinoma of apocrine sweat glands in a capuchin, a squamous cell carcinoma in a white-lipped tamarin, and a mammary adenocarcinoma in a male squirrel monkey (Brunnert *et al.,* 1990b; Cameron and Conroy, 1976; Khan *et al.,* 1999; McClure, 1980; Richter and Buyukmihci, 1979; Waggie *et al.,* 2000).

Reported benign epithelial tumors in Old World species include basal cell tumors in rhesus, Japanese and cynomolgus macaques, papilloma in a rhesus macaque, apocrine cystadenoma in a macaque, trichoepithelioma in a Barbary ape (*Macaca sylvanus*), epidermal inclusion cyst in a rhesus, and multiple papilliferous cystadenoma in a cynomolgus macaque (Beniashvili, 1989; Brack and Martin, 1984; Kim and Palazzo, 1978; Lowenstine, 1986; McClure, 1980; Seibold and Wolf, 1973; Yanai *et al.,* 1995c). Malignant neoplasms include squamous cell carcinomas in baboons, cynomolgus and rhesus macaques, and a malignant basal cell tumor in a rhesus macaque (Knightly *et al.,* 1996; Brack and Martin, 1984; Hubbard *et al.,* 1983; Hubbard *et al.,* 1984; Lowenstine, 1986; McClure, 1980; Migaki *et al.,* 1971; Morin *et al.,* 1980). Mammary gland tumors include fibroadenomas in macaques and carcinomas in rhesus macaques and a baboon (Beniashvili, 1989; Cohen *et al.,* 2001; Eydelloth and Swindle, 1983; Hubbard *et al.,* 1984; Lowenstine, 1986; Smith *et al.,* 2000; Tekeli and Ford, 1980).

Benign mesenchymal tumors include lipomas in macaques, an African green monkey, and baboons, and fibromas in pigtailed and rhesus macaques (Beniashvili, 1989; Chapman, Jr., 1968; Lowenstine, 1986; McClure, 1980; Seibold and Wolf, 1973; Squire *et al.,* 1978). Malignant mesenchymal tumors include a fibrosarcoma with metastasis, extraosseous osteosarcoma, and hemangiosarcoma in rhesus macaques (Gliatto *et al.,* 1990; Myers, Jr. *et al.,* 2001; Todd *et al.,* 1973).

Mast cell tumors have been reported in a baboon and two rhesus macaques (Colgin and Moeller, 1996; Seibold and Wolf, 1973) and a melanocytoma was described in a rhesus macaque (Frazier *et al.,* 1993).

Musculoskeletal system

Disorders of joints are some of the most common musculoskeletal lesions in nonhuman primates and include primary and secondary degenerative changes, infectious and metabolic disease, trauma, immune-mediated disease, deformity, crystal arthropathies, and neoplasms. This section will cover primary osteoarthritis and spondyloarthropathies.

Osteoarthritis

Primary osteoarthritis has been reported in cynomolgus, rhesus and pigtailed macaques, baboons, and owl monkeys (Carlson *et al.*, 1994; Chateauvert *et al.*, 1989; Rothschild, 1993; Rothschild *et al.*, 1999; Rothschild and Woods, 1992). This degenerative joint disease is characterized primarily by progressive destruction of articular cartilage matrix, bone abnormalities and minor synovial changes. It increases in prevalence and severity with advancing age and can affect the spine, knee, hip, elbow, wrist and interphalangeal joints. Female rhesus macaques showed a higher frequency of disease than did males in one study (DeRousseau, 1988). There was no correlation between articular lesions of the knee and factors of gender and weight in a study of cynomolgus macaques (Carlson *et al.*, 1996). Increased parity in rhesus macaque females was associated with increased frequency in another study (Chateauvert *et al.*, 1989). Caging did not affect the incidence of osteoarthritis; however, caged monkeys had less severe disease than did their free ranging counterparts (Chateauvert *et al.*, 1989). Clinical signs may be absent or include decreased range of motion, decreased activity, gait abnormalities, crepitus, and joint enlargement and deformity. Radiographs often reveal narrowing of the joint space, increased thickness of the subchondral bone, subchondral bone cysts and osteophyte formation at the joint periphery. Gross softening, fibrillation, erosion, and ulceration of the articular cartilage may be observed as well as eburnation of subchondral bone and osteophytosis. Key microscopic features include fibrillation, clefting and erosion of the cartilage matrix, chondrocyte necrosis, proliferation of chondrocytes, and subchondral bone proliferation (Carlson *et al.*, 1994; Chateauvert *et al.*, 1989). Hyperplasia of the synovial lining cells and a minimal inflammatory infiltrate composed of few lymphocytes and plasma cells may be present (Carlson *et al.*, 1994).

The pathogenesis of this disorder is unknown and it is not clear if the initiating changes occur primarily in the articular cartilage or in the subchondral bone (Bailey and Mansell, 1997). Both structures are important for synovial joint health and function and defects in either are potential causes of lesions in the other.

Spondyloarthropathies

In the human literature, spondyloarthropathies are inflammatory diseases often affecting the sacroiliac, spinal and peripheral joints. They are characterized by arthritis and inflammation at the insertions of ligaments, tendons or joint capsules to bone. There is a strong association between the *HLA-B27* gene and an absence of rheumatoid factor in humans. A series of skeletal studies documented spondyloarthropathy in Old World and New World monkeys including *Papio, Macaca,* and *Callithrix* (Rothschild, 1993; Rothschild *et al.*, 1997; Rothschild and Woods, 1992). Bony erosion or fusion of joints was observed and affected joints included the sacroiliac, ankle, knee, metacarpophalangeal, interphalangeal of the hand, carpal, elbow, shoulder, and intervertebral joints. Clinical signs may be absent or include lameness and mildly swollen joints. A stiffened gait, joint contracture, muscle atrophy, and kyphosis occur with disease progression. Grossly, there is articular cartilage and subchondral bone erosion and a thickened synovium partially overlying the cartilaginous surface and present at the leading edge of the erosions (Pritzker and Kessler, 1998). Joint fusion and osteophytosis of the vertebral bodies may be observed with chronic disease. Microscopically, with active disease, an inflammatory infiltrate composed of neutrophils, lymphocytes, macrophages and plasma cells in varying concentrations may be present. There is proliferation of the synovial lining cells and erosion of the articular cartilage and subchondral bone. Subsequently ankylosis may develop.

Reactive arthritis is considered to be a subset of the spondyloarthropathies in humans and follows infection with enteric or urogenital pathogens such as *Yersinia, Salmonella, Campylobacter, Shigella* and *Chlamydia.* Typically, bacterial agents are not cultured from affected joints. Reactive arthritis was reported following a natural outbreak of shigellosis in 13 of 33 rhesus macaques (Urvater *et al.*, 2000). Clinically, one to two months after infection the animals presented with partial to non-weight bearing lameness. The most commonly affected joints were the stifle, elbow, coxofemoral and interphalangeal joints. Joint effusion varied from a minimal to moderate degree. Neutrophilia, with a left shift, was present during the diarrheic phase. A mature neutrophilia was observed during the arthritic stage. Synovial fluid analysis revealed mature neutrophils. Bacteria were not apparent microscopically and cultures were negative. Progression of clinical signs to severe muscle atrophy with joint contracture occurred in some animals.

Multisystemic diseases

Nutritional

Vitamin C deficiency

Nonhuman primates have a dietary requirement for Vitamin C. Outbreaks of vitamin C deficiency occur occasionally due to defects in manufacturing or handling of commercial feeds (Blackwell *et al.*, 1974; Demaray *et al.*, 1978; Eisele *et al.*, 1992; Kessler, 1980; Ratterree *et al.*, 1990). There are species dependent differences in the manifestation of disease. Squirrel monkeys, particularly juvenile and subadults, develop subcutaneous and subperiosteal hemorrhages of the head, leading to ossification and exostoses. Cephalhematomas result in enlargement of the head (described as "turban head") and deformation of facial features (Blackwell *et al.*, 1974). Affected animals may also exhibit weakness and gingival hemorrhages (Ratterree *et al.*, 1990). Rhesus macaques with scurvy develop changes in the long bones, teeth and ribs. Presenting clinical signs are lameness, reluctance to move and weakness (Ratterree *et al.*, 1990). Animals may exhibit swelling of the wrists, periorbital hemorrhage, cutaneous bruising, anemia and loose teeth. Lesions include metaphyseal fractures, periosteal hemorrhages and gingival hemorrhage (Roberts, 1993; Shaw *et al.*, 1945).

The basis of the lesions, in both species, is defects in collagen and intercellular cement formation resulting in increased vascular fragility and defects in bone formation. Nonhuman primates lack L-gulono-gamma-lactone oxidase and thus are unable to synthesize L-ascorbic acid. Ascorbic acid is required for the synthesis of hydroxyproline, an important component of collagen (Jones *et al.*, 1997).

Vitamin E deficiency

Several Vitamin E related diseases have been reported. In callitrichids, a syndrome of anemia and skeletal muscle necrosis associated with low serum levels of vitamin E has been described. A Heinz body hemolytic anemia, concurrent with skeletal muscle myopathy, has been described in a colony of common marmosets (*Callithrix jacchus*) and resulted in illness and death in over half of the 348 animals in the study (Chalmers *et al.*, 1983). Vitamin E was not detected in the sera of affected monkeys. A similar syndrome was reported in rufiventer marmosets (*Saguinus labiatus*) (Baskin *et al.*, 1983). Anemia, lumbosacral and rear limb muscle atrophy, skeletal muscle necrosis, and generalized steatitis were found in these animals. Serum levels of vitamin E were low despite normal dietary levels and adequate diet intake. Neither parenteral nor oral Vitamin E supplementation resulted in raised serum levels but did ameliorate disease in one group of animals (Baskin *et al.*, 1983).

Vitamin E responsive anemia occurs in owl monkeys and is restricted to specific karyotypes of this species (King, 1993). The disease is characterized by hemolytic anemia and skeletal muscle necrosis (Bronson, 1980; Chalifoux *et al.*, 1981). Additional findings attributed to anemia are centrilobular hepatic necrosis and cerebral ring hemorrhages (King, 1993). Although serum vitamin E levels were not low in affected animals, administration of parenteral vitamin E and selenium was effective in treating the disease (Sehgal *et al.*, 1980).

Cardiomyopathy attributed to Vitamin E deficiency resulted in death of seven of 14 Gelada baboons. Grossly, the hearts appeared pale and streaked. Microscopic findings included myocyte necrosis and fibrosis (Liu *et al.*, 1984). Supplementation with dietary Vitamin E prevented recurrence of disease within the remaining group (Liu *et al.*, 1985).

Miscellaneous multisystemic diseases

Fatal fasting syndrome

Fatal fasting syndrome, also known as fatal fatty liver syndrome, has been reported in four species of macaques (*Macaca fascicularis, M. mulatta, M. arctoides,* and *M. radiata*) and one African green monkey (Bronson *et al.*, 1982; Christe and Valverde, 1999; Laber-Laird *et al.*, 1987). Obese, adult female macaques appear predisposed (Bronson *et al.*, 1982; Laber-Laird *et al.*, 1987). Clinically, a brief period of anorexia, precipitous

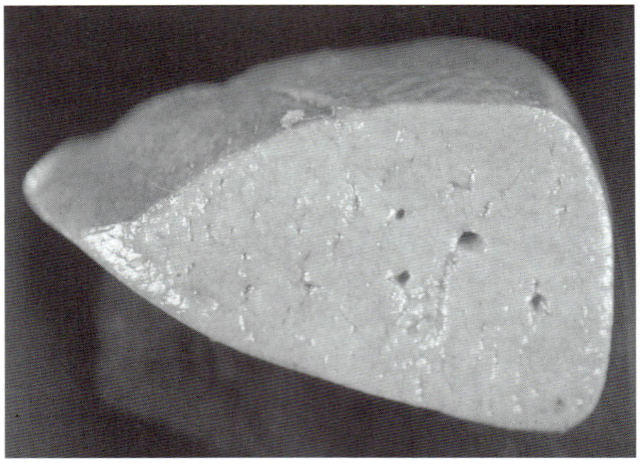

Figure 4.18 Liver, hepatic lipidosis, fatal fasting syndrome, rhesus macaque. Note the smooth, tense capsule, rounded margins and the greasy appearance of this liver section.

weight loss of up to 30% and sudden death are seen. Serum biochemical profiles often reveal azotemia. At necropsy, most animals have abundant stores of subcutaneous and visceral fat despite recent weight loss. Dull white granular flecks corresponding to foci of necrosis are often dispersed throughout the abdominal fat. The liver is enlarged, pale and friable (Figure 4.18). The kidneys are frequently pale and soft and the gastrointestinal tract is often empty. Microscopically, hepatocytes (Figure 4.19) and proximal renal tubular epithelium (Figure 4.20) exhibit fatty vacuolar change (Bronson *et al.*, 1982; Laber-Laird *et al.*, 1987). Renal tubular necrosis may be present. Pancreatic lesions are described and include small foci of necrosis, acinar ectasia, acinar cell attenuation and decreased zymogen

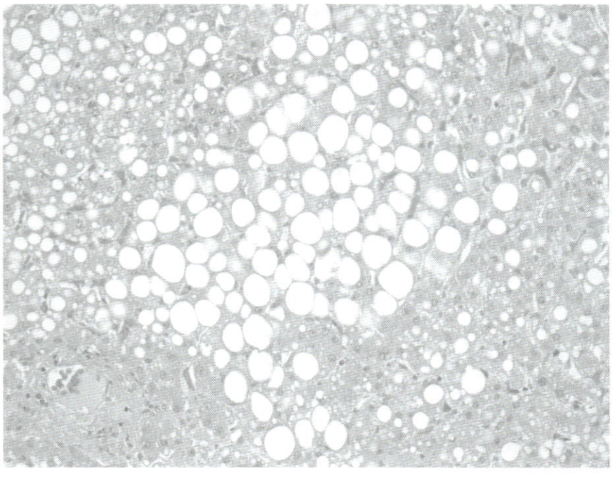

Figure 4.19 Liver, hepatic lipidosis, fatal fasting syndrome, rhesus macaque. Hepatocytes contain discrete, variably sized, round cytoplasmic vacuoles that occasionally displace the nucleus. H&E, original magnification 200×.

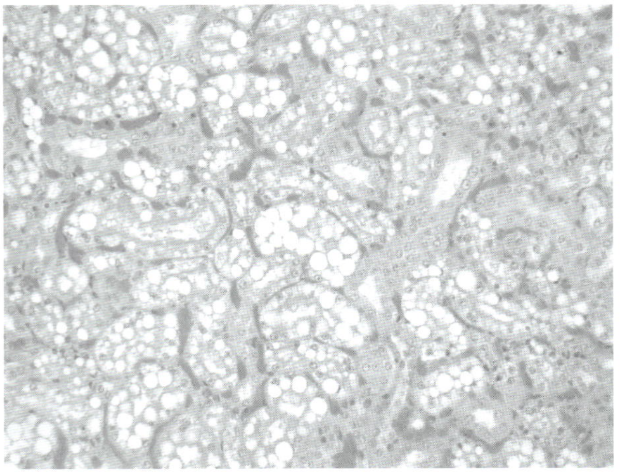

Figure 4.20 Kidney, fatal fasting syndrome, rhesus macaque. Note the cytoplasmic vacuolation of the proximal tubular epithelial cells. H&E, original magnification 100×.

content (Bronson *et al.*, 1982; Laber-Laird *et al.*, 1987). Ten of 19 animals in one study had grossly normal thyroid glands but demonstrated diffusely distended follicles on microscopic examination (Laber-Laird *et al.*, 1987). Association with a hypothyroid state was not determined.

The occurrence of this syndrome does not appear to be associated with any one disease or husbandry practice. Stressful situations such as a housing change may induce suboptimal feed intake (Bronson *et al.*, 1982; Laber-Laird *et al.*, 1987). The pathogenesis is not understood but the rapid weight loss indicates a negative energy balance that presumably initiates fatty acid mobilization from body fat depots with ensuing fatty change in the liver and kidney. Blood chemistry profiles are more consistent with renal malfunction rather than liver failure. This syndrome has similarities to hyperlipemia in ponies and hepatic lipidosis in anorexic obese cats and guinea pigs (cited from Gliatto and Bronson, 1993a).

Systemic amyloidosis

Systemic amyloidosis is a relatively common disease of several species of nonhuman primates. It is one of a group of diverse amyloid diseases characterized by the depositioning in tissue of insoluble protein fibrils in a beta sheet configuration. Amyloidoses are classified by the precursor protein deposited in tissues. In systemic or reactive amyloidosis, amyloid A (AA) fibrils are derived from serum AA, an acute phase reactant protein produced by the liver. The disease has been described in a number of nonhuman primate species including rhesus macaques, pigtail macaques, chimpanzees, baboons, squirrel monkeys, cynomolgus macaques, a mangabey, and a stumptailed macaque (Banks and Bullock, 1967; Blanchard *et al.,* 1986; Doepel *et al.,* 1984; Hubbard *et al.,* 2001; Hubbard *et al.,* 2002; Jennings *et al.,* 1995; McClure, 1984; Slattum *et al.,* 1989b). Retrospective studies involving larger numbers of individuals have been published on rhesus macaques (Blanchard *et al.,* 1986; Doepel *et al.,* 1984; Rodger *et al.,* 1980), pigtail macaques (Ellsworth *et al.,* 1992; Slattum *et al.,* 1989a; Slattum *et al.,* 1989b) and chimpanzees (Hubbard *et al.,* 2001). A bimodal age distribution is seen in both pigtail and rhesus macaques; larger numbers of cases are seen in juvenile to subadult animals under four years of age and mature adults ten years and older (Blanchard *et al.,* 1986; Slattum *et al.,* 1989a). There is no sex predisposition.

The clinical signs of generalized amyloidosis are related to the organs of deposition and include weight loss, diarrhea, hepatomegaly and splenomegaly. Clinical disease is generally a reflection of protein losing enteropathy, induced by amyloid deposition in the small intestine. Laboratory findings are inconsistent but hypoalbuminemia, hypergammaglobulinemia and an albumin:globulin (A:G) ratio of one or less are suggestive (McClure, 1984). Hepatic enzyme levels are generally unaffected even in the presence of severe amyloid deposition.

Amyloid is most frequently found in the gastrointestinal tract, liver and spleen. In severe cases it may also be found in lymph nodes, adrenals, thyroid, gallbladder and kidney. Gross appearance is variable. There may be inconspicuous to massive enlargement of the liver. In severe cases, the liver may comprise up to 14% of the body weight. The liver appears red to pale tan and is firm to waxy in texture (Figure 4.21). The spleen is enlarged, pale red and waxy. Deposition in the intestine is generally not grossly visible but may present as pallor and thickening of the mucosa. Deposition in other organs is not grossly visible.

Histologically, amyloid appears as a homogeneous to slightly fibrillar, extracellular eosinophilic material. In the liver, amyloid deposition begins in the space of Disse and extends along hepatic cords and into sinusoids. Progressive deposition leads to hepatic cord atrophy and almost complete effacement of normal architecture (Figure 4.22). In the gastrointestinal tract, the most severe accumulation is in the lamina propria of the small intestine (Figure 4.23). Severely affected cases may show deposition of lesser amounts of amyloid in the stomach and colon. Amyloid is deposited predominantly in the interstitium at the corticomedullary junction of the adrenal gland (Figure 4.24). In the kidney, the majority of the deposition is in the medullary interstitium; the glomeruli are rarely involved (Blanchard *et al.,* 1986; McClure, 1984; Slattum *et al.,* 1989b).

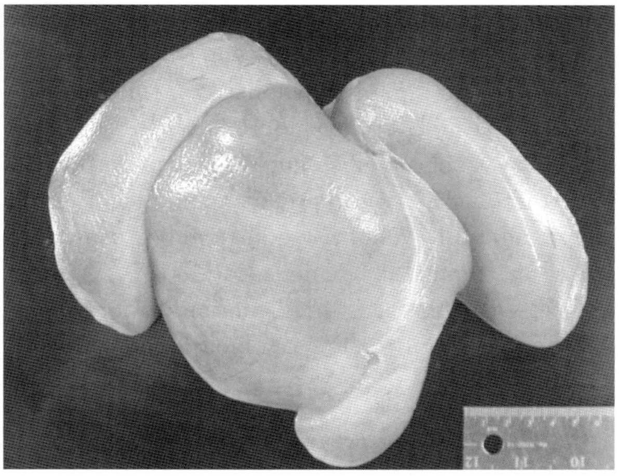

Figure 4.21 Liver, systemic amyloidosis, rhesus macaque. The markedly enlarged liver appears pale, waxy, dull and dry. Note the rounded hepatic margins.

The pathogenesis of generalized amyloidosis is poorly understood. Factors associated with the development of amyloidosis in nonhuman primates include chronic enterocolitis, osteoarthritis, chronic vascular catheterization, and retroperitoneal fibromatosis associated with Type D retroviral infection (Blanchard *et al.,* 1986; Doepel *et al.,* 1984; Ellsworth *et al.,* 1992; McClure, 1984; Slattum *et al.,* 1989a). In humans, generalized amyloidosis is associated with conditions

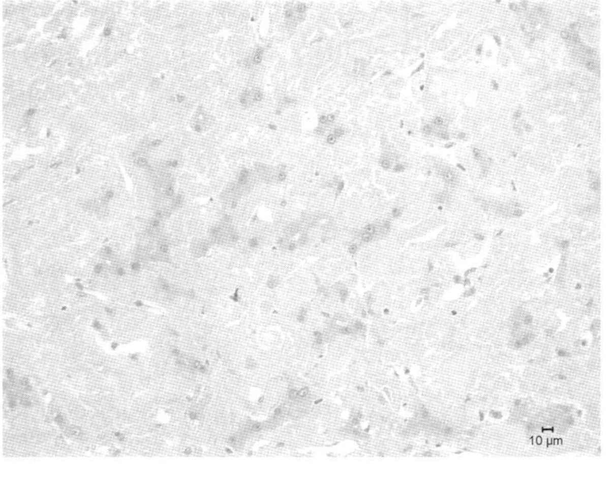

Figure 4.22 Liver, systemic amyloidosis, rhesus macaque. Note the abundant amyloid deposition and marked attenuation of hepatic cords. H&E, original magnification 200×.

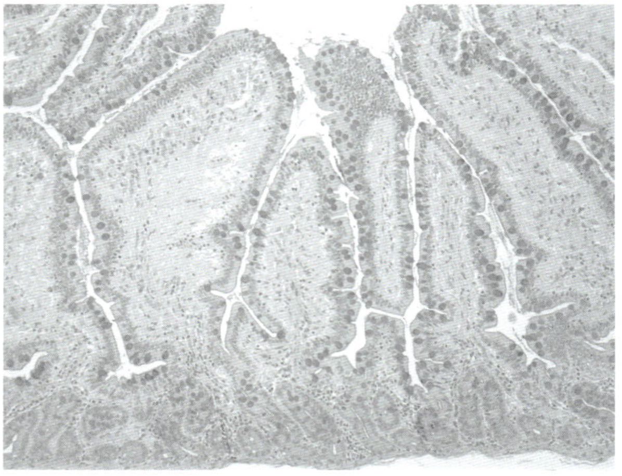

Figure 4.23 Small intestine, systemic amyloidosis, rhesus macaque. The lamina propria is expanded by diffuse deposition of amorphous acellular material (amyloid). H&E, original magnification 100×.

of underlying inflammation, such as rheumatoid arthritis, chronic sepsis, Crohn's disease and familial Mediterranean fever. In both man and nonhuman primates, serum AA is highly elevated in patients with active inflammation and generalized amyloidosis (Doepel *et al.,* 1981; Gillmore *et al.,* 2001). The factors leading to the development of generalized amyloidosis in a smaller proportion of patients with chronic inflammatory disease are unclear. They may relate to genetically determined differences in the precursor protein, impaired processing of circulating AA and or the presence of enhancing cofactors (reviewed in Cunnane, 2001).

Figure 4.24 Adrenal gland, systemic amyloidosis, rhesus macaque. Abundant amorphous amyloid is deposited at the corticomedullary junction. H&E, original magnification 200x.

Acknowledgement

This work was supported in part by Public Health Service Award RR00163 from the National Center for Research Resources.

Correspondence

Any correspondence should be directed to Anne Lewis, Division of Animal Resources, Oregon National Primate Research Center, Oregon Health and Science University, 505 NW 185th Avenue, Beaverton, OR 97006, USA. Email: lewisann@ohsu.edu

References

Adams, M.R. and Bond, M.G. (1979). *Lab. Anim. Sci.* 29(5), 674–676.

Adler, R.R., Moore, P.F., Schmucker, D.L. and Lowenstine, L.J. (1993). In Jones, T.C., Mohr, U. and Hunt, R.D. (eds) *Monographs on Pathology of Laboratory Animals: Nonhuman Primates II,* pp. 81-86. Berlin and New York: Springer-Verlag.

Ami, Y., Suzaki, Y. and Goto, N. (1993). *J. Vet. Med. Sci.* 55(1), 7–11.

Amin, H.K. (1974). *J. Comp. Pathol.* 84(2), 161–167.

Baer, J.F., Gibson, S.V., Weller, R.E., Buschbom, R.L. and Leathers, C.W. (1992). *Lab. Anim. Sci.* 42(5), 463–466.

Bailey, A.J. and Mansell, J.P. (1997). *Gerontology* 43(5), 296–304.

Banks, K.L. and Bullock, B.C. (1967). *J. Am. Vet. Med. Assoc.* 151(7), 839–842.

Baskerville, A., Cook, R.W., Dennis, M.J., Cranage, M.P. and Greenaway, P.J. (1992). *J. Comp. Pathol.* 107(1), 49–57.

Baskerville, M., Baskerville, A. and Manktelow, B.W. (1984). *J. Comp. Pathol.* 94(3), 329–338.

Baskin, G.B. and Hubbard, G.B. (1980). *Vet. Pathol.* 17(1), 100–102.

Baskin, G.B., Ratterree, M., Davison, B.B., Falkenstein, K.P., Clarke, M.R., England, J.D., Vanier, M.T., Luzi, P., Rafi, M.A. and Wenger, D.A. (1998). *Lab. Anim. Sci.* 48(1), 476–482.

Baskin, G.B., Wolf, R.H., Worth, C.L., Soike, K., Gibson, S.V. and Bieri, J.G. (1983). *Lab. Anim. Sci.* 33(1), 74–80.

Beniashvili, D.Sh. (1989). *J. Med. Primatol.* 18(6), 423–437.

Benjamin, S.A. and Lang, C.M. (1969). *J. Am. Vet. Med. Assoc.* 155(1), 1236–1240.

Bertens, A.P., Helmond, F.A. and Hein, P.R. (1982). *Lab. Anim.* 16(3), 281–284.

Berthe, J., Barneon, G., Richer, G. and Mazue, G. (1980). *Lab. Anim. Sci.* 30(4), Pt 1, 703–705.

Betton, G.R. (1984). *Vet. Pathol.* 21(2), 193–197.
Birkebak, T.A., Wang, N.P. and Weyhrich, J. (1996). *J. Med. Primatol.* 25(5), 367–369.
Blackwell, C.A., Manning, P.J., Hutchinson, T.C. and Fisk, S.K. (1974). *Lab. Anim. Sci.* 24(3), 541–544.
Blanchard, J.L., Baskin, G.B. and Watson, E.A. (1986). *Vet. Pathol.* 23(4), 425–430.
Brack, M. (1988). *Vet. Pathol.* 25(4), 270–276.
Brack, M. (1995). *Lab. Anim.* 29(1), 54–58.
Brack, M. (1998). *J. Med. Primatol.* 27(6), 319–324.
Brack, M. and Martin, D.P. (1984). *J. Med. Primatol.* 13(3), 159–164.
Brack, M. and Rothe, H. (1981). *Vet. Pathol.* 18, Suppl 6, 45–54.
Brack, M. and Weber, M. (1995). *Nephron* 69(3), 286–292.
Bronson, R.T. (1980). *J. Neurol. Sci.* 47(1), 105–109.
Bronson, R.T., O'Connell, M., Klepper–Kilgore, N., Chalifoux, L.V. and Sehgal, P. (1982). *Lab. Anim. Sci.* 32(2), 187–192.
Bronson, R.T. and Schoene, W.C. (1980). *J. Neuropathol. Exp. Neurol.* 39(2), 181–196.
Brown, R.J. and Kupper, J.L. (1972). *Lab. Anim. Sci.* 22(5), 741–742.
Brunnert, S.R., Herron, A.J. and Altman, N.H. (1990a). *Vet. Pathol.* 27(5), 372–374.
Brunnert, S.R., Herron, A.J. and Altman, N.H. (1990b). *Vet. Pathol.* 27(2), 126–128.
Bunton, T.E. (1986). *Vet. Pathol.* 23(4), 431–438.
Bunton, T.E. (1993) In Jones, T.C., Mohr, U. and Hunt, R.D. (eds) *Nonhuman Primates II,* pp. 118–122. Berlin Springer-Verlag.
Bunton, T.E. and Lollini, L. (1983). *J. Med. Primatol.* 12(2), 106–111.
Cameron, A.M. and Conroy, J.D. (1976). *J. Med. Primatol.* 5(1), 56–59.
Carlson, C.S., Loeser, R.F., Jayo, M.J., Weaver, D.S., Adams, M.R. and Jerome, C.P. (1994). *Journal of Orthopaedic Research* 12(3), 331–339.
Carlson, C.S., Loeser, R.F., Purser, C.B., Gardin, J.F. and Jerome, C.P. (1996). *Journal of Bone and Mineral Research* 11(9), 1209–1217.
Carvalho, K.S., Bodensteiner, J.B., Connolly, P.J. and Garg, B.P. (2001). *J. Child. Neurol.* 16(8), 574–580.
Chalifoux, L.V. (1993a) In Jones, T.C., Mohr, U. and Hunt, R.D. (eds) *Monographs on Pathology of Laboratory Animals: Nonhuman Primates II,* pp. 155–160. Berlin and New York: Springer-Verlag.
Chalifoux, L.V. (1993b) In Jones, T.C., Mohr, U. and Hunt, R.D. (eds) *Monographs on Pathology of Laboratory Animals: Nonhuman Primates II,* pp. 150–154. Berlin and New York: Springer–Verlag.
Chalifoux, L.V., Brieland, J.K. and King, N.W. (1985). *Dig. Dis. Sci.* 30(12) Suppl, 54S–58S.
Chalifoux, L.V. and Bronson, R.T. (1981). *Gastroenterology* 80(5), Pt 1, 942–946.
Chalifoux, L.V., Bronson, R.T., Sehgal, P., Blake, B.J. and King, N.W. (1981). *Vet. Pathol.* 18(6), 23–37.
Chalifoux, L.V., King, N.W. Jr. and Johnson, L.D. (1993) In Jones, T.C., Mohr, U. and Hunt, R.D. (eds) *Monographs on Pathology of Laboratory Animals: Nonhuman Primates II,* pp. 87–93. Berlin and New York: Springer-Verlag.
Chalmers, D.T., Murgatroyd, L.B. and Wadsworth, P.F. (1983). *Lab. Anim.* 17(4), 270–279.
Chapman, W.L. Jr. (1968). *J. Am. Vet. Med. Assoc.* 153(7), 872–878.
Chapman, W.L., Jr., Crowell, W.A. and Isaac, W. (1973). *Lab. Anim. Sci.* 23(3), 434–442.
Chateauvert, J., Grynpas, M.D., Kessler, M.J. and Pritzker, K.P.H. (1989). *J. Rheumatol.* 17(1), 73–83.
Christe, K.L. and Valverde, C.R. (1999). *Contemp. Top. Lab Anim. Sci.* 38(4), 12–15.
Clapp, N.K., Littlefield, L.G. and Lushbaugh, C.C. (1982). *Gastroenterology* 83(2), 519.
Clapp, N.K., Lushbaugh, C.C., Humason, G.L., Gangaware, B.L. and Henke, M.A. (1985a). *Dig. Dis. Sci.* 30(12), 107S–113S.
Clapp, N.K., Lushbaugh, C.C., Humason, G.L., Gangaware, B.L., Henke, M.A. and McArthur, A.H. (1985b). *Prog. Clin. Biol. Res.* 186, 247–261.
Clarkson, T.B. (1998). *Lab. Anim. Sci.* 48(6), 569–572.
Clarkson, T.B., Hamm, T.E., Bullock, B.C. and Lehner, N.D. (1976a). *Primates Med.* 9, 66–89.
Clarkson, T.B., Lehner, N.D., Bullock, B.C., Lofland, H.B. and Wagner, W.D. (1976b). *Primates Med.* 9, 90–144.
Coe, C.L., Lemieux, A.M., Rier, S.E., Uno, H. and Zimbric, M.L. (1998). *J. Gerontol. A Biol. Sci. Med. Sci.* 53(1), M3–M7.
Cohen, M., Saidla, J.E. and Schlafer, D.H. (2001). *J. Med. Primatol.* 30(2), 121–126.
Colgin, L.M. and Moeller, R.B. (1996). *Lab. Anim. Sci.* 46(1), 123–124.
Cork, L.C. and Adams, R.J. (1993) In Jones, T.C., Mohr, U. and Hunt, R.D. (eds) *Monographs on Pathology of Laboratory Animals: Nonhuman Primate II,* pp. 188–192. Berlin and New York: Springer-Verlag.
Cork, L.C. and Walker, L.C. (1993) In Jones, T.C., Mohr, U. and Hunt, R.D. (eds) *Monographs on Pathology of Laboratory Animals: Nonhuman Primates II,* pp. 173–183. Berlin and New York: Springer-Verlag.
Cork, L.C., Masters, C., Beyreuther, K. and Price, D.L. (1990). *Am. J. Pathol.* 137(6), 1383–1392.
Crews, L.M., Kerber, W.T. and Feinman, H. (1967). *Pathol. Vet.* 4(2), 157–161.
Cunnane, G. (2001). *Lancet* 358(9275), 4–5.
D'Hooghe, T.M. (1997). *Fertil. Steril.* 68(4), 613–625.
D'Hooghe, T.M., Bambra, C.S., Cornillie, F.J., Isahakia, M. and Koninckx, P.R. (1991). *Biol. Reprod.* 45(3), 411–416.
D'Hooghe, T.M., Bambra, C.S., De, J., I, Lauweryns, J.M. and Koninckx, P.R. (1996a). *Acta. Obstet. Gynecol. Scand.* 75(2), 98–101.
D'Hooghe, T.M., Bambra, C.S., Isahakia, M. and Koninckx, P.R. (1992). *Fertil. Steril.* 58(2), 409–412.

D'Hooghe, T.M., Bambra, C.S., Raeymaekers, B.M. and Koninckx, P.R. (1996b). *Hum. Reprod.* 11(9), 2022–2025.

Davis, J.A., Banks, R.E. and Young, D. (1988). *Lab. Anim. Sci.* 38(3), 312–315.

Davis, K.J., Bell, R.C., Wilhelmsen, C.L., Langford, M.J. and Montali, R.J. (1994). *Vet. Pathol.* 31(4), 479–480.

de Koning, E.J., Bodkin, N.L., Hansen, B.C. and Clark, A. (1993). *Diabetologia* 36(5), 378–384.

Demaray, S.Y., Altman, N.H. and Ferrell, T.L. (1978). *Lab. Anim. Sci.* 28(4), 457–460.

DePaoli, A. and McClure, H.M. (1982). *Vet. Pathol. Suppl* 7, 104–125.

DeRousseau, C.J. (1988). *Contributions to Primatology: Osteoarthritis in Rhesus Monkeys and Gibbons* 25, 1–144. Basel: S. Karger.

Dick, E.J. Jr., Hubbard, G.B., Martin, L.J. and Leland, M.M. (2003). *J. Med. Primatol.* 32(1), 39–47.

DiGiacomo, R.F. (1977). *Vet. Pathol.* 14(6), 539–546.

DiGiacomo, R.F., Hooks, J.J., Sulima, M.P., Gibbs, C.J. Jr. and Gajdusek, D.C. (1977). *J. Am. Vet. Med. Assoc.* 171(9), 859–861.

DiGiacomo, R.F. and McCann, T.O. (1970), *Am. J. Obstet. Gynecol.* 108(4), 538–542.

Doepel, F.M., Glorioso, J.C., Newcomer, C.E., Skinner, M. and Abrams, G.D. (1981). *Lab. Invest.* 45(1), 7–13.

Doepel, F.M., Ringler, D.H. and Petkus, A.R. (1984). *Lab. Anim. Sci.* 34(5), 494–496.

DuFrain, R.J. (1985). *Cancer Genet.Cytogenet.* 14(1–2), 83–87.

Eisele, P.H., Morgan, J.P., Line, A.S. and Anderson, J.H. (1992). *Lab. Anim. Sci.* 42(3), 245–249.

Ellsworth, L., Farley, S., DiGiacomo, R.F. and Tsai, C.C. (1992). *Lab. Anim. Sci.* 42(4), 352–355.

Engle, E.T. and Stout, A.P. (1940) *Am. J. Cancer* 39, 334–337.

Eydelloth, R.S. and Swindle, M.M. (1983). *J. Med. Primatol.* 12(2), 101–105.

Fanton, J.W. and Golden, J.G. (1991). *Radiat.Res.* 126(2), 141–146.

Fanton, J.W., Hubbard, G.B., Witt, W.M. and Wood, D.H. (1984). *J. Am. Vet. Med. Assoc.* 185(11), 1377–1378.

Fanton, J.W., Hubbard, G.B. and Wood, D.H. (1986). *Am. J. Vet. Res.* 47(7), 1537–1541.

Fikes, J.D., O'Sullivan, M.G., Bain, F.T., Jayo, M.J., Harber, E.S. and Carlson, C.S. (1996). *Vet. Pathol.* 33(2), 171–175.

Fincham, J.E., Benade, A.J., Kruger, M., Smuts, C.M., Gobregts, E., Chalton, D.O. and Kritchevsky, D. (1998). *Nutrition* 14(1), 17–22.

Fincham, J.E., Quack, G., Wulfroth, P. and Benade, A.J. (1996). *Arzneimittelforschung.* 46(5), 519–525.

Fiori, M.G., Zanoni, R. and Nunzi, M.G. (1994). *J. Comp. Pathol.* 110(4), 413–417.

Ford, E.W., Roberts, J.A. and Southers, J.L. (1998) In Bennett, B.T., Abee, C.R. and Henrickson, R. (eds) *Nonhuman Primates in Biomedical Research: Diseases,* pp. 312–361. San Diego: Academic Press.

Fox, J.G., Handt, L., Xu, S., Shen, Z., Dewhirst, F.E., Paster, B.J., Dangler, C.A., Lodge, K., Motzel, S. and Klein, H. (2001). *J. Med. Microbiol.* 50(5), 421–429.

Frazier, K.S., Herron, A.J., Hines, M.E. and Altman, N.H. (1993). *Vet. Pathol.* 30(3), 306–308.

Gearing, M., Rebeck, G.W., Hyman, B.T., Tigges, J. and Mirra, S.S. (1994). *Proc. Natl. Acad. Sci. U.S.A* 91(20), 9382–9386.

Geula, C., Nagykery, N. and Wu, C.K. (2002). *Acta. Neuropathol. (Berl)* 103(1), 48–58.

Giddens, W.E. Jr., Seifert, R. and Boyce, J.T. (1979). In Bowden, D.M. (eds) *Aging in Nonhuman Primates,* pp. 324–334. New York: Von Nostrand Reinhold.

Gillett, C.S., Ringler, D.H. and Pond, C.L. (1984). *Lab. Anim. Sci.* 34(1), 91–93.

Gillmore, J.D., Lovat, L.B., Persey, M.R., Pepys, M.B. and Hawkins, P.N. (2001) *Lancet* 358(9275), 24–29.

Giudice, L.C., Tazuke, S.I. and Swiersz, L. (1998). *J. Reprod. Med.* 43(3), 252–262.

Gliatto, J.M., Bree, M.P. and Mello, N.K. (1990). *J. Med. Primatol.* 19(5), 507–513.

Gliatto, J.M. and Bronson, R.T. (1993a). In Jones, T.C., Mohr, U. and Hunt, R.D. (eds) *Monographs on Pathology of Laboratory Animals,* pp. 198–202. Springer-Verlag.

Gliatto, J.M. and Bronson, R.T. (1993b). In Jones, T.C., Mohr, U. and Hunt, R.D. (eds) *Monographs on Pathology of Laboratory Animals: Nonhuman Primates II,* pp. 183–187. Berlin and New York: Springer-Verlag.

Gozalo, A., Dagle, G.E., Montoya, E., Weller, R.E. and Malaga, C.A. (1992a). *J. Med. Primatol.* 21(5), 279–284.

Gozalo, A., Nolan, T. and Montoya, E. (1992b). *J. Med. Primatol.* 21(1), 39–41.

Guhad, F.A., Farah, I.O., Chai, D.C. and Logan–Henfrey, L.L. (1997). *Contemp. Top. Lab. Anim. Sci.* 36(3), 76–77.

Hadfield, R.M., Yudkin, P.L., Coe, C.L., Scheffler, J., Uno, H., Barlow, D.H., Kemnitz, J.W. and Kennedy, S.H. (1997). *Hum. Reprod. Update.* 3(2), 109–115.

Harber, E.S., O'Sullivan, M.G., Jayo, M.J. and Carlson, C.S. (1996). *Vet. Pathol.* 33(4), 431–434.

Hartig, W., Bruckner, G., Schmidt, C., Brauer, K., Bodewitz, G., Turner, J.D. and Bigl, V. (1997). *Brain Res.* 751(2), 315–322.

Herring, J.M., Reyes, M., Dochterman, L.W. and Bennett, B.T. (1990). *Lab. Anim. Sci.* 40(6), 645–647.

Hertig, A.T., MacKey, J.J., Feeley, G. and Kampschmidt, K. (1983). *Am. J. Obstet. Gynecol.* 145(8), 968–980.

Hisaw, F.L. and Hisaw, F.L.J. (1958). *Cancer* 11, 810–816.

HogenEsch, H., Broerse, J.J. and Zurcher, C. (1992). *Vet. Pathol.* 29(4), 359–361.

Howard, C.F. Jr. (1978). *Diabetes* 27(4), 357–364.

Howard, C.F. Jr. (1986). *Diabetologia* 29(5), 301–306.

Howard, C.F. Jr. and Van Bueren, A. (1986). *Diabetes* 35(2), 165–171.

Hubbard, G.B. (2001). *Veterinary Clin. North Am. Exot. Anim. Pract.* 4(2), 573–583.

Hubbard, G.B., Eason, R.L. and Wood, D.H. (1985). *Vet. Pathol.* 22(1), 88–90.

Hubbard, G.B., Lee, D.R., Steele, K.E., Lee, S., Binhazim, A.A. and Brasky, K.M. (2001). *J. Med. Primatol.* 30(5), 260–267.

Hubbard, G.B., Steele, K.E., Davis, K.J., III and Leland, M.M. (2002). *J. Med. Primatol.* 31(2), 84–90.

Hubbard, G.B., Wood, D.H. and Butcher, W.I. (1984). *Vet. Pathol.* 21(5), 531–533.

Hubbard, G.B., Wood, D.H. and Fanton, J.W. (1983). *Lab. Anim. Sci.* 33(5), 469–472.

Huneke, R.B., Foltz, C.J., VandeWoude, S., Mandrell, T.D. and Garman, R.H. (1996). *J. Med. Primatol.* 25(6), 404–413.

Hunt, R.D., Van Zwieten, M.J., Baggs, R.B., Sehgal, P.K., King, N.W. Roach, S.M. and Blake, B.J. (1976). *Lab. Anim. Sci.* 26(6), Pt 2, 1088–1092.

Hunt, R.D., Blake, B.J. and Chalifoux, L.V. (1993). In Jones, T.C., Mohr, U. and Hunt, R.D. (eds) *Monographs on Pathology of Laboratory Animals, Nonhuman Primates II,* pp. 164–168. Berlin and New York: Springer-Verlag.

Jackson, C.B. (1996). *Contemp. Top. Lab. Anim. Sci.* 35(5), 60–62.

Jaikaran, E.T. and Clark, A. (2001). *Biochim. Biophys. Acta* 1537(3), 179–203.

Jennings, V.M., Dillehay, D.L. and Webb, S.K. (1995). *Lab Anim. Sci.* 45(6), 680–682.

Johnson, E.H., Morgenstern, S.E., Perham, J.M. and Barthold, S.W. (1996a). *J. Med. Primatol.* 25(6), 435–438.

Johnson, L.D., Ausman, L.M., Sehgal, P.K. and King, N.W. Jr. (1996b). *Gastroenterology* 110(1), 102–115.

Johnson, W.D., Hughes, H.C., Lang, C.M. and Stenger, V.G. (1978). *Lab. Anim. Sci.* 28(1), 81–84.

Jones, T.C., Hunt, R.D. and King, N.W. (1997). *Veterinary Pathology,* 6th edition, 1–1392. Baltimore: Williams & Wilkins.

Juan-Salles, C., Marco, A., Ramos-Vara, J.A., Resendes, A., Verges, J., Valls, X. and Montesinos, A. (2002). *Primates.* 43(3), 179–190.

Kaspareit, J., Friderichs–Gromoll, S., Buse, E., Korte, R. and Vogel, F. (2001). *Exp. Toxicol. Pathol.* 53(4), 267–269.

Kast, A., Peil, H. and Weisse, I. (1994a). *Lab. Anim.* 28(1), 80–89.

Kast, A., Peil, H. and Weisse, I. (1994b). *Lab. Anim.* 28(1), 80–89.

Kerrick, G.P. and Brownstein, D.G. (2000). *Contemp. Top. Lab. Anim. Sci.* 39(4), 40–42.

Kessler, M.J. (1980). *J. Med. Primatol.* 9(5), 314–318.

Kessler, M.J., Kupper, J.L., Brown, R.J. and O'Neill, T.P. (1979). *J. Med. Primatol.* 8(6), 377–381.

Khan, K.N., Silverman, L., Logan, A. and Harris, R.K. (1999). *J. Vet. Diag. Invest* 11, *No. 5,* 478–480.

Kim, J.C. and Palazzo, M.C. (1978). *J. Med. Primatol.* 7(3), 189–191.

King, C.S., Streett, J.W., Moody, K.D. and Barthold, S.W. (1996). *J. Med. Primatol.* 25(5), 361–366.

King, N.W. (1993) In Jones, T.C., Mohr, U. and Hunt, R.D. (eds) *Monographs on Pathology of Laboratory Animals: Nonhuman Primates II,* pp. 226–232. Berlin and New York: Springer–Verlag.

King, N.W. Jr., Baggs, R.B., Hunt, R.D., Van Zwieten, M.J. and MacKey, J.J. (1976). *Lab. Anim. Sci.* 26(6), Pt 2, 1093–1103.

Kirkwood, J.K., Pearson, G.R. and Epstein, M.A. (1986). *J. Comp. Pathol.* 96(5), 507–515.

Klumpp, S.A., McClure, H.M. and Clarkson, T.B. (1993). In Jones, T.C., Mohr, U. and Hunt, R.D. (eds) *Monographs on Pathology of Laboratory Animals: Nonhuman Primates II,* pp. 105–117. Berlin and New York: Springer–Verlag.

Knightly, F.A., Lamberski, N., Harmon, B.G. and Latimer, K.S. (1996). *Am. Assoc. Zoo. Vet. Annual Conf. Proc.,* pp. 500–502.

Laber-Laird, K.E., Jokinen, M.P. and Lehner, N.D. (1987). *Lab. Anim. Sci.* 37(2), 205–209.

Lapin, B.A. (1982). *J. Med. Primatol.* 11(6), 327–341.

Lembo, T.M., Tinkey, P.T., Cromeens, D.M., Gray, K.N. and Price, R.E. (1997). *J. Med. Primatol.* 26(5), 229–232.

Lewis, R.W., Kim, J.C., Irani, D. and Roberts, J.A. (1981). *Prostate* 2(1), 51–70.

Lindberg, B.S. and Busch, C. (1984). *Ups. J. Med. Sci.* 89(2), 129–134.

Line, S.W., Ihrke, P.J. and Prahalada, S. (1984). *Lab. Anim. Sci.* 34(6), 616–618.

Liu, S.K., Dolensek, E.P. and Tappe, J.P. (1985). *Heart Vessels Suppl* 1, 288–293.

Liu, S.K., Dolensek, E.P., Tappe, J.P., Stover, J. and Adams, C.R. (1984). *J. Am. Vet. Med. Assoc.* 185(11), 1347–1350.

Lowenstine, L.J. (1986). In Benirschke, K. (eds) *Primates: the Road to Sulf-Sustaining Populations,* pp. 781–814. New York and Berlin : Springer-Verlag.

Lowenstine, L.J. (2003). *Toxicol. Pathol.* 31 Suppl, 92–102.

Lunn, S.F. (1980). *Vet. Rec.* 106(18–20), 414.

Lushbaugh, C., Humason, G. and Clapp, N. (1985). *Dig. Dis. Sci.* 30(12) Suppl, 119S–125S.

Lushbaugh, C.C., Humason, G.L., Swartzendruber, D.C., Richter, C.B. and Gengozian, N. (1978). *Primates Med.* 10, 119–134.

MacKenzie, W.F. and Casey, H.W. (1975). *Am. J. Pathol.* 80(2), 341–344.

Madara, J.L., Podolsky, D.K., King, N.W. Sehgal, P.K., Moore, R. and Winter, H.S. (1985). *Gastroenterology* 88(1) Pt 1, 13–19.

Majeed, S.K. and Gopinath, C. (1980). *Lab. Anim.* 14(4), 363–365.

Marini, R.P., Esteves, M.I., Dangler, C.A. and Fox, J.G. (1999). *Contemp. Top. Lab. Anim. Sci.* 38(4), 42–48.

Martin, C.B. Jr., Misenhimer, H.R. and Ramsey, E.M. (1970). *Lab. Anim. Care* 20(4), Pt 1, 686–692.

Martin, L.J., Pardo, C.A., Cork, L.C. and Price, D.L. (1994). *Am. J. Pathol.* 145(6), 1358–1381.

McCann, T.O. and Myers, R.E. (1970). *Am. J. Obstet. Gynecol.* 106(4), 516–523.

McClure, H.M. (1973). *Am. J. Phys. Anthropol.* 38(2), 425–429.

McClure, H.M. (1975). In Bourne, G.H. (eds) *The Rhesus Monkey: Management, Reproduction, and Pathology,* pp. 369–398. New York: Academic Press.

McClure, H.M. (1979). In Andrews, E.J., Ward, B.C. and Altman, N.H. (eds) *Spontaneous Animal Models of Human Disease,* pp. 215–218. New York: Academic Press.

McClure, H.M. (1980). In Montali, R.J. and Migaki, G. (eds) *The Comparative Pathology of Zoo Animals,* pp. 549–565. Washington, D.C: Smithsonian Institution Press.

McClure, H.M. (1984). *Adv. Vet. Sci. Comp. Med.* 28, 267–304.

McClure, H.M., Amin, M., Anderson, D., Keene, T.E., Klumpp, S., Phillip, A., Insel, T., Karr, J. and Petros, J.A. (1999). *J. Urol.* 161(4S) Supplement, no. April, 128–138.

McIntosh, G.H., Giesecke, R., Wilson, D.F. and Goss, A.N. (1985). *Vet. Pathol.* 22(1), 86–88.

Migaki, G., Digiacomo, R. and Garner, F.M. (1971). *Lab. Anim. Sci.* 21(3), 410–411.

Montesdeoca, D., Roberts, J. and Donnelly, T.M. (1992). *Lab. Animal* 21(9), 18–21.

Moody, K.D., Weir, E.C., Morgenstern, S.E. and Barthold, S.W. (1996). *Lab. Anim. Sci.* 46(1), 120–122.

Moore, C.M., Hubbard, G.B., Leland, M.M., Dunn, B.G. and Best, R.G. (2003). *J. Med. Primatol.* 32(1), 48–56.

Moore, C.M., McKeand, J., Witte, S.M., Hubbard, G.B., Rogers, J. and Leland, M.M. (1998). *Am. J. Primatol.* 46(4), 323–332.

Morin, M.L., Renquist, D.M. and Allen, A.M. (1980). *Lab. Anim. Sci.* 30(1), 110–112.

Morris, T.H. (1994). *J. Med. Primatol.* 23(5), 317–318.

Mueller-Heubach, E. and Battelli, A. (1981). *J. Med. Primatol.* 10(4-5), 265–268.

Myers, D.D. Jr., Dysko, R.C., Chrisp, C.E. and Decoster, J.L. (2001). *J. Med. Primatol.* 30(2), 127–130.

Myers, R.E. (1972). *J. Reprod. Med.* 9(4), 171–198.

Nakamura, S., Nakayama, H., Goto, N., Ono, F., Sakakibara, I. and Yoshikawa, Y. (1998). *J. Med. Primatol.* 27(5), 244–252.

O'Brien, T.D., Butler, P.C., Westermark, P. and Johnson, K.H. (1993). *Vet. Pathol.* 30(4), 317–332.

O'Brien, T.D., Wagner, J.D., Litwak, K.N., Carlson, C.S., Cefalu, W.T., Jordan, K., Johnson, K.H. and Butler, P.C. (1996). *Vet. Pathol.* 33(5), 479–485.

O'Gara, R.W. and Adamson, R.H. (1972). In T-W-Fiennes, N. (eds) *Pathology of Simian Primates. Part I: General Pathology,* pp. 190-238. Basel: S. Karger.

O'Sullivan, M.G. and Carlson, C.S. (2001). *J. Comp. Pathol.* 124(2-3), 212–215.

Ohagi, S., Nishi, M., Bell, G.I., Ensinck, J.W. and Steiner, D.F. (1991). *Diabetologia* 34(8), 555–558.

Palotay, J.L. and Howard, C.F. Jr. (1982). *Vet. Pathol.* Suppl 7, 181–192.

Peters, A. (2002). *Prog. Brain Res.* 136, 455–465.

Plesker, R., Coulibaly, C. and Hetzel, U. (2002). *J. Vet. Med. A. Physiol. Pathol. Clin. Med.* 49(8), 428–429.

Podolsky, D.K., Madara, J.L., King, N., Sehgal, P., Moore, R. and Winter, H.S. (1985). *Gastroenterology* 88(1) Pt 1, 20–25.

Poduri, A., Gearing, M., Rebeck, G.W., Mirra, S.S., Tigges, J. and Hyman, B.T. (1994). *Am. J. Pathol.* 144(6), 1183–1187.

Pond, C.L., Newcomer, C.E. and Anver, M.R. (1982). *Vet. Pathol. Suppl* 7, 126–133.

Pritzker, K.P.H. and Kessler, M.J. (1998). In Bennett, B.T., Abee, C.R. and Henrickson, R. (eds) *Nonhuman Primates in Biomedical Research – Diseases,* pp. 415-459. Academic Press.

Ratterree, M.S., Didier, P.J., Blanchard, J.L., Clarke, M.R. and Schaeffer, D. (1990). *Lab. Anim. Sci.* 40(2), 165–168.

Richter, C.B. and Buyukmihci, N. (1979). *Vet. Pathol.* 16(2), 263–265.

Richter, C.B., Lushbaugh, C.C. and Swartzendruber, D.C. (1980). In Montali, R.J. and Migaki, G. (eds) *The Comparative Pathology of Zoo Animals,* pp. 567–571. Washington, DC: Smithsonian Institution Press.

Rier, S.E., Martin, D.C., Bowman, R.E., Dmowski, W.P. and Becker, J.L. (1993). *Fundam. Appl. Toxicol.* 21(4), 433–441.

Roberts, E.D. (1993). In Jones, T.C., Mohr, U. and Hunt, R.D. (eds) *Monographs on Pathology of Laboratory Animals: Nonhuman Primates I,* pp. 202–205. Berlin and New York: Springer-Verlag.

Roberts, J.A. (1972). In T-W-Fiennes, N. (eds) *Pathology of Simian Primates. Part I: General Pathology,* pp. 878–888. Basel: S. Karger.

Rodger, R.F., Bronson, R.T., McIntyre, K.W. and Nicolosi, R.J. (1980). *J. Am. Vet. Med. Assoc.* 177(9), 863–866.

Rodriguez, N.A., Garcia, K.D., Fortman, J.D., Hewett, T.A., Bunte, R.M. and Bennett, B.T. (2002). *J. Med. Primatol.* 31(2), 74–83.

Rohovsky, M.W., Fox, J.G. and Chalifoux, L.V. (1977). *Lab. Anim. Sci.* 27(2), 280–281.

Rothschild, B.M. (1993). *J. Med. Primatol.* 22(5), 313–316.

Rothschild, B.M. and Woods, R.J. (1992). *Seminars in Arthritis and Rheumatism* 21(5), 306–316.

Rothschild, B.M., Hong, N. and Turnquist, J.E. (1997). *Clin. Exp. Rheumatol.* 15(1), 45–51.

Rothschild, B.M., Hong, N. and Turnquist, J.E. (1999). *Semin. Arthritis Rheum.* 29(2), 100–111.

Rouse, R., Bronson, R.T. and Sehgal, P.K. (1981). *J. Med. Primatol.* 10(4-5), 199–204.

Rubio, C.A. and Hubbard, G.B. (1998). *Anticancer Res.* 18(2A), 1143–1147.

Rubio, C.A. and Hubbard, G.B. (2001). *In Vivo* 15(1), 109–116.

Rubio, C.A. and Hubbard, G.B. (2002). *In Vivo* 16(3), 191–195.

Ruch, T.C. (1959a) *Diseases of Laboratory Primates,* pp. 1–598. Philadelphia: W.B. Saunders Company.

Ruch, T.C. (1959b) *Diseases of Laboratory Primates,* pp. 443–471. Philadelphia: W.B. Saunders Company.

Ruch, T.C. (1959c). In Ruch, T.C. (ed.) *Diseases of Laboratory Primates,* pp. 529–567. Philadelphia: W.B. Saunders Company.

Sani, S., Traul, D., Klink, A., Niaraki, N., Gonzalo-Ruiz, A., Wu, C.K. and Geula, C. (2003). *Acta. Neuropathol. (Berl)* 105(2), 145–156.

Santoro, M.G. (2000). *Biochem. Pharmacol.* 59(1), 55–63.

Sasseville, V.G., Pauley, D.R., Spaulding, G.L., Chalifoux, L.V., Lee-Parritz, D. and Simon, M.A. (1993). *J. Med. Primatol.* 22(4), 272–275.

Schultz, C., Dehghani, F., Hubbard, G.B., Thal, D.R., Struckhoff, G., Braak, E. and Braak, H. (2000a). *J. Neuropathol. Exp. Neurol.* 59(1), 39-52.

Schultz, C., Hubbard, G.B., Rub, U., Braak, E. and Braak, H. (2000b). *Neurobiol. Aging* 21(6), 905–912.

Schultz, C., Dick, E.J., Cox, A.B., Hubbard, G.B., Braak, E. and Braak, H. (2001). *Neurobiol. Aging* 22(4), 677–682.

Scott, W.J., Fradkin, R. and Wilson, J.G. (1975). *J. Med. Primatol.* 4(3), 204–206.

Scully, R.E., Young, R.H. and Clement, P.B. (1998). In Rosai, J. and Sobin, L.H. (eds) *Tumors of the Ovary, Maldeveloped Gonads, Fallopian Tube, and Broad Ligament,* pp. 267–284. Washington, D.C: Armed Forces Institute of Pathology.

Sehgal, P.K., Bronson, R.T., Brady, P.S., McIntyre, K.W. and Elliott, M.W. (1980). *Lab. Anim. Sci.* 30(1), 92–98.

Seibold, H.R. and Wolf, R.H. (1973). *Lab. Anim. Sci.* 23(4), 533–539.

Seli, E., Berkkanoglu, M. and Arici, A. (2003). *Obstet. Gynecol .Clin. North Am.* 30(1), 41–61.

Selkoe, D.J., Bell, D.S., Podlisny, M.B., Price, D.L. and Cork, L.C. (1987). *Science* 235 (4791), 873–877.

Shaw, J.H., Phillips, P.H. and Elvehjam, C.A. (1945). *J. Nutr.* 20, 365–372.

Sheffield, W.D., Squire, R.A. and Strandberg, J.D. (1981). *Vet. Pathol.* 18(3), 326–334.

Sheldon, W.G. and Gleiser, C.A. (1971). *Vet. Pathol.* 8(1), 16–18.

Siddiqi, Z.A. and Peters, A. (1999). *J. Neuropathol. Exp. Neurol.* 58(9), 903–920.

Simkins, M.D. and Bullock, B.C. (1995). *Lab. Anim. Sci.* 45(4), 470.

Skelton-Stroud, P.N. and Glaister, J.R. (1994). *Lab. Anim.* 28(3), 265–269.

Skelton-Stroud, P.N. and Ishmael, J. (1985). *Vet. Pathol.* 22(2), 141–146.

Slattum, M.M., Rosenkranz, S.L., DiGiacomo, R.F., Tsai, C.C. and Giddens, W.E. Jr. (1989a). *Lab. Anim. Sci.* 39(6), 560–566.

Slattum, M.M., Tsai, C.C., DiGiacomo, R.F. and Giddens, W.E., Jr. (1989b). *Lab. Anim. Sci.* 39(6), 567–570.

Sloane, J.A., Pietropaolo, M.F., Rosene, D.L., Moss, M.B., Peters, A., Kemper, T. and Abraham, C.R. (1997). *Acta Neuropathol. (Berl)* 94(5), 471–478.

Smith, A.W., Casey, H.W., LaCroix, J.T. and Johnson, D.K. (1969). *J. Am. Vet. Med. Assoc.* 155(7), 1241–1244.

Smith, J.M., Stump, K.C., Rao, S.S. and DeTolla, L.J. (2000). *Contemp. Top. Lab. Anim. Sci.* 39(4), 88.

Soave, O.A. (1978). *Lab. Anim. Sci.* 28(3), 331–334.

Splitter, G.A., Kirk, J.H., Mac Kenzie, W.F. and Rawlings, C.A. (1972a). *Vet. Pathol.* 9(4), 249–262.

Splitter, G.A., Pryor, W.H. Jr. and Casey, H.W. (1972b). *J. Am. Vet. Med. Assoc.* 161(6), 710–713.

Squire, R.A., Goodman, G.D., Valerio, M.G., Fredrickson, T.N., Strandberg, J.D., Lingeman, C.H., Levitt, M.H., Harshbarger, J.C. and Dawe, C.J. (1978). In Benirschke, K., Garner, F.M. and Jones, T.C. (eds) *Pathology of Laboratory Animals,* pp. 1051–1283. Berlin and New York: Springer-Verlag.

Stam, J. (2003). *Adv. Neurol.* 92, 225–232.

Steiner, M.S., Couch, R.C., Raghow, S. and Stauffer, D. (1999). *J. Urol.* 162(4), 1454-1461.

Strickland, H.L. and Bond, M.G. (1983). *Lab. Anim. Sci.* 33(6), 589–592.

Strozier, L.M., McClure, H.M., Keeling, M.E. and Cummins, L.B. (1972). *J. Am. Vet. Med. Assoc.* 161(6), 704–706.

Struble, R.G., Price, D.L. Jr., Cork, L.C. and Price, D.L. (1985). *Brain Res.* 361(1-2), 267–275.

Suzuki, J., Gotoh, S., Miwa, N., Terao, K. and Nakayama, H. (2000). *J. Med. Primatol.* 29(2), 88-94.

Swindle, M.M., Adams, R.J. and Craft, C.F. (1981). *J. Med. Primatol.* 10(4–5), 269–273.

Swindle, M.M., Blum, J.R., Lima, S.D. and Weiss, J.L. (1985). *Circulation* 71(1), 146–153.

Tekeli, S. and Ford, T.M. (1980). *Vet. Pathol.* 17(4), 502–504.

Todd, G.C., Griffing, W.J. and Koenig, G.R. (1973). *Vet. Pathol.* 10(4), 342–346.

Toyosawa, K., Okimoto, K., Koujitani, T. and Kikawa, E. (2000). *Vet. Pathol.* 37(2), 186–188.

Tsai, C.C. and Giddens, W.E. Jr. (1985). *Lab. Anim. Sci.* 35(1), 85–88.

Uno, H., Alsum, P.B., Dong, S., Richardson, R., Zimbric, M.L., Thieme, C.S. and Houser, W.D. (1996). *Neurobiol. Aging* 17(2), 275–281.

Uno, H., Alsum, P., Zimbric, M.L., Houser, W.D., Thomson, J.A. and Kemnitz, J.W. (1998). *Am. J. Primatol.* 44(1), 19–27.

Urvater, J.A., McAdam, S.N., Loehrke, J.H., Allen, T.M., Moran, J.L., Rowell, T.J., Rojo, S., Lopez de Castro, J.A.,

Taurog, J.D. and Watkins, D.I. (2000). *Immunogenetics* 51(4–5), 314–325.
Valverde, C.R., Tarara, R.P., Griffey, S.M. and Roberts, J.A. (2000). *Comp. Med.* 50(5), 540–544.
van der Linden, P.J.Q. (1997). *Front Biosci.* 2, e48–e52.
Vinatier, D., Orazi, G., Cosson, M. and Dufour, P. (2001). *Eur. J. Obstet. Gynecol. Reprod. Biol.* 96(1), 21–34.
Waggie, K.S., Tolwani, R.J. and Lyons, D.M. (2000). *Vet. Pathol.* 37(5), 505–507.
Wakui, S., Furusato, M., Kato, H., Nomura, Y., Kano, Y. and Aizawa, S. (1989). *Vet. Pathol.* 26(5), 447–448.
Walker, L.C. (1997). *Brain Res. Brain Res. Rev.* 25(1), 70–84.
Walker, L.C., Kitt, C.A., Struble, R.G., Wagster, M.V., Price, D.L. and Cork, L.C. (1988). *Neurobiol. Aging* 9(5–6), 657–666.
Walker, L.C., Masters, C., Beyreuther, K. and Price, D.L. (1990). *Acta Neuropathol. (Berl)* 80(4), 381–387.
Weller, R.E. (1994). In Baer, J.F., Weller, R.E. and Kakoma, I. (eds) *Aotus: the Owl Monkey,* pp. 177–215. San Diego: Academic Press.
Westermark, P., Engstrom, U., Johnson, K.H., Westermark, G.T. and Betsholtz, C. (1990). *Proc. Natl. Acad. Sci. U.S.A.* 87(13), 5036–5040.
Willwohl, D., Kettner, M., Braak, H., Hubbard, G.B., Dick, E.J. Jr., Cox, A.B. and Schultz, C. (2002). *Acta Neuropathol. (Berl)* 103(3), 276–280.
Wood, J.D., Peck, O.C., Tefend, K.S., Rodriguez, M., Rodriguez, M., Hernandez, C., Stonerook, M.J. and Sharma, H.M. (1998). *Dig. Dis. Sci.* 43(7), 1443–1453.
Wood, J.D., Peck, O.C., Tefend, K.S., Stonerook, M.J., Caniano, D.A., Mutabagani, K.H., Lhotak, S. and Sharma, H.M. (2000). *Dig. Dis. Sci.* 45(2), 385–393.
Yager, J.A. and Scott, D.W. (1993). In Jubb, K.V.F., Kennedy, P.C. and Palmer, N. (eds) *Pathology of Domestic Animals,* pp. 531–738. San Diego and New York: Academic Press.
Yanai, T., Teranishi, M., Manabe, S., Takaoka, M., Yamoto, T., Matsunuma, N. and Goto, N. (1992). *Vet. Pathol.* 29(6), 569–571.
Yanai, T., Masegi, T., Tomita, A., Kudo, T., Yamazoe, K., Iwasaki, T., Kimura, N., Katou, A., Kotera, S. and Ueda, K. (1995a). *Vet. Pathol.* 32(1), 57–59.
Yanai, T., Wakabayashi, S., Masegi, T., Ishikawa, K., Yamazoe, K., Iwasaki, T. and Ueda, K. (1995b). *J. Comp Pathol.* 112(2), 127–131.
Yanai, T., Wakabayashi, S., Masegi, T., Iwasaki, T., Yamazoe, K., Ishikawa, K. and Ueda, K. (1995c). *Vet. Pathol.* 32(3), 318–320.

CHAPTER 5

Common Viral Infections of Laboratory Primates

Nicholas W. Lerche
California National Primate Research Center,
University of California, Davis, CA, USA

Introduction

It has long been recognized that microbial agents are potential interfering factors in animal research and the ability of pathogenic agents to disrupt studies by increasing morbidity and mortality is well documented (Fortmeyer, 1982; Pakes *et al.,* 1984; Van der Logt, 1993). More recently there has been a re-emphasis on the distinction between infection and disease and a growing recognition of the potential for subclinical infections to adversely affect the results of animal research through more subtle effects on complex biological processes (Van der Logt, 1993; Lerche and Osborn, 2002). These concepts have been most fully developed and accepted for rodents and other small animal experimental models, but are equally applicable to laboratory primates (Lerche and Osborn, 2002). Only relatively recently, however, has the term "specific pathogen-free" (SPF) been used with regard to populations of laboratory primates in recognition of this fact (Lerche *et al.,* 1994; Desrosiers, 1997).

Laboratory primate species are natural hosts to a wide variety of viral agents, representing a broad range of pathogenic potential from subclinical infection to fatal disease. Many of these viruses establish persistent infections, while others produce transient infections but are relatively common in primate populations maintained for research. This chapter will describe some of the common viral infections of laboratory primates and their potential impact on research protocols.

Retroviruses

Retroviruses, family *Retroviridae,* are single-stranded, enveloped RNA viruses. All retroviruses possess an enzyme, reverse transcriptase, that allows retrograde transfer of genetic information from RNA to DNA, and retrovirus replication involves the integration of a DNA "provirus" into the host genome (Lowenstine and Lerche, 1988). Simian retroviruses establish lifelong persistent infections in the primate host, and exhibit a broad spectrum of pathogenic potential ranging from highly pathogenic to nonpathogenic depending on various host, virus and environmental factors. Latent or subclinical infections are common, and

The Laboratory Primate
Copyright 2005 Elsevier
ISBN 0-1208-0261-9

procedures associated with experimental protocols may lead to virus reactivation and disease (Schroder *et al.*, 2000; Lerche and Osborn, 2002).

Simian type D retrovirus

Simian type D retroviruses (SRV) are a group of closely related viruses that, to date, have been isolated almost exclusively from Asian monkeys of the genus *Macaca*. These isolates are all related to the prototypic type D retrovirus, Mason-Pfizer monkey virus (MPMV), isolated in 1969 from a mammary carcinoma of a rhesus monkey (Chopra, 1970). SRV has been isolated from many macaque species including rhesus *(Macaca mulatta)*, cynomolgus (*M. fascicularis*) and pig-tailed (*M. nemestrina*) monkeys, species commonly used in biomedical research (Lerche, 1992; Gardner *et al.*, 1994). Five distinct serotypes are currently recognized (Marx *et al.*, 1985; Lerche, unpublished data). Within these distinct serotypes, there may be considerable genetic variation among isolates (Marracci *et al.*, 1995). All macaque species appear susceptible to infection with all serotypes, but some general serotype-host species associations have been recognized. Cynomolgus *(Macaca fasciularis)* and pig-tailed *(M. nemestrina)* macaques are predominantly infected with serotype 2 (SRV-2) viruses, while in rhesus macaques *(M. mulatta)* SRV-1 is the predominant serotype (Lerche, 1992). Prevalence of SRV infection is variable, ranging from complete absence in some populations, to hyper-endemic in others where infection rates may exceed 50% (Lerche *et al.*, 1987; Daniel *et al.*, 1988; Lerche *et al.*, 1994). Geographic origin of animals, as well as management and husbandry practices, are factors influencing the level of SRV endemicity in macaque populations maintained for research (Lerche *et al.*, 1986, 1987; Lerche, 1992; Gardner *et al.*, 1994).

SRV has a broad cellular tropism, including both lymphoid and non-lymphoid tissues. SRV can be demonstrated in many tissues and organs by immunohistochemical staining, *in situ* hybridization, or polymerase chain reaction (PCR) (Maul *et al.*, 1988; Lackner *et al.*, 1988; 1990). Virus can be isolated from many body fluids, including saliva, urine, blood, lacrimal secretions, cerebrospinal fluid and breast milk (Lerche, 1992; Gardner *et al.*, 1994). High titers of virus can occur in saliva (Lerche *et al.*, 1986).

Transmission of SRV occurs horizontally, either through direct contact between infected and susceptible animals, or indirectly through contact with contaminated instruments or equipment (e.g. tattoo needles, transfer boxes, dental instruments, or gavage tubes) (Lerche, 1992). A major mode of transmission is via contact with virus shed in saliva, either during mutual grooming or aggressive interactions involving biting and scratching (Lerche *et al.*, 1986). Transplacental transmission has also been documented in females that are viremic during pregnancy (Tsai *et al.*, 1987).

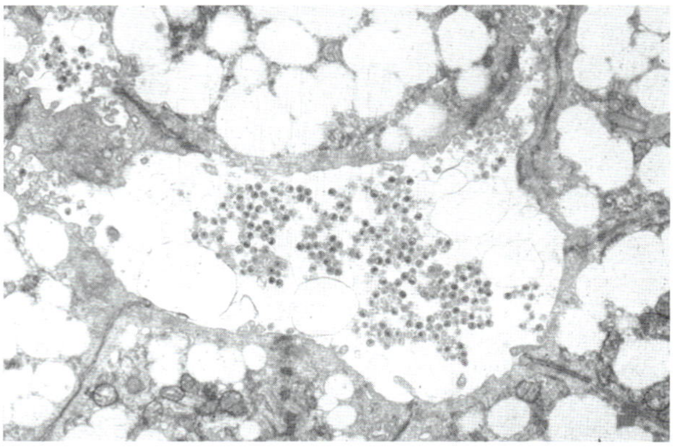

Figure 5.1 Simian type D retrovirus virions in a duct lumen in the parotid salivary gland of an SRV infected rhesus macaque.

SRV can elicit a broad spectrum of clinical and pathologic manifestations, ranging from a subclinical carrier state to rapidly fatal immunosuppressive disease (Lerche, 1992; Gardner *et al.*, 1994). Many clinical variations exist between these two extremes. SRV infected macaques can appear clinically healthy for prolonged periods (Lerche, 1992; Guzman *et al.*, 1999). Common clinical and laboratory findings in SRV-infected macaques include diarrhea, weight loss, splenomegaly, lymphadenopathy, anemia, neutropenia, lymphopenia and occasional neoplastic disease, including cutaneous fibrosarcoma and retroperitoneal fibromatosis (RF) (Lerche *et al.*, 1987; Marx *et al.*, 1985; Marx and Lowenstine, 1987; Tsai *et al.*, 1987; Guzman *et al.*, 1999). In SRV-infected cynomolgus macaques, but not in other species of macaque, rare B-cell lymphomas have been reported (Paramastri *et al.*, 2002; Lerche, unpublished data). There is now accumulating evidence that the fibromatosis and lymphomas are due to two different herpesvirus cofactors secondary to immunosuppression of SRV-infection in macaques. The etiologic agent of RF is thought to be a macaque rhadinovirus (gamma-herpesvirus), the macaque counter-part of human Kaposi's sarcoma virus (human herpesvirus-8, HHV-8) (Rose *et al.*, 1997), while the rare B-cell lymphomas observed in SRV-infected cynomolgus are likely due to co-infection with an EBV-like lymphocryptovirus (Paramastri *et al.*, 2002). Clinical findings identified in a retrospective case review of SRV-infected rhesus and cynomolgus macaques are presented in Table 5.1. In both species 20–30% of infected animals had no clinical abnormalities (Lerche and Osborn, 2002). While the interval between infection and the onset of clinical disease can be variable, true subclinical carriers have been identified that remain healthy, virus culture or PCR positive, and seronegative for many years (Lerche *et al.,* 1986; Moazad and Thouless, 1993). This true carrier state allows for SRV maintenance in small populations.

Histopathologic findings in SRV infected macaques are most consistent in lymphoid tissues. SRV infection is marked by early lymphoid hyperplasia, frequently at "ectopic" sites such as kidney or bone marrow (Osborn *et al.,* 1984; Guzman *et al.,* 1999). Later stages of infection are characterized by varying degrees of lymphoid depletion involving loss of both T and B lymphocyte populations (Maul *et al.*, 1988).

Lesions are also commonly noted in the gastrointestinal tract of SRV-infected macaques, and include goblet cell hyperplasia, villous blunting, and enteritis (Osborn *et al.,* 1984; Guzman *et al.,* 1999). SRV antigens can be demonstrated in tissues throughout the gastrointestinal tract, and lesions can be observed in the absence of any other detectable enteric pathogens, suggesting a possible primary pathogenic role for SRV in gastrointestinal lesions (Lackner *et al.,* 1990). Other histopathology in SRV-infected macaques is attributable to specific opportunistic pathogens, including cytomegalovirus, *Cryptosporidium parvum, Candida spp., Pneumocystis carinii*, and others. (Osborn *et al.,* 1984). Chronic enteritis and amyloidosis has also been observed in long-term SRV infections of rhesus and pig-tailed macaques (Osborn *et al.,* 1984; Lerche, unpublished data).

Development of neutralizing antibodies is a major component of the host immune response to SRV infection. Animals exhibiting strong antibody responses frequently become aviremic, while a significant proportion of viremic animals are seronegative. In general, less severe histologic lesions and clinical disease are seen in seropositive, nonviremic macaques compared with viremic animals with or without detectable antibody (Lerche *et al.,* 1987; Guzman, 1999).

Hematologic abnormalities are a prominent feature of SRV infection, with anemia and/or neutropenia being the most commonly observed in both experimental and naturally acquired infections (Fine *et al.,* 1975; Lerche *et al.,* 1987; Letvin *et al.,* 1984; MacKenzie *et al.,* 1986).

TABLE 5.1: Clinical findings in SRV-infected* cynomolgus and rhesus macaques: a retrospective review of 62 cases at the CNPRC

Clinical signs	Cynomolgus macaques (N = 27)	Rhesus macaques (N = 35)
Clinically healthy	8 (31)[a]	8 (23)
Diarrhea	8 (31)	19 (54)
Splenomegaly	8 (31)	12 (34)
Lymphadenopathy	5 (19)	27 (77)
Weight loss (>10%)	4 (15)	19 (54)
Anemia (Hct <30%)	2 (8)	19 (54)
Neutropenia (<1,500/ul)	1 (4)	4 (11)
Thrombocytopenia	1 (4)	3 (9)
Abdominal mass (RF)	1 (4)	0 (0)
Fibrosarcoma (cutaneous RF)	0 (0)	2 (16)

[a]No. animals (%).
*Infection confirmed by immunoblot or culture/PCR.
Adapted from Lerche and Osborn (2002).

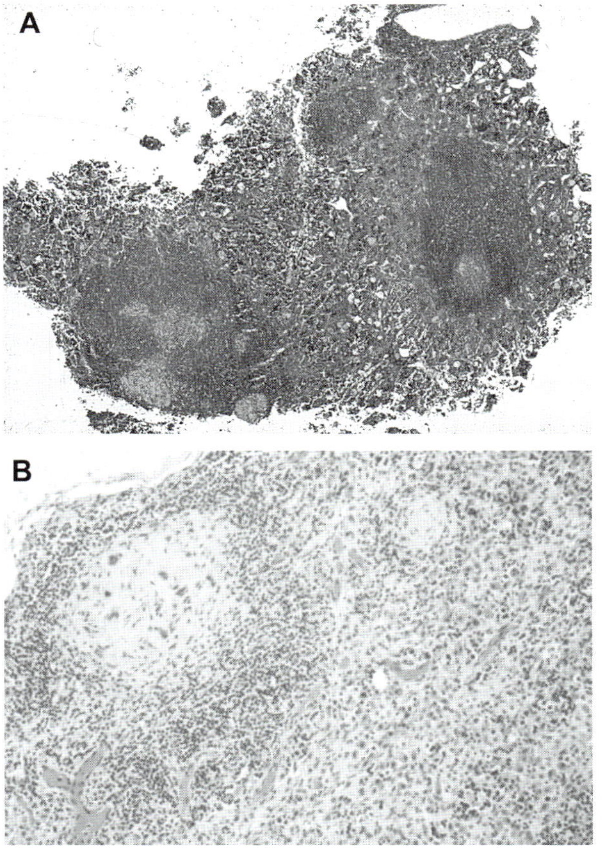

Figure 5.2 A: Lymphoid hyperplasia in "ectopic" foci in the bone marrow of an SRV infected rhesus macaque. B: SRV-associated lymphoid depletion in peripheral lymph node of a rhesus macaque.

Paradoxical hyperplastic bone marrow has been described in animals with profound anemia and neutropenia, and hypoplastic marrow has been described in late-stage SRV infection (MacKenzie *et al.,* 1986). The mechanisms responsible for anemia and granulocytopenia are not known, although direct viral injury to hematopoietic stem cells has been postulated as a contributing factor (MacKenzie *et al.,* 1986; Yoshioka, 2000). SRV can be isolated from bone marrow and, at least *in vitro,* SRV appears capable of inhibiting granulocytic precursor cell differentiation (Yoshioka, 2000). Persistent co-infection with simian parvovirus (SPV) secondary to SRV-induced immune suppression may be responsible for a large proportion of the severe anemia observed in SRV-infected macaques (O'Sullivan *et al.,* 1994, 1996).

Due to its unique epidemiologic, clinical, and pathological features, SRV has been the most prominent simian virus with regard to compromising primate biomedical research protocols (Lerche and Osborn, 2002). Little is currently known about the more subtle effects of SRV on cell surface markers and cytokine expression. By analogy with other retroviruses, such effects are certainly to be expected but the exact nature and scope of these effects remains to be determined.

Simian T-lymphotropic virus

Simian T-lymphotropic virus (STLV) is a C-type member of the oncornavirus subgroup of retroviruses. STLV was identified in the early 1980s, shortly after the initial isolation of HTLV-I, and represents the nonhuman primate counterpart of human T-lymphotropic viruses types I and II (HTLV-I, -II) (Miyoshi *et al.,* 1983). This group of genetically related human and nonhuman primate viruses is often referred to collectively as primate T-lymphotropic viruses (PTLVs). The natural hosts of STLV are many species of African and Asian monkeys and apes, and STLV is endemic in many populations of wild and captive nonhuman primates (Daniel *et al.,* 1988; Lowenstine *et al.,* 1992; Otsyula *et al.,* 1996). To date, three distinct subtypes of STLV (STLV-1, -2, -3) have been described in various nonhuman primate species (Meertens and Gessain, 2003). Estimates of STLV seroprevalence in populations of laboratory primates housed in the USA have ranged from 3–12% (Daniel *et al.,* 1988; Lerche *et al.,* 1994).

STLV is a highly cell-associated virus, with a cellular tropism primarily for CD4+ T-lymphocytes *in vivo* (Fultz, 1994; Gabet *et al.,* 2003). Natural transmission requires transfer of lymphocytes from infected to susceptible animals, primarily by transfer of infected cells in semen or cervical secretions to mucosal surfaces during breeding, or in breast milk during nursing of infants on infected dams (Fultz, 1994).

The vast majority of STLV infections are clinically silent. Only a very small proportion of infected animals develop T-cell lymphoma or lymphoproliferative disease (LPD), and usually only after a prolonged period of infection (Lowenstine and Lerche, 1988; Fultz, 1994). STLV-related disease in African primate species, including African green monkeys, baboons, and gorillas, is well documented (Hubbard *et al.,* 1996; Lowenstine and Lerche, 1988). Persistent lymphocytosis with atypical lymphocytes having lobulate or convoluted nuclei ("flower cells"), similar to the preleukemic state of adult T-cell leukemia in humans was observed in three African green monkeys naturally infected with STLV-1 (Noda *et al.,* 1986). Infiltration of leukemic cells may also be found in other tissues, such as the lungs.

Despite comparable STLV infection rates, the association with lymphoproliferative disease or T-cell

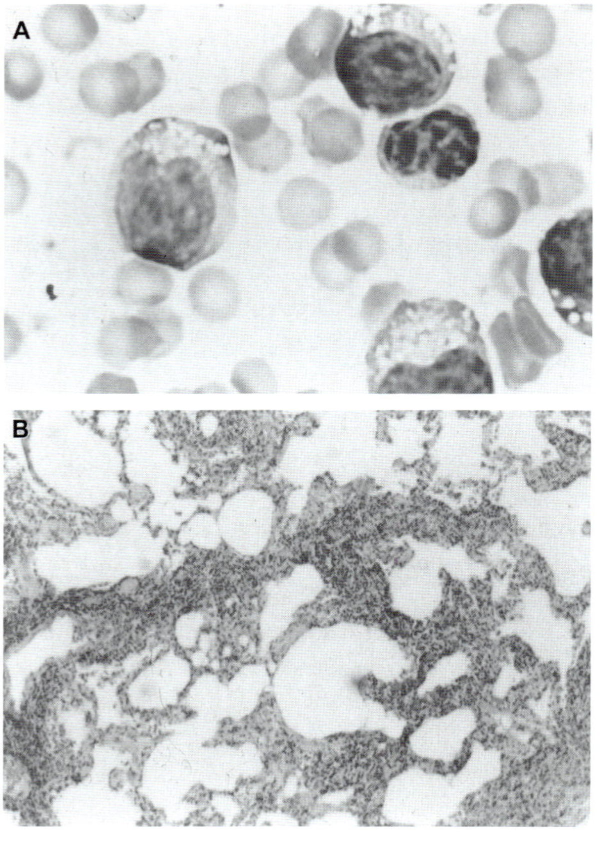

Figure 5.3 A: Characteristic atypical lymphocytes with lobulated nuclei in STLV-infected baboons. B: Infiltration of leukemic cells in the lung of an STLV-infected baboon. [Photos courtesy of Gene Hubbard and Jonathan Allan, Southwest Foundation for Biomedical Research]

leukemia/lymphoma is not well established in Asian macaques (Daniel *et al.,* 1988; Richards *et al.,* 1998). Cross-species transmission of STLV-1 of rhesus macaque origin resulted in a dramatic increase in the incidence of T-cell leukemia/lymphoma in baboons, a rate much higher than that observed in baboons naturally infected with baboon strains of STLV-1 (Voevodin *et al.,* 1996). No counterpart to the progressive neurologic disease, tropical spastic paraparesis (TSP)/HTLV-associated myelopathy (HAM) seen in some HTLV-infected humans, has yet been recognized in nonhuman primates (Fultz, 1994).

Altered cytokine profiles have been reported in healthy human HTLV carriers, including increased expression of IFN-γ, TNF-α, IL-1α, and IL-6 (Lal *et al.,* 1991; Shimamoto *et al.,* 1996; Carvalho *et al.,* 2001). Altered cytokine profiles have also been observed in STLV-infection of nonhuman primates. Several STLV-transformed cell lines have been found to constitutively release TNF-β, GM-CSF, β-FGF, and IL-6 (Lazo and Bailer, 1996). In addition, peripheral blood mononuclear cells (PBMC) from healthy STLV-infected rhesus macaques were found to release significantly higher levels of IFN-γ *in vitro,* in the absence of any stimulation, compared to PBMC from STLV-negative rhesus matched for age and sex (Yee and Lerche, 2003; Lerche, unpublished data).

Development of T-cell LPD or T-cell leukemia/lymphoma, as well as altered cytokine profiles in STLV-infected primates represent potential confounding variables in research utilizing laboratory primates.

Simian immunodeficiency virus

Simian immunodeficiency viruses (SIV) are a group of genetically related viruses belonging to the lentivirus subgroup of retroviruses. The natural hosts of SIV include many species of African monkeys and the chimpanzee *(Pan troglodytes)* (Gardner *et al.,* 1994; Peeters *et al.,* 2002). Prevalence of infection in various populations of captive and wild African primates may exceed 50% (Gardner *et al.,* 1994). Infection in these natural host species, however, is almost always subclinical (Lowenstine *et al.,* 1986; Gardner *et al.,* 1994). Asian macaques are susceptible to SIV infection but naturally acquired (i.e. non-experimental) infections in macaques are an artifact of captivity, the result of cross-species transmission from African species through direct contact with infected animals or their tissues or body fluids (Khan *et al.,* 1991; Lowenstine *et al.,* 1992; Gardner *et al.,* 1994). SIV is not a naturally occurring infection of wild macaques in their countries of origin (Lowenstine *et al.,* 1986; Ohta *et al.,* 1988). Extensive screening of captive macaque populations in the USA has demonstrated that SIV is absent from contemporary colonies and recently imported animals (Lerche *et al.,* 1994). Once introduced into macaque populations, however, SIV can cause epidemics of severe immunodeficiency disease (Lowenstine *et al.,* 1992). Various strains of SIV are highly pathogenic in macaques and produce a severe immune deficiency disease characterized by a profound decline in CD4+ T-lymphocytes, hematologic abnormalities (e.g. anemia, lymphopenia), opportunistic infections and lymphoma, similar to human acquired immunodeficiency syndrome (AIDS) (Gardner *et al.,* 1994). Experimental infection of rhesus macaques is now the premier animal model for AIDS-related research (Johnson and Hirsch, 1992; Gardner *et al.,* 1994).

SIV rarely causes overt disease or histologic lesions in natural host species, African monkeys and apes. However, alterations of cytokine profiles and cell surface markers have been reported, including dysregulation of protein kinases (e.g. ROR2), involved in T-cell signaling pathways (Bostik *et al.,* 2001) and increased production of IL-10 in SIV-infected sooty mangabeys *(Cercocbus atys)* (Giavedoni *et al.,* 2000). Altered cytokine expression in SIV-infected macaques include increased production of INF-γ, IFN-α/β, IL-12, and IL-18 (Giavedoni *et al.,* 2000; Zou *et al.,* 1997). Altered cell surface markers, including increased expression of CD69, an early T-cell activation marker, and dysregulation of protein tyrosine kinases involved in T-cell signaling pathways, have also been reported in SIV-infected macaques (Bostik *et al.,* 2001; Giavedoni *et al.,* 2000; Zou *et al.,* 1997). Continued utilization of the SIV/macaque model for AIDS-related research will undoubtedly identify additional SIV related alterations of cytokine and chemokine expression and cell signaling pathways, as well as the mechanisms underlying these changes.

Simian foamy virus

Foamy viruses, members of the spumavirus subgroup of retroviruses, have a very broad host range, and are able to infect a wide variety of vertebrate species and replicate in nearly all types of cultured cells (Falcone *et al.,* 1999; Morozov and Legaye, 1998). Foamy virus infection is highly prevalent in virtually all species of nonhuman primates, approaching 100% in adult animals in many populations. At least 12 serotypes or genetic variants are currently recognized (Lowenstine and Lerche, 1988).

SFV appears to be nonpathogenic in all primate hosts. SFV establishes latent infection in many different tissues, while replication appears to be restricted to the mucosae of the naso- and oro-pharynx (Falcone *et al.,* 1999). SFV appears to produce no lesions *in vivo.* However, reactivation of latent virus in cultures of primary cells from nonhuman primates produces a highly cytolytic infection, with rapid cell death (Feldman *et al.,* 1997). As a result of this propensity, SFV is often considered more of a nuisance virus than a pathogen. Recent studies, however, have revealed that human infection with SFV is relatively common among persons occupationally exposed to nonhuman primates, although no disease has been reported in association with these infections (Heneine *et al.,* 1998).

Because of its nonpathogenic nature and ability to infect a wide variety of tissues and cells, SFV is considered as a potential candidate viral vector for gene therapy. Preexisting SFV infection would preclude such studies in nonhuman primates and SFV has now been included as a targeted agent for establishing specific pathogen-free (SPF) breeding colonies of macaques (Lerche, unpublished).

Little is known regarding the more subtle effects of SFV infection that might adversely affect primate research. Unlike other retrovirus infections, SFV appears not to induce an increased IFN-γ response (Falcone *et al.,* 1999). SFV infection does, however, alter cell surface markers *in vitro*, inducing increased major histocompatibility complex (MHC) class I expression (Colas *et al.,* 1995).

Herpesviruses

Herpesviruses are double-stranded DNA viruses comprising three subfamilies, the alpha-, beta- and gamma-herpesviruses. Herpesviruses establish life-long persistent infections that are mostly latent or subclinical in their respective natural host species. Predisposing factors, such as immunosuppression or cross-species transmission can give rise to overt disease.

Alpha-herpesviruses

The simian alpha-herpesviruses are neurotropic viruses that establish latent infections in ganglia of the CNS. One group of simian alpha-herpesviruses is genetically related to human herpes simplex virus and species-specific variants have been identified in most species of laboratory primates (Eberle and Hilliard, 1995).

B virus

B virus (also known as *Cercopithecine herpesvirus-1, Herpesvirus simiae*, or Herpes B) is highly endemic in most populations of macaques (genus *Macaca*) (Kessler and Hilliard, 1990; Weigler *et al.,* 1990). There is evidence that different species of macaques harbor distinct genetic variants of B virus (Smith *et al.,* 1998; Ohsawa *et al.,* 2002). B virus is mostly nonpathogenic in its natural hosts but primary infection may manifest as conjunctivitis, or as characteristic vesicular lesions in the oral cavity, particularly on the tongue or at the mucocutaneous boundary of the lips where virus replication occurs in epithelial cells (Weigler, 1991 for review). Rare disseminated infection with massive necrosis in multiple organs has also been reported (Simon *et al.,* 1993; Carlson *et al.,* 1997). Following primary replication in epithelial tissues, B virus enters sensory and autonomic nerve endings and makes its way to the CNS by axonal transport to the nucleus of nerve cell bodies where the virus becomes latent in lumbosacral and/or trigeminal ganglia of the spinal cord (Boulter, 1975; Weigler *et al.,* 1995). Reactivation of latent virus and shedding is sporadic and may be induced by "stress" or by immunosuppressive drugs (Chellman *et al.,* 1992; Weigler *et al.,* 1993; Weigler *et al.,* 1995). Transmission occurs by direct contact with virus shed in oral or genital secretions (Zwartouw and Boulter, 1984; Lees *et al.,* 1991; Weigler *et al.,* 1993). B virus is highly pathogenic in humans and non-macaque species of primates, causing severe and often fatal encephalomyelitis (Loomis *et al.,* 1981; Weigler, 1992).

Simian Agent 8

Herpesvirus papio type 2 (HVP-2) in baboons *(Papio spp.)* and Simian Agent 8 (SA8) in African green monkeys *(Chlorocebus aethiops)* share genetic relatedness and similar biology with B virus, including the establishment of latency in sensory ganglia. Both viruses appear to be highly endemic in their respective host populations and most infections are clinically silent (Eberle *et al.,* 1997; Plesker and Coulibaly, 2002). An outbreak of vesicular disease in a baboon colony, affecting the oral and genital mucosa and originally attributed to SA8, was later determined to be due to HVP-2 (Levin *et al.,* 1986; Levin *et al.,* 1988; Eberle *et al.,* 1995). Primary SA8 infection in a group of African green monkeys was associated with transient vesicular stomatitis in young animals (Plesker and Coulibaly, 2002). To date, neither HVP-2 nor SA8 has been recognized as a human pathogen.

Simian varicella virus

Simian varicella virus (SVV) comprises a distinct group of alpha-herpesviruses of Old World monkeys, genetically related to human varicella-zoster virus (VZV) (Gray and Gusick, 1996). Naturally occurring infections and epizootics of SVV have been reported in African green monkeys *(Chlorocebus aethiops)*, Patas monkeys *(Erythrocebus patas)* and several species of macaques *(genus Macaca)* (Soike, 1992; Treuting *et al.*, 1998). Various strain designations for SVV isolates include Delta herpesvirus, Medical Lake macaque virus and Liverpool vervet virus (Padovan and Cantrell, 1986). Transmission of SVV is thought to be by exposure to droplet aerosols of respiratory secretions of actively infected animals (Mahalingam *et al.*, 2002). Disease caused by SVV infection includes fever, anorexia, lethargy and a vesicular rash on the face, abdomen and extremities (Soike, 1992). Disseminated infection may result in life-threatening encephalitis, pneumonia and hepatitis (Padovan and Cantrell, 1986). The vesicles develop from degeneration of the Malpigian layer of the dermis, and are characterized, histologically, by the formation of syncytial giant cells and eosinophilic intranuclear inclusion bodies (Padovan and Cantrell, 1986). Focal necrosis and hemorrhage may also occur in visceral organs, including the lung, spleen, liver and pancreas, with intranuclear inclusion bodies present in all affected tissues (Soike, 1992).

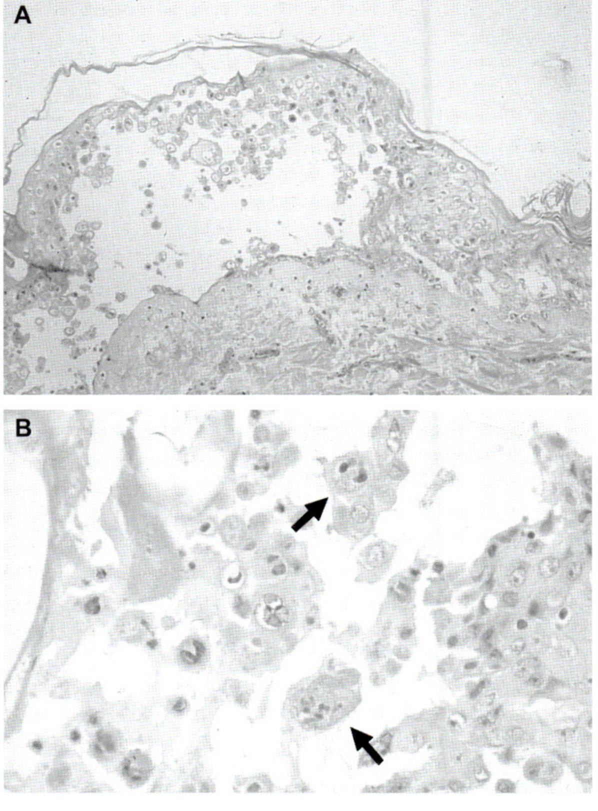

Figure 5.4 A: Vesicular lesion of SVV showing dermal necrosis. B: Presence of multinucleated giant cells. [Photos courtesy of Linda J. Lowenstine, University of California, Davis]

SVV infection is severe in African green monkeys and Patas monkeys, resembling the disseminated VZV disease seen in immunosuppressed humans, and mortality may exceed 50% (Padovan and Cantrell, 1986; Soike, 1992). SVV-induced disease appears milder in macaques although fatal cases have been reported (Padovan and Cantrell, 1986; Treuting *et al.*, 1998). In contrast to human VZV infection, SVV viral DNA was detectable in peripheral blood mononuclear cells of African green monkeys for several months following experimental primary SVV infection (Mahalingam *et al.*, 2002). In animals surviving primary infection, SVV establishes latent infection of the sensory ganglia (Mahalingam *et al.*, 1991; White *et al.*, 2002). Documentation of reactivation of latent SVV infection is limited but is presumed to be the source of infection for susceptible monkeys and an important cause of sporadic SVV epizootics (Soike, 1992; Gray and Gusick, 1996; Mahalingam *et al.*, 2002).

Beta-herpesviruses

Cytomegalovirus

Cytomegalovirus (CMV) is a ubiquitous virus that establishes persistent infection in humans and many, if not all, species of laboratory primates. Simian CMV has been isolated from common marmosets, baboons, African green monkeys, chimpanzees, and rhesus and other macaque species (Asher *et al.*, 1974; Eizuru *et al.*, 1989). CMV strains appear to be species-specific and isolates from various host species are genetically distinct (Eizuru *et al.*, 1989). Natural cross-species transmission of CMV has not been documented. CMV infection is extremely common, with up to 100% infection rates in animals over one year of age maintained in breeding colonies (Swack and Hsuing, 1982; Kessler *et al.*, 1989; Vogel *et al.*, 1994; Andrade *et al.*, 2003). Transmission of CMV is by direct contact with virus shed intermittently in urine or saliva of persistently infected animals, and natural transmission can occur across oral and genital mucosae. A strong humoral and cell-mediated immune response is evoked by CMV infection that is protective against disease but incapable of eliminating reservoirs of persistent virus within host tissues (Lockridge *et al.*, 1999). CMV is nonpathogenic in immunocompetent hosts but is a serious opportunistic infection of immunocompromised primates, including those co-infected with immunosuppressive retroviruses or receiving immunosuppressive medication (Ohtaki *et al.,* 1988; Baskin, 1987; Lockridge *et al.,* 1999; Sequar *et al.,* 2002). Characteristic intranuclear or intracytoplasmic "owl's eye" inclusion bodies can be found in cytomegalic cells in all infected tissues often associated with tissue necrosis and neutrophilic infiltrates (Baskin, 1987). CMV-associated retinitis has been reported in two macaques co-infected with simian immunodeficiency virus (SIV) (Conway *et al.,* 1990).

Maternal-fetal transmission of CMV in nonhuman primates has not been documented. Experimentally infected fetal rhesus macaques show severe developmental brain defects including microcephaly, lissencephaly, ventricular dilation, leptomeningitis and encephalitis, as well as intrauterine growth restriction and disseminated CMV disease (London *et al.,* 1986; Tarantal *et al.,* 1998). Similar fetal defects are observed following human primary CMV infection during pregnancy. Most nonhuman primates are infected with CMV long before reaching sexual maturity and, as a result, primary infection during pregnancy and associated fetal abnormalities are extremely rare.

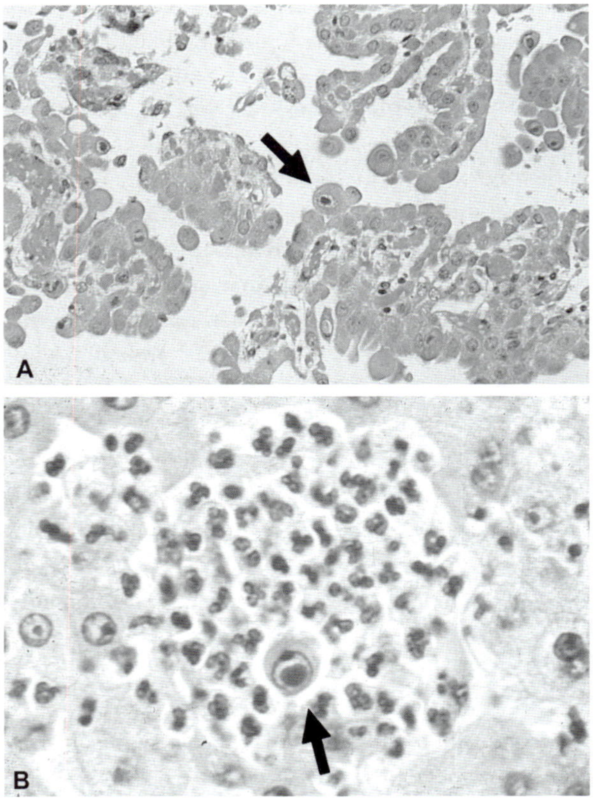

Figure 5.5 Characteristic CMV "owl's eye" intracytoplasmic or intranuclear inclusion bodies in cytomegalic cells in brain (choroids plexus) and liver of a rhesus macaque. [Photos courtesy of Peter Barry, University of California, Davis]

Gamma-herpesviruses

Lymphocryptoviruses

Gamma-1-herpesviruses include the Lymphocryptoviruses (LCV), oncogenic lymphotropic herpesviruses related to human Epstein-Barr virus (EBV). EBV-like lymphocryptoviruses have been identified in many species of Old World and New World primates. LCV infection is common in laboratory primates maintained in breeding colonies with antibody prevalence exceeding 95% in animals over two years of age (Fujimoto and Honjo, 1991; Jenson *et al.,* 2000). LCV has been etiologically linked to post-transplant lymphoproliferative disease in cynomolgus macaques undergoing experimental renal allografts (Schmidtko *et al.,* 2002) and to B-cell lymphomas in rhesus macaques with SIV- or SRV-induced immunodeficiency (Feichtinger *et al.,* 1992; Paramastri *et al.,* 2002). Recently, a novel LCV has been isolated from spontaneous B-cell lymphomas of otherwise healthy and immunocompetent common marmosets *(Callithrix jacchus)* (Cho *et al.,* 2001). Clinical signs in affected marmosets included weight loss, anorexia, diarrhea, and a firm palpable mass, detectable in the mid-portion of the abdomen. Some animals had peripheral lymphadenopathy with atypical lymphocytes present in the peripheral blood (Ramer *et al.,* 2000). At necropsy, a consistent finding was enlarged mesenteric lymph nodes, with the lymph node architecture replaced by heterogeneous sheets of neoplastic round cells. Similar neoplastic cell infiltrates were found in the proximal colon, as well as in the jejunum, ileum, and duodenum. In some affected marmosets, neoplastic cells were found in the liver, kidney and lungs (Ramer *et al.,* 2000).

Rhadinovirus

A second subfamily, the Gamma-2-herpesviruses, include the Rhadinoviruses (genus *Rhadinovirus*). These viruses are genetically related to the prototype rhadinovirus, *Herpesvirus saimiri* of squirrel monkeys *(Saimiri sciureus)* and the recently identified human herpesvirus-8 (HHV-8). Since the discovery of HHV-8, related viruses have been identified in many species of Old World primates, including baboons, macaques, African green monkeys and chimpanzees (Desrosiers *et al.,* 1997; Damania and Desrosiers, 2001;

Greensill *et al.*, 2000; Lacoste *et al.*, 2001). Two distinct lineages of gamma-2 herpesviruses have been identified in Old World primates (Bosch *et al.*, 1998; Greensill *et al.*, 2000; Schulz *et al.*, 2000). Limited serologic surveys indicate that rhadinovirus infections are common in rhesus macaques and other laboratory primate species (Desrosiers *et al.*, 1997).

Rhadinovirus infections appear mild to subclinical in immune competent natural hosts, although cross-species transmission or infection in immunodeficient hosts may result in disease. Cross-species transmission of *Herpesvirus saimiri* to other New World monkeys results in T-cell lymphoma (Melendez *et al.*, 1969). *H. saimiri* is also capable of transforming T-cells of humans and Old World monkeys *in vitro*. Experimental infection of immunocompetent macaques with rhesus rhadinovirus (RRV), a B-lymphotropic rhadinovirus, resulted in transient peripheral lymphadenopathy and seroconversion (Mansfield *et al.*, 1999). Similar experimental infections in rhesus macaques co-infected with simian immunodeficiency virus (SIV) resulted in the development of lymphoproliferative disease resembling the multicentric plasma cell variant of Castleman's disease, characterized by persistent lymphadenopathy, hepato-splenomegaly, and hypergammaglobulinemia (Wong *et al.*, 1999). Rhadinoviruses have also been implicated in the etiology of retroperitoneal fibromatosis (RF), a rapidly fatal vascular fibroproliferative neoplasm of macaques that is strongly associated with infection with an immunosuppressive type D retrovirus, SRV-2 (Rose *et al.*, 1997; Bosch *et al.*, 1999). RF can occur both in localized form, or in a progressive form that develops rapidly to fill the abdominal cavity and encase the intestines and other viscera. The lesions of RF may be either proliferative, involving spindle-shaped cells of vascular smooth muscle origin, or sclerotic, in which fibroblasts appear to be the major cell type involved (Giddens *et al.*, 1985). The proliferative lesion of RF shares some similarities with human Kaposi's sarcoma (KS), a tumor etiologically linked to infection with the human rhadinovirus human herpesvirus-8 (HHV-8). Several properties of rhadinoviruses may affect cellular and immune function and play a role in the pathogenesis of RF in macaques and KS in humans. The genome of RRV has been found to encode a functional homologue of interleukin-6 (IL-6) that may potentiate normal cellular IL-6 signaling to B lymphocytes (Kaleeba *et al.*, 1999). Elevated levels of circulating IL-6, a cytokine that plays a central role in KS, were found in SRV-infected macaques with persistent rhadinovirus viremia (Bosch *et al.*, 1999). In addition, the R1 gene of RRV has been shown to be an oncogene, with homology to the saimiri transforming gene of *Herpesvirus saimiri* and the K1 gene of HHV-8 (Damania *et al.*, 1999), and the protein expressed by R1 is capable of signal transduction that elicits B-lymphocyte activation (Damania *et al.*, 2000).

Parvoviruses

Parvoviruses are small (18-26 nm in diameter) non-enveloped, single-stranded DNA viruses of the family *Parvoviridae*. Parvoviruses require rapidly dividing cells for replication. This requirement is the basis for many aspects of parvoviral pathogenesis in tissues with high cellular turnover, including tissues of the developing fetus, gastrointestinal tract, and bone marrow (Bultman *et al.*, 2003).

Simian parvovirus

Simian parvoviruses (SPV) are recently identified members of the *Erythrovirus* genus of parvoviruses, genetically related to human parvovirus B19 (O'Sullivan *et al.*, 1994; Brown and Young, 1997; Green *et al.*, 2000). Unknown prior to 1992, SPV was discovered during an investigation of an outbreak of severe anemia in cynomolgus macaques *(Macaca fascicularis)* (O'Sullivan *et al.*, 1994). Genetically related but distinct simian parvoviruses have now been identified in cynomolgus, pig-tailed *(M. nemestrina)* and rhesus *(M. mulatta)* macaques (Brown and Young, 1997; Green *et al.*, 2000). Genetic analysis of these isolates has shown them to be 65–80% genetically similar to human B19, and the nonhuman primate isolates appear to be equally divergent from each other as they are from the human parvovirus (O'Sullivan *et al.*, 1994; Brown *et al.*, 1995; Brown and Young, 1997).

Based on limited serological surveys, exposure to SPV appears to be common, with an estimated seroprevalence of from 35–50% in adult macaques maintained in research colonies (Brown and Young, 1997). The precise mode of transmission remains unclear but direct contact, and possibly droplet aerosols, have been implicated (O'Sullivan *et al.*, 1996). Infection in immunocompetent animals appears to be mild and self-limiting and infection likely often goes undetected. Experimental infection of adult, immunocompetent, seronegative cynomolgus macaques by the intravenous and intranasal routes resulted in mild clinical signs, including anorexia and weight loss in some animals (O'Sullivan *et al.*, 1997). A transient viremia was observed in these animals, peaking at 10–12 days post-inoculation coincident with the appearance of antibodies and the transient disappearance of reticulocytes from the peripheral blood (O'Sullivan *et al.,* 1997). Animals with pre-existing SPV antibody could not be reinfected. SPV-associated severe anemia is usually observed in animals with some pre-disposing factor. The cynomolgus macaques, in which SPV infection was initially identified, were all found to be co-infected with simian type D retrovirus (SRV), a virus known to induce immunosuppression (O'Sullivan *et al.,* 1994). Subsequent outbreaks of SPV-related anemia were also observed in populations with a high prevalence of SRV infection and among animals receiving experimental drug treatments that may have contributed to persistent SPV infection (O'Sullivan *et al.,* 1996). Similarly, SPV variants, identified in outbreaks of severe anemia in pig-tailed and rhesus macaques, were found in animals experimentally infected with SIV/HIV (SHIV), a lentiviral chimera construct known to be immunosuppressive (Green *et al.,* 2000).

In immunocompromised animals, SPV infection may present with sudden onset of nonspecific clinical signs including lethargy, weakness and pallor, and severe, progressive normocytic normochromic anemia. Hematocrit values may drop below 10% (O'Sullivan *et al.,* 1994; Green *et al.,* 2000). Early in infection, a regenerative response may be present, but later the anemia is characteristically non-regenerative. Examination of bone marrow shows depletion of erythroid lineages (O'Sullivan *et al.,* 1994; Forseman *et al.,* 1999). In affected animals, intranuclear inclusion bodies, with margination of chromatin, may be observed in cells of the erythroid lineage (O'Sullivan *et al.,* 1994; 1997). Electron micrographs of these inclusion bodies show parvovirus approximately 24 nm in size and occasional parvovirus arrays (O'Sullivan *et al.,* 1994). Although dyserythropoiesis, characterized by morphologically abnormal erythroid cells with bizarre nuclear forms including blebbing, appendages, and multilobulation was observed in the initial report describing the discovery of SPV (O'Sullivan *et al.,* 1994), this is not a typical finding in SPV infections and may be related to the dual SPV/SRV infection present in those animals (O'Sullivan, 2003, personal communication).

In human beings, B19 infection during pregnancy is associated with non-immune fetal hydrops and fetal losses (Bultman *et al.,* 2003). Experimental infection of pregnant cynomolgus monkeys results in fetal infection that may culminate in severe anemia and fetal hydrops (O'Sullivan *et al.,* 1999), and appears to be an excellent animal model of B19 infection of human fetuses (O'Sullivan, 2003, personal communication).

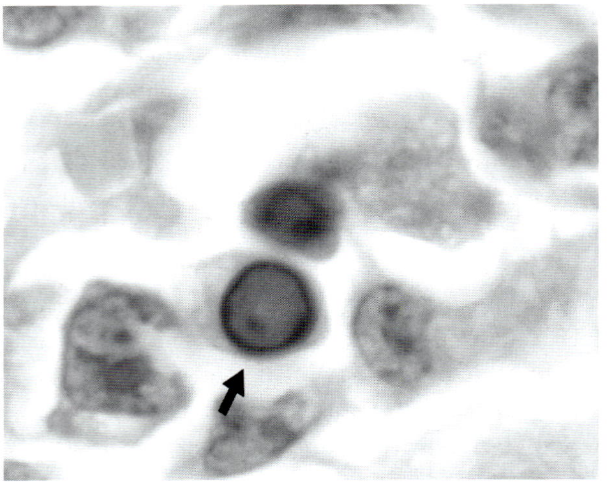

Figure 5.6 Simian parvovirus intranuclear inclusion bodies in erythroid precursor cells in bone marrow of a cynomolgus macaque. [Photo courtesy of M. Gerard O'Sullivan, 3M Corporation, Minneapolis, MN]

Simian parvovirus infection appears to be an excellent experimental model system to study human parvovirus pathogenesis. As with other viruses, undetected or unintended SPV infection in laboratory primates can represent a potentially serious confounding variable in experimental protocols, resulting in compromised research results (Schroder *et al.,* 2000; Lerche and Osborn, 2002).

Polyomaviruses

Simian virus 40

Simian Virus 40 (SV40) is a nonenveloped, double stranded DNA virus in the *Polyomavirus* genus of the family *Papovaviridae*, which also includes the BK and JC polyomaviruses of humans. SV40 persists as a latent infection in kidney, CNS and circulating lymphocytes and transmission is thought to occur through direct contact with virus shed in urine (Simon *et al.,* 1999). SV40 infection is common in various species of macaques, although seronegative populations have been reported (Kessler *et al.,* 1989). SV40 infection is non-pathogenic in immunocompetent hosts, but primary infection is associated with interstitial nephritis. Reactivation of latent infection in immunocompromised hosts, such as those co-infected with SIV, is etiologically linked to progressive multifocal leukoencephalopathy (PML), a progressive demyelinating disease of the CNS secondary to SV40 infection of oligodendrocytes

(Horvath *et al.*, 1992). Characteristic lesions of PML include multiple, often confluent, foci of demyelination in the subcortical white matter and brain stem. A marked reactive astrocytosis is often present, and large inclusions are frequently seen in oligodendrocytes (Chretien *et al.*, 2000).

A meningoencephalitis (ME) primarily affecting gray matter of the cerebrum and brain stem has been described as a second CNS manifestation of SV40 infection, distinct from PML. (Simon *et al.*, 1999). In SV40-associated ME, meninges are thickened by edema and cellular infiltrates around vessels. A distinctive feature of SV40 related ME is the presence of cells with enlarged nuclei containing intranuclear inclusions (Simon *et al.*, 1999). In the immunocompromised host, it appears that primary SV40 related CNS disease manifests as ME, whereas reactivation of SV40 in the CNS manifests as PML (Simon *et al.*, 1999). Primary SV40 infection has also been associated with interstitial pneumonia and rare nephritis characterized by renal tubular necrosis (Sheffield *et al.*, 1980).

Integrated polyomavirus DNA may be oncogenic, as inoculation of hamsters and other non-primate species with SV40 induces tumor formation (Mutti *et al.*, 1998).

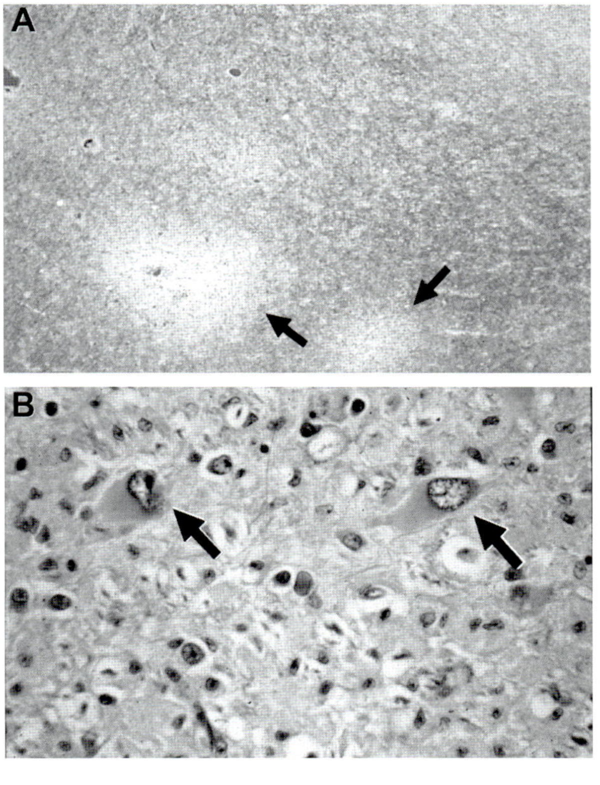

Figure 5.7 SV40 associated PML, A: Foci of demyelination in white matter. B: Intranuclear inclusion bodies in oligodendrocytes. [Photos courtesy of Linda J. Lowenstine, University of California, Davis]

The zoonotic potential of SV40 remains undetermined. SV40 was an undetected contaminant of live oral polio vaccine prepared in rhesus macaque kidney cell cultures in the 1950s and early 1960s. As a result, millions of vaccine recipients were exposed to SV40 infection. Recently, SV40-specific nucleic acids have been detected, by genetic amplification (polymerase chain reaction, PCR), in human tumors, specifically ependymoma, mesothelioma, and certain types of lymphoma (Mutti *et al.*, 1998).

Simian agent 12

Simian Agent 12 (SA12), a polyomavirus related to, but distinct from, SV40 has been isolated from Chacma baboons *(Papio ursinus)* (Valis *et al.*, 1977), and a third virus, distinct from both SV40 and SA12, was isolated from olive baboons (*P. anubis*) (Gardner *et al.*, 1989). Neither of these baboon viruses has been associated with clinical disease but seroprevalence studies indicate that infection with these viruses is common (15–25%) in baboon populations surveyed (Valis *et al.*, 1997; Gardner *et al.*, 1989).

Correspondence

Any correspondence should be directed to Nicholas Lerche, California National Primate Research Center, University of California, Davis, CA, USA. Email: nwlerche@ucdavis.edu

References

Andrade, M.R., Yee, J.L., Barry, P., Spinner, A., Leite, J.P., Roberts, J.A. and Lerche, N.W. (2003). *Amer. J. Primatol.* 59, 123–128.

Asher, D.M., Gibbs, C.J. Jr., Lang, D.J., Gajdusek, D.C. and Chanock, R.M. (1974). *Proc. Soc. Exp. Biol. Med.* 145, 794–801.

Baskin, G.B. (1987). *Am. J. Pathol.* 129, 345–352.

Bosch, M.I., Strand, K.B. and Rose, T.M. (1998). *J. Virol.* 72, 8458–8459.

Bosch, M.I., Harper, E., Schmidt, A., Strand, K.B., Thormahlen, S., Thouless, M.E. and Wang, Y. (1999). *J. Gen. Virol.* 80, 467–475.

Bostik, P., Wu, P., Dodd, G.L., Villinger, F., Mayne, A.E., Bostik, V., Grimm, B.D., Robinson, D, Kung, H-J. and Ansari, A.A. (2001). *J. Virol.* 75, 11298–11306.

Boulter, E.A. (1975). *J. Biol. Stand.* 3, 279–280.

Brown, K.E. and Young, N.S. (1997). *Rev. Med. Virol.* 7, 211–218.

Brown, K.E., Green, S.W., O'Sullivan, M.G. and Young, N.S. (1995). *Virol.* 210, 314–322.

Bultman, B.D., Klinger, K., Sotlar, K., Bock, C.T. and Kandolf, R. (2003). *Virchows Arch.* 442, 8–17.

Carlson, C.S., O'Sullivan, M.G., Jayo, M.J., Anderson, D.K., Harber, E.S., Jerome, W.G., Bullock, B.C. and Heberling, R.L. (1997). *Vet. Pathol.* 34, 405–414.

Carvalho, E.M., Bacellar, O., Porto, A.F., Braga, S., Galvao-Castro, B., and Neva, F (2001). *J. Acq. Immuno. Syndromes* 27, 1–6.

Chellman, G.J., Lukas, V.S., Eugui, E.M., Altera, K.P., Almquist, S.J. and Hilliard, J.K. (1992). *Lab. Anim. Sci.* 42, 146–151.

Cho, Y-G., Ramer J., Rivailler, P., Quink, C., Garber, R.L., Beier, D.R. and Wang, F. (2001). *PNAS USA* 98, 1224–1229.

Chopra, H.C. (1970). *Proc. Amer. Soc. Cancer Res.* 11, 16–18.

Chretien, F., Boche, D., de la Grandmaison, G.L., Ereau, T., Mikol, J., Hurtrel, M., Hurtrel, B. and Gray, F. (2000). *Acta Neuropathol.* 100, 332–336.

Colas, S., Bourge, J.F., Wybier, J., Chelbi-Alix, M.K., Paul, P. and Emanoil-Ravier, R. (1995). *J. Gen. Virol.* 76, 661–667.

Conway, M.D., Didier, P., Fairburn, B., Soike, K.F., Martin, L., Murphey-Corb, M., Meiners, N. and Insler, M.S. (1990). *Curr. Eye Res.* 9, 759–770.

Damania, B., Li, M., Choi, J-K., Alexander, L., Jung, J.U. and Desrosiers, R.C. (1999). *J. Virol.* 73, 5123–5131.

Damania, B., DeMaria, M., Jung, J.U. and Desrosiers, R.C. (2000). *J. Virol.* 74, 2721–2730.

Damania, B. and Desrosiers, R.C. (2001). *Philos. Trans. R. Soc. Lond. B Biol. Sci.* 29, 595–604.

Daniel, M.D., Letvin, N.L., Sehgal, P.K., Schmidt, D.K., Silva, D.P., Solomon, K.R., Hodi, F.S., Ringler, D.F., Hunt, R.D. and King, N.W. (1988). *Int. J. Cancer* 41, 601–608.

Desrosiers, R.C. (1997). *AIDS Res. Hum. Retroviruses* 13, 5–6.

Desrosiers, R.C., Sasseville, V.G., Czajak, S.C., Zhang, X., Mansfield, K.G., Kaur, A., Johnson, R.P., Lackner, A.A. and Jung, J.U. (1997). *J. Virol.* 71, 9764–9769.

Eberle, R., Black, D.H., Blewett, E.L. and White, G.L. (1997). *Lab. Anim. Sci.* 47, 256–262.

Eberle, R., Black, D.H., Lipper, S. and Hilliard, J.K. (1995). *Arch. Virol.* 140, 529–545.

Eberle, R. and Hilliard, J. (1995). *Infect. Agents Dis.* 55–70.

Eizuru, Y., Tsuchiya, K., Mori, R. and Minamishima, Y. (1989). *Arch. Virol.* 107, 65–75.

Faucher, S., Dimock, K. and Wright, K.E. (2002). *Virus Res.* 90, 63–75.

Falcone, V., Leupold, J., Clotten, J., Urbanyi, E., Herchenroder, O., Spatz, W., Volk, B., Bohm, N., Toniolo, A., Neumann-Haefelin, D. and Schweizer, M. (1999). *Virol.* 257, 7–14.

Feichtinger, H., Kaaya, E., Putkonen, P., Li, S.L., Ekman, M., Gendelman, R., Biberfeld, G. and Biberfeld, P. (1992). *AIDS Res. Hum. Retroviruses* 8, 339–348.

Feldman, G., Fickenscher, J., Bodemer, W., Sparing, M., Neblein, T., Hunsmann, G. and Dittmer, U. (1997). *Virol.* 229, 106–112.

Felsenfeld, A.D. and Schmidt, N.J. (1977). *Infect. Immun.* 15, 807–812.

Fine, D.L., Landon, J.C., Pienta, R.J., Kubicek, M.T., Valerio, M.G., Loeb, W.F. and Chopra, H.C. (1975). *J. Nat. Cancer. Instit.* 54, 651–658.

Fortmeyer, H.P. (1982). In Bartosek, I., Guaitani, A. and Pacei, E. (eds) *Animals in Toxicology Research,* pp 13–32. Raven Press, New York.

Forseman, L., Narayan, O. and Pinson, D. (1999). *Contemp. Topics* 38, 20–22.

Fujimoto, K. and Honjo, S. (1991). *J. Med. Primatol.* 20, 42–45.

Fultz, P. (1994). In Levy, J.A. (ed.) *The Retroviridae,* vol. 3, pp 111–131. Plenum Press, New York.

Gabet, A.S., Gessain, A. and Wattel, E. (2003). *Int. J. Cancer* 107, 74–83.

Gardner, M.B., Endres, M., Barry, P. (1994). In Levy, J.A. (ed.) *The Retroviridae*, vol. 3, pp 133–176. Plenum Press, New York.

Gardner, S.D., Knowles, W.A., Hand, J.F. and Porter, A.A. (1989). *Archiv. Virol.* 105, 223–233.

Giavedoni, L.D., Velasquillo, M.C., Parodi, L.M., Hubbard, G.B. and Hodara, V.L. (2000). *J. Virol.* 74, 1648–1657.

Giddens, W.E. Jr., Tsai, C.C., Morton, W.R., Ochs, H.D., Knitter, G.H. and Blakley, G.A. (1985). *Amer. J. Pathol.* 119, 253–263.

Gray, W.L. and Gusick N.J. (1996). *Virol.* 224, 161–166.

Green, S.W., Malkovska, I., O'Sullivan, M.G. and Brown, K.E. (2000). *Virol.* 269, 105–112.

Greensill, J. and Schulz, T.F. (2000). *AIDS* 14, Suppl. 3, S11–S19.

Greensill, J., Sheldon, J.A., Renwick, N.M., Beer, B.E., Norley, S., Goudsmit, J. and Schulz, T.F. (2000). *J. Virol.* 74, 1572–1577.

Guzman, R.E., Kwelin, E.L., Zimmerman, T.E. (1999). *Toxicol. Pathol.* 27, 672–677.

Hayashi, K., Chen, H-L., Yanai, H., Koirala, T.R.,Ohara, N., Teramoto, N., Oka, T., Yoshino, T., Takahashi, K., Miyamoto, K., Fujimoto, K., Yoshikawa, Y. and Akagi, T. (1999). *Lab. Invest.* 79, 823–835.

Heneine, W., Switzer, W.M., Sundstrom, P., Brown, J., Vedapuri, S., Schable, C.A., Khan, A.S., Lerche, N.W., Schweizer, M., Neumann-Haefelin, D., Chapman, L.E. and Folks, T.M. (1998). *Nature Medicine* 4, 403–407.

Horvath, C.J., Simon, M.A., Bergsagel, D.J., Pauley, D.R., King, N.W., Garcea, R.L. and Ringler, D.J. (1992). *Amer. J. Pathol.* 140, 1431–1440.

Hubbard, G.B., Mone, J.P., Allan, J.S., Davis, K.J. 3rd, Leland, M.M., Banks, P.M. and Smir, B. (1993). *Lab. Anim. Sci.* 43, 301–309.

Jenson, H.B., Ench, Y., Gao, S.J., Rice, K., Carey, D., Kennedy, R.C., Arrand, J.R. and Mackett, M. (2000). *J. Infect. Dis.* 181, 1462–1466.

Johnson, P.R. and Hirsch, V.M. (1992). *AIDS Res Hum Retroviruses* 8, 367–372.

Kaleeba, J.A.R., Bergquam, E.P. and Wong, S.W. (1999). *J. Virol.* 73, 6177–6181.

Kessler, M.J. and Hilliard, J.K. (1990). *J. Med. Primatol.* 19, 155–160.

Kessler, M.T., London, W.T., Madden, D.L., Dambrosia, J.M., Hilliard, J.K., Soike, K.F. and Rawlins, R.G. (1989). *Puerto Rican Health Sci.* 8, 95–97.

Khan, A.S., Galvin, T.A., Lowenstine, L.J., Jennings, M.B., Gardner, M.B. and Buckler, C.E. (1991). *J. Virol.* 65, 7061–7065.

Kuhn, E.M., Stolte, N, Matz-Rensing, K., Mach, M., Stahl-Henning, C., Hunsmann, G. and Kaup, F.J. (1999). *Vet. Pathol.* 36, 51–56.

Lackner, A.A., Rodriguez, M.A., Bush, C.E., Munn, R.J., Kwang, K-S., Moore, P.F., Osborn, K.G., Marx, P.A., Gardner, M.B. and Lowenstine, L.J. (1988). *J. Virol.* 62, 2134–2142.

Lackner, A.A., Moore, P.F., Marx, P.A., Munn, R.J., Gardner, M.B. and Lowenstine, L.J. (1990). *J. Med. Primatol.* 19, 339–349.

Lacoste, V., Mauclere, P., Dubreuil, G., Lewis, J., Georges-Courbot, M-C. and Gessain, A. (2001). *Genome Res.* 11, 1511–1519.

Lal, R.B. and Rudolph, D.L. (1991). *Blood* 78, 571–574.

Lazo, A. and Bailer, R.T. (1996). *J. Med. Primatol.* 25, 257–266.

Lees, D.N., Baskerville, A., Cropper, L.M. and Brown, D.W. (1991). *Lab. Anim. Sci.* 41, 360–364.

Lerche, N.W., Yee, J.L. and Jennings, M.B. (1994). *Lab. Anim. Sci.* 44, 217–221.

Lerche, N.W., Osborn, K.G., Marx, P.A., Prahalada, S., Maul, D.H., Lowenstine, L.J., Munn, R.J., Bryant, M.L., Henrickson, R.V., Arthur, L.O., Gilden, R.V., Barker, C.S., Hunter, E. and Gardner, M.B. (1986). *J. Nat. Cancer. Instit.* 77, 489–495.

Lerche, N.W., Marx, P.A., Osborn, K.G., Maul, D.H., Lowenstine, L.J., Bleviss, M.L., Moody, P., Henrickson, R.V. and Gardner, M.B. (1987). *J. Nat. Cancer. Instit.* 79, 847–854.

Lerche, N.W., Yee, J.L. and Jennings, M.B. (1994). *Lab. Anim. Sci.* 44, 217–221.

Lerche, N.W. and Osborn, K.G. (2002). *Toxicol. Pathol.* 31 (Suppl.), 103–110.

Lerche, N.W. (1992). In Matano, S., Tuttle, R., Ishida, H. and Goodman, M. (eds) *Topics in Primatology,* vol. 3, pp 439–447. University of Tokyo Press.

Lerche, N.W., Cotterman, R., Dobson, M.D., Yee, J.L., Rosenthal, A.N. and Heneine, W. (1997). *Lab. Anim. Sci.* 47, 263–268.

Letvin, N.L., Daniel, M.D., Sehgal, P.K., Chalifoux, L.V., King, N.W., Hunt, R.D., Aldrich, W.R., Holley, K., Schmidt, D.K. and Desrosiers, R.C. (1984). *J. Virol.* 52, 683–686.

Levin, J.L., Hilliard, J.K., Geisler, C.A., Butler, T.M. and Goodwin, W.J. (1986). *Lab. Anim. Sci.* 36, 551.

Levin, J.L., Hilliard, J.K., Lipper, S.L., Butler, T.M. and Goodwin, W.J. (1988). *Lab. Anim. Sci.* 38, 394–397.

Lockridge, K.M., Sequar, G., Zhou, S.S., Yue, Y., Mandell, C.P. and Barry, P.A. (1999). *J. Virol.* 73, 9576–9583.

Lockridge, K.M., Zhou, S.S., Kravitz, R.H., Johnson, J.L., Sawai, E.T., Blewett, E.L. and Barry, P.A. (2000). *Virol.* 268, 272–280.

London, W.T., Martinez, A.J., Houff, S.A., Waller, W.L., Curfman, B.L., Traub, R.G. and Sever, J.L. (1986). *Teratol.* 33, 323–331.

Loomis M.R., O'Neill, T., Bush, M. and Montali, R.J. (1981). *J. Amer. Vet. Med. Assoc.* 179, 1236–1239.

Lowenstine, L.J. and Lerche, N.W. (1988). *J. Zoo Anim. Med.* 19, 168–187.

Lowenstine, L.J., Pedersen, N.C., Higgins, J., Pallis, K.C., Uyeda, A., Marx, P., Lerche, N.W., Munn, R.J. and Gardner, M.B. (1986). *Int. J. Cancer* 38, 563–574.

Lowenstine, L.J., Lerche, N.W., Yee, J.L., Oyeda, A., Jennings, M.B., Munn, R.J., McClure, H.M., Anderson, D.C., Fultz, P.N. and Gardner, M.B. (1992). *J. Med. Primatol.* 21, 1–14.

MacKenzie, M., Lowenstine, L.J., Lalchandani, R., Lerche, N.W., Osborn, K.G., Spinner, A., Bleviss, M., Henrickson, R.V. and Gardner, M.B. (1986). *Lab. Anim. Sci.* 36, 59–64.

Mahalingam, R., Smith, D., Wellish, M., Wolf, W. Dueland, A.N., Cohrs, R., Soike, K and Gilden, D. (1991). *Proc. Nat. Acad. Sci, USA* 88, 2750–2752.

Mahalingam, R., Clarke, P., Wellish, M., Dueland, A.N., Soike, K.F., Gilden, D.H., and Cohrs, R. (1992). *Virol.* 188, 193–197.

Mahalingam, R., Traina-Dorge, V., Wellish, M., Smith, J. and Gilden, D.H. (2002). *J. Virol.* 76, 8548–8550.

Mansfield, K.G., Westmoreland, S.V., DeBakker, C.D., Czajak, S., Lackner, A.A. and Desrosiers, R.C. (1999). *J. Virol.* 73, 10320–10328.

Marracci, G.H., Kelley, R.D., Pilcher, K.Y., Crabtree, L, Shiigi, S.M., Avery, N., Leo, G., Webb, M.C., Hallick, L.M., Axthelm, M.K. and Machida, C.A. (1995). *J. Virol.* 69, 2621–2628.

Marx, P.A. and Lowenstine, L.J. (1987). *Cancer Surv.* 6, 101–114.

Marx, P.A., Bryant, M.L., Osborn, K.G., Maul, D.H., Lerche, N.W., Lowenstine, L.J., Kluge, J.D., Zaiss, C.P., Henricksons, R.V., Shiigi, S.M., Wilson, B.J., Malley, A., Olson, L.C., McNulty, W.P., Arthur, L.O., Gilden, R.V., Barker, C.S., Hunter, E., Munn, R. J., Heidecker, G. and Gardner, M.B. (1985). *J. Virol.* 56, 571–578.

Maul, D.H, Zaiss, C.P., MacKenzie, M.R., Shiigi, S.M., Marx, P.A. and Gardner, M.B. (1988). *J. Virol.* 62, 1768–1773.

Meertens, L. and Gessain, A. (2003). *J. Virol.* 77, 782–789.

Melendez, L.V., Hunt R.D., Daniel, M.D., Garcia, F.G. and Fraser, C.E. (1969). *Lab. Anim. Care* 19, 378–386.

Miyoshi, I., Fujishita, M., Taguchi, T. Matsubayashi, K., Miya, N. and Tanioka, Y. (1983). *Int. J. Cancer* 32, 333–336.

Moazad, T.C. and Thouless, M.E. (1993). *J. Med. Primatol.* 22, 382–389.

Morozov, V.A. and Lagaye, S. (1998). *Lancet* 351, 1705–1711.

Mutti, L., Carbone, M., Giordano, G.G. and Giordano, A. (1998). *Monaldi Arch Chest Dis.* 53, 198–201.

Nigida, S.M., Falk, L.A., Wolfe, L.G. and Deinhardt, F. (1979). *Lab. Anim. Sci.* 29, 53–60.

Noda, Y., Isikawa, K., Sasauaua, A., Honjos, S., Mori, S., Tsujimoto, H. and Hayami, M. (1986). *Jpn. J. Cancer Res.* 77, 1227–1234.

Ohta, Y., Tsujimoto, J., Ishikawa, K., Kodama, T., Morikawa, S., Nakai, M., Honjo, S. and Hayami, M. (1988). *Int. J.Cancer* 41, 115–122.

Ohtaki, S., Hondo, R., Kodama, H. and Kurata, T. (1988). *Acta Patholog. Jpn.* 36, 1553–1563.

Osborn, K.G., Prahalada, S., Lowenstine, L.J., Gardner, M.B., Maul, D.H. and Henrickson, R.V. (1984). *Am. J. Pathol.* 114, 94–103.

Ohsawa, K., Black, D.H., Torii, R., Sato, H. and Eberle, R. (2002). *Comp. Med.* 52, 555–559.

O'Sullivan, M.G., Anderson, D.C., Fikes, J.D., Bain, F.T., Carlson, C.S., Green, S.W., Young, N.S. and Brown, K.E. (1994). *J. Clin. Invest.* 93, 1571–1576.

O'Sullivan, M.G., Anderson, D.K., Lund, J.E., Brown, W.P., Green, S.W., Young, N.S. and Brown, K.E. (1996). *Lab. Anim. Sci.* 46, 291–297.

O'Sullivan, M.G., Anderson, D.K., Goodrich, J.A., Tulli, H., Green, S.W., Young, N.S. and Brown, K.E. (1997). *J. Virol.* 71, 4517–4521.

O'Sullivan, M.G., Feeney, D.A., Jarvis, J.B., Veille, J.C., Block, W.A., Young, N.S., Brown, K.E. (2000). Program and Abstracts, 8th Parvovirus Workshop, p 153.

Otsyula, M., Yee, J.L., Jennings, M., Suleman, M., Gettie, A., Tarara, R., Ishihakia, M., Marx, P. and Lerche, N.W. (1996). *Ann. Trop. Med. Parasitol.* 90, 65–70.

Padovan, D. and Cantrell, C.A. (1986). *Lab. Anim. Sci.* 36, 7–13.

Pakes, S.P., Lu, Y-S. and Mevnier, P.C. (1984). In Fox, J.G., Cohen, B.J. and Loew, F.M. (eds) *Laboratory Animal Medicine,* pp 649–666. Academic Press, New York.

Palmer, A.E. (1987). *J. Med. Primatol.* 16, 99–130.

Paramastri, Y.A., Wallace, J.M., Sallenge, K.J., Wilkinson, L.M., Malarkey, D.E. and Cline, J.M. (2002). *Vet. Pathol.* 39, 399–402.

Peeters, M, Courgnaud, V., Abela, B., Auzel, P., Pourrut, X., Bibollett-Ruche, F., Loul, S., Liegeois, F., Butel, C., Koulagna, D., Mpoudi-Ngole, E., Shaw, G.M., Hahn, B. and Delaporte, E. (2002). *Emerg. Infect. Dis.* 8, 451–457.

Plesker, R. and Coulibaly, C. (2002). *Primate Report* 63, 27–31.

Ramer, J.C., Garber, R.L., Steele, K.E., Boyson, J.F., O'Rourke, C. and Thomson, J.A. (2000). *Lab. Anim. Sci.* 50, 59–68.

Rangan, S.R., Martin, L.N., Bozelka, E., Wang, N. and Gormus, B.J. (1986). *Int. J. Cancer* 38, 425–432.

Rau, P., Jiang, H. and Wang, F. (2000). *J. Clin. Microbiol.* 38, 3219–3225.

Richards, A.L., Giri, A., Iskandriati, D., Pamungkas, J., Sie, A., Rosen, L., Anthony, R.L. and Franchini, G. (1998). *J. Acquired Immune Deficiency Syndromes and Human Retrovirology* 19, 542–545.

Rivadeneira, E.D., Ferrari, M.G., Jarrett, R.G., Armstrong, A.A., Markham, P., Birkebak, T., Takemoto, S., Johnson-Delaney, C., Pecon-Slattery, J., Clark, E.A. and Franchini, G. (1999). *Blood* 94, 2090–2101.

Rose, T.M., Strand, K.B., Schultz, E.R., Schaefer, G., Rankin, G.W., Thouless, M.E., Tsai, C.C. and Bosch, M.L. (1997). *J. Virol.* 71, 4138–4144.

Rose, T.M., Ryan, J.T., Schultz, E.R., Raden, B.W. and Tsai, C.C. (2003). *J. Virol.* 77, 5084–5097.

Sakakibara, I., Sugimoto, Y., Sasagawa, A., Honjo, S., Tsujimoto, H., Nakamura, H. and Hayami, M. (1986). *J. Med. Primatol.* 15, 311–318.

Schmidtko, J., Wang, R., Wu, C.L., Mauiyyedi, S., Harris, N.L., Della Pella, P., Brousaides, N., Zagachin, L., Ferry, J.A., Wang, F., Kawai, T., Sachs, D.H., Cosimi, B.A. and Colvin, R.B. (2002) *Transplantation* 15, 1431–1439.

Schroder M.A., Fisk, S.K. and Lerche, N.W. (2000). *Contemp. Topis Lab. Anim. Sci.* 39, 16–23.

Schultz, E.R., Rankin, G.W. Jr., Blanc, M-P., Raden, B.W., Tsai, C-C. and Rose, T.M. (2000). *J. Virol.* 74, 4919–4928.

Sequar, G., Britt, W.J., Lakeman, F.D., Lockridge, K.M., Tarara, R.P., Canfield, D.R., Zhou, S.S., Gardner, M.B., and Barry, P.A. (2002). *J. Virol.* 76, 7661–7671.

Sheffield W.D., Strandberg, J.D., Braun, L., Shah, K. and Kalter, S.S. (1980). *J. Infect. Dis.* 142, 618–622.

Shimamoto, Y., Funai, N., Watanabe, M., and Suga, K. (1996). *Int. J. Hematol.* 64, 111–118.

Simon, M.A., Daniel, M.D., Lee-Parritz, D., King, N.W. and Ringler, D.J. (1993). *Lab. Anim. Sci.* 43, 545–550.

Simon, M.A., Ilynski, P.O., Baskin, G.B., Knight, H.Y., Pauley, D.R. and Lackner, A.A. (1999). *Amer. J. Pathol.* 154, 437–446.

Smith, A.L., Black, D.H. and Eberle, R. (1998). *J. Virol.* 72, 9224–9232.

Soike, K.F. (1992). *Annals of the New York Academy of Science* 653, 323–333.

Strand, K., Harper, E., Thormahlen, S., Thouless, M.E., Tsai, C-C., Rose, T. and Bosch, M.L. (2000). *J. Clin. Virol.* 16, 253–269.

Swack, N.S. and Hsuing, G.D. (1982). *J. Med. Primatol.* 11, 169–177.

Tarantal, A.F., Salamat, M.S., Britt, W.J., Luciw, P.A., Hendrickx, A.G. and Barry, P.A. (1998). *J. Infect. Dis.* 177, 446–450.

Treuting, P.M., Johnson-Delaney, C., and Birkebak, T.A. (1998). *Lab. Anim. Sci.* 48, 384–386.

Tsai, C.C., Warner, T.F.C., Uno, H. and Giddens, W.E. Jr. (1987). *Am. J. Pathol.* 120, 30–37.

Tsai, C.C., Follis, K.E., Snyder, K., Windsor, S., Thouless, M.E., Kuller, L. and Morton, W.R. (1990). *J. Med. Primatol.* 19, 203–216.

Valis, J.D., Newell, N., Reissig, M., Malherbe, H., Kaschila, V.R. and Shah, K.V. (1977). *Infect. Immun.* 18, 247–252.

Van der Logt, J.T.M. (1993). *Aging Clin. Exp. Res.* 5, 317–323.

Voevodin, A., Samilchuk, E., Schatzl, H., Boeri, E., Franchini. G. (1996). *J. Virol.* 70, 1633–1639.

Vogel, P., Weigler, B.J., Kerr, H., Hendrickx, A.G., Barry, P.A. (1994). *Lab. Anim. Sci.* 44, 25–30.

Weigler, B.J., Roberts, J.A., Hird, D.W., Lerche, N.W. and Hilliard, J.K. (1990). *Lab. Anim. Sci.* 40, 257–261.

Weigler, B.J. (1992). *Clin. Infect. Dis.* 14, 555–567.

Weigler, B.J., Hird, D.W., Hilliard, J.K., Lerche, N.W., Roberts, J.A. and Scott, L.M. (1993). *J. Infect. Dis.* 167, 257–263.

Weigler, B.J., Scinicariello, F. and Hilliard, J.K. (1995). *J. Infect. Dis.* 171, 1139–1143.

White, T.M., Mahalingam, R., Traina-Dorge, V. and Gilden, D.H. (2002). *J. Neurovirol.* 8, 191–205.

Wutzler, P., Meerbach, A., Farber, I., Wolf, H. and Scheibner, K. (1995). *Arch. Virol.* 140, 1979–1995.

Wong, S.W., Bergquam, E.P., Swanson, R.M., Lee, F.W., Shiigi, S.M., Avery, N.A., Fanton, J.W. and Axthelm, M.K. (1999). *J. Exp. Med.* 190, 827–840.

Yee, J.L. and Lerche, N.W. (2003). *AIDS Res. Hum. Retovir.* 19 Supp., 565.

Yoshioka, S.T. (2000). Unpublished MS Thesis, University of California, Davis.

Zou, W., Lackner, A.A., Simon, M., Durand-Gasselin, I., Galanaud, P., Desrosiers, R.C. and Emilie, D. (1997). *J. Virol.* 71, 1227–1236.

Zwartouw, H.T. and Boulter, E.A. (1984). *Lab. Anim.* 18, 65–70.

Zwartouw, H.T., MacArthur, J.A., Boulter, E.A., Seamer, J.H., Marston, J.H. and Chamove, A.S. (1984). *Lab. Anim.* 18, 125–130.

Chapter 6

Modeling Parasitic Diseases in Nonhuman Primates: Malaria, Chagas' Disease, and Filariasis

Mario T. Philipp[1] and Jeanette E. Purcell[2]
Divisions of Bacteriology and Parasitology[1] and Veterinary Medicine,[2] Tulane National Primate Research Center, Tulane University Health Sciences Center, Covington, Louisiana, USA

Introduction

Numerous successful attempts have been made at modeling parasitic diseases of humans in nonhuman primates. Those we have selected to portray are the models for malaria, Chagas' disease, and lymphatic filariasis, in part because of the sheer relevance of these diseases in terms of prevalence, morbidity, or both. In addition, our choice is influenced by the achievements attained and/or potential offered by these models in understanding disease pathogenesis, developing vaccines, or improving chemotherapy. Whenever possible, we have focused on models that make use of the rhesus macaque, *Macaca mulatta,* as this monkey species has emerged as the one preferentially used in studies of the pathogenesis of infectious diseases.

Nonhuman primate models of malaria

Malaria continues to be an enormous threat to human health worldwide. Yearly incidence varies between 300 and 500 million clinical cases, with up to 0.5% mortality, mostly in children (Artavanis-Tsakonas *et al.,* 2003). *Plasmodium falciparum, P. malariae, P. vivax,* and *P. ovale,* all infect humans but severe manifestations and deaths are caused largely by *P. falciparum,* the most virulent of the human plasmodia (Trampuz *et al.,* 2003; Stauffer and Fischer, 2003). This is perhaps the reason why most studies on malaria in nonhuman primates have attempted to model falciparum malaria, either with *P. falciparum* itself or with primate malaria species that mimic that disease in humans.

The Laboratory Primate
Copyright 2005 Elsevier
ISBN 0-1208-0261-9

Models of malaria pathogenesis

In the last ten years, two manifestations of pathogenesis of malaria have received most of the attention of researchers working with nonhuman primates: cerebral malaria, and the complications associated with malaria in pregnancy.

Cerebral malaria

Cerebral malaria is the most morbid complication of severe falciparum malaria. It is largely prevalent in children and non immune adults and is often fatal (Warrell, 1999). Clinically it is sometimes preceded by headache, irritability, and confusion. In non immune individuals, coma may ensue within hours (Warrell, 1999). Seizures, and other signs such as nystagmus, disconjugate gaze, papilledema, retinal hemorrhages and altered respiration, are fairly common. Results of imaging studies are consistent with cerebral edema or ischemia (Chandy, 2003). This agrees with the characteristic pathologic finding at autopsy, namely vessels in the cerebral vascular bed plugged with parasitized erythrocytes, petechial hemorrhages and edema (de Souza and Riley, 2002). Two hypotheses of the pathogenesis of cerebral malaria have guided most of the research. The first is the mechanical hypothesis, whereby selective cytoadherence and sequestration of parasitized erythrocytes in cerebral venules cause obstruction to blood flow in the brain, leading to coma and death. The second is the toxin/cytokine hypothesis, whereby the secretion of a "toxin" by the parasite results in the dysregulated production of pro-inflammatory cytokines such as TNF-alpha and other mediators (reviewed in Warrell, 1997). Possible parasitized erythrocytes' ligands include the erythrocyte band 3, sequestrin, and the PfEMP-1 family of proteins (*P. falciparum* erythrocyte membrane protein 1), found concentrated in the electron-dense "knobs", long described as the putative link between red cells and the endothelium. The venous endothelium receptors, thus far incriminated, include thrombospondin and CD36, which are constitutively expressed, and ICAM-1 (intracellular adhesion molecule 1) and E-selectin, which can be upregulated in response to inflammatory cytokines such as TNF-alpha (Warrell, 1997). Unraveling the pathogenesis of such a complex syndrome is a true challenge to which efforts made using nonhuman primate models have contributed significantly.

There have been numerous attempts at modeling cerebral malaria in nonhuman primates and both human and monkey *Plasmodium* species have been used. These include *P. coatneyi*, *P. knowlesi*, and *P. fragile*, among the monkey malarias, and *P. falciparum*. Rhesus macaques have proven to be the best host to mimic features of severe cerebral malaria upon infection with monkey malaria species. Unlike with *Macaca fascicularis*, in which infection with monkey malaria species is benign, these organisms cause an often lethal infection in *Macaca mulatta* (de Souza and Riley, 2002). Four of five rhesus macaques that were given an intravenous inoculation with a virulent strain of *P. knowlesi*, became parasitemic within a week. As with severe malaria in the human, they progressively manifested fever, loss of appetite and fatigue, terminating with coma and death (Ibimoye *et al.*, 1993). Neuropathological findings included brain edema, accumulation of parasitized red blood cells and macrophages in the brain microvasculature, and increased accumulation of fibroblasts and collagen deposits in the extracellular spaces adjacent to the parasite-filled microvessels (Ibimoye *et al.*, 1993). The overall picture of microvascular cramming of parasitized red cells closely resembles the human situation. On the other hand, the apparent sequestration of macrophages differs from the human (Ibimoye *et al.*, 1993). Moreover, the rather prevalent edema formation in this model has been attributed to breakdown of the blood-brain barrier, elicited perhaps by the production of kinines, or other vasoactive amines, in a manner that is rare in human cerebral malaria (de Souza and Riley, 2002).

Rhesus monkeys infected with *P. fragile* also show promise as a model of cerebral malaria. In a limited (two-animal) experiment, it was demonstrated that 11 days post-inoculation (PI) the animals became comatose (Fujioka *et al.*, 1994). Cerebral microvessels from both animals exhibited sequestered parasitized erythrocytes, albeit in different amounts per animal. Unlike the *P. knowlesi*/rhesus model, no cerebral edema was evident. Electron microscopy revealed characteristic electron-dense knobs on parasitized red blood cells, adhering to the microvessel endothelium (Fujioka *et al.*, 1994). Importantly, using immunofluorescent labeling, expression of CD36, ICAM-1, and thrombospondin was detected in cerebral microvessels of infected animals, but not in the corresponding microvessels of an uninfected control rhesus.

Similar knob-associated cytoadherence and upregulation of endothelial receptors were obtained with the rhesus-*P. coatneyi* model (Aikawa *et al.*, 1992), both supporting the "mechanical" hypothesis in the pathogenesis of cerebral malaria. Brain cytokine expression profiles were also assessed in this model (Tongren *et al.*, 2000).

Brains of four animals, infected with *P. coatneyi,* were examined post-mortem, 11 days PI. Cytokine, adhesion molecule, and inducible nitric oxide synthetase (iNOS) mRNA expression was assessed in tissue specimens from midbrain, cerebellum, and cortex and white matter of the cerebrum, by reverse transcriptase-polymerase chain reaction (RT-PCR). The expression levels of TNF-alpha, IFN-gamma, IL-1beta, ICAM-1, and iNOS were highest in the cerebellum of infected animals. Interestingly, this correlated well with an early finding in this model, namely, the preferential accumulation of sequestered parasitized red blood cells in this region of the brain (Sein *et al.,* 1993). Infected animals also had elevated TNF-alpha expression levels in the cortex, and IL-1beta expression levels in the cortex, white matter, and midbrain. This model thus offers the opportunity to explore not only the mechanical hypothesis, but also the toxin/cytokine hypothesis of cerebral malaria pathogenesis.

Malaria in pregnancy

Pregnancy enhances malarial morbidity in the infected mother and may affect fetal and infant health and survival (McGregor, 1984; Bruce-Chwatt, 1983; Menendez, 1995). It is estimated that between 75,000 and 200,000 infant deaths per year are associated with malaria that was acquired during pregnancy (Steketee *et al.,* 2001). Low birth weight (LBW) from prematurity and/or intrauterine growth retardation (IUGR), intrauterine exposure to the parasite and congenital infection, as well as maternal anemia may all contribute to this outcome. Neither the pathophysiologic mechanisms underlying the enhanced maternal malarial morbidity, infant LBW and IUGR, nor the role of the parasite's presence in the placenta are known. *Plasmodium coatneyi* infections of rhesus monkey dams have been shown to faithfully model the maternal malarial syndrome (Davison *et al.,* 1998, 2000). *P. coatneyi* is a natural parasite of *Macaca fascicularis.* While the infection is benign in this macaque species, *P. coatneyi* causes death in rhesus monkeys in about a third of the infected animals with high parasitemias (Coatney, 1971). Inoculation of four rhesus dams with *P. coatneyi* blood-stage parasites, during the first trimester, led, in three of the animals, to abortions which correlated with peak parasitemias at seven to ten days post-inoculation. Six dams given smaller inocula (10^6 organisms rather than 10^3) also became parasitemic, yet were able to complete their pregnancy. In all cases the infants that were delivered weighed significantly less than infants of uninfected mothers (Davison *et al.,* 1998). Two of the infants exhibited IUGR, one symmetrical and the other asymmetrical, the former dying five days post-delivery. An initial assessment of the placentas of infected dams revealed several changes associated with infection and related to pregnancy outcome. LBW and IUGR were significantly associated with high parasite pigment scores and elevated numbers of activated macrophages in the intervillous space (Davison *et al.,* 2000). Accumulation of the parasite pigment hemozoin, a by-product of hemoglobin metabolism by *Plasmodia,* is an indirect measure of parasitemia, late in gestation. This is because fibrin, where most hemozoin is deposited, is not a major component of the first-trimester placenta. Numbers of activated macrophages, quantified immunohistochemically by the activation marker LN5, is a measure of local inflammation. Interestingly, placental damage as assessed by scoring distinct fibrinoid lesions, infarcts, chorionic plate thromboses, and syncytiotrophoblast disruption, did not correlate with LBW, IURG, or early infant mortality (Davison *et al.,* 2000). Thus, in this first approach to defining the pathophysiology of fetal LBW and IUGR in malarious rhesus dams, it appears that elevated late parasite burdens and placental inflammation seem to be required. Clearly, this model has much to offer as a research tool to understand the pathogenesis of the maternal malarial syndrome.

Models to investigate malaria vaccines and antimalarials

Malaria vaccines

Malaria ranks first among all other parasitic diseases in terms of efforts devoted to its prevention by immunological means. Pre-clinical testing of malaria vaccines, i.e. evaluation of safety, immunogenicity and efficacy immediately prior to clinical trials in humans, is done in nonhuman primates. Although certain malaria vaccine studies have made use of rhesus macaques (see below), the World Health Organization has recommended the New World monkey genera *Aotus* and *Saimiri* as the genera of choice for pre-clinical trials of malaria vaccines (WHO/OMS Memorandum, 1988). Accordingly, numerous pre-clinical assessments of the efficacy of human malaria subunit vaccines have been done in these New World monkeys, mostly in *Aotus* species (Enders *et al.,* 1992; Herrera *et al.,* 1992; Kumar *et al.,* 1995; Chang *et al.,* 1996; Gysin *et al.,* 1996; Burghaus *et al.,* 1996; Moreno *et al.,* 1999; Egan *et al.,* 2000; Jones *et al.,* 2001; Stowers, *et al.,* 2001, 2002ab; Hisaeda *et al.,* 2002; Baruch *et al.,* 2002;

Rivera *et al.,* 2002; Rosas *et al.,* 2002; Purmova *et al.,* 2002; Cubillos *et al.,* 2002; Perlaza, *et al.,* 2003) but also in *Saimiri* (Collins *et al.,* 1989, 1990, 1999, 2000a; Bonnefoy *et al.,* 1994; Yang *et al.,* 1999).

The main reason for this selection, which has elicited heated debate, both for it (Stowers *et al.,* 2001), and against it (Heppner *et al.,* 2001), is the susceptibility of these animals to infection with the human malariae *P. falciparum, P. vivax,* and *P. malariae.* The complexity of the malaria life cycle, and the numerous difficulties encountered thus far in developing an effective subunit vaccine, has dictated the strategy of approaching malaria immunoprophylaxis in a comprehensive fashion (Carvalho *et al.,* 2002). Six parasite developmental stages are being considered as targets: sporozoites, the form in which the parasite enters the human host; the liver stage, which is the first multiplicative phase; the merozoite, which is the stage that is able to invade and reinvade the red blood cell; the infected erythrocyte itself; parasite toxins; and the sexual stages. Since all of these stages of all of the malaria parasite species that infect humans may be accessed in *Aotus* and *Saimiri* monkeys, this comprehensive approach to malaria vaccine development further underscores the need for these two primate models.

The *Aotus* and *Saimiri* species that are susceptible to *P. falciparum, P. vivax,* and *P. malariae* are *Aotus lemurinus griseimembra* and *A. lemurinus lemurinus* from Panama and Colombia, *A. nancymai* and *A. vociferans* from Peru, and *A. azarae boliviensis* from Bolivia, as well as *Saimiri sciureus* from Panama, Colombia and Guyana, and *S. boliviensis* and *S. peruviensis* (Collins *et al.,* 2002b). *P. ovale* is not infectious to these New World species (Collins *et al.,* 2002b), and not all strains of the other three malaria species of humans are infectious either. However, once a parasite isolate has been adapted to an *Aotus* or *Saimiri* species, it also has been found to be infectious to other species from these genera (Herrera *et al.,* 2002; Collins *et al.,* 2002b). This process of adaptation requires splenectomy (Taylor *et al.,* 1979; Weller *et al.,* 1992; Collins *et al.,* 1994), for although, in most cases, susceptibility to sporozoite infection is readily achieved, eusplenic animals do not show a detectable infection with blood stages of malaria (Collins *et al.,* 1977, 1996ab; Gramzinski *et al.,* 1999; Zapata *et al.,* 2002). Fortunately, once adaptation of the parasite has been accomplished, eusplenic animals may be used in some cases (Herrera *et al.,* 2002).

Chiefly on the basis of their phylogenetic kinship with the human (Miyamoto *et al.,* 1988), chimpanzees (*Pan troglodytes*) also have been employed in experimental malaria vaccine efficacy trials (Daubersies *et al.,* 2000). However, several reasons discourage their use for this purpose. First, chimpanzees need to be splenectomized to increase susceptibility to human malaria infection. Second, compared to monkeys they are difficult to house and manage humanely. Third, and perhaps most importantly, in areas such as Gabon which, together with the Republic of Congo, are estimated to harbor the majority of the world's chimpanzees, the combined effects of commercial deforestation and hunting and the spread of Ebola hemorrhagic fever, compounded with the chimpanzee's slow reproductive cycle, have diminished the chimpanzee population by over 50% between 1983 and 2000 (Walsh *et al.,* 2003; Tutin, 2001). Thus it has been recommended to elevate this species to "critically endangered" status (Walsh *et al.,* 2003).

Evidence of immunologically relevant similarities between *Aotus* and humans should mitigate the potential consequences of a decision to follow that recommendation. When T-cell receptor alpha chain cDNA segments of *A. nancymai* monkeys were analyzed, even though the *Aotus* gene sequences diverged more from the human sequences than those of the chimpanzee or the rhesus macaque, none of the 29 gene segments analyzed lacked a human counterpart (Favre *et al.,* 1998). In most cases the identity of amino acid sequences between corresponding *Aotus* and human genes was greater than 80%. This marked conservation of gene sequences, which indicates a close structural relationship of *Aotus* and human TcR, further substantiates the validity of using *Aotus* monkeys as models for the evaluation of malaria subunit vaccines (Favre *et al.,* 1998).

Rhesus macaques and plasmodium species that infect rhesus monkeys have been used to assess efficacy of malaria vaccines in cases where there is evidence that the monkey and human malaria antigens have immunologically relevant structural similarities (Millet *et al.,* 1995; Kocken *et al.,* 1999), or when a novel immunization regimen requires preliminary testing (Bhardwaj *et al.,* 2002; Rogers *et al.,* 2001, 2002; Kumar *et al.,* 2001). These, as well as the testing of safety and/or immunogenicity of subunit vaccines, or DNA vaccines derived from human malaria species (Coban *et al.,* 2004; Kumar *et al.,* 2002; Shankar *et al.,* 2001; Angov *et al.,* 2003; Wang *et al.,* 1998; Tine *et al.,* 1996) are sensible applications of rhesus models to malaria vaccine development.

Antimalarials

The chief driving forces to develop new drugs to prevent and cure malaria have been: the evolution of widespread resistance to chloroquine, initially (1950s) by *P. falciparum* and later (1980s) by *P. vivax* (Wellems and Plowe, 2001); the subsequent appearance of localized foci of resistance to most other antimalarials, and the toxicity of these other drugs (Kumar *et al.*, 2003). Nonhuman primate models have played, and will continue to play, an important role in this process, most crucially with *P. vivax* which, unlike *P. falciparum* (Trager and Jensen, 1979), has not yet been grown in routine continuous culture.

Chloroquine resistant strains of both *P. falciparum* and *P. vivax* have been adapted to *Aotus* and *Saimiri* species (e.g. Schmidt, 1979; Sullivan *et al.*, 1999; Collins *et al.*, 2000b), thus setting the scene to investigate new therapeutics applicable to such strains. As with vaccine studies, the fact that all of the mammalian developmental stages of *P. vivax* and *P. falciparum* are accessible in New World monkeys makes *Aotus* and *Saimiri* species essential for drug efficacy studies. However, new drugs are often assessed also in Old World monkeys, specifically the rhesus macaque, using monkey malaria species such *P. cynomolgy* and/or *P. fragile* (Edstein *et al.*, 1994; Puri and Dutta, 2003). Application examples range from studies of drug efficacy against specific developmental stages (Obaldia *et al.*, 1997), drugs that may reverse chloroquine resistance (Oduola *et al.*, 1998), drugs with therapeutic effects against specific disease manifestations, such as cerebral malaria (Tripathi *et al.*, 1997), drug toxicity (Petras *et al.*, 1997) and many others.

From the preceding examples it is apparent that nonhuman primate models of malaria will continue to provide invaluable information on the pathogenesis, immunoprophylaxis and treatment of this important disease.

Nonhuman primate models of Chagas' disease

Chagas' disease in humans

No effective drugs to cure, or vaccines to prevent, Chagas' disease are available and the pathogenesis of this disease is poorly understood (Cerecetto and Gonzalez, 2002). Therein lies the importance of research on American trypanosomiasis, Chagas' disease. It affects between 16 and 18 million people, largely in Central and South America, although cases have been reported in the U.S. (Hagar and Rahimtoola, 1991; Leiby *et al.*, 1997). Many more (*ca.* 100 million) are at risk of infection with *Trypanosoma cruzi*, the disease's etiologic agent (http://www.who.int/ctd/chagas). *T. cruzi* is transmitted to humans by blood-sucking reduviid bugs of the genus *Triatoma*. During the acute phase of Chagas' disease, which may last from a few weeks to several months, the organism disseminates extracellularly (the amastigote form), and resides and multiplies intracellularly (the trypomastigote, replicative form). A local inflammatory lesion, called a chagoma, may develop at the site of entry of the parasite. Histologically, the chagoma shows mononuclear cell infiltration, interstitial edema, and intracellular aggregates of amastigotes in cells of the subcutaneous tissue and muscle. Biopsy specimens from enlarged lymph nodes show hyperplasia, and amastigotes may be present in reticular cells. Severe clinical signs of acute Chagas' disease are rare, as only 5–20% of infected individuals manifest the high fever, prostration, and chills that characterize severe acute disease. More commonly, no symptoms, or a transient flu-like syndrome characterizes this phase of the infection (Añez *et al.*, 1999). A clinically silent "indeterminate" phase follows, with sub-patent parasitemias that contrast with the high levels of circulating organisms that characterize the acute stage. Eventually, after many years of quiescence, the truly morbid, chronic phase of Chagas' disease ensues. Up to 40% of patients go on to develop cardiomyopathy, megacolon, and/or megaesophagus syndrome, and nerve conduction disorders of the heart and gastro-intestinal tract. There is high mortality, usually of heart failure, in the more serious cases (Umezawa *et al.*, 2001).

Models of pathogenesis of Chagas' disease

Many New World monkey species are naturally susceptible to *T. cruzi,* and several Old World primate species, that are incidentally exposed to *T. cruzi* transmission, may become infected with this organism. However, most systematic assessments of the potential of nonhuman primates, as models to investigate the pathogenesis of Chagas' disease, have been performed in two monkey species: the brown capucine monkey, *Cebus apella*, and the rhesus macaque.

Early studies in rhesus, albeit in a small number of animals ($n = 4$), indicated that, as with humans, parasites were readily detectable in the circulation within the first 40 weeks of infection, but not thereafter, even though the animals remained infected, as determined by subculture (Seah *et al.*, 1974). In a further experiment, one animal of a group of four rhesus that were infected with a virulent strain of *T. cruzi*, died with emaciation after the development of megaesophagus, 166 days PI (Marsden *et al.*, 1976). Later, a more comprehensive set of experiments firmly established the rhesus macaque as a workable model to investigate pathogenesis of Chagas' disease. Thirteen animals were infected subcutaneously with a Colombian strain of *T. cruzi*. Two animals were available as uninfected controls. Parasitemia appeared as early as 13 days PI, and lasted in some animals up to day 59 PI (Bonecini-Almeida *et al.*, 1990). Subsequently, parasites were demonstrated only by hemoculture and/or xenodiagnosis. The animals appeared apathetic during the first three weeks PI, and manifested diminished appetite and an average loss of weight of 6%. Low fever (1°C above normal values) was detected in three animals. A chagoma, appearing as a papular erythematous lesion, appeared in nine of the infected animals at the site of inoculation. Afferent lymph nodes were enlarged in all of the animals (Bonecini-Almeida *et al.*, 1990). Mild electrocardiographic anomalies, similar to those observed in non-lethal human acute Chagas' myocarditis also were detected in nine animals. These anomalies uniformly disappeared by the fourth month PI. Both leukocyte and lymphocyte levels were significantly increased in the peripheral circulation from the fifth week PI until the 15th week. Circulating specific IgM and IgG antibodies were observed from week two PI. IgM antibodies subsided by month nine PI, but IgG levels persisted long thereafter (Bonecini-Almeida *et al.*, 1990). Myocarditis and myositis, characterized by multiple foci of lympho-histiocytic inflammatory infiltrates, were present in animals sacrificed on days 41 (n = 2), 70 (n = 1), and 76 (n = 1) PI but not in the one animal sacrificed 39 months PI. These outcomes thus agreed, respectively, with features of both the acute and indeterminate phases of Chagas' disease.

The chronic phase was investigated in a group of seven surviving animals from the study described above (Bonecini-Almeida *et al.*, 1990). At the time the evaluation of these animals was initiated, they had been infected for an average of 16.7 years (Carvalho *et al.*, 2003). Subpatent parasitemia was detected in all of the animals, using hemoculture (n = 2), artificial xenodiagnosis (n = 3), and polymerase chain reaction (n = 6). In two assessments that were performed at an interval of 16 months, no cardiac or gastrointestinal radiologic alterations were observed, using contrast radiography of the gastrointestinal tract, either with regard to relative heart/chest size, or diameter of esophagus and colon (Carvalho *et al.*, 2003). However, significant electrocardiographic abnormalities were detected in three of six animals. Only one of six infected monkeys showed an echocardiographic abnormality. These results indicate that rhesus monkeys, experimentally infected with *T. cruzi*, not only reproduce the acute phase of Chagas' disease, but also develop chronic chagasic cardiomyopathy (Carvalho *et al.*, 2003).

Overall, the results obtained with cebus monkeys are similar to those achieved with rhesus, both during the acute and the chronic phases of infection (Rosner *et al.*, 1988; Falasca *et al.*, 1990; Malchiodi *et al.*, 1993; Riarte *et al.*, 1995). They will not be described here, except for one study in which it was demonstrated that *C. apella* animals manifest some of the gastrointestinal complications of chronic Chagas' disease (Falasca *et al.*, 1992). In a study that involved 53 animals, 14 were inoculated with different strains of *T. cruzi* (CA1, n = 10; Colombian, n = 4; and Tulahuen, n = 4), and 35 were used as uninfected controls. Barium contrast X-rays studies of the gastrointestinal tract were performed at baseline and at one and three years PI. No changes in diameter, length or motility of the colon and esophagus were observed in the control animals. However, three of the infected animals had an enlarged colon and at post-mortem (between 21 and 67 months PI) lesions were seen in nine of the 12 assessed animals, either in the colon alone (n = 4) or in both colon and esophagus (n = 5) (Falasca *et al.*, 1992). Over a five-year period, the animals infected with the Tulahuen and Colombian strains of *T. cruzi* (but not those infected with the CA1 strain) showed significantly decreased peak gastric acid output, as compared to uninfected controls.

From studies in rhesus and in cebus monkeys, it is readily apparent that the pathogenesis of Chagas' disease may be reliably investigated in these models. Although efforts to control the triatomine vector using pyrethroid insecticides have been very successful – the World Health Organization has estimated that transmission of Chagas' disease has ceased in the majority of endemic areas of Brazil and Argentina (Dias *et al.*, 2002; Cerecetto *et al.*, 2002) – the fact that vaccine development is very incipient, and that much still needs to be done to improve chemotherapy (Cerecetto and Gonzalez, 2002), indicates that monkey models that faithfully reproduce Chagas' disease will be very useful.

Chagas' disease immunoprophylaxis and chemotherapy

Antigens that elicit protective immune responses against lethal challenge with *T. cruzi* have been recently identified (Garg and Tarleton, 2002; Schnapp *et al.*, 2002) with the aid of experiments in mice. Similarly, trials with new drugs for Chagas' disease are being actively performed in murine models (Caffrey *et al.*, 2000; Zuccotto *et al.*, 2001; Stoppani, 1999). To our knowledge, however, these results have not yet been significantly translated into pre-clinical trials in non-human primates.

The two drugs of most widespread clinical use, Nifurtimox and Benznidazole, are effective in acute Chagas' disease, but may not be effective in the chronic phase. Their efficacy varies from strain to strain of *T. cruzi*, they are toxic, have generated resistance, and their supply is often problematic (Stoppani, 1999; Cerecetto and Gonzalez, 2002). It is expected that the dramatic need for better drugs, and for vaccines against Chagas' disease, will propel further advances in drug and vaccine research and eventually bring this field of research to the point where pre-clinical tests in non-human primates are apposite.

Nonhuman primate models of lymphatic filariasis

Three species of filariae cause lymphatic filariasis in humans: *Wuchereria bancrofti*, *Brugia malayi*, and *Brugia timori*. The most prevalent is *W. bancrofti*, as it is estimated to infect about 90% of the 129 million people thought to harbor lymphatic filariae (Michael *et al.*, 2001). Attempts have been made, with some success, at establishing *W. bancrofti* in several species of nonhuman primates (Misra *et al.*, 1997; Latendresse *et al.*, 1987; Campbell *et al.*, 1987; Palmieri *et al.*, 1982, 1983; Sucharit *et al.*, 1982; Dissanaike and Mak, 1978, 1980) but, to our knowledge, these models have been applied only rarely to the study of the pathogenesis of bancroftian filariasis (Latendresse *et al.*, 1987). In contrast, *B. malayi* has been used to generate an important nonhuman primate model of lymphatic filariasis in the rhesus monkey (see below). This is possibly because this parasite is more cosmopolitan with regard to host species than *W. bancrofti*. It may also be grown in rodents (Ash and Riley, 1970) but, unfortunately, rodents of no species are able to mimic the spectrum of clinical signs and concomitant inflammatory responses observed in human filariasis, when infected with *B. malayi* (Philipp *et al.*, 1984).

Lymphatic filariasis in humans

Lymphatic filariasis is a chronic debilitating disease that manifests clinically as recurrent adenolymphangitis, fever, lymphedema, hydrocele, or elephantiasis (Kumaraswami, 2000). Adult worms reside in the afferent lymphatic ducts or lymph-node sinuses, where they cause lymphatic dilation, obstruction, and inflammation. Pathologic changes in chronic lymphatic filariasis result chiefly from inflammatory damage to the lymphatics. Infiltration with plasma cells, eosinophils, and macrophages in, and around the infected vessels are all histologic hallmarks of lymphatic involvement. Proliferation of the endothelium and connective tissue may also be present, sometimes with damage to lymph valves. Lymphedema may be evident clinically. The pathogenesis paradigm of lymphatic filariasis is that its pathology relates, in part, to misdirected immune responses. In the lymphatics, local immune responses directed toward dead or dying adult worms are believed to cause the granulomatous and proliferative processes that precede lymphatic obstruction (Dreyer *et al.*, 2000). Immunologically, a spectrum of responses has been described in endemic areas (Maizels *et al.*, 1995; Nutman, 2001). On one extreme there are clinically asymptomatic individuals with circulating microfilariae, the first-stage larvae released by the female adult worms, yet their peripheral blood mononuclear cells (PBMC) fail to proliferate or produce certain cytokines in response to parasite antigen (Maizels *et al.*, 1995; Nutman, 2001). On the other extreme there are individuals with no detectable circulating microfilariae but with a relatively high prevalence of lymphatic obstruction and whose PBMC proliferate and produce IL-2 and IFN-gamma (Maizels *et al.*, 1995; Nutman, 2001).

Filariasis pathogenesis in the rhesus monkey

Given the clinical and immunological complexity of lymphatic filariasis, modeling this disease in experimental animals has been a challenge. In the seventies, it was established that the rhesus monkey was susceptible to inoculation with third stage (infective) larvae of *B. malayi* (Wong *et al.,* 1969, 1977; el-Bihari and Ewert, 1970). Animals were shown to become microfilaremic, eosinophilia was a common finding, and intermittent fever, that correlated with lymphadenopathy, was frequently observed (Wong *et al.,* 1977). Several years later, features of the early course of *B. malayi* infection were investigated in the rhesus monkey, in a study of the histopathological, lymphoscintigraphical and immunological changes (Dennis *et al.,* 1998). Five of the ten animals in the study were inoculated with infective larvae in the lower right leg and five were sham-inoculated controls. All of the animals given larvae became microfilaremic between weeks 10 and 12 PI. As with humans, in whom microfilariae often cease to circulate, three animals later became amicrofilaremic, at weeks 17, 26, and 27 PI, respectively. Inguinal nodes at the side of inoculation were more severely affected than nodes on the contralateral side. The right-side inguinal lymph nodes of all infected animals were moderately to severely enlarged beginning at week 5 PI. Peak enlargement was observed between weeks 10 and 16 PI with a gradual decrease in size until week 33 PI, when all nodes were of normal size again. Lymph-node histology, assessed in biopsy tissue from two infected animals at week 10 PI, indicated that the right-side inguinal nodes were hyperplastic with increased size and number of germinal centers, and increased mitotic activity of lymphoblasts. There was a moderate to severe eosinophilic lymphadenitis in the paracortical and medullary areas. Connective tissue surrounding the node was moderately inflamed, with perivascular infiltration of lymphocytes, plasma cells, and eosinophils. Portions of parasitic larvae were also present. The right and left nodes from the control animal were normal.

At week 24 PI, the right and left inguinal nodes from the five infected and the five control animals were biopsied and assessed histologically. The left inguinal lymph nodes from all infected animals were essentially unaffected and could not be differentiated histologically from the right and left nodes of control animals. Histological changes in infected nodes ranged from minimal accumulations of eosinophils and mast cells in medullary areas to nodes with mild lymphoid hyperplasia. Most nodes had a slight increase in mast cells over control nodes. The histopathologic changes, observed in the right inguinal nodes of infected animals at week 10 PI, were severe compared to those observed at week 24 PI. These changes apparently did not correlate with peak microfilarial densities that occurred between 20 and 24 weeks PI (Dennis *et al.,* 1998).

Perhaps the most revealing results were obtained from lymphoscintigraphic measurements. Isotopic lymphoscintigraphy permits the indirect measurement of lymph flow by timing the accumulation of a radioactive compound (usually a derivative of 99mTc) in a lymph node whose afferent ducts are under investigation (Szuba *et al.,* 2003). Lymphoscintigraphy was performed at 7 and 15 weeks PI. Significant impairment of lymph flow rate was detected between the popliteal and the inguinal lymph nodes of the right leg of all of the infected animals at week 7 PI, and of three of these animals at week 15 PI. Compared to control uninfected animals, the rate of lymph flow also was diminished in the left leg of inoculated animals, but much less markedly than in the right leg. These experiments showed that pathophysiological alterations leading to partial lymphatic obstruction can occur very early in the infection process (Dennis *et al.,* 1998).

Lymphatic pathology, and its correlation with cellular immune responsiveness and parasite burden, also were investigated in the *B. malayi*/rhesus model in animals that had been infected for a much longer time (Giambartolomei *et al.,* 1998). Nine animals received multiple inoculations with a small number of third-stage larvae at monthly intervals over a 48-month period. Six of these animals had been inoculated once with 200 larvae, two years before the initiation of the "trickle" inoculation protocol (Giambartolomei *et al.,* 1998). There were also three sham-inoculated controls. The inoculation regimen strongly influenced immunological responsiveness and clinical outcome. The three animals that underwent the dual inoculation protocol had lymphedema, whereas the other six infected animals did not. All of the infected, but clinically normal, animals had PBMCs that were essentially non-responsive to *B. malayi* antigen, in that they proliferated only marginally and had minimally increased transcription of IL-2 and IFN-gamma mRNA. In contrast, PBMCs from animals with lymphedema proliferated vigorously and evidenced significantly higher IL-2 and IFN-gamma transcription levels (Giambartolomei *et al.,* 1998). Correlating with this non-responsiveness, and perhaps contributing to the cause of it, was the finding of a significantly lower proportion of T-cells bearing the IL-2-receptor, and a significantly diminished expression of the CD80 co-stimulatory ligand among PBMCs of the non-responding animals (Giambartolomei *et al.,* 1998, 2001). Interestingly, the average level of microfilariae per ml of blood was 50 times higher in the group of nonresponder animals than in the responders. Five of those six animals, but only one of the three responder animals, showed detectable microfilaremia.

Most recently, evidence has been obtained that bacteria of the genus *Wolbachia,* which are endosymbiotic in lymphatic and other filarial species, play a role in the inflammatory pathogenesis of filariasis (Taylor, 2003). In the rhesus monkey chronically infected with *B. malayi,* two out of three animals that exhibited lymphedema also had circulating antibodies that recognized a surface antigen of *Wolbachia*. This antibody was not present in any of nine infected animals with no lymphedema or in the two uninfected controls (Punkosdy *et al.,* 2001). Thus, the polarization of immune responsiveness, its correlation with lymphatic pathology, inverse correlation with microfilaremia, and the possible involvement of *Wolbachia* in filariasis pathogenesis observed in humans with lymphatic filariasis also is observed in the rhesus macaque infected with *B. malayi* (Giambartolomei *et al.,* 1998, 2001). We believe this model holds promise as a tool to help understand the pathogenesis of lymphatic filariasis.

Chemotherapy and immunoprophylaxis of lymphatic filariasis

In 1997 the World Health Organization issued the Global Program to Eliminate Lymphatic Filariasis (reviewed in Molyneux and Taylor, 2001). The program is currently based on the application of albendazole, combined either with diethylcarbamazine (DEC) or ivermectin, as a means to interrupt transmission of lymphatic filariasis. The drugs target primarily the microfilariae, whose level in the peripheral circulation is drastically reduced by their administration in a single yearly dose over a five-to-six-year period (Karam and Ottesen, 2000). Two of these drugs, DEC and ivermectin, had previously been tested against lymphatic filariae in nonhuman primates: ivermectin in the silvered leaf monkey (*Presbytis cristatus*) (Kurniawan *et al.,* 1992; Mak *et al.,* 1987, 1988), a natural host for these parasites, and DEC in both the leaf and rhesus monkeys (Kurniawan *et al.,* 1992; Wong *et al.,* 1977). As the filariasis eradication program expands, these animals could be of use again to elucidate potential side effects that may still emerge.

The incipient development of the field of filariasis vaccines research currently precludes the need for nonhuman primates for preclinical trials. This situation may soon change, however, as the Filarial Genome Project, which was initiated in 1994 under the auspices of the WHO, continues to progress (Williams *et al.,* 2000). This program comprises an expressed sequence tag (EST) gene discovery component, combined with genome sequencing and chromosome physical mapping (Williams *et al.,* 2000). The chosen lymphatic filaria is *B. malayi,* and over 22,000 ESTs have been sequenced thus far (Blaxter *et al.,* 2002). Analysis of the chemotherapeutic and immunoprohylactic potential of these gene products may soon yield new targets for drug and vaccine development.

Concluding remarks

The examples provided in this chapter should help to illustrate the usefulness of nonhuman primates in modeling parasitic diseases. Much of the progress made in our understanding of the pathogenesis of cerebral malaria, chronic Chagas' disease or lymphatic filariasis, has been due to the use of these models and should lead to the design of better drugs and vaccines. These, too, may now be thoroughly tested both for safety and efficacy, in nonhuman primate models.

Acknowledgements

This work was supported in part by grant RR00164 from the National Center for Research Resources, National Institutes of Health. Secretarial help from Avery Maclean is gratefully acknowledged.

Correspondence

Any correspondence should be directed to Mario Philipp, Division of Bacteriology and Parasitology, Tulane National Primate Research Center, Tulane University Health Sciences Center, Covington, Louisiana, USA. Email: philipp@tpc.tulane.edu

References

Aikawa, M., Brown, A.E., Smith, C.D., Tegoshi, T., Howard, R.J., Hasler, T.H., Ito, Y., Collins, W.E. and Webster, H.K. (1992). *Mem. Inst. Oswaldo Cruz.* 87 Suppl 3, 443–447.

Añez, N., Carrasco, H., Parada, H., Crisante, G., Rojas, A., Gonzalez, N., Ramirez, J.L., Guevara, P., Rivero, C., Borges, R. and Scorza, J.V. (1999). *Am. J. Trop. Med. Hyg.* 60, 215–222.

Angov, E., Aufiero, B.M., Turgeon, A.M., Van Handenhove, M., Ockenhouse, C.F., Kester, K.E.,

Walsh, D.S., McBride, J.S., Dubois, M.C., Cohen, J., Haynes, J.D., Eckels, K.H., Heppner, D.G., Ballou, W.R., Diggs, C.L. and Lyon, J.A. (2003). *Mol. Biochem. Parasitol.* 128, 195–204.

Artavanis-Tsakonas, K., Tongren, J.E. and Riley, E.M. (2003). *Clin. Exp. Immunol.* 133, 145–152.

Ash, L.R. and Riley, J.M. (1970). *J. Parasitol.* 56, 969–73.

Baruch, D.I., Gamain, B., Barnwell, J.W., Sullivan, J.S., Stowers, A., Galland, G.G., Miller, L.H. and Collins, W.E. (2002). *Proc. Natl. Acad. Sci. USA* 99, 3860–3865.

Bhardwaj, D., Hora, B., Singh, N., Puri, S.K., Lalitha, P., Rupa, P. and Chauhan, V.S. (2002). *FEMS Immunol. Med. Microbiol.* 34, 33–43.

Blaxter, M., Daub, J., Guiliano, D., Parkinson, J. and Whitton, C. (2002). *Trans. R. Soc. Trop. Med. Hyg.* 96, 7–17.

Bonecini-Almeida Mda, G., Galvao-Castro, B., Pessoa, M.H., Pirmez, C. and Laranja, F. (1990). *Mem. Inst. Oswaldo Cruz.* 85, 163–171.

Bonnefoy, S., Gysin, J., Blisnick, T., Guillotte, M., Carcy, B., Pereira da Silva, L. and Mercereau-Puijalon, O. (1994). *Vaccine* 12, 32–40.

Bruce-Chwatt, L.J. (1983). *Br. Med. J. (Clin Res Ed),* 286, 1457–1458.

Burghaus, P.A., Wellde, B.T., Hall, T., Richards, R.L., Egan, A.F., Riley, E.M., Ballou, W.R. and Holder, A.A. (1996). *Infect. Immun.* 64, 3614–3619.

Caffrey, C.R., Scory, S. and Steverding, D. (2000). *Curr. Drug. Targets* 1, 155–162.

Campbell, J.R., Marwoto, H.A., Tirtokusumo, S., Masbar, S., Rusch, J.T., Purnomo, and Trenggono, B. (1987). *Lab. Anim. Sci.* 37, 502–504.

Carvalho, C.M., Andrade, M.C., Xavier, S.S., Mangia, R.H., Britto, C.C., Jansen, A.M., Fernandes, O., Lannes-Vieira, J. and Bonecini-Almeida, M.G. (2003). *Am. J. Trop. Med. Hyg.* 68, 683–691.

Carvalho, L.J., Daniel-Ribeiro, C.T. and Goto, H. (2002). *Scand. J. Immunol.* 56, 327–343.

Cerecetto, H. and Gonzalez, M. (2002). *Curr. Top. Med. Chem.* 2, 1187–1213.

Chandy, J. (2003). *Infect. Med.* 20, 53–58.

Chang, S.P., Case, S.E., Gosnell, W.L., Hashimoto, A., Kramer, K.J., Tam, L.Q., Hashiro, C.Q., Nikaido, C.M., Gibson, H.L., Lee-Ng, C.T., Barr, P.J., Yokota, B.T. and Hut, G.S. (1996). *Infect. Immun.* 64, 253–261.

Coatney, G.R. and National Institute of Allergy and Infectious Diseases (U.S.) (1971). *The Primate Malarias,* U.S. National Institute of Allergy and Infectious Diseases; Supt. of Docs., U.S. Govt. Print. Off., Washington, Bethesda, MD.

Coban, C., Philipp, M.T., Purcell, J.E., Keister, D.B., Okulate, M., Martin, D.S. and Kumar, N. (2004). *Infect. Immun.* 72, 253–259.

Collins, W.E., Galland, G.G., Sullivan, J.S., Morris, C.L. and Richardson, B.B. (1996b). *Am. J. Trop. Med. Hyg.* 54, 380–385.

Collins, W.E., Galland, G.G., Sullivan, J.S., Morris, C.L., Richardson, B.B. and Roberts, J.M. (1996a). *Am. J. Trop. Med. Hyg.* 54, 372–379.

Collins, W.E., Kaslow, D.C., Sullivan, J.S., Morris, C.L., Galland, G.G., Yang, C., Saekhou, A.M., Xiao, L. and Lal, A.A. (1999). *Am. J. Trop. Med. Hyg.* 60, 350–356.

Collins, W.E., Lobel, H.O., McClure, H., Strobert, E., Galland, G.G., Taylor, F., Barreto, A.L., Roberto, R.R., Skinner, J.C., Adams, S. and *et al.* (1994). *Am. J. Trop. Med. Hyg.* 50, 28–32.

Collins, W.E., Nussenzweig, R.S., Ballou, W.R., Ruebush, T.K., 2nd, Nardin, E.H., Chulay, J.D., Majarian, W.R., Young, J.F., Wasserman, G.F., Bathurst, I. and *et al.* (1989). *Am. J. Trop. Med. Hyg.* 40, 455–464.

Collins, W.E., Nussenzweig, R.S., Ruebush, T.K., 2nd, Bathurst, I.C., Nardin, E.H., Gibson, H.L., Campbell, G.H., Barr, P.J., Broderson, J.R., Skinner, J.C. and *et al.* (1990). *Am. J. Trop. Med. Hyg.* 43, 576–583.

Collins, W.E., Sullivan, J.S., Fryauff, D.J., Kendall, J., Jennings, V., Galland, G.G. and Morris, C.L. (2000a). *Am. J. Trop. Med. Hyg.* 62, 491–495.

Collins, W.E., Walduck, A., Sullivan, J.S., Andrews, K., Stowers, A., Morris, C.L., Jennings, V., Yang, C., Kendall, J., Lin, Q., Martin, L.B., Diggs, C. and Saul, A. (2000b). *Am. J. Trop. Med. Hyg.* 62, 466–479.

Collins, W.E., Warren, M., Skinner, J.C., Richardson, B.B. and Kearse, T.S. (1977). *J. Parasitol.* 63, 57–61.

Cubillos, M., Salazar, L.M., Torres, L. and Patarroyo, M.E. (2002). *Biochimie* 84, 1181–1188.

Daubersies, P., Thomas, A.W., Millet, P., Brahimi, K., Langermans, J.A., Ollomo, B., BenMohamed, L., Slierendregt, B., Eling, W., Van Belkum, A., Dubreuil, G., Meis, J.F., Guerin-Marchand, C., Cayphas, S., Cohen, J., Gras-Masse, H., Druilhe, P. and Mohamed, L.B. (2000). *Nat. Med.* 6, 1258–1263.

Davison, B.B., Cogswell, F.B., Baskin, G.B., Falkenstein, K.P., Henson, E.W., Tarantal, A.F. and Krogstad, D.J. (1998). *Am. J.Trop. Med. Hyg.* 59, 189–201.

Davison, B.B., Cogswell, F.B., Baskin, G.B., Falkenstein, K.P., Henson, E.W. and Krogstad, D.J. (2000). *Am. J. Trop. Med. Hyg.* 63, 158–173.

de Souza, J.B. and Riley, E.M. (2002). *Microbes Infect.* 4, 291–300.

Dennis, V.A., Lasater, B.L., Blanchard, J.L., Lowrie, R.C., Jr. and Campeau, R.J. (1998). *Exp. Parasitol.* 89, 143–152.

Dias, J.C., Silveira, A.C. and Schofield, C.J. (2002). *Mem. Inst. Oswaldo Cruz.* 97, 603–612.

Dissanaike, A.S. and Mak, J.W. (1978). *Southeast Asian J. Trop. Med. Public Health* 9, 451–452.

Dissanaike, A.S. and Mak, J.W. (1980). *J. Helminthol.* 54, 117–122.

Dreyer, G., Noroes, J., Figueredo-Silva, J. and Piessens, W.F. (2000). *Parasitol. Today* 16, 544–548.

Edstein, M.D., Corcoran, K.D., Shanks, G.D., Ngampochjana, M., Hansukjariya, P., Sattabongkot, J.,

Webster, H.K. and Rieckmann, K.H. (1994). *Am. J. Trop. Med. Hyg.* 50, 181–186.
Egan, A.F., Blackman, M.J. and Kaslow, D.C. (2000). *Infect. Immun.* 68, 1418–1427.
el-Bihari, S. and Ewert, A. (1971). *J. Parasitol.* 57, 1170–1174.
Enders, B., Hundt, E. and Knapp, B. (1992). *Mem. Inst. Oswaldo Cruz.* 87(3), 413–422.
Falasca, C.A., Gili, M., Grana, D., Gomez, E., Zoppi, J. and Mareso, E. (1990). *Rev. Inst. Med. Trop. Sao Paulo* 32, 151–161.
Falasca, C.A., Merlo, A.B., Gomez, E., Grana, D., Malateste, C. and Mareso, E. (1992). *Rev. Inst. Med. Trop. Sao Paulo* 34, 489–498.
Favre, N., Daubenberger, C., Marfurt, J., Moreno, A., Patarroyo, M. and Pluschke, G. (1998). *Immunogenetics* 48, 253–259.
Fujioka, H., Millet, P., Maeno, Y., Nakazawa, S., Ito, Y., Howard, R.J., Collins, W.E. and Aikawa, M. (1994). *Exp. Parasitol.* 78, 371–376.
Garg, N. and Tarleton, R.L. (2002). *Infect. Immun.* 70, 5547–5555.
Giambartolomei, G.H., Lasater, B.L., Villinger, F. and Dennis, V.A. (1998). *Exp. Parasitol.* 90, 77–85.
Giambartolomei, G.H., Lasater, B.L., Villinger, F. and Dennis, V.A. (2001). *Acta Trop.* 78, 67–71.
Gramzinski, R.A., Obaldia, N., 3rd, Jones, T.R., Rossan, R.N., Collins, W.E., Garrett, D.O., Lal, A.A. and Hoffman, S.L. (1999). *Am. J. Trop. Med. Hyg.* 61, 19–25.
Gysin, J., Moisson, P., Pereira da Silva, L. and Druilhe, P. (1996). *Res. Immunol.* 147, 397–401.
Hagar, J.M. and Rahimtoola, S.H. (1991). *N. Engl. J. Med.* 325, 763–768.
Heppner, D.G., Cummings, J.F., Ockenhouse, C., Kester, K.E., Lyon, J.A. and Gordon, D.M. (2001). *Trends Parasitol.* 17, 419–425.
Herrera, S., Herrera, M.A., Certa, U., Corredor, A. and Guerrero, R. (1992). *Mem. Inst. Oswaldo Cruz.* 87(3), 423–428.
Herrera, S., Perlaza, B.L., Bonelo, A. and Arevalo-Herrera, M. (2002). *Int. J. Parasitol.* 32, 1625–1635.
Hisaeda, H., Saul, A., Reece, J.J., Kennedy, M.C., Long, C.A., Miller, L.H. and Stowers, A.W. (2002). *J. Infect. Dis.* 185, 657–664.
Ibimoye, M.O., Howard, C.V., Sibbons, P., Hasan, M. and van Velzen, D. (1993). *J. Comp. Pathol.* 108, 303–310.
Jones, T.R., Narum, D.L., Gozalo, A.S., Aguiar, J., Fuhrmann, S.R., Liang, H., Haynes, J.D., Moch, J.K., Lucas, C., Luu, T., Magill, A.J., Hoffman, S.L. and Sim, B.K. (2001). *J. Infect Dis.* 183, 303–312.
Karam, M. and Ottesen, E. (2000). *Med. Trop. (Mars).* 60, 291–296.
Kocken, C.H., Dubbeld, M.A., Van Der Wel, A., Pronk, J.T., Waters, A.P., Langermans, J.A. and Thomas, A.W. (1999). *Infect. Immun.* 67, 43–49.
Kumar, A., Weiss, W., Tine, J.A., Hoffman, S.L. and Rogers, W.O. (2001). *J. Immunol. Methods* 247, 49–60.
Kumar, A., Katiyar, S.B., Agarwal, A. and Chauhan, P.M. (2003). *Curr. Med. Chem.* 10, 1137–1150.
Kumar, S., Yadava, A., Keister, D.B., Tian, J.H., Ohl, M., Perdue-Greenfield, K.A., Miller, L.H. and Kaslow, D.C. (1995). *Mol. Med.* 1, 325–332.
Kumar, S., Villinger, F., Oakley, M., Aguiar, J.C., Jones, T.R., Hedstrom, R.C., Gowda, K., Chute, J., Stowers, A., Kaslow, D.C., Thomas, E.K., Tine, J., Klinman, D., Hoffman, S.L. and Weiss, W.W. (2002). *Immunol. Lett.* 81, 13–24.
Kumaraswami, V. (2000). *Lymphatic Filariasis.* Imperial College Press, London.
Kurniawan, A., Sartono, E., Setiowati, W.E., Supali, T., Wibowo, H., Purnomo, Rukmono, B. and Partono, F. (1992). *Southeast Asian J. Trop. Med. Public Health* 23, 288–292.
Latendresse, J.R., Campbell, J.R., Tirtokusumo, S. and Marwoto, H.A. (1987). *J. Comp. Pathol.* 97, 653–665.
Leiby, D.A., Read, E.J., Lenes, B.A., Yund, A.J., Stumpf, R.J., Kirchhoff, L.V. and Dodd, R.Y. (1997). *J. Infect. Dis.* 176, 1047–1052.
Maizels, R.M., Sartono, E., Kurniawan, A., Partono, F., Selkirk, M. and Yazdanbakhsh, M. (1995). *Parasitol. Today* 11, 50–56.
Mak, J.W., Lam, P.L., Rain, A.N. and Suresh, K. (1987). *J. Helminthol.* 61, 311–314.
Mak, J.W., Lam, P.L., Rain, A.N. and Suresh, K. (1988). *Parasitol. Res.* 74, 383–385.
Malchiodi, E.L., Carbonetto, C.H., Grana, D., Eiguchi de Palmero, K., Chiaramonte, M.G., Falasca, C.A. and Margni, R.A. (1993). *Trop. Med. Parasitol.* 44, 86–90.
Marsden, P.D., Seah, S.K., Draper, C.C., Pettitt, L.E., Miles, M.A. and Voller, A. (1976). *Trans. R. Soc. Trop. Med. Hyg.* 70, 247–251.
McGregor, I.A. (1984). *Am. J. Trop. Med. Hyg.* 33, 517–525.
Menendez, C. (1995). *Parasitol. Today* 11, 178–183.
Michael, E. (2001) *Lymphatic Filariasis.* Imperial College Press, London.
Millet, P., Grady, K.K., Olsen, M., Galland, G.G., Sullivan, J.S., Morris, C.L., Richardson, B.B., Collins, W.E. and Hunter, R.L. (1995). *Am. J. Trop. Med. Hyg.* 52, 328–335.
Misra, S., Tyagi, K. and Chatterjee, R.K. (1997). *Exp. Parasitol.* 86, 155–157.
Miyamoto, M.M., Koop, B.F., Slightom, J.L., Goodman, M. and Tennant, M.R. (1988). *Proc. Natl. Acad. Sci. USA.* 85, 7627–7631.
Molyneux, D.H. and Taylor, M.J. (2001). *Curr. Opin. Infect. Dis.* 14, 155–159.
Moreno, C.A., Rodriguez, R., Oliveira, G.A., Ferreira, V., Nussenzweig, R.S., Moya Castro, Z.R., Calvo-Calle, J.M. and Nardin, E. (1999). *Vaccine* 18, 89–99.
Nutman, T.B. (2001). *Curr. Opin. Infect. Dis.* 14, 539–546.

Obaldia, N., 3rd, Rossan, R.N., Cooper, R.D., Kyle, D.E., Nuzum, E.O., Rieckmann, K. H. and Shanks, G.D. (1997). *Am. J. Trop. Med. Hyg.* 56, 508–510.

Oduola, A.M., Sowunmi, A., Milhous, W.K., Brewer, T.G., Kyle, D. E., Gerena, L., Rossan, R. N., Salako, L. A. and Schuster, B. G. (1998). *Am. J. Trop. Med. Hyg.* 58, 625–629.

Palmieri, J.R., Connor, D.H., Purnomo, Dennis, D.T. and Marwoto, H. (1982). *J. Helminthol.* 56, 243–245.

Palmieri, J.R., Connor, D.H., Purnomo and Marwoto, H. A. (1983). *Am. J. Pathol.* 112, 383–386.

Perlaza, B.L., Zapata, C., Valencia, A.Z., Hurtado, S., Quintero, G., Sauzet, J.P., Brahimi, K., Blanc, C., Arevalo-Herrera, M., Druilhe, P. and Herrera, S. (2003). *Eur. J. Immunol.* 33, 1321–1327.

Petras, J.M., Kyle, D.E., Gettayacamin, M., Young, G.D., Bauman, R.A., Webster, H.K., Corcoran, K.D., Peggins, J.O., Vane, M.A. and Brewer, T.G. (1997). *Am. J. Trop. Med. Hyg.* 56, 390–396.

Philipp, M., Worms, M.J., Maizels, R.M. and Ogilvie, B.M. (1984). *Contemp. Top Immunobiol.* 12, 275–321.

Punkosdy, G.A., Dennis, V.A., Lasater, B.L., Tzertzinis, G., Foster, J.M. and Lammie, P.J. (2001). *J. Infect. Dis.* 184, 385–389.

Puri, S.K. and Dutta, G.P. (2003). *Acta Trop.* 86, 35–40.

Purmova, J., Salazar, L.M., Espejo, F., Torres, M.H., Cubillos, M., Torres, E., Lopez, Y., Rodriguez, R. and Patarroyo, M.E. (2002). *Biochim. Biophys. Acta.* 1571, 27–33.

Riarte, A., Sinagra, A., Lauricella, M., Bolomo, N., Moreno, M., Cossio, P., Arana, R. and Segura, E.L. (1995). *Mem. Inst. Oswaldo Cruz.* 90, 733–740.

Rivera, Z., Granados, G., Pinto, M., Varon, D., Carvajal, C., Chaves, F., Calvo, J., Rodriguez, R., Guzman, F. and Patarroyo, M.E. (2002). *J. Pept. Res.* 59, 62–70.

Rogers, W.O., Baird, J.K., Kumar, A., Tine, J.A., Weiss, W., Aguiar, J.C., Gowda, K., Gwadz, R., Kumar, S., Gold, M. and Hoffman, S.L. (2001). *Infect. Immun.* 69, 5565–5572.

Rogers, W.O., Weiss, W.R., Kumar, A., Aguiar, J.C., Tine, J.A., Gwadz, R., Harre, J.G., Gowda, K., Rathore, D., Kumar, S. and Hoffman, S.L. (2002). *Infect. Immun.* 70, 4329–4335.

Rosas, J.E., Pedraz, J.L., Hernandez, R.M., Gascon, A.R., Igartua, M., Guzman, F., Rodriguez, R., Cortes, J. and Patarroyo, M.E. (2002). *Vaccine* 20, 1707–1710.

Rosner, J.M., Schinini, A., Rovira, T., de Arias, A., Velasquez, G., Idalia Monzon, M., Maldonado, M., Ferro, E.A. and Galeano, R. (1988). *Trop. Med. Parasitol.* 39, 51–55.

Schmidt, L.H. (1979). *Am. J. Trop. Med. Hyg.* 28, 793–807.

Schnapp, A.R., Eickhoff, C.S., Sizemore, D., Curtiss, R., 3rd and Hoft, D. F. (2002). *Infect. Immun.* 70, 5065–5074.

Seah, S.K., Marsden, P.D., Voller, A. and Pettitt, L.E. (1974). *Trans. R. Soc. Trop. Med. Hyg.* 68, 63–69.

Sein, K.K., Maeno, Y., Thuc, H.V., Anh, T.K. and Aikawa, M. (1993). *Am. J. Trop. Med. Hyg.* 48, 504–511.

Shankar, P., Schlom, J. and Hodge, J.W. (2001). *Vaccine* 20, 744–755.

Stauffer, W. and Fischer, P.R. (2003). *Clin. Infect. Dis.* 37, 1340–1348.

Steketee, R.W., Nahlen, B.L., Parise, M.E. and Menendez, C. (2001). *Am. J. Trop. Med. Hyg.* 64, 28–35.

Stoppani, A.O. (1999). *Medicina (B Aires)* 59(2), 147–165.

Stowers, A.W., Cioce, V., Shimp, R.L., Lawson, M., Hui, G., Muratova, O., Kaslow, D.C., Robinson, R., Long, C.A. and Miller, L.H. (2001). *Infect. Immun.* 69, 1536–1546.

Stowers, A.W., Chen Lh, L.H., Zhang, Y., Kennedy, M.C., Zou, L., Lambert, L., Rice, T.J., Kaslow, D.C., Saul, A., Long, C.A., Meade, H. and Miller, L.H. (2002a). *Proc. Natl. Acad. Sci. USA* 99, 339–344.

Stowers, A.W., Kennedy, M.C., Keegan, B.P., Saul, A., Long, C.A. and Miller, L.H. (2002b). *Infect. Immun.* 70, 6961–6967.

Sucharit, S., Harinasuta, C. and Choochote, W. (1982). *Am. J. Trop. Med. Hyg.* 31, 599–601.

Sullivan, J.S., Morris, C.L., Richardson, B.B., Galland, G.G., Jennings, V.M., Kendall, J. and Collins, W.E. (1999). *J. Parasitol.* 85, 672–677.

Szuba, A., Shin, W.S., Strauss, H.W. and Rockson, S. (2003). *J. Nucl. Med.* 44, 43–57.

Taylor, D.W. and Siddiqui, W.A. (1979). *J. Parasitol.* 65, 267–271.

Taylor, M.J. (2003). *Ann. N.Y. Acad. Sci.* 990, 444–449.

Tine, J.A., Lanar, D.E., Smith, D.M., Wellde, B.T., Schultheiss, P., Ware, L.A., Kauffman, E.B., Wirtz, R.A., De Taisne, C., Hui, G.S., Chang, S.P., Church, P., Hollingdale, M.R., Kaslow, D.C., Hoffman, S., Guito, K.P., Ballou, W.R., Sadoff, J.C. and Paoletti, E. (1996). *Infect. Immun.* 64, 3833–3844.

Tongren, J.E., Yang, C., Collins, W.E., Sullivan, J.S., Lal, A.A. and Xiao, L. (2000). *Am. J. Trop. Med. Hyg.* 62, 530–534.

Trager, W. and Jensen, J. (1979). *Science.* 193, 673–675.

Trampuz, A., Jereb, M., Muzlovic, I. and Prabhu, R.M. (2003). *Crit. Care* 7, 315–323.

Tripathi, R., Vishwakarma, R.A. and Dutta, G.P. (1997). *Exp. Parasitol.* 87, 290–292.

Tutin, C.E. (2001). *Reprod. Fertil. Dev.* 13, 469–476.

Umezawa, E.S., Stolf, A.M., Corbett, C.E. and Shikanai-Yasuda, M.A. (2001). *Lancet* 357, 797–799.

Walsh, P.D., Abernethy, K.A., Bermejo, M., Beyers, R., De Wachter, P., Akou, M.E., Huijbregts, B., Mambounga, D.I., Toham, A.K., Kilbourn, A.M., Lahm, S.A., Latour, S., Maisels, F., Mbina, C., Mihindou, Y., Obiang, S.N., Effa, E.N., Starkey, M.P., Telfer, P., Thibault, M., Tutin, C.E., White, L.J. and Wilkie, D.S. (2003). *Nature* 422, 611–614.

Wang, R., Doolan, D.L., Charoenvit, Y., Hedstrom, R.C., Gardner, M.J., Hobart, P., Tine, J., Sedegah, M.,

Fallarme, V., Sacci, J. B., Jr., Kaur, M., Klinman, D.M., Hoffman, S.L. and Weiss, W.R. (1998). *Infect Immun.* 66, 4193–4202.

Warrell, D. (1997). *Ann. Trop. Med. Parasit.* 91, 875–884.

Warrell, D. (1999). *Parassitologia* 41, 287–294.

Wellems, T.E. and Plowe, C.V. (2001). *J Infect. Dis.* 184, 770–776.

Weller, R.E., Baer, J.F., Valentine, N.B., Buschbom, R.L., Ragan, H.A. and Malaga, C.A. (1992). *Am. J. Trop. Med. Hyg.* 46, 366–370.

Williams, S.A., Lizotte-Waniewski, M.R., Foster, J., Guiliano, D., Daub, J., Scott, A.L., Slatko, B. and Blaxter, M.L. (2000). *Int. J. Parasitol.* 30, 411–419.

Wong, M.M., Fredericks, H.J. and Ramachandran, C.P. (1969). *Bull. World Health Organ.* 40, 493–501.

Wong, M. M., Guest, M. F., Lim, K. C. and Sivanandam, S. (1977). *Southeast Asian J. Trop. Med. Public Health* 8, 265–273.

Yang, C., Collins, W.E., Sullivan, J.S., Kaslow, D.C., Xiao, L. and Lal, A.A. (1999). *Infect. Immun.* 67, 342–349.

Zapata, J.C., Perlaza, B.L., Hurtado, S., Quintero, G.E., Jurado, D., Gonzalez, I., Druilhe, P., Arevalo-Herrera, M. and Herrera, S. (2002). *J. Parasitol.* 88, 723–729.

Zuccotto, F., Zvelebil, M., Brun, R., Chowdhury, S.F., Di Lucrezia, R., Leal, I., Maes, L., Ruiz-Perez, L.M., Gonzalez Pacanowska, D. and Gilbert, I.H. (2001). *Eur. J. Med. Chem.* 36, 395–405.

CHAPTER 7

Reproduction: Definition of a Primate Model of Female Fertility

Almuth Einspanier[1,2] and Mauvis A. Gore[3]
[1] Institute of Physical Biochemistry, Veterinary Faculty, University of Leipzig, Leipzig, Germany
[2] Dept. of Reproductive Biology, German Primate Centre, Göttingen, Germany
[3] University Marine Biological Station, University of London and University of Glasgow, Millport, Scotland

Introduction

There is a wide variety of animal models to study human disease, many of which are rodents and knockout animals (Levallet *et al.,* 1999; Tsika, 1994). The popularity of rodent models is due to their easy breeding, their husbandry and their very short generation time. Why then do we need primate models? Non-human primates were studied to understand the basics of human physiology, due to their relative genetic and physiological similarity (Hearn *et al.,* 1978). *Macaca mulatta* in particular are in great demand and the number bred in captivity is not currently keeping up with the demand (Sunday, 2003).

A model for human female reproduction has to relate to human females, with regard to fertility and infertility, ovulation, implantation, parturition, lactation, amenorrhoea, aging, genes, and endocrine pattern, and differences should be defined so they can be considered in the interpretation of results.

In this chapter we present the characteristics that make non-human primates good models for human female reproduction, the differences between primate species and the effect these differences may have on the model. The classification of primates in the paper was according to Groves (2001).

Basics of primate reproduction

Why do we need a model for reproduction research? As reproduction involves the transfer of genes from one generation to another, it has a long-term effect on future generations. Over the reproductive life, females are subject to cyclic ovarian changes, pregnancy and lactation

The Laboratory Primate
ISBN 0-1208-0261-9

until menopause occurs at the end of the reproductive life. Reproductive outcome is influenced by the age of mating pairs, group composition, social status within the group, nutritional status, body weight, external factors (e.g. noise, disturbance) and stress (Johnson *et al.*, 1991; Poole and Evans, 1982; Tardiff and Jaquish, 1994; Jaquish *et al.*, 1996; Lunn, 1983). Each of these phases will be discussed to provide the background to female primate reproduction.

Normal ovarian cycles are approximately one month in duration for catarrhines (Old World monkeys). Cycle lengths and menstrual flow within and between species do vary, however (Harvey *et al.*, 1987; Martin, 1990). Cycle length can range from 22 days (d) in *Semnopithecus entellus* to 36 d in *Miopithecus talapoin* and *Pan troglodytes*, while sexual maturity varies from 25–46 months in *Chlorocebus aethiops* to 22–187 months in *Pan troglodytes*. By comparison, cycle length in *Homo sapiens* ranges from 18–53 d (Martin, 1990). Most of the differences in the cycle length occur in the follicular phase (Rowell, 1972). Variations in the cycle have important effects on the reproduction rate of females, including length of the postpartum amenorrhea, the number of cycles before conceiving, gestation length, and seasonality. The interbirth interval differs widely between species, from the relatively short period of 152 d in *Tarsius spectrum*, 157 d in *Callithrix jacchus* and 242 d in *Saguinus fuscicollis*, to an intermediate period of 360 d in *M. mulatta*, 365 d in *C. aethiops*, 414 d in *Saimiri sciureus*, 420 d in *Papio cynocephalus*, 511 d in *Lemur catta*, to a long interval of 1095 d in *Ateles fusciceps*, 1460 d in *Gorilla gorilla* and 1825 d in *Pan troglodytes* (Harvey *et al.*, 1987).

Seasonality, in the species where it occurs, is related to a number of endocrine and environmental factors, such as food resources. For example, seasonal reproduction in *C. aethiops* is strongly correlated with the availability of acacia flowers, which are rich in flavanoids that have estrogenic properties (Whitten, 1983,1984). In *M. mulatta*, rainfall appears to trigger reproductive synchrony (Vandenbergh, 1973; Drickamer, 1974).

Ovulation is spontaneous in primates and does not have to be induced by copulation. *Callithrix* and *Saguinus* are the only primate species (haplorrhine platyrrhines) with multiple ovulations and which regularly produce two (range of 1–4) offspring. The number of preovulatory follicles varies between 1 and 4 and relates to the number of offspring born (e.g. Hearn *et al.*, 1978). Most primates have a single follicle ovulation per cycle, followed by a singleton birth. Associated with spontaneous ovulation is a highly developed social system, where males interact with females regularly, providing the opportunity for fertilisation. This points to the importance, in primates in reproduction, of social behaviour which will be discussed later.

Menstruation takes place regularly in catarrhines, but in very few platyrrhines (New World monkeys), such as *Cebus apella* and perhaps Ateles. There are few, if any, external markers of cyclicity and there is no skin swelling (Dixson, 1998). Why menstruation occurs at all has been discussed widely and three hypotheses suggested (Strassmann, 1997). The first hypothesis argues it is a defence against sperm-borne pathogens and the second that it is a signal of fertility. The most recent hypothesis suggests that endometrial cyclicity saves energy, and that vaginal bleeding is a side effect when there is too much blood for efficient reabsorption by the body. More work is still needed in this area.

TABLE 7.1: General reproductive information

	Pavian (*Papio* spp.)	Marmoset monkey (*Callithrix jacchus*)	Cynomolgus monkey (*Macaca fascicularis*)	Women (*Homo sapiens*)
Weight of adult females	8–15 kg	0.3–0.4 kg	3–7 kg	50–70 kg
Seasonality	No	No	Yes	No
Generation Interval (age of puberty)	5–6 years	1.5–2 years	4–6 years	>14 years
Duration of ovarian cycle	35 days	28 days	28 days	28 days
Period of pregnancy	185 days	144 days	165 days	9 months
Anovulatory phase during lactation	1–2 years	No	1 year	~3 months
Offspring per year	0.8	5	1	1

Fertility

Onset of maturity

There are a number of factors that can affect fertility, but the interaction between genes and environmental factors are key in the onset of fertility (Bronson, 1989). The onset of female sexual maturity varies between families and can differ considerably between species (Rowe, 1996), as, for example, 10 months in *Eulemur fulvus* but 24 months in *E. macaco;* 6.7 months in *Galago senegalensis* and 18–24 months for *Otolemur crassicaudatus*; 8.5 months for *Callimico goeldii*, 12 months for *Callithrix jacchus*, 18 months for *Saguinus oedipus* but 24 months for *C. pygmaea*. For *Callicebus moloch*, onset is at 30 months and 84 months for *Cebus apella*. There is a range of times of onset of sexual maturity in macaques, from 35 months in *Macaca nemestrina,* 42 months in *M. fuscata,* 42–48 months in *M. mulatta, M. sylvanus* and *M. radiata*, to 51.6 months in *M. fascicularis*. In the species *Papio*, onset is at only 38 months in *P. ursinus*, but 51 months in *P. cynocephalus* and *P. hamadryas* and 54 months in *P. anubis*. In other cercopithecoids, onset is at 36 months in *Cercocebus atys* and *Erythrocebus patas* and 46.5 months in *Semnopithecus entellus*. Of the hominoids, sexual maturity is at 108 months in *Hylobates lar* and 135 months in *Pan troglodytes*. In the hunter-gatherer *Homo sapiens*, menarche is at 16.9 years, while it is 10.5–15.5 years in Western Europeans (Dixson, 1998).

In terms of non-human primate models for the human ovarian cycle, a review by diZerega and Hodgen (1981) noted that folliculogenesis in captive *M. mulatta* and *M. fascicularis* was uniform in response to, and virtually indistinguishable from, the menstrual/ovarian cycle in women. Similar results were noted by Zeleznik (1993) for humans and primates in follicular growth.

Gestation

Gestation represents a great cost to females in time and energy for fetal development (Robbins, 1983). Pregnant mammals have 17–32% higher energy requirements and food intake than non-reproducing females. Gestation length varies from less than 90 d in cheirogaleids, to between 100 and 200 d in lemurids, lorisids, galagonids, cebidae, cercopithecins and colobins, and 210–260 d in the hominids (Harvey *et al.*, 1987). Slowing fetal growth rate can be important to females where protein or minerals are limited in their diet, as might be the case for frugivorous or folivorous primates. Most fruit and leaves are low in available protein. Mammalian gestation periods and birth weights increase as power functions of maternal weight. When species with delayed implantation or fertilisation are excluded from the analysis, primate fetuses grow 76% more slowly than other mammals. Most primates produce singletons, with only the *Callithrix* and *Saguinus* species regularly producing twins, and neonatal weight is highly correlated with maternal body weight (see Harvey *et al.*, 1987, Fig. 16.3). Jaquish *et al.* (1996) postulated that fetal loss in the *Callithrix* may represent a mechanism by which females can adjust for intrauterine crowding or excessive energy demands by selectively expelling or reabsorbing an individual conceptus. This has been supported by Nubbemeyer *et al.* (1997).

Estrus versus menstruation

Much discussion of the terms estrus and menstruation has taken place in relation to non-human primates but a key point is that, where sexual receptivity is relatively independent of hormonal control in primates, estrus does not occur (Dixson, 1998). Martin (1990) suggested referring rather to the pregnancy cycle under natural conditions, because long periods of estrous cycling represent a failure of male-female reproductive interaction. Martin also argues that menstruation refers to the discharge of uterine blood at regular intervals, whereas estrus is a regular but very limited period of intense sexual receptivity. For example, lemuroids, lorids and *Tarsius* exhibit estrus, but not obvious menstruation, through limited periods of sexual receptivity during the peri-ovulatory stage of the ovarian cycle (Martin, 1990). Matings, in a wild troop of *Lemur catta,* have been observed to take place within a 12 day period of females exhibiting estrus, leading to a highly synchronised birthing period (Jolly, 1967). Catarrhine primates do not exhibit estrus, but they do show external signs of bleeding.

Behavioural signs of reproductive activity

Beach (1976) categorised female mammalian sexuality into three phases: attractiveness, proceptivity and receptivity. Sexual attractiveness refers to the non-behavioural effects of ovarian hormones on genital appearance and odours that cause a sexual stimulus in males. Proceptivity is sexual invitational and affiliative behaviour displayed by females to attract and support sexual interactions with males. Some proceptive behaviours are not necessarily related only to sexual invitation. For example, lip-smacking in female *M. mulatta* is both a proceptive behaviour and one used in attracting the attention of infants. Similarly, quadrupedal stances and mounting in many macaques can also be used to show hierarchical rank. Receptivity is the acceptance by females of males and copulation with intravaginal ejaculation. A detailed review of interspecific variation in sexual signalling and behaviour among primates has been compiled by Hrdy and Whitten (1987; see Table 30.1).

Physical signs

Many catarrhine species display a coloured swelling, called sexual skin, that changes in character over the cycle. In most species this is largely in the anogenital area but, in *Theropithecus gelada*, it is on the chest. *M. mulatta* females can show wrinkled swellings on their rumps, shoulders and brow, with older females exhibiting a much deeper red than younger females (Gore, pers. obs.). The swellings tend to change from pink to red, relating to estrogen levels. Swelling and deepening of the colour occurs in the follicular phase, with diminishing edema and colour fading taking place during the luteal phase (Dixson, 1998). These effects are mediated by estrogens that also have behavioural effects like increased motoric activity, less food intake, irritability and proceptivity. Changes in vaginal cytology and skin swelling are also induced by estrogens. Sexual skin is not found in all catarrhine species, as noted by Dixson (1998; see Table 7.8). For example, *Pan troglodytes* do have very pronounced, bright swellings, but *Gorilla gorilla* and *Pongo* species do not. Gorilla females show a very slight tumescence of the labia at mid-cycle when they are highly proceptive. Hapalines, however, do not menstruate and show no obvious external changes.

As Dixson (1990) noted, visual communication in monkeys and apes plays an important role in reproductive behaviour and it is often used in proceptive gestures in female primates, combined with specific facial expressions. Attractivity and proceptivity fluctuate during the ovarian cycle and are extremely variable among primate species (Dixson, 1983). In many primates, female sexual attractivity is demonstrated by both behavioural and non-behavioural cues, such as visual or chemical cues.

During both precopulation and copulation, behavioural and physiological responses can be observed in a number of female species (see Table 5.9 in Dixson, 1998). Behaviours range from head-shake, looking back at, and clutching the leg of, the mounted male in females of several catarrhine species, to body tension and spasm with vocalisation in female *Pan troglodytes*. Physiological responses include contraction of the uterine, vaginal, anal and pelvic muscles, increased heart and respiratory rate and vaginal secretions. Reproductive behaviour occurs mainly at the time of ovulation and as a consequence of neuroendocrine and endocrine changes. For example, female *Callithrix* present their genital area and may arch their backs for males to signal sexual receptivity and to stimulate the males to copulate (Hershkovitz, 1977, Wissman,1999).

Social contraception

Behaviour can also have a negative effect on reproduction. Social suppression occurs in platyrrhines through socio-endocrinology (Abbott and Hearn, 1978; Abbott *et al.,* 1993). There is usually one dominant female as breeder in a social group, with the non-breeders as social subordinates and helpers. Social suppression is behaviour-dependent and results mainly in a contraceptive effect, although endocrine parameters are not necessarily suppressed (Ziegler and Sousa, 2002). There is a relationship between the social status in the group and ovarian competence (Smith *et al.,* 1997), but this reproductive suppression does not affect the process of sexual maturation in platyrrhines (Ziegler *et al.,* 1987).

Orgasm

A primate response that has been neglected until recently is the orgasm. Non-human primate females may experience orgasms under specific circumstances. Female copulatory orgasms have a variable occurrence and the proximate causation is poorly understood (see review in Dixson, 1998). A study of *M. fuscata* reported that, of 240 copulations of 68 different heterosexual pairings of 16 males and 26 females, female orgasms were recorded 80 times (Troisi and Carosi, 1998). Orgasms were not related to female age or dominance, but with longer lasting copulations with a higher number of mounts and pelvic thrusts. After statistically controlling for the level of physical stimulation experienced by females, during copulation, the highest frequency of female orgasms was noted in pairs of high-ranking males and low-ranking females, and least among pairs of low-ranking males and high-ranking females. The conclusion was that proximate mechanisms controlling orgasmic threshold were more responsive to social stimuli and less constrained by physiological limitations.

Endocrinology and reproduction

As mentioned above, the sociosexual behaviour is related to the female ovarian cycle. Hormones have an effect upon proceptivity and the male sexual solicitations of the female are related to ovulation (Dixson, 1987). Dixson postulated that proceptivity functions as a proximate stimulus for enhancing female sexual attractivity. This attractivity is based on hormonal changes, an example being that the sexual skin reflects a change in estrogen and progestin levels. Male primates investigate the female's body and test her urine for information on her ovarian/hormonal status and therefore her likely reproductive success.

The term hypothalamic-pituitary-ovary-uterus axis well describes the hierarchy in female reproduction. External factors, such as mating partner, stress or light can influence the secretion pattern of the gonadotropin releasing factors, and these induce the secretion of gonadotropins from the pituitary. The gonadotropins, FSH and LH, are controlled by the pulsatile secretion of the gonadotropin releasing hormone that stimulates steroids and other hormones within the ovary and induce follicle growth, ovulation and luteinisation. The ovarian cycle is characterised by hormones typical for the follicular and luteal phases, and that can be identified in body fluids and secretions.

The main hormones involved

The major hormones of the ovary are estrogens, progesterone, relaxin, growth hormones, prostaglandins and inhibin. Estrogens are secreted mainly during the follicular phase by the preovulatory follicle, and reach maximal plasma levels around ovulation. They induce changes in behaviour, vaginal epithelium, peri-anal skin colour and skin swelling. In general, all the estrogen induced changes make the female more attractive to the male. Female proceptivity, however, decreases when progesterone levels increase in the luteal phase, suggesting a negative effect of progesterone on the occurrence of proceptive behaviour (Carosi *et al.*, 1999; Heistermann *et al.*, 1996; Graham *et al.*, 1973; Wildt *et al.*, 1977). This phase is also associated with a significant decline in the frequency of male ejaculations. In platyrrhines, proceptivity can be directly inhibited by administering progesterone (Kendrick and Dixson, 1985).

Estrogens positively increase LH-secretion, and both peak around the time of ovulation and decline thereafter (*Callithrix*: Hearn *et al.*, 1975; Harlow *et al.*, 1984; *Aotus*: Bonney and Setchell, 1980; *Cebus*: Nagle and Denari, 1983). Ovulation can be detected by determining the [E^2 + LH] preovulatory rise of estradiol and LH in blood or urine. This is important in general primate management, assisted reproduction and pregnancy duration.

Hormone levels are much higher in the platyrrhine than in catarrhine species. An example of this is the circulating steroids during the cycle and pregnancy in *Callithrix* (Chambers and Hearn, 1979; Hodges *et al.*, 1983). In humans and most catarrhines, ovulation is preceded by a clear increase in urinary or serum estrogen levels (Mitchell *et al.*, 1982; Setchell *et al.*, 1980), although accurate timing of ovulation is difficult to achieve through urine samples. This is due to a time lag in the metabolic pathway. However, *Callithrix* do not show a clear estrogen pattern in the blood or skin swelling, or a clear change in vaginal cytology. A good predictor of ovulation in *Callithrix* is the detection of urinary estrogen metabolites, such as estrone-3-sulphate that can be used for establishing cycle status (e.g. Hodges and Eastman, 1984; Eastman *et al.*, 1984) as well as being a direct assay for conjugated estrogens (Hodges and Eastman, 1984). However, co-analysis of creatinine is required to compensate for variations in urine concentrations and volume.

Hormone detection

Hormone analysis of urine or feces allows conceptive and non-conceptive ovarian cycles in captive and free-living primates to be monitored non-invasively. This does, however, require information on urinary metabolites, metabolic pathways and temporal relationships between changes, in order to interpret the results obtained. A diurnal variation of the excretion patterns of fecal steroids in the marmoset monkeys has been obtained in this way (Sousa and Ziegler, 1998) and needs to be considered in the interpretation of data.

Non-invasive monitoring of conceptive and non-conceptive ovarian cycles can be achieved by measuring immunoreactive E1C, estradiol-17β, pregnandiol (Pd), LH and progesterone in urine and feces. Alternatively, hormone detection in the blood is an invasive procedure and it can be obtained by direct hormone analysis without information on metabolites. This involves blood sampling. In general, hormone analysis is mainly used for monitoring cycles, and a single sample does not provide the information necessary, except in the case of pregnancy markers such as chorion gonadotropin (CG) or relaxin (RLX).

After ovulation, the new corpus luteum or corpora lutea have formed. These produce progesterone which is the dominant hormone of the luteal phase and, therefore, a good indicator that ovulation and luteinisation have occurred. Luteal phase and early pregnancy can also be confirmed by the presence of RLX (O'Byrne and Steinetz, 1976; Steinetz *et al.*, 1995; Einspanier *et al.*, 1999) or CG (Ottobre *et al.*, 1984; Norman *et al.*, 1984; Diamond *et al.*, 1987; Steinetz *et al.*, 1992). RLX is produced by the uterus and placenta and the levels of RLX and CG increase significantly during implantation and early pregnancy, making RLX and CG early pregnancy markers. Both hormones can be detected in serum samples, from both catarrhine and platyrrhine monkeys, and a single sample can provide information about the pregnancy status of the individual primate. There is a difference between RLX content in a normal pregnancy and in pregnancy failure, where a progesterone decline occurs much later (Einspanier *et al.*, 1999). RLX is an indicator, therefore, of the placental situation and fetal health. In general, the RLX serum concentration provides information on the pathology and physiology of pregnancy.

Hormones and social status

Hormonal changes can also offer information on social status. In acyclic, subordinate *Callithrix*, LH secretion is reduced with an apparent lack of pulsatile LH-secretion (Abbott, 1986, 1987). This seems to be suppressed by the endogenous secretion of GnRH from the hypothalamus. Subordinate females have suppressed pituitary LH secretion and a reduced pituitary response to GnRH. This situation can be reversed very rapidly and is entirely dependent upon social status (Abbott *et al.*, 1986, 1987). For example, the removal of subordinates from their social group results in a rapid elevation of their plasma LH, leading to ovulation. In general, platyrrhines can experience both a physiological and a behavioural block, leading to an extreme social control over female reproduction (Abbott *et al.*, 1993). In a study on *Callithrix*, subordinate daughters did ovulate, but none became pregnant as a result of a behavioural block by the dominant female (French and Stribly, 1987). This social suppression is also common in *Theropithecus gelada* (Dunbar, 1989) and *Miopithecus* (Keverne, 1976). Elevated plasma levels of prolactin or cortisol are also associated with suppressed reproductive function in subordinate female primates (e.g. Kaplan *et al.*, 1986). However, cortisol levels are known to be responsive to certain female reproductive conditions. For example, an exogenous estrogen treatment causes cortisol elevation in humans (Doe *et al.*, 1960; Altemus *et al.*, 2001), *Papio* (Pepe *et al.*, 1982) and *Saimiri* (Coe *et al.*, 1986). Periovulatory elevated cortisol levels can also be used as a predictor of ovulation in humans (Genazzani *et al.*, 1975), *Saguinus* (Ziegler *et al.*, 1995) and *Callithrix* (Saltzman *et al.*, 1997).

External factors influencing reproduction

Social systems

Primates have highly differentiated social systems and behaviour. Differences in social systems of small-bodied platyrrhines, such as *Aotus*, *Callicebus*, *Saimiri*, *Callimico*, *Saguinus*, *Leontopithecus* and *Callithrix*, have been found to be closely associated with the ontogenetic requirements of their young (Garber and Leigh, 1997). Factors such as competition in predation and foraging cannot explain the variation in infant development and subadult growth rate. In hapalines, such as *Aotus* and *Callicebus*, mothers have a high reproductive output but limit their investment per offspring and depend on extra-maternal helpers for infant survival. *Saimiri* mothers, however, have long inter-birth intervals with resulting low reproductive output. Interestingly, some hapalines are able to have multiple pregnancies and high reproductive efficiency.

Nutrition

Periods of chronic or severe under-nutrition can result in the suppression of reproductive hormone secretion. Studies have now shown that the period of under-nutrition can be very brief and that the metabolic cue is related to caloric intake rather than time of under-nutrition (Cameron, 1996). It is also independent of changes in body mass or composition, intake of a specific nutrient, plasma glucose, insulin levels, the taste or smell of food, or the physical process of ingestion. In *Theropithecus gelada* and *Papio cynocephalus*, food availability relates to plant nutritional quality (Dunbar *et al.,* 2002). It is clear that the ability of food intake to stimulate activity of the reproductive axis would be an evolutionary advantage. Fertility would be quickly maximised when adequate food resources become available. In the human, there is an important connection between ease of labour and a hunter-gatherer diet as well as a tendency towards difficult labour for people with an agriculture-based diet (Gebbie, 1981; Roy, 2003). Interestingly, there is no evidence that a rapid suppression of reproductive hormones, with short-term fasting, causes infertility (Cameron, 1996). Specific nutrients may be required, as in hapalines, which have a high vitamin C requirement that is suggested to have an influence on reproduction (Wissmann, 1999). Under-nutrition blocks the hypothalamic generator and suppresses LH pulsatility (Thalabard, 1992). Lactation, in terms of energy, is a costly activity for female mammals. The cost of lactation can be expected to increase as the infant grows and its gross energy requirements increase. For example, Lee (1987) pointed out that it is important for a mother to assess the demand versus the needs of her infant to allow her to alter suckling frequency and thus her fertility.

Just as female primates, of different species, vary in their strategy for gestation, the milk produced for lactation varies in nutritional quality and therefore cost to the mother. Data on the composition of primate milk is limited, but *Miopithecus* milk has a relatively high percentage of protein (2.1%) compared to lemuroids (1.9%) and baboons (1.5%). The percentage fat in the milk of lemurs (2.3%) and *Miopithecus* (3.0%) is considerably less than that of baboons (4.6%) (Robbins, 1983). Weaning can take place as early as 40 days (d) in *Microcebus murinus* and 45 d in *Galago demidovii*, to a range between 55–90 d in *Callithrix*, between 180 and 450 d in cercopithecoids, generally longer in *Colobus* (360–597 d), and longest in hominids of 410–2110 d (Harvey *et al.,* 1987).

Birth sex ratio

How do parents divide investment between sons and daughters, and is there any evidence that mothers bias investment? Would this be influenced by the physical condition of mother? Studies on birth sex ratios (BSR) have provided a range of results supporting diverse and conflicting evolutionary arguments. This variation between studies suggests that factors that include population density, adult sex ratio and competition for mating, influence BSR of high and low ranking mothers. Physiological mechanisms influencing BSR are still unknown so the null hypothesis, that BSR does not vary predictably with maternal rank, remains the most plausible explanation at present. As there is no correlation between maternal condition, birth sex ratio and maternal care, additional factors probably have been overlooked or alternative adaptive explanations are required for particular species or populations, (Brown, 2001).

Diseases

Sexually transmitted diseases (STDs) are known to exist in wild and domesticated animals. For example, Lockhart and colleagues (1966) reported over 200 sexually transmitted parasites among 27 orders of host species. Host species can counter STDs through immune defence and behaviour pattern. Reviews by Donovan (2000a, 2000b) suggest that although humans employ behavioural defence strategies to avoid STDs, efforts to control the spread of STDs in humans continue (Holmes *et al.,* 1994). Little is known, however, about behavioural defences animals use to reduce risk of STDs. Comparative tests revealed no support for genital inspection as a means to identify and avoid infected individuals. Genital inspection was more common in males than females, and was not correlated with mating promiscuity. Primate species with increased promiscuity were not more likely to display genital self-grooming following mating, nor were these males or females more likely to urinate immediately after mating (suggested to flush micro-organisms from urethra and surrounding genital areas). Pre-copulatory behavioural defences to STDs, e.g. mate choice, are unlikely to be fully effective (Nunn, 2003).

Infertility

Cycles and follicles

Some of the main problems concerning infertility, such as irregular cycles and anovulation, are of interest in studies of human infertility. Menstrual cycles can be irregular in primates and environmental factors can influence the follicular phase in particular (Dixson, 1998). For example, female *Papio anubis*, subjected to isolation from their social group, tended to have longer follicular phases (Rowell, 1970). In an experiment with *P. anubis*, Rowell observed that it was social change rather than social context that influenced cycle length. Female primates have a narrow window of time for conception. Females that fail to become pregnant remain acyclic for the season (Keverne, 1987). Anovulatory and irregular cycles can occur in *M. mulatta* outside of the breeding season, and concurrent with low gonadotrophin and estradiol secretion (Koering, 1986). In *M. mulatta*, it was found that the dominant follicle can still remain susceptible to the atretogenic effects of 17β-estradiol even by day 8 in the cycle (Dierschke *et al.*, 1994a, 1994b). Furthermore, irregular cycles and anovulation have been recorded in the *Callithrix jacchus* colony at the German Primate Centre, during the seasonal changes from winter to spring and autumn to winter (Einspanier, pers. obs.). This suggests that these pathologies are based on environmental effects (e.g. light, temperature, nutrition) that affect the hypothalamus-pituitary-axis. In the Canadian Inuit community, for example, birth seasonality, effect of photoperiod and social change have been ascribed (Condon, 1991) to modernisation and decline in traditional life style, including nutritional changes.

Luteinised Unruptured Follicles (LUFs) were induced and these appeared to be similar to spontaneous LUFs in human females, suggesting a useful model may be available to study this aspect of infertility in women. A good review on the morphological correlates of follicle maturation and atresia in primates has been provided by Koering (1987) and a model suggesting how dominant follicles come to ovulate by Gore *et al.* (1995, 1997).

Polycystic ovaries

A model for polycystic ovaries in primates has been developed. Work on *M. nemestrina* showed that FSH levels during pregnancy were similar to the concentrations in the early follicular phase and that FSH may initiate follicular growth during pregnancy. High levels of estrogen did not suppress FSH secretion and ovarian histology revealed extensive growth of follicles and the corpus luteum during gestation. The ovary had high numbers of pre-antral, small antral and medium Graafian follicles. Some follicles were cystic and atretic, and the ovary resembled human polycystic ovary syndrome (Chandrashekar *et al.*, 1987).

Implantation and pregnancy loss

Implantation failure and early pregnancy loss are important topics in human fertility. To investigate these, there is a need for an animal model with the same type of implantation and similar Corpus luteum (CL) function. Primates are the closest species for such a model, but when comparing the different conception rates and, therefore, efficiency, one has to choose a species very carefully. For example, *Papio* has a 47% conception rate with an average 60% spontaneous abortion rate (Kuehl *et al.*, 1992). Similar rates are found in *M. mulatta* (Ghosh and Sengupta, 1992) and in humans (Boklage *et al.*, 1990). *Callithrix jacchus*, however, has a 90% efficiency rate (Nubbemeyer *et al.*, 1997). The differences may be a result of evolutionary trends and these could help us to understand reproductive strategies, suggesting that one should not depend on only one model.

Social contraception

Social contraception often results in dominant females raising more offspring than their subordinates (Abbott, 1989). Social suppression of ovulation is most extreme in the hapalines. When the social suppression is removed, the individual female begins to cycle, but she ceases if she subsequently becomes subordinate again. This suppression is not related to a low body weight or high levels of cortisol or prolactin measured in the blood. It is a result of insufficient gonadotrophic stimulation from the anterior pituitary gland, which is a result of low levels of gonadotrophin releasing hormone (GnRH) from the hypothalamus.

A study of *Saguinus oedipus* showed the dramatic effect of reproductive suppression in females (Savage *et al.*, 1988). Behavioural observations and analyses of hormonal samples of young female *S. oedipus* were taken when the subjects were with their family groups, isolated from conspecifics or paired with an unrelated male conspecific. The results showed that reproductive suppression in the subjects occurred when they were in a family group. Once removed and isolated, the females displayed increased scent marking behaviour and hormonal levels. When paired with a male, hormonal levels rose sharply and the female began ovarian cycling, along with sociosexual behaviour and increased scent marking. A change in social structure of *Leontopithecus chrysomelas* has been shown to take place when the dominant female is contracepted for longer than one year (Vleeshouwer *et al.*, 2003). A previously subordinate female may then begin to ovulate and conceive.

Social and physiological constraints result in a higher number of miscarriages in *Papio cynocephalus* (Wasser, 1995). Wasser noted that dominant females conceived more easily at lower progesterone levels than low ranking females. Dominant females have higher estrogen levels during conceptive cycles. Estrogen increases the progesterone receptor density in the epithelial layer of the endometrium, resulting in a better-prepared endometrium and therefore higher implantation rates in dominant females. However, estrogen has little effect on the progesterone receptor density in the stroma and myometrium, where they are needed to sustain a pregnancy. As a result, the opposite effect is found in subordinate females, whereby they have difficulty in conceiving but are better able to maintain a pregnancy.

Aging

The reproductive value for free-ranging *M. mulatta* on Cayo Santiago illustrates a decline in reproduction with age (Dunbar, 1989). The highest number of female offspring born to an average female peaked at 4.5–6.4 years of age. Age at first breeding is closely associated with age at maturity, but it ranges from 10.8 months in *Galago senegalensis* and 11.5 months in *Microcebus murinus*, to 30 months in *Lemur catta*, 24.1 months in *Saguinus fuscicolis*, 42 months in *Cebus apella*, 43.3 months in *M. mulatta*, 47.7 months in *Chlorocebus aethiops* and 73 months in *P. cynocephalus*, to the later 118.2 months in *Gorilla gorilla* and 138 months in *Pan troglodytes* (Harvey *et al.*, 1987). Markers of aging, including ovarian cyclicity, hormonal level and sexual behaviour in *M. nemestrina* suggested that there are two dimensions of reproductive function, behavioural and physiological (Short *et al.*, 1989). In a study of *M. fuscata* females aged 3–32 years, a decline in ovarian function occurred at 21–25 years, accompanied by a slight increase in plasma LH during the breeding season. Menopause onset occurred at around 27 years (Nozaki *et al.*, 1995). Over a ten-year period, a decline in hypothalamus-pituitary-ovarian activity occurs in aging *Callithrix jacchus*, leading to declining steroid and peptide hormone concentrations, an increase in apoptotic indices in follicles and fewer regular cycles (A. Einspanier pers. obs.).

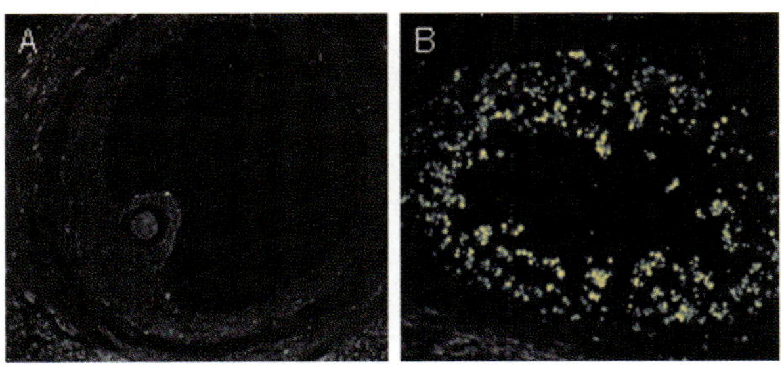

Figure 7.1 A: Preovulatory follicle from a four year old female *Callithrix jacchus*. No signs of apoptosis.
B: Preovulatory follicle from a 12 year old *Callithrix jacchus*. Granulosa cells show a high degree of apoptosis.

Gilardi and colleagues (1997) studied hormonal and menstrual data in *M. mulatta* and categorised individuals as pre-, peri- or post-menopausal. In the peri-menopausal individuals, a long follicular phase and a lack of patterned pregnanediol-3-glucuronide dynamics were noted, as well as breakthrough bleeding in 3% of the *M. mulatta*. The post-menopausal individuals were amenorrheic with low urinary estrogen conjugate levels. They also showed a decline in ovarian function in the third decade of life, paralleling the menstrual and hormonal pattern related to the menopausal transition period in women. In the review of Bellino and Wise (2003) the topic of menopause in primate models is extensively discussed.

Aging in the spiral artery of the ovary results in morphological changes in both *M. fuscata* and women (Shimada *et al.*, 1993). The resulting change in blood flow of the spiral artery may influence the ovary in relation to menopause. A loss of innervation of sympathetic and vasoactive intestinal peptide (VIP) nerves in senescent ovaries indicated that the decline in ovarian function with age, in *M. mulatto*, was related to reduced neural input to the endocrine component of the gland (Schultea *et al.*, 1992). Innervation of a subpopulation of primordial follicles by VIPergic fibres was observed in *M. mulatta* (Dissen *et al.*, 1993). The VIPergic fibres appeared to target and enclose selected follicles only, resulting in deprivation of innervation in other follicles. As the phenomenon is restricted to VIPergic fibres and is not in neuropeptide Y (NPY) or tyrosine hydroxylase (TH) fibres, it suggests that the pattern contributes to follicular selection at an early stage of development.

Pavelka and Fedigan (1991) examined menopause, as a life history trait rather than the endocrinology, by comparing lifespan instead of life expectancy between

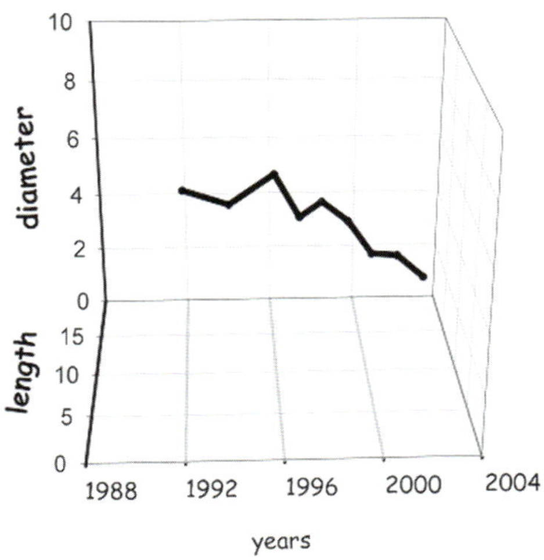

Figure 7.2 Aging in a Callithrix jaachus 12 year old female in 2004. The curve shows the decline in both length and diameter of the uterus with age.

humans and non-human primates. They suggested a non-adaptive pleiotropy perspective or the adaptive grandmother hypothesis. Prior tests of the grandmother hypothesis suggested that post-reproductive female (PR) *M. fuscata* do not significantly improve survivorship of their descendents. However, not all PR are grandmothers and not all grandmothers are PRs. Pavelka and colleagues assessed availability and the adaptive value of repro of PR mothers and grandmothers. The presence of a mother, irrespective of reproductive status, was associated with improved reproduction in her adult daughters while presence of a PR grandmother was associated with improved survival of grandchildren to age one year. While improved maternal investment did not appear to be the primary explanation for reproductive termination, the few PRs with unweaned grandchildren available seem to have a positive influence on the survival of the offspring (Pavelka *et al.*, 2002).

Summary

For human reproduction and corresponding areas, such as stem cell research and its possible use, there is an urgent need for primate models to study the physiology and to develop treatments where necessary. Each model has its benefits and limitations, however, and these need to be acknowledged when data are presented and adapted to the human situation. The use of a number of different primate models would be ideal, but the choice of primate species and number is often dictated by cost and availability.

Primates have both similarities with, and interesting differences to, human female reproduction and fertility. Nutrition is a key factor in both humans and primates in all aspects of reproduction, and particularly fertility, gestation and lactation. Environmental factors such as exogenous estrogen and its effect on fertility is a topical area in human research. The reasons behind menstruation, orgasm, menopause and birth sex ratio are also hotly debated in research on humans and might benefit from research in these areas on non-human primates. Contraception can result from primate social behaviour and social change and this could provide insight into human infertility. Similarly, aging is a key topic in human fertility research, with new assisted reproduction methods and current discussions on cloning. Sexually transmitted diseases are a major reproductive health concern for their effect on fertility, as well as general health, and the research on primate STDs may allow useful models to be developed. Finally, many primates have an earlier onset of maturity and shorter reproductive cycles, allowing more reproductive cycles and quicker results than in human subjects.

Glossary

BSR	birth sex ratio
CG	chorion gonadotropin
CL	corpus luteum
FSH	follicle stimulating hormone
GnRH	gonadotrophic releasing hormone
LH	luteinising hormone
LUF	luteinised unruptured follicle
PR	post-reproductive female
RLX	relaxin
STD	sexually transmitted disease

Primates

Strepsirrhine	lemuroid, lorid, Daubentoniid primates
Haplorrhine	tarsier, simians
Platyrrhine	New World monkeys
Catarrhine	simians: Old World monkeys and hominoids
Hapaline	*Callithrix, Saguinus, Leontopithecus, Callimico* species
Galago senegalensis	northern lesser bush baby
Galago demidoff	Demidoff's bush baby, previously *G. demodovii*
Otolemur crassicaudatus	thick-tailed greater bush baby
Microcebus murinus	grey mouse lemur
Lemur catta	ring-tailed lemur
Eulemur fulvus	brown lemur
Eulemur macaco	black lemur
Daubentonia	aye-aye
Tarsius spectrum	spectral tarsier
Callimico goeldii	Goeldi's monkey
Callithrix jacchus	common marmoset
Callithrix pygmaea	pygmy marmoset, previously *Cebuella*
Saguinus oedipus	cotton-top tamarin
Saguinus fuscicollis	saddleback tamarin
Leontopithecus chrysomelas	golden-headed lion tamarin
Aotus	night or owl monkey
Callicebus molloch	dusky titi monkey
Cebus apella	tufted or brown capuchin monkey
Saimiri sciureus	common squirrel monkey

Ateles fusciceps	brown-headed spider monkey
Macaca mulatta	rhesus macaque
Macaca nemestrina	pig-tailed macaque
Macaca fuscata	Japanese macaque
Macaca sylvanus	barbary macaque
Macaca radiata	bonnet macaque
Macaca fascicularis	long-tailed or crab-eating macaque
Papio cynocephalus	olive baboon
Papio ursinus	chacma baboon
Papio hamadryas	hamadryas baboon
Papio anubis	yellow baboon
Theropithecus gelada	gelada baboon
Cercocebus atys	sooty magabey
Miopithecus talapoin	dwarf guenon or southern talapoin monkey, previously *Cercopithecus*
Erythrocebus patas	patas monkey
Chlorocebus aethiops	vervet, grivet or green monkey, previously *Cercopithecus*
Semnopithecus entellus	hanuman langur, previously *Presbytis*
Hylobates lar	white handed gibbon
Pongo	orangutan
Gorilla gorilla	gorilla
Pan troglodytes	chimpanzee

Correspondence

Any correspondence should be directed to Almuth Einspanier, Institute of Physical Biochemistry, Veterinary Faculty, University of Leipzig, Germany.
Email : einspanier@vetmed.uni-leipzig.de

References

Abbott, D.H. (1986). *Primatologia* 2, 1–16.

Abbott, D.H. (1987). *J. Zool. Lond.* 213, 455–470.

Abbott, D.H. (1989). In Standen, V. and Foley, R.A. (eds) *Comparative Socioecology: The Behavioural Ecology of Humans and Other Mammals*, pp 285–304. Blackwell Scientific Publications, Oxford.

Abbott, D.H. and Hearn, J.P. (1978). *J. Reprod. Fertil.* 53(1), 155–166.

Abbott, D.H., Barrett, J. and George, L.M. (1993). In Rylands, A.B. (ed.) *Marmosets and Tamarins: Systematics, Behaviour and Ecology*, pp 152–163. Oxford University Press, Oxford.

Altemus, M., Roca, C., Galliven, E., Romanos, C. and Deuster, P. (2001). *J. Clin. Endocrinol. Metab.* 86(6), 2525–2530.

Beach, F.A. (1976). *Horm. Behav.* 7, 105–138.

Bellino, F.L. and Wise, P.M. (2003). *Biol. Reprod.* 68, 10–18.

Boklage, C.E. (1990). *Int. J. Fertil.* 35(2), 75, 79–80, 81–94.

Bonney, R.C. and Setchell, K.D. (1980). *J. Steroid Biochem.* 12: 417–421.

Bronson, F.H. (1989). *Mammalian Reproductive Biology*. The University of Chicago Press, Chicago.

Brown, G.R. (2001). *Anim. Behav.* 61(4): 683–694.

Cameron, J.L. (1996). *Rev. Reprod.* 1: 117–126.

Carosi, M, Heistermann, M. and Visalberghi, E. (1999). *Horm. Behav.* 36: 252–265.

Chambers, P.L. and Hearn, J.P. (1979). *J. Reprod. Fertil.* 56(1): 23–32.

Chandrashekar, V., Kierschke, D.J. and Wolf, R.C. (1987). *Am. J. Primat.* 13, 145–153.

Coe, C.L., Murai, J.T., Wiener, S.G., Levine, S. and Siiteri, P.K. (1986). *J. Endocrinol.* 118, 435–440.

Condon, R.G. (1991). *Hum. Ecol.* 19(3), 287–321.

Diamond, E.J., Aksel S., Hazelton, J.M., Wiebe, R.M. and Abee, C.R. (1987). *J. Reprod. Fertil.* 80(2), 373–381.

Dierschke, D.J., Chaffin, L. and Hutz, R.J. (1994a). *Trends in Endocrine Metab.* 5(5), 215–219.

Dierschke, D.J., Golos, T.G., Durning, M. and Hutz, R.J. (1994b). *Am. J. Primat.* 34, 261–273.

Dissen, G.A., Dees, W.L. and Ojeda, S.R. (1993). In Adashi, E.Y. and Leung, P. (eds) *The Ovary*, pp 1–19. Raven Press Ltd., New York.

Dixson, A.F. (1983). *Adv. Study Behav.* 13, 63–106.

Dixson, A.F. (1987). *Physiol. Behav.* 39(4), 495–459.

Dixson, A.F. (1990). *Folia Primatol.* 54(1–2), 46–56.

Dixson, A.F. (1998). *Primate Sexuality: Comparative Studies of the Prosimians, Monkeys, Apes, and Human Beings*. Oxford University Press, Oxford.

diZerega, G.S. and Hodgen, G.D. (1981). *Endocrine Rev.* 2(1), 27–49.

Doe, R.P., Zinneman, H.H., Flink E.B. and Ulstrom, R.A. (1960). *J. Clin. Endocrinol. Metab.* 20, 1484–1492.

Donovan, B. (2000a). *Sexually Transmitted Infections* 76, 7–12.

Donovan, B. (2000b). *Sexually Transmitted Infections* 76, 88–93.

Drickamer, L.C. (1974). *Folia Primatol.* 21, 61–80.

Dunbar R.I.M. (1989). *Folia Primatol.* 53(1–4), 235–246.

Dunbar, R.I.M., Hannah-Stewart, L. and Dunbar, P. (2002). *Anim. Behav.* 64(5), 801–805.

Eastman, S.A., Makawiti, D.W., Collins, W.P. and Hodges, J.K. (1984). *J. Endocrinol.* 102(1), 19–26.

Einspanier, A., Nubbemeyer, R., Schlote, S., Schumacher, M., Ivell, R., Fuhrmann, K. and Marten, A. (1999). *Biol. Reprod.* 61(2), 512–520.

French, J.A. and Stribly, J.A. (1987). In Solomon, N.G. and French, J.F. (eds) *Cooperative Breeding in Mammals*, pp 34–75. Cambridge University Press, Cambridge.

Garber, P.A. and Leigh, S.R. (1997). *Folia Primatol.* 68, 1–22.

Gebbie, D. (1981). *Reproductive Anthropology-Descent Through Woman*. Wiley, New York.

Genazzani, A.R., Lemarchand-Beraud, T., Aubert, M.L. and Felber, J.P. (1975). *J. Clin. Endocrinol. Metab.* 41(3), 431–437.

Ghosh, D. and Sengupta, J. (1992). *Acta Endocrinol. (Copenh).* 127(2), 168–173.

Gilardi, K.V.K., Shideler, S.E., Valverde, C.R., Roberts, J.A. and Lasley, B.L. (1997). *Biol. Reprod.* 57, 335–340.

Gore, M.A., Nayudu, P.L., Vlaisavljevic, V. and Thomas, N. (1995). *Human Reprod.* 10, 2313–2319.

Gore, M.A., Nayudu, P.L. and Vlaisavljevic, V. (1997). *Human Reprod.* 12, 101–107.

Graham, C.E., Keeling, M., Chapman, C., Cummins, L.B. and Haynie, J. (1973). *Am. J. Phys. Anthropol.* 38, 211–216.

Groves, C. (2001). *Primate Taxonomy.* Smithsonian Institution Press, London.

Harlow, C.R., Hearn, J.P. and Hodges, J.K. (1984). *J. Endocrinol.* 103(1), 17–24.

Harvey, P.H., Martin, R.D. and Clutton-Brock, T.H. (1987). In Smuts, B.B., Cheney, D.L., Seyfarth, R.M., Wrangham, R.W. and Struhsacker, T.T. (eds) *Primate Societies*, pp 181–196. University of Chicago Press, Chicago.

Hearn, J.P., Lunn, S.F., Burden, F.J. and Pilcher, M.M. (1975). *Lab. Animals* 9(2), 125–134.

Hearn, J.P., Abbott, D.H., Chambers, P.C., Hodges, J.K. and Lunn, S.F. (1978). *Primates Med.* 10, 40–49.

Heistermann, M., Möhle, U., Vervaecke, H., van Elsacker, L. and Hodges, J.K. (1996). *Biol. Reprod.* 55, 844–853.

Hershkovitz, P. (1977). *Living New World Monkeys (Platyrrhini)*, 1, 440–475. University of Chicago Press, Chicago.

Hodges, J.K. and Eastman, S.A.K. (1984). *Am. J. of Primatol.* 6, 187–197.

Hodges, J.K., Brand, H., Henderson, C. and Kelly, R.W. (1983). *J. Reprod. Fertil.* 67(1), 73–82.

Holmes, K.K., Sparling, P.F., Mardh, P.A., Lemon, S.M., Stamm, W.E., Piot, P. and Wasserheit, J.N. (1994). *Sexually Transmitted Diseases.* McGraw-Hill, New York.

Hrdy, S.B. and Whitten, P.L. (1987). In Smuts, B.B., Cheney, D.L., Seyfarth, R.M., Wrangham, R.W. and Struhsacker, T.T. (eds) *Primate Societies*, pp 370–384. University of Chicago Press, London.

Jaquish, C.E., Cheverud, J.M. and Tardiff, S.D. (1996). *J. Hered.* 87(1), 74–77.

Jolly, A. (1967). In Altmann, S.A. (ed.) *Breeding Synchrony in Wild Lemur catta. Social Communication Among Primates*, pp 3–14. University of Chicago, Chicago.

Johnson, E.O., Kamilaris, T.C., Carter, S., Gold, P.W. and Chrousos, G.P. (1991). *Am. J. Primatol* 25, 191–201.

Kaplan, J.R., Adams, M.R., Koritnik, D.R., Rose, J.C. and Manuck, S.B. (1986). *Am. J. Primatol.* 11,181–193.

Kendrick, K.M. and Dixson, A.F. (1985). *Neuroendocrinology* 41(6), 449–453.

Keverne, E.B. (1976). *J. Soc. Cosmet. Chem.* 27(6), 257–269.

Keverne, E.B. (1987). *J. Zool. Lond.* 213, 395–408.

Koering, M.J. (1986). In Dukelow, W.R. and Erwin, J. (eds) *Comparative Primate Biology*, pp 215–262. New York.

Koering, M.J. (1987). In Stouffer, R.L. (ed.) *The Primate Ovary*, pp 3–24. New York, Plenum Press.

Kuehl, T.J., Kang, I.S. and Siler-Khodr, T.M. (1992). *Am. J. Primatol.* 28, 41–48.

Lee, P.C. (1987). *Journal of Zoology* 213, 409–422.

Levallet, J., Pakarinen, P. and Huhtaniemi, I.T. (1999). *Arch. Med. Res.* 30(6), 486–494.

Lockhart, A.B., Thrall, P.H. and Antonovics, J. (1966). *Laboratory Animal Science* 71, 415–471.

Lunn, S.F. (1983). *Folia Primatol.* 28, 41–48.

Martin, R.D. (1990). *Primate Origins and Evolution, a Phylogenetic Reconstruction.* Princeton University Press, Princeton.

Mitchell, B.F., Seron-Ferre, M. and Jaffe, R.B. (1982). *Endocrinology* 111(6), 1837–1842.

Nagle, C.A. and Denari, J.H. (1983). In Hearn, J. (ed.) *Reproduction in the New World Primates: New Model in Medical Science,* pp. 39–67. MTP Press, Lancaster/Boston.

Norman, R.L., Lindstrom, S.A., Bangsberg, D., Ellinwood, W.E., Gliessman, R. and Spiess, H.G. (1984). *Endocrinology* 115(1), 261–266.

Nozaki, M., Mitsunaga, F. and Shimizu, K. (1995). *Biology of Reproduction* 52, 1250–1257.

Nubbemeyer, R., Heistermann, M., Oerke, A.K. and Hodges, J.K. (1997). *J. Med. Primatol.* 26, 139–146.

Nunn, C.L. (2003). *Animal Behaviour* 66(1), 37–48.

O'Byrne, E.M. and Steinetz, B.G. (1976). *Proc. Soc. Exp. Biol. Med.* 152(2), 272–276.

Ottobre, J.S., Nixon, W.E. and Stouffer, R.L. (1984). *Biology of Reproduction* 31(5), 1000–1006.

Pavelka, M.S.M. and Fedigan, L.M. (1991). *Yearbook of Physical Anthropology* 34, 13–38.

Pavelka, M.S.M., Fedigan, L.M. and Zohar, S. (2002). *Animal Behaviour* 64(3), 407–414.

Pepe, G.J., Johnson, D.K. and Albrecht, E.D. (1982). *Steroids* 39(5), 471–477.

Poole, T.B. and Evans, R.G. (1982). *Lab. Animal* 16, 88–97.

Robbins, C.T. (1983). *Wildlife Feeding and Nutrition.* Academic Press, London.

Roy, R.P. (2003). *Obstetrics & Gynecology* 101(2), 397–401.

Rowe, N. (1996). *The Pictorial Guide to the Living Primate.* Pogonias Press, New York.

Rowell, T.E. (1970). *J. Reprod. Fertil.* 21, 133–141.

Rowell, T.E. (1972). *Adv. Study Behav.* 105, 69–105.

Saltzman, W., Schultz-Darken, N.J., Wegner, F.H., Wittwer, D.J. and Abbott, D.H. (1997). *Horm. Behav.* 33(1), 58–74.

Savage, A., Ziegler, T.E. and Snowden, C.T. (1988). *American Journal of Primatology* 14, 345–359.

Schultea, T.D., Dees, W.L. and Ojeda, S.R. (1992). *Biology of Reproduction* 47, 760–767.

Setchell, K.D., Bull, R. and Adlercreutz, H. (1980). *J. Steroid Biochem.* 12, 375–384.

Shimada, T., Morita, T., Nagai, K., Sato, F., Mori, H. and Campbell, G.R. (1993). *Horm. Res.* 39 (suppl. 1), 9–15.

Short, R., England, N., Bridson, W.E. and Bowden, D.M. (1989). *J. Gerontology* 44(5), B131–138.

Smith, T.E., Schaffner, C.M. and French, J.A. (1997). *Horm Behav.* 31(2),159–168.

Sousa, M.B. and Ziegler, T.E. (1998). *Am. J. Primatol.* 46(2),105–117.

Steinetz, B.G., Randolph, C. and Mahoney, C.J. (1992). *Endocrinology* 130(6), 3601–3607.

Steinetz, B.G., Randolph, C. and Mahoney, C.J. (1995). *Biology of Reproduction* 53(4), 834–839.

Strassmann, B.I. (1997). *Evolutionary Anthropology* 6, 157–164.

Sunday, S.O. (2003). *Scotland on Sunday*. Edinburgh: 23.

Tardiff, S.D. and Jaquish, C.E. (1994). *Ann NY Acad. Sci.* 709, 214–215.

Thalabard, J.C. (1992). *J. Biosoc. Sci.* 24(3), 289–300.

Troisi, A. and Carosi, M. (1998). *Animal Behaviour* 56(5), 1261–1266.

Tsika, R.W. (1994). *Exerc. Sport. Sci. Rev.* 22, 361–388.

Vandenbergh, J.G. (1973). Environmental influences on breeding in rhesus monkeys. Symposium of the IVth International Congress of Primatology. Karger, Basel.

Vleeshouwer, K.D., Leus, K. and Van Elsacker, L. (2003). *Animal Welfare* 12(2), 251–268.

Wasser, S.K. (1995). *Nature* 376(6537), 219–220.

Wildt, D.E., Doyle, U., Stone, S.C. and Harrison, R.M. (1977). *Primates* 18, 216–270.

Wissmann, M.A. (1999). *Husbandry and Nutrition* 2(1), 209–240.

Whitten, P.L. (1983). *Am. J. Phys. Anthro.* 60, 269–270.

Whitten, P.L. (1984). *Am. J. Phys. Anthro.* 63, 234.

Zeleznik, A.J. (1993). In Adashi, E.Y. and Leung, P. (eds) *The Ovary*, pp 41–55. New York, Raven Press.

Ziegler, T.E., Savage, A., Scheffler, G. and Snowdon, C.T. (1987). *Biol. Reprod.* 37, 618–627.

Ziegler, T.E., Scheffler, G. and Snowdon, C.T. (1995). *Horm. Behav.* 29(3), 407–424.

Ziegler, T.E. and Sousa, M.B. (2002). *Horm. Behav.* 42(3), 356–367.

Chapter 8

Male Reproduction and Fertilization

Richard M. Harrison
and H. Michael Kubisch
Tulane National Primate Research Center, Covington, Louisiana, USA

Introduction

There are many similarities in the anatomy and physiology of the male reproductive system between the general classifications of nonhuman primates, i.e., New World monkeys, Old World monkeys, and the apes. This chapter will provide a general overview in each area of male reproduction using the most common nonhuman primate models and will indicate any differences between these models and other classifications. Descriptions of various manipulations of the system will not be presented in this chapter.

Control of male reproduction

Endocrinology

Overview

Knowledge of the function of the hypothalamic-pituitary-testicular axis is the key to the understanding of reproduction in the male primate. The interrelationship of these tissues is essential for both the endocrine

The Laboratory Primate
Copyright 2005 Elsevier
ISBN 0-1208-0261-9

and exocrine functions of the testes. In brief, gonadotropin-releasing hormone (GnRH), a decapeptide, is synthesized in the hypothalamus and is secreted in pulses to the anterior pituitary by way of the portal vessels in the pituitary stalk. In early reports GnRH is referred to as LHRH, since those studies showed an increase in luteinizing hormone (LH) following administration of the hormone. In the anterior pituitary the gonadotropic cells are stimulated by the GnRH to secrete LH and follicle-stimulating hormone (FSH). Note that these hormones are identified by the names given for their functions in the female but are chemically identical in both genders. These hormones enter the general circulation and LH stimulates Leydig cells in the interstitial spaces of the testes to secrete testosterone. LH receptors were demonstrated on rhesus (*Macaca mulatta*) and African green monkey (*Chlorocebus aethiops*) Leydig cells (Davies *et al.*, 1979a; 1979b) and GnRH was shown to stimulate LH secretion in male rhesus monkeys (Ferin *et al.*, 1974, Toivola *et al.*, 1978), in cynomologus monkeys (*M. fascicularis*) (Mori and Hafez, 1973), and in baboons (Koyama, 1976). FSH stimulates the Sertoli cells in the seminiferous tubules to produce androgen binding protein (ABP), inhibin, and activin. A pulse generator in the hypothalamus controls the frequency and amplitude of the GnRH pulses. It appears that, in monkeys that have a seasonal breeding period, the frequency of the GnRH pulses does not change between the recrudescent phase and the in-season period but the amplitude of the pulses increases in the breeding season (Wickings *et al.*, 1986).

The hypothalamic-pituitary-testicular axis has feedback controls at two levels to maintain testosterone secretion. Testosterone from the Leydig cells has a negative feedback on the hypothalamus to decrease amplitude and frequency of the GnRH pulses. Estradiol, converted from testosterone by aromatase, has a negative feedback on the anterior pituitary to make the gonadotropic cells less sensitive to the GnRH stimulation. These two processes decrease the amount of LH secretion and thereby reduce testosterone levels. Inhibin and activin from the Sertoli cells have, respectively, negative and positive feedback on the anterior pituitary to maintain FSH levels.

Endocrinology in the pre-adult

Fetal endocrinology

It has been shown that the hypothalamic-pituitary-testicular axis is functional in the male rhesus fetus as early as 100 days of gestation. Studies in gonadectomized fetuses proved that testosterone or its metabolite, dihydrotestosterone, were both equally sufficient for feedback regulation of LH in the developing male rhesus fetus (Resko and Ellinwood, 1985). The timing of this functionality was demonstrated in earlier reports. Male rhesus fetus testosterone values in umbilical artery plasma at 60 days gestation averaged 2100 pg/ml. This decreased to 1000 pg/ml by 100 days of gestation and then increased to approximately 1500 pg/ml by 160 days of gestation (Resko, 1970). Male rhesus monkey testicular tissue, collected at approximately 45 and 60 days of gestation (from male rhesus monkeys) was capable of converting ^{14}C-pregnenolone to testosterone and androstenedione. Plasma testosterone and androstenedione levels in the male fetuses increased from 66 and 43 ng/100 ml of plasma at fetal age of 100 days to 102 and 140 ng/100 ml of plasma at 150 days of gestation respectively (Resko, 1970). The prenatal development of the ductus epididymides is a response to androgens secreted by the fetal testes (Alexander, 1972). There is little differentiation of the cuboidal epithelium of the epididymis up to 130 days of gestation. At this time the secretion of androgens induces differentiation of the cells so that they resemble those found in adults, including sterocilia cells in the corpus and cauda epididymides. When the androgen levels drop after birth the epithelial cells regress to an undifferentiated state until the androgens increase at puberty.

Endocrinology in the neonate and juvenile monkey

In the rhesus monkeys, levels of the androgens decreased sharply after birth and remained low until puberty (Resko, 1970). FSH levels were reported as 2.4 ± 0.8 ng/ml in juvenile rhesus males, 6.4 ± 1.8 ng/ml in pubertal males, and 16.1 ± 1.8 ng/ml in adults, while LH levels were not detectable in juveniles but rose to 16.2 ± 3.1 ng/ml in pubertal males (Arslan *et al.*, 1986). Juvenile African green monkeys did show LH responses to GnRH but the responses were related to the age of the monkeys and testosterone secretion did not reach adult-like levels until full maturity (Brady *et al.*, 1985). Male cynomologus monkeys aged 1.5–2.5 years had significantly less plasma inhibin and androgen-binding protein, a protein secreted by the Sertoli cells, than did adult males (Keeping *et al.*, 1990). Rhesus males showed measurable levels of inhibin-B at age one week, an increase concomitant with a neonatal rise in plasma FSH. They then decreased, but still at detectable levels, until one year of age. Adult values are higher than any values before one year of age (Winters and Plant, 1999).

A study (Ginther *et al.,* 2002) using neonatal and pubertal cotton-top tamarins (*Saguinus oedipus oedipus*) found that pubertal LH secretion began at week 37. Testosterone surges followed at weeks 41–44 and the conversion of testosterone to dihydrotestosterone (DHT) occurred at about 49 weeks. These endocrine changes were preceded by rapid weight gain, completed by week 24, and followed by rapid testicular growth, completed by about 76 weeks of age. In marmosets, the pubertal rise in testosterone was found at 50–60 weeks of age (Kelnar *et al.,* 2002). At this age, marmosets (*Callithrix jacchus*) were judged to have sufficiently mature testes to be used for safety evaluation (Jackson and Edmunds, 1984).

The quiescent state of the testes, prior to puberty, is due to a limited secretion of FSH and LH from the anterior pituitary. This limitation of gonadotropic hormones is occasioned by the interruption of pulsatile secretion of GnRH from the hypothalamus at this stage of development. Stimulation of GnRH neurons within the hypothalamus of juvenile monkeys (Plant *et al.,* 1989) did result in the onset of precocious puberty. The hypothalamic-pituitary-Leydig cell axis was fully activated and spermatogenesis was initiated. This would suggest that the limiting component for puberty is the stimulation of the neurons that control the GnRH pulse generator.

The role of inhibin in the negative feedback system for the secretion of FSH in chimpanzees, prior to the onset of pubertal testicular growth (Po), has been reported (Marson *et al.,* 1993). Serum inhibin, LH, and testosterone begin to increase six months prior to an increase in FSH and Po. At this time, inhibin was related to testosterone levels but not the gonadotropins. After Po, inhibin was still related to testosterone but inversely related to gonadotropins and primarily to FSH. This indicated that the negative feedback system for FSH by inhibin is not functional until the onset of puberty.

As puberty approaches, the GnRH pulse generator is reactivated resulting in increased gonadotropin levels, increases in growth rate, and full development of the reproductive system. At this same time blood leptin levels also increase. The suggestion that leptin may be a signal that triggers the onset of puberty has been shown to be incorrect (Barker-Gibbs *et al.,* 2002) but may indicate a state at which puberty can proceed if other critical control mechanisms are all operational (Mann and Plant, 2002).

Endocrinology in the adult

In the adult human and nonhuman male primate GnRH, secreted in the hypothalamus, stimulates cells in the anterior pituitary to secrete LH and FSH. LH stimulates the Leydig cells, located in the interstitial spaces of the testes to secrete testosterone and FSH stimulates the Sertoli cells in the seminiferous tubules which are involved in spermatogenesis. For any hormone to have an effect on specific tissues there must be specific receptors for that hormone in the tissues. A study conducted in *M. mulatta, M. fascicularis, M. nemestrina,* and *Papio cynocephalus* (Berman and Sairam, 1984) found no differences in the binding of ^{125}I-labeled FSH to a particular fraction of testicular homogenates from these four species. In the rhesus monkeys, the FSH receptor was more abundant than the LH receptor. Testosterone levels apparently modulate the release of LH and FSH that occurs in response to changing GnRH pulse frequencies in the adult *M. fascicularis* (Adams *et al.,* 1988). LH response to GnRH is not affected by pulse frequency but FSH responded at a slow frequency when plasma levels were low.

Inhibin

In rhesus monkeys the testicular regulation of the secretion of FSH is primarily by a negative feedback action of inhibin B acting at the pituitary level (Plant, 1994). The inhibin-β_B subunit peptides and the corresponding mRNA have been localized in the Sertoli cells of the primate testes from three species of macaques, *M. mulatta, M. fascicularis, M. arctoides* (Vliegen *et al.,* 1993; Zhang *et al.,* 1997). Since the secretion of inhibin B is stimulated by gonadotropin and, considering the parallel concentrations of inhibin B, LH, and FSH from birth to adulthood (Ramaswamy and Plant, 2001), the role of LH, as well as the role of FSH in this regulation, is important (Majumdar *et al.,* 1997). A study in juvenile and adult rhesus monkeys (Ramaswamy *et al.,* 2003) concluded that inhibin B secretion by the monkey testes is regulated by the inhibitory action of LH and the stimulatory action of FSH. The action of LH is probably indirect and mediated by testosterone inhibition of inhibin-β_B gene expression.

Testes

Testosterone from the Leydig cells diffuses into the general circulation and into the seminiferous tubules. One report by Dierschke *et al.* (1975) provided evidence of a mechanism by which blood flowing into the testes had a significantly higher concentration of testosterone than blood in the femoral artery. Blood was collected simultaneously from the distal testicular artery and the femoral artery and the testosterone levels were 5.10 ± 1.15 and 4.47 ± 1.00 ng/ml (p,0.02) respectively. This suggests a local transfer of testosterone from the venous vessels of the pampiniform plexus to the testicular artery.

Accessory reproductive glands

Sperm exit the testes via the ductuli efferentes and enter the epididymis, which is composed of three sections that differ morphologically and functionally. These differences are involved in the maturation of the sperm. Sperm pass in sequence through the caput epididymis, the corpus epididymis, the cauda epididymis, and then into the vas deferens. In the cynomolgus monkey (*M. fascicularis*), sperm differ morphologically in the location of the cytoplasmic droplet (Mahony *et al.*, 1994). From caput to corpus to cauda, sperm steadily exhibited a more distal cytoplasmic droplet. When examined for motility, caput sperm were not progressively motile, corpus sperm had poor duration of motility, and cauda sperm showed progressive motility of good duration. In the rhesus monkeys, sperm transport through the epididymis takes approximately five days (Amann *et al.*, 1976). These authors found that the average production of spermatozoa per testis in the rhesus monkey during the breeding season was 547 ± 69 × 10^6. The amount of androgen receptors and the form of the androgen are important to the functioning of these sections of the epididymis. A study in adult rhesus monkeys (Roselli *et al.*, 1991) found that both androgen receptors and 5α-reductase activity were higher in the caput and corpus of the epididymis compared to the ductuli efferentes and cauda epididymis. 5α-reductase is the enzyme that converts testosterone to dihydrotestosterone (DHT), a biologically more active form of androgen. These results suggest that the maturation changes that occur in sperm during their passage through the epididymis occur in the caput and corpus epididymis and that sperm in the cauda are fully mature and capable of fertilization. Similar studies in the vervet monkey (*Cercopithecus aethiops*) (van der Horst *et al.*, 1999) found that only when the spermatozoa had reached the cauda epididymis and vas deferens did they attain the full vigor to swim rapidly, and with progressive motility.

Two accessory glands of the male reproductive tract that are involved in the production of semen, are the seminal vesicles and the prostate. The seminal vesicles in *Loris tardigradus lydekerianus* vary in weight during the year, with a peak prior to the breeding season (Ramakrishna and Prasad, 1967) suggesting stimulation by testosterone. In *Daubentinia* the seminal vesicles are absent (Roberts, 1972). The sizes of the seminal vesicles vary greatly among the primates. The prostate is probably responsive to dihydrotestosterone (DHT) since this is the hormone responsible for the differentiation of the fetal organ. Testosterone may be converted to DHT by 5α-reductase in the target tissues. The administration of testosterone influences the uptake of zinc by the prostate (Schoones *et al.*, 1969).

Factors affecting male reproduction

Seasonality

A limited number of studies have been conducted on the effects of seasonality on male reproduction in male nonhuman primates. Some of the studies have been conducted on captive populations in environmental conditions different from their native habitats. Under different conditions, however, some species have shown definitive seasonal patterns in many parameters associated with breeding. The factors that influence these patterns may include temperature, light, humidity, and food sources. A problem with the definition of seasonality comes from the observation that some species will have some births in all months but the majority of births occur within a limited period (Kriewaldt and Hendrickx, 1968). Because of this, reports of a birth season in a species that others report as being non-seasonal do occur.

Rhesus monkeys have been reported to be seasonal breeders with definite periods for mating and consequently a period for births, both in the laboratory and free-ranging (Wickings and Nieschlag, 1980; Gordon *et al.*, 1976; Drickamer, 1974; Zamboni *et al.*, 1974). At the Tulane National Primate Research Center in Covington, Louisiana, the early breeding season for rhesus monkeys is August to September, prime-breeding season is November to January, and late breeding season is February to April (Phillippi-Falkenstein and Harrison, 2003). Other species that have been reported to have seasonal breeding periods are *Saimiri sciureus* (Coe and Roseblum, 1978; Mendoza *et al.*, 1978), *Cercopithecus mitis kolbi* (Omar and De Vos, 1971), *M. fuscata* (Kawai *et al.*, 1967; Matsubayashi, 1974), and *M. sylvana* (Roberts, 1978). Those reported as nonseasonal breeders include *Aotus trivagatus* (Dixson *et al.,* 1980); *Ateles belzebuth* and *A. geoffroyi* (Klein, 1971); *M. fascicularis* (Mahone and Dukelow, 1979); *M. radiata* (Murty *et al.,* 1980); *M. arctoides* (Slob *et al.,* 1979); *M. nemestrina* (Bernstein *et al.,* 1978); and *Pan troglodytes* (Graham and Hodgen, 1979). In species that are seasonal breeders there may be changes in testosterone levels, as indicated earlier, as well as changes in body weights, testicular size, and semen parameters (Wickings and Nieschlag, 1980). Rhesus monkeys have testosterone levels that peak in the beginning of the breeding season with values 400% greater than in the periods of sexual inactivity (Gordon *et al.,* 1976). A similar pattern is found for DHT which peaks in October to December in males maintained in constant laboratory conditions (Wickings and Nieschlag, 1980). The response of rhesus monkeys to electroejaculation is also decreased in the non-breeding season (Zamboni *et al.,* 1974; Wickings and Nieschlag, 1980). Paradoxically, rhesus males were found to lose weight during the breeding season and to regain this weight in the four months preceeding the next breeding season (Bernstein *et al.,* 1989).

In the Southern Hemisphere, the breeding season for rhesus is from March to August (Bielert and Vandenberg, 1981) and at this time the sex skin color in males is most intense.

Male squirrel monkeys in social groups showed elevations in testosterone during the breeding season (Schiml *et al.,* 1999), but males housed in individual cages had body weight peaks and cortisol peaks prior to the breeding season and testosterone peaks followed the breeding season. The individually housed males were not in full view of each other and had no physical contact with other monkeys. They were housed in a separate building from the socially housed monkeys. The squirrel monkeys usually show a very definitive breeding season. In a semi-natural environment at the Monkey Jungle near Miami, Florida, the breeding is confined to the period from January to April (DuMond and Hutchinson, 1967). In November and December before the breeding season, the males develop a "fatted" condition. An increase in body weight and an increase in pelage of the upper torsos, shoulders, and arms characterize this condition. Spermatogenesis was evident in all seminiferous tubules in testicular tissue collected during the breeding season and regressed to a basal layer of Sertoli cells and spermatogonia with little mitotic activity in the non-breeding season. There are also seasonal changes in thyroid hormones in male squirrel monkeys with highest concentrations prior to the breeding season (Kaack *et al.,* 1980).

Bonnet macaques (*M. radiata*) showed a definite seasonal pattern (Click, 1979). Copulations peaked during October and November, months with higher humidity and precipitation, lower temperature and sunshine. Testicular sizes and serum testosterone levels increased and peaked during the breeding season.

Japanese macaques (*M. fuscata*) have seasonal breeding seasons but the timing of these seasons is affected by latitude, rainfall, temperature, and social factors (Kawai *et al.,* 1967).

The testes of male crowned lemurs (*Lemur coronatus*) reach their maximum size in late December, at the time females enter their breeding season, and regress by late March (Kappeler, 1987).

The Southern black howler monkey (*Alouatta caraya*) has the highest fecal testosterone levels in May and June, the same months in which sperm concentration, in collected semen samples, is the highest (Moreland *et al.,* 2001).

Male capuchin monkeys (*Cebus apella*) in captivity show a seasonal variation in testicular volume relative to the breeding season (de B. vaz Guimaraes *et al.,* 2003).

Spermatogenesis

A detailed review of spermatogenesis by de Rooij and Russell (2000) is recommended for readers that are not familiar with the terminology. The process of spermatogenesis is best understood as three continuous phases: (1) stem cell renewal, (2) proliferation of the germ cells, and (3) spermiogenesis, or the maturation of the spermatids to spermatozoa. The process of spermatogenesis becomes functional with the completion of puberty. In the cynomolgus monkey (*M. fascicularis*), the first spermatocytes appeared at 3–4 years of age at a body weight of 3.24 ± 0.15 kg and full spermatogenesis was attained between 3.6 years and 4.3 years at a body weight of 3.5–3.8 kg (Dang and Meusy-Dessolle, 1984).

Stem cell renewal

The stem cells, or spermatogonia, are the beginning of the germ cells line. All spermatozoa are derived from these stem cells. They are diploid undifferentiated cells that replicate by mitosis. It is generally considered that there are two types of stem cells in the higher primates: dark type A (Ad) and pale type (Ap) spermatogonia (Plant and Marshall, 2001). The Ad spermatogonia are believed to provide a reserve population of stem cells. The Ap cells actively divide and are the renewing stem cells (Schulze, 1979). These cells divide by mitosis and produce two identical daughter cells. In rhesus monkeys,

10.5 days (Clermont, 1969; de Rooij *et al.,* 1986; Rosiepen *et al.,* 1997) after this first replication, half of the cells will replicate themselves and the other half will produce the B type spermatogonia, the first generation of differentiated spermatogonia. This is referred to as the spermatogenic cycle.

Proliferation of germ cells

In Old World monkeys these B type spermatogonia undergo three mitotic divisions producing sequential generations of B_2, B_3, and B_4 spermatogonia. The B_4 spermatocytes undergo another mitotic division to produce primary spermatocytes. These primary spermatocytes undergo meiosis to produce secondary spermatocytes after the first meiotic division and spermatids after the second division. Hence each primary spermatocyte eventually produces four haploid spermatids.

The final stage of spermatogenesis is the morphological changes of the spermatids to the highly differentiated fertilizable mature spermatozoa. The changes include the loss of most of the cytoplasm, the development of the middle piece that contains the mitochondria, the formation of the tail, and the acrosomal cap. Studies conducted in India (Moudgal and Sairam, 1998) show that, in primates, FSH promotes quantitative spermatogenesis and has a critical role in regulating spermiogenesis.

Cycles of spermatogenesis

The 10.5-day cycle of the Ap spermatogonia in the rhesus monkeys means that there are new type B spermatogonia produced at that interval in a continuous manner. Since all parts of any seminiferous tubule are not synchronized, there is a continuous production of spermatozoa. The onset of puberty in the rhesus monkeys is associated with a rapid and substantial increase in Sertoli cells and this is followed by amplification of stem spermatogonia (Marshal and Plant, 1996). The spermatogenic capacity of the testis is dependent on the number of Ap spermatogonia, which, in turn, depends on the expanding population of Sertoli cells since each Sertoli cell can support only a limited number of germ cells (Orth *et al.,* 1988; Sharpe, 1994). In the adult rhesus testes, the population of Ap spermatogonia in the adult is a magnitude greater than in the juvenile testis (Marshall and Plant, 1996).

Duration of spermatogenesis in the rhesus monkey is 48 days, or 4.5 spermatogenic cycles (de Rooij *et al.,* 1986). In man there is only one generation of B type spermatogonia but the length of spermatogenesis is around 74 days (Heller and Clermont, 1964). The spermatogenic cycle length in *M. fascicularis* is 10.16 ± 0.44 days (Aslam *et al.,* 1951).

Chimpanzee spermatogenesis has been reported to be most similar to the human in both structure and function (Smithwick and Young, 1996a). The spermatogenic cycle in the chimpanzee is approximately 14 days. There are six stages and the time from Ap spermatogonia to mature spermatids is approximately 62.5 days (Smithwick and Young, 1996b).

In the common marmoset (*Callithrix jacchus*) the length of the spermatogenic cycle was estimated to be ten days and the duration of spermatogenesis to be 37 days (Millar *et al.,* 2000). This species was reported to convert type A spermatogonia to type B spermatogonia at a rate several-fold higher than reported for other primate (Weinbauer *et al.,* 2001).

Hormonal control of spermatogenesis

The roles of FSH and LH have not been strictly defined in relation to the initiation and maintenance of spermatogenesis. FSH may not be required for the initiation of spermatogenesis and some studies suggest that spermatogenesis can be maintained with LH (Plant and Marshall, 2001). It is, however, recognized that the normal physiological circumstances, where both FSH and LH are present to stimulate their respective target tissues, results in maximal sperm production in men (Steinberger, 1971; Nieschlag *et al.,* 1999).

Administration of FSH to adult monkeys stimulated spermatogenesis (van Alphen *et al.,* 1988). *M. fascicularis* responded to the FSH injections to a greater degree than *M. mulatta,* based on the increase in A type spermatogonia. In *M. fascicularis,* both types of A spermatogonia are present at birth (Kluin *et al.,* 1983) and a few B types are present by the end of the first year. At puberty these spermatogonia begin differentiation and a large number of spermatocytes and spermatids are present.

In male chimpanzees, the emission of first ejaculates and the apparent onset of activity of the accessory sex glands occurred approximately four months after the onset of pubertal testicular growth (Po) (Marson *et al.,* 1993). Sperm were not found in any semen samples until nine months after Po. Sperm counts, motility and viability increased over time but sperm morphology did not change.

Three adult male cynomolgus monkeys (*M. fascicularis*) were used to study the effects of mild testicular hyperthermia on spermatogenesis (Lue *et al.*, 2002). The scrota containing the testes were immersed in a 43° C water bath for 30 minutes once daily for six consecutive days. Serum inhibin B levels declined in all monkeys two weeks after treatment. Two monkeys were azoospermic by six or eight weeks and the third monkey sperm count was 10% of pretreatment levels. Full recovery was noted by 12 weeks after the first treatment. This species is reported to have a high efficiency of spermatogenesis (Zhengwei *et al.*, 1997) where the conversion of primary spermatocytes to spermatids is 3.94 ± 0.19. The maximum theoretical value is 4.

Fertilization

Sperm chemotaxis

Before a sperm can fertilize an ovum it has to undertake the arduous task of traveling to the oviduct. The fact that oocytes may release signals that are able to attract sperm has been well documented in marine species with external fertilization, such as sea urchins (Cosson, 1990). However, finding evidence for sperm chemotaxis in mammals, as well as detecting the nature of the signals and the pathways they elicit, has proven much more difficult. One of the first compounds shown to have chemotactic properties *in vitro* was follicular fluid, although these early observations have not always been conclusive because of the difficulties inherent in trying to distinguish chemotaxis from sperm accumulations caused by other factors (Schwartz *et al.*, 1957; Ralt *et al.*, 1991; Makler *et al.*, 1995)

The idea of chemotaxis in mammals has gained new support with recent discoveries of olfactory receptors on human sperm cells. It had been known for some time that olfactory receptor proteins are found not only in human testicular tissues but also in the membrane of mature dog spermatozoa (Parmentier *et al.*, 1992; Vanderhaeghen *et al.*, 1993). Recently the odorant receptor hOR17-4, which belongs to a family of nearly 1000 olfactory receptors that are primarily expressed in the sensory neurons of the nose, has been cloned from human testicular tissue (Spehr *et al.*, 2003). Through heterologous expression of this gene in a human kidney cell line it was possible to identify several agonists. One of these, bourgeonal, was not only able to induce chemotactic behavior in sperm but was also able to produce some of the changes in Ca^{2+} usually associated with sperm modifications seen *in vivo*.

Sperm capacitation and hyperactivation

Early attempts to use ejaculated sperm for *in vitro* fertilization of oocytes met with little success until the work of two groups in 1951 (Austin, 1951; Chang, 1951). They almost simultaneously discovered that sperm had to remain in the female reproductive tract for some time before it gained the capacity to fertilize an oocyte. The modifications that take place in sperm during this period were termed capacitation. Capacitation appears to be restricted to mammals and is a sort of "priming" that allows sperm rapidly to undergo the necessary changes for fertilization of the oocyte. Capacitation is not accompanied by visible morphological changes. However, there are modifications in the composition of membrane surface antigens, including depletion of a sperm surface coating protein, ESP13.2. This has recently been found, in the rhesus macaque, to change membrane phospholipids, and increase intracellular Ca^{2+} (Yudin *et al.*, 2003).

Capacitation of rhesus and cynomolgus sperm can be induced *in vitro* by exposure to the cyclic nucleotide analogue dbcAMP and caffeine (Boatman and Bavister, 1984; VandeVoort *et al.*, 1992, 1994). Results of a recent study have shown that the exposure of cynomolgus sperm to these reagents results in increased tyrosine phosphorylation of sperm tail proteins, which may be one of the mediators of the changes in motility associated with capacitation in this species (Mahony and Gwathmey, 1999). These changes in motility are referred to as hyperactivation and were first reported by Yanagimachi (1969). Other factors that may regulate hyperactivation include increased levels of cAMP as well as changes in the intracellular level of Ca^{2+} in the flagellum (Aoki *et al.*, 1999; Ho and Suarez, 2001). The purpose of hyperactivation is most likely to aid sperm in penetrating oviductal mucus and the cumulus cell matrix. Whether or not hyperactivation and capacitation are coupled, *in vivo*, has not been resolved. A number of studies using various sperm treatments have shown that, at least *in vitro*, the two processes can be uncoupled (Boatman and Robbins, 1991; Ho *et al.*, 2002). It is interesting to note that, in humans, only capacitated sperm appear to be chemotactically responsive to follicular fluid (Cohen-Dayag *et al.*, 1995).

Sperm binding to the zona pellucida

The oocyte is surrounded by a mass of cumulus cells which are embedded in an extracellular matrix that is rich in hyaluronic acid and contain various proteins. Sperm passes through this matrix with the aid of a membrane-bound hyaluronidase (Lin *et al.,* 1994). The presence of one of these, termed PH-20, has been confirmed in macaque and cynomolgus sperm (Lin *et al.,* 1993, 1994; Cherr *et al.,* 1996).

The zona pellucida of the oocyte contains three glycoproteins, ZP1, ZP2 and ZP3 (or ZPA, ZPB and ZPC) that are crucial for the binding of sperm. The genes encoding these have been cloned in a variety of species including cynomolgus monkey and baboon (Harris and Piersen, 2003) and they have been localized histochemically on the zona pellucida in cynomolgus monkey oocytes (Eberspaecher *et al.,* 2001). Binding to the zona appears to be completed in two phases, an initial loose attachment that is quickly followed by a more tight binding (Hartmann *et al.,* 1972). Evidence gained mostly from work in the mouse suggests that ZP3 is the protein primarily responsible for binding of sperm cells and the induction of the acrosome reaction while ZP1 acts as a secondary receptor (Florman and Wassarman, 1985, Bleil *et al.,* 1988; Saling 1991). All three ZP homologues have been cloned in the bonnet monkey and a recent study, in which recombinant ZP proteins that had originated from bonnet monkey were used in sperm binding assays, similarly showed the ZP3 homologue to be the primary receptor for capacitated sperm (Gahlay *et al.,* 2002).

Several monosaccharides located on ZP3 have been implicated as the binding epitopes including mannose, α-galactose and β-N-acetylglucosamine (Bleil and Wassarman, 1988; Miller *et al.,* 1992; Johnston *et al.,* 1998; Tulsiani *et al.,* 1992). Much of this evidence relies on the use of synthetic oligosaccharides in competition assays *in vitro*, which allow for basic binding studies, but these may not necessarily reflect the affinities found *in vivo*.

Conversely the receptor on the sperm membrane that mediates sperm binding to ZP3 in primates remains to be identified. Several putative receptors have been isolated from mouse or pig sperm that include zonadhesin, proacrosin and spermadhesins (reviewed by Jansen *et al.,* 2001). The receptor that has been studied most extensively is β1,4-galactosyltransferase (Miller *et al.,* 1992). Support for its role in sperm binding comes from studies with genetically engineered mice, where over expression has led to increased binding to ZP3 (Youakim *et al.,* 1994). However, it is likely that β1,4-galactosyltransferase may act jointly with other molecules because male mice, in which β1,4-galactosyltransferase expression has been ablated, remain fertile, although they exhibit altered sperm binding characteristics and slower penetration of the zona (Lu and Shur 1997).

The acrosome reaction

When a sperm cell makes contact with the zona pellucida it undergoes morphological changes which are termed the acrosome reaction. The acrosome is a large secretory vesicle, located at the anterior part of sperm cells, surrounded by a membrane and containing a number of enzymes which include hyaluronidase and acrosin. In rhesus sperm, the acrosome reaction can be observed as quickly as 11 seconds after tight binding to the zona and is associated with further changes in the pattern of tail movements (Tollner *et al.,* 2003).

During the acrosome reaction, the acrosomal membrane is modified by fusion with the overlaying sperm plasma membrane resulting in the release of acrosomal enzymes. The molecular events inside sperm that trigger the acrosome reaction are not well understood. However, there is evidence that ZP3 binding activates two pathways, one leading to the opening of a channel which produces a transient influx of Ca^{2+}, the other to an internal rise in pH. Both mechanisms appear necessary for the sustained release of Ca^{2+} from an intracellular Ca^{2+} pool (reviewed by Florman *et al.,* 1998). The acrosome reaction also leads to the release and activation of acrosin, a proteolytic enzyme that is most likely one of the enzymes involved in the digestion of the zona pellucida necessary to facilitate the transport of sperm into the perivitelline space (Tesarik *et al.,* 1990). It is interesting to note that an inhibitor of acrosin has been isolated from oviductal fluid in the rhesus monkey (Stambaugh *et al.,* 1974).

While most work in this area has been conducted in sperm of animal species commonly used in the laboratory or humans, some of the enzymes released during the acrosome reaction have been isolated from nonhuman primates, including the rhesus monkey and chimpanzee (Stambaugh and Buckley, 1970; Srivasta *et al.,* 1981).

Oocyte-sperm fusion and oocyte activation

Mature mammalian oocytes are arrested at metaphase II (MII). This developmental stage is maintained by a maturation promoting factor (MPF), a protein complex consisting of a kinase termed $p34^{cdc2}$ and cyclin B. The primary function of MPF is to promote spindle assembly, chromatin condensation and the breakdown of the nuclear envelope. The activity of MPF appears to be dependent on the presence of a cytostatic factor (CSF) which, in turn, requires regulatory activity of two genes, the c-mos proto-oncogene and Emi1 (Pal *et al.,* 1994, Colledge *et al.,* 1994, Reimann and Jackson, 2002).

After penetrating the zona pellucida, the sperm enters the perivitelline space and binds to the oocyte membrane, usually along the equatorial segment of the sperm head. Binding is mediated by a complex of egg membrane proteins, probably including integrin that binds to complementary membrane molecules on sperm (Bronson and Fusi, 1996). Fusion with the sperm results in an oscillating increase in Ca^{2+} in the oocyte which, in turn, triggers the release of enzyme from cortical granules that are located below the oocyte membrane. These enzymes are released into the perivitelline space where they induce changes in the zona pellucida that prevent further passage of sperm (Bleil and Wassarman, 1981). The precise mechanism by which sperm binding elicits the Ca^{2+} response, remains unclear but there is strong evidence that the trigger is some factor that is released by the sperm into the oocyte cytoplasm. This is supported by observations that rhesus sperm extracts, when injected into oocytes, can elicit similar calcium oscillations to those observed during fertilization (Meng and Wolf, 1997).

Fusion of the sperm with the oocyte is followed by decondensation of the sperm head during which some of the proteins, used to tightly "package" its chromosomes, are replaced by oocyte-derived histones. This is followed by formation of a pronuclear membrane and the male pronucleus. Following entry of the sperm tail into the oocyte, sperm-derived mitochondria, which have been marked by ubiquitin tags during spermatogenesis, are selectively degraded by oocyte-derived factors, a phenomenon that has been observed in several species including rhesus monkeys (Sutovsky *et al.,* 1999).

How the Ca^{2+} oscillations in the oocytes are "translated" into the series of modifications that are observed during activation remains largely unknown. It is interesting to note, in this context, that the number and amplitude of sperm-induced Ca^{2+} oscillations appear to be able to regulate differentially the initiation, completion, and, possibly, the temporal order of most, if not all, of the events that occur during activation (Lawrence *et al.,* 1998; Nixon *et al.,* 2002; Ducibella *et al.,* 2002). One mediator of the Ca^{2+} signal appears to be Ca^{2+}/calmodulin-dependent kinase II (CaMKII). It has been shown to undergo transient increases in response to Ca^{2+} oscillation while, conversely, CaMKII inhibition prevents oocyte activation (Markoulaki *et al.,* 2003). CaMKII appears to be involved in the regulation of a variety of targets including cyclin degradation, secretion-promoting proteins and control of oocyte calcium channels.

One important consequence of oocyte activation is the inactivation of MPF through degradation of its cyclin B subunit. Similarly, the activity of mitogen-activated protein kinase (MAP), which is involved in regulating the cell cycle and is responsible for maintenance of chromatin in its condensed state, is inactivated (Moos *et al.,* 1995). Both steps are necessary for the oocyte to resume meiosis.

During oocyte activation the oocyte enters anaphase II, during which the second polar body is extruded, rendering the oocyte haploid. A pronuclear membrane surrounds the chromosomes, leading to the formation of the female pronucleus. Replication of male and female-derived DNA occurs synchronously, the pronuclear membranes disappear and the zygote initiates the first mitotic cell division.

Pronuclear migration and proper chromosomal segregation depend on the presence and function of a centriole. In most mammals, with the exception of mice and possibly other rodents, centrioles appear to be exclusively contributed by the sperm, while oocytes contain only a rudimentary similar structure (Sathananthan *et al.,* 1991; Palermo *et al.,* 1994). The centriole initially forms the sperm aster which is necessary to guide the female pronucleus toward its male counterpart. It is replicated during the pronuclear stage and, subsequently, organizes the bipolar assembly of the mitotic spindle. In humans, defects in centrioles have been linked to reduced sperm motility and developmental arrest of embryos (Williamson *et al.,* 1984; Ryder *et al.,* 1990; Asch *et al.,* 1995).

In vitro fertilization

In vitro fertilization in nonhuman primates has been successful in several species including squirrel monkeys (Gould *et al.,* 1973; Kuehl and Dukelow, 1975), rhesus monkeys (Bavister *et al.,* 1983, Wolf *et al.,* 1989), cynomolgus monkeys (Balmaceda *et al.,* 1984), pigtail macaques (Cranfield *et al.,* 1989), baboons (Kraemer *et al.,* 1979; Irsigler *et al.,* 1984), marmosets (Lopata *et al.,* 1988), chimpanzees (Gould, 1983), and lowland gorilla (Pope *et al.,* 1997), although rates of *in vitro* development in early experiments were often low. Development to the eight-cell stage in rhesus embryos generated by *in vitro* fertilization was first reported by Bavister *et al.* (1983), who subsequently provided the first report of the birth of a rhesus infant following transfer of cleavage stage embryos generated by *in vitro* fertilization (Bavister *et al.,* 1984). The outcomes and pregnancy rates, after transfer of rhesus embryos, are often difficult to compare between different groups working with primates because of variations in experimental procedures. In most cases, however, early stage embryos have been transferred into an oviduct under laparotomy and pregnancy rates, reported as the percentage of female recipients becoming pregnant, have been variable. Early studies exclusively utilized fresh embryos. For example, Hodgen (1983) reported four births of singletons following transfer of *in vivo*-derived rhesus embryos into 11 recipients, while Bavister *et al.* (1983) obtained one live birth after transfer of 22 *in vitro* generated rhesus embryos into 11 recipient females. More recently, Schramm and Paprocki (2000) reported the birth of a rhesus infant following *in vitro* maturation/fertilization of rhesus oocytes and subsequent transfer of fresh embryos into two recipients.

Subsequently, several reports have demonstrated the viability of embryos following cryopreservation. Transfer of frozen-thawed early stage embryos into four recipients resulted in the birth of one and three sets of twins (of which one was stillborn), respectively (Wolf *et al.,* 1989; Lanzendorf *et al.,* 1990). Similarly, following vitrification at the blastocyst stage, transfer of embryos into three recipients resulted in the birth of one set of twins (Yeoman *et al.,* 2001).

The generation of rhesus offspring has also been reported following oocytes and embryo manipulations. Generally, while such manipulations reduce the viability of embryos and significantly retard their developmental competence, there is evidence that infants can be obtained following a variety of manipulative techniques. For example, Meng *et al.* (1997) reported the birth of two infants following nuclear transfer of embryonic blastomere nuclei into enucleated oocytes and embryo transfers into nine recipients. Similarly, offspring have been generated following embryo dissociation in an attempt to generate identical animals (1 offspring/13 recipients, Chan *et al.,* 2000a), by intracytoplasmic injection of sperm carrying exogenous DNA (1 offspring/7 transfers, Chan *et al.,* 2000b), and retroviral gene transfer (3 offspring/20 transfers, one being transgenic, Chan *et al.,* 2001; 2 offspring/2 transfers, Wolfgang *et al.,* 2001). While these results demonstrate that such manipulations result in reduction in pregnancy rates, they, nevertheless, demonstrate the feasibility of manipulating preimplantation rhesus embryos and obtaining live infants from the transfer of such embryos.

Attempts have also been made to transfer IVF-derived rhesus embryos into pigtail macaques to avoid the limitations imposed by seasonal breeding patterns of rhesus monkeys. These efforts have resulted in at least one birth proving the feasibility of such heterospecific pregnancies (Kubisch and Harrison, unpublished).

Senescence

No definitive longitudinal studies on ageing in male nonhuman primates have been conducted. The decrease in specific binding of testosterone to plasma protein has been correlated to ageing in chimpanzees (McCormack, 1971). Studies in rhesus males, however, found that the decrease in sexual performance by ageing males was not related to either free or bound serum testosterone levels (Phoenix and Chambers, 1982).

Correspondence

Any correspondence should be directed to Richard Harrison, Tulane National Primate Research Center, Covington, Louisiana, USA. Email: harrison@tulane.edu

References

Adams, L.A., Clifton, K., Bremmer, W.J. and Steiner, R.A. (1988). *Biol. Reprod.* 38, 156–162.

Alexander, N.J. (1972). *Am. J. Anat.* 135, 119–134.

Amann, R.P., Johnson, L., Thompson, D.L. Jr. and Pickett, B.W. (1976). *Biol. Reprod.* 15, 586–592.

Aoki, F., Sakai, S. and Kohmoto, K. (1999). *Mol. Reprod. Dev.* 53, 77–83.

Arslan, M., Mahmood, S., Khurshid, S., Naqvi, S.M.S., Afzal, M.A. and Baig, S.M. (1986). *J. Med. Primatol.* 15, 351–359.

Asch, R., Simerly, C., Ord, T., Ord, V.A. and Schatten, G. (1995). *Hum. Reprod.* 10, 1897–1906.

Aslam, H., Rosiepen, G., Krishnamurty, H., Arslan, M., Clemen, G. and Austin, C.R. (1951). *Austr. J. Sci. Res. Series* B4, 581–589.

Austin, C.R. (1951). *Austr. J. Sci. Res. Series* B4, 581–589.

Balmaceda, J.P., Pool, T.B., Arana, J.B., Heitman, T.S. and Asch, R.H. (1984). *Fertil. Steril.* 42, 791–795.

Barker-Gibbs, M.L., Sahu, A., Pohl, C.R. and Plant, T.M. (2002). *J. Clin. Endocrinol. Metab.* 87, 4976–4983.

Bavister, B.D., Boatman, D.E., Leibfried, L., Loose, M. and Vernon, M.W. (1983). *Biol. Reprod.* 28, 983–999.

Bavister, B.D., Boatman, D.E., Collins, K., Dierschke, D.J. and Eisele, S.G. (1984). *Proc. Natl. Acad. Sci. (USA)* 81, 2218–2222.

Bernstein, I.S., Gordon, T.P., Rose, R.M. and Peterson, M.S. (1978). *Behav. Biol.* 24, 400–404.

Bernstein, I.S., Weed, J.L., Judge, P.G. and Ruehlmann, T.E. (1989). *Am. J. Primatol.* 18, 251–257.

Berman, M.I. and Sairam, M.R. (1984). *J. Reprod. Fertil.* 70, 463–471

Bielert, C. and Vandenberg, J.G. (1981). *J. Reprod. Fertil.* 62, 229–233.

Bleil, J.D. and Wassarman, P.M. (1981). *Dev. Biol.* 86, 189–197.

Bleil, J.D., Greve, J.M. and Wassarman, P.M. (1988). *Dev. Biol.* 128, 376–385.

Boatman, D.E. and Bavister, B.D. (1984). *J. Reprod. Fertil.* 71, 357–366.

Boatman D.E., Robbins, R.S. (1991). *Biol. Reprod.* 44, 806–813.

Brady, A.G., Koritnik, D.R. and Stevens, S.W. (1985). *J. Med. Primatol.* 14, 293–304.

Bronson, R.A. and Fusi, F.M. (1996). *Molec. Hum. Reprod.* 2, 153–168.

Chan, A.W., Dominko, T., Luetjens, C.M., Neuber, E., Martinovich, C., Hewitson. L, Simerly, C.R. and Schatten, G.P. (2000a). *Science* 287, 317–319.

Chan, A.W., Luetjens, C.M., Dominko, T., Ramalho-Santos, J., Simerly, C.R., Hewitson, L. and Schatten, G. (2000b). *Mol. Hum. Reprod.* 6, 26–33.

Chan, A.W., Chong, K.Y., Martinovich, C., Simerly, C. and Schatten, G. (2001). *Science* 291, 309–312.

Chang, M.C. (1951). *Nature* 168, 697.

Cherr, G.N., Meyers, S.A., Yudin, A.I., VandeVoort, C.A., Myles, D.G., Primakoff, P. and Overstreet, J.W. (1996). *Dev. Biol.* 175, 142–153.

Clermont, Y. (1969). *Am. J. Anat.* 126, 57–72.

Click, B.B. (1979). *Folia Primatol.* 32, 268–289.

Coe, C.L. and Rosenblum, L.A. (1978). *Folia Primatol.* 29, 19–42.

Cohen-Dayag, A., Tur-Kaspa, Y., Dor, J., Mashiach, S. and Eisenbach, M. (1995). *Proc. Natl. Acad. Sci. (USA)* 92, 11039–11043.

Colledge, W.H., Carlton, M.B., Udy, G.B., Evans, M.J. (1994). *Nature* 370, 65–68.

Cosson, M.P. (1990). In Gagnon, C. (ed.) *Controls of Sperm Motility: Biological and Clinical Aspects,* pp 103–135. CRC Press, Boca Raton.

Cranfield, M.R., Schaffer, N., Bavister, B.D., Berger, N., Boatman, D.E., Kempske, S., Miner, N., Panos, M., Adams, J. and Morgan, P.M. (1989). *Zoo Biol. Suppl.* 1, 33–46.

Dang, D.C. and Meusy-Dessolle, N. (1984). *Arch. Androl. (Suppl.)* 12, 43–51.

Davies, T.F., Walsh, P.C., Hodgen, G.D., Dufau, M.L. and Catt, K.J. (1979a). *J. Clin. Endrocrinol. Metab.* 13, 1–10.

Davies, T.F., Walsh, P.C., Hodgen, G.D., Dufau, M.L. and Catt, K.J. (1979b). *J. Clin. Endocrinol. Metab.* 48, 680–685.

de B. Vaz Guimaraes, M.A., Alvarenga de Pliveira, C. and Campanarut Barnade, R. (2003). *Folia Primatol.* 74, 54–56.

de Rooij, D.G., van Alphen, M.M.A. and van de Kant, H.J.G. (1986). *Biol. Reprod.* 35, 587–591.

de Rooij, D.G. and Russell, L.D. (2000). *J. Androl.* 21, 776–798.

Dierschke, D.J., Walsh, S.W., Mapletoft, J., Robinson, J.A. and Ginther, O.J. (1975). *Proc. Soc. Exp. Biol. Med.* 148, 236–242.

Dixson, A.F., Martin, R.D., Bonney, R.C. and Fleming, D. (1980). In Anand Kumar (ed.) *Non-Human Primate Models for Study of Human Reproduction,* pp 61-68. Karger Verlag, Basel.

Drickamer, L.C. (1974). *Folia Primatol.* 21, 61–80.

Ducibella, T., Huneau, D., Angelichio, E., Xu, Z., Schultz, R., Kopf, G., Fissore, R., Madoux, S. and Ozil, J.-P. (2002). *Dev. Biol.* 250, 280–291.

DuMond, F.V. and Hutchinson, T.C. (1967). *Science* 158, 1067–1070.

Eberspaecher, U., Becker, A., Bringmann, P., van der Merwe, L. and Donner, P. (2001). *Cell Tissue Res.* 303, 277–287.

Ferin, M., Warren, M., Dyrenfurth, I., Vande Wiele, R.L. and White, W.F. (1974). *J. Clin. Endrocrinol. Metab.* 38, 231–237.

Florman, H.M. and Wassarman, P.M. (1985). *Cell* 41, 313–324.

Florman, H.M., Arnoult, C., Kazam, I.G., Li, C. and O'Toole, C.M.B. (1998). *Biol. Reprod.* 59, 12–16.

Gahlay, G.K., Srivastava, N., Govind, C.K. and Gupta, S.K. (2002). *J. Reprod. Immunol.* 53, 67–77.

Ginther, A.J., Carlson, A.A., Ziegler, T.E. and Snowdon, C.T. (2002). *Biol. Reprod.* 66, 282–290.

Gordon, T.P., Rose, R.M. and Bernstein, I.S. (1976). *Horm. Behav.* 7, 229–243.

Gould, K.G. (1983). *Fertil. Steril.* 40, 378–383.

Gould, K.G., Cline, E.M. and Williams, W.L. (1973). *Fertil. Steril.* 24, 260–268.

Graham, C.E. and Hodgen, G.D. (1979). *J. Med. Primatol.* 8, 265–272.

Harris, J.D. and Piersen, C.E. (2003). *Mol. Reprod. Dev.* 65, 237–244.

Hartmann, J.F., Gwatkin, R.B. and Hutchison, C.F. (1972). *Proc. Natl. Acad. Sci. (USA)* 69, 2767–2769.

Heller, C.G. and Clermont, Y. (1964). *Res. Prog. Horm. Res.* 20, 545–575.

Ho, H.-C. and Suarez, S.S. (2001). *Biol. Reprod.* 65, 1606–1615.

Ho, H.-C., Granish, K.A and Suarez, S.S. (2002). *Dev. Biol.* 250, 208–217.

Hodgen, G.D. (1983). *JAMA*. 250, 2167–2171.
Irsigler, U.M., Van der Merwe, J.V. and Botes, A.D. (1984). *S. Afr. Med. J.* 66, 447–450.
Jackson, M.R. and Edmunds, J.G. (1984). *Lab. Anim.* 18, 173–178.
Jansen, S., Ekhlasi-Hundrieser, M. and Topfer-Petersen, E. (2001). *Cells Tissue Organs* 168, 82–92.
Johnston, D.S., Wright, W.W., Shaper, J.H., Hokke, C.H., Van den Eijnden, D.H. and Joziasse, D.H. (1998). *J. Biol. Chem.* 273, 1888–1895.
Kaack, B., Walker, M. and Walker, L. (1980). *Arch. Androl.* 4, 133–136.
Kawai, M., Azuma, S. and Yoshiba, K. (1967). *Primates* 8, 35–74.
Kappeler, P.M. (1987). *Am. J. Primatol.* 12, 487–503.
Keeping, H.S., Winters, S.J., Attardi, B. and Troen, P. (1990). *Endocrinology* 126, 2858–2867.
Kelnar, C.J., McKinnell, C., Walker, M., Morris, K.D., Wallace, W.H., Saunders, P.T., Fraser, H.M. and Sharpe, R.M. (2002). *Hum. Reprod.* 17, 1367–1378.
Klein, I.L. (1971). *Folia Primatol.* 15, 233–243.
Kluin, P.M., Kramer, M.F. and de Rooij, D.G. (1983). *Int. J. Androl.* 6, 25–43.
Koyama, T. (1976). *Folia Endocrinol. Jap.* 52, 881–897.
Kraemer, D.C., Flow, B.L., Schriver, M.D., Kinney, G.M. and Pennycook, J.E. (1979). *Theriogenology* 11, 51–56.
Kriewaldt, F.H. and Hendrickx. (1968). *Lab. Anim. Care* 18, 361–370.
Kuehl, T.J. and Dukelow, W.R. (1975). *J. Med. Primatol.* 4, 23–31.
Lanzendorf, S.E., Zelinski-Wooten, M.B., Stouffer, R.L. and Wolf, D.P. (1990). *Biol. Reprod.* 42, 703–711.
Lawrence, Y., Ozil, J.-P. and Swann, K. (1998). *Biochem. J.* 335, 335–342.
Lin, Y., Kimmel, L.H., Myles, D.G. and Primakoff, P. (1993). *Proc. Natl. Acad. Sci. (USA)* 90, 10071–10075.
Lin, Y., Mahan, K., Lathrop, W.F., Myles, D.G. and Primakoff, P. (1994). *J. Cell. Biol.* 125, 1157–1163.
Lopata, A., Summers, P.M. and Hearn, J.P. (1988). *Fertil. Steril.* 50, 503–509.
Lu, Q. and Shur, B.D. (1997). *Development* 124, 4121–4131.
Lue, Y.H., Laughli, L.S., Swerfloff, S., Hikim, A.P., Overstreet, J.W. and Wang, C. (2002). *J. Androl.* 23, 799–805.
Mahone, J.P. and Dukelow, W.R. (1979). *J. Med. Primatol.* 8, 179–183.
Mahony, M.C. and Gwathmey, T. (1999). *Biol. Reprod.* 60, 1239–1243.
Mahony, M.C., Fan, Q., Cartwright, S. and Hodgen, G.D. (1994). *Am. J. Reprod. Immunol.* 312, 133–140.
Majumdar, S.S., Winters, S.J., Plant, T.M. (1997). *Endocrinology* 133, 1363–1373.
Makler, A., Stoller, J., Reichler, A., Blumenfeld, Z. and Yoffe, N. (1995). *Fertil. Steril.* 63, 1077–1082.
Mann, D.R. and Plant, T.M. (2002). *Semin. Reprod. Med.* 20, 93–102.
Markoulaki, S., Matson, S., Abbotta, A.L. and Ducibella, T. (2003). *Dev. Biol.* 258, 464–474.
Marshall, G.R. and Plant, T.M. (1996). *Biol. Reprod.* 54, 1192–1199.
Marson, J., Fraser, H.M., Wickings, E.J., Cooper, R.W., Jouannet, P. and Meuris, S. (1993). *Biol. Reprod.* 48, 490–494.
Matsubayashi, K. (1974). *Lab. Anim.* 8, 253–255.
McCormack, S.A. (1971). *Endocrinology* 89, 1171–1177.
Mendoza, S.P., Lowe, E.L., Resko, J.A. and Levine, S. (1978). *Physiol. Behav.* 20, 515-522.
Meng, L. and Wolf, D.P. (1997). *Biol. Reprod.* 12, 1062–1068.
Meng, L., Ely, J.J., Stouffer, R.L. and Wolf, D.P. (1997). *Biol. Reprod.* 57, 454–459.
Miller, D.J., Macek, M.B. and Shur, B.D. (1992). *Nature* 357, 589–593.
Moos, J., Visconti, P.E., Moore, G.D., Schultz, R.M. and Kopf, G.S. (1995). *Biol. Reprod.* 53, 692–699.
Millar, M.R., Sharpe, R.M., Weinbauer, G.F., Frase, H.M. and Saunder, P.T. (2000). *Int. J. Androl.* 23, 266–277.
Moreland, R.B., Richardson M.E., Lamberski, N. and Long, J.L. (2001). *J. Androl.* 22, 395–403.
Mori, J. and Hafez, E.S.E. (1973). *J. Reprod. Fertil.* 34, 155–157.
Moudgal, N.R. and Sairam, M.R. (1998). *Hum. Reprod.* 13, 916–919.
Murty, G.S.R.C., Ramasharma, K., Mukku, V.R., Srinath, B.R. and Moudgal, N.R. (1980). In Anand Kumar, T.C. (ed.) *Non-Human Primate Models for Study of Human Reproduction*, pp 50-54. Karger Verlag, Basel.
Nieschlag, E., Simoni, M., Gromoll, J. and Weinbauer, G.F. (1999). *Clin. Endocrinol.* 51, 139–146.
Nixon, V., Levasseur, M., McDougall, A. and Jones, K. (2002). *Curr. Biol.* 12, 746–750.
Omar, A. and De Vos, A. (1971). *Folia Primatol.* 16, 206–215.
Orth, J.M., Gunsalalus, G.L. and Lamperti, A.A. (1988). *Endocrinology* 122, 787–794.
Pal, S.K., Torry, D., Serta, R., Crowell, R.C., Seibel, M.M., Cooper, G.M. and Kiessling, A.A. (1994). *Fertil. Steril.* 61, 496–503.
Palermo, G., Munne, S. and Cohen, J. (1994). *Hum. Reprod.* 9, 1220–1225.
Parmentier, M., Libert, F., Schurmans, S., Schiffmann, S., Lefort, A., Eggerickx, D., Ledent, C., Mollereau, C., Gerard, C. and Perret, J. (1992). *Nature* 355, 453–455.
Paterson, M., Wilson, M.R., Morris, K.D., van Duin, M. and Aitken, R.J. (1998). *Am. J. Reprod. Immunol.* 40, 198–209.
Phoenix, C.H. and Chambers, K.C. (1982). *Int. J. Primatol.* 3, 221–229.
Phillippi-Falkenstein, K. and Harrison, R.M. (2003). *Am. J. Primatol.* 60, 23–28.

Plant, T.M. (1994). In Burger, H.G., deKretser, D.M., Findlay, J., Petraglia, F. and Robertson, D.M. (eds) *Frontiers in Endocrinology,* Vol. 3, pp 135-143. Ares-Serono Symposia, Rome.

Plant, T.M., Gay, V.L., Marsall, G.R. and Arslan, M. (1989). *Proc. Natl. Acad. Sci. USA* 86, 2506–2510.

Plant, T.M. and Marshall, G.R. (2001). *Endocr. Rev.* 22, 764–786.

Pope, C.E., Dresser, B.L., Chin, N.W., Lui, J.H., Loskutoff, N.M., Behnke, E.J., Brown, C., McRae, M.A., Sinoway, C.E., Campbell, M.K., Cameron, K.N., Owens, O.M., Johnson, C.A., Evans, R.R. and Cedars, M.I. (1997). *Am. J. Primatol.* 41, 247–260.

Ralt, D., Goldenberg, M., Fetterolf, F.P., Thompson, D., Dor, J., Mashiach, S., Garbers, D.L. and Eisenbach, M. (1991). *Proc. Natl. Acad. Sci. (USA)* 88, 2840–2844.

Ramakrishna, P.A. and Prasad, M.R.N. (1967). *Folia Primatol.* 5, 176–189.

Ramaswamy, S. and Plant, T.M. (2001). *Mol. Cell. Endocrinol.* 180, 93–101.

Ramaswamy, S., Marshall, G.R., Pohl, C.R., Friedman, R.L. and Plant, T.M. (2003). *Endocrinology* 144, 1175–1185.

Reimann, J.D. and Jackson, P.K. (2002). *Nature* 416, 850–854.

Resko, J.A. (1970). *Endocrinology* 87, 680–687.

Resko, J.A. and Ellinwood, W.E. (1985). *Biol. Reprod.* 33, 346–352.

Resko, J.A., Malley, A., Begley, D. and Hess, D.L. (1973). *J. Reprod. Fertil.* 48, 181–183.

Roberts, J.A. (1972). In Fiennes, R.N.T-W. (ed.) *Pathology of Simian Primates, Part I,* pp 878-888. Karger, Basel.

Roberts, M.S. (1978). *Folia Primatol.* 29, 229–235.

Roselli, C.E., West, N.B. and Brenner, R.M. (1991). *Biol. Reprod.* 44, 739–745.

Rosiepen, G., Arslan, M., Clemen, G., Nieschlag, E. and Weibauer, G.F. (1997). *Cell Tissue Res.* 288, 365–369.

Ryder, T.A., Mobberley, M.A. and Hughes, L. (1990). *Fertil. Steril.* 53, 556–560.

Saling, P.M (1991). *Biol. Reprod.* 44, 246–251.

Sathananthan, A.H., Kola, I., Osborne, J., Trounson, A., Ng, S.C., Bongso, A. and Ratnam, S.S. (1991). *Proc. Natl. Acad. Sci. (USA)* 88, 4806–4810.

Schiml, P.A., Mendoza, S.P., Saltzman, W., Lyons, D.M. and Mason, W.A. (1999). *Am. J. Primatol.* 47, 93–103.

Schramm, R.D. and Paprocki, A.M. (2000). *Hum. Reprod.* 15, 2411–2414.

Schulze, C. (1979). *Cell Tissue Res.* 198, 191–199.

Schwartz, R., Brook, W. and Zinsser, H.H. (1957). *Fertil. Steril.* 9, 300-308.

Schoones, R., DeKlerk, J.N. and Murphy, G.P. (1969). *Invest. Urol.* 6, 476–484.

Sharpe, R.M. (1994). In Knobil, E. and Neill, J.D. (eds) *The Physiology of Reproduction,* pp 1363–1434. Raven Press Ltd, New York.

Slob, A.K., Ooms, M.P. and Vreeburg, J.T.M. (1979). *Biol. Reprod.* 20, 981–984.

Smithwick, E.B. and Young, L.G. (1996a). *Tissue Cell* 28, 137–148.

Smithwick, E.B. and Young, L.G. (1996b). *Tissue Cell* 28, 357–366.

Spehr, M., Gisslemann, G., Poplawski, A., Riffell, J., Wetzel, C.H., Zimmer, R.K. and Hatt, H. (2003). *Science* 299, 2054–2057.

Srivasta, P.N., Farooqui, A.S. and Gould, K.G. (1981). *Biol. Reprod.* 25, 363–369.

Stambaugh, R. and Buckley, J. (1970). *Biol. Reprod.* 3, 275–282.

Stambaugh, R., Seitz, H.M. and Mastroianni, L. (1974). *Fertil. Steril.* 25, 352–357.

Steinberger, E. (1971). *Physiological Rev.* 51, 1–22.

Sutovsky, P., Moreno, R.D., Ramalho-Santos, J., Dominko, T., Simerly, C. and Schatten, G. (1999). *Nature* 402, 371–372.

Tesarik, J., Drahorad, J., Testart, J. and Mendoza, C. (1990). *Development* 110, 391–400.

Toivola, P.T.K., Bridson, W.E. and Robinson, J.A. (1978). *Endocrinology* 102, 1815–1821.

Tollner, T.L., Yudin, A.I., Cherr, G.N. and Overstreet, J.W. (2003). *Biol. Reprod.* 68, 664–672.

Tulsiani, T.R.P, Nagdas, S.K., Cornwall, G.A. and Orgebin-Crist, M.C. (1992). *Biol. Reprod.* 46, 93–100.

van Alphen, M.M.A., van de Kant and de Rooij, D.G. (1988). *Endocrinology* 123, 1449–1455.

Vanderhaeghen, P., Schurmans, S., Vassart, G. and Parmentier, M. (1993). *J. Cell Biol.* 123, 1441–1452.

Van der Horst, G., Seier, J.V., Spinks, A.C. and Hendrickx, S. (1999). *Int. J. Androl.* 22, 197–207.

VandeVoort, C.A., Tollner, T.L. and Overstreet, J.W. (1992). *J. Androl.* 13, 428–432.

VandeVoort, C.A., Tollner, T.L. and Overstreet, J.W. (1994). *Mol. Reprod. Dev.* 37, 299–304.

Vliegen, M.K., Schlatt, S., Weinbauer, G.F., Bergmann, M., Groome, N.P. and Nieschlag, E. (1993). *Cell Tissue Res.* 273, 261–268.

Weinbauer, G.G., Aslam, H., Krishnamurthy, H., Brinkworth, M.H., Einspanier, A. and Hodges, J.K. (2001). *Biol. Reprod.* 64, 120–126.

Wickings, E.J. and Nieschlag, E. (1980). *Inter. J. Androl.* 3, 87–104.

Wickings, E.J., Marshall, G.R. and Nieschlag, E. (1986). In Dukelow, W.R. and Erwin, J. (eds) *Comparative Primate Biology. Vol. 3. Reproduction and Development,* pp 149-170. Alan R. Liss, Inc., New York.

Williamson, R.A., Koehler, K.K., Smith, W.D. and Stenchever, M.A. (1984). *Fertil. Steril.* 41, 103–107.

Winters, J. and Plant, T.M. (1999). *Endocrinology* 140, 5497–5504.

Wolf, D.P., Vandevoort, C.A., Meyer-Haas, G.R., Zelinski-Wooten, M.B., Hess, D.L., Baughman, W.L. and Stouffer, R.L. (1989). *Biol. Reprod.* 41, 335–346.

Wolfgang, M.J., Eisele, S.G., Browne, M.A., Schotzko, M.L., Garthwaite, M.A., Durning, M., Ramezani, A.,

Hawley, R.G., Thomson, J.A. and Golos, T.G. (2001). *Proc. Natl. Acad. Sci. (USA)* 98, 10728–10732.
Yanagimachi R. (1969). *J. Reprod. Fertil.* 18, 275–286.
Yeoman, R.R., Gerami-Naini, B., Mitalipov, S., Nusser, K.D., Widmann-Browning, A.A. and Wolf, D.P. (2001). *Hum. Reprod.* 16, 1965–1969.
Youakim, A., Hathaway, H.J., Miller, D.J., Gong, X. and Shur, B.D. (1994). *J. Cell Biol.* 126, 1573–1583.
Yudin, A.I., Tollner, T.L., Li, M.-W., Treece, C.A., Overstreet, J.W. and Cherr, G.N. (2003). *Biol. Reprod.* (electronic publication ahead of print).
Zamboni, L., Conaway, C.H. and Van Pelt, L. (1974). *Biol. Reprod.* 11, 251–267.
Zhang, T., Zhou, H.M. and Liu, Y.X. (1997). *Mol. Hum. Reprod.* 3, 223–231.
Zhengwei, Y., McLachlan, R.I., Bremmer, W.J. and Wreford, N.G. (1997). *J. Andrrol.* 18, 681–687.

CHAPTER 9

Primate Natural History and Social Behavior: Implications for Laboratory Housing

Corrine K. Lutz[1] and Melinda A. Novak[1,2]
[1]New England Primate Research Center, Harvard Medical School, Southborough, MA, USA
[2]Department of Psychology, University of Massachusetts, Amherst, MA, USA

Introduction

Creating appropriate environments for housing nonhuman primates is a challenge in the laboratory. Recent regulations call for the consideration of an animal's psychological well-being in addition to their physical health when housing nonhuman primates (Shalev, 1999). However, what well-being entails is still under debate. Novak and Suomi (1988) described four interrelated measures of well-being in captive nonhuman primates. In addition to assessing an animal's physical health and the range of species-typical behaviors displayed, physiological reactions to stress and the ability of the animal to cope with environmental challenges were also considered important. Since 1988, other criteria have been proposed to assess psychological well-being of captive primates including measures of reproductive success and the quality of parental care (Snowdon, 1991).

Some people argue that captive primates should exhibit the full range of behaviors displayed by their wild counterparts. However, captive environments do not replicate what a primate might experience in the wild, nor are all species-typical behaviors desirable in captivity. For example, affiliative and exploratory behavior may be more desirable than aggressive behavior in the laboratory environment. Therefore, the goal of housing primates in captivity should be to promote the display of species-typical behaviors suitable for the given conditions, while eliminating abnormal or problem behaviors.

Two different approaches for housing primates in captivity may attain this same goal. One approach is of homology, creating an environment that duplicates the experiences in the wild, and the second is of analogy, adapting the important experiences in the wild to the laboratory environment. The aim of the homologous approach is to create in the laboratory an environment that simulates nature and is therefore "naturalistic."

The Laboratory Primate
ISBN 0-1208-0261-9

Examples of this may include allowing the animals to live in species-appropriate groups in outdoor environments, or in naturalistic indoor environments. Although this approach is common in many zoos, it is not practical in most laboratory settings due to space, cost, and experimental constraints. Therefore, the analogous approach may be more appropriate. This approach focuses on the behavioral outcome rather than the environment, itself. The attempt here is to create laboratory conditions that allow an animal to exhibit species-typical behaviors and fulfill its behavioral "needs." This approach may be more feasible in the laboratory situation, because the same behavioral outcome may be accomplished by a number of different means. For example, foraging may be recreated with a variety of devices that require that the animal expend some effort to obtain food. Regardless of the approach taken, the ultimate objective of environmental enrichment is to increase appropriate species-typical behaviors while decreasing levels of abnormal behavior in laboratory primates.

There is a general consensus that social companionship is an inherent aspect of psychological well-being in nonhuman primates (Novak and Suomi, 1991; Reinhardt and Reinhardt, 2000). However, historically, individual housing has been a standard in many laboratories because of concerns ranging from disease transmission, to injury from fights, to easy access for treatment or sampling. Therefore, the purpose of this chapter is to examine the effects of social and asocial housing conditions in the laboratory on both species-typical and abnormal behaviors in laboratory primates. Rhesus macaques (*Macaca mulatta*), a species commonly used in the laboratory, will be used as a model to demonstrate how knowledge of an animal's natural history and behavior in the wild can be translated into appropriate laboratory housing. We will first describe the natural history of rhesus monkeys and then will assess how rearing and housing conditions in the laboratory can affect the behavior and psychological well-being of laboratory primates.

Rhesus macaque natural history

Rhesus macaques (*Macaca mulatta*) are social animals, typically living in large multimale troops ranging in size between 8 and 85 individuals (Teas *et al.*, 1980; Melnick *et al.*, 1984; Lindburg, 1971). Despite the wide range, the average group size is around 20–30 animals (Teas *et al.*, 1980; Southwick *et al.*, 1965; Lindburg, 1971). The troop is essentially a closed society, and troop members react very aggressively to unfamiliar monkeys (Southwick *et al.*, 1965). The troops are not homogeneous. Instead, they can be divided into matrilineal subgroups with dominance hierarchies (Berman, 1983). At any given time, there may be smaller, more fluid groups such as infant play groups or sexual consorts for breeding (Southwick *et al.*, 1965).

Rhesus monkeys have the widest range of habitats of all species of macaques. Their range extends from Afghanistan eastward to South and Southeast Asia (Teas *et al.*, 1980) and their habitats vary from mangrove swamps to mountain forests and from crowded cities to remote forest areas (Teas *et al.*, 1980; Southwick *et al.*, 1965). Although rhesus monkeys live in many different habitats, they are usually found in areas with disturbed vegetation. In fact, they appear to spend more time in disturbed areas than would be expected. These disturbed sites tend to be associated with human activity (Goldstein and Richard, 1989; Richard *et al.*, 1989). Rhesus monkeys have adapted well and thrive in areas associated with human activity such as villages, towns, railway stations, temples, and along roads and canals. They can survive and reproduce well under crowded and competitive conditions, situations that would normally be considered aversive. Rhesus monkeys appear to have the appropriate temperament and behavioral variability to adapt to close association with humans in these disturbed areas (Teas *et al.*, 1980).

Under free-ranging conditions, rhesus monkeys spend nearly 50% of their time moving to food sites and foraging for food (Teas *et al.*, 1980; Goldstein and Richard, 1989). Rhesus monkeys have an extremely varied diet and have been observed to consume over 100 species of plants (Lindburg, 1971), selectively feeding on shoots, fruits, and seeds. Rhesus monkeys also raid agricultural crops and occasionally eat eggs, insects, and small animals. This widely varied diet may contribute to their ability to flourish in very different environments.

Rhesus monkeys are also highly social organisms and one important activity that is thought to facilitate social relationships is grooming (Teas *et al.*, 1980). Monkeys spend a significant amount of time grooming one another and adult females groom more than any other age/sex class (Lindburg, 1971; Teas *et al.*, 1980). Long bouts of reciprocal grooming are also observed in male/female consort pairs (Lindburg, 1971). This disparity in grooming relationships based on age, sex, and circumstance suggests that social grooming is an important mechanism for both creating and maintaining social relationships.

As with most monkey species, rhesus monkeys have a long period of infant dependency and an extended mother–infant relationship. Although the time spent interacting with the mother declines during the first year of life, approximately half of the infants still nurse at one year of age and spend about 15% of their time in contact with their mother. However, this close maternal contact is greatly reduced after the birth of a sibling (DeVinney *et al.,* 2001). Weaning occurs shortly before or at the birth of the next infant, which typically occurs around one year of age (Lindburg, 1971; DeVinney *et al.,* 2001). After the first year, female infants continue to maintain close ties with their mother, whereas males join playgroups (Lindburg, 1971). Female rhesus monkeys typically remain in their natal group and associate closely with their mothers and female relatives (matriline) throughout their lives, whereas most adolescent males leave their natal group before reaching sexual maturity. Emigrating males lead a solitary life or join other peripheralized males until they successfully enter a new troop. Emigration is a time of significant danger for young adolescent males. Because rhesus society is closed to outsiders, strangers are faced with aggression and possible death (Southwick *et al.,* 1974). In fact, it is estimated that 40–50% of adolescent males do not survive this transfer (Berard, 1989; Dittus, 1979). Despite the considerable risks, some males transfer to new troops more than once during adult life (Koford, 1965; Melnick *et al.,* 1984).

The troop provides the structure and the rules by which monkeys move, forage, and interact with each other and with outsiders. However, the troop is also made up of individuals, and extensive evidence has shown that individual members of a troop differ widely in their reactions to environmental events. Although the underlying basis for these differences is not fully understood, much of the research has focused on genetic-based temperaments (Suomi, 2000).

Converging findings from both laboratory and field suggest that monkeys differ in their emotional responsiveness. About 20% of the rhesus monkey population appears to be quite reactive to novel events. This reactivity or inhibition is manifested by heightened and prolonged activation of the hypothalamic-pituitary-adrenal (HPA) axis and by behavioral responses including fear and withdrawal. In contrast, the remaining members of the population show only mild activation of the HPA axis and only brief responses of wariness or caution in response to novel stimuli (Suomi, 1991). In nature, monkeys with a high reactive temperament show more emotional responses to maternal disruption (i.e., during consort pairing when mothers frequently rebuff contact from their infants; Berman *et al.,* 1994). Emotionality is also linked to the timing of emigration with high reactive adolescent males leaving their natal troop at a significantly later age than less reactive individuals (Suomi *et al.,* 1992).

Individual rhesus monkeys also vary with respect to their level of social competence and risk taking behavior. Some monkeys (~5% of the population) appear to take risks that can result in injury or death. They engage in severe and inappropriately targeted aggression and do not moderate their aggressiveness with appeasement or affiliative behavior. A hallmark of this temperament is the presence of low levels of serotonin in brain as measured by the metabolite, 5-HIAA, 5-hydroxyindoleacetic acid. In nature, low levels of 5-HIAA in male rhesus monkeys are associated with extreme aggression, earlier emigration from their natal troop compared to other adolescent males (Mehlman *et al.,* 1995) and increased risk of mortality (Higley *et al.,* 1996). Females with low CNS serotonin are more likely to be the targets of aggression and have an increased risk of illness and infant death (Westergaard *et al.,* 2003).

Two general conclusions can be drawn from our descriptions of the life history of rhesus monkeys. First, free-ranging monkeys live in extraordinarily complex social environments. Thus, social housing would appear to be crucial for maintaining rhesus monkeys in captivity. However, there are important nuances to this view, an understanding of which is necessary for considering how social housing is to be achieved in captivity. It is likely that the social needs of rhesus monkeys vary by sex. Females live in large kin groups throughout their lives whereas males emigrate and occasionally become solitary. Rhesus monkeys are not necessarily friendly to other monkeys. In nature, rhesus monkey troops are like "closed societies" and troop members react very aggressively to strangers. Finally, some male monkeys with low CNS serotonin levels show extreme aggression and are ultimately forced out of their natal troop. These factors need to be taken into account when forming and monitoring social groups in captivity.

In addition to a complex social life, rhesus monkeys also live in a demanding physical environment where they must actively seek food and find suitable resting sites. Both movement and exploration are important behavioral responses that relate to survival and reproductive success in nature. Thus, exposure to novel stimuli (i.e., environmental enrichment) would appear to be essential for housing monkeys in captivity. However, this view must be adjusted because some monkeys show heightened stress responses to novel events. Careful selection of novel stimuli and an assessment of monkey reactions may be required to minimize negative reactions.

Laboratory environment and abnormal behavior

In contrast to the social and environmental complexity in the wild, the laboratory environment is often viewed as stark, lacking in variety, and with few opportunities for social interaction. Although the laboratory environment provides for an animal's physical needs (e.g., food or veterinary care), it is often viewed as being inadequate for satisfying an animal's psychological or behavioral needs (e.g., social contact). Although not all laboratory environments are the same, one possible outcome of housing monkeys in laboratory environments is the development of abnormal behavior. Extensive evidence suggests that both the severity and frequency of abnormal behavior are correlated with the quality of the laboratory environment. (*See also Chapter14, by Victor Reinhardt.*) Of particular concern is individual cage housing because it can lead to the development of severely abnormal behavior. For example, in two surveys of individually housed rhesus macaques, 5–11% had a record of self-inflicted wounding (Bayne *et al.,* 1995; Lutz *et al.,* 2003) while 84% of females and 92.5% of males exhibited at least one type of abnormal behavior (Lutz *et al.,* 2003). The development of abnormal behavior was significantly affected by the age of first exposure to individual housing. Those individually housed as juveniles were more likely to engage in self-directed biting and self-inflicted wounding compared to those individually housed later as adolescents or adults (Lutz *et al.,* 2003). Although the laboratory environment does not replicate conditions the animals experience in the wild, some experiences can be simulated in meaningful ways for the animals, promoting a behavioral repertoire that closely parallels that of animals in the wild. At the same time, the display of abnormal behavior can be eliminated or reduced either by preventing the behavior from occurring in the first place, or by utilizing a form of intervention once the behavior occurs.

Rearing

Extensive research conducted decades ago has shown that an animal's early rearing experiences can have a significant impact on behavioral development with effects that extend into and persist through adulthood (Harlow and Harlow, 1962a, 1962b). Thus, the first year of life in rhesus monkeys is crucial for the attainment of normal social skills. Several different variations in early rearing have been studied ranging from extreme deprivation, where all social contact is eliminated, to nuclear family environments. Although extreme rearing conditions such as isolation rearing are no longer practiced today, isolation rearing will be described, because it stands at one end of a continuum of possible rearing conditions for infant monkeys.

Total isolation

Total social isolation was a rearing condition in which an infant monkey was separated from its mother at birth and housed alone in a completely enclosed cage. The cage contained a light, provisions for food, water, and cleaning, and constant white noise to mask external sounds (Sackett, 1968). The subjects were housed in this condition during infancy for durations ranging up to one year.

Isolation rearing had devastating effects on social and exploratory behavior. Upon removal from isolation, isolated infants exhibited abnormal behaviors rarely seen in animals born in the wild (Harlow and Harlow, 1962a). When placed with stimulus animals, the isolates were fearful and exhibited behaviors such as crouching, rocking, self-clasping, and self-orality, while the socially-raised control animals paced, were piloerected, and cooed (Mitchell *et al.,* 1966). Compared to controls, isolates had high levels of fear, flight, and hostility and low levels of social sex and social play (Mitchell *et al.,* 1966). Isolates also showed more immobility and disturbed behavior such as rocking, crouching, and wall-hugging (Mitchell, 1968).

Duration and timing of isolation played an important role in the animal's resulting behavior. In comparison to the six-month isolates, infants isolated for 12 months were more fearful, less playful, and showed the least amount of sexual behavior (Mitchell, 1966). However, the age at which isolation occurred was also a significant factor in the resulting behaviors. Infants isolated for six months between the ages of six and 12 months (six months late) spent more time in social interactions than infants isolated between the ages of one and six months (six months early). They also spent more time in nonsocial play and exploration and were less likely to show immobility or disturbance (Mitchell, 1968). The emerging picture for rhesus monkeys was that early isolation had profound effects, retarding the development of normal social behavior and facilitating the development of bizarre, stereotypical, highly abnormal behaviors. The length of the isolation period was positively correlated with the amount of fearful and withdrawn behavior exhibited and negatively correlated with the levels of activity and social contact. These effects persisted throughout development (Sackett, 1967).

Although isolation rearing had devastating effects on rhesus infants, the severity of these effects varied both as a function of sex and species. Female rhesus monkeys appeared to be less affected by isolation rearing than males. When later tested in early adulthood, females were more active and exhibited more exploratory behavior than males (Sackett, 1972). Isolation rearing did not impact other macaque species in quite the same way. Pigtailed macaques (*Macaca nemestrina*) reared in isolation showed significantly lower levels of abnormal behavior and higher levels of social behavior than rhesus macaques (Sackett *et al.,* 1976).

Early attempts to rehabilitate isolate-reared monkeys met with failure. When isolates were socialized by same-age or adult stimulus animals, they reacted with fearful behavior and flight (Mitchell *et al.,* 1966). However, rehabilitation with younger "therapist" monkeys proved to be more successful in socializing former isolates. Monkeys reared in isolation for six months were allowed to interact with younger (three months of age), socially normal, "therapist" monkeys for approximately eight hours per week. Upon first exposure, the reaction of the isolates was to rock and self-clasp in a corner. The younger therapist monkey would typically cling to the isolate as it would to a mother or to a surrogate. As a result, the isolates gradually accepted the social contact. This contact ultimately resulted in a decrease in self-directed behavior and an increase in exploration. Furthermore, the development of the social behavior paralleled that of the younger therapist monkeys (Suomi *et al.,* 1974). Similar results were obtained with 12-month isolates using therapists that were ten months younger than the isolate subjects. As in the previous study, self-directed behaviors decreased whereas species-specific social behaviors increased in the isolates during therapy (Novak and Harlow, 1975). However, when the "rehabilitated" isolates were tested with normal age-mates, further treatment was required before age-appropriate social behaviors emerged (Novak, 1979). These studies demonstrate the effectiveness of "younger monkey therapy" in reversing some of the consequences of isolation rearing and serve to underscore the importance of social contact in the development of a normal behavioral repertoire.

Partial isolation

Although total isolation was utilized solely for research purposes, a less severe form of isolation, partial isolation, has occasionally been used in laboratory environments. In this condition, the infant is raised alone in a wire cage with visual and auditory, but not physical access to other monkeys. This rearing procedure may be employed in situations where mothers reject infants or in research protocols requiring infants to be physically separated from conspecifics. Although the animal is only isolated from actual physical contact (i.e., monkeys can see, hear, and communicate with other monkeys in nearby cages), severe behavioral problems are associated with this rearing condition.

Early studies revealed that partial isolates exhibited many of the same behavioral abnormalities as observed in total isolates, including various forms of stereotypic behavior, self-biting and wounding (Cross and Harlow, 1965). Social behavior was also impacted. Monkeys reared in partial isolation failed to show appropriate sexual behavior. This effect was observed in all males and in almost all females. When impregnated, the females proved to be inadequate mothers (Harlow and Harlow, 1962b), although their maternal skills improved with the birth of a second infant (Ruppenthal *et al.,* 1976). In comparison to control animals raised in the wild for the first two years of life, adult partial isolates showed less locomotion and exploration and higher levels of abnormal behavior than wild-born controls. Rocking and self-biting were only observed in the partial isolates (Suomi *et al.,* 1971). Similarly, catatonic contracture (a bizarre behavioral pattern in which an animal's arm slowly floats upward as if it were not a part of the monkey's own body) was observed only in monkeys reared in isolation (Cross and Harlow, 1965). In general, the behavioral effects of partial isolation were less severe than those produced by total isolation rearing. The rank ordering of rearing condition from less severe to more severe (i.e., partial isolates vs. six-month isolates vs. one-year isolates) was reflected in the monkeys' subsequent behavior at four years of age (Sackett, 1967).

Rearing with other species

Rearing conditions in which infants are deprived of contact with conspecifics have a devastating effect on behavior. However, contact with another species can help ameliorate some of the effects of this deprivation. Mason and Kenney (1974) separated infant rhesus monkeys from their mothers, age-mates, or cloth surrogates, and housed them with dogs. A control group of monkeys was separated and housed with wooden hobbyhorses. All of the dog-reared infants formed strong attachments to their dog surrogates. In comparison to monkeys reared with inanimate hobbyhorses, dog-reared infants were more visually responsive when presented with pictures and novelty (Wood *et al.*, 1979). Dog-reared monkeys were also more active and showed a significant decline in defecation and urination during social testing in comparison to those reared with a hobbyhorse. Despite these improvements, both groups failed to develop the social skills necessary for normal social interaction (Capitanio, 1984).

Peer-only rearing

In this housing condition, infants are reared together and have continuous access to their partner or group. Monkeys reared in this manner acquire most of the species-typical behavior seen in normally reared monkeys. However, this rearing condition also yields greater emotionality in infancy as manifested by high levels of clinging behavior and low levels of play (Chamove, 1973). The number of infants in a group can play a role in the development of behavior. Infants housed in pairs spent most of their time clinging to each other. Normal infantile sexual behavior was absent, and the monkeys were sexually inadequate as adults. When group size was increased to four animals, clinging behavior was much lower. Nonetheless, regardless of group size, play behavior in peer-reared monkeys was lower than that of normally reared monkeys (Chamove *et al.*, 1973).

Surrogate-peer rearing

An alternative to pairing infants with each other is surrogate-peer rearing. In this rearing condition, the infant is housed with an inanimate surrogate, which serves as the attachment figure. The infants are also exposed to same-age peers, or playmates, for approximately 30 minutes to two hours per day. Because clinging behavior is directed to the inanimate surrogate, play behavior is directed towards their peers. In contrast to peer-only rearing, clinging behavior is low and interactions with other infants consist mostly of play and environmental exploration. During the first year of life, surrogate-peer reared infants tended to show somewhat higher levels of disturbance behavior than the mother-peer reared subjects, a difference that tends to disappear by the end of the first year of life (Hansen, 1966). Monkeys reared with surrogates and peers develop normal, species-typical behavior and behave normally as adults. As in mother-reared monkeys, females showed higher levels of social contact than males. Female surrogate-peer reared monkeys gave birth normally and showed appropriate maternal behavior. Surrogate-peer rearing during the first year of life appears to result in virtually normal behavior (Novak *et al.*, 1992).

Mother-only rearing

Although appropriate contact with peers has been shown to be important for normal social development, infants have been successfully raised with the mother in the absence of peers. Mother-only reared infants were observed to develop normally, displaying normal social behavior without any manifestation of abnormal behavior. However, juvenile mother-only reared monkeys tended to be more aggressive in their interactions with peers than monkeys reared with both the mother and peers, suggesting that early exposure to peers may be important in the development of affiliative interactions (Alexander and Harlow, 1965).

Mother-peer rearing

In most situations, the optimal environment for the development of normal social behavior in infant rhesus monkeys is one that allows the infant to remain with its mother in a species-typical social group that includes other infants (Shalev, 1999). Many different social environments have been designed to achieve these goals. One early example of social housing called the "playpen" consisted of a pen surround by four living cages. Each living cage housed a single mother and infant. Only the infants had free access to both the living unit and the playpen. In comparison to surrogate-peer-reared infants, the mother-peer-reared infants were more socially responsive and developed complex play behavior at an earlier age (Harlow and Harlow, 1962b). This arrangement was subsequently modified into a "nuclear family" arrangement in which a large playpen area was surrounded by floor to ceiling pens consisting of an adult male and female rhesus monkey with their offspring. Only infants less than two years old had access to the playpen environment (Ruppenthal *et al.*, 1974). Since the 1960s, numerous configurations of mother-peer rearing have been employed, ranging from single male harem groups (Kessler *et al.*, 1985) to multi-male, multi-female groups (Suomi *et al.*, 1996). Rearing monkeys in these kinds of environments not only fosters adequate social development but also promotes breeding success.

Housing

The previous section focused on infancy, a stage of life in which animals appear to be most vulnerable to environmental events and to social experiences. In contrast, this section is concerned with housing rhesus monkeys after the first year of life. A growing literature once again indicates that some form of social housing is probably optimal for most juvenile, adolescent, and adult monkeys. Monkeys housed in social groups exhibit complex social behavior such as grooming (Suomi *et al.,* 1996) and play (Brown and Dixson, 2000). However, this rich social environment may come at a cost of less desirable behaviors such as aggression (Line *et al.,* 1990).

Social groups

One of the most significant challenges of social housing is in group formation. Because rhesus monkeys are known to be highly reactive to strangers, adding an individual to an already-formed group can result in severe aggression and possible death to the stranger (Southwick *et al.,* 1974). Therefore, care must be taken when forming or reconstructing groups. Two different strategies have been used to form groups. One involves a gradual process of introductions whereas the other strategy consists of placing strangers together all at once. Westergaard *et al.* (1999) tested these two methods for forming groups of rhesus macaques. The gradual process involved a staged strategy in which small groups of animals were introduced gradually over a period of weeks. The rapid strategy involved introducing all of the animals on the same day. Resulting wounding rates were higher when group members were released simultaneously in comparison to those released incrementally in subgroups. However, the gradual process is not guaranteed to be successful for all animals in all situations. A group of older monkeys that had been individually housed for an average of 13.1 years was formed utilizing the gradual process (Line *et al.,* 1990). For this strategy, the animals were first introduced as pairs in separate cages. Then in a larger, divided, cage, the males were grouped together on one side, and the females were grouped together on the other side. The final procedure consisted of combining the males and the females together into one large group. Despite the use of a gradual procedure, 10/13 monkeys sustained injuries and one female died. The group was later split into two smaller groups, one of higher-ranking animals and one of lower-ranking animals, resulting in less overall aggression. Multi-male, multi-female groups appear to be most successful when younger females and older males are used; groups of all older animals tend to show more aggression. However, age appears to play less of a role when single-male harem groups are formed (Schapiro *et al.,* 1994). Visual barriers may also aid in safe group formation. When visual barriers were placed in the corrals containing large numbers of males and females, wounding rates were lower than when the corrals had no visual barriers (Westergaard *et al.,* 1999).

Although housing monkeys in groups approximating the composition of those found in nature may be the best way to promote psychological well-being of most animals, housing animals in large groups may not be feasible for many research protocols or laboratory spaces. Group housing limits access to individual animals for treatment and may make testing difficult. Therefore, alternative housing methods need to be addressed.

Pair housing

An alternative method of social housing is pair housing in which two compatible monkeys are housed in the same cage. The benefits of this housing condition are that animals have social contact, but at the same time, researchers have easy access to the animals. Researchers have had relatively high success rates when pairing rhesus monkeys. When 21 pairs of rhesus macaque females were tested, three were separated due to fighting, resulting in an 86% success rate (Eaton *et al.,* 1994). Of the successful pairs, there was a significant preference to be in close proximity to one another, especially at night. Pair-housed females spent much of their time social grooming. Similarly, paired juveniles spent a lot of time in socially-directed behavior (social grooming and social play) in addition to maintaining proximity with each other during periods of feeding and inactivity (Schapiro and Bloomsmith, 1994).

Pair housing substantially reduces the risk of acquiring abnormal behavior. Pair-housed monkeys had lower levels of abnormal behavior including eye-poking, self-biting, somersaulting, and hair-pulling compared to individually housed monkeys (Reinhardt *et al.,* 1988; Eaton *et al.,* 1994). Housing did not affect several health measures. There were no differences in physical health in pair-housed vs. individually housed controls; nor did health vary significantly as a function of dominance rank within the pair (Eaton *et al.,* 1994).

In fact, the presence of a preferred companion appeared to modulate stress in female rhesus monkeys when removed from a social group (Gust *et al.,* 1994). As with group-housed animals, an optimal procedure for forming successful pairs is to first give them a familiarization period prior to pairing. They can then be paired when a clear-cut rank relationship has been formed. Using this strategy, Reinhardt (1989) formed five male pairs and all were successful.

Limited contact

In some cases, pairing may not be possible due to restrictions of some experimental protocols or lack of same-sex partners. When this situation arises, a viable alternative is the use of grooming-contact bars (Crockett *et al.,* 1997). In this housing condition, the two members of a pair are housed in separate cages, but the cages are divided by widely spaced bars that permit physical contact, but prevent the animals from entering the other animal's cage. This caging system was tested with adult longtailed macaques (*Macaca fascicularis*). The female/female and male/female pairs were the most successful, with 100% compatibility. Male/male pairs were somewhat less compatible, with an 89% success rate. The animals remained in contact approximately 12% of the time and in the nine male/female pairs that were housed long-term, 12% of their time was spent social grooming.

Individual housing

Individual housing without grooming-contact bars is the most restrictive kind of social environment. Monkeys have visual and auditory access to conspecifics but do not experience physical contact. Although less devastating than the isolation rearing of infant monkeys, individual housing at a later age has also been associated with a form of "social deprivation syndrome" consisting of numerous abnormal activities (Goosen, 1981). The extent of abnormal behavior depends largely on two factors – the age at which monkeys are first exposed to individual cage housing and the overall length of the individual cage housing period (Lutz *et al.,* 2003). The development of abnormal behavior is negatively correlated with the age of first exposure. Younger monkeys are more vulnerable and therefore are more likely to develop both stereotypic and severely abnormal behavior. The development of abnormal behavior is also positively linked to the overall time spent in individual cage housing. The longer monkeys are housed in individual cages, the greater the likelihood that they will develop both stereotypic and severely abnormal behavior (Lutz *et al.,* 2003).

Conclusions

The behavior of captive primates is affected by the environment in which they live. An ideal environment is one that promotes species-typical behavior, either by replicating the natural environment or by creating an artificial environment containing the relevant factors. Social contact has been demonstrated to be one important variable necessary for promoting species-typical behaviors while reducing the levels of abnormal behavior. This is apparent when viewing the different rearing and housing conditions and their behavioral outcomes. A monkey reared in isolation exhibits a far different behavioral repertoire than that of a monkey reared with its mother in a social group. Unfortunately not all laboratory environments, husbandry procedures or research protocols allow for housing animals in groups similar to those found in the wild. Therefore, one must create the best possible environment under the given circumstances. A number of environments might suffice depending on the setting and the research protocol. For example, if infants have to be reared apart from their mothers, then surrogate-peer rearing of infants is probably an acceptable alternative. Researchers and facility managers need to be aware of the impact of housing conditions on behavior and psychological well-being in order to make informed choices for maintaining the animals in their care. A thorough knowledge of the behavior of primates in natural habitats can help identify relevant strategies for captive housing and at the same time can indicate potential problems and pitfalls to avoid.

Correspondence

Any correspondence should be directed to Dr. Melinda Novak, Department of Psychology, Tobin Hall, 135 Hicks Way, University of Massachussetts, Amherst, MA, 01003-9271, USA. Supported by NCCR grants RR11122 and RR00168.

References

Alexander, B.K. and Harlow, H.F. (1965). *Zool. Jb. Abt. Allgemeine Zool. Physiol. der Tiere* 71, 489–508.

Bayne, K., Haines, M., Dexter, S., Woodman, D. and Evans, C. (1995). *Lab. Animal* 24, 40–44.

Berard, J.D. (1989). *Puerto Rico Health Sciences Journal* 8, 61–64.

Berman, C.M. (1983). In Hinde, R.A. (ed.) *Primate Social Relationships: An Integrated Approach*, pp 132–134. Sunderland, MA: Sinauer Associates, Inc.

Berman, C.M., Rasmussen, K.L.R. and Suomi, S.J. (1994). *Child Development* 65, 1028–1041.

Brown, G.R. and Dixson, A.F. (2000). *Primates* 41, 63–77.

Capitanio, J.P. (1984). *Journal of Comparative Psychology* 98, 35–44.

Chamove, A.S. (1973). *Behaviour* 47, 48–66.

Chamove, A.S., Rosenblum, L.A. and Harlow, H.F. (1973). *Animal Behaviour* 21, 316–325.

Crockett, C.M., Bellanca, R.U., Bowers, C.L. and Bowden, D.M. (1997). *Contemporary Topics* 36, 53–60.

Cross, H.A. and Harlow, H.F. (1965). *Journal of Experimental Research in Personality* 1, 39–49.

DeVinney, B.J., Berman, C.M. and Rasmussen, K.L.R. (2001). *American Journal of Primatology* 54, 193–210.

Dittus W.P.J. (1979). *Behaviour* 69, 265–302.

Eaton, G.G., Kelley, S.T., Axthelm, M.K., Iliff-Sizemore, S.A. and Shiigi, S.M. (1994). *American Journal of Primatology* 33, 89–99.

Goldstein, S.J. and Richard, A.F. (1989). *International Journal of Primatology* 10, 531–567.

Goosen, C. (1981). *Biological Psychiatry* 16, 697–716.

Gust, D.A., Gordon, T.P., Brodie, A.R. and McClure, H.M. (1994). *Physiology and Behavior* 55, 681–684.

Hansen, E.W. (1966). *Behaviour* 27, 107–149.

Harlow, H.F. and Harlow, M.K. (1962a). *Scientific American* 207, 136–146.

Harlow, H.F. and Harlow, M.K. (1962b). *Bulletin of the Menninger Clinic* 26, 213–224.

Higley, J.D., Mehlman, P.T., Higley, S.B., Fernald, B., Vickers, J., Lindell, S.G., Taub, D.M. Suomi, S.J. and Linnoila, M. (1996). *Archives of General Psychiatry* 53, 537–543.

Kessler, M.J., London, W.T., Rawlins, R.G., Gonzalez, J., Martinez, H.S. and Sanchez, J. (1985). *Journal of Medical Primatology* 14, 91–98.

Koford, C.B. (1965). In DeVore, I. (ed.) *Primate Behavior Field Studies of Monkeys and Apes*, pp 160–174. New York: Holt, Rinehart & Winston.

Lindburg, D.G. (1971). In Rosenblum, L.A. (ed.) *Primate Behavior Developments in Field and Laboratory Research, Vol. 2*, pp 1–106. New York: Academic Press.

Line, S.W., Morgan, K.N., Roberts, J.A. and Markowitz, H. (1990). *Laboratory Primate Newsletter* 29, 8–12.

Lutz, C., Well, A. and Novak, M. (2003). *American Journal of Primatology* 60, 1–15.

Mason, W.A. and Kenney, M.D. (1974). *Science* 183, 1209–1211.

Mehlman, P.T., Higley, J.D., Faucher, I., Lilly, A.A., Taub, D.M., Vickers, J., Suomi, S.J. and Linnoila, M. (1995). *American Journal of Psychiatry* 152, 907–913.

Melnick, D.J., Pearl, M.C. and Richard, A.F. (1984). *American Journal of Primatology* 7, 229–243.

Mitchell, G.D. (1968). *Folia Primatologica* 8, 132–147.

Mitchell, G.D., Raymond, E.J., Ruppenthal, G.C. and Harlow, H.F. (1966). *Psychological Reports* 18, 567–580.

Novak, M.A. (1979). *Developmental Psychology* 15, 50–61.

Novak, M.A. and Harlow, H.F. (1975). *Developmental Psychology* 11, 453–465.

Novak, M.A., O'Neill, P. and Suomi, S.J. (1992). *American Journal of Primatology* 28, 125–138.

Novak, M.A. and Suomi, S.J. (1988). *American Psychologist* 43, 765–773.

Novak, M.A. and Suomi, S.J. (1991). *Laboratory Animal Science* 41, 308–314.

Reinhardt, V. (1989). *American Journal of Primatology* 17, 243–248.

Reinhardt, V., Houser, D., Eisele, S., Cowley, D. and Vertein, R. (1988). *American Journal of Primatology* 14, 135–140.

Reinhardt, V. and Reinhardt, A. (2000). *Lab. Animal* 29, 34–41.

Richard, A.F., Goldstein, S.J. and Dewar, R.E. (1989). *International Journal of Primatology* 10, 569–594.

Ruppenthal, G.C., Arling, G.L., Harlow, H.F., Sackett, G.P. and Suomi, S.J. (1976). *Journal of Abnormal Psychology* 85, 341–349.

Ruppenthal, G.C., Harlow, M.K., Eisele, C.D., Harlow, H.F. and Suomi, S.J. (1974). *Child Development* 45, 670–682.

Sackett, G.P. (1967). *Journal of Comparative and Physiological Psychology* 64, 363–365.

Sackett, G.P. (1968). In Fox, M.W. (ed.) *Abnormal Behavior in Animals*, pp 293–331. Philadelphia: Saunders.

Sackett, G.P. (1972). *Developmental Psychology* 6, 260–270.

Sackett, G.P., Holm, R.A. and Ruppenthal, G.C. (1976). *Developmental Psychology* 12, 283–288.

Schapiro, S.J. and Bloomsmith, M.A. (1994). *American Journal of Primatology* 32, 159–170.

Schapiro, S.J., Lee-Parritz, D.E., Taylor, L.L., Watson, L., Bloomsmith, M.A. and Petto, A. (1994). *Laboratory Animal Science* 44, 229–234.

Shalev, M. (1999). *Lab. Animal* 28, 16–19.

Snowdon, C.T. (1991). In Novak, M.A. and Petto, A.J. (eds) *Through the Looking Glass*, pp 103–115. Washington, DC: American Psychological Association.

Southwick, C.H., Beg, M.A. and Siddiqi, M.R. (1965). In DeVore, I. (ed.) *Primate Behavior Field Studies of Monkeys and Apes*, pp 111–159. New York: Holt, Rinehart & Winston.

Southwick, C.H., Siddiqi, M.F., Farooqui, M.Y. and Pal, B.C. (1974). In Holloway, R.L. (ed.) *Primate Aggression, Territoriality, and Xenophobia: A Comparative Perspective*, pp 185–209. New York: Academic Press.

Suomi, S.J. (1991). In Brauth, S., Hall, W. and Dooling, R. (eds) *Plasticity of Development*, pp 27–56. Cambridge, MA: MIT Press.

Suomi, S.J. (2000). In Balter, L. and Tamis-Lamode, C. (eds.) *Child Psychology: A Handbook of Contemporary Issues*, pp 510–525. New York: Taylor & Francis.

Suomi, S.J., Harlow, H.F. and Kimball, S.D. (1971). *Psychological Reports* 29, 1171–1177.

Suomi, S.J., Harlow, H.F. and Novak, M.A. (1974). *Journal of Human Evolution* 3, 527–534.

Suomi, S.J., Novak, M.A. and Well, A. (1996). *Developmental Psychology* 32, 1116–1128.

Suomi, S.J., Rasmussen, K.L.R. and Higley, J.D. (1992). In McAnamey, E.R., Kriepe, R.E., Orr, D.P. and Comerci, G.D. (eds.) *Textbook of Adolescent Medicine*, pp 135–139. Philadelphia: Saunders.

Teas, J., Richie, T., Taylor, H. and Southwick, C. (1980). In Lindburg, D.G. (ed.) *The Macaques Studies in Ecology, Behavior and Evolution,* pp 247–262. New York: Van Nostrand Reinhold Company.

Westergaard, G.C., Izard, M.K., Drake, J.H., Suomi, S.J. and Higley, J.D. (1999). *American Journal of Primatology* 49, 339–347.

Westergaard, G.C., Cleveland, A., Trenkle, M.K., Lussier, I.D. and Higley, J.D. (2003). *Journal of Medical Primatology* 32, 95–104.

Wood, B.S., Mason, W.A. and Kenney, M.D. (1979). *Journal of Comparative and Physiological Psychology* 93, 368–377.

PART 2

Primate Management

Contents

Figure 10.2 Examples of different housing systems.

factors are important: Temperature: 24°C (25–29); Humidity: 60%; Light: 12 hours light/dark.

Other important factors for callitrichid well-being are smell, noise and view. Since marmosets and tamarins communicate through odour, cleaning should not be so intensive that it removes the family smell of the home cage. Marmosets and tamarins vocalise a lot and it is attractive for them to hear other animals. The monkeys are usually interested in the voices of the personnel and might accept music of low volume. Marmosets and tamarins are nosy and therefore will always try to see as much as they can. Neighbouring groups should be visually separated by foliage, a hanging screen or a curtain, in order to avoid stress. However, single housed animals or young peer groups should be able to see each other.

Handling

Handling of animals is inevitable for quarantine, transport, health checks, routine examinations and experimental procedures. In general, no callitrichid species likes to be handled but can be trained to co-operate by receiving rewards like Nutri-Cal® or marshmallows. As a rule, handling should only be done by skilled persons, in a calm and careful way, to avoid injuries to both animal and handler.

All handling assumes trapping of the animals first. This can be done in several ways depending on cage design and group size. In larger cages and groups, there should be an area which can be used to separate individuals or, as mentioned before, the sleeping box. If there is a passage or corridor that the group has to use, this can also be used for separation of individuals. Animals can also be trained to enter a smaller cage or their sleeping box which will then be closed. Alternatively individuals have to be caught with a net.

Identification of animals

It is necessary to identify animals individually. Microchip, tattoo, chain collar, cutting of hair at the tail or colouring of ear tufts can easily be maintained. Microchips, usually placed under the soft skin in the neck, can get lost by allogrooming or may have to be removed in the case of MRI. Animals have to be restrained for reading both the microchip and the tattoo, since the numbers are not easily visible due to pigmentation of the skin or hair growing. Whilst chain collars have to be checked frequently for the right size, in order to avoid injuries to the animals, cutting or colouring of hair also needs to be repeated on a regular basis.

Feeding and nutrition

General consideration

The nutritional status has a major influence on the growth, reproduction and longevity of nonhuman primates as well as their ability to resist pathogenic and other environmental stress (Knapka *et al.,* 1995). Due to their small body size, limited gut volume, and rapid rate of food passage (Garber, 1986) callitrichids require a diet high in nutritional quality and available energy. To develop an appropriate diet, the following items should be considered:

- information from feeding in the wild;
- information from published nutrient requirements;
- food available at the facility;
- food preference of animals.

Callitrichids are omnivorous, consuming both insects and plant items such as flowers and fruits. Garber (1992) reported that marmosets spend 30–70% of their feeding time on exudates like gum arabic, with their specially adapted lower anterior dentition and V-shaped mandible. Because of their "long-tusked" dentition, tamarins are called opportunistic exudate eaters which benefit from the bark gouging of marmosets (Ferrari and Martins, 1992). The National Research Council (NRC) recommendations (2003) state that diet of tamarins should consist of 45% insects, 35% fruits, 10% exudates, 7% nectar and 3% seeds, whereas marmosets eat 45% exudates, 16% fruits, 39% insects. Gum arabic provides the marmosets with a source of complex polysaccharides and carbohydrates, certain minerals, especially calcium (Carroll, 1997), and tannin. Trials, in different callitrichid species, with two different diets with or without gum arabic, showed that the addition of gum slowed the gut passage rate in marmosets, without depressing the digestive efficiency (Power and Oftedal, 1996). Marmosets are able to ferment tree exudates with the help of micro-organisms in their bigger cecum (Ferarri and Martins, 1992).

The feeding schedule should be orientated to the activity pattern of Callitrichidae. Active periods occur early in the morning, at noon and during early afternoon. Housing conditions are also known to influence the feeding modus like indoors/outdoors group or single housing. When primates are group housed, it is important to ensure that the lowest ranking individual has sufficient access to food and water. Food should be offered in feeding bowls well above floor level. In larger groups several bowls should be offered. Feeding bowls and water bottles have to be cleaned every day.

Pellets represent a combination of all essential nutritive compounds. They are best fed *ad libitum* and should add up to 50% of the total diet. Marmosets eat an average of 20–25 g pellets/day. Many facilities moisten the marmoset pellets with milk or juice, feeding this as "porridge" in the morning. Different commercial pellets are available (SSNIFF, SDS, MAZURI, PMI) and are listed for comparison of ingredients in Table 10.1.

A standard feeding regime in many colonies is a high calory vitamin porridge for breakfast, fruits, vegetables, and a protein source for lunch, and probably insects in the afternoon. Varying extras are offered, like cooked rice, noodles, potatoes, rusks, hardboiled eggs, cottage cheese, boiled chicken, cat food, raisins, sunflower seeds, dried figs or dates, peanuts, carob, grasshoppers, mealworms and waxworms during the week. The feeding intervals should be 4.5 to 6.5 hours. It is difficult to determine how much food an individual consumes on average. On a dry matter basis, an active adult animal consumes approximately 5% of its bodyweight per day.

The following is an overview of special nutrient requirements for nonhuman primates (NHP) with emphasis on callitrichid nutrition (NRC, 2003).

Energy, fat and protein

Energy is required to support the basic life functions. The daily energy intake must be sufficient to meet requirements for basal metabolism and activity. The average amount for callitrichids is 3.9 to 4.2 kcal/g diet. Adult cotton-top tamarins consume, on average, 152 kcal/kg LBW and lactating females up to 260 kcal/kg LBW (Kirkwood and Underwood, 1984).

Fat is an important resource for energy and commercial diets contain about 9% of the essential fatty acids. Dry skin and hair loss are evidence of deficiencies of unsaturated fatty acids in the diet. Experimental diets with high saturated fat and cholesterol concentrations lead to arterioscleroses and a higher incidence of arterial aneurysm (McIntosh *et al.*, 1987).

Observations in the field, and also in the laboratory, have shown that callitrichids have a big demand for high quality proteins with up to 25% originating from an animal source. Flurer and Zucker (1988) fixed the daily protein requirements between 2.4 and 3.47 g/kg LBW. Ausman *et al.* (1986) indicated that soy protein is half as effective as lactalbumin and also reduces the iron resorption. Essential amino acid requirements for nonhuman primates have not been established (NRC, 2003) but, in 1998, Flurer and Zucker reported that arginine and histidine are essential amino acids in adult *Callithrix jacchus*. Pathological findings due to protein deficiencies are alopecia, facial oedema, diarrhoea, fatty liver syndrome and anaemia.

TABLE 10.1: Comparison of different commercial pellets available				
Nutrients	**Units**	**Mazuri marmoset pellets**	**Ssniff marmoset pellets**	**SDS trio munch**
Crude Protein	%	25.40	26.00	23.80
Crude Oil	%	7.50	7.00	5.30
Crude Fibre	%	3.70	2.50	4.90
Ash	%	10.50	6.00	6.30
Calcium	%	2.16	1.00	1.12
Phosphorus	%	1.46	0.70	0.90
Sodium	%	0.33	0.20	0.29
Magnesium	%	0.29	0.20	0.15
Potassium	%	0.81	0.70	0.69
Water	%	10.00		
N.F.E.	%	42.90		49.70
Met. Energy	%	12.30	14.90	12.40
Vitamins				
A	IU/kg diet	30142.00	18000.00	33462.00
D3	IU/kg diet	11640.00	3000.00	11000.00
E	mg/kg diet	105.60	120.00	123.20
B1	mg/kg diet	27.70	18.00	16.70
B2	mg/kg diet	18.20	24.00	13.50
B6	mg/kg diet	14.10	18.00	11.40
B12	mg/kg diet	39.40	100.00	25.00
Biotin	µg/kg diet	398.00	500.00	220.00
Panthotenic Acid	mg/kg diet	37.30	50.00	25.30
Choline	mg/kg diet	1951.00	1600.00	1070.00
Folic Acid	mg/kg diet	10.20	7.00	5.20
Nicotinic Acid	mg/kg diet	92.70	70.00	45.50
K3	mg/kg diet	5.30	6.00	5.58
Inositol	mg/kg diet	1649.00	60.00	1510.00
Ascorbic Acid	mg/kg diet	2966.00	3500.00	400.00
Trace Minerals				
Manganese (mg)	mg/kg diet	85.00	90.00	74.00
Copper (mg)	mg/kg diet	18.00	14.00	12.00
Zinc (mg)	mg/kg diet	71.00	90.00	89.00
Iodine (mg)	mg/kg diet	3.38	2.00	1.05
Iron (mg)	mg/kg diet	358.00	260.00	152.00
Selenium (mg)	mg/kg diet	0.23	0.20	0.25
Cobalt (mg)	mg/kg diet	2.02	2.00	2.02
Fluorine (mg)	mg/kg diet	54.00		13.00
Amino Acids				
Lysine	%	1.43	1.40	1.23
Methionine	%	0.38	0.40	0.53
Phenylalanine	%	1.04	1.30	1.29
Histidine	%	0.57	0.70	0.55

(Continued)

TABLE 10.1 *(Continued)*

Nutrients	Units	Mazuri marmoset pellets	Ssniff marmoset pellets	SDS trio munch
Tryptophan	%	0.22	0.40	0.21
Threonine	%	0.93	1.00	0.83
Isoleucine	%	0.94	1.10	1.00
Leucine	%	1.82	2.00	2.16
Valine	%	1.18	1.20	1.18
Arginine	%	1.70	1.80	1.13
Tyrosine	%	0.73	0.80	0.83
Cystine	%	0.36	0.40	0.26

Carbohydrates and fibre

Carbohydrates provide about 40% of the metabolised energy in the diet. Crude fibre concentrations in commercial diets vary between 2 and 8% but the addition of 5 to 10% is recommended. Clapp and Tardiff (1985) described a diet for marmosets consisting of 4.2 to 10% fibre, while Power and Oftedal (1996) suggested an addition of 16% of total fibre to the diet. An increase in dietary fibre increases faecal volume and the digestive passage through the gastrointestinal tract, thus reducing the time for digestion.

Vitamins and minerals

Supplementation of fat soluble vitamins must be carried out carefully due to their toxicity in higher concentrations. The NRC recommendation for Vitamin A is 10,000 to 15,000 IU/kg diet. Like all other primates, callitrichids are dependent on an external supply of Vitamin C of 15 mg/kg metabolic bodyweight. With the lack of UV B light, callitrichids can only utilise Vitamin D_3 and so the daily requirement for marmosets is 110 IU/100 g LBW and 33 IU/400 g LBW for tamarins (Knapka *et al.*, 1995). In 1997 Power *et al.*

Figure 10. 3 Marmoset food at the German Primate Centre.

collected blood samples from 18 free ranging cotton-top tamarins. The blood values ranged from 25.5 to 120 ng/ml. Juveniles had higher serum concentrations than adults and pregnant females lower than nonpregnant. The high circulating levels of vitamin D metabolites in captive New World primates are hypothesised to be necessary for their health status because of the low binding affinity of their vitamin D receptor. Serum concentrations below 50 ng/ml may indicate suboptimal vitamin D status. Vitamin B_{12} is important for the function of the gastrointestinal cells, bone marrow and nerve cells. To date, there are no reliable data about the daily requirements of callitrichids. Recommendations for other vitamins are: Retinol: 171 g/kg LBW, Vitamin E + Se: 0.1–0.16 mg/kg LBW and Vitamin K: 2 μg/kg LBW.

Quantitative mineral requirements of nonhuman primates are poorly defined (NRC, 2003) but the following minerals are the minimum suggested for inclusion in a primate diet: Calcium: 0.5%, Phosphorus: 0.3 to 0.4%, Potassium: 0.24 to 1.1%, Sodium: 0.25 to 0.65%, Magnesium: 0.08%, Chloride: 0.27 to 0.62%, Iron: 0.018%, Copper: 12–20 mg/kg diet, Iodine: 2.2 mg/kg, Manganese: 70–100 mg/kg, Zinc: 150 mg/kg, Selenium: 0.01–0.02 mg/kg, Chromium: 150 μg/day.

Environmental enrichment

Why and how

In the wild, callitrichids have a huge home range. They live in social groups which can comprise up to 15 family members. They spend about 50% of their day on foraging. In order to compensate for this high activity of the animals in the wild, it is necessary to provide their environment in captivity with some enrichment devices. A starting point for environmental enrichment is to reproduce some of the main features of their natural habitat and to create opportunities for captive animals to develop skills they might need in the wild. Enrichment can be offered through several ways, like food, play or encounters.

Diet and foraging

Food can become more interesting when fruits and vegetables are offered as large pieces. These food items should ideally be distributed throughout the cage, even at places where animals have to find a way to reach it, e.g. in a container with drilled holes or hanging on a chain from the ceiling. Another possibility is to offer live insects, like crickets and mealworms. Vignes *et al.* (1992) described a mealworm feeder as a foraging enrichment device, made out of 500 ml water bottles with holes of 0.5 cm diameter and hung horizontally.

For species of the genera Callithrix and Cebuella, the mode of feeding of gum arabic can be altered to provide environmental enrichment. Gum is usually provided as a powder to be mixed with water, or as crumbles of different sizes, but gum powder, mixed with water to a thick fluid, can also be painted on clean branches and shelves. More recently, Ventura and Buchanan-Smith (2003) introduced artificial gum trees to stimulate the peculiar feeding skills of marmosets, and De Rosa *et al.* (2002) have observed the use of puzzle feeders.

Cage and furnishing

In order to encourage play and explorative behaviour, toys can be provided easily in the form of available laboratory material that will be used and "destroyed" by the monkeys. Paper rolls, cardboard boxes, plastic tubes or wooden blocks represent perfect toys for marmosets. Majolo and Buchanan-Smith (2003) introduced different novel objects for enrichment such as film cases containing a marble, or a cup containing ten small plastic test tubes. A foraging tree was made from PVC pipe cut into sections and connected with T-shaped PVC tubes (Byron, 2001). Toys that are built for cages have to be checked for safety to avoid injuries.

Social environment

Free ranging callitrichids are usually living as monogamous groups, but can also be encountered as polyandrous or multi-male and multi-female groups. If there is a need for single housing, there should at least be the possibility of visual and olfactory contact with other conspecifics. Different sex pairs can be housed together, e.g. with a vasectomised male. Same sex pairs have to be introduced by giving them visual contact first and the new homecage should be the cage of the subordinate animal. Females are more aggressive than males (Scott, L., personal communications) and female-female pairs do not represent natural group compositions and are therefore less stable, with allogrooming being rarely observed (Majolo and Buchanan-Smith, 2003).

For further information on environmental enrichment see Heath and Libretto (1993), Kitchen and Martin (1996), Poole (1990) and Schoenfeld (1989).

Breeding

Basic information on reproductive biology

Reproduction in callitrichids is characterised by several peculiarities: females show ovarian cycles all year round with a high rate of fecundity, males copulate throughout the cycle and, even during pregnancy, with a higher frequency around the time of ovulation (Kendrick and Dixson, 1983). Callitrichids are the only simian primates with multiple ovulation. Births usually comprise twins, but increasingly triplets or even quadruplets in captivity. Shortly after birth, and despite lactation, callitrichids ovulate (8 to 18 days post-partum) and can conceive again. In cotton-top tamarins, Ziegler *et al.* (1987) determined different post-partum ovulation periods depending on litter size: 27.3 ± 1 days after the birth of twins and only 16 ± 0.75 days after giving birth to a singleton. Information on the length of the ovarian cycle and pregnancy, as well as the time of occurence of post-partum ovulation, is given in Table 10.2.

Another characteristic of callitrichids is the inhibition of reproduction in sub-dominant females. A group usually consists of only one breeding pair. The presence of a dominant female (normally the mother) prevents lower ranking females (the daughters) from reproducing. However, when an unrelated animal is introduced to a group, polygyny or polyandry are increasingly observed. In the common marmoset, up to 50% of daughters can ovulate while living in their natal family, but sexual behaviour with the fathers does not occur (Abbott, 1994; Hubrecht, 1989; Saltzmann *et al.*, 1997). In cotton-top tamarins, none of the daughters ovulates in the presence of the mother (French *et al.*, 1984; Tardif, 1984; Ziegler *et al.*, 1987).

Pregnancy and birth

Pregnancy can be detected by abdominal palpation, ultrasonography with a 7.5–10 MHz probe (Jaquish *et al.*, 1995) or by measuring hormones (e.g. progesterone) in urine, faeces or blood (Harlow *et al.*, 1984). Implantation of the early embryos is superficial. The placenta consists of two discoid parts connected by vascular anastomosese (*Placenta bidiscoidalis*) and is permeable to antibodies (*Placenta haemochorialis*). Thus twins are immunological blood chimers. At conception the litter may comprise a higher number than will be born, since marmosets are capable of resorbing individual embryos without disturbance of the development of the others (Jaquish *et al.*, 1995). Resorption can only occur until the end of the embryonic period. Death of the foetus in later gestation stages might either cause abortion of the whole litter or mummification of the dead foetus.

Birth usually takes place at night. Observations in multiparous females show that the delivery occurs almost always at the same time (Layne, D., personal communications). If the female shows signs of labour during the day something is usually wrong and veterinary examination necessary and a probable caesarian indicated.

Methods for monitoring the reproductive status

For many experimental studies, and also colony management, it is important to know about the reproductive status of the females. Monitoring can be performed using invasive and non-invasive methods.

TABLE 10.2: Reproductive data of callitrichid species kept in laboratories

	Callithrix jacchus	Saguinus fuscicollis	Saguinus oedipus
Cycle length (days)	28	26	21
Gestation period (days)	140–145	150–155	180–185
Post-partum ovulation (days)	10	17–18	17–18

Hormone measurements in blood, urine or faeces are indirect methods since hormone values reflect the function of the reproductive organs. Ultrasonography provides direct results through an immediate view of the ovaries and uterus (Tarantal and Hendrickx, 1988). To determine the time of ovulation, all sample collection for hormone analysis, needs to be performed at least twice, but preferably three times, a week.

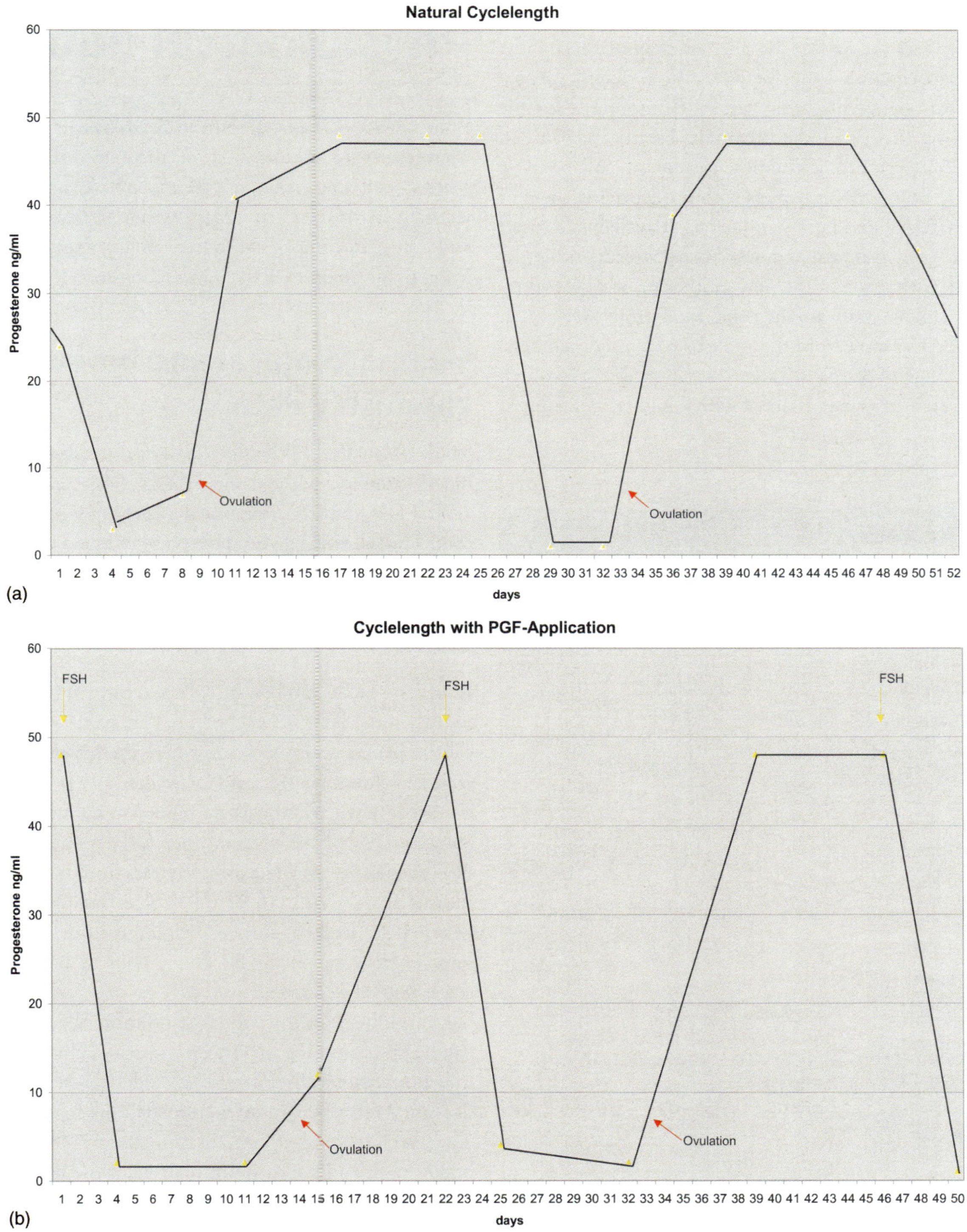

Figure 10. 4a,b Progesterone profile of a common marmoset with natural cycle and with $PGF_{2\alpha}$-application.

Breeding management and raising of offspring

New pairs should only be made up of fully grown animals in good condition, not younger than 24 months. In immature females, pregnancy may restrict normal growth and very young mothers more often fail to rear their first offspring. It is advisable to place the new pair away from their natal groups in order to prevent stress and the suppression of ovarian cycles by the mother.

Infants are carried by all members of the family, but participation varies between the different species. This co-operative rearing system is important for the breeding success of the offspring. Hearn and Burden (1979) developed a rotation system of collaborative rearing of marmoset triplets without depriving them of the maternal and family influence. If available, a foster mother can also raise the new-borns, after marking the babies with urine of the new mother prior to presentation. Infants start to eat solid food from week 3–4, stealing the food from the mouths of family members. At the age of 90 days, infants are completely weaned. Twin fights can be observed with a higher incidence at the age of eight months. Offspring should be removed from their family at the age of eighteen months.

Irreversible control of fertility can be performed by sterilisation or castration of the male or female. Pregnancy can be prevented by injecting the female with Prostaglandin $F_{2\alpha}$ every three weeks or implanting melengestrol-acetate between the shoulder blades. This MGA implant can last for two years (Moehle *et al.*, 1999).

TABLE 10.3: Physiological data of common marmoset and cotton-top tamarin

		Common marmoset	Cotton-top tamarin
Bodyweight	g	350–450	550–650
Sexual maturity	months	24	24
Estrus lengths	days	28	21
Gestation period	days	144	180–185
Post-partum ovulation	days	10	17–18
Birthweight	g	26–32	40–45
Weaning age	days	60–90	60–90
Life span	years	13	15
Body temperature	Celsius	36.8–38.6	39.3–40.1
AF	/min		
HF	/min	194–242	
Daily pellets intake	g	20	25
Urine/24 hours	ml	16.99	

Source: Fortman *et al.* (2002); Savage (1995).

Physiological data

Body mass differs significantly between species. Marmosets are the smallest of these four laboratory species, weighing between 350 and 400 g, and moustached tamarins are the largest, weighing between 650 and 700 g. In comparison to free ranging animals, those in captivity are up to 20% bigger (Kingston, 1969).

Haematological and blood chemistry data

Many New World Primates are trained to allow routine blood sampling without anaesthesia. Some values like AST, LDH and CK are elevated after Ketamine application. The following gives an overview of blood chemistry values of different callitrichid species.

Experimental procedures

Urine and faeces can best be collected early in the morning when the light is switched on. A container or a mat can be placed under the cage where the sample will be collected. Anzenberger and Gossweiler (1993) described a procedure where animals are trained to go into a small compartment of the cage and pee for a reward. Hearn (1975) collected 24-hour urine samples from marmosets, measuring 3–47 ml (16.99 ml). Faeces samples can be coloured by feeding nutrient colour. If urine or faeces have to be sampled for 24 hours, animals have to be single housed in a metabolism cage for this period.

Intramuscular injections, of no more than 0.2 ml, can be given into the quadriceps muscle. Subcutaneous injections are ideally introduced under the skin of the dorsum, with a maximum volume of 2 ml on each side. The average blood volume of animals is 7–8% of the bodyweight and a maximum of 10% of the blood volume can be taken in 14 days (0.7 ml/100 g LBW) without expecting health problems. If more is required, up to 15% fluid has to be substituted. Blood samples

TABLE 10.4: Normative haematological values of marmosets and tamarins

		Common marmoset	Cotton-top tamarin	Moustached tamarin	Saddle-back tamarin
WBC	x 1000/Ul	6.1 ± 2.2	11.2 ± 5.2	12.3 ± 2.8	8.7 ± 4.06
RBC	x 1000000/Ul	5.6 ± 0.78	6.3 ± 0.61	6.06 ± 0.65	5.39 ± 1.02
HGB	mg/dl	15.0 ± 1.4	15.9 ± 1.7	14.3 ± 1.9	14 ± 2.5
HCT	%	44.6 ± 7.1	47.9 ± 5	48.2 ± 6.5	44.4 ± 6.6
MCH	mg/dl	25.8 ± 2.7	25.4 ± 1.5	24.2 ± 0.7	26.4 ± 3.3
MCHC	g/dl	34.2 ± 4.5	33.1 ± 2.3	30.1 ± 2.6	33.1 ± 3.9
MCV	fl	74.3 ± 10.9	76.3 ± 5.4	78.2 ± 5.7	78.9 ± 7.6
SEGS	x 1000/Ul	3.2 ± 1.5	7.03 ± 4.5	5.1 ± 1.8	8.2 ± 4.5
BANDS	x 1000/Ul	0.17 ± 0.08	0.33 ± 1.7	0.08 ± 0.01	
Lymphocytes	x 1000/Ul	3.0 ± 1.6	3.3 ± 1.7	6.4 ± 2.5	1.9 ± 0.92
Monocytes	x 1000/Ul	0.25 ± 0.18	0.54 ± 0.44	0.85 ± 0.52	0.3 ± 0.12
Eosinophils	x 1000/Ul	0.23 ± 0.14	0.21 ± 0.17	0.39 ± 0.26	0.28 ± 0.17
Basophils	x 1000/Ul	0.16 ± 0.15	0.1 ± 0.06	0.22 ± 0.2	0.18 ± 0.08
NRBC	/100 WBC	3.0 ± 2.0	1.0 ± 1	3.0 ± 2	9 ± 19
Platelets	x 1000/Ul	609 ± 200	361 ± 74	840 ± 142	546 ± 113

Source: Abou-Madi (1999); Fortman *et al.* (2002); Savage (1995).

can be taken from alert animals from the femoral or lateral tail vein with a 1–2 ml syringe and a 25–26 gauge needle.

Hearn (1977) developed a restraining device to allow a single person to carry out routine procedures. Intravenous injections or fluid administration can be given into the Vena saphena, V. cephalica or V. coccygea, using a 24–26 gauge catheter. If the animal's veins are collapsed, fluid can be given intraosseally, directly into the bone-marrow of the tibia or femur, using a 20 G disposable intraosseus infusion needle with a T-handle. This needle can also be used to collect bone marrow from the trochanteric fossa, iliac crest or proximal humerus. Vascular access ports can also be implanted into marmosets to deliver compounds intravenously, e.g. the V. femoralis (Dalton, 1985).

Oral gavage can easily be maintained. Semen can be collected by penile stimulation (Kuederling *et al.*, 2000) or electroejaculation (Cui *et al.,* 1991). Osmotic pumps can be placed subcutaneously between the shoulder blades, or intraperitoneally (Fortman *et al.*, 2002) to deliver drugs continuously for up to four weeks without restraining animals. Geretschläger *et al.* (1987) described a method to collect cerebrospinal fluid from anaesthetised marmosets with a 25 gauge scalp vein set, Venofix®, and a 1 ml syringe, collecting 0.1 ml per puncture from the cisterna magna.

Veterinary care

Health management

Health checks should be made by care takers during feeding schedules, once or, preferably, twice daily, but it should definitely be the first duty, in the morning, to see if an animal is down, in labour or healthy. Observations should be made of behaviour and general condition of the animals, such as appetite, attentiveness, locomotion, bodyweight, coat quality, faecal quality, stress between group members and how the animals go to the food, which might be quite difficult in larger groups. If necessary, the veterinarian should be consulted to decide if the animal requires a proper physical examination, including body temperature, rectal and vaginal swab, faecal sample, blood sample, palpation of the abdomen, auscultation of heart and

TABLE 10.5: Normative blood chemistry values of marmosets and tamarins					
		Common marmoset	Cotton-top tamarin	Moustached tamarin	Saddle-back tamarin
Glucose	mg/dl	177 ± 65	179 ± 82	117 ± 63	173 ± 66
BUN	mg/dl	19 + 5	15 ± 8	13 ± 5	14 ± 5
Creatinine	mg/dl	0.7 ± 0.2	0.7 ± 0.3	0.7 ± 0.4	0.5 ± 0.2
Uric Acid	mg/dl	0.5 ± 0.2	1.0 ± 0.7		0.8 ± 0.1
Calcium	mg/dl	9.5 ± 1.1	8.9 ± 0.9	8.7 ± 1.2	8.9 ± 0.9
Phosporus	mg/dl	5.3 ± 1.9	4.8 ± 1.5	8.0 ± 3.0	5.2 ± 1.1
Sodium	mEq/l	147 ± 8	150 ± 8	154 ± 7	154 ± 1
Potassium	mEq/l	4.9 ± 2.6	4.0 ± 0.8	4.9 ± 1.6	3.4 ± 0.7
Chloride	mEq/l	103 ± 11	104 ± 8	104 ± 8	110 ± 1
Iron	mg/dl	129 ± 1	127 ± 73		
Magnesium	mg/dl		2.4 ± 0		
Cholesterol	mg/dl	176 ± 73	121 ± 42	106 ± 79	65 ± 12
Triglyceride	mg/dl	160 ± 43	69 ± 32		80 ± 0
Total Proteins	mg/dl	6.8 ± 1.0	6.6 ± 0.7	6.5 ± 0.7	7.5 ± 1.0
Albumin	mg/dl	5.1 ± 0.6	3.8 ± 0.5	3.5 ± 1.0	4.2 ± 0.2
Globulin	mg/dl	1.7 ± 0.5	2.8 ± 0.5	2.3 ± 1.3	2.5 ± 0.1
AST	I.U./l	112 ± 112	157 ± 56	56 ± 85	491 ± 892
ALT	I.U./l	13 ± 24	38 ± 41	7 ± 14	26 ± 32
GGT	I.U./l		21 ± 21		
Total Bilirubin	mg/dl	0.2 ± 0.3	0.2 ± 0.2	0.1 ± 0.1	0.3 ± 0.4
Direct Bilirubin	mg/dl	0.0 ± 0.0	0.0 ± 0.1		
Indir. Bilirubin	mg/dl	0.1 ± 0.0	0.2 ± 0.1		
Amylase	μg/l	337–1523	575 ± 400	496 ± 0	
alk. Phosphatase	I.U./l	125 ± 64	184 ± 110	358 ± 341	129 ± 68
LDH	I.U./l	551 ± 429	460 ± 319	594 ± 326	390 ± 226
CK	I.U./l	543 ± 0	645 ± 706		
Lipase	I.U./l		40 ± 16		
CO_2	mMol/l		18.1 ± 8.3	24 ± 0	11.7 ± 3.5
Cortisol	μg/dl		570 ± 0		

Source: Abou-Madi (1999); Fortman *et al.* (2002); Savage (1995).

lungs, ultrasound and x-ray if indicated. Sick animals should be kept warm (e.g. heating lamp or pad) and, if necessary, the individual should be isolated from its group for intensive care.

Anaesthesia

Drug of choice for mild anaesthesia is Ketamine (50 mg/kg), with a maximum of 25 mg/animal because of its myotoxicity (Davy *et al.*, 1987). Saffan® is also very safe (12–18 mg/kg). For longer surgery, a combination of Saffan (18 mg/kg) and Valium® (0.25 mg/animal) is reliable in common marmosets, and Ketamine (25 mg/kg) + Midazolam (25 mg/kg) for cotton-top tamarins. The combinations, with an inhalation narcotic (e.g. Isoflurane) for longer lasting surgery is very effective either with a modified tubus (2.0 mm) or a face mask.

Quarantine

Most imported animals arrive in a stressed condition, dehydrated and underfed and should therefore be rested (Deinhardt, 1967). Quarantine duration depends on national regulations but should be a minimum

TABLE 10.6: Drugs recommended for Callitrichidae

Drug	Dosage mg/kg	Application Route	
Analgetics			
Aspirin	20	PO	TID
Flunixin	10	IM	SID
Buprenorphine	0.01	IM/IV	TID
Oxymorphone	0.075	IM/IV	TID
Antibiotics			
Amikacin	2.5	IM	SID
Amoxicillin	10	IM	SID/QOD
Ampicillin	5	PO/IM	BID
Cefazolin	25	IM	BID
Cefotaxime	75–100	IM	TID
Chloramphenicol	25–50	IM/PO	BID
Clindamycin	11	PO	SID
Doxycycline	8	PO	SID
Enrofloxacin	2.5–5	PO	SID
Erythromycin	40–75	IM/PO	BID
Gentamycin	2	IM	BID
Kanamycin	7.5	IM	BID
Neomycin	10	PO	SID
Oxytetracycline	10	IM	QOD
Trimethoprim/ Sulfadiazine	12	IM	QOD
Antimycotic			
Amphotericin B	0.25–1.0	IV	SID
Griseofulvin	20	PO	SID
Ketoconazol	20–30	PO	TID
Antiparasitic			
Albendazol	25	PO	SID
Fenbedazol	50	PO	SID
Ivermectin	0.2	IM/PO	SID
Metronidazol	35–50	PO	BID
Paramomycine	50	PO	BID
Miscellaneous			
Dexamethasone	0.25–1.0	IV/IM/PO	SID
Furosemide	2	IV/IM/PO	BID
Kaolin	0.5–1.0 ml	PO	SID
Oxytocin	1–2 IU	IV/IM	SID
Prednisolone	10	IV	SID
Baypamune		SC	QOD
Atropine	0.05	IV	
Epinephrine	0.2–0.4	IV	

of 30 days. Animals should be totally isolated from the rest of the colony. Two days after arrival, each animal should undergo a general health examination including a tuberculin-test. This examination is repeated at the end of the quarantine period, and can be done either with alert or anaesthetised animals, depending on the animal.

Diseases

Only a short overview can be given about the most common diseases in callitrichids, and for further details see Bennett *et al.* (1995, 1998), Potkay (1992) and Savage (1995). The occurrence of diseases can differ a lot with different housing conditions, and whether animals are imported from the wild or bred in a laboratory, housed outdoors/indoors or behind a barrier system with restricted access of personnel.

Viral diseases

Herpes simplex or hominis can be transmitted to all callitrichids from humans with an active Herpes infection, e.g. via saliva. It produces severe ulceration in the facial area, oral cavity and oesophagus and can lead to death within 48 hours with an incubation period of seven days. Many facilities therefore recommend that people with cold sores should not work with the animals. The reservoir host for Herpes tamarinus is the Squirrel monkey. It is transmitted via saliva, bite wounds, capture nets and gloves. After an incubation period of 7–10 days, the animals develop multiple ulcerations of lips, eyes and oesophagus and nasal discharge, apathy and anorexia, death occurring from between 2 and 3 days. Callitrichids and cebids should therefore not be housed together (King, 1967). Herpes saguinus was described by Melendez (1971) as not leading to clinical symptoms. Cytomegalovirus (CMV), which occurs very often as a latent infection in Old World Primates (OWP), seems not to be relevant in marmosets, and has been isolated from the salivary gland of tamarins without clinical symptoms (Nigida *et al.*, 1979). Herpes saimiri is latent in Squirrel monkeys without symptoms but leads to malignant lymphomas in callitrichids and Old World Primates. Herpes ateles is also known to produce lymphomas. Inoculation with Varicella zoster evokes delayed antibody titres but no clinical symptoms. Epstein-Barr-Virus is an established animal model in callitrichids but it does not occur naturally. Ramer *et al.* (2000) described a syndrome of weight loss, loss of appetite, diarrhoea and palpable abdominal mass with grossly large mesenteric lymph nodes. Cho *et al.* (2001) isolated, from these spontaneous B cell lymphomas, an EBV related Lymphocryptovirus called CalHV-3. It is the third Herpes virus isolated in Callitrichidae.

Monkeypox is a zoonosis, transmitted through direct contact with lesions on the skin and mucosa of the oral cavity. Smallpox, vaccinia and monkeypox infection can be fatal diseases in marmosets characterised by cutaneus erythemateous papules on the tail, hands and feet, anogenital region and abdomen. They are also associated with weight loss (Gough *et al.,* 1982), and can even be lethal.

The role in gastroenteritis of Rotavirus, which occurs more often in captivity than in the wild, has not been verified (Kalter, 1982). Mansfield *et al.* (2001) and Thomson and Scheffler (1996), isolated Coronavirus from animals with watery diarrhoea and an acute *Escherichia coli* infection.

Para-influenza Virus Type I produces symptoms from nasal discharge to pulmonary lesions whereas Type II and III have been isolated from cotton-top tamarins without symptoms (Murphy *et al.,* 1972). Paramyxovirus saguinus, in combination with gastroenterocolitis and a high mortality rate, in cotton-top tamarins have been described. Clinical symptoms include apathy, kachexie, diarrhoea and death within 24 hours.

Spread as an aerosol, measles is a very contagious disease with high morbidity and mortality up to 95%. This Morbillivirus infection starts with swollen eyelids, fever, nasal discharge, facial oedema and exanthema, ending up with interstitial pneumonia. Secondary complications include septicaemia, abortion, metritis, severe dysentery and disseminated intravascular coagulopathy. Human measles vaccine and gamma globulin can be used prophylactically.

Christe *et al.* (2002) immunised juvenile rhesus monkeys successfully with a canine distemper vaccine. The incubation period for Hepatitis A Virus, a picornavirus, is 30–40 days but animals do not develop clinical symptoms. It is of more importance as an anthropozoonosis, transmitted via urine or faeces, and personnel should be vaccinated. Callitrichid Hepatitis Virus (CHV), transmitted by mice, is closely related to the murine Lymphocytic Choriomeningitis Virus (LCMV). Clinical findings are anorexia, lethargy and dyspnoea and levels of aspartate aminotransferase, alkaline phosphatase and bilirubin are elevated. Outbreaks in different colonies have shown that it is connected with a high incidence of morbidity and mortality (Asper *et al.,* 2001; Montali *et al.,* 1993).

Bacterial diseases

Bordetella bronchiseptica causes death in juvenile marmosets and tamarins and clinical symptoms are mucopurulent nasal discharge, fever and pneumonia. Antibiotic treatment with Doxycycline works well. Vaccination is indicated if there is a manifest infection in the colony (Brack *et al.,* 1997). Campylobacter spp. is one of the most frequently isolated organisms from NHP with diarrhoea, or even from asymptomatic animals. Campylobacter jejuni is a cause of diarrhoea and enterocolitis in tamarins (Paul-Murphy, 1993) producing yellowish, soft mucoid faeces that can also contain occult blood. Faecal-oral route is the primary mode of transmission. It is of higher prevalence in wild caught animals and is a sign of poor hygiene in captivity (Gozalo *et al.,* 1991). Erythromycin is the antibiotic of choice (Johnson *et al.,* 2001). Brack *et al.* (1998) reported cases of Erysipelothrix insidiosa septicaemia in red bellied tamarins and common marmosets. Pathological lesions were gastrointestinal haemorrhages, hepatitis and myocarditis. A vaccination with porcine Erysipelothrix insisdiosa vaccine terminated the infection. Pathogenic strains of Escherichia coli are an important cause of diarrhoea although not well documented as a cause of diarrhoea in NHP. The enteropathogenic (EPEC) strain induces watery, non inflammatory, non bloody diarrhoea while the EHEC (enterohaemorrhagic) strain induces life-threatening haemorrhagic diarrhoea due to the production of a shiga-like toxin, associated with anaemia and neutrophilic leucocytosis (Thomson and Scheffler, 1996). Animals become anorexic, inactive, lethargic and develop a recognised clinical dehydration (Mansfield *et al.,* 2001). Treatments are Enrofloxacine and supportive fluids.

Klebsiella pneumoniae is an opportunistic pathogen that is very common in callitrichids, leading either to sudden death, without prior clinical signs, or pneumonia, septicaemia, peritonitis, lymph node abscessation and enteritis (Berendt *et al.,* 1978). The strains can quickly develop multi drug resistance due to plasmid-transfer.

Bronchopneumonia can be related to a Pseudomonas aeruginosa infection, with associated conditions including endocarditis, myocarditis, empyema and septicaemia (Deinhardt, 1967).

Salmonella spp. can be manifested as gastroenteritis with watery diarrhoea, anorexia and fever with severe dehydration. Transmission is via the faecal–oral route or contaminated food and rodents or insects can also be vectors (Savage, 1995).

Shigellosis doesn't seem to play such as an important role in New World Primates as it does in Old World Primates, but Potkay (1992) mentioned Shigellosisas as an important pathogen in callitrichids, seen in concurrent infections with Salmonella.

In comparison to Old World Primates, New World Primates are not very sensitive to infections with Mycobacterium tuberculosis. Michel and Huchzermeyer (1998) described a case of an anthropozoonosis in a common marmoset, kept as a pet in South Africa, with loss of condition and palpable mass in the abdomen, identified as an abscessed mesenteric lymph node.

Yersinia pseudotuberculosis is of great importance in facilities with outdoor housing. This enterobacterium is spread by rodents. Infected animals develop diarrhoea, ulcerative enterocolitis and mesenteric lymphadenitis, associated with hepatosplenic necrosis (McClure *et al.,* 1986). A polyvalent vaccine should be administered in colonies at higher risk

Parasitic diseases

The Acanthocephalan Prosthenorchis elegans penetrates into the wall of small and large intestines, mainly the lower ileum and caecum as far as the serosa, resulting in ulceration, necrosis, perforation and peritonitis. Trichospirura leptostoma is a spiroid nematode that inhabits the pancreatic duct of common marmosets. Animals with high parasitosis have moderate to severe fibrosis in the pancreas (Hawkins *et al.* 1997). Clinical symptoms are weight loss and increased faecal volume. 50 mg/kg Fenbendazol SID for 14 days is the most effective treatment.

Severe pterigodermatitis (Rictularia nycticebus) is manifested by diarrhoea, weakness, hypoproteinaemia and anaemia. This spiroid is attached to the small intestines and may be treated with Mebendazol or Ivermectine. Cockroaches are the reservoir hosts and should be eliminated (Potkay, 1992).

Giardia lamblia is a flagellate protozoan found world-wide, infecting humans and animals. Clinical signs range from none to watery bloody mucoid diarrhoea, associated with abdominal cramps, bloating, anorexia and nausea (Kalishman *et al*, 1996). Other protozoal infections are Balantidium spp. and Entamoeba spp. Entamoeba histolytica produces cysts in the liver, whereas E. dispar is apathogenic but can also result in diarrhoea, anorexia, weakness, abdominal pain and nausea. Natural infections are very uncommon. Drug of choice is Metronidazol or Paramomycinsulfate.

New World Primates are very sensitive to infections with Toxoplasma gondii. Transmission is mostly by ingestion of sporulated oocysts shed by felidae, or diaplazentar (Potkay, 1992). Death occurs after 5–6 days with non-specific clinical symptoms like anorexia, weakness, fever, coughing, dyspnoe , leukopenia and abortion.

Filariasis can be overwhelming with thrombus-like occlusions of small vessels in the lung, heart muscle and liver.

Mycotic diseases

There are not many case reports of mycotic infections in callitrichids. They can be observed after long and repeated antibiotic treatment. Candida sp. is a normal inhabitant of mucous membranes and skin (Savage, 1995). Juan-Salles *et al.* (1998) described a case of intestinal Cryptococcosis in a common marmoset with fibrinonecortizing enteritis.

Non-infectious diseases

Geula *et al.* (2002) reported that 60% of marmosets, older than seven years, have deposits of β-amyloid in the brain, liver and kidney, without having clinical symptoms.

Dental abscesses, recognised by a typical swelling beneath the eye, are very common and not only in aged monkeys. The oral cavity should be carefully examined for cracking, splitting or looseness of teeth, exposed pulp cavity or peridontitis. Antibiotics should be administered prior to tooth extraction under general anaesthesia.

In 1988 Brack carried out a retrospective study of all Callitrichidae at the German Primate Centre, revealing that 91% of tamarins and marmosets, older than six months, had an IgM-mediated mesangioproliferative glomerulonephropathy. Clinical signs can be proteinuria and haematuria. Eitner *et al.* (2001) considered that it is an IgA-mediated nephropathy.

Hemosiderosis is the deposition of the iron pigment, hemosiderin, in the liver, and is a common finding in many New World Primates (Miller *et al.,* 1997). They found 100% incidence of hemosiderosis in animals with Wasting Marmoset Syndrome. There is also evidence of immune dysfunction and increased susceptibility to infection in affected individuals (DeSousa, 1989). Spelman *et al.* (1989) hypothesised that the clinical disease associated with hemosiderosis in captive lemurs is caused by excessive dietary iron intake, high dietary ascorbic acid and low amounts of tannin. Sergejew *et al.* (2000) treated iron overloaded marmosets successfully with different chelations.

Osteomyelitis can occur after bite wounds on the digits or tail. Disinfection with poviodine and antibiotic administration are the best prophylactic measures. Amputation of digits may be necessary.

Wasting Marmoset Syndrome (WMS) is a disease of poorly understood aetiology. It may be related to stress, malnutrition (e.g. protein deficiency, too much fruit in the diet), parasitic, bacterial or viral infections or colitis. Clinical symptoms are rapid weight loss (>30% of LBW/week), dull fur, alopecia, particularly on the tail and breast, muscle atrophy, diarrhoea or obstipation colitis. Clinical pathology includes normochromic anaemia, thrombocytosis, hypoproteinaemia, hypoalbuminaemia, and elevated alkaline phosphatase levels (Logan and Khan, 1996; Tucker, 1984). Lewis *et al.* (1987) compared the faecal microflora of healthy marmosets with animals who developed WMS. The latter showed an increase of bacteriodes and fewer lactobacilli. The value of dietary lactobacilli in NHP is a largely unexplored area. A chronic inflammation may be the primary insult in the development of WMS in many marmosets. Treatment with antibiotics and supportive care with fluids, immune modulators, Vitamin D_3 and Calcium, Ensure®, Nutri-Cal® may help.

There are very few cases of Diabetes mellitus in NWP (Howard and Yasudu, 1990). Metabolic Bone Disease is mostly associated with a diet low in Vitamin D_3 and calcium and the lack of sunlight, followed by osteomalacia and secondary hyperparathyroidism (Hatt and Sainsbury, 1998). Clinical findings are bone fractures. Multiparous females are of higher risk because of their high demand for Calcium and D_3 during pregnancy and lactation.

Nasopharyngeal Squamous Cell Carcinoma was described in two related marmoset colonies by Betton (1983) and McIntosh *et al.* (1985), and characterised by conjunctivitis, mucoid nasal discharge and exophthalmos.

Gozalo *et al.* (1993) described a case of a renal hemangiosarcoma in a moustached tamarin.

In captivity, cotton-top tamarins develop colon adenocarcinoma spontaneously, with a high incidence of up to 35% of adult animals aged 5–10 years. If chronic diarrhoea and weight loss are present, colon carcinoma should be considered. Animals respond to Sulfazalazine treatment given for more than two months (Clapp, 1993).

Dystocia is frequently observed in callitrichids and not only in primiparous animals. Caesarean section is indicated if the female shows signs of labour for more than one hour. It is also possible that the females deliver one or two babies without problems and the last baby has to be delivered via c-section. Retentio secundinaria has been observed very rarely. A case of Placenta previa is described by Lunn (1980).

Abbreviations

BID	=	Twice Daily
CITES	=	Convention on International Trade in Endangered Species on Wild Fauna and Flora
IM	=	Intramuscularly
IU	=	International Units
IV	=	Intravenously
LBW	=	Live Bodyweight
MRI	=	Magnetic Resonance Imaging
NCR	=	National Council Research
NHP	=	Nonhuman Primates
NWM	=	New World Monkeys
NWP	=	New World Primates
OWM	=	Old World Monkeys
OWP	=	Old World Primates
PO	=	Orally
QOD	=	Every Other Day
SID	=	Once Daily
SC	=	Subcutaneously
TID	=	Three Times Daily
WMS	=	Wasting Marmoset Syndrome

Correspondence

Any correspondence should be directed to Susanne Rensing, Department of Animal Health, Covance Laboratories, Kesselfeld 29, D-48163, Muenster, Germany.

References

Abbott, D.H. (1984). *Am. J. Primatol.* 6, 169–186.

Abou-Madi, N. (1999). In Sodaro, V. and Saunders, N. (eds) *Callitrichid Husbandry Manuel,* pp 127–141. Neotropical Primate Taxon Advisory Group.

Anzenberger, G. and Gossweiler, H. (1993). *Am. J. Primatol.* 31, 223–230.

Asper, M., Hofmann, P., Osmann, C., Funk, J., Metzger, C., Bruns, M., Kaup, F.J., Schmitz, H. and Guenther, S. (2001). *Virology* 284, 203– 213.

Ausman, L.M., Gallina, D.L., Hayes, K.C. and Hegsted, D.M. (1986). *Am. J. Clin. Nutr.* 43, 112–127.

Bennett, B.T., Abee, C.R. and Henrickson, R. (1995). In Bennett, B.T. (ed.) *Nonhuman Primates in Biomedical Research: I Biology and Management*. Academic Press, San Diego, USA.

Bennett, B.T., Abee, C.R. and Henrickson, R. (1998). In Bennett B.T. (ed.) *Nonhuman Primates in Biomedical Research, II. Diseases*. Academic Press, San Diego, USA.

Berendt, R.F., Knutsen, G.L. and Powanda, M.C. (1978). *Infect. Immun.* 22, 275–281.

Betton, G.R. (1983). *Vet. Pathol.* 21, 193–197.

Brack, M. (1988). *Vet. Pathol.* 25, 270–276.

Brack, M., Reimbrodt, R., Tschaepe, H. Gilhaus, H. and Rensing, S. (1997). *Erkrank. Zootiere* 38, 317–321.

Brack, M., Rensing, S. and Gatesman, T.J. (1998). *Infect. Disease Review* 1, 15–19.

Byron, J.K. and Bodri, M.S. (2001). *Lab. Animal* 30, 42–48.

Carpenter, J.W., Mashima, T.Y. and Rupiper, D.J. (1996). In Carpenter, J.W. (eds) *Exotic Animal Formulary*, Greystone Publications, Kansas, USA.

Carroll, J.B. (1997). In Pryce, C., Scott, L. and Schnell, C. (eds) *Handbook: Marmosets and Tamarins in Biology and Biomedical Research*, pp 102–109. DSSD Imagery, Salisbury, UK.

Cho, Y., Ramer, J., Rivailler, P., Quink, C., Garber, R.L., Beier, D.R. and Wang, F. (2001). *Proc. Natl. Acad. Sci. USA* 98, 1224–1229.

Christe, K., McChesney, M.B., Spinner, A., Rosenthal, A.N., Allen, P.C., Valverde, C.R., Roberts, J.A. and Lerche, N.W. (2002). *Comp. Med.* 52, 467–472.

Clapp, N.K. and Tardiff, S.D. (1985). *Dig. Dis Sci.* 30, 17–23.

Clapp, N.K. (1993). *A Primate Model for the Study of Colitis and Colonic Cancer: the Cotton-top Tamarin (Saguinus oedipus)*. CRC Press, Boca Raton, Florida, USA.

Coimbra-Filho, A.F. and Mittermeier, R. (1981). In Coimbra-Filho, A.F. and Mittermeier, R.A. (eds) *Ecology and Behavior of Neotropical Primates*, Vol. 1, p 496. Rio de Janeiro: Academia Brasileira de Ciencias.

Cui, K.H., Flaherty, S.P., Newble, C.D., Guerin, M.V., Napier, A.J. and Matthews, C.D. (1991). *J. Androl.* 12, 214–220.

Dalton, M.J. (1985). *Lab. Anim.* 14, 21–22.

Davy, C.W., Trennery, P.N., Edmunds, J.G., Altman, J.F.B. and Eichler, D.A. (1987). *Lab. Anim.* 21, 60–67.

Deinhardt, F. (1967). *Lab. Anim. Care* 17, 11–29.

De Rosa, C., Vitale, A. and Puopolo, M. (2002) *Lab. Anim.* 37, 100–107.

DeSousa, M. (1989). *Clin. Exp. Immunol.* 75, 1–6.

Dietz, J.M. and Baker, A.J. (1993). *Anim. Behav.* 46, 1067–1078.

Eitner, F., Ostendorf, T., Kitahara, M., Rensing, S., Hofmann, P., Maetz-Rensing, K., Kaup, F.J., Groene, H.J. and Floege, J. (2001). *Kidney Blood Pressure Res.* 24, 332.

Ferrari, S.F. and Martins, E.S. (1992). *Am. J. Phys. Anthropol.* 88, 97–103.

Flurer, C.I., Krommer, G. and Zucker, H. (1988). *Lab. Anim. Sci.* 38, 183–186.

Flurer, C.I. and Zucker, H. (1989). *Z. Ernaehrungswiss.* 28, 49–55.

Fortman, J.D., Hewtt, T.A. and Bennett, B.T. (2002). In Fortman, J.D. (ed.) *The Laboratory Nonhuman Primate*. CRC Press, Boca Raton, Florida, USA.

French, J.A., Abbott, D.H. and Snowdon, C.T. (1984). *Am. J. Primatol.* 6, 155–167.

French, J.A., Inglett, B.J. and Dethlefs, T.M. (1989). *Am. J. Primatol.* 18, 73–86.

Garber, P.A. (1986). *Am. J. Phys. Anthropol.* 69, 202.

Garber, P.A. (1992). *Am. J. Phys. Anthropol.* 88, 469–482.

Geretschläger, E., Russ, H., Mihatsch, W. and Przuntek, H. (1987). *Lab. Anim.* 21, 91–94.

Geula, C., Nagykery, N. and Wu, C. (2002). *Acta Neuropathol.* 103, 48–58.

Gozalo, A., Block, K., Montoya, E., Moro, J. and Escamilla, J. (1991). *Primatol. Today*, 675–676.

Gozalo, A., Chavera, A., Dagle, G.E., Montoya, E. and Weller, R.E. (1993). *J. Med. Primatol.* 22, 431–432.

Gough, A.W., Barsoum, N.J., Gracon, S.I., Mitchell, L. and Sturgess, J. (1982). *Lab. Anim. Sci.* 32, 87–90.

Harlow, C.R., Hearn, J.P. and Hodges, J.K. (1984). *J. Endocrinol.* 103, 17–24.

Hatt, J.M. and Sainsbury, A.W. (1998). *Vet. Rec.* 143, 78–80.

Hawkins, J.V., Clapp, N.K., Carson, R.L., Henke, M.A., McCracken, M.D., Faulkner, C.T. and Patton, S. (1997). *Contemp. Top. Lab. Anim. Sci.* 36, 52–55.

Hearn, J.P. (1977). *Lab. Anim.* 11, 261–262.

Hearn, J.P. and Burden, F.J. (1979). *Lab. Anim.* 13, 131–133.

Hearn, J.P., Lunn, S.F., Burden, F.J. and Pilcher, M.M. (1975). *Lab. Anim.* 9, 125–134.

Heath, M. and Libretto, S.E. (1993). *Anim. Technol.* 44, 163–173.

Hershkovitz, P. (1977). In Hershkovitz, P. (ed.) *Living New World Primates (Platyrrhini), with an introduction to Primates, Vol. I*, pp 397–911. The University of Chicago Press, USA.

Howard, C.F. and Yasuda, M. (1990). *J. Med. Primatiol.* 19, 609–625.

Hubrecht, R.C. (1989). *Primates* 30, 423–432.

Jaquish, C.E., Toal, R.L., Tardiff, S.D. and Carson, R.L. (1995). *Am. J. Primatol.* 36, 259–275.

Johnson, L.D., Ausman, L.M., Rolland, R.M., Chalifoux, L.V. and Russel, R.G. (2001). *Comp. Med.* 51, 257–261.

Juan Salles, C., Marco, A. and Domingo, M. (1998). *J. Med. Primatol.* 27, 298–302.

Kalishman, J., Paul-Murphy, J., Scheffler, J. and Thomson, J.A. (1996). *Lab. Anim. Sci.* 46, 116–119.

Kalter, S.S. (1982). *Vet. Pathol.* 19 (Suppl. 7), 33–43.

Kendrick, K.M. and Dixson, A. (1983). *Physiol. Behav.* 30, 732–742.

King, N.W., Hunt, R.D., Daniel, M.D. and Melendez, L.V. (1967). *Lab. Anim. Care* 17, 413–423.

Kingston, W.R. (1969). *Lab. Anim. Handbooks* 4, 243–250.

Kirkwood, J.K. and Underwood, S.J. (1984). *Folia Primatol.* 42, 180.

Kitchen, A.M. and Martin, A.A. (1996). *Lab. Anim.* 30, 317–326.

Kleiman, D.G. (1978). In Kleiman, D.G. (ed.) *The Biology and Conservation of the Callitrichidae.* Smithsonian Institution Press, Washington DC, USA.

Knapka, J.J., Barnard, D.E., Bayne, K.A.L., Lewis, S.M., Marriott, B.M. and Oftedal, O.T. (1995). In Bennet, B.T., Abee, C.R. and Henrickson, R. (eds) *Nonhuman Primates in Biomedical Research: I Biology and Management,* pp 211–248. Academic Press, San Diego.

Kuederling, I., Schneiders, A., Sonksen, J., Nayudu, P. and Hodges, J.K. (2000). *Am. J. Primatol.* 52, 149–154.

Lewis, D.H., Stein, F.J., Sis, R.F. and McMurray, M.N. (1987). *Lab. Anim. Sci.* 37, 103–105.

Logan, A.C. and Khan, K.N.M. (1996). *Tox. Pathol.* 6, 707–709.

Lunn, S.F. (1980). *Vet. Rec.* 106, 414.

Majolo, B. and Buchanan-Smith, H.M. (2003). *Lab. Animal Europe* 3, 25–33.

Mansfield, K.G., Lin, K., Xia, D., Newman, J.V., Schauer, D.B., MacKey, J., Lackner, A.A. and Carville, A. (2001). *J. Infect. Dis.* 184, 803–807.

McClure, H.M., Brodie, A.R., Anderson, D.C. and Swenson, R.B. (1986). In Benirschke, K. (ed.) *Primates: The Road to Self-sustaining Populations,* pp 531–556. New York: Springer-Verlag.

McIntosh, G.H., Giesecke, R., Wilson, D.F. and Goss, A.N. (1985). *Vet. Pathol.* 22, 86–88.

McIntosh, G.H., McMurchie, E.J., James, M., Lawson, C.A., Bulman, F.H. and Charnock, J.S. (1987). *Arteriosclerosis* 7, 159–165.

Melendez, L.V., Daniel, M.D., Barahona, H.H., Fraser, C.E., Hunt, R.D. and Garcia, F.G. (1971). *Lab. Anim. Sci.* 21, 1050–1054.

Michel A.L. and Huchzermeyer, H.F.A.K. (1998). *J. S. Afr. Vet. Ass.* 69, 64–65.

Miller, G.F., Barnard, D.E., Woodward, R.A., Flynn, B.M. and Bulte, J.W.M. (1997). *Lab. Anim. Sci.* 47, 138–142.

Moehle, U., Heistermann, M., Einspanier, A. and Hodges, K. (1999). *J. Med. Primatol.* 28, 36–47.

Montali, R.J., Scanga, C.A., Pernikoff, D., Wessner, D.R., Ward, R. and Holmes, K.V. (1993). *J. Inf. Diseases* 167, 946–950.

Murphy, B.L., Maynard, J.E., Krushak, D.H. and Berquist, K.R. (1972). *Lab. Anim. Sci.* 22, 339–343.

National Research Council, Committee on Animal Nutrition (2003). In *Nutrient Requirements of Nonhuman Primates.* National Academy of Science, Washington DC, USA.

Nigida, S.M., Falk, L.A., Wolfe, L.G. and Deinhardt, F. (1979). *Lab. Anim. Sci.* 29, 53–60.

Paul-Murphy, J. (1993). In Fowler, M.E. (ed.) *Zoo and Wild Animal Medicine: Current Therapy 3.* pp 344–351. WP Saunders. Philadelphia, USA.

Poole, T.B. (1990). *Anim. Technol.* 41, 81–86.

Poole, T., Hubrecht, R. and Kirkwood, J.K. (1999). In Poole, T. and English, P (eds) *The UFAW Handbook on the Care and Management of Laboratory Animals,* pp 559–573. Blackwell Science, Oxford, UK.

Potkay, S. (1992). *J. Med. Primatol.* 21, 189–236.

Power, M.L. and Oftedal, O.T. (1996). *Am. J. Primatol.* 40, 131–144.

Power, M.L., Oftedal, O.T., Savage, A., Blumer, E.S., Soto, L.H., Chen, T.C. and Hollick, M.F. (1997). *Zoo Biology* 16, 39–46.

Ramer, J.C., Garber, R.L., Steele, K.E., Boysen, J.F., O'Rourke, C. and Thomson, J.A. (2000). *Lab. Anim. Sci.* 50, 59–68.

Rothe, H., Wolters, H.J. and Hearn, J.P. (1978). In *Biology and Behaviour of Marmosets.* Proceedings of the Marmoset Workshop Göttingen, Germany, September, 2–5, 1977. Mercke-Druck, Göttingen, Germany.

Rylands, A.B., Coimbra–Filho, A.F. and Mittermeier, R.A. (1993). In Rylands, A.B. (ed.) *Marmosets and Tamarins,* pp 11–77. Oxford University Press, UK.

Saltzmann, W., Schultz-Darken, N.J. and Abbott, D.H. (1997). *Am. J. Primatol.* 41, 159–177.

Savage, A. (1995). In Savage, A. (ed.) *The Cotton-top Tamarin Husbandry Manuel.* Roger Williams Park Zoo, Providence, RI, USA.

Schoenfeld, D. (1989). *Am. J. Primatol.* (Suppl.1): 45–51.

Sergejew, T., Forgiarini, P. and Schnebli, H. (2000). *Br. J. Hematol.* 110, 985–992.

Spelman, L.H., Osborn, K.G. and Anderson, M.P. (1989). *Zoo Biology* 8, 239–251.

Tarantal, A.F. and Hendrickx, A.G. (1988). *J. Med. Primatol.* 17, 105–112.

Tardif, S.D. (1984). *Am. J. Primatol.* 6, 199–209.

Thomson, J.A. and Scheffler, J.J. (1996). *Lab. Animal. Sci.* 46, 275–279.

Tucker, M.J. (1984). *Lab. Anim.* 18, 351–358.

Ventura, R. and Buchanan-Smith, H.M. (2003). *Int. J. Primatol.* 24, 399–413.

Vignes, S., Newman, J.D. and Roberts, R.L. (1992). *Contemp. Top. Lab. Anim. Sci.* 40, 26–29.

Ziegler, T.E., Savage, A., Scheffler, G. and Snowdon, C.T. (1987). *Biol. Reprod.* 37, 618–627.

CHAPTER 11

Management of Old World Primates

Keiji Terao
Tsukuba Primate Center for Medical Science, National Institute of Infectious Diseases, 1-Hachimandai, Tsukuba, Ibaragi 305-0843, Japan

Introduction

Macaques, especially rhesus and cynomolgus monkeys, are widely used in biomedical research and have become the most common laboratory primates internationally. The care and management of macaques requires well-designed facilities, equipment and standard operating procedures supported by well-trained and experienced personnel. This chapter describes practical procedures of care and management of macaques to maintain healthy breeding laboratory primates.

Housing

The type of caging has the most profound influence on the life of primates maintained for research. The approach to housing has changed considerably throughout the years. While, historically, issues of sanitation, infection and experimental control have governed cage design, current philosophies are strongly influenced by welfare considerations, particularly psychological welfare. The factors influencing the choice of housing also include the type of facility, research requirements, location and climate.

Single or pair housing

Cages utilized in single or pair housing are typically constructed of stainless steel mesh or wire (Figures 11.1, 11.2). Aluminium is also used but lacks the strength of stainless steel and galvanized metal and, although the least expensive, it is not often used today. Galvanized metal has the shortest lifespan, followed by aluminium, due to interaction with sanitizing materials and temperatures, while stainless steel lasts indefinitely.

Cage sizes are determined in some countries by local legislation and in others by professional judgement. Minimum cage sizes vary internationally but some recommendations are provided in Table 11.1 for Old World Primates (modified from IPS, 1988).

It is essential that floors and walls of animal rooms are constructed from materials that are non-toxic and water resistant. Floors should slope towards a trench

The Laboratory Primate

ISBN 0-1208-0261-9

Figure 11.1 Individual cages for female breeders of cynomolgus monkey. Left: 40 cm wide, 60 cm deep and 60 cm high for infant and juvenile. Right: 50 cm wide, 80 cm deep and 80 cm high for breeders.

which, in turn, inclines towards a drain, so that waste can be flushed effectively.

Communal housing

Communal housing involving more than three animals can be indoors, indoors-outdoors or outdoors, depending on the particular situation and location of each facility. Indoor communal housing is usually achieved by either compartmentalizing an animal room with panels of wire mesh, or utilizing entire animal rooms. For indoor-outdoor housing, the indoor communal cages are linked to kennel type outdoor runs and both sections have to be large enough to accommodate all animals at the same time. Types of outdoor housing range from cages constructed from completely enclosed structures, made from galvanized wire mesh panels, to open topped corrals. In the latter case, it is important to ensure that no animal can escape from the corral and this is done either with a moat or by constructing the

Figure 11.2 Indoor individual housing system.

TABLE 11.1: Recommended cage sizes							
Weight (kg)	Cercopithecoids (e.g. *Papio, Macaca, Cercopithecus*) relating to C	Floor area/animal (sq m)			Min. cage height (cm)		
		A	B	C	A	B	C
1–3	Up to 4 kg	0.28	0.35	0.60	77	75	1.00
3–5	Up to 4 kg	0.40	0.50	0.60	77	80	1.00
5–7	Up to 6 kg	0.40	0.70	0.80	77	85	1.10
7–9	Over 6 kg	0.40	0.90	1.40	77	90	1.50
9–15	–	0.56	1.10	–	82	125	–
15–25	–	0.74	1.50	–	92	125	–

A: US Public Health Service recommendations (1985).
B: Council of Europe guidelines (1986).
C: Royal Society UFAW guidelines (1987).

corral walls from solid galvanized panels that incline at about 15° inwards.

The floors in communal housing can be constructed from wire mesh, concrete, gravel or soil. The latter three are generally utilized in outdoor housing. Outdoor housing must include shelters to provide for complete protection from the elements, and such shelters must be able to accommodate all of the animals in that enclosure. Depending on the species and geographic location, corrals may also have to incorporate heated areas.

Size of communal housing, as well as the type of the enrichment devices provided, will depend on the species kept. Terrestrial species, for example, will require more floor space, while more arboreal species need climbing structures and vertical space.

The Tsukuba experience

Function

The Tsukuba Primate Center, in Japan, is an indoor facility housing Cynomolgus monkeys as breeding pairs in a closed colony system. Breeding has progressed through successive generations and a computer program provides daily schedules of mating, selection of partners, pregnancy diagnosis, management of pregnant females, weaning, regular body weighing, physical examination and medical treatments.

The facility and operating procedures

The animal facility is completely isolated from the staff area, and each of the breeding, rearing, holding and quarantine areas functions, independently. Each has its own ventilation system and staff to prevent the spread of infectious diseases among the different animal populations. The animal rooms have no windows and each is air conditioned with a constant filtrated fresh air supply, 24 hours/day throughout the year. All air is exhausted from the facility to the outside, after filtration with deodorant and HEPA filters. The relative air pressure is higher in the animal rooms, creating a one way-air flow to prevent the spread of aerosol-borne diseases. The air temperature of animal rooms is maintained at around $25 \pm 5°C$ and the relative humidity at $60 \pm 10\%$, with a photoperiod of 12-hours from 7:00am to 7:00pm. Infants are reared by their mothers and, after weaning, two infants of the same body size and age are housed together until two years of age. Food trays are washed every day, to remove waste, and floors are washed daily with water to remove excreta and waste, followed by application of disinfectant.

Personnel are required to change their clothes to overalls when they enter the animal rooms and shower when they exit. In addition to overalls, staff are required to wear a vinyl apron, rubber boots, arm cover, rubber gloves and a face guard (Figure 11.3). Used uniforms are autoclaved, washed and re-distributed. Sick personnel, or those whose family members suffer from infectious diseases, are prohibited from entering the animal facility until they recover.

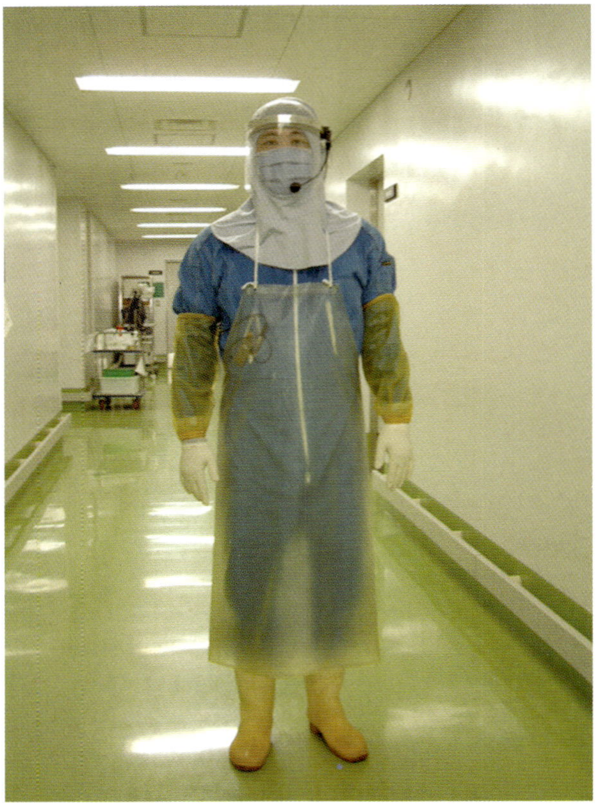

Figure 11.3 Overall uniform in animal rooms.

Feeding

Adult monkeys are each fed 100 g of fresh apples in the morning and 70 g of commercially prepared monkey diet in the afternoon. The tap water is supplied by automatic watering devices, and the water supply line flushed at least once a week to reduce disease hazards from the water supply. Diet and fruits, for the monkeys, are kept in a cold room in the food preparation area and distributed every morning in a separate tray for each animal room.

Health and microbiological monitoring

A daily individual health check of the monkeys is important to detect signs of disease and/or menstruation. The personnel carefully check for normal activity, bleeding or injury on the body surface, stool consistency, occurrence of exanthema and nasal discharge, presence of menstrual bleeding and any other clinical signs. All observations are recorded electronically and downloaded onto the host computer via a computer terminal in each animal facility (Figure 11.4). Food consumption of each monkey is also recorded.

Figure 11.4 Daily observation and record keeping through portable recorders.

All animals are monitored at least once every two years to obtain sero-epidemiological data for microbiological management, and also for the establishment of each monkey's baseline values for biological, hematological and biochemical variables in order to be able to diagnose when these are abnormal in any individual. The following investigations are performed (Figure 11.5):

1. General conditions by palpation, auscultation, internal and intra oral examination.
2. Body weight and temperature.
3. Haematology, biochemistry, microbiology and virology.
4. Tuberculin skin testing.

All data obtained from these health examinations are stored in the individual's record files and regularly analysed for changes, with time, of biological and physiological parameters. All of the short- and long-term monitoring procedures that are established in the facility are to ensure early detection of infectious disease, injury or other pathology and the validation or modification, when required, of management procedures. The examination and recording of the pathology of every spontaneous death is important as these could also directly affect procedures for the short- or long-term management of the section and/or the facility.

The purpose of microbiological management is to minimize the risk of the zoonotic, epizootic and zooanthroponosic infections between animals and personnel or among animals. The microbiological investigations that are conducted include testing for Tubercle bacilli, Herpes B (HBV), simian varicella virus (SSV), SIV and measles virus. These are undertaken every alternate year to ensure the maintenance of specific pathogen free (SPF) status.

Results of monitoring

Tables 11.2, 11.3, 11.4 and 11.5 summarize baseline values established in our primate center during the past 20 years. Since these values are influenced by age, environmental condition, diet, housing or other factors, it is necessary to establish these for every facility. Indeed, the longitudinal monitoring of these parameters in our facility led to the reduction of the calorie content of our diet to decrease both the frequency of obesity and the serum triglyceride levels in our breeders (Figure 11.6).

Standardized procedures assure reliable and reproducible results since it is well established that the laboratory environment influences the psychological and physiological condition of the animals which, in turn, affect research results.

Acknowledgements

I wish to thank Drs. I. Sakakibara and T. Yoshida for preparing data and to Drs. R. Mukai, T. Yoshida, T. Sankai, K. Fujimoto, F. Ono, N. Ageyama and Mr. H. Ohto for their cooperation.

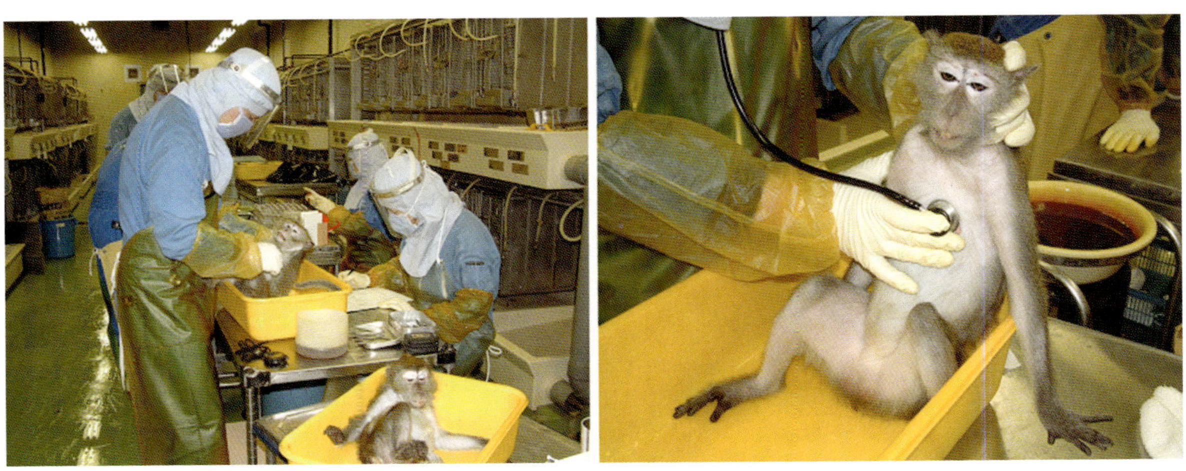

Figure 11.5 Regular health examination.

TABLE 11.2: Abbreviation and unit of physiological parameters		
Hematological parameter	**Abbreviation**	**Unit**
White blood cell count	WBC	$\times10^2$/l
Red blood cell count	RBC	$\times10^4$/l
Hemoglobin concentration	HGB	g/dl
Hematocrit value	HCT	%
Mean corpuscular volume	MCV	fl
Mean corpuscular hemoglobin	MCH	pg
Mean corpuscular hemoglobin concentration	MCHC	g/dl
Platelet count	PLT	$\times10^4$/l
WBC-small cell ratio	W-SCR	%
WBC-large cell ratio	W-LCR	%
WBC-small cel count	W-SCC	$\times10^2$/l
WBC-large cell count	W-LCC	$\times10^2$/l
RBC distribution width	RDW-SD	fl
PLT distribution width	PDW	fl
Mean platelet volume	MPV	fl
Serum biochemical parameter		
Glutamic oxaloacetic transaminase activity	GOT	IU/l
Glutamic pyruvic transaminase activity	GPT	IU/l
Total protein concentration	TP	g/dl
Albumin concentration	ALB	g/dl
Albumin-globulin ratio	A/G	
Blood urea nitrogen	BUN	mg/dl
Glucose concentration	GLU	mg/dl
Total cholesterol concentration	T-CHO	mg/dl
Free cholesterol concentration	F-CHO	mg/dl
Triglyceride concentration	TG	mg/dl
Alkaline phosphatase activity	ALP	IU/l

TABLE 11.3: Normative hematological values in laboratory-bred cynomolgus monkeys at different ages

Item	Sex	Age in year								
		1	2	3–4	5–6	7–8	9–10	11–12	13–14	15
WBC	F	99 ± 38	103 ± 39	106 ± 33	91 ± 30	87 ± 30	93 ± 27	88 ± 26	111 ± 27	88 ± 30
	M	89 ± 28	98 ± 34	93 ± 16	86 ± 25	93 ± 40	82 ± 31	97 ± 46	76 ± 33	96 ± 30
RBC	F	576 ± 85	603 ± 56	560 ± 60	597 ± 69	576 ± 68	561 ± 55	548 ± 69	533 ± 52	545 ± 74
	M	554 ± 62	547 ± 54	591 ± 81	562 ± 67	556 ± 80	528 ± 64	509 ± 63*	528 ± 41	526 ± 41
HGB	F	11.9 ± 1.1	12.0 ± 0.7	11.7 ± 1.1	12.1 ± 0.9	12.1 ± 1.0	12.3 ± 1.2	13.0 ± 0.9**	13.5 ± 0.5**	13.3 ± 1.6**
	M	11.7 ± 1.2	11.6 ± 0.8	13.0 ± 0.9**	12.2 ± 1.0	12.4 ± 1.0**	12.8 ± 1.2**	12.8 ± 1.6**	13.0 ± 1.0**	13.8 ± 1.1**
HCT	F	37.3 ± 4.2	39.5 ± 3.2*	37.1 ± 3.9	39.7 ± 3.7*	38.5 ± 3.4	38.2 ± 3.7	40.0 ± 4.5*	39.5 ± 3.1	40.0 ± 5.5*
	M	35.6 ± 3.9	35.9 ± 3.0	39.9 ± 3.9**	37.1 ± 4.1	37.2 ± 4.4	37.0 ± 3.7	37.9 ± 4.7*	39.3 ± 3.1**	40.5 ± 3.6**
MCV	F	65.2 ± 4.9	65.7 ± 2.5	66.2 ± 3.2	66.8 ± 4.7	67.2 ± 5.0	68.3 ± 6.2*	73.3 ± 3.7**	74.2 ± 3.6**	73.6 ± 5.8**
	M	64.4 ± 3.7	66.0 ± 4.7	68.0 ± 5.5*	66.2 ± 3.9	67.4 ± 4.9**	70.6 ± 6.5**	74.5 ± 4.6**	74.4 ± 3.5**	76.9 ± 2.6**
MCH	F	21.0 ± 2.9	19.9 ± 1.3	21.0 ± 1.5	20.4 ± 2.0	21.2 ± 2.4	22.1 ± 2.6	23.9 ± 1.9**	25.5 ± 2.2**	24.5 ± 2.4**
	M	21.3 ± 1.9	21.4 ± 2.5	22.4 ± 3.3	21.8 ± 1.8	22.6 ± 2.7*	24.5 ± 3.3**	25.4 ± 3.2**	24.9 ± 2.6**	26.3 ± 2.1**
MCHC	F	32.2 ± 2.4	30.3 ± 1.4**	31.7 ± 1.7	30.6 ± 2.6**	31.5 ± 1.9	32.3 ± 1.6	32.5 ± 2.0	34.4 ± 1.8**	33.3 ± 1.8*
	M	33.1 ± 1.7	32.5 ± 2.1	32.8 ± 2.5	33.0 ± 2.2	33.6 ± 2.4	34.6 ± 2.5*	33.9 ± 2.7	33.4 ± 3.5	34.2 ± 2.8
PLT	F	36.1 ± 11.4	43.5 ± 12.3**	39.3 ± 8.8	40.1 ± 9.4	48.3 ± 12.5**	42.3 ± 12.3*	45.6 ± 6.9**	42.2 ± 3.2	44.8 ± 10.6**
	M	34.2 ± 8.5	39.0 ± 9.8*	41.3 ± 8.0*	33.8 ± 8.7	33.8 ± 8.3	37.4 ± 8.8	39.9 ± 6.7*	37.6 ± 10.5	42.3 ± 14.1**
W-SCR	F	40.5 ± 16.5	44.1 ± 14.1	37.0 ± 15.4	40.8 ± 13.1	38.2 ± 15.3	38.7 ± 12.9	38.7 ± 15.4	39.7 ± 18.5	39.5 ± 15.8
	M	46.1 ± 13.5	44.1 ± 12.1	47.0 ± 12.8	42.7 ± 13.6	40.9 ± 13.7	39.5 ± 15.2	41.1 ± 15.1	33.1 ± 10.8**	33.6 ± 14.2**
W-LCR	F	59.5 ± 16.5	55.9 ± 14.1	63.0 ± 15.4	59.2 ± 13.1	61.8 ± 15.3	61.3 ± 12.9	61.3 ± 15.4	60.3 ± 18.5	60.5 ± 15.8
	M	53.9 ± 13.5	55.9 ± 12.1	53.0 ± 12.8	57.3 ± 13.6	59.1 ± 13.7	60.5 ± 15.2	58.9 ± 15.1	66.9 ± 10.8**	66.4 ± 14.2**
W-SCC	F	41 ± 27	44 ± 17	37 ± 15	36 ± 13	31 ± 13*	34 ± 12	33 ± 14	43 ± 21	33 ± 15
	M	40 ± 14	41 ± 12	44 ± 15	36 ± 14	38 ± 20	30 ± 13*	37 ± 16	25 ± 13**	30 ± 11
W-LCC	F	58 ± 25	59 ± 33	69 ± 34	55 ± 25	56 ± 30	58 ± 25	56 ± 28	68 ± 31	55 ± 29
	M	49 ± 23	57 ± 30	49 ± 15	50 ± 20	56 ± 29	52 ± 29	60 ± 39	51 ± 25	65 ± 31
RDW-SD	F	32.4 ± 2.4	32.5 ± 2.1	32.0 ± 2.0	33.0 ± 2.6	32.7 ± 2.3	33.5 ± 3.0	35.1 ± 2.2**	35.3 ± 2.5**	35.2 ± 3.4**
	M	30.8 ± 2.3	32.3 ± 3.0*	32.8 ± 2.3*	31.9 ± 2.1	32.1 ± 2.3*	32.5 ± 3.0**	34. ± 11.5**	33.9 ± 1.3**	34.4 ± 1.7**
PDW	F	11.2 ± 1.8	10.9 ± 2.4	11.0 ± 2.0	10.7 ± 2.0	10.1 ± 1.6*	10.6 ± 1.1	10.3 ± 1.1	11.2 ± 1.8	10.6 ± 1.4
	M	10.6 ± 2.0	9.9 ± 1.6	10.0 ± 0.8	11.0 ± 1.6	10.7 ± 1.8	10.3 ± 1.6	10.0 ± 1.3	11.5 ± 1.7	10.6 ± 1.6
MPV	F	10.2 ± 1.0	9.9 ± 1.2	10.2 ± 1.1	10.0 ± 1.2	9.7 ± 1.0*	10.0 ± 0.7	10.1 ± 0.7	10.5 ± 1.1	10.1 ± 0.8
	M	9.8 ± 0.9	9.5 ± 1.0	9.6 ± 0.7	10.2 ± 0.9	10.0 ± 1.1	10.0 ± 0.9	9.8 ± 0.7	10.7 ± 0.9**	10.2 ± 0.9

1. M: Male, F: Female.
2. Mean ± S.D. Asterisks indicate significant differences vs. the one year old group (*P<0.05, **P<0.01).

TABLE 11.4: Normative serum biochemistry parameters in laboratory-bred cynomolgus monkeys at different ages

Item	Sex	Age in year								
		1	2	3	4	5	6	7	8–9	10
GOT	F	32.0 ± 6.8	35.3 ± 10.5	32.0 ± 6.4	32.7 ± 11.1	28.7 ± 7.1	32.5 ± 5.7	26.0 ± 5.6	29.2 ± 8.3	30.1 ± 9.5
	M	31.7 ± 7.2	32.9 ± 9.0	35.0 ± 6.5	33.9 ± 10.3	31.8 ± 9.8	37.1 ± 6.5	40.0 ± 26.1	29.6 ± 9.9	30.3 ± 7.6
GPT	F	21.5 ± 7.7	28.3 ± 15.7	24.4 ± 8.8	21.8 ± 11.9	26.7 ± 17.7	26.5 ± 7.5	23.6 ± 7.5	35.7 ± 29.8	31.3 ± 26.1
	M	22.6 ± 7.6	24.7 ± 8.4	34.5 ± 24.8	28.6 ± 32.1	21.0 ± 12.0	31.6 ± 14.1	36.4 ± 23.0	25.7 ± 8.7	27.6 ± 23.9
TP	F	6.69 ± 0.37	6.75 ± 0.38	7.16 ± 0.24	6.96 ± 0.45	6.88 ± 0.58	7.02 ± 0.43	6.85 ± 0.37	7.05 ± 0.79	6.56 ± 0.65
	M	6.60 ± 0.38	6.61 ± 0.36	6.90 ± 0.41	7.17 ± 0.47	7.22 ± 0.43	7.48 ± 0.38	7.38 ± 0.47	6.96 ± 0.53	7.18 ± 0.36
ALB	F	4.10 ± 0.15	4.02 ± 0.16	4.16 ± 0.11	4.04 ± 0.24	4.00 ± 0.26	3.94 ± 0.23	3.95 ± 0.26	3.91 ± 0.33	3.70 ± 0.25
	M	4.06 ± 0.21	4.02 ± 0.19	4.09 ± 0.19	4.13 ± 0.20	4.12 ± 0.14	4.11 ± 0.17	4.12 ± 0.15	4.05 ± 0.22	3.95 ± 0.11
A/G	F	1.59 ± 0.21	1.47 ± 0.11	1.39 ± 0.18	1.40 ± 0.13	1.39 ± 0.14	1.28 ± 0.10	1.36 ± 0.11	1.24 ± 0.13	1.30 ± 0.17
	M	1.60 ± 0.13	1.55 ± 0.14	1.45 ± 0.17	1.36 ± 0.14	1.33 ± 0.13	1.22 ± 0.11	1.27 ± 0.10	1.39 ± 0.14	1.22 ± 0.11
BUN	F	19.1 ± 2.6	19.1 ± 4.3	16.2 ± 1.2	19.3 ± 3.2	17.5 ± 4.0	18.7 ± 3.8	15.1 ± 3.1	15.4 ± 3.6	15.7 ± 4.2
	M	19.2 ± 2.6	18.4 ± 3.1	18.2 ± 4.3	18.2 ± 3.2	18.0 ± 3.3	19.6 ± 2.1	20.6 ± 4.1	18.9 ± 2.3	18.2 ± 3.9
GLU	F	38 ± 10	48 ± 20	47 ± 20	36 ± 20	50 ± 28	43 ± 12	35 ± 11	33 ± 12	38 ± 15
	M	39 ± 8	52 ± 16	40 ± 17	46 ± 27	55 ± 34	86 ± 42	42 ± 17	52 ± 41	48 ± 24
T-CHO	F	104 ± 25	112 ± 20	108 ± 16	109 ± 22	112 ± 36	127 ± 36	121 ± 32	110 ± 29	102 ± 32
	M	102 ± 21	117 ± 21	113 ± 17	118 ± 41	100 ± 28	84 ± 21	109 ± 24	93 ± 10	91 ± 22
F-CHO	F	14.1 ± 6.1	19.8 ± 4.5	18.3 ± 3.9	20.3 ± 5.4	22.4 ± 8.2	26.5 ± 8.9	23.2 ± 5.9	22.9 ± 5.8	22.4 ± 8.5
	M	14.4 ± 4.9	19.9 ± 4.9	19.9 ± 4.5	20.0 ± 6.8	18.1 ± 5.3	15.7 ± 5.5	23.9 ± 5.2	18.1 ± 3.8	18.6 ± 3.6
TG	F	48 ± 15	51 ± 22	51 ± 17	72 ± 24	75 ± 43	92 ± 81	102 ± 21	73 ± 32	93 ± 27
	M	53 ± 24	49 ± 14	56 ± 18	48 ± 15	72 ± 31	85 ± 43	98 ± 42	76 ± 46	99 ± 60
ALP	F	56 ± 13	47 ± 11	34 ± 9	21 ± 10	14 ± 6	7 ± 1	8 ± 4	9 ± 2	8 ± 3
	M	54 ± 15	49 ± 11	43 ± 12	39 ± 8	25 ± 11	14 ± 4	8 ± 2	7 ± 1	6 ± 2

TABLE 11.5: Mean body weight of laboratory-bred cynomolgus monkeys at different ages

Female monkeys					
Age in week	0	1	2	3	6
Mean(g)	324.8	324.7	343.8	368	452.1
S.D.	53.3	61.1	63.4	66.8	81
Number of monkeys	1509	1237	1219	1245	1126
Age in week	9	12	15	16	17
Mean(g)	534.8	613	691.1	712.5	736.1
S.D.	88.9	99.1	101.8	111.2	100.5
Number of monkeys	1111	1091	1195	697	844
Age in week	18	19	20	21	22
Mean(g)	751.9	768.9	786.1	794.4	810.6
S.D.	97	97	97.2	101	104.1
Number of monkeys	1449	1213	1311	1499	1417
Age in week	23	24	25	26	27
Mean(g)	827.3	843.3	861.5	887	898.5
S.D.	105.7	110.7	116.1	120	125.7
Number of monkeys	1360	1389	1106	1113	1071
Age in week	28	29–31	34–36	38–42	48–52
Mean(g)	929.8	960.6	1083.2	1170.6	1311.1
S.D.	139.8	149.6	163.8	172.9	214.2
Number of monkeys	950	2315	1463	1510	745
Age in year	1.2	1.5	1.9	2.3	2.7
Mean(g)	1460.4	1692.8	1935.4	2090.9	2267.4
S.D.	227.1	247.3	272.2	333.3	391.5
Number of monkeys	730	580	759	801	710
Age in year	3.1	3.5	3.8	4.8	5.8
Mean(g)	2475.2	2659.2	2744.6	2976.1	3185.7
S.D.	420	490.4	502.7	606.6	677.9
Number of monkeys	704	600	490	508	493
Age in year	6.7	7.7	8.7	9.5	
Mean(g)	3387.2	3565.9	3802.2	3832.7	
S.D.	778.3	839.7	997.7	1021	
Number of monkeys	420	352	237	181	
Male Monkeys					
Age in week	0	1	2	3	6
Mean(g)	351.1	349	370.3	395.9	482.1
S.D.	59	61.6	67.2	72.7	89.8
Number of monkeys	1504	1183	1166	1184	1128
Age in week	9	12	15	16	17
Mean(g)	571	655.8	741.2	761.9	780.7
S.D.	103.2	113.7	114.8	116	112.7
Number of monkeys	1095	1074	1246	879	1061

(Continued)

TABLE 11.5 *(Continued)*

Male monkeys *(Continued)*					
Age in week	18	19	20	21	22
Mean(g)	798.1	813.1	825.4	841.3	860.9
S.D.	108.5	109.4	116.5	114.2	115.5
Number of monkeys	1578	1400	1416	1530	1461
Age in week	23	24	25	26	27
Mean(g)	881.4	903.7	923.4	947.5	971
S.D.	119.6	125.5	132.9	134.3	141.5
Number of monkeys	1341	1271	1052	1005	911
Age in week	28	29–31	34–36	38–42	48–52
Mean(g)	998.5	1042.5	1155.8	1251.5	1420.9
S.D.	145.1	152.4	168.9	185.2	196
Number of monkeys	838	1956	1313	1301	672
Age in year	1.2	1.5	1.9	2.3	2.7
Mean(g)	1558	1789.1	2027.9	2236.7	2443.7
S.D.	222.4	289.2	339.4	407.6	464.1
Number of monkeys	756	591	872	716	598
Age in year	3.1	3.5	3.8	4.8	5.8
Mean(g)	2687.2	3014.3	3320.5	3917.7	4608.3
S.D.	576.7	654.6	796.3	1250	1650.2
Number of monkeys	569	463	332	189	161
Age in year	6.7	7.7	8.7	9.5	
Mean(g)	5258.8	5831	6110.1	6305.3	
S.D.	1514.2	1143.8	1188.9	1036.4	
Number of monkeys	167	146	124	79	

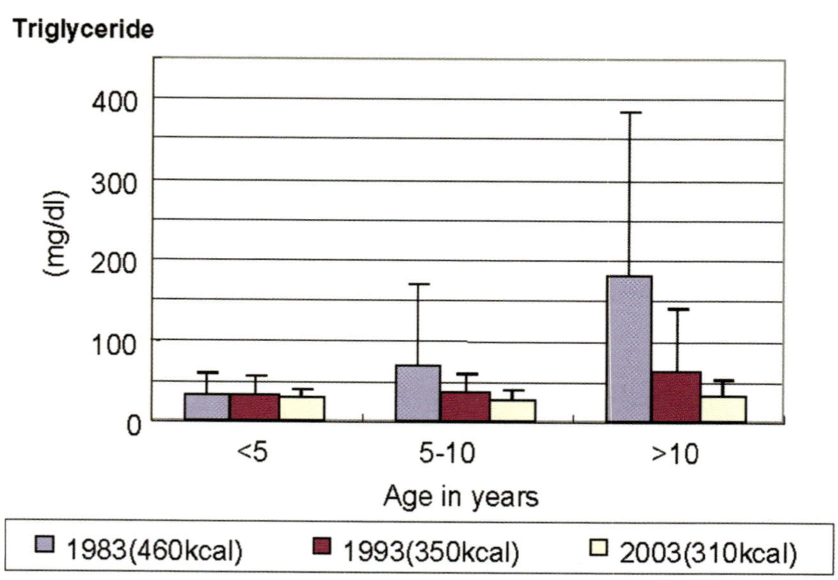

Figure 11.6 Changes in serum triglyceride levels in relation to decrease in calorie intake. Calorie intake was decreased from 460 kcal to 350 kcal in 1982 and from 350 kcal to 310 kcal in 2001.

Correspondence

Any correspondence should be directed to Keiji Terao, Tsukuba Primate Center for Medical Science, National Institute of Infectious Diseases, 1-Hachimandai, Tsukuba, Ibaragi 305-0843, Japan. Email: terao@nih.go.jp

References

Honjo, S. (1985). *J.Med.Primatol.* 14, 75–89.

Honjo, S., Cho, F. and Terao, K. (1984). In Hendrickx, A.G. (ed.) *Advances in Veterinary Science and Comparative Medicine,* Vol. 28. Research on Nonhuman Primates, pp 51–80. Academic Press, New York.

Chapter 12

Vervet Monkey Breeding

Jürgen Seier
Primate Unit, MRC, PO Box 19070, Tygerberg 7505,
South Africa

Introduction: breeding biology

The vervet or African green monkey (*Chlorocebus aethiops*) is one of the two African nonhuman primates utilized in biomedical research. Vervet monkeys are widely distributed throughout sub-saharan Africa and live in small multi-male, multi-female groups which generally consist of up to 50 individuals. Vervet monkeys have been described as marginal seasonal breeders in the wild, but propagate throughout the year in captivity (Eley, 1992; Kushner *et al.*, 1982; Seier, 1986). Usually only one infant is born in the wild, per year, and will remain with the mother for about 12 months (Lee, 1987). When the infant is 3–6 months old, the mother intensifies rejection, thereby beginning the weaning process (Fairbanks and McGuire, 1987). It was estimated that, in some regions, the abortion rate in the wild is about 26% (Turner *et al.*, 1987) and that about 30–57% of infants die during the first year of life (Cheney *et al.*, 1988; Eley, 1992). Vervet monkey females exhibit what is known as "aunting" behaviour, in that they also care for the offspring of other females. The males, however, take no part in the rearing of infants (Gartlan, 1969).

Breeding and rearing systems in captivity

Multi-male, multi-female groups and harems

The most biological system of breeding vervet monkeys is in heterosexual groups. However, some behavioural characteristics of vervet monkeys may positively or negatively affect the success of social housing. A few of these key factors are that:

- males can be aggressively challenged by coalitions of females (Fairbanks and McGuire, 1987; Hector and Raleigh, 1992);
- subordinate males are more likely than dominant males to attack inappropriate targets such as females or immature animals (Raleigh and McGuire, 1990);
- lower ranking females have lower fecundity and higher fetal wastage and neonatal losses (Fairbanks and McGuire, 1984; Turner *et al.* 1987).

These factors have the potential to induce tension, decrease productivity and increase morbidity and

The Laboratory Primate
Copyright 2005 Elsevier
ISBN 0-1208-0261-9

mortality in group situations. Considerable fighting between females was observed in harem groups made up of single males and 5–10 females maintained in outdoor communal cages (Else, 1985). Of ten original groups, three had to be disbanded and, occasionally, it was found necessary to disband and reform the remaining groups. Although just under 90% of females in these groups still conceived annually, the fighting was associated with impaired productivity reflected by a high infant mortality rate of 21%, and a low annual production rate of 67%.

Smaller harem groups, consisting of one male and two to six females, (Kushner *et al.*, 1982) appear to function better. When maintained in this way, in indoor communal cages, with pregnant females removed from the group before birth and returned two weeks after delivery, the numbers of stillbirths and abortions amounted to a low 19% with infant mortality at only 14% within the first 150 days. The birth interval was 357 days, infants were reared in peer groups and no particular behavioural problems were reported in this situation.

Pair breeding

Socially a lot less complex and with a considerably lower potential for conflict, pair breeding is another useful system of propagating vervet monkeys in captivity. Since vervet monkey troops, in the wild, disband only when fewer than two adults remain in a group (Isbell *et al.*, 1991), it seemed logical that pair breeding could still approximate the biological condition. Although the fecundity of even low ranking females should be high, due to the fewer compatibility problems associated with the uncomplicated social structure of housing in pairs, about 86% of females conceive annually in this system (Seier, 1986), no more than in a group situation (Else, 1985). Furthermore, placing the female into the home cage of the male, rather than the reverse, was found to be preferable (Seier, 1986), while removing the female from the male, for pregnancy and lactation, only slightly increased the inter-birth-interval to 373.8 days (Seier, 1986) compared to 357 days in harem systems (Kushner *et al.*, 1982). Although the lower potential for conflict and trauma, in pair breeding, has not resulted in decreased rates of stillbirths and abortions, from the 16.7%–28% reported (Johnson *et al.*, 1973; Seier, 1986), infant mortality has been reduced to about 4.2% (Seier, 1986). However, unless pairs are housed together with other juveniles, this system would be less suitable for infant rearing, due to the lack of access to peers for play and normal development.

The menstrual cycle

Female vervet monkeys have no cyclic perineal swelling and menstruation is scant (Seier *et al.*, 1991), often only detectable by vaginal swab. This makes timed matings considerably more difficult than in baboons or macaques. The length of the menstrual cycle is about 30–32 days, 18 days of which is the duration of the luteal phase (Eley *et al.*, 1989; Hess *et al.*, 1979). Menstruation lasts for 2.5–5.0 days (Cho *et al.*, 2002; Johnson *et al.*, 1973; Seier *et al.*, 1991).

Mating, conception, pregnancy and birth

In captivity, mating takes place throughout the cycle and pregnancy and not just periovulatory (Rowell, 1971), but the frequency of copulations appears to be low in this species (Eley, 1992). In a pair breeding situation, 76 females required an average of 39.8 days of exposure to the male to conceive, irrespective of the stage of the menstrual cycle (Seier, 1986).

Implantation bleeding is not a reliable marker for conception. It has been reported to occur for 1–5 days in only 10 out of 42 pregnancies, between 13 and 29 days after mid-cycle timed mating (Hess *et al.*, 1979) and, in another study, for one day in 33% of pregnancies (Johnson *et al.*, 1973). Pregnancy can be determined using an electronic linear scanner but, in vervet monkeys, it was found necessary to insert an index finger into the rectum to elevate the uterus toward the abdominal wall (Seier *et al.*, 2000) before it was possible to obtain images. Key diagnostic features are provided in Table 12.1.

Earlier signs, such as an endometrial line swelling, or an endometrial "pregnancy" ring, could not be used reliably and consistently to diagnose pregnancy in vervet monkeys. A simple, yet effective way of determining pregnancy is by rectal palpation (Seier, 1986). When performed by an experienced person, this method enables pregnancy diagnosis 25–30 days after breeding (Hess *et al.*, 1979), or 15–21 days post conception (Seier, 1986) with prior knowledge of each female's non-gravid uterine size and shape.

TABLE 12.1: Key diagnostic features for dating vervet monkey pregnancy ultrasonographically

	Mean	Range	±SD	n
Menstrual age at first definite diagnosis of pregnancy (days)	33.1	30–35	1.48	20
Ventro-dorsal diameter of gestational cavity at first diagnosis (mm)	2	1–4	0.80	20
Menstrual age at first visualization of the yolk sac (days)	38.0	32–41	3.10	20
Size of yolk sac at first visualization (mm)	3.3	3–4	0.40	20
Menstrual age at first detectable heart beat (days)	45.5	44–48	1.73	4
Menstrual age of first measurable embryo (days)	35.0	–	–	1
Size of first measurable embryo (mm)	2	–	–	1
Biparietal diameter at 9 weeks (mm)	10	9–11	0.80	4
Biparietal diameter at 12 weeks (mm)	20	17–22	0.21	4

Source: From Seier, 2000; with permission from Blackwell Munksgaard.

Table 12.2 provides the changes in the uteri of 26 females and 103 pregnancies.

The gestation period in vervet monkeys is between 157 and 167 days (Johnson *et al.*, 1973; Hess *et al.*, 1979) and the rate of abortions and stillbirths ranges from 16.7 to 28% (Seier, 1986; Johnson *et al.*, 1973; Kushner *et al.*, 1982).

When comparing captive bred with wild caught individuals (Seier, 1986), the former experienced a higher abortion rate but slightly lower stillbirth rate (Table 12.3). However the higher rate of abortions in the captive bred females was due to a few habitual aborters.

The age of females at first birth is about 36–41 months (Fairbanks and McGuire, 1984; Seier 1986), so is obviously the age range to start breeding with females. Males are sexually mature at about 46 months of age (Hiyaoka *et al.*, 1990). Although the first pregnancy is sometimes considered to be high risk (Eley, 1992), others found no trend of pregnancy outcome in relation to successive pregnancies (Lambrecht *et al.*, 1999; Seier, 1986).

Some seasonality in the number of births was observed but basically they occurred every month of the year in captivity (Hess *et al.*, 1979; Seier, 1986). Most births occur either late in the afternoon, at night or early in the morning and the birth process takes about 15 to 20 minutes (Seier, 1986). The rate of twinning is reported to be about 0.4% (Pollack and Raleigh, 1994), but no female has successfully raised twins. About 24–49 days after birth, the uterus has completed involution (Seier, 1986) and lactational amenorrhea lasts about three to six months (Eley, 1992). The average birth weight is 364 and 352 g for males and females respectively (Seier, 1986), while others report male and female birth weights as 343 and 318 g

TABLE 12.2: Changes in uterine size and consistency during the first 70 days of gestation

Stage (days)	Mean		Mean		Shapes and Consistencies
	Width (cm)	Range (cm)	Length (cm)	Range (cm)	
Non-gravid	1.5	0.5–2.5	2.8	1.0–5.5	Club shaped, hard
8–14	2.4	2.0–3.0	4.3	3.0–6.0	Fundus globular
22–28	2.7	2.0–3.5	5.2	4.0–7.0	Globular soft, sometimes tightly inflated like balloon
36–42	3.3	2.5–4.5	6.3	4.0–9.0	Round-oval, feels like tightly inflated balloon
50–56	4.1	3.0–5.5	7.8	5.0–11.0	Still oval balloonlike
64–70	5.3	4.0–7.0	10.8	7.0–14.0	Diffuse, difficult to palpate

Source: Modified from Seier, 1986; with permission from Blackwell Munksgaard.

TABLE 12.3: Fetal wastage rate (%)		
	Abortions	Stillbirths
Captive bred females	18.2	4.9
Wild caught females	10.2	6.4
Total	14.0	5.7

Source: Modified from Seier, 1986; with permission from Blackwell Munksgaard.

respectively (Cho *et al.*, 2002). The placenta is bi-lobed and weighs about 74.6 g (Seier, 1986).

Timed matings

The timing of matings to produce dated pregnancies is considerably more difficult in vervet monkeys than in other Old World species. Due to the lack of cyclic perineal swelling and scant menstruation, there are no external markers to schedule mating.

The only reliable visual aid is swabbing for menstrual blood by inserting a cotton tip applicator into the vagina (Seier *et al.*, 1991; Johnson *et al.*, 1973). The first day of menstruation is day one of the menstrual cycle. Various timed mating strategies have produced different results: Johnson *et al.* (1973) mated on cycle day 10 for a period of 6 days, which required 4.8 to 11.4 matings per conception, resulting in a conception rate of 20.8 and 8.7%. Hess *et al.* (1979) mated during midcycle, either exposing females to the males each day for 8 h for 2–3 days or every other day for three days or, as in a third technique, for two days before midcycle. The first two methods achieved a conception rate (conception/mating) of 45.8% and the latter 21.8%. In the author's experience mating continuously from cycle day 7 to 22 achieved a conception rate of 45% (Seier *et al.* 2000), similar to Hess *et al.* (1979). Cho *et al.* (2002) mated on and after the day of ovulation, and achieved a conception rate of 48.9%.

It appears that exposing the female for a longer period of time to the male, fails to produce more conceptions.

Growth and development

Neonatal and infant mortality rates of 4.2% to 21.0% have been reported, depending on the rearing system used. Data on the physical development of infant vervet monkeys is provided in Table 12.4. The first permanent molars appear at about 14 months (Seier, 1986).

Correspondence

Any correspondence should be directed to Jürgen Seier, Primate Unit, MRC, PO Box 19070, Tygerberg 7505, South Africa. Email: jurgen.seier@mrc.ac.za

TABLE 12.4: Physical development of 57 infant vervet monkeys until completion of deciduous dentition

Days	Weight	Dentition	Pelage
1–7	406	Lower incisors	Fur black, face pink
8–14	414	Upper central incisors	Fur still black, face white-pink
15–21	486	Upper lateral incisors	Most face white, pelage lighter, first green appears at base of tail
22–28	523	–	Fur lighter, often grey-black
29–35	552	–	First animals develop white brow band, face dark, pelage lighter and thicker
36–42	595	–	More white brow bands, faces more furry and dark
43–49	618	Upper and lower canines	Green fur on arms, base of tail, legs and shoulders, most have white brow bands
64–70	760	Upper and lower first molars	–
113–119	1016	Lower second set of molars	Most individuals have adult fur
134–140	1012	Upper second set of molars	

Source: Modified from Seier, 1986; with permission from Blackwell Munksgaard.

References

Cheney, D.L., Seyfarth, R.M., Andelman, S.J. and Lee, P.C. (1988). In Clutton-Brock T.H. (ed.) *Reproductive Success: Studies of Individual Variation in Contrasting Breeding Systems,* pp 384–402. University of Chicago Press.

Cho, F., Hiyaoka, A., Suzuki, M.T. and Honjo, S. (2002). *Exp. Anim.* 51, 343–351.

Else, J.G. (1985). *Lab. Anim. Sci.* 35, 373–375.

Eley, R.M. (1992). *Utafiti* 4, 1–33.

Eley, R.M., Tarara R.P., Worthman C.M. and Else J.G. (1989). *Am. J. Primatol.* 17, 1–10.

Fairbanks, L.A. and McGuire, M.T. (1984). *Am. J. Primatol.* 7, 27–38.

Fairbanks, L.A. and McGuire, M.T. (1987). *Int. J. Primatol.* 8, 351–366.

Gartlan, J.S. (1969). *J. Reprod. Fert.* 6, 137–150.

Hector, A.K. and Raleigh, M.J. (1992). *Am. J. Primatol.* 26, 77–87.

Hess, D.L., Hendrickx, A.G. and Stabenfeldt, G.H. (1979). *J. Med. Primatol.* 8, 273–281.

Hiyaoka, A., Yoshida, T., Cho, F. and Goto, N. (1990). *Exp. Anim.* 39, 345–352.

Isbell, L.A., Cheney, D.L. and Seyfarth, R.M. (1991). *Am. J. Primatol.* 25, 57–65.

Johnson, P.T., Valerio, D.A. and Thompson, G.E. (1973). *Lab. An. Sci.* 33, 355–359.

Kushner, H., Kraft-Schreyer, N., Angelakos, E.T. and Wudarski, E.M. (1982). *J. Med. Primatol.* 11, 77–84.

Lamprechts, C., Seier, J.V. and Mdhluli, C. (1999). *Proceedings of the International Joint Meeting of the 12th ICLAS General Assembly & Conference and 7th FELASA Symposium. Lab. Anim.,* 170–172.

Lee, P.C. (1987). *J. Zool. Lond.* 213, 409–422.

Pollack, D.B. and Raleigh, M.J. (1994). *Am. J. Primatol.* 32, 57–60.

Raleigh, M.J. and McGuire, M.T. (1990). In Ziegler, T.E. and Bercovitch, F.B. (eds) *Socioendocrinology of Primate Reproduction,* pp 95–111. Wiley-Liss, New York.

Rowell, T.E. (1971). *Anim. Behav.* 19, 625–645.

Seier, J.V. (1986). *J. Med. Primatol.* 15, 339–349.

Seier, J.V., Venter F.S., Fincham J.E. and Taljaard J.J.F. (1991). *J. Med. Primatol.* 20, 1–5.

Seier, J.V., van der Horst G., de Kock M. and Chwalisz K. (2000). *J. Med. Primatol.* 29, 70–75.

Turner, T.R.,Whitten P.L., Jolly, C.J. and Else, J.G. (1987). *Am. J. Primatol.* 12, 197–200.

CHAPTER 13

Nutrition and Nutritional Diseases

Sherry M. Lewis and Charlotte E. Hotchkiss
The Bionetics Corporation, National Center for Toxicological Research, FDA, Jefferson, AR 72079, USA

Duane E. Ullrey
Professor Emeritus, Departments of Animal Science and Fisheries and Wildlife, Michigan State University, East Lansing, MI 48824, USA

Introduction

The order Primates includes prosimians, New World monkeys, Old World monkeys, apes, and humans. Taxonomists have proposed that extant species and subspecies of primates number over 300 and 600, respectively (Groves, 2001). Of these, 28 species were found in U.S. National Institutes of Health (NIH) National Primate Research Centers in 1998 (NRC, 2003). In 2003, 15 species (Table 13.1) were specifically identified in facilities affiliated with the NIH National Center for Research Resources (www.ncrr.nih.gov).

Some of these and additional species are found in commercial laboratories. In these various laboratories, primates serve as human surrogates in studies of normal physiology and in research concerned with diagnosis, prevention and treatment of disease. This chapter on nutrition has been included in *The Laboratory Primate* to provide insight to those involved in primate research.

Very few primate species have been the subject of nutrition research. The National Academies of Science/National Research Council (NRC, 2003) found that energy requirements of fewer than 20 species have been studied, and protein, mineral, and vitamin requirements have been studied in less than 10. Much has been made of presumed differences between New World and Old World monkeys. However, the NRC concluded that such generalizations are unwarranted because so few representatives of either group have been studied, and it is more appropriate to draw distinctions between individual species when consistent differences are evident. The nutrients required by nonhuman primates are described in the next section, followed by

The Laboratory Primate
Copyright 2005 Elsevier
ISBN 0-1208-0261-9

TABLE 13.1: Species of nonhuman primates reported in NIH National Primate Research Centers and NCRR affiliated institutions in 2003 (www.ncrr.nih.gov)

Family	Common name	Scientific name
Callitrichidae	Common marmoset	*Callithrix jacchus*
Cebidae	Owl monkey	*Aotus trivirgatus*
	Red-bellied titi	*Callicebus moloch*
	Tufted brown capuchin	*Cebus apella*
	Squirrel monkey	*Saimiri sciureus*
Cercopithecidae	Sooty mangabey	*Cercocebus atys*
	African green (vervet) monkey	*Chlorocebus aethiops*
	Patas monkey	*Erythrocebus patas*
	Cynomolgus macaque	*Macaca fascicularis*
	Japanese macaque	*Macaca fuscata*
	Pigtailed macaque	*Macaca nemestrina*
	Rhesus macaque	*Macaca mulatta*
	Baboon	*Papio* spp.
	Yellow baboon	*Papio cynocephalus*
Pongidae	Chimpanzee	*Pan troglodytes*

discussions on diet formulation, and possible dietary contaminants.

Nutrient requirements

Specific nutrient requirements have not been established for the various species of laboratory primates. Requirement estimates have been proposed, based upon studied groups of primates, and apply to primates fed diets comprised of components with high nutrient bioavailability and an apparent metabolizable energy of 4 kcal·g^{-1} of dietary dry matter (NRC, 2003). Table 13.2 has been adapted from the *Nutrient Requirements of Nonhuman Primates (2nd Revised Edition, 2003)* and lists estimated dietary nutrient concentrations (dry matter basis) proposed as adequate in diets containing conventional feed ingredients and intended for post-weaning primates. Estimates of nutrient bioavailability and dietary sources of various nutrients are also presented.

Energy

The energy in food, although not a nutrient *per se,* is basic to all physiological functions. It is conventionally measured in units of heat, either as kilocalories (sometimes called the large calorie or Calorie) or kilojoules because the calorie and joule are inconveniently small. One kilocalorie (Cal) is equal to 4.184 kilojoules (kJ).

When the energy in food is released by complete oxidation to carbon dioxide and water, it is known as gross energy (GE) or heat of combustion. The average GE of carbohydrates, proteins, and fats has been estimated to be 4.1, 5.6, and 9.4 kcal·g^{-1}, respectively (Mayes, 1996). Common foods contain, on average, about 4.4 kcal GE·g^{-1} of dry matter (DM).

Apparent digestible energy (DE) of a food is the GE in food minus the GE in feces when this food is eaten (NRC, 1981). It is not a constant but is affected by diet composition, amount of food consumed per unit of time, and the digestive efficiency of the consuming animal. Dietary fiber components, particularly cellulose and hemicelluloses, are poorly utilized by animals with simple gastrointestinal systems, whereas lignin is not utilized at all. Some nonhuman primates (e.g., Colobinae and *Alouatta* spp.) have the capacity to utilize substantial amounts of cellulose and hemicelluloses owing to the presence of gastrointestinal compartments harboring fiber-fermenting microorganisms (NRC, 2003).

Apparent metabolizable energy (ME) of a food is equal to food GE, minus fecal and urinary GE, and minus GE of gaseous products of microbial fermentation such as methane. Gaseous GE loss is often disregarded when calculating food ME for animals with

TABLE 13.2: Nutrient concentrations, per kg dietary dry matter, of conventional laboratory diets adequate to meet the needs of laboratory housed post-weanling nonhuman primates

Nutrient	Concentration[1,2]	Nutrient bioavailability[2]	Nutrient source or supplement[2]
Crude Protein, %	15–22	63–88% natural or purified diets	corn, wheat, soybean meal, dried whey, oats, corn gluten meal, fish meal, alfalfa meal
Essential n–3 fatty acids, %	0.5	in feedstuffs, 95–99%	soybean oil, flaxseed oil, marine oils
Essential n–6 fatty acids, %	2	in feedstuffs, 95–99%	corn oil, soybean oil, peanut oil, cottonseed oil
Neutral–detergent fiber, %	10–30	dependent upon source, intake level, gut adaptation, low	natural fibers endogenous to diet components
Acid–detergent fiber, %	5–15	dependent upon source, intake level, gut adaptation, low	natural fibers endogenous to diet components
Vitamin A, IU kg^{-1}	8,000	in feedstuffs, 5–50%; organic, good	vitamin A acetate or palmitate
Vitamin D_3, IU·kg^{-1}	2,500	in feedstuffs, approximately 50%; organic, good	D activated animal sterol
Vitamin E, mg·kg^{-1}	100	in feedstuffs, 20–75%; organic/synthetic, 85%	DL-α–tocopheryl acetate (all-*rac*-α–tocopheryl acetate), D-α–tocopheryl acetate (*RRR*-α-tocopheryl acetate)
Vitamin K, mg·kg^{-1}	0.5	in feedstuffs, 10–70%; menadione, 60%	menadione dimethylpyrimidinol bisulfite, menadione nicotinamide bisulfite
Thiamin (B_1), mg·kg^{-1}	3.0	in feedstuffs, high	thiamin mononitrate, thiamin hydrochloride
Riboflavin (B_2), mg·kg^{-1}	4.0	in feedstuffs, approximately 95%	riboflavin
Pantothenic acid, mg·kg^{-1}	12.0	in feedstuffs, 50–60%	D-calcium pantothenate
Niacin, mg·kg^{-1}	25.0	in grains, low; organic/synthetic, high	nicotinic acid, nicotinamide
Pyridoxine (B_6), mg·kg^{-1}	4.0	in feedstuffs, 40–60%	pyridoxine hydrochloride
Biotin, mg·kg^{-1}	0.2	in feedstuffs, 20–85%; synthetic, good	D-biotin
Folic acid, mg·kg^{-1}	4.0	natural, approximately 50%; synthetic, approximately 85%	pteroylmonoglutamate
Vitamin B_{12}, mg·kg^{-1}	0.03	devoid in plant feedstuffs; organic/synthetic, high	cyanocobalamin
Choline, mg·kg^{-1}	750	in feedstuffs, 60–90%	choline chloride, choline bitartrate
Vitamin C, mg·kg^{-1}	200	in feedstuffs, 80–90%	L-ascorbyl-2-poylphosphate

(Continued)

TABLE 13.2 (*Continued*)

Nutrient	Concentration[1,2]	Nutrient bioavailability[2]	Nutrient source or supplement[2]
Calcium, %	0.8	lowered by phytate and oxalate; inorganic salts, high	dicalcium phosphate, calcium carbonate
Phosphorus (total), %	0.6	phytate bound, negligible; inorganic salts, high	dicalcium phosphate, monocalcium phosphate
Magnesium, %	0.08	magnesium oxides, approximately 75%	magnesium oxide, magnesium sulfate
Potassium, %	0.4	in feedstuffs, >85%; organic/inorganic salts, high	potassium chloride, potassium carbonate and sulfate
Sodium, %	0.2	in feedstuffs, 60–90%; inorganic salts, high	sodium chloride, sodium sulfate
Chloride, %	0.2	in feedstuffs, approximately 85%; inorganic salts, high	sodium chloride, calcium chloride
Iron, $mg \cdot kg^{-1}$	100	in feedstuffs, 40–50%; organic/inorganic salts, high	ferrous sulfate monohydrate
Copper, $mg \cdot kg^{-1}$	20	inorganic salts, high; lowered by high zinc	cupric sulfate pentahydrate
Manganese, $mg \cdot kg^{-1}$	20	organic salts, 30–35%; inorganic salts, 60–75%	manganous sulfate monohydrate
Zinc, $mg \cdot kg^{-1}$	100	organic/inorganic salts, high; lowered by phytate	zinc sulfate monohydrate
Iodine, $mg \cdot kg^{-1}$	0.35	inorganic salts, high	ethylene diamine dihydroiodide, calcium iodate
Selenium, $mg \cdot kg^{-1}$	0.30	in feedstuffs, 60–90%; inorganic salts, high	sodium selenite
Chromium (trivalent), $mg \cdot kg^{-1}$	0.2	inorganic salts, high	chromic potassium sulfate, chromic chloride

[1]Concentrations based upon highly digestible diets containing 4 $kcal \cdot g^{-1}$ ME, dry matter basis (NRC, 2003).

[2]Data adapted from NRC, 2003; Ammerman *et al.*, 1995; NRC, 1989; McDowell, 1989; Underwood, 1977; Rogers, 1979.

simple gastrointestinal systems (Lloyd *et al.,* 1978). The heat increment (or thermogenic effect) of nutrient digestion, absorption, intermediary metabolism, and microbial fermentation is generally a lost benefit to animals housed in thermoneutral environments, although it may be useful in maintaining body temperature in a cold environment. ME concentrations of diets have sometimes been estimated, rather than directly determined, by assigning mathematical constants to dietary carbohydrate, protein, and fat (Merrill and Watt, 1955). The physiological fuel values, 4 $kcal{\cdot}g^{-1}$ for carbohydrates and protein and 9 $kcal{\cdot}g^{-1}$ for fat, give reasonable approximations of apparent ME for many diets consumed by simple-stomached species. Fiber components are not generally considered energy sources unless substantial gut microbial fermentation is a factor.

Maintenance energy requirements

The ME intake required for maintenance must provide sufficient energy to meet basal metabolism (support of basic life functions), thermoregulation, and activity costs (Lloyd *et al.,* 1978; Robbins, 1993; Torun *et al.,* 1996). Kleiber (1975) used the concept of metabolic body size as a power function of bodyweight (BW^n) to compare the basal metabolic rate (BMR) of fasting adult animals ranging in weight from mice (0.021 kg) to cattle (600 kg), and including macaques (4.2 kg) and chimpanzees (38 kg). He concluded that metabolic body size could be represented by $BW_{kg}^{0.75}$ and BMR could be expressed as 70 $kcal{\cdot}d^{-1} \times BW_{kg}^{0.75}$. Theoretically, BMR should be measured when animals are in a post-absorptive state, stress-free, and housed in a thermoneutral environment (Curtis, 1983). Prediction of maintenance energy requirements must consider gender, age, reproductive status, dietary intake, activity, health status, body composition and environmental circumstances such as temperature, relative humidity and wind. A review of published data suggests that maintenance energy requirements of adult nonhuman primate species, in a laboratory setting, range widely (NRC, 2003). The average maintenance ME requirement of two important laboratory primate species, adult rhesus and cynomolgus macaques, was 109 kcal $ME{\cdot}BW_{kg}^{0.75}{\cdot}d^{-1}$ or $1{\cdot}55 \times$ BMR.

Growth energy requirements

Dietary ME requirements for growth of nonhuman primates are difficult to estimate because they differ so widely among species (NRC, 2003). The BMRs of growing animals may be up to four times those of adults (Clarke *et al.,* 1977; Scott, 1986; NRC, 2003). A value of 5.6 kcal $ME{\cdot}g^{-1}$ of expected BW gain has been used as a rule-of-thumb to estimate energy requirements for growth for several species (NRC, 2003). This value is intermediate between a theoretical maximum of 9 $kcal{\cdot}g^{-1}$ for fat deposition and a low of 1.5–3.5 $kcal{\cdot}g^{-1}$ of BW gain for animals that accumulate minimal fat during neonatal growth (Robbins, 1993).

Gestation and lactation energy requirements

During pregnancy, additional energy is required to support maternal and fetal development, and gestational energy requirements may be 30% above maintenance (Clarke *et al.,* 1977; NRC, 2003). Lactation energy costs equal the energy content of milk and energy costs of milk production; thus, the energy requirements for lactation may exceed those for rapid growth. Daily milk-energy (GE) output, proportional to a power function of mass, averages 124 $kcal{\cdot}BW_{kg}^{0.75}$ for mammals (Brody, 1945; NRC, 2003). Nonhuman primates have typical daily milk yields of 45–70 $g{\cdot}BW_{kg}^{0.75}$ (Oftedal, 1984).

Energy deficiencies and excesses

Clinical signs of energy and other nutrient deficiencies are presented in Table 13.3.

The primary consequence of energy excess is obesity, which is associated with decreased life span, increased incidence and severity of degenerative diseases, and earlier onset and higher incidence of neoplasia (McCay, 1935; Weindruch and Walford, 1988; Yu *et al.,* 1985; Masoro, 1996). Rhesus (*M. mulatta*), pigtail (*M. nemestrina*), celebes (*M. nigra*), and squirrel monkeys (*Saimiri sciurius*) are nonhuman primate models that have been used in aging studies (Hansen *et al.,* 1981; Ingram *et al.,* 1990; Kemnitz *et al.,* 1993; Lane *et al.,* 1996). The effects of dietary ME restriction upon aging, obesity, and signs of diabetes in adult rhesus macaques (*M. mulatta*) indicate how complex these issues are (Hansen and Bodkin, 1993; Bodkin *et al.,* 1995; NRC, 2003).

Carbohydrates and fiber

Carbohydrates (50–80% of plant dry matter) are primary sources of dietary energy and provide >40% of ME in the diets of most laboratory primate species (Asp, 1994). Cereal starches and soluble sugars, such as sucrose, lactose, fructose, and glucose are commonly used in research diets (Knapka *et al.,* 1995; NRC, 2003).

TABLE 13.3: Signs of nutrient deficiency	
Nutrient	**Clinical and pathologic signs of deficiency**
Energy	↓ activity, ↓ limb growth, ↓ abdominal fat, ↑ insulin sensitivity, ↓ gonadotropins, ↑drug elimination time
Protein	Alopecia, brittle skin, facial edema, anorexia, ↓ skeletal growth, ↓ cranial growth during fetal development, ↑ cortisol, ↓ T3 and T4, anemia, altered lipid metabolism, enzyme activity, neural cytochemistry, and iron metabolism; behavioral changes, ↓ ability to thermoregulate in newborns, villous atrophy, fatty liver, may contribute to marmoset wasting syndrome
Fat	Dry, scaly skin, alopecia, visual loss, behavioral changes, anemia, bone marrow hyperplasia
Vitamin B1 (thiamin)	↓ body weight, anorexia, apathy, cachexia, muscle weakness, myocardial necrosis, EKG abnormalities, pulmonary congestion (beri-beri); cavitary necrosis in brain, ataxia, dementia, paralysis, convulsions, ophthalmoplagia (Wernicke's encephalopathy)
Vitamin B2 (riboflavin):	↓ body weight, weakness, alopecia, lack of vigor, dermatitis, gingivitis, digestive disturbances blindness, sudden death, normocytic hypochromic anemia, epithelial atrophy of intestine and bladder, ataxia due to demyelination of peripheral nerves, hypoalbuminemia, fatty liver, abnormal tryptophan metabolism, adrenal cortical atrophy or hemorrhage
Vitamin B6 (pyridoxine)	↓ body weight, anorexia, poor growth, apathy, weakness, alopecia, depigmentation of the hair, dermatitis, hyperirritability, ataxia, tremors, neuropathy, convulsions, hypochromic microcytic anemia with polychromasia and nucleated red blood cells, lymphopenia, fatty liver, hepatic necrosis and cirrhosis, oxaluria, arteriosclerosis, neural degeneration of the cerebral cortex
Vitamin B12 (cobalamin)	Blindness, spastic paralysis of the hind limbs and tail, general weakness, apathy, death, demyelination, axon loss Note: megaloblastic (pernicious) anemia not reported in nonhuman primates
Folic acid	Megaloblastic anemia, ↓ hemoglobin, ↓ RBC, ↑ MCV, poikilocytosis, hypersegmented neutrophils, leukopenia, thrombocytopenia; bone marrow contains giant metamyelocytes, megaloblastic erythroid precursors, and may be hemorrhagic; stomatitis, gingivitis, anorexia, diarrhea, ↓ body weight, lethargy, weakness, alopecia, scaly dermatitis, petechial hemorrhages, edema, boils, depressed immunocompetence, atresic and cystic ovarian follicles with depletion of granulosa cells, megablastosis, multinucleation, and impairment of orderly proliferation and maturation in cervicovaginal epithelium
Niacin	↓ body weight, alopecia, abnormal skin pigmentation on the face, phalangeal joints, and perineum, anemia, diarrhea, chronic atrophic gastritis, atrophic necrotizing enterocolitis, reduced concentrations of erythrocyte pyridine nucleotides
Biotin	Scaly dermatitis on hands and feet, alopecia, depigmentation of the hair, increased susceptibility to infection
Pantothenic acid	↓ growth, diarrhea, cachexia, depigmentation of the hair, alopecia, ataxia, anemia
Vitamin C (ascorbic acid)	Weakness, depression, reluctance to move, diaphyseal swellings, physeal fractures, bruising, bleeding gums, loose teeth, anemia, and subperiosteal hemorrhages, cephalohemotomas (especially squirrel monkeys)
Choline	↓ body weight, lethargy, alopecia, decreased blood lipids, fatty livers, portal cirrhosis, portal hypertension

(Continued)

TABLE 13.3 (*Continued*)	
Nutrient	**Clinical and pathologic signs of deficiency**
Vitamin A	Squamous metaplasia of respiratory epithelium, increased susceptibility to infection, anorexia, diarrhea, weakness, ↓ growth, xerophthalmia, night blindness, corneal destruction, retinal degeneration
Vitamin D	↓ intestinal calcium absorption, hypocalcemia, ↓ bone mineralization (rickets/osteomalacia), ↓ bone density, skeletal deformity, bone fracture, subperiosteal resorption in the hands, loss of lamina dura of the tooth sockets
Vitamin E	Muscular dystrophy (white muscle disease), anemia, multinucleated red cell precursors, orthochromatophilic normoblasts, ↑ erythrocyte fragility, hemolysis, granulocytosis, retinal degeneration, atherosclerosis, pulmonary edema, pancreatic atrophy, steatitis, chronic gastritis, respiratory distress, death
Vitamin K	↑ clotting time
Calcium	Acute: tremors, muscle fasciculations, cardiac arrhythmias, anorexia, nausea, vomiting, hypocalcemia. Chronic: ↓ mobility, bowing of the long bones, hypocalcemia, hyperphosphatemia, ↑ alkaline phosphatase, ↓ bone mineralization, soft tissue mineralization
Phosphorus	Hypophosphatemia, ↑ alkaline phosphatase
Magnesium	Hypocalcemia, tremors, muscle fasciculations, anorexia, nausea, vomiting, cardiovascular changes
Iron	Microcytic, hypochromic anemia, ↓ serum iron, ↑ total iron-binding capacity, ↓ liver and bone marrow iron stores
Copper	↓ growth, alopecia, depigmentation of the hair (achromotrichia), anemia, bone disorders, ↑ plasma cholesterol, GI disturbances, spinal cord lesions. Otolemur species develop progressive cholinergic axonomal dystrophy, cholinergic denervation, generalized gliosis, and later extensive cerebral amyloidosis
Zinc	Alopecia, dermal lesions, parakeratosis and thickening of the tongue (squirrel monkeys), ↓ neutrophil chemotaxis, ↓ mitogen responses, ↓ immunoglobulin production; anorexia during pregnancy, stillbirths, abortions, delivery complications, low birth weights; in the neonates: altered myelination, taste dysfunction, ↓ growth, lethargy, apathy, hypoactivity, ↓ alkaline phosphatase, ↓ ALT, ↓ immune function, hypochromic microcytic anemia, delayed skeletal maturation and bone mineralization
Selenium	↓ body weight, listlessness, alopecia, cardiac and skeletal muscle degeneration, hepatic necrosis, nephrosis
Iodine	↓ T4, ↑ TSH, thyroid hyperplasia; in fetal development ↓ growth, weakness, alopecia, ↓ brain weight, morphologic changes in brain, delayed bone maturation, prominent abdomen
Manganese	↑ clasping, ↑ clinging, ↓ righting response
Chromium	Impaired glucose tolerance, corneal lesions
Fluorine	↑ susceptibility to dental caries

See text for references.

The end-products of digestion of starch and most soluble sugars are glucose, fructose, and galactose. The non-starch polysaccharides, cellulose and hemicelluloses, cannot be digested by endogenous mammalian enzymes, although hemicelluloses can be partially hydrolyzed in the acid stomach (NRC, 2003). Anaerobic microbial fermentation is required for effective use of cellulose and hemicelluloses, and short-chain, volatile fatty acids are the energy yielding end-products of that fermentation. The ME potential of organic acids is about 3 kcal·g^{-1} (Souci *et al.,* 1994).

Typical laboratory diets provide between 4 and 14% crude fiber (CF) to nonhuman primates. CF determinations underestimate plant structural fiber components (Van Soest, 1983; Knapka *et al.,* 1995; NRC, 2003), but a sequential detergent system of fiber fractionation more accurately separates soluble cell contents from insoluble fiber in plant cell walls (Robertson and Van Soest, 1981). After removal of soluble cell components, the residual insoluble neutral-detergent fiber (NDF) includes primarily cellulose, hemicelluloses, and lignin (NRC, 2003). Acid-detergent fiber (ADF) is primarily cellulose and lignin. The quantity of hemicelluloses may be estimated by subtraction of ADF from NDF. When chimpanzees (*Pan troglodytes*) were fed diets containing either 14 or 34% NDF, the higher fiber concentration decreased gut transit time and diet digestibility (Milton and Demment, 1988).

Certain fibers with a high cation-exchange capacity can reduce absorption of calcium, copper, iron, and zinc (Renan and van Rensburg, 1980; Klevay *et al.,* 1981; Kriek *et al.,* 1982; Schneeman, 1990).

Various fiber sources and quantities have been shown to increase, decrease, or leave unchanged, serum lipid and cholesterol concentrations and atherosclerosis incidence in rhesus (*Macaca mulatta*) and vervet or green (*Chlorocebus aethiops*) monkeys (Heine *et al.,* 1984; Kritchevsky *et al.,* 1986, 1988).

While a fiber requirement has not been established for nonhuman primates, there is evidence of a beneficial role for dietary fiber among nonhuman primates whose gastrointestinal tracts are specialized for foregut or hindgut microbial fermentation (Stevens and Hume, 1995; Edwards, 1995; Edwards and Ullrey, 1999). Concentrations of 10–30% NDF and 5–15% ADF are recommended in diets of selected primate species to promote gut health (NRC, 2003).

Nonhuman primates, particularly rhesus macaques, are susceptible to gastric dilatation (bloat), which may be fatal (Kim *et al.,* 1978; Elwell and DePaoli, 1978; Boyce and Miller, 1980; Bennett *et al.,* 1980; Holmberg *et al.,* 1982). Bloat has been linked to a variety of causes, among which are rapid fermentation of carbohydrates and insufficient dietary fiber. Bloat generally occurs upon refeeding after a prolonged fast, and *Clostridium perfringens* may be involved in the disease process (Bennett *et al.,* 1980). Some New World monkeys, especially infants, are susceptible to potentially-fatal hypoglycemia after prolonged fasting (Brady *et al.,* 1990; Mann, 1968).

Protein

Requirements for dietary protein are greatly influenced by digestibility and protein quality (amounts and proportions of essential amino acids which cannot be synthesized in the animal body and must be supplied by diet). The biological value (BV) of a protein is a measure of the presence of essential amino acids in proportion to needs (Mitchell and Block, 1946). The essential amino acid in lowest concentration in relation to need is the "limiting amino acid." Another measure of quality is relative nutritional value (RNV), established by comparing animal response to a novel or test protein with the response to a reference protein at varying dietary concentrations (NRC, 2003).

Natural-ingredient diets generally combine plant protein sources (e.g., grains, grain by-products, soybean meal), and sometimes animal protein sources (e.g., meat and fish meals, milk and dairy by-products), in proportions designed to ensure that essential amino acids will not be limiting. If needed, specific amino acids, such as methionine or lysine, may be added. Insects and insect meals also can be natural sources of dietary protein. Protein in purified and semipurified diets frequently is provided by casein, soy isolates, or lactalbumin, and, if needed, amino acid mixtures (NRC, 2003).

Grain proteins can be limiting in the essential amino acid, lysine, whereas legume proteins can be limiting in methionine. Soy protein, limiting in methionine, had 50% the potency (RNV) of casein or lactalbumin for growth of infant and young squirrel and cebus monkeys (Ausman *et al.,* 1979, 1985, 1986; Samonds and Hegsted, 1973). The addition of methionine to the soy protein diet resulted in a RNV for nitrogen balance equivalent to that of the casein diet (Ausman *et al.,* 1986). Nitrogen balance studies with growing cebus monkeys demonstrated that the RNV of casein was 60 to 70% that of lactalbumin due to limited concentrations of the sulfur-containing amino acids, methionine and cysteine, in casein. Additions of threonine, methionine, and lysine improved the value of a wheat gluten-containing diet for growing cebus monkeys, suggesting multiple amino acid deficiencies (Ausman and Hegsted, 1980).

When male vervet monkeys (*Cercopithecus aethiops*, now *Chlorocebus aethiops*) were fed amino acid mixtures, omitting tryptophan (an amino acid deficient in corn) heightened aggression (Chamberlain *et al.*, 1987). The aggressive behavior was subsequently shown to be inversely correlated with the concentrations of tryptophan and 5-hydroxyindoleacetic acid in cerebrospinal fluid (Young *et al.*, 1989).

Adult marmosets (*Callithrix jacchus*) may have a requirement for dietary arginine and histidine (Flurer and Zucker, 1985) although some primates, notably adult humans, may be able to synthesize these amino acids at a rate sufficient to meet needs.

Infant rhesus monkeys fed a commercial formula low in phenylalanine up to 70 days of age developed lethargy, anemia, anorexia, diarrhea, dermatitis and edema (Kerr *et al.*, 1969). Supplementation of the formula with phenylalanine ameliorated all but the dermatitis.

Infant monkeys fed soy-based, human-infant formulas lacking supplemental taurine exhibited depressed growth and an altered glycine to taurine ratio in conjugated bile acids (Hays *et al.*, 1980). A loss in visual acuity and retinal degeneration was demonstrated in infant rhesus monkeys fed a taurine-free diet (Sturman *et al.*, 1984). While synthesized in liver and brain of adult nonhuman primates, taurine synthesis may be inadequate in young or preterm animals (Hayes, 1985); thus, supplementation of the young might be required during the first year of life (NRC, 2003).

Experimental protein deficiency in nonhuman primates has been used to model the human diseases of kwashiorkor (Deo *et al.*, 1965), tropical sprue (Mehta *et al.*, 1979), mucoid vasculopathy (Sandhyamani, 1992), tropical splenomegaly syndrome, and endomyocardial fibrosis (Sezi, 1996). Clinical signs of protein deficiency are presented in Table 13.3.

Overall, dietary protein requirements are increased during pregnancy and lactation, and during periods of stress and illness. Protein requirements decrease as growth rates decline and animals mature, and appear to be similar to those predicted from studies of other mammals (NRC, 2003). Requirements for juvenile to adult primates, expressed as grams of protein per kilogram of BW per day, range from 0.59 g $BW_{kg}^{-1} \cdot d^{-1}$ for adult humans to 4.3 g$\cdot BW_{kg}^{-1} \cdot d^{-1}$ for juvenile squirrel monkeys; most adult nonhuman primates required less than 3.0 g$\cdot BW_{kg}^{-1} \cdot d^{-1}$ (NRC, 2003). Protein concentrations needed to support adult maintenance were 4.6–7.5% of ME calories or 6.4–8% of dietary dry matter (NRC, 2003).

Fat and essential fatty acids

Most commercial nonhuman primate diets provide 4–9% crude fat. The most common dietary fats are triacylglycerols of plant origin. Besides energy, dietary fats provide essential fatty acids (EFA) which function to: (1) promote normal growth, organ development, and reproductive function, (2) prevent or alleviate skin abnormalities, (3) maintain normal ratios of polyunsaturated fatty acids required for synthesis of tissue lipids, cellular membranes, and red blood cell integrity, and (4) provide for the normal absorption and utilization of cholesterol and fat-soluble vitamins (Reisbick *et al.*, 1990; Knapka *et al.*, 1995; NRC, 2003).

The polyunsaturated EFAs, α-linolenic acid (C18:3 n-3) and linoleic acid (C18:2 n-6), cannot be manufactured by the body and must be included in the nonhuman primate diet. The requirements for these fatty acids change during various life-stages, and depot stores in adults may sustain needs for short periods (Innis, 1991; NRC, 2003). The n-3 fatty acids, in particular the long-chain docosahexaenoic and eicosapentaenoic acids, appear critical for normal fetal brain development, but these can be transferred across the placenta subsequent to elongation and desaturation of dietary α-linolenic acid by tissues of the mother (Greiner *et al.*, 1996). Rhesus infants raised on an n-3 fatty acid-deficient diet showed reduced visual acuity by four weeks of age (Neuringer *et al.*, 1984). Both n-3 and n-6 fatty acids provide protection against coronary heart disease (Rudel *et al.*, 1995). It is recommended that 0.5% (by weight) of dietary DM (approximately 1% of ME) be present as n-3 fatty acids to support normal development and maintenance of the brain and nervous system (NRC, 2003). To avoid a deficiency of n-6 fatty acids, 2% of linoleic acid, by weight, or 4% of ME is recommended in dietary DM (NRC, 2003). Corn oil in the diet (supplying about 2% of ME) prevented linoleic acid deficiency in young adult rhesus monkeys, indicating that the requirements of adults may be lower than those of infants (Greenberg, 1970). Diets supplemented with high concentrations of unsaturated fatty acids may require increased concentrations of vitamin E due to its destruction during fatty acid peroxidation (NRC, 2003).

Oils derived from a variety of seeds (corn, cottonseed, soybeans) contain 50% or more linoleic acid (Rogers, 1979). Hydrogenated vegetable fat contains approximately 30%. Lard, butter, and beef tallow contain between 2 and 10% linoleic acid. Soybean oil contains about 15% saturated fatty acids, 23% monounsaturated fatty acids, 51% linoleic, and 7% α-linolenic acids.

Flaxseed oil contains about 5% linoleic acid and 20% α-linolenic acid.

Nonhuman primates have been studied extensively as models for atherosclerosis. The development of this disease is affected by individual and species differences, types of dietary fat, other dietary components, and non-dietary factors such as stress. Clinical signs of fat deficiency or excess are presented in Table 13.3.

Vitamins

Vitamins are complex organic molecules, required in small amounts to maintain normal physiologic functions and health. They must be consumed or synthesized in optimal amounts to support life-stages such as growth, maintenance, reproduction, and aging. Vitamins are classified according to their solubility in fat (A, D, E and K) or water (B-complex, and vitamin C). The fat-soluble vitamins cannot be synthesized by the body (except for cutaneous biosynthesis of vitamin D during exposure to the sun) and must be supplied by the diet (or by microbial synthesis in the case of vitamin K). They are absorbed by mechanisms associated with absorption of fat. B-complex vitamins function as enzymatic cofactors, cannot be synthesized by the body, and must be supplied by the diet or by microbial synthesis in the gut. Deficiency of a single vitamin of the B-complex is rare. Definite syndromes are, however, characteristic of specific vitamin deficiencies. Quantitative vitamin requirements for only a few species of nonhuman primates are known. Even among the most studied vitamins, minimum and maximum requirements are not well established. However, vitamin concentrations in natural-ingredient diets have been proposed that are expected to meet needs for post-weaning growth, maintenance, and reproduction of nonhuman primate species used in biomedical research (NRC, 2003). Signs of vitamin deficiencies are presented in Table 13.3. Additionally, vitamin bioavailabilities can be affected by interactions with other nutrients, compounds in the diet, or environmental factors. Table 13.4 provides a partial list of potential interactions and their subsequent effects.

Fat-soluble vitamins

Vitamin A

Vitamin A is a term referring to derivatives of β-ionone that have the biological activity of all-*trans*-retinol. Vitamin A may be found in foods of animal origin, or may result from metabolism of carotenoids found in plant foods. Vitamin A is the primary signaling molecule in the visual process in which 11-*cis*-retinal, a component of the visual pigment rhodopsin in rod cells and iodopsin in cone cells, is converted by light to all-*trans*-retinal, eliciting a response in the visual cortex of the brain (NRC, 2003). Vitamin A also participates in cell-to-cell communication, cellular differentiation, embryologic development, spermatogenesis, and the immune response. Vitamin A induces mucous cell differentiation, alters keratin expression, and is required for normal mucociliary differentiation (Huang *et al.*, 1994). Deficiency results in squamous metaplasia of the respiratory tract epithelium (Miller *et al.*, 1993); without protective mucus, the respiratory tract is prone to infection. Other early signs are loss of appetite, diarrhea, weakness, and retarded growth. With prolonged deficiency, xerophthalmia, night blindness, corneal damage, and retinal degeneration can occur (Knapka *et al.*, 1995). Vitamin A in excess can be toxic; signs include skin dryness and pigmentation, alopecia, anorexia, weakness, leukopenia, hypoplastic anemia, enlarged liver and spleen, hepatocellular damage, bleeding lips and gums, stiffness in joints, and pruritis (Knapka *et al.*, 1995; Hendrickx *et al.*, 2000). Vitamin A is a known teratogen. All-*trans* and 13-*cis* retinoic acid have teratogenic effects similar to those of retinol, including embryolethality, skeletal abnormalities, and craniofacial malformations consisting of external ear defects, mandibular hypoplasia, cleft palate, and temporal bone deformities, hypoplasia of the thymus, and cardiac malformations in cynomolgus monkeys (Hendrickx and Hummler, 1992; Hummler *et al.*, 1990). Many of these defects can be traced to abnormalities of the embryonic hindbrain and its associated neural crest cells which contribute to the first and second pharyngeal arches (Hendrickx *et al.*, 2000).

Vitamin D

Provitamin D_2 (ergosterol), found in plants and fungi, can be converted to vitamin D_2 by solar irradiation in the portion of ultraviolet-B (UVB) that reaches the earth's surface (290–315 nm). Provitamin D_3 (7-dehydrocholesterol), found in the skin of animals, is converted by UVB to previtamin D_3, followed by thermal conversion to vitamin D_3. Vitamin D produced in the skin, or absorbed from the diet, is hydroxylated in the liver to 25(OH)D, the major circulating form used to assess vitamin D status. Under the influence of parathyroid hormone (PTH), 25(OH)D undergoes 1α-hydroxylation in the kidney to $1,25(OH)_2D$, the active form of the vitamin responsible for calcium and

TABLE 13.4: Reported nutrient-nutrient or nutrient-environmental interactions and subsequent effects among nonruminant adult mammals

Nutrient	Interaction with nutrient/compound/environmental event	Effects of nutrient/compound/environmental event interaction
Vitamin A	oxygen, ↑ diet processing temperature	↓ biopotency[1,2,3]
	moisture in feeds, free choline chloride, trace minerals, and ↓ pH	↓ stability[1]
	dietary fat, protein, and vitamin E	↑ absorption, utilization[2]
Vitamin D_3	oxygen, ↑ diet processing temperature	↓ biopotency[1,3]
	↓ calcium and phosphorus, or imbalance in ratio	↑ rickets and osteomalacia[3]
Vitamin E	oxygen, ↑ diet processing temperature	↓ biopotency[1]
	moisture in feeds, trace minerals, and organic acids ↑	↓ stability[1]
	↑ polyunsaturated fatty acids	↑ requirement[2,3]
Vitamin K	oxygen, ↑ temperature	↓ biopotency[1]
	moisture in feeds, free choline chloride, trace minerals, and alkalinity	↓ stability[1]
	↑ vitamin E intake	vitamin K-responsive hemorrhagic signs[2]
Thiamin (vitamin B_1)	↓ folic acid	↓ absorption[1]
	↑ temperature	Maillard reaction, ↓ biopotency[1]
Riboflavin (vitamin B_2)	aflatoxins, alcohol	↓ absorption, utilization[1]
	↑ energy expenditure	↑ requirement[2]
	ultraviolet light exposure	↓ stability[3]
Pantothenic acid	↑ temperature and moisture	↓ biopotency[1]
Niacin	↓ tryptophan	↓ niacin synthesis[1,2,3]
	lime treatment of cereal grains	↑ availability of bound niacin[2]
Pyridoxine (vitamin B_6)	↑ diet processing temperature; freezing fruits and vegetables	Maillard reaction, ↓ biopotency[1,2]
	↑ protein intake	↑ requirement[2]
	carbonates and mineral oxides	↓ biopotency[1]
Biotin	oxygen	↓availability[1]
	avidin (in egg albumin)	binds biotin, ↓availability[1]
	bound in wheat	↓availability[2]
Folic acid	↑ diet processing temperature, oxygen, and ultraviolet light	Maillard reaction or molecule cleavage, ↓ biopotency[1,2]
	↓ vitamin B_{12}	↓ utilization, ↓ thymidine synthesis (↓ DNA)[1]
Vitamin B_{12}	↓ cobalt	↓ cobalamin synthesis, ↓ intrinsic factor synthesis = ↑ pernicious anemia[1,2]
Choline	hygroscopic in crystalline form	stress agent to vitamins in vitamin-mineral premix[1]
	alkaline treatment of grain oils	↓ availability[1]

(Continued)

TABLE 13.4 (Continued)

Nutrient	Interaction with nutrient/compound/environmental event	Effects of nutrient/compound/environmental event interaction
Vitamin C	↑ temperature, light, oxidation, pelleting, and extrusion	↓ biopotency[1,2,3]
	↑ copper, iron, zinc, and pectin	↓ absorption[1]
	reducing agent properties	↓ autooxidation of other vitamins in premixes[1]
Calcium	bound by phytate and oxalate, aluminum	↓ availability[1,2]
	vitamin D	↑ active absorption[1]
	phosphorus	optimal ratio with calcium, 1:1 to 2.2:1[1,3]
	copper, zinc, manganese, fluorine, silicon, and boron	essential for proper utilization[2]
	↑ protein intake	↓ tubular calcium resorption[2]
Cobalt	inorganic sulfate	↓ liver stores[1]
	molybdenum	↑ liver stores[1]
	↓ iron	↑ absorption[4]
Copper	bound by phytate, ↑ anionic molybdenum and sulfur	↓ absorption and utilization[1,3,4]
	↑ zinc, iron, cadmium	↓ absorption[4], retention, and anemia[1,2]
	ascorbic acid, sucrose, fructose, and cysteine	↓ availability[1,4]
Iodine	goitrogens (in kale, cabbage, turnips, soybean meal, peanuts)	↓ availability[1]
	fluoride, cobalt, manganese, and nitrate, ↓ selenium	↓ availability[1]
	↑ calcium	antithyroid effect[1]
Iron	ascorbic acid, cysteine, histidine, lysine, and organic acids	↑ absorption[1,2,4]
	fructose and sorbitol	↑ absorption[4]
	↑ calcium, phosphorus, cobalt, and zinc	↓ absorption[1,4], ↓ ferritin incorporation[1]
	↓ copper	↓ iron(III)-transferrin formation[4]
	↑ nickel	↑ liver stores, ↑ hematopoiesis[1]
	sucrose	↓ absorption, ↓ hemoglobin[1]
Magnesium	animal fat, hydrogenated whole oil	↓ availability[1]
	↑ phosphorus, calcium	↓ absorption[1]
Manganese	↑ phosphorus	↓ availability[1]
	↑ iron, cobalt	↓ absorption[1,4]
	↑ calcium	↓ absorption and retention[4]
	vitamin K	prothombin synthesis[4]
Phosphorus	bound by phytate, ↑aluminum hydroxide, and magnesium	↓ availability[1,2]
	↑ iron	↓ utilization[3]
Potassium	↑ lean body mass	↑ requirement[2]
Selenium	vitamin E	prevention of disorders (liver necrosis, white muscle disease)[1]

(Continued)

TABLE 13.4 *(Continued)*

Nutrient	Interaction with nutrient/compound/ environmental event	Effects of nutrient/ compound/environmental event interaction
	high copper	↓ glutathione peroxidase activity in blood, liver, testes, and kidneys[1], ↓ toxicity[4]
	methionine, inorganic sulfur, and ↑ protein intake	↓ absorption, utilization[1,4]
	↑ protein intake and arsenic	↓ decreased toxicity[3]
	ascorbic acid	↑ absorption of selenite and selenate[1]
	antioxidants	↑ selenium utilization[1,4]
	cadmium	↓ cadmium toxicity[4]
Zinc	bound by phytate (inositol hexaphosphate) in oil seeds, cereals	↓ availability[1], ↑ requirement[2,4]
	↑ calcium and inorganic phosphorus	↓ availability[1,4]
	↓ calcium	↑ release of skeletal zinc[4]
	ascorbate	↑ availability[1]
	iron, copper, cobalt, cadmium, lead, tin, and chromium	↓ absorption and utilization[1,4]

[1]Data adapted from Ammerman *et al.*, 1995.
[2]Data adapted from NRC, 1989.
[3] Data adapted from McDonald *et al.*, 1978.
[4]Data adapted from Underwood, 1977.

phosphorus homeostasis. $1,25(OH)_2D$ binds to the nuclear vitamin D receptor (VDR) and retinoic acid X receptor (RXR) to form a complex that attaches to vitamin D-responsive elements of deoxyribonucleic acid (DNA) and regulates transcription. In the intestine, this vitamin D metabolite enhances calcium absorption; in bone, PTH and $1,25(OH)_2D$ increase both osteoclastic and osteoblastic activity, mobilizing calcium stores in response to hypocalcemia or increasing bone mass, depending on the overall calcium status and the timing and route of administration. In young animals, vitamin D deficiency causes rickets, characterized by hypertrophy of the epiphyseal plates, seen as bulges at the ends of long bones and at costochondral junctions. In older animals, vitamin D deficiency results in osteomalacia during remodeling. The clinical result is diminished bone density, skeletal deformities, bone fractures, subperiosteal resorption, and disappearance of the lamina dura of the tooth sockets (Yamaguchi *et al.*, 1986).

Animals maintained indoors are not exposed to the UV radiation necessary to synthesize vitamin D in the skin, and therefore must receive vitamin D in the diet. The relative potencies of vitamins D_2 and D_3 are not the same in all species. In studied Old World monkeys, D_2 is 2- to 3-fold less active than D_3, but both are able to prevent rickets and osteomalacia. In studied New World monkeys (certain *Cebus, Saimiri, Saguinus,* and *Lagothrix* species), although vitamin D_2 was converted to $25(OH)D_2$ with only 3-fold less efficiency than D_3, D_2 did not prevent osteomalacia, even when fed at high doses (Marx *et al.,* 1989) suggesting that these New World monkeys require dietary vitamin D_3. $1,25(OH)_2D_3$ and $25(OH)D_3$ are competitively bound to intracellular vitamin D binding protein (IDBP) in the hsp-70 family in certain New World monkeys (Gacad *et al.*, 1997; Gacad and Adams, 1998). This results in high circulating concentrations of $1,25(OH)_2D_3$ that can be present even in animals with clinical signs of rickets (Yamaguchi *et al.,* 1986; Shinki *et al.,* 1983). IDBP also binds other steroid hormones, as reflected by high circulating levels of cortisol, testosterone, progesterone, and estrogen (McCamant *et al.*, 1987; Gacad and Adams, 1992). New World owl monkeys (*Aotus sp.*) express low levels of IDBP and do not exhibit resistance to vitamin D. If vitamin D resistance evolved as a protective mechanism to prevent toxicity, nocturnal owl monkeys which receive little sun exposure would not require as much protection. Recently, another protein, vitamin D response element binding protein (VDRE-BP), has been found in certain New World monkeys, which binds directly to DNA, preventing stimulation by vitamin D (Chen *et al.,* 2000). The relative importance of the effects of IDBP and VDRE-BP in resistance to vitamin D has not been determined. Vitamin D toxicity is characterized by hypercalcemia. Squirrel monkeys subjected to experimental hypervitaminosis D_3 exhibited hypercalcemia, hyperphosphatemia, uremia, and death with only minor nephrocalcinosis (Hunt *et al.,* 1969). Rhesus monkeys developed significant nephrocalcinosis, and capuchins exhibited mineralization of the kidneys, aorta, lungs, myocardium, stomach, and various arteries and arterioles (Hunt *et al.,* 1969, 1972). Although primates (especially studied New World monkeys) are relatively resistant to hypervitaminosis D, it has been reported that histoplasmosis granulomas in an owl monkey enhanced conversion of dietary vitamin D to $1,25(OH)_2D_3$, resulting in hypercalcemia (Weller *et al.,* 1990).

Vitamin E

Vitamin E (α-tocopherol and closely related compounds) acts to prevent free radical damage to polyunsaturated fatty acids (PUFA) by reacting with the radicals to form a tocopheroxyl radical. Vitamin E is restored by reactions with hydrogen donors, such as vitamin C or glutathione. Selenium-containing glutathione peroxidases use reducing equivalents from glutathione to catabolize destructive hydroperoxides. Because of these interactions, the amount of vitamin E needed to maintain good health varies with the amount of PUFA, selenium, and antioxidants in the diet. Malabsorption syndromes such as those caused by pancreatitis may cause vitamin E deficiency (NRC, 2003; Juan-Salles *et al.,* 2000). The major effects of vitamin E deficiency are muscular dystrophy and anemia (McIntosh *et al.,* 1987; Boonjawat *et al.,* 1979). Vitamin E-deficient muscular dystrophy is associated with increased urinary excretion of creatinine and allantoin, and associated weakness and weight loss. Classic white muscle disease, with pale patches of myocardium, has been reported in vitamin E-deficient baboons (Liu *et al.,* 1984). Vitamin E-responsive anemia is due both to ineffective erythropoesis, characterized by the presence of multinucleated red cell precursors and orthochromataphilic normoblasts in bone marrow and peripheral blood, as well as to increased erythrocyte fragility and hemolysis (Santiyanont *et al.,* 1977; McIntosh *et al.,* 1987). Owl monkeys (*Aotus* sp.) exhibit greater lipid peroxidation of red blood cell membranes compared to squirrel or cebus monkeys, which makes their erythrocytes more susceptible to hemolysis (Brady *et al.,* 1982). Some owl monkeys develop vitamin E-responsive anemia despite apparently normal circulating concentrations of vitamin E (Brady *et al.,* 1982). Vitamin E is also needed for normal neurologic function (Muller, 1986), and a deficiency also can cause retinal degeneration. Because of the antioxidant effect on PUFA and inhibition of platelet aggregation, vitamin E may influence development of atherosclerosis (Verlangieri and Bush, 1992; McIntosh *et al.,* 1987). Other signs of deficiency include pulmonary edema, pancreatic atrophy, steatitis, chronic gastritis, respiratory distress, granulocytosis, and death (Wixson and Griffith, 1986).

Vitamin K

Vitamin K is a cofactor for γ-glutamyl carboxylase, an enzyme involved in the carboxylation of glutamate (Gla residues). The primary form in higher plants is phylloquinone (vitamin K_1). Bacteria produce a variety of menaquinones (vitamin K_2). Vitamin K is essential for blood clotting; in the liver vitamin K functions in the synthesis of several clotting factors (II, VII, IX, and X) and in extrahepatic tissues in the synthesis of bone-related proteins (osteocalcin and matrix Gla protein). Because vitamin K is synthesized by intestinal bacteria, natural deficiencies are rare. Oral antibiotic treatment in conjunction with a vitamin K-deficient diet can increase clotting times (Hill *et al.,* 1964). Vitamin K may have potential for prevention of osteoporosis by helping preserve bone mineral density (BMD) and reducing the incidence of osteoporotic fractures (Iwamoto *et al.,* 2001; Shiraki *et al.,* 2000). However, vitamin K deficiency has been induced in rhesus monkeys with warfarin (as measured by increased prothrombin time), but no changes in BMD or bone metabolism, as measured by markers of bone turnover, were detected (Binkley *et al.,* 2000). Matrix Gla protein is present in blood vessels as well as in bone, and may inhibit vascular calcification (Schurgers *et al.,* 2001; Shearer, 2000). The potential of vitamin K for treatment of atherosclerosis has not been studied in nonhuman primates.

Water-soluble vitamins

Thiamin

Vitamin B_1, as the coenzyme thiamin pyrophosphate, is involved in the oxidative decarboxylation of α-ketoacids, and functions in transketolase reactions. Thiamin is critical for decarboxylation of pyruvate in preparation for entry into the tricarboxylic acid cycle (NRC, 2003); impaired energy metabolism in thiamin deficiency may be related to the clinical signs of weight loss, anorexia, apathy, cachexia, and muscle weakness. The pathologic lesions produced by thiamin deficiency are primarily in the myocardium and the nervous system. Necrosis of myocardial fibers with electrocardiographic abnormalities are associated with signs of cardiac insufficiency (Waisman and McCall, 1944). Central nervous system lesions characterized by cavitary necrosis of the striatum and a microvacuolar periventricular lesion of the brain stem result in ataxia, dementia, paralysis, convulsions, and ophthalmoplagia characteristic of Wernicke's encephalopathy (Waisman and McCall, 1944; Rinehart *et al.,* 1949; Blank *et al.,* 1975; Witt and Goldman-Rakic 1983a,b).

Riboflavin

Vitamin B_2 is required for the synthesis of flavin mononucleotide (FMN) and flavin adenine dinucleotide (FAD), coenzymes essential to carbohydrate, fat, and protein metabolism (Knapka *et al.,* 1995). A deficiency causes generalized signs of weight loss, weakness, alopecia, and lack of vigor. Squirrel monkeys and baboons also exhibit digestive disturbances (Day *et al.,* 1935; Foy *et al.,* 1964; Foy *et al.,* 1972; Peretti and Baird, 1975; Foy and Kondi, 1984). Normocytic, hypochromic anemia is prominent in the rhesus monkey (*Macaca mulatta*) (Day *et al.,* 1935) and baboon (*Papio anubis*) (Foy *et al.,* 1964), but not in the capuchin (*Cebus albifrons*) (Mann *et al.,* 1952; Mann, 1968) or squirrel monkey (*Saimiri sciureus*) (Peretti and Baird, 1975). Dermatitis characterized by seborrhea or small red lesions on the face that progress to large, nodular, scabby lesions over the entire body was seen in the rhesus monkey, baboon, and capuchin (Waisman, 1944; Cooperman *et al.,* 1945; Mann, 1968; Foy and Kondi, 1984). Other lesions reported were gingivitis, epithelial atrophy of intestine and bladder, ataxia due to demyelination of peripheral nerves, hypoalbuminemia, blindness, fatty liver, abnormal tryptophan metabolism, adrenal cortical atrophy or hemorrhage, and sudden death (Greenberg and Moon, 1963; Foy *et al.,* 1964; Mann *et al.,* 1952, Mann, 1968).

Pyridoxine

Vitamin B_6 in the form of pyridoxine, pyridoxal, or pyridoxamine, is required for the synthesis of amino acids and in glycogen and lipid metabolism (Knapka *et al.,* 1995; NRC, 2003). Due to altered protein metabolism, deficiency causes alopecia, dermatitis, poor growth, apathy, weight loss, hyperirritability, hypochromic microcytic anemia, leukopenia, fatty liver, hepatic necrosis and cirrhosis, and oxaluria (McCall *et al.,* 1946). In macaques, but not capuchins, pyridoxine deficiency also causes widespread arteriosclerosis (Mann, 1968; Kuzuya, 1993). Deficiency signs such as neural degeneration of the cerebral cortex, ataxia, tremors, neuropathy, and convulsions may be related to the fact that pyridoxine increases the rate of serotonin synthesis in rhesus monkey brain (Hartvig, 1995).

Vitamin B_{12}

B_{12} (cobalamin), as part of two mammalian coenzymes, is involved in regeneration of folic acid and in nucleic acid metabolism. It is found only in animal products and microorganisms; herbivores receiving adequate dietary cobalt obtain adequate amounts through foregut or hindgut microbial fermentation, sometimes in conjunction with coprophagy. Baboons fed a vegetarian diet had depleted cobalamin stores, but were resistant to clinical deficiency (Siddons, 1974a). In humans, cobalamin deficiency causes megaloblastic anemia, but overt anemia has not been reported in nonhuman primates. After 33–45 months, deficiency may cause neurologic abnormalities, including blindness, spastic paralysis of the hind limbs and tail, general weakness, apathy, and death due to demyelination and axon loss (Agamanolis *et al.,* 1976; Chester *et al.,* 1980).

Folic acid

Folates are involved in the metabolism of single-carbon compounds and nucleotides required for synthesis of DNA, ribonucleic acid (RNA), and protein (Knapka *et al.,* 1995). The classic sign of folic acid deficiency in all primates studied is megaloblastic anemia, characterized by decreased hemoglobin and red cell counts with increased mean corpuscular volume and poikilocytosis. Leukopenia, thrombocytopenia, and intramedullary hemolysis in the bone marrow also occur (Rasmussen *et al.,* 1979). More generalized signs of deficiency are stomatitis, gingivitis, diarrhea, weight loss, lethargy, alopecia, asthenia, scaly dermatitis, and depressed immunocompetence (Dreizen *et al.,* 1970; Siddons, 1974b). Deficiency affects reproductive function and causes atresic and cystic ovarian follicles with depletion of granulosa cells concurrent with megablastosis, multinucleation, and impairment of orderly proliferation and maturation in cervicovaginal epithelium

(Mohanty and Das, 1982). Supplementation with folate beyond amounts in the standard diet improved hematologic and folate status, maternal weight gain during pregnancy, and infant birth weight in squirrel monkeys (Rasmussen *et al.,* 1980).

Niacin

Niacin, nicotinic acid, is a component of the coenzymes nicotinamide adenine dinucleotide (NAD) and nicotinamide adenine dinucleotide phosphate (NADP), required for oxidation-reduction and dehydrogenase pathways (Knapka *et al.,* 1995; NRC, 2003). The ability to synthesize niacin from tryptophan has been demonstrated in some primate species, and a spontaneous deficiency on a protein-adequate diet seems unlikely. Signs of experimental deficiency include weight loss, alopecia, anemia, diarrhea, chronic atrophic gastritis, atrophic necrotizing enterocolitis, reduced concentrations of erythrocyte pyridine nucleotides, and abnormal skin pigmentation on the face, phalanges, and perineum (Tappan *et al.,* 1952).

Biotin

Biotin is a cofactor for carboxylation and decarboxylation enzymes involved in lipogenesis, gluconeogenesis, and protein synthesis (Knapka *et al.,* 1995). Deficiency causes scaly dermatitis on hands and feet, alopecia, depigmentation of the hair, and increased susceptibility to infection. Spontaneous deficiency is rare because biotin is synthesized by intestinal bacteria. Deficiency can be induced experimentally by feeding raw egg white, which contains avidin, a strong binder of biotin, or by feeding sulfa drugs, which inhibit synthesis by intestinal microflora.

Pantothenic acid

Pantothenic acid is part of coenzyme A, which serves as a cofactor in the tricarboxylic acid cycle, in fatty acid synthesis and degradation, and in the formation of acetylcholine in nervous tissue (NRC, 2003). Deficiency causes growth retardation, anemia, diarrhea, cachexia, depigmentation of the hair, alopecia, and ataxia (McCall *et al.,* 1946).

Vitamin C

Ascorbic acid is a cofactor in many enzymatic reactions, including those involved in the hydroxylation of proline or lysine in the formation of collagen. Studied primate species (with the exception of some prosimians) lack the enzyme gulonolactone oxidase, which is needed to synthesize ascorbic acid, and therefore must receive vitamin C in the diet. Vitamin C deficiency causes scurvy. In young animals, clinical signs are related to failure in formation and cross-linking of the organic matrix of developing bone. Bone growth and bone strength are impaired, and affected primates exhibit weakness, depression, reluctance to move, diaphyseal swellings, and epiphyseal fractures (Eisele *et al.,* 1992). At any age, defective collagen synthesis is associated with increased capillary permeability, resulting in bruising, bleeding gums, and subperiosteal hemorrhages (Machlin *et al.,* 1979). Cephalohematomas have been seen in vitamin C-deficient squirrel monkeys and capuchin monkeys (Lehner *et al.,* 1968; Demaray *et al.,* 1978; Kessler, 1980; Ratterree *et al.,* 1990; Borda *et al.,* 1996). Anemia is common due both to blood loss and the role of vitamin C in iron and folic acid metabolism (Eisele *et al.,* 1992); the anemia may be microcytic to macrocytic and hypochromic to normochromic. Periodontal ligaments are weakened, gums necrose, alveolar bone is destroyed, and teeth are lost in scorbutic animals (Anonymous, 1981). Young animals may require more ascorbate than mature monkeys, and stress increases requirements (Tillotson and O'Connor, 1980) by increasing the metabolism of ascorbic acid to CO_2, which is then exhaled (Flurer *et al.,* 1990). Flurer and Zucker (1987) found that tamarins (*Saguinus fuscicollis*) had lower plasma ascorbate concentrations than marmosets (*Callithrix jacchus*). Whether this finding was a species difference or due to a failure of the tamarins to adapt to their housing conditions was unclear (Flurer *et al.,* 1990). Certain experimental treatments, such as administration of oral contraceptives, may also increase dietary requirements for ascorbic acid (Weininger and King, 1982).

Despite the addition of vitamin C to commercial primate diets, spontaneous scurvy was common in the past (Ratterree *et al.,* 1990; Eisele, 1992). Manufacturing errors, use of diets containing unstable vitamin C forms, improper storage, and soaking diets in water can result in inadequate dietary vitamin C (Demaray *et al.,* 1978). The recent availability of ascorbyl-2-polyphosphate, a stable and biologically active form of vitamin C, has the potential to eliminate scurvy as a practical problem.

Choline

Choline is an essential component of cell membranes, ensuring cellular integrity and normal signaling functions (NRC, 2003). Since it can be manufactured endogenously, disease due to deficiency does not normally occur unless the diet is low in protein or methionine and high in fat. Low choline intakes have resulted in fatty livers in several primate species, and occasionally, decreased blood lipids, alopecia, weight loss, lassitude, portal cirrhosis, and portal hypertension (Wilgram *et al.,* 1958; Cueto *et al.,* 1967; Patek *et al.,* 1975).

Carnitine

While carnitine is not essential, it functions in fatty acid transport into mitochondria, and it has been suggested that dietary supplementation with carnitine may be beneficial in management of liver impairment, diabetes, kwashiorkor, hypopituitarism, adrenal insufficiency, pregnancy, physical exertion, and fasting disorders (Knapka *et al.,* 1995). Cholinergic axonal dystrophy in *Otolemur* species has been associated with low plasma, liver, and muscle carnitine (Schmechel *et al.,* 1996).

Minerals

Minerals, inorganic elements required in a variety of metabolic roles, may be divided arbitrarily into two groups: (1) macrominerals which are required in amounts greater than 100 $mg{\cdot}d^{-1}$ (dietary concentrations commonly expressed in percentage units), and (2) microminerals or trace elements which are required in amounts less than 100 $mg{\cdot}d^{-1}$ (dietary concentrations commonly expressed in parts per million [ppm or $mg{\cdot}kg^{-1}$] or parts per billion [ppb or $\mu g{\cdot}kg^{-1}$]) (Mayes, 1996). Essential macrominerals include calcium, phosphorus, magnesium, potassium, sodium, chlorine, and sulfur; essential microminerals include iron, copper, manganese, zinc, iodine, selenium, chromium, and cobalt (as part of vitamin B_{12}, cobalamin) (NRC, 2003). Other trace elements, e.g., fluorine, molybdenum, silicon, boron, nickel, and tin may also be required. Mineral requirements of laboratory primates are commonly met by the contributions of primary dietary ingredients plus specific mineral additions. Signs of mineral deficiencies are presented in Table 13.3. Potential mineral interactions with other nutrients, compounds in the diet, or environmental factors and their subsequent effects are listed in Table 13. 4.

Calcium

Calcium, along with phosphorus, comprises the hydroxyapatite crystals responsible for bone rigidity. In addition to this structural role, calcium is essential for metabolic functions involving membrane transport and activation of cellular proteins. Calcium-requiring processes include cell movement, muscle contraction, nerve transmission, glandular secretion, blood clotting, and cell division (NRC, 2003). More than 98% of the body's calcium is contained in bones and teeth, providing a reservoir that can be drawn upon during short-term dietary deficiencies. Therefore, reductions in circulating calcium severe enough to cause metabolic complications such as cardiac arrhythmias are generally due to disease processes, rather than deficient dietary calcium, especially in adults. Wild monkeys may not always ingest adequate calcium. When mature cynomolgus monkeys were brought into captivity and fed diets containing 0.24–0.6% calcium, bone mass increased significantly (Jerome *et al.,* 1997; Jayo *et al.,* 1998; Brommage, *et al.,* 1999). This did not occur in captive-bred rhesus fed a 1.0% calcium diet (Keller *et al.,* 2000). The effects of calcium deficiency are age-dependent. In growing animals, inadequate mineralization of the organic matrix of bone causes rickets – widening and irregularities of physeal growth plates, enlarged joints and costochondral junctions, and bowed and fractured bones. In adult animals, the net loss of mineral from previously mineralized bone during remodeling leaves a surfeit of organic matrix and is called osteomalacia (soft bone). Bone cortices are thinner, there are losses of bone density, and fractures result. While 0.14% calcium maintained bone mass for two years in mature cynomolgus macaques (Hotchkiss *et al.,* 2001), young rhesus (two years old initially) fed 0.15% calcium exhibited decreased bone density and accelerated epiphyseal fusion (Griffiths *et al.,* 1975). PTH released in response to a low-calcium, high-phosphorus diet may result in secondary hyperparathyroidism in which hypertrophic organic matrix replaces normally mineralized bone, particularly of the face. Lemurs at the Cincinnati Zoo developed hypocalcemia, hyperphosphatemia, increased alkaline phosphatase activity, impaired mobility, bowing of the long bones, poorly mineralized skeletons, and soft tissue mineralization, characteristic of nutritional secondary hyperparathyroidism when allowed free choice among an assortment of foods (Tomson *et al.,* 1978). Several factors, other than dietary calcium concentration, affect the amount of calcium available to the body. Vitamin D is needed for intestinal calcium absorption. In humans, high dietary protein and sodium will increase urinary calcium excretion without a corresponding increase in intestinal calcium absorption (Knapka *et al.,* 1995); however, in the presence of adequate dietary calcium, increased protein intakes did not adversely affect bone mineral density in rhesus monkeys (Grynpas *et al.*, 1993). Lactation greatly increases the demand for calcium. Bone mineral content was significantly reduced in cynomolgus macaques (*Macaca fascicularis*) during lactation but was restored after weaning when the diets had sufficient calcium (Lees *et al.*, 1998). Restricted physical activity results in bone resorption and loss of body calcium (Lueken *et al.*, 1993). Renal disease and neoplasia also can have profound effects on calcium metabolism.

Phosphorus

Phosphorus, the second major mineral component of bone, is also a component of proteins, nucleic acids, and phospholipids. As phosphate, it helps maintain osmotic and acid-base balance of body fluids and plays a vital role in metabolic energy transfers and gluconeogenesis (NRC, 2003). Phosphorus deficiency results in decreased plasma inorganic phosphorus concentrations, while plasma alkaline phosphatase activity increases as bone is resorbed in an attempt to meet soft tissue metabolic demands. In various species, alterations in dietary calcium:phosphorus ratios have pronounced effects upon bone metabolism, but some adult primates appear resistant to these effects (Knapka *et al.*, 1995).

Magnesium

While the majority of magnesium in the body is in the skeleton, it is also essential for several hundred metabolic reactions (NRC, 2003). Magnesium forms a complex with adenosine-5′-triphosphate (ATP), which interacts with adenylate cyclase to form cyclic adenosine-5′-monophosphate (cAMP). Magnesium deficiency results in hypocalcemia and neuromuscular (tremors and muscle fasciculations), gastrointestinal (anorexia, nausea, and vomiting), and cardiovascular (arrhythmias) abnormalities, perhaps related to the interactions between cAMP and cellular calcium influx during cellular activation by hormones such as parathyroid hormone (PTH) (Dunn, 1971; NRC, 2003).

Sodium

Sodium is the primary extracellular cation, and is important in regulating body water. While most diets contain adequate amounts of sodium, losses due to vomiting or diarrhea may require replacement by oral or parenteral electrolyte solutions. Loss of sodium, due to furosemide treatment, induced salt hunger in rhesus monkeys (Schulkin *et al.*, 1984). Because of the association of increased salt intake with high blood pressure in humans, several nonhuman primate species have been evaluated as experimental surrogates. Increased salt intake in African green monkeys increased blood pressure on average, but there was significant individual variation such that individuals with high initial blood pressures showed a greater response to increases in dietary sodium (Srinivasan *et al.*, 1984). Spider monkeys and hamadryas baboons also exhibited increases in blood pressure, but rhesus monkeys did not, when a diet containing 3% NaCl was fed (Cherchovich *et al.*, 1976; Srinivasan *et al.*, 1980; Srinivansan *et al.*, 1984).

Potassium

Potassium is the major intracellular cation, and is involved with cell membrane transport, acid-base balance, and carbohydrate metabolism (Knapka *et al.*, 1995). It is found in high concentrations in both plant and animal tissues, making a deficiency rare (NRC, 2003). Potassium requirements may vary depending upon species, life stage, and diet composition.

Chlorine

Chloride is the major anion involved in osmotic regulation in association with sodium and potassium, and is a factor in membrane transport. It is secreted as hydrochloric acid in gastric secretions. Although a dietary deficiency of chloride has not been reported in nonhuman primates, significant amounts may be lost through vomiting, and may require replacement through oral or parenteral electrolyte administration.

Sulfur

Sulfur is a component of sulfur-containing amino acids in proteins, as well as the vitamins biotin and thiamin. A deficiency of sulfur cannot be distinguished from deficiencies of sulfur-containing amino acids or these vitamins (Knapka *et al.*, 1995). High intakes of sulfur-containing amino acid residues (cystine, cysteine, methionine, and taurine) may increase renal loss of calcium (NRC, 2003).

Iron

A major role of iron is concerned with oxygen transport in heme in the protein hemoglobin in red blood cells. Iron is an essential component of myoglobin, ferritin, and cytochromes, and acts as an enzyme cofactor. Iron deficiency results in microcytic, hypochromic anemia with decreased serum iron, increased total iron-binding capacity, and decreased liver and bone marrow iron stores (Knapka *et al.,* 1995; NRC, 2003). It can be quickly reversed by administration of iron, marked by a rapid increase in reticulocyte count. Iron deficiency may occur secondary to blood loss from trauma, ulcers, parasitism, menstruation, or endometriosis (Wixson and Griffith, 1986). A primary dietary iron deficiency is unusual, but other ingested materials such as phytates, plant polyphenols, soil, or calcium, copper, manganese, and zinc may influence iron absorption and metabolism (Fitch *et al.,* 1964; Rosenberg and Solomons, 1982; Ashmead and Christy, 1985).

Iron overload (hemosiderosis) can cause liver damage (Nath *et al.,* 1972). Cebus monkeys (Wolfe *et al.,* 1989) and marmosets (Sergejew *et al.,* 2000) are useful models for iron chelation therapy following iron overload. Some prosimians have exhibited increased liver iron stores compared to certain other primates (Schmechel *et al.,* 1996), and they may be more susceptible to hemosiderosis under certain husbandry conditions (Spelman *et al.,* 1989; Miller *et al.,* 1997).

Copper

Copper is needed for hemoglobin formation, copper-dependent enzymes, and regulation of cholesterol metabolism (Knapka *et al.,* 1995). A deficiency is associated with anemia, poor growth, bone disorders, depigmentation of the hair (achromotrichia), increased plasma cholesterol, gastrointestinal disturbances, and spinal cord lesions (Knapka *et al.,* 1995). *Otolemur* species have developed progressive cholinergic axonomal dystrophy, cholinergic denervation, generalized gliosis, and later, extensive cerebral amyloidosis associated with low tissue copper stores and carnitine deficiency (Schmechel *et al.,* 1996). Young rhesus monkeys kept in galvanized caging developed achromotrichia, alopecia, weakness, and microcytic anemia related to decreased plasma copper and increased plasma and liver zinc (Stevens *et al.,* 1977; Obeck, 1978; Wagner *et al.,* 1985).

Zinc

Over 200 zinc enzymes are known (NRC, 2003). Zinc activates enzymes of protein metabolism, is part of carbonate dehydratase and carbon dioxide transfer, and is important for immune function (Knapka *et al.,* 1995). While spontaneous zinc deficiency is not common, experimental deficiency causes alopecia and dermal lesions on the extremities, face and abdomen in rhesus and bonnet macaques (Swenerton and Hurley, 1980), squirrel monkeys (Barney *et al.,* 1967; Macapinlac *et al.,* 1967) and moustached tamarins (Chadwick *et al.,* 1979). Parakeratosis and thickening of the dorsal surface of the tongue were present in the squirrel monkeys (Barney *et al.,* 1967). Rhesus monkeys also exhibit depressed immune responses, as evidenced by decreased neutrophil chemotaxis, impaired mitogen responses, and inhibition of immunoglobulin production (Vruwink *et al.,* 1991). In pregnant rhesus monkeys, zinc deficiency caused anorexia, stillbirths, abortions, delivery complications, low birth weights (Golub *et al.,* 1984a,b), altered myelination in the offspring (Liu *et al.,* 1992), taste dysfunction, impaired growth, and reduced dietary efficiency (Golub *et al.,* 1984c). In infant rhesus monkeys, zinc deficiency caused apathy, lethargy, depressed immune function, decreased plasma alkaline phosphatase activity, hypochromic microcytic anemia, delayed skeletal maturation, and impaired bone mineralization (Leek *et al.,* 1984; Golub *et al.,* 1985; Haynes *et al.,* 1985; Leek *et al.,* 1988). Zinc overload can interfere with absorption of copper in infant rhesus monkeys (Lonnerdal *et al.,* 1999). Elevated serum zinc was associated with light-induced seizures in Senagalese baboons (*Papio papio*) which responded to zinc-chelation therapy (Alley *et al.,* 1981).

Selenium

Selenium incorporated into selenoproteins is primarily associated with cysteine. Four of these selenoproteins are glutathione peroxidases, which protect cells from peroxidative damage. Because of similar metabolic functions, there is an interaction between requirements for selenium and vitamin E, but both are required (NRC, 2003). In squirrel monkeys, signs of selenium deficiency are weight loss, listlessness, alopecia, cardiac and skeletal muscle degeneration, hepatic necrosis, and nephrosis (Muth *et al.,* 1971). However, in rhesus monkeys cardiomyopathic lesions were seen only with concurrent protein deficiency (Butler *et al.,* 1988). Selenium toxicity signs in cynomolgus monkeys can be characterized as dermatologic (dermatitis, hyperkeratosis, xerosis), gastrointestinal (constipation, anorexia, hemorrhagic diarrhea, emesis, gastroenteritis), central nervous system-related (lethargy and hypothermia), and reproductive (disturbances in menstruation, fetal growth retardation, and fetal death) (Cukierski *et al.,* 1989; Tarantal *et al.,* 1991). Toxicity of selenium varies with the amounts and chemical forms of the mineral ingested and to some extent species. Limited evidence suggests that monkeys are more susceptible to selenium in drinking water than rats or hamsters (Underwood, 1977).

Cobalt

Cobalt is incorporated into vitamin B_{12} during production by microorganisms, but a deficiency of elemental cobalt has not been reported in nonhuman primates (NRC, 2003).

Iodine

Iodine is an essential component of the thyroid hormones, thyroxine (T4) and triiodothyronine (T3), which are involved in regulation of growth and metabolic rate (NRC, 2003). Iodine deficiency in pregnant marmosets reduced plasma thyroxine, increased plasma TSH, and produced thyroid hyperplasia in both mothers and infants. The infants were small and weak with reduced hair growth, and had decreased fetal brain weight with morphologic changes (Mano *et al.,* 1987). A surviving infant exhibited stunted growth, delayed bone maturation, and a prominent abdomen, all signs of hypothyroidism.

Manganese

Manganese is necessary for enzyme activation, phosphate transferases and decarboxylases in particular (Knapka *et al.,* 1995). In general, deficiency causes poor reproductive success, poor growth, and leg deformities; however the only changes reported in nonhuman primates were altered behavioral development, with manganese-deficient infant rhesus monkeys showing an abnormal increase in clasping and clinging responses in conjunction with a decreased righting response (Riopelle and Hubbard, 1977). Experimental manganese toxicity in nonhuman primates resulted in similar neurologic signs (Pentshew *et al.,* 1963). High concentrations of dietary iron can reduce manganese stores (Hurley *et al.,* 1983).

Molybdenum

Molybdenum is essential for purine metabolism and is a component of xanthine oxidase. No deficiency or toxicity has been reported in nonhuman primates.

Chromium

Chromium is involved in carbohydrate, lipid, protein, and nucleic acid metabolism (NRC, 2003). As part of the oligopeptide chromodulin, chromium potentiates stimulation of the insulin receptor by insulin; a deficiency was found to cause impaired glucose tolerance in squirrel monkeys (Davidson *et al.,* 1967; Davidson and Blackwell, 1968). Irreversible corneal lesions have also been reported in chromium-deficient squirrel monkeys (Martin *et al.,* 1972).

Fluorine

Fluoride is incorporated into the mineral structure of bones and teeth as fluorohydroxyapatite. While it has not been demonstrated to be essential, fluoride supplementation decreases the incidence of dental caries in cynomolgus monkeys (Cohen and Bowen, 1966; Bowen, 1973). Excess fluoride can cause mottling of teeth. Because of hard tissue incorporation, fluoride accumulates in the body throughout life. Maximal concentrations of approximately 1200 ppm of fluoride in bone mineral increases bone strength; greater concentrations cause brittle bones with increased risk of fracture despite high bone mineral content (Turner, 1992).

Water

Captive primates generally share with humans a common municipal water supply; consequently, established water quality standards for humans are likely to ensure satisfactory water quality for nonhuman primates (NRC, 2003). Although quantitative data on water consumption are available for few species of nonhuman primates, maintenance of body water balance is the homeostatic objective. Diets high in fiber, salt, sodium bicarbonate, or protein increase water intake (Harris and Van Horn, 1992). Thus, water requirements vary with diet composition and intake, as well as with the effects of activity and environment and the need to dissipate body heat (NRC, 2003). Due to the complexity of factors affecting water requirements, potable water should be provided *ad libitum* to caged laboratory animals. Under group-housing conditions, competition for water supplies may require special measures to ensure individual access (Weisbard and Goy, 1976). Based on water intake data from studies of adult pig-tailed macaques, adult rhesus macaques, and adult humans, daily water needs approximate 1 $ml{\cdot}kcal^{-1}$ ME expenditure (NRC, 2003). Smaller species of nonhuman primates may have greater water requirements due to larger surface areas per unit of mass; 161 observations of adult common marmosets (*Callithrix jaccus*) demonstrated an average water intake of 11.7 $ml{\cdot}d^{-1}$ (Lunn, 1989).

Nonhuman primate diet formulations

Natural-ingredient diets

Natural-ingredient diets, most commonly fed to nonhuman primates, are composed primarily of grains and by-products of human food processing. Diets manufactured from natural ingredients generally do not pose health risks to the animal, but may affect experimental results (Rao and Knapka, 1987). Ingredient variability due to plant variety, soil or weather conditions, and harvesting or storage procedures may influence nutrient concentrations (Knapka *et al.,* 1995).

Most commercial primate diets are prepared by extrusion (steam-moistened feed forced through small openings at the end of a high-pressure, high-temperature chamber), forming a low density biscuit that tends to be more palatable than pellets and is easy for primates to handle (NRC, 2003). Processing by extrusion, pelleting, or baking can destroy heat labile vitamins such as vitamins A, D, E, K, C, thiamin, and folate. Vitamin C, as ascorbic acid, is the most labile vitamin and >50% can be destroyed by extrusion (Lovell and Lim, 1978). Stable forms of vitamins are available but not all commercial sources are equal in this respect. Use of L-ascorbyl-2-polyphosphate, a form of vitamin C with full biologic activity and stable to oxidation, has increased the safe primate diet storage interval from 90 to 180 days (Machlin *et al.,* 1979; NRC, 2003). If possible, primate diets should be stored in air-conditioned areas or coolers in which temperatures are below 21° C (70° F) and relative humidity is below 60% (Knapka *et al.,* 1995).

Extruded laboratory primate diets generally provide between 2.5 and 3.5 kcal ME·g^{-1} and are designed to be nutritionally complete. While these diets may meet the nutrient requirements of a variety of nonhuman primates, no single diet meets the requirements for all laboratory species. Specialty formulations may be required for certain species during specific life phases, for gastrointestinal health, or for discrete protocol requirements (Knapka *et al.,* 1995; NRC, 2003).

Purified diets

Purified diets are formulated with highly refined ingredients such that each component of the diet provides a single nutrient or class of nutrients to the diet, e.g., specific carbohydrates and fatty acids, crystalline amino acids, isolated fibers, vitamins, and minerals (Knapka *et al.,* 1995). These diets typically are not affected by the batch-to-batch product variation seen with natural-ingredient diets; the composition of purified diets allows for addition of nutrients, test compounds, or manipulations requiring various dosages of a chemical or nutrient for research purposes. Purified diets are expensive and may be less palatable to nonhuman primates.

Liquid purified diets also can be used in primate research. These diets have been used in strictly controlled intake studies with animals of all ages. Canned liquid diets, formulated for human use, provide 1 kcal ME·ml^{-1} and have been used successfully to rear nonhuman primate infants and adolescent rhesus macaques and baboons during growth studies (Hansen and Jen, 1979; Rutenberg and Coelho, 1988; NRC, 2003). Canned, high-moisture solid diets are frequently fed to nonhuman primates, and many are specialized formulations for specific species such as marmosets and tamarins, designed to provide 3.5–4.2 kcal ME·g^{-1}, which helps control "marmoset wasting syndrome" (Knapka *et al.,* 1995; NRC, 2003). Product information usually can be found on manufacturers' websites.

Live prey and supplemental foods

The provision of live prey for captive primates can promote foraging activity and contribute to environmental enrichment (Knapka *et al.,* 1995; NRC, 2003). Beetles, caterpillars, grasshoppers, ants, crickets, mealworms, wax-moth larvae, and other invertebrates also may be an important source of nutrients for obligate insectivores. Wax-moth larvae contain about 26% crude fat; crickets and mealworms contain approximately 20% crude protein; however, due to their high moisture content and high cost, insects are not a practical replacement for a complete diet for omnivores or facultative insectivores (Knapka *et al.,* 1995). High-calcium diets (also fortified with other nutrients) have been formulated for crickets, mealworms, and wax-moth larvae to be fed for 1–2 days before the insects themselves become a meal for insectivorous primates. The insects, plus their gut contents, provide a more nutritionally complete meal than insects not receiving these supplements (Strzelewicz *et al.,* 1985; Allen and Oftedal, 1989). Risks associated with feeding certain live prey include consumption of pathogenic nematodes by callitrichids feeding on cockroaches and hepatitis infection of callitrichids by lymphocytic choriomeningitis virus when laboratory mouse pups are fed (NRC, 2003).

When succulent fruits or vegetables are provided as treats, environmental enrichment, or as supplements to a balanced diet, care must be taken that moisture in these items does not excessively dilute nutrients in the balanced diet. For instance, high-moisture vegetables or fruits, fed in conjunction with high-moisture canned or gel diets, can limit the ability of a small primate (marmosets and tamarins) to consume sufficient DM to meet nutrient and energy needs (Barnard *et al.*, 1988). Cultivated fruits (bananas, oranges, apples, and grapes) are not comparable in composition to wild fruits, which generally are higher in indigestible fiber and DM (NRC, 2003). When succulent foods are provided in unlimited amounts to captive primates, more palatable treats may be substituted for nutritionally complete formulated diets, and nutritional imbalances may result.

Food contaminants

Plants introduce many naturally occurring chemicals into diets of nonhuman primates. Most harmful chemicals and natural toxicants, except phytic acid and perhaps oxalic acid, are at a sufficiently low concentration that biological impact is generally negligible. Natural ingredient diets may contain endogenous mineral contaminants derived from soil, e.g., arsenic, lead, mercury, and selenium. The toxicity of some environmentally-acquired contaminants may be exacerbated by harvest, processing, and storage conditions (Rao and Knapka, 1987). Nitrates in excess can give rise to carcinogenic nitrosamines. Pesticide and herbicide residues also may be present if approved practices are not followed. Purified diets generally do not contain toxic endogeneous dietary or microbial contaminants due to their refined nature.

Phytoestrogens

Most constitutive, natural plant chemicals are classified as: flavonoids, phenolic acids, phenylpropanoids, coumarins, cyclitols, isothiocyanates, catechins, simple phenols, monoterpenes, sesquiterpenes, amino acids, and anthraquinones (NRC, 1996); several are metabolized with significant biological effects. Biologically active phytoestrogen flavonoids, plant compounds structurally and/or functionally similar to ovarian and placental estrogenic metabolites include: isoflavonoids, lignans and some flavones, flavanones, chalcones, coumestans, and stilbenes (Wiseman, 2000; Whitten and Patisaul, 2001).

Soybean and alfalfa meals and other legumes contribute phytoestrogens to animal diets although soybean meal is the primary contributor of the isoflavone glycosides, genistin and daidzin, which are hydrolyzed by bacteria in the large intestine to their respective aglycones genistein (4′,5,7-trihydroxyisoflavone) and daidzein (4′,7-dihydroxyisoflavone) (Wiseman, 2000; Yang and Bittner, 2002). The most significant isoflavanoid is equol, a metabolite of daidzein. Coumestrol is the best-known coumestan and the isoflavonoid with the highest estrogenic potency. Alfalfa is one of the richest sources of coumestans (Boettger-Tong *et al.*, 1998).

Isoflavonoids found in legumes, especially in soybeans and soybean-based products can contain as much as 0.2–1.6 $mg{\cdot}g^{-1}$ isoflavones on a dry weight basis (Whitten and Patisaul, 2001). Total isoflavonoid concentration of soy protein isolate can range between 0.62 and 0.99 $mg{\cdot}g^{-1}$ (Anderson and Wolf, 1995). While purified diets are generally isoflavone-free (Brown and Setchell, 2001), some may contain soybean protein isolates.

Effects of metabolized dietary isoflavones, which increase serum and urinary concentrations, are known in humans (Adlercreutz *et al.*, 1991, 1993) and rodents (Boettger-Tong *et al.*, 1998; Brown and Setchell, 2001). These highly bioavailable compounds have led to high steady-state serum isoflavone concentrations in adult rats (2613 ± 873 ng/mL) and mice (2338 ± 531 ng/mL), which exceed an endogenous estrogen level by 30,000–60,000-fold (Brown and Setchell, 2001). Because phytoestrogen compounds have the potential to modulate genotypic and phenotypic expression among animal models studied to date, it is possible that nonhuman primates may also be affected by these compounds. Primate diets have the potential to deliver large daily doses of isoflavones during the lifespans of animals, including the *in utero* period.

Mycotoxins

Mycotoxins are naturally occurring products of fungal growth that develop on grains under field, harvest, or storage conditions and vary geographically. Contamination by one or several mycotoxins is common as a single fungus can generate several mycotoxins or several mycotoxin-producing fungi may infect the same plant (NRC, 1996). Contamination by two toxigenic species of *Aspergillus, A. flavus* and *A. parasiticus* which produce hepatocarcinogenic aflatoxins, appears ubiquitous. Corn and other grains, peanuts, and cottonseed meal are common media for growth of aflatoxin-producing fungi. *A. flavus* produces aflatoxins B_1 and B_2, whereas *A. parasiticus* produces aflatoxins B_1, B_2, G_1 and G_2 (Pitt *et al.* 1993). While all four aflatoxins are toxic and believed to be carcinogenic in animals, B_1 is the most prevalent and the most potent, causing hepatocellular adenomas and carcinomas and colon tumors in rats (NRC, 1996). Environmental stressors (e.g., drought or insect attack) cause corn crops to become particularly susceptible to *A. flavus* growth (NRC, 1996). The median levels of aflatoxins in corn range from <0.1 to 80 $ng{\cdot}kg^{-1}$; aflatoxin B_1 is the only mycotoxin for which the Food and Drug Administration (FDA) has set levels of contamination at which products are regarded adulterated (Riley *et al.*, 1993). Ochratoxin A, produced predominantly by *Aspergillus ochraceus* and *Penicillium verrucosum*, occurs worldwide in many grains (barley and wheat) and it is implicated in urinary tumorigenesis in humans and rodents (NRC, 1996).

While aflatoxin contamination is of great concern, the fumonisins are rapidly becoming the focus of increased research interest. Cereal grains infected by the fungal genus *Fusarium* produce estrogenic compounds (mycoestrogens) which are highly stable compounds that can be ingested, inhaled, or absorbed through the skin (Whitten and Patisaul, 2001; Yang and Bittner, 2002). Zearalenone, an estrogenic mycotoxin, is produced by *Fusarium graminearium* and *F. culmorum* which are commonly found in plants, soil, and stored grains such as barley, corn, rice, oats, and rye (Park *et al.*, 1996; Boettger-Tong *et al.*, 1998). Zearalenone, has been associated with mammary tumorigenesis (Shier *et al.*, 2001) and may be uterotropic (Sheehan *et al.*, 1984). When 500 cereal samples from 19 countries were analyzed for zearalenone, over 40% had concentrations as high as .045 $\mu g{\cdot}g^{-1}$ cereal (Tanaka *et al.*, 1988) compared to concentrations of 21 $\mu g{\cdot}g^{-1}$ that have been reported in moldy corn in the United States (Park *et al.*, 1996).

Fusarium moniliforme, another toxigenic fungi, is a ubiquitous contaminant in corn which produces the water-soluble fumonisins B_1 (FB_1) and B_2 (FB_2) and fusarin C and the T_2 toxin (Riley *et al.*, 1993; NRC, 1996; Voss *et al.*, 2001). FB_1 does not cross the placenta and is not teratogenic *in vivo* in rats, mice, or rabbits, but is embryotoxic at maternally toxic doses (Voss *et al.*, 2001). Fumonisin B_1 is carcinogenic in rats (renal carcinogen) and mice (hepatocarcinogen) when included in the diet at concentrations of 50 ppm or greater. Unlike aflatoxin, which produces heritable genetic damage, fumonisins inhibit ceramide synthetase, a key enzyme in *de novo* sphingolipid synthesis, which increases intracellular sphinganine (Howard *et al.*, 2001). Cellular apoptosis is induced; susbsequent cell proliferation could potentially give rise to spontaneous or chemically induced tumors. Vervet monkeys *(Cercopithecus aethiops,* now *Chlorocebus aethiops)* fed varying levels of *Fusarium verticillioides* culture material daily for 13.5 years, exhibited altered liver and kidney function, as well as adversely affected cholesterol, creatine kinase, and blood parameters (Gelderblom *et al.*, 2001). Thus, a diversity of lesions was induced by the feeding of *F. verticillioides*. The magnitude of the effects of short- and long-term feeding of diets containing these natural contaminants is clearly undefined and warrants further research.

Correspondence

Any correspondence should be directed to Duane Ullrey, Professor Emeritus, Departments of Animal Science and Fisheries and Wildlife, Michigan State University, East Lansing, MI 48824, USA. Email: dsullrey@aol.com; ullrey@msu.edu

References

Adlercreutz, H., Honjo, H., Higashi, A., Fotsis, T., Hamalainen, E., Hasegawa, T. and Okada, H. (1991). *Am. J. Clin. Nutr.* 54, 1093–1100.

Adlercreutz, H., Fotsis, T., Lampe, J., Wahala, K., Makela, T., Brunow, G. and Hase, T. (1993). *Scand. J. Clin. Lab. Invest. Suppl.* 215, 5–18.

Agamanolis, D.P., Chester, E.M., Victor, M., Kark, J.A., Hines, J.D. and Harris, J.W. (1976). *Neurology* 26, 905–914.

Allen, M.E. and Oftedal, O.T. (1989). *J. Zoo Wildl. Med.* 20, 26–33.

Alley, M.C., Killam, E.K. and Fisher, G.L. (1981). *J. Pharmacol. Exp. Therapeut.* 217, 138–146.

Ammerman, C.B., Baker, D.H., Lewis, A.J. (1995). *Bioavailability of Nutrients for Animals, Amino Acids, Minerals, and Vitamins*, pp 95–431. San Diego: Academic Press.

Anderson, R.L. and Wolf, W.J. (1995). *J. Nutr.* 125, 581S–588S.

Anonymous. (1981). *JAMA* 246, 730–731.

Ashmead, D. and Christy, H. (1985). *Anim. Nutr. Health* 8, 10–13.

Asp, N.G. (1994). *Am. J. Clin. Nutr.* 59, 679S–681S.

Ausman, L.M. and Gallina, D.L. (1979) *Primates in Nutritional Research*, pp 39–57. New York: Academic Press.

Ausman, L.M. and Hegsted, D.M. (1980). *Am. J. Clin. Nutr.* 33, 2551–2558.

Ausman, L.M., Gallina, D.L. and Nicolosi, R.J. (1985). In Rosenblum, L.A. and Coe, C.L. (eds) *Handbook of Squirrel Monkey Research.* New York: Plenum Press.

Ausman, L.M., Gallina, D.L., Hayes, K.C. and Hegsted, D.M. (1986). *Am. J. Clin. Nutr,* 43, 112–127.

Barnard, D., Knapka, J. and Renquist, D. (1988). *Lab. Anim. Sci.* 38, 282–288.

Barney, G.H., Macapinlac, M.P., Pearson, W.N. and Darby, W.J. (1967). *J. Nutr.* 93, 511–517.

Bennett, B.T., Cuasay, L., Welsh, T.J., Beluhan, F.Z. and Schofield, L. (1980). *Lab. Anim. Sci.* 30, 241–244.

Binkley, N., Krueger, D., Engelke, J. and Suttie, J.W. (2000). *J. Bone Miner. Res.* 15 (Suppl. 1), S425.

Blank, N.K., Vick, N.A. and Schulman, S. (1975). *Acta. Neuropath.* 31, 137–150.

Bodkin, N.L., Ortmeyer, H.K. and Hansen, B.C. (1995). *J. Gerontol. Biol. Sci. Med. Sci.* 50, B142–B147.

Boettger-Tong, H., Murthy, L., Chiappetta, C., Kirkland, J.L., Goodwin, B., Adlercreutz, H., Stancel, G.M. and Makela, S. (1998). *Environ. Health Perspect.* 106, 369–373.

Boonjawat, J., Wilairat, P. and Vimokesant, S.L. (1979). *Am. J. Clin. Nutr.* 32, 2065–2067.

Borda, J.T., Patino, E.M., Ruiz, J.C. and Sanchez-Negrette, M. (1996). *Lab. Prim. Newslet.* 35, 5–6.

Bowen, W.H. (1973). *Brit. Dent. J.* 135, 489–493.

Boyce, L. and Miller, C. (1980). *Vet. Med. Small Anim. Clinician* 75, 130–131.

Brady, P.S., Sehgal, P.K. and Hayes, K.C. (1982). *Am. J. Vet. Res.* 43, 1489–1491.

Brady, A.G., Williams, L.E. and Abee, C.R. (1990). *Lab. Anim. Sci.* 40, 262–265.

Brody, A. (1945). In Brody, S. (ed.) *Bioenergetics and Growth,* pp 352–403. New York: Reinhold.

Brommage, R., Hotchkiss, C.E., Lees, C.J., Stancill, M.W., Hock, J.M. and Jerome, C.P. (1999). *J. Clin. Endocrinol. Metab.* 84, 3757–3763.

Brown, N.M. and Setchell, K.D. (2001). *Lab. Invest.* 81, 735–747.

Butler, J.A., Whanger, P.D. and Patton, N.M. (1988). *J. Am. Coll. Nutr.* 7, 43–56.

Chadwick, D.P., May, J.C. and Lorenz, D. (1979). *Lab. Anim. Sci.* 29, 482–485.

Chamberlain, B., Ervin, F.R., Pihl, R.O. and Young, S.N. (1987). *Pharmacol. Biochem. Behav.* 28, 503–510.

Chen, H., Hu, B., Allegretto, E.A. and Adams, J.S. (2000). *J. Biol. Chem.* 275, 35557–35564.

Cherchovich, G.M., Capek, K., Jefremova, Z., Pohlova, I. and Jelinek, J. (1976). *J. Appl. Physiol.* 40, 601–604.

Chester, E.M., Agamanolis, D.P., Harris, J.W., Victor, M., Hines, J.D. and Kark, J.A. (1980). *Acta. Neurologica. Scand.* 61, 9–26.

Clarke, H.E., Coates, M.E., Eva, J.K., Ford, D.J., Milner, C.K., O'Donoghue, P.N., Scott, P.P. and Ward, R.J. (1977). *Lab. Anim.* 11, 1–28.

Cohen, B. and Bowen, W.H. (1966). *Brit. Dent. J.* 121, 269–276.

Cooperman, J.M., Waisman, H.A., McCall, K.B. and Elvehjem, C.A. (1945). *J. Nutr.* 30, 45–57.

Cueto, J., Tajen, N., Gilbert, E. and Currie, R.A. (1967). *Annals. Surg.* 166, 19–28.

Cukierski, M.J., Willhite, C.C., Lasley, B.L., Hendrie, T.A., Book, S.A., Cox, D.N. and Hendrickx, A.G. (1989). *Fund. Appl. Toxicol.* 13, 26–39.

Curtis, S.E. (1983). In *Environmental Management in Animal Agriculture,* pp 79–96. Ames: Iowa State University Press.

Davidson, I.W.F., Lang, C.M. and Blackwell, W.L. (1967). *Diabetes* 16, 395–401.

Davidson, I.W.F. and Blackwell, W.L. (1968). *Proc. Soc. Exp. Biol. Med.* 127, 66–70.

Day, P.L., Langston, W.C. and Shukers, C.F. (1935). *J. Nutr.* 9, 637–644.

Demaray, S.Y., Altman, N.H. and Ferrell, T.L. (1978). *Lab. Anim. Sci.* 28, 457–460.

Deo, M.G., Sood, S.K. and Ramalingaswami, V. (1965). *Archs. Path.* 80, 14.

Dreizen, S., Levy, B.M. and Bernick, S. (1970). *J. Dent. Res.* 49, 616–620.

Dunn, M.J. (1971). *Clin. Sci.* 41, 333–344.

Edwards, M.S. (1995). In *Comparative Adaptations to Folivory in Primates* (Ph.D. Dissertation). E. Lansing: Michigan State University Press.

Edwards, M.S. and Ullrey, D.E. (1999). *Zoo Biol.* 18, 537–549.

Eisele, P.H., Morgan, J.P., Line, A.S. and Anderson, J.H. (1992). *Lab. Anim. Sci.* 42, 245–249.

Elwell, M.R. and DePaoli, A. (1978). *J. Am. Vet. Med. Assoc.* 173, 1235–1236.

Fitch, C.D., Harville, W.E., Dinning, J.S. and Porter, F.S. (1964). *Proc. Soc. Exp. Biol. Med.* 116, 130–133.

Flurer, C.I. and Zucker, H. (1985). *Primates* 26, 479–490.

Flurer, C.I. and Zucker, H. (1987). *Int. J. Vitamin Nutr. Res.* 57, 297–298.

Flurer, C.I., Geyer, G., Berg, D. and Rambeck, W.A. (1990). *Z. Ernahrungswiss.* 29, 192–196.

Foy, H., Kondi, A. and Mbaya, V. (1964). *Brit. J. Nutr.* 18, 307–318.

Foy, H., Kondi, A. and Verjee, Z.H. (1972). *J. Nutr.* 102, 571–582.

Foy, H. and Kondi, A. (1984). *J. Natl. Cancer Inst.* 72, 941–948.

Gacad, M.A. and Adams, J.S. (1992). *Endocrinology* 131, 2581–2587.

Gacad, M.A., Chen, H., Arbelle, J.E., LeBon, T. and Adams, J.S. (1997). *J. Biol. Chem.* 272, 8433–8440.

Gacad, M.A. and Adams, J.S. (1998). *J. Clin. Endocrinol. Metab.* 83, 1264–1267.

Gelderblom, W.C., Seier, J.V., Snijman, P.W., Van Schalkwyk, D.J., Shephard, G.S. and Marasas, W.F. (2001). *Environ. Health Perspect.* 109 (Suppl. 2), 267–276.

Golub, M.S., Gershwin, M.E., Hurley, L.S., Baly, D.L. and Hendrickx, A.G. (1984a). *Am. J. Clin. Nutr.* 39, 265–280.

Golub, M.S., Gershwin, M.E., Hurley, L.S., Baly, D.L. and Hendrickx, A.G. (1984b). *Am. J. Clin. Nutr.* 39, 879–887.

Golub, M.S., Gershwin, M.E., Hurley, L.S., Saito, B.S. and Hendrickx, A.G. (1984c). *Am. J. Clin. Nutr.* 40(6), 1192–1202.

Golub, M.S., Gershwin, M.E., Hurley, L.S., Hendrickx, A.G. and Saito, W.Y. (1985). *Am. J. Clin. Nutr.* 42, 1229–1239.

Greenberg, L.D. and Moon, H.D. (1963). *Fed. Proc.* 22, 592.

Greenberg, L.D. (1970). In Harris, R.S. (ed.) *Feeding and Nutrition of Nonhuman Primate,* pp 117–142. New York: Academic Press.

Greiner, S.R.C., Zhang, Q., Goodman, K.J., Giussani, D.A., Nathanielsz, P.W. and Brenna, J.T. (1996). *J. Lipid Res.* 37, 2675–2686.

Griffiths, H.J., Hunt, R.D., Zimmerman, R.E. and Finberg, H. (1975). *Invest. Radiol.* 10, 263–268.

Groves, C. (2001) *Primate Taxonomy,* Washington, D.C.: Smithsonian Institution Press.

Grynpas, M.D., Hancock, R.G., Greenwood, C., Turnquist, J. and Kessler, M.J. (1993). *Calc. Tissue Int.* 52, 399–405.

Hansen, B.C. and Jen, K.C. (1979). In Hayes, K.C. (ed.) *Primates in Nutritional Research,* pp 59–71. London: Academic Press.

Hansen, B.C., Jen, K.L. and Kribbs, P. (1981). *Physiol. Behav.* 26, 479–486.

Hansen, B.C. and Bodkin, N.L. (1993). *Diabetes* 42, 1809–1814.

Harris, J.B. and Horn, H.H.V. (1992). In *Dairy Production Guide (Circular 1017).* Florida Cooperative Extension Service.

Hartvig, P., Lindner, K.J., Bjurling, P., Laengstrom, B. and Tedroff, J. (1995). *J. Neur. Transmiss. Gen. Sec.* 102, 91–97.

Hayes, K.C., Stephan, Z.F. and Sturman, J.A. (1980). *J. Nutr.* 110, 2058–2064.

Hayes, K.C. (1985). *Nutr. Rev.* 43, 65–70.

Haynes, D.C., Gershwin, M.E., Golub, M.S., Cheung, A.T., Hurley, L.S. and Hendrickx, A.G. (1985). *Am. J. Clin. Nutr.* 42, 252–262.

Heine, R.J., Schouten, J.A., van Gent, C.M., Havekes, L.M., Koopman, P.A. and van der Veen, E.A. (1984). *Ann. Nutr. Metab.* 28, 201–206.

Hendrickx, A.G. and Hummler, H. (1992). *Teratology* 45, 65–74.

Hendrickx, A.G., Peterson, P., Hartmann, D. and Hummler, H. (2000). *Repro. Toxicol.* 14, 311–323.

Hill, R.B., Schendel, H.E., Rama Rao, P.B. and Johnson, B.C. (1964). *J. Nutr.* 84.

Holmberg, C.A., Leininger, R., Wheeldon, E., Slater, D., Henrickson, R. and Anderson, J. (1982). *Vet. Pathol.* 19(Suppl. 7), 163–170.

Hotchkiss, C.E., Stavisky, R., Nowak, J., Brommage, R., Lees, C.J. and Kaplan, J. (2001). *Bone* 29, 7–15.

Howard, P.C., Warbritton, A., Voss, K.A., Lorentzen, R.J., Thurman, J.D., Kovach, R.M. and Bucci, T.J. (2001). *Environ. Health Perspect.* 109(Suppl 2), 309–314.

Huang, T.H., Ann, D.K., Zhang, Y.J., Chang, A.T., Crabb, J.W. and Wu, R. (1994). *Am. J. Resp. Cell Mol. Biol.* 10, 192–201.

Hummler, H., Korte, R. and Hendrickx, A.G. (1990). *Teratology* 42, 263–272.

Hunt, R.D., Garcia, F.G. and Hegsted, D.M. (1969). *Am. J. Clin. Nutr.* 22, 358–366.

Hunt, R.D., Garcia, F.G. and Walsh, R.J. (1972). *J. Nutr.* 102, 975–986.

Hurley, L.S., Keen, C.L. and Lonnerdal, B. (1983). *Fed. Proc.* 42, 1735–1739.

Ingram, D.K., Cutler, R.G., Weindruch, R., Renquist, D.M., Knapka, J.J., April, M., Belcher, C.T., Clark, M.A., Hatcherson, C.D., Marriott, B.M. *et al.* (1990). *J. Gerontol.* 45, B148–B163.

Innis, S.M. (1991). *Prog. Lipid Res.* 30, 39–103.

Iwamoto, J., Takeda, T. and Ichimura, S. (2001). *J. Orthoped. Sci.* 6, 487–492.

Jayo, M.J., Register, T.C. and Carlson, C.S. (1998). *Bone* 23, 361–366.

Jerome, C.P., Power, R.A., Obasanjo, I.O., Register, T.C., Guidry, M., Carlson, C.S. and Weaver, D.S. (1997). *Bone* 20, 355–364.

Juan-Salles, C., Prats, N., Ruiz, J.M., Valls, X., Gine, J., Garner, M.M., Verges, J. and Marco, A. (2000). *J. Comp. Pathol.* 123, 202–206.

Keller, E.T., Binkley, N., Stebler, B.A., Hall, D.M., Johnston, G.M., Zhang, J. and Ershler, W.B. (2000). *Bone* 26, 55–62.

Kemnitz, J.W., Weindruch, R., Roecker, E.B., Crawford, K., Kaufman, P.L. and Ershler, W.B. (1993). *J. Gerontol.* 48, B17–B26.

Kerr, G.R., Chamove, A.S. and Harlow, H.F. (1969). *J. Pediatr.* 75, 833–837.

Kessler, M.J. (1980). *J. Med. Primatol.* 9, 314–318.

Kim, J.C., Abee, C.R. and Wolf, R.H. (1978). *Vet. Med. Small Anim. Clinician* 73, 1303–1306.

Kleiber, M. (1975). *The Fire of Life: An Introduction to Animal Energetics,* 2nd revised ed., pp 3–8. New York: R.E. Krieger Publ. Co.

Klevay, L.M., Mendelsohn, D., Vanderwatt, J.J., Davidson, L.M. and Kritchevsky, D. (1981). *S. Afr. Med. J.* 59, 605–609.

Knapka, J.J., Barnard, D.E., Bayne, K.A.L., Lewis, S.M., Marriot, B.M. and Oftedal, O.T. (1995). In Bennett, B.T., Abee, C.R., Henrickson, R. (eds) *Nonhuman Primates in Biomedical Research: Biology and Management,* pp 211–248. New York: Academic Press.

Kriek, N.P., Sly, M.R., du Bruyn, D.B., de Klerk, W.A., Renan, M.J., Van Schalkwyk, D.J. and Van Rensburg, S.J. (1982). *Br. J. Exp. Pathol.* 63, 254–268.

Kritchevsky, D., Davison, L.M., Goodman, G.T., Tepper, S.A. and Mendelsohn, D. (1986). *Lipids* 21, 338–341.

Kritchevsky, D., Davidson, L.M., Scott, D.A., Van der Watt, J.J. and Mendelsohn, D. (1988). *Lipids* 23, 164–168.

Kuzuya, F. (1993). *Nagoya. J. Med. Sci.* 55, 1–9.

Lane, M.A., Baer, D.J., Rumpler, W.V., Weindruch, R., Ingram, D.K., Tilmont, E.M., Cutler, R.G. and Roth, G.S. (1996). *Proc. Natl. Acad. Sci. USA* 93, 4159–4164.

Leek, J.C., Vogler, J.B., Gershwin, M.E., Golub, M.S., Hurley, L.S. and Hendrickx, A.G. (1984). *Am. J. Clin. Nutr.* 40, 1203–1212.

Leek, J.C., Keen, C.L., Vogler, J.B., Golub, M.S., Hurley, L.S., Hendrickx, A.G. and Gershwin, M.E. (1988). *Am. J. Clin. Nutr.* 47, 889–895.

Lees, C.J., Jerome, C.P., Register, T.C. and Carlson, C.S. (1998). *J. Clin. Endocrinol. Metab.* 83, 4298–4302.

Lehner, N.D., Bullock, B.C. and Clarkson, T.B. (1968). *Proc. Soc. Exp. Biol. Med.* 128, 512–514.

Liu, S.K., Dolensek, E.P., Tappe, J.P., Stover, J. and Adams, C.R. (1984). *J. Am. Vet. Med. Assoc.* 185, 1347–1350.

Liu, H., Oteiza, P.I., Gershwin, M.E., Golub, M.S. and Keen, C.L. (1992). *Biological Trace Element Res.* 34, 55–66.

Lloyd, L.E., McDonald, B.E. and Crampton, E.W. (1978). *Fundamentals of Nutrition,* pp 396–438. San Francisco: W.H. Freeman and Co.

Lonnerdal, B., Jayawickrama, L. and Lien, E.L. (1999). *Am. J. Clin. Nutr.* 69, 490–496.

Lovell, R.T. and Lim, C. (1978). *Trans. Am. Fish. Soc.* 107, 321–325.

Lueken, S.A., Arnaud, S.B., Taylor, A.K. and Baylink, D.J. (1993). *J. Bone Miner. Res.* 8, 1433–1438.

Lunn, S.F. (1989). *Lab. Anim.* 23, 353–356.

Macapinlac, M.P., Barnye, G.H., Pearson, W.N. and Darby, W.J. (1967). *J. Nutr.* 93, 499–510.

Machlin, L.J., Garcia, F., Kuenzig, W. and Brin, M. (1979). *Am. J. Clin. Nutr.* 32, 325–331.

Mann, G.V., Watson, P.L., McNally, A. and Goddard, J. (1952). *J. Nutr.* 47, 225–241.

Mann, G.V. (1968). *Vitamins and Hormones* 26, 465–485.

Mano, M.T., Potter, B.J., Belling, G.B., Chavadej, J. and Hetzel, B.S. (1987). *J. Neurol. Sci.* 79, 287–300.

Martin, G.D., Stanley, J.A. and Davidson, I.W.F. (1972). *Invest. Ophthalmol.* 11, 153–158.

Marx, S.J., Jones, G., Weinstein, R.S., Chrousos, G.P. and Renquist, D.M. (1989). *J. Clin. Endocrinol. Metab.* 69, 1282–1290.

Masoro, E.J. (1996). *Toxicol. Pathol.* 24, 738–741.

Mayes, P.A. (1996). In Murray, R.K., Granner, D.K., Mayes, P.A. and Rodwell, V.W. (eds) *Harper's Biochemistry,* pp 625–634. Stamford, CT: Appleton & Lange.

McCall, K.B., Waisman, H.A., Elvehjem, C.A. and Jones, E.S. (1946). *J. Nutr.* 31, 685–697.

McCamant, S.K., Klosterman, L.L., Goldman, E.S., Murai, J.T. and Siiteri, P.K. (1987). *Steroids* 50, 549–557.

McCay, C.M., Crowell, M.F. and Maynard, L.A. (1935). *J. Nutr.* 10, 63–79.

McDonald, P., Edwards, R.A., Greenhalgh, J.F.D. (1978). *Animal Nutrition*, 2nd ed., pp 59–112. Singapore: Huntsmen Group Ltd.

McDowell, L.R. (1989). *Vitamins in Animal Nutrition, Comparative Aspects to Human Nutrition*, pp 10–387. San Diego: Academic Press.

McIntosh, G.H., Bulman, F.H., Looker, J.W., Russell, G.R. and James, M. (1987). *J. Nutr. Sci. Vitaminol.* 33, 299–312.

Mehta, S.K., Chakravarti, R.N., Nain, C.K., Das, K.C., Bhagwat, A.G. and Mehta, S. (1979). *Israel J. Med. Sci.* 15, 348–355.

Merrill, A.L. and Watt, B.K. (1955). *Energy Value of Foods, Basis and Derivation*, Agr. Res. Serv., U.S. Dept. Agr., Agr. Hndbk. No. 74, pp 24–43. Washington, D.C.: U.S. Government Printing Office.

Miller, L.A., Cheng, L.Z. and Wu, R. (1993). *Cancer Res.* 53, 2527–2533.

Miller, G.F., Barnard, D.E., Woodward, R.A., Flynn, B.M. and Bulte, J.W. (1997). *Lab. Anim. Sci.* 47, 138–142.

Milton, K. and Demment, M.W. (1988). *J. Nutr.* 118, 1082–1088.

Mitchell, H.H. and Block, R.J. (1946). *J. Biol. Chem.* 163, 599–620.

Mohanty, D. and Das, K.C. (1982). *J. Nutr.* 112, 1565–1576.

Muller, D.P. (1986). *Postgrad. Med. J.* 62, 107–112.

Muth, O.H., Weswig, P.H., Whanger, P.D. and Oldfield, J.E. (1971). *Am. J. Vet. Res.* 32, 1603–1605.

Nath, I., Sood, S.K. and Nayak, N.C. (1972). *J. Pathol.* 106, 103–111.

National Research Council. (1981). *Nutritional Energetics of Domestic Animals and Glossary of Energy Terms.* 2nd revised ed., pp 9–22. Washington, D.C.: National Academy Press.

National Research Council. (1989). *Recommended Dietary Allowances*, pp 86–261. Washington, D.C.: National Academy Press.

National Research Council. (1996). *Carcinogens and Anticarcinogens in the Human Diet, A Comparison of Naturally Occurring and Synthetic Substances*, pp 35–126. Washington, D.C.: National Academy Press.

National Research Council. (2003). *Nutrient Requirements of Nonhuman Primates,* 2nd revised ed., pp 41–194. Washington, D.C.: National Academy Press.

Neuringer, M., Connor, W.E., Van Petten, C. and Barstad, L. (1984). *J. Clin. Invest.* 73, 272–276.

Obeck, D.K. (1978). *Lab. Anim. Sci.* 28, 698–704.

Oftedal, O.T. (1984). *Symp. Zool. Soc. Lond.* 51, 33–85.

Park, J.J., Smalley, E.B. and Chu, F.S. (1996). *Appl. Environ. Microbiol.* 62, 1642–1648.

Patek, A.J., Bowry, S. and Hayes, K.C. (1975). *Proc. Soc. Exp. Biol. Med.* 148, 370–374.

Pentshew, A., Ebner, F.F. and Kovatch, R.M. (1963). *J. Neuropathol. Exp. Neurol.* 22, 488–499.
Peretti, P.O. and Baird, M. (1975). *J. Nutr. Sci. Vitaminol.* 21, 199–206.
Pitt, J.I., Hocking, A.D., Bhudhasamai, K., Miscamble, B.F., Wheeler, K.A. and Tanboon–Ek, P. (1993). *Int. J. Food Microbiol.* 20, 211–226.
Rao, G.N. and Knapka, J.J. (1987). *Fundam. Appl. Toxicol.* 9, 329–338.
Rasmussen, K.M., Thenen, S.W. and Hayes, K.C. (1979). *Am. J. Clin. Nutr.* 32, 2508–2518.
Rasmussen, K.M., Thenen, S.W. and Hayes, K.C. (1980). *J. Med. Primatol.* 9, 169–184.
Ratterree, M.S., Didier, P.J., Blanchard, J.L., Clarke, M.R. and Schaeffer, D. (1990). *Lab. Anim. Sci.* 40, 165–168.
Reisbick, S., Neuringer, M., Hasnain, R. and Connor, W.E. (1990). *Physiol. Behav.* 47, 315–323.
Renan, M.J. and van Rensburg, S.J. (1980) *Phys. Med. Biol.* 25, 433–444.
Riley, R.T., Norred, W.P. and Bacon, C.W. (1993). *Annu. Rev. Nutr.* 13, 167–189.
Rinehart, J.F., Friedman, M. and Greenberg, L.D. (1949). *Arch. Pathol.* 48, 129–139.
Riopelle, A. and Hubbard, D.G. (1977). *J. Orthomolecular Psych.* 4, 327–333.
Robbins, R.C. (1993) *Wildlife Feeding and Nutrition,* pp 114–174. San Diego: Academic Press.
Robertson, J.B. and Van Soest, P.J. (1981). In James, W.P.T. and Theander, O. (eds) *The Analysis of Dietary Fiber in Food*, pp 123–158. New York: Marcel Dekker.
Rogers, A.E. (1979). In Baker, H.J., Lindsey, J.R. and Weisbroth, S.H. (eds) *The Laboratory Rat*, Vol. 1, pp 123–152. San Diego: Academic Press.
Rosenberg, I.H. and Solomons, N.W. (1982). *Am. J. Clin. Nutr.* 35, 781–782.
Rudel, L.L., Parks, J.S. and Sawyer, J.K. (1995). *Arterioscler. Thromb. Vasc. Biol.* 15, 2101–2110.
Rutenberg, G.W. and Coelho, A.M., Jr. (1988). *Am. J. Phys. Anthropol.* 75, 529–539.
Samonds, K.W. and Hegsted, D.M. (1973). *Am. J. Clin. Nutr.* 26, 30–40.
Sandhyamani, S. (1992). *Int. Angiol.* 11, 256–260.
Santiyanont, R., Yaipimol, C. and Wilairat, P. (1977). *J. Nutr.* 107, 2026–2030.
Schmechel, D.E., Burkhart, D.S., Ange, R. and Izard, M.K. (1996). *Exp. Neurol.* 142, 111–127.
Schneeman, B.O. (1990). *Adv. Exp. Med. Biol.* 270, 37–42.
Schulkin, J., Leibman, D., Ehrman, R.N., Norton, N.W. and Ternes, J.W. (1984). *Behavioral Neurosci.* 94, 753–756.
Schurgers, L.J., Dissel, P.E., Spronk, H.M., Soute, B.A., Dhore, C.R., Cleutjens, J.P. and Vermeer, C. (2001). *Z. Kardiol.* 90(Suppl. 3), 57–63.
Scott, M.L. (1986). *Nutrition in Humans and Selected Animal Species,* pp 12–78. New York: John Wiley & Sons.
Sergejew, T., Forgiarini, P. and Schnebli, H.P. (2000). *Brit. J. Haematol.* 110, 985–992.
Sezi, C.L. (1996). *East African Med. J.* 73, S24–S28.
Shearer, M.J. (2000). *Curr. Opin. Clin. Nutr. Metabol. Care* 3, 433–438.
Sheehan, D.M., Branham, W.S., Medlock, K.L. and Shanmugasundaram, E.R. (1984). *Teratology* 29, 383–392.
Shier, W.T., Shier, A.C., Xie, W. and Mirocha, C.J. (2001). *Toxicon.* 39, 1435–1438.
Shinki, T., Shiina, Y., Takahashi, N., Tanioka, Y., Koizumi, H. and Suda, T. (1983). *Biochem. Biophys. Res. Comm.* 114, 452–457.
Shiraki, M., Shiraki, Y., Aoki, C. and Miura, M. (2000). *J. Bone Miner. Res.* 15, 515–521.
Siddons, R.C. (1974a). *Brit. J. Nutr.* 32, 219–228.
Siddons, R.C. (1974b). *Brit. J. Nutr.* 32, 579–587.
Souci, S.W., Faichmann, W. and Kraut, H. (1994). In *Food Composition and Nutrition Tables,* 5th ed. Boca Raton: CRC Press.
Spelman, L.H., Osborn, K.G. and Anderson, M.P. (1989). *Zoo Biol.* 8, 239–251.
Srinivasan, S.R., Berenson, G.S., Radhakrishnamurthy, B., Dalferes, E.R., Underwood, D. and Foster, T.A. (1980). *Am. J. Clin. Nutr.* 33, 561–569.
Srinivasan, S.R., Dalferes, E.R.J., Wolf, R.H., Radhakrishnamurthy, B., Foster, T.A. and Berenson, G.S. (1984). *Am. J. Clin. Nutr.* 39, 792–796.
Stevens, M.D., MacKenzie, W.F. and Anand, V.D. (1977). *Vet. Pathol.* 14, 508–509.
Stevens, C.E. and Hume, I.D. (1995) *Comparative Physiology of the Vertebrate Digestive System*, pp 1–10. Cambridge, U.K.: Cambridge University Press.
Strzelewicz, M.A.,Ullrey, D.E., Schafer, S.F. and Bacon, J.P. (1985). *J. Zoo Anim. Med.* 16, 25.
Sturman, J.A., Wen, G.Y., Wisniewski, H.M. and Neuringer, M.D. (1984). *Int. J. Develop. Neurosci.* 2, 121–129.
Swenerton, H. and Hurley, L.S. (1980). *J. Nutr.* 110, 575–583.
Tanaka, T.A. (1988). *J. Agric. Food Chem.* 36, 979–983.
Tappan, D.V., Lewis, U.J., Register, U.D. and Elvehjem, C.A. (1952). *J. Nutr.* 40, 75–85.
Tarantal, A.F., Willhite, C.C., Lasley, B.L., Murphy, C.J., Miller, C.J., Cukierski, M.J., Book, S.A. and Hendrickx, A.G. (1991). *Fund. Appl. Toxicol.* 16, 147–160.
Tillotson, J.A. and O'Connor, R. (1980). *Int. J. Vit. Nutr. Res.* 50, 171–178.
Tomson, F.N. and Lotshaw, R.R. (1978). *J. Am. Vet. Med. Assoc.* 173, 1103–1106.
Torun, B., Davies, P.S.W., Livingstone, M.B.E., Paolisso, M., Sackett, R., Spurr, G.B. and Guzman, M.P.E.D. (1996). *European J. Clin. Nutr.* 50, S37–S81.
Turner, C.H., Akhter, M.P. and Heaney, R.P. (1992). *J. Orthop. Res.* 10, 581–587.

Underwood, E.J. (1977). *Trace Elements in Human and Animal Nutrition,* 4th ed., pp 25–346. New York: Academic Press.

Van Soest, P.J. (1983). *Nutritional Ecology of the Ruminant,* 2nd ed., pp 76–69. Ithaca: Cornell University Press.

Verlangieri, A.J. and Bush, M.J. (1992). *J. Am. Coll. Nutr.* 11, 131–138.

Voss, K.A., Riley, R.T., Norred, W.P., Bacon, C.W., Meredith, F.I., Howard, P.C., Plattner, R.D., Collins, T.F., Hansen, D.K. and Porter, J.K. (2001). *Environ. Health Perspect.* 109 (Suppl. 2), 259–266.

Vruwink, K.G., Fletcher, M.P., Keen, C.L., Golub, M.S., Hendrickx, A.G. and Gershwin, M.E. (1991). *J. Immunol.* 146, 244–249.

Wagner, J.L., Jones, C., Hinkle, D.K. and Renquist, D.L. (1985). *Lab. Anim. Sci.* 35, 531.

Waisman, H.A. and McCall, K.B. (1944). *Arch. Biochem.* 4, 265–279.

Waisman, H.A. (1944). *Proc. Soc. Exp. Biol. Med.* 55, 69–71.

Weindruch, R. and Walford, R.L. (1988). *The Retardation of Aging and Disease by Dietary Restriction.* Springfield, IL: Charles C. Thomas.

Weininger, J. and King, C. (1982). *Am. J. Clin. Nutr.* 35, 1408–1416.

Weisbard, C. and Goy, R.W. (1976). *Folia. Primatol. (Basel)* 25, 95–121.

Weller, R.E., Dagle, G.E., Malaga, C.A. and Baer, J.F. (1990). *J. Med. Primatol.* 19, 675–680.

Whitten, P.L. and Patisaul, H.B. (2001). *Environ. Health Perspect.* 109(Suppl. 1), 5–20.

Wilgram, G.F., Colin, C., Lucas, C. and Best, C.H. (1958). *J. Exp. Med.* 108, 361–371.

Wiseman, H. (2000). In Wiseman, H., Goldfarb, P., Ridgway, T. and Wiseman, A. (eds) *Biomolecular Free Radical Toxicity: Causes and Prevention.* West Sussex, England: John Wiley & Sons.

Witt, E.D. and Goldman–Rakic, P.S. (1983a). *Annals. Neurol.* 13, 376–395.

Witt, E.D. and Goldman-Rakic, P.S. (1983b). *Annals. Neurol.* 13, 396–401.

Wixson, S.K. and Griffith, J.W. (1986). *Lab. Anim. Sci.* 36, 231–236.

Wolfe, L.C., Nicolosi, R.J., Renaud, M.M., Finger, J., Hegsted, M., Peter, H. and Nathan, D.G. (1989). *Brit. J. Haematol.* 72, 456–461.

Yamaguchi, A., Kohno, Y., Yamazaki, T., Takahashi, N., Shinku, T., Horiuchi, N., Suda, T., Koizumi, H., Tanioka, Y. and Yoshiki, S. (1986). *Calc. Tissue Int.* 39, 22–27.

Yang, C.Z. and Bittner, G.D. (2002). *Lab. Anim. (NY).* 31, 43–48.

Young, S.N., Ervin, F.R., Pihl, R.O. and Finn, P. (1989). *Psychopharmacology (Berl).* 98, 508–511.

Yu, B.P., Masoro, E.J. and McMahan, C.A. (1985). *J. Gerontol.* 40, 657–670.

Chapter 14

Environmental Enrichment and Refinement of Handling Procedures

Viktor Reinhardt
Animal Welfare Institute, Washington, DC 20027, USA

Introduction

"Proper housing and management of animal facilities are essential to animal well-being" and "to the quality of research data. Animals should be housed with a goal of maximizing species-specific behaviors and minimizing stress-induced behaviors" (National Research Council, 1996, pp. 21, 22). Traditionally, primates have been subjected to housing and handling conditions that were *not* in line with these principles:

- Single-housing in barren cages was the norm (Figure 14.1) in spite of the fact that primates are intelligent social animals who *need* appropriate stimulation to develop and behave normally and experience a state of ease. Stimuli-deprived subjects often show physiological and clinical signs of distress (Shively *et al.*, 1989; Schapiro and Bushong, 1994; Lilly *et al.*, 1999; Schapiro *et al.*, 2000) and engage in behavioral pathologies such as trichotillomania and self-mutilation (Jorgensen *et al.*, 1998; Tully, 2002) which in human primates reflect severe emotional disturbance (Seaman, 1971; Yaroshevsky, 1975; Simeon *et al.*, 1992).
- Enforced restraint during husbandry and research-related procedures was condoned regardless of the fact that primates respond with distress which affects not only their well-being but also the quality of research data (Figure 14.2; Ives and Dack, 1956; Berendt and Williams, 1971; Bush *et al.*, 1977; Albrecht *et al.*, 1978; Streett and Jonas, 1982; Fuller *et al.*, 1984; Hayashi and Moberg, 1987; Landi *et al.*, 1990; Schnell and Wood, 1993).

The animals' plight was so obvious that the USA, the country with the greatest use of nonhuman primates [more than 55,000 animals; USDA, 2001], amended its animal welfare law in 1985 to promulgate:

1. standards to govern the humane handling, care and treatment of primates, and
2. requirements to promote the well-being of primates (Animal Welfare Act, 1985).

The Laboratory Primate
Copyright 2005 Elsevier
ISBN 0-1208-0261-9

Figure 14.1 "The common practice of housing rhesus monkeys, singly, calls for special attention" (National Research Council, 1998, p. 99). "Solitary confinement is a severe punishment even for monkeys" (Sokol, 1993, p. 12).

This law stipulates that:

1. the "handling of all animals shall be done ... in a manner that does not cause ... behavioral stress ... or unnecessary discomfort" (USDA, 1991, §2.38,f);
2. research facilities have to develop and follow an "*environmental enhancement plan*" addressing the "*social needs*" of nonhuman primates whose primary enclosures "must [*sic*] be enriched by providing means of expressing non injurious species-typical activities" (USDA, 1991, 3.81,a,b).

These rules led to the development of:

- *environmental enrichment* strategies to promote the expression of non-injurious social behaviors [*animate enrichment*], foraging behaviors [*feeding enrichment*] and object-oriented behaviors [*inanimate enrichment*], and
- *training protocols* to obtain the animals' cooperation during procedures.

The focus of this chapter is published material that includes *empirical evidence* or *scientific data* relevant to environmental enrichment and training of laboratory primates.

Environmental enrichment

Environmental enrichment is the provision of stimuli that promote the expression of species-appropriate activities.

Animate enrichment

Compatible companionship addresses the *social needs* of primates and provides ever-changing stimulation. "Social companions are the one 'enrichment device' to which habituation never occurs" (de Waal, 1992, p. 86).

Group-housing

Living in a group allows the individual to engage in a great variety of social behaviors. The constant distraction resulting from companionship may account for the fact that the incidence of stereotyped activities declines when single-caged subjects are transferred to group-housing arrangements (Missakian, 1972; Goosen, 1988; Bayne *et al.*, 1991; Schapiro *et al.*, 1996; Bloomsmith *et al.*, 1998; Alexander and Fontenot, 2003).

Group formation

Introducing unfamiliar *adult* primates into the same enclosure is accompanied by dominance-determining interactions which may jeopardize the animals' safety.

Rhesus macaques pose the greatest challenge during attempts at forming new groups, introducing strangers into existing groups, or reintroducing separated members back into their home group. Regardless of whether the animals are of the same or opposite sex, carefully pre-familiarized or not, are introduced at the same time or in stages, the result is usually dramatic with individual animals suffering distress and life-threatening injuries resulting from harassment and fighting (Southwick, 1967; Bernstein *et al.*, 1974; Bernstein

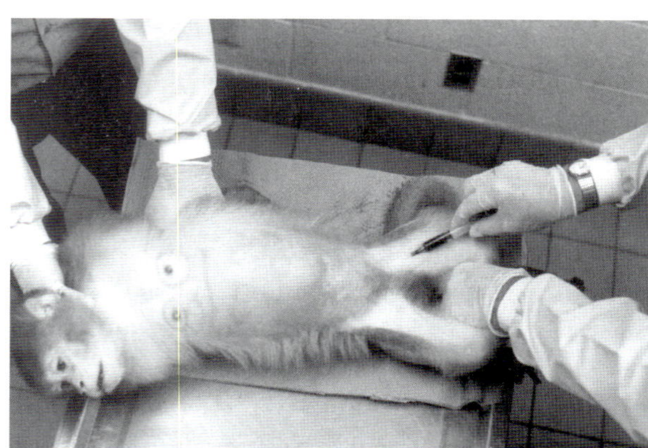

Figure 14.2 "Restraint of nonhuman primates may introduce major variables to complicate experimentation or clinical testing" (Streett and Jonas, 1982, p. 263). "Of 56 physically restrained primates, 30 (54%) experienced severe metabolic acidosis" (Bush *et al.*, 1977, p. 866).

and Gordon, 1977; Goo and Sassenrath, 1980; Jensen *et al.*, 1980; Kaplan *et al.*, 1980; Line *et al.*, 1990a; Gust *et al.*, 1991; Clarke and Blanchard, 1994; Conlee *et al.*, 1996). Group formations of pre-familiarized male long-tailed and unfamiliar pig-tailed macaques of both sexes were also accompanied by aggression but there was no serious wounding (Clarke *et al.*, 1995; Gust *et al.*, 1996; Asvestas and Reininger, 1999; Hartner *et al.*, 2000).

The establishment of heterosexual groups of partially familiarized vervets and baboons can be accompanied by dangerous fighting and harassment (Else, 1985; Else *et al.*, 1986). Published information on isosexual group formation of these two species is lacking. Chimpanzees can be introduced as heterosexual or isosexual female groups without undue risk if the animals are first familiarized with each other (Fritz and Fritz, 1979; McDonald, 1994; Bloomsmith *et al.*, 1998; McNary, 1992). Wounding can be a major problem when groups of males are formed (Alford *et al.*, 1995; Bloomsmith and Lambeth, 1996; Bloomsmith and Baker, 2001; Fritz and Howell, 2001; Seres *et al.*, 2001).

For capuchin monkeys "no foolproof recipe is yet available. Overall, group formation is a stressful procedure both for the animals and the care-givers, and although cumulative experience may help to reduce the risks of failure, the outcome can never be predicted" (Visalberghi and Anderson, 1999, p. 606). Nevertheless, in a study of heterosexual groups of capuchins, aggressive interactions were absent when the familiarized, carefully pre-selected subjects were brought together in a large exercise cage (Wolff and Ruppert, 1991).

The "formation of same-sex groups of squirrel monkeys is rarely accompanied by injurious aggression. Once unisexual groups have stabilized, formation of larger heterosexual groups generally proceeds smoothly" (Mendoza, 1991, p. 347; cf. Vermeer, 1997). Severe aggression is commonplace when heterosexual groups are established without first introducing same-sex animals to one another (King and Norwood, 1989; Mendoza, 1991). "The only real problem encountered with the social housing of marmosets is the keeping of groups of non-breeding individuals. Same sex individuals kept together may fight and often one individual is bullied and needs to be removed" (Poole, 1990, p. 83).

Introducing unfamiliar *immature* primates into the same enclosure usually is uneventful (macaques: Goosen *et al.*, 1984; Schapiro *et al.*, 1994; chimpanzees: Alford *et al.*, 1995; Bloomsmith and Lambeth, 1996; baboons: Marks *et al.*, 2000). The addition of an experienced adult helps in the harmonious maintenance of such kindergartens (Figure 14.3).

Figure 14.3 Rhesus kindergarten under the supervision of a mature female.

Group management

The constraints set by confinement and research can make the group-housing of primates a challenging endeavor.

Group members must be able to keep social distances and break visual contacts to avoid disputes. When there is a conflict, subordinates must be able to escape otherwise they can become the target of persistent aggression. Therefore, a generous space allocation is a basic condition to control undue aggression within a group (Southwick, 1967; Elton, 1979; Howell *et al.*, 1993; Boyce *et al.*, 1998). (See also chapter 9, by Lutz and Novak.) The provision of cover and blinds allowing visual seclusion, and the installation of structures opening up the vertical dimension of the enclosure are effective means to mitigate social tensions and minimize overt aggression (Erwin, 1977; Maple and Finlay, 1987; Williams *et al.*, 1988; Mendoza, 1991; Ricker *et al.*, 1995; Neveu and Deputte, 1996; Maninger *et al.*, 1998; Nakamichi and Asanuma, 1998; Westergaard *et al.*, 1999; McCormack and Megna, 2001).

Social incompatibility problems often can be resolved by carefully removing certain individuals (Reinhardt *et al.*, 1987; Judge *et al.*, 1994), changing routine husbandry procedures (Bloomsmith *et al.*, 1988; Zimmermann and Feistner, 1996; Boccia and Hijazi, 1998; Catlow *et al.*, 1998; Chamove, 2001) or training specific group members to refrain from aggression (Bloomsmith *et al.*, 1994; Schapiro *et al.*, 2002).

The social structure of a primate group can easily be destabilized if an animal is removed permanently or taken out for some time and subsequently returned. If this happens, rank relationships are rearranged, a process that bears a high risk of social distress and serious aggression (Erwin, 1979; Kessler *et al.*, 1985;

Coe and Scheffler, 1989; Cohen *et al.*, 1992; Ha *et al.*, 1999). When research protocols require that individuals be removed for extended time periods, pair-housing offers a practicable alternative to group-housing.

Pair-housing

Even after long-term housing with the same partner, paired companions spend about the same amount of time engaged in social behaviors as wild animals do (Ranheim and Reinhardt, 1989; Line *et al.*, 1990b; Brent, 1992). This suggests that pair-housing sufficiently addresses the animals' social needs (Figures 14.4 and 14.5). Sharing a cage with a conspecific not only permits social interactions but it also has a therapeutic effect. In macaques, stereotypies decrease (Goosen, 1988; Eaton *et al.*, 1994; Schapiro *et al.*, 1996) and self-injurious behavior can be eliminated altogether when single-caged animals are transferred to compatible pair-housing (Reinhardt and Rossell, 2001).

Figure 14.4 "To enhance the life-style of a primate, one of the most effective, but often overlooked, improvement is pair housing" (Rosenberg and Kesel, 1994, p. 469).

Figure 14.5 "A compatible conspecific probably provides more appropriate stimulation to a captive primate than any other potential environmental enrichment factor" (International Primatological Society, 1993, p. 11). Here, two male long-tailed macaques engrossed in social grooming (photo by Richard Lynch).

Pair formation

Successful iso-sexual [to avoid uncontrolled breeding] pair-formation of *adults* has been reported in female baboons (Jerome and Szostak, 1987), female and male chimpanzees (Fritz, 1994; Brent *et al.*, 1997), male bonnet macaques (Taylor and Laudenslager, 1998), female and male long-tailed macaques (Line *et al.*, 1990b; Crockett *et al.*, 1994; Asvestas, 1998; Lynch, 1998; Seelig, 1998; Watson, 2002), female pig-tailed macaques (Byrum and St. Claire, 1998), male and female rhesus macaques (Eaton *et al.*, 1994), female and male stump-tailed macaques (Reinhardt, 1994b), female and male owl monkeys (Weed and Watson, 1998), female marmosets (Majolo *et al.*, 2001) and male squirrel monkeys (Gwinn, 1996).

Experts had warned that "especially when new pairs are formed and dominance relationships are being established, there is a strong likelihood that the veterinarian will be kept quite busy suturing wounds" (Coe, 1991, p. 79; cf. Line, 1987; Rosenberg and Kesel, 1994). Indeed, it would be unethical and contradict basic ethological principles to put two strange primates in a cage and wait for the predictable, possibly injurious fight over dominance. If, however, potential partners are paired only under the conditions in which:

1. they have established a rank relationship during a brief non-contact familiarization, and
2. the pairing takes place in an environment that is new to both,

the two animals do not need to fight over dominance nor do they have a reason to show territorial antagonism

(Reinhardt, 1989). When these precautions were applied in rhesus macaques, the species that has the reputation of being particularly intolerant and vicious, pair formation was accompanied by fighting in two of 77 female pairs and in none of 20 male pairs tested. Only two (1%) of the 194 animals were injured during pairing and required veterinary care (Reinhardt, 1994a).

Pairs of *adults and juveniles* can be established, without undue risks, by directly introducing unfamiliar partners (Redican and Mitchell, 1973; Reinhardt, 1994a; Majolo *et al.,* 2001). Pair-formation of nonfamiliarized *juveniles* is also relatively safe (Brandt and Mitchell, 1973; Reinhardt, 1994a).

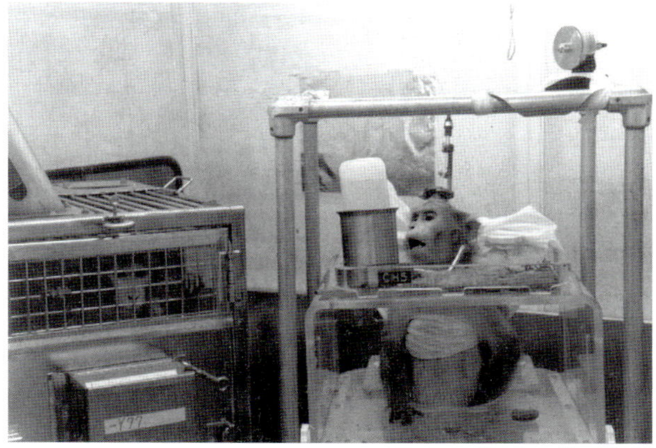

Figure 14.6 "The presence of another animal of the same species has a protective effect under stress" (Bovard, 1959, p. 269).

Pair management

Clinical records do not support the belief that "social pairing is associated with high health risks to monkeys" (Morgan *et al.,* 1998, p. 168). In a small sample of rhesus macaques, no difference was found in rates of clinical morbidity between pair- and single-housed animals (Eaton *et al.,* 1994). In two large rhesus colonies the incidence of veterinary intervention was *lower* in pairs than in single-caged individuals (Reinhardt, 1990; Schapiro and Bushong, 1994; Schapiro *et al.,* 1997).

Single- and pair-housed animals do not differ in their cortisol levels, and dominant partners have concentrations that are equivalent to those of their subordinate counterparts (Coe *et al.,* 1982; Reinhardt *et al.,* 1991; Schapiro *et al.,* 1993). Rather than being a source of stress, the presence of the companion buffers stress during alarming situations (Mason, 1960; Gonzalez *et al.,* 1982; Coelho *et al.,* 1991; Gust *et al.,* 1994; Gwinn, 1996; Gerber *et al.,* 2002).

Pair-housing does not interfere with routine husbandry procedures and common experimental procedures (Reinhardt and Reinhardt, 2001). Metabolic and controlled food-intake studies can be done in the pair's homecage by separating the two partners with a special cage divider allowing continuous communication (Reinhardt and Reinhardt, 2001). Whenever a partner has to be removed for experimental reasons the companion is brought along in a mobile cage to provide psychological support (Figure 14.6). When primates are hospitalized there is often no reason for keeping them alone (Lindburg and Coe, 1995). In a study with long-tailed macaques, partners were returned on the same day of a surgical procedure to their cage mates. "Change in hierarchy status, self-traumatic events, weight loss or diarrhea did not occur in any of these 15 animals, and the incision sites healed unremarkably. The animals ate and drank normally, and received their postoperative treatments without problem" (Murray *et al.,* 2002, pp. 112–123).

Aggressive conflicts between pair-housed macaques can be forestalled by the following:

1. Allow a new pair to develop a stable relationship by keeping the two animals continuously together for several weeks.
2. Paired partners need *social space* to cope with permanent confinement. At a minimum, allocate twice the floor area that is legally required for single-housing. Double cages can be created by removing the dividing panels of twin modules or by interconnecting adjacent cages with a short tunnel (Bellinger *et al.,* 1992; Baskerville, 1999; Reinhardt and Reinhardt, 2001).
3. A privacy panel, with a passage hole close to the back wall of the cage, diminishes squabbles over food. It also helps the subordinate partner to get out of the dominant's sight during potential conflicts. This avoids antagonism while fostering affiliation (Reinhardt and Reinhardt, 2001).
4. Horizontal, rather than vertical, arrangement of the pair's double cage eliminates competition over access to the preferred upper section of the cage.
5. Provide each half of the cage with a raised resting surface, a food receptacle and a water spout to prevent competition over basic resources.
6. Check each pair on a daily basis to make sure that the animals' well-being is not jeopardized by unnoticed incompatibility.
7. Do not keep animals together if: (a) their rank relationship becomes equivocal (e.g., bidirectional threatening), (b) one of them persistently shows signs of depression (e.g., crouching in a corner), (c) one of them

monopolizes food, or (d) one inflicts on the other an injury that requires medical treatment. It is advisable to permanently separate a pair under conditions a, b and c. Condition d is often associated with a dominance reversal and permits reunion shortly after wound treatment.

8. If partners have to be housed in different rooms for more than one week, do not simply re-unite them in the home cage thereafter but give them the opportunity to briefly recognize each other across a temporary transparent barrier. This makes it unnecessary for the two to treat each other as strangers, ready to fight over dominance.
9. Never threaten or scare paired animals. This could excite them so much that they redirect their aggressive tension toward each other.

Human interaction

Positive interaction with humans is an important enrichment option contributing to the alleviation of behavioral pathologies (Bayne *et al.*, 1993; Choi, 1993). It "is essential for the well-being of the animal, data validity, and ease of handling" (Wolfle, 1987, p. 1220). A trust relationship between humans and their nonhuman primate charges is the very foundation of scientifically sound research methodology (Reinhardt, 2003a). Animal personnel and investigators should visit the animals in their charge on a regular basis at times *other* than their normal routines and *outside* the experimental context (Figure 14.7). Such friendly encounters are highlights for the imprisoned animals who gradually overcome their angst and develop a trustful relationship with humans. "Nonhuman primates are quick to forget, or perhaps forgive, the momentary fear or resentment they feel towards a human being who has just subjected them to an unpleasant experience if a strong bond of trust already exists with that person" (Mahoney, 1992, p. 35).

Feeding enrichment

Feeding enrichment is the attempt to emulate aspects of species-typical foraging.

Primates voluntarily work for food even when identical food is freely available (Murphy, 1976; Anderson and Chamove, 1984; Evans *et al.*, 1989; Line *et al.*, 1989a; Reinhardt, 1994c; Taylor, 2002). This shows that they have a *need* to get actively involved in the food acquisition process. In the laboratory, this need is usually not met because the food is offered in such a way that no, or only very little, effort is required to search for, retrieve and process it.

Feeding on substrate

Scattering food on a substrate rather than throwing it on the bare floor or placing it in food boxes is the easiest way of promoting food searching activities (Figure 14.8; Chamove and Anderson, 1979; DiGregorio, 1990; Byrne and Suomi, 1991; Fragaszy and Adams-Curtis, 1991; Poenisch, 1992; Boinski *et al.*, 1994; Beck, 1995; Baker, 1997; Goodwin, 1997; Taylor *et al.*, 1997). The animals tend to be less aggressive and food-competitive and engage less often in behavioral disorders when spending more time gathering food and having the food spread out more evenly (Anderson and Chamove, 1984; Boccia, 1989; Baker, 1997; Chamove, 2001).

Figure 14.7 "Personnel should routinely devote some time to positive interactions with animals" (National Research Council, 1998, p. 88).

Figure 14.8 Macaques foraging on woodchip litter scattered with sunflower seeds (photo by Bushmitz Moshe).

Whole produce

Produce of the season introduces variety into the standard diet (Figure 14.9). Individually housed chimpanzees increased the time contacting food from 5 to 33 minutes/h when one ear of corn was added every other day to their portion of biscuits (Nadler *et al.*, 1992). Group-housed rhesus macaques, who received two ears of corn per animal once every week, spent an average of 47 minutes/h husking corn ears, chewing husk and eating corn (Beirise and Reinhardt, 1992).

Using cage structures to promote foraging

Primate enclosures consist of metal mesh, bars and/or chain-link fencing. Presenting the food behind this material induces skilful food gathering behavior (Figure 14.9). Distributing the daily biscuit ration on the mesh ceiling of the cage, rather than in the food box, increased the average foraging time from less than 1 to 22 minutes in pair-housed rhesus macaques (Reinhardt, 1993a). Lemurs forage about as much as their wild counterparts do when their food is distributed on the mesh roof of their cage (Britt, 1998). A more labor-intensive feeding enrichment option is the floor of the cage itself. Trays filled with a food mixture are attached to the underside of the cage or placed on the drop pan. This type of food presentation can provide macaques, baboons and chimpanzees with hours of food acquisition behavior (Bryant *et al.*, 1988; Mahoney, 1992; Spector *et al.*, 1994).

Food puzzles are created by remounting the ordinary food boxes away from the access holes directly onto the mesh wall (Figure 14.10). Offering the daily biscuit ration in two food puzzles increased the amount of time pair-housed rhesus macaques spent gathering food from less than 1 minute to 42 minutes. The monkeys did not hoard biscuits when they had to work to obtain them. As a consequence, there was practically no spoilage with the puzzles compared to the food boxes (Reinhardt, 1993b; cf. Markowitz, 1979; Murchison, 1994, 1995; Rosenblum and Andrews, 1995; Bertrand *et al.*, 1999). The usefulness of the food puzzle as primary feeder has been confirmed in stump-tailed (Reinhardt, 1993c), Japanese (Yanagihara *et al.*, 1994) and long-tailed macaques (Reinhardt and Garza-Schmidt, 2000). Different cage modules may require different design modifications to make the removal of biscuits from the box a more time-consuming activity. For example, replacing the large circular opening of a standard feeder with four small openings increased the time pig-tailed macaques spent removing their biscuit ration from less than 1 minute to 7 minutes (Murchison, 1995).

Figure 14.9 It would be a waste of time to chop fruits and vegetables for the animals; they have all the time needed to process and retrieve the food themselves.

Figure 14.10 The mesh of the cage can readily serve as a foraging "device" prompting skilful foraging behavior.

Foraging devices

Puzzle feeders require that the animal manoeuvres food items from behind a barrier to an access hole. Group-housed chimpanzees foraged from puzzles filled with peanuts, treats or dried fruit for 4–10 minutes/h (Bloomstrand *et al.,* 1986; Brent and Eichberg, 1991; Perret *et al.,* 1998). In group-housed Japanese macaques, puzzles filled with treats "proved to be a learning success, but they were best used sporadically to prevent the animals from becoming bored" (Goodwin, 1997, p. 514). Paired tamarins increased the time spent contacting food from 2 minutes to 23 minutes/h when their standard ration was placed in a puzzle rather than in food dishes (Glick-Bauer, 1997). Manipulation of a puzzle, by single-caged rhesus macaques, was associated with a reduction in stereotypies, "but this effect was transient, occurring only during the first hour after the puzzle feeder was filled with treats. Puzzle feeder manipulation had no effect on self-injurious behavior, and in fact, some monkeys with this disorder actually bit themselves while extracting peanuts" (Novak *et al.,* 1998, p. 226). Foraging time increased from 5 to 22 minutes when single-caged pig-tailed macaques had their daily biscuit ration offered in a puzzle rather than in the food box (Murchison, 1994).

Probe feeders demand cognitive and manipulative skills to extract food from a hidden source. Group-housed chimpanzees poked sticks through the holes of a food loaded polyvinyl chloride [PVC] pipe and pushed the content to an open end of the pipe during about 5–30 minute observations (Gilloux *et al.,* 1992). They were engrossed in sham "termite fishing" for about 10–30 minutes when they had access to a pipe feeder filled with sticky food stuff (Maki *et al.,* 1989). The presence of honey in a bottle to be "fished" with artificial tools elicited foraging activities during about 15–30 minutes of observation in pair-housed chimpanzees (Celli *et al.,* 2003). Group-housed tamarins spent approximately 8–30 minutes extracting food from a bamboo pipe feeder (Steen, 1995). Social-housed marmosets engaged in foraging for about 15–30 minutes when they had access to hollow, subdivided hardwood dowels filled with liquid gum (McGrew *et al.,* 1986). They quickly lost interest in such feeders when exposed to them on a daily basis (Roberts *et al.,* 1999). When the standard mixture of biscuits and produce was hidden in compartmentalized PVC pipes with different sized holes, rather than placed in food boxes, the amount of time group-housed capuchins foraged increased from an average of 18 to 24 minutes/h (Hayes, 1990).

Substrate feeders prompt the searching for edible items in a lawn-like substrate. Single-caged long-tailed macaques foraged for about 8–30-minute observations when they had access to a fleece pad sprinkled with morsels of food (Lam *et al.,* 1991). Single-caged rhesus macaques showed a temporary decline in stereotypical activities in the presence of turf boards baited with tidbits (Bayne *et al.,* 1992a). Long-tailed macaques manipulated such boards for approximately 2–30 minutes but this "did not have a detectable effect on the amount of time the subjects engaged in stereotypies" (Lutz and Farrow, 1996, p. 76). Social-housed chimpanzees spent about 6–30 minutes foraging from a grass container scattered with sunflower seeds. No effect on abnormal behaviors was noted (Lambeth and Bloomsmith, 1994).

Inanimate enrichment

Inanimate enrichment stimulates primates to explore and manipulate interesting objects or make use of the arboreal dimension of the enclosure.

Interesting objects

Toys exert a strong novelty effect, but primates are too intelligent not to lose interest quickly in such unresponsive objects unless these are presented on a regular basis for only short periods of time, are frequently exchanged with other objects or are made of destructible material.

Single-caged chimpanzees spent about 3 minutes playing with a durable rubber toy that was put into their cages once every day for a 10-minute period (Brent *et al.,* 1989). The same type of gadget led to a rapid habituation in long-tailed and pig-tailed macaques (Crockett *et al.,* 1989). Rhesus macaques who had access to two different durable toys, for an 8-week period, showed a decline in behavioral disorders and spent about 10% of the time contacting the toys. When these were removed, behavioral disorders increased to levels *exceeding* those observed before the animals had been exposed to the toys (Bayne *et al.,* 1992b). Single-caged pig-tailed macaques engaged less in behavioral disorders while playing with a collection of five different durable toys, but toy use became negligible by the end of five weeks (Kessel and Brent, 1998).

Group-housed chimpanzees resorted to increased abnormal behavior in the presence of durable toys that, after a brief novelty phase, no longer evoked any interest (Paquette and Prescott, 1988). When they were given hard-rubber toys or wrapping paper on separate days twice a week, interaction time with the paper was about 16 minutes/h, and with the rubber toys was 6 minutes/h.

The use of the durable rubber toy declined to zero while the use of the destructible paper remained constant over the course of four months (Pruetz and Bloomsmith, 1992). The greater attraction of objects that change in shape, size and texture during destruction has also been confirmed in single-caged chimpanzees (Brent and Stone, 1998). Group-housed rhesus macaques who received one cardboard box once a week spent approximately 33 minutes/h tearing the box apart and manipulating and chewing shredded pieces (Beirise and Reinhardt, 1992).

Gnawing sticks are branch segments cut from dead deciduous trees (Figure 14.11). Because of gradual wear and dehydration, the sticks retain an ever-changing stimulatory effect. Caged macaques use gnawing sticks approximately 4% of the time (Line and Morgan, 1991; Reinhardt and Reinhardt, 2001).

Mirrors elicit curiosity behavior (Figure 14.12). During a 2-week period, individually housed pig-tailed macaques contacted a mirror hung in front of the cage 12–18 times/h. Long-tailed macaques showed little interest in the beginning, but contact rates reached those of pig-tailed macaques at the end of the study (O'Neill *et al.,* 1997). Rhesus macaques lost interest very quickly (O'Neill *et al.,* 1997; cf. Goode *et al.,* 1998).

Visual and auditory stimuli

Single-caged rhesus macaques, exposed to video programs for five consecutive days, viewed the monitor about 3% of the time (Schapiro and Bloomsmith, 1995). Chimpanzees, who were shown television programs daily for a 2-week period, watched the monitor about 20% of the time during the second hour of the programs (Brent *et al.,* 1989). Operant conditioning studies

Figure 14.11 There is a general consensus that wooden objects provide inexpensive, long-term and effective stimulation for laboratory primates without causing health and hygienic problems (Eckert *et al.,* 2000).

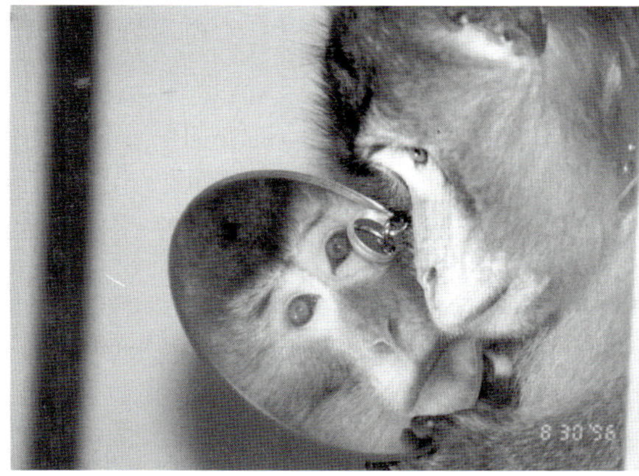

Figure 14.12 Mirrors elicit curiosity behavior (photo by Richard Lynch).

indicate that "TV is not a valued commodity" for rhesus macaques (Harris *et al*, 1999, p. 48). Windows offer more entertainment (Figure 14.13). Pairs of long-tailed macaques, who were transferred regularly for 1 1/2-hour periods to a playroom, spent about 67% of the time looking out the windows (Lynch and Baker, 2000).

Single-caged baboons did not change their behavior when radio music was available but their heart rate was lowered, indicating that they were calmer when they could listen to the music (Brent and Weaver, 1996). Studies with group-housed rhesus macaques and chimpanzees also suggest that music can have a calming effect, decreasing aggression while increasing affiliative interactions (Novak and Drewsen, 1989; Howell *et al.,* 2002).

Elevated structures

Large cages are usually outfitted with raised structures that are heavily used by the animals. For example, group-housed rhesus macaques and chimpanzees may

Figure 14.13 Windows offer very attractive entertainment (photo by Richard Lynch).

spend more than three-quarters of the day on high platforms and climbing frames (Ochiai and Matsuzawa, 1999). Marmoset pairs stopped showing startle responses after their home cage was furnished with perches. By increasing possibilities of social distancing in the arboreal dimension, the perches led to a significant decrease in aggression (Kitchen and Martin, 1996). Similar observations were made in mangabeys and in Japanese macaques (Neveu and Deputte, 1996; Nakamichi and Asanuma, 1998).

Small cages often lack elevated structures because of the extreme spatial restriction. To be useful, such structures must be installed in such a way that an animal can:

- retreat on it to the back of the cage in alarming situations;
- sit on it in front of the cage and have visual control of the activities in the room (Figure 14.14);
- rest comfortably on it without touching the ceiling of the enclosure;
- use the space beneath it for free postural adjustments (Figure 14.15).

Adult singly caged rhesus males perched, on average, 28% of the time on PVC pipes to which they had continual access for one year (Reinhardt and Reinhardt, 2001; cf. Watson, 1991; Woodbeck and Reinhardt, 1991).

Caged macaques clearly prefer perches over swings, probably because swings permit relaxed posturing rather than unstable balancing (Kopecky and Reinhardt, 1991). They may manipulate swings but show little inclination to actually use them for swinging (Dexter and Bayne, 1994).

Figure 14.14 Primates try to escape upward to avoid terrestrial threats. "Therefore, these animals might perceive the presence of humans above them as particularly threatening" (National Research Council, 1998, p. 118). In light of these ethological considerations, bottom-row caging is an inadequate housing arrangement not only for marmosets and tamarins (National Research Council, 1998) but for primates in general.

Figure 14.15 "Perches or shelves should be provided in all cages" (European Commission, 2002, p. 88). Perches can easily be installed in cages with movable back wall (Schmidt *et al.*, 1989; Reinhardt and Reinhardt, 2001; Martin *et al.*, 2002; photo by Lisa Knowles).

Exercise pen and space

Exercise pens offer single-caged primates periodic access to relatively large enclosures that are equipped in such a way that the subjects are stimulated to take advantage of the enhanced space.

Long-tailed macaques were transferred on a daily basis for one hour to a *playpen* almost four times as big as their home cages. Of the many activities available those that consistently captured the monkeys' attention were foraging in woodchips, viewing a companion in the adjacent playpen and tearing apart telephone directories. Stereotypical behaviors and self-directed aggression were virtually absent (cf. Leu *et al.*, 1993) but reappeared once the animals returned to their home cages (Bryant *et al.*, 1988). Decreased self-biting and hair-pulling was noted in rhesus macaques who were permitted access to an exercise cage for several hours daily (Storey *et al.,* 2000). Baboons who were transferred to an *activity cage* more than four times the size of their home cage on two consecutive days once every month used enrichment objects about 24% of the time and showed an amelioration of behavioral disorders that was maintained when they were back in their home cages (Kessel and Brent, 1995). Japanese macaques were released every four days for 24 hours into a seven times larger *recreation cage* that contained the same toys as the home cage. Time spent using the toys increased from 9% in the home cage to 19% in the recreation cage, and time spent engrossed in stereotypies decreased from 6% to 2%. The animals retained their enhanced interest in the toys upon being returned to their small home cages, presumably because they could always see another monkey playing in the recreation cage (Tustin *et al.,* 1996).

Even though space has no stimulatory value, primates show positive behavioral changes in large versus small cages. When individual rhesus macaques were each tested in small, empty cages and in six-times larger empty pens for 90-minute observations, subjects engaged in stereotyped locomotion 20% of the time in the small, but only 7% of the time in the large enclosure (Paulk *et al.,* 1977). A small increment of cage size above the legal requirement is unlikely to have an influence on the caged subject (Bayne and McCully, 1989; Line *et al.,* 1989b, 1990c; Crockett *et al.,* 1993, 1995, 2000; Gaspari *et al.,* 2000) unless the space is structured. Paired marmosets, for example, showed a significant increase in perching, plus a significant decrease in stereotypic and aggressive behavior, after being transferred from their small cages, equipped with a nest box and two branches, to double-size cages that were also outfitted with a nest box and two branches (Kitchen and Martin, 1996). Tamarins stopped stereotypic head bobbing after being transferred from a small perch-equipped cage to a three-times larger perch-equipped cage (Box and Rohrhuber, 1993).

Water

Water can elicit a broad spectrum of activities (Bercovitch and Kessler, 1993). Group-housed rhesus macaques, who had access to a trough with water for ten days, spent about 30% of the time drinking, splashing, submerging arms or legs and exploring the trough (Parks and Novak, 1993). For some species, water may not be an adequate enrichment option. Group-housed marmosets, for example, had little interest in a water bath "other than to use it as a toilet" (Hazlewood, 2001, p. 150).

Training for cooperation during procedures

Training is a reward-based human-animal interaction sequence aiming at the voluntary cooperation of the animal during a procedure. Successful training minimizes or eliminates stress reactions (Elvidge *et al.,* 1976; Turkkan, 1990; Luttrell *et al.,* 1994; Bentson *et al.,* 2003; Reinhardt, 2003b), provides valuable distraction from boredom (Hediger, 1964; Hogan, 1991; Laule, 1992; Martin, 1996; Goodwin, 1997; Philipp, 1997; Kobert, 1998) and increases the safety of the handling personnel (Figures 14.16a,b,c and 14.17).

Technique

1. Acquire first-hand knowledge of the ethological characteristics of the animal you are going to train.
2. Establish a relationship based on trust and respect with the trainee.
3. Be confident, patient, gentle and firm.
4. Break the training program into small steps with specific goals.
5. Never terminate a session before the goal of the particular training step is achieved.
6. Reward the trainee with food treats and/or praise whenever he/she meets your expectation, e.g., touches the target or allows you to hold a limb.
7. Never punish the trainee.
8. If the trainee does not meet your expectation, withhold the reward.
9. Be consistent.

Application

Successful training protocols have been described for the following procedures:

Blood collection	rhesus macaques: Reinhardt, 2003b; Phillippi-Falkenstein and Clarke, 1992 stump-tailed macaques: Reinhardt, 2003b chimpanzees: Laule and Whittaker, 2001
Presentation of vascular access ports	rhesus macaques: Grant and Doudet, 2003

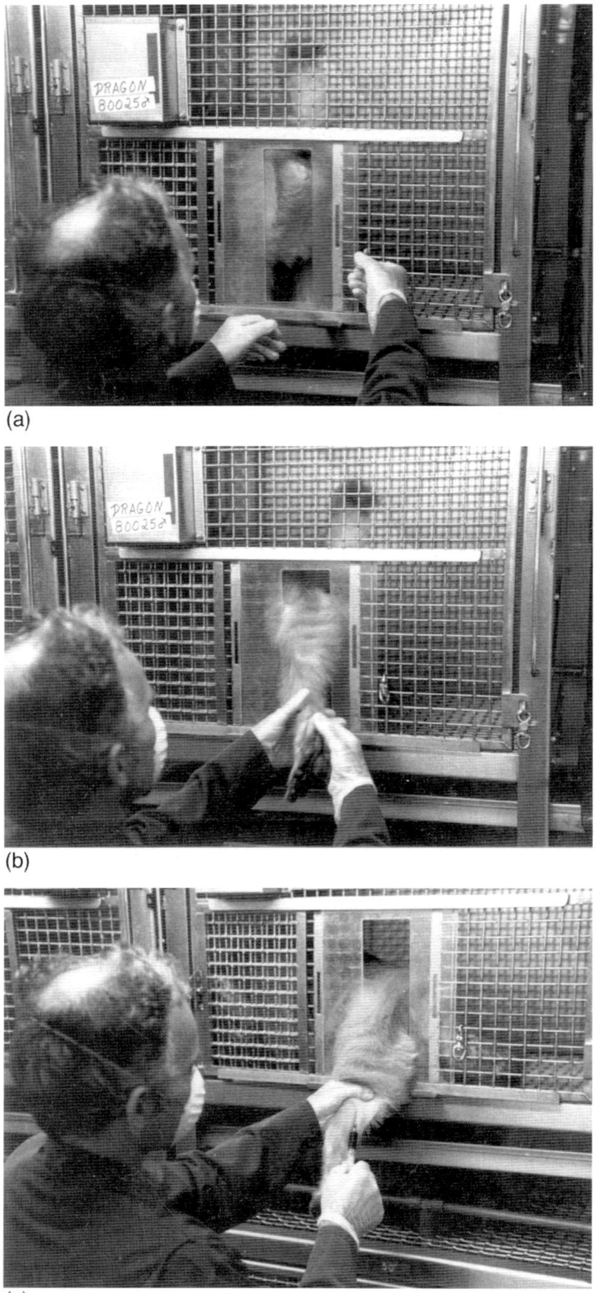

(a) (b) (c)

Figure 14.16a,b,c "Procedures that reduce reliance on forced restraint … are less stressful for animals and staff, safer for both, and generally more efficient" (National Research Council, 1998, p. 45). Here, a male rhesus macaque cooperates during in-home cage blood collection. The animal is in control of the situation and, therefore, has no reason for showing aggressive defense reactions triggered by fear.

Injection	chimpanzees: Spragg, 1940 baboons: Levison *et al.,* 1964 long-tailed macaques: Nelms *et al.,* 2001
Blood pressure measurement	baboons: Turkkan *et al.,* 1989
Topical treatment	stump-tailed macaques: Reinhardt and Cowley, 1990
Oral drug administration	baboons: Turkkan *et al.,* 1989 Pig-tailed macaques: Crouthamel and Sackett 2004
Urine collection	vervets: Kelley and Bramblett, 1981 tamarins: Ziegler *et al.,* 1987 marmosets: Anzenberger and Gossweiler, 1993; McKinley *et al.,* 2003 chimpanzees: Laule *et al.,* 1996; Lambeth *et al.,* 2000
Saliva collection	squirrel monkeys: Tiefenbacher *et al.,* 2003
Vaginal swabbing	stump-tailed macaques: Bunyak *et al.,* 1992
Weighing	marmosets: McKinley *et al.,* 2003
Capture	long-tailed macaques: Clarke *et al.,* 1988 chimpanzees: Kessel-Davenport and Gutierrez, 1994 rhesus macaques: Luttrell *et al.,* 1994

Training primates to cooperate, rather than resist, during procedures is not necessarily a very time-consuming undertaking:

- In order to train a heterosexual group of 45 rhesus macaques to cooperate during the routine one-by-one capture procedure, a cumulative total of 15 hours [20 minutes per animal] was invested. Conventionally, the animals had been caught using nets. This process could take 60 minutes or longer, and incidents of acute diarrhea, rectal prolapse and laceration were common. Once the group was trained, the monkeys no longer showed signs of distress and the capture procedure was accomplished in less than 20 minutes (Luttrell *et al.,* 1994).
- An average cumulative total of 52 minutes of training was invested per male/female pair of marmosets to reliably obtain urine samples of both partners (McKinley *et al.,* 2003).
- To get the cooperation during home cage blood collection of adult rhesus and stump-tailed macaques,

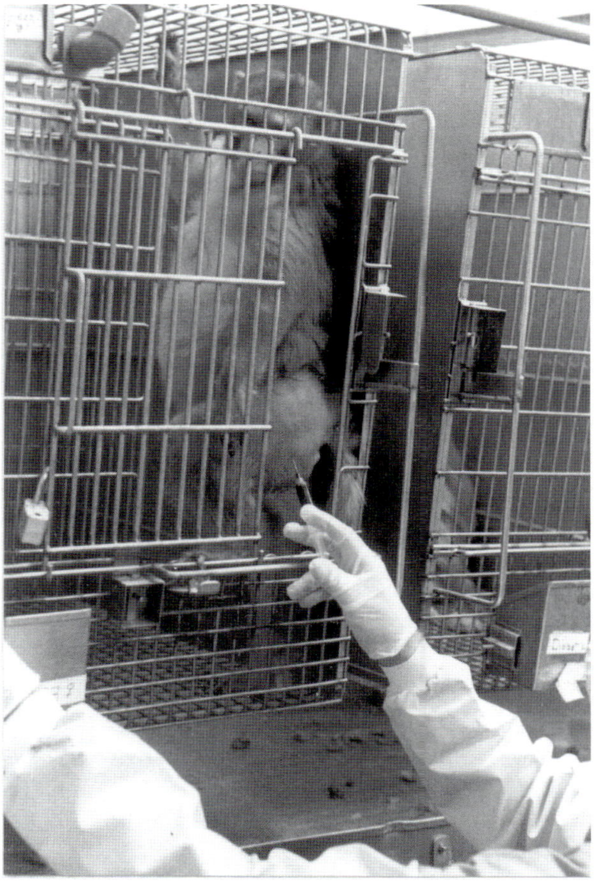

Figure 14.17 "The least distressing method of handling is to train the animal to co-operate in routine procedures. Advantage should be taken of the animal's ability to learn" (Home Office, 1989, p. 18). Here a female rhesus macaque cooperating during in-home cage injection.

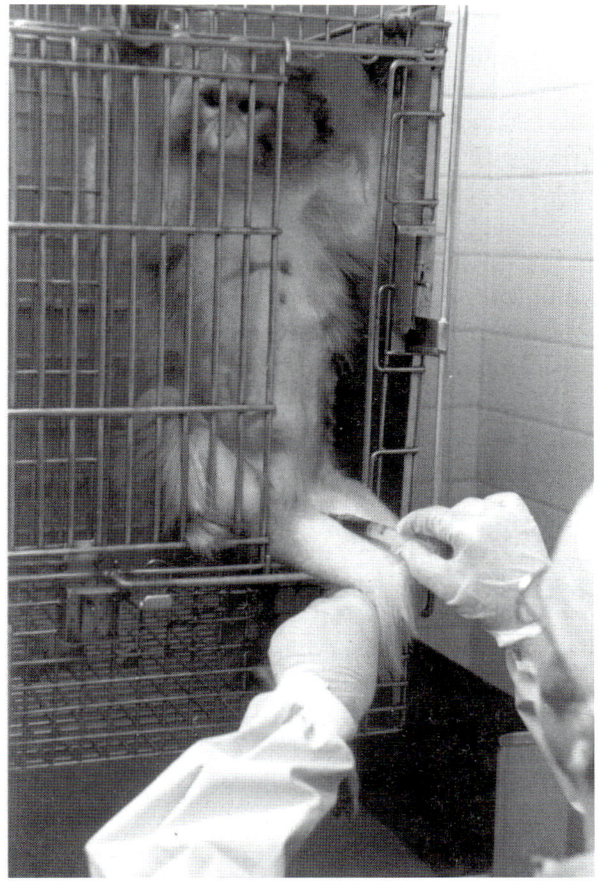

Figure 14.18 The typical, significant cortisol response to traditional blood collection involving enforced restraint is absent in rhesus macaques trained to voluntarily cooperate (Reinhardt, 2003).

of both sexes, less than one hour of training was required per animal (Reinhardt, 2003b). This initial time investment quickly paid off: the subject no longer had to be controlled by a second person, and research data no longer were confounded by avoidable stress reactions (Figure 14.18).

Conclusion

Compatible companionship is the best choice of environmental enrichment, addressing the animals' need for social contact and providing ever-changing stimulation for the expression of species-appropriate activities. Housing the animals in pairs is the easiest way to achieve social enrichment and can even be implemented with species stigmatized as being particularly aggressive. Pair-housing does not interfere with routine husbandry procedures and common research protocols but it creates a condition in which the animals can experience optimal behavioral health and social well-being. Almost as important as conspecific companionship, is a trust-based relationship with the attending personnel, *including* the investigator. It is a fundamental condition to protect laboratory primates from stress resulting from fear-inducing procedures.

For animals kept in pens, woodchip litter is a hygienically acceptable substrate on which the food can be scattered, thereby promoting foraging activities. For caged animals, the standard diet can readily be presented in such a way that time-consuming, skillful foraging behavior is necessary to retrieve the food. Foraging devices have the disadvantage that they are expensive and require extra time to load and clean them.

Manipulable objects have limited stimulatory value unless they change their shape and texture through wear. Television programs and videotapes may offer valued environmental enrichment for personnel but it is questionable if this holds true for the animals themselves.

Elevated resting surfaces are *essential* for laboratory primates to access the quasi-safe arboreal dimension. Space has little stimulatory value but it becomes an extremely important enrichment factor if it is structured. Periodic release into large, furnished cages, gives single-caged primates a temporary relief from their boring, small living quarters.

Environmental enrichment may distract primates sufficiently so that they engage less in stereotypical activities, but, in most instances, it is not a cure for such behavior patterns.

Taking advantage of the learning capability of primates, by training them to cooperate during procedures, not only provides mental stimulation but also refines research methodology by minimizing stress responses.

Acknowledgements

I am very thankful to Lynn and Mike Noel and my wife Annie for editing the draft of this manuscript.

Correspondence

Any correspondence should be directed to Viktor Reinhardt, Animal Welfare Institute, Washington, DC 20027. Email: shasta@siskiyou.net

References

Albrecht, E.D., Nightingale, M.S. and Townsley, J.D. (1978). *J. Endocrinol.* 77, 425–426.

Alexander, S. and Fontenot, M.B. (2003). *AALAS Official Program*, 141.

Alford, P.L., Bloomsmith, M.A., Keeling, M.E. and Beck, T.F. (1995). *Zoo. Biol.* 14, 347–359.

Anderson, J.R. and Chamove, A.S. (1984). In *UFAW Standards in Laboratory Animal Management. Proceedings of a Symposium,* pp. 253–256. The Universities Federation for Animal Welfare, Potters Bar, UK.

Animal Welfare Act, Pub. L. No. 99–198 (1985).

Anzenberger, G. and Gossweiler, H. (1993). *Am. J. Primatol.* 31, 223–230.

Asvestas, C. (1998). *Laboratory Primate Newsl.* 37(3), 5.

Asvestas, C. and Reininger, M. (1999). *Lab. Primate Newsl.* 38(3), 14.

Baker, K.C. (1997). *Zoo. Biol.* 16, 225–236.

Baskerville, M. (1999). In Poole, T. and English, P. (eds) *The UFAW Handbook on the Care and Management of Laboratory Animals, Seventh Edition,* pp. 611–635. Blackwell Science, Oxford, UK.

Bayne, K. and McCully, C. (1989). *Lab. Anim.* 18, 25–28.

Bayne, K., Dexter, S.L. and Suomi, S.J. (1991). *Lab. Primate Newsl.* 30(2), 9–12.

Bayne, K., Dexter, S.L., Mainzer H, McCully, C., Campbell, G. and Yamada, F. (1992a). *Anim. Welfare* 1, 39–53.

Bayne, K., Hurst, J.K. and Dexter, S.L. (1992b). *Lab. Anim. Sci.* 42, 38–45.

Bayne, K., Dexter, S.L. and Strange, G.M. (1993). *Cont. Top. Lab. Anim. Sci.* 32(2), 6–9.

Beck, R.P.A. (1995). *RATEL* 22(4), 112–126.

Beirise, J.H. and Reinhardt, V. (1992). *Lab. Primate Newsl.* 31(1), 7–8.

Bellinger, L.L., Hill, E.G. and Wiggs, R.B. (1992). *Cont. Top. Lab. Anim. Sci.* 31(3), 10–12.

Bentson, K.L., Capitanio, J.P. and Mendoza, S.P. (2003). *J. Med. Primatol.* 32, 148–160.

Berendt, R. and Williams, T.D. (1971). *Lab. Anim. Sci.* 21, 502–509.

Bercovitch, F. and Kessler, M.J. (1993). *Humane Innov. Altern. Anim. Exper.* 7, 435–439.

Bernstein, I.S. and Gordon, T.P. (1977). *Lab. Anim. Sci.* 27, 532–540.

Bernstein, I.S., Gordon, T.P. and Rose, R.M. (1974). In Holloway, R.L. (ed.) *Primate Aggression, Territoriality, and Xenophobia,* pp. 211–240. Academic Press, New York, NY.

Bertrand, F., Seguin, Y., Chauvier, F. and Blanquié, J.P. (1999). *Folia Primatol.* 70, 207.

Bloomsmith, M.A. and Lambeth, S.P. (1996). *AZA Regional Conference Proceedings*, 449–452.

Bloomsmith, M.A. and Baker, K.C. (2001). In Brent, L. (ed.) *Special Topics in Primatology Volume 2 – The Care and Management of Captive Chimpanzees,* pp. 204–241. The American Society of Primatologists, San Antonio, TX.

Bloomsmith, M.A., Alford, P.L. and Maple, T.L. (1988). *Am. J. Primatolol.* 16, 155–164.

Bloomsmith, M.A., Laule, G.E., Alford, P.L. and Thurston, R.H. (1994). *Zoo. Biol.* 13, 557–566.

Bloomsmith, M.A., Baker, K.C., Ross, S.K. and Lambeth, S.P. (1998). *Am. J. Primatolol.* 45, 171.

Bloomstrand, M., Riddle, K., Alford, P.L. and Maple, T.L. (1986). *Zoo. Biol.* 5, 293–300.

Boccia, M.L. (1989). *Lab. Primate Newsl.* 28(2), 18–19.

Boccia, M.L. and Hijazi, A.S. (1998). *Lab. Primate Newsl.* 37(3), 1–5.

Boinski, S., Noon, C., Stans, S., Samudio, R., Sammarco, P. and Hayes, A. (1994). *Lab. Primate Newsl.* 33(4), 1–4.

Bovard, E.W. (1959). *Psychol. Rev.* 66, 267–277.

Box, H.O. and Rohrhumber, B. (1993). *Anim. Technol.* 44, 19–30.

Boyce, W.T., O'Neill–Wagner, P.L., Price, C.S., Haines, M.C. and Suomi, S.J. (1998). *Health Psychol.* 17, 285–289.

Brandt, E.M. and Mitchell, G. (1973). *Dev. Psychol.* 2, 222–228.

Brent, L. (1992). *Anim. Welfare* 1, 161–170.

Brent, L. and Eichberg, J.W. (1991). *Zoo. Biol.* 10, 353–360.

Brent, L. and Weaver, D. (1996). *J. Med. Primatol.* 25, 370–374.

Brent, L. and Stone, A.M. (1998). *J. Appl. Anim. Welfare Sci.* 1, 5–14.

Brent, L., Lee, D.R. and Eichberg, J.W. (1989). *Am. J. Primatolol.* 19(Suppl. 1), 65–70.

Brent, L., Kessel, A.L. and Barrera, H. (1997). *Zoo. Biol.* 16, 335–342.

Britt, A. (1998). *Zoo. Biol.* 17, 379–392.

Bryant, C.E., Rupniak, N.M.J. and Iversen, S.D. (1988). *J. Med. Primatol.* 17, 257–269.

Bunyak, S.C., Harvey, N.C., Rhine, R.J. and Wilson, M.I. (1982). *Am. J. Primatolol.* 2, 201–204.

Bush, M., Custer, R., Smeller, J. and Bush, L.M. (1977). *J. Am. Vet. Med. Assoc.* 171, 866–869.

Byrne, G.D. and Suomi, S.J. (1991). *Am. J. Primatolol.* 23, 141–151.

Byrum, R. and St. Claire, M. (1998). *Lab. Primate Newsl.* 37(4), 1.

Catlow, G., Ryan, P.M. and Young, R.J. (1998). In Hare, V.J. and Worley, E. (eds) *Proceedings of the Third International Conference on Environmental Enrichment*, pp. 209–217. The Shape of Enrichment, San Diego, CA.

Celli, M.L., Tomonagaa, M., Udonob, T., Teramotob, M. and Naganob, K. (2003). *Appl. Anim. Behav. Sci.* 81, 171–182.

Chamove, A.S. (2001). *Austral. Primatol.* 14(3), 16–19.

Chamove, A.S. and Anderson, J.R. (1979). *Anim. Technol.* 30, 69–74.

Choi, G.C. (1993). *WARDS Newsl.* 4, 3–7, 13.

Clarke, A.S., Mason, W.A. and Moberg, G.P. (1988). *Lab. Anim. Sci.* 38, 305–309.

Clarke, A.S., Czekala, N.M. and Lindburg, D.G. (1995). *Primates* 36, 41–46.

Clarke, M.R. and Blanchard, J.L. (1994). *Lab. Primate Newsl.* 33(2), 5–8.

Coe, C.L. (1991). In Novak, M.A. and Petto, A.J. (eds) *Through the Looking Glass. Issues of Psychological Well-being in Captive Nonhuman Primates,* pp. 78–92. American Psychological Association, Washington, DC.

Coe, C.L. and Scheffler, J. (1989). *Zoo. Biol. Suppl.* 1, 89–99.

Coe, C.L., Franklin, D., Smith, E.R. and Levine, S. (1982). *Physiol. & Behav.* 29, 1051–1057.

Coelho, A.M., Carey, K.D. and Shade, R.E. (1991). *Am. J. Primatol.* 23, 257–267.

Cohen, S., Kaplan, J.R., Cunnick, J.E., Manuck, S.B. and Rabin, B.S. (1992). *Psychol. Sci.* 3, 301–304.

Conlee, K.M., Lilly, A.A. and Taub, D.M. (1996). XVIth Congress of the International Primatological Society/XIXth Conference of the American Society of Primatologists, Madison Abstract No. 671.

Crockett, C.M., Bielitzki, J.T., Carey, A. and Velez, A. (1989). *Lab. Primate Newsl.* 28(2), 21–22.

Crockett, C.M., Bowers, C.L., Sackett, G.P. and Bowden, D.M. (1993). *Am. J. Primatolol.* 30, 55–74.

Crockett, C.M., Bowers, C.L., Bowden, D.M. and Sackett, G.P. (1994). *Am. J. Primatolol.* 32, 73–94.

Crockett, C.M., Bowers, C.L., Shimoji, M., Leu, M., Bowden, D.M. and Sackett, G.P. (1995). *J. Comp. Psychol.* 109, 368–383.

Crockett, C.M., Shimoji, M. and Bowden, D.M. (2000). *Am. J. Primatol* 52, 63–80.

Crouthamel, B. and Sackett, G. (2004). *Lab. Primate Newsl.* 43(1), 5–6.

de Waal, F.B.M. (1992). In Erwin, J. and Landon, J.C. (eds) *Chimpanzee Conservation and Public Health: Environments for the Future,* pp. 83–87. Diagnon/Bioqual, Rockville, MD.

Dexter, S.L. and Bayne, K. (1994). *Lab. Primate Newsl.* 33(2), 9–12.

DiGregorio, G. (1990). *Dissert. Abstr. Int.* B50(11), 5353.

Eaton, G.G., Kelley, S.T., Axthelm, M.K., Iliff-Sizemore, S.A. and Shiigi, S.M. (1994). *Am. J. Primatolol.* 33, 89–99.

Eckert, K., Niemeyer, C., Anonymous, Rogers, R.W., Seier, J., Ingersoll, B., Barklay, L., Brinkman, C., Oliver, S., Buckmaster, C., Knowles, L. and Pyle, S. (2000). *Lab. Primate Newsl.* 39(3), 1–4.

Else, J.G. (1985). *Lab. Anim. Sci.* 35, 373–375.

Else, J.G., Tarara, R., Suleman, M.A. and Eley, R.M. (1986). *Lab. Anim. Sci.* 36, 168–172.

Elton, R.H. (1979). In Erwin, J., Maple, T. and Mitchell, G. (eds) *Captivity and Behavior*, pp. 125–139. Van Nostrand Reinhold, New York, NY.

Elvidge, H., Challis, J.R.G., Robinson, J.S., Roper, C. and Thorburn, G.D. (1976). *J. Endocrinol.* 70, 325–326.

Erwin, J. (1977). *Lab. Anim. Sci.* 27, 541–547.

Erwin, J. (1979). In Erwin, J., Maple, T. and Mitchell, G. (eds) *Captivity and Behavior,* pp. 139–171. Van Nostrand, New York, NY.

European Commission (2002). *The Welfare of Non-human Primates* – Report of the Scientific Committee on Animal Health and Animal Welfare. European Commission, Strasbourg, France.

Evans, H.L., Taylor, J.D., Ernst, J. and Graefe, J.F. (1989). *Lab. Anim. Sci.* 39, 318–323.

Fragaszy, D.M. and Adams-Curtis, L.E. (1991). In Box, H.O. (ed.) *Primate Responses to Environmental Change,* pp. 247–264. Chapman and Hall, New York, NY.

Fritz, J. (1994). *Lab. Primate Newsl.* 33(1), 5–7.

Fritz, P. and Fritz, J. (1979). *J. Med. Primatol.* 8, 202–221.

Fritz, J. and Howell, S. (2001). In Brent, L. (ed.) *Special Topics in Primatology Volume 2 – The Care and Management of Captive Chimpanzees,* pp. 172–203. The American Society of Primatologists, San Antonio, TX.

Fuller, G.B., Hobson, W.C., Reyes, F.I., Winter, J.S.D. and Faiman, C. (1984). *Proc. Soc. Exp. Biol. Med.* 175, 487–490.

Gaspari, F., Perretta, G. and Schino, G. (2000). *Folia Primatol.* 71, 291.

Gerber, P., Schnell, C.R. and Anzenberger, G. (2002). *Primates* 43, 201–216.

Gilloux, I., Gurnell, J. and Shepherdson, D. (1992). *Anim. Welfare* 1, 279–289.

Glick-Bauer, M. (1997). *Lab. Primate Newsl.* 36(1), 1–3.

Gonzalez, C.A., Coe, C.L. and Levine, S. (1982). *Psychoneuroendocrin.* 7, 209–216.

Goo, G.P. and Sassenrath, E.N. (1980). *J. Med. Primatol.* 9, 325–334.

Goode, T.L., McPherson, H., Hughes, J., Presti, P., Conboy, T., Smith, S., Bone, A., Zimmerman, W., Holder, D., Klein, H. (1998). *Cont. Top. Lab. Anim. Sci.* 37(4), 100.

Goodwin, J. (1997). *AZA Regional Conference Proceedings*, 510–515.

Goosen, C. (1984). *Standards in Laboratory Animal Management. Proceedings of a Symposium,* pp. 245–249. The Universities Federation for Animal Welfare, Potters Bar, UK.

Goosen, C. (1988). In Beijnen, A.C. and Solleveld, H. (eds) *New Developments in Biosciences: Their Implication for Laboratory Animal Science*, pp. 67–70. Martinus Nijhoff Publishers, Dordrecht, Netherlands.

Grant, J.L. and Doudet, D.J. (2003). *Lab. Primate Newsl.* 42(1), 1–3.

Gust, D.A., Gordon, T.P., Wilson, M.E., Ahmed-Ansari, A., Brodie, A.R. and McClure, H.M. (1991). *Brain Behav. Immun.* 5, 296–307.

Gust, D.A., Gordon, T.P., Brodie, A.R. and McClure, H.M. (1994). *Physiol. Behav.* 55, 681–684.

Gust, D.A., Gordon, T.P., Wilson, M.E., Brodie, A.R., Ahmed-Ansari, A. and McClure, H.M. (1996). *Am. J. Primatolol.* 39, 263–273.

Gwinn, L.A. (1996). *Cont. Top. Lab. Anim. Sci.* 35(4), 61.

Ha, J.C., Robinette, R.L. and Sackett, G.P. (1999). *Am. J. Primatolol.* 47, 153–163.

Harris, L.D., Briand, E.J., Orth, R. and Galbicka, G. (1999). *Cont. Top. Lab. Anim. Sci.* 38(2), 48–53.

Hartner, M.K., Hall, J., Penderghest, J., White, E., Watson, S. and Clark, L. (2000). *Cont. Top. Lab. Anim. Sci.* 39(4), 67.

Hayashi, K.T. and Moberg, G.P. (1987). *Am. J. Primatol.* 12, 263–273.

Hayes, S.L. (1990). *Lab. Anim. Sci.* 40, 515–519.

Hazlewood, S.J. (2001). *Anim. Technol.* 52, 149–152.

Hediger, H. (1964). *Wild Animals in Captivity.* Dover Publications, New York, NY.

Hogan, M. (1991). *AZA Annual Conference Proceedings*, 629.

Home Office (1989). Animals (Scientific Procedures) Act 1986. *Code of Practice for the Housing and Care of Animals Used in Scientific Procedures.* Her Majesty's Stationery Office, London, UK.

Howell, S.M., Matevia, M., Fritz, J., Nash, L. and Maki, S. (1993). *Anim. Welfare* 2, 153–163.

Howell, S., Roeder, E., Nelson, C., Fritz, J. and Schwandt, M. (2002). *Am. J. Primatolol.* 57, 83–84.

International Primatological Society. (1993). *Primate Report* 35, 3–29.

Ives, M. and Dack, G.M. (1956). *J. Lab. Clin. Med.* 47, 723–729.

Jensen, G.D., Blanton, F.L. and Gribble, D.H. (1980). *Experimental Gerontology* 15, 399–406.

Jerome, C.P. and Szostak, L. (1987). *Lab. Anim. Sci.* 37, 508–509.

Jorgensen, M.J., Kinsey, J.H. and Novak, M.A. (1998). *Am. J. Primatolol.* 45, 187.

Judge, P.G., de Waal, B.M., Paul, K.S. and Gordon, T.P. (1994). *Lab. Anim. Sci.* 44, 344–350.

Kaplan, J.R., Manning, P. and Zucker, E. (1980). *Lab. Anim. Sci.* 30, 565–570.

Kelley, T.M. and Bramblett, C.A. (1981). *Am. J. Primatolol.* 1, 95–97.

Kessel, A.L. and Brent, L. (1995). *Cont. Top. Lab. Anim. Sci.* 34(6), 74–79.

Kessel, A.L. and Brent, L. (1998). *J. Appl. Anim. Welfare Sci.* 1, 227–234.

Kessel-Davenport, A.L. and Gutierrez, T. (1994). *The Newsletter* 6(2), 1–2.

Kessler, M.J., London, W.T., Rawlins, R.G., Gonzales, J., Martines, H.S. and Sanches, J. (1985). *J. Med. Primatol.* 13, 91–98.

King, J.E. and Norwood, V.R. (1989). In Segal, E.F. (ed.) *Housing, Care and Psychological Wellbeing of Captive and Laboratory Primates,* pp. 102–114. Noyes Publications, Park Ridge, NJ.

Kitchen, A.M. and Martin, A.A. (1996). *Lab. Animals* 30, 317–326.

Kobert, M.G. (1998). In Hare, V.J. and Worley, E. (eds) *Proceedings of the Third International Conference on Environmental Enrichment,* pp. 230–236. The Shape of Enrichment, San Diego, CA.

Kopecky, J. and Reinhardt, V. (1991). *Lab. Primate Newsl.* 30(2), 5–6.

Lam, K., Rupniak, N.M.J. and Iversen, S.D. (1991). *J. Med. Primatol.* 20, 104–109.

Lambeth, S.P. and Bloomsmith, M.A. (1994). *Anim. Welfare* 3, 13–24.

Lambeth, S.P., Perlman, J.E. and Schapiro, S.J. (2000). *Am. J. Primatolol.* 51 (Suppl. 1), 79–80.

Landi, M.S., Kissinger, J.T., Campbell, S.A., Kenney, C.A. and Jenkins, E.L. (1990). *J. Am. Coll. Toxicol.* 9, 517–523.

Laule, G.E. (1992). *Enrichment* 1(2), 11–12.

Laule, G.E., Thurston, R.H., Alford, P.L. and Bloomsmith, M.A. (1996). *Zoo Biol.* 15, 587–591.

Laule, G. and Whittaker, M. (2001). In Brent, L. (ed.) *Special Topics in Primatology Volume 2 – The Care and Management of Captive Chimpanzees,* pp. 242–266. The American Society of Primatologists, San Antonio, TX.

Leu, M., Crockett, C.M., Bowers, C.L. and Bowden, D.M. (1993). *Am. J. Primatol.* 30, 327.

Levison, P.K., Fester, C.B., Nieman, W.H. and Findley, J.D. (1964). *J. Exp. Anal. Behav.* 7, 253–254.

Lilly, A.A., Mehlman, P.T. and Higley, J. (1999). *Am. J. Primatol.* 48, 197–223.

Lindburg, D.G. and Coe, J. (1995). In Gibbons, E.F., Durrant, B.S. and Demarest, A.J. (eds) *Conservation of*

Endangered Species in Captivity, pp. 553–570. SUNY Press, Albany, NY.

Line, S.W. (1987). *J. Am.Vet. Med. Assoc.* 190, 854–859.

Line, S.W., Markowitz, H., Morgan, K.N. and Strong, S. (1989a). In Driscoll, J.W. (ed.) *Animal Care and Use in Behavioral Research: Regulation, Issues, and Applications,* pp. 103–117. Animal Welfare Information Center National Agricultural Library, Beltsville, MD.

Line, S.W., Morgan, K.N., Markowitz, H. and Strong, S. (1989b). *Am. J. Vet. Res.* 40, 1523–1526.

Line, S.W., Morgan, K.N., Roberts, J.A. and Markowitz, H. (1990a). *Lab. Primate Newsl.* 29(1), 8–12.

Line, S.W., Morgan, K.N., Markowitz, H., Roberts, J. and Riddell, M. (1990b). *Lab. Primate Newsl.* 29(4), 1–5.

Line, S.W., Morgan, K.N., Markowitz, H. and Strong, S. (1990c). *Am. J. Primatol.* 20, 107–113.

Line, S.W. and Morgan, K.N. (1991). *Lab. Anim. Sci.* 41, 365–369.

Luttrell, L., Acker, L., Urben, M. and Reinhardt, V. (1994). *Anim. Welfare* 3, 135–140.

Lutz, C.K. and Farrow, R.A. (1996). *Cont. Top. Lab. Anim. Sci.* 35(3), 75–78.

Lynch, R. (1998). *Lab. Primate Newsl.* 37(1), 4–5.

Lynch, R. and Baker, D. (2000). *Lab. Primate Newsl.* 39(1), 12.

Mahoney, C.J. (1992). *Lab Anim.* 21(5), 27, 29, 32–37.

Majolo, B., Buchanan-Smith, H.M. and Morris, K. (2001). *Primate Eye* 73, 12–13.

Maki, S., Alford, P.L., Bloomsmith, M.A. and Franklin, J. (1989). *Am. J. Primatol. Suppl.* 1, 71–78.

Maninger, N., Kim, J.H. and Ruppenthal, G.C. (1998). *Am. J. Primatolol.* 45, 193–194.

Maple, T.L. and Finlay, T.W. (1987). *Appl. Anim. Behav. Sci.* 18, 5–18.

Markowitz, H. (1979). In Erwin, J., Maple, T. and Mitchell, G. (eds) *Captivity and Behavior,* pp. 217–238. Van Nostrand Reinhold, New York, NY.

Marks, D., Kelly, J., Rice, T., Ames, S., Marr, R., Westfall, J., Lloyd, J. and Torres, C. (2000). *Lab. Primate Newsl.* 39(4), 9–10.

Martin, D.P., Gilberto, T., Burns, C. and Pautler, H.C. (2002). *Cont. Top. Lab. Anim. Sci.* 41(5), 47–49.

Martin, S. (1996). *AZA Regional Conference Proceedings,* 139–143.

Mason, W.A. (1960). *J. Abnorm. Soc. Psychol.* 60, 100.

McCormack, K. and Megna, N.L. (2001). *Am. J. Primatolol.* 54(Suppl. 1), 50–51.

McDonald, S. (1994). *Int. Zoo Yearbook* 33, 235–247.

McGrew, W.C., Brennan, J.A. and Russel, J. (1986). *Zoo Biol.* 5, 45–50.

McKinley, J., Buchanan-Smith, H.M., Bassett, L. and Morris, K. (2003). *J. Appl. Anim. Welfare Sci.* 6, 209–220.

McNary, J.K. (1992). In Fulk, R. and Garland, C. (eds) *The Care and Management of Chimpanzees (Pan troglodytes) in captive environments* (R), pp. 88–100. North Carolina Zoological Society, Asheboro, NC.

Mendoza, S.P. (1991). *Lab. Anim. Sci.* 41, 344–349.

Missakian, E.A. (1972). *J. Person. Soc. Psychol.* 21, 131–134.

Morgan, K.N., Line, S.W. and Markowitz, H. (1998). In Shepherdson, D.H., Mellen, J.D. and Hutchins, M. (eds) *Second Nature – Environmental Enrichment for Captive Animals,* pp. 153–171. Smithsonian Institution Press, Washington, DC.

Murchison, M.A. (1994). *Lab. Primate Newsl.* 33(1), 7–8.

Murchison, M.A. (1995). *Lab. Primate Newsl.* 34(1), 1–2.

Murphy, D.E. (1976). *Int. Zoo News* 137(23.5), 24–26.

Murray, L., Hartner, M. and Clark, L.P. (2002). *Cont. Top. Lab. Anim. Sci.* 41(4), 112–113.

Nadler, R.D., Herndon, J.G., Metz, B., Ferrer, A.C. and Erwin, J. (1992). In Erwin, J. and Landon, J.C. (eds) *Chimpanzee Conservation and Public Health: Environments for the Future,* pp. 137–145. Diagnon/Bioqual, Rockville, MD.

Nakamichi, M. and Asanuma, K. (1998). *Percept. Motor Skills* 87, 707–714.

National Research Council (1996). *Guide for the Care and Use of Laboratory Animals,* 7th edition. National Academy Press, Washington, DC.

National Research Council (1998). *The Psychological Well-Being of Nonhuman Primates.* National Academy Press, Washington, DC.

Neveu, H. and Deputte, B.L. (1996). *Am. J. Primatolol.* 38, 175–185.

Nelms, R., Davis, B.K., Tansey, G. and Raber, J.M. (2001). *AALAS Official Program,* 97–98.

Novak, M.A. and Drewsen, K.H. (1989). In Segal, E.F. (ed.) *Housing, Care and Psychological Wellbeing of Captive and Laboratory Primates,* pp. 161–182. Noyes Publications, Park Ridge, NJ.

Novak, M.A., Kinsey, J.H., Jorgensen, M.J. and Hazen, T.J. (1998). *Am. J. Primatolol.* 46, 213–227.

Ochiai, T. and Matsuzawa, T. (1999). *Primate Res.* 15, 289–296.

O'Neill, P.L., Wright, A.C. and Weed, J.L. (1997). *AZA Regional Conference Proceedings,* 95–101.

Paquette, D. and Prescott, J. (1988). *Zoo Biol.* 7, 15–23.

Parks, K.A. and Novak, M.A. (1993). *Am. J. Primatolol.* 29, 13–25.

Paulk, H.H., Dienske, H. and Ribbens, L.G. (1977). *J. Abnorm. Psychol.* 86, 87–92.

Perret, K., Büchner, S. and Adler, H.J. (1998). *Zool. Garten* 68, 95–111.

Philipp, C. (1997). *AZA Regional Conference Proceedings,* 102–105.

Phillippi-Falkenstein, K. and Clarke, M.R. (1992). *Lab. Anim. Sci.* 42, 83–85.

Poenisch, T. (1992). *The Newsletter* 4(1), 1.

Poole, T.B. (1990). *Anim. Technol.* 41, 81–86.

Pruetz, J.D. and Bloomsmith, M.A. (1992). *Anim. Welfare* 1, 127–137.

Ranheim, S. and Reinhardt, V. (1989). *Lab. Primate Newsl.* 28(3), 1–2.

Redican, W.K. and Mitchell, G. (1973). *Am. J. Phys. Anthropol.* 38, 523–526.
Reinhardt, V. (1989). *Am. J. Primatol.* 17, 243–248.
Reinhardt, V. (1990). *Lab. Primate Newsl.* 29(3), 7–11.
Reinhardt, V. (1993a). *Anim. Welfare* 2, 165–172.
Reinhardt, V. (1993b). *Zoo Biol.* 12, 307–312.
Reinhardt, V. (1993c). *Folia Primatol.* 61, 47–51.
Reinhardt, V. (1994a). *J. Med. Primatol.* 23, 426–431.
Reinhardt, V. (1994b). *Animal Technology* 5, 37–41.
Reinhardt, V. (1994c). *Primates* 35, 95–98.
Reinhardt, V. (2003a). *J. Appl. Anim. Welfare Sci.* 6, 123–130.
Reinhardt, V. (2003b). *J. Appl. Anim. Welfare Sci.* 6, 189–197.
Reinhardt, V. and Cowley, D. (1990). *Lab. Primate Newsl.* 29(4), 9–10.
Reinhardt, V. and Garza-Schmidt, M. (2000). *Lab. Primate Newsl.* 39(2), 8–10.
Reinhardt, V. and Reinhardt, A. (2001). *Environmental Enrichment for Caged Rhesus Macaques (Macaca mulatta) – Photographic Documentation and Literature Review* (second edition). Animal Welfare Institute, Washington, DC.
Reinhardt, V. and Rossell, M, (2001). *J. Appl. Anim. Welfare Sci.* 4, 285–294.
Reinhardt, V., Reinhardt, A., Eisele, S., Houser, W.D. and Wolf, J. (1987). *Appl. Anim. Behav. Sci.* 18, 371–377.
Reinhardt, V., Cowley, D. and Eisele, S. (1991). *J. Exper. Anim. Sci.* 34, 73–76.
Ricker, R.B., Williams, L.E., Brady, A.G., Gibson, S.V. and Abee, C.R. (1995). *Cont. Top. Lab. Anim. Sci.* 34(4), 55.
Roberts, R.L., Roytburd, L.A. and Newman, J.D. (1999). *Contemp.Topics Lab. Anim. Sci.* 38(5), 27–31.
Rosenberg, D.P. and Kesel, M.L. (1994). In Rollin, B.E. and Kesel, M.L. (eds) *The Experimental Animal in Biomedical Research. Volume II, Care, Husbandry, and Well-Being – An Overview by Species,* pp. 457–483. CPR Press, Boca Raton, FL.
Rosenblum, L.A. and Andrews, M.W. (1995). In Bennett, B.T., Abee, C.R. and Henrickson, R. (eds) *Nonhuman primates in Biomedial Research. Biology and Management* (B.T.), pp. 101–112. Academic Press, New York, NY.
Schapiro, S.J. and Bloomsmith, M.A. (1995). *Am. J. Primatolol.* 35, 89–101.
Schapiro, S.J. and Bushong, D. (1994). *Anim. Welfare* 3, 25–36.
Schapiro, S.J., Bloomsmith, M.A., Kessel, A.L. and Shively, C.A. (1993). *Appl. Anim. Behav. Sci.* 37, 251–263.
Schapiro, S.J., Lee-Parritz, D.E., Taylor, L.L., Watson, L.M., Bloomsmith, M.A. and Petto, A.J. (1994). *Lab. Anim. Sci.* 44, 229–234.
Schapiro, S.J., Bloomsmith, M.A., Porter, L.M. and Suarez, S.A. (1996). *Appl. Anim. Behav. Sci.* 48, 158–172.
Schapiro, S.J., Nehete, P.N., Perlman, J.E. and Sastry, K.J. (1997). *Am. J. Primatolol.* 42, 147.
Schapiro, S.J., Nehete, P.N., Perlman, J.E. and Sastry, K.J. (2000). *Appl. Anim. Behav. Sci.* 68, 67–84.
Schapiro, S.J., Bloomsmith, M.A. and Laule, G.E. (2002). *XIXth Congress of the International Primatological Society, Abstracts,* pp. 181–182. Mammalogical Society of China, Beijing, China.
Schmidt, E.M., Dold, G.M. and McIntosh, J.S. (1989). *Lab. Anim. Sci.* 39, 166–167.
Schnell, C.R. and Wood, J.M. (1993). *Proceedings of the Fifth Federation of European Laboratory Animal Science Associations Symposium,* 107–111.
Seaman, W.B. (1971). *Hospital Pract.* 6(9), 146–147.
Seelig, D. (1998). *Lab. Primate Newsl.* 37(3), 14–16.
Seres, M., Aureli, F. and deWaal, F.B.M. (2001). *Zoo Biol.* 50, 501–515.
Shively, C.A., Clarkson, T.B. and Kaplan, J.R. (1989). *Atherosclerosis* 77, 69–76.
Simeon, D., Stanley, B. and Frances, A. (1992). *Am. J. Psychiatry* 149, 221–226.
Sokol, K.A. (1993). *Lab Animal* 22(5), 40–45.
Southwick, C.H. (1967). *Behaviour* 28, 182–209.
Spector, M., Kowalczky, M.A., Fortman, J.D. and Bennett, B.T. (1994). *Cont. Top. Lab. Anim. Sci.* 33(5), 54–55.
Spragg, S.D.S. (1940). *Comp. Psychol. Monogr.* 15, 1–132.
Steen, Z. (1995). *Int. Zoo News* 42, 284–298.
Storey, P.L., Turner, P.V. and Tremblay, J.L. (2000). *Contemp. Topics Lab. Anim. Sci.* 39(1), 14–16.
Streett, J.W. and Jonas, A.M. (1982). *Lab. Anim. Sci.* 32, 263–266.
Taylor, T.D. (2002). *Shape of Enrichment* 11(2), 1–3.
Taylor, W.J. and Laudenslager, M.L. (1998). *Lab. Anim.* 27(4), 28–31.
Taylor, W.J., Brown, D.A., Lucas-Awad, J. and Laudenslager, M.L. (1997). *Lab. Primate Newsl.* 36(3), 1–3.
Tiefenbacher, S., Lee, B., Meyer, J.S. and Spealman R.D. (2003). Noninvasive technicque for the repeated sampling of salivary free cortisol in awake, unrestrained squirrel monkeys. *American Journal of Primatology* 60, 69–75.
Tully, L.A., Jenne, M. and Coleman, K. (2002). *Cont. Top. Lab. Anim. Sci.* 41(4), 7.
Turkkan, J.S. (1990). *J. Med. Primatol.* 19, 455–466.
Turkkan, J.S., Ator, N.A., Brady, J.V. and Craven, K.A. (1989). In Segal, E.F. (eds) *Housing, Care and Psychological Wellbeing of Captive and Laboratory Primates,* pp. 305–322. Noyes Publications, Park Ridge, NJ.
Tustin, G.W., Williams, L.E. and Brady, A.G. (1996). *Lab. Primate Newsl.* 35(1), 5–7.
USDA (1991). *Fed. Reg.* 56(32), 6426–6505.
USDA (2001). *Animal Welfare Report – Fiscal Year 2000.* U.S. Department of Agriculture – Animal Care, Riverdale, MD.
Vermeer, J. (1997). *Int. Zoo News* 44, 146–149.
Visalberghi, E. and Anderson, J.R. (1999). In *The UFAW Handbook on the Care and Management of Laboratory Animals Seventh Edition* (Poole T. and English, P., eds), pp. 601–610. Blackwell Science, Oxford, UK.
Watson, D.S.B. (1991). *Lab. Anim. Sci.* 41, 378–379.

Watson, L.M. (2002). *Lab. Primate Newsl.* 41(2), 6–9.
Weed, J.L. and Watson, L.M. (1998). *Am. J. Primatol.* 45, 212.
Westergaard, G.C., Izard, M.K., Drake, J.D., Suomi, S.J. and Higley, J.D. (1999). *Am. J. Primatol.* 49, 339–347.
Williams, L.E., Abee, C.R., Barnes, S.R. and Ricker, R.B. (1988). *Lab. Anim. Sci.* 38, 289–291.
Wolff, A. and Ruppert, G. (1991). *Lab. Anim.* 20(2), 36–39.
Wolfle, T.L. (1987). *J. Am. Vet. Med. Assoc.* 191, 1219–1221.
Woodbeck, T. and Reinhardt, V. (1991). *Lab. Primate Newsl.* 30(4), 11–12.
Yanagihara, Y., Matsubayashi, K. and Matsuzawa, T. (1994). *Primate Res.* 10, 95–104.
Yaroshevsky, F. (1975). *Can. Psych. Asso. J.* 20, 443–446.
Ziegler, T.E., Bridson, W.E., Snowdon, C.T. and Eman, S. (1987). *Am. J. Primatolol.* 12, 127–140.
Zimmermann, A. and Feistner, A.T.C. (1996). *Dodo* 32, 67–75.

CHAPTER 15

Development of Specific Pathogen Free Nonhuman Primate Colonies

Keith Mansfield
New England Primate Research Center, Harvard Medical School, One Pine Hill Drive, Southborough, MA 01772, USA

Introduction

In contrast to many other laboratory animal species, demand for nonhuman primates has increased in recent years and there currently exists a significant shortage of such animals for utilization in biomedical research. Well-defined research animals are required by a number of disciplines, including programs investigating infectious diseases, vaccine development, neurosciences, transplantation and reproductive biology. Genetically defined animals, free of common infectious agents that may confound research or potentially infect human handlers, are in high demand. Their poor availability forces delays in critical research representing a significant impediment to international biomedical research objectives (Desrosiers, 1997; Roberts *et al.*, 2000). Thus breeding production of nonhuman primates has taken on increasing importance and requires an understanding of normal primate biology, husbandry and infectious disease control.

Historical perspectives on specific pathogen free primate colonies

In the past 15 years, the Acquired Immunodeficiency Syndrome (AIDS) pandemic has been the driving force

The Laboratory Primate
Copyright 2005 Elsevier
ISBN 0-1208-0261-9

behind the development of nonhuman primate colonies, specific pathogen free (SPF) of a number of viral agents (Desrosiers, 1997). According to the World Health Organization, the Human Immunodeficiency Virus (HIV) is recognized as the leading infectious cause of human morbidity and mortality worldwide. Greater than 30 million individuals have contracted this agent with the vast majority of infections occurring in sub-Saharan Africa. Development of effective strategies for control, prevention and treatment remains a significant biomedical objective that must be addressed as we move into the next millennium.

SIV-infected rhesus macaques represent one important tool with which to advance our knowledge of this disease and these animals have been used extensively to investigate aspects of viral pathogenesis and host immunity. For reasons of biosafety, and to eliminate confounding variables, such investigations are ideally carried out in animals that are free of a number of viral agents commonly found in conventional colonies, including Simian Retrovirus type D (SRV-D), Simian T-lymphotropic Virus (STLV-1), Simian Immunodeficiency Virus (SIV) and B Virus (BV) (Desrosiers, 1997; Holmes *et al.*, 1995). In addition to these requirements, investigators in pursuit of their research goals increasingly request animals of defined major histocompatibility (MHC) genotype or animals free of additional infectious agents in pursuit of their research goals. MHC defined macaques have been utilized to investigate cytotoxic T lymphocyte (CTL) responses during SIV infection through the use of tetramer technology (Sauermann, 2001). Animals free of additional agents have been utilized to investigate novel vaccine strategies and the pathogenesis of opportunistic infections during AIDS (Desrosiers *et al.*, 1997; Kaur *et al.*, 2002; Mansfield *et al.*, 1997b; Wang *et al.*, 2001). Such requirements are likely to increase with further refinements of the model.

The establishment of SPF colonies has proven difficult and current production from domestic sources is not adequate to meet demand from private and academic institutions. Such animals are costly and not readily available. The limited availability of well-defined research animals is an impediment to the continued progress and success of animal model based research. In recognition of these obstacles, the National Council on Research Resources (NCRR) and Office of AIDS Research (OAR) of the National Institutes of Health (NIH) have taken a lead role in the funding and development of SPF colonies, many of which have been located within the National Primate Research Center (NPRC) program.

Definition of specific pathogen free status

Consistent use of terminology to define the virologic status of colonies and individual animals is imperative to avoid confusion and misunderstanding. Specific pathogen free macaques may be defined as offspring arising from breeding programs in which selected agents have been eliminated through an extensive "test and remove" strategy of founder animals. SPF founders are animals obtained from conventional colonies and which have been repeatedly tested for the target agents and housed in separation from other virally infected monkeys. SPF founders may also be known as level 1 SPF animals and offspring as level 2 SPF. By definition SPF animals have been housed either individually or only in contact with animals of similar virologic status. "Conventional" animals are those in which no testing and segregation strategy has been employed. A "seronegative" designation is given to animals that arise from conventional colonies, lack antibodies to a particular agent and have not been housed with other potentially infected animals. An animal lacking antibodies to a particular agent but housed with other infected animals should also be considered conventional. Thus, understanding the current virologic status of an individual animal requires an appreciation not only of current virologic test results but also previous housing history and virologic status of all contact animals. An unambiguous animal identification system and detailed housing and medical records are prerequisites to the development of SPF colonies.

SPF target viruses for macaque colonies

The SPF target viruses may differ from colony to colony and will be dependent on the nonhuman primate species and expected utilization. The minimum target viruses for macaque colonies, supported through the current NCRR/OAR program, are BV, STLV, SIV and SRV-D.

Simian immunodeficiency virus

The Simian immunodeficiency viruses are a group of closely related primate lentiviruses that cause minimal disease in their natural African nonhuman primate host but progressive immunodeficiency when transmitted to Asian macaques (Baskin *et al.*, 1988 Lackner *et al.*, 1994). They have been investigated extensively as an animal model of human HIV infection (Joag, 2000; Johnson, 2002). Investigation of spontaneously occurring immunodeficiency in colony macaques, housed at several NPRCs in the 1970s and 1980s, revealed that these animals had become infected with SIV (Lowenstine *et al.*, 1992; Mansfield *et al.*, 1995). Presumably this had resulted from cross species transmission between captive animals. While an antibody negative viremic state exists, a simple test and removal strategy, using standard whole virus ELISA, was successful in eliminating this virus from colonies (Daniel *et al.*, 1988). Although SIV is a target virus of most SPF colonies, spontaneous SIV infection of macaques has not been recognized in domestic colonies since the mid 1980s.

Simian retrovirus type D

SRV-D is the principle cause of immunodeficiency in domestically bred macaques and is a significant cause of morbidity and mortality when present in captive colonies. While extensive effort has been devoted to the study of SIV, relatively little work has been conducted on SRV-D. The virus readily infects T cells (CD4 and CD8), B cells, macrophages and epithelial cells (Lackner *et al.*, 1988). SRV-D may be isolated from saliva of healthy carrier animals and biting, with inoculation of saliva or blood, is the most likely method of horizontal transmission (Lerche *et al.*, 1986). Mother to infant transmission may occur in the perinatal and postnatal period and clinical disease is characterized by chronic diarrhea and acquisition of opportunistic infections. Retroperitoneal fibromatosis and Noma are clinical conditions associated with chronic SRV-D infection. As with SIV, a test and remove strategy has been successful in eliminating SRV-D from colonies in which it occurs as an enzootic infection (Daniel *et al.*, 1988; Schroder *et al.*, 2000). An antibody negative viremic state occurs, occasionally following *in utero* or perinatal transmission. While this antibody negative state may complicate testing strategies based on the detection of host immune responses, such animals generally present with clinical disease early in life and can be eliminated by maintaining a high index of clinical suspicion and routinely attempting viral SRV-D isolation of ill animals. Alternatively, molecular based assays, such as PCR or routine viral isolation, can be incorporated in the initial screening program but may not be warranted in all situations (Liska *et al.*, 1997). The cost effectiveness of additional testing would be dependent on the prevalence of SRV-D in the source colony and the age at which candidate SPF founders are selected.

B virus

BV is an alphaherpes virus that causes widespread infection of all species of macaques kept in captivity. Infection of the natural host generally results in a self-limiting clinical disease with the establishment of latency following primary infection (Weigler *et al.*, 1993). The virus is readily transmitted, with most animals seroconverting by two years of age (Weigler *et al.*, 1990; Weigler *et al.*, 1993). While of limited consequence in the natural host, zoonotic transmission of BV to human handlers or other species may result in severe clinical disease and often death. Although a rare occurrence, with fewer than 40 fatal cases described in literature since 1932, many facilities will not accept B virus antibody positive animals for reasons of biosafety (Cohen *et al.*, 2002; Holmes *et al.*, 1995).

BV has been the most problematic to eliminate from SPF colonies (Ward *et al.*, 2000; Ward and Hilliard, 2002). The reasons for this are most likely multi-factorial. A phenomenon of suspected delayed seroconversion has been recognized clinically in SPF macaque colonies since their inception. Delayed seroconversion can be defined as occurring when animals develop BV antibodies after repeatedly testing negative for BV for periods of greater than one year (Freifeld *et al.*, 1995; Ward and Hilliard, 1994). The phenomenon is poorly understood but one hypothesis is that animals that are infected early in life may establish latency without an adequate antibody response. These animals may then develop a primary humoral immune response during periods of viral reactivation. Regardless of their SPF status, it is common policy to treat all macaques as if they potentially harbor B virus and take appropriate precautions (Cohen *et al.*, 2002).

Simian T lymphotropic virus

STLV refers to a group of related type C retroviruses known to infect a number of nonhuman primate species. These agents are closely related to Human T-lymphotropic Virus type-1 of man (Franchini and

Reitz, Jr., 1994). The incidence of STLV infection may vary greatly from colony to colony with reported seroprevalence rates of 0–20% (Daniel *et al.*, 1988; Ishikawa *et al.*, 1987). Serologic surveys indicate an increasing prevalence with age and, while the mechanism of natural transmission is unknown, parenteral and sexual routes are suspected of being of greater importance than perinatal transmission (Ishikawa *et al.*, 1987). Seronegative virus-positive animals are uncommon. STLV appears to have limited health consequences in immunologically normal macaques but has been associated with lymphoproliferative disorders such as lymphoma in AIDS (Homma *et al.*, 1984). STLV has limited zoonotic potential but has been selected as a target SPF virus due to its potential as a confounding variable in immunologic studies, particularly those investigating other simian retroviruses. Commercially available ELISA reagents, to detect antibodies to HTLV-1, also detect antibodies to STLV and can be used in a "test and remove" strategy (Daniel *et al.*, 1988). All positive tests should be verified, at a reference laboratory, by Western blot as the test used in this setting may lack specificity.

SPF target agents in non-macaque primate colonies

Historically most efforts have focused on the development of SPF rhesus macaque colonies due to their widespread use in biomedical research programs and the presence of significant spontaneously occurring infectious agents in this species. In some sense, all nonhuman primate colonies used in biomedical research are maintained SPF of certain common infectious agents, such as *Mycobacterium tuberculosis*. However, in comparison to macaques, development of colonies free of ubiquitous viral agents has lagged in other nonhuman primate species. Recently NCRR/NIH has funded the establishment of a baboon resource colony free of specific viral agents including STLV-1, SIV, herpes papionis virus-2 (HPV-2), simian virus 40 (SV40), simian foamy virus (SFV) and cytomegalovirus (CMV) (Table 15.1). Development of this colony has followed techniques established in the SPF rhesus macaque program and is based on early separation of offspring and extensive serologic evaluation of candidate animals.

TABLE 15.1: Potential SPF target agents in non-macaque primate species

Species	Potential target agent
Papio anubis	Cytomegalovirus
	Simian T-lymphotropic virus
	Simian immunodeficiency virus
	Herpes papionis virus 2
	Papovavirus
African green monkey	Simian immunodeficiency virus
	Simian agent 8
	Simian virus 40
Squirrel monkey	Herpes virus saimiri
	GB virus A variant(s)
	Herpes tamarinus
Callithrix jacchus	GB virus A variant(s)
	Callitrichine herpesvirus 3

While New World primates harbor a number of viral agents that could potentially impact on research programs, considerably less work has been conducted in the establishment of SPF colonies. Examples of potential SPF target agents include *Herpesvirus saimiri* of squirrel monkeys, GB virus A variants of callitrichidae and gammaherpes virus infection of common marmosets (Bukh and Apgar, 1997; Cho *et al.*, 2001; Jung *et al.*, 1999). Development of such colonies would require significant economic support and would be dependent on demonstration of significant need by the research community. Techniques established in macaques would likely be successful in these species.

Viral testing

Detection of viral specific antibody responses, and a "test and remove" strategy, form the foundation for establishing nonhuman primate colonies free of target agents. Screening and confirmational tests are utilized in most SPF derivation algorithms. The ideal screening test would be inexpensive, sensitive and specific, and technically simple to perform. Antibody-detection based ELISA procedures fulfill these criteria and have been used extensively in the establishment of SPF colonies. Serologic testing by ELISA should be backed up by additional assays to confirm positive tests and investigate indeterminant values. Confirmational testing may include cELISA, Western blot, PCR and viral isolation and are available at a number of reference

TABLE 15.2: NCRR/NIH supported reference laboratories

National B Virus Resource Center
Viral Immunology Center Georgia State University 50 Decatur Street Atlanta, Georgia 30303 Phone: 404-651-0808 Laboratory Director: Dr. Julia Hilliard
Simian Retrovirus Laboratory
California National Primate Research Center Road 98 at Hutchinson University of California Davis, California 95616 Phone: 503-752-8242 Laboratory Director: Dr. Nicholas Lerche

laboratories, including two supported by NCRR/NIH (Table 15.2).

When utilizing screening assays in a population, to diagnose enzootic viral infections, it is important to understand the concept of test sensitivity and specificity (Gardner *et al.*, 2000). Sensitivity is defined as the proportion of subjects with a disease that test positive for that disease. It is a measure of a test's ability to detect animals that are truly positive or affected by a condition such as a viral infection. Specificity is defined as the proportion of subjects without the disease who have a negative test. Both may be calculated by comparing test results to those obtained with a gold standard and are inherent properties of the test regardless of the population tested. In ELISA assays, the sensitivity and specificity of the test may be affected by altering such parameters as the dilution of serum or conjugate, incubation times or calculation of cutoff values.

While the sensitivity and specificity are inherent properties of the test, the false positive and negative rates are influenced by the prevalence of the disease in a population and directly impact decisions regarding the derivation and management of SPF colonies. When SPF colonies are initially formed, the prevalence of viral infection is high and this results in a low false positive rate and a relatively high false negative rate (Figure 15.1). In other words, during the initial phase of SPF colony derivation, one should be suspicious that negative results may be positive but can be relatively certain that positive results are truly positive. In contrast, as the colony matures and the prevalence of disease decreases, the reverse becomes true: the false positive rate will increase and the false negative rate decrease. In mature SPF colonies, most positive screening tests will be false positives and considerably more effort should be devoted to confirmational testing of these animals.

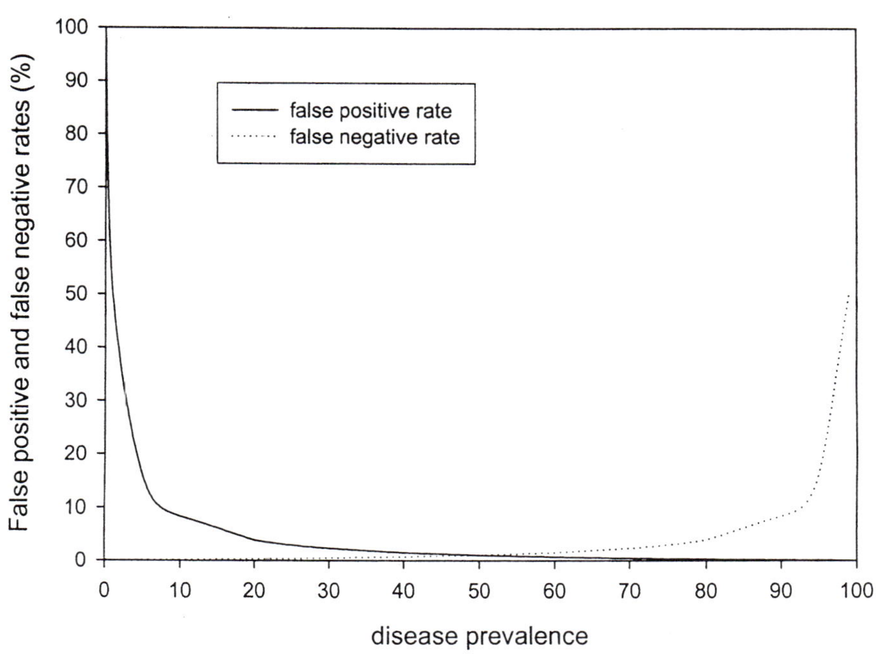

Figure 15.1 Test sensitivity and specificity. Changes in false positive and false negative rates are observed as disease prevalence changes from 0.0% to 99.0%. Even with high test sensitivity and specificity (set at 99%), low disease prevalence results in a high false positive rate.

On-site screening of samples by ELISA provides some advantages over sending all samples off to reference laboratories for analysis. On-site testing is less expensive and has a shorter turn-around time. With proper planning, tests can be run within 24–48 hours of collection and can greatly expedite colony management decisions. Delays in obtaining results may be extensive, particularly if samples must be sent from the country of origin for testing, and may adversely impact on husbandry decisions.

ELISA plates may be made for the four target viruses, utilizing purified whole virus preparations (Table 15.3) (Daniel *et al.*, 1988). HSV-1 or SA8/HPV-2 may be used in lieu of BV for biosafety reasons (Blewett *et al.*, 1999; Takano *et al.*, 2001). HSV-1 and HTLV-1 plates are also available commercially and can be adapted for use in a surrogate testing program for BV and STLV-1 respectively. SRV-D and SIV plates are not currently available commercially and must be prepared from purified virus. Antibody testing for BV may lack sensitivity in certain instances. Delayed seroconversion of SPF founder animals has previously been described and represents a problem in the establishment of SPF colonies as does delayed or poor antibody response in viral positive animals (Ward and Hilliard, 1994). Whether such animals shed virus and remain a threat to other colony animals and humans, remains unknown.

TABLE 15.3: Techniques for the production of viral antigen and Enzyme Linked Immunosorbent Assays (ELISA)

Viral purification
Cells for virus production grown in 600 ml T-150 flasks
Specific cell lines utilized support growth of target virus
Harvested by centrifugation 72 hrs after splitting
Cell-free supernatants pelleted by ultracentrifugation (48,000 g at 4 c for 3 hr)
Purified by Sepharose 4B column chromatography
ELISA plates coated with viral antigen diluted 1:1,000 in Triton X-100-PBS and blocked with 0.3% BSA
Antigen concentration dependent on viral antigen
ELISA
Test serum at 1:10–1:100 dilution (1 hr)
1:100 dilution of goat anti-human IgG(Fc) conjugated with alkaline phosphatase (1 hr) (Kirkegaard & Perry Laboratories)
Developed with p-Nitrophenylphosphate substrate (30 min)
Absorbance read at 410 nm with Dynatech ELISA reader

Specific pathogen free animal derivation strategies

The optimal testing strategies may vary, depending on the founding population and target SPF agents, and should be tailored to the individual facility (Ward and Hilliard, 1994, 2002). If initial serologic surveys indicate that the source colony is free of a target virus, considerably fewer resources need be devoted to this pathogen but rather refocused on the remaining agents. A potential testing strategy for macaque target agents is outlined in Figure 15.2. Juvenile animals, between 8 and 12 months of age make the best SPF candidate animals because maternal immunity offers a level of protection and most animals do not seroconvert to the target viruses until 2–3 years of age. While animals may be taken at an earlier age, this increases husbandry requirements and may result in increased health problems and behavioral disorders. Animals are selected at 8–12 months when they are weaned from natal groups. At this time, blood samples are obtained while routine preventative health care is performed. Animals are individually housed until initial virologic testing is available and seropositive animals are returned to the conventional colony. Seronegative animals are placed in small age matched peer groups of 3–4 animals. These small peer groups allow for normal behavioral development but also limit the potential spread of infectious agents. In general, a 1:8 male to female ratio is used in the harvest and selection of candidates. If detailed pedigrees are available, selection should be based on maintaining the genetic diversity present within the founder population. Following formation of peer groups, animals are examined quarterly and blood drawn for serologic assessment. Indeterminate or positive ELISA tests obtained at this stage should be confirmed at a reference laboratory. If seroconversion is detected, the entire group should be removed to the conventional colony. Quarterly testing for retroviruses should continue for a minimum of two years and animals placed in breeding harems at 3–3.5 years of age. Because of difficulties in eliminating BV, quarterly testing for this agent should continue indefinitely.

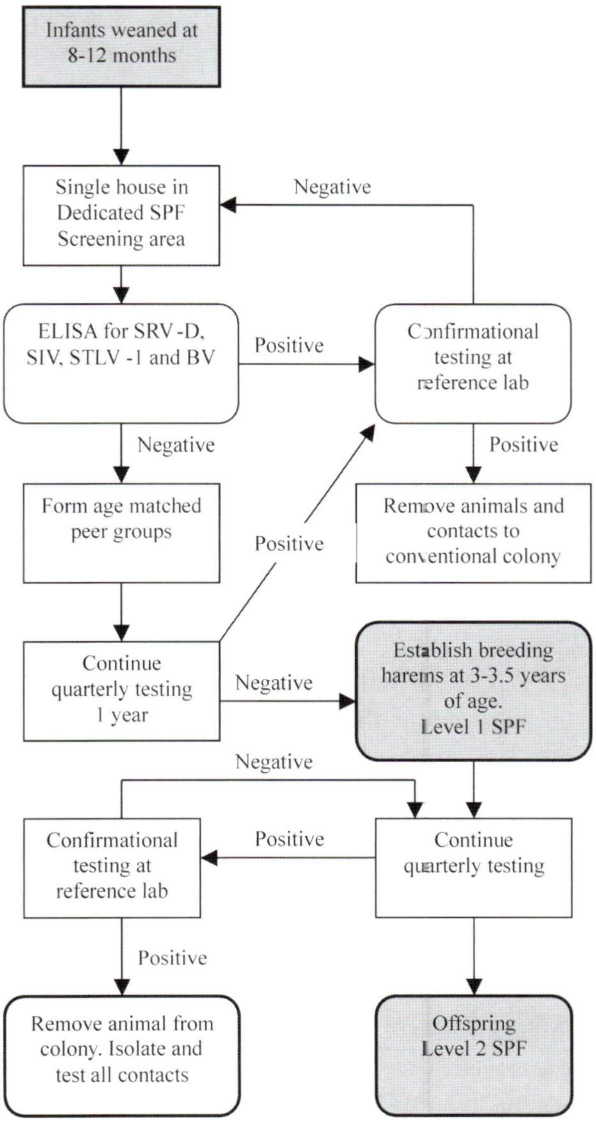

Figure 15.2 Testing algorithm for derivation of specific pathogen free macaques.

Animal housing configurations

Selection of housing configuration may directly affect program success and is an important component of SPF colony management. Common strategies for housing of macaque species include time mating, corral breeding and harem breeding configurations. Each offers distinct advantages and disadvantages. Corral breeding is a common configuration used in areas of the world in which year round climatic conditions allow housing of animals outdoors. Corrals generally contain from 15–20 breeding males and 100–125 adult females. Corrals have reduced fixed and variable costs including reduced construction, labor and supply costs. Reproductive efficiency may be better than in other housing configurations. Unfortunately corral housing presents several distinct disadvantages from the standpoint of establishing SPF colonies. The main disadvantage is that pathogen containment is more difficult in large group settings. A break in SPF status of one animal puts all contact animals at risk. When this occurs in breeding harems, potential contacts can be limited to 8–10 animals housed together. Testing of the affected group can be intensified or the entire group can be eliminated. A break in SPF status in corral housing may put 150–200 animals at risk and a subsequent "test and remove" strategy is difficult or impossible to perform.

Breeding harems are another commonly used housing configuration and consist of one male and 8–10 adult females. Animals are housed in indoor pens or outdoor "corn cribs" and their smaller size effectively contains breaks in SPF status. Breeding harems also allow for accurate construction of breeding pedigrees and enhance the ability to observe sick or injured animals. Smaller group size increases the ability to observe and treat injured animals and to perform routine preventative health care.

While time mating offers similar advantages to breeding harems in pathogen containment and allows the production of timed pregnancies for research purposes, it is labor intensive, results in high caging costs, and often reduces reproductive performance.

Veterinary care program

A strong preventative health care program is critical to the success of SPF colonies. Such programs not only allow the tracking of target SPF agents but also the control of other potential primate pathogens that impact on colony health. Components of a preventative health program should include quarterly routine physical examination of all animals and banking of biological samples (Table 15.4). At this time, intradermal tuberculin testing and phlebotomy for viral testing can be performed. Serum and, ideally, DNA should be obtained and stored. Vaccinations based on risk assessment for measles, rabies and tetanus may be administered. Standard operating procedures should be developed and followed and individual animal medical records maintained. Individual animal medical records

TABLE 15.4: Preventative health care program for rhesus macaque SPF colonies

Quarterly Physical Examination
Blood draw, weight and serum storage
Tuberculin test
Animal records
Vaccination program: measles, rabies, and tetanus
Serum and DNA banks (<1 year)
Separation of SPF colony from conventional and indigenous primates
Disease surveillance program
Diagnostic program of diseased animals
Complete necropsy and histologic examination

should, at least, track serologic testing results, housing history, reproductive performance, genealogy and clinical information.

Disease surveillance is an important component of the preventative health program. A rigorous diagnostic program for ill or sick animals should be developed to help in the identification of potential disease outbreaks that may impact on colony health. A production-animal herd-health approach should govern the decision making process. All animals that have died or are euthanased for poor health should be necropsied and have tissue examined histologically. Results should be entered in a computerized database to allow tracking of disease trends. Necropsies not only allow identification of cause of death in individual animals but also surveillance of the colony for primate pathogens and identification of subclinical disease.

Strict separation of SPF colony animals from conventional colony animals, indigenous primates and other wild or feral populations must also be maintained. Indigenous primates may represent a potential exogenous source of the SPF target agents. Wild or feral populations such as rodents and dogs may transmit such important pathogens as leptospirosis, encephalomyocarditis virus or rabies virus. If conventional colonies are to be maintained at the same facility, care should be taken to prevent transmission of agents directly, or by fomites, in common use areas such as clinic or procedure rooms. Once colonies are established, introduction of animals from other sources should be discouraged. If introduction of animals is required to maintain genetic diversity, extended quarantines and repeated testing for a period of at least one year are recommended. Introduction of males will have the greatest and most immediate impact on increasing the genetic diversity of the colony.

Expanded SPF programs

In addition to these four SPF target viruses, nonhuman primates may be infected with a variety of other viral, bacterial and parasitic agents. Such agents may adversely impact on colony health or represent a significant zoonotic risk. Pathogen free animals may also be required for specific research protocols. Several facilities have developed or are developing an "expanded" SPF program in which animals are free of other agents in addition to the original four target viruses (Table 15.5). Such expanded programs should be based on the present and anticipated research needs of facilities being supplied by the breeding program. In establishing new colonies, it may be educational to screen a subset of the source colony for additional agents. If the seroprevalence of such additional agents is low or non-existent, it may be cost effective to include these agents in the SPF program.

Viral agents

With refinement of the SIV AIDS model, investigators have requested animals free of additional agents to study the impact of concurrent infections on disease

TABLE 15.5: Additional SPF target agents in macaque species

Agent	Zoonotic potential
Viral	
Rhesus Rhadinovirus (RRV)	Unknown
Lymphocryptovirus (LCV)	Unknown
Rhesus cytomeglovirus (RhCMV)	Unknown
Simian foamy virus (SFV)	Yes
Simian virus 40 (SV40)	Yes
Bacterial	
Helicobacter pylori	Yes
Enteropathogenic E. coli	Yes
Parasitic	
Cryptosporidia parvum	Yes
Enterocytozoon bieneusi	Yes

pathogenesis and to investigate novel antigen delivery systems. Recently, expanded SPF programs have been established at several facilities. In addition to BV, STLV-1, SRV-D and SIV, these facilities have targeted Simian virus 40 (SV40), lymphocryptovirus (LCV), rhesus rhadinovirus (RRV), rhesus CMV and SFV. All have been recognized to cause widespread infection in most rhesus macaque colonies. In particular, colonies free of SV40 and SFV may be required for some experimental protocols and would represent a valuable SPF asset. Furthermore, in some cases (e.g. simian foamy viruses) these agents may pose a newly recognized zoonotic risk to personnel (Heneine *et al.*, 1998). In response to these demands, re-derivation of macaques by cesarean section from existing colonies has been undertaken at some facilities. Following re-derivation, animals are hand reared in isolation from existing colonies and tested for the presence of additional viral and non-viral pathogens.

To perform this testing, additional ELISA procedures have been developed using purified virus for SV40, RRV, SFV and rhesus CMV. Testing for SV40 and SFV is also available at some commercial laboratories (Desrosiers *et al.*, 1997; Kaur *et al.*, 2002; Wang *et al.*, 2001). Testing for LCV was initially performed by PCR on DNA isolated from peripheral blood mononuclear cells and, more recently, a peptide ELISA for antibody testing has also been developed utilizing two peptides representing the carboxyterminal domains of the rhesus LCV VCAp18 (Rao *et al.*, 2000).

Parasitic agents

Parasitic agents are frequent pathogens found in most nonhuman primate colonies. In both conventional and SPF colonies, common parasites include *Cryptosporidia parvum, Enterocytozoon bieneusi, Giardia lamblia, Trichomonas, Entameoba* and *Strongyloides stercoralis*. These agents cause varying degrees of morbidity in immunologically normal animals and may result in severe disease and mortality in immunosuppressed animals. Furthermore, rhesus macaques may serve as animal models for some of these agents and a number of investigators, actively investigating aspects of disease pathogenesis and host immune response utilize these models. The derivation process for the expanded viral targets colony may also eliminate these parasitic agents. Serologic evidence indicates that *Cryptosporidia* has been excluded from animals re-derived from existing SPF colonies. Similarly, molecular data from our laboratory suggests that *E. bieneusi* may also be absent (Mansfield *et al.*, 1997a).

Bacterial agents

Bacterial pathogens are arguably the major cause of morbidity and mortality in nonhuman primate colonies. Important microbes include *Shigella flexneri, Klebsiella pneumoniae, Campylobacter jejuni, Yersinia pseudotuberculosis, Mycobacterium tuberculosis* and enteropathogenic *E. coli*. Despite this fact, advances in understanding the epidemiology and pathogenesis of spontaneous bacterial diseases in nonhuman primates has lagged behind other areas of investigative primatology. The significance of many bacterial pathogens to colony health, and their effect on experimental research, are poorly understood. *Shigella* and *Campylobacter* are well-recognized causes of clinical disease in Old World species of nonhuman primates and their exclusion from colonies would likely result in an overall decrease in enterocolitis, a frequent cause of morbidity and mortality in captive colonies. Likewise, enteropathogenic *E. coli* has been frequently identified in both normal and immunocompromised rhesus macaques and may be associated with severe diarrhea in infant and juvenile animals (Mansfield *et al.*, 2001). The total elimination of all bacterial pathogens would be difficult and require that animals be raised in a barrier type facility. A more targeted approach might be warranted such as the elimination of *Helicobacter pylori* for groups interested in utilizing a primate model of this bacterial infection (Solnick *et al.*, 1999).

Summary recommendations

The requirements and utilization of nonhuman primates, SPF for various target agents, will likely further increase in the coming years. For macaque species, the minimum target agents should include BV, SIV, SRV-D and STLV-1. Consideration should be given to the inclusion of other agents such as RRV, CMV, LCV, SV40 and SFV based on perceived research requirements and seroprevalence in the source colony. On-site serologic testing has many advantages including cost effectiveness and rapid turn-around time but should be backed up by confirmational testing at a reference laboratory.

For macaques, candidate SPF animals should be selected at 8–12 months of age and initially housed in small peer groups. These animals should be tested for the target agents on a quarterly basis. During this candidate

TABLE 15.6: Recommendations for the establishment of SPF macaque breeding colonies

Minimum SPF target viruses: BV, SIV, STLV and SRV-D
Initial serologic surveys
On-site serologic testing capabilities/ Off-site confirmational testing
Candidate SPF animal selected at 6–12 months of age
Initially housed in small peer groups and tested quarterly for target viruses for two years
Breeding harems established at 3–3.5 years of age and begin testing for retroviruses annually and BV quarterly
Rigorous preventative health care and diagnostic program in place
Strict separation of SPF colonies from conventional and indigenous primates
Maintenance of genetic diversity.

selection phase, animals that seroconvert and all contacts should be removed from the colony. Breeding harems can be established at 3–3.5 years of age. Harems provide the best opportunity at pathogen containment early in SPF colony development. Strict separation of SPF animals from conventional and indigenous primates should be maintained as such animals represent a potential source for breaks in SPF status. A vigorous preventative healthcare program must be in place. This program should include quarterly physical exams, blood and DNA banking, viral screens, vaccinations and individual animal medical records. SPF animals should be managed to preserve genetic diversity and maintained as out-bred colonies. Detailed pedigrees and breeding records should be kept and the use of software for the management of genetic, pedigree and demographic data considered.

Acknowledgements

This work was supported by a grant from the National Institutes of Health, National Council of Research Resources P51 RR000168.

Correspondence

Any correspondence should be directed to Keith Mansfield, New England Primate Research Center, Harvard Medical School, One Pine Hill Drive, Southborough, MA 01772, USA. Email: keith mansfield @hms.harvard.edu

References

Baskin, G.B., Murphey-Corb, M., Watson, E.A. and Martin, L.N. (1988). *Vet. Pathol.* 25, 456–467.

Blewett, E.L., Saliki, J.T. and Eberle, R. (1999). *J. Virol. Methods* 77, 59–67.

Bukh, J. and Apgar, C.L. (1997). *Virology* 229, 429–436.

Cho, Y., Ramer, J., Rivailler, P., Quink, C., Garber, R.L., Beier, D.R. and Wang, F. (2001). *Proc. Natl. Acad. Sci. U.S.A.* 98, 1224–1229.

Cohen, J.I., Davenport, D.S., Stewart, J.A., Deitchman, S., Hilliard, J.K. and Chapman, L.E. (2002). *Clin. Infect. Dis.* 35, 1191–1203.

Daniel, M.D., Letvin, N.L., Sehgal, P.K., Schmidt, D.K., Silva, D.P., Solomon, K.R., Hodi, F.S., Jr., Ringler, D.J., Hunt, R.D., King, N.W. and Desrosiers, R.C. (1988). *Int. J. Cancer* 41, 601–608.

Desrosiers, R.C. (1997). *AIDS Res. Hum. Retroviruses* 13, 5–6.

Desrosiers, R.C., Sasseville, V.G., Czajak, S.C., Zhang, X., Mansfield, K.G., Kaur, A., Johnson, R.P., Lackner, A.A. and Jung, J.U. (1997). *J. Virol.* 71, 9764–9769.

Franchini, G. and Reitz, M.S., Jr. (1994). *AIDS Res. Hum. Retroviruses* 10, 1047–1060.

Freifeld, A.G., Hilliard, J., Southers, J., Murray, M., Savarese, B., Schmitt, J.M. and Straus, S.E. (1995). *J. Infect. Dis.* 171, 1031–1034.

Gardner, I.A., Stryhn, H., Lind, P. and Collins, M.T. (2000). *Prev. Vet. Med.* 45, 107–122.

Heneine, W., Switzer, W.M., Sandstrom, P., Brown, J., Vedapuri, S., Schable, C.A., Khan, A.S., Lerche, N.W., Schweizer, M., Neumann-Haefelin, D., Chapman, L.E. and Folks, T.M. (1998). *Nat. Med.* 4, 403–407.

Holmes, G.P., Chapman, L.E., Stewart, J.A., Straus, S.E., Hilliard, J.K. and Davenport, D.S. (1995). *Clin. Infect. Dis.* 20, 421–439.

Homma, T., Kanki, P.J., King, N.W., Jr., Hunt, R.D., O'Connell, M.J., Letvin, N.L., Daniel, M.D., Desrosiers, R.C., Yang, C.S. and Essex, M. (1984). *Science* 225, 716–718.

Ishikawa, K., Fukasawa, M., Tsujimoto, H., Else, J.G., Isahakia, M., Ubhi, N.K., Ishida, T., Takenaka, O., Kawamoto, Y. and Shotake, T. (1987). *Int. J. Cancer* 40, 233–239.

Joag, S.V. (2000). *Microbes. Infect.* 2, 223–229.

Johnson, R.P. (2002). *Vaccine* 20, 1985–1987.

Jung, J.U., Choi, J.K., Ensser, A. and Biesinger, B. (1999). *Semin. Cancer Biol.* 9, 231–239.

Kaur, A., Daniel, M.D., Hempel, D., Lee-Parritz, D., Hirsch, M.S. and Johnson, R.P. (1996). *J. Virol.* 70, 7725–7733.

Kaur, A., Hale, C.L., Noren, B., Kassis, N., Simon, M.A. and Johnson, R.P. (2002). *J. Virol.* 76, 3646–3658.

Lackner, A.A., Rodriguez, M.H., Bush, C.E., Munn, R.J., Kwang, H.S., Moore, P.F., Osborn, K.G., Marx, P.A.,

Gardner, M.B. and Lowenstine, L.J. (1988). *J. Virol.* 62, 2134–2142.
Lackner, A.A., Vogel, P., Ramos, R.A., Kluge, J.D. and Marthas, M. (1994). *Am. J. Pathol.* 145, 428–439.
Lerche, N.W., Osborn, K.G., Marx, P.A., Prahalada, S., Maul, D.H., Lowenstine, L.J., Munn, R.J., Bryant, M.L., Henrickson, R.V. and Arthur, L.O. (1986). *J. Natl. Cancer Inst.* 77, 489–496.
Liska, V., Lerche, N.W. and Ruprecht, R.M. (1997). *AIDS Res. Hum. Retroviruses* 13, 433–437.
Lowenstine, L.J., Lerche, N.W., Yee, J.L., Uyeda, A., Jennings, M.B., Munn, R.J., McClure, H.M., Anderson, D.C., Fultz, P.N. and Gardner, M.B. (1992). *J. Med. Primatol.* 21, 1–14.
Mansfield, K.G., Carville, A., Shvetz, D., MacKey, J., Tzipori, S. and Lackner, A.A. (1997a) *Am. J. Pathol.* 150, 1395–1405.
Mansfield, K.G., Carville, A., Shvetz, D., MacKey, J., Tzipori, S. and Lackner, A.A. (1997b). *Am. J. Pathol.* 150, 1395–1405.
Mansfield, K.G., Lerche, N.W., Gardner, M.B. and Lackner, A.A. (1995). *J. Med. Primatol* 24, 116–122.
Mansfield, K.G., Lin, K.C., Newman, J., Schauer, D., MacKey, J., Lackner, A.A. and Carville, A. (2001). *J. Clin. Microbiol.* 39, 971–976.
Rao, P., Jiang, H. and Wang, F. (2000). *J. Clin. Microbiol.* 38, 3219–3225.
Roberts, J.A., Smith, D.G. and Hendrickx, A. (2000). *Science* 287, 1591.
Sauermann, U. (2001). *Curr. Mol. Med.* 1, 515–522.
Schroder, M.A., Fisk, S.K. and Lerche, N.W. (2000). *Contemp. Top. Lab. Anim. Sci.* 39, 16–23.
Solnick, J.V., Canfield, D.R., Yang, S. and Parsonnet, J. (1999). *Lab. Anim. Sci.* 49, 197–201.
Takano, J., Narita, T., Fujimoto, K., Mukai, R. and Yamada, A. (2001). *Exp. Anim.* 50, 345–347.
Wang, F., Rivailler, P., Rao, P. and Cho, Y. (2001). *Philos. Trans. R. Soc. Lond Biol. Sci.* 356, 489–497.
Ward, J.A. and Hilliard, J.K. (1994). *Lab. Anim. Sci.* 44, 222–228.
Ward, J.A. and Hilliard, J.K. (2002). *Contemp.Top. Lab. Anim. Sci.* 41, 36–41.
Ward, J.A., Hilliard, J.K. and Pearson, S. (2000). *Comp. Med.* 50, 317–322.
Weigler, B.J., Roberts, J.A., Hird, D.W., Lerche, N.W. and Hilliard, J.K. (1990). *Lab. Anim. Sci.* 40, 257–261.
Weigler, B.J., Hird, D.W., Hilliard, J.K., Lerche, N.W., Roberts, J.A. and Scott, L.M. (1993). *J. Infect. Dis.* 167, 257–263.

CHAPTER 16

Medical Care

James Mahoney
New York University School of Medicine, Sanctuary Support Program, LLC, 11 Rea Court, Monroe, New York 10950, USA

Animal health monitoring and surveillance

A definition of what constitutes medical care

Detecting the first subtle signs of illness in a nonhuman primate, as with any animal, depends on recognizing it, not only in terms of its species, or even its age and sex, but also its individuality as a unique being. This is the "art" in support of the "science" of veterinary medicine and animal management, the practical hands-on versus the purely erudite approach. The image of the individual animal conjured up in the mind's eye of the veterinarian and, even more importantly, the caregiver, is critical to good medical care.

Careful daily observation of animals by caregivers and recording normal, as well as abnormal, findings in them is the basis of good animal management. Yet the earliest, subtle signs of ill health in nonhuman primates can be easily missed if caregivers do not take, or are not given, adequate time to perform their daily observations. Clinical rounds carried out regularly by veterinarians, in company with the caregivers/animal technicians of each animal area, must become a religiously observed routine. To place the onus of observing and reporting illness in the animals on the caregiver alone is to invite disaster. The inconvenience of having to disrobe contaminated outer work clothes and don fresh "clean" garments, in order to meet the biosafety rules of entering office areas or clinical laboratories, can be enough to make all but the most dedicated put off, until later, the reporting of critical information concerning animals to supervisors or clinicians. In the process, this information may not be passed on in a timely fashion and even may be totally forgotten. The physical structure and layout of animal facilities, in relation to the veterinarians' and supervisors' offices and clinical laboratory centers, can also make the difference between an efficient operation and one doomed to frequent failure. Even a short flight of stairs can impede the transfer of such information.

The Laboratory Primate

ISBN 0-1208-0261-9

Management of the stable colony

Body weight

Body weight is perhaps the single most important criterion of normal health and development. Except for the normal increase in body weight seen in young animals, or during late pregnancy, body weight should fluctuate up or down no more than 5% over periods of months or even years. A sudden change in weight of even 5%, say over a period of a few days or weeks, should be investigated to find the cause.

A significant decline in body weight (10% or more), unaccompanied by any specific symptoms, is indicative of malnutrition or food deprivation, as may occur with the competition of social housing, persistent parasitism, low-grade chronic pneumonia, heart or kidney disease, diabetes mellitus and, never forget, tuberculosis. A marked short-term increase in body weight, or a more progressive increase over time, may be indicative of inappropriate diet, overeating and/or lack of exercise, or diabetes mellitus or unrecognized pregnancy.

Animals should be weighed on the occasion of every instance of tranquilization. Hand reared infants should be weighed daily during the first four to six weeks of life and at least weekly thereafter for some time. Mother reared infants should be weighed at least every two weeks in early life, taking into account stress or danger that might be caused by infant separation, and then monthly thereafter, especially if a reliable visual assessment of normal development cannot be made on a frequent (i.e., daily) basis. In adults, body weights should be obtained at least every three months, especially of socially housed primates, in order to identify feeding problems quickly. Automated weighing scales can be incorporated into tunnel systems through which group-housed primates are trained to move, one at a time, for treat rewards or routine feeding.

Physical examination

Each primate in the stable must be physically examined at least once a year, and preferably every six months. Physical examination should include the following:

- Assessment of general appearance and body condition; examination of the hair and skin (integumentary system), oral cavity, teeth and dental formula, and superficial lymph nodes (tonsils, submandibular, cervical, axillary and inguinal lymph nodes); auscultation of the chest for heart and lung sounds; deep palpation of the abdomen to determine the size of the spleen and liver and presence of abdominal masses, etc.; examination of the external and internal urogenital systems, the ears, eyes, nose and throat and the head, trunk and limbs. An electrocardiogram should be obtained at least once a year.

Such examination may reveal conditions that would otherwise go unnoticed until they become major problems. This is particularly so for early detection of orodental disease (especially tooth root abscesses and gingivitis, common in New World species and major causes of poor thriving in elderly primates of all species), heart disease, abdominal or intestinal masses, and uterine pathology (particularly endometriosis and endometrial hyperplasia, common causes of premature death in aged macaques). In group-housed primates, especially those maintained in large outdoor areas, hands-on physical examination may reveal bite wounds and other integumentary and skeletal injuries that have gone totally unrealized.

Hematology

Complete blood counts should be obtained at least every six months, and as often as animals are anesthetized for other purposes, in order to build up substantial individual, as well as colony, baseline hemograms.

Moderate and even severe anemia, caused by acute or chronic injury, chronic undernourishment or outright food deprivation and low-grade generalized infection may go unnoticed until blood examination. While normal red blood cell (RBC) parameters tend to lie within a fairly narrow range among individual animals of a particular species (although marked interspecies variation is seen), total white blood cell counts (WBC) and leucocyte differentials show a great deal of individual variation (Table 16.1). For example, a total WBC of 16.3×10^9/L may be within normal limits for one individual, whereas in another it would be a clear indication of leucocytosis, as in response to infection. Likewise, the ratio of neutrophils to lymphocytes, within certain limits, may be quite variable among individuals of a species, or it may change within normal limits in relation to an animal's age, or during pregnancy.

Serum chemistry profiles

As would be expected, levels of electrolytes are fairly consistent across species, but most of the other serum chemistry parameters vary markedly, as can be seen

TABLE 16.1: Comparative hemogram profiles in four species of laboratory primates

International Species Information System: Physiological Data Reference Values – 2002
Edited by J. Andrew Teare, D.V.M.
The Rhesus Monkey (*M. mulatta*), Common Marmoset (*C. jacchus*),
Squirrel Monkey (*S. sciureus*), and Chimpanzee (*P. troglodytes*).
Standard International Units
(Combined Sexes and Ages)

Test	Rhesus monkey	Marmoset	Squirrel monkey	Chimpanzee	Units
	Mean ± S.D.	Mean ± S.D.	Mean ± S.D.	Mean ± S.D.	
WBC	9.969 ± 3.603	7.214 ± 3.031	8.623 ± 4.016	10.60 ± 4.777	10^9/L
RBC	5.42 ± 0.48	6.09 ± 1.01	6.98 ± 0.76	5.49 ± 0.67	10^{12}/L
Hgb	122 ± 12	152 ± 20	133 ± 14	143 ± 17	g/L
Hct	0.397 ± 0.047	0.467 ± 0.078	0.415 ± 0.045	0.422 ± 0.049	L/L
MCV	71.8 ± 3.2	76 ± 10.0	59.8 ± 5.1	79.9 ± 6.1	fL
MCH	22.4 ± 1.4	25.2 ± 3.1	19.2 ± 1.6	26.2 ± 2.2	pg/cell
MCHC	312 ± 1.7	332 ± 37	323 ± 19.2	328 ± 16	g/L
Platelets	0.468 ± 0.850	0.599 ± 0.2820	0.326 ± 0.1130	0.260 ± 0.065	10^{12}/L
Neutros	4.772 ± 3.280	3.540 ± 2.075	4.449 ± 3.021	7.041 ± 4.431	10^9/L
Lymphs	4.872 ± 2.889	3.724 ± 1.824	3.349 ± 2.209	2.912 ± 1.657	10^9/L
Monos	0.292 ± 0.246	0.255 ± 0.265	0.358 ± 0.371	0.364 ± 0.276	10^9/L
Eos	0.467 ± 0.370	0.183 ± 0.137	0.272 ± 0.396	0.227 ± 0.187	10^9/L
Basos	0.119 ± 0.055	0.126 ± 0.130	0.097 ± 0.070	0.092 ± 0.059	10^9/L
Neutro Bands	0.157 ± 0.115	0.175 ± 0.239	0.278 ± 0.536	0.427 ± 1.026	10^9/L

WBC: White Blood Cells
RBC: Red Blood Cells
Hgb: Hemoglobin
Hct: Hematocrit
MCV: Mean Corpuscular Volume
MCH: Mean Corpuscular Hemoglobin
MCHC: Mean Corpuscular Hemoglobin Concentration
Neutros: Segmented Neutrophils
Lymphs: Lymphocytes
Monos: Mononuclear Leucocytes
Eos: Eosinophilic Leucocytes
Basos: Basophilic Leucocytes
Neutro Bands: Neutrophilic Band Cells
By kind permission of the International Species Information System.

from data presented in Table 16.2 from four of the most commonly used laboratory primates, the rhesus monkey (*Macaca mulatta*), common marmoset (*Callithrix jacchus*), squirrel monkey (*Saimiri sciureus*) and chimpanzee (*Pan troglodytes*). Part of this variation is related to age, sex and physiologic status (e.g., pregnancy), while some is undoubtedly due to variations in composition of laboratory diet, degree of water deprivation prior to anesthesia and blood sampling, ambient temperature, and even reference standards of the reporting laboratory. Some degree of the variation may be related to chronic stress in the subjects, however.

TABLE 16.2: Comparative serum chemistry profiles in four species of laboratory primates

International Species Information System: Physiological Data Reference Values – 2002
Edited by J. Andrew Teare, D.V.M.
The Rhesus Monkey *(M. mulatta)*, Common Marmoset *(C. jacchus)*,
Squirrel Monkey *(S. sciureus)*, and Chimpanzee *(P. troglodytes)*.
Standard International Units
(Combined Sexes and Ages)

Test	Rhesus Monkey Mean ± S.D.	Marmoset Mean ± S.D.	Squirrel Monkey Mean ± S.D.	Chimpanzee Mean ± S.D.	Units
Glucose	2.664 ± 1.277	10.60 ± 3.663	4.940 ± 2.220	4.662 ± 1.388	mMol/L
BUN	8.211 ± .7140	8.211 ± .7140	10.35 ± 4.284	3.570 ± 1.071	mMol/L
Creatinine	133 ± 9	44 ± 18	71 ± 27	88 ± 27	μMol/L
Total Prot	68 ± 3	66 ± 7	67 ± 6	73 ± 6	g/L
Cholesterol	5.646 ± 1.709	4.222 ± 1.450	4.999 ± 1.450	5.076 ± 0.191	mMol/L
Tot Bilirub	3 ± 2	5 ± 3	9 ± 7	5 ± 3	μMol/L
Alk Phosph	163 ± 52	107 ± 67	439 ± 407	184 ± 133	U/L
Cholesterol	5.646 ± 1.709	4.222 ± 1.450	4.999 ± 1.450	5.076 ± 0.191	mMol/L
Calcium	–	2.48 ± 0.08	2.20 ± 0.25	2.33 ± 0.15	mMol/L
Phosphorus	1.16 ± 0.61	1.81 ± 0.94	1.65 ± 0.68	1.32 ± 0.39	mMol/L
Magnesium	–	0.782 ± 0.000	0.782 ± 0.000	0.745 ± 0.111	mMol/L
Sodium	148 ± 3	149 ± 6	148 ± 4	140 ± 4	mMol/L
Potassium	3.7 ± 0.6	4.1 ± 2.0	4.2 ± 0.9	4.0 ± 0.5	mMol/L
Chloride	111 ± 3	106 ± 8	115 ± 7	103 ± 4	mMol/L
Albumin	37 ± 2	41 ± 7	39 ± 5	42 ± 0	g/L
Globulin	31 ± 4	24 ± 5	28 ± 7	35 ± 6	g/L
Amylase	65.31 ± 14.9	64.75 ± 53.28	39.41 ± 49.21	6.845 ± 3.885	U/L
Triglycer	1.424 ± 0.497	2.667 ± 1.435	0.8249 ± 0.542	1.119 ± 0.485	mMol/L
ALT (SGOT)	26 ± 6	30 ± 42	112 ± 79	30 ± 12	U/L
AST (SGPT)	38 ± 12	130 ± 59	140 ± 75	23 ± 11	U/L
LDH	509 ± 136	511 ± 311	498 ± 554	398 ± 205	U/L
CPK	–	312 ± 171	838 ± 1130	229 ± 215	U/L
GGT	32 ± 7	6 ± 5	23 ± 24	29 ± 11	U/L

Triglycer	Triglyceride
ALT (SGOT)	Alanine Aminotransferase
AST (SGPT)	Aspartate Aminotransferase
LDH	Lactic Dehydrogenase
CPK	Creatine Phosphokinase
GGT	Gamma Glutamyltransferase

By kind permission of the International Species Information System.

Fasting levels of glucose are significantly higher in the marmoset, as well as in the gibbon (*Hylobates spp*), especially in the latter above 20 years of age (personal observations), than in man or other species of nonhuman primates.

The triglycerides tend to be significantly lower in the squirrel monkey and gibbon, and higher in the marmoset, than in man, but within normal human range in most other species of nonhuman primates.

Alkaline phosphatase is significantly higher in most species of nonhuman primates, especially in the squirrel monkey, than it is in man. It is at its normal highest in all species in the young and during the latter half of pregnancy, returning to baseline levels in the dam within 24 hours after placental separation reflecting rapid juvenile and fetal bone growth.

The aminotransferase enzymes, alanine (ALT/SGPT) and aspartate (AST/SGOT), tend to have significantly broader "normal" ranges in nonhuman primates than in man.

Creatine phosphokinase (CPK) and lactic dehydrogenase (LDH) also tend to be significantly higher in nonhuman primates than in man.

Triglycerides tend to be significantly lower in the squirrel monkey and gibbon.

Urinalysis

The importance of urinalysis in the direct and differential diagnosis of disease in nonhuman primates should not be overlooked. Urine should be tested every six months, and no less than annually. The sample is centrifuged and the supernatant observed for color and clarity, measured spectrometrically for specific gravity, and then tested by chemical paper strip for chemical characteristics (pH, glucose, RBCs and WBCs, hemoglobin, ketones, protein and urobilinogen). The sediment is examined microscopically for the presence of RBCs and WBCs, epithelial cells, spermatozoa, bacteria and fungal threads, the various types of casts (e.g., fatty, granular, hyaline and waxy), crystals and mucus.

Urine for bacteriological examination must be obtained by catheterization of the urethra under sedation, using sterile technique. It should be remembered that urethral catheterization carries a risk of bacteremia developing, and prophylactic systemic treatment with a long-acting penicillin is strongly recommended.

Except for singly caged animals, obtaining freshly voided urine samples for analysis from identifiable individual primates presents problems in most husbandry systems, especially if the animals are housed in social groups in large areas. A common method for obtaining urine and other biologic materials is to remove primates from their home cages and place them in metabolic cages over night. These cages usually have solid metal sides, at least in the lower sections of the walls, and a funnel device beneath the barred floor that allows voided urine to collect in a container placed underneath. In more traditional laboratory settings, metabolic cages are usually contained in special areas, away from immediate contact with the home cage or general housing area, often in otherwise empty rooms. This degree of separation is nothing short of extreme isolation, a stressful and terrifying experience for most primates. Innovative cage design can alleviate this problem. Cages permitting single animal occupancy can be incorporated into social-housing systems, for example, and small species such as marmosets and tamarins can be confined in their familiar sleep boxes over night for urine collection.

Parasite testing and control

In well-established indoor colonies, testing for enteric parasites should be performed in staggered fashion at least once annually in each animal. During quarantine and in the first one to two years following quarantine, fecal tests should be performed more frequently (please see the section on quarantine below). Fecal testing of individual animals on a frequent, routine basis in large outdoor colonies, especially if maintained on earth or grass, may be particularly difficult. Reliance may have to be made on frequent random sampling from unidentified animals. On finding reliable evidence of particular species of parasites, one would assume that infection is widespread and the entire troop should be treated. However, much is to be said for routine anthelmintic treatment, carried out at three to four month intervals, regardless of whether parasitic infection is demonstrated.

Fecal samples must be obtained fresh from the cage substratum or, if possible, directly from the rectum. Protozoal parasites, such as *Balantidium coli, Entemeba hystolitica, Giardia* and unidentified flagellates, are best seen by microscopical examination of wet smears in saline solution. Although wet smears are the simplest and fastest method of identifying ova, larvae and adult forms of metazoan parasites, they are the least reliable; a negative result cannot necessarily be taken as negative. The more complicated and time consuming flotation and, better still, fecal concentration techniques, using zinc sulfate or sodium nitrate solution, are the preferred methods of testing.

With appropriate anthelmintic treatment, most species of nematodes, cestodes and trematodes in primates housed under indoor conditions, or on concrete in outdoor areas, can be brought under control and finally eradicated. Certain types of enteric parasites, however, are difficult and even impossible to eradicate from a colony. The protozoan, *Balantidium coli* and the nematode, *Enterobius vermicularis* (pinworm), which resist high ambient temperatures and desiccation in their encysted larval or egg forms, are examples. As in the human being (Neva, 1986), Strongyloidiasis (threadworm disease), caused by *Strongyloides spp,* is particularly difficult to eradicate in nonhuman primates. This is in part because of the asexual, parthenogenic form of reproduction that occurs deep within the intestinal wall of the host, but more because of the autoinfectious, as opposed to the indirect fecal-soil-skin penetration route of reinfection. In rhesus monkeys maintained under semi-free conditions in the tropics, Strongyloidiasis is said to have no significant clinical effects. (Kessler *et al.,* 1984). Personal experience with chimpanzees in indoor housing, however, suggests that Strongyloidiasis, persisting for several years, can lead to intestinal abscess formation, recurring and progressively worsening bouts of localized peritonitis, intestinal obstruction, and in some cases, fatal septicemia.

Acanthocephaliasis (*Prosthenorchis elegans*, or Thorny-headed worm), a particularly devastating disease of marmosets and tamarins, depends on the cockroach as an intermediary host. Control of this insect under laboratory conditions may be difficult in some locations.

Pulmonary acariasis or lung mite disease (caused by the arthropod, *Pneumonyssus simicola*), is common in wild-born macaques. Although it may produce chronic coughing, its presence is often not realized until X-ray examination of the lung fields for routine TB surveillance or at necropsy. Its route of transmission is not well understood, and it is probably untreatable.

Tuberculosis

Tuberculosis is not the scourge it once was for primate colonies. Vast improvements since the late 1960s and early 1970s in capture techniques, pre-shipment health screening, air transport of wild-caught primates from their countries of origin, and a greater reliance on domestic bred primates for research, reared under strict conditions of medical control, have diminished the potential for epidemics. Yet the importance of maintaining a vigilant eye on the disease cannot be over-emphasized.

Tuberculosis in different primate species

Marmosets and tamarins are regarded as highly resistant to tuberculosis and most colonies elect not to perform screening tests on them (Hearn, 1995). The larger New World species, such as the squirrel, capuchin and spider monkey, however, are considered susceptible and routine testing is mandatory. Rhesus monkeys are highly susceptible to the disease. The interval between acquiring the bacillus and death in this species can be as short as six weeks. In confined, indoor housing, the disease spreads extremely rapidly. This is because primary lung lesions do not become encapsulated, as they do in man and chimpanzees, for example, and the coughing of affected animals readily sheds the organism into the air. Wild-caught cynomolgus or crab-eating macaques were considered less likely to be infected with, and more resistant to, tuberculosis than feral rhesus monkeys, making them the larger primate of choice in Europe in the 1960s–70s.

The development of disease in baboons is somewhat slower than in the rhesus macaque. Infection of wild troops with bovine TB has been described (Sapolsky & Else, 1987). Tuberculosis develops slowly in chimpanzees, as it does in man, because of the encapsulation of thoracic (mediastinal) and other lymph nodes. In the absence of a strict policy of frequent periodic testing, nonetheless, the disease may spread widely in this species before its presence is recognized.

Tuberculin testing

The test for tuberculosis in primates depends on a delayed cellular hypersensitivity response to killed *Mycobacterium tuberculosis* bacilli in the tuberculin test material.

There is no universal agreement on which type of tuberculin is best suited for use in nonhuman primates. Mammalian or Old tuberculin (human isolates) is the type mandated in the United States; Purified Protein Derivative (P.P.D.) is the type favored in Europe. In either case, 0.1 ml of tuberculin is used for injection via a syringe needle preferably of 29 g, and certainly no larger than 27 g size, no matter the size or species of the individual, or the route of administration.

In stable colony animals, TB tests should be performed every three months, and never less than every six months. Early detection of disease is increased by testing animals throughout the colony on a staggered timetable, rather than all at one time.

Tests are usually performed intrapalpebrally (IP), in the upper eyelid, except in smaller species, or in infants (<one-year old) of larger species. It should be noted that this is not an intradermal site, but rather an *inter*dermal location (i.e., between the outer and inner dermal layers of the eyelid). Intradermal (ID) inoculation is reserved more for small species, or for young juveniles or individuals known to have marked nonspecific or traumatic reactions if the eyelid is used, making interpretation of results difficult. The abdominal skin, midline below the xiphisternum, is the most

convenient ID site for observation. More rarely, ID testing can be performed on the inner aspect of the forearm, which is especially suited to trained animals, such as chimpanzees, so avoiding the need for chemical restraint.

Animals should be observed for IP reactions at very close quarters, at a distance not exceeding 3 meters or so. Needless to say, interpreting eyelid tests long distance through binoculars in primates maintained in large outdoor corrals or indoor pens is not ideal. A drawback with ID tests is that interpretation of results depends on gentle palpation and rolling of the skin at the injection site between thumb and finger, in order to assess any degree of thickening and induration. This is not only potentially dangerous for the examiner, but more than a momentary manipulation of the skin has a tendency to cause swelling and hyperemia of the injection site. This may make interpretation all but impossible.

IP test reactions are scored on a 6-point scale based on the appearance and extent of swelling and/or erythema, ranging from 0 to 5 (Hall *et al.*, unpublished data) (Table 16.3). In contrast, ID reactions are based on the degree of induration and erythema surrounding the injection site. A zone of induration exceeding 1.0 cm in diameter is taken as a highly suspicious or positive reaction.

It is important to read tests daily for three consecutive days. A reaction that increases markedly in intensity over the 72-hour period is cause for concern (e.g., from a 3^+ to a 3^{+++}). Two individuals should be responsible for reading test results; the caregiver familiar with the individual animal and the attending veterinarian. Tuberculosis is far too important a disease to leave the responsibility for its detection on the shoulders of one person alone.

Not only do false negative and false positive results occur, but, at best, interpretation of results is subjective, rather than objective. Difficulty in interpretation of results is particularly likely in recently acquired primates, especially if they originate from the wild, where personal familiarity with individual animals is lacking (please see the section on quarantine below).

False negative results may be obtained during the incubation period, before immune recognition of the antigen has developed, and in advanced stages of disease, in the presence of overwhelming levels of antigen, when anergy (immune exhaustion) comes into play.

False positive results can occur because of trauma (needle bruising, hemorrhage, infection of injection site by microbial skin contaminants, or pre-existing trauma to the eyelid); individual non-specific reactions (slight residual reaction at 72 hours), especially likely to cause the animal to rub and scratch the site; immunological cross-reaction with atypical mycobacterial species, which are common in soil and on plant bedding materials such as straw, and even cross-reaction with avian tuberculosis.

In the event of obtaining an indefinite or somewhat suspicious result (e.g., a score of 3^+ to 3^{+++} at 72 hours in a stable colony animal, or even a 3^+ in a quarantine situation), the course of action then taken will be governed to some extent by the species in question

TABLE 16.3: Interpretation standards for intra-palpebral tuberculin testing in nonhuman primates

Results of tests are graded at 24h, 48h, and 72h on a scale of 0 to 5, as follows:	
0	NO CHANGE observable.
$1^0 - 1^{+++}$	BRUISING (Hint, to slight, moderate or marked, due to needle trauma, self-inflicted injury, etc. – may last beyond 72 hrs – not part of the tuberculin reaction).
$2^0 - 2^{+++}$	ERYTHEMA (Hint, to slight, moderate or marked degrees of pink).
$3^0 - 3^{+++}$	SWELLING with/without erythema (Hint, to slight, moderate or marked. Eyelid not drooped).
4	SWELLING + ERYTHEMA (Swelling to degree producing drooping of eyelid and distinct asymmetry between eyes).
5	EYELID CLOSURE, ABSCESSATION, NECROSIS, etc.

Hall, A.S. (Oregon Regional Primate Research Center), Henrickson, R.V. and Anderson, J. (California Regional Primate Research Center) (Unpublished data, 1976) for use in the rhesus monkey. Adapted and modified by Mahoney, C.J., for use in macaques, baboons, gibbons, chimpanzees and gorillas.

(e.g., rhesus monkey versus chimpanzee) and history of the animal(s) in question, as well as the living conditions. At the very least, the animal(s) with a well-established history of negative results should be re-tested two weeks later, using the opposite eyelid as well as an abdominal skin site for comparison. If such animals exhibit an indefinite or suspicious reaction a second time, they should be immediately isolated, and re-tested with serial dilutions of tuberculin (e.g., a 1:1, 1:2, and 1:3 dilution in saline) ID (not IP). Chest X-rays, in both lateral and antero-posterior planes, should be taken for comparison with previously obtained radiographs (e.g., during initial quarantine). It would be wise to re-test all other animals in contact or sharing the same common airspace, and place them under temporary quarantine. Serious consideration should be given at this time to obtaining stomach and tracheal/ bronchial washings for culture from the suspect animals.

A clear-cut numerical test score of 4 or 5, where the eyelid droops to the point of near closure or is abscessed at any stage during the 72-hour test period in stable colony or quarantine animals, is cause for serious concern.

In the event of obtaining highly suspicious results during the first or second test, animals should be isolated immediately for further rigorous evaluation or culled by performing euthanasia. All in-contact animals should be radiographed immediately and evaluated for hilar (mediastinal) lymph node enlargement, and re-tested within two weeks, even if their X-ray findings are negative.

Reactive animals that have been released from quarantine in the recent past should be separated immediately from other animals in the same room or common area and placed in isolation. Other primates that shared the same quarantine period with TB-suspect animals should be immediately identified and re-tested to be sure that infection may not have already spread. It should be noted that, in rare instances, TB may appear in rhesus monkeys even after a 3-month, unremarkable quarantine period. A chest X-ray should be obtained and evaluated for comparison with radiographs taken in quarantine (please see section on quarantine).

Treatment of tuberculosis in nonhuman primates

In all but chimpanzees and other endangered species of apes, including gibbons and siamangs, nonhuman primates known or strongly suspected to be infected with tuberculosis should be euthanized without delay. This is to limit or arrest the spread of disease. Necropsy examination of subjects must be carried out with extreme caution, not only in respect to using sterile technique, but also in the observance of strict biohazard control for human safety. Under no circumstances can dispersal of mycobacteria be allowed through environmental control systems, such as positive-negative directional airflow systems, air conditioning and humidifier systems, or by prevailing wind.

Treatment of tuberculosis, even in great apes, must be undertaken with the strictest precautions and the realization of the enormity of the consequences if the disease spreads to other animals or personnel.

In known-infected animals, a triple therapeutic treatment, lasting 9–12 months, consists of the oral antibiotics, isoniazid (10 mg/Kg SID) and rifampin (10 mg/Kg SID), and prophylactic use of pyridoxine (vitamin B_6) (50 mg SID) to lessen the risk of toxic neuritis and liver damage occurring. Liver function should be monitored on a periodic basis (serum ALT, AST and GGT), and chest X-rays obtained at the outset and end of treatment for evidence of hilar lymphadenopathy. In tuberculin reactive subjects with no evidence of tubercle lesions, treatment can be limited to isoniazid/ pyridoxine, but the same liver and lung monitoring must be followed.

Record keeping and data retrieval/analysis

Cumbersome systems of record keeping, failure to review past records of animals prior to planned work, reliance on memory (although there is an important place for this, as part of the historical continuity of the colony), are all apt to lead to serious oversights.

Open to prominent display should be information vital to individual animals, as in posting notices as reminders that a particular animal requires a muscle relaxant, such as acepromazine or diazepam, before being given an anesthetic agent such as ketamine hydrochloride.

All data relating to health should be stored in individual animal files. The importance of keeping records is so that current information can be immediately compared with easily retrieved past information. Systems of recording and storing data that do not permit easy review should be avoided. For example, body weights are best presented in graphic form, or listed numerically in vertical display so that increases and decreases in weight over time are striking to the eye. In contrast, such information as hematological data or blood chemistry results are best logged in a horizontal format across the page, so that the eye can slip easily from one value to the next for each given profile.

Management of quarantine and isolation

Quarantine procedures

Primate importers commonly employ a 31- or 42-day quarantine period before delivery of wild-caught animals to the buyer. However, it should be clearly understood that such a short term covers only the incubation period for most of the bacterial and viral respiratory and enteric diseases. It most certainly does not cover tuberculosis in any species, nor such diseases as hepatitis B in chimpanzees, for example, both of which may have up to a six-month incubation. Research facilities in particular often prefer to be responsible for the full quarantine care and surveillance of newly arrived animals, especially if they originate from the wild. National, state and local law may require facilities to be officially registered as quarantine sites, however.

It is recommended that quarantine isolation be no less than three months for any species of nonhuman primate. For some species, such as marmosets and tamarins, a strong case can be made for extending the quarantine period to nine months, or even longer. Newly arrived animals may introduce to the resident colony latent viruses that they might be carrying, but to which they, themselves, show no ill effect. Conversely, marmosets and tamarins recently released from quarantine may fall prey to pathogens resident in the stable colony, such as viruses that might be implicated in the disease known as wasting syndrome, or bacteria such as *Campylobacter spp.,* a common cause of enteritis. There is evidence that *Shigella flexneri* is not carried by wild-caught rhesus monkeys, but is contracted by them only when they enter an infected stable colony (Mahoney, personal observations).

Tuberculin tests, blood sampling for hematology and serum chemistries, urinalysis and fecal examination for parasites, with appropriate anthelmintic treatment, should be performed no less than twice during a preliminary 31-day quarantine (at the beginning and end). During a full three-month quarantine, tuberculin tests and blood sampling should be carried out at two-week intervals, although the frequency can be reduced to monthly intervals if quarantine is extended to six months. An entry and exit physical examination, to include chest X-ray (lateral and antero-posterior views) for tuberculosis monitoring, fecal parasite and bacteriological examination for *Shigella*, *Salmonella*, *Klebsiela* and *Campylobacter spp.,* and urinalysis must be performed before animals can be released from quarantine to enter the stable colony.

In the event of infectious disease breaking out in one or more newly arrived primates (such as respiratory or enteric infection), the quarantine period must begin again and be extended for *all* animals in the group after final disappearance of symptoms.

Serum must be obtained at the beginning and end of the quarantine period for serological testing for pathogens, especially viruses likely to be found in the particular primate species (e.g., B-virus, simian respiratory viruses, filoviruses, etc., in macaques; HIV and cytomegalovirus, the hepatitis viruses A, B and C, and HIV, etc., in chimpanzees).

An aliquot of serum, obtained at the beginning and end of quarantine, should also be banked in the frozen state for future reference in the event of disease occurring.

Blood typing

In those species in which blood typing and transfusion techniques are well established, such as in macaques, baboons, gibbons and chimpanzees, all newly acquired animals should be blood typed during quarantine.

Nonhuman primates carry both human and simian type blood groups. Unlike the simian blood groups, however, the ABO and other human-type blood groups in monkeys are not expressed on the surface of the red blood cells, as they are in the anthropoid apes, but in the saliva. In order to fully type blood groups in monkeys, therefore, fresh saliva, as well as serum and whole blood must be tested (Socha and Rufié, 1983). (Please see section below on blood transfusion).

Vaccination procedures

Consideration should be given to vaccination of primates against measles, especially if they are maintained in indoor housing, in close contact with personnel.

As protection against tetanus and rabies there is a strong case for using killed virus preparations for vaccination of primates who have access to the outdoors in endemic areas. Although there is no hard evidence, there is a general feeling among primate experts that combined vaccine preparations against diphtheria, pertussis and tetanus should not be given as they may produce serious and even fatal side-effects.

Psychological stress and disease in newly arrived primates

Psychological stress can occur in primates under any circumstances in the laboratory/captive setting. Procedures to ensure humane care and psychological and environmental enrichment are designed to address this important issue.

The psychological stress that primates, newly arrived in a laboratory, may suffer is of a particular type, however. Primates obtained from the wild, and even those born and reared under semi-free captive conditions, such as islands, corrals, or even large indoor or indoor-outdoor pens, may have little or no experience of human beings at close range.

One cannot help be struck by the utter silence on entering a quarantine room housing new arrivals, the characteristic chatter among the animals strangely absent. Not only must caregivers be aware of this stressful situation for the animals, but they must also realize that detecting early signs of ill-health in them will be that much more difficult because of the animals' withdrawn behavior.

Personnel health monitoring and surveillance policies

The successful medical care of laboratory primates is dependent on institutional programs of employee health monitoring and disease surveillance: they are inextricably related and one cannot be achieved without the other.

Pre-employment and employee health policies

There is no universally accepted policy for the hiring of new employees to work with nonhuman primates. Rules that apply to employment seekers may differ markedly from the regulations governing employee health policies. In the United Kingdom, for example, the standard policy in former years was not to hire new recruits to work with primates unless they had a tuberculin-positive skin test with negative chest X-ray findings, or had been inoculated with BCG vaccine. Negative reactors were deemed to be at too high a risk of contracting tuberculosis from infected wild-caught monkeys being brought into laboratories in those days. In the United States, on the other hand, candidates were not hired if they demonstrated a positive tuberculin skin test, regardless of whether they had a negative chest X-ray; this policy still stands.

Yet, procedures followed for employees having had negative tuberculin tests in the past but now showing a positive skin test are removed from contact with primates and investigated further for evidence of active tuberculosis (i.e., radiographic evidence and follow up of personal and family medical histories by appropriate medical authorities).

Personnel must be tested for tuberculosis semi-annually. Those having converted or who have been vaccinated with BCG (which has not been proved to protect against TB) should receive chest X-rays every two years or so, according to the standard practice of regional medical authorities.

While candidates must provide proof of sero-negativity to hepatitis-B antigen, as a precondition for employment, they cannot be questioned about their HIV infectious status.

All employees dealing with nonhuman primates should be vaccinated against hepatitis A and B (especially if working with chimpanzees or other Great Apes), measles, mumps and rubella. Because of the risk of being bitten or scratched, they should also be vaccinated against rabies and tetanus, especially if dealing with primates maintained in outdoor conditions.

Pre-employment aliquots of serum should be collected and stored at –20°C for future reference from all employees working directly with nonhuman primates or who handle their biological materials. Ideally, serum aliquots should be obtained and stored annually thereafter.

Biosafety and prevention of zoonoses

Prevention of zoonotic infections being contracted by personnel depends on rigorous personal hygiene and strict programs of vaccination protection.

Frequent hand washing when dealing with nonhuman primates and their biologics is perhaps the single most important personal precaution that can be taken, even when disposable gloves and other protective garments are worn. Gloves must be changed frequently during the course of the working day. Next in line of defense is facial protection in the form of disposable surgeon's masks. These, too, must be changed frequently throughout the working period, and at all costs, personnel must avoid touching their facemasks with gloved hands (as in response to a facial itch) which may result in direct contamination of the nose and lips.

Effectively designed, transparent face shields prevent splashing and aerosol contamination of the eyes, especially important as protection against the rare but almost invariably fatal contraction of *Herpesvirus simiae* from macaques.

Head and foot covering, in the form of disposable surgeon's hats and booties, as well as a long-sleeved top and trousers which are laundered daily on site, complete the garb for dealing with non-infectious animals. Personnel must not be allowed to leave animal areas without discarding worn, potentially soiled clothing or donning appropriate fresh outer clothing.

Biosafety level-II conditions, enforced in the study of non-airborne spread of infectious agents (B-virus, HIV and SIV, the human and simian immunodeficiency viruses), require dressing in entire body covering, composed of jump suit, hood and the other components described above.

Biosafety level-III, required for dealing with airborne infectious agents, deals also with environmental control, such as hepafilter systems.

First aid and critical care

Fluid and electrolyte replacement therapy

Rehydration and electrolyte replacement therapy are often the single most important requirements of first aid and critical care, not to mention emergency treatment in primates. Acute, life-threatening dehydration and electrolyte imbalance occur from a wide variety of pathologic states, including persistent vomiting and/or diarrhea, hemorrhage, traumatic and septic shock, hypoglycemia, diabetic coma, hyperthermia, and renal, respiratory, pancreatic and liver disease.

Treatment may be by oral (*per os* or PO), subcutaneous (SC), intraperitoneal (IP) or intravenous (IV) routes of administration. Oral solutions, used for human infants or athletes, can be offered the conscious primate to voluntarily lap or can be administered, with caution, to the unconscious or semi-conscious subject by naso-gastric or oro-gastric tube. The most commonly used parenteral solutions are dextrose (final concentration ranging from 2.5% to 5.0%, either in water or 0.9% saline), Ringer's lactate, or some combination of the two.

Because of the difficulty or inability to monitor fluctuations in fluid/electrolyte and acid/ base balance in the tissues, blood rehydration therapy in primates is usually a hit or miss affair, at best, and based more on experience than calculation. This shortcoming is compounded by the need, for personnel safety reasons, to handle all but the smallest or youngest primates under sedation or anesthesia. In order to minimize the period of chemical restraint, fluid therapy must often be given in bolus form, by whatever route, rather than by the preferable slow administration over a prolonged period of time, as is the standard method of treatment in human patients.

Nonetheless, certain guidelines for fluid therapy can be followed. Oral solutions must never be given parenterally, not only out of consideration for sterility but also because of their inappropriate chemical composition; SC fluids should not exceed a 2.5% final concentration of dextrose in order to maintain physiological osmolarity; in the event that IV administration cannot be readily achieved, because of difficulty in "raising" a vein in hypotensive states, consideration should be given, but, initially, to PO or SC therapy to avoid unnecessary loss of critical time or added stress to the patient, followed later by renewed attempts at IV administration, if deemed feasible; SC fluid therapy should consist of multiple boluses, not to exceed 5 ml each, in very small species, or 100 ml in large animals such as adult chimpanzees; IV fluids can be given at an approximate ratio of 15–30 ml/kg over a period of 30–45 min in small species (e.g., a 500-gm adult marmoset), or 1 to 2 hours in larger species, such as an adult chimpanzee weighing 50 kg; IP administration of IV fluids should be considered if large volumes cannot be given IV (again, not to exceed 2.5% final concentration if dextrose is added). Caution must be used to assure that injection is intraperitoneal and not SC or submuscular), and that perforation of the intestine is avoided.

Shock treatment

The term "shock" covers a multitude of physiological states that are poorly defined and not well understood (such as traumatic, hemorrhagic, hypovolemic, surgical and anaphylactic shock, etc.).

Besides oral or parenteral fluids, one of the main weapons in the arsenal of shock treatments is the family of corticosteroids (Rosenberg and Kesel, 1995). Among a wide range of physiologic actions, they improve capillary blood flow in damaged tissues, and promote mobilization of liver glycogen stores. Because of their anti-inflammatory effects, they must be used with caution, and covered by an appropriate antibiotic if any risk of infection exists (opportunistic or otherwise). The most commonly used agents are methylprednisolone, at a dosage of 1–1.5 mg/kg (12–36 hours' activity) and dexamethasone, at 0.5–2.0 mg/kg IM (36–72 hours' activity), dosage depending on the severity of the condition and the duration of anticipated short-term use. A step-down dosage regimen must be employed when terminating long-term or high-dose steroid treatment, to avoid drug withdrawal complications.

Blood transfusion

The clinical need for blood transfusion in any species of mammal is governed by the degree and rapidity of onset of anemia or outright blood loss that has occurred. In acute hemorrhage in primates, such as that following infliction of severe bite wounds, surgical accident or obstetrical calamity, a sudden drop in hematocrit to around 22% may require emergency transfusion. In more chronic conditions, however, which have developed over a period of several weeks or months, such as in chronic parasitism, the body can adjust hemodynamically to hematocrits as low as 17% or 16% without immediate threat to life. It must be borne in mind, however, that the animal's true packed cell volume may be masked by any state of shock. Circulating blood may be lacking in areas such as the alimentary tract, giving the impression of hypovolemia, or the hematocrit may be falsely elevated because of dehydration or vasoconstriction in the intestine, skin or other tissues.

All species of nonhuman primates demonstrate a variety of blood groups of both the human type (e.g., A-B-O, M-N and Rh systems) and simian type (e.g., V-A-B, C-E-F systems) (Erskin and Socha, 1978; Socha and Rufié, 1983; Socha *et al.*, 1995; Blanché *et al.*, 1997). To avoid transfusion reactions due to the possible presence of antibodies in the recipient and donor's sera, blood for transfusion must be of the correct type and, except in the direst emergency, should be cross-matched with the recipient's blood beforehand. In monkeys, these antibodies are usually directed towards simian-type antigens. Unlike the anthropoid apes, however, Old World monkeys do not have A or B antigens on their red cell surfaces, but only in the saliva and various tissues (e.g., endothelial tissue). They do have potent anti-A and/or anti-B agglutinins in their sera, however (an important consideration in organ/tissue transplantation research).

When possible, washed erythrocytes, resuspended in physiologic saline, should be used in preference to whole blood, especially if less than an ideal match exists between the donor and recipient. This reduces the level of any hemolyzing antibodies that might be present in either the donor or recipient's blood. In anticipation of the need for multiple transfusions, consideration should be given to performing the transfusions close together, to reduce the risk of the recipient's developing hemolyzing antibodies against the donor red cells.

Needless to say, extreme caution must be exercised to avoid inadvertent transmission of infectious agents from donor to recipient, such as experimentally acquired viruses (e.g., simian immunodeficiency virus in monkeys, or hepatitis and AIDS viruses in chimpanzees) or spontaneous disease such as SRV, herpes-B and CMV.

Body temperature control

Animals requiring first aid and critical care, because of the nature of their condition, are often at serious risk of losing crucial body temperature. Because of the relatively larger ratio of body surface area to body weight, this is especially likely in small species or neonates and infants (a mouse loses body temperature at a much faster rate than an elephant!).

Although electrical heating pads may be used with the greatest caution, they should preferably not used at all, because of the high risk of causing burns. Water circulating pads are preferable. A continuous cycle of warming towels in a clothes drier is an effective method of providing safe heat to weakened newborn infants. Immersion of easily handled infants and small species up to the neck in constantly replenished warm water is another effective method.

Wound and bone repair

Depending on the extent and severity of wound damage or bone fracture, it is often advisable to attend to general aspects of first aid to physiologically stabilize the patient (i.e., by attending to fluid needs, treatment of shock, maintenance of body temperature and prevention of infection) before attempting immediate, extensive repair of tissues or bone. This is not so, of course, in the case of uncontrolled or heavy hemorrhage, which must be attended to rapidly.

Nonhuman primates rarely tolerate plaster casts. As in other types of animals, if casts are unavoidably necessary, the cast material must always be extended to the limit, or beyond the distal end of the affected appendage, in order to avoid constriction of terminal blood flow and serious tissue swelling (with the risk of gangrene setting in).

Intramedullary pinning of fractured bones is the preferred method of approach. Even in small species, such as pygmy marmosets, high-grade stainless steel trocars, from appropriate-sized intraspinal catheters, can serve as effective intramedullary pins. Wiring and bone plates can also be used in certain cases. In spite of all efforts to employ medical solutions to wound repair and bone setting, primates often repeatedly remove dressings and other support, yet the lesions heal with surprising perfection by secondary intention.

Emergency animal care

Definition of an emergency

For the purpose of this discussion, a medical emergency is defined as any condition which, if not eliminated or alleviated within a time period ranging from, perhaps, as short as a few seconds to a maximum of around 20 minutes, depending on the nature of the crisis, will lead to severe impairment, injury or death of an animal. In extreme cases, there may be little or no time to organize an immediate full-scale medical response to the emergency. Caregivers, finding themselves alone or in less than adequate numbers, may have to take initial steps on their own, until they achieve a certain degree of stabilization of the animal. Then, and maybe only then, will they be free to take the time necessary to call for additional help, including that of the veterinarian.

Under no circumstances must personnel expose themselves to physical danger when responding to an emergency situation. Except when dealing with small species, such as marmosets and tamarins, or infants of larger species, it must be determined that the animal in question is sufficiently unconscious or restrained (physically or chemically) before any attempt is made to open primary enclosure doors.

True medical emergencies are rare, but herein lies a commonly faced problem. Without frequent practice sessions, animal care personnel and attending veterinarians may find themselves inadequately organized as a team to respond rapidly to a crisis.

Well-stocked emergency drug and medical supply boxes must be readily available at key locations throughout the animal facility. These should contain only the materials and drugs necessary to get personnel through the first critical moments of a crisis, and should not contain "the cupboard and the kitchen sink"!

Table 16.4 lists some of the most important items of an Emergency Medical Box.

As with human emergencies, such as in vehicular crashes, one of the most important conditions to ensure is that the nonhuman primate has free airflow to the lungs. This may require no more than straighteing out a kink in the neck, or rolling the animal over, either by hand or with the aid of a rod or stick-like device, from a dorsal supine position into ventral or lateral recumbency, so that it can begin to breath spontaneously. This one simple action may save the animal's life and give the technician/caregiver time needed to secure help.

Anesthetic/surgical crises

With adequate pre-anesthetic and pre-surgical planning, pre-arranged assignment of tasks to specifically dedicated personnel, and constant, close observation of the patient throughout any surgical procedure, many *potential* problems will be eliminated before they even arise. The patient's responses to anesthesia, such as heart rate and ECG, respiratory rate and depth, blood oxygenation and pCO_2 levels, intravenous drip rates, body temperature and the like, must be constantly monitored and adjusted accordingly. Before any surgical procedure with a known or suspected high risk of uncontrollable hemorrhage is undertaken, adequate amounts of pre-typed and cross-matched blood, or blood volume expanders, must be available *at close hand.*

The greatest risk in surgery and anesthesia, however, is most likely to occur during an emergency procedure. Not only may the animal patient already be in a precarious physiologic state, but also an adequate number of specifically trained staff may not be available, at short notice, to perform the various tasks required. The surgeon(s) must then assume the prominent role of constant oversight, distracting him/ her from the central task. Before undertaking

TABLE 16.4: Emergency medical requirements for laboratory primates

Anesthetics/Tranquilizing Agents		
	Ketamine/Telezol	
	Atropine (Salivation control/Cardiac stimulant)	
Emergency Drugs		
	Epinephrine (Cardiac Stimulant)	
	Lidocaine	
	Respiratory stimulant (e.g., Dopram)	
	Local Anesthetic Spray (for laryngeal block)	
Preparation Materials		
	Razors and/or clippers	
	Sterile & Non-sterile gauze	
	Alcohol wipes *or* 70% alcohol solution	
IV Setup	*IV/SC/IP Fluids*:	Saline (0.9%), 5% dextrose saline or Ringer's lactate (250 ml–1000 ml bags).*
	IV Lines:	Microdrip (60 drops per ml)
		Macrodrip (15 drops per ml)*
	IV Catheters:	Indwell catheters (25 g–18 g)*
		Butterfly catheters (25 g–18 g)*
Syringes and Hypodermic Needles		
	Syringes (Luer lock):	(1 ml–60 ml)*
	Needles:	(25 g–16 g)*
Sundry Supplies and Materials		
	Adhesive tape to fix I.V. lines/catheters	
	Tongue depressors/ Cotton-tipped Applicators	
	Sharps Container (for sharp instruments/supplies disposal)	
	Surgical jelly	
Instruments & Sundry Equipment		
	Stethoscope	
	Laryngoscope:	Variable sized blades*
	Endotracheal Tubes:	Variable sizes*
	Mouth gag	
	Amboy bag (Insuflator bag)	
	Torch/Penlight	
	Bandage Scissors	
	Scalpel Blades	
	Simple suture pack (e.g., for hemorrhage control)	
Paper Work		
	Physical Examination Data Sheets/Anesthetic Log, etc.	

* Depending on species/size of animal.

emergency surgery, careful consideration must be given to whether it might be better to only stabilize the animal initially, and postpone full surgical intervention until later.

Post-anesthesia/tranquilization and post-surgical crises

Unlike the days of barbiturate, morphine and phencyclidine use, rarely nowadays do nonhuman primates die as a *direct* consequence of anesthetic or tranquilizing agents. Ketamine HCl, Telazole® (tiletamine HCl/zolazepam HCl) and gas anesthetics, such as isoflurane, have a wide margin of safety. Yet, sudden unexpected death, unrelated to direct effects of the anesthetic agent, can occur during post-surgical recovery.

The commonest cause is obstruction to free tracheal airflow, resulting from compression of the nostrils or a kink of the neck, accumulation of saliva or mucus in the pharynx, or regurgitation of stomach contents into the mouth, pharynx or glottis. Gross obesity and compression of the chest by the mass of abdominal contents pressing against the diaphragm are other common causes of post-anesthetic compromise. Animals should be placed in ventral recumbency during post-surgical/post-anesthetic recovery, with limbs and neck stretched out. At all costs, they must not be allowed to crawl, semi-conscious, into corners of a cage, where their necks may become bent, chin compressed against chest, or roll onto their backs, a position which allows saliva and mucous secretions to accumulate in the back of the throat. Hypothermia may become a serious issue if the animal is slow to recover from anesthesia and regain muscle activity.

It is essential to provide continuous surveillance and close observation of animals during the post-surgical recovery phase; failure to do so amounts to nothing less than negligence.

Gastric torsion/dilatation (bloat); intestinal torsion (volvulus) and intussusception

One of the common causes of emergency in nonhuman primates is gastric torsion and dilatation, also known as bloat, which can occur in all species, but is especially common in macaques and baboons.

Early symptoms, such as hyperactivity, hypersensitivity, agitation, guarded gait and abdominal rigidity, are often not recognized. By the time the condition is discovered the patient may already be comatose, severely acidotic cyanotic and apneic.

The cause(s) of bloat is not well understood, although lack of exercise, feeding only once, rather than divided times daily, and sudden changes in diet, are predisposing factors. Emergency treatment includes relieving gastric gas distension by large-bore stomach tube, anti-shock (corticosteroid) therapy, large doses of broad spectrum antibiotics aimed particularly at Clostridia and other anaerobes, and parenteral fluid therapy (please see section above on first aid and critical care).

Volvulus and intussusception of the intestine are, similarly, causes of sudden death if not detected early. In these cases, emergency abdominal surgery, along with other quickly applied supportive measures, are required. In both gastric torsion and intestinal volvulus, death is due to interruption of the mesenteric blood supply, which results in a rapid production of bacterial endotoxins.

Hyperthermia/hypothermia

Hyperthermia due to sunstroke or failure of environmental control systems is a not-uncommon cause of emergencies and death in nonhuman primates. Unless absolutely essential, anesthetizing primates during hot weather conditions should be avoided and undertaken only if special precautions are in place to guard against overheating in direct sunlight or unventilated shade. The danger occurs during the post-tranquilization recovery phase, when brain-stem regulatory systems may be dysfunctional. Failure of thermostatic controls in indoor-housing heating systems is an ever-present risk in all animal facilities.

Vigorous attempts to reduce body temperature must be undertaken immediately. These include continuous showering of the animal's body with cold water, application of ice packs, especially to the spine, and anti-shock drug therapy (e.g., administration of corticosteroids). Intravenous fluid therapy should be instituted once other stabilizing procedures have been put in place. Rectal body temperatures should be monitored frequently (e.g., every 15 min).

Hypothermia is less a cause of emergency or sudden death in primates than hyperthermia, although it may contribute significantly to other types of illness. It is most likely to occur in abandoned neonates, or following Cesarean delivery. Frost bite of the tail and digits, and subsequent tetanus infection, are not uncommon causes of death in primates housed outdoors, especially if they come into contact with cold metal perches or cage wire.

Obstetrical and neonatal emergencies

Uterine hemorrhage

Placenta abruptio (premature separation of the placenta) and placenta praevia (implantation of the placenta low in the uterine segment or over the internal os of the cervix) account for approximately 2% of pregnancy-related emergencies in macaques (Mahoney *et al.,* 1979) and other species.

Second stage labor in nonhuman primates is usually not marked by significant blood loss. In fact, vaginal bleeding may not even be seen before the birth of the infant. Heavy vaginal hemorrhage, especially of sudden onset, is therefore an ominous sign and its cause must be immediately investigated. Not only may the life of the infant be threatened but also that of the dam.

Surgical intervention (hysterotomy), with careful haste, is the order of the day if there is to be any hope of saving not only the fetus but also the mother. The mother, if she survives the acute hemorrhage, will almost certainly require emergency blood transfusion with typed and cross-matched blood. Under no circumstances should labor be induced by giving IM boluses of oxytocin, because this risks causing uterine rupture and fatal crushing of the fetal head during the ensuing uterine spasm.

Dystocia due to fetal malpresentation/malposition

In macaques, baboons and chimpanzees, dystocia because of breech birth is the commonest category of fetal death. In many cases, however, the mother dies as well because of uterine atony, physical exhaustion and shock if her plight is not soon discovered and corrective measures applied (i.e., manual or surgical delivery of the dead fetus, intravenous hydration, anti-shock therapy). Breech dystocia is rarely a problem for mother or neonates in small, multitocus species like marmosets and tamarins.

Suffocation of the neonate

Suffocation of the newborn because of failure of the chorio-amniotic membranes to fully tear away from the infant's face occurs rarely. Two cases in chimpanzees, and one in a rhesus monkey birth, have been seen by the author over a 33-year period. Nonetheless, the condition requires immediate action. In one chimpanzee case, the mother, while holding the suffocating baby to breast, was stimulated to jump and prance by threatening her with a fire extinguisher. The membranes tore as a consequence (the infant was named "Shake," as a way of reminding staff what to do in any future such case).

Trauma/bite wounds

Rarely are bite wounds in primates the direct cause of emergencies, but the physical pounding the victim receives, especially in gang fights, by other members of a social group can cause acute physical and psychological stress. Group fights among rhesus monkeys and other macaques, and even among diminutive marmosets and tamarins, often result in multiple bite wounds, each of which might be insignificant, but collectively amount to severe injury. Extensive mincing and crushing of muscle tissue can occur through repeated, non-penetrating bite wounds in macaques or by wrist and feet pummeling in chimpanzees, leading to onset of fatal shock.

Concluding remarks

This chapter has sought to stress the importance of knowing animals as individuals, if their medical care and needs are to be adequately served. It stresses the practical, rather than the ideal textbook approach. Although mainly directed towards the laboratory primate, the author hopes that it may also serve those dealing with primates in sanctuary and zoological park settings.

Correspondence

Any correspondence should be directed to James Mahoney, New York University School of Medicine, Sanctuary Support Program, LLC, 11 Rea Court, Monroe, New York 10950, USA. Email: rocher@frontiernet.net

References

Blanché, A., Klein, J. and Socha, W.W. (1997). *Molecular Biology and Evolution of Blood Groups and MHC Antigens in Primates.* Springer, Berlin, Heidelberg and New York.

Erskine, A.G. and Socha, W.W. (1978). *The Principles and Practice of Blood Grouping,* second edition, pp 207–221. The C.V. Mosby Company, St. Louis.

Hearn, J.P. (1995). In Rollin, B.E. and Kesel, M.L. (eds) *The Experimental Animal in Biomedical Research*, Vol. II, Care, Husbandry and Well-Being, An Overview by Species, pp 483–494. CRC Press, Boca Raton.

Holmberg, C.A., Henrickson, R.V., Malaga, C., Schneider, R. and Gribble, D. (1982). *Vet. Path.* 19 (Suppl 7), 9–16.

Kessler, M.J., Yarborough, B., Rawlins, R.G. and Bernard, J. (1984). Intestinal parasites of the free-ranging Cayo Santiago rhesus monkeys (*Macaca mulatta*). *J. Med. Primatol.* 13, 57–66.

Lee, D.R. and Guhard, F. (2001). In Brent, L. (ed.) *The Care and Management of Captive Chimpanzees*, pp 83–117, Vol. 2 of the series "Special Topics in Primatology". Series ed: Wallace, J., Amer. Soc. Primatol. San Antonio, TX.

Mahoney, C.J., Eisele, S. and Capriolo, M. (1979). In Ruppenthal, G.C. (ed.) *Nursery Care of Nonhuman Primates,* pp 3–20. Plenum Press, New York.

Mahoney, C.J. (1995). In Rollin, B.E. and Kesel, M.L. (eds) *The Experimental Animal in Biomedical Research*, Vol. II, Care, Husbandry and Well-Being, An Overview by Species, pp 495–510. CRC Press, Boca Raton.

Neva, F.A. (1986). *J. Infect. Dis.* 153, 397–406.

Rosenberg. D.P. and Kesel, M.L. (1995). In Rollin, B.E. and Kesel, M.L. (eds) *The Experimental Animal in Biomedical Research,* Vol. II, Care, Husbandry and Well-Being, An Overview by Species, pp 457–482. CRC Press, Boca Raton.

Sapolsky, R.M. and Else, J.G. (1987). *J. Med. Primatol.* 16, 229–236.

Sesline, D.H., Schwart, L.W., Osburn, B.I., Thoeel, C.O., Terrell, T., Anderson, J.H. and Henrickson, R.V. (1975). *J. Amer. Vet. Assn.* 167, 639–645.

Socha, W.W., Blancher, A. and Moor-Jankowski, J. (1995). *J. Med. Primatol.* 24, 282–305.

Socha, W.W. and Rufié, J. (1983). *Blood Groups of Primates: Theory, Practice and Evolutionary Meaning.* Alan R. Liss, Inc., New York.

Teare, J.A. (2002). *Reference Ranges for Physiological Values in Captive Wildlife.* Electronic database: International Species Information System, Apple Valley, Minnesota, USA.

Chapter 17

Factors Affecting the Choice of Species

Heinz Weber
Dr. med. vet, Klingental 7, CH-4058 Basel, Switzerland

Introduction

Four main areas of research in primates

Research in primates can be divided into four main areas:

- Basic or fundamental research on biological characteristics of the species.
- Both fundamental and applied research of pathogenesis and course of disease as well as treatment that cannot be studied in human patients for ethical reasons.
- Testing of pharmaceutical compounds for unwanted (toxic) effects.
- Production and safety evaluation of vaccines.

Current usage

In Europe, up to 90% of primates used for experimental purposes are estimated to be in the latter two areas of experimental use of primates, namely in industrial drug research and in the development and safety testing of vaccines (Council of Europe, 2003). In the USA it appears that only about 40% are used for this purpose, probably due to the important contributions to many different areas of fundamental research performed at the many Regional Primate Centres (Dukelow, 1999).

A breakdown of species used is difficult to obtain because this is usually not requested for the annual reports of the national authorities. However, it seems that the majority of primates used in Europe are cynomolgus monkeys followed by rhesus monkeys, marmosets and squirrel monkeys as well as some baboons and tupaias. Chimpanzees have not been used for experimental purposes since 2000 (Animal Procedures Committee, 2002). In contrast, in the USA, rhesus monkeys account for about 60% of all primates used. This is probably based on the historical background in that they were abundantly available from India before the export ban in 1977. Squirrel monkeys are also used more than in Europe and research on chimpanzees is substantial, again probably due to Regional Primate Research Centres.

The Laboratory Primate

ISBN 0-1208-0261-9

Factors affecting choice

Ethics and licensing

Non-human primates hold a special position in the animal species used for biological and medical research, and this forces scientists to think carefully if they really want to embark on a project requiring a non-human primate. Ethical considerations have led to guidelines for the use of laboratory animals for scientific purposes. The use of primates, considered as highly developed with regards to cognitive functions, is subject to more scrutiny than other laboratory animals. An important factor, affecting the choice of species, is the ethical requirement that a species should be chosen with the lowest capacity of pain perception and suffering. Additionally, for a long time primates have been obtained from the wild, from various geographical regions, and only recently have they been bred in order to obtain animals with a known background.

Regulations for obtaining licences for experiments on vertebrate animals, generally, require that:

- the method applied be appropriate to obtain the results envisaged;
- the knowledge obtained is new and needed for the benefit of humans and animals;
- the goal cannot be achieved by a non-animal experiment;
- animals of lower development level, with regards to pain perception and distress susceptibility, should be used;
- pain and distress should be minimized;
- animals should be cared for and housed so that their basic needs are covered.

The close similarity of non-human primates to humans may, in many cases, justify their use for investigations of pathological processes that cannot be studied in patients, and for the development of medicines and vaccines. However, the advanced state of development of these species, their limited availability and high care requirements must lead to careful consideration of other models and other species first.

In a recently published report to the British Home Office (Animal Procedures Committee, 2002) on present and future use of primates, it is concluded that, although the choice of species should be governed by scientific considerations, it is affected by many other more practical considerations. In this chapter it is intended to discuss some of these factors that influence the choice of a specific non-human primate species.

Availability

Based on the references published by the University of Washington, DC, Primate Information Center, in the last printed volume of Current Primate References (1999) in a vast field of biological and medical research, approximately 6,600 citations relating to more than 150 species and subspecies (excluding fossile and putative primates) are quoted. The most frequently mentioned species are rhesus monkeys (*Macaca mulatta*), cynomolgus monkeys (*Macaca fascicularis*), pigtail macaques (*Macaca nemestrina*), vervets or Green monkeys (*Chlorocebus aethiops*), baboons (*Papio papio* and *Papio anubis*), common marmosets (*Callithrix jacchus*) and chimpanzees (*Pan troglodytes*). Of the prosimians, *Microcebus* sp. are mentioned most. Many of the citations, especially those on chimpanzees, cover ethological and ecological field studies and publications on nature conservation rather than biomedical experimentation. Chimpanzees are listed as endangered species, and their capture from the wild is unacceptable (Buhjinsky, 1992).

The number of citations does not reflect the actual number of animals required for experimental purposes. In the USA it has been estimated that about 57,000 primates have been used in 2000 and 13,000 in 1999 by NIH grantees (Robinson, 2002). In the member states of the Council of Europe, the figures have remained around 8,000 to 9,000 used in experiments annually in recent years (Animal Procedures Committee, 2002; Council of Europe, 2003; European Commission, 2002; Hunsmann, 2003). In the USA and Europe, primates make up for 0.5 and 0.1%, respectively, of all laboratory animals used. In South Africa about 120 vervets and 210 baboons are used annually (Seier, 2003). The interpretation of the figures is not easy, due to variable data collection in the individual countries. They do not necessarily cover actual needs of primate numbers because, in some countries, animals kept over years for repeated non-terminal tests are counted every year as "animals in experimentation". A breakdown into species is not available from all countries. According to a survey among European pharmaceutical companies (Weber, 1997) and a report of the British Animal Procedures Committee (2002), 60 to 70% are macaques, chiefly cynomolgus, fewer rhesus, 30% marmosets, 10% squirrel monkeys and some other species. Baboons, at present, are only used in France, Germany and Spain.

Purpose-bred animals v. wild-caught

In European countries only purpose-bred animals are permitted to be used in experiments, unless sound justification is presented for the use of wild-caught animals. Whether island-bred animals can be considered as "purpose-bred" is debatable, but if they are, animals from the Caribbean islands would also be acceptable because it may become difficult to draw a line between Mauritius and Barbados and smaller islands like Kayo Santiago. In the USA, breeding establishments have been sponsored to produce SPF animals and to supply primate species whose export from countries of origin has been banned.

From the ethical and scientific point of view, the use of purpose-bred animals is recommended. Only these can provide animals of known age, life history and health status. However, the time required to breed colonies to meet increased demands for groups of defined animals, within narrow age ranges, is not to be underestimated because of low productivity and slow development of most primate species (see Table 17.1). Considering, for instance, that macaque females cannot be expected to give birth to a first infant before the age of three years, a minimum of five to six years has to be taken into account, from the birth of a captive-born mother until her first two-year-old offspring is available for research purposes. The shift to cynomolgus monkeys, after the export ban for rhesus monkeys from India, and the search for other alternative species resulted in the establishment of breeding colonies for the most frequently required primates. Currently, purpose-bred cynomolgus and rhesus macaques and marmosets are available in large numbers. Cynomolgus monkeys are commercially available from the Philippines, Indonesia, Vietnam and Mauritius and rhesus monkeys from China. Smaller colonies also exist in Israel and Europe. A number of commercial and in-house marmoset breeding colonies can satisfy 100% of the needs of biomedical research (Hunsmann, 2003). Both purpose-bred and wild-caught vervets are available from the Carribean area (Ervin and Palmour, 2003). Thanks to the support of the National Centre for Research Resources, SPF rhesus monkeys, other macaques, baboons, squirrel monkeys and other species are available for NIH grantees in the USA (Robinson and Beattie, 2003). Purpose-bred vervets and baboons are also available in South Africa although the export from these countries is practically impossible (Seier, 2003). A large breeding colony of squirrel monkeys exists in French Guyana.It is estimated that about 80% of all non-human primates used in research are purpose-bred, but this could increase to meet future needs (Animal Procedures Committee, 2002; Research Resources Information Center, 2002).

Numbers required

In fundamental research on the pathogenesis and course of spontaneous or induced disease models such as Alzheimer, Parkinsonism, Arteriosclerosis, Diabetes etc.,

TABLE 17.1: Comparison of recommended floor areas for two animals in m²

Species	Floor areas (m²)				
	USA*	CoE (1986)	UFAW	International workshop Berlin 1993	Recent proposals
marmosets	0.3	0.25	0.25	1.0	CoE not finalized
squirrel monkeys	0.56	0.35	0.5	not mentioned	CoE not finalized
vervets	0.8	0.7	1.6	2.0	2.0[1]
rhesus monkeys	1.12	1.1	2.8	3.0	2.0[1]
baboons	1.41	1.5	2.8	not mentioned	7.0**
chimpanzees	4.66	not mentioned	not mentioned	not mentioned	CoE not mentioned
dogs	–	5.6	–	5.6 with exercise area	

Data sources: International Primatological Society (1993); Poole (1993); Deputte *et al.* (2003[1])

* The values for USA are calculated by adding the recommended values for single animals.

** The floor area of 7 m² has been calculated on the basis of the same principles as that for macaques, i.e. four steps along each side.

the group size of animals can be low. In longitudinal studies cited in this research category, it is not uncommon to find that the experiments had been performed on one to five animals. When testing the efficacy of new pharmaceuticals, large numbers of monkeys in various dose groups would be preferred. In experiments for the safety evaluation of new pharmaceuticals, a minimum of three animals per sex and dose and three dose groups, in addition to a control group, are required for non-rodent species, i.e. at least 24 animals per test (OECD, 1998). In neurovirulence tests, that are required for testing seed lots of live vaccines against measles, mumps, varicella and yellow fever, groups of at least 11 animals, in addition to a control group of four macaques or vervets, are necessary. For oral live polio vaccines, a reference group has to be included (European Pharmacopoeia, Council of Europe, 2001). The animals should be at least 1.5 kg and, depending on the species and the growth rate of the colony, these animals are estimated to be about 18 to 24 months of age.

Large groups of animals in narrow weight ranges can only be supplied by large breeding establishments if capture from non-endangered natural populations is to be avoided.

Problems related to source

Though animals obtained from breeding colonies in the countries of origin are usually cheaper than those produced in Europe or the USA, problems with transportation have to be considered. Many air transport companies have stopped transporting primates for experimental purposes, under the pressure of animal rights groups, or only accept direct transport into the country of the user in the airline's home country. Additionally, insecurities about the health status of such animals has resulted in the imposition of a quarantine period (European Commission, 2002). Such obligations prolong the time from ordering the animals until a project can be started.

Co-operation

When working with conscious animals, one important aspect is the acceptance of a test animal to co-operate with the experimenter. Anxiety and fear of being handled, and objection to restraint, can cause distress. Contrary to most of the commonly used laboratory animal species, primates have not had the chance of being "domesticated" and there have not been enough generations of purpose-bred animals to be able to select animals that are less aggressive towards humans. Though primates can be trained to come forward and accept minor manipulations, such as topical applications, vaginal swabbing and intravenous blood sampling and injections (Reinhardt and Seelig, 1998), in our experience they will always remain wary towards humans. Whereas dogs and swine can easily be trained to lie on a table or in a hammock, without further restraint, and accept blood-pressure measurements or electrocardiogram recording, most primates will have to be restrained in some way or other. This wariness towards humans, or other perceived enemies, seems to be transmitted to infants by those mothers who, themselves, were descendants of animals from the wild.

From personal experience with macaques and squirrel monkeys, hand-reared infants that have been rejected by their mothers in their first days of life, or after hysterectomy for experimental reasons, usually accept humans readily without much aggression. It was impressive, for example, to see how the care staff of van Wagenen in New Haven in 1970 could carry adult hand-reared female rhesus monkeys around as if they were children. Similar observations have been made of female baboons by Scheuber (2003) in Munich. However, adult males sooner or later become unreliable and, since hand-rearing is not a feasible way of obtaining compliant monkeys, non-human primates will probably be less co-operative animals than dogs for a long time.

Ease of handling of a specific species may be a matter of experience and the quality of the animal care staff, but co-operative behaviour may differ between species. Amongst macaques, *Macaca arctoides* females have been said to be more co-operative than rhesus and cynomolgus females (personal communications). Adolescent baboons are also comparatively docile. Years ago we were warned by Kummer in Zürich against trying to use talapoin monkeys (*Miopithecus talapoin*) as experimental animals. Defence reactions against manipulation can result in the infliction of severe wounds to the handler, and the larger the animal, the greater the risk. Working with chimpanzees, the nearest genetic relatives to humans, should be restricted to specialized institutions anyway.

Smaller species may be said to be less prone to stress from handling, but this could just be due to their size making handling easier. In the 1980s we used squirrel and rhesus monkeys for non-invasive blood pressure measuring, in restraint chairs, and gained the impression that the rhesus monkeys supported an eight hour session better than squirrel monkeys. Though no cardiovascular effects were observed, the squirrel monkeys became fidgety after four hours and lost weight if the procedure was performed at weekly intervals. Unless blood sampling is required, blood pressure and pulse frequency measurements can alternatively be performed in marmosets, with radio-telemetry, to avoid handling. Their small size makes them more favourable for such an experimental set up (Schnell and Gerber, 1977).

Compatibility and group stability

Primates should be housed in social groups. Group size and composition will be determined by the experimental requirements and species specific characteristics (International Primatological Society, 1993). Except for marmosets, tamarins, owl monkeys and prosimians, the most traditionally used species live in fairly large groups in the wild (Williams and Bernstein, 1995). The confined conditions required for housing animals in experiments, and the necessity in most cases to maintain single sex groups, differ significantly from natural conditions. Group stability, therefore, will depend strongly on conflict coping behaviour, reconciliation patterns of the species and the ability to cope with disturbances such as periodic transient isolation of an individual animal from the group. From our experience, squirrel monkeys living under natural conditions in fairly large groups, often with males and females segregated (Williams and Glasgow, 2000), usually pose no great problems. However, signs of antagonism may be discrete and, thus, require careful observation to detect mobbing against individual animals within a group.

Vervets seem to pose more problems for group housing although multi-male and multi-female groups can be maintained communally, if they were reared together from weaning (Seier, 2001). For males that had to be singly housed for particular studies, Seier developed an amicable system of rotating a mobile cage, containing a female, between the singly housed individuals (Seier, 1996). Behavioural patterns of three macaque species were investigated under semi-natural conditions and it was concluded that group stability could be ranked in the descending order of *M. tonkeana* – *M. arctoides* – *M. fascicularis* – *M. mulatta* (Thierry, 2000). Others have experienced a higher incidence of bite wounds and have found incompatible adult male cynomolgus monkeys more difficult to pair than rhesus monkeys (Bürge, 2003, pers. com.). However, differences in behaviour may not only be species specific, but could also be related to rearing conditions during breeding. Macaques are often separated at the age of six months and reared in peer groups where acquisition of full social competence may be compromised.

Life cycle and growth

Puberty

The choice of a model for studying menopause, osteoporosis, degenerative processes resembling Alzheimer's or Parkinson's disease may be determined by the time phases associated with growth, sexual maturation, life span and the occurrence of ageing processes in each primate species. Most primate species have slower development and a longer life span than other laboratory animals. In Table 17.2 the ages at which animals reach sexual maturity, reproductive performance, adult body size and life span are given for the commonly

TABLE 17. 2: Development of commonly used primate species

Species	Annual reproductive performance per female (no. of newborn/year)	Earliest reproductive age (years)	Adult body weight* (kg)	Approximate life span potential (age in years)
marmosets	1.5–2.8	1.5	0.2–0.3	12
squirrel monkeys	0.65	2.5–3	0.6–1.2	20
vervets	1.0–1.5	3	2–6	25
rhesus monkeys	0.6–0.8	3	4–11	30
baboons	0.8–0.85	4–5	12–24	40
chimpanzees		10	30–60	50

Data sources: Napier and Napier (1967); Williams and Glasgow (2000); Röder and Timmermann (2000).

* Lower weight limits applicable to young adults and females, upper limits to older males.

used primate species. The data are derived from Napier and Napier (1967), Hendricks and Dukelow (1995), Williams and Glasgow (2000) and Röder and Timmermann (2002).

In sub-chronic and chronic toxicity tests, of three, six or 12 months duration, regulatory agencies usually require that the duration of the experiment includes maturation of the sexual organs. Although the start of a sub-chronic toxicity study is set at the age of four to six months for dogs (OECD, 1998), the trend is rather to begin toxicity tests with older animals of six to eight months of age. Applying the same principles to non human primates would mean that macaques should be at least 2.5 years, squirrel monkeys 1.5 years and marmosets one year old in order to involve the period of puberty in the shorter tests.

Reproduction

Age is also a determining factor for the choice of species in reproductive studies. After the thalidomide (Contergan) disaster, a great deal of interest was focused on primates in which practically identical effects to those in humans could be produced. Primates remain of interest for teratogenesis testing, complementary to the traditional species, the rat and the rabbit. For practical reasons, species with regular ovulation throughout the year, with short gestation periods and multiple foetuses, and with the possibility for timed mating, would be preferred. The larger primate species all have the disadvantage that the females cannot be used before they are three to four years old (see Table 17.2). Experience has also shown that primipar animals should not be used because pregnancy rate is lowest in young primipar animals, better in "prime" animals and highest in prime secundipar animals (Philippi-Falkenstein and Harrison, 2003). Of the macaques, rhesus monkeys have the disadvantage of seasonal anovulatory cycles in contrast to other macaques and baboons (Dukelow, 1975; Grauwiler and Brüggemann, 1972). *Callithrix jacchus* reach maturity at an earlier age, have no seasonal variation, but the lack of menstrual cycles renders timing of mating difficult. Female marmosets also have a shorter gestation time and frequently bear twins. Though smaller than macaques, hysterotomy can also be performed repeatedly in marmosets to avoid loss of foetuses due to abortion (Siddall, 1978). Recently, however, there have been reservations about the advantages of marmosets over macaques for reproductive toxicology studies due to differences between marmosets and humans. These include multiovulation, twin placental discs supplied by blood vessels from both twins and functionality of spermatogenesis in marmosets (Zülke and Weinbauer, 2003).

Life span

The long life span of most primate species could preclude their use for life-time studies on the pharmacology and toxicology of ageing. Tests for assessing the cancerogenic potential of new chemicals are usually performed on mice, rats and, occasionally, hamsters because it is possible to observe the major part of the approximately two year life span of animals in these rodent species (Meyer and Svendson, 2003). The life span of primates significantly exceeds that of rodents and only in marmosets is it comparable to the life span of dogs. Some primate species appear, spontaneously, to develop changes similar to those observed in aged humans, e.g., senile plaques, menopause and osteoporosis. However, some of these changes appear at a relatively high age only attained by animals in captivity. Bone loss, increasing with age, occurs in rhesus monkeys older than 10 to 11 years and slightly earlier in cynomolgus monkeys (Colman *et al.,* 1999, also for literature references concerning further species). Hormonal changes, associated with amenorrhea and other climacteric conditions, appear in macaques above 22–25 years of age (Hodgen *et al.,* 1977; Hendricks and Dukelow, 1995). Senile plaques and cerebral amyloidosis, indicative of Alzheimer's disease, have been described in cynomolgus monkeys above 20 to 22 years of age, but similar histopathological changes have been reported in other species including other Old and New World monkeys and prosimians with a shorter life span (Nakamura *et al.,* 1998). Although, in rhesus monkeys, an age related neuronal loss in the substantia nigra appears to be gradual, a marked difference exists between younger adults and those above 20 years of age (Siddiqi *et al.,* 1999). Gradual age associated changes in brain composition, measured by magnetic resonance imaging (MRI), have also been reported (Andersen *et al.,*1999) for three age groups (5–8; 12–17 and 21–27 years) of rhesus monkeys.

Primates are excellent models for studying the course and underlying mechanisms of spontaneous disease processes occurring in humans. However, because of the irregular onset, not sufficient numbers of animals are available at a given time. There is also a limited availability of aged primates, so induced models, especially of ageing processes, like the model produced by Ridley and Baker (2002) for Alzheimer's disease in marmosets, could be of interest.

Size

The size of the species can influence selection for research. Smaller species can have the advantage, sometimes, of being cheaper, easier to handle, and requiring less space and food. For the pharmaceutical and chemical industries, the smaller amount of substance needed for the development of pharmaceuticals, or in safety testing of other chemicals, may be a factor contributing to the cost. This is one of the reasons why, in Europe at least, there is an increasing use of the marmoset for toxicity testing (Animal Procedures Committees, Part 1, 2002; Siddall, 1978). Small size, however, limits the amount of body fluids that may be sampled at any one time, or at frequent intervals. Recommendations for application volumes and sample sizes are to be found in a publication of a joint working group of the European Federation of Pharmaceutical Industries and the European Center for the Validation of Alternative Methods (Diehl *et al.,* 2001).

There are also reasons for choosing larger rather than smaller species as, for example, in xenotransplantation studies when organ sizes of recipient and donor should be matched (Kozlowski *et al.,* 1999; Whiting and Ardehali, 2003). Another advantage of using larger models would be the possibility of using the same techniques and equipment used for investigations in humans. This could be the case, for instance, in lung function tests on macaques (Fozard and Büscher, 2000; Fozard *et al.,* 2003) or in experiments on visually based behavioural tests using rhesus monkeys (Henn *et al.,* 1992; Hess *et al.,* 1992, Newsome and Stein-Avilez, 1999).

Health status

In contrast to the traditional laboratory animals, especially mice and rats, primates are still often silent carriers of potentially pathogenic organisms, some of which can cause fatal disease when transmitted to humans. Similarly, most primates are very susceptible to some human diseases. It is not within the scope of this chapter to review microbiological problems encountered with primates extensively, but to point out some major risks which should be avoided when using primates in the laboratory. An extensive list of facultative pathogenic organisms has been published by a working group of the German Society for Laboratory Animal Science (GV-SOLAS, 1980). Important micro-organisms that can be found in the commonly used primate species, especially from the wild, are listed in a report of the Federation of European Animals Science Associations (FELASA, 1999) together with recommendations for a monitoring system. The primary intention of the report was to persuade users and suppliers to supply information on the health status of the animals. This would facilitate interpretation of results and the taking of necessary precautions against infection of existing colonies or humans handling the animals. A second intention was to encourage breeders and suppliers to take appropriate steps to eliminate unwanted micro-organisms.

Infectious agents causing overt clinical disease, such as malaria (Stokes *et al.,* 1983), in a carrier animal are of less concern than those persisting in the host animal without causing disease but, when transmitted to other species, such as humans, can lead to fatal outbreaks. Thus macaques can be silent carriers of Herpes B virus (*Cercopithecine herpesvirus 1*) and vervets have been incriminated in having caused the fatal Marburg disease in humans in 1967 (Hennessen, 1970). Vervets and mangabeys are carriers of SIV (Baskin, 2003; Kalter *et al.,* 1997), apes (and in one case also a vervet) may be carriers of hepatitis B virus (Heckel *et al.,* 2001; Thung *et al.,* 1981), while squirrel monkeys may host *Plasmodium falciparum* (Gozalo *et al.,* 1997) and *Trypanosoma cruzi,* the parasite causing Chagas' disease in humans (Ndao *et al.,* 2000).

In view of xenotransplantation of tissues, and possibly whole organs, from animals, a new emerging cause of concern are the exogenous and endogenous retroviruses (Brown *et al.,* 1998; Mukai *et al.,* 1992). In addition to pigs, of which genetically modified breeds have been created, baboons are also being exploited. Because of their close proximity to humans, it is expected that, analogous to the genetically modified swine, hyperacute rejection can be circumvented. However, primates may harbour an array of exogenous retroviruses and baboons, in particular, an additional endogenous retrovirus which may be genetically transmitted.

Not much is known, in primates, about the interference of micro-organisms with experimental results, apart from the potential disease outbreaks under immunosuppression (Lednicky *et al.,* 1998), or when transferred to another species such as the transmission of *saimiriine herpesviruses 1 and 2* from squirrel monkeys to callithricids (Potkay, 1992). Health aspects are of particular importance for the production and safety testing of vaccines. Animals used for neurovirulence test serum samples, taken from test animals at the time of inoculation and from the control animals up to 10 days after the test animals have been killed, must be negative for the wild type of the virus to be tested and for measles virus. Monkeys used for the preparation of kidney cell cultures, for the culture and control of inactivated polio vaccine, must ensure no contamination with simian viruses such as SIV, Simian Virus 40, filoviruses, herpes B virus and yellow fever virus (European Pharmacopoeia, Council of Europe, 2001).

Because of these problems, biomedical research institutions would prefer purpose-bred animals of known health status and background information on the history of the animals which is necessary for an unbiased interpretation of results. Unfortunately, eradication of infectious agents in breeding populations, that have been initiated with animals from the wild, is often not easy and embryo transfer is not very practical as a method of getting rid of infectious agents in primate colonies. Intermittent shedding of B-virus from silent carriers could not be excluded in a colony of macaques with animals positive for B-virus (Weigler *et al.,* 1993). In order to avoid the risk of further fatal outbreaks of disease in personnel handling the animals, efforts were made to create B-virus free colonies by separating infants from their mothers before they could be infected by bites inflicted by positive animals (Hilliard and Ward, 1999). By careful selection of sero-negative animals, various in-house and commercial breeders of macaques have succeeded in creating colonies that are free of herpes B and retroviruses (Pamungkas and Sajuthi, 2003). In addition, it is possible to obtain herpes B and Ebola-Reston free cynomolgus monkeys from suppliers on the island of Mauritius. Vervets free of a number of pathogens, including hepatitis, STLV and SV 40 can also be obtained from the Caribbean islands. The reason appears to be that both species were brought over to the islands, in the 17th century, by sailors who had obtained them as pets in the countries of origin, at an age before they were infected during conflicts in the group (Baulu *et al.,* 2001; Ervin and Palmour, 2003; Stanley, 2003). Vervets from Barbados are also preferred, by the vaccine industry, because of their low incidence of foamy viruses that are undesirable for kidney cell culturing of polio virus (personal communication).

Breeding colonies in the country of origin are at risk of re-infection with pathogenic agents, indigenous in the wild primate population, and need to prevent contact with feral animals (Miranda *et al.,* 1999). Careful monitoring of the health status of animals is required, especially when breeders replenish or expand their breeding stock with wild conspecifics. In the case of herpes B, for instance, delayed seroconversion and intermittent shedding of virus may lead to late detection of infected animals (Weigler *et al.,*1993). Though breeding colonies outside the countries of origin do not risk infection with primate specific pathogenic organisms from the wild conspecifics, they too, need to take necessary precautions to avoid infection of undesired endemic micro-organisms, either through humans or other vectors. Animal welfare organisations exert pressure to provide outdoor enclosures. While there is no doubt that these have significant advantages in terms of environmental enrichment, they are open to wild rodents, cats, foxes and birds. It is, therefore, not surprising that occasionally contamination with toxoplasmosis, leptospirosis, or pseudotuberculosis is encountered (Bacciarini *et al.,* 2001; Cunningham *et al.,* 1992; Plesker and Claros, 1992; Vore *et al.,* 2001). Recently, Echinococcosis of gorillas, baboons and cynomolgus monkeys has posed problems in zoos and primate centres with open air corrals (Blankenburg *et al.,* 2002; Rehmann, 2002; Rehmann *et al.,* 2003). On the other hand, no contamination with Leishmaniosis has been observed clinically in a colony of baboons in the Mediterranean area of France, although the infection is endemic in the area (Bray, 1985; Dubreuil, 2002).

It should be clear that it is preferable to obtain animals from breeders and suppliers that monitor their primate colonies reliably. Should it be necessary, however, to obtain animals from other sources, there has to be a long quarantine period before animals can be released for an experiment. Twelve weeks is estimated to be a minimum time and is demanded by some national legislations (OIE, 1990). Apart from the risk of humans acquiring diseases from non-human primates, the opposite risk should not be forgotten. Their closeness to humans renders primates susceptible to a number of pathogenic organisms normally found in humans. Old World primates and apes are known to be very susceptible to tuberculosis of all types, and the fact that tuberculosis seems to be increasing in recent years (Grange and Zumla, 1999) should make those responsible for primate houses mindful about preventive measures against infection from personnel.

Species specificity, subspecies and infra-species differences

In an article on the founding principle of animal models, Schiffer (2002) discusses the value of animal models on the basis of evolution and points out that "the phylogenetic relatedness of humans with other animals holds that a shared genetic background allows for shared vulnerabilities to disease". This would explain why some animal models can serve as true homologues to human conditions, as, for example, the chimpanzee which has

98.7% homology of DNA sequences with the human. However, since the phylogenetic relationship between animals and humans is not linear but branching, leading to genetic diversity and patterns of vulnerability, animal models usually fail to match identically with the human condition. Nevertheless, focusing on why certain species are vulnerable to certain diseases, and others not, may offer us new insight into underlying (genetic) mechanisms. One must ask why African primates are carriers of SIV without succumbing to the infection and why macaques develop the disease. Non-human primates offer the possibility of studying diseases or degenerative processes that occur in humans but rarely in non-primate species. Some of these, such as ageing diseases, may be spontaneous in non-human primates while others may be induced. Even spontaneous models are rarely restricted to one or two specific primate species. Exceptions would be a very localized group of photosensitive epileptic baboons (Meldrum *et al.*, 1975) and the squirrel and owl monkeys which are susceptible to the malaria types *Plasmodium falciparum* and *Plasmodium vivax* (Gallond, 2000). Spontaneous diseases have the disadvantage that they do not readily lend themselves to treatment schedules due to a lack of standardization. Non-human primates are, however, valuable in gaining new insight into the pathogenesis and the mechanisms of the diseases and so trigger ideas for new therapies.

Because of lack of standardization in spontaneous diseases, far greater use is made of induced models (Meyer and Svendson, 2003) for which more animals are available at a given time. This is typically the case for the testing of pharmaceuticals where only restricted use is made of primate species. The general assumption that primates are better models than other species has been criticized (Howard and Pollard, 1983; McGuire *et al.*, 1983; Miczek and Gold, 1983). In a review article on models for post-angioplastic restenosis, Lafont and Faxon (1998) consider the use of non-human primates as rather impractical, although the pathological changes seen in these species (macaques, baboons and chimpanzees) would make them nearly ideal for investigating arteriosclerosis. Indeed, since useful animal models for disease in humans can be found in a large variety of animal species, the justification of using a non-human primate model may become delicate. Nevertheless, it may well be that, with increasing knowledge of the genetics of primate species and the genetical basis of human disease, the interest in using primates in research and development will grow (Animal Procedures Committee, 2002).

Regulations

Neurovirulence tests for vaccines for human use, and drug dependency tests for neurologically active pharmaceutical compounds, are the only fields where the choice of primates to be used are regulated. The European Pharmacopoeia Council of Europe (2001) recommends the use of macaques or vervets for the neurovirulence test for measles, mumps, poliomyelitis, varicella and yellow fever vaccines. A replacement of primates by polio susceptible transgenic mice for the neurovirulence test for polio vaccines is being evaluated in a WHO collaborative study (Dragunsky *et al.*, 2003). Occasionally some national authorities, as in Canada and Japan, may require tests in non-human primates for the evaluation of the drug abuse potential of new drugs in the preclinical phase (Kelly and Römer, 1991).

Inter and intraspecies variations in pharmaceutical use

As previously mentioned, the majority of primates are used in the pharmaceutical industry and, within the industry, most primates are required for safety evaluation tests. Primates are used less frequently in the early phase of pharmaceutical characterization of new compounds. If they are, it is usually as models to verify activities found in other tests when cognitive functions, cardiovascular or immunological mechanisms are of interest. For the safety testing of pharmaceutical products, national authorities demand only that the tests should be performed on two species, one being a non-rodent. Normally the dog is used as the second species, but their routine use has been contested (Zbinden, 1993). It has been proposed that a species should be selected either because a compound has proven to be active in the species in pharmacological tests or because pharmacokinetic and metabolic data prove the suitability of the species. Although it has long been postulated that primates have more similarities to humans than most other animal species in drug kinetics and metabolism (Yanagita, 1973), it was realized that this cannot be generalized. A number of studies have compared pharmacokinetics and metabolism of pharmaceuticals in humans with results obtained in apes, simians, prosimians and other laboratory animals (Caldwell *et al.*, 1979; Kiechel *et al.*, 1975; Weber and Madörin, 1977; Smith and Williams, 1972). Marked differences were found in the metabolic pathways and in the quantitative recovery of unchanged drug and its metabolites not only between primates, including humans, and non-primate species, but also between different primate species tested. Thus, some pathways were common to both man and non-primate species, others to man and various primate species and yet others only to individual primate species. In a comparative study on six different metabolic pathways occurring in primate species, including humans, Smith and Williams (1972) could show that the formation of N^1-glucuronide of sulphadimethoxine, and the O-methylation of 3,5-diiodo-4-hydroxybenzoic acid, appear to be metabolic reactions largely restricted to man and other primates. Glutamine conjugation of phenylacetic acid and indolacetic acid occurs in apes and in both New and Old World monkeys, but not in prosimians or non-primate species. N^4-methylation of sulphadimethoxine occurs in humans, primates and non-primates but with marked quantitative differences within the simian species tested. Caldwell and others (1979) compared the metabolism of the analgesic, meperidine, in vervets, patas, Mona monkeys and a mangabey with that in humans and rats. Though the overall recovery of unchanged drug and its metabolites was most comparable between man and vervets, the relative amounts of unchanged meperidine and metabolites that was excreted, varied between the four species. The complete data suggested that the mangabey would have been the best, and the vervet the least suitable, species whilst the rat differed significantly from the human situation. Based on analytical investigations of metabolization of the beta blocker, pindolol (Kiechel *et al.*, 1975), Weber and Madörin (1977) pointed out that the pattern of the unchanged compound excreted in the urine of six animal species was qualitatively identical to that in humans, baboons and the rhesus monkey although it differed significantly in the squirrel monkey, the dog, the cat and the rat. However, when the percentage of unchanged compound responsible for its activity was taken into account, the baboon and the dog came quantitatively closer to humans. A further retrospective study (Weber, 1992) on the metabolism of five compounds used in clinical testing on rats, dogs and rhesus monkeys, illustrated (Table 17.3) that, except for one compound where the strong emetic effect precluded the use of dogs for repeated dose toxicity tests, the rhesus money showed no overall clear advantage over dogs. When seeking a species most closely resembling the human situation, all factors of compound availability (absorption, distribution, metabolism and elimination) have to be taken into account (Smyth and Hottendorf, 1980).

Toxicologists will, therefore, carefully balance the necessity of using a primate against the advantages of being able to compare effects with historical data available with the traditional animal species. However, with the development of new biotech compounds, peptides and cytokines, etc. non-human primates may become indispensable when a new drug has to be tested in a multi-dosage study, because of greater immunological closeness to humans. The immune reaction of the traditionally used beagle dog may be too marked for the use of this species in chronic toxicity tests. As mentioned before, a smaller species than macaques would be preferred and increasing use appears to be made of marmosets in Europe (Smith *et al.*, 2001).

Apart from differences between species and distinct subspecies, a different origin of the same species may give rise to converse results in comparable experiments. This is not surprising in view of the often vast geographical area of distribution of some of the species in which the development of subpopulations, normally hindered by migration between neighbouring colonies, cannot be avoided. Genetically homogeneous animals may only be expected in island populations. Examples of this are the cynomolgus macaques on the island of Mauritius and the vervets on the Caribbean islands, that are supposed to have originated from a limited number of animals brought in by sailors and immigrants from the countries of origin (Baulu *et al.*, 2002; Stanley, 2003). The rhesus colony on Cayo Santiago will also have become rather homogenous since its foundation in 1938/39, although the original founder animals were collected from seven different districts in India (Rawlins and Kessler, 1986). The homogeneity of the animals obtained from purpose or captive-bred monkeys will depend on the origin of the founder animals, or of the area from which breeding stock is replaced (Williams-Blangero and Vandenberg, 2003).

The differences in external phenotypic characteristics between rhesus monkeys of Indian, Chinese and Burmese origin, would suggest marked intra-species differences in physiological characteristics which, under different circumstances, may have led to differentiation into subspecies. Not unexpectedly, such differences have been reported chiefly in the field of immunology (Williams-Blangero and Vandenberg, 2003).

TABLE 17.3: Comparability of metabolic and pharmacokinetic parameters of five compounds in rhesus monkeys, beagle dogs and rats with those in humans after oral application

Compound	Parameter	Comparability to humans		
		Primate	Dog	Rat
Pindolol	Metabolic pattern	++	–	–
(indolyl-isopropyl- propranolol)	Percent unchanged compound	+	+	–
	Peak value	++	+	–
	Time to reach peak	++	++	+
	Elimination (pathway and time)	+	++	–
Pazindol	Metabolic pattern	–	++	+
(imidazo-isoindol)	Percent unchanged compound	+	++	++
	Peak value	++	+	+
	Time to reach peak	–	++	++
	Elimination (pathway and time)	++	++	–
Bromocriptin	Metabolic pattern	++	not tested	+
(ergot-derivative)	Percent unchanged compound	++	not tested	++
	Peak value	–	not tested	+
	Time to reach peak	++	not tested	–
	Elimination (pathway and time)	++	not tested	++
Proquazone	Metabolic pattern	+	++	+
(phenylquinazolinone)	Percent unchanged compound	++	+	–
	Peak value	–	+	+
	Time to reach peak	–	+	+
	Elimination (pathway and time)	–	++	+
Ketotifen	Metabolic pattern	+	+	–
(benzoheptathiophen)	Percent unchanged compound	+	–	++
	Peak value	+	++	++
	Time to reach peak	+	++	++
	Elimination (pathway and time)	++	++	–

Close resemblance: ++; comparable: +; marked differences: –

From: H. Weber: "Survey on the use of primates in the European pharmaceutical industry" (Primate Report 49: 11–18 (1997)).

Differences have also been reported, between animals of Chinese and Indian origin, in the susceptibility of rhesus monkeys to SIV infections, the Indian appearing to develop SIV much more rapidly than those of Chinese origin (Marthas *et al.*, 2003; Warren, 2002). The strain of viruses used, as well as methodological factors such as the route of administration, could also be contributing to such differences in reactivity. Significant differences in the rejection of kidney transplants were also apparent between cynomolgus monkeys originating from the Philippines and a group from Mauritius (Menninger *et al.*, 2002).

Thus it is of importance to know the source of the species, and the conditions under which they were used, in order to compare and interpret results of experiments at different institutions (GV-SOLAS, 1985).

Conclusions

It is important to select a model species that will help you understand the underlying mechanisms and repair processes of a disease and the integration of an isolated function within a complex system. Where useful disease models can be found in other animal species, these will be favoured for ethical and practical reasons. However, non-human primates offer opportunities, in a number of areas, to study the course and mechanisms of many human diseases that are not reproducible in other species. Due to their close genetic similarity to humans, non-human primates are also the species of choice for testing new therapeutic agents that closely resemble human proteins and peptides. When using non-human primates in biological and medical research, however, it is important to consider that they differ from other laboratory animal species with respect to uniformity, domestication, life expectancy and health aspects, and there is often a lack of obtainable historical background data. Marked physiological behavioural and immunological differences may exist, not only between species but also within a species from different geographical areas. The characterization of the animals and their source is essential for the interpretation of results from different institutions.

Acknowledgements

I should like to thank all of my colleagues, not cited in the chapter, especially T. Bürge and R. Pfister from Novartis, P. Thomann of Zurich University, E. Rommel, consultant, and I. Kohler from ZLB Bioplasma AG, who helped me with personal information and recommendations and drew my attention to pertinent literature.

Correspondence

Any correspondence should be directed to Heinz Weber, Dr. med. vet, Klingental 7, CH-4058 Basel, Switzerland. Email: heinz.weber@freesurf.ch

References

Anderson, A.H., Zhang Z., Zhang M., Gash, D.M. and Avison, M.J. (1999). *Brain Research* 829, 90–98.

Animal Procedures Committee (2002). In Banner, M. (ed.) *The Use of Primates under the Animals (Scientific Procedures) Act (1986),* pp 1–39. London: Home Office.

Bacciarini, L.N., Voellm, J., Janovski, M., Kappler, A. and Groene, A. (2001). *Europ. J. Vet. Pathol.* 7, 67–73.

Baskin, G. (2003). In Institute of Laboratory Animal Research. National Research Council (ed.) *International Perspectives. The Future of Nonhuman Primate Resources,* pp 143–148. Washington, DC: National Academies Press.

Baulu, J., Evans, G. and Sutton, C. (2001). *Lab. Primate Newsl.* 41, 4–6.

Blankenburg, A., Maetz-Rensing, A., Dinkel, A., Sauermann, U. and Kaup, F-J. (2002). In Erken, A.H.M., Dorrestein, G.M. and Dollinger, P. (eds) *Proceedings of the European Association of Zoo and Wildlife Veterinarians and European Wildlife Disease Association Meeting, Heidelberg 2002,* pp 367–370. Van Setten Kwadraat, Houton NL.

Bray, R.S. (1985). In Chang, K.P. and Bray, R.S. (eds) *Leishmaniasis,* pp 479–481. Amsterdam: Elsevier Science Publishers.

Brown, J., Mathews, A.L., Sandstrom, P.A. and Chapman, L.E. (1998). *Trends in Microbiology* 6, 411–415.

Buhjinski, T.M. (1996). *African Primates* 2, 5–9.

Caldwell, J., Notarianni, L.J., Smith, R.L., Fafunso, M.A., French, M.R., Dawson, P. and Bassir, O. (1979). *Toxicology and Applied Pharmacology* 48, 273–278.

Captive Care Breeding Committee of the International Primatological Society (1993). In Poole, T.B. and Schwibbe M. (eds) *International Guidelines for the Acquisition, Care and Breeding of Nonhuman Primates,* pp 3–27. Goettingen: Erich Goeltze GmbH.

Colman, K.J., Lane, M.A., Binkley, N., Wegner F.H. and Kemnitz J.W. (1999). *Bone* 24, 17–23.

Council of Europe (1986). *European Convention for the Protection of Vertebrate Animals used for Experimental and other Scientific Purposes. Explanatory report.* Strasbourg: Council of Europe Press.

Council of Europe (2001). *European Pharmacopoeia, Fourth Edition,* pp 2171–2226. Strasbourg: Council of Europe.

Council of Europe (2003). www.coe.int/animalwelfare

Cunningham, A.A., Buxton, D. and Thomson, K.M. (1992). *Journal of Comparative Pathology* 107, 207–219.

Deputte, J.R., Anderson, D., Gosset, D., Koene, P., Ludes, J. and Herrenschmidt, N. (2005). In preparation.

Diehl, K.H., Hull, R., Morton, D., Pfister, R., Rabemampianina, Y., Smith, D., Vidal, J.M. and van de Vorstenbosch, C. (2001). *J. Appl. Toxicol.* 21, 15–23.

Dragunsky, E., Nomura, T., Karpinski, K., Furesz, J., Wood, D.J., Pervikov, Y., Abe, S., Kurata, T., Vanloocke, O., Karganova, G., Taffs, R., Heath, A., Ivshina, A. and Levenbook, I. (2003). *Bulletin of the World Health Organisation* 81, 251–260.

Dubreuil, G. (2002). Personal communication.

Dukelow W.R. (1975). *Lab. Primate Newsl.* 14 (4), 1–4.

Dukelow W.R. (1999). *Am. J. Primatol.* 49, 129–132.

Ervin, F. and Palmour, R. (2003). *International Perspectives. The Future of Nonhuman Primate Resources,* pp 49–53. ILAR, National Research Council, Washington, DC: The National Academies Press.

European Commission (2002). *The Welfare of Non-human Primates Used in Research. Report of the Scientific Committee on Animal Health and Welfare. Directorate C,* Brussels.

FELASA Working Group on Non-human Primate Health (1999). *Lab. Animals* 33 (Suppl.1), S1:3–S1:18.

Fozard, J.R. and Büscher, H. (2000). *Pulm. Pharmacol. Ther.* 14, 289–295.

Fozard, J.R., Büscher, H., Farr, D.C. and Mazzoni, L. (2004). *Autonomic and Autocoid Pharmacology,* 24, 9–16.

Gallond G.G. (2000). *ILAR Journal* 41, 37–43.

Gozalo, A., Lucas, C., Cachay, M., Montoya, E., Ballou, W.R., Wooster, M.T. and Watts, D.M. (1997). *J. Med. Primatol.* 26, 204–206.

Grange, J.M. and Zumla, A. (1999). *The Lancet* 353, 996.

Grauwiler, J. and Brüggemann, S. (1972). In Goldsmith, E.I. and Moor-Jankowski, J. (eds) *Medical Primatology*, Part 3, pp 250–258. Basel: Karger.

GV-SOLAS Working Committee for SPF and Gnotobiotic Laboratory Animals (1980). In Rossbach, W. (ed.) *Obligate and facultative pathogenic germs of primates, with Supplement: Classification of Primates.* Basel: Emminger.

GV-SOLAS Working Committee for the Biological Characterization of Laboratory Animals (1985). *Lab. Animals* 19, 106–108.

Heckel, J-O., Rietschel, W. and Hufert, F.T. (2001). *J. Med. Primatol.* 30, 14–19.

Hendrickx, A.G. and Dukelow, W.R. (1995). In Bennet, B.T., Abee, C.R. and Hendrickson R. (eds) *Nonhuman Primates in Biomedical Research, Biology and Management,* pp 147–191. San Diego: Academic Press.

Henn, V., Straumann, D., Hess, B.J.M., Halswanter, T.H. and Kawachi, N. (1992). *Ann. N Y Acad. Sci.* 656, 166–180.

Hess, B.J.M., Van Opstal, A.J., Straumann, D. and Hepp, K. (1992). *Vision Res.* 32, 1647–1654.

Hennessen, W. (1970). In Balner, H. and Beveridge, W.I.B. (eds) *Infections and Immunosuppression in Subhuman Primates,* pp 35–38. Copenhagen: Munksgard.

Hilliard, J.K. and Ward, J.A. (1999). *Lab. Animal Sci.* 49, 144–148.

Hodgen, G.D., Goodmann, A.L., O'Connor, A. and Johnson, D.K. (1977). *Am. J. Obstet. Gynecol.* 127, 581–584.

Howard, J.L. and Pollard, G.T. (1983). In Miczek K.A. (ed.) *Ethopharmacology: Primate Models of Neuropsychiatric Disorders,* pp 307–312. New York: Alan R. Liss.

Hunsmann, G. (ed.) (2003) *International Perspectives; The Future of Nonhuman Primate Resources,* pp 63–68. Institute of Laboratory Animal Research, National Research Council. Washington, DC: National Academies Press.

Kalter, S.S., Heberling, R.L., Cooke, A.W., Barry, J.D., Tian, P.Y. and Northam, W.J. (1997). *Lab. Anim. Sci.* 47, 461–467.

Kelly, P.H. and Römer, D. (1991). In Hess, R. (ed.) *Arzneimitteltoxikologie: Anforderungen – Verfahren – Bedeutung,* pp 384–392. Stuttgart: Georg Thieme.

Kiechel, J.R., Niklaus, P., Schreier, E. and Wagner, H. (1975). *Xenobiotica* 12, 741–754.

Kozlowski, T., Shimizu, A., Lambrigts, D., Yamada, K., Fuchimoto, Y., Glaser, R., Monroy, R., Xu, Y., Awwad, M., Colvin, R.B., Cosimi, A.B., Robson, S.C., Fishman, J., Spitzer, T.R., Cooper, D.K.C. and Sachs, D.H. (1999). *Transplantation* 67, 18–30.

Lafont A. and Faxon, D. (1998). *Cardiovasc. Res.* 39, 50–59.

Lednicky, J.A., Arrington, A.S., Stewart, A.R., Min Dal, X., Wong, C., Jafar, S., Murphey-Corb, M. and Butel, J. (1998). *J. Virol.* 72, 3980–3990.

Marthas, M.L., Lu, D., Penedo, M.C.T., Hendrickx, A.G. and Miller, C.J. (2003). *International Perspectives: The Future of Nonhuman Primate Resources.* Institute of Laboratory Animal Research, National Research Council, pp 128–130. Washington DC: National Academies Press.

McGuire, M.T., Brammer, G.L. and Raleigh, M.J. (1983). In Miczek, A. (ed.) *Ethopharmacology: Primate Models of Neuropsychiatric Disorders,* pp 313–328. New York: Alan R. Liss.

Meldrum, B., Anlezark, G. and Trimble, M. (1975). *Eur. J. Pharmacol.* 32, 203–213.

Menninger, K., Wieczorek, G., Riesen, S., Kunkler, A., Audet, M., Blancher, A., Schuurman, H.J., Quesniaux, V. and Bigaud, M. (2002). *Transplant Proc.* 34, 2887–2888.

Meyer, O. and Svendsen, O. (2003). In Hau, J. and Van Hoosier Jr., G.L. (eds) *Handbook of Laboratory Science, Second Edition*, Vol. III, pp 11–40. Boca Raton, London, New York, Washington, DC: CRC Press.

Miczek, K.A. and Gold, L.H. (1983). In Miczek, K.A. (ed.) *Ethopharmacology: Primate Models of Neuropsychiatric Disorder,* pp 138–153. New York: Alan R. Liss.

Miranda, M.E., Ksiazek, T.G., Retuya, T.J., Khan, A.S., Sanchez, A., Fulhorst, C.F., Rollin, P.E., Calaor, A.B., Manalo, D.L., Roces, M.C., Dayrit, M.M. and Peters, C.J. (1999). *J. Infect. Dis.* 179 (Suppl. 1), S115–S119.

Mukai, R., Narita, T., Kobayashi, R., Fujimoto, K., Takasaka, M. and Honjo, S. (1992). In Matano, S., Tuttle, R.H., Ishida H. and Goodman, M. (eds) *Topics in Primatology, Vol 3,* pp 399–408. Tokyo: Tokyo University Press.

Nakamura, S., Nakayama, H., Goto, N., Ono, F., Sakakibara, I. and Yoshikawa, Y. (1998). *J. Med. Primatol.* 27, 244–252.

Napier, J.R. and Napier, P.H. (1967). *A Handbook of Living Primates.* New York: Academic Press.

Ndao, M., Kelly, N., Normandin, D., MacLean, J.D., Whiteman, A., Kokoskin, E., Arevalo, I. and Ward, B.J. (2000). *Comparative Medicine* 50, 658–665.

Newsome, W.T. and Stein-Aviles, J.A. (1999). *ILAR Journal* 40, 78–91.

OECD (1998). OECD 409 Guidelines for the Testing of Chemicals; Repeated Dose 90-Day Oral Toxicity Study in Non-Rodents.

OIE/Office International des Epizooties (1999). *International Animal Health Code,* pp 48–53. Paris: O.I.E.

Pamungkas, J. and Sajuthi, D. (2003). *International Perspectives: The Future of Nonhuman Primate Resources,* p 20. Institute of Laboratory Animal Research, National Research Council. Washington, DC: National Academies Press.

Philippi-Falkenstein, K. and Harrison, R.M. (2003). *Am. J. Primatol.* 60, 23–28.

Plesker, R. and Claros, M. (1992). *J. Vet. Med. B.* 39, 201–208.

Poole, T.B., Costa, P., Netto, W.J., Schwarz, K., Wechster, B. and Whittaker, D. (1993). In Donoghue, P.N. (ed.) *The Accomodation of Laboratory Animals in Accordance with Animal Welfare Requirements,* pp 81–86. Bonn: Bundesministerium für Ernährung, Landwirtschaft und Forsten.

Potkay, S. (1992). *Journal of Medical Primatology* 21, 189–236.

Rawlins, R.G. and Kessler, M.J. (1986). In Rawlins, R.G. and Kessler, M.J. (eds) *The Cayo Santiago Macaques,* pp 13–45. Albany: State University of New York Press.

Rehmann, P. (2002) "Alveolaere Echinokkose bei Primaten aus dem Zoologischen Garten Basel". Thesis, University Berne.

Rehmann, P., Groene, A., Lawrenz, A., Pagan, O., Gottstein, B. and Bacciarini, L.N. (2003). *J. Comp. Pathol.* 129, 85–88.

Reinhardt, V. and Seelig, D. (1998). *Environmental Enhancement for Caged Macaques. A Photographic Documentation.* Washington, DC: Animal Welfare Institute.

Ridley, R.M. and Baker, H.F. (2002). *7th EMRG Workshop October 14th–16th 2002,* p 50. European Marmoset Research Group. Göttingen: DPZ.

Robinson, J. (2002). In Research Resources Information Center (ed.) *Rhesus Monkey Demands in Biomedical Research. A Workshop Report,* p 7. National Academy of Sciences, Washington, DC.

Robinson, J. and Beattie, G. (2003). In Institute of Laboratory Animal Research. National Research Council (ed.) *International Perspectives. The Future of Nonhuman Primate Resources,* pp 72–80. Washington, DC: National Academies Press.

Röder, I.L. and Timmermann, J.A. (2002). *Lab. Animals* 36, 221–242.

Scheuber, H-P. (2003). Personal communication.

Schiffer, S.P. (2002). *Comparative Medicine* 52, 305–306.

Schnell, C.R. and Gerber, P. (1997). *Primate Report* 49, 61–70.

Seier, J. (1996). *J. Med. Primatol.* 25, 64–68.

Seier, J. (2001). *SAALAS Newsletter* 2, 4–8.

Seier, J. (2003). In Seier, J. (ed.) *International Perspectives; The Future of Nonhuman Primate Resources,* pp 16–19. Institute of Laboratory Animal Research, National Research Council. Washington, DC: National Academies Press.

Siddall, R.A. (1978). *Primates in Medicine* 10, 215–224.

Siddiqi, Z., Kemper, T.L. and Killiany, R. (1999). *J. Neuropathol. Exp. Neurol.* 58, 959–971.

Smith, D., Trennery, P., Farningham, D. and Klapwiy, J. (2001). *Lab. Animals* 35, 117–130.

Smith, R.L. and Williams, R.T. (1974). *Journal of Medical Primatology* 3, 138–152.

Smyth, R.D. and Hottendorf, G.H. (1980). *Toxicology and Applied Pharmacology* 53, 179–195.

Stanley, M.A. (2003). *International Perspectives; The Future of Nonhuman Primate Resources,* pp 46–48. Institute of Laboratory Animal Research, National Research Council, Washington, DC: National Academies Press.

Stokes, W.S., Donovan, J.C., Montrey, R.D., Thompson, W.L., Wannemacher Jr, R.W. and Rozmiarek, H. (1983). *Lab. Anim. Sci.* 33, 81–85.

Thierry, B. (2000). In Aureli, F. and de Waal, F.B.M.(eds) *Natural Conflict Resolution,* pp 106–128. Berkeley: University of Berkeley Press.

Thung, S.N., Gerber, M.A., Purcell, R.H., London,W.T., Mihalik, K.B. and Popper, H. (1981). *Am. J. Pathol.* 105, 328–332.

Vore, S.J., Peele, P.D., Barrow, P.A., Bradfield, J.F. and Pryor, W.H. (2001). *J. Med. Primatol.* 30, 20–25.

Warren, J. (2002). *Rhesus Monkey Demands in Biomedical Research. A Workshop Report,* pp 9–10. Research Resources Information Center, Bethesda: National Academy of Sciences.

Weber, H. (1997). *Primate Report* 49, 11–18.

Weber, H. and Madörin, M. (1977). In Prasad, M.R.N. and Anand Kumar, T.C. (eds) *Use of Non-human Primates in Biomedical Research,* pp 380–390. New Delhi: Indian National Science Academy.

Weigler, B.J., Hird, D.W., Hilliard, J.K., Lerche, N.W., Roberts, J.A. and Scott, L.M. (1993). *J. Infect Dis.* 167, 257–263.

Whiting, D. and Ardehali, A. (2003). In Hau, J. and Van Hoosier, G.L. (eds) *Handbook of Laboratory Animal Science. Second Edition,* pp 81–94. Boca Raton, London, New York and Washington, DC: CRC Press.

Williams, L.E. and Bernstein, I.S. (1995). In Taylor Bennet, B., Abee C.R. and Hendrickson, R. (ed.) *Nonhuman Primates In Biomedical Research. Biology and Management,* pp 77–100. San Diego: Academic Press.

Williams, L. and Glasgow, M. (2000). *ILAR Journal* 41, 26–36.

Williams-Blangero, S. and Vandenberg, J.L. (2003). *International Perspectives; The Future of Nonhuman Primate Resources,* pp 114–121. Institute of Laboratory Animal Research, National Research Council, Washington, DC: National Academies Press.

Yanagita, T. (1973). In Spiegel, A. (ed.) *The Laboratory Animals in Drug Testing. 5th ICLA Symposium, Hannover,* pp 177–184. Stuttgart: Fischer.

Zbinden, G. (1993). *Regul. Toxicol. Pharmacol.* 85–94.

Zülke, U. and Weinbauer, G. (2003). *Toxicol. Pathol.* 31 (Suppl.), 123–127.

Part 3

Research Techniques and Procedures

Contents

CHAPTER 18

Anaesthesia

Steve Unwin
Chester Zoo, North of England Zoological Society,
Chester CH2 1LH, UK

Introduction

The purpose of this chapter is not only to review various chemical restraint regimes for non-human primates used in a laboratory setting, but also to suggest improved methods of anaesthesia for specific species. It aims to outline what is anaesthetic best practice, indicate instances and procedures where anaesthesia may not even be necessary, and also to investigate how each anaesthetic regime works. Specific investigations for various anaesthetic regimes are highlighted and boxed within the text. Section 1 deals specifically with anaesthesia, while Section 2 deals with drug administration and sample collection. Readers are directed to the primate anaesthetic chapter by Horne (2001) for a thorough overview of primate anaesthesia.

Section 1: Anaesthesia

Basic considerations

Is anaesthesia necessary?

When first considering an anaesthetic procedure, a standard set of questions should always be asked. Anaesthesia can be physiologically stressful on any animal, and a decision to sedate or anaesthetise

The Laboratory Primate

ISBN 0-1208-0261-9

an animal must not be taken lightly. However, with anaesthesia, the animal is immobile, risk of disease transmission (from bites and scratches) is also reduced, and the quality of samples will be enhanced. Ethical and welfare considerations must also be taken into account. Thus:

- Could the process you are investigating be conducted with the animal unanaethetised? Would the animal remain stress free in this situation?
- Is the animal trainable to provide what is required in a non-stressful way, such as presenting an arm for blood sampling?
- If manual restraint is to be considered, will the stress induced in the animal be more damaging than an anaesthetic? Have health and safety issues been considered for staff? Is there ease of access to the animal for sample collection? (a big problem if it is struggling inside a net). Is the process repeatable in the future without sedation?
- If anaesthesia is considered the best process, what regime will be employed? Will there be an effect of any anaesthetic regime on samples/experimental parameters?

Animal and environmental parameters, such as size, age and health status of the animal, whether it is part of a social group or isolated, how long and how frequent the procedure will need to be repeated, etc., will impact on the decision whether to anaesthetise or not.

Preparing the animal

Pre-anaesthetic fasting

Primates being anaesthetised for elective surgical procedures should be fasted. Except for marmosets and tamarins, this should be for at least 12 hours beforehand (remove water 6 hours beforehand). In these smaller species, 6 hours fasting is usually sufficient.

Analgesia

This should be considered for all procedures that would be considered painful in humans. Not only is this ethically sound but, after a painful procedure, the animal will recover a lot faster from an anaesthetic, with pain relief as part of the anaesthetic plan.

Sedation or general anaesthesia

Whichever would be more appropriate for the situation.

Emergency drugs and protocol

At every facility a plan must be in place as to what action needs to be taken if an anaesthetic does not go as planned. The standard **A**irway **B**reathing **C**irculation **D**rug therapy protocol can be employed, or a variant of it. Emergency drugs such as adrenaline, atropine and doxapram should be readily available, either with doses already drawn up into labelled syringes for that animal, or with dose charts readily to hand. Each facility will need to adapt any emergency anaesthetic complication protocol to their own situation.

Animal restraint while under anaesthetic

Some method of physical restraint is mandatory, especially in the larger monkey and ape species.

Route and method of anaesthetic induction and maintenance

This will be based on the animal's species, size, age, and state of health.

Induction using injectable agent

Most anaesthetic induction agents are given via intramuscular or intravenous injection. Care is needed to ensure that primates do not injure themselves either following the administration of injectable anaesthetics or during the recovery phase (see below). If an animal is remotely injected with anaesthetic (such as with a blowpipe or dart gun), they should be on the ground when darted so they are not injured if they fall. Preferably, all monkeys should be induced in crush cages. This process is quick and accurate. Enclosure design should accommodate the use of various sized crush cages. When animals are released after dosing with an injectable anaesthetic, for example, they invariably sit high up on a perch or cling high up on the sides of the cage or pen, and in a large cage may fall heavily when they lose consciousness. This can be prevented by using a small cage for anaesthetic induction, or by supporting the animal as it succumbs.

Induction using inhalation agent

In small species, such as marmosets and tamarins, or in juvenile Old World monkeys, either manual restraint or an induction box can be employed and the animals masked down with inhalation agent. This can be more stressful to the individual concerned, particularly with an agent such as isoflurane which has an unpleasant odour and is a respiratory irritant. Sevoflurane does not have these negative aspects and should be considered in animals too sick for an injectable agent, or those more amenable to handling. Little is yet published on the use of this anaesthetic in non-human primates.

Anaesthetic maintenance

Whatever the method of induction, placement of an endotracheal tube, to maintain a patent airway, is recommended in all general anaesthetic cases, even if inhalation anaesthetics are not to be administered. As intubation does not present any particular problems in primates, a well fitting face mask can alternatively be employed by experienced anaesthetists. Endotracheal intubation allows precise delivery of inhaled anaesthetic agents. Intubation in small non-human primates (less than 1 kg), is straightforward, using commercially available equipment and careful positioning of the animal (Morris *et al.*, 1997).

In all primates, tracheal length is relatively short, so endotracheal tubes should only be inserted to just past the larynx, to prevent bronchiole intubation and consequent over inflation of one or other lung.

In chimpanzees, laryngeal masks could be used to minimise tracheal irritation. The larynx should be sprayed with local anaesthetic before intubation, particularly following induction with ketamine alone, as the laryngeal reflexes will still be functioning. Very young or very sick primates can be induced using gas anaesthesia or intravenous propafol.

Anaesthetic monitoring

During induction

Most injectable anaesthetics have a time to maximal effect of at least 10 minutes. For staff and animal safety, it is recommended that remote monitoring (i.e. Respirations) only be carried out for 15 minutes post induction for *any* intramuscular injectable regime. This will allow maximal effect to occur, without the potential of external stimuli causing the animal to 'fight' the anaesthetic.

During procedure

The primary purpose of monitoring is to maximise the safety of an anaesthetic procedure, particularly by avoiding excessive anaesthetic depth. The depth of anaesthesia is often gauged by the rate and depth of respiration, heart rate, and degree of jaw tension. The palpebral and pedal reflexes may also be used, depending on the anaesthetic used. While these can all be useful in determining if the animal is 'too light', they provide little information about an animal that is 'too deep' (Horne, 2001). Hypoxia and hypothermia are the two main causes of anaesthetic emergencies. Continuous monitoring of oxygen-haemoglobin, with pulse oximetry, to maintain above 85% saturation and of temperature, to detect hypoxia and hypothermia, represent *a bare minimum* that must be undertaken. Anaesthetic emergencies can also occur because of hypoventilation (measure end tidal CO_2 via capnography), hypotension (measure blood pressure) and cardiac arrythmias (measure cardiac electrical activity via ECG). Careful monitoring of all these parameters allows for continuous assessment of cardiopulmonary function and provides early warning, should problems arise (Horne, 2001).

A note on body temperature

Hypothermia may be a significant problem during surgery on primates, unless care is taken to keep animals warm during anaesthetic induction and the surgical preparation time. For example, Ketamine has been shown to produce a fall in rectal temperature to 35°C within 10 minutes at an ambient temperature of 30°C. Induction in the home cage means that the animal is lying on metal mesh or bars when it loses consciousness. Any drop in temperature may then be exacerbated by clipping extensive areas of hair and by the application of large volumes of rapidly evaporating skin preparation solutions (Baskerville, 1999). Intraoperatively, monkeys should have their body temperature monitored and maintained by use of a circulating heating blanket, heat pads and/or heat lamps. Warm intravenous fluids are also helpful in preventing hypothermia. Important for long-term anaesthesia is the placement of an indwelling catheter that allows for administration of fluids and emergency drugs. Sterile technique for catheter placement is imperative (Hrapkiewicz *et al.*, 1998). Latex gloves can be filled with hot water to provide short-term hot water bottles. These, however, must be removed and replaced regularly, and removed totally before recovery. Hypothermia is indicated at temperatures below 36°C.

The stress of being handled may also result in a rise in body temperature to, at *most*, 39.5°C. Any temperature above this, or one which doesn't reduce, usually indicates an underlying disease process.

Two commonly used anaesthetic combinations, ketamine acepromazine and tiletamine-zolazepam, were compared to identify their effects upon body temperature in cynomolgus macaques (Lopez *et al.*, 2002). Thirty cynomolgus macaques, previously implanted with subcutaneous telemetry devices, were allocated into two groups of 15 animals. Baseline temperature data were collected for 3 days before administering anaesthesia to establish normal diurnal temperature patterns for each monkey. Each group was then anaesthetised with either ketamine-acepromazine or tiletamine-zolazepam, and their body temperatures were recorded at 15-min intervals. Both groups had marked decreases in body temperature, with the greatest decreases in the tiletamine-zolazepam group. In addition, both groups had notable post-anaesthesia elevations in body temperature that often lasted for more than 24 h post induction.

Anaesthetic recovery

Recovery must be conducted in as calm and quiet environment as possible. Keep the animal warm, and away from draughts. Extubation of the endotracheal tube should not occur until the animal can swallow. A string can be tied to the tube so it can be removed remotely. Providing a favourite toy for the animal may be appropriate to help keep it calm until it is sure of its surroundings. During recovery, primates try to climb upwards as soon as they recover consciousness. Fortunately, they rapidly regain the ability to cling with their hands and feet, and support their own weight. They usually then stay quietly clinging onto something until they have fully recovered.

A note on euthanasia

For animal welfare and staff safety reasons, this must *always* be conducted when the animal is fully anaesthetised. A phenobarbitone overdose intravenously (femoral vein) is the most common regime. This substance is extremely irritating for the animal when injected perivascularly as it causes rapid onset tissue necrosis.

Summarised anaesthetic information and dose rates are presented in Tables 18.2, 18.3 and 18.4 at the end of this section. Suggested analgesic dose rates are presented in Table 18.5.

How do anaesthetics work?

Pre-anaesthetic medications

Analgesics should be given for invasive procedures that are known to cause pain in other mammals. Buprenorphine (Temgesic) is recommended for control of acute or chronic visceral pain but can cause sedation. Butorphanol (Torbugesic) is recommended for mild postoperative discomfort. Both these drugs are opiate based preparations. Note however that primates have more of an agonist effect at the Mu receptor (compared to other mammals) which may cause severe respiratory depression when using butorphanol, especially in combination with anaesthetic agents (Ligouri *et al.*, 1996). The respiratory depressant effects of opioids and Benzodiazepines (such as diazepam, see below, and medazolam) are synergistic in humans. Benzodiazepines (for example diazepam) alone have minimal respiratory depressant effects but, combined with even small doses of opioids, have been shown to cause respiratory collapse in some humans. Thus, these two classes of drug should not be mixed in non-human primates. Oral diazepam, used as a pre-induction agent at 1 mg/kg at about one hour pre-induction, provides amnesia. However, clinical sedation when using this drug is highly variable between individuals, and species. Note that diazapam can also be given intravenously, to reduce seizures, but is not effective in primates when given intramuscularly (Horne, 2001). Benzodiazepines have minimal effects on cardiopulmonary function and are generally considered very safe compounds to use in primates.

Injectable anaesthetics

Ketamine

Ketamine used alone as an injectable anaesthesia agent has long been the mainstay of primate anaesthetics. Ketamine would be the induction agent of choice if pre-anaesthetic fasting is not possible (for example if an animal has a fight wound requiring immediate attention), as gagging reflexes are maintained.

Ketamine belongs to the cyclohexamine class of drug and, like PCP, is a non-competitive NMDA receptor antagonist which is short acting and has been used as a dissociative anaesthetic as well as a research tool in psychosis (Shiigi and Casey, 1999). As a dissociative agent, higher doses produce a state of catalepsy and profound analgesia. Cardiovascular function is slightly stimulated, rather than depressed, when ketamine is used, and there is respiratory depression only at high doses (Baskerville, 1999). The pharyngeal and laryngeal reflexes are well maintained, except at very high doses, which is an advantage since primates often have to be anaesthetised when food has not been withheld beforehand (thus increasing the risk of emesis and reflux, possibly leading to aspiration). The drug may also be given on repeated occasions, though some tolerance may develop. As ketamine is frequently used for restraint for taking blood samples it should be noted that high doses decrease the leucocyte and erythrocyte values. Animals given ketamine are often adequately anaesthetised for placement of an endotracheal tube and can then be maintained by inhalation anaesthesia. If a monkey is too light to be intubated on the initial dose of ketamine, a small intravenous bolus of ketamine (at a quarter the original dose) can be given to achieve intubation for gaseous anaesthesia, or administration of oxygen alone.

Ketamine alone is not satisfactory for major surgery since it does not provide adequate analgesia. Muscle tone is increased during ketamine anaesthesia, making it unsuitable for use alone for invasive surgery. It can be mixed with other anaesthetic drugs to improve the rating of anaesthetic. The most common combination is with medetomidine (see Table 18.4).

Anaesthetic Research Box

Ghaly *et al.* (2001) examined the effect of incremental hypnotic ketamine dosages on somatosensory and neural motor volleys, recorded epidurally in response to transcranial magnetic stimulation. Their findings reflected the maintenance of a state of neural excitability under ketamine induced anaesthesia. This helps to explain the lack of muscle relaxation and the anecdotal tolerance to this drug in a number of species.

Leopold *et al.* (2002) used optokinetic responses and functional magnetic resonance imaging (fMRI) to examine visual processing in monkeys whose conscious state was modulated by low doses (1–2 mg/kg) of ketamine. They found that, despite the animal's dissociated state and despite specific influences of ketamine on the oculomotor system, optokinetic nystagmus (OKN) could be reliably elicited with large, moving visual patterns. Responses were horizontally bidirectional for monocular stimulation, indicating that ketamine did not eliminate cortical processing of the motion stimulus. Also, results from fMRI directly demonstrated that the cortical blood oxygenation level-dependent response to visual patterns was preserved at the same ketamine doses used to elicit OKN. Finally, in the ketamine-anaesthetised state, perceptually bistable motion stimuli produced patterns of spontaneously alternating OKN that normally would be tightly coupled to perceptual changes. Taken together these results demonstrate that, after ketamine administration, cortical circuits continue to process visual patterns in a dose-dependent manner, despite the animal's behavioural dissociation.
While perceptual experience is difficult to evaluate, under these conditions, oculomotor patterns revealed that the brain not only registers, but also acts upon, its sensory input, employing it to drive a sensorimotor loop and even responding to a sensory conflict by engaging in spontaneous perception-related state changes. This is an important finding, as ketamine alone does not provide amnesia, thus creating increased stress in the individual in future anaesthetics. The experimenters concluded that the ketamine-anaesthetised monkey preparation offers a safe and viable paradigm for the behavioural and electrophysiological investigation of issues related to conscious perception and anaesthesia, as well as neural mechanisms of basic sensory processing.

Ketamine: Medetomidine combination

This combination is the most common induction regime of non-human primates in the UK. It provides an ideal injectable combination in primates for minor short-term procedures (30–45 mins). Metidomidine can be reversed using atipamezole (see Table 18.4).

Medetomidine is an α-2 agonist, thus produces profound sedation and analgesia by virtue of its ability to modulate neurotransmitter release in noradrenergic and serotonergic pathways of the spinal cord and brain (Horne, 2001). There is slight hypotension on administration of this drug, and this can readily be controlled by IV fluid administration. Note however that it is unsafe to use medetomidine alone, particularly in great apes, because they can readily be roused, even from a very deep sedation (see oral use of medetomidine in the chimpanzee section). Thus, a medetomidine/ketamine combination is used (see Table 18.4 for dose rates).

A recent study, in a laboratory setting, compared ketamine alone with ketamine plus medetomidine for balanced anaesthesia and assessed the repeated intramuscular use of ketamine and its potential for tissue damage (Sun *et al.*, 2003). The combination of ketamine and medetomidine was tested in newly arrived macaques undergoing a period of quarantine in an animal facility. Results indicated that the medetomidine and ketamine combination induced a deeper, more level plane of anaesthesia of longer duration than did ketamine alone. Furthermore, use of the medetomidine-reversing agent, atipamezole, permitted more rapid recovery. In addition, a preliminary study in adult rats was undertaken to assess tissue damage induced by intramuscular injection of ketamine versus the combination of ketamine and medetomidine. Histological evaluation of tissue inflammation and muscle necrosis, in rats, indicated that the lower dose of ketamine, afforded by combination with medetomidine, caused markedly less damage to muscle tissue at injection sites. This combination of ketamine and medetomidine has long been the combination of choice for injectable primate anaesthesia in many zoological settings.

Anaesthetic Research Box

The cardiorespiratory effects, effectiveness, and reversibility of two injectable anaesthetic combinations were compared in captive patas monkeys (Erythrocebus patas) (Kalema-Zikusoka *et al.*, 2003). Seven patas monkeys were hand-injected with medetomidine (40 microg/kg, i.m.), butorphanol (0.4 mg/kg., i.m.), and ketamine (3.0 mg/kg., i.m.), and seven were injected with the same dosages of medetomidine and butorphanol plus midazolam (0.3 mg/kg, i.m.). Heart rates decreased in monkeys in both treatment groups and were lower than those previously recorded in patas monkeys anaesthetised with either ketamine or ketamine and isoflurane. Mean arterial pressures were highest in ketamine-treated monkeys but remained within normal limits for both groups. End tidal CO_2 values increased gradually over time in both groups and were above physiologic norms after 20 min. Respiratory rates were similar between groups and remained constant throughout the procedures. Despite adequate ventilation parameters, initial low percent oxygen-hemoglobin saturation (SpO_2) values were suggestive of severe hypoxemia. It was not clear whether these were accurate readings or an artifact of medetomidine-induced peripheral vasoconstriction. Oxygen supplementation restored SpO_2 values to normal (>94%) in both groups. Both combinations effectively produced a state of light anaesthesia, although spontaneous recoveries occurred after 30 min in three ketamine-treated monkeys. All monkeys were given i.m. atipamezole (0.2 mg/kg) and naloxone (0.02 mg/kg), whereas midazolam-treated monkeys also received flumazenil (0.02 mg/kg, i.v.), which resulted in faster (median recovery time = 5 min) and more complete recoveries in this group. Both combinations were considered safe to use when supplemented with oxygen, although the midazolam combination provided a longer anaesthetic period and was more fully reversible.

Tiletamine-zolazapam (Telazol/Zoletil) is a 1:1 combination of the cyclohexamine, tiletamine, and the benzodiazepine, zolazapam. It has been widely used in primate medicine, particularly in field situations, either alone or combined with an α-2 agonist (see Table 18.4 for dose rates). It is a good combination for anaesthetic novices, as side effects are minimal, and the therapeutic index is wide, so potential overdosing is unlikely to cause long-term problems. Tiletamine is more potent than ketamine and thus small volumes of zoletil are required for injection. For example, in the author's experience, zoletil at 3–4 mg/kg provides excellent immobilisation in chimpanzees for 30–40 minutes, but full recovery can take a couple of hours. Zoletil immobilised primates tend to maintain stable cardiopulmonary parameters (Horne, 2001).

Propafol (see Tables 18.2 and 18.4), given intravenously through an indwelling catheter provides a smooth, rapid (less than 30 seconds) induction of anaesthesia and can be used to maintain anaesthesia by intermittent bolus administration – which would be required every 5–10 minutes. Monkeys maintained on propafol should be intubated, as propafol may induce a short period of apnea. Monkeys recover from propafol smoothly and rapidly (Hrapkiewicz *et al.*, 1998).

Inhalation anaesthetics

All of the common inhalation agents have been used in primates, with isoflurane currently being the most popular. Although generally acknowledged as being much safer than halothane, Isoflurane does have potent vasodilatory properties that can cause severe hypotension if not used carefully (Horne, 2001). Sevoflurane (widely used in humans, but so far only used experimentally in non-human primates) is not as soluble in the blood as isoflurane, while maintaining a 50% greater alveolar concentration, translating clinically to much faster induction and recovery times. In a calm patient, mask induction and intubation can be achieved in as little as 5 minutes (compared to 10 minutes for isoflurane). Whatever inhalation agent is used, they all cause cardiovascular depression, although, unlike halothane, this is very minimal with sevoflurane and isoflurane, and all depress the respiratory system.

Anaesthetic Research Box

The effects of sevoflurane on cerebral metabolism and hemodynamics have been studied in rhesus monkeys (Yoshikawa *et al.*, 1997). At 3.0% sevoflurane, regional cerebral blood flow (CBF) increased significantly in response to the increase in the mean arterial pressure, suggesting the inhibition of autoregulation of CBF. However, regional CBF/CMR O_2 ratio was not significantly different among the cerebral regions with each condition. It could be concluded that CBF, during up to 3.0% sevoflurane anaesthesia might become dependent on the cerebral perfusion pressure and the changes in regional CBFs varied among the regions On the other hand, the ratio of oxygen consumption and delivery was well maintained throughout the brain regions.

Group specific regimes

Marmosets and tamarins

Induction recommendation: Ketamine:medetomidine combination or mask induction with Isoflurane or sevoflurane.

The following sedative and anaesthetic agents have proven suitable for marmosets. Ketamine hydrochloride at 10–15 mg/kg, by intramuscular injection, can be used for short-term restraint for minor, non-painful procedures, such as fitting identity collars or identity chips, but muscle relaxation is often poor. A ketamine/medetomidine mixture (3 mg/kg ketamine with 50 ug/kg medetomidine given intramuscularly) has the advantage that the anaesthesia can be reversed by the administration of atipamezole intravenously or intramuscularly (Poole *et al.*, 1999). This must only be done at least 30 minutes after induction, to minimise residual ketamine effects. Inhalation agents such as Isoflurane or sevoflurane can be used as induction agents as well as to maintain anaesthesia: these are well tolerated by the animals who recover rapidly. When using Isoflurane as an induction agent, with the animal in manual restraint, note that it has an unpleasent smell, and can be irritating to the airways. A well fitting face mask or, preferably, intubation should be used whatever anaesthetic regime is used, in order to maintain a patent airway if emergency resuscitation is required.

Anaesthetic Research Note

A safe means of anaesthetising common marmosets (Callithrix jacchus) for a study using magnetic resonance imaging (MRI) to investigate cerebral ischaemia was required (Whelan *et al.*, 1999). Continuous infusion of alphaxalone/alphadalone was used to anaesthetize 37 marmosets for non-recovery and recovery experiments. This was found to give safe, reliable anaesthesia when coupled with pulse oximetry and electrocardiographic (ECG) monitoring.

Owl monkeys and other similar sized species

Induction recommendation: Ketamine:medetomidine combination.

For blood sampling and extensive physical examination etc., light sedation with ketamine HCl (10–15 mg/kg) IM is recommended (Eckhert, 1999).

Ideally, these animals should be placed in a crush cage and hand injected with the anaesthetic. They should then be released from the crush, and observed from a distance for the anaesthetic to take effect. Handling during the induction period should be discouraged as this causes the animal to 'fight' the anaesthetic.

When anaesthesia is required in squirrel monkeys, ketamine hydrochloride (10 mg/kg) is generally sufficient for examinations, blood sampling and minor surgical procedures. Isoflurane (1–2%), administered as an inhalant via intubation, should be used in addition to ketamine for more invasive or extensive surgical procedures. Administration of an analgesic, such as buprenorphine (0.01–0.03 mg/kg twice daily via IM injection), is recommended following major surgical procedures. Animals should be held in small transport cages and closely monitored during recovery from anaesthesia (Mendoza, 1999).

Cynomolgus monkeys were anaesthetised with either intramuscular ketamine (10 mg/kg) or intramuscular ketamine (2 mg/kg) and medetomidine (50 microg/kg) (Young *et al.*, 1999). Various physiological measurements were made once the animals were safe to handle and again 10 min later. Cardiovascular and respiratory function were well maintained, with both regimens, but the heart rate was lower, and arterial-alveolar carbon dioxide gradient was higher in the animals that received medetomidine. In those animals that received medetomidine, atipamezole was given to reverse the medetomidine but there was no difference in recovery times between the two regimens. (This is highly

Anaesthetic agent	Pharmacology	Pharmacokinetics	Contraindications	Adverse effects
Ketamine	GABA inhibitor. Effective to Stage II anaesthesia. No change or increased muscle tone. Increased cardiac output due to increased sympathetic tone. Retains pinnal, pedal, photic, laryngeal, corneal and pharayngeal reflexes.	Well distributed to all tissues– highest levels in brain, liver, lung and fat. Increasing the dose will increase anaesthetic duration but not the intensity.	Hypersensitivity, major surgery (when used alone), hypertension, heart failure, arterial aneurysms.	Respiratory depression following high doses, emesis, vocalisation, erratic and prolonged recovery, dyspnoea, spastic jerky movements, convulsions, muscular tremors, hypertonicity, opisthotonos and cardiac arrest. Eyes remain open– use Lacrilube® or similar.
Medetomidine	Alpha adrenergic receptor $\alpha 2{:}\alpha 1$ sensitivity factor of 1620:1 (cf xylazine at 10:1). Sedation, decreased GI and endocrine secretions, peripheral and cardiac vasoconstriction, bradycardia, respiratory depression, diuresis, hypothermia, analgesia, muscle relaxation, blanched or cyanotic mucous membranes and anxiolytic effects.	Onset of effect – 5 mins post IV, 10–15 mins post IM. Can be absorbed via the oral mucosa.	Animal MUST BE CALM BEFORE INDUCTION WITH MEDETOMIDINE. Cardiac disease, respiratory disorders, liver or kidney disease, shock, severe debilitation, stress due to heat, cold or fatigue.	An extension of the pharmacological effects – bradycardia, occasional AV blocks, decreased respiration, hypothermia, urination, vomiting, hyperglycaemia, pain on injection. Rare effects have also been reported including prolonged sedation, paradoxical excitation, hypersensitivity, apnoea and death from circulatory failure.
Atipamezole	Medetomidine reversal agent–inhibits $\alpha 2$ adrenergic receptors.	Peak effect in 10 mins. Rapidly metabolised by the liver.	None. Caution in pregnant or lactating animals.	Vomiting, diarrhoea, hypersalivation, tremors and brief excitation/ apprehensiveness. Because reversal can occur rapidly, care should be exercised as animals recovering from sedation and analgesia may exhibit apprehensive or aggressive behaviours. Reversal should only be undertaken when staff safety is assured.

Zolazepam/ Tiletamine	Retains reflexes as for Ketamine.	Onset of action is variable, but fairly rapid – 6–10 mins.	Pancreatic disease, severe cardiac or pulmonary disease.	Hypersalivation in some species (use atropine to counter at induction), transient apnoea, erratic and/or prolonged recovery.
Propafol	Short acting hypnotic – method of action is not well understood.	Onset usually less than 1 min post IV injection. Its short duration of action (10 mins) is due to rapid redistribution from the CNS to other tissues.	Hypersensitivity. Caution in pregnant or lactating animals.	High incidence of apnoea with resultant cyanosis if propafol is given too rapidly. Give slowly (23% of calculated dose every 30 seconds until desired effect. SINGLE USE ONLY. Extravasation of injection is not irritating nor does it cause tissue sloughing.
Diazepam	CNS depressant (anxiolytic, sedative, skeletal muscle relaxant, anticonvulsant).	Peak levels 30 mins to 2 hrs following oral administration. Slower following IM injection.	Too rapid injection of IV diazepam, in small animals or neonates, may cause cardiotoxicity secondary to the propylene glycol in the preparation. Hypersensitivity. Caution in pregnant or lactating animals.	Very variable effects.
Isoflurane	MAY interfere with the function of nerve cells in the brain. Induces CNS depression, depression of body temperature regulation centres, increased cerebral blood flow, repiratory depression, hypotension, vasodilation, and myocardial depression (less so than with halothane) and muscular relaxation.	Rapid absorption from the alveoli to the brain. Most (99.8%) is eliminated via the lungs. The remainder is metabolised in the liver.	Malignant hyperthermia. Use with caution in animals with CSF or head injury and during pregnancy (may be fetotoxic).	Hypotension (secondary to vasodilation) is considered to be dose related. Dose dependent respiratory depression, GI upset, cardiac arrythmias.

unusual when compared to other primate species). Anaesthesia was not entirely reliable with medetomidine/ketamine and the researchers recommend caution when using this mixture in cynomolgus monkeys.

Anaesthetic Research Box

Results have confirmed that ketamine is a suitable anaesthetic agent to immobilise male cynomolgus monkeys in experimental studies (short- and long-term studies) aimed at elucidating hormonal changes (Malaivijitnond *et al.*, 1998). There were no statistically significant differences in serum cortisol, testosterone, ILH and Bio LH values between the first blood sample (before the ketamine injection) and sequential blood samples in monkeys.

Macaques and baboons (Old World monkeys)

Induction recommendation: Ketamine:medetomidine combination or Zoletil:Medetomidine combination.

Sedation and general anaesthesia are of particular importance in Old World primates because of the frequent need to provide chemical restraint for safety reasons. General anaesthesia also presents additional difficulties in species which cannot be manually restrained for induction. Injectable anaesthetics should be used for induction, to avoid the stress of induction by inhalation, and anaesthesia can be maintained at 1–1.5% isoflurane or sevoflurane in oxygen.

Ketamine is the most commonly used agent, both for chemical restraint and for induction of anaesthesia in all but very young animals. It can be given intramuscularly, in small volumes, and has a wide safety margin. The ability to bite is lost at low doses and at an early stage of induction. Doses of 5–10 mg/kg produce effective sedation for handling and close examination and for minor procedures, such as taking blood samples or passing a gavage tube. Doses of 10–25 mg/kg give sufficient depth of anaesthesia for minor surgical procedures. Due to the increased muscle tone when using ketamine alone, analgesia, with good muscle relaxation, can be produced by combining ketamine (5–10 mg/kg) with medetomidine (50 ug/kg), an alpha 2 agonist. This combination also has a wide therapeutic index.

Anaesthetic Research Box

A detailed anaesthetic technique for baboons (Papio anubis), undergoing heterotopic abdominal cardiac xenotransplantation, has recently been described (Santerre *et al.*, 2001). Twenty-two baboons served as transplant recipients. Donors were either crossbred farm pigs (Sus scrofa) (n = 4) or transgenic pigs (Sus scroefa) (n = 18) expressing human complement regulatory proteins on the endothelium. Intra-operative management was complicated by the physiological consequences of infrarenal abdominal aortic cross-clamping, in addition to the immunological sequelae related to cross-species transplantation. In choosing anaesthetics for this procedure, the researchers considered the need for maximal cardiac stability, throughout a long surgical procedure that required abdominal aortic cross-clamping, to facilitate the implantation of an oversized porcine cardiac graft. Baboons received a balanced anaesthetic consisting of inhaled isoflurane in oxygen, intravenous fentanyl and intravenous pancuronium. The pharmacological techniques employed were found to be safe and reliable and were well tolerated by our recipients without any significant side-effects.

To anaesthetise adult rhesus monkeys, to obtain high resolution 3D MR images, Fowler *et al.* (2001), demonstrated the usefulness of intravenous propofol immediately followed by conversion to inhalation anaesthesia and stereotactic intracranial surgery with the head frame 'in situ'. There was minimal morbidity while achieving a high degree of precision for the stereotactic targeting.

Panadero *et al.* (2000) investigated anaesthetic techniques for experimental otoneurologic surgery in macaques. Induction consisted of an intramuscular mixture of ketamine, midazolam and atropine. The surgical procedure was performed under intubated general anaesthesia, after propofol (1.5 mg/kg) administration, and maintained with nitrous oxide and halotane. The mixture of ketamine, midazolam and atropine produced a deep anaesthesia in 17 minutes, permitting safe animal handling. Atraumatic nasotracheal intubation, without muscle relaxing agents, was easily achieved in all animals. Anaesthesia was adequately maintained with nitrous oxide and halotane. Animals did not present any relevant incidents during surgery and were extubated 2.5 minutes after cessation of gas administration. No relevant surgical complications occurred post operatively.

The experimenters concluded that this anaesthetic technique provided optimal restraint and anaesthesia for experimental otoneurosurgical procedures with primates. It offered a quick recovery and avoided the use of muscular relaxing agents for intubation, and thus could be safely used in other kinds of surgical procedures.

Anaesthesia Research Box

Du Ploy *et al.* (1998) investigated the stability of cardiodynamics, and some blood parameters, during a slow, continuous infusion of a combination of ketamine and diazepam. Contractility (dP/dt), myocardial relaxation (Tln), left ventricular end-diastolic pressure (LVEDP), left ventricular systolic pressure (LVSP), arterial blood pressure and certain blood parameters were assessed in 3 male and 3 female juvenile baboons (Papio ursinus). Anaesthesia was induced with 15 mg/kg ketamine i.m. and maintained with a continuous i.v. infusion (40–60 ml/h) of ketamine and diazepam. The mixture consisted of 2 ml ketamine (100 mg/ml), 2 ml diazepam (5 mg/ml) and 50 ml saline. A period of 10 min was allowed for preparation of the animals, after which lead II of the ECG, femoral artery blood pressure and left ventricular pressure were recorded at 15-min intervals for a period of 2 h with a total duration of anaesthesia of 195 min. Arterial blood samples were analysed, at 30-min intervals, for blood gases, electrolytes, glucose and insulin. Left ventricular parameters were derived from the left ventricular pressure curve. Tln, LVSP and LVEDP showed small fluctuations. Contractility decreased ($p < 0.037$) at the 195-min interval. No arrhythmias or ECG changes were seen, while blood pressure decreased gradually. Serum calcium concentration decreased, and blood glucose levels increased gradually over time. Anaesthesia and analgesia were sufficient and no other drugs were necessary. The animals appeared sedated and dazed 60–80 min after the procedure. It was concluded that a continuous infusion of a combination of ketamine and diazepam, for a duration of 150 min, can provide stable anaesthesia for cardiodynamic measurements.

Wild red howler monkeys (Alouatta seniculus) were translocated during the flooding of the forest at a hydroelectric dam site in French Guiana (Vie *et al.*, 1998). For a variety of minor clinical procedures, 96 monkeys were anaesthetised with various intramuscular injections of combinations of medetomidine and ketamine. The howler population was composed of healthy animals (42 males and 54 females) of various ages. Medetomidine (150 micrograms/kg) associated with ketamine (4 mg/kg) gave the best results and was used on 63 animals. The injection rapidly resulted in complete immobilisation with good to excellent myorelaxation. The induction stage was quiet, with absence of both corneal and pedal withdrawal reflexes, in 57 animals after 10 minutes. Six animals required an additional injection. Rectal temperature and respiratory and heart rates decreased during anaesthesia, whereas relative oxyhemoglobin saturation increased. One death occurred during anaesthesia. One abortion and one death also occurred the day following anaesthesia but were more probably a result of capture stress. Atipamezole, given IM, 45 min after the anaesthetic injection, at a dose of five times the medetomidine dose led to standing recovery in 4.5 min. Spontaneous recovery, after an average of 45 min, occurred in 17 animals before the atipamezole injection. Total recovery time was shorter in young animals. Medetomidine/ketamine induced good myorelaxation and provided considerably shortened immobilisation duration, which are two notable advantages for field studies. This combination was recommended for short procedures, including minor surgery, in red howler monkeys.

Anaesthetic Research Box

The influence of thiopentone intravenous infusion, or halothane inhalation, on the results of radiorenography was evaluated, in six chacma baboons (Papio ursinus), using 99mTc-diethyltriamine-pentaacetic acid (DTPA) as scanning agent (Dormehl *et al.*, 1984). The renogram parameters, which depend on the condition of the cardiovascular system, differed significantly for the two anaesthetic agents. Since, in baboon studies, anaesthesia is necessary for the duration of renogram acquisition, it is imperative to standardise an experimental procedure which will leave the cardiovascular system relatively stable. From this investigation it seemed most appropriate to use a constant intravenous infusion of thiopentone as an anaesthetic.

A major advantage with medetomidine is that its effects are reversible with atipamezole. The sedative effects of medetomidine, and a medetomidine-midazolam combination, in Japanese macaques, and the antagonism of medetomidine-midazolam with atipamezole have been evaluated. Medetomidine (120 μg kg^{-1}) alone, or a medetomidine (30 μg kg^{-1}) plus midazolam

(0.3 mg kg^{-1}) mixture, were injected intramuscularly in the hind limb of 12 animals (n = 6 for each group) and their effects, particularly behavioural changes, response to external stimuli, sedative onset time, time to lateral recumbency and time in lateral recumbency, were monitored for 120 minutes.

Another group (n = 7) were given medetomidine-midazolam and injected 30 minutes later with atipamezole (120 µg kg^{-1}). Behavioural changes and responses to external stimuli were assessed as before. Animals given medetomidine became sedated but could be aroused by external stimuli. Despite the lower (25%) dose of medetomidine involved, the effects of medetomidine-midazolam were more marked. Macaques, given this combination, became sedated in 4 ± 2 minutes (mean ± SD) and remained unresponsive to external stimuli for at least 60 minutes. Five out of six macaques became laterally recumbent for 74 ± 37 minutes. Intramuscular atipamezole effectively reversed sedation, shortening the arousal and total recovery time. The recovery from sedation was rapid and smooth, being completed 19 ± 11 minutes after antagonism. It was concluded that the medetomidine-midazolam combination provided useful chemical restraint and may prove useful in macaques undergoing some experimental, diagnostic or therapeutic procedures. Thus it was shown that the use of atipamezole, as an antagonist, increases the value of this technique in macaques.

Chimpanzees

Induction recommendation: Ketamine:medetomidine combination or Zoletil.

Chimpanzees should be appropriately fasted for food and water prior to scheduled procedures requiring sedation or anaesthesia. Following sedation or anaesthesia, food and water should continue to be withheld until the animal has regained full consciousness. Selection of an anaesthetic is often based on the nature of recovery of individual animals; some exhibit severe motor movements and apparent hallucinations, and should be monitored carefully. Recovery cages, that restrict movement and permit access to the animal, should be used. Animals should not be returned to their home cage until they are fully recovered. Veterinary and care staff should learn the response of each animal to various procedures and sedatives and be aware that other chimpanzees, within the compound or in an adjacent cage, might attack a sedated animal. Recognition of pain in chimpanzees, and the need for analgesics, is best achieved by care-givers who know species-typical chimpanzee behaviours, and who are sufficiently familiar with the individual to recognise subtle changes in behaviour and mood (Fritz *et al.*, 1999).

Diazepam has been used for many years as a pre-requisite to immobilising chimpanzees One suggested regime has been 5 mg per animal for juveniles and 10–15 mg per animal for adults, for 5 nights prior to immobilisation with ketamine. However, this regime produces a 'calmer' animal, rather than drowsiness. An alternative to pre-induction sedation in chimpanzees is training them to accept hand injections.

Staff must be aware that apparently sedated animals can awaken quickly and attack with little notice (Fritz *et al.*, 1999; Sanderson, pers. comm.). This is particularly common with a ketamine:medetomidine combination.

For major surgical procedures in chimpanzees, properly administered inhalant anaesthetics, such as isoflurane or sevoflurane, are recommended, and offer many advantages over other anaesthetics, especially as they provide ready access to the respiratory system if the animal stops breathing. IV catheterisation should also be considered as standard procedure for all surgery. Six adult female chimpanzees (Pan troglodytes) were anaesthetised for the placement of intrauterine contraceptive devices, microchips for identification, routine blood sampling, and physical measurements. Anaesthesia was induced with medetomidine, in combination with ketamine, administered by intramuscular injection with a projectile syringe. Induction was smooth and rapid but five of the animals were insufficiently relaxed for orotracheal intubation. The plane of anaesthesia was deepened by administering isoflurane delivered in oxygen and nitrous oxide, and general anaesthesia was maintained for up to 74 minutes. The action of medetomidine was reversed with atipamezole, at the end of each procedure, and the animals recovered smoothly and uneventfully.

Ketamine has analgesic properties in low doses and produces anaesthesia in high doses. It does not produce good muscle relaxation, which is a common reason for using it in combination with other drugs such as medetomidine or xylazine.

Chimpanzees can develop tolerance to ketamine, especially if they must be frequently immobilised. In addition, there are times when the injection appears to have no effect and it must be considered that the dose had been injected into fat or muscle fascia and in these cases there might be a prolonged recovery period.

Telazol/Zolitle, a 1:1 mixture of cyclohexamine tiletamine (2–3 times more potent than ketamine), and the benzodiazepine zolazapam, are used for diagnostic procedures and short anaesthesia, when deeper anaesthesia and fewer muscular contractions are needed than those achieved with ketamine alone. Note however that recovery time is extended beyond that of ketamine.

Butorphanol is not utilised in combination with other anaesthetics due to commonly seen respiratory depression. However, used alone for relatively short periods of sedation, the respiratory depression has not been observed (Fritz *et al.,* 1999).

Anaesthetic Research Box

Five chimpanzees (Pan troglodytes) initially received oral droperidol sedation (1.25 mg for a juvenile chimpanzee, body wt = 18.5 kg, and 2.5 mg for adults, body wt >20 kg, range: 18.5–71 kg) followed by transmucosal carfentanil administration at 2.0 microg/kg (Kearns *et al.*, 2000). This preinduction regimen was developed to produce heavy sedation, or even light anaesthesia, in order to eliminate the need for, or at least minimise the stress of, darting with tiletamine/zolazepam at 3 mg/kg i.m. This study was designed to assess the safety and efficacy of transmucosal carfentanil. Once each animal was unresponsive to external stimuli, or at approximately 25 min (range 24–34 min) after carfentanil administration, naltrexone and tiletamine/zolazepam (N/T/Z) were combined into one intramuscular injection for anaesthetic induction. Naltrexone was administered at 100 times the carfentanil dose in milligrams. For comparison, two chimpanzees received only droperidol, 2.5 mg p.o., followed by tiletamine/zolazepam, 3 mg/kg i.m. The preinduction period for all animals receiving carfentanil was characterised as smooth, with chimpanzees becoming gradually less active and less responsive to external stimuli. Two animals became very heavily sedated at 24 and 35 min, respectively, and were hand injected with N/T/Z. The other three chimpanzees became sternally recumbent but retained some response to stimuli, and N/T/Z was administered, by remote injection, with minimal response. Rectal body temperatures, pulse and respiratory rates, arterial oxygen haemoglobin saturation, and arterial blood gases were measured at initial contact (t = 0 min) and at 10-min intervals thereafter. Respiratory depression was present in all chimpanzees, regardless of protocol. Mean hemoglobin saturation was 91% for both groups. Mean partial pressure of oxygen and arterial values for carfentanil-treated and control animals were 64.4, 7.6 and 63.5 6.0 at t = 0, respectively. Only the partial arterial pressure of carbon dioxide, ($Paco_2$) and pH showed significant differences between treated and control animals. Mean $Paco_2$ was greater and mean pH lower for the carfentanil-treated group compared with the controls at t = 0 (58.9 3.7 and 50.3 3.1 for $Paco_2$ and 7.33 0.02 and 7.40 0.30 for pH, respectively). The results of this study suggest that oral droperidol, followed by transmucosal carfentanil, can be used effectively as a premedication regimen to produce profound sedation. This limits the stress of darting during parenteral anaesthetic induction with tiletamine/zolazepam in chimpanzees. The main side effect of respiratory depression appears to be adequately managed by reversing the carfentanil at the time of induction.

TABLE 18.1: Physiologic parameters of non-human primates

	Rhesus	Baboon
Body temperature	37–39°C	39°C
Heart rate	120–180/min	150/min
Respiratory rate	32–50/min	35/min
Tidal volume	21 ml	50 ml

TABLE 18.2: Restraint/Preanesthesia regimes for non-human primates

Indication and drugs	Dosage and route of administration	
Atropine	0.02–0.05 mg/kg	IM, SQ
Ketamine	see following table	
Diazepam, C-IV (Valium®)	0.5–1.0 mg/kg	IM, IV

TABLE 18.3: General anaesthesia regimes used in non-human primates – those in bold recommended by the author (see text)

Indication and drugs	Dosage and route of administration	
Sodium Pentobarbital, C-II	20–30 mg/kg	IV
Sodium Thiopental, C-III (2.5%)	15–20 mg/kg	IV
Thiamylal Sodium, C-III (2.5%) (Surital®, Bio-Tal®)	25 mg/kg	IV
Ketamine (Ketaset®, Vetalar®)	**see following table**	
Ketamine/Diazepam:		
Ketamine	5–25 mg/kg ; 10 mg/kg	IM
Diazepam (Valium®)	0.5 mg/kg; 7.5 mg/kg	IM/IV
Ketamine/Xylazine:		
Ketamine	7 mg/kg	IM
Xylazine (Rompun®)	0.6–1.0 mg/kg	IM
Ketamine/ medetomidine:		
Ketamine	**5–7.5 mg/kg**	**IM**
Medetomidine (Domitor/Zalapine®)	**0.033–0.075 mg/kg**	
Propafol	2–4 mg/kg; baboons for induction	IV
	2.5–5 mg/kg; macaques	IV
	5–10 mg/kg	IV
Tiletamine + zolazapam (Zolitel/ Telazol®)	**1–15 mg/kg (higher dose rate for smaller species)**	**IM**
	2–6 mg/kg	
Tiletamine + zolazapam/ medetomidine:		
Tiletamine + zolazapam	**1.25 mg/kg (Apes)**	**IM**
Medetomidine:	**0.03–0.04 mg/kg (Apes)**	**IM**
Methoxyflurane (Metofane®)	To effect	IH
Halothane (Fluothane®)	To effect	IH
Isoflurane	**To effect**	**IH**
Sevoflurane	**To effect**	**IH**
Halothane/Nitrous Oxide (50% O_2 + 50% N_2O)	To effect	IH

Reverse Medetomidine with Atipamizole at 0.1–0.25 mg/kg (equal volume to medetomidine 1 mg/mL)

NB. Erythrocebus patas may require a higher dose of xylazine. Cercopithecus sp require a much lower dose of barbiturates.

TABLE 18.4: Analgesia suggestions for non- human primates

Indication and drugs	Dosage and route of administration	
Morphine, C-II	0.5–2.0 mg/kg q4h	SC IM IV
Oxymorphone, C-II		
Old World Primates	0.15 mg/kg	SC IM IV
New World Primates	0.075 mg/kg	SC IM IV
Flunixin meglumine	**0.3–1 mg/kg q12-24h**	**SC, IV**
Meloxicam (Metacam®)		
Meperidine, C-II (Demerol®)	2–4 mg/kg	IM
Pentazocine, C-IV (Talwin®) not to exceed total dose of 60 mg	1.5–3.0 mg/kg q3-4h	IM SC
Buprenorphine (Temgesic®)	**0.01–0.03 mg/kg q8-12h**	**IM, IV**
Acetylsalicytic Acid (Aspirin)	10–20 mg/kg q6h	PO
Acetaminophen	10 mg/kg q8h	PO
	5–10 mg/kg q6h	PO
Butorphanol tartrate (Torbugesic®)	**0.025 mg/kg q3-6h**	**IM**

TABLE 18.5: Suggested initial dosages of ketamine for non-human primate species

Species	Preanaesthesia (mg/kg)	Anaesthesia (mg/kg)
Aotus trivirgatus (Owl)	10.0–12.0	20.0–25.0
Cebus sp. (Capuchin)	13.0–15.0	25.0–30.0
Cercopithecus aethiops (African green)	10.0–12.0	25.0–30.0
Erythrocebus patas (Patas)	3.0–5.0	5.0–7.5
Macaca fascicularis (Cynomolgus)	12.0–15.0	20.0–25.0
Macaca fuscata (Japanese)	5.0	
Macaca mulatta (Rhesus)	5.0–10.0	20.0–25.0
Macaca nemestrina (Pig-tailed)	5.0–7.5	20.0–25.0
Macaca radiata (Bonnet)	12.0–15.0	25.0–30.0
Macaca arctoides (Stump-tailed)	5.0–7.5	20.0–25.0
Papio anubis (Baboon)	5.0–7.5	10.0–15.0
Papio cynocephalus (Baboon)	5.0–7.5	7.5–10.0
Papio papio (Baboon)	10.0–12.0	
Saimiri sciureus (Squirrel)	12.0–15.0	25.0–30.0
Pan troglodytes (Chimpanzee)	5.0–7.5	10.0–15.0

Section 2: Drug administration and sample collection

Methods of drug administration

Manual restraint

Callitrichids do not like being handled or restrained, but they can be trained to co-operate in procedures if the operator uses food treats. Individuals soon become used to handling and, in anticipation of a reward, put up little resistance. For injections, they can be netted and held around the shoulders by a competent handler. It should not be necessary to use a restraining device (Poole *et al.,* 1999). Manual restraint should not be considered for any species over 5 kg, or in any animal not used to being handled.

Oral

Oral drug application should be done on a voluntary basis to avoid stressing the animals by catching them. Small doses of fluids may be injected into locusts or mealworms or trickled onto small pieces of crackers to be offered to the individual by hand. It is also recommended that the monkeys be trained to lick tasty fluids such as sweet vanilla sauce or fruit juice from a syringe so that, subsequently, drugs such as antihelmintics, antibiotics or oral anaesthetics can be administered in the same way. This method is especially useful when chronic administration of a drug is required. Special training methods for voluntary drug intake or working in Skinner boxes etc. have been developed for primate species such as macaques, squirrel monkeys and common marmosets (Erkert, 1999).

Oral gavage for Old World monkeys should only be considered when accurate oral dosing is required for research purposes. Unless the animal is well trained, this should be conducted either intranasally or orally when the animal is sedated, to allow accurate placement of the gavage tube (i.e. so it isn't directed into the trachea). In all circumstances, the throat must be palpated, and the tube's progress down the throat visualised before dosing.

Anaesthetic Research Box

Tips on administering unpalatable oral medications such as antibiotics and sedatives.

- Flagyl with lemon juice and honey
- Use sodas – e.g. Fanta
- Inject drug into a passionfruit – need to make a second hole to prevent bursting
- Mango milkshakes
- Clindamycin mixed to a paste in avocado
- Chocolate
- Cerelac
- Make use of pediatric preparations where can
- Put medication on fur – will often lick it off
- Consider nebulisation – especially for small primates
- Give medicated feed before main feed, and give treats to others if in a group situation to prevent under dosing.

Injectable

In chimpanzees and larger monkeys, which are difficult to handle, and where the animal is untrained or too ill to move to the front of the cage to take medication from a spoon or cup, alternative methods of administering analgesics, sedatives or medication-filled syringes must be used. A blow pipe is preferred to a compressed air, CO_2 or percussion-type gun which are recommended for emergency situations only, such as for capturing escaped animals. Blow pipes require minimal maintenance, are silent, and the projectile syringe imparts less trauma on impact (Fritz *et al.,* 1999). Their maximum range is, at most, 10 metres. Smaller species can be placed in a crush cage or net, and hand injected.

Inhalation

Inhalation of medications via nasal sprays is also a possibility. This is supported by the author's experience, where an asthmatic orangutan was trained to self-medicate with a flixotide facemask inhaler. Pediatric anaesthesiologists make little syringe tips for syringes to convert a liquid anaesthetic to an aerosol for children. A combination could be introduced via this method to sedate the patient enough to allow hand injection or to place a gas anaesthesia mask. This possibility of application needs to be further explored.

Procedures

Withdrawal of body fluids, infusing of materials and catheterisation

Blood samples can easily be obtained from percutaneous venipuncture of the femoral vein or artery of monkeys. They are usually sedated for this procedure, and the femoral triangle must be cleaned with alcohol and clipped if needed. If an arterial sample is taken, direct pressure must be applied for a minimum of 3–5 minutes to obtain adequate hemostasis. Other collection sites include the cephalic and saphenous veins. These sites must always be clipped to avoid contamination. When numerous blood samples are required, placement of an indwelling catheter is preferable to multiple venipunctures. Catheters increase the efficiency and ease of blood sample collection, and reduce stress and pain to the animal. Monkeys can be trained to offer their arms or legs for blood collection with positive reinforcement, but this requires dedicated staff and a considerable amount of time. Surgical placement of vascular access ports and tether systems allow for long-term blood collection and chronic dosing (Hrapkiewicz *et al.,* 1998). The saphenous vein provides an excellent, easily accessible site to place a catheter.

In small species, such as marmosets and tamarins, blood samples can be taken from the femoral vein using a short 0.4–0.5 mm diameter (25–27 gauge) needle with a 1 or 2.5 ml syringe.

A useful rule of thumb is that single blood samples of up to 0.5 ml per 100 g body weight can safely be taken. If repeated blood samples have to be taken, the animal should be given an iron supplement and no more than 0.86 ml per week, or 15% of total blood volume in any month, taken from animals with a weight of 350 g (Poole *et al.,*1999).

In owl monkeys, blood samples of up to 4 ml can be collected from the femoral vein, which runs slightly medial and parallel to the artery. After anaesthetising the animal, the inside of the upper thigh is shaved along the midline and the skin disinfected with alcohol. Under visual guidance, or by palpating the artery with a gloved fingertip, the cannula should be inserted about 1 mm medial to the vein at an acute angle to penetrate it. To avoid haemolytic reactions and rapid coagulation, the use of wide lumen needles (0.9 mm) and heparinised or EDTA-containing syringes are recommended. Immediately after withdrawing the needle, the vein should be compressed briefly to prevent the formation of a haematoma (Erkert, 1999). Blood samples can be collected from squirrel monkeys under manual restraint alone (Mendoza, 1999).

Old World primates can be trained to present a limb for blood sampling and, where repeated sampling is required, this is much less stressful for the animal than being repeatedly crushed up and sedated. Since both ketamine anaesthesia and stress affect haematological results, samples from unstressed conscious animals are likely to be more meaningful when these parameters are being investigated (Baskerville, 1999). A useful chimpanzee blood collection technique involves using a butterfly catheter, with a vacutainer attachment, for femoral or caudal tibial veins. This is rapid and very effective and prevents needle slipping if the animal moves.

Urine collection can be via cystocentesis (anaesthetic required), urinary catheter placement (anaesthetic required), or by free catch (collection pan under the cage).

Surgery

Some of the routine surgeries performed on NHPs include finger and tail amputations and laceration repair. Use of a subcuticular suturing pattern is recommended when closing incisions. This method usually prevents monkeys from pulling sutures out and eliminates the need for suture removal. Postoperatively, monkeys can be placed in a recovery cage (or in their home cages if they're housed singly). A heat source such as a heat lamp should be provided to help prevent hypothermia. They should be observed frequently until they are sitting in an upright position. Monkeys with incisions should be monitored for wound dehiscence, often caused by cage mates grooming. For social reasons, however, reintroduction to the group should not be delayed.

Correspondence

Any correspondence should be directed to Steve Unwin, Veterinary Officer, Chester Zoo, North of England Zoological Society, Chester CH2 1LH, UK.

References

Adams, W.A., Robinson, K.J., Jones, R.S. and Sanderson, S. (2003). *Vet. Rec.* 152(1), 18–20.

Baskerville, M. (1999). In Poole, T. and English, P. (eds) *The UFAW Handbook on the Care and Management of*

Laboratory Animals (7th ed.) *Terrestrial Vertebrates,* Vol. 1, pp 611–635. Blackwell Science, Oxford.

Dormehl, I.C., Jacobs, D.J., du Plessis, M. and Goosen, D.J. (1984). *J. Med. Primatol.* 13(1), 5–10.

Du Plooy, W.J., Schutte, P.J., Still, J., Hay, L. and Kahler, C.P. (1998). *J. S. Afr. Vet. Assoc.* 69(1), 18–21.

Erkert, H.G.E. (1999). In Poole, T. and English, P. (eds) *The UFAW Handbook on the Care and Management of Laboratory Animals* (7th ed.) *Terrestrial Vertebrates,* Vol.1, pp 574-590. Blackwell Science, Oxford.

Fowler, K.A., Huerkamp, M.J., Pullium, J.K. and Subramanian T. (2001). *Brain Res. Protoc.* 7(2), 87–93.

Fowler, K.A. (2003). *Zoo and Wild Animal Medicine* (5th ed.) W. B. Saunders, Missouri.

Fritz, J., Wolfe, T.C., Howell, S. (1999). In Poole, T. and English (eds) *The UFAW Handbook on the Care and Management of Laboratory Animals* (7th ed.) *Terrestrial Vertebrates,* Vol. 1, pp 643–658. Blackwell Science, Oxford.

Ghaly, R.F., Ham, J.H. and Lee, J.J. (2001). *Neurol. Res.* 23(8), 881–886.

Horne, W.A. (2001). *Veterinary Clin. North Am. Exot. Anim. Pract.* 4(1), 239–266.

Hrapkiewicz, K., Medina, L. and Holmes, D.D. (1998). *Clinical Medicine of Small Mammals and Primates: An Introduction* (2nd ed.) Manson Publishing/The Veterinary Press, London.

Kalema-Zikusoka, G., Horne, W.A., Levine, J. and Loomis, M.R. (2003). *J. Zoo. Wildl. Med.* 34(1), 47–52.

Kearns, K.S., Swenson, B. and Ramsay, E.C. (2000). *J. Zoo. Wildl. Med.* 31(2), 185–189.

Leopold, D.A., Plettenberg, H.K. and Logothetis, N.K. (2002). *Exp. Brain Res.* 143(3), 359–372.

Ligouri, A., Morse, W.H. and Bergman, J. (1996). J. *Pharmacol. Exp. Ther.* 227: 462.

Lopez, K.R., Gibbs, P.H. and Reed, D.S. (2002). *Contemp. Top Lab. Anim. Sci.* 41(2), 47–50.

Malaivijitnond, S., Takenaka, O., Sankai, T., Yoshida, T., Cho, F. and Yoshikawa, Y. (1998). *Lab. Anim. Sci.* 48(3), 270–274.

Mendoza, S.P. (1999). In Poole, T. and English, P. (eds) *The UFAW Handbook on the Care and Management of Laboratory Animals* (7th ed.) *Terrestrial Vertebrates,* Vol. 1, pp 591-600. Blackwell Science, Oxford.

Morris, T.H., Jackson, R.K., Acker, W.R., Spencer, C.K. and Drag, M.D. (1997). *Lab. Anim.* 31(2), 157–162.

Panadero, A., Saiz-Sapena, N., Cervera-Paz, F.J. and Manrique, M. (2000). *Rev. Med. Univ. Navarra.* 44(4), 12–18.

Plumb, D.C. (1999). *Veterinary Drug Handbook* (3rd ed.) Iowa State University Press.

Poole, T., Hubrecht, R. and Kirkwood, J.K. (1999). In Poole, T. and English, P. (eds) *The UFAW Handbook on the Care and Management of Laboratory Animals* (7th ed.) *Terrestrial Vertebrates,* Vol. 1, pp 559–573. Blackwell Science, Oxford.

Santerre, D., Chen, R.H., Kadner, A., Lee-Parritz, D. and Adams, D.H. (2001). *Vet. Res. Commun.* 25(4), 251–259.

Shiigi, Y. and Casey, D.E. (1999). *Psychopharmacology (Berl).* 146(1), 67–72.

Sun, F.J., Wright, D.E., and Pinson, D.M. (2003). *Contemp. Top Lab. Anim. Sci.* 42(4), 32–37.

Miyabe, T., Nishimura, R., Mochizuki, M., Sasaki, N. and Mastubayashi, K. (2001).*Veterinary Anaesthesia and Analgesia* 28(3), 168.

Vie, J.C., De Thoisy, B., Fournier, P., Fournier-Chambrillon, C., Genty, C. and Keravec, J. (1998). *Am. J. Primatol.* 45(4), 399–410.

Whelan, G., James, M.F., Samson, N.A. and Wood, N.I. (1999). *Lab. Anim.* 33(1), 24–29.

Yoshikawa, T., Ochiai, R., Kaneko, T., Takeda, J., Fukushima, K., Tsukada, H., Seki, C. and Kakiuchi, T. (1997). *Masui.* 46(2), 237–243.

Young, S.S., Schilling, A.M., Skeans, S. and Ritacco, G. (1999). *Lab. Anim.* 33(2), 162–168.

Chapter 19

Rigid Endoscopy

John W. Fanton
Oregon National Primate Research Center, Beaverton, OR 97006, USA

Introduction

The advent of minimally invasive surgery has resulted in improvements for nonhuman primates (NHP) for both research applications and clinical care. By definition, minimally invasive surgery is the use of small instruments to penetrate body cavities, gaining access to organs and viscera without the requirement of making large surgical incisions. Rigid endoscopy within the abdomen is termed laparoscopy, and thoracoscopy when utilized in the thorax. Despite the attendant reduction in postoperative morbidity resulting from smaller body wall incisions, the physiologic consequences of the actual procedure itself can be profound, possibly even lethal. In order to facilitate visualization of target organs, both laparoscopy and thoracoscopy require that a working space be created within each respective body cavity. This is accomplished by insufflation of the abdomen with a gas under pressure. In the case of thoracoscopy, penetration of the thorax itself creates pneumothorax, resulting in the lungs falling away from the thoracic wall. The physiologic consequences of these situations will be discussed later.

Laparoscopy has been utilized for over three decades in NHP, predominantly for examination of the reproductive tract (Dukelow *et al.,* 1971) (Jewett and Dukelow, 1971, 1972) (Dukelow *et al.,* 1973; Graham *et al.,* 1973; Rawson and Dukelow, 1973). Conversely, thoracoscopy is a relatively new technique in veterinary medicine, and NHP in particular (Miura *et al.,* 2000; Bohm, Jr. *et al.,* 2000).

Minimally invasive surgery has almost unlimited application potential in NHP. While gynecologic surgery continues to be one of the most common applications, the upper abdominal and thoracic viscera are readily accessible due to technologic innovations, and a number of common research applications. One of the most significant enhancements has been the introduction of video equipment for all types of endoscopy. Compact video cameras are attached to the viewing port of the telescope and project the image upon a video monitor, allowing multiple participants, all of whom have an excellent view of the surgical site. This one factor has been highly significant in changing the role of laparoscopy from one of observation to one of sophisticated manipulation. Much more complex surgical procedures are now feasible because the number of surgeons who may participate has multiplied.

A second reason for the more extensive implementation of rigid endoscopy is the refinement in laparoscopic equipment. Vast improvements have occurred, mainly oriented around the diminishing size of the instruments which penetrate the abdominal and thoracic wall. Smaller incisions result in reduced morbidity, a basic premise of the concept of minimally invasive surgery. As more

The Laboratory Primate
Copyright 2005 Elsevier
ISBN 0-1208-0261-9

refinements in equipment occur, and the equipment becomes more affordable to scientists and health care providers, rigid endoscopy will undergo an expanding role in the practice of NHP medicine and research.

In human surgery, predictions are that in the future as much as 80% of all abdominal surgery will be accomplished laparoscopically (Schauer, 2000). There has been a corresponding increase of rigid endoscopy in both animal health practices and in research applications. In the author's practice some common procedures, such as oophorectomy and salpingo-oophorectomy, have transitioned over the past five years from traditional laparotomy to being exclusively managed with laparoscopy.

Laparoscopy

Physiologic responses to laparoscopy

One of the fundamental procedures in laparoscopy is the introduction of pneumoperitoneum, or insufflation of the abdominal cavity with gas. This greatly enhances visualization of the viscera and provides a working space. However, insufflation with carbon dioxide (CO_2) results in absorption of the gas into the body, where it becomes biologically active, the end result of which is hypercarbia and acidemia. A second negative effect results from the creation of positive pressure within the abdominal cavity. This creates a reduction in lower body venous return, decreasing cardiac output and causing a reflex tachycardia. Additionally, cardiac arrhythmias may result from direct sympathetic stimulation and catecholamine release, and bradycardia may result from peritoneal stretching. One complication which may occur is regurgitation of gastric contents, attributable to both the increased intraabdominal pressure and Trendelenburg (tilted with head down) positioning to enhance exposure of the lower abdominal viscera. These findings are in agreement with findings for human laparoscopic surgery (Schauer, 2000).

A variety of gases have been used for insufflation, including air, nitrous oxide, nitrogen, oxygen, helium and argon. However, the gas most frequently employed for insufflation has been CO_2 (Schauer, 2000). In addition to being noncombustible, CO_2 is relatively inexpensive, both of which are distinct advantages. Perhaps most important, CO_2 is rapidly absorbed by the body and therefore is much less likely to form a gas embolism, a complication to be avoided at all costs.

Abdominal insufflation causes a reduction in oxygenation directly attributable to reduced lung volume and diaphragmatic displacement. The significance of this is indicated by the fact that, in human laparoscopy patients, the functional residual capacity is reduced 20–25% simply by Trendelenburg positioning, and another 20% simply by general anesthesia. When combined with the diaphragmatic compression arising from pneumoperitoneum, significant perturbations in pulmonary function and acid base balance can occur. An anesthetist is required for each laparoscopy case, allowing minute by minute monitoring and corrective action.

A dedicated anesthetist is a requirement for laparoscopy cases involving NHP. Some earlier reports of NHP laparoscopy include descriptions of anesthesia with injectable agents, and without airway management. With the great strides in animal anesthesia over the past 20 years, gas anesthesia has become a standard methodology. Airway management with orotracheal intubation, and monitoring of end tidal CO_2 now offer the laparoscopist a much improved methodology. Due to rapid transperitoneal CO_2 absorption, combined with hemodynamic and ventilatory impairment, subjects may be predisposed to developing cardiac arrhythmias (Benitez and Edelman, 2000). Monitoring of end tidal CO_2 is considered by some to be mandatory (Benitez and Edelman, 2000). Mechanical ventilation of the patient during pneumoperitoneum also is highly recommended, if not mandatory. The combination of CO_2 monitoring and mechanical ventilation support allows the laparoscopist to make rapid adjustments in ventilatory rate and tidal volume when required. In the author's experience, rhesus macaques undergoing routine laparoscopy procedures may produce end tidal CO_2 values in excess of 55–60 mm Hg if mechanical ventilation is not performed. Mechanical ventilation will reverse these adverse effects, but the anesthetist must be careful to not over ventilate and produce hypocarbia.

When laparoscopy was originally introduced to the NHP research community, injectable anesthesia was more common than gas anesthesia and comprehensive airway support, with positive pressure ventilation, was not considered a prerequisite. Although thousands of laparoscopic surgeries have been conducted with injectable anesthesia, current standards of practice dictate that intubation and airway support should be the accepted methodology. While many different injectable agents may provide quite adequate anesthesia, either alone or in combinations, airway management is critical to patient health. With the advent of more modern monitoring devices, specifically expired CO_2 monitors, patients can be maintained in a very stable physiologic state throughout laparoscopic procedures.

Equipment

Although the basic equipment has been described in the literature many times over the past 20 years (Harrison, 1980), there have been improvements from original techniques (Dierschke and Clark, 1976). Changes in modern laparoscopy equipment have spurred the development of quite sophisticated procedures which are now commonplace in human surgery.

Insufflation of the abdomen is accomplished with a Verres needle, a blunt tipped device which allows passage of carbon dioxide gas into the abdomen without causing organ trauma. Rubber or plastic tubing connects the Verres needle to the insufflator, which can be of either a mechanical or electronic design. Insufflators provide gas to inflate the abdomen, creating a pneumoperitoneum which, in turn, provides a large open space within the abdomen to allow visualization of the abdominal viscera. Mechanical insufflators typically meter the insufflation gas at a single predetermined rate. Conversely, electronic insufflators function on demand and will alter the flow rate in response to low intraabdominal pressure. Electronic units meter the gas into the abdomen at variable rates, based on the intraabdominal pressure. These insufflators typically provide flow rates of 1–15 liters per minute, which is a significant advantage during procedures wherein gas leakage through accessory ports may be a factor.

The main viewing instrument, or telescope, is a rigid metal tube containing multiple nonadjustable focusing lenses, and can vary in diameter from 2.5–10 mm. Obviously, the smaller the telescope, the less morbidity from body wall penetration. This is of special significance in working with smaller NHP. A 10 mm telescope would be unacceptable to use in a marmoset and is considered quite large, even in macaques. The downside of a smaller telescope is that fewer light fibers are available than with larger equipment and the field of view is narrower. However, this is one area of tremendous improvement in the past 20 years. The difference in the projected image between a modern 10 mm and 5 mm telescope is barely noticeable. One other choice which the laparoscopist must make is whether an offset objective lens is required. Telescopes are available with an offset angulation of the objective lens of 10–30 degrees, allowing the effect of viewing around a corner. This can be of benefit for observation of inaccessible areas of major organs. However, a zero offset view is usually less disorienting for the novice laparoscopist, and for many

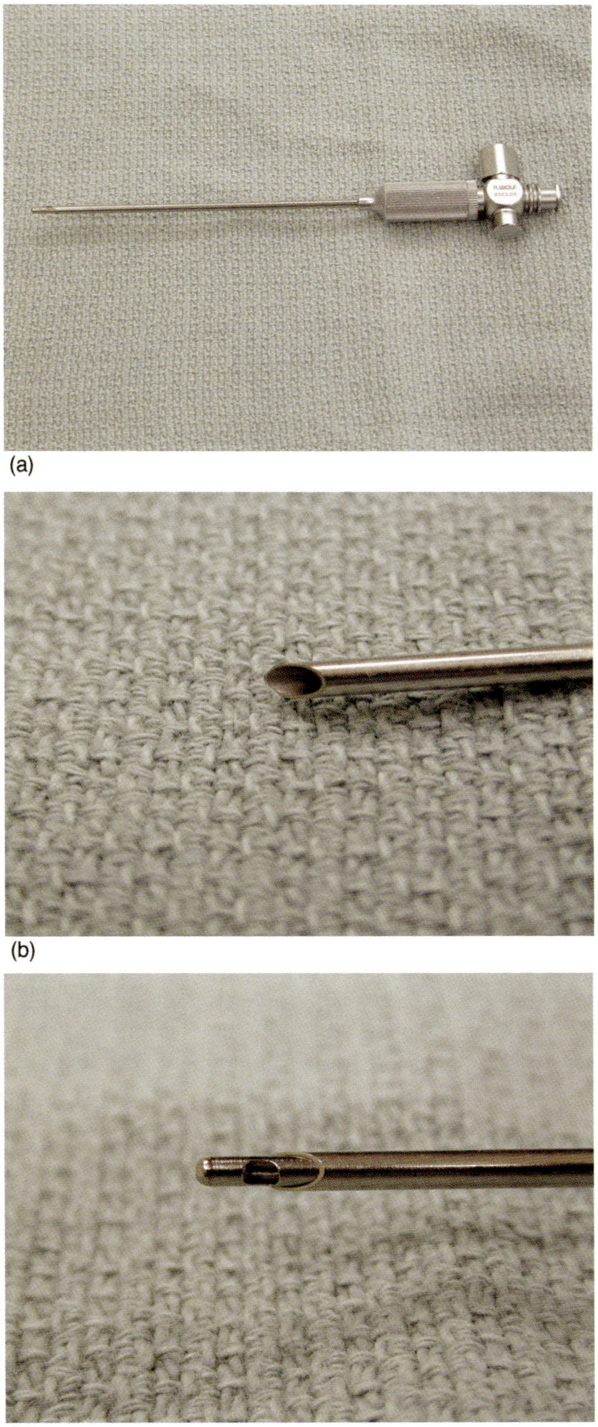

(a) (b) (c)

Figure 19.1 (a) An example of a Verres needle. The sharp tip of a Verres needle is an external sheath which is normally retracted over a blunt tipped needle. (b) Here the external sheath has been pushed down showing the sharp tip. (c) Here the sheath has been allowed to slide back into place, allowing the blunt tipped needle to protrude beyond the sharp tip. This is the normal position of the sheath, which is controlled by a spring inside the needle shaft. The blunt tip is to prevent organ damage after insertion of the Verres into the abdomen.

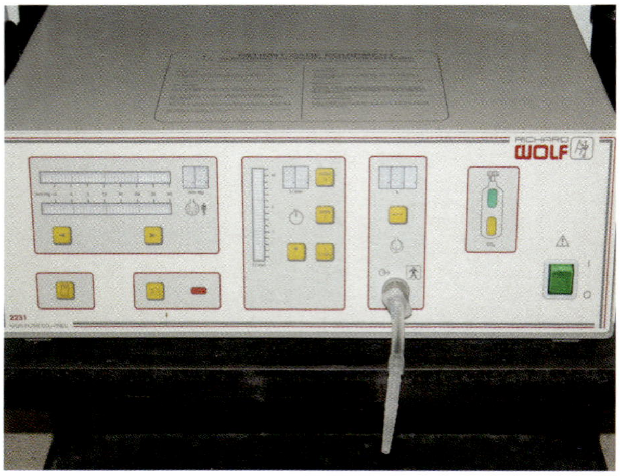

Figure 19.2 An example of an electronic insufflator which will provide gas flows of from 1 to 15 liters per minute, and automatically responds to loss of intraabdominal pressure by increasing the flow rate. This device will maintain intraabdominal pressure at a prescribed limit, typically 15–18 mm Hg.

of the more common laparoscopic procedures. Following abdominal insufflation, telescopes are inserted into the abdomen through cannulas of slightly larger diameter. Cannulas have a trocar which protrudes down the shaft and provides a sharp point for puncturing through the body wall. Once placed, the trocar is removed and the telescope is inserted through the cannula into the body cavity for viewing. Accessory ports or cannulas function on the same principle for manipulating instruments.

Attached to the telescope is a flexible fiberoptic light cable which delivers the illuminating light to the instrument. As with all endoscopic equipment, light cables can vary in diameter, and it is of some importance to match the light cable to the telescope. Directing a large number of light fibers from a large diameter light cable into a smaller diameter telescope can result in heat accumulation at various points within the viewing device, so connecting light cables and viewing instruments of disparate diameters is not recommended.

Light sources provide the illumination for the procedures. Light sources have traditionally contained halogen light bulbs, but xenon light bulbs have become the industry standard. Xenon light sources almost instantly attain maximum brightness when switched on, while halogen light sources require a warm-up period before maximum brightness is attained. Also, xenon bulbs produce light with a more natural color spectrum and white balance, aiding with photography. All light sources create significant heat, so they should be positioned in a ventilated area.

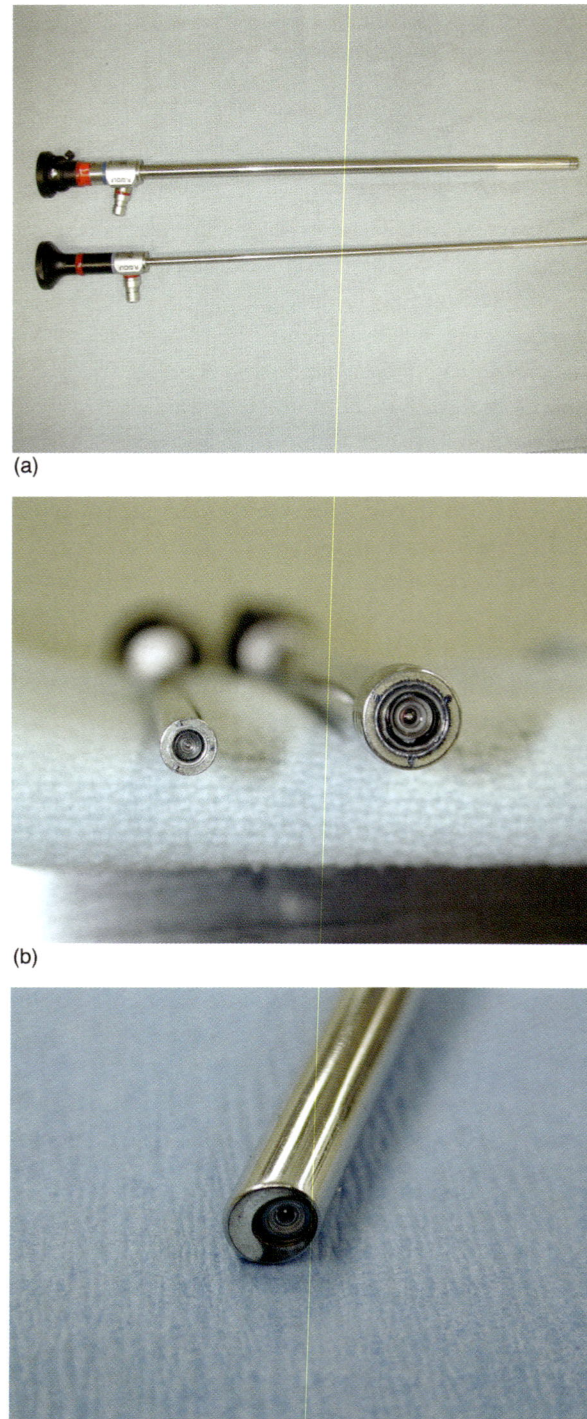

Figure 19.3 Viewing telescopes. Telescopes may have a diameter of from 2.5–10 mm. (a) Pictured here are a 5 mm and 10 mm zero offset telescope. These instruments may have as many as 28 focusing lenses to provide a high resolution image. (b) Note the circumferential position of the light fibers around the zero degree offset objective lenses of these telescopes. (c) This is an example of a 4 mm arthroscope with a 30 degree offset. This device may be used for either laparoscopy or thoracoscopy in small nonhuman primates.

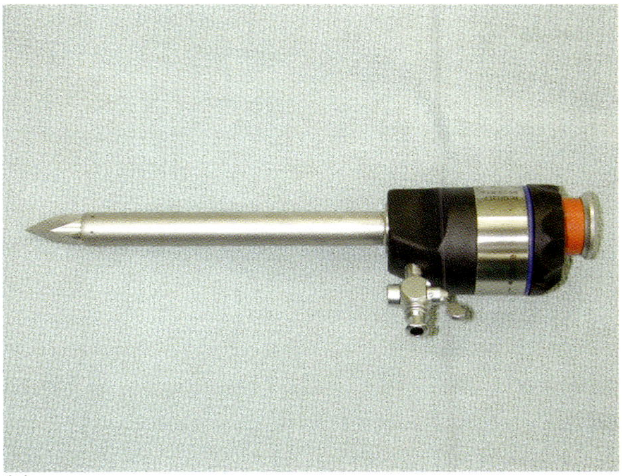

(a)

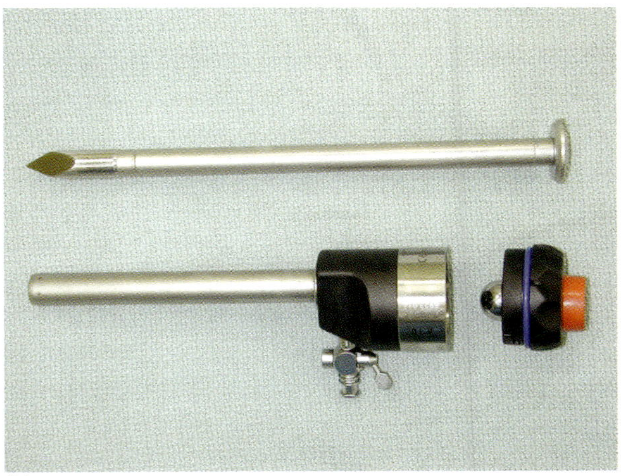

(b)

Figure 19.4 A trocar and cannula for telescope insertion. After removal of the Verres needle and trocar insertion, continued insufflation is accomplished though the side port valve on the cannula. All cannulas and ports have a one way valve which prevents loss of insufflation gas when the port is not in use.

The most notable enhancement for all forms of endoscopy is the widespread implementation of video equipment. The common practice today is to attach a compact video camera to the viewing telescope. The camera connects to a video control unit which processes the image and sends it to a video monitor, allowing all surgeons and technical staff to simultaneously view the procedure in real time. An example of a video control box is shown in Figure 19.6b. Video cameras may be either one chip or three chip systems. One chip cameras process the RGB (red, green, and blue) signals on just one chip, whereas three chip cameras have a single chip to process each of these colors. Three chip systems are also capable of producing much higher resolutions than one chip camera. Resolutions exceeding 800 lines per inch are possible with three chip systems. The only disadvantage of three chip cameras is expense, costing approximately 2–3 times more than a one chip camera.

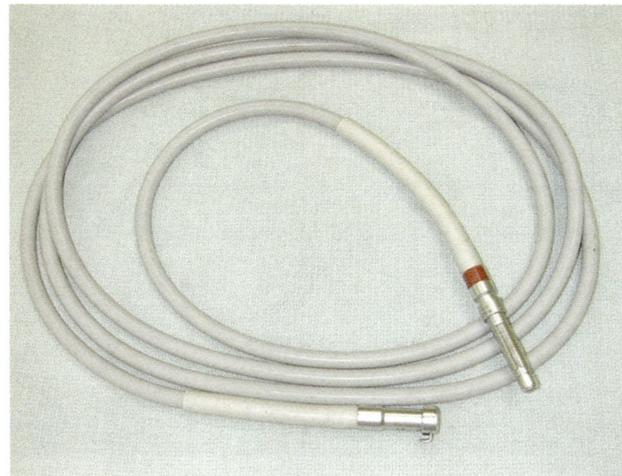

(a)

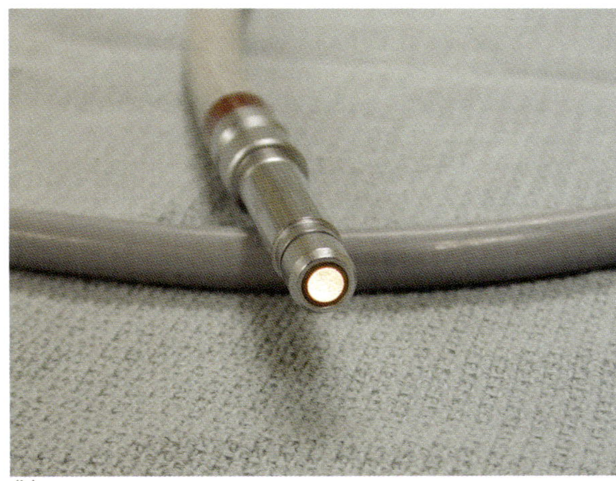

(b)

(c)

Figure 19.5 The light cable. (a) These are quite flexible and efficiently carry light from the light source to the telescope. (b) This is how the tip of a light cable should appear when the light fibers are intact. (c) If the cable is damaged, a loss of light carrying capacity will occur. The cable will appear dirty, as depicted here.

(a)
(b)
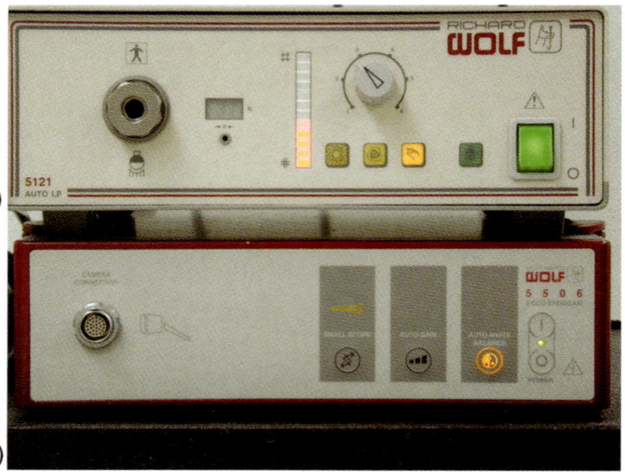

Figure 19.6 (a) A xenon light source. Most modern light sources have xenon light bulbs, which produce a natural white light, and are ready for use instantly after being switched on. Older halogen light sources require a short period of time to warm up in order to provide a natural spectrum of light. (b) The camera control system for a 3 chip camera. The camera cable is passed off the surgery table and plugged into this unit. The video signal processing is performed by this unit, which then sends it to the video monitor.

Video monitors are the last link in the video system and are the limiting factor for resolution. Most video monitors have a maximum resolution of approximately 700 lines per inch, and cameras; with resolution capability greater than this, will be limited to this level. While a standard television set may be attached to the camera in some systems, it will yield a poor resolution picture and is not recommended.

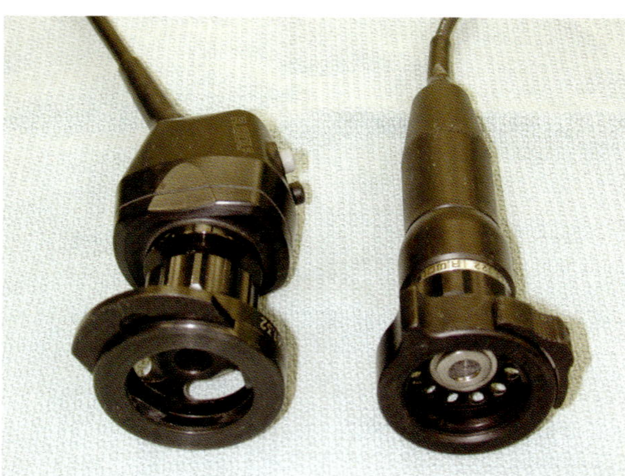

Figure 19.7 The heart of the video assisted endoscopic surgery system is the video camera which is attached to the end of the telescope and sends the image to a video monitor. Cameras may be either 1 chip or 3 chip systems. Pictured is a 1 chip camera (right), and a 3 chip camera (left). A 3 chip camera provides superior resolution and color balance.

One additional piece of equipment in the video system is a recorder. Most scientists and clinicians desire video recordings of the procedures undertaken. The most simple method for this is to connect a standard videotape recorder to the camera control unit. However, more capability is provided by attaching a digital video recorder, allowing more comprehensive processing of the video signal. Digital movies may be constructed with this technology, and single frames may be edited on a computer. All of the surgical photographs in this chapter were produced by this method, using a simple digital camcorder and inexpensive computer software.

There is an almost infinite variety of accessories available for rigid endoscopy. Tissue manipulation is managed with many different types of grasping forceps, scissors, pre-tied ligatures and stapling instruments.

Most are available as either reusable or disposable versions, at significantly different prices. Many instruments can be connected to electrocautery generators, allowing for the projection of electrocautery capability inside the abdomen, a very valuable process which allows cauterization of bleeding vasculature and bloodless dissection of tissue with the use of a cutting current.

The shafts of these instruments are insulated to prevent inadvertent electrocautery of the adjacent tissues within the abdomen or thorax.

Working instruments are inserted into the abdomen through accessory ports or cannulas, which are placed in a wide variety of locations, depending on the target organ or procedure to be performed. A general rule of endoscopy is to create a triangle whenever possible, with the viewing telescope at the apex, and the manipulating instruments forming the base of the triangle.

This allows the best viewing angle in relation to the working instruments. This principle holds true for both laparoscopy and thoracoscopy, whenever possible. The diameter of these accessory ports is dependent on the manipulating instruments. The most commonly used manipulating instruments for macaque sized NHP are 5 mm diameter. However, if stapling instruments are required for a procedure, 11 mm accessory ports are typically required, as most staplers are only available in 10 mm diameter. Both viewing ports and accessory ports are available in a wide variety of configurations. Some have smooth cannulas to facilitate insertion, while others have shafts with raised edges like a screw and must be twisted during insertion, providing improved security against inadvertent extraction.

Almost all laparoscopy instruments used in NHP research were created for human surgical applications. One disadvantage to be overcome, as a result of this, is the length of the instrumentation. Grasping instruments,

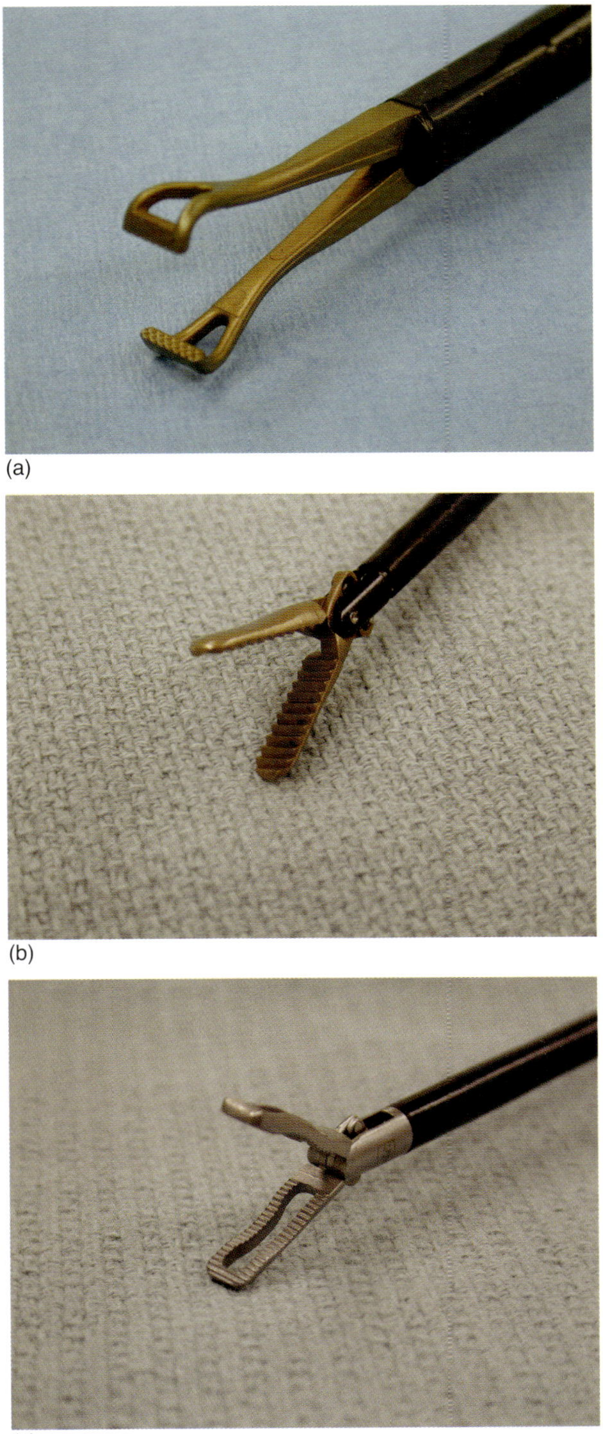

(a)

(b)

(c)

Figure 19.8 Examples of grasping instruments which may be used to manipulate tissues inside the abdomen or thorax. All instruments have rotating shafts which may be controlled near the handle by the surgeon. Grasping forceps may have smooth or serrated tips, and locking or nonlocking handles.

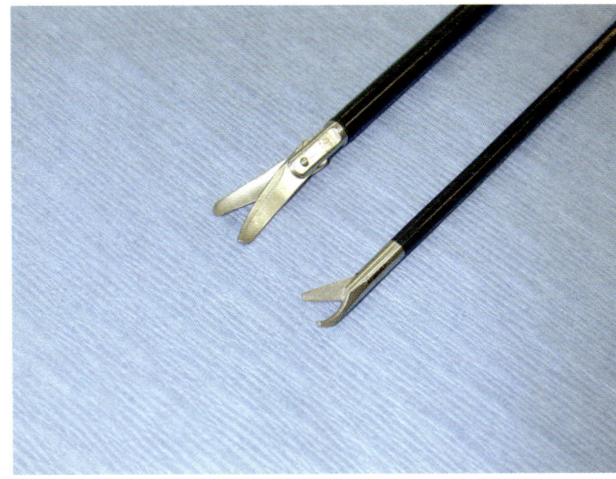

Figure 19.9 Examples of small Metzenbaum scissors and hooked scissors. Hooked scissors offer an advantage when cutting tissues because the tissue cannot slip out of the jaws of the instrument. This is the preferred instrument for performing laparoscopic oophorectomy in nonhuman primates.

scissors, and stapling devices are typically available in lengths of 20–30 cm. Most of the length of these devices is outside the body of the patient, which can result in awkward working conditions for the novice laparoscopist. Any disadvantages presented by working with these instruments will be overcome by practice and repeated use. An alternative to using such longer instruments is to acquire arthroscopy instruments which are considerably shorter and may be more easily manipulated by novices. A disadvantage of such instruments is that the tips may not be suitable for the desired procedure.

Another equipment item, which will aid with endoscopy procedures, is an easily adjustable tilting surgery table. Patient positioning is often a critical prerequisite for successful endoscopic surgery. Most modern veterinary surgical tables have tilting mechanisms but those with simple lever systems, which require the least manipulation, will facilitate the procedure.

The equipment described above is readily available, but can be quite expensive. A complete set of high quality instrumentation may cost in excess of $25,000. However, quite sophisticated instrument sets can be accumulated for much less.

Below is a listing of the major components utilized by the author to perform in excess of 4000 endoscopic procedures over the past five years. These procedures have been predominantly in the gynecologic category, but almost all of the equipment listed can also be used for thoracoscopic work. The equipment items listed are available from a variety of different equipment manufacturers, also listed below. This equipment list should be considered the minimum requirement for performing the surgical procedures described.

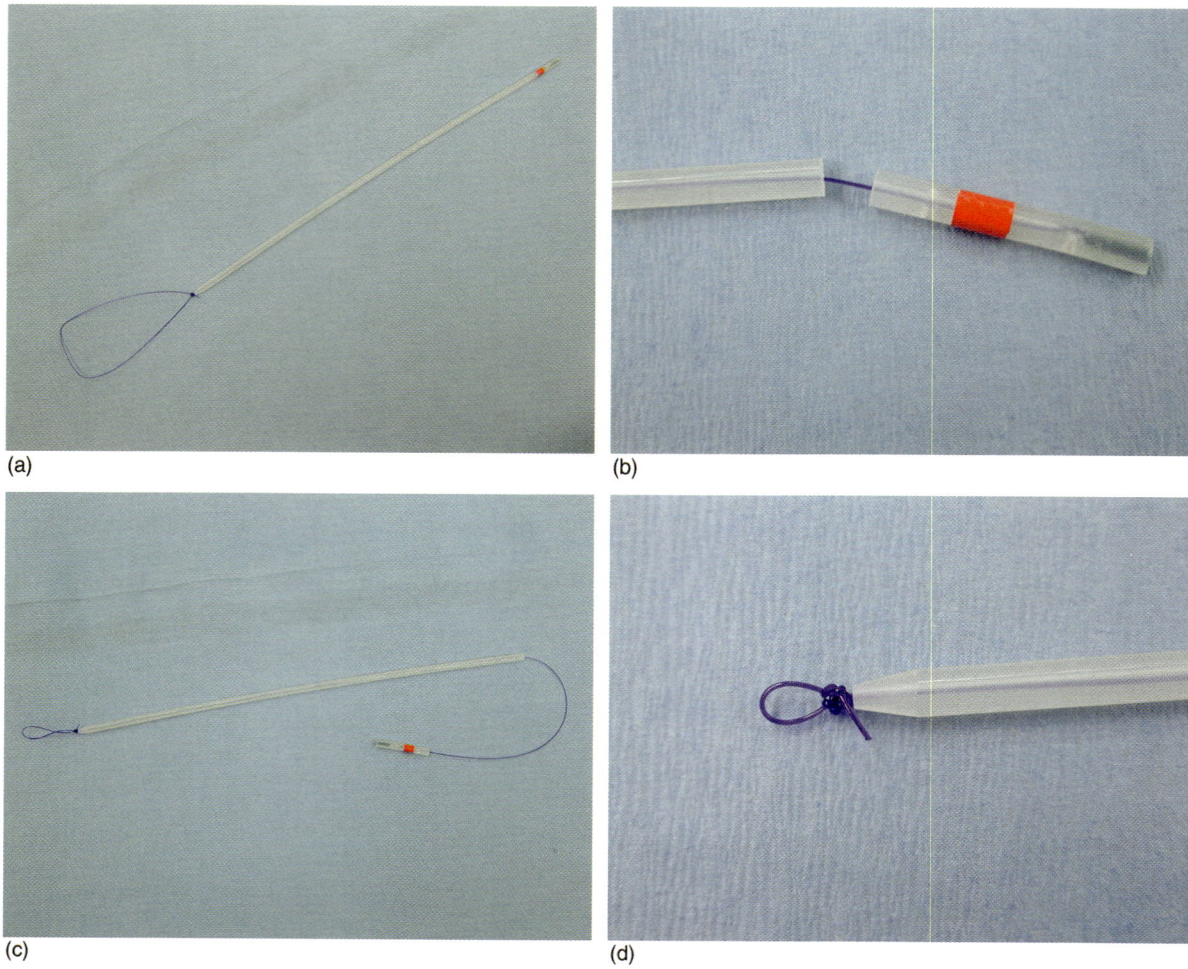

Figure 19.10 (a) An Endoloop™, or pre-tied ligature. The ligature is inserted through an accessory port, the external tip is broken from the shaft (b) and used to pull the ligature tight (c and d). These allow ligation of tissues to be accomplished from outside the abdomen or thorax.

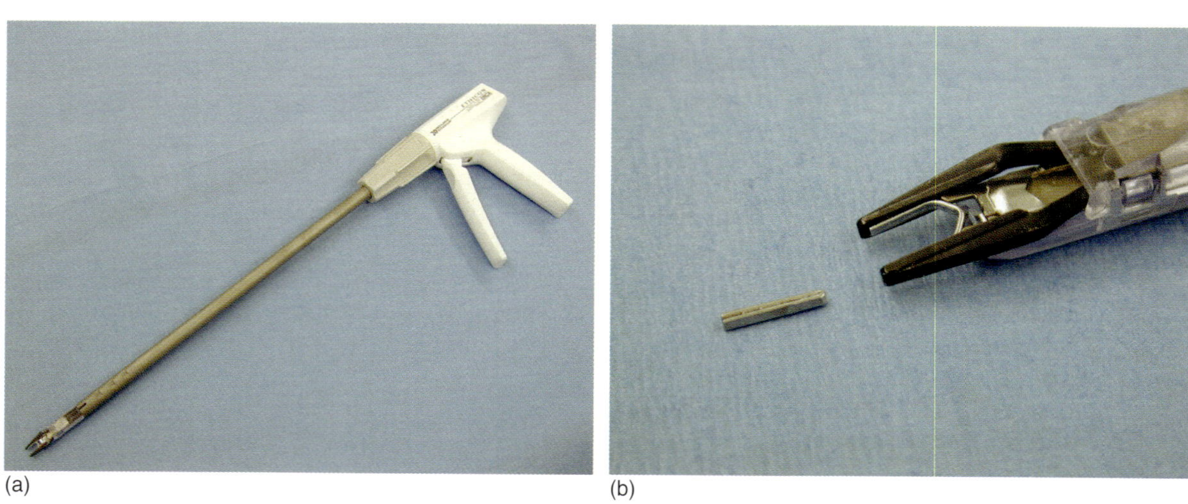

Figure 19.11 Staplers are available in a wide variety of shapes and sizes. They may be used for ligation of vasculature inside a body cavity, and are quite useful. One disadvantage is that most stapling instruments are only available in 10 mm diameter, requiring larger accessory ports than are normally used in nonhuman primates.

(a)

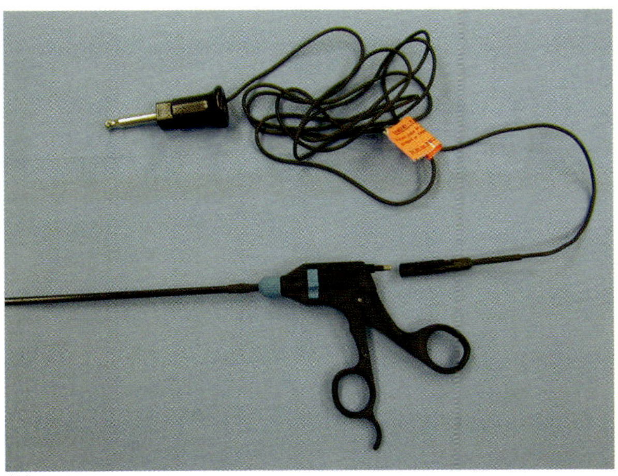

(b)

Figure 19.12 Some instruments have connectors to allow electrocautery inside a body cavity. This is an important technique for hemorrhage control in both laparoscopy and thoracoscopy. The electrode on the handle of the instrument is connected to a special cable which is then passed off the surgery table and connected to a standard electrocautery generator. Both monopolar and bipolar electrocautery may be implemented.

Equipment Set

High Resolution Color Video Monitor – Sony Model PVM-1343MD
High-Flow CO_2 insufflator – R. Wolf Model 2231.501
High intensity Xenon light source – R. Wolf Model 5121.012
Video camera control unit – R. Wolf Model 5506.751
Three chip CCD endoscopic camera – R. Wolf Model 5506.962
Light cable for 5 mm telescope
Verres needle
Viewing telescope, 5 mm diameter, 0° lens offset
6 mm introducer sheath and trocar
5 mm accessory port and trocar (2)
Grasping forceps – 24 inch and 15 inch lengths
Hooked scissors – 24 inch length

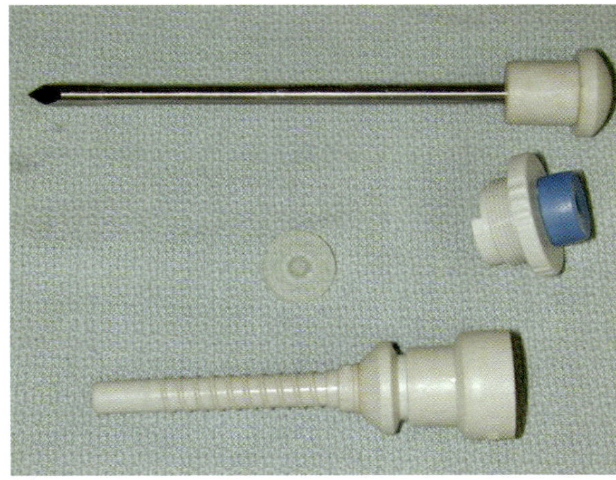

(a)

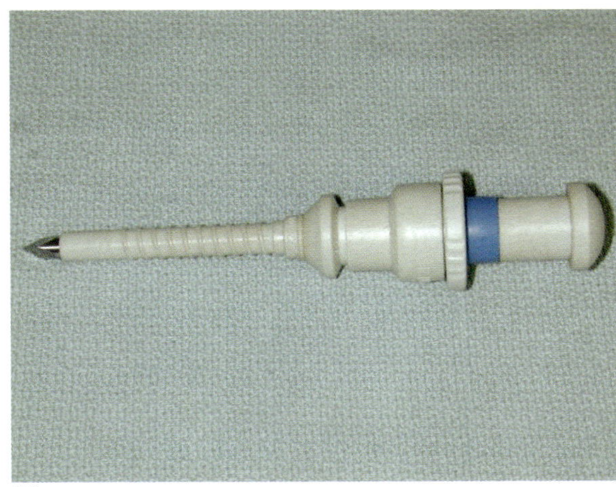

(b)

Figure 19.13 An example of a 5 mm diameter accessory port. The small white plastic disc is a diaphragm which prevents loss of insufflation gases when an instrument is not inserted through the port.

Irrigation and aspiration cannula
CO_2 source – "E" size tank with connectors to insufflator

Equipment Sources

Richard Wolf Medical Instruments Corp.
353 CorporateWoods Parkway
Vernon Hills, Illinois 60061, U.S.A.

Karl Storz Veterinary Endoscopy
175 Cremona Drive
Goleta, California 93117, U.S.A.

Ethicon Endo-Surgery
4545 Creek Road
Cincinnati, Ohio 45242, U.S.A.

U.S. Surgical
150 Glover Avenue
Norwalk, Connecticut 06856, U.S.A.

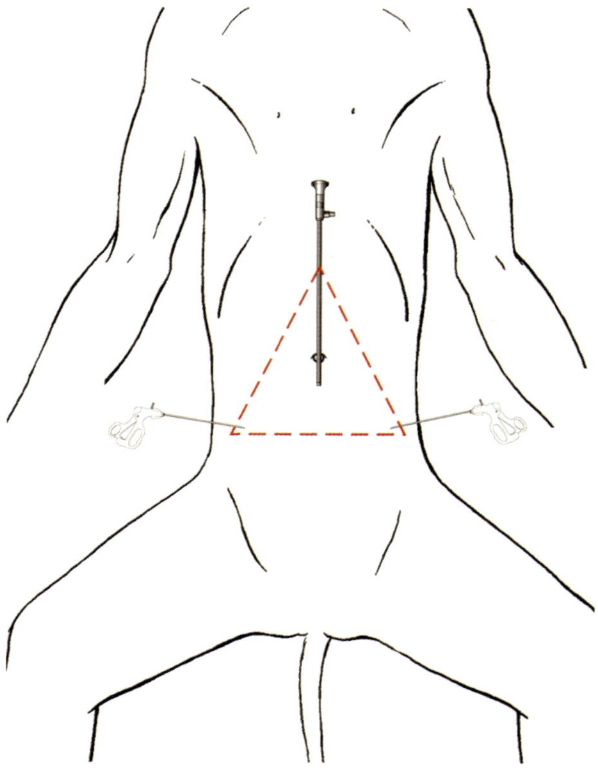

Figure 19.14 To enhance viewing, the telescope and the working instruments should be positioned to create a triangle whenever possible. This applies to both laparoscopy and thoracoscopy.

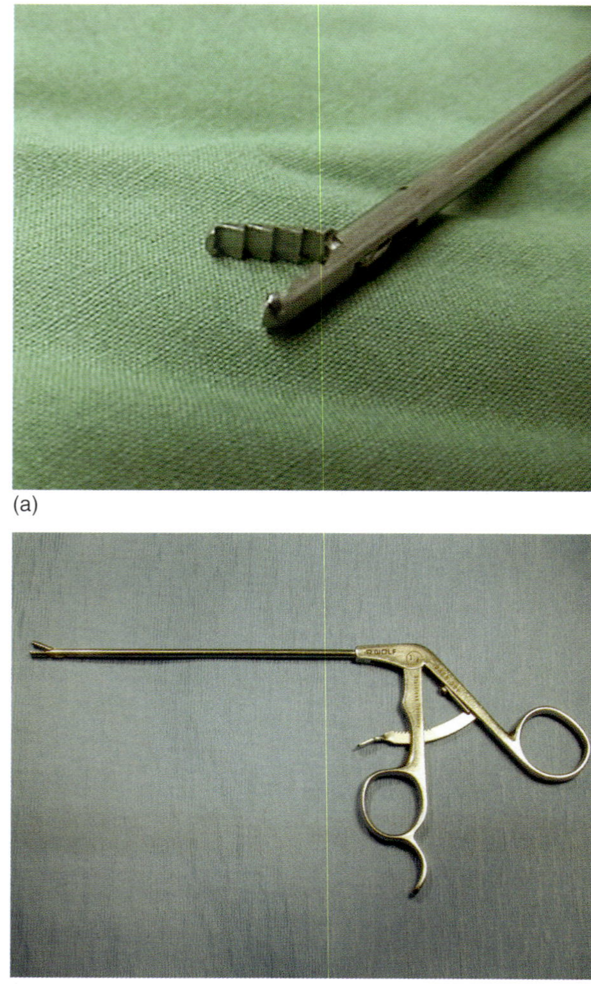

(a)

(b)

Figure 19.15 An example of arthroscopy forceps. The forceps shown are 15 cm in length and are used regularly by the author for gynecologic manipulations in rhesus monkeys. The only disadvantage is the smooth surface of the grasping tips may not provide adequate traction for some procedures.

Olympus America, Inc.
2 Corporate Center Drive
Melville, New York 11747, U.S.A.

Preoperative care and preparation

Preoperative examination of the patient is mandatory to evaluate general health and to assess any special facts about body conformation, such as obesity, which may complicate the procedure. Also, it is important to note scars from prior abdominal surgery. Previous abdominal surgery does not disqualify the patient from laparoscopy, but the surgeon should exercise additional care during the insertion of trocars and cannulas to avoid accidental organ perforation, especially loops of bowel which may be adhered to the anterior body wall. Although this is not a significant problem in macaques, it has been reported in chimpanzees (Graham, 1976).

Postoperative care

Postoperative analgesia is required for all methods of body cavity-penetrating endoscopic surgery. Pain management protocols should be tailored to individual patient requirements to ensure adequate levels of post-operative comfort. Standard analgesia employed by the author for postoperative pain, in both laparoscopy and thoracoscopy, is based upon the administration of opioid agonists for several days. Both oxymorphone (0.15 mg/kg IM) and hydromorphone (0.15–0.25 mg/kg IM) have been utilized with good results. Hydromorphone has the advantage in that it can be administered orally, negating the requirement for repeated injections. Hydromorphone also is quite economical in comparison with others drugs in this class. Other agents which may be used are buprenorphine (0.03 mg/kg IM), an opioid agonist/antagonist which may provide prolonged relief of up to 8 hours in some NHP. Also to be considered are nonsteroidal anti inflammatory agents, such as ketoprofen, which have been utilized to provide

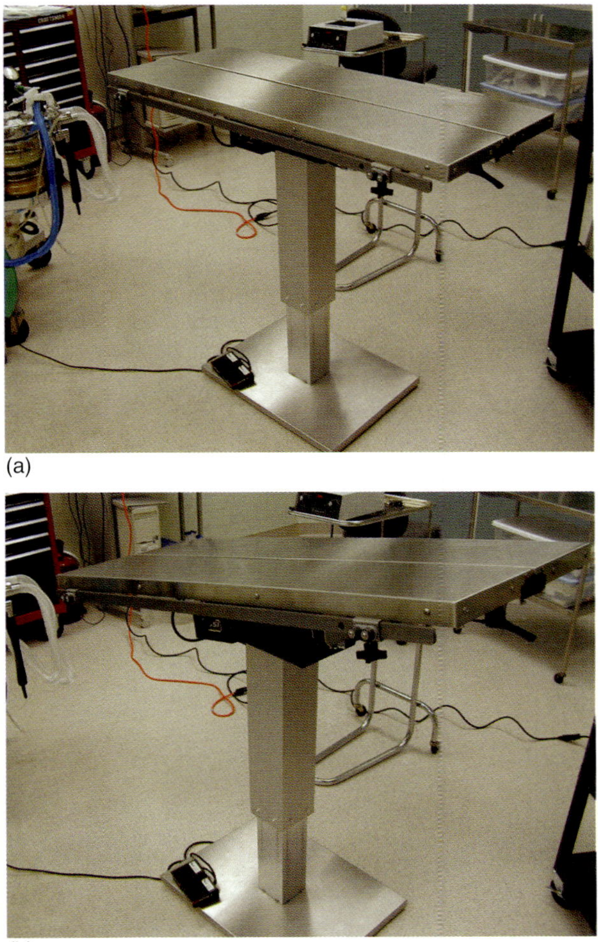

(a)

(b)

Figure 19.16 An example of a tilt table which can be operated by the surgeon without assistance. The lever mechanism which controls the tilt angle protrudes from the back of the table and is covered by the sterile drapes during the surgery, allowing the surgeon to manipulate the tilt angle without contamination of the sterile field.

adequate analgesia for postlaparoscopic pain (Perret-Gentil *et al.*, 2000).

Personnel

One of the most critical elements of minimally invasive surgery is teamwork. Videoendoscopy allows multiple operators working simultaneously in the same small body cavity. Choreography between the surgeon and assistants is essential for streamlining the process, reducing surgical time and producing an acceptable outcome. Assistant surgeons and ancillary staff are just as important during laparoscopic procedures as for any traditional surgery. The anesthetist plays a central role and should not be tasked with ancillary duties during these procedures. The anesthetist, more than any other staff member, must be attentive to patient status and often must adjust ventilator settings to maintain the patient in homeostasis.

Laparoscopic procedures

Ovarian follicle aspiration

This is one of the most commonly performed NHP laparoscopic procedures during the past decade. Monkeys to be utilized as oocyte donors typically will undergo a standard protocol for controlled ovarian stimulation. An example of one such protocol is as follows: during the first four days of menstruation, daily injections of a gonadotropin releasing hormone (GnRH) antagonist are initiated and continued for ten days. Starting on day two of these GnRH treatments, follicle stimulating hormone is administered twice per day and, on days 7, 8 and 9, luteinizing hormone is also injected twice daily. Transabdominal ultrasound examination of the ovaries is performed three days prior to the planned surgery date to verify that both the number and rate of follicle growth has increased. If adequate stimulation has occurred, a single injection of HCG is administered at approximately 27–32 hours preoperatively.

As with all laparoscopic surgery, fasting for at least 14 hours is recommended to reduce the possibility of regurgitation during the period of pneumoperitoneum. In the author's practice, induction of anesthesia is routinely performed with an intramuscular injection of ketamine HCl (10–20 mg/kg IM). Patient preparation then proceeds with removal of hair from the ventral abdomen, typically with an electric clipper. Additional patient preparation includes intravenous catheterization for administration of crystalloid fluids to maintain blood pressure and volume. Peripheral vascular access is most often obtained in the antecubital fossa, or even more peripherally in the distal cephalic vein. Saphenous vein catheterization is also an acceptable alternative. Following induction anesthesia, maintenance anesthesia is provided by isoflurane gas vaporized in 100% oxygen and delivered via cuffed orotracheal tube. Other volatile gas agents such as halothane and sevoflurane are quite acceptable alternatives for maintenance anesthesia. Atropine sulfate is utilized (0.04 mg/kg IV) to prevent bradycardia. After sterile skin preparation and draping, a 5 mm skin incision is created approximately 1–2 cm superior to the umbilicus. This is then used as an access point for the Verres needle which is inserted through the fascia and muscle into the abdomen, after which it is connected to the insufflator with appropriately sized tubing. The surgeon may facilitate insertion of the Verres needle by gently grasping the skin near the incision and tenting the skin upwards to provide counter traction.

To verify that the Verres needle has penetrated into the abdominal cavity, the surgeon should place a drop of saline on the needle hub. If the saline slowly gravitates downward into the needle and disappears, the needle tip is within the cavity. The insufflation gas is infused initially at approximately 1 liter/minute until intraabdominal pressure of 15–18 mm Hg is attained. Use of high rates of insufflation will typically cause expansion of the omentum within the abdomen. Although this is not a pathologic event, it will initially obscure the view through the telescope when it is inserted. Manipulating the omentum with the telescope tip often easily results in deflation.

The Verres needle is then removed, after which the trocar and cannula for the viewing telescope are inserted

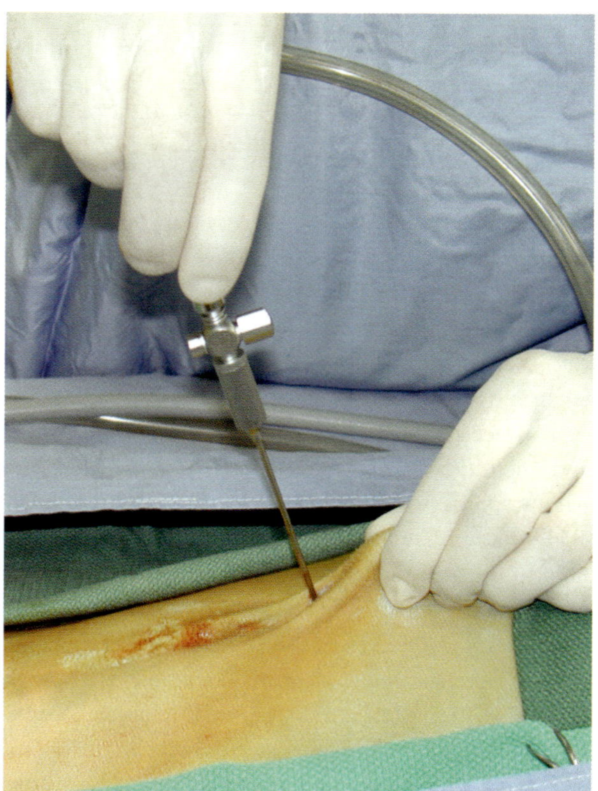

Figure 19.17 The insertion process for a Verres needle. After a small skin incision with a scalpel, one hand is used to pull up on the skin and provide counter traction to needle insertion. When the needle has successfully entered the abdomen, the sheath will pull back over the sharp tip to protect the abdominal viscera.

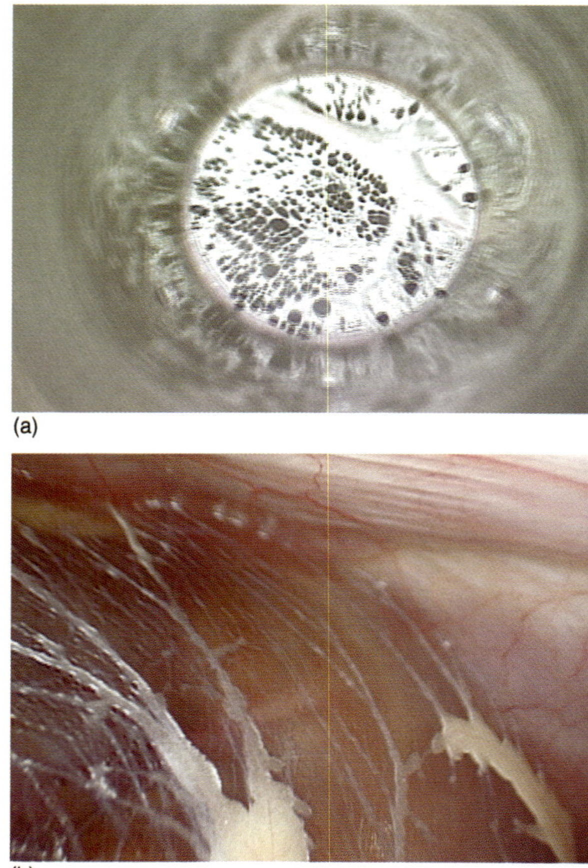

(a)

(b)

Figure 19.18 When insufflation is performed too rapidly, using flow rates in excess of 1 liter per minute, the omentum often will inflate like a balloon within the abdomen. (a) This is an example of how the initial view when inserting the telescope through the cannula is obscured in this situation. (b) The omentum is inflated against the body wall, and must be manipulated with the tip of the telescope to cause it to deflate.

at the same site. The trocar is a larger diameter device than the Verres needle, and considerable force may be required to penetrate into the abdomen. There is danger in this procedure, in that trocar trauma is one of the most common complications of laparoscopic surgery. It is possible that, if excessive force is utilized, the trocar may suddenly penetrate and the operator will not be able to stop before it impacts on the organs lying beneath. Trocar insertions can be well controlled by the surgeon holding his arms tightly against his body, leaning over the patient on the table and using the force of the entire body for trocar insertion, rather than working with arms extended. The telescope is then inserted for the initial visualization of the reproductive organs, facilitated by placing the patient in Trendelenburg positioning. This is standard methodology for almost every gynecologic procedure. Accessory cannulas are

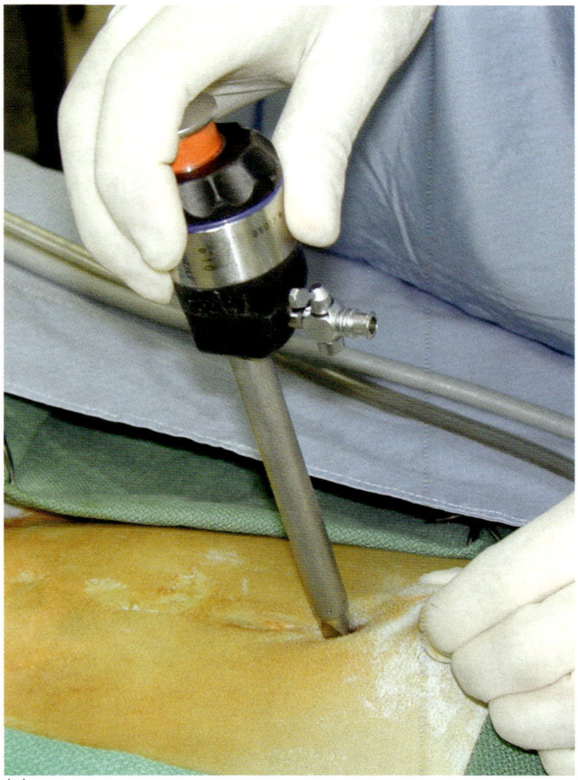

(a)

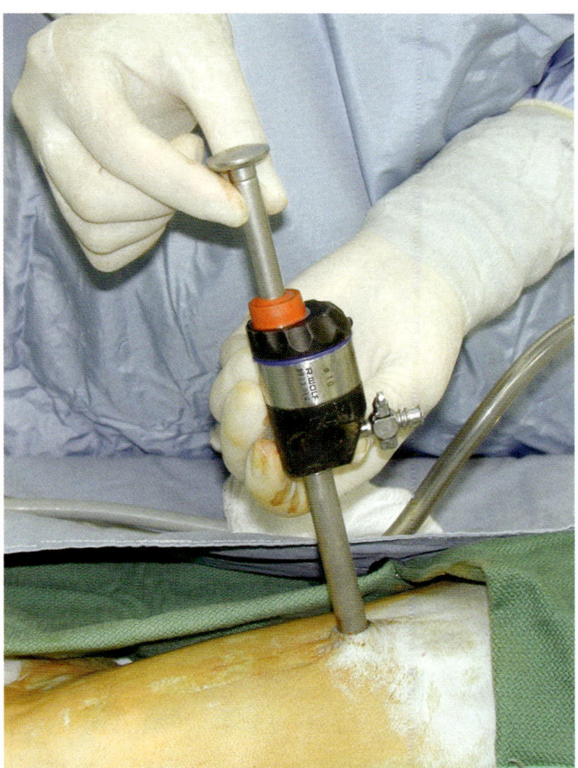

(b)

Figure 19.19 This is the insertion process for the telescope cannula. The Verres is removed and the trocar and cannula are inserted at the same location. Counter traction may be required even if insufflation is adequate. Care must be taken to not accidentally inset the trocar too far and damage the viscera.

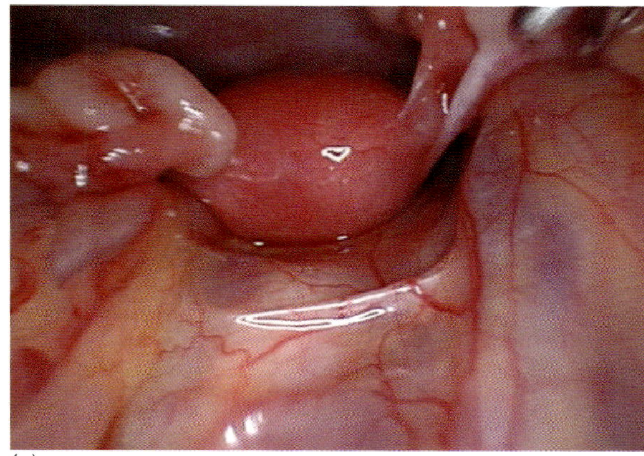

(a)

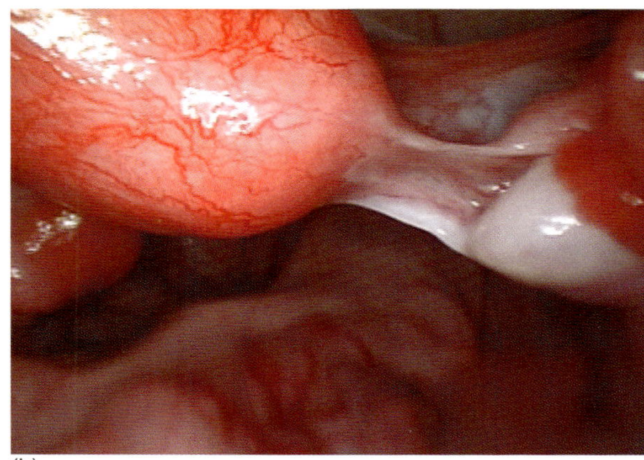

(b)

Figure 19.20 The uteroovarian ligament (large arrow). In macaques it is a thin white colored ligament running between the uterine fundus and the ovary. This is one of the two the standard places to place grasping instruments for safely manipulating the ovary. Care should be exercised to not include the isthmus of the oviduct (small arrow), which lies in close approximation.

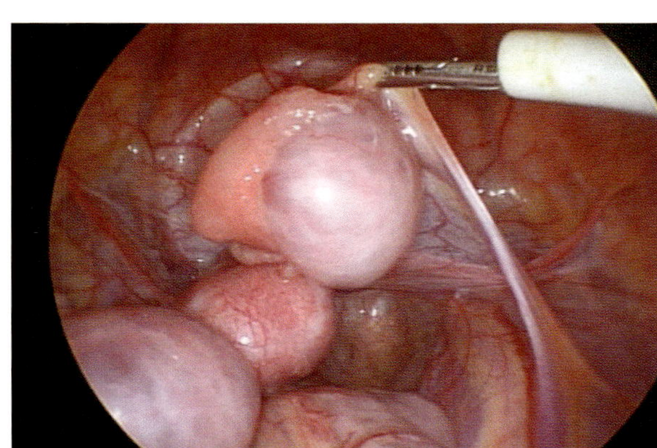

Figure 19.21 An example of a hyperstimulated ovary with multiple gravid follicles, stabilized by a grasping forceps placed on the infundibulopelvic ligament.

inserted in a variety of positions, with the most common being in a wide caudal paramedian location. This allows the insertion of forceps for manipulating the uterus, ovaries, urinary bladder, descending colon, ureters and omentum. The accessory ports can be placed in a position offset from each other to allow a greater range of motion to aid in organ manipulation. For oocyte harvest, grasping forceps are utilized to mobilize the ovaries by securing either the uteroovarian ligament or the infundibulopelvic ligament. The ovary is manipulated into a variety of positions using the forceps, being careful not to cause iatrogenic damage to the follicles. The novice laparoscopist may require two forceps for stabilizing the ovary, placing it in tension between the two instruments. As experience is gained, the operator will be able to stabilize the ovary with a single forceps negating the requirement for multiple accessory ports. The apparatus for follicular aspiration has been described (Jewett and Dukelow, 1973; Dukelow, 1980). The basic procedure consists of stabilizing the ovary with forceps, inserting a needle into the follicles

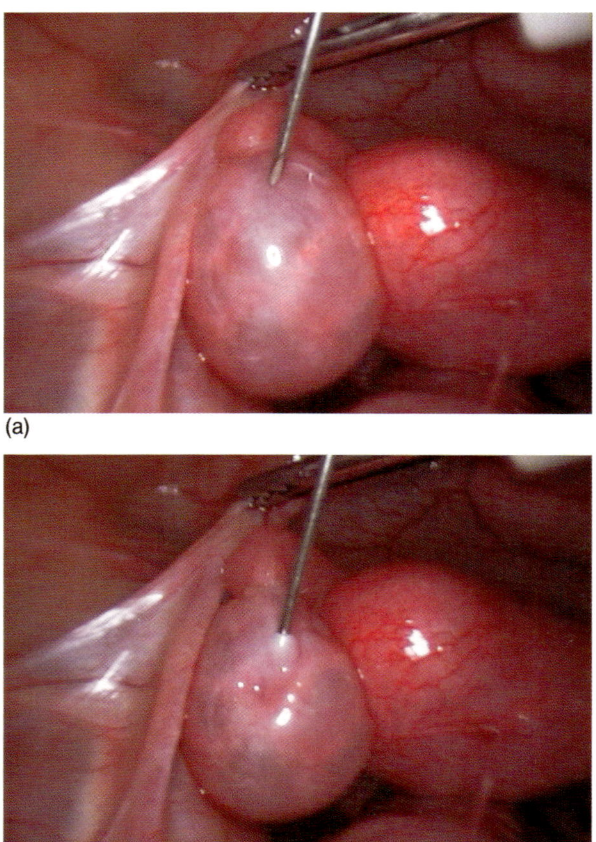

(a)

(b)

Figure 19.22 Follicle aspiration. (a) The appearance of the ovary and follicle immediately prior to needle insertion and aspiration. (b) The follicular fluid has been aspirated, resulting in a flattened or dimpled surface on the ovary.

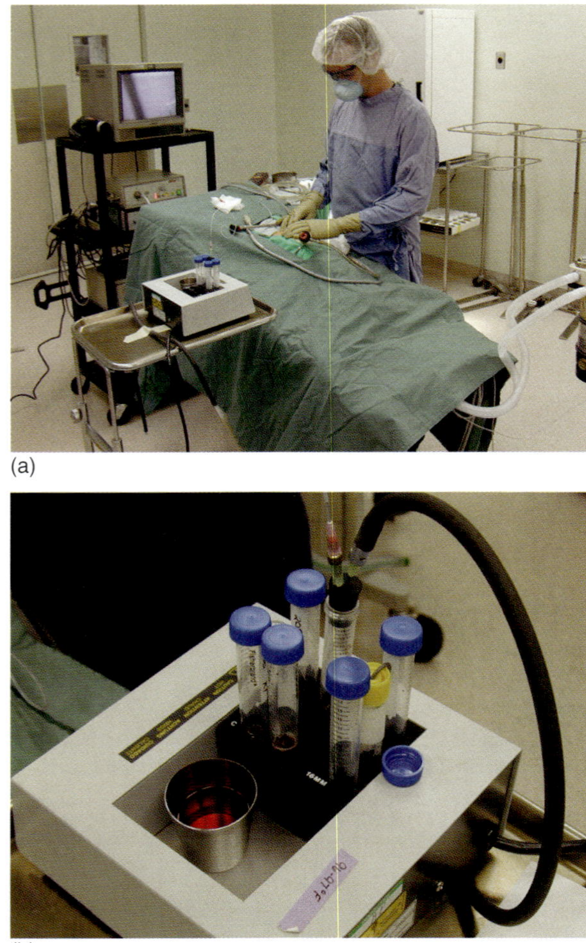

(a)

(b)

Figure 19.23 The system for follicle aspiration as used by the author. Teflon tubing runs from the aspiration needle to the collection tubes, and the whole system is maintained under constant vacuum. The collection tubes are maintained in a warming tray at 36°C during the procedure.

and applying suction to remove the follicular fluid and oocytes.

This may consist of a simple syringe and needle combination, or the needle may be attached to tubing which empties into a reservoir. The latter system is typically maintained under continuous vacuum to facilitate the process. A simplified reservoir system has been utilized by the author for over 2000 such procedures in macaques, mainly rhesus monkeys. The aspiration apparatus consists of a 3 inch long, 22 gauge needle attached to Teflon tubing. The tubing runs across the surgical table to the collection reservoir, usually a 15 ml conical specimen tube, which is maintained under 90–120 mm Hg continuous vacuum.

The recipient tubes contain 2–3 ml of heparinized Talp-Hepes solution, a buffered form of Tyrodes media. The Talp-Hepes media is used for oocyte transport or other procedures which may take place outside of the incubators or processing rooms. Several tubes are used for each ovary, to separate that portion of the aspirate which may be contaminated with blood.

Despite extreme care being exercised by the surgeon, to prevent follicular trauma during the aspiration process, hemorrhage may occur. This is much more common when the surgeon is inexperienced. The twofold result is the presence of blood within the aspirate fluid, and blood remaining in the abdominal cavity at the conclusion of the aspiration procedure. The aspirate fluid may be readily filtered to remove erythrocytes. A standard of surgical practice for all abdominal procedures is to remove blood, which has collected, by thorough lavage and suction of the abdomen with warmed saline. An aspiration cannula is the commonly used instrument for this purpose. The lavage process is important because blood left in the abdomen may induce adhesion formation, a complication which is best avoided for further laparoscopic surgeries.

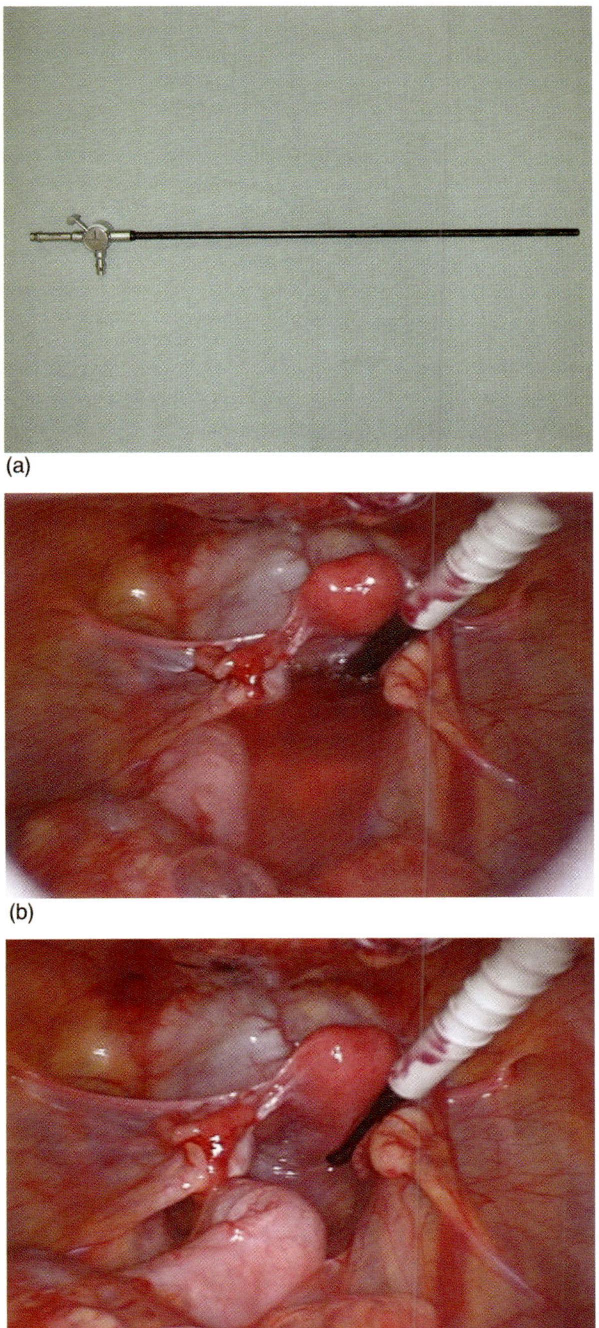

(a)

(b)

(c)

Figure 19.24 (a) An aspiration cannula used for lavage of the abdomen. This one has a three-way valve, allowing saline to be flushed in and then aspirated with one handed operation. (b and c) Lavage of the abdomen with warmed saline after a laparoscopic surgery. Any blood which collects in the abdomen should be lavaged and removed, to preclude adhesion formation.

Laparoscopic incisions are closed with synthetic absorbable suture material. The author uses 4–0 monofilament synthetic absorbable suture, initially closing the rectus fascia with a simple interrupted pattern, followed by skin closure using the same suture material in a continuous intradermal pattern.

After receipt of the oocytes in the laboratory, hyaluronidase is added to the Talp-Hepes solution to remove the cumulus and, using a microscope, the oocytes are separated and placed in an incubator at 37°C and 5% CO_2. In the author's experience with rhesus macaques, a total of 35–55 oocytes are typically harvested per aspiration procedure, when the reaction to hormonal stimulation has been good. However, it is not unusual to retrieve much larger numbers of oocytes, with as many as 150 being retrieved on many occasions.

At the author's facility, follicular aspiration as described has produced oocytes for use in a successful *in vitro* fertilization program, with rhesus monkey infants being born from both standard *in vitro* fertilization and a variety of intracytoplasmic sperm injection methods.

Laparoscopic embryo transplantation

Embryo transfers are timed events, with the initial screening of potential recipients based on menstrual activity. The pool of recipients is bled daily, for analysis of estrogen and progesterone levels, starting on day 8 after the last menstruation. In rhesus monkeys, most of the recipients can be expected to ovulate by day 12 of the cycle, with estrogen levels expected to peak at 250–450 pg/ml. Resting progesterone levels will be 0.1–0.2 ng/ml, and will typically peak at >1 ng/ml within 2–3 days of the estrogen peak. Embryo transfers are timed to occur during this peak, with variance

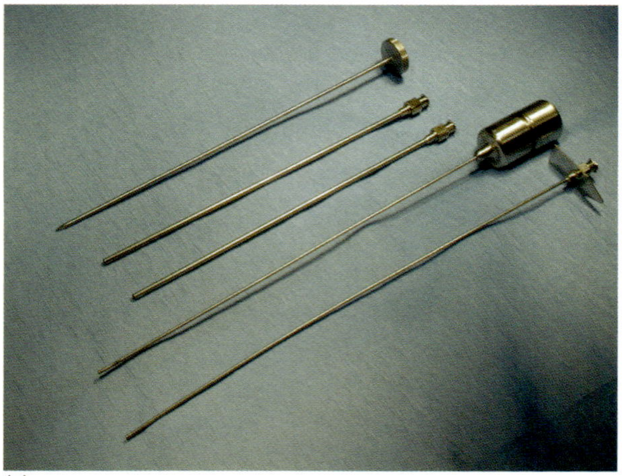

(a)

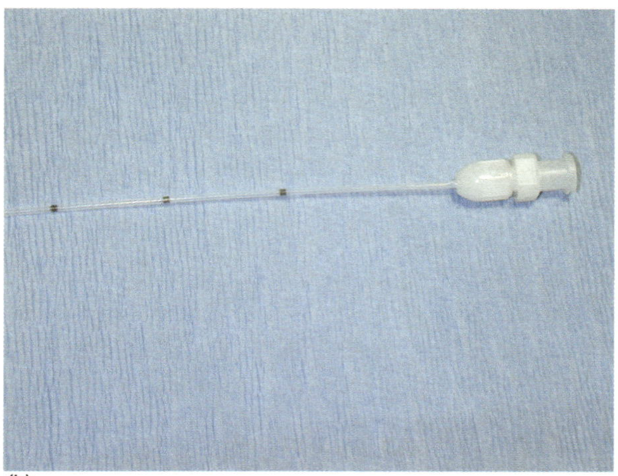

(b)

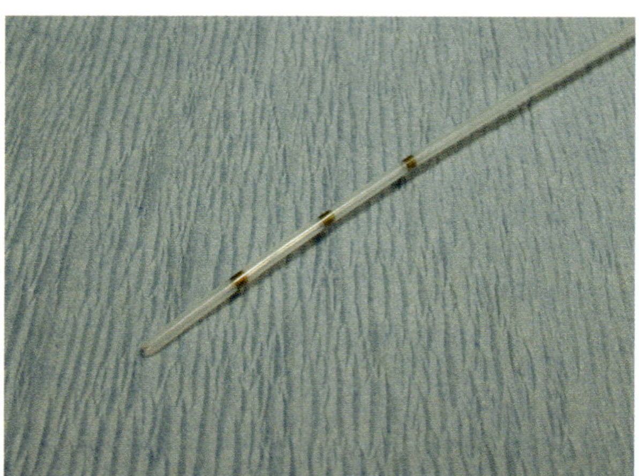

(c)

Figure 19.25 (a) An equipment set for intrafallopian embryo transfers. Instruments include self retaining graspers, cannulas for insertion into the abdomen, and the cannula and catheter for insertion into the oviduct. (b and c) The catheter used for laparoscopic embryo transfers. Note the black markings on the distal and proximal tips, to aid in judging depth of placement.

allowed for the age of the embryos. At two weeks after embryo transfer, daily blood samples are again taken to determine if pregnancy has been achieved. In pregnant rhesus monkeys, estrogen will typically remain at approximately 300 pg/ml, and progesterone will remain at 2–7 ng/ml.

Laparoscopic transfer of embryos directly into the oviducts of squirrel monkeys has been described (Dukelow, 1980), using micropipettes as the delivery device. The instrument set, used by the author, includes a variety of cannulas, grasping forceps, and embryo delivery catheters. Insufflation and telescope positioning are identical to the process for follicle aspiration.

Next, a 3 mm cannula is inserted transabdominally caudal and lateral to the telescope, and a self-retaining 2.5 mm grasping forceps is inserted through the cannula to grasp the fimbria. This grasping forceps contains a single small tooth, which is atraumatic to the fimbria, but any small grasping forceps would suffice for this purpose. The forceps is placed on the edge of the fimbria, and the grasping forceps is slowly withdrawn from the abdomen to place the oviduct under tension, reducing the tight flexures and corners of the normal oviduct. Next, a second 3 mm cannula is used to penetrate the abdomen, contralateral to the ovary and oviduct, and a 2 mm smooth tipped cannula with stylet is inserted through the cannula and directly through the os of the fimbria into the oviduct. The stylet is then removed. During this process, a second operator is preparing the embryos, which are aspirated into a special 24 cm Teflon catheter. After attaching a 1 ml syringe, the embryos are aspirated into the catheter tip using a low power dissecting microscope for magnification. To aid in rapid location of the embryos inside the catheter, they are positioned between two small air bubbles near the catheter tip, which are visible grossly. While the surgeon maintains tension on the Fallopian tube and fimbria, the assistant inserts the catheter through the intrafallopian cannula and into the oviduct to the desired depth.

The cannula is retracted until the catheter can be visualized entering the fimbria, and the embryos are deposited with gentle syringe pressure. Black lines on the outside of the catheter, one cm apart, allow accurate assessment of depth of placement. The typical intrafallopian depth for embryo placement is 2–4 cm, which is easily measured by these markings. While depth of placement is most critically affected by anatomical variation of the Fallopian tube, a general guideline is to deposit mature embryos more deeply in the oviducts, attempting to approximate embryo location which would occur during normal fertilization. The apparatus is then removed,

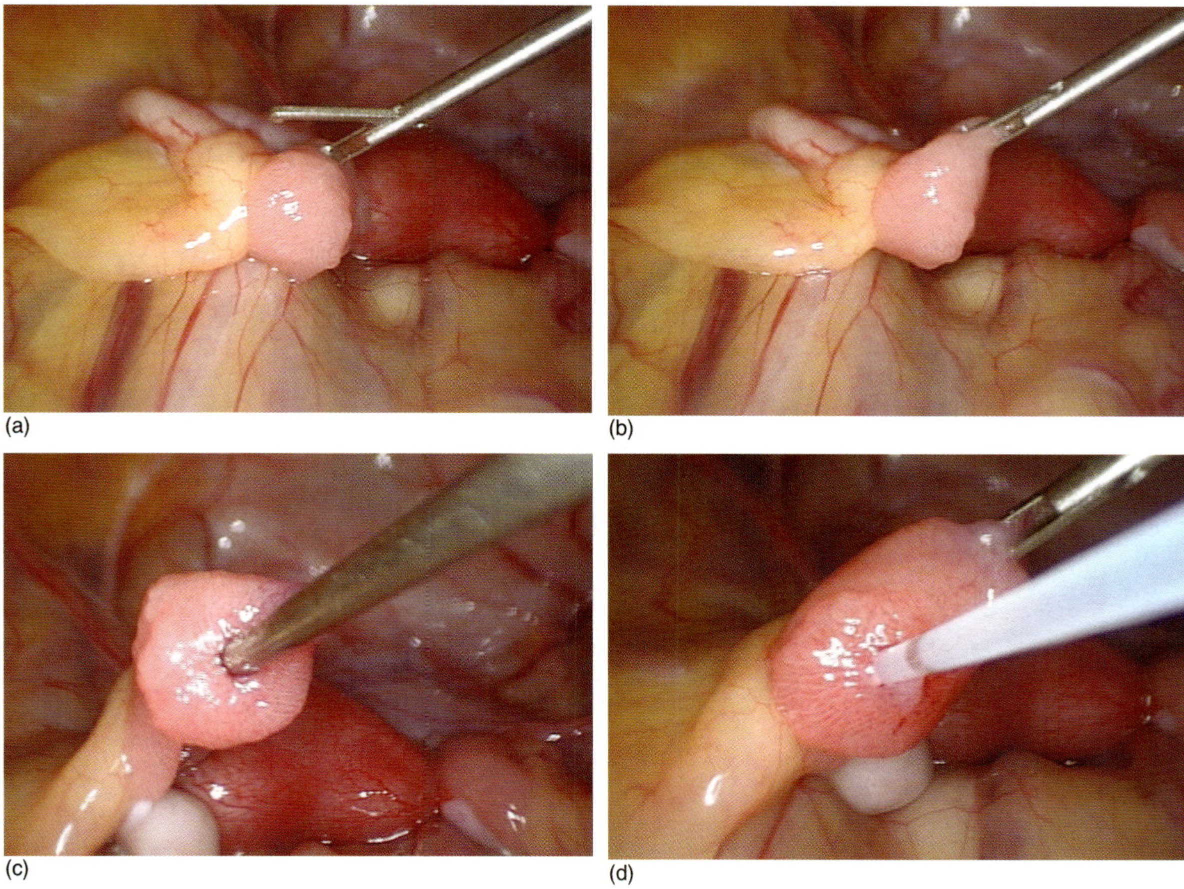

Figure 19.26 Laparoscopic embryo transfer in progress. (a and b) The grasping forceps is attached to the fimbria and places it in traction, followed by (c) insertion of the guide cannula. The last step is insertion of the catheter (d) to the desired depth, using the centimeter markings on the side of the catheter to indicate depth of the catheter tip.

the abdomen deflated, and standard closure of the incisions is performed. Although this procedure can be performed by one operator, it is greatly facilitated by an assistant surgeon.

Oophorectomy and salpingoophorectomy

For these two procedures an assistant surgeon is mandatory. Prior to the surgery, the surgeon should review the medical record for any prior gynecologic surgical procedures. Prior manipulation of the ovaries and supporting ligaments may have resulted in fibrosis and shortening of the uteroovarian ligament, impairing mobilization of the ovaries and increasing the difficulty of this procedure significantly. Anesthesia, patient positioning, and placement of the viewing telescope are identical to the procedures for oocyte harvest, with the exception that bilateral accessory port placement is required. The ovary is stabilized by grasping the uteroovarian ligament with a locking forceps. Hooked tip scissors, with insulated tips for electrocoagulation, are introduced through the contralateral accessory port and utilized to dissect the desired tissues. For oophorectomy, the assistant maintains traction on the uteroovarian ligament, close to the ovary, while the surgeon divides and coagulates the ovarian tissue away from the oviduct, infundibulopelvic ligament, and fimbria.

Care must be taken to prevent overly ambitious use of the electrocautery, as it can cause stricture of the oviduct or fimbria, resulting in hydrosalpinx. Also, the ovary itself may be damaged. If the procedure is properly performed, the ovary will be isolated and may be removed via an accessory port.

Salpingoophorectomy is technically easier to perform as the oviduct and fimbria are resected with the ovary. The careful dissection technique of separating the oviduct from the ovary is not required. The ovary is stabilized with single forceps on the infundibulopelvic ligament and hooked scissors are used to cut and coagulate directly through the uteroovarian ligament and oviduct to the level of the broad ligament,

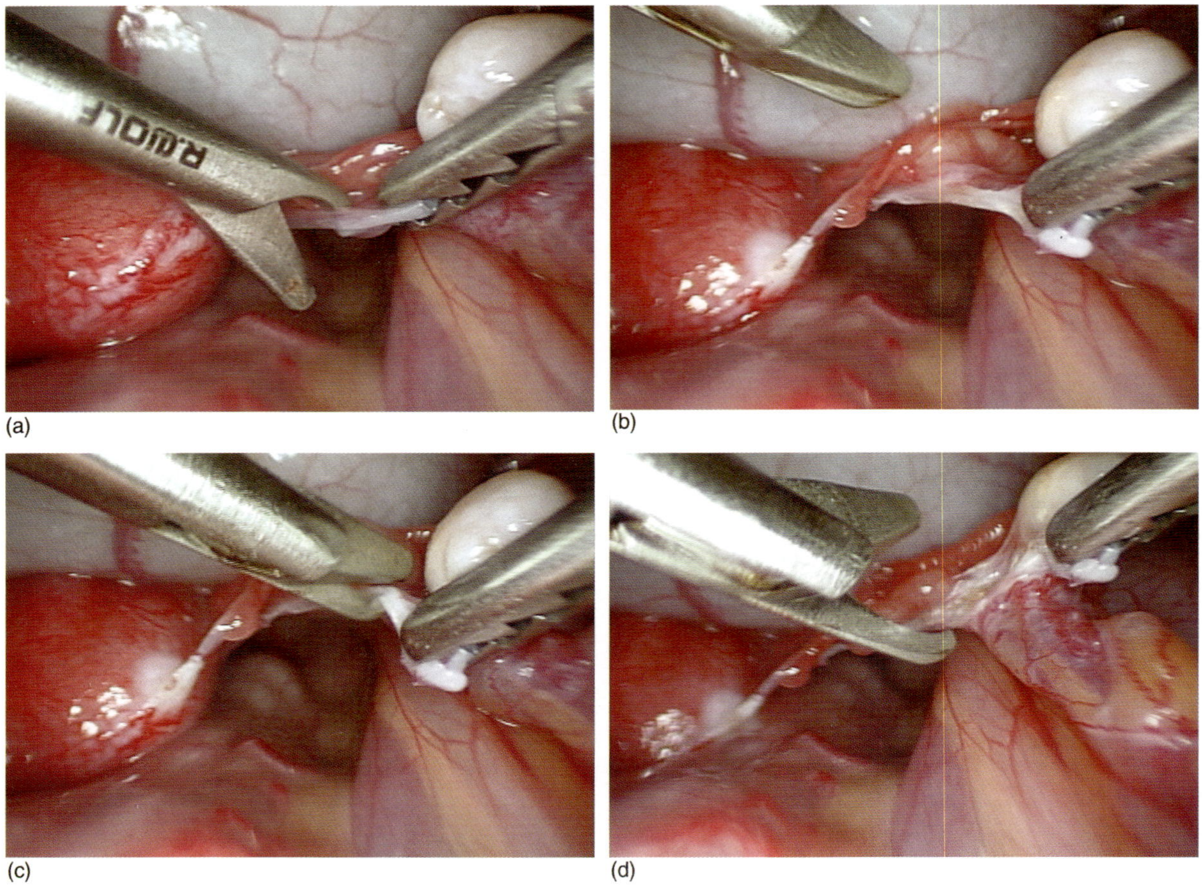

Figure 19.27 Laparoscopic oophorectomy. (a) A hooked scissors in position to transect the uteroovarian ligament. (b) Immediately after the ligament has been cut. (c and d) Continuing the dissection of the ovary from the oviduct, with tissues being cut and cauterized simultaneously with the hooked scissors. Care must be exercised to avoid damage to the uteroovarian vasculature.

greatly simplifying the procedure. A second technique involves the use of pre-tied ligatures which are slipped over the ovary and cinched down tightly to ligate the uteroovarian vasculature, the oviduct and infundibulopelvic ligament. Scissors are then inserted and used to transect these tissues, distal to the ligature. The resected specimens are removed from the abdomen through one of the accessory ports. This second technique is dependent on the length of the uteroovarian ligaments. If they are fibrosed, from prior surgery, or are excessively short, placing a ligature as described may be quite difficult.

Hepatic biopsy

A blood clotting profile should be performed, preoperatively, as significant intraabdominal hemorrhage may occur even with a normal patient. In patients with clotting deficiencies hemorrhage may be life threatening.

After anesthesia has been induced, the viewing telescope is inserted near the umbilicus, very similar to the previously discussed gynecologic procedures. The telescope is oriented in a cephalad direction towards the liver. The patient should be positioned in reverse Trendelenburg, to allow gravity to pull the intestines and spleen away from the liver, enhancing visualization. One or two paramedian accessory ports are placed in a standard triangulated position. A 5 mm oval cupped biopsy forceps is inserted through an accessory port and directed towards the liver. Under visualization, the jaws of the forceps are tightly closed over the edge of the liver and a small biopsy is removed. Even if they are new and quite sharp, it is difficult to cut the hepatic tissue with these biopsy forceps. Therefore traction, combined with a twisting or rocking motion of the forceps, will aid in obtaining the sample.

A second technique is to use a true-cut biopsy punch to obtain the sample.

The biopsy instrument is inserted trans-abdominally through a small 1–2 mm skin subcostal incision. An accessory port is required only if grasping forceps are deemed necessary to position the liver. In most cases,

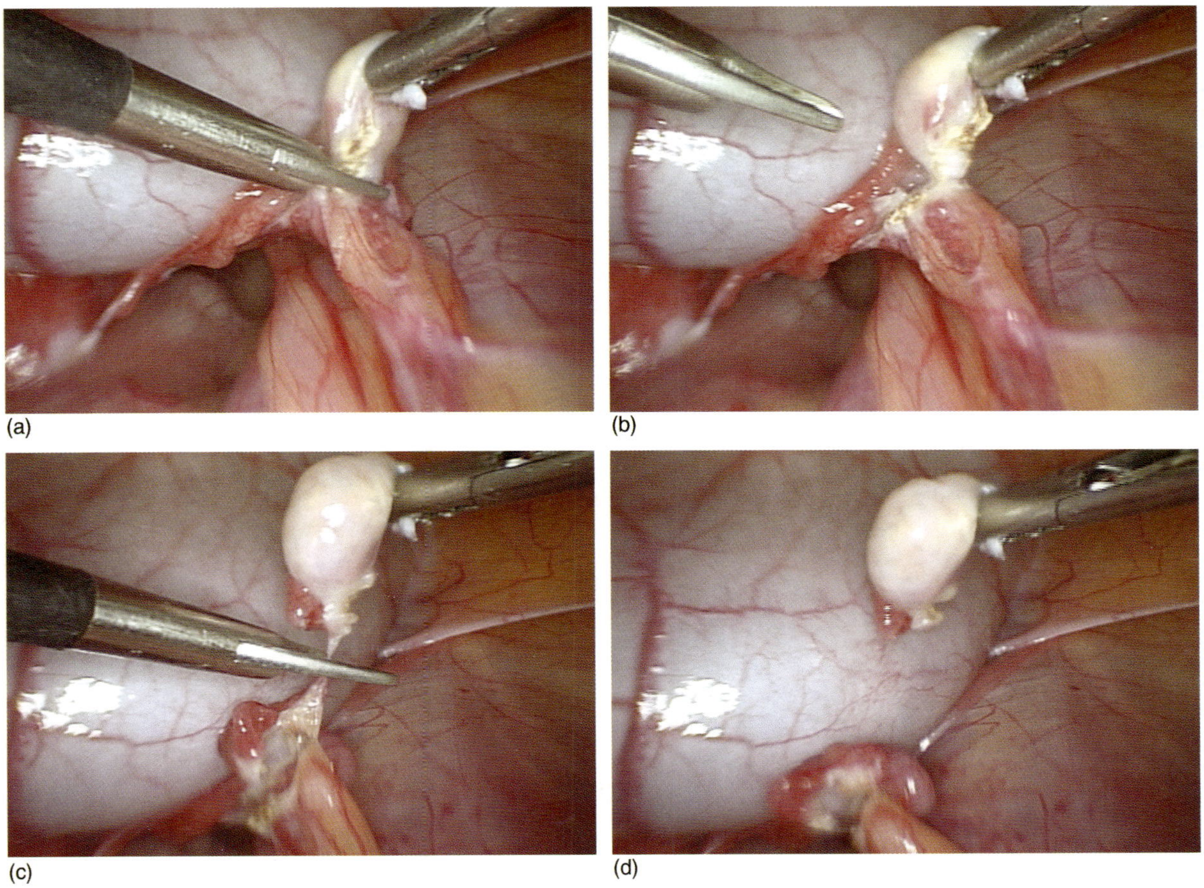

Figure 19.28 Laparoscopic oophorectomy. (a and b) The uteroovarian ligament is placed in traction to allow the oviduct to be spared. (c and d) The ovary has been excised and may be removed through an accessory port.

the liver will be suspended and exposed very adequately for this procedure, with patient positioning in a slight feet down orientation. This causes the intestinal viscera to fall away and expose the liver and stomach. After the biopsy is removed, the biopsy site must be observed to assure that hemorrhage does not continue. Hemorrhage may be controlled by placing pressure on the biopsy site with a probe, or the tip of a grasping instrument. This will usually control bleeding within 2–3 minutes.

Splenic biopsy

Trocar placement for the telescope is on the ventral midline immediately over, or caudal to, the umbilicus. Accessory ports should be positioned in the left wide paramedian region and a second accessory port triangulated in the low subxyphoid region. Biopsy of the spleen is often best accomplished with stapling devices which can be used to place a line of secure staples across the tip of the spleen, after which the isolated tissue is resected with scissors. Alternatively, a True-Cut biopsy punch can be used in exactly the same manner as for hepatic biopsy. Close attention must be paid to post biopsy hemorrhage, which can be controlled with pressure on the biopsy site. If biopsy specimens are not adequate, complete splenectomy may be accomplished using vascular stapling instruments to control all splenic vasculature prior to resection.

Intestinal biopsy

The small intestine may be biopsied using 4–0 Endoloop™ ligatures (Perret-Gentil, 2000). Viewing telescope and accessory port placement are as described above for hepatic biopsy, with two accessory ports required. The Endoloop™ is inserted through one of the ports, and a grasping forceps through the other port. The grasper is inserted through the ligature loop to grasp the antimesenteric border of the intestine. A small piece of the intestine is placed in traction and the Endoloop™ is secured around it, after which the full thickness biopsy is resected with hooked scissors. Constriction of the intestinal lumen, by approximately 20%, has been reported with this technique (Perret-Gentil, 2000). However, morbidity was much lower than accomplishing intestinal biopsy with traditional laparotomy.

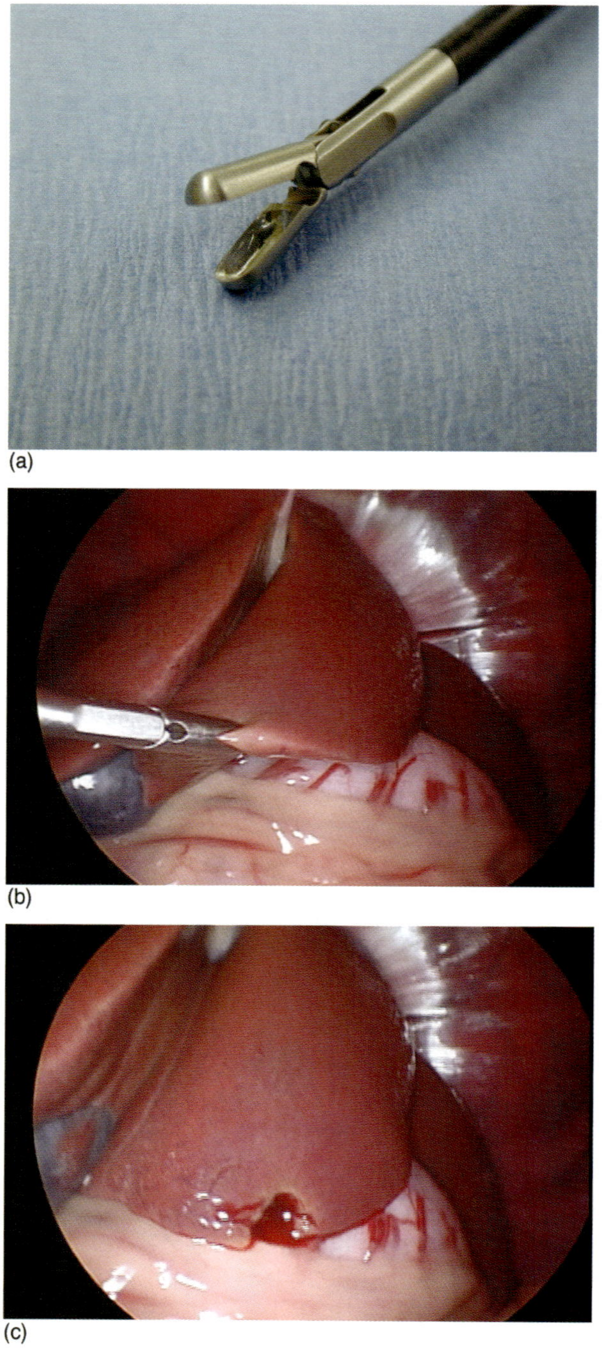

Figure 19.29 (a) Cup tipped forceps used for excising small biopsy specimens from soft tissues. (b) The forceps are in place on the caudal edge of the liver in a rhesus monkey. The forceps are then tightly closed and rocked slightly to dislodge the biopsy specimen. (c) Immediately after the hepatic biopsy has been removed. Note the minimal hemorrhage. If hemorrhage continues, any type of forceps may be used to place pressure on the biopsy site for a few minutes.

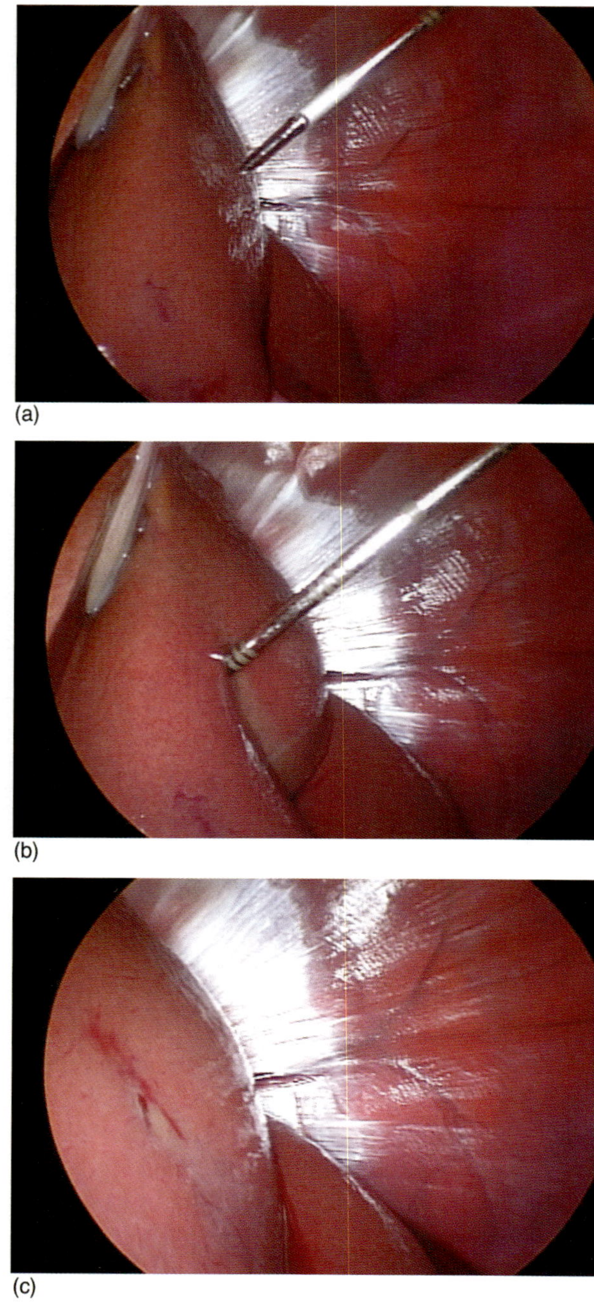

Figure 19.30 (a) A True-Cut biopsy punch inserted through the upper left abdominal wall and positioned for a hepatic biopsy. (b) The biopsy punch has been inserted into the liver. (c) After the biopsy punch has been removed, minimal hemorrhage occurs from the biopsy site (arrow). If hemorrhage does occur, an instrument may be used to place pressure on the biopsy site until clotting occurs.

Mesenteric lymph node excision

Telescope and accessory port placement is identical to the procedure for intestinal biopsy (Perret-Gentil, 2000). After selection of the lymph node to be excised, it is placed in traction. An Endoloop™ is placed over the node and the vasculature is ligated, after which the node is resected with hooked scissors. Care must be taken to preserve blood supply to adjacent bowel.

Specimen removal

Tissues to be retrieved from within the abdomen must first be completely resected. The most simple method of retrieval is to extract the specimen through one of the accessory ports. If the tissue is too large for this, the accessory port may be removed and the specimen extracted directly through the incision.

For more friable tissues, which may fragment upon retrieval, a specimen bag may be used. This is simply a sterile plastic bag, which may be closed by pulling on a string around the top. The bag is inserted into the abdomen through an accessory port and the tissues are inserted into the bag. The opening of the bag is closed and the bag is then retrieved in the same fashion as tissues. The difference is that more traction may be placed on the bag, literally squeezing it through an accessory port. Specimen bags are commercially available from a variety of medical supply firms. However, properly sterilized single use plastic food bags may also suffice in some circumstances.

Ultimately, the method of tissue retrieval depends on the importance of specimen morphology. If deforming a specimen is not possible by pulling it through a small diameter accessory port, then the only option is to remove a port and lengthen the incision adequately to accommodate the specimen.

Thoracoscopy

Thoracoscopy is the use of a rigid endoscope within the thorax, and is sometimes distinguished from video-assisted thoracic surgery (VATS), which employs a video camera attached to the viewing telescope in exactly the same manner as the laparoscopic procedures described above. These are minimally invasive procedures for the examination of the pleural cavity and thoracic viscera. Traditional thoracotomy incisions are often quite long and result in considerable postoperative pain. Most thoracoscopic procedures require only three small incisions through the thoracic wall, minimizing the trauma associated with the procedure.

Equipment

The same basic equipment utilized for laparoscopic surgery can also be used for many thoracic procedures. The major difference is that CO_2 insufflation is often not required, as the rib cage maintains the thoracic cavity in an expanded state. Telescopes used for thoracoscopic surgery in macaques, and larger NHP species, are 5 mm in diameter. Telescopes, with a 20–30 degree angled field of view, are often required for proper examination of thoracic viscera, unlike laparoscopic procedures. Again, an arthroscope, with an angled field of view, may provide the perfect combination of narrow diameter and optical capability. Cannulas for thoracoscopy are either open or closed. Open cannulas do not maintain positive pressure within the pleural cavity, while closed cannulas have a valve for creation of a controlled pneumothorax, or even tension pneumothorax. Open cannulas are preferred for longer duration procedures. Cannula diameter is mandated by the size of the manipulating instruments. As with laparoscopy, much of the available stapling equipment has a diameter of 10 mm, thus mandating the use of 11 mm cannulas.

Patient preparation

As with laparoscopy in NHP, anesthesia at a surgical plane is a requirement. Endotracheal intubation is mandatory as pneumothorax is induced in the operated hemithorax, and ventilatory support is required to maintain expired CO_2 levels in the normal range. One lung ventilation is often performed with selective intubation of either right or left lungs, allowing improved exposure. When one-lung ventilation is used, the need for monitoring end-tidal CO_2 levels is quite important. Physiologic monitoring should be identical to those standards recommended for laparoscopy.

Patient positioning is in either dorsal or lateral recumbency. There are two basic endoscopic approaches to the thorax, either transdiaphragmatic or intercostal. The transdiaphragmatic approach allows visualization of each hemithorax. For a transdiaphragmatic approach, the patient is positioned in dorsal recumbency. A 5 mm skin incision is performed just caudal to the xyphoid cartilage. Then, a screw-in cannula may be inserted from a subxyphoid position and directed cranially.

The cannula is partially inserted and then the telescope is placed inside the cannula for the final entry of the thoracic cavity. At the point of penetration into the thorax, the telescope is advanced into the thoracic cavity, allowing visualization of the viscera and preventing iatrogenic trauma. If the mediastinum remains intact, only one hemithorax can be examined at this time. For examination of the contralateral hemithorax, the cannula and telescope must be partially retrieved and then directed across the midline to the other side. It is preferred to use a 0 degree telescope for this initial examination.

After initial examination, two other cannulas, for instrument utilization, are placed under thoracoscopic visualization through the intercostal spaces nearest the pathology or organ to be treated. As a general rule, the cannulas should be placed as far dorsal as possible to allow maximum mobility of the instruments.

Alternatively, the instruments can be inserted directly through the chest wall, without using a cannula, by making a small skin incision and penetrating the chest wall with a hemostatic forceps like a Rochester or Kelly. The tips of the forceps are spread apart to separate the tissues widely enough to allow passage of the instruments. This intercostal approach allows visualization only of one side of the thoracic cavity but provides better visualization of the dorsal part of the lung and cavity.

The patient is placed in lateral recumbency and one-lung ventilation is often used for this approach, rather than for trans-diaphragmatic approach. The telescope is typically placed through the 10th intercostal space, while the accessory cannulas are inserted through the 7th or 8th intercostal spaces for introduction of instruments. The three cannulas are placed in triangle fashion, very similar to the preferred arrangement for most abdominal endoscopy.

Exploration of the thoracic cavity begins with visualization of the chest wall, and ribs and intercostal muscles may also be evaluated. The parietal pleura can be examined in its entirety, and the intercostal vasculature is prominent. Other structures which may be approached are the cranial vena cava, phrenic nerves, thoracic inlet, and internal thoracic arteries. A lung retractor may be introduced to enhance visualization of the thoracic inlet. The pericardium and mediastinal lymph nodes may be observed. If the diaphragm is to be examined, the telescope must be inserted through one of the accessory cannulas to provide a proper focal distance. The dorsal part of the lung parenchyma is more easily examined with a 30 degree telescope, and the lungs may be retracted to allow visualization of the hilus.

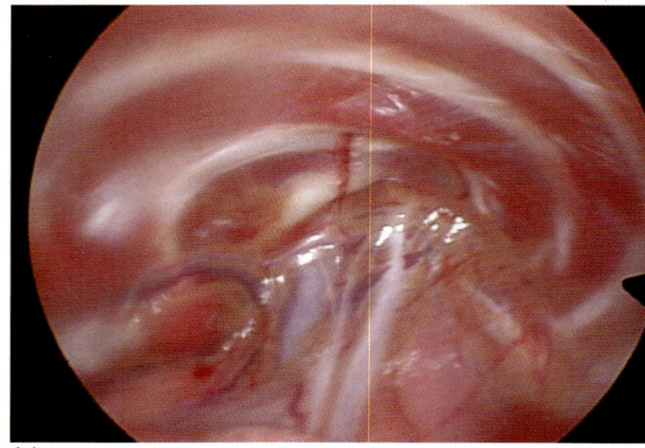

(a)

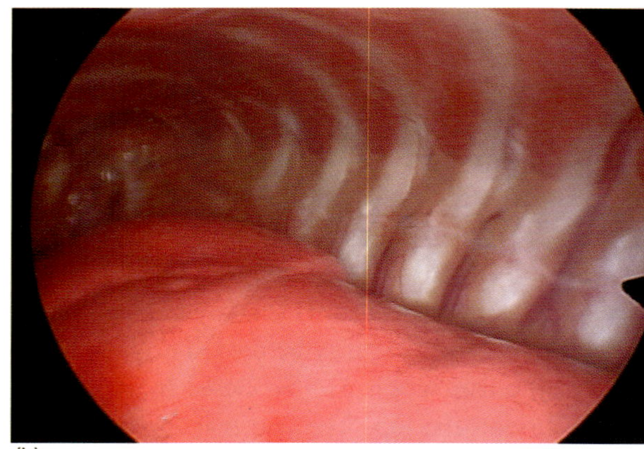

(b)

Figure 19.31 (a) Thoracoscopic view of the dorsal thorax, using a 9th intercostal space placement of a 5 mm telescope without a cannula. Pneumothorax created during the telescope insertion allows the lung to collapse and provide good visualization. (b) The telescope has been turned to show the prominent intercostal vasculature, and provides a good view of the collapsed lung.

Closure of incisions

The incisions created for thoracoscopic surgery are quite small and do not require closure in multiple layers. Absorbable sutures in a simple interrupted pattern are adequate for closure of the subcutaneous spaces, while skin closure may be effected with either intradermal sutures of absorbable material, or simple interrupted skin sutures of monofilament nylon. At the conclusion of any thoracoscopic procedure, the pneumothorax must be corrected. Often the most simple method is the insertion of an intravenous catheter through an intercostal space away from the surgical sites. The catheter is attached to a three-way stopcock which is then connected to a 30–60 ml syringe. After all cannula and telescope incisions have been closed, the syringe and stopcock are used in conjunction to evacuate the pneumothorax. Rarely is a large diameter thoracostomy tube required to correct the iatrogenic pneumothorax. However, if the lung has been biopsied in such a manner as to allow leakage of air, or if the skin closure has been inadequate, there will be recurrent pneumothorax and a thoracostomy tube may be required for several hours or days to correct the problem. If this is necessary, the thoracostomy tube should be placed in the affected hemithorax, preferably in the 9th intercostal space, and should be connected to a Heimlich valve or some other form of self-evacuating one-way valve.

Thoracoscopic procedures

Lung biopsy

Biopsy of the lung requires a pre-tied ligature (Endoloop™) which eliminates the requirement for knot tying inside the cavity. The Endoloop™ in introduced through one of the accessory cannulas, and a grasping forceps is inserted through the other cannula and is projected through the loop to grasp the tip of the lung for biopsy. The lung parenchyma is pulled back inside the loop and the loop is pulled tight. Scissors are introduced through the other accessory port and are used to remove the biopsy specimen, after which they can be used to cut the long ends of the ligature. In humans, perioperative morbidity and postoperative length of hospitalization were significantly less with VATS versus open thoracotomy for this procedure (Landreneau, 1999).

Thymic biopsy

This procedure has been well described for macaques (Bohm, 2000). The telescope and cannula placement are different for this specific procedure. The telescope is inserted through the 6th intercostal space, while the accessory cannulas are placed through the 3rd and 5th

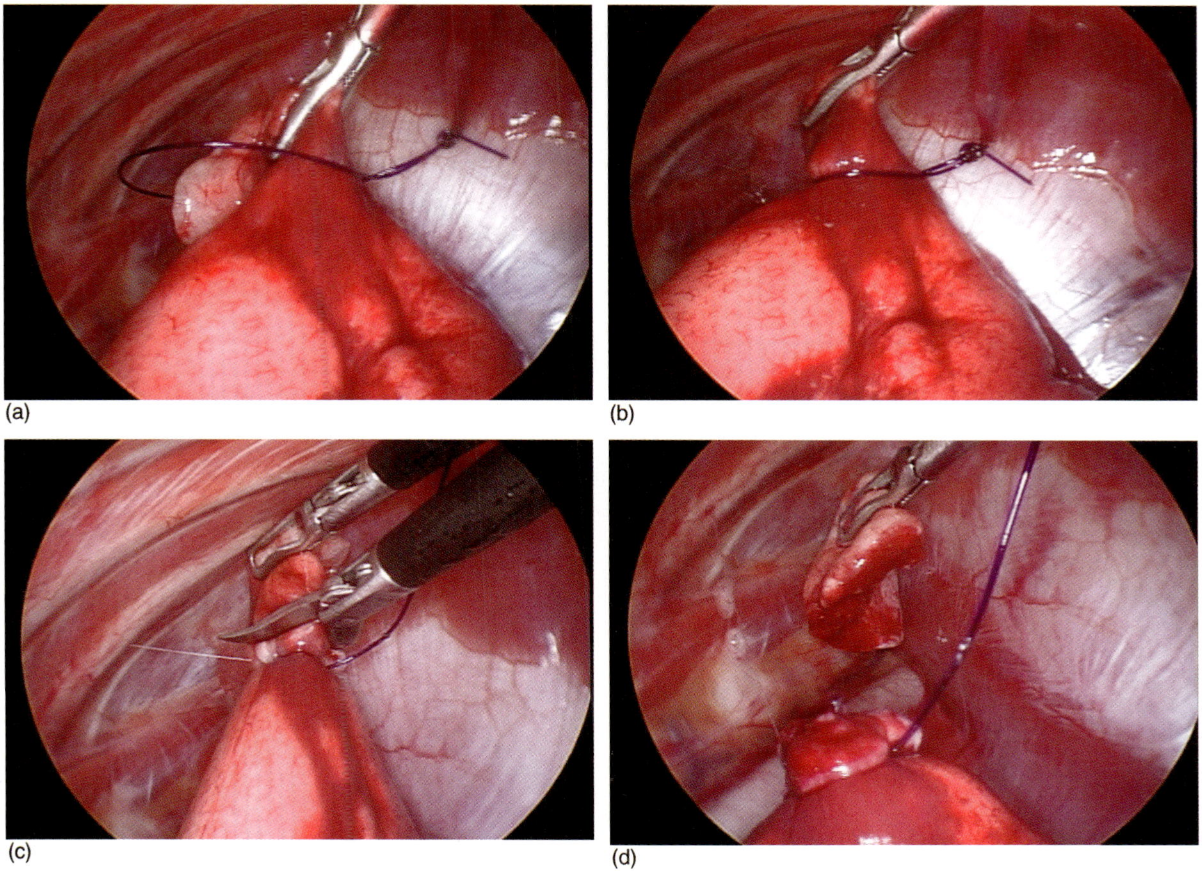

Figure 19.32 Lung biopsy. (a) The viewing telescope has been inserted at the 4th intercostal space and is directed toward the diaphragm. A standard grasping forceps is inserted through the 10th intercostal space, without a cannula. The forceps is being used to elevate the caudal lung lobe. (b) An Endoloop™ is being used for ligation of the selected tissues. (c) The lung specimen has been ligated with the Endoloop™, and a scissors has been inserted to resect the ligated tissues. (d) The lung biopsy has been resected and is ready for extraction, after which the ligature is cut and the incisions repaired.

intercostal spaces, and pneumothorax at 6 mmHg is maintained throughout the procedure. The biopsy specimen is obtained by direct excision with a Metzenbaum scissors and is removed through an accessory port.

Summary comments

Over the past decade, the advent of video-assisted rigid endoscopy has revolutionized how surgical procedures are performed. It has become a standard methodology for many human surgical procedures due to the decreased level of morbidity, and is also rapidly gaining acceptance within the veterinary community for both clinical and research applications. Continuing improvements in equipment will result in even more widespread utilization, to the benefit of the patient. Using video-assisted rigid endoscopy, the only limiting factor on the scope of procedures possible is the imagination of the operator.

Correspondence

Any correspondence should be directed to John Fanton, Oregon National Primate Research Center, Beaverton, OR 97006, USA.

References

Benitez, L.D. and Edelman, D.S. (2000). In Eubanks, W.S., Swanstrom L.L. and Soper, N.J. (eds) *Anesthesia Concerns in Surgical Endoscopy,* pp 48–56. Lippincott Williams & Wilkins, Philadelphia.

Bohm, R.P., Jr., Rockar, R.A., Ratterree, M.S., Blanchard, J.L., Harouse, J., Gettie, A. and Cheng-Mayer, C. (2000). *Contemp. Top. Lab. Anim. Sci.* 39, 24–26.

Dierschke, D.J. and Clark, J.R. (1976). *J. Med. Primatol.* 5, 100–110.

Dukelow, W.R. (1980). In Wildt, D.E. and Harrison, R.M. (eds) *Laparoscopy in Small Animals and Ancillary Techniques,* pp 95–105. Williams & Wilkins, Baltimore.

Dukelow, W.R., Jarosz, S.J., Jewett, D.A. and Harrison, R.M. (1971). *Lab. Anim. Sci.* 21, 594–597.

Dukelow, W.R., Jewett, D.A. and Rawson, J.M. (1973). *Am. J. Phys. Anthropol.* 38, 207–209.

Graham, C. E. (1976). *J. Med. Primatol.* 5, 111–123.

Graham, C.E., Keeling M., Chapman C., Cummins L.B. and Haynie, J. (1973). *Am. J. Phys. Anthropol.* 38, 211–215.

Harrison, R.M. (1980). In Wildt, D.E. and Harrison, R.M. (eds) *Laparoscopy in Monkeys and Apes,* pp 73–93. Williams & Wilkins, Baltimore.

Jewett, D.A. and Dukelow, W.R. (1971). *Lab. Prim. News.* 10, 16–17.

Jewett, D.A. and Dukelow, W.R. (1972). *J. Reprod. Fertil.* 31, 287–290.

Jewett, D.A. and Dukelow, W.R. (1973). *J. Med. Primatol.* 2, 108–113.

Landreneau, R.J. and Mack, M.J. (1999). In Eubanks, W.S., Swanstrom, L.L. and Soper, N.J. (eds) *Thoracoscopic Resection of Pulmonary Parenchymal Lesions,* pp 471–479. Lippincott Williams & Wilkins, Philadelphia.

Miura, T., Shimada, T., Tanaka, K., Chujo, M. and Uchida, Y. (2000). *J. Thorac. Cardiovasc. Surg.* 120, 437–447.

Perret-Gentil, M.I., Sinanan, M.N., Dennis, M.B., Anderson, D.M., Pasieka, H.B., Weyhrich, J.T. and Birkeban, T.A. (2000). *J. Invest. Surg.* 13, 181–195.

Rawson, J.M. and Dukelow, W.R. (1973) *J. Reprod. Fertil.* 34, 187–190.

Schauer, P.R. (2000). In Eubanks, W.S., Swanstrom, L.L. and Soper, N.J. (eds) *Physiologic Consequences of Laparoscopic Surgery,* pp 22–37. Lippincott Williams & Wilkins, Philadelphia.

CHAPTER 20

Ultrasound Imaging in Rhesus (*Macaca mulatta*) and Long-tailed (*Macaca fascicularis*) Macaques: Reproductive and Research Applications

Alice F. Tarantal
California National Primate Research Center and Department of Pediatrics, University of California, Pedrick and Hutchison Roads, Davis, California 95616-8542, USA

Section 1: Introduction

Ultrasonography is a routine imaging modality used for colony management and experimental protocols at many nonhuman primate facilities. Investigations using the monkey model have incorporated standard two-dimensional imaging and pulsed and color Doppler for assessing the cardiovascular and urinary systems, whole organ transplants and vascular grafts, and to obtain ocular and transcranial measures, to name a few (see Section 7). Reproductive applications of ultrasound in macaque breeding colonies can provide an efficient method for pregnancy detection, monitoring the fetus during gestation, and for routine assessments of breeding females. With limited cost investment compared to other imaging modalities, ultrasound can be readily incorporated as a routine procedure. For example,

The Laboratory Primate
Copyright 2005 Elsevier
ISBN 0-1208-0261-9

routine screening can identify animals with reproductive pathology, confirm gestational age for animals with known breeding history, predict gestational age for those animals where mating information is not available, and monitor the conceptus as a part of experimental and colony-based protocols. In addition, ultrasonography plays a major role in the identification of animals with obstetrical complications and is invaluable as an adjunct for interventional procedures to avoid invasive, labor-intensive, and costly surgical approaches.

The intent of this chapter is to highlight some of the more common uses of ultrasound in macaques, with a focus on reproduction. While many publications have documented the use of ultrasound in new world species (Corradini *et al.*, 1998; Jaquish *et al.*, 1996; Kuederling and Heistermann, 1997; Mohle *et al.*, 1999; Narita *et al.*, 1988; Nubbemeyer *et al.*, 1997; Tardif *et al.*, 1998; Windle *et al.*, 1999), baboons (Brans *et al.*, 1990; Cseh *et al.*, 2001; 2002; Devonald *et al.*, 1996; Ervin *et al.*, 1998; Farine *et al.*, 1988; Hankins *et al.*, 1990; Hennessy *et al.*, 1999; Herring *et al.*, 1992; Jin *et al.*, 1996; Santolaya-Forgas *et al.*, 1997), vervets (Seier *et al.*, 2000), and the great apes (Hatasaka *et al.*, 1997; Yeager *et al.*, 1981), this chapter will primarily address applications in rhesus (*Macaca mulatta*) and long-tailed or cynomolgus (*Macaca fascicularis*) macaques. Section 7 addresses other non-reproductive applications of ultrasound that are primarily related to experimental procedures. As with other sections in this chapter, Section 7 is not intended to cover all applications or published studies, but to highlight some of the uses and investigations performed.

Section 2: Equipment and scanning techniques

Transabdominal imaging in rhesus and long-tailed macaques requires high frequency transducers (also called scanheads; ≥7.5 MHz), because of the depth of anatomical structures characteristic of these species. With a transabdominal approach, the uterine focal zone is typically 2–3 cm in most animals, hence the need for high frequency transducers that can provide high quality images at shallow depths. In addition, because of the overall size of the animals, a transducer with a relatively small footprint is required, similar to imaging requirements in human pediatrics. In the human clinical setting, endovaginal scanning is a standard technique because of improved anatomical detail when compared to transabdominal imaging. However, this approach is not necessary in these species because of the depth of the structures of interest and ease in obtaining high quality images when using the transabdominal approach. In addition to two- and three-dimensional imaging, pulsed and color Doppler aid in assessing the vascularity of normal and aberrant structures, and the blood flow hemodynamics for diagnostic and experimental purposes.

When beginning the examination, it is first necessary to ensure good contact between the transducer and abdominal surface; thus, shaving hair and applying a sufficient quantity of good quality acoustic gel is essential. Crucial in the imaging process is an understanding of anatomy, the anatomical relationships of the structures under investigation, and the relationship of these structures to the image viewed on the monitor of the ultrasound imaging system. An understanding of normal anatomical variation is also important, as well as establishing baseline data for individual animals. For example, as discussed below, uterine position and flexion can vary and benign structures, such as mesenteric cysts, are routinely observed. In addition, while a distended urinary bladder is necessary for imaging human reproductive structures, this is not a requirement for macaques. Poor image quality due to the presence of air from intervening bowel can be addressed by gentle pressure and guiding the impeding structures to an alternate location. Similar to humans, a distended urinary bladder can distort anatomy and result in misdiagnoses, as, for example, when assessing the location of placental discs in relation to the uterine cervix (see comments, Section 4).

Standard imaging of anatomical structures includes serial transverse and sagittal scans to ensure complete analysis of the organ of interest. When assessing the uterus, for example, this will ensure that pathologic changes or an early pregnancy are not missed. Evaluating abdominal and pelvic structures such as the adrenals, kidneys, spleen, gastrointestinal tract, pancreas, and hepatobiliary system are also important in order to obtain all relevant anatomical information. This is particularly important when sonographically assessing animals for the presence of a mass or abdominal adhesions, which can involve multiple structures (Tarantal, 1992).

Efforts should also be made to maintain the lowest scanning outputs and equipment settings to obtain the best quality images. Limiting the examination period to only the essential time required to obtain the necessary information will minimize any potential for heating that can occur during the ultrasound imaging process. While there are extensive reviews in the literature that support the safety of diagnostic ultrasound, it is important to avoid any effects that could potentially result from either heating or cavitation, both of which are established biologic effects of ultrasound at diagnostic levels (Barnett *et al.,* 2000; Fowlkes and Holland, 2000). Focusing for long periods of time on any anatomical structures, particularly in the fetus, should thus be avoided (Tarantal, 1998).

Animals are routinely scanned under ketamine (10 mg/kg) or telazol (5–8 mg/kg). Telazol is recommended when procedures are planned that will require immobilization for greater than 15–20 minutes. Animals can also be scanned in restraint chairs once trained (Golub and Anderson, 1986), or hand-held in a supine position on an examination table by experienced animal handlers. This approach may be required for evaluations where frequent ketamine administration would be prohibitive, for evaluations where it is essential that fetal sleep:wake cycles are not disturbed, or if the potential effects of ketamine on the developing brain need to be avoided (Olney, 2002). Ketamine and telazol readily cross the placenta, thus immobilizing the fetus as well as the dam. If monkeys are restrained by hand or in chairs, care must be taken to ensure that they are comfortably positioned and in a manner such that they are unable to bite or scratch the hand of the individual scanning.

Section 3: Nongravid animals

Uterine anatomy

The macaque uterus is "pear-shaped," simplex, and the uterine fundus may have a shallow groove representing the fusion of the paramesonephric ducts and formation of the uterovaginal primordium during development. Anatomical variations observed in humans also occur in monkeys because the uterus develops in a similar manner. If the right and left paramesonephric ducts do not completely fuse, then a bicornuate uterus can form. Fusion may also be incomplete and the lumen may fail to completely develop (see human embryology text for further information). Unlike humans, the macaque uterus is typically flattened ventrodorsally, and the cervix of the rhesus and long-tailed macaque is typically thick-walled and bulbous, with a serpentine cervical canal and prominent colliculi and several blind passages (Hill, 1974) which can be imaged sonographically (Figure 20.1).

Sonographic reproductive evaluations

Sonographic analyses of breeding females are routinely performed to detect pregnancy (Tarantal and Hendrickx, 1988a; Tarantal *et al.,* 1997b), to identify any uterine changes that may diminish fecundity (Tarantal, 1992), and to assess reproductive maturation (Golub *et al.,* 2003). The goals of these analyses are to identify animals with pathology or other factors that could alter reproductive potential, and to ensure that pregnancy is proceeding normally. The macaque uterus is typically found in the midline, although the uterine body may be highly mobile and flexed either to the right, left, ventral (anteflexed), or dorsal (retroflexed) in relation to the cervix. The uterus may be found in other locations due to abdominal or pelvic adhesions. A central

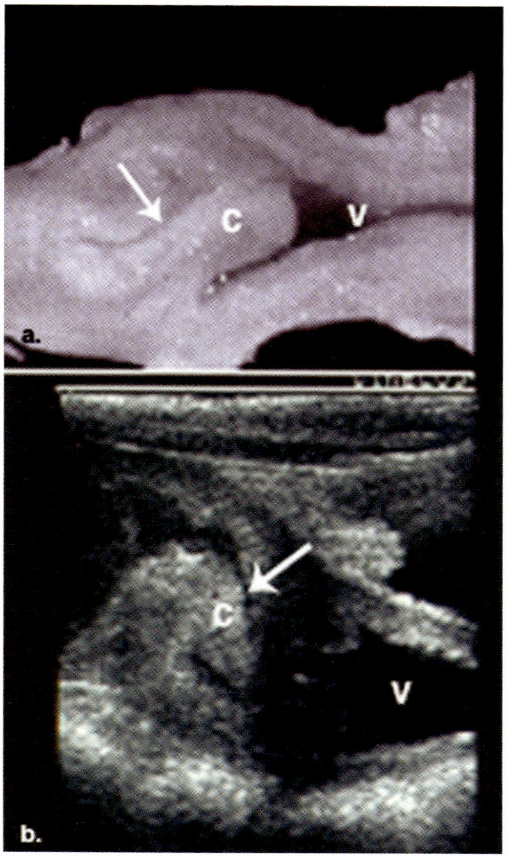

Figure 20.1 **(a)** Uterine cervix of a long-tailed macaque. Note colliculus (c) and tortuous cervical canal (arrow). **(b)** Sagittal scan of the uterine cervix showing a prominent colliculus (c) and tortuous cervical canal (arrow). v = vaginal canal.

linear echo is typically observed sonographically and represents the uterine cavity (Figure 20.2). The uterine or endometrial cavity echo is a useful landmark for identifying the uterus, detecting pregnancy, and assessing uterine pathology.

For uterine imaging, serial sagittal (longitudinal) scans, right-to-left and left-to-right, and transverse scans from the fundus cranially to the cervix caudally, are required to ensure that a complete evaluation is performed. It may also be necessary to angle the plane of section obliquely in order to obtain a complete view from the fundus to the cervix in some animals. The uterus is usually located in the midline between the bladder and rectum, but the uterus can have various right-to-left and ventral-to-dorsal positions depending on the degree of bladder and rectal distention, normal anatomical variation, and pelvic adhesions. Other uterine parameters assessed sonographically include size, shape, contour, and the appearance and thickness of the endometrium and myometrium. The macaque uterus is typically smooth with a rounded contour. Minor degrees of contour irregularity may be the only indication of the presence of uterine pathology (see below),

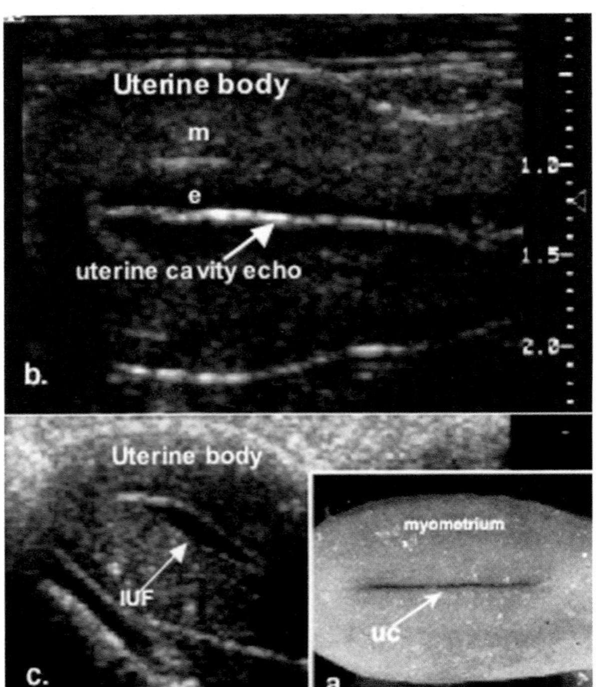

Figure 20.2 **(a)** Insert on lower right shows longitudinal section through the uterine body at gross examination. Note uterine cavity (uc). **(b)** Longitudinal scan of an adult rhesus monkey uterus. Note linear uterine cavity echo, endometrium (e), and myometrium (m). Uterine fundus to left of image and cm cursors noted on right. **(c)** Intrauterine fluid (IUF) noted within the uterine cavity of a nongravid rhesus monkey.

or may be the result of prior surgical procedures. Thus, historical data on individual animals is useful to review prior to imaging. Some females may have an elongated lower uterine segment with the upper portion of the uterine body pedunculated (Figure 20.3).

The uterine body may be anteflexed, although the macaque uterus is not generally anteversed (tilted ventrally) as typically observed in humans. Studies have shown that while the size of the uterine body in relation to the cervix changes with puberty, the uterine body size does not change significantly during the breeding season or with the stage of the menstrual cycle (Tarantal, 1992). Echogenic changes characteristic of both the proliferative and secretory phase of the menstrual cycle have been documented (Foster *et al.,* 1992; Morgan *et al.,* 1987; Tarantal, 1992), and are similar to findings in humans. As noted above, routine collection of baseline data for individual animals is important for identifying any anatomical changes or irregularities that may arise over time and with aging.

Evaluations of the uterus include measures of the total uterine length which is obtained from a midline sagittal scan using the cursors of the imaging system which are placed at the fundus and the caudal aspect of the cervix. As previously reported, sonographic measures of the mean total uterine length are 50.2 ± 7.2 mm for mature rhesus and 44.5 ± 6.9 mm for mature long-tailed macaques (Tarantal, 1992). Length measures are obtained with a longitudinal scan, and measures of width and height of the uterine body are obtained from a transverse scan at the widest portion. The overall

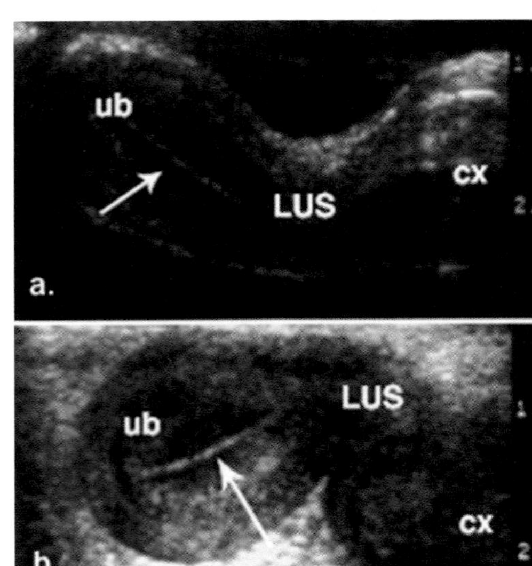

Figure 20.3 **(a)** Longitudinal image of a rhesus macaque uterus with an elongated lower uterine segment (LUS). ub = uterine body, cx = cervix. **(b)** Retroflexed uterine body (ub). Note uterine cavity echo (arrow).

appearance and echogenicity of the endometrium and myometrium are also evaluated, and the width of the endometrium is assessed by measuring from the endometrial/myometrial border to the uterine cavity echo. The uterine cavity is important to evaluate for evidence of echogenic foci, fluid, or irregular contour. Although small volumes of intrauterine and extrauterine fluid may be observed during the follicular phase in nongravid animals, these findings differ from those associated with impending or recent abortion. The presence of large volumes of echogenic intrauterine fluid may also be indicative of hematometra subsequent to a spontaneous abortion (Figure 20.4) or post-delivery.

Finally, assessments of uterine texture (homogeneous, heterogeneous), and identification of the presence of localized areas of increased or decreased echogenicity completes the examination.

Abnormalities and pathology

Although ultrasound has proven useful for the diagnosis and evaluation of a wide variety of uterine abnormalities, differentiating benign from malignant structures may be difficult. Generally, the more unusual the echo pattern and shape of the mass, the greater likelihood for malignancy. Errors in sonographic evaluation of pelvic masses can be attributed to poor scanning technique, lesions that are below the scanning resolution of the equipment used, or misinterpretation of bowel for a pelvic mass. Features assessed sonographically include: (1) confirming the presence or absence of a mass; (2) determining the location, origin, and anatomical relationship(s); (3) evaluating size, internal consistency, and contour; (4) assessing involvement of the mass with other structures and organs, and (5) identifying the presence or absence of ascites and other lesions.

Location of a pelvic mass may be uterine or adnexal/extrauterine, bilateral, or unilateral. Adnexal masses are those that are confined to the pelvis and involve the oviduct and/or ovaries. The location of the mass should be documented in at least two scan planes because the mass may appear separate from the uterus in one plane, whereas a second scan plane may suggest the mass is an extension of the uterus or other anatomical structure(s). A fully distended bladder can also displace an adnexal mass out of the pelvis and into the abdominal cavity. If the mass is very large, it may also distort normal pelvic landmarks and make anatomical identification difficult (Figure 20.5). The uterine cavity echo is helpful in locating the uterine body in these cases.

Assessments of an identified mass include measures of length and width, evaluation of internal consistency (cystic, complex, or solid), and degree of border definition (well-defined, moderately-defined, poorly-defined). For fluid-containing structures, assessing shape, internal echogenicity and presence of debris is helpful in the diagnosis. If a fluid-filled cystic structure is found, it may represent an ovarian, paraovarian, or mesenteric cyst; hydrosalpinx; cystadenoma; or endometrioma. Septated structures may be endometriomas or carcinomas; generally the borders of endometriomas are smooth whereas carcinomas are irregular. A complex (cystic and solid) structure may be an endometrioma, abscess, cystadenoma, ectopic pregnancy or a teratoma, all of which have been reported in macaques (Ami *et al.*, 1993; Baskin *et al.*, 2002; Beniashvili, 1989; Birkebak *et al.*, 1996; Golub *et al.*, 2003; Lowenstine, 2003; Plesker *et al.*, 2002; Rippy *et al.*, 1996; Tarantal, 1992). A solid mass could also be an endometrioma, leiomyoma (fibroid), or carcinoma. Whether a tumor has spread to other areas may be difficult to determine, although the presence of ascites is suggestive. Once a pelvic mass is delineated, other areas should be extensively examined including the cul-de-sac, pericolic and perihepatic spaces for fluid, liver parenchyma for the presence of metastatic foci, peritoneal surfaces and omentum for evidence of tumor implants, and the kidneys for obstructive uropathy (Figure 20.6).

A fluid-filled mass suspected to be an endometrioma can be confirmed by ultrasound-guided aspiration

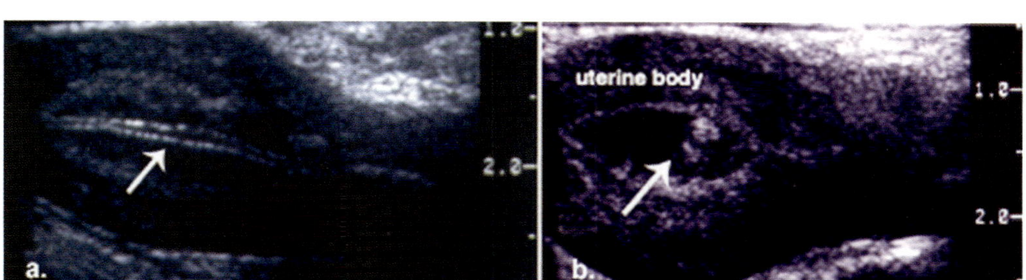

Figure 20.4 **(a)** Intrauterine fluid noted (arrow) in a nongravid rhesus monkey. **(b)** Ultrasound examination post-abortion. Note enlarged uterine cavity and presence of blood (arrow).

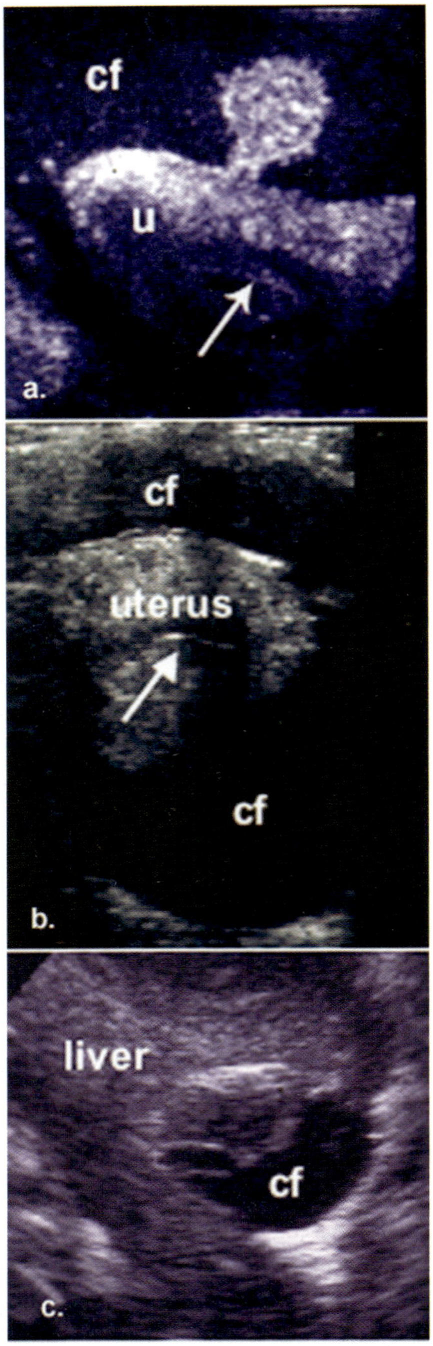

Figure 20.5 **(a)** Oblique image showing the uterus (u) and uterine cavity (arrow) with a large endometrioma. Note "chocolate fluid" (cf) surrounding the uterus. **(b)** Transverse scan through the uterine body of a rhesus macaque; note fluid (cf) surrounding the uterus. Uterine cavity echo noted (arrow). **(c)** Endometrioma associated with the left lateral aspect of the liver.

of "chocolate fluid" (Figure 20.5). While the majority of endometriomas are found on the uterine body or within the adnexa, they have also been found on the lateral margin of other organs such as the liver and in surgical scars. In addition, while endometriosis is known to be associated with infertility (Buyalos and Agarwal, 2000; Giudice *et al.*, 2002), endometriomas concurrent with a viable pregnancy have been reported (Tarantal, 1992).

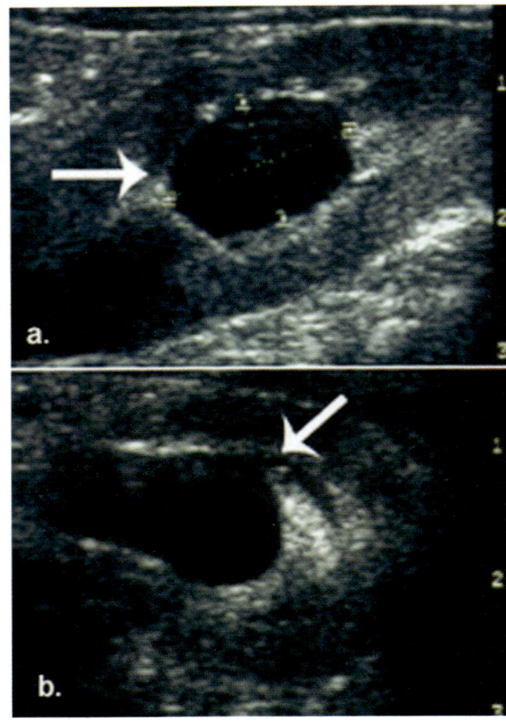

Figure 20.6 Hydronephrosis resulting from endometriosis shown in both **(a)** longitudinal and **(b)** transverse scans of the left kidney.

Other pathologic conditions that can be identified sonographically include adenomyosis, ovarian and cervical cysts, and leiomyomas. Leiomyomas are the most common benign neoplasm in women, are usually multiple, and may be accompanied by discomfort and uterine bleeding. They may be classified as submucosal, intramural, or subserosal. Leiomyomas may undergo secondary changes such as degeneration, calcification, hemorrhage, or necrosis. The sonographic appearance depends on location, presence or absence of secondary changes, and relative amounts of stromal and muscular constituents, and can, therefore, have a wide variety of appearances. They may appear hypoechoic in relation to the uterus, are most often identified by a deformation of uterine contour, and can obscure the uterine cavity. Fibroids are stimulated by estrogens and, therefore, may grow rapidly during anovulatory cycles and pregnancy.

Characteristic features of uterine abnormalities, such as leiomyoma, carcinoma, adenomyosis, endometritis, and endometrial hyperplasia, include alterations in uterine size, contour and echogenicity and, therefore, definitive diagnosis by ultrasound may be difficult. For example, although diffuse uterine enlargement with preserved myometrial and uterine cavity appearance is

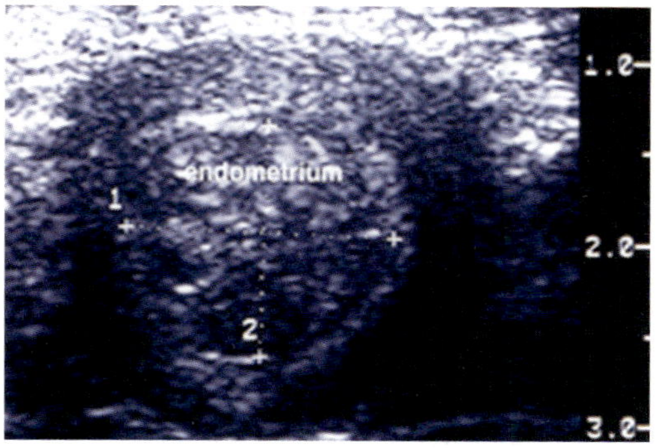

Figure 20.7 Endometrial hyperplasia. Transverse section of the uterine body of a rhesus monkey where the thickness of the endometrium is more than twice the width typically observed. Placement of cursors at the endometrial/myometrial border indicate the total width of the endometrium is approximately 1.2 cm.

suggestive of adenomyosis, these characteristic features may also be observed with endometritis or uterine fibroids. While uterine biopsy may be helpful to confirm the diagnosis, careful sonographic evaluations can limit the possibilities. Endometrial hyperplasia, another relatively common finding in monkeys, is suggested when the endometrial thickness from the endometrial/myometrial border to the uterine cavity is greater than 5 mm (Figure 20.7), whereas typical endometrial thickness ranges from 2–4 mm (usually 3 mm).

Endometrial proliferation can involve dilated uterine glands which may be observed sonographically by the appearance of endometrial "striations" and a heterogenous appearance, similar to findings in early pregnancy (see below). The degree of proliferation can be mild or severe, and can persist as an asymptomatic condition. Many cases are found incidentally on routine ultrasound examinations, and may increase in older animals, similar to humans.

Published studies that focused on 224 female macaques through five breeding seasons revealed that the most frequent abnormalities encountered were uterine/adnexal adhesions (31%), irregular uterine shape and contour (11%), and endometriosis (4%). Other documented findings included endometrial hyperplasia, endometritis, adenomyosis, uterine fibroids, and mesenteric, ovarian, and cervical cysts. Irregularities of uterine contour were also associated with a ruptured uterus and adhesions and, in some cases, the result of surgical procedures and prior uterine implants. Uterine echogenic foci were also noted indicating retained sutures and focal connective tissue changes (Figure 20.8).

Section 4: Gravid animals

Early pregnancy and pregnancy detection

Sonographic features have been established and reported for pregnancy detection during the peri-implantation and post-implantation periods in both

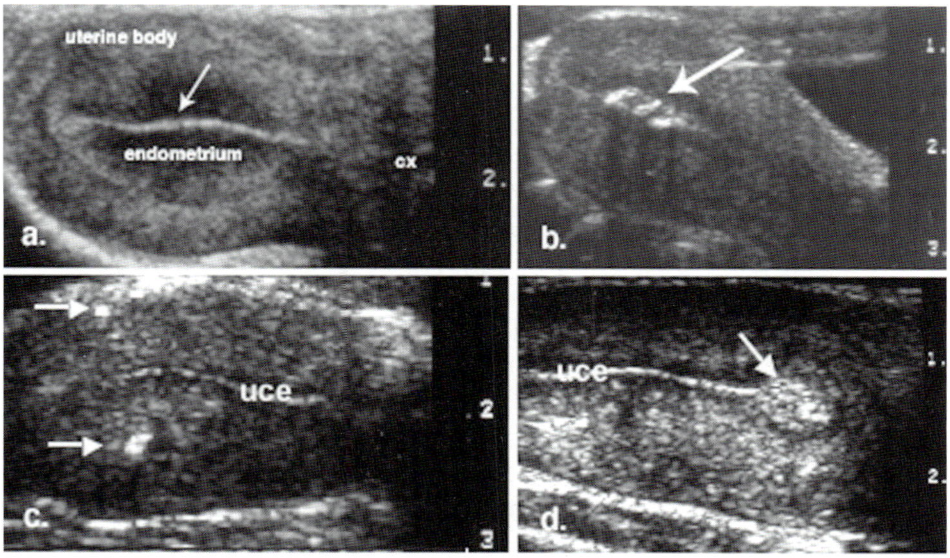

Figure 20.8 Uterine imaging of rhesus monkeys which shows normal anatomy including the uterine cavity echo (arrow) **(a)**. In **(b)** echogenic retained suture material (arrow) with shadowing is noted, in **(c)** echogenic foci within the myometrium (arrows), and polyp in the lower uterine segment (arrow) in **(d)** cx = cervix, uce = uterine cavity echo.

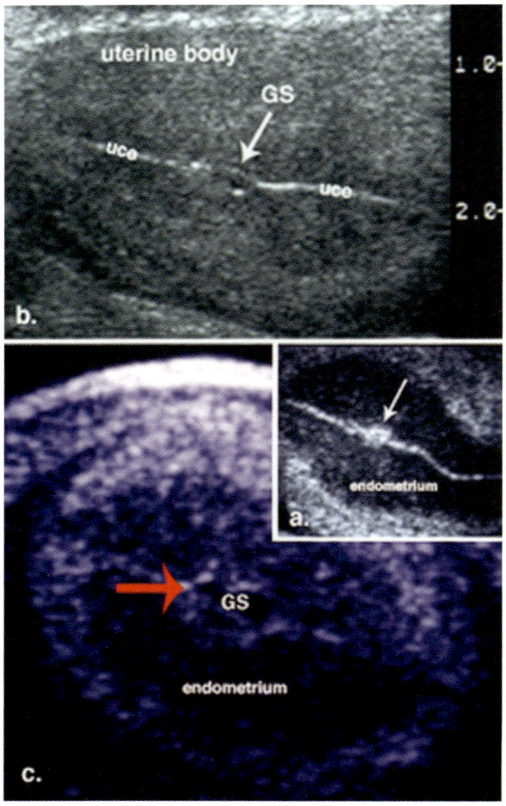

Figure 20.9 Uterine images showing early signs of pregnancy. Longitudinal scan shown in insert **(a)**; arrow points to early gestational sac (GS) formation. By 14 **(b)** and 16 **(c)** days gestation the GS is fluid-filled. uce = uterine cavity echo.

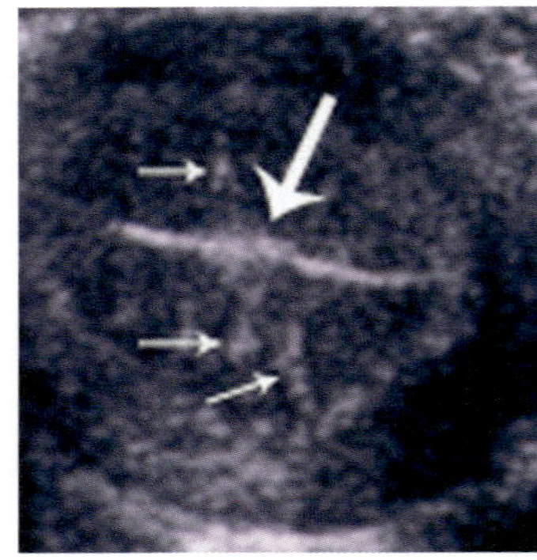

Figure 20.10 Oblique transverse image of uterine body shows endometrial striations (small arrows) and early GS formation (arrow) indicating a pregnancy at approximately 12 days gestation. The striations typically are observed sonographically radiating from the epithelial plaque.

species (Conrad *et al.,* 1989; Shimizu, 1988; Tarantal *et al.,* 1997b; Tarantal and Hendrickx, 1988a–d). These investigations have shown that ultrasonography provides reliable evidence of pregnancy during this early period of development, and can be used to gain greater insight into the developmental events associated with pregnancy in macaques.

The sonographic features that confirm pregnancy during the early peri-implantation period are based on established endometrial changes and a focal, uterine cavity hyperechoic thickening, consistent with epithelial plaque and early gestational sac formation (Tarantal *et al.,* 1997b) (Figure 20.9). These sonographic findings are similar to changes described histologically (Enders, 1991), which are both generalized (endometrial) and localized (epithelial plaque). The generalized findings represent the effects of hormonal stimulation which results in thickening of the endometrium, increased glandular secretion and tortuosity, and stromal edema (Figure 20.10). The localized response is unique to the monkey and is associated with the implanting blastocyst and plaque formation (Ramsey *et al.,* 1976). Both features are essential for positive identification of viable pregnancies at 12–14 days gestation or within approximately 2–3 days of implantation (Tarantal *et al.,* 1997b). Documenting both of these anatomical features is crucial because some animals can show endometrial changes such as thickening and striations without development of a viable pregnancy. These animals may be exhibiting early pregnancy loss, hormonal stimulation unassociated with pregnancy, or endometritis. Because of the high rate of early pregnancy loss in human and nonhuman primates, follow-up evaluations are essential to confirm the pregnancy is viable (Figure 20.11).

It is important to note that the sonographic appearance of early pregnancy in monkeys differs from humans. This is because, in humans, the embryo becomes completely embedded within the uterine stroma, whereas in the monkey, the blastocyst remains superficially attached (Ramsey *et al.,* 1976). As it grows, it adheres to the side opposite the initial attachment site forming the location of the future secondary placental disc; placental discs are typically bidiscoid in macaques (~80%) (King, 1993; Myers, 1972). Thus, with superficial implantation, the gestational sac develops as a fluid-filled sac within the uterine cavity rather than as a "double sac" as described in humans (Nyberg *et al.,* 1983) where interstitial implantation occurs.

Typically, most macaque embryos will implant in the upper one-third of the uterus, although implantation can occur eccentrically within the fundus or in the lower uterine segment (Figure 20.12).

Once pregnancy is well-established, published sonographic developmental guidelines assist in ensuring that the pregnancy is developing normally (Table 20.1). For example, the formation of a gestational sac, the appearance of the yolk sac and the

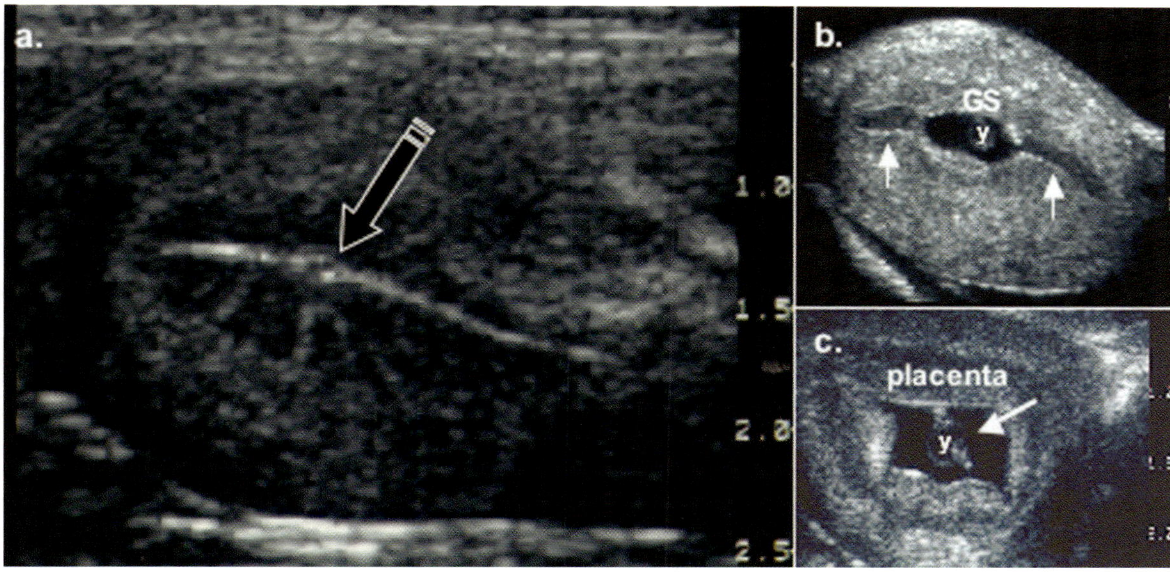

Figure 20.11 Normal developmental features observed at 12 days gestation **(a)**; by 18 days gestation **(b)** the yolk sac (y) is evident within the GS. Note implantation bleeding (arrows). By 25 days gestation **(c)** the yolk sac (y) and embryo (arrow) with a beating heart is readily imaged.

embryo, and the initial detection of cardiac activity should all be observed within the first four weeks of pregnancy. Thus, these guidelines are useful for confirming that the development of the conceptus is within the normal range, and for assessing animals that may be at risk for pregnancy loss. These features are particularly important to document prior to assigning animals to experimental studies.

Standardized measures of the conceptus during early gestation include the mean gestational sac diameter (measured in the short and long axis – length, width, height – with the mean value calculated), and the greatest length (GL) of the embryo, once it is visible (Figure 20.13) (Tarantal and Hendrickx, 1988a).

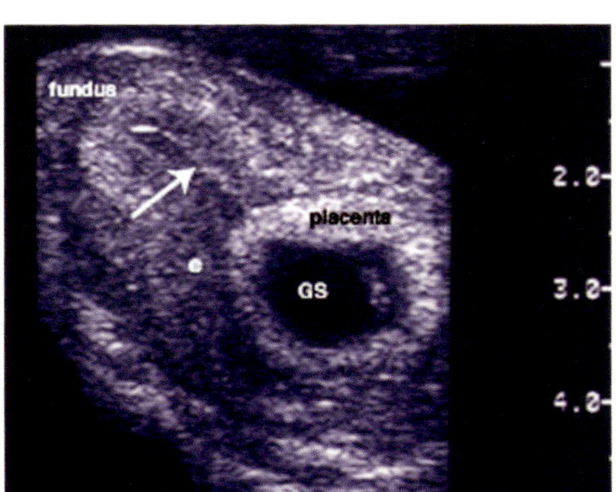

Figure 20.12 Embryos can implant at a variety of intrauterine sites. Note GS in the lower aspect of the uterine body; arrow indicates uterine cavity echo. e = endometrium.

Cardiac activity should be evident by approximately 25 days gestation.

In monkeys, the GL is measured from a midline sagittal section, ensuring that the natural curvature of the embryo is taken into consideration. If the natural C-shape is not appreciated, then the measure will be artificially shortened. Cursors are placed at the highest point of the crown and at the base of the tail; the tail should not be included in the GL measurement. Growth charts for the gestational sac mean dimension and the GL have been previously published (Tarantal

TABLE 20.1: Sonographic developmental guidelines in early pregnancy for rhesus and long-tailed macaques (12 to 25 days gestation)

Gestational age (days)	Sonographic features
12–14	• Uterine endometrial striations • Endometrial plaque formation • Initial sign of developing gestational sac
15–16	• Gestational sac formation with fluid
18–20	• Yolk sac • Growth of gestational sac • Placental disc(s) evident
20–25	• Embryo identified • Cardiac activity observed • Growth of gestational sac

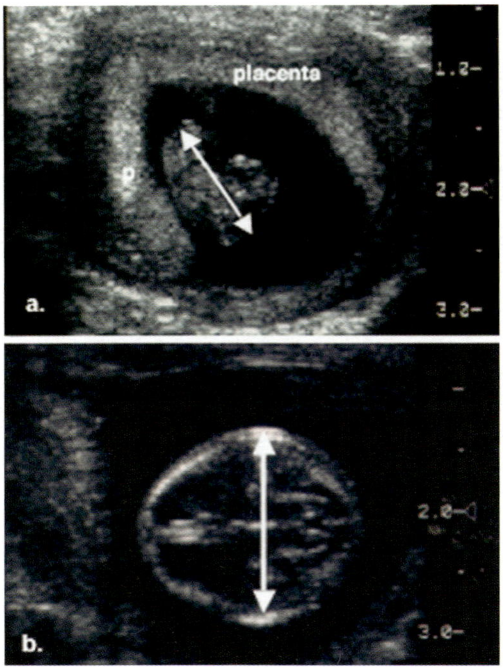

Figure 20.13 Image of a macaque embryo at 30 days gestation **(a)**. Note the greatest length (GL) is measured by placing cursors at the crown and base of the tail. The normal curvature of the embryo at these early stages of gestation must be imaged in order to obtain an accurate measure. p = placenta. In the early second trimester, the biparietal diameter (BPD) is measured by obtaining a transverse section through the skull at the level of the parietal bones and developing thalamus **(b)**. The cursors are placed at the leading edge to leading edge of the skull margin closest to the transducer.

and Hendrickx, 1988a,b), and are not included herein. As the conceptus approaches the second trimester, measures of defined anatomical structures improve accuracy (see Section 5).

Twins versus singletons

The incidence of live-born twins in rhesus and long-tailed monkeys is rare (Figure 20.14), with a twinning rate of approximately 0.1% based on the literature (Bercovitch *et al.*, 2002; Geissmann, 1989; Resuello, 1987; Tarantal and Hendrickx, 1988c). Sonographic studies have, however, indicated that the twinning rate may be greater in rhesus monkeys than reported because of the appearance of twin gestational sacs, one with a viable embryo and one anembryonic (Tarantal and Hendrickx, 1988d) similar to "vanishing twins" observed in humans (Landy and Keith, 1998). While in humans the majority of cases result in singleton births, a high rate of loss occurs in monkeys. Complete abortions occur early in gestation, or placental abruptions and subsequent

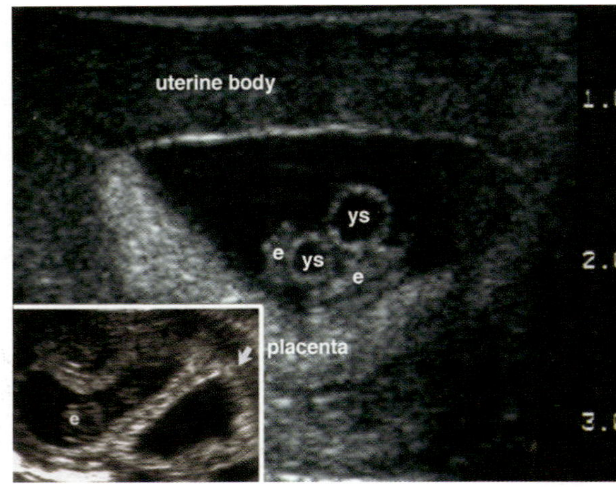

Figure 20.14 Twin yolk sacs (ys) noted, both associated with viable embryos (e). Twin GS, one with a viable embryo (e) and one anembryonic (arrow) shown in insert, left.

abortion of all products of conception in the second trimester (Tarantal and Hendrickx, 1989a). A small subpopulation of animals have delivered normal, healthy singletons at term.

Implantation bleeding or "placental sign"

Macaques have a menstrual cycle similar to humans, typically encompassing 28–30 days (Hartman, 1932; Jewett and Dukelow, 1972). Vaginal bleeding is a common finding in macaques in early pregnancy, and can occur at the same time menses is expected, thus precluding the use of this finding to diagnose pregnancy. Bleeding can last from a few days to over a month, with an average of 24 days reported (Hartman, 1932). Implantation bleeding can be observed sonographically and clearly distinguished from the intrauterine bleeding and hematoma formation that occurs with "threatened" or impending abortion (Figure 20.15). Echogenic hematomas are not observed with implantation bleeding, whereas discrete hypoechoic regions can be imaged cranial and caudal to or surrounding the developing gestational sac. Any vaginal and/or intrauterine bleeding that occurs after approximately 50 days gestation should be pursued since this may be an indication of a placental abruption, either retroplacental or subchorionic, which can place both the fetus and dam at risk (see below). Findings associated with placental abruption contrast early pregnancy loss or implantation bleeding both of which are normal for these species and typically have few hematologic complications.

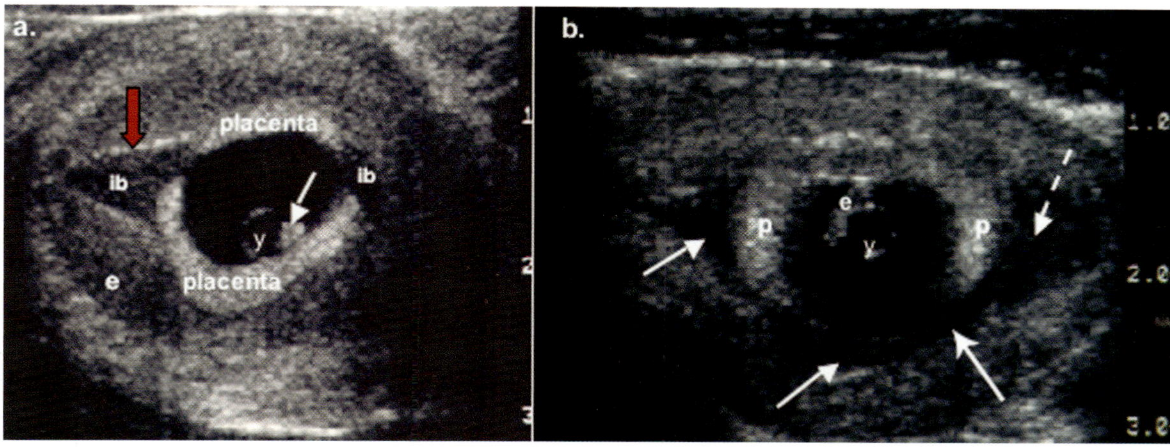

Figure 20.15 Implantation bleeding versus impending abortion. **(a)** Oblique longitudinal section showing signs of intrauterine implantation bleeding (ib) in early gestation; note hypoechoic region (large arrow) cranial to the GS. y = yolk sac, embryo (arrow), e = endometrium. These findings differ when compared to animals at a similar gestational age with signs of threatened or impending abortion **(b)** where blood surrounds the GS and separates it from the uterine cavity (arrows). In these cases, the GS will frequently appear very round rather than ovoid, and hematomas will form (dashed arrow). Heavy vaginal bleeding is usually associated with these findings. p = placenta.

Pregnancy loss

The established sonographic guidelines shown in Table 20.1 for early pregnancy are useful for monitoring developmental status and accurately identifying signs of early embryonic demise. Spontaneous loss can result in resorption, early embryonic death, and/or abortion. Signs of impending abortion include the presence of heavy intrauterine bleeding and echogenic hematomas, as described above, with or without the presence of a viable embryo. Signs of early embryonic demise and/or resorption are evident when the size of the gestational sac does not correlate with the known gestational age, and when sequential examinations do not reveal any change in size or evidence of further development. By 18–20 days gestation, the yolk sac should be observed as a 3 mm circular structure within the developing gestational sac (Figure 20.11) and if not found could be an indication of early embryonic loss, with subsequent formation of an anembryonic gestational sac. Anembryonic gestational sacs can persist for extended periods of time, in some cases for 3–4 months, prior to spontaneous abortion of all products of conception.

Obstetrical problems

Sonographic monitoring for obstetrical problems is useful from a colony management perspective and for experimental studies. Such monitoring provides the opportunity to clinically manage the dam and fetus, and also preserve the integrity of studies. The number and location of placental discs, and changes in echogenicity with placental "aging" are particularly important to include. Those animals identified with either a marginal or complete previa during the second trimester are further evaluated later in gestation, and cesarean-sections performed if the placenta is in close proximity or completely covering the cervix (Figure 20.16). Follow-up examinations are important because the uterus continues to grow during gestation and it is possible that the placenta may "migrate" away from the cervical os. Therefore, it is important to perform examinations at select times during the later stages of gestation to monitor and confirm the diagnosis. Diagnosis of this condition requires accurate identification of the location of the placenta in relation to the cervix on longitudinal scans. A distended urinary bladder can alter this relationship and, thus, lead to misinterpretation.

Animals that display retroplacental or subchorionic hemorrhage require close monitoring to determine if the condition resolves, or if surgical removal of all products of conception is required (Figure 20.17). The most severe conditions are those with concealed hemorrhages where no vaginal bleeding is observed. Continued surveillance of individual animals for repeat incidence with subsequent pregnancies is important for colony management and to remove females from reproductive studies that have repeated pregnancy loss. Prior unpublished studies have suggested that approximately 60% of animals identified with a placental abruption showed decidual proliferation sonographically. Thus, animals

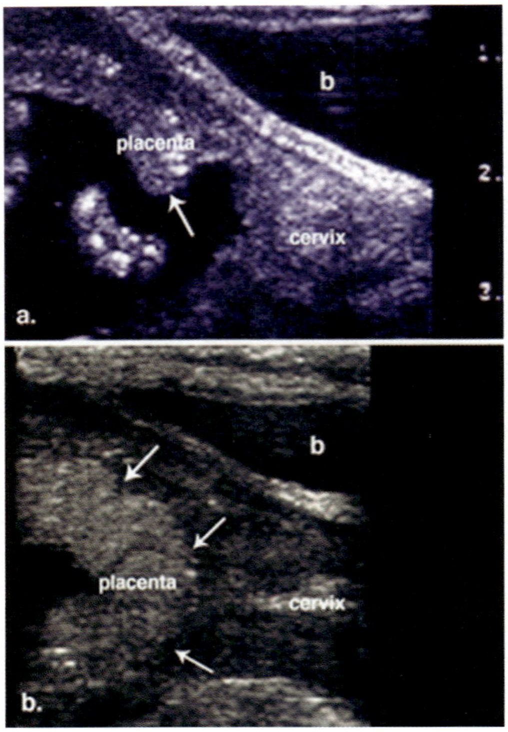

Figure 20.16 Longitudinal uterine scans to evaluate placental location. **(a)** Margin of the placenta is in close proximity to the cervix (arrow). b = urinary bladder. **(b)** Complete placenta previa shown. Note placental disc is completely covering the uterine cervix (arrows). Here, further growth of the uterus will not change this anatomical relationship.

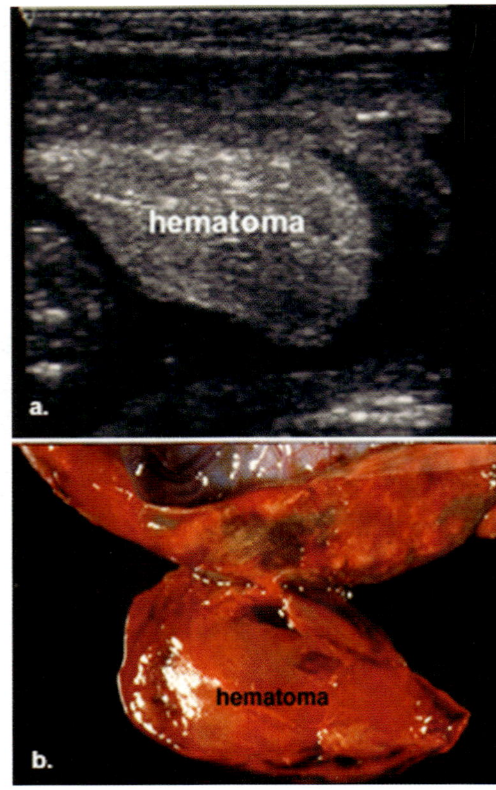

Figure 20.17 Placental abruption. **(a)** Rhesus macaque scanned in the second trimester with evidence of a placental abruption; note echogenic hematoma. **(b)** Note hematoma attached to the margin of the primary placental disc upon gross examination.

identified with proliferative decidua early in gestation can be monitored more frequently to ensure that, if a placental abruption does occur, it can be identified early.

For time-mated females that are overdue and have not delivered by the anticipated time period (165 ± 10 days gestation), sequential monitoring at select time points to determine the need for cesarean-section is a reasonable management practice. Sonographic monitoring should be balanced with the frequency of ketamine administration. Thus, it is important that the maximum information be obtained at each examination. If an ultrasound examination is scheduled at approximately 160 days gestation, information on placental location, cervical softening, and cervical length (measured from the cranial aspect of the internal os to caudal aspect of the external os), and fetal position and status provide important information. If delivery has not occurred by 170 days gestation, reassessment of these parameters is indicated. Delivery can be anticipated within 24–48 hours if the cervix is found to be completely dilated and the fetal head is engaged (Figure 20.18). Some animals may show one or multiple days of vaginal bleeding prior to delivery, and these findings should not be viewed as a complication necessitating surgery or frequent ultrasound examinations. In addition, the delivery history of individual animals should be taken into consideration; if animals have a history of late delivery of viable, healthy offspring, then this should be factored into the decision-making process. A lack of sonographically-observed cervical softening or shortening and advanced gestational age (175 days gestation, if time-mated) suggests surgical intervention may be required. Signs of fetal distress such as gasping and chronically slow heart rates (<120 bpm) unrelated to acute changes that may be the result of uterine contractions or maternal position suggests that immediate delivery of the infant by cesarean-section is indicated. The timing of the observations in relation to ketamine administration should also be considered. Thus, information on fetal status and cervical changes, in combination with the animal's prior delivery history, provides the information necessary to optimize management practices.

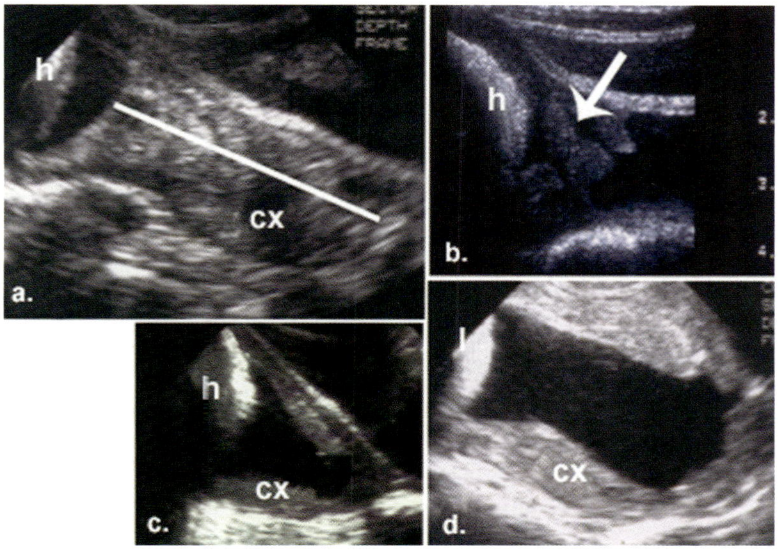

Figure 20.18 **(a)** Sonographic evaluation of a term rhesus monkey indicates the cervix (cx) (internal and external os) is closed. Line shown indicates cervical length. h = fetal head. **(b)** Arrow indicating shortening and softening of cervix with colliculi apparent (arrow). **(c, d)** Cervix is dilated and fetus is in cephalic presentation.

Section 5: Fetal development

Growth and growth charts

Standardized growth charts using both predicted and mean values have been available for both species for many years, and changes during development, variations in growth patterns, and differences when comparing different species have been previously reported (Tarantal and Gargosky, 1995; Tarantal and Hendrickx, 1988a–d; Tarantal *et al.,* 1997a). Normative prenatal growth parameters include multiple head (biparietal [BPD] and occipitofrontal diameters, head area and circumference [HC]), abdominal (area and circumference [AC]), and limb measures (humerus, femur lengths [FL]) in addition to early gestational biometrics (mean gestational sac dimensions, GL, yolk sac) (Tarantal and Hendrickx, 1988a,b). Growth rates are similar for rhesus and long-tailed macaques through the second trimester at which time the period of growth acceleration begins (~110 days gestation; beginning of the third trimester) and the species diverge (Tarantal and Gargosky, 1995; Tarantal and Hendrickx, 1988b). The rhesus monkey fetus shows a rapid increase in size compared to the fetal long-tailed monkey, with both species showing a characteristic decline in cranial growth during the later stages of the third trimester similar to the human fetus.

Initially, the viability of the conceptus is established, and a fetal heart rate obtained to determine if it is within normal limits for the gestational age. The gestational sac mean dimension is the earliest measurement obtained and is assessed before the embryo can be imaged. Once the embryo is visible, the GL is used to assess growth because it is a more reliable and less variable measure when compared to the growing gestational sac (see Figure 20.13). At the end of the embryonic period (~50 days gestation), the BPD and FL are measured to increase accuracy. Standardized scan plans for these measures are based on established anatomical landmarks. For the head, an ovoid, transverse section at the level of the thalamus is preferred (Figure 20.13). Measures are taken from leading-edge to leading-edge of the parietal bone of the developing skull, as previously reported (Tarantal and Hendrickx, 1988b). For FL, the longest view of the ossified portion of the long bone is used for the measurement. These are established methods that have been used for several decades.

From a colony perspective, one numerical value is most efficient for predicting or confirming gestational age rather than using mean values or ranges of values. Therefore, predicted gestational age (pGA) charts with single values identified for each gestational age and parameter have been generated using regression analysis (Tarantal and Hendrickx, 1988b). However, it is essential when using such predicted charts that two measures be combined to enhance accuracy (Tarantal and Hendrickx, 1988b). For example, in the early second trimester, one can use both the pGA for BPD and the pGA for FL, each based on the respective measurement. If the BPD measurement reflects a pGA of 60 days, and the pGA for the FL measurement is 64 days gestation, then 60 days + 64 days = 124 divided by 2 measures = 62 days; therefore the pGA for the fetus on the day of the observation is 62 days gestation. Depending on the manner in which the animals are bred, acceptable ranges from the established breeding dates are ±2 days in early gestation, ±5 days mid-gestation, and up to 1–2 weeks later in gestation (see comments, below) (Tarantal and Hendrickx, 1988b). For example, if the fetus described above was the result of a timed mating and the gestational age at the time of the ultrasound examination was 60 days, a pGA of 62 days is within the normal range. If intrauterine growth restriction (IUGR) is suspected, adding measures of HC and AC may prove informative. As discussed above, it is recommended that the best use of these tables is to combine information from multiple measurements. Thus, Tables 20.2–20.5 have been included for BPD, HC, AC, and FL measures for both species.

As noted above, it is important to consider the gestational time period when predicting or confirming gestational age. The earlier gestational time points are the most accurate if the time of mating is not known; prior to 100 days gestation is recommended, although the earlier the observation the more accurate the pGA. It is also essential to keep in mind that attempting to predict or confirm gestational age late in gestation is difficult because of the manner in which the fetus grows. This is primarily because of the large variation in growth and the "growth spurt" that occurs after 110 days gestation, and the fact that the growth of the fetal head does not change significantly at the later stages of development (Tarantal and Gargosky, 1995). From an experimental perspective, more detail on growth parameters may be useful, thus Tables 20.6–20.9 include 95% confidence intervals, means ± standard deviations, and ranges of values for BPD, HC, AC, and FL for both species.

The methods used for systematically assessing developmental anatomy is dependent upon the age of the conceptus. In early gestation, the overall appearance and shape of the embryo, imaging of the amnion, amniotic fluid, chorionic jelly, and yolk sac all aid in ensuring development is proceeding normally (Figure 20.19). As the fetus grows and matures, transverse, oblique, and coronal sections through the fetal skull allow one to evaluate intracranial structures such as cerebral and cerebellar hemispheres; third, fourth, and lateral ventricles; choroid plexus; midbrain; cerebral vasculature; cranial base; and meninges; and other structures such as eyes and orbital cavities; nose, lip, maxilla and mandible; palate; and profile (Figure 20.20). Such examinations are important for documenting when cranial abnormalities occur (Figure 20.21) (Tarantal and Hendrickx, 1987; Tarantal *et al.,* 1994a; 1999b).

The axial and appendicular skeleton includes imaging of ossification centers of the vertebrae; clavicles; scapulae; ribs; humerus, radius/ulna, metacarpals and digits (fingers); and femur, tibia/fibula, metatarsals, and digits (toes) (Figure 20.22). Assessment of the tail, particularly in long-tailed macaques, may be informative.

For the thoracic and abdominal cavities, imaging the thymus, heart and great vessels; lungs; liver, kidneys and adrenals; stomach and intestine; presence of meconium; and urinary bladder can all be readily accomplished (Figure 20.23), and aid in the identification of abnormalities (Cukierski *et al.,* 1986).

In human and nonhuman primates, fetal echocardiography typically includes 4- and 5-chamber views, right and left ventricular outflow tracts, long- and short-axis views, aortic outflow tract, velocities across the mitral, tricuspid, pulmonic, aortic valves, venous flow, and right and left ventricular output (Tarantal, unpublished). Readers are referred to human fetal echocardiography texts for further information.

Similar to humans, amniotic fluid volumes are highly variable, although extremes such as oligohydramnios and polyhydramnios can be detected. Amniotic fluid flocculence may vary and typically represents sloughed fetal cells; increased flocculence may be suggestive of fetal distress. Conditions such as amniotic bands can also be identified (Tarantal and Hendrickx, 1987). The placental location and number of placental discs, placental and umbilical vessels, and the umbilical cord insertion site are readily imaged and can be routinely evaluated. Pulsed and color Doppler are also useful for assessing velocity waveforms in the umbilical and fetal circulations (Nimrod *et al.,* 1996; Panigel *et al.,* 1993; Ragavendra and Tarantal, 2002; Schmiedl *et al.,* 1998; Simpson *et al.,* 1997; Tarantal *et al.,* 1995b). Such information may be helpful in identifying changes in blood flow with fetal growth restriction (Tarantal *et al.,* 1995b).

Fetal sex determination

The earliest indication of sexual differentiation sonographically in monkeys is at approximately 50 days gestation (late first trimester) (Tarantal and Hendrickx, 1988b,c). At this developmental time point, differences in the position of the phallus in relation to the perineum/pelvis are observed (Prahalada *et al.,* 1997). For males, the phallus, which develops in both males

TABLE 20.2: Predicted values for biparietal diameter (BPD) for rhesus (Mm[a]) and long-tailed (Mf[b]) macaques from the late first to third trimesters (50–165 days gestation)

Age	Mm	Mf	Age	Mm	Mf	Age	Mm	Mf
50	12.4	11.8	**89**	31.0	30.0	**128**	44.0	41.4
51	12.9	12.3	**90**	31.4	30.3	**129**	44.3	41.5
52	13.4	12.9	**91**	31.9	30.7	**130**	44.5	41.7
53	13.9	13.4	**92**	32.3	31.1	**131**	44.7	41.9
54	14.4	13.9	**93**	32.7	31.5	**132**	44.9	42.1
55	15.0	14.5	**94**	33.1	31.8	**133**	45.1	42.3
56	15.5	15.0	**95**	33.5	32.2	**134**	45.3	42.5
57	16.0	15.5	**96**	33.9	32.5	**135**	45.5	42.6
58	16.5	16.0	**97**	34.3	32.9	**136**	45.7	42.8
59	17.0	16.5	**98**	34.7	33.2	**137**	45.9	42.9
60	17.5	17.0	**99**	35.0	33.6	**138**	46.0	43.1
61	18.0	17.5	**100**	35.4	33.9	**139**	46.2	43.2
62	18.5	18.0	**101**	35.8	34.2	**140**	46.4	43.4
63	19.0	18.5	**102**	36.2	34.5	**141**	46.5	43.5
64	19.5	19.0	**103**	36.5	34.9	**142**	46.7	43.7
65	20.0	19.5	**104**	36.9	35.2	**143**	46.8	43.8
66	20.5	20.0	**105**	37.2	35.5	**144**	46.9	43.9
67	21.0	20.5	**106**	37.6	35.8	**145**	47.1	44.0
68	21.5	21.0	**107**	37.9	36.1	**146**	47.2	44.1
69	22.0	21.4	**108**	38.3	36.4	**147**	47.3	44.2
70	22.4	21.9	**109**	38.6	36.7	**148**	47.4	44.4
71	22.9	22.4	**110**	38.9	37.0	**149**	47.5	44.4
72	23.4	22.8	**111**	39.3	37.3	**150**	47.6	44.5
73	23.9	23.3	**112**	39.6	37.5	**151**	47.7	44.6
74	24.3	23.7	**113**	39.9	37.8	**152**	47.8	44.7
75	24.8	24.2	**114**	40.2	38.1	**153**	47.8	44.8
76	25.3	24.6	**115**	40.5	38.3	**154**	47.9	44.9
77	25.7	25.1	**116**	40.8	38.6	**155**	48.0	44.9
78	26.2	25.5	**117**	41.1	38.9	**156**	48.0	45.0
79	26.7	25.9	**118**	41.4	39.1	**157**	48.1	45.0
80	27.1	26.3	**119**	41.7	39.4	**158**	48.1	45.1
81	27.6	26.8	**120**	42.0	39.6	**159**	48.1	45.1
82	28.0	27.2	**121**	42.3	39.8	**160**	48.2	45.2
83	28.4	27.6	**122**	42.5	40.1	**161**	48.2	45.2
84	28.9	28.0	**123**	42.8	40.3	**162**	48.2	45.2
85	29.3	28.4	**124**	43.0	40.5	**163**	48.2	45.3
86	29.8	28.8	**125**	43.3	40.7	**164**	48.3	45.3
87	30.2	29.2	**126**	43.5	40.9	**165**	48.3	45.3
88	30.6	29.6	**127**	43.8	41.2			

[a]Mm: $BPD = -14.582 + (5.320 \times 10^{-1})\,age + (1.000 \times 10^{-3})\,age^2 + (-9.267 \times 10^{-6})\,age^3$ [$R^2 = 0.98$]

[b]Mf: $BPD = -20.048 + (7.370 \times 10^{-1})\,age + (2.000 \times 10^{-3})\,age^2 + (-1.801 \times 10^{-6})\,age^3$ [$R^2 = 0.97$]

TABLE 20.3: Predicted values for head circumference (HC) for rhesus (Mm[a]) and long-tailed (Mf[b]) macaques from the late first to third trimesters (50–165 days gestation)

Age	Mm	Mf	Age	Mm	Mf	Age	Mm	Mf
50	49.6	46.4	**89**	120.7	115.4	**128**	174.9	163.9
51	51.4	48.2	**90**	122.4	117.0	**129**	175.9	164.7
52	53.2	50.0	**91**	124.1	118.6	**130**	176.7	165.4
53	55.1	51.8	**92**	125.8	120.2	**131**	177.6	166.1
54	56.9	53.6	**93**	127.5	121.7	**132**	178.4	166.7
55	58.7	55.5	**94**	129.2	123.3	**133**	179.2	167.4
56	60.5	57.3	**95**	130.8	124.8	**134**	180.0	167.9
57	62.4	59.1	**96**	132.5	126.4	**135**	180.7	168.5
58	64.2	60.9	**97**	134.1	127.9	**136**	181.4	169.0
59	66.0	62.8	**98**	135.7	129.3	**137**	182.1	169.5
60	67.9	64.6	**99**	137.3	130.8	**138**	182.7	170.0
61	69.7	66.4	**100**	138.9	132.3	**139**	183.3	170.4
62	71.6	68.2	**101**	140.5	133.7	**140**	183.9	170.8
63	73.4	70.1	**102**	142.0	135.1	**141**	184.4	171.1
64	75.3	71.9	**103**	143.6	136.5	**142**	184.9	171.4
65	77.2	73.7	**104**	145.1	137.9	**143**	185.4	171.7
66	79.0	75.5	**105**	146.6	139.2	**144**	185.8	171.9
67	80.9	77.3	**106**	148.0	140.6	**145**	186.2	172.1
68	82.7	79.1	**107**	149.5	141.9	**146**	186.5	172.3
69	84.6	80.9	**108**	150.9	143.2	**147**	186.8	172.4
70	86.4	82.7	**109**	152.4	144.5	**148**	187.1	172.5
71	88.3	84.5	**110**	153.8	145.7	**149**	187.3	172.6
72	90.1	86.3	**111**	155.2	146.9	**150**	187.5	172.6
73	92.0	88.1	**112**	156.5	148.1	**151**	187.7	172.6
74	93.8	89.9	**113**	157.8	149.3	**152**	187.8	172.6
75	95.6	91.6	**114**	159.2	150.5	**153**	187.9	172.6
76	97.5	93.4	**115**	160.4	151.6	**154**	187.9	172.6
77	99.3	95.1	**116**	161.7	152.7	**155**	187.9	172.6
78	101.1	96.9	**117**	163.0	153.8	**156**	187.9	172.6
79	102.9	98.6	**118**	164.2	154.9	**157**	187.9	172.6
80	104.8	100.3	**119**	165.4	155.9	**158**	187.9	172.6
81	106.6	102.1	**120**	166.5	156.9	**159**	187.9	172.6
82	108.4	103.8	**121**	167.7	157.9	**160**	187.9	172.6
83	110.1	105.5	**122**	168.8	158.8	**161**	187.9	172.6
84	111.9	107.1	**123**	169.9	159.8	**162**	187.9	172.6
85	113.7	108.8	**124**	171.0	160.6	**163**	187.9	172.6
86	115.5	110.5	**125**	172.0	161.5	**164**	187.9	172.6
87	117.2	112.1	**126**	173.0	162.3	**165**	187.9	172.6
88	118.9	113.8	**127**	174.0	163.2			

[a]Mm: $HC = -21.117 + (8.35 \times 10^{-1})\ age + (1.600 \times 10^{-2})\ age^2 + (-7.888 \times 10^{-5})\ age^3$ [$R^2 = 0.97$]

[b]Mf: $HC = -30.846 + (1.096)\ age + (1.300 \times 10^{-2})\ age^2 + (-7.234 \times 10^{-5})\ age^3$ [$R^2 = 0.98$]

TABLE 20.4: Predicted values for abdominal circumference (AC) for rhesus (Mm[a]) and long-tailed (Mf[b]) macaques from the late first to third trimesters (50–165 days gestation)

Age	Mm	Mf	Age	Mm	Mf	Age	Mm	Mf
50	34.2	33.5	**89**	103.0	96.3	**128**	142.8	131.1
51	36.4	35.6	**90**	104.3	97.5	**129**	143.6	131.7
52	38.7	37.7	**91**	105.6	98.7	**130**	144.4	132.4
53	40.9	39.7	**92**	106.8	99.8	**131**	145.2	133.0
54	43.1	41.7	**93**	108.1	100.9	**132**	145.9	133.6
55	45.2	43.6	**94**	109.3	102.0	**133**	146.7	134.2
56	47.4	45.6	**95**	110.5	103.1	**134**	147.5	134.8
57	49.5	47.5	**96**	111.7	104.2	**135**	148.3	135.4
58	51.5	49.4	**97**	112.9	105.3	**136**	149.0	136.0
59	53.6	51.2	**98**	114.1	106.3	**137**	149.8	136.6
60	55.6	53.1	**99**	115.2	107.3	**138**	150.5	137.2
61	57.6	54.9	**100**	116.3	108.3	**139**	151.3	137.8
62	59.5	56.7	**101**	117.4	109.3	**140**	152.0	138.3
63	61.5	58.4	**102**	118.5	110.3	**141**	152.8	138.9
64	63.4	60.2	**103**	119.6	111.3	**142**	153.5	139.4
65	65.2	61.9	**104**	120.7	112.2	**143**	154.3	140.0
66	67.1	63.6	**105**	121.8	113.2	**144**	155.0	140.5
67	68.9	65.3	**106**	122.8	114.1	**145**	155.7	141.1
68	70.7	66.9	**107**	123.8	115.0	**146**	156.5	141.6
69	72.5	68.5	**108**	124.8	115.9	**147**	157.2	142.1
70	74.2	70.1	**109**	125.8	116.7	**148**	157.9	142.6
71	75.9	71.7	**110**	126.8	117.6	**149**	158.7	143.2
72	77.6	73.2	**111**	127.8	118.5	**150**	159.4	143.7
73	79.3	74.8	**112**	128.8	119.3	**151**	160.1	144.2
74	80.9	76.3	**113**	129.7	120.1	**152**	160.8	144.7
75	82.6	77.7	**114**	130.7	120.9	**153**	161.6	145.2
76	84.2	79.2	**115**	131.6	121.7	**154**	162.3	145.7
77	85.7	80.6	**116**	132.5	122.5	**155**	163.0	146.2
78	87.3	82.1	**117**	133.4	123.3	**156**	163.8	146.7
79	88.8	83.5	**118**	134.3	124.0	**157**	164.5	147.2
80	90.3	84.8	**119**	135.2	124.8	**158**	166.0	147.7
81	91.8	86.2	**120**	136.1	125.5	**159**	166.2	148.2
82	93.3	87.5	**121**	136.9	126.3	**160**	166.7	148.6
83	94.7	88.8	**122**	137.8	127.0	**161**	167.5	149.1
84	96.1	90.1	**123**	138.6	127.7	**162**	168.2	149.6
85	97.6	91.4	**124**	139.5	128.4	**163**	169.0	150.1
86	98.9	92.7	**125**	140.3	129.1	**164**	169.7	150.6
87	100.3	93.9	**126**	141.1	129.7	**165**	170.5	151.1
88	101.6	95.1	**127**	142.0	130.4			

[a]Mm: $AC = -125.714 + (4.233)\,age + (-2.300 \times 10^{-2})\,age^2 + (5.159 \times 10^{-5})\,age^3$ $[R^2 = 0.94]$

[b]Mf: $AC = -110.796 + (3.781)\,age + (-2.000 \times 10^{-2})\,age^2 + (4.001 \times 10^{-5})\,age^3$ $[R^2 = 0.96]$

TABLE 20.5: Predicted values for femur length (FL) for rhesus (Mm[a]) and long-tailed (Mf[b]) macaques from the late first to third trimesters (50–165 days gestation)

Age	Mm	Mf	Age	Mm	Mf	Age	Mm	Mf
50	1.7	2.3	**89**	19.5	18.4	**128**	34.8	32.0
51	2.2	2.8	**90**	19.9	18.8	**129**	35.2	32.3
52	2.6	3.2	**91**	20.4	19.2	**130**	35.5	32.6
53	3.1	3.6	**92**	20.8	19.5	**131**	35.8	32.9
54	3.6	4.0	**93**	21.2	19.9	**132**	36.1	33.2
55	4.0	4.5	**94**	21.7	20.3	**133**	36.5	33.5
56	4.5	4.9	**95**	22.1	20.7	**134**	36.8	33.8
57	4.9	5.3	**96**	22.5	21.1	**135**	37.1	34.0
58	5.4	5.7	**97**	22.9	21.4	**136**	37.4	34.3
59	5.9	6.2	**98**	23.4	21.8	**137**	37.7	34.6
60	6.3	6.6	**99**	23.8	22.2	**138**	38.0	34.9
61	6.8	7.0	**100**	24.2	22.5	**139**	38.3	35.1
62	7.2	7.4	**101**	24.6	22.9	**140**	38.6	35.4
63	7.7	7.8	**102**	25.0	23.3	**141**	38.8	35.7
64	8.2	8.3	**103**	25.5	23.6	**142**	39.1	35.9
65	8.6	8.7	**104**	25.9	24.0	**143**	39.4	36.2
66	9.1	9.1	**105**	26.3	24.4	**144**	39.7	36.5
67	9.5	9.5	**106**	26.7	24.7	**145**	39.9	36.7
68	10.0	9.9	**107**	27.1	25.1	**146**	40.2	37.0
69	10.5	10.3	**108**	27.5	25.4	**147**	40.4	37.2
70	10.9	10.8	**109**	27.9	25.8	**148**	40.7	37.5
71	11.4	11.2	**110**	28.3	26.1	**149**	40.9	37.7
72	11.8	11.6	**111**	28.7	26.5	**150**	41.2	37.9
73	12.3	12.0	**112**	29.0	26.8	**151**	41.4	38.2
74	12.8	12.4	**113**	29.4	27.2	**152**	41.6	38.4
75	13.2	12.8	**114**	29.8	27.5	**153**	41.9	38.6
76	13.7	13.2	**115**	30.2	27.8	**154**	42.1	38.8
77	14.1	13.6	**116**	30.6	28.2	**155**	42.3	39.1
78	14.6	14.0	**117**	30.9	28.5	**156**	42.5	39.3
79	15.0	14.4	**118**	31.3	28.8	**157**	42.7	39.5
80	15.5	14.8	**119**	31.7	29.2	**158**	42.9	39.7
81	15.9	15.2	**120**	32.0	29.5	**159**	43.1	39.9
82	16.4	15.6	**121**	32.4	29.8	**160**	43.3	40.1
83	16.8	16.0	**122**	32.8	30.1	**161**	43.5	40.3
84	17.3	16.4	**123**	33.1	30.4	**162**	43.7	40.5
85	17.7	16.8	**124**	33.5	30.8	**163**	43.8	40.7
86	18.2	17.2	**125**	33.8	31.1	**164**	44.0	40.9
87	18.6	17.6	**126**	34.2	31.4	**165**	44.2	41.0
88	19.1	18.0	**127**	34.5	31.7			

[a]Mm: $FL = -19.386 + (3.620 \times 10^{-1})\ age + (2.000 \times 10^{-3})\ age^2 + (-9.243 \times 10^{-6})\ age^3$ [$R^2 = 0.98$]

[b]Mf: $FL = -18.916 + (4.100 \times 10^{-1})\ age + (1.000 \times 10^{-3})\ age^2 + (-5.051 \times 10^{-6})\ age^3$ [$R^2 = 0.96$]

TABLE 20.6: 95% confidence intervals for biparietal diameter (BPD) for rhesus (Mm) and long-tailed (Mf) macaques (N = 10–100) from the late first to the third trimesters. (Mean ± standard deviation, *Range*)

Age	Mm	Mf	Age	Mm	Mf	Age	Mm	Mf
50–52	12.5–13.1 (12.8 ± 0.9) *11–14*	11.9–12.6 (12.3 ± 1.2) *10–15*	**89–91**	31.4–31.8 (31.6 ± 1.4) *28–35*	30.1–31.3 (30.7 ± 1.8) *27–36*	**128–130**	43.9–45.3 (44.6 ± 1.7) *40–47*	40.7–43.4 (42.1 ± 2.6) *39–49*
53–55	13.3–14.5 (13.9 ± 1.4) *11–16*	13.4–14.3 (13.9 ± 1.2) *12–16*	**92–94**	31.6–32.3 (32.0 ± 1.3) *30–35*	30.8–32.3 (31.6 ± 1.6) *28–34*	**131–133**	44.8–45.8 (15.3 ± 1.3) *43–47*	40.3–43.1 (41.7 ± 1.8) *40–45*
56–58	15.7–16.7 (16.2 ± 1.5) *14–19*	14.8–16.4 (15.6 ± 1.8) *13–19*	**95–97**	33.2–34.2 (33.7 ± 1.7) *31–36*	31.8–33.3 (32.5 ± 1.5) *30–36*	**134–136**	44.8–45.8 (45.3 ± 1.0) *44–47*	41.2–45.1 (43.1 ± 2.1) *41–47*
59–61	17.4–18.3 (17.8 ± 1.7) *14–20*	16.2–17.2 (16.7 ± 1.2) *14–19*	**98–100**	34.4–35.5 (34.9 ± 1.8) *32–43*	32.1–33.6 (32.9 ± 1.7) *30–35*	**137–139**	45.2–46.8 (46.0 ± 1.3) *44–48*	41.5–44.5 (43.0 ± 1.2) *42–45*
62–64	18.8–19.7 (19.2 ± 1.6) *16–22*	16.6–20.1 (18.3 ± 2.2) *13–20*	**101–103**	35.6–36.7 (36.2 ± 1.7) *32–40*	33.3–35.1 (34.2 ± 2.1) *30–38*	**140–142**	45.4–47.0 (46.2 ± 1.5) *43–48*	41.5–43.5 (42.5 ± 1.4) *40–44*
65–67	20.0–21.0 (20.5 ± 1.2) *17–23*	19.1–20.4 (19.8 ± 1.2) *18–22*	**104–106**	36.5–37.5 (37.0 ± 1.5) *35–40*	34.9–36.8 (35.9 ± 1.7) *32–39*	**143–145**	46.1–48.1 (47.1 ± 1.7) *45–50*	41.8–44.6 (43.2 ± 2.0) *41–48*
68–70	21.0–22.2 (21.5 ± 1.6) *18–26*	21.4–22.3 (21.8 ± 1.2) *19–24*	**107–109**	37.2–38.5 (37.8 ± 1.6) *35–42*	35.3–37.8 (36.6 ± 2.3) *32–40*	**146–148**	46.2–48.1 (47.1 ± 1.6) *45–50*	43.4–45.7 (44.5 ± 1.8) *42–48*

(Continued)

TABLE 20.6 (*Continued*)

Age	Mm	Mf	Age	Mm	Mf	Age	Mm	Mf
71–73	23.1–24.1 (23.6 ± 1.1) *20–26*	21.9–23.5 (22.7 ± 1.6) *20–25*	**110–112**	38.8–40.3 (39.6 ± 1.8) *36–43*	35.2–37.3 (36.2 ± 1.8) *34–39*	**149–151**	47.0–48.2 (47.6 ± 1.1) *46–50*	44.0–46.8 (45.4 ± 2.2) *42–50*
74–76	25.0–26.0 (25.5 ± 1.6) *22–28*	23.9- 24.6 (24.3 ± 0.5) *24–25*	**113–115**	39.6–40.8 (40.2 ± 1.5) *38–43*	37.4–40.0 (38.7 ± 2.1) *36–43*	**152–154**	46.5–48.3 (47.4 ± 1.8) *44–51*	42.7–45.3 (44.0 ± 1.7) *42–48*
77–79	25.7–26.4 (26.1 ± 1.1) *24–29*	25.7–27.3 (26.5 ± 1.5) *24–29*	**116–118**	41.0–42.0 (41.5 ± 1.3) *40–44*	37.3–40.4 (38.9 ± 1.7) *36–41*	**155–157**	46.8–47.9 (47.4 ± 1.0) *44–50*	44.8–48.1 (46.4 ± 2.1) *44–50*
80–82	27.1–28.0 (27.5 ± 1.8) *23–33*	25.8–26.9 (26.3 ± 1.3) *23–28*	**119–121**	41.6–42.7 (42.1 ± 1.7) *39–46*	39.8–41.6 (40.7 ± 1.8) *37–44*	**158–160**	47.5–49.0 (48.3 ± 1.5) *46–52*	43.3–45.8 (44.6 ± 1.6) *42–46*
83–85	28.7–29.3 (29.0 ± 1.5) *26–33*	27.7–29.2 (28.4 ± 1.7) *24–32*	**122–124**	42.0–43.6 (42.8 ± 1.8) *40–47*	38.7–41.0 (39.8 ± 1.9) *37–43*	**161–163**	47.0–48.6 (47.8 ± 0.8) *47–49*	42.5–44.1 (43.3 ± 0.5) *43–44*
86–88	30.2–30.6 (30.4 ± 1.3) *28–34*	27.8–29.7 (28.7 ± 2.3) *21–32*	**125–127**	43.1–44.6 (43.9 ± 1.4) *41–46*	39.0–41.4 (40.2 ± 1.6) *38–44*	**164–166**	47.6–49.0 (48.3 ± 1.0) *46–51*	43.6–47.0 (45.3 ± 1.8) *43–48*

TABLE 20.7: 95% confidence intervals for head circumference (HC) for rhesus (Mm) and long-tailed (Mf) macaques (N = 10–100) from the late first to the third trimesters. (Mean ± standard deviation, *Range*)

Age	Mm	Mf	Age	Mm	Mf	Age	Mm	Mf
50–52	49.1–52.4 (50.7 ± 3.2) *43–56*	46.9–49.6 (48.3 ± 4.1) *41–57*	**86–88**	116.4–118.4 (117.4 ± 5.2) *106–129*	110.5–117.7 (114.1 ± 7.0) *101–126*	**122–124**	164.3–172.2 (168.3 ± 7.1) *158–179*	152.9–161.3 (157.1 ± 6.3) *148–168*
53–55	52.4–57.9 (55.1 ± 5.4) *45–63*	51.0–54.4 (52.7 ± 4.1) *47–62*	**89–91**	122.4–124.3 (123.4 ± 5.1) *111–134*	114.1–117.9 (116.0 ± 4.5) *107–125*	**125–127**	170.0–176.2 (173.1 ± 4.4) *162–177*	151.8–164.2 (158.0 ± 8.1) *148–172*
56–58	61.2–67.3 (64.3 ± 6.3) *54–76*	55.8–62.3 (59.1 ± 6.7) *48–70*	**92–94**	124.0–126.9 (125.5 ± 4.9) *113–135*	119.1–124.4 (121.8 ± 3.5) *118–130*	**128–130**	174.7–179.9 (177.3 ± 5.6) *169–189*	159.6–172.9 (166.2 ± 11.0) *154–187*
59–61	64.6–68.8 (66.7 ± 5.8) *55–76*	61.2–66.3 (63.7 ± 5.2) *54–75*	**95–97**	130.8–134.9 (132.8 ± 5.9) *124–144*	123.4–128.6 (126.0 ± 4.4) *120–136*	**131–133**	175.1–180.6 (177.8 ± 5.1) *172–188*	154.2–169.1 (161.6 ± 8.9) *152–177*
62–64	72.1–75.8 (73.9 ± 5.3) *63–83*	60.1–77.3 (68.7 ± 9.3) *49–77*	**98–100**	134.0–140.7 (137.3 ± 8.4) *126–169*	124.7–130.7 (127.7 ± 5.7) *114–134*	**134–136**	176.6–181.4 (179.0 ± 4.2) *174–187*	158.0–184.0 (171.0 ± 8.2) *163–179*
65–67	73.0–81.6 (77.3 ± 8.9) *49–92*	73.0–77.8 (75.4 ± 4.1) *69–83*	**101–103**	140.9–145.4 (143.1 ± 6.0) *129–152*	132.2–140.0 (136.1 ± 6.4) *126–145*	**137–139**	179.5–188.1 (183.8 ± 5.6) *177–194*	158.0–176.0 (167.0 ± 3.6) *164–171*

(Continued)

TABLE 20.7 (*Continued*)

Age	Mm	Mf	Age	Mm	Mf	Age	Mm	Mf
68–70	81.0–86.0	80.4–83.8	**104–106**	144.1–147.7	131.1–144.6	**140–142**	179.4–187.2	165.4–176.4
	(83.5 ± 7.3)	(82.1 ± 4.0)		(145.9 ± 4.9)	(137.9 ± 8.8)		(183.3 ± 5.4)	(170.9 ± 7.1)
	70–108	*69–87*		*137–157*	*126–158*		*177–193*	*162–181*
71–73	86.7–92.8	86.3–91.6	**107–109**	144.5–151.3	139.8–154.2	**143–145**	182.4–193.8	167.9–183.9
	(89.8 ± 5.7)	(88.9 ± 4.6)		(147.9 ± 7.0)	(147.0 ± 7.8)		(188.1 ± 8.0)	(175.9 ± 12.0)
	76–98	*80–97*		*139–161*	*137–158*		*178–202*	*163–198*
74–76	95.4–100.1	91.5–97.6	**110–112**	154–160.3	141.0–150.8	**146–148**	180.9–190.0	169.0–174.5
	(97.7 ± 7.3)	(94.5 ± 1.9)		(157.4 ± 8.5)	(145.9 ± 6.4)		(185.4 ± 6.4)	(171.8 ± 3.6)
	82–108	*93–97*		*141–176*	*139–156*		*175–193*	*169–180*
77–79	98.5–102.1	96.5–105.3	**113–115**	156.3–161.4	145.5–158.1	**149–151**	184.9–191.2	168.2–184.4
	(100.3 ± 5.1)	(100.9 ± 5.7)		(158.9 ± 6.4)	(151.8 ± 9.4)		(188.0 ± 4.4)	(176.3 ± 8.8)
	92–117	*94–113*		*148–176*	*139–173*		*181–194*	*166–194*
80–82	104.8–108.1	97.9–102.3	**116–118**	159.5–165.8	148.5–159.6	**152–154**	182.9–195.8	167.5–175.5
	(106.5 ± 6.8)	(100.1 ± 4.6)		(162.7 ± 6.7)	(154.0 ± 4.5)		(189.4 ± 7.7)	(171.5 ± 3.8)
	87–124	*90–110*		*152–172*	*149–160*		*176–200*	*166–176*
83–85	110.4–112.7	103.7–109.1	**119–121**	164.4–168.6	156.2–167.0	**155–158**	178.7–186.9	165.5–182.5
	(111.6 ± 5.4)	(106.4 ± 4.4)		(166.5 ± 6.3)	(161.6 ± 9.0)		(182.8 ± 3.3)	(169.0 ± 3.1)
	96–121	*101–114*		*154–177*	*151–182*		*180–188*	*164–174*

TABLE 20.8: 95% confidence intervals for abdominal circumference (AC) for rhesus (Mm) and long-tailed (Mf) macaques (N = 10–100) from the late first to third trimesters. (Mean ± standard deviation, *Range*)

Age	Mm	Mf	Age	Mm	Mf	Age	Mm	Mf
50–52	37.8–41.3 (39.6 ± 3.1) *35–44*	36.5–39.1 (37.8 ± 3.3) *32–47*	**86–88**	100.3–104.3 (102.3 ± 10.3) *89–176*	91.5–100.3 (95.9 ± 8.0) *82–109*	**122–124**	134.7–145.0 (139.9 ± 9.0) *128–164*	117.1–136.6 (126.8 ± 9.3) *110–137*
53–55	42.2–46.5 (44.4 ± 3.8) *39–50*	37.8–42.8 (40.3 ± 4.9) *27–46*	**89–91**	105.0–107.5 (106.2 ± 6.7) *81–128*	94.8–100.4 (97.6 ± 6.4) *86–109*	**125–127**	132.5–147.3 (139.9 ± 8.0) *131–150*	121.5–131.5 (126.5 ± 6.0) *117–133*
56–58	47.6–53.1 (50.4 ± 5.3) *41–59*	43.6–50.9 (47.3 ± 5.4) *39–57*	**92–94**	104.0–108.9 (106.4 ± 7.7) *92–127*	95.2–106.4 (100.8 ± 7.3) *92–114*	**128–130**	141.8–149.1 (145.5 ± 7.6) *134–157*	127.6–140.4 (134.0 ± 9.5) *125–160*
59–61	53.5–58.7 (56.1 ± 8.4) *44–74*	45.0–50.0 (47.5 ± 3.9) *41–54*	**95–97**	107.9–112.9 (110.4 ± 6.9) *100–124*	99.0–110.2 (104.6 ± 9.3) *91–120*	**131–133**	142.7–163.9 (153.3 ± 12.7) *135–176*	116.8–133.2 (125.0 ± 8.9) *109–131*
62–64	57.5–61.2 (59.4 ± 4.9) *46–65*	47.0–65.0 (56.0 ± 8.6) *41–67*	**98–100**	112.5–119.4 (116.0 ± 8.2) *103–139*	104.9–109.7 (107.3 ± 5.2) *102–117*	**134–136**	135.7–152.6 (144.2 ± 12.6) *130–165*	
65–67	61.3–67.2 (64.2 ± 6.3) *52–77*	58.4–63.2 (60.8 ± 3.3) *56–68*	**101–103**	115.6–122.0 (118.8 ± 8.1) *102–133*	108.3–112.1 (110.2 ± 3.7) *102–116*	**137–139**	140.7–155.3 (148.0 ± 8.8) *135–163*	

(Continued)

TABLE 20.8 *(Continued)*

Age	Mm	Mf	Age	Mm	Mf	Age	Mm	Mf
68–70	68.3–72.7	66.5–70.7	**104–106**	120.4–125.9	107.3–120.9	**140–142**	140.0–159.7	124.5–163.9
	(70.5 ± 6.5)	(68.6 ± 4.9)		(123.2 ± 7.4)	(114.1 ± 9.6)		(149.9 ± 10.7)	(144.2 ± 15.9)
	59–88	*54–77*		*108–137*	*97–131*		*137–166*	*126–167*
71–73	72.6–78.7	69.9–75.3	**107–109**	115.4–122.5	112.9–126.9	**143–145**	147.2–172.5	129.2–149.6
	(75.6 ± 5.8)	(72.6 ± 3.7)		(118.9 ± 6.7)	(119.9 ± 8.4)		(159.9 ± 15.1)	(139.4 ± 12.2)
	64–86	*68–78*		*112–133*	*107–133*		*143–189*	*127–160*
74–76	80.8–85.7		**110–112**	123.9–130.5	109.2–127.4	**146–148**	147.2–172.8	137.9–145.3
	(83.2 ± 7.3)			(127.2 ± 6.5)	(118.3 ± 9.9)		(160.0 ± 10.3)	(141.6 ± 4.4)
	63–93	*73–80*		*114–140*	*103–131*		*149–171*	*132–145*
77–79	82.8–86.7	81.4–89.8	**113–115**	125.0–131.8	112.3–126.7	**149–151**	137.5–168.5	143.5–157.4
	(84.7 ± 5.6)	(85.6 ± 5.9)		(128.4 ± 7.9)	(119.5 ± 8.6)		(153.0 ± 6.3)	(150.4 ± 5.6)
	76–98	*77–96*		*115–142*	*109–135*		*148–160*	*141–155*
80–82	90.2–94.6	82.4–88.5	**116–118**	125.4–137.6	108.9–124.7	**152–155**	141.3–176.7	137.7–149.1
	(92.4 ± 8.9)	(85.4 ± 6.5)		(131.5 ± 11.8)	(116.8 ± 6.4)		(159.0 ± 11.1)	(143.3 ± 4.6)
	74–109	*73–99*		*115–160*	*107–122*		*145–172*	*136–148*
83–85	94.1–96.8	89.4–93.7	**119–121**	135.6–141.2	123.7–134.7			
	(95.5 ± 6.2)	(91.5 ± 3.6)		(138.4 ± 8.7)	(129.2 ± 8.1)			
	81–110	*86–98*		*128–155*	*116–140*			

TABLE 20.9: 95% confidence intervals for femur length (FL) for rhesus (Mm) and long-tailed (Mf) macaques from the late first to third trimesters . (Mean ± standard deviation, *Range*)

Age	Mm	Mf	Age	Mm	Mf	Age	Mm	Mf
50–52	1.9–5.1 (3.5 ± 1.5) *2–6*	2.5–3.5 (3.0 ± 1.2) *2–6*	**89–91**	19.7–20.2 (20.0 ± 1.3) *17–23*	18.3–19.6 (18.9 ± 2.0) *16–24*	**128–130**	33.9–35.9 (34.9 ± 2.3) *30–39*	31.3–34.3 (32.8 ± 2.8) *28–38*
53–55	3.0–4.7 (3.9 ± 0.9) *3–5*	3.5–4.5 (4.0 ± 1.0) *2–6*	**92–94**	20.7–21.6 (21.1 ± 1.6) *18–24*	19.5–20.9 (20.2 ± 1.7) *17–23*	**131–133**	35.8–37.4 (36.6 ± 1.9) *33–41*	27.2–33.4 (30.3 ± 3.4) *27–36*
56–58	4.1–5.5 (4.8 ± 1.3) *3–8*	4.4–5.9 (5.1 ± 1.4) *3–8*	**95–97**	22.0–23.0 (22.5 ± 1.7) *19–26*	20.4–22.2 (21.3 ± 1.8) *18–24*	**134–136**	35.8–37.2 (36.5 ± 1.3) *35–39*	29.4–35.8 (32.6 ± 2.6) *30–36*
59–61	5.6–6.5 (6.0 ± 1.3) *4–8*	5.7–7.5 (6.6 ± 1.8) *4–9*	**98–100**	23.3–24.0 (23.6 ± 1.3) *21–27*	21.6–23.3 (22.4 ± 1.9) *19–26*	**137–139**	37.1–39.8 (38.5 ± 2.0) *36–42*	32.5–38.2 (35.3 ± 1.2) *34–36*
62–64	6.2–7.5 (6.8 ± 1.5) *4–9*	6.7–9.5 (8.1 ± 2.0) *3–10*	**101–103**	24.9–26.1 (25.5 ± 1.7) *23–29*	22.2–23.9 (23.1 ± 1.9) *20–27*	**140–142**	36.7–39.9 (38.3 ± 2.7) *34–44*	33.4–37.0 (35.2 ± 1.7) *33–38*
65–67	8.4–9.6 (9.0 ± 1.5) *5–13*	7.9–9.9 (8.9 ± 1.8) *7–14*	**104–106**	26.1–27.3 (26.7 ± 1.7) *23–31*	23.1–26.5 (24.8 ± 3.0) *20–29*	**143–145**	38.7–41.7 (40.2 ± 2.7) *37–45*	34.4–37.5 (35.9 ± 2.4) *33–42*
68–70	9.7–10.4 (10.1 ± 1.1) *8–13*	9.7–10.7 (10.2 ± 1.3) *7–14*	**107–109**	26.2–27.3 (26.8 ± 1.4) *23–30*	25.0–27.0 (26.0 ± 2.0) *22–29*	**146–148**	39.2–41.6 (40.4 ± 2.1) *36–43*	34.7–40.2 (37.4 ± 3.5) *33–42*

(Continued)

TABLE 20.9 (*Continued*)

Age	Mm	Mf	Age	Mm	Mf	Age	Mm	Mf
71–73	11.5–12.3	11.0–12.5	**110–112**	27.9–29.0	23.8–27.1	**149–151**	39.7–41.7	38.3–41.1
	(11.9 ± 0.9)	(11.8 ± 1.5)		(28.4 ± 1.4)	(25.5 ± 2.8)		(40.7 ± 1.8)	(39.7 ± 1.9)
	9–13	*9–14*		*26–31*	*22–31*		*37–44*	*37–42*
74–76	13.2–14.1	11.8–13.1	**113–115**	28.7–30.0	25.9–29.6	**152–154**	40.4–43.0	35.8–40.7
	(13.6 ± 1.4)	(12.4 ± 0.9)		(29.3 ± 1.7)	(27.7 ± 2.8)		(41.7 ± 2.4)	(38.3 ± 3.0)
	10–16	*11–14*		*26–33*	*23–31*		*38–46*	*34–44*
77–79	13.7–14.8	13.9–15.9	**116–118**	30.6–31.8	24.5–28.9	**155–157**	40.5–43.0	39.0–44.1
	(14.3 ± 1.7)	(14.9 ± 1.8)		(31.2 ± 1.5)	(26.7 ± 2.4)		(41.8 ± 2.2)	(41.6 ± 2.8)
	11–19	*13–20*		*29–33*	*25–31*		*36–45*	*35–45*
80–82	15.7–16.4	13.9–15.4	**119–121**	31.7–32.9	29.1–32.0	**158–160**	41.8–44.5	35.9–44.1
	(16.0 ± 1.5)	(14.6 ± 1.7)		(32.3 ± 1.8)	(30.5 ± 2.8)		(43.1 ± 2.5)	(40.0 ± 2.6)
	12–21	*12–18*		*29–37*	*26–35*		*39–47*	*37–43*
83–85	16.9–17.5	15.8–17.5	**122–124**	32.1–33.4	28.8–31.8	**161–163**	39.0–45.0	33.7–42.3
	(17.2 ± 1.5)	(16.6 ± 1.8)		(32.7 ± 1.5)	(30.3 ± 2.5)		(42.0 ± 3.2)	(38.0 ± 1.7)
	13–21	*14–22*		*30–36*	*27–35*		*37–46*	*36–39*
86–88	18.6–19.1	16.7–18.1	**125–127**	34.3–35.5	28.9–33.7	**164–166**	43.8–46.2	35.0–45.0
	(18.8 ± 1.4)	(17.4 ± 1.7)		(34.9 ± 1.1)	(31.3 ± 3.1)		(45.0 ± 1.0)	(40.0 ± 2.0)
	16–22	*15–22*		*33–36*	*28–36*		*44–46*	*38–42*

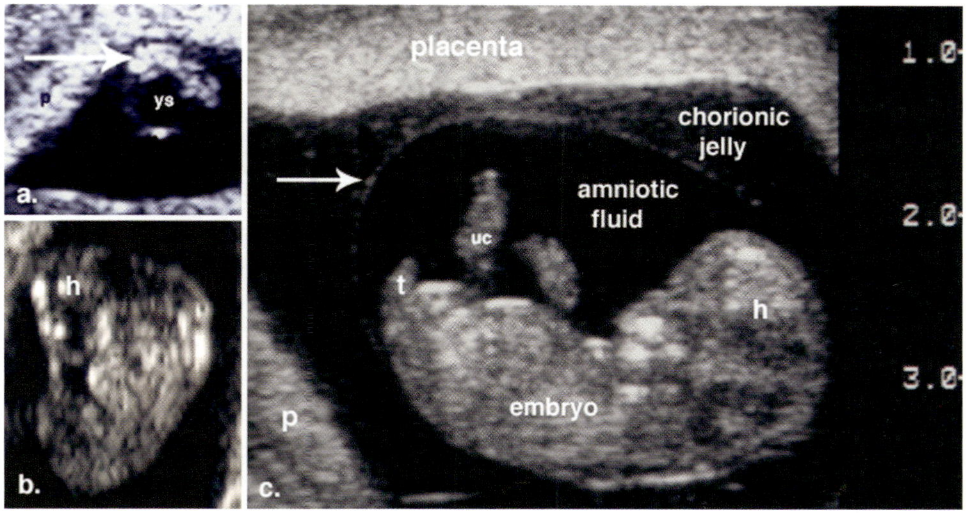

Figure 20.19 Embryonic development. **(a)** At 27 days gestation, the embryo (arrow) is closely associated with the yolk sac (ys). **(b)** By 32 days gestation, the C-shape of the embryo is evident. h = head. **(c)** By 50 days gestation, the embryo has grown significantly. The amnion (arrow) and chorion have not yet fused, and the chorionic jelly appears more echogenic and viscous when compared to the amniotic fluid. h = head, t = tail, uc = umbilical cord, p = placenta.

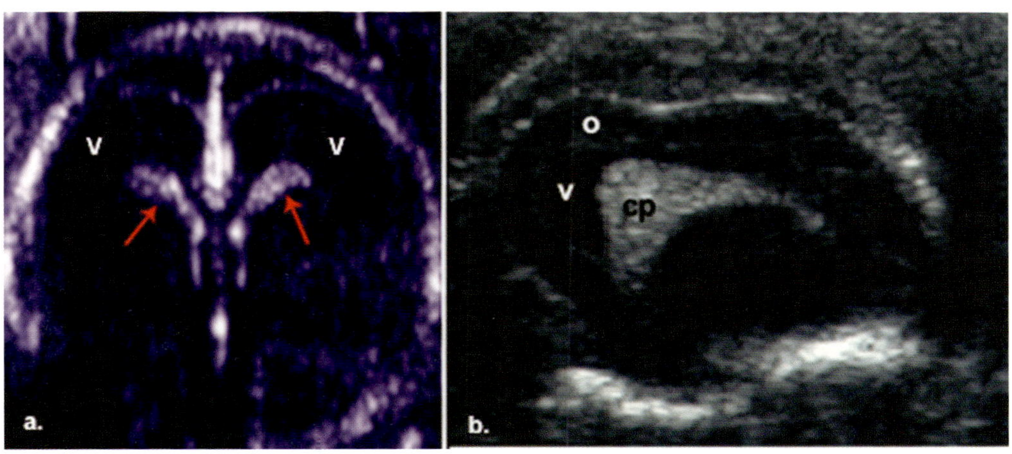

Figure 20.20 Fetal intracranial anatomy. **(a)** Coronal scan showing the ossified skull, choroid plexus (small arrows) and lateral ventricles (v) of a second trimester fetus. **(b)** Oblique sagittal section showing the choroid plexus (cp), lateral ventricle (v), and occipital lobe (o).

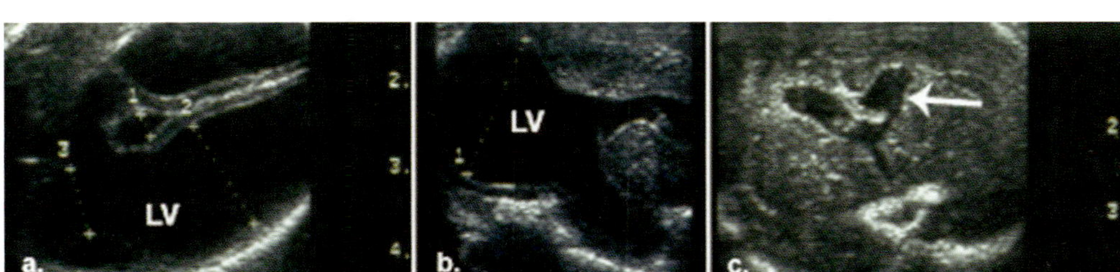

Figure 20.21 Second trimester fetus with hydrocephalus shown in transverse **(a)**, sagittal **(b)**, and coronal **(c)** scans. Note lateral ventricles (LV) (arrow in **c**). Cursors in **(a)** indicate the 3rd ventricle has a width of 0.3 cm, and the lateral ventricle ranges from 0.8–1.4 cm.

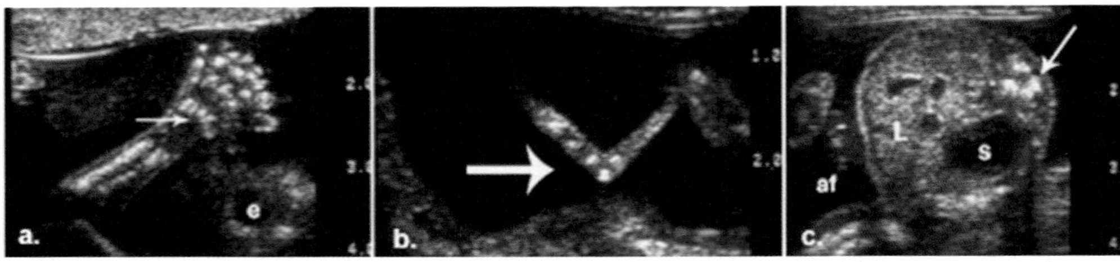

Figure 20.22 **(a)** Left arm and hand with imaging of the ossification centers of the radius and ulna, metacarpals (arrow), and digits. e = eye. **(b)** Tail abnormality noted in a second trimester fetal long-tailed macaque (arrow). **(c)** Transverse section through an early second trimester fetal abdomen shows ossification centers of the vertebrae (arrow). L = liver, s = stomach, af = amniotic fluid.

and females, is cranially oriented while in the female, the phallus is positioned more caudally. By 50–60 days gestation, a distinct midline raphe indicating fusion of the labioscrotal folds and formation of the scrotum is a defining feature for the identification of male fetuses (see Figure 20.24). As previously reported, by 70 days gestation all external structures are evident, and clearly identify the fetus as male or female (Tarantal and Hendrickx, 1988b,c) unless genital ambiguity occurs (Prahalada *et al.,* 1997). Thus, the effects of exposure to androgenic agents during the critical window of external sexual differentiation can be readily monitored and documented.

If identifying the sex of the fetus is essential prior to the period of sexual differentiation, then other methods can be used to determine gender. Previous reports focused on techniques such as chorionic villus sampling (CVS) or placental biopsy (Tarantal, 1990), with analysis of collected specimens using a PCR-based assay for the rhesus Y chromosome. CVS can be readily performed under ultrasound guidance at 18–40 days gestation. With this procedure, it is important

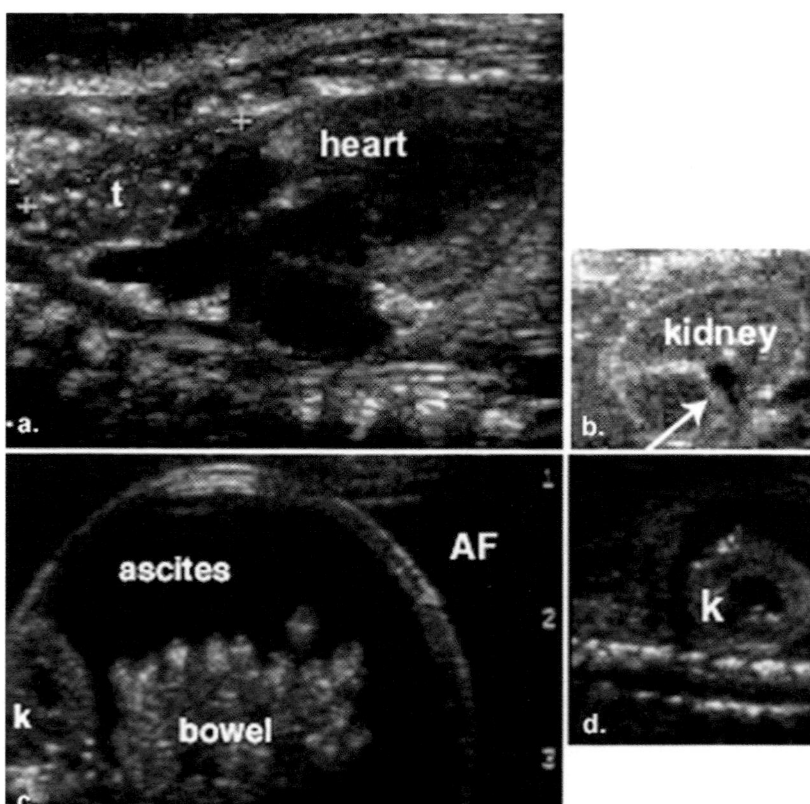

Figure 20.23 **(a)** Note heart and thymus (t). Cursors indicate the length of the second trimester fetal thymus. **(b)** Longitudinal section of normal fetal kidney with hilum noted (arrow). **(c, d)** Fetal ascites and hydronephrosis noted in a second trimester fetus. k = kidney, AF = amniotic fluid.

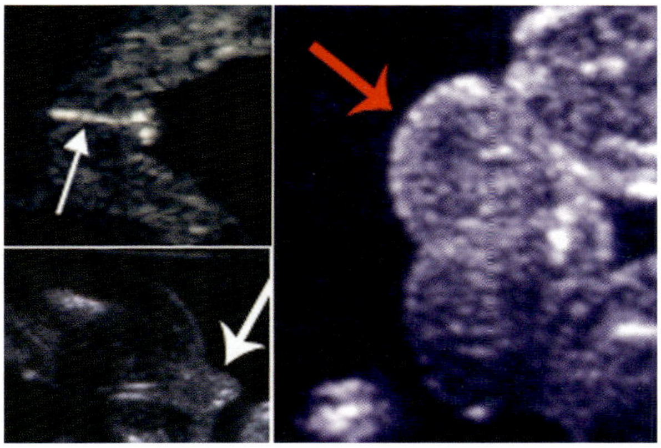

Figure 20.24 Male fetuses can be identified with 100% accuracy once the midline raphe (upper left panel; arrow) and the scrotum (right panel; arrow) have formed. Female fetuses are identified by the clitoris (left lower panel; arrow).

to ensure that aspiration of chorionic villi is confirmed microscopically prior to sample analysis. When obtaining chorionic villi, care needs to be taken to guide the tip of the aspirating needle into the body of the placental disc, avoiding the site of umbilical cord insertion or placental disc margins, and ensuring that the needle is not advanced into the chorionic jelly, amniotic fluid, yolk sac, or the embryo. The sex of the embryo can also be identified using less invasive measures through the evaluation of maternal blood samples and quantitative PCR-based assays (Jimenez and Tarantal, 2003).

Fetal physiology

Similar to human fetuses, the monkey fetus is very active *in utero*. By the second trimester, vigorous whole body movements are less frequently observed and more selective activities such as darting eye movements, those associated with the mouth and oral cavity, and extension and flexion of the limbs and head can be observed, provided the dam has not been anesthetized. The macaque Biophysical Profile (mBPP) is used in awake animals (either hand- or chair-restrained) for evaluating the development of *in utero* fetal activity patterns (Golub *et al.*, 1992). The mBPP incorporates a 20-minute observation period during which time the following characteristics are quantitated and evaluated: fetal heart rates (obtained every 5 minutes during the 20 minute observation period), fetal movements (gross body movements; trunk and/or limbs), fetal breathing movements (chest and abdominal excursions), muscle tone (extension and/or flexion of the extremities and/or spine), and whole body startle. Observations are initiated during the second trimester and can be performed periodically until term delivery. Data are typically analyzed as frequency of occurrence rather than by a categorized score assignment.

Section 6: Ultrasound-guided procedures

The use of ultrasound-guided procedures has been described previously, and the application to both the gravid and nongravid macaque established (Tarantal, 1990; 1993). The primary advantage of these techniques is the elimination of the need for surgical procedures with the associated time, cost, potential trauma, and increased risk factors for adhesions, endometriosis, and alterations in pelvic anatomy. With the limitations on the number of surgical procedures that can be performed in laboratory-housed macaques, these techniques will continue to serve a wider application, and further enhance the health and well-being of these valuable research animals. Some examples are discussed below. All procedures are performed under aseptic conditions, with a standard surgical preparation of the animal, and the use of sterile sheaths and other sterile supplies (needles, syringes, gloves). The key to success is a good understanding of anatomy and anatomical relationships, constant monitoring of the path of the advancing needle tip, and performing the procedure in a relatively short period of time. The approach should be initially identified and the traversing needle and needle tip imaged at all times.

Many ultrasound-guided procedures are routinely used for reproductive-related research applications and colony management purposes such as intrauterine insemination (Tarantal *et al.*, 1990b), recovery of cervical mucus (VandeVoort *et al.*, 1989), endometrial or myometrial biopsy (Tarantal, 1992), and follicular aspiration after hormonal stimulation (VandeVoort and Tarantal, 1991; 2001). Nonsurgical ultrasound-guided methods have been very successful because they can be performed quickly, safely, and with great efficiency. Other reproductive applications in male monkeys include ultrasound-guided biopsy of the prostate (Kamischke *et al.*, 1997; 1999), and gonadal transplant of germ cells (Schlatt *et al.*, 1999).

Ultrasound is also used extensively as an adjunct for guiding the delivery of agents into various fetal compartments (intra-chorionic and intra-amniotic, intraperitoneal, intramuscular, intravenous, intracardiac,

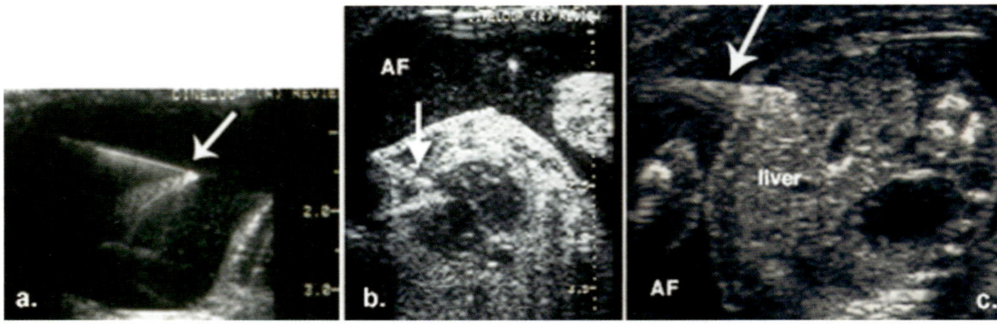

Figure 20.25 A variety of fetal interventional procedures can be performed safely under ultrasound-guidance. **(a)** Amniocentesis is routinely performed. Note needle and needle tip (arrow). **(b)** Fetal blood samples are collected routinely by cardiocentesis. Note needle within the left ventricle (arrow). **(c)** Oblique section through the fetal abdomen in the early second trimester. Note needle (arrow) within the fetal liver. AF = amniotic fluid.

intracranial, intrapulmonary, intraportal, intrahepatic, intrarenal) (Figure 20.25) and for collecting fetal specimens (amniotic fluid, blood, liver, muscle, urine, chorionic villi) (Tarantal, 1990; 1993). Transient catheter placement has also been used as a method for monitoring intrauterine tissue temperature, delivery of exogenous substances, or for sample collection (Tarantal *et al.*, 1993a).

Many fetal monkey models have been developed using ultrasound-related techniques and procedures. These models have focused on human fetal diseases and dynamically explored mechanisms of pathogenesis and corrective therapies. Importantly, these models have allowed the investigation of crucial questions that cannot be ethically explored in humans. Included are studies on fetal growth restriction and regulation (Tarantal and Gargosky, 1995; Tarantal *et al.*, 1995b; 1997a), the prenatal transplantation of stem cells (Cowan *et al.*, 1996; 2001; Crombleholme *et al.*, 1988; Duncan *et al.*, 1991; Harrison *et al.*, 1989; Tarantal *et al.*, 2000), *in utero* growth factor therapy (Coulter *et al.*, 1996; Gilbert *et al.*, 1997; Goetzman *et al.*, 1993; Plopper *et al.*, 1992; Read *et al.*, 1989; Tarantal *et al.*, 1997a; Tarantal and Cowan, 1999), fetal gene transfer/gene therapy (Tarantal *et al.*, 2001b,c), obstructive renal disease and dysplasia (Matsell *et al.*, 2002; Matsell and Tarantal, 2002; Tarantal *et al.*, 2001a), and for the induction of viral and parasitic infections and pathogenesis by direct inoculation of fetuses *in utero* (Barr *et al.*, 1994; Chang *et al.*, 2003; Ho *et al.*, 1997; Lane *et al.*, 1996; Tarantal *et al.*, 1995b; 1998; 1999b). Monkey fetuses have also been treated with antiviral agents in order to assess transplacental transport, effects of these agents on growth, development, and hematopoiesis, and efficacy in reversing various aspects of disease (Castillo *et al.*, 2002; Tarantal *et al.*, 1994b; 1997a; 1999a; 2003). Readers are referred to the above publications for further details.

Other applications include routine biopsy of abdominal organs such as the liver and kidney (infant to aged). Additional applications include intraportal or intrahepatic injections of cells or other exogenous agents for experimental purposes (Tarantal, unpublished). If an organ or structure can be imaged and the path readily identified, access can be achieved under direct ultrasound guidance.

Section 7: Other ultrasound imaging applications

As noted above, ultrasound is a well-established and routine imaging modality for nonhuman primates. There are several human ultrasound texts that provide good reference material, although there are some anatomical differences when comparing humans and monkeys. Understanding anatomy, anatomical relationships, and anatomical variations; identification of good acoustic windows; balancing settings to avoid artifacts; and obtaining multiple images in different scan planes are all components for obtaining the best quality images and correctly interpreting the findings. Some examples are presented below, which highlight studies where ultrasound has been used with rhesus and long-tailed macaques either for diagnostic or experimental purposes. This is not intended to be a complete list of all published reports and studies with these species.

General abdominal imaging from the diaphragm to the pelvis in monkeys is similar to humans, and includes evaluations of the stomach and bowel, hepatobiliary system, pancreas, adrenals, kidneys, and the spleen. Vessels are readily identified and typically provide a constant relationship to most abdominal organs, and can be used as landmarks. The aorta enters the abdomen through the diaphragm, passes dorsal to the liver, gives off its major branches as celiac, superior mesenteric, and inferior mesenteric arteries, then bifurcates into right and left common iliacs. The aorta runs longitudinally and is identified sonographically by characteristic pulsations when compared to the larger and more thin-walled inferior vena cava. The aorta may appear tortuous with age, and the walls may become thickened; care should be taken to ensure that tortuous vessels are not confused with a dissecting aneurysm. Human vascular anatomy is similar when compared to monkeys, and includes anatomical variations in branching patterns. Generally, the celiac artery has a short trunk that splits into common hepatic, splenic, and left gastric branches. The common hepatic gives rise to the hepatic proper (right and left hepatics, cystic) and gastroduodenal arteries. The common iliacs (right and left) course laterally with branches deviating medially (internal iliacs) and laterally (external iliac to femoral). The inferior vena cava forms from the right and left iliac veins at the same level as the aortic bifurcation, which is generally near the site of the umbilicus. The renal veins are found at the same level as the renal arteries, and may have a slightly larger diameter and thinner vessel wall. The inferior vena cava and renal veins are located ventral to the arteries and the left renal vein passes ventral to the aorta. Hepatic veins are associated with the inferior vena cava and appear sonographically as "slits" in the hepatic parenchyma. However, the portal venous system, which is formed from the splenic, superior, and inferior mesenteric veins, is similar to humans as it divides into intrahepatic branches which are more numerous than hepatic veins, and the vessel walls more echogenic when compared to hepatic veins.

Size, shape, contour, and texture are used to assess the liver and gallbladder (Figure 20.26). Pathologic changes may be diffuse, focal, or perihepatic. Comparing the anatomy of humans with macaques, the ligamentum teres is generally lodged more deeply in the monkey, the lobes of the liver may be more irregular with the quadrate lobe narrower than humans, and the portal vessels somewhat larger with less branching.

The kidneys are readily identified because of their location. The left kidney is frequently lower in the abdomen than the right kidney (Hartman, 1933). Macaque kidneys are also somewhat flattened, with the cephalic pole slightly smaller and sharper than the caudal pole. Left kidney volumes have been estimated sonographically and reported to be smaller than right kidney volumes, similar to humans (Gaschen *et al.*, 2000). In healthy animals, benign renal cysts are commonly observed, similar to findings in humans (Figure 20.27). Ultrasound is also useful for assessing the urinary tract, such as for the diagnosis and management of eosinophilic cystitis (Tarantal, unpublished).

Echocardiographic profiles include ductal patency/closure in hand-held newborns, and structural analyses using 4- and 5-chamber views. The best images are obtained using a transducer with a small footprint, and with the animals in left lateral recumbency. A standard human echocardiography text is useful for reference. Cardiovascular reference values have been published for healthy rhesus monkeys under ketamine (Korcarz *et al.*, 1997), and studies have been performed to assess diet-induced coronary heart disease (Williams *et al.*, 1991), experimentally-induced myocarditis (Lima *et al.*, 1986), and spontaneous mitral valve prolapse (Swindle *et al.*, 1985). The effects of ketamine and transesophageal echocardiography on left ventricular, systemic arterial, and baroreflex responses have also been explored (Koenig *et al.*, 2001), and age-related changes in vascular and myocardial stiffness (Vaitkevicius *et al.*, 2001), and cardiac damage as a result of total body irradiation (Wondergem *et al.*, 1999) reported. Echocontrast agents have also been studied for both

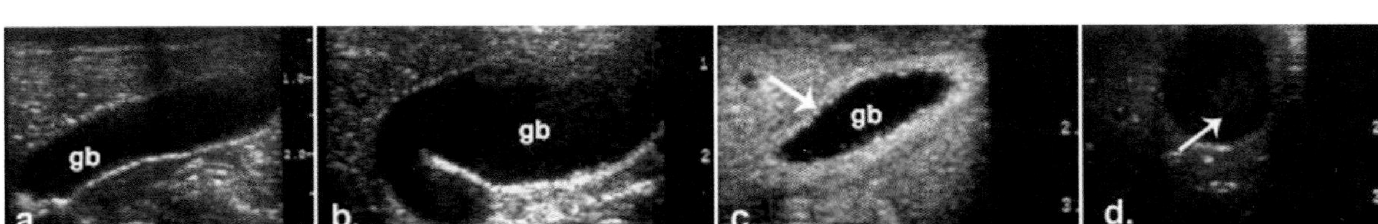

Figure 20.26 Images of the gallbladder (gb) in rhesus monkeys include normal findings **(a, b),** and animals with diffuse thickening of the gallbladder wall **(c)** and sludge within the lumen **(d)**.

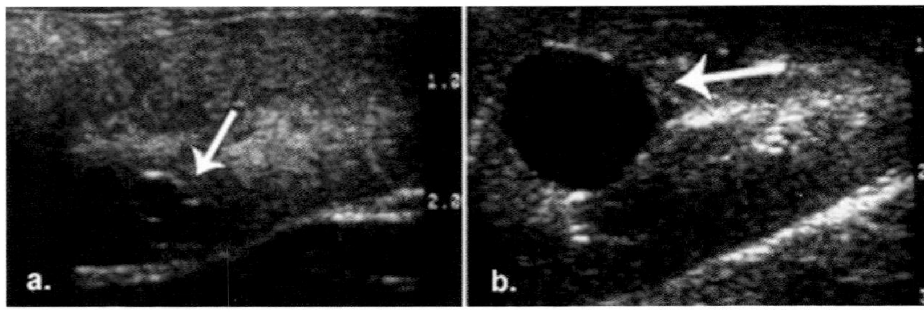

Figure 20.27 Benign renal cysts can appear small with an irregular shape **(a)** or large with a smooth contour **(b)**.

safety and efficiency in macaques prior to human application (Bommer *et al.*, 1998; Grauer *et al.*, 1996; Kwan *et al.*, 1994; Ostensen *et al.*, 1992; Simpson *et al.*, 1998).

Ultrasound imaging has been used for assessing the placement and patency of femoral catheters (Fig. 20.28), and for monitoring vascular grafts and whole organ transplants. One study focused on the evaluation of rapamycin monotherapy for preventing disease in cynomolgus recipients of orthotopic aortic allografts (Dambrin *et al.*, 2004). Recent sonographic studies have also shown luminal narrowing of arterial allografts with increased levels of circulating C-reactive protein (Fitzgerald *et al.*, 2004). Other vascular applications include techniques for identification of venous thrombus formation and resolution (Fowlkes *et al.*, 1998). Monitoring kidney transplants has been successfully accomplished for many years using pulsed, color, and power Doppler (Gaschen and Schuurman, 2002; 2001a–c; Gaschen *et al.*, 2001a,b; Mychaliska *et al.*, 1997). Observations include assessments of the size of the transplant, appearance of the corticomedullary border, and evaluation of the collecting system and urinary bladder. Ultrasound has also been useful for identifying diminished organ perfusion and tissue necrosis (Tarantal, unpublished). A scoring system has been used for evaluating transplanted kidneys where a score of *1* = uniform flow throughout the kidney; *2* = diminished flow with mild peripheral cortical hypoperfusion; *3* = low or no cortical flow; and *4* = no visible parenchymal flow except in the central vessels. Similar to applications in human subjects, the resistive index (RI) has proven informative for assessing changes associated with rejection. Systematic evaluation of transplanted kidneys begins with a two-dimensional imaging examination to assess anatomy, followed by more functional assessments. Finally, ultrasound-guided urine collection and biopsy of the kidney is performed with care taken to avoid traversing vascular structures and the ureter insertion site.

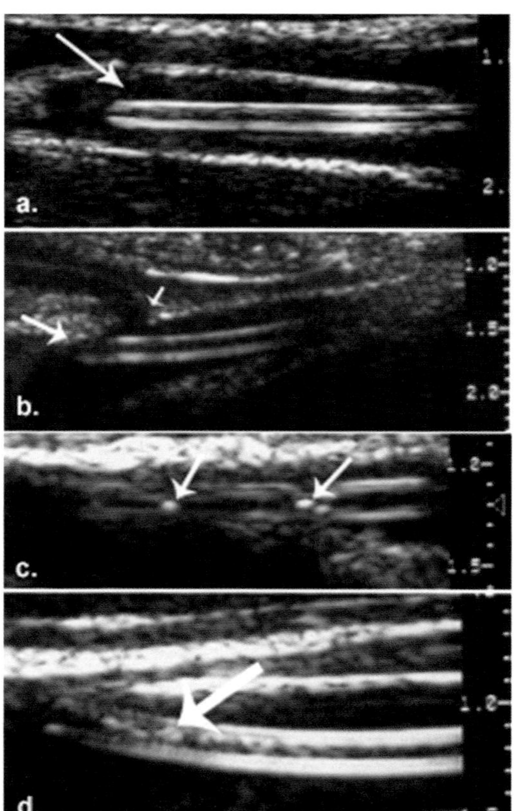

Figure 20.28 Note tip of the catheter (arrow) within the vessel **(a)** and a catheter with the tip near a bifurcation (small arrow) and in close proximity to the vessel wall **(b)**. With time, echogenic foci are frequently noted within the catheter possibly due to microthrombi (arrows) **(c, d)**.

Monitoring intracranial flow patterns has also been reported in macaques (Hennerici *et al.*, 1989), and ultrasound has also been used as an adjunct for the safe placement and monitoring of intracranial probes, for reducing some of the uncertainties associated with sterotaxic surgeries (Glimcher *et al.*, 2001; Tokuno *et al.*, 2000), to assess accommodative ocular biometric changes (Vilupuru and Glasser, 2003) and whether ambient lighting at night compromises the functional integrity of vision-dependent mechanisms in infant monkeys (Smith *et al.*, 2003), and to examine the relationship of ocular components to refraction during adult life (Fernandes *et al.,* 2003). Color Doppler has also been used to evaluate flow in the retrobulbar circulation (Netland *et al.,* 1997).

The biologic effects of ultrasound have been extensively investigated in the monkey model, both from a developmental perspective (Tarantal and Hendrickx, 1989b,c; Tarantal *et al.,* 1993a–d; 1995a), and to explore the potential effects of exposure on structures such as the lung (Tarantal and Canfield, 1994). Whereas these studies have focused primarily on the potential adverse effects of heating and cavitation, more recently cavitation has been used as a method for improved gene delivery (Miller *et al.,* 2002).

In conclusion, while the monkey will continue to serve as one of the premier animal models for addressing a wide variety of biomedical issues, ultrasound can play a major role in colony management, and for the conduct of studies that have direct relevance and translation to the human clinical setting.

Acknowledgement

Many of the studies described above were supported by grants funded through the National Institutes of Health (NIH) and the base operating grant for the California National Primate Research Center (#RR00169).

Correspondence

Any correspondence should be directed to Alice Tarantal, California National Primate Research Center, University of California, Pedrick and Hutchison Roads, Davis, California 95616-8542, USA. Email: aftarantal @ucdavis.edu

References

Ami, Y., Suzaki, Y. and Goto N. (1993). *J. Vet. Med. Sci.* 55, 7–11.

Barnett, S.B., Ter Haar, G.R., Ziskin, M.C., Rott, H.D., Duck, F.A. and Maeda, K. (2000). *Ultrasound Med. Biol.* 26, 355–366.

Barr, B.C., Conrad, P.A., Sverlow, K.W., Tarantal, A.F. and Hendrickx, A.G. (1994). *Lab. Invest.* 71, 236–242.

Baskin, G.B., Smith, S.M. and Marx, P.A. (2002). *Vet. Pathol.* 39, 572–575.

Beniashvili, D. (1989). *J. Med. Primatol.* 18, 423–437.

Bercovitch, F.B., Widdig, A., Berard, J.D., Nürnberg, P., Kessler, M.J., Schmidtke, J., Trefilov, A. and Krawczak, M. (2002). *Am. J. Primatol.* 57, 31–34.

Birkebak, T.A., Wang, N.P. and Weyhrich J. (1996). *J. Med. Primatol.* 25, 367–369.

Bommer, W.J., Tarantal, A.F., Yamamoto, R., Pulido, G., Liberty, S. and Dhond, M.R. (1998). *Proceedings of the 48th Annual Meeting of the American College of Cardiology.*

Brans, Y.W., Kuehl, T.J., Hayashi, R.H. and Reyes, P. (1990). *J. Med. Primatol.* 19, 641–649.

Buyalos, R.P. and Agarwal, S.K. (2000). *Curr. Opin. Obstet. Gynecol.* 12, 377–381.

Castillo, A.B., Tarantal, A.F., Watnik, M.R, and Martin, R.B. (2002). *J. Orthop. Res.* 20, 1185–1189.

Chang, W.L.W., Tarantal, A.F., Zhou, S.-S., Borowsky, A.D., Barthold, S.W. and Barry, P.A. (2003). *J. Virol.* 76, 9493–9504.

Conrad, S.H., Sackett, G.P. and Burbacher, T.M. (1989). *J. Med. Primatol.* 18,143–154.

Corradini, P., Recabarren, M., Seron-Ferre, M. and Parraguez, V.H. (1998). *J. Med. Primatol.* 27, 287–292.

Coulter, C.L., Read, L.C., Carr, B.R., Tarantal, A.F. and Styne, D.M. (1996). *J. Clin. Endocrinol. Metab.* 81, 1254–1260.

Cowan, M.J., Tarantal, A.F., Capper, J., Harrison, M. and Garovoy, F. (1996). *Bone Marrow Transplant* 17, 1157–1165.

Cowan, M.J., Chou, S.-H. and Tarantal, A.F. (2001). *Stem Cells from Cord Blood, In Utero Stem Cell Development, and Transplantation-Inclusive Gene Therapy, Ernst Schering Research Foundation Workshop #33.* Springer-Verlag, New York, pp 145–196.

Crombleholme, T.M., Longaker, M.T., Slotnick, N.R., Tarantal, A.F., Golbus, M.S., Zanjani, E.D. and Harrison, M.R. (1988). *Surg. Forum* 39, 564–566.

Cseh, S., Corselli, J., Chan, P. and Bailey, L. (2001). *Reprod. Nutr. Dev.* 41, 531–534.

Cseh, S., Corselli, J., Chan, P. and Bailey, L. (2002). *Anim. Reprod. Sci.* 70, 287–293.

Cukierski, M.A., Tarantal, A.F. and Hendrickx, A.G. (1986). *J. Med. Primatol.* 15, 227–234.

Dambrin, C., Klupp, J., Birsan, T., Luna, J., Suzuki, T., Lam, T., Stahr, P., Hausen, B., Christians, U.U., Fitzgerald, P., Berry, G. and Morris, R. (2003). *Circulation* 107, 2369–2374.

Devonald, K.J., Harewood, W.J., Ellwood, D.A. and Phippard, A.F. (1996). *J. Med. Primatol.* 25, 339–345.

Duncan, B.W., Harrison, M.R., Zanjani, E.D., Tarantal, A.F., Adzick, N.S., Bradley, S.M., Longaker, M.T., Jennings, R.W., Roberts, L.J., and Bigler, M.E. (1991) *Transplant Proc.* 23, 841–813.

Enders, A.C. (1991). *Placenta* 12, 309–325.

Ervin, M.G., Seidner, S.R., Leland, M.M., Ikegami, M. and Jobe, A.H. (1998). *Am. J. Physiol.* 274, R1169–1176

Farine, D., MacCarter, G.D., Timor-Tritch, I.E., Yeh, M.N. and Stark, R.I. (1988). *J. Med. Primatol.* 17, 215–221.

Fernandes, A., Bradley, D.V., Tigges, M. and Herndon, J.G. (2003). *Invest. Ophthalmol. Vis. Sci.* 44, 2373–2380.

Fitzgerald, J.T., Sena, M.J., Johnson, J.R., Griffey, S.M., Tarantal, A.F., Barry, P.A., McChesney, M.B., Vandewalker, K., Ramsamooj, R. and Perez, R.V. *Transplantation* 78, 367–374.

Foster, W.G., Stals, S.I. and McMahon, A. (1992). *J. Med. Primatol.* 21, 30–34.

Fowlkes, J.B. and Holland, C.K. (2000). *J. Ultrasound Med.* 19, 69–72.

Fowlkes, J.B., Strieter, R.M., Downing, L.J., Brown, S.L., Saluja, A., Salles-Cunha, S., Kadell, A.M., Wrobleski, S.K. and Wakefield, T.W. (1998). *Ultrasound Med. Biol.* 24, 1175–1182.

Gaschen, L. and Schuurman, H.J. (2002). *Invest. Radiol.* 37, 376–380.

Gaschen, L. and Schuurman, H.J. (2001a). *J. Med. Primatol.* 30, 88–93.

Gaschen, L. and Schuurman, H.J. (2001b). *Invest. Radiol.* 36, 335–340.

Gaschen, L. and Schuurman, H.J. (2001c). *Br. J. Radiol.* 74, 411–419.

Gaschen, L., Menninger, K. and Schuurman, H.J. (2000). *J. Med. Primatol.* 29, 76–84.

Gaschen, L., Audett, M., Menninger, K. and Schuurman, H.J. (2001a). *J. Med. Primatol.* 30, 46–55.

Gaschen, L., Kunkler, A., Menninger, K. and Schuurman, H.J. (2001b). *Vet. Radiol. Ultrasound.* 42, 259–264.

Geissmann, T. (1989). Il Sedicesimo, Firenze.

Gilbert, W.M., Eby-Wilkens, E. and Tarantal, A.F. (1997). *Obstet. and Gynecol.* 89, 462–465.

Giudice, L.C., Telles, T.L., Lobo, S. and Kao, L. (2002). *Ann. NY Acad. Sci.* 955, 252–264.

Glimcher, P.W., Ciaramitaro, V.M., Platt, M.L., Bayer, H.M., Brown, M.A. and Handel, A. (2001). *J. Neurosci. Methods* 108, 131–144.

Goetzman, B.W., Read, L.C., Plopper, C.G., Tarantal, A.F., George-Nascimento, C., Merritt, T.A., Whitsett, J.A. and Styne, D. (1993). *Ped. Res.* 35, 30–36.

Golub, M.S. and Anderson, J.H. (1986). *Lab. Anim. Sci.* 36, 507–511.

Golub, M.S., Hogrefe, C.E., Germann, S.L., Lasley, B.L., Natarajan, K. and Tarantal, A.F. (2003). *Tox. Sci.* 74, 103–113.

Golub, M.S., Tarantal, A.F., Gershwin, M.E., Keen, C.L., Hendrickx, A.G. and Lonnerdal, B. (1992). *Am. J. Clin. Nutr.* 55, 734–740.

Grauer, S.E., Xu., J., Pantely, G.A., Ge, S., Gong, Z., Shiota, T., Zhou, X. and Sahn, D.J. (1996). *Acad. Radiol.* 3(2), S405–406.

Hankins, G.D., Lowery, C.L. Jr., Scott, R.T., Morrow, W.R., Carey, K.D., Leland, M.M. and Colvin, E.V. (1990). *Am. J. Obstet. Gynecol.* 163, 728–732.

Harrison, M.R., Slotnick, N.R., Crumbleholme, T.M., Golbus, M.S., Tarantal, A.F. and Zanjani, E.D. (1989). *Lancet* 2, 1425–1428.

Hartman, C.G. (1932). *Contributions to Embryology* 134 (23), 1–85.

Hartman, C.G. (1933). *The Anatomy of the Rhesus Monkey.* William and Wilkins Co., New York.

Hatasaka, H.H., Schaffer, N.E., Chenette, P.E., Kowalski, W., Hecht, B.R., Meehan, T.P., Wentz, A.C., Valle, R.F., Chatterton, R.T. and Jeyendran, R.S. (1997). *J. Assist. Reprod. Genet.* 14, 102–110.

Hennerici, M., Burrig, K.F. and Daffertshofer, M. (1989). *Hypertension* 13, 315–321.

Hennessy, A., Gillin, A.G., Duggin, G.G., Horvath, J.S. and Tiller, D.J. (1999). *Clin. Exp. Pharmacol. Physiol.* 26, 849–852.

Herring, J.M., Fortman, J.D., Anderson, R.J. and Bennett, B.T. (1992). *Lab. Anim. Sci.* 41, 602–605.

Hill, W.C.O. (1974). *Primates Comparative Anatomy and Taxonomy. VII. Cynopithecinae.* John Wiley & Sons, New York.

Ho, M.S., Barr, B.C., Tarantal, A.F., Lai, L.T., Hendrickx, A.G., Marsh, A.E., Sverlow, K.W., Packham, A.E., and Conrad, P.A. (1997). *J. Clin. Microbiol.* 35, 1740–1745.

Jaquish, C.E., Tardif, S.D., Toal, R.L. and Carson, R.L. (1996). *J. Med. Primatol.* 25, 57–63.

Jewett, D.A. and Dukelow, W.R. (1972). *Primates* 13, 327–330.

Jimenez, D.F. and Tarantal, A.F. (2003). *J. Med. Primatol.* 32, 315–319.

Jin, B., Turner, L., Crawford, B., Birrell, A. and Handelsman, D.J. (1996). *J. Androl.* 17, 342–352.

Kamischke, A., Behre, H.M., Weinbauer, G.F. and Nieschlag, E. (1997). *J. Urol.* 157, 2340–2344.

Kamischke, A., Weinbauer, G.F., Semjonow, A., Lerchl, A., Richter, K.D. and Nieschlag, E. (1999). *J. Androl.* 20, 601–610.

King, B.F. (1993). *J. Exp. Zoo.* 266, 528–540.

Koenig, S.C., Ludwig, D.A., Reister, C., Fanton, J.W., Ewert, D. and Convertino, V.A. (2001). *Comp. Med.* 51, 513–517.

Korcarz, C.E., Padrid, P.A., Shroff, S.G., Weinert, L. and Lang, R.M. (1997). *J. Med. Primatol.* 26, 287–298.

Kuederling. I. and Heistermann, M. (1997). *J. Med. Primatol.* 26, 299–306.

Kwan, O.L., Tarantal A.F., Kessler D., Cotter B., Dalton N., Ross J. and DeMaria, A. (1994). *Proceedings of the American Society of Echocardiography.*

Landy, H.J. and Keith, L.G. (1998). *Hum. Reprod. Update* 4, 177–183.

Lane, J.H., Tarantal, A.F., Pauley, D., Marthas, M., Miller, C.J. and Lackner, A.A. (1996). *Am. J. Pathol.* 149, 1097–1104.

Lima, J.A., Szarfman, A., Lima, S.D., Adams, R.J., Russell, R.J., Cheever, A., Trischmann, T. and Weiss J.L. (1986). *Circulation* 73, 172–179.

Lowenstine, L.J. (2003). *Tox. Pathol.* 31, 92–102.

Matsell, D.G. and Tarantal A.F. (2002). *Ped. Nephrol.* 17, 470–476.

Matsell, D.G., Mok, A. and Tarantal, A.F. (2002). *Kidn. Intern.* 61, 1263–1269.

Miller, D.L., Pislaru, S.V. and Greenleaf, J.E. (2002). *Somat. Cell Mol. Genet.* 27, 115–134.

Mohle, U., Heistermann, M., Einspanier, A. and Hodges, J.K. (1999). *J. Med. Primatol.* 28, 36–47

Morgan, P.M., Hutz, R.J., Kraus, E.M. and Bavister, B.D. (1987). *Biol. Reprod.* 36, 463–469.

Mychaliska, G.B., Rice, H.E., Tarantal, A.F., Stock, P.G., Capper, J., Garovoy, M.R., Olson, J.L., Cowan, M.J. and Harrison, M.R. (1997). *J. Pediatr. Surg.* 32, 976–981.

Myers, R.E. (1972). *J. Repro. Med.* 9, 171–198.

Narita, H., Hamano, M. and Cho, F. (1988). *Jikken Dobutsu* 37, 393–397.

Netland, P.A., Siegner, S.W. and Harris A. (1997). *Invest. Ophthalmol. Vis. Sci.* 38, 2655–2661.

Nimrod, C., Simpson, N., Hafner, T., de Vermette, R., Fournier, J., Coady, L. and Baccanale, C. (1996). *J. Med. Primatol.* 25, 106–111.

Nubbemeyer, R., Heistermann, M., Oerke, A.K. and Hodges, J.K. (1997). *J. Med. Primatol.* 26, 139–146.

Nyberg, D.A., Laing, F.C., Filly, R.A., Uri-Simmons, M. and Jeffrey, R.B. Jr. (1983). *Radiol.* 146, 755–759.

Olney, J.W. (2002). *Neurotoxicol.* 23, 659–668.

Ostensen, J., Hede, R., Myreng, Y., Ege, T. and Holtz E. (1992). *Acta Physiol. Scand.* 144, 307–315.

Panigel, M., Dixon, T., Constantinidis, I., Sheppard, S., Swenson, R., McLure, H., Campbell, W.E., Huddleston, J., Polliotti, B. and Nahmias, A. (1993). *J. Med. Primatol.* 22, 393–399.

Plesker, R., Coulibaly, C. and Hetzel, U. (2002). *J. Vet. Med.* 49, 428–429.

Plopper, C.G., St. George, J.A., Read, L.C., Nishio, S.J., Weir, A.J., Edwards, L., Tarantal, A.F., Pinkerton, K.E., Merritt, T.A., Whitsett, J.A., George-Nascimento, C.G. and Styne, D. (1992). *Am. J. Physiol. Lung Cell Molec. Phys.* 262, L313–L321.

Prahalada, S., Tarantal, A.F., Harris, G.S., Ellsworth, K.P., Clarke, A., Skiles, G.L., MacKenzie, K.I., Kruk, L.F., Ablin, D.S., Cukierski, M.A., Peter, C.P., van Zwieten, M.J. and Hendrickx, A.G. (1997). *Teratology* 55, 119–131.

Ragavendra, N. and Tarantal, A.F. (2002). *Placenta* 22, 200–205.

Ramsey, E.M., Houston, M.L. and Harris, J.W.S. (1976). *Am. J. Obstet. Gynecol.* 124, 647–652.

Read, L.C., Tarantal, A.F. and George-Nascimento, C. (1989). *Acta Paed. Scand. (Suppl)* 351, 97–103.

Resuello, R.G. (1987). *Vet. Rec.* 121, 155.

Rippy, M.K., Lee, D.R., Pearson, S.L., Bernal, J.C. and Kuehl, T.J. (1996). *J. Med. Primatol.* 25, 346–355.

Santolaya-Forgas, J., Vengalil, S., Meyer, W. and Fortman J. (1997). *Am. J. Primatol.* 43, 323–328.

Schlatt, S., Rosiepen, G., Weinbauer, G.F., Rolf, C., Brook, P.F. and Nieschlag, E. (1999). *Hum. Reprod.* 14, 144–150.

Schmiedl, U.P., Komarniski, K., Winter, T.C., Luna, J.A., Cyr, D.R., Ruppentha, G. and Schlief R. (1998). *J. Ultrasound. Med.* 17, 75–80.

Seier, J.V., van der Horst, G., de Kock M. and Chwalisz, K. (2000). *J. Med. Primatol.* 29, 70–75.

Shimizu, K. (1988). *J. Med. Primatol.* 17, 247–256.

Simpson, N.A., Nimrod, C., De Vermette, R. and Fournier, J. (1997). *Placenta* 18, 287–293.

Simpson, N.A., Nimrod, C., De Vermette, R., Leblanc, C. and Fournier, J. (1998). *Ultrasound Obstet. Gynecol.* 11, 204–208.

Smith, E.L. 3rd, Hung, L.F., Kee, C.S., Qiao-Grider, Y. and Ramamirtham, R. (2003). *Optom. Vis. Sci.* 80, 374–382.

Swindle, M.M., Blum, J.R., Lima, S.D., and Weiss, J.L. (1985). *Circulation* 71, 146–153.

Tarantal, A.F. (1990). *J. Med. Primatol.* 19, 47–58.

Tarantal, A.F. (1992). *J. Med. Primatol.* 21, 308–315.

Tarantal, A.F. (1993). *Am. J. Primatol.* 29, 209–219.

Tarantal, A.F. (1998). *Progress in Obstetrics and Gynecologic Sonography Series, Safety of Diagnostic Ultrasound.* Chapter 5, Parthenon Publishing, pp 39–51.

Tarantal, A.F. and Canfield, D.R. (1994) *Ultrasound Med. Biol.* 20, 65–72.

Tarantal, A.F., Castillo, A., Ekert, J.E., Bischofberger, N. and Martin, R.B. (2002). *J AIDS* 29, 207–220.

Tarantal, A.F., Chu, F., O'Brien, W.D. Jr. and Hendrickx, A.G. (1993a). *J. Ultrasound Med.* 5, 285–295.

Tarantal, A.F. and Cowan. M.J. (1999). *Cytokine* 11, 290–300.

Tarantal, A.F. and Gargosky, S.E. (1995). *Growth Regulation* 5, 190–198.

Tarantal, A.F., Gargosky, S.E., O'Brien, W.D. Jr. and Hendrickx, A.G. (1995a). *Ultrasound Med. Biol.* 21, 1073–1081.

Tarantal, A.F., Goldstein, O., Barley, F. and Cowan, M.J. (2000). *Transplantation* 69, 1818–1823.

Tarantal, A.F., Han, V.K.M., Cochrum, K.C., Mok, A., daSilva, M. and Matsell, D.G. (2001a). *Kidn. Intern.* 59, 446–456.

Tarantal, A.F. and Hendrickx, A.G. (1987). *J. Med. Primatol.* 16, 291–299.

Tarantal, A.F. and Hendrickx, A.G. (1988a). *J. Med. Primatol.* 17, 105–112.

Tarantal, A.F. and Hendrickx, A.G. (1988b). *Am. J. Primatol.* 15, 309–323.

Tarantal, A.F. and Hendrickx, A.G. (1988c). *Anat. Rec.* 222, 177–184.

Tarantal, A.F. and Hendrickx, A.G. (1988d). *Non-human Primates – Developmental Biology and Toxicology.* Ueberreuter Wissenschaft, Berlin, pp 91–99.

Tarantal, A.F. and Hendrickx, A.G. (1989a). *Am. J. Primatol.* 18, 166A.

Tarantal, A.F. and Hendrickx, A.G. (1989b). *Teratology* 39, 137–147.

Tarantal, A.F. and Hendrickx, A.G. (1989c). *Teratology* 39, 149–162.

Tarantal, A.F., Hendrickx, A.G., Matlin, S.A., Lasley, B.L., Gu, Q.–Q., Thomas, C.A.A., Vince, P.M. and Van Look, P.F.A. (1993b). *Contraception* 47, 307–316.

Tarantal, A.F., Hendrickx, A.G. and O'Rahilly, R. (1990a). *J. Ultrasound Med.* 9, 614–615.

Tarantal, A.F., Hunter, M.K. and Gargosky, S.E. (1997a). *Endocrinology* 138, 3349–3358.

Tarantal, A.F., Laughlin, L.S., Dieter, J., Tieu, J., Hendrickx, A.G., Overstreet, J.W. and Lasley, B.L. (1997b). *Early Preg: Biol. and Med.* 4, 281–290.

Tarantal, A.F., Lee, C.I., Ekert, J.E., McDonald, R., Kohn, D.B., Plopper, C.G., Case, S.S. and Bunnell, B.A. (2001b). *Molec. Ther.* 4, 614–621.

Tarantal, A.F., Marthas, M.L., Gargosky, S.E., Otsyula, M., McChesney, M.B., Miller, C.J. and Hendrickx, A.G. (1995b). *J. AIDS Hum. Retrovirol.* 10, 129–138.

Tarantal, A.F., Marthas, M.L., McChesney, M.B., Miller, C.J., Lerche, N.W., Pedersen, N.C. and Hendrickx, A.G. (1993c). *Ped. AIDS and HIV Inf: Fetus to Adol.* 4, 373–380.

Tarantal, A.F, Marthas, M.L., Shaw, J.–P., Cundy, K. and Bischofberger, N. (1999a). *J. AIDS Hum. Retrovirol.* 20, 323–333.

Tarantal, A.F., O'Brien, W.D., Jr. and Hendrickx, A.G. (1993d). *Teratology* 47, 159–170.

Tarantal, A.F., O'Rourke, J.P., Case, S.S., Newbound, G.C., Li, J., Lee, C.I., Kohn D.B. and Bunnell B.A. (2001c). *Molec. Ther.* 3, 128–138.

Tarantal, A.F., Otiang-a-Owiti, G. and Hendrickx, A.G. (1994a). *J. Med. Primatol.* 23, 319–324.

Tarantal, A.F., Salamat, M.S., Britt, W.J., Luciw, P., Hendrickx, A.G. and Barry, P.A. (1999b). *J. Inf. Dis.* 177, 446–450.

Tarantal, A.F., Spanggord, R.J. and Hendrickx, A.G. (1994b). *Ped. AIDS HIV Inf: Fetus to Adol.* 5, 10–19.

Tarantal, A.F., VandeVoort, C.A. and Overstreet, J.W. (1990b). *J. Med. Primatol.* 19, 447–453.

Tardif, S.D., Jaquish, C.E., Toal, R.L., Layne, D.G. and Power, R.A. (1998). *J. Med. Primatol.* 27, 28–32.

Tokuno, H., Hatanaka, N., Takada, M. and Nambu A. (2000). *Neurosci. Res.* 36, 335–338.

Vaitkevicius, P.V., Lane, M., Spurgeon, H., Ingram, D.K., Roth, G.S., Egan, J.J., Vasan, S., Wagle, D.R., Ultrich, P., Brines, M., Wuerth, J.P., Cerami, A., and Lakatta, E.G. (2001). *Proc. Natl. Acad. Sci. USA* 98, 1171–1175.

VandeVoort, C.A. and Tarantal, A.F. (1991). *J. Med. Primatol.* 20, 110–116.

VandeVoort, C.A. and Tarantal, A.F. (2001). *J. Med. Primatol.* 30, 304–307.

VandeVoort, C.A., Tollner, T.L., Tarantal, A.F. and Overstreet, J.W. (1989). *Gamete Res.* 24, 1–5.

Vilupuru, A.S., and Glasser, A. (2003). *Optom. Vis. Sci.* 80, 383–394.

Williams, J.K., Anthony, M.S., and Clarkson, T.B. (1991). *Arch. Pathol. Lab.* 115, 784–790.

Windle, C.P., Baker, H.F., Ridley, R.M., Oerke, A.K., and Martin R.D. (1999). *J. Med. Primatol.* 28, 73–83.

Wondergem, J., Persons, K., Zurcher, C., Frolich, M., Leer, J.W., and Broerse, J. (1999). *Radiother. Oncol.* 53, 67–75.

Yeager, C.H., O'Grady, J.P., Esra, G., Thomas, W., Kramer, L., and Gardner, H. (1981). *J. Am. Vet. Med. Assoc.* 179, 1309–1310.

CHAPTER 21

Functional Magnetic Resonance Imaging in Conscious Marmoset Monkeys: Methods and Applications in Neuroscience Research

Craig F. Ferris[1] and Charles T. Snowdon[2]
[1]Center for Comparative Neuroimaging, University of Massachusetts Medical School, Worcester, Massachusetts, USA
[2]Department of Psychology, University of Wisconsin, Madison, Wisconsin, USA

Introduction

Understanding how the brain functions is an important part of experimental research with nonhuman primates. However, nonhuman primates present difficulties for experimental research that are not found with other taxa such as rodents. Primates have long lives and long generation times relative to rodents. As a result of this, the supply of nonhuman primates in research is greatly restricted. This means that researchers must use relatively small numbers of primates and, ideally, use primates in multiple studies. Furthermore, all scientists working with nonhuman primates must be concerned with the physical health and psychological well-being of the nonhuman primates. Psychological well-being is required by law for primates but not for rodents. Primates are long lived, hard to acquire, expensive to maintain, etc.

Additionally, because primate behavior and the primate brain are more complex and more human-like than

The Laboratory Primate

ISBN 0-1208-0261-9

other animal models, a basic understanding of the primate nervous system is critical in biomedical research using nonhuman primates. An important question to consider is how can researchers balance the value of research with nonhuman primates with the limited supply and the ethical issues associated with primate research?

Several methods used in neuroscience research are shown in Figure 21.1. The methods are plotted in terms of time-scale on the x-axis, relative size on the y-axis and degree of invasiveness, indicated by the gray scale of each block. Two methods: magnetic encephalography (MEG) and event-related potential (ERP), involving scalp electrode recording, MEG+ERP and functional magnetic resonance imaging (fMRI), appear as non-invasive methods. While fMRI occupies a niche that overlaps somewhat, in temporal and spatial resolution, with Positron Emission Tomography (PET), although it does not involve the use of radioactive materials (Neil, 1993), it also overlaps with MEG+ERP methods. However, fMRI covers a broader range of both spatial resolution and time course than either PET or MEG+ ERP but does not provide the long-lasting effects of lesion methods nor the spatial resolution of single unit recording and patch clamp methods (Cohen and Brookheimer, 1994). Nonetheless fMRI methods can be used effectively to understand the organization of functional systems within the brain, in response to specific stimuli.

A further virtue of fMRI is that it provides a view of neuronal activity throughout the entire brain. So rather than focusing on one or two areas at a time, as is done with lesion or single unit recording methods, a researcher can examine activity of the entire brain in response to a set of stimuli. Furthermore, the procedure is not terminal. The subject does not need to be sacrificed, when the research is over, to locate the electrode site or determine the degree of the lesion. It also allows the researcher to make use of the longevity of nonhuman primates by doing repeated studies on the same individuals. Longitudinal and developmental studies of brain function become possible with fMRI.

These virtues of fMRI have made it a method of choice in studies of human brain function with an explosion of increasingly detailed and sophisticated studies appearing over the last decade. What then is the need for doing fMRI research with nonhuman primates when so much is being done with human primates? In our opinion, there are four important answers to this question. First, with nonhuman primates it is possible to have control over the animal's entire development. A researcher can know precisely what previous experience a monkey has had, to what environmental conditions it was exposed and what its social history was like. A much greater knowledge of, and control over, previous experience reduces noise in data, increases the precision and improves the interpretation of results. Second, certain procedures of interest, such as studying brain changes during conditioning to either positive or negative emotional stimuli, can be carried out with nonhuman primates that might not be possible with humans. Third, fMRI studies with humans can be greatly

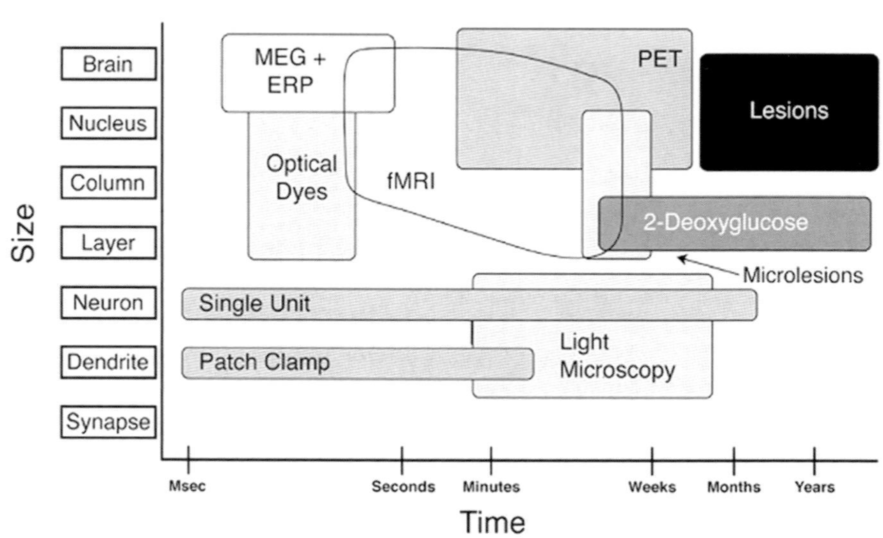

Figure 21.1 Schematic diagram of methods used to understand brain function. The x-axis plots increasing time, the y-axis increasing amount of brain tissue and shading indicates increasing degree of invasiveness. (PET = positron emission tomography, EGG + ERP = electroencephalography plus evoked potential). Adapted from Cohen and Brookheimer (1994).

enhanced by previous research with nonhuman species which could suggest which brain areas are of greatest interest for studying a particular problem. Indeed, fMRI studies of nonhuman primates can provide a guide in determining the relationship between results obtained with invasive methods, such as lesions or single unit recording methods, and brain activation as measured by fMRI, and these could provide validation for the results from humans. Finally, nonhuman primates vary greatly in social structure, mating systems and cognitive ability. Comparative research with carefully selected species, therefore, can help determine which animal models are the most useful in understanding various components of human behavior and brain function.

In the remainder of this chapter we shall give a basic explanation of how fMRI works, raise several cautionary points about using fMRI with awake, conscious primates and, using our own research with common marmosets, illustrate some of the exciting insights that this method provides about the relationship of brain and behavior.

What is fMRI and how does it work?

fMRI indirectly detects neural activity in different parts of the brain by comparing contrast in the magnetic resonance (MR) signal intensity prior to and following, stimulation. What is responsible for the change in signal intensity? Areas of the brain with increased synaptic and neuronal activity require increased levels of oxygen to sustain this activity. Enhanced neuronal activity is accompanied by an increase in metabolism, concomitant with changes in cerebral blood flow and volume to the area of activation (Sokolloff *et al.,* 1977; Fox *et al.,* 1986; Belliveau *et al.,* 1990; Ramsey *et al.,* 1996). As a result, there is a reactive hyperemia or increase in capillary blood flow in brain regions that are active (Ogawa *et al.,* 1990; Ogawa *et al.,* 1992; Logothetis *et al.,* 1999). The enhanced blood flow usually exceeds the metabolic demand, exposing the active brain area to blood high in oxygenated hemoglobin. Hemoglobin, highly saturated with oxygen, increases the MR signal intensity. Conversely, hemoglobin low in oxygen diminishes the MR signal. This relationship between metabolism, hemoglobin saturation, blood flow and MR signal intensity is shown in Figure 21.2 and is called the **blood oxygen level dependent** technique for functional imaging (Ogawa *et al.,* 1990).

How does oxygenated hemoglobin affect the signal? Mobile protons, associated with hydrogen atoms in water and fat, are the primary source of MR signal. With its single charged proton, the hydrogen nucleus spins, creating a surrounding electromagnetic field with the characteristics of a magnetic dipole (think of a tiny bar magnet). When placed in an external magnetic field, quantum mechanics tells us this hydrogen nucleus can have two "spin" states or energy levels. Most of the nuclei prefer the lower energy state rather than the higher. When electromagnetic radiation in the radiofrequency

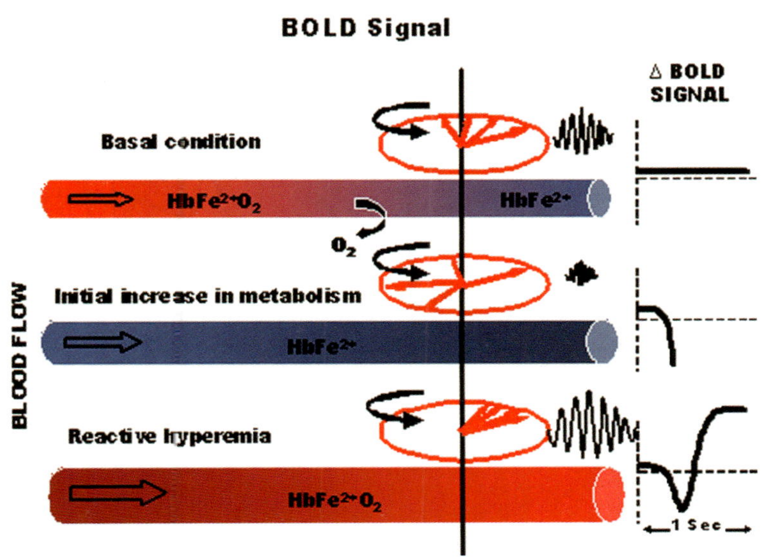

Figure 21.2 Schematic diagram showing conditions contributing to blood oxygen level dependent signal changes.

range (MHz) is applied, nuclei can absorb energy and be elevated to the higher energy state. The energy given off, as these nuclei "relax" back into their lower state, is the source of the MR signal.

A more intuitive, but less accurate, description of nuclear magnetic resonance is provided by classical mechanics. Think of each hydrogen nucleus as a spinning top with a random orientation in a nonmagnetic environment as shown in Figure 21.3. When placed into a magnetic field, the hydrogen nuclei align in parallel with the field lines of the magnet. The net magnetization, from all of these aligned nuclei, is shown as a vector parallel to main external magnetic field B_0. These aligned nuclei are the source of potential energy for the MR signal. Stronger magnetic fields recruit more hydrogen protons for imaging. The difference in hydrogen proton density between tissues like water and fat are one reason for signal contrast in neuroanatomical imaging.

The spinning hydrogen nuclei "wobble" or show an angular precessional frequency, when placed in the main magnetic field, as depicted by the red circles in Figure 21.3. This precessional frequency is directly related to field strength of the magnet. Animal magnets, with field strengths of 4.7, 7.0 and 9.4 Tesla, create precessional frequencies for hydrogen nuclei of 200, 300 and 400 MHz, respectively. Brief radiofrequency (RF) pulses can be used to create a magnetic field, B_1, perpendicular to the main field, B_0, as shown in Figure 21.4. If these RF pulses have the same frequency as the precessing hydrogen nuclei, the resonant energy is absorbed, flipping the nuclei into the transverse x-y plane where they continue to precess. When initially flipped, all of the hydrogen nuclei precess at the same frequency and are said to be in-phase. This "in-phase" precession in the transverse plane sets up an oscillating MR signal that can be picked up by a special antenna called an RF receiver or probe. However, this oscillating signal rapidly decays, lasting only 20–50 msec. Interactions between these nuclei and inhomogeneties, in their surrounding microenvironment, cause some to precess slower than others resulting in "de-phasing."

So how does the technique work to enhance the MR signal? Deoxygenated hemoglobin is paramagnetic and creates its own micromagnetic field, promoting de-phasing and reducing MR signal (see Figure 21.2). Oxygenated hemoglobin, on the other hand, has very little magnetic susceptibility. Active brain areas with enhanced blood flow are high in oxygenated hemoglobin, promoting phase coherence and a stronger MR signal.

Following an RF pulse, hydrogen nuclei undergo two simultaneous events. They precess in the transverse plane, generating MR signal for as long as they remain in-phase, and they relax back into the orientation of the main magnetic field, returning to their lowest energy state. Both time courses are exponential functions. The time constant for relaxation is called T_1 while the time constant for de-phasing is called T_2. T_1 is measured in seconds while T_2 is measured in milliseconds. A third time constant called T_2* is shorter than T_2 and reflects de-phasing, caused by conditions in the microenvironment

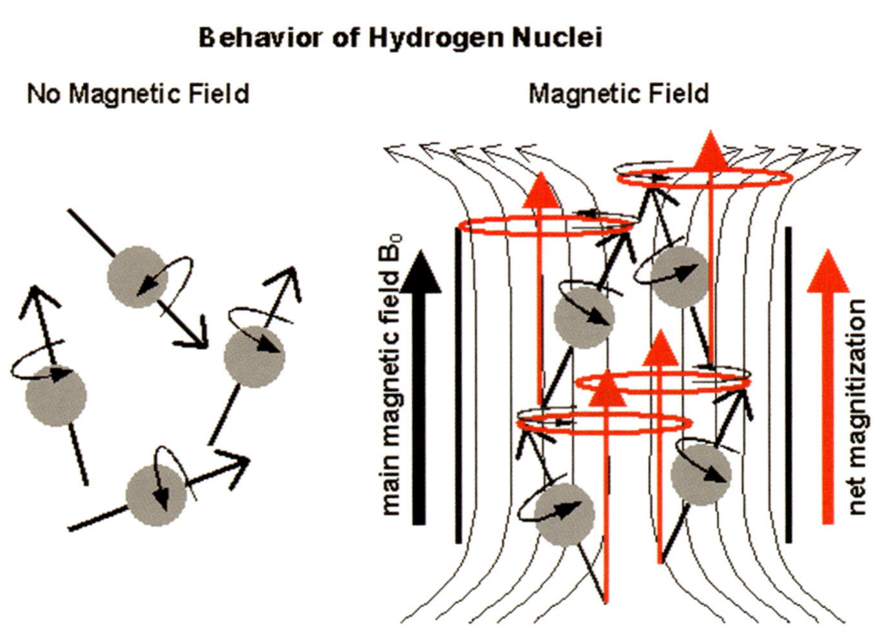

Figure 21.3 Schematic diagram showing the behavior of hydrogen nuclei in magnetic and nonmagnetic environments.

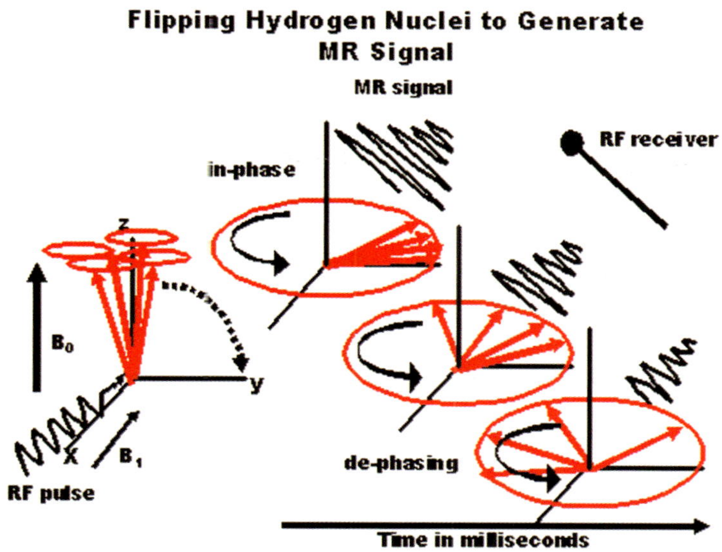

Figure 21.4 Schematic diagram showing behavior of hydrogen nuclei flipped into the transverse x-y plane with a radiofrequency (RF) pulse. B_0 is the main magnetic field and B_1 is the magnetic field created in the transverse plane by the RF pulse.

contributing to magnetic susceptibility. Changes in T_2* are the bases behind signal changes in functional imaging. It is important to note, as long as there is net magnetization in the transverse plane, there is potential for generating a series of MR signals. Is there a way to bring the hydrogen nuclei back into phase or have them "echo" repeatedly until the system comes back to equilibrium? Yes, spin echo and gradient echo pulse sequences have been developed to do just this.

In a spin echo pulse sequence, the initial 90° pulse flips the hydrogen nuclei into the transverse plane as shown in Figure 21.5. Afterward, a refocusing 180° pulse is applied, reversing the direction of the precessional spins. The faster spinning hydrogen nuclei now have been turned around and are facing the slower spinning nuclei. The MR signal waxes and wanes as the faster nuclei catch up and pass the slower nuclei. The 180° refocusing pulse can be applied several times while there is transverse magnetization. Usually after 1.5 to 2.5 seconds, the sequence is repeated (time to repeat or T_R) and another 90° pulse is applied to flip the hydrogen nuclei into the transverse plane and the echo series

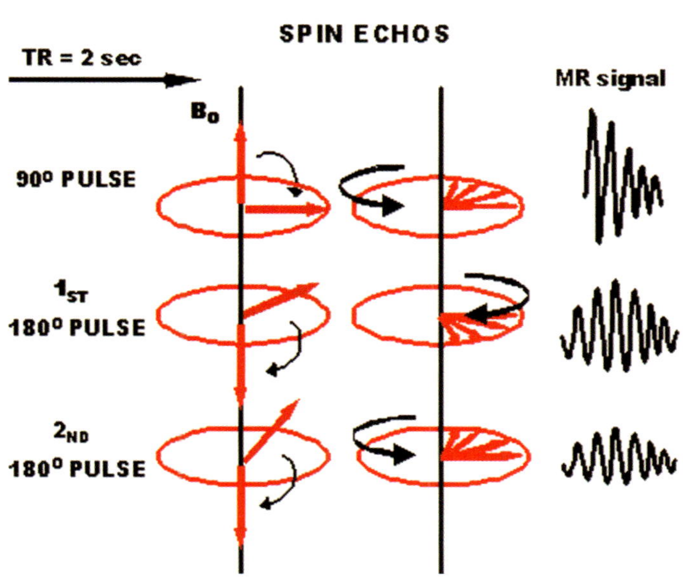

Figure 21.5 Schematic diagram showing the behavior of precessing hydrogen nuclei in a spin echo pulse sequence. Net magnetization (red arrows in the main magnetic field B_0) slowly relaxes back to equilibrium over successive 180° pulses following the initial 90° pulse. TR is the repetition time between 90° pulses.

begins again. One of the major advantages of fast spin echo pulse technique, particularly in imaging fully conscious animals, is its tolerance to artifacts from physiologic motion (e.g. cerebrospinal fluid movement) or magnetic susceptibility (e.g. distorted signal at air/liquid interfaces) (see Figure 21.11). However, magnetic susceptibility caused by deoxygenated hemoglobin is a key component in the signal. Hence, the fast spin echo technique is less sensitive to changes in signal than the gradient echo technique. This problem of sensitivity can be addressed with higher field strengths and better RF electronics, as will be discussed later.

In a gradient echo pulse sequence, the initial RF pulse only partially flips the hydrogen nuclei into the transverse plan (Figure 21.6). Since these nuclei are only partially flipped, they rapidly lose their transverse magnetization as they return to equilibrium in the main magnetic field. Hence a series of short T_Rs, usually between 50 and 75 msec, are necessary in order to collect multiple MR signals. Indeed, the purpose of the gradient echo technique is to increase the speed of the scan. The rephasing event required for echo generation is accomplished by reversing the local magnetic field with special gradient coils. Gradient echo is particularly sensitive to magnetic susceptibility and is commonly used in functional imaging. However, this major advantage is also its major disadvantage because of the susceptibility artifacts, most noticeably at air/tissue interfaces associated with sinuses. Gradient echo is also very sensitive to shimming (improving the field homogeneity) making it difficult to collect undistorted slices across the entire brain.

Magnetic resonance imaging requires considerable hardware. During an experiment, the participant is actually within five concentric rings of magnets. First, there is the basic magnet or spectrometer for generating the main magnetic field. Animal magnets can range from 2.0 to 11.7 Tesla. The greater the field strength, the more energy is used and the greater the heat produced. Most machines are supercooled to about 4° Kelvin through a jacket of circulating liquid helium surrounding the magnet. MRI creates large and strong magnetic fields and needs to be carefully shielded. People with pacemakers, or carrying metallic implants, should stay far away from an MRI.

Real living bodies placed within a magnet create nonhomogeneous magnetic fields and, to correct for these, the second ring of magnets serves as tunable shims to adjust to the most homogeneous field possible. A system that only uses the plane of magnetization of the MRI machine would not be able to detect in three dimensions. Therefore the third ring of a MRI machine is a set of gradient coils that can be used to detect changes in all three dimensions (x, y and z coordinates). In order to produce the RF pulses, that temporarily depolarize the protons, an RF transmitter coil is needed. Often this can be located directly above the organ of interest. Thus, if one is interested in the brain,

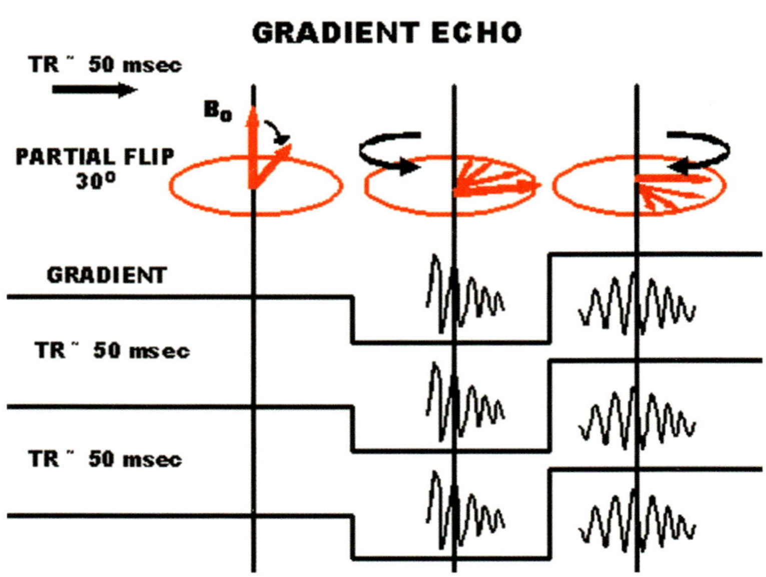

Figure 21.6 Schematic diagram showing the behavior of precessing hydrogen nuclei during successive gradient echo pulse sequences. Net magnetization (red arrows) in the main magnetic field (B_0) is partially flipped into the transverse plane at an angle of 30°. During the short-lived event the gradient coils are used to reverse the magnetic field (black horizontal line) to refocus the precessing nuclei.

the RF coil need only be large enough to cover the extent of the brain, not the entire body (see Figure 21.7). The RF transmitter is the 4th ring of the MRI machine. Finally, there is an RF receiving coil to detect changes in magnetic activity after a pulse.

All of this equipment is controlled via a workstation that can control the operation of each of the parts of the fMRI, including shimming, controlling pulse structure and pulse sequences and simultaneously recording the data from the RF receiver coil. In a typical study, the subject is placed at the ideal location in the magnetic field for the area being imaged. Shimming is carried out to minimize magnetic non-homogeneities. After this there is usually a period of anatomical data collection so that subsequent functional data can be localized with accuracy. Now we are ready to consider imaging nonhuman primates.

Problems associated with fMRI in nonhuman animals

Fully conscious animals have not been used routinely in fMRI studies because of technical problems associated with motion artifacts. Any minor head movement distorts the image and may also create a change in signal intensity that can be mistaken for stimulus-associated changes in brain activity (Hajnal *et al.*, 1994). In addition to head movement, motion outside the field of view caused by respiration, swallowing and muscle contractions in the face and neck are other major sources of motion artifacts (Yetkin *et al.*, 1996; Birn *et al.*, 1998). It is for these reasons that most fMRI studies in animals have used general anesthetics to immobilize the animal. However, anesthetics preclude the study of brain activity involving cognition and emotion. Furthermore, anesthetics depress neuronal activity reducing the signal (Lahti *et al.*, 1999; Brevard *et al.*, 2003). In addition, a monkey that has been immobilized to eliminate motion artifacts, must also be sufficiently relaxed to attend to the stimuli of interest and to give neural responses that are relatively uncontaminated by stress. Monkey brains are much smaller than human brains and therefore have a proportionally greater amount of the brain in contact with air and fluids, creating potential susceptibility artifacts. Finally, fMRI reflects changes in blood oxygen levels relative to some baseline, so the selection of appropriate stimuli, including baseline control stimuli, is critical.

These problems can be circumvented by addressing the following issues:

1. Selection of a species that is not only interesting for studying cognitive and emotional responses but is also one that is easy to work with and readily habituated for fMRI work.

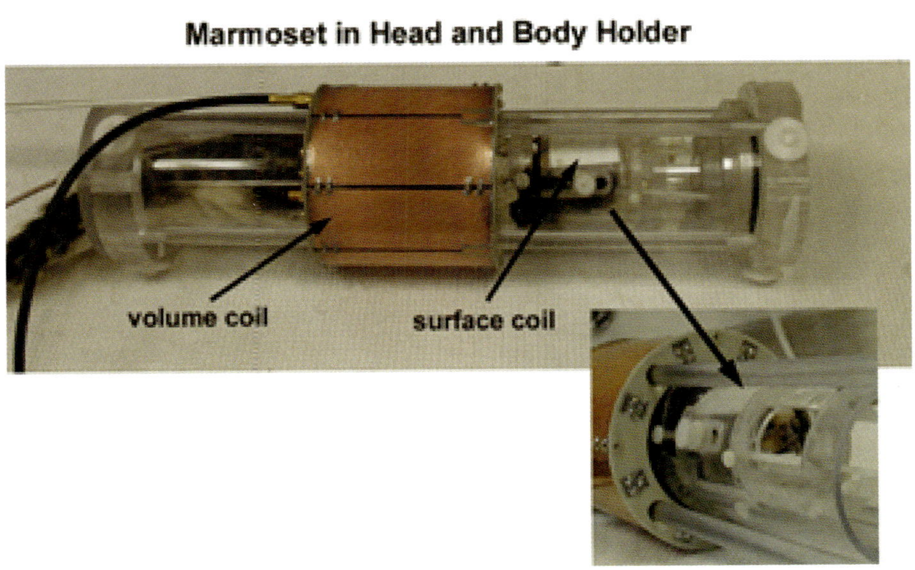

Figure 21.7 Composite photograph showing positioning of a common marmoset in a dual coil restrainer. Once secured the volume coil is pushed along the guide rails and positioned over the surface coil. The volume coil is used to send RF pulses while the surface coil is used to receive RF signal.

2. Development of a system for immobilizing awake, conscious monkeys and assessing that motion artifacts do not occur.
3. Discovering a method of anesthesia that allows a monkey to be anesthetized long enough to be placed in the immobilization apparatus, but to recover full brain function by the time of imaging.
4. Developing the ideal set of pulse sequences to provide good spatial and temporal resolution while minimizing susceptibility artifacts that can occur at boundaries between brain tissue and air or cerebrospinal fluid.
5. Selecting stimuli that allow investigators to communicate with a monkey while it is being imaged in order to get clear answers to research questions.

We describe below the solutions we have found to each of these five issues.

Selection of species and habituation

We have chosen to work with common marmosets (*Callithrix jacchus*) for several reasons. These are small (<500 g) monkeys that breed well in captivity and are quite easy to handle, especially when compared with macaques and baboons. They exhibit high levels of curiosity and exploration of novel objects (Day *et al.*, 2003) compared with related species. They habituate readily to a variety of laboratory procedures.

Marmosets are cooperative breeders with an affiliative family structure similar to humans, i.e. mother, father and related siblings usually numbering 6–8 individuals per family. Marmosets are socially monogamous and breeding adults form close pair bonds, although animals will solicit and copulate with other animals when the mate is not present. Marmosets have a short developmental time course of approximately 20 months from birth to reproductive maturity. Older siblings can leave the family after 18–20 months if suitable mates are found. Managers of captive colonies can create new breeding groups at 20 months. As a result of this relatively short generation time, researchers can study the behavior of individuals across several generations. Maintaining and breeding these monkeys in small family groups is quite cost-effective.

Marmosets routinely give birth to dizygotic twins, allowing experimental designs where one twin can serve as the control for the other. Recent studies (reviewed in Snowdon, 2001) have demonstrated high levels of cognitive ability in marmosets and related species, with some skills in social cognition equaling only those of great apes. Due to their small size, they easily fit into the small bore of ultra-high field MRI spectrometers making it possible to study developmental changes in their brain anatomy, chemistry and function over the course of their life.

Dual coil RF electronics and restrainer for marmosets

The holding device developed for imaging an awake marmoset is shown in Figure 21.7 (Insight Neuroimaging Systems, Worcester Massachusetts, USA). Under a short acting anesthetic the marmoset is placed into a cloth body suit that prevents the monkey from using its limbs which might disrupt the imaging session. Once in the body suit, the monkey is placed into a head restrainer with a built-in saddle-shaped surface coil. A plastic semicircular headpiece, with blunted ear supports that fit into the ear canals, is positioned over the ears. The head is placed into the cylindrical head holder with the animal's canines secured over a bite bar and ears positioned inside the head holder with adjustable screws fitted into lateral sleeves. The head holder is secured to a center post at the front of the chassis. The body of the animal is placed into the body restrainer. The body restrainer "floats" down the center of the chassis, connecting at the front and rear-end plates and buffered by rubber gaskets. This novel design isolates all of the body movement from the head restrainer and minimizes motion artifact. The body restrainer is designed to allow for unrestricted respiration with minimal movement. Once the animal is positioned in the body holder, the volume coil is placed over the head restrainer and locked into position. With this dual coil design the volume coil is used to send RF pulses for enhanced field homogeneity and linearity while the surface coil receives RF signals for increased signal-to-noise ratio. This procedure is completely noninvasive, requires no surgery (e.g. skull implants to immobilize the head) and takes approximately 10 min to set up. The holder is designed to fit into 20 G/cm gradient set with an internal diameter of 12 cm.

Acclimating marmoset monkeys to the imaging protocol

Prior to any functional imaging of the monkeys, they undergo a three to five day habituation period where they are anesthetized, placed in the body suit and head restraint and held for several minutes after recovery from anesthesia so that they become habituated to the procedure. Immediately after a habituation or an imaging trial, the monkey is removed from the head coil and body restraint within 3 min, is fed a highly preferred food and immediately placed with its cage mate.

Marmoset males ($n = 5$) were habituated to the imaging restraint over three consecutive days in a "simulated" imaging session involving anesthesia, anesthesia recovery and restraint for a time period of 90–120 min. Shown in Figure 21.8 are plasma cortisol data for baseline and restraint on the third day of habituation trials. These data are compared to plasma cortisol levels from normal, male marmosets from a stable peer group (midday baseline) and on the third day of the psychosocial stress of an unstable peer group formation (Saltzman *et al.,* 1994). Levels of cortisol can go as high as 400–500 μg/dl when animals are exposed to the psychosocial stress of unstable peer groups (Johnson *et al.,* 1996). These comparative measures of stress-induced cortisol show that the stress from the imaging procedure is in the range of normal baselines variation and 2–3 fold less than levels seen with psychosocial stress. Behavioral notes during the habituation process indicated that animals did adapt well. On the initial day of training, marmosets would sometimes vocalize, keep eyes closed tightly and move their tails. By the third day, these same animals would remain quiet with eyes open, looking out at their surroundings.

Anesthesia and recovery prior to imaging

Since animals must be lightly anesthetized to be secured in the restraining device, there is concern that animals are not fully conscious during the functional imaging study. To address this problem we chose the anesthetic Domitor that can be rapidly reversed with Antiseden. Once the marmoset is in its body suit and its head has been fully secured in the head holder, we inject the monkey with a dose of Antisedan to reverse the anesthesia. An example of typical physiological data, confirming the reversal from anesthesia in the magnet prior to imaging, is shown in Figure 21.9. However, it should be noted that the physiological measures of anesthesia reversal, such as EEG, EKG, and respiration, shown in Figure 21.9 do not measure the perceptual and cognitive level of alertness in the animal. In other words, are the animals still "dopey" from the anesthetic? To address this question we investigate all marmosets that are to be used in the study, for their recovery from anesthesia. Animals are anesthetized and awakened with the Domitor/Antiseden regimen and timed for how long it takes to demonstrate normal perceptual and motor function. For example, they are expected to perform normal motor activity attending to, tracking and reaching for moving food rewards.

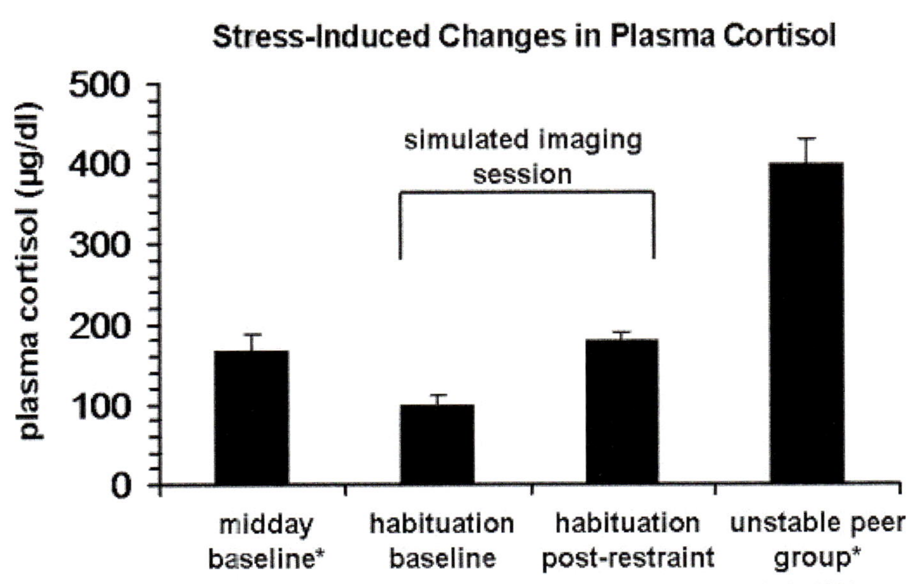

Figure 21.8 Shown are the change in plasma cortisol levels in marmoset monkeys prior to and following three days of habituation in a simulated imaging environment. These data are compared to those of Saltzman *et al.*, 1994 looking at the level of stress associated with social change in marmosets.

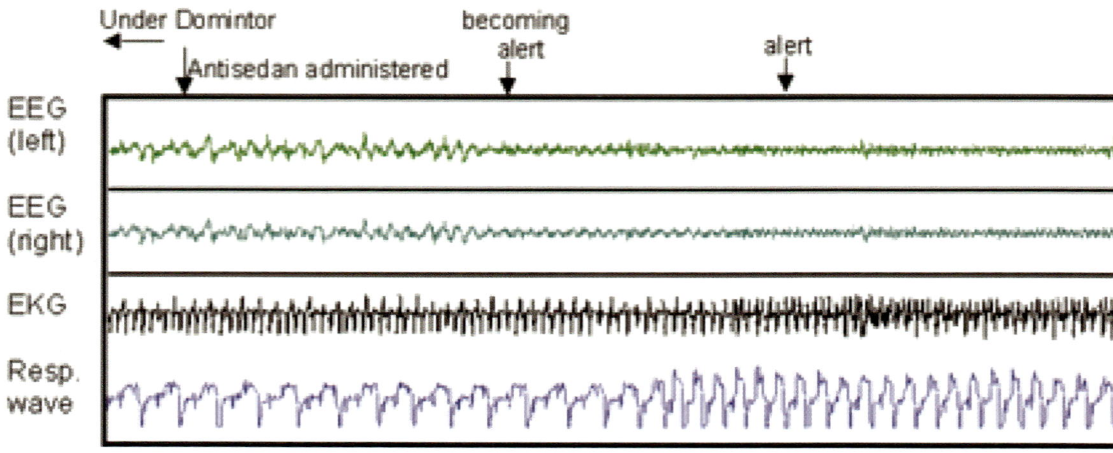

Figure 21.9 Shown are data from EEG, EKG, and respiratory waveforms of an animal coming out of anesthesia while in the restrainer and in the magnet. The animal was initially anesthetized with Domintor (0.02 mg/kg, IM) and put into the restrainer and in the magnet. The anesthesia was reversed by administering Antisedan (0.05 mg/kg, IM) and the animal was aroused within ~ 2 mins as assessed by the different physiological parameters. The slow and large amplitude EEG waves are characteristic of an animal being under anesthetics, whereas the fast and low amplitude EEG waves are characteristic of a conscious animal.

Using these criteria, we found considerable individual variation, with animals ranging in recovery time from 12–40 min after administering the drug that reverses the anesthesia. Hence, for each monkey we use in the magnet, we first obtain a precise history of their response latency to full recovery from Domitor/Antiseden. During a session, we delay acquisition of the first functional images until after the recovery time shown in the studies performed for each animal previously.

Assessing motion artifact

Each of the steps taken so far has been done with the intent of minimizing movement while assuring that animals are awake and fully conscious and not overly stressed by the procedure. However, assessment of motion artifact is still necessary for every imaging session with every marmoset. Any trials where there is any noticeable motion artifact must be discarded. Shown in Figure 21.10 are anatomical images, from an awake marmoset, collected at various times over a single imaging session that lasted just over two hours. Voxel-by-voxel subtractions were performed between the images taken at each of the three time points. The resulting images show no "ghosts" or edge effects indicative of movement. These images were taken on a 9.4T scanner with a bird-cage coil. The absence of any movement in these images assures that, even at very high magnet strength, high quality repeatable images can be obtained.

Minimizing physiological noise and susceptibility artifact

Spin echo vs Gradient echo

There are several methodological considerations when imaging fully conscious animals. Given the exceptional signal-to-noise ratio with the dual coil electronics, shown in Figure 21.7, we have examined the feasibility of using the fast spin echo technique for imaging. Fast spin echo is much less sensitive than the conventional gradient echo used in functional imaging. However, it has advantages, as noted earlier, that can compensate for its relative lack of sensitivity.

Figure 21.11 shows data collected in awake rats during separate imaging sessions, one using a fast spin echo sequence (30 mm Field of View, 64 × 64 data matrix, 1.0 mm slice, 56 ms TEeff, 2.48 s TR, NEX =1) and the other using a gradient echo (15 ms TEeff, 240 ms TR 30° pulse angle, NEX = 1). Sixty data sets, of 18 slices each, were collected at 10 sec intervals over a 10 min imaging session. Visual inspection of all 60 data sets revealed little or no distortion in any image in the spin echo sequence. However, in several of the gradient echo data sets, there was an ostensible "wave pattern" of field inhomogeneity. This artifact was most likely caused by disturbances in magnetic field homogeneity due to respiration, swallowing and muscle contractions in the face and neck (Yetkin *et al.*, 1996; Birn *et al.*, 1998).

For the same reason that spin echo sequences are less sensitive to this type of motion artifact, they are also

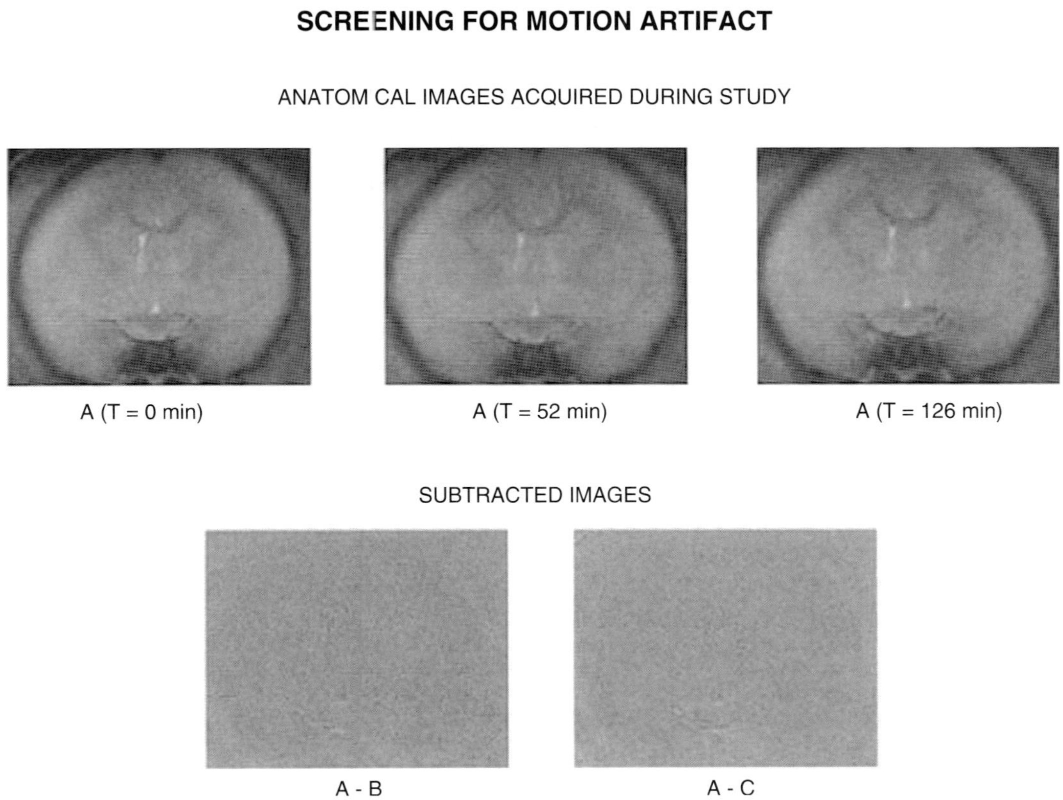

Figure 21.10 Shown are anatomical images collected over time from the same brain slice. Voxel by voxel subtraction of these images show no image evidence there was no movement of the head during the imaging study (Ferris *et al.*, 2001).

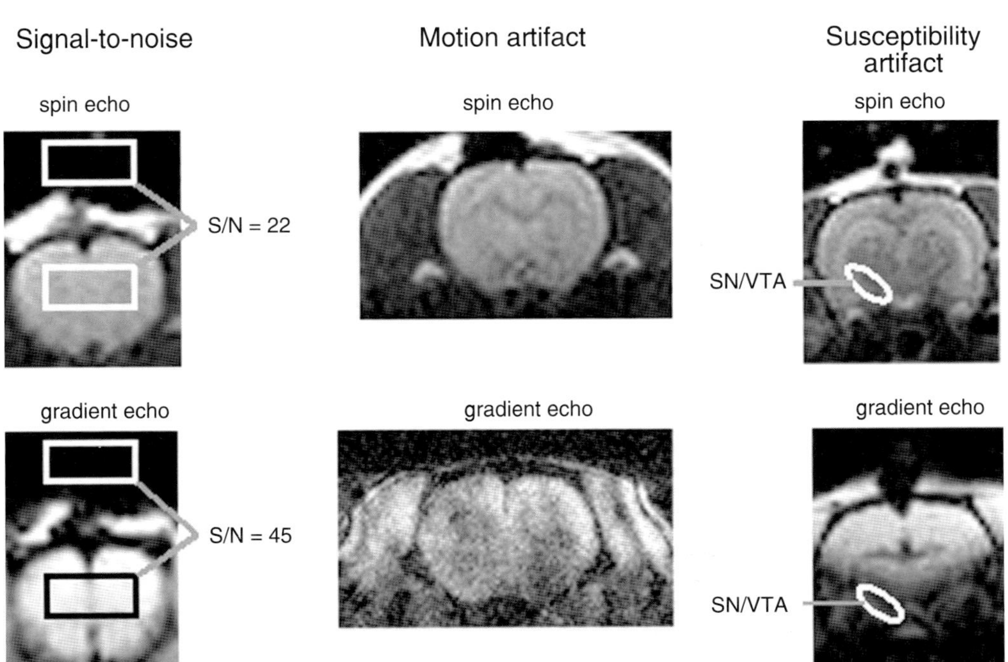

Figure 21.11 Shown are MR images highlighting the advantages and disadvantages of spin echo and gradient echo pulse sequences. All images were collected from the same animal over the same imaging session. Susceptibility artifact is very pronounced in the substantia nigra (SN) and ventral tegmental area (VTA) (Ludwig *et al.*, 2003) S/N = Signal-to-Noise ratio.

less sensitive to potential changes in signal intensity as shown in these T_2^* weighted images. Gradient echo sequences have twice the signal-to-noise ratio as spin echo, making gradient echo sequences the better method for detecting small changes in signal intensity. This lack of sensitivity notwithstanding, the spin echo technique is more than sensitive enough to pick up signal changes in fully conscious marmosets presented with odors, as shown later.

A second compelling reason for choosing between spin echo and gradient echo is the susceptibility artifact that limits visualization and analysis of brain areas near air cavities or ventricles, such as the amygdala or temporal cortex. As shown in Figure 21.11, the critical dopaminergic pathways originating in the ventral tegmental area and substantia nigra, that are central to any study looking at motivation, cannot be imaged with the gradient echo pulse technique due to their proximity to air sinuses. This problem is resolved by use of the spin echo technique.

Temporal and spatial resolution

The position of a typical 18 slice data set, covering much of the marmoset brain, is presented in Figure 21.12. Each slice is 1 mm thick and contiguous with no space between slices. A 19th asymmetrical slice was collected at the level of olfactory bulbs to assess activation of the olfactory system. During functional imaging these slices can be collected in 10 sec using a data matrix of 64×64 in a field of view of 3.0 cm giving a functional resolution of $468 \times 468 \times 1000$ μm. This, in plane resolution of *ca.* 500 μm^2, can be reduced to 250 μm^2 if we double the data matrix to 128×128 and extend the acquisition time to 20 sec. The high resolution neuroanatomical data, used for activational maps, have a data matrix of 250×250 with a 125×125 μm in plane resolution. Shown in Figure 21.13 is a formalin fixed marmoset brain imaged on a 9.4T/89 mm spectrometer with a 512×512 data matrix and field of view of 3.0 cm.

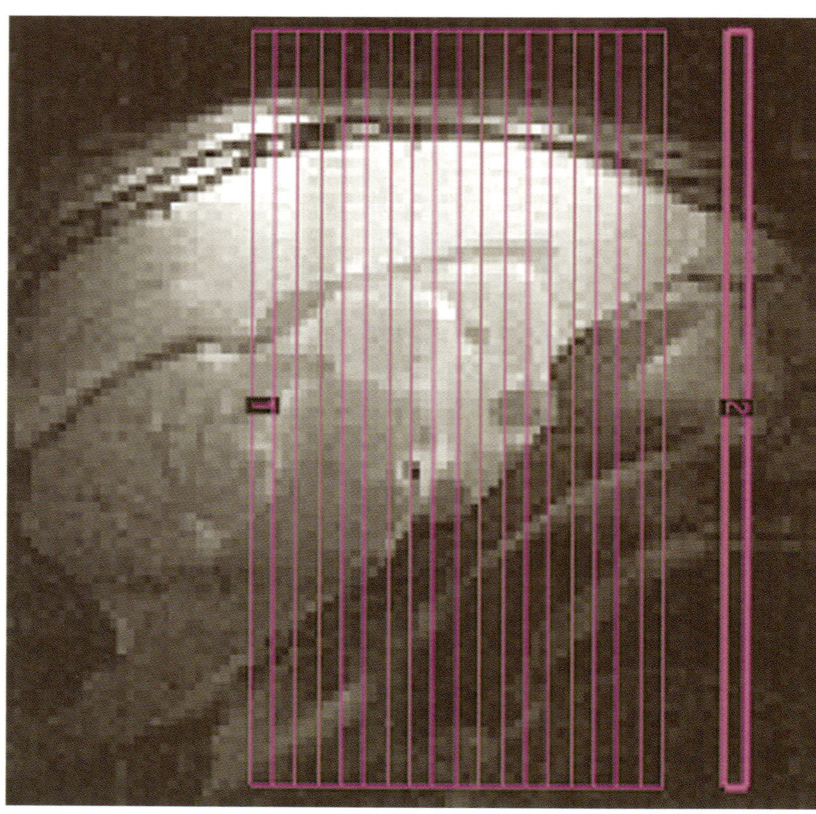

Figure 21.12 Scout image showing a sagittal view of the marmoset brain and slice selection prior to a functional imaging study. One millimeter thick slices are contiguous with an asymmetric slice (far right) positioned at the level of the olfactory bulbs.

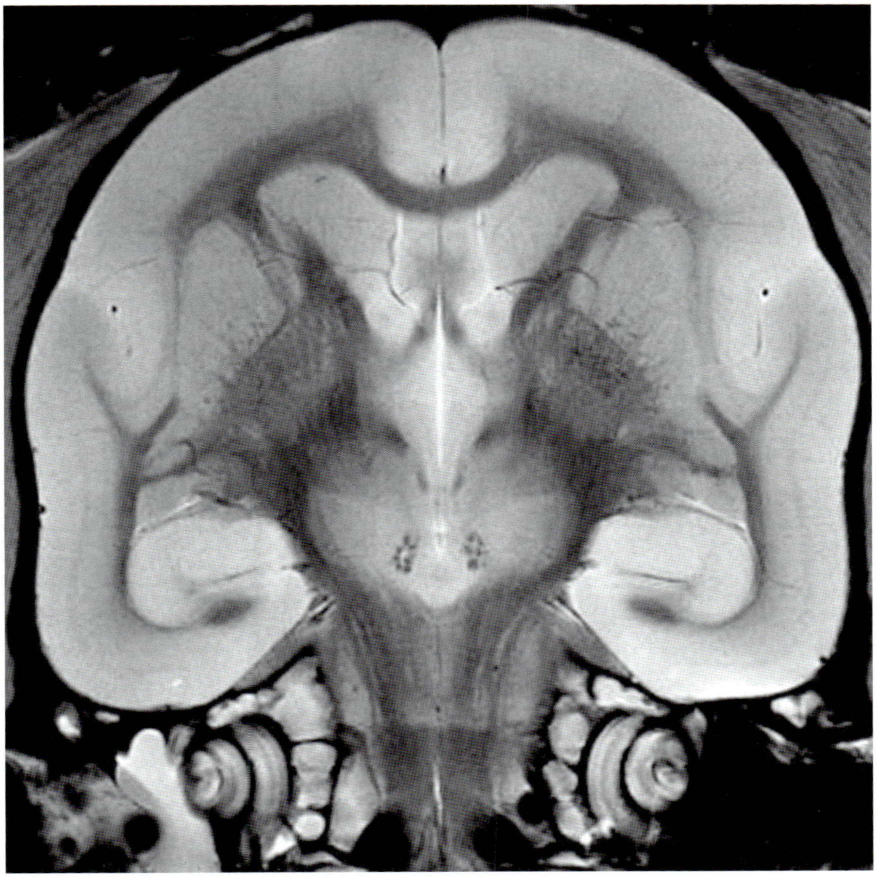

Figure 21.13 High resolution neuroanatomical image of a formalin fixed marmoset brain taken on a 9.4T spectrometer.

The in-plane resolution is *ca.* 60 μm². Anatomical data sets, with image quality similar to this, can be taken using a 9.4T microimager and used to develop a 3D digital MRI atlas of the marmoset for data registration and analysis.

Selection of stimuli and experimental design

In our initial work we chose to work with chemical stimuli. Chemical signals offered several advantages over other modalities. The bore of a 9.4T magnet is quite small with the gradient set in place, and the bore on the first machine we used was quite long, making it difficult to present visual images and to observe whether the marmosets were attending to the images. The magnet with its RF pulses produces considerable noise and our method of immobilization currently uses ear bars so the use of auditory stimuli was precluded, at least initially. Fortunately, common marmosets make extensive use of scents for communication, especially reproductive status (Smith and Abbott, 1998) and individual signatures (Epple, 1974). We were able to place scent marks on small wooden blocks attached to long rods and insert these into the magnet so that the stimulus was at the level of the marmoset's nose and mouth. Since marmosets in the wild mark on wood substrates, this approximates the natural presentation of olfactory stimuli.

But what are the appropriate controls for evaluating the effects of an olfactory stimulus? As noted above, fMRI measures changes in brain activity, both increases and decreases, but these are always relative to something else. Functional MRI cannot measure absolute levels of neural or synaptic activity. The statistical analyses of fMRI results can indicate no change in activity relative to some baseline, increased activation (or positive signal) and decreased activation (or negative). Because synaptic inhibition is an active process involving the consumption of energy, it is important not to confuse a negative signal with neural inhibition.

With respect to chemical signals there are three important controls. First, chemical signals are typically dissolved and stored in some sort of vehicle, so data involving response to a particular stimulus must be evaluated relative to the vehicle in which the signal is presented. Second, every type of stimulus presented will produce activation in the basic sensory areas that process these incoming stimuli. In order to know which brain areas are activated by the cognitively or emotionally relevant aspect of a signal, the basic sensory areas, activated by all stimuli of that modality type, must be identified. Thus, for inhaled chemical signals, the use of an arbitrary odor with no known biological significance should activate the basic sensory areas involved in processing olfactory stimuli, allowing these areas to be ignored when analyzing the biological effects of the stimulus of interest.

Third, a no stimulus baseline is important. Olfactory cues may not dissipate immediately when the stimulus is withdrawn and so there is a potential for carry-over effects from previous stimulation. Use of a no stimulus baseline period, before presentation of any control or experimental stimulus, allows a period to evaluate any carry-over effects from the previous stimulus. In work we have done so far, we have found no difference in activation between the vehicle control and the no stimulus baseline, so either can be used as the basis for evaluating the effects of biologically significant odors.

A typical fMRI design, that works well with auditory and visual signals, is a boxcar design where brief periods of stimulation are alternated with periods of control stimuli. By presenting several repetitions of this stimulus pattern, and tightly linking activation patterns to experimental versus control stimuli, the brain areas activated by the experimental stimuli can be identified. Because of the long decay time of olfactory signals, we have used a much longer period for stimulation, typically 7 min followed by a 10 min washout period, than would be typical of studies using stimuli in other modalities.

Data reduction and analysis

Data reduction is by no means a simple task. Based on the field of view, the resolution power of the magnet, the pulse sequences used and the thickness of brain slices selected, the data obtained refer to small volumes of brain tissue. The smallest volume that can be resolved, given the imaging parameters, is a "voxel" which is analogous to a "pixel" in a two-dimensional space. The activity of each voxel, in response to an experimental stimulus, is evaluated against the activity observed with the control stimulus. Different thresholds can be set for determining if the activity, in response to the experimental stimulus, is significantly above (positive) or below (negative) the activity in the control condition. In human studies a typical response is an elevation of 0.5–1% above control. However, we have found robust responses in marmosets with increased activity of 3–5%. Once voxels with significant changes in activation have been identified, these must then be registered against the brain anatomy of the subject in order to determine which areas are active. This means that some additional steps must be taken during an imaging session. We acquire an anatomical data set at the start and end of each functional imaging session, both as a way of evaluating whether motion artifact has occurred or not, and to provide the anatomical basis for evaluating which voxels have been activated.

Based on prior research, using lesions, single unit recording or other methods, one may already have specific hypotheses about which areas should be activated by specific types of stimuli. However, one of the virtues of the fMRI method is that the entire brain can be imaged at once. Thus, in addition to hypothesis testing, one can also do exploratory research to determine which brain areas are simultaneously activated by an experimental stimulus, leading to inferences about connections between different areas. One can test multiple hypotheses simultaneously. For example, we could predict that the odor from an ovulating female should have an effect on brain areas known to be directly associated with sexual arousal and, at the same time, test hypotheses about the activation of the dopaminergic reward system and explore for brain areas involved in mate recognition, by comparing responses to odors from a mate versus an unfamiliar female. Because the same individuals can be studied repeatedly, exploratory research can lead to specific hypotheses that can be tested later on the same individuals.

Applications in neuroscience research

Emotional states

Fear, anger, hunger and sexual arousal are examples of emotional states that are fertile areas of investigation using fMRI. One can collect a library of vocalizations, smells and visual images, with proven ethological significance in the marmoset's natural habitat as well as in the semi-natural environment of the laboratory setting. These stimuli can be used to communicate with the animal in the magnet. As an example of how stimuli can be used with functional brain imaging, we present some of our research with common marmosets (Ferris *et al.*, 2001; Ferris *et al.*, 2004). Common marmoset females scent mark frequently and the quality of the mark changes with the ovulatory cycle. Males can detect these differences (Smith and Abbott, 1998). Lesion studies on male common marmosets have shown that destruction of the anterior hypothalamus and medial preoptic regions lead to impaired sexual arousal and copulation (Dixson and Lloyd, 1988). However, we do not know if there is any effect of scent marks on arousal in these two areas. We tested four males, presenting them with scents from novel ovulating females, from ovariectomized females and vehicle control. Both scent marks produced elevated neural activity in the anterior hypothalamus and medial preoptic areas compared to the vehicle control. However, activation was significantly greater in response to odors from the ovulating female (Figure 21.14). Two of the males were virgins with no prior sexual experience. However, they demonstrated similar responses to those of sexually experienced males, suggesting that this response is independent of previous sexual experience.

Because we also gathered data from the entire brain, we could also test not only a specific hypothesis about the activation of two specific brain areas, we could also look at patterns of positive and negative in several different brain areas, including those involved in the reward system, in general emotional processing (not specifically sexual) and in areas affecting stimulus evaluation and decision making (Figure 15, Ferris *et al.*, 2004).

Several other studies are possible with the same sample. Can males learn to associate a previously neutral stimulus with the opportunity for mating and, if so, how do neural responses to the stimulus change from pre- to post-conditioning? Since marmosets do form close pair bonds with a specific female and can mate with that female throughout the ovulatory cycle and during pregnancy, do males show different neural responses to odors from their mates versus unfamiliar females, and are they sexually responsive to their mate's odors regardless of reproductive state?

Brain/environment interactions in development

There are myriad examples in animal studies showing that early emotional or environmental insult can affect brain development with long-term neurobiological and behavioral consequences. Insights into the etiology of

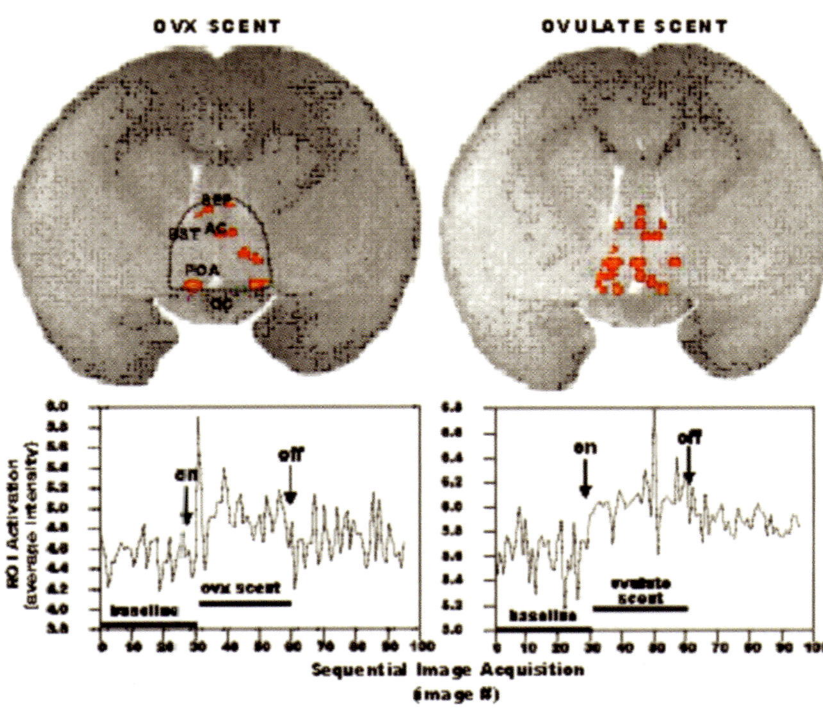

Figure 21.14 BOLD signal changes in the preoptic area of male marmosets exposed to the scent from a peri-ovulatory (ovulate scent) or the scent from an ovariectomized (OVX) female. The dome-shaped boundary shown in the left image outlines the region of interest (ROI). The average changes in signal intensity in response to OVX scent and ovulatory scent in the ROI are shown in the time course data below. (Abr: SEP, septum; BST, bed nucleus of the stria terminalis; POA, preoptic area; OC, optic chiasm; AC, anterior commissure).

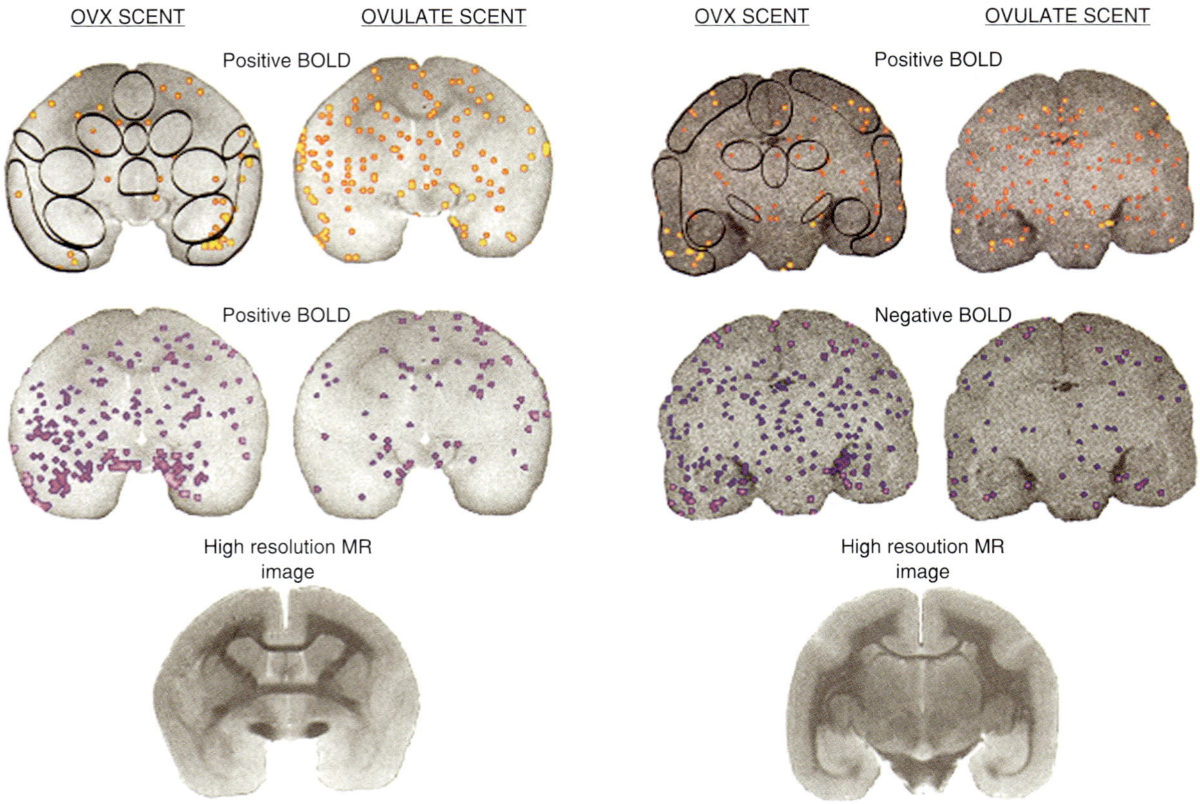

Figure 21.15 Significant changes in positive and negative BOLD signal are shown for presentation of OVX scent and ovulatory scent. Rather than show positive and negative BOLD for each scent as a single composite they are presented individually one above the other, respectively. Regions of interest are shown in the circumscribed areas in the upper right coronal section. A high resolution image taken from a formalin fixed marmoset brain is presented to shown neuroanatomical detail (Ferris *et al.*, 2004).

mental illness may be gleaned by longitudinal studies on marmosets, examining the interaction between a vulnerable gene pool and a stressful environment at critical times in development. Since marmosets generally have twins, one twin can serve as a control for the other. Since fMRI is non-invasive, and can be used to study the same animal over the course of its life, it is possible to observe developmental changes in neuroanatomy, brain activity and brain chemistry (spectroscopy).

Drugs effects on brain activity

Two other obvious applications of fMRI are studying changes in brain activity in response to acute and prolonged exposure to psychotherapeutic drugs and drugs of addiction. For example, many psychotropic drugs cause a prompt increase in brain levels of neurotransmitters. Nonetheless, patients typically require weeks of treatment before reporting an improvement in their condition. This would suggest that drug efficacy for the treatment of mental illness is due to secondary changes in the neurochemical signals and pathways that are slowly affected by the continuous exposure to the psychotropic agent. Functional MRI could help to resolve this mechanism of action. In the case of drugs of addiction, marmosets can be trained to self-administer cocaine, after which they are withdrawn from the drug and later reinstated in response to conditioned cues. It is feasible to image the brain during each of these different phases of cocaine addiction.

Testing cognitive performance

Since animals will readily respond to peripheral stimulation while in the magnet, marmosets (or monkeys more generally) may be used in studies of cognitive performance. Several recent studies have shown that marmosets and related species have complex cognitive skills in areas such as imitation, social transmission of information about food, novel foraging tasks and cooperation. Although studies requiring an operant response, such as bar pressing, are difficult to do without inducing motion artifact, eye movements directed toward one object or another could be monitored to test cognitive performance. Imaging studies on conscious rhesus monkeys (*Macaca mulatta*) (Logothetis *et al.*, 1999) show that cognitive testing is feasible and opens the area of cognitive neuroscience to investigation with fMRI in marmosets.

Correspondence

Any correspondence should be directed to Craig F. Ferris, Department of Psychiatry, University of Massachusetts Medical School, 55 Lake Ave North, Worcester, MA 01655. Tel: 508 856 5530. Fax: 508 856 6426. E-mail: craig.ferris@umassmed.edu

Acknowledgements

This work was funded by a grant from the National Institute of Mental Health, Program in Behavioral and Integrative Neuroscience MH58700, and by RR00167 to the Wisconsin Primate Research Center.

References

Belliveau, J.W., Kennedy, D.N. Jr., McKinstry, R.C., Buchbinder, B.R., Weisskoff, R.M. Cohen, M.S., Vevea, J.M., Brady, T.J. and Rosen, B.R. (1990). *Science* 254, 716–719.

Birn, R.M., Bandettini, P.A., Cox, R.W., Jesmanowicz, A. and Shaker, R. (1998). *Magn. Reson. Med.* 40, 55–60.

Brevard, M.E., Duong, T.Q., King, J.A. and Ferris, C.F. (2003). *Magn. Reson. Imaging* 21, 995–1001.

Cohen, M.S. and Brookheimer, S.Y. (1994). *Trends Neurol. Sci.* 17, 268–277.

Day, R.L., Coe, R.L., Kendal, J.R. and Laland, K.N. (2003). *Anim. Behav.* 65, 559–571.

Dixson, A.F. and Lloyd, S.A.C. (1988). *Symp. Zool. Soc. Lond.* 60, 81–117.

Epple, G. (1974). Studies on olfactory communication in South American primates. *Ann. NY. Acad. Sci.* 237, 261–278.

Ferris, C.F., Snowdon, C.T., King, J.A., Duong, T.K., Ziegler, T.E., Ugurbil, K., Ludwig, R., Schultz-Darken, N.J., Wu, Z. Olson, D.P., Sullivan, J.M. Jr., Tannenbaum, P.L. and Vaughan, J.T. (2001). *Neuro Report* 12, 2231–2236.

Ferris, C.F., Snowdon, C.T., King, J.A., Sullivan J.M. Jr., Ziegler, T.E., Ludwig, R., Schultz-Darken, N.J., Wu, Z., Olson, D.P., Tannenbaum, P.L., Einspanier, A., Vaughan, J.T. and Duong, T.Q. (2004). *J. Magn. Reson. Imaging* 19, 168–175.

Fox, P.T., Mintum, M.A., Raichle, M.E., Miezin, F.M., Allman, J.M. and Van Essen, D.C. (1986). *Nature* 323, 806–809.

Hajnal, J.V., Myers, R., Oatridge, A., Schwieso, J.E., Young, I.R. and Bydder, G.M. (1994). *Magn. Reson. Med.* 31, 283–291.

Johnson, E.O., Kamilaris, T.C., Carter, C.S., Calogero, A.E., Gold, P.W. and Chrousos, G.P. (1996). *Biol. Psychiatry* 40, 317–337.

Lahti, K., Ferris, C.F., Fuhai, L., Sotak, C. and King, J.A. (1998). *J. Neurosci. Methods* 82, 75–83.

Logothetis, N.K., Guggenberger, H., Peled, S. and Pauls, J. (1999). *Nat. Neurosci.* 2, 555–562.

Ludwig, R., Bogdanov, G., King, J.A., Allard, A. and Ferris, C.F. (2004). A dual RF resonator system for high-field functional magnetic-resonance imaging of small animals. *Journal of Neuroscience Methods* 132, 125–135.

Ogawa, S., Lee, T.M., Nayak, A.S. and Glynn, P. (1990). *Mag. Reson. Med.* 14, 68–78.

Ogawa, S., Tank, D.W., Menon, R., Ellermann, J.M., Kim, S.-G., Merkle, H. and Ugurbil, K. (1992). *Proc. Natl. Acad. Sci. U.S.A.* 89, 5951–5955.

Ramsey, N.F., Kirkby, B.S., Van Gelderen, P., Berman, K.F., Duyn, J.H., Frank, J.A., Mattay, V.S., Van Horn, J.D., Esposito, G., Moonen, C.T.W. and Weinberger, D.R. (1996). *J. Cereb. Blood Flow Metab.* 16, 755–764.

Saltzman, W., Schultz-Darken, N.J., Scheffler, G., Wegner, F.H., Abbott, D.H. (1994). *Physiol. Behav.* 56, 801–810.

Smith, T.E. and Abbott, D.H. (1998). Behavioral discrimination between circumgenital odor from peri-ovulatory dominant and anovulatory female common marmosets (*Callithrix jacchus*). *Am. J. Primatol.* 46, 265–284.

Snowdon, C.T. (2001). *Anim. Cogn.* 4, 247–257.

Sokolloff, L., Reivich, M., Kennedy, C., Des Rosiers, M.H., Patlak, C.S., Pettigrew, K.D., Sakurada, O. and Shinohara, M. (1977). *J. Neurochem.* 28, 897–916.

Yetkin, F.Z., Haughton, V.M., Cox, R.W., Hyde, J., Birn, R.M., Wong, E.C. and Prost, R. (1996). *Am. J. Neuroradiol.* 17, 1005–1009.

CHAPTER 22

Radiographic Imaging of Nonhuman Primates

Celia R. Valverde[*] and Kari L. Christe[†]
Sacramento Veterinary Surgical Services Inc.[*] and
California Regional Primate Research Center,[†]
University of California, Davis, USA

Introduction

A high quality radiograph requires optimal radiographic density, adequate contrast, minimal patient motion and proper patient positioning (Owens, 1999). High quality radiographs must be obtained for accurate interpretation and assessment. A radiograph exposure technique chart is essential for consistent high-quality radiograph production. Most poor quality radiographs are due to operator error or inadequate techniques (Ewers, 2000; Stender, 1990). Since radiograph exposure technique charts are based on anatomic thickness (in centimeters), formulating a chart for small primates is relatively easy. As thickness decreases it becomes easier to visualize the different densities within that thickness (Ferron, 1967). However, as mass increases radiograph exposure technique charts must be tailored precisely to the anatomical study (Ferron, 1967). In addition, short exposure times (1/60 or 1/120 seconds), fast film screen systems and effective patient restraint, eliminate most motion artifacts. Nonhuman primates require either physical or chemical restraint for proper patient positioning. Unfortunately, this restraint may also alter the radiographic technique or quality.

Thoracic radiograph

Anatomic features

The thoracic cavity in the rhesus monkey extends from the thoracic inlet to the 14th thoracolumbar segment dorsally and just caudal to the 7th sternebrae ventrally (Silverman and Morgan, 1980a). The adult rhesus macaque thorax is wedge-shaped, deep sagitally, and broader dorsally than ventrally. The macaque's thorax is more elongated craniocaudally than the dog's thorax.

The costodiaphragmatic angle is sharp; the right costodiaphragmatic angle is sometimes located more caudally than the left (Silverman and Morgan, 1980a).

The mediastinum is complete. The thymus is bilobed in the macaque; therefore, the thymic "sail sign" may be visualized on both sides of the mediastinum in young animals. The mediastinum in the adult macaque is concave bilaterally on the frontal radiograph and is similar to the cat's mediastinum. Irregularities of the paravertebral mediastinal stripe on the frontal projections are considered an abnormal finding and are usually caused by mediastinal lymphadenopathy (Silverman and Morgan, 1980a).

The Laboratory Primate
Copyright 2005 Elsevier
ISBN 0-1208-0261-9

Radiographic technique

All routine thoracic examinations of nonhuman primates should be performed on sedated patients, restrained in the upright position by an acceptable restraint device. Radiograph positioners have been well described in the literature (Ferron, 1966; Silverman and Morgan, 1980a; Silverman *et al.,* 1983).

A routine thoracic study should include erect lateral and ventrodorsal views. The relatively greater depth of the costodiaphragmatic recesses in macaques, as compared with those in human beings and dogs, increases the importance of the lateral projection. The caudal extent of the costodiaphragmatic recesses cannot be fully evaluated on the frontal projections due to the superimposition of the cranial abdominal organs.

Respiratory system

Anatomic features

Trachea and major airways

The trachea of the adult rhesus monkey is flattened transversely and is approximately 9.5–11 cm long and 0.9 cm wide. There are approximately 27 cartilaginous C-shaped tracheal rings. The dorsal portion of the trachea is incomplete and is formed by an elastic membrane. The right bronchus extends in a near straight line from the trachea, whereas the left bronchus comes off at a distinct angle (Silverman and Morgan, 1980a).

Dorsal tracheal deviation on the lateral projections and irregularity of the paravertebral mediastinal stripe on the frontal projections are the most accurate signs of hilar lymphadenopathy (Silverman and Morgan, 1980a).

Lung

The macaque lung most closely resembles the dog lung in that they both have a thick pleura, minimal interlobular septation, and similar arrangement of vessels. Pulmonary lobation in the rhesus monkey is similar to that in the dog and cat. The left lung is divided into two lobes: cranial and caudal. The cranial lobe is subdivided into cranial and caudal segments. The right lung is divided into four lobes: cranial, middle, caudal, and accessory. Bronchi in macaque lungs are developed to at least three generations before the first bronchiole is reached. The bronchi open into short (2–3 mm) terminal bronchioles that are only developed to a single generation and lined by pseudostratified columnar epithelium.

Frequently there is no intermediate bronchiole in rhesus macaque. Other macaque species have terminal bronchioles that are developed to 3–4 generations and lined by simple cuboidal epithelium (Silverman and Morgan, 1980a).

The pulmonary parenchyma of macaques has an accentuation of the interstitial densities resulting in diminution of the distinctiveness of pulmonary bronchovascular markings. It is important not to overemphasize the significance of the interstitial pattern and diagnose interstitial pulmonary disease. The accentuation of the interstitial pattern may reflect a normal pulmonary parenchyma of nonhuman primates, or may be secondary to pulmonary acariasis (*Pneumonyssus siminicola*) (Silverman and Morgan, 1980a).

Angulating the x-ray beam 15 degrees caudally, for the ventrodorsal projection, can enhance evaluation of the caudal portions of the lungs.

Artifactual densities or normal densities can be confused with thoracic pathology, e.g. obliteration of the cardiac silhouette by pericardial or sternal fat deposits and the presence of nipple shadows resembling "coin lesions."

Special imaging techniques

Planimetry is a radiographic technique utilized to determine the total lung capacity in nonhuman primates. The technique has been described in anesthetized baboons (Harris *et al.,* 1971). The authors reported an excellent correlation between total lung capacity and radiographic lung area. This correlation was better for young than aged baboons (Bennett *et al.,* 1995).

Diseases/Disorders

Air sacculitis

Air sacs are not found in all species of nonhuman primates. When present, the air sac acts as a resonator modifying sound (Hilloowala, 1976). New World primates such as howler monkeys (*Allouatta sp*), titi (*Callicebus sp*) and owl monkeys (*Aotus sp*) have a varying degree of laryngeal air sac development. Old World primates have cervical laryngeal air sacs that communicate with the laryngeal lumen via a central ostium at the base of the epiglottis (Lowenstine, 2003).

Great apes have well-developed laryngeal air sacs which are large in the siamang, gibbons, orangutans, and gorillas. These air sacs have paired openings from the lateral saccules and extend along the ventral neck, beneath the clavicles and into the axilla. In orangutans and gorillas, the apparatus extends along the thoracic wall. Air sacculitis is often an extension of chronic respiratory infections. Air sacculitis due to gram negative organisms such as *Pseudomonas* is common, especially in orangutans. Radiographic signs may include increased soft tissue density of the cervical and intermandibular areas, thickening of the mucosa of the perilaryngeal air sacs and the presence of exudate (Baer *et al.,* 1994).

Pneumonia

Bronchointerstitial pneumonia is one of the major causes of morbidity and mortality in nonhuman primate species. The differential diagnosis for interstitial pneumonia is quite extensive and includes bacterial (*Streptococcus sp, Staphylococcus sp, Pseudomonas aeruginosa, Klebsiella pneumoniae, Bordetella sp, Francisella tularensis, Proteus mirabilis*), parasitic (*Filarioides sp, Filariopsis sp*), protozoal (*Toxoplasma gondii, Pneumocystis carinii*), fungal (*Histoplasma capsulatum, Coccidiomyocoides immitis*), and viral agents (measles, other paramyxoviruses, influenza viruses).

Tuberculosis

Tuberculosis is the most common disease producing mediastinal lymphadenopathy in nonhuman primates. Irregularity of the paravertebral mediastinal stripe on the frontal projections and dorsal trachea deviation on the lateral projections are the most accurate signs of hilar lymphadenopathy. In contrast to human beings, pulmonary tuberculosis in nonhuman primates rarely shows any nodular calcification. Most commonly the tubercles undergo extensive caseation necrosis. Extra-pulmonary sites of infection may not be readily apparent on radiographs. Differential diagnosis of pulmonary tuberculosis includes lung mite infestation (*Pneumonyssus simicola*), nocardiosis, and deep mycosis such as histoplasmosis, or coccidioidomycosis (Silverman and Morgan, 1980a).

Cocciodiomycosis

Radiographic findings of cocciodiomycosis include the formation of numerous multiloculated cavities that arise from conducting airways and evidence of severe diffuse pulmonary parenchymal disease. Serial thoracic radiographs provide an excellent indication of the extent of infection and the progress of the disease (Converse *et al.,* 1962; Breznock *et al.,* 1975).

Pulmonary acariasis

Lung mite (*Pneumonyssus simicola*) infestation is usually a subclinical disorder seen in virtually all wild and imported wild-caught rhesus. It may persist in subsequent generations of captive-born group housed macaques. Treatment with ivermectin can eliminate the infection, but the lesions of chronic bronchiolitis, bronchiectasis, and pigmentation may persist as an incidental finding (Lowenstine, 2003). The typical radiographic manifestation of pulmonary acariasis in rhesus macaques consists of an increased nonstructured interstitial pulmonary density, increased thickness and density of the bronchial walls, increased peribronchial density, pleural thickening, pleural adhesions, indistinct pulmonary vasculature and cavitary pulmonary lesions. Follow-up radiographs usually demonstrate minimal or no progression of the peribronchial and interstitial densities in uncomplicated cases. Pleural thickening is a prominent radiographic finding; the interlobar fissures are thickened and irregular, differing from their normal thin, smooth appearance. Pleural adhesions are common. Cavitary pulmonary disease consisting of bullae and subpleural blebs can be extensive. The most frequent complication of pulmonary acariasis is pneumothorax. Because, of the complete separation of the mediastinum of the macaque, unilateral pneumothorax occurs frequently (Silverman and Morgan, 1980a).

Cardiovascular system

Anatomic features

The cardiovascular system of most nonhuman primates is similar in anatomy and physiology to that of human beings, and the domestic dog (Bonakdarpour *et al.,* 1967). The major mediastinal branches of the aortic arch in the macaque and dog are the brachiocephalic branch and right and left subclavian arteries. The vertebral and carotid arteries originate from the brachiocephalic trunk.

Special imaging techniques

Angiography

Angiography is a radiographic contrast study using iodinated contrast media and sequential radiographs to visualize the cardiovascular system (heart, arteries, and veins). With the advent of ultrasonography and color Doppler, contrast procedures to examine the cardiovascular system have been reduced to the diagnosis of the more complex congenital defects and associated conditions. Angiography, in conjunction with pressure measurements, is essential in examining cardiac lesions to determine the severity of the functional and anatomical defects.

Selective angiography is the placement of an indwelling catheter, near the suspected lesion, for delivery of a contrast media. Selective angiography requires specialized equipment (e.g. catheters, contrast material, an automated contrast injector, a rapid film recording device, and fluoroscopy). Nonselective angiography is less accurate in demonstrating lesions than selective angiography. Its use is limited, although it can be adequately performed with routine radiographic equipment (Owens, 1999).

Arterial angiography can be performed, usually through catheterization of the femoral or carotid artery through a small surgical incision (cutdown). A catheter can be advanced to the area of interest. The advancement of the catheter can be monitored using fluoroscopic technique or through serial injections of an iodinated contrast material. Lateral and ventrodorsal projections should be evaluated. The assessment of the arterial, capillary and venous phases of angiography can be performed.

Angiographic studies have been used to provide a comparison between nonhuman primate and human vascular anatomic variations and collateral pathways, such as the angiographic anatomy of the external carotid artery system in the rhesus monkey (TerBrugge *et al.,* 1989) and the coronary artery distribution in bonnet monkeys (*Macaca radiata*) (Buss *et al.,* 1982). Placental radioangiography, as a method for the study of uteroplacental blood flow, was described in the pregnant rhesus monkey (Kato *et al.,* 1987). Clinical applications of the arterial angiogram have been described for the diagnosis of dissecting arterial aneurysm secondary to arteriosclerosis (Boorman *et al.,* 1976), congenital arteriovenous fistula in a rhesus macaque (Rosenberg *et al.,* 1983), and for the diagnosis of atrial septal defect in a Sumatran orangutan (Greenberg *et al.,* 1999).

Diseases/Disorders

Cardiomyopathy

This is a major cause of morbidity and mortality in the owl monkeys (*Aotus sp*). A prominent feature of this condition is ventricular dilation and myocardial hypertrophy, primarily of the left ventricular chamber, left ventricular free wall and the interventricular septum. Thoracic radiographs are used as an ancillary diagnostic tool. Findings might include cardiomegaly, vascular dilation and pleural and/or pericardial effusion. Echocardiography is the imaging modality of choice. Differential diagnosis includes viral myocarditis (Cocksackie B virus, encephalomyocarditis caused by picornaviruses) and parasitic infection (*Toxoplasma gondii* in New World primates, particularly squirrel monkeys; *Trypanosoma cruzi* in Southern American monkeys) (Baer *et al.,* 1994).

Chronic myocardial interstitial fibrosis is a common finding in many species of primates, especially great apes in which sudden cardiac death is common (Lowenstine, 2003).

Atherosclerosis

Nonhuman primates are used extensively as experimental models for atherosclerosis. Atherosclerosis can be readily induced in many species. Naturally occurring atherosclerosis is seen in captive primates and is also reported in free-ranging animals, but is usually mild and limited to the formation of fatty streaks (Lowenstine, 2003). Atherosclerosis has been associated with coronary artery disease, cardiomyopathy, dissecting aortic aneurysms, and congestive heart failure.

Aortic aneurysm

Spontaneously occurring aortic aneurysms have been described in the gorilla and squirrel monkey (*Saimiri sp*), howler (*Allouatta sp*), capuchin (*Cebus sp*), patas (*Erythrocebus patas*), African green (*Chlorocebus aethiops*), spider monkeys (*Atelles sp*) and pygmy chimpanzee (*Pan paniscus*). Diagnosis of aortic aneurysm can be made on the basis of a radiographic examination, sonography, abdominal palpation and ascultation of bruits. Definitive diagnosis may require angiography or computerized tomography (Baer *et al.,* 1994; Lowenstine, 2003).

Abdominal radiograph

All routine abdominal examinations of nonhuman primates should be performed on sedated or anesthetized patients. A routine abdominal study should include lateral and ventrodorsal views. A foam pad placed underneath the animal is helpful with correct positioning and comfort since, nonhuman primates have prominent sacral bones with protruding tails.

Abdominal organs are difficult to visualize on nonhuman primate abdominal radiographs. The lack of intra-abdominal fat reduces organ resolution and gastrointestinal contents obscure organ visualization. The abdomen is also the thickest part of the body and therefore, a higher kVp must be used to attain adequate radiation penetration. Moreover, motion artifact can present a major problem when imaging the gastrointestinal tract.

Gastrointestinal system: Esophagus

Special imaging techniques

Esophagography

An esophagram is performed using positive contrast media to evaluate esophageal location and morphology. Fluoroscopy is the optimum technique for the assessment of esophageal motility and function. One should obtain lateral and ventrodorsal projections of the cervical region and thorax, and a right ventrodorsal oblique projection of the thorax. Sedation and anesthesia are not recommended because esophageal motility will be altered. Esophagrams can outline potential strictures, esophageal foreign bodies, megaesophagus, and trauma to the esophagus.

Stomach

Anatomic features

Tarsiers, New World and Old World monkeys have a flask-shaped simple stomach. The rhesus stomach is positioned lower and more horizontally than in humans, due to the shape and size of their liver (Hartman *et al.*, 1933). The stomach structure of colobines (*Colobus* and *Presbytis*) differs from any other primate and resembles that found in ungulates, with a pseudoruminant anterior fermentation area in a large multichambered stomach. The stomach is large and sacculated, though not truly compartmentalized. The sacculations are produced by reduced longitudinal muscle bands. The size of the stomach and colon tend to be proportionally larger in folivorous nonhuman primates. These animals also tend to be larger in stature to accommodate for their sizeable gastrointestinal tract. The colobine stomach may constitute up to a quarter of the adult body weight and up to half for a semi-weaned infant. The Pongidae stomach is indistinguishable from human beings.

Special imaging techniques

Gastrography

Gastrography is a radiographic contrast study of the stomach using negative, positive, or double contrast techniques for evaluation of gastric morphology, function and the rate of gastric emptying. If gastrointestinal perforation is suspected barium should not be used; an iodinated contrast medium should be used instead. Full strength iodinated contrast (240-300 mg Iodine/ml) should be diluted 50:50 with water (Wallack, 2003). Ideally, no sedation or anesthesia should be used in a functional study, although it can be used for a morphologic evaluation of the stomach.

Pneumogastrogram

- Obtain survey abdominal radiographs.
- Using an orogastric tube, infuse sufficient gas to make the stomach tympanic.
- Obtain left and right laterals, dorsoventral and ventrodorsal radiographic views.
- Upon completion of the study, the gastric distention should be relieved by passing an orogastric tube.

Positive contrast gastrogram

- Obtain survey abdominal radiographs.
- Using an orogastric tube, infuse micropulverized barium sulfate 30% weight/volume, dose range 0.5–3 ml/kg (Owens, 1999).
- If barium is used, the abdomen must be penetrated by the x-rays, requiring higher kVp than for skeletal or mammographic images.
- Obtain left and right laterals, dorsoventral and ventrodorsal radiographic views at 1, 15, 30 minutes and 1, 2 and 3 hours. Additionally, obtain hourly radiographs until gastric emptying is nearing completion.

Double contrast gastrogram

- Using an orogastric tube infuse barium sulfate.
- A high density barium of 100% weight/volume is preferable (Wallack, 2003). Using the dog weight as a guideline, dose range is 1.5 ml/kg (>40 kg) to 3 ml/kg (<80 kg) (Wallack 2003).
- Use a dose range 0.5-1 ml/kg if a 30 to 50% weight/volume barium sulfate.
- Infuse sufficient gas to make the stomach tympanic.
- Carefully, roll the animal 360°.
- Obtain left and right laterals, dorsoventral and ventrodorsal radiographic views.
- Upon completion of the study, the gastric distention should be relieved by passing an orogastric tube.

Pneumoperitoneography can be used in combination with gastrointestinal positive contrast studies; both the mucosal and serosal surfaces are visualized (Silverman *et al.*, 1975).

Diseases/Disorders

Trichobezoars are relatively common in many species of primates (Keller *et al.*, 1982; Lapin, 1963; Mook, 2002).

Trichobezoars are formed from plant fibers, hair, and mucus. Clinical signs may be subtle such as anorexia, vomiting, weight loss and occasionally the stool may contain hair.

Gastrointestinal ulceration occurs most commonly in newly captured, stressed or geriatric nonhuman primates (e.g. African Green monkeys, squirrel monkeys, macaques). Nonhuman primates on NSAID therapy are at risk of developing gastrointestinal ulcers.

Acute gastric dilation/bloating is seen almost exclusively in indoor-housed macaques and baboons, but has also been reported in callitrichids (Cicmanec, 1977). Clinical signs are enlarged abdomen differentiated from colonic distention, discomfort, distress and shock. Microcardia and a narrowed caudal vena cava are observed due to hypovolemia and decreased venous return. Death ensues unless the animal is treated immediately. The etiology is unknown though it often associated with management changes in feeding. *Clostridium perfringens* in the feed has also been associated with acute gastric distention (Newton *et al.*, 1971).

Small and large intestines

Anatomic features

Prosimians have no sigmoid flexure, taenia (bands) or haustra (saccules) in the colon. Also, the length of the cecum, relative to that of the colon and rectum, is far greater than in other primate groups. New World monkeys have no sigmoid flexure or haustra in the colon. Old World monkeys are mostly hindgut fermenters. They have a cecum but lack a vermiform appendix; the colon has taenia, haustra, and a sigmoid flexure. Hominids have a colon well developed with taenia, haustra, and a cecum with a vermiform appendix. Valvulae conventes are present in some higher primates and humans (Goldberg *et al.*, 1982).

Special imaging techniques

Upper gastrointestinal study (UGI)

A UGI is a radiographic study using positive contrast media to provide functional and morphological evaluation of the small intestine. Barium sulfate suspension of 30% weight/volume provides adequate mucosal coating and contrast for radiographic visualization. Iodinated gastrointestinal contrast media should be used in animals with suspected perforation. However, ionic and nonionic iodinated media are less sensitive for the detection of mucosal lesions and may not demonstrate a perforation or a fistula. In addition, nonionic iodine contrast is hypertonic, drawing fluid into the intestinal lumen, diluting the contrast media further, and dehydrating the patient. Also, iodine has a faster transit time than barium so lateral and ventrodorsal projections should be made more frequently (usually every 10–30 minutes) until the contrast is seen in the large intestine (Owens, 1999). The transit time to the colon is variable, depending on the animal's age, temperature, problem and the type and length of anesthesia, making interpretation difficult.

- Obtain survey lateral and ventrodorsal abdominal radiographs.
- Using an orogastric tube, infuse barium sulfate 30% weight/volume.
- Dose ranges are derived from guidelines from the domestic dog and cat. Owens *et al.* (1999) recommended a dose range 1.5–3 ml/kg. Wallack, 2003 recommended 10 ml/kg (>20 kg) to 15 ml/kg (<20 kg).
- Obtain sequential radiographs at 5, 30 minutes, and hourly thereafter, until the contrast is seen in the large intestine (canine protocol). The recommended radiographic time sequence for the feline is 5, 10, 30 minutes and every 30 minutes thereafter, until the contrast is seen in the large intestine. The feline protocol may be more applicable to small New World species that have high metabolic rates.
- For iodinated contrast examination:
  - Dose ranges are based on domestic dog and cat.
  - Use Iohexol 600–875 mg Iodine/kg. Dilute the amount with water to obtain a whole volume of 10 ml/kg (Wallack, 2003).

Barium enema

This is an abdominal radiographic study, using retrograde infusion of negative contrast, positive contrast or double contrast, for the morphologic assessment of the large intestine.

- Thoroughly cleanse the large intestine with oral laxatives and enemas.
- Obtain survey abdominal radiographs.
- Anesthesia as needed.
- Place a balloon catheter cranial to the anal sphincter that completely occludes the anal canal.
- Place the animal in left lateral recumbency.
- Infuse 7–30 ml/kg of micropulverized barium sulfate rectally. A 10–20% weight/weight or 20–25% weight/ volume concentration is recommended. The starting volume should be 7–15 ml/kg (Wallack, 2003).

- Obtain lateral and ventrodorsal abdominal radiographic views.
- Evacuate the barium.
- Place the animal in right lateral recumbency and infuse air to redistend the colon.
- Obtain lateral and ventrodorsal abdominal radiographic views.
- Deflate the balloon and remove the catheter.

Pneumocolon

Pneumocolon is used to discriminate between a normal gas filled large intestine and a gas dilated small intestine and to identify large bowel strictures and intraluminal and intramural masses.

- Through colonic cleansing with enemas is optional since this study is often only performed to verify the position of the colon.
- Obtain survey right lateral and ventrodorsal abdominal radiographs.
- Anesthesia as needed.
- Place the animal in right lateral recumbency.
- Insert a lubricated catheter into the rectum.
- Administer approximately 1–3 ml/kg of air into the rectum and colon (Wallack, 2003).
- Repeat a right lateral abdominal radiograph to verify if sufficient air has been administered to adequately delineate the colon.
- If enough air is present, then obtain ventrodorsal and oblique abdominal radiographic views.
- If there is insufficient air, then repeat administration.

Double contrast barium enema

Double contrast barium enema is used to evaluate the colonic mucosa, wall thickness or identify strictures and/or intraluminal and intramural masses.

- If performing a barium enema, first complete the procedures for barium enema, then follow the steps below.
- If not performing a barium enema first, then obtain survey right lateral and ventrodorsal abdominal radiographs.
- Administer 4–6 ml/kg of micropulverized barium sulfate suspension rectally. A 10–20% weight/weight or 20–25% weight/volume concentration is recommended on the dog and cat (Wallack, 2003).
- Elevate the patient's torso to encourage barium drainage from the colon.
- Place the animal in right lateral recumbency.
- Insert a lubricated catheter into the rectum.
- Administer approximately 7–30 ml/kg of air into the rectum and colon. A starting volume of 7–11 ml/kg is recommended (Wallack, 2003).
- Repeat a right lateral abdominal radiograph to verify if sufficient air has been administered to adequately delineate the colon.
- If enough air is present then obtain ventrodorsal and oblique abdominal radiographic views.
- If there is insufficient air, then repeat administration.

Diseases/Disorders

A **linear foreign body** may cause the intestines to appear "plicated" on abdominal radiographs. An intestinal linear foreign body injury was reported in a cynomolgus macaque; the ulceration, perforations and septic peritonitis were attributed to the ingestion of sisal rope used for environmental enrichment (Hahn, 2000).

Intestinal perforation can occur as a result of intestinal foreign body, intestinal neoplasia or abdominal trauma.

Rocks in the cecum are common radiographic findings of nonhuman primates housed in outdoor corrals with gravel or a rocky substrate.

Enterocolitis is the most common cause of morbidity and mortality in nonhuman primate species. The differential diagnosis includes bacterial (*Campylobacter sp, Salmonella sp, Shigella sp, Yersinia sp, enteropathogenic E. coli, Clostridium sp, Mycobacterium avium-intracellulare complex*), parasitic (*Giardia, Balantidium coli, Cryptosporidium sp, Entamoeba histolytica, Trichuris trichiura, Strongyoides,* Strongyles and cestodes), fungal (*Candida albicans*), and viral agents (rotaviruses, coronaviruses, paramyxovirus).

Marmoset wasting syndrome is a poorly understood disease of callitrichids. The etiology is probably multifactorial, including low dietary protein and zinc content, feeding behavior, gastrointestinal tract infection and malabsorption. Clinical signs are nonspecific but include alopecia, diarrhea, chronic colitis, weight loss and anemia. It has been postulated that their dietary protein requirement is greater than that of other primates.

Megacolon of cynomolgus macaques (*M. fascicularis*) has been reported. Indoor-housed cynomolgus macaques appeared bloated but not in pain. Gastrointestinal signs included diarrhea, mucus in the stool, anorexia, and failure to pass stool, with repeated episodes of extreme abdominal distension and accumulation of gas and feces in abnormally enlarged colons. Intra-abdominal adhesions were noted in all animals. Most of the animals had a history of a prior obstetric surgery. Abdominal radiographs confirmed abnormal dilation and positioning of the colon. Conservative management with stool softeners was palliative; a partial colectomy was curative (Eisele *et al.*, 1991).

Diverticulosis is commonly seen in geriatric nonhuman primates. It is most severe in the descending and sigmoid colon. Severe diverticulosis can lead to diverticulitis and ultimately rupture through the intestinal wall causing focal or generalized peritonitis.

Intestinal adenocarcinoma is the most common malignant neoplasm in rhesus macaques and a significant cause of morbidity and mortality in the geriatric population (Valverde *et al.,* 2000). Typical presenting signs are weight loss, microcytic anemia and fecal occult blood positive. Anorexia, constipation and obstruction occur quite late in the disease course. Metastases to the liver and mesenteric lymph nodes can occur. Diagnosis is achieved by abdominal palpation, ultrasonography, contrast radiography, or exploratory laparotomy. Naturally occurring colonic adenocarcinoma is highly prevalent in the cotton-top tamarin (*Saguinus oedipus*) (Lowenstine, 2003).

Rectal prolapse is commonly seen in animals that are stressed or have recurrent gastroenteritis. Rectal prolapse can self-reduce or be manually reduced. Nonreducible or severely traumatized rectal prolapses require surgical resection.

Liver

Anatomic features

The rhesus liver is primitive with a left and right lateral lobe placed dorsally, and a single large ventrally placed central lobe (Hartman and Straus, 1933). The caudal border of the liver does not extend beyond the costal arches and the caudal liver margins should be sharp.

Diseases/Disorders

Hepatitis The differential diagnosis for hepatitis is quite extensive and includes bacterial, viral (Hepatitis A, B, C, D, E, G, TT, Callitrichid hepatitis due to lymphocytic choriomeningitis virus in callitrichids and owl monkey, GBV-A-like flavivirus of tamarins) or parasitic (schistosomiasis, trematodiasis).

Amyloidosis appears more disseminated in Asian macaques then in African or New World nonhuman primates. Macaques are prone to systemic amyloidosis following chronic inflammatory processes (i.e., retroviral infections or chronically implanted devices (Lowenstine, 2003). Systemic amyloidosis usually involves the liver and spleen and may or may not include the gastrointestinal tract or kidneys. Animals with severe amyloidosis are usually debilitated and cachectic with sparse, dry hair coats and marked hepatomegaly. Animals with disseminated amyloidosis can live for extended periods of time without severe clinical problems but the condition is irreversible.

Hepatic lipidosis/fatal fasting syndrome has been reported in four macaque species (*M. arctoides, M. fascicularis, M. mulatta, and M. radiate*) and one African Green monkey (*Chlorocebus atheiops*) (Christe and Valverde, 1999). The disease is characterized by rapid weight loss (0.1 kg/day) in obese animals, with anorexia and lethargy often being the only presenting clinical signs. The liver grossly appears enlarged with rounded edges, pale and friable. The kidneys may also be enlarged, pale and soft.

Pancreas

Anatomic features

The macaque pancreas is located just behind the stomach along the transverse axis (Hartman and Straus, 1933).

Diseases/Disorders

Pancreatitis is uncommonly diagnosed antemortem. However, it should be included in the differential list for anorexia and abdominal pain. Pancreatitis is more commonly noted in obese and diabetic monkeys. Radiographic findings may include an increased soft tissue radiopacity and diminished contrast in the right cranial abdomen, the stomach antrum is displaced to left, the proximal duodenum is displaced to right, the transverse colon is displaced caudally and focal mineralization of pancreas may be seen (uncommon).

Peritoneal cavity

Special imaging techniques

Peritoneography

Peritoneography is a radiographic contrast study of the peritoneal cavity using negative or positive contrast media to outline the peritoneal surface of the diaphragm, abdominal wall and serosal surfaces of the abdominal viscera. It is especially useful in assessing the integrity of the diaphragm or abdominal wall for congenital or acquired hernias. It enables the evaluation of the liver lobes and masses associated with the liver and adrenals.

- Evacuate the urinary bladder.
- Provide anesthesia.
- After the skin is surgically scrubbed and prepped; place the animal in a supine position.
- Insert a 19 ga, 1½ needle vertically along the midline, half way between the umbilicus and the pubis.

Proper needle placement is confirmed by aspirating for air, urine or blood (manifestations of improper needle placement). Also, moving the tip of the needle gently from side to side ensures that the needle does not enter the retroperitonium or a solid abdominal organ.

- For a positive contrast study:
  - Inject approximately 200–400 mg iodine/ml of an iodinated contrast at a dosage of 1.1 ml/kg. Use 2.2 ml/kg if abdominal fluid is present (Wallack, 2003)
  - A nonionic contrast is recommended.
  - Barium must never be used for this study.
  - Place the animal prone on a radiographic table turning gently from side to side for about 2 minutes to ensure that the contrast is distributed throughout the peritoneal cavity.
  - Raise the head 35° or hold the animal upright to allow the contrast agent to flow ventrally over the inguinal rings to confirm or identify an inguinal hernia (James *et al.,* 1975). Alternately, place the animal in trendlenburg allowing the contrast agent to flow cranially highlighting a diaphragmatic hernia.
- For a negative contrast study:
  - Infuse CO_2 at a sufficient dose to make the abdomen tympanic. Nitrous oxide or carbon dioxide is preferable to oxygen and room air because of the increased solubility.
  - After positioning, allow 2–3 min so that the gas within the peritoneal cavity has sufficient time to rise.
  - When finished, place the animal in left lateral recumbency so potential CO_2 emboli will localize in the right atrium and be filtered by the lungs decreasing the risk of air embolization and mortality.
  - Obtain lateral and ventrodorsal abdominal radiographic views. Other radiographic projections to consider include, anterioposterior erect, lateral erect, right and left decubitus, recumbent dorsoventral, the opposite lateral and a recumbent lateral using horizontal beam depending on the suspected diagnosis.

Contrast agents are rapidly absorbed from the peritoneal cavity and excreted by the kidneys; a radiograph at 45 minutes post-injection can also demonstrate the renal collecting system. Pneumoperitoneography has shown no serious side effects and the CO_2 is usually absorbed into the body tissues within 1–2 hours (Hoffman *et al.*) Pulmonary function must be monitored during the procedure. If the gas media used was room air, it should be removed at the end of the procedure.

Diseases/Disorders

Diaphragmatic hernias cause a cranial displacement of the diaphragm. They are not related to normal anatomical openings, but are usually, defects in one or more leaflets and the central tendon. Diaphragmatic hernias have been found in many different species such as golden lion tamarins, squirrel monkeys, and a pregnant baboon (Bush *et al.,* 1980; Hendrickx and Gasser, 1967; T-W-Fiennes, 1972).

Inguinal hernias are relatively common. They are most common in males and may progress to an inguinoscrotal hernia. Omentum is most commonly herniated, although intestines may herniate as well. Using positive contrast peritonography on 100 human patients, the accuracy of the herniogram was 97% compared to only 59% with physical exam (James *et al.,* 1975).

Perineal hernia is the result of weakened uterine and perineal ligaments from multiple births and a vertical posture. The bladder frequently herniates into a subcutaneous position in the perineal region, usually during the last trimester of pregnancy. It will often reoccur with each pregnancy.

Peritonitis\Ascites\Hemoabdomen are commonly associated with ruptured hollow viscera. Radiological signs of peritonitis include localized or general loss of abdominal detail and failure to visualize the serosal surfaces of the abdominal viscera. Free abdominal gas is usually associated with rupture of gastrointestinal tract; but a small amount of gas is usually undetectable. If gas is present, the serosal surface of the bowel may be more distinct. Also, since gas tends to rise, it may be visible radiographically adjacent to the diaphragm; the highest point in abdomen is the left side, between the diaphragm and the liver. In chronic peritonitis there may be marked abdominal effusion.

Spleen

Anatomic features

The spleen lies in the upper left quadrant of the abdomen just under the diaphragm. It is triangular in shape and more elongated than a human spleen.

Urinary system

Anatomic features

Nonhuman primate kidneys are flattened and slightly irregular in outline. The cranial pole is sharper than the caudal pole and the left kidney is slightly larger then the right. Both kidneys are firmly attached to the dorsal abdominal wall near the thoracolumbar junction

(Silverman and Morgan, 1980b). The left kidney is located more caudally and ventrally than the right, similar to the renal orientation in the dog and cat, but opposite to the human (Gray, 1959).

Nonhuman primates have unipyramidal kidneys, apart from spider monkeys (*Ateles sp*). Spider monkeys have multipapillary kidneys like humans, pigs, cattle and elephants. It is difficult, therefore, to opacify the renal collecting system of the spider monkey.

Survey radiographs do not identify the renal location or contour well in nonhuman primates, due to the paucity of abdominal fat.

Special imaging techniques

Intravenous pyelography (IVP)

Intravenous pyelography is also known as Excretory Urography (EU), or Intravenous Urography (IVU). It is a radiographic study done after intravenous injection of an iodinated contrast media to visualize the renal structure and the collecting system. Sequential radiographs highlight the renal vasculature (vascular phase), parenchyma (nephrogram phase), renal collecting ducts and ureters (pyelogram phase) as the contrast media is excreted by the kidneys and passes through the renal collecting ducts and ureters into the bladder. An IVP must not be performed in a dehydrated or severely renal compromised patient.

- Animal preparation includes a 12–24 hour fast and appropriate cleansing enemas to evacuate the gastrointestinal tract.
- The animal should be well hydrated.
- Anesthetize as needed.
- Obtain survey radiographs.
- Rapidly inject 880 mg Iodine/kg of an iodinated contrast media intravenously. The dosage should be increased 10% for patients with elevated BUN and creatinine (Wallack, 2003).
- Obtain ventrodorsal and lateral abdominal radiographs immediately after injection of contrast media, and at 0, 5, 15 and 30 minutes. Animals with poor renal function might require additional radiographs taken at 45, 60, 90 and 180 minutes if urinary excretion is delayed.
- The nephogram phase shows dense opacification of both kidneys with complete visualization of the renal contours on the ventrodorsal view.
- Caudial abdominal compression can provide better visualization of the renal collecting system and proximal ureters by delaying drainage of the contrast medium. Tightly bind the caudal abdomen of the animal with an elastic bandaging material to produce ureteral compression. Larger primates with broad or pendulous abdomens, such as chimpanzees, are more difficult to compress (Silverman and Morgan, 1980b).
- The pylogram phase shows the renal pelvis and ureters, seen clearly on ventrodorsal compression radiographs. If not, recheck radiographs at 30 minutes.
- At 15 minutes post-injection, ventrodorsal and lateral radiographs are retaken. If the resulting radiographs show complete filling of the renal pelvis and ureteral dilation proximal to the compression band, compression is released. The major disadvantages of the compression technique are: inconsistent ureteral compression, distortion of the renal pelvis and potential discomfort.
- The draining phase: radiographs (dorsoventral and lateral) are repeated five minutes after removing the compression band. Due to a rich uterine vascular supply, uterine opacification by the contrast media can be observed.

Cystography

Cystography is a contrast study of the urinary bladder used to evaluate its position, morphology, integrity, distensibility, wall thickness and for intraluminal or intramural lesions (tumor, calculi, polyps).

Positive contrast cystography is optimal for evaluating bladder wall integrity and bladder position. Mural masses may be seen if the projection is tangential to the lesion.

- Fast the animal 12–24 hours to evacuate the gastrointestinal tract. Enemas are recommended to empty the colon and rectum.
- Anesthetize as needed.
- Obtain survey radiographs.
- Aseptically catheterize and empty the bladder. Also, flush and remove any possible blood clots.
- Infuse diluted iodinated contrast media (1 part contrast to 4 parts saline) until the bladder is distended (dose range: 2–4 mls/kg). Infuse the contrast when palpating the bladder to avoid over-distension (Owens, 1999).
- Obtain lateral and ventrodorsal oblique radiographic views.

Double contrast cystography provides the best mucosal detail and is optimal for the assessment of urinary calculi and intramural masses.

- Fast the animal 12–24 hours to evacuate the gastrointestinal tract. Enemas are recommended to empty the colon and rectum.

- Anesthetize as needed.
- Obtain survey radiographs.
- Aseptically catheterize and empty the bladder. Also, flush and remove any possible blood clots.
- Infuse a small volume of iodinated contrast media into the bladder (dose range: 1–2 ml for small primates and 2–10 mls for larger species).
- Rotate the patient to coat the bladder mucosa with contrast media.
- Infuse sufficient gas to distend bladder, while palpating to avoid overdistension (dose range: 2–4 ml/kg) (Owens, 1999). Soluble gases, e.g. carbon dioxide and nitrous oxide are preferred, although room air can be used.
- Obtain lateral and ventrodorsal oblique radiographic views.

Pneumocystography highlights the bladder wall, but is the least preferred technique due to the risk of air embolism.

- Use preparation as described in the previous studies.
- Infuse gas (soluble gas is preferred) sufficient to distend the bladder (approximately 2–4 mls/kg) (Owens, 1999).
- Obtain lateral and ventrodorsal oblique radiographic views.

Urethrography

A positive contrast study of the urethra that evaluates the location, morphology and integrity of the urethra.

- Evacuate the colon and rectum with an enema.
- Anesthetize as needed.
- Obtain survey radiographs.
- For retrograde urethrogram, place a balloon-tip catheter in the distal urethra, and inflate the balloon.
  - Infuse iodinated contrast media (dose range: 5–10 ml). Take a lateral projection radiograph at the end of the injection (Owens, 1999).
- A voiding urethrogram can be performed after a cystogram by applying pressure to the urinary bladder with a radiolucent paddle or wooden spoon (Owens and Biery, 1999).

Diseases/Disorders

Renal cysts of varying sizes and numbers are commonly noted in nonhuman primates. Until they compose or compress more than 50% of the renal tissue, or are within the renal medulla, they are generally not clinically significant.

Glomerulonephritis is seen fairly frequently in all nonhuman primates but the callitrichids are especially prone to develop IgM/IgA nephropathy. Immune complexes directed toward parasitic and dietary antigens are hypothesized to play a role (Lowenstine, 2003). Glomerulonephritis has been a major cause of death and is a frequent cause of end stage renal disease in New World primates, especially in owl monkeys, squirrel monkeys, and other cebids. Owl monkeys have an increased incidence and severity of glomerulonephritis with advancing age (Jones *et al.,* 1993a).

Hydronephrosis is a dilatation of the renal pelvis and calyces, leading to flattening of the renal papillae and atrophy of the renal cortex. Hydronephrosis is a frequent result of urinary tract obstruction (e.g., sequelae of endometriosis, tumors, renoliths).

Urolithiasis. The renal pelvis and calyces are sites for formation and accumulation of calculi. Renoliths may be well tolerated, but in some cases they lead to severe hydronephrosis. In the advent of pyelonephritis, renoliths can become a nidus of infection. Urolithiasis is rare in nonhuman primates while nephrocalcinosis is more common (Faltas, 2000; Silverman and Morgan, 1980b).

Obstructed bladder occasionally occurs in rhesus males during the breeding season when they have retrograde ejaculation and the coagulum obstructs the urethra or bladder. This is corrected by urinary catheterization or an emergency cystotomy.

Reproductive System

Prostate

Anatomic features

The nonhuman primate prostate does not form a ring encircling the urethra but lies solely on the posterior and lateral surfaces. It consists of two histologically distinct divisions, the cranial and caudal lobes (Hartman and Straus, 1933). The cranial prostate supplies the semen coagulation component, while the caudal prostate is homologous to the human prostate. Only the marmoset (*Callithrix jacchus*) and the orangutans (*Pongo pygmaeus*) have a single lobed prostate (Harrison and Lewis, 1986).

Uterus/Vagina

Anatomic features

A common characteristic of all prosimians is a bicornuate uterus. The rhesus macaque has a unicornuate uterus that is slightly flattened anteriorposteriorly. In many primate species, the cervical canal is not straight, but has varying degrees of tortuosity (Hafez and Jaszczak, 1972).

Special imaging techniques

Hysterosalpingography The fallopian tube patency can be assessed with a hysterosalpingography despite the tortuosity of the primate cervix. Once anesthetized, a blunt 18 ga. needle or catheter is placed in the cervical canal and manipulated until it passes through to the uterine cavity, and a positive contrast medium (approximately 1 ml) is injected. Good quality hysterosalpingograms can be obtained and radiopaque material can be visualized in both fallopian tubes in 75% of monkeys (Bennett *et al.,* 1995).

Vaginography Vaginography is a radiographic study using a retrograde infusion of positive contrast media, performed to evaluate vagina, cervix and urethra including strictures, fistulas, masses, or ectopic ureters.

- A balloon-tip catheter is inflated inside the vestibule.
- Undiluted iodinated contrast media is infused until the vagina is adequately filled.
- A lateral radiographic view of the pelvis is obtained.

Pelvimetry Pelvimetry of squirrel monkeys is a good predictor of perinatal mortality (Bennett *et al.,* 1995). Lateral and anterioposterior pelvic radiographs of Bolivian squirrel monkeys were obtained 6 months postpartum. The pelvic inlet, midpelvis and pelvic outlet dimensions were measured and compared, based on the outcome of parturition. It was concluded that the size of the pelvis was a determinant of the outcome of pregnancy. A 10% reduction in pelvic outlet size resulted in a 20% reduction in the area of the birth canal. The size difference was only 0.17 cm, which is a minor difference to be detected by a physical measurement.

Diseases/Disorders

Pregnancy detection Pregnancy is always abdominal in nonhuman primates (Abitbol, 1993). Radiographs can be used to confirm pregnancy, especially in the later stages of pregnancy when the skeleton calcifies. Bone maturation and ossification can be used as growth standards and age indicators in experimental animals whose exact age is unknown. However, significant acceleration of bone maturation in nonhuman primates is observed, compared to humans. For example, the onset of wrist ossification of the rhesus macaque is approximately 120 days of gestation compared to a chimpanzee fetus in the last four weeks of gestation or a full-term newborn human (Michejda, 1980).

Endometriosis This is one of the most common reproductive disorders in Old World primates. It is defined clinically as the presence of both endometrial glands and stroma outside of the uterine cavity. The ectopic endometrial tissue responds to cyclic hormonal activity and proliferates, necroses and sloughs just as it would inside the uterus. This tissue either forms a cystic structure (endometrioma or "chocolate" blood-filled cysts) or sloughs into the peritoneal cavity where it can cause peritonitis or abdominal adhesions.

Uterine leiomyomas These smooth muscle tumors are common in all species of nonhuman primates (Lowenstine, 2003).

Neurologic system

Anatomic features

The thoracolumbar spine of the rhesus macaque consists of 20 segments, the anticlinal region being near the 10th thoracic segment. The thoracic spine is slightly kyphotic and the caudal lumbar spine is slightly lordotic. There are 13 thoracic vertebrae, all bearing ribs. There are 7 sternebrae.

Special imaging techniques

Myelography

In myelography the spinal cord is outlined by a nonionic contrast media injected into the subarachnoid space.

- Obtain survey spinal column radiographs with the animal under general anesthesia.
- Perform aseptical spinal puncture of the subarachnoid space of either the cysterna magna or at the caudal lumbar spine (L5–L6) using an appropriate size spinal needle.
- Collect the cerebrospinal fluid for cytological analysis, serology titers, and culture sensitivity, if clinically indicated.

- Slowly inject a nonionic iodinated contrast medium (e.g., iohexol, iopamidol) at a dose to fill the subarachnoid space (dose range 0.25–0.5 ml/kg) (Owens, 1999). Iodine concentration: 200–300 mg iodine/ml at a dosage of 0.45 ml/kg (Wallack, 2003).
- A test injection with a small fraction of the total dose of contrast is given to verify the correct needle placement in the subarachnoid space. After the test dose is injected, a radiograph can be obtained with the needle still in place.
- After the total dose of contrast is administered, obtain lateral and ventrodorsal radiographs of the spine. Additional projections, such as oblique and stressed positions (extended and flexed lateral, traction view), are taken as needed.
- Depending on the flow of the contrast media, the body may be tilted to aid in moving the contrast and for better filling of the subarachnoid space at a specific site.

Epidurography

Epidurography is a radiographic contrast investigation of the epidural space using positive contrast medium to assess the cauda equina and proximal portions of the nerve roots. The main indication for epidurography is in the assessment of the lumbosacral region of the vertebral column or to confirm placement of epidural catheters (De Weert *et al.,* 1995).

- Obtain survey spinal column radiographs with the animal under general anesthesia.
- With the animal in sternal or lateral recumbency, place a spinal needle aseptically into the floor of the spinal canal through the lumbosacral (L7-S1) or coccigeal interacuate space (S3-Co1 or Co1-Co2). The L7-S1 intervertebral space can be located on the midline at an intersection of a line connecting the iliac crests.
- To identify the epidural space, attach a 3-6 ml glass syringe to the spinal needle to assess for loss of resistance. Loss of resistance is identified by slowly advancing the spinal needle while gently tapping on the glass syringe plunger until air resistance is no longer encountered.
- Inject a nonionic iodinated contrast media at a dose to fill the epidural space (dose 0.15 ml/kg). Iodine concentration: 200–300 mg Iodine/ml at a dose of 5 ml (Wallack, 2003).
- Remove the needle and obtain radiographs including lateral, and ventrodorsal or dorsoventral projections.

Discography

In discography, positive contrast medium is injected into the nucleous pulposus of the intervertebral disc. Normally, only a very small amount can be introduced. If there is damage of the annulus fibrosus, more contrast can be injected and the leakage will be evident on subsequent radiographs. This technique is usually reserved for diagnosis of protrusion of the L7 intervertebral disc.

Musculoskeletal

Skeletal radiographic procedures are performed similarly to the methods described for dogs (Morgan *et al.,* 1975). The radiographic changes of bone are a reflection of the underlying disease process. Radiographic abnormalities include alterations of size, shape, contour and radiopacity.

Appendicular skeletal maturation

Longitudinal radiographic evaluation of skeletal maturation in rhesus monkeys demonstrated that the majority of appendicular ossification centers were identified, radiographically, by 175 days of age and that physeal closure was complete at 7.2 years in females and 7.3 years in male rhesus monkeys (Silverman *et al.,* 1983). The order of physeal closure was similar, but not identical, in both sexes.

Special imaging techniques

Arthrography

Arthrography is a radiographic contrast study of a joint used to delineate joint margins and to assess the articular cartilage, intraarticular ligaments, tendons, meniscus, and joint capsule.

- General anesthesia is required.
- Obtain survey radiographs of the joint, including lateral and caudocranial projections.
- Perform an aseptic articular puncture. Remove as much joint fluid as possible. Analysis of the synovial fluid, serology titers and culture and sensitivity is performed if clinically indicated.
- Inject a nonionic contrast medium for a positive contrast study (preferable).
- Recommended dosage based on a study of the shoulder joint on dogs ranging from 23 to 44 kg (Wallack, 2003).

  - Iopamidol at a concentration of 100 mg Iodine/ml. Dose: 0.4 ml/kg (Wallack, 2003).
  - M-S Diatrizoate, M-S Ioxaglate and Iopromide at a concentration of 140 mg Iodine/ml. Dose: 1.5–4 ml/joint (Wallack, 2003).
- Dilute contrast with isotonic sterile saline to achieve recommended iodine concentration (Wallack, 2003).
- Remove the needle and manipulate the joint to ensure uniform filling of the joint.
- Obtain radiographs of the joint, including lateral, caudocranial and lateral oblique projections.

Diseases/Disorders

Noma or cancrum oris, derived from the Greek noun "to devour," is an acute gangrenous process which most frequently affects the oral cavity, particularly the gingiva, cheeks and lips, often producing bone denudation, sequestration and extensive facial disfigurement (Adams, 1980; Lackner *et al.,* 1993). The occurrence of noma is highly suggestive of infection with an immunosuppressive type D retrovirus. Radiographic findings might include osteonecrosis, osteomyelitis, and bone sequestration.

"**Potts' disease"** has been described in nonhuman primates as tuberculous lesions involving the vertebrae and adjacent spinal cord (Jones *et al.,* 1993a; Jones *et al.,* 1993b).

Calcinosis circumscripta is a deposition of amorphous calcium salts in the subcutaneous tissue and skin, usually on the extremities, and over bony prominences. Occasionally the nodules ulcerate and drain a chalky white semisolid substance. A histologic examination is required to differentiate types of mineralization (e.g. ossification or calcification). Radiographically, there is diffuse soft tissue calcification (Line *et al.,* 1984).

Metabolic bone disease. Radiographic lesions do not occur until late in the course of metabolic bone disease (MBD), after 40% of the osseous mineral has been absorbed. Nonetheless, radiography is frequently the key to diagnosis. Typical findings include marked osseous demineralization, decreased cortical density, thinning of the cortices, increased trabecular pattern and folding fractures in long, weight supporting bones. The lamina dura dentis disappears naturally with skeletal maturity and in the early in the course of MBD in young animals (Fowler, 1978).

Nutritional secondary hyperparathyroidism is the excessive production of parathyroid hormone as a response to hypocalcemia, resulting in calcium resorption from bone. Under special dietary deficiencies, calcium resorption becomes detrimental to the integrity of the bone. The ultimate result is rickets in the young animal and osteomalacea in the adult. It presents as a progressive nutritional disease characterized by hyperphophatemia, hypocalcemia, increased alkaline phosphatase, soft tissue mineralization, impaired locomotion, and poor skeletal mineralization (Tomson *et al.,* 1978; Snyder *et al.,* 1978). Radiographic findings include generalized areas of decreased bone mineralization in the axial and appendicular skeleton, and a wide variety of pathological fractures. It might include thin bone cortices of the long bones, bony deformities caused by folding and stress fractures of long bones, and increased soft tissue swelling at areas of tendinous attachment, e.g. patella, tuber calcis, and olecranon (Martin, 1978).

Rickets is a failure of mineralization of osteoid or cartilaginous bone matrix in young growing animals. Radiographically, there is a widening of the radiolucent epiphyseal plate, bowing of long bones, and widening of the metaphysis ("cupping").

Osteomalacia is characterized by softening of bone and a decrease in bone density caused by insufficient mineralization of osteoid in the adult bone. It is a common result of metabolic bone disease. Radiographically, there is loss of bone density, thinning of cortices, coarsened trabecular pattern, mottled radiolucent areas, folding fractures, and bowed long bones.

Vitamin D deficiency. New World primates require dietary vitamin D3 to maintain normal skeletal mineral homeostasis. Failure to supply vitamin D3 to growing animals results in rickets, and osteomalacia in adults. Osteodystrophia fibrosa is considered to be a descriptive form of these conditions (Potkay, 1992).

Fibrous osteodystrophy is a condition that may be seen as a result of mineral imbalance or osteoporosis. There is an osteoclastic resorption of osteoid being replaced by a highly cellular connective tissue. Bones of the face and mandible are affected most frequently. It is a common manifestation of nutritional secondary hyperparathyroidism in New World primates such as spider and woolly monkeys. The mandible may become soft and pliable ("rubber jaw"). Radiographically, there is a marked decrease in bone density at the craniofacial bones.

Osteoporosis is a condition in which resorption of osteoid overbalances the deposition of new bone. Osteoporosis may be observed by protein malnutrition, hyperadrenocorticism (Cushing's disease), and iatrogenically by the exogenous administration of corticosteroids. Radiographic signs are detected only in advanced disease; bones are light, brittle and fragile. In young growing animals the sequence of cartilaginous transformation to bone may be delayed. Cancellous bone is primarily involved, non-weight bearing trabeculae being the first to be resorbed. Radiographic signs include thinning of the cortices with a corresponding increase in the medullary space.

Vitamin C deficiency or **Scurvy.** Most severe bone and joint changes associated with scurvy result from deficiency during periods of rapid bone growth. In the growth plate of long bones there is disruption of the normal maturation process. Radiographic signs include a transverse methaphyseal "white line" attributed to the thickening of the provisional zone of calcification on the growth plate and the "scurvy line", a transverse zone of rarefaction shaftward to the "white line" (Eisele *et al.,* 1992, Ratterree *et al.,* 1990).

Additional radiographic findings were epiphyseal separations and displacements, peripheral metaphyseal clefts, cortical thinning, enlarged costochondral junctions, subperiosteal hemorrhages and physeal fractures (Eisele *et al.,* 1992; Martin, 1978; Ratterree *et al.,* 1990). Subsequent radiographs demonstrate angular deformities associated with physeal fractures. Subperiosteal cranial hemorrhages, or cephalohematomas, is a well known manifestation of vitamin C deficiency in squirrel monkeys. Subsequently, hyperostosis occurs due to deposition of new bone by the periosteum (Demaray *et al.,* 1978; Blackwell *et al.,* 1974).

Fluoroscopy

Fluoroscopy provides real time radiographic viewing of moving anatomic structures. It can evaluate respiratory function or assess motility and function of the pharynx, esophagus, stomach and bowel. Fluoroscopy is quite useful in interventional studies involving directed aspirates, biopsies, and catheter placements. Serial jejunal biopsies (bacterial culture, virus isolation, IgA levels, administration of therapeutic or experimental agents directly into the proximal small intestines, by-passing acid secretions) can be obtained with a steerable catheter and fluoroscopy (Ford *et al.,* 1989). The advantage of this technique is the non-invasive nature of the procedure vs. operative alternatives.

Nuclear imaging

Nuclear imaging is a modality that provides diagnostic information on the functional status of an organ or body part. Radiopharmaceutical drugs that emit radioactive gamma rays are administered into the patient and, after the radionuclide has been deposited in the organ or tissue of interest, a gamma scintillation camera is used to detect ionizing gamma rays emitted from the animal's body. The radionuclide distribution is usually recorded as an image on x-ray film. Examples of scintigraphy studies are: thyroid, brain and bone scintigraphy. Scintigraphy imaging can also be used to observe and evaluate the presence of portosystemic shunts, gastric emptying, renal function, pulmonary perfusion, and for pulmonary ventilation studies (Owens, 1999).

Correspondence

Any correspondence should be directed to Dr. Kari L. Christe, California National Primate Research Center, University of California, Davis, One Shields Avenue, Davis, CA 95616-8542, USA. Phone: (530) 752-0447; Fax: (530) 752-2880; Email: cnprc_info@primate.ucdavis.edu

References

Abitbol, M.M. (1993). *Am. J. Phys. Anthropol.* 91, 367–378.

Adams, R.J. (1980). In Montali, R.J. and Migaki, G. (eds) *The Comparative Pathology of Zoo Animals,* pp 77–85. Smithsonian Institution Press, Washington.

Baer, J.F., Weller, R. E. and Kakoma, I. (1994). *Aotus: The Owl Monkey.* Academic Press, San Diego.

Bennett, B.T., Abee, C.R. and Henrickson, R. (1995). *Nonhuman Primates in Biomedical Research.* Academic Press, San Diego.

Blackwell, C.A., Manning, P.J., Hutchinson, T.C. and Fisk S.K. (1974). *Lab. Anim. Sci.* 24(3), 541–544.

Bonakdarpour, A., Lynch, P.R., Lapayowker, M.S. and Stauffer, H.M. (1967). *Invest. Radiol.* 2, 432–441.

Boorman, G.A., Silverman, S. and Anderson, J.H. (1976). *Lab. Anim. Sci.* 26, 942–947.

Breznock, A.W., Henrickson, R.V., Silverman, S. and Schwartz, L.W. (1975). *J. Am. Vet. Med. Assoc.* 167, 657–661.

Bush, M., Montali, R.J., Kleiman, D.G., Randolph, J., Abramowitz, M.D. and Evans, R.F. (1980). *J. Am. Vet. Med. Assoc.* 177, 858–862.

Buss, D.D., Hyde, D.M., Poulos, P.W. Jr. (1982). *Anat. Rec.* 203(3), 411–417.

Christe, K.L. and Valverde, C.R. (1999). *Contemp. Top. Lab. Anim. Sci.* 38, 12–15.

Cicmanec, J.L. (1977). In Kleiman, D.G. (ed.) *The Biology and Conservation of the Callitrichidae,* pp 331–336. Smithsonian Institution Press, Washington.

Converse, J.L., Lowe, E.P., Castleberry, M.W., Blundell, G.P. and Besember A.R. (1962). *J. of Bacteriology* 83, 871–878.

Demaray, S.Y., Altman, N.H., Ferrell, T.L. (1978). *Lab. Anim. Sci.* 28(4), 457–460.

De Weert, T.M., Golub, M.S. and Kaaekuahiwi, M.A. (1995). *Lab. Anim. Sci.* 45, 94–97.

Eisele, P.H., Markovits, J.E. and Paul-Murphy, J.R. (1991). *Lab. Anim. Sci.* 41(5), 436–441.

Eisele, P.H., Morgan, J.P., Line, A.S. and Anderson, J.H. (1992). *Lab. Anim. Sci.* 42, 245–249.

Ewers, R.S. and Hofmann-Parisot, M. (2000). *Vet. Re.* 147, 7–11.

Faltas, N.H. (2000). *Contemp. Top. Lab. Anim. Sci.* 39, 18–19.

Ferron, R.R. (1966). *Lab. Anim. Care* 16, 459–464.

Ferron, R.R. (1967). *Lab. Anim. Care* 17, 594–600.

Ford, E., Anderson, J., Cox, K., Dencer, M. and Cello, S. (1989). *Lab. Anim. Sci.* 39, 609–612.

Fowler, M.E. (1978). *Zoo and Wild Animal Medicine,* pp 55–76. Saunders, Philadelphia.

Goldberg, H.I., Gould, R., Rosenquist, J., Royal, S., Owen, R.L. and Silverman, S. (1982). *Radiology* 142, 53–58.

Gray, H. (1959). In Goss, C.M. (ed.) *Anatomy of the Human Body,* pp 1326–1327. Lea & Febiger, Philadelphia.

Greenberg, M.J., Janssen, D.L., Jamieson, S.W., Rothman, A., Frankville, D.D., Cooper, S.D., Kriett, J.M., Adsit, P.K., Shima, A.L., Morris, P.J. and Sutherland-Smith, M. (1999). *J. Zoo. Wildl. Med.* 30, 256–261.

Hafez, E.S.A. and Jaszczak, S. (1972). *Primates* 13, 297.

Hahn, N.E., Lau, D., Eckert, K. and Markowitz, H. (2000). *Comp. Med.* 50, 556–558.

Harrison, M.R. and Lewis R.W. (1986). *The Male Reproductive Tract and its Fluids. Comparative Primate Biology, Vol 3: Reproduction and Development,* pp 101–148. Alan R. Liss Inc.

Harris, T.R., Pratt, P.C. and Kilburn, K.H. (1971). *Am. J. Med.* 50, 756–763.

Hartman, C.G. and Straus, W.L. (1933) *The Anatomy of the Rhesus Monkey* (Macaca mulatta). The Williams & Wilkins Company, Baltimore.

Hendrickx, A.G. and Gasser, R.F. (1967). *Folia Primatol.* (Basel) 7, 66–74.

Hilloowala, R.A. (1976). *Acta. Anat.* (Basel) 95, 260–278.

Hoffman, R.A., Silverman, S. (1975). *Lab. Anim. Sci.* 25(5), 609–613.

James, A.E., Jr., Heller, R.M., Jr., Bush, M., Gray, C.W. and Oh, K.S. (1975). *J. Med. Primatol.* 4, 114–119.

Jones, T.C., Mohr, U. and Hunt, R.D. (1993a). *Nonhuman Primates I.* Springer-Verlag. Distribution rights for North America Canada and Mexico by ILSI, Berlin; New York, Washington, DC.

Jones, T.C., Mohr, U. and Hunt, R.D. (1993b). *Nonhuman primates II.* Springer-Verlag. Distribution rights for North America Canada and Mexico by ILSI, Berlin; New York, Washington, DC.

Kato, T., Yasue, T., Shoji, Y., Shimabukuro, S., Ito, Y., Goto, S., Motooka, S., Uno, T. and Ojima, A. (1987). *Acta. Pathol. Jpn.* 37, 361–373.

Keller, G.L., Kramer, L., Butler, W.B. and Knapke, F.B., Jr. (1982). *J. Zoo. Anim. Med.* 13, 148–151.

Lackner, A.A., and Schiodt, M.S. (1993). In Jones, T.C., Mohr, U. and Hunt, R.D. (eds) *Nonhuman Primates I. Monographs on Pathology of Laboratory Animals,* pp 70–3. Springer-Verlag. Distribution rights for North America Canada and Mexico by ILSI, Berlin; New York, Washington, DC.

Lapin, B.A. (1963). *Comparative Pathology in Monkeys.* Charles C. Thomas, Springfield, Ill.

Line, S.W., Ihrke, P.J. and Prahalada, S. (1984). *Lab. Anim. Sci.* 34, 616–618.

Lowenstine, L.J. (2003). *Toxicol Pathol* 31 Suppl, 92–102.

Martin, D.P. (1978). In Fowler, M.E. (ed.) *Zoo and Wild Animal Medicine,* pp 525–552. Saunders, Philadelphia.

Michejda, M. (1980). *Dev. Biol. Stand.* 45, 45–50.

Mook, D.M. (2002). *Comp. Med.* 52, 560–562.

Morgan, J.P., Silverman, S., Zontine, W.J. (1975). *Techniques of Veterinary Radiology.* Veterinary Radiology Associates, Davis.

Newton, W.M., Beamer, P.D. and Rhoades, H.E. (1971). *Lab Anim Sci.* 21(2), 193–196.

Owens, J.M. and Biery, D.N. (1999). *Radiographic Interpretation for the Small Animal Clinician.* Williams & Wilkins, Baltimore.

Owens, J.M., Biery, D.N. and Tennant, J. (1982). *Radiographic Interpretation for the Small Animal Clinician.* Ralston Purina Company, St. Louis, Mo.

Potkay, S. (1992). *J. Med. Primatol.* 21, 189–236.

Ratterree, M.S., Didier, P.J., Blanchard, J.L., Clarke, M.R. and Schaeffer, D. (1990). *Lab. Anim. Sci.* 40, 165–168.

Rosenberg, D.P., Link, D.P. and Prahalada, S. (1983). *Lab. Anim. Sci.* 33, 183–186.

Silverman, S. (1975). *Lab. Anim. Sci.* 25, 748–752.

Silverman, S. and Morgan, J.P. (1980a). *Am. J. Vet. Res.* 41, 1704–1719.

Silverman, S. and Morgan, J.P. (1980b). *Vet. Radiology* 21, 213–223.

Silverman, S., Morgan, J.P., Ferron, R., McNulty, W. Merten, D. (1983). *Vet. Radiology* 24, 25–34.

Snyder, S.B., Mdahl, J.L., Law, D.H., Froelich J.W. (1978). In Montali, R.J., Migaki, G. (eds) *The Comparative Pathology of Zoo Animals.* pp 51–57.

Stender, H.S. (1990). *Bildqualitat in der Rontgendiagnostik,* pp 65–78. Arzte-Verlag, Koln Deutscher.

TerBrugge, K., Lasjaunias, P., Chiu, M.C., Marotta, T.R., Tatton, W. and Glen, J. (1989). *Am. J. Neuroradiol.* 10, 1203–1208.

Tomson, F.N., Keller, G.L. and Knapke, F.B. (1978). In Montali, R.J. and Migaki, G. (eds) *The Comparative Pathology of Zoo Animals,* pp 59–64.

T-W-Fiennes, R.N. (1972). *Pathology of Simian Primates.* S. Karger, Basel, New York.

Valverde, C.R., Tarara, R.P., Griffey, S.M. and Roberts, J.A. (2000). *Comp Med* 50, 540–544.

Wallack, S.T. (2003). The Handbook of Veterinary Contrast Radiography. Seth T. Wallack, Solana Beach, CA.

CHAPTER 23

Imaging: Positron Emission Tomography (PET)

Svetlana Chefer
National Institute on Drug Abuse, National Institute of Health, Department of Health and Human Services, 5500 Nathan Shock Drive, Baltimore, Maryland 21224, USA

Introduction

Positron emission tomography (PET) and single photon emission computed tomography (SPECT) are radionuclide imaging methods providing non-invasive measurements of three-dimensional distribution of radio-labeled compounds in living organisms. Radionuclide imaging is an *in vivo* analog to tissue dissection, autoradiography, and other techniques that involve either sectioning and imaging or counting excised tissue samples taken from animals into which a compound, labeled with radioisotopes, has been introduced. However, instead of film, an array of scintillation detectors is used to obtain images.

Imaging technologies are playing a growing role in research involving non-human primates, enabling large, expensive laboratory animals to be studied non-invasively as well as reducing the number of smaller laboratory animals required for an experiment. The technology allows both the temporal and the spatial biodistribution of a radiotracer to be determined in a living organism (Figure 23.1). Since the same animal can be used repeatedly, the variability caused by inter-animal differences is removed. Moreover, the ability to study an animal more than once allows each animal to serve as its own control, improving the statistical power of the study and permitting interventional strategies to be followed over time.

Based on the radiotracer principle developed by George De Hevesey at the beginning of the 20th century (De Hevesy and Chiewitz, 1935), these methods conceptually consist of three components: (i) scanners that provide tomographic images of the concentration of radioactivity in the body; (ii) ligands labeled with radioactive atoms; and (iii) mathematical models that describe the *in vivo* behavior of radioligands and allow the physiological process under study to be quantified from the image.

The Laboratory Primate
Copyright 2005 Elsevier
ISBN 0-1208-0261-9

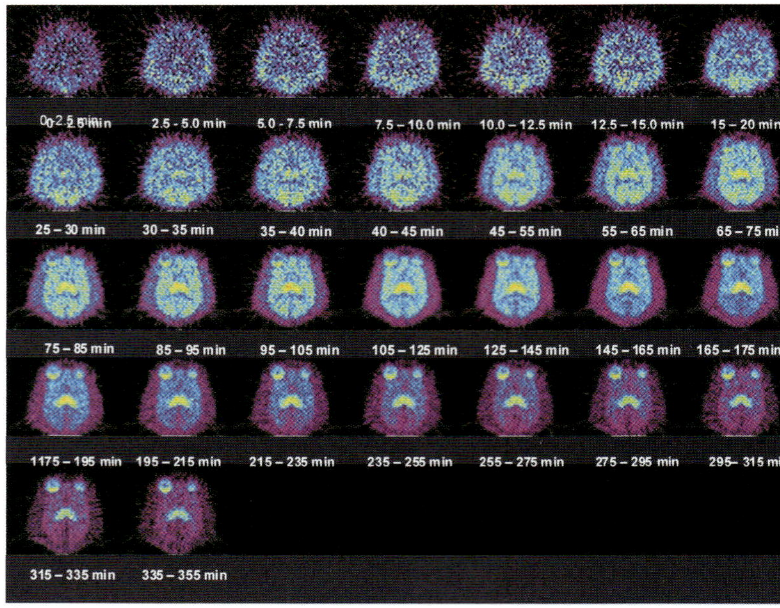

A.

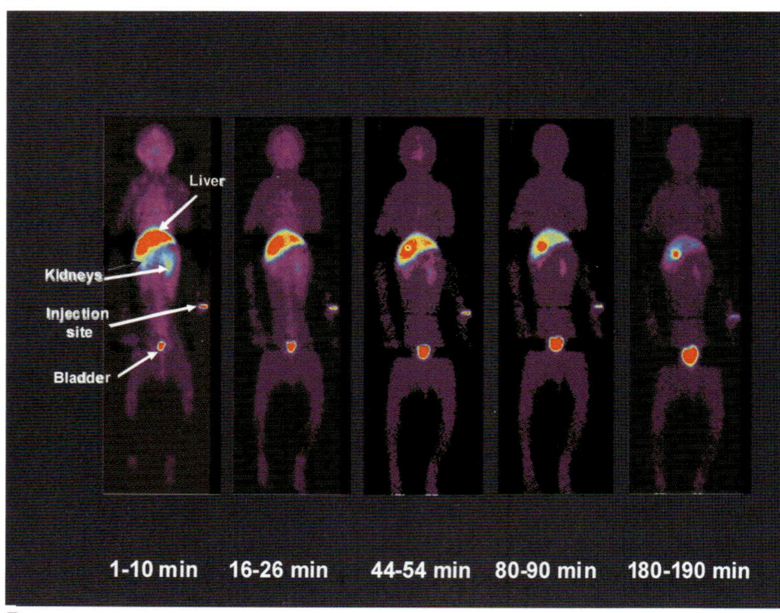

B.

Figure 23.1 PET images of Rhesus monkeys' brain (A) and whole body (B) acquired in two animals showing the accumulation and distribution of 2- [^{18}F]F-A-85380 (a ligand for $\alpha4\beta2^*$ subtype of nicotinic acetylcholine receptors (nAChRs)) over time. **A.** Following intravenous bolus injection of 2- [^{18}F]F-A-85380 at a dose 0.8 mCi/kg, the radiotracer is delivered through the major blood vessels (0–2.5 min), distributed through the brain (2.5–105 min) and sequestered in the thalamus (>105 min), the region with the highest density of $\alpha4\beta2^*$ nAChRs, while tracer in other parts of the brain slowly clears. **B.** Most of the radioaligand immediately after the injection (dose 2.3 mCi/kg) is accumulated in the liver. Over time the radiotracer is cleared from the body and collected in the urine (urinary bladder). Images are obtained using ECAT HR+ clinical tomograph at Neuroimaging Research Branch, IRP, NIDA, NIH.

Principles of emission computed tomography

Physics and instrumentation

The principles for detecting the emerging radiations and forming images are similar for single photon emitters and positron emitters, but the underlying physics and the instruments employed are different. In the case of radiotracers labeled with a single photon-emitting isotope, such as iodine-123 or technetium-99m, nuclear decay results in the emission of uncorrelated photons in the 80–350 keV energy range. Gamma cameras use the principal of mechanical collimation to detect and localize the photons (Figure 23.2).

A lead collimator, with many holes, is used to allow photons traveling in a preferential direction to continue on to interact with a scintillation crystal. Most of the photons, not traveling in the preferential direction of the collimator holes, are stopped by the lead septa. This dramatically reduces the detected gamma ray event rate per unit of injected dose of radioactivity (or sensitivity of the system). The photons with this directionality are recorded at each of many projections around the organ being imaged. The detector typically comprises a thin crystal slab of thallium-activated sodium iodide (NaI(TI)) which absorbs the gamma ray and, in response, emits visible light protons which are converted to an electrical signal by an array of photomultiplier tubes (PMTs). The signals produced by the PMT array indicate the position of interaction and energy of the gamma ray. This information is used to produce a planar image of the distribution of radioactivity in the body. In order to obtain tomographic images, the camera head is rotated around the body to obtain multiple views. These are combined to reconstruct tomographic images of the radioactivity distribution.

Systems that create images from positron-emitting tracers, such as carbon-11, nitrogen-13, oxygen-15, and fluorine-18, use the principle of electronic collimation. These systems take advantage of the fact that positrons (positive electron antiparticles) emitted from radionuclides travel a short distance, typically about 2 mm to 7 mm range. They lose the kinetic energy during this flight and then interact with orbital electrons through a process known as pair annihilation. In annihilation,

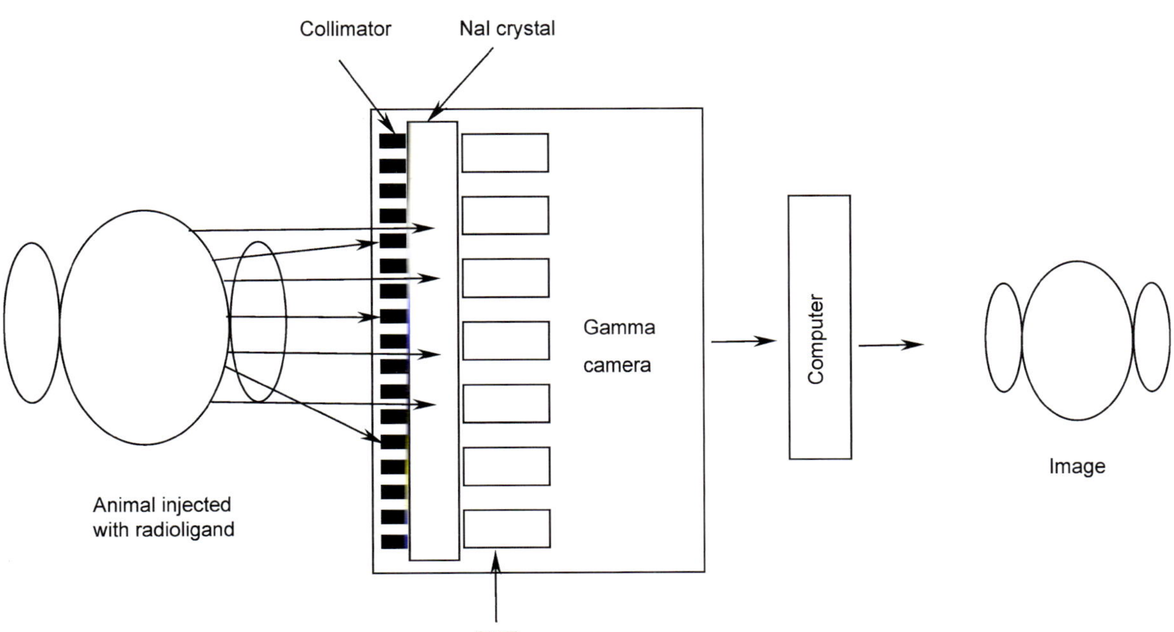

Figure 23.2 Basic component of a single photon imaging device used for SPECT. Shown are the collimator, sodium iodide crystal detector behind the collimator and photomultiplier tubes. Photons or gamma rays pass through the parallel holes in the collimator to the crystal and are registered by PMTs and electronic circuitry behind the crystal. As the head rotates around the subject, images are collected at each scanning angle. Scan profiles for these images are used as an input for the SPECT reconstructed algorithm to obtain tomographic images of the distribution of radioactivity in the body.

the mass of the electron and positron pair is usually converted into energy in the form of photons that are equal in energy to the rest mass of an electron (511 keV each), but travel in almost exactly opposite directions, very close to 180 degrees apart. The photons emitted can penetrate tissues at nearly the speed of light.

The detection device consists of a cylindrical array of scintillation crystals, commonly bismuth germanate (BGO) or a continuous sheet of sodium iodide scintillator, which has good stopping power for the relatively energetic annihilation photons. The detectors are coupled with the PMTs, which convert the light photons into electrons and amplify the signal by a factor of approximately 10^6. Most of the PET cameras have detectors arranged in a series of rings to image the whole organ at one time. Thus, the scintillation detector converts the interaction of a 511 keV gamma ray, in the scintillator, into a robust current pulse than can be detected and processed by relatively standard electronics. Detection of photon pairs is used to measure both the location and the amount of radioactivity in the field of view (FOV) of the scanner. The two annihilation photons are detected by two opposing radiation detectors connected by an electronic coincidence circuit (Figure 23.3). Thus, if the two detectors record an event within a predetermined time window (so-called coincidence, the time window for coincidence detection is typically 5–20 ns), then the pair annihilation must have taken place somewhere in a straight line between the two detectors. These events are registered and then processed by an algorithm that allows an image to be reconstructed by computer. Images are then created as the intersection of the lines by using planes cutting across the object of interest. Since electronic circuitry has provided the direction of the emission, the system is said to use electronic collimation, in contrast to the mechanical lead collimator with hole used in conventional SPECT imaging. Because the direction of an incident gamma ray is defined electronically, PET has at least one order of magnitude advantage in sensitivity against SPECT (sensitivity is defined as the ability of the system to detect as much as possible coincidence per unit of radioactivity in the FOV).

The data acquired by SPECT and PET scanners are projections of the radiotracer distribution along many lines of sight. Therefore, the reconstruction process can be compared to drawing the lines. Using mathematical reconstruction methods, similar to those used in X-ray and CT imaging, and appropriate correction factors for phenomena such as gamma rays attenuation and scatter, it is then possible to compute cross-sectional images through the object that quantitatively reflect the concentration of the radionuclide inside the object. The intensity of each point or pixel in the reconstructed image is proportional to the concentration of radioactivity at the corresponding location in the body.

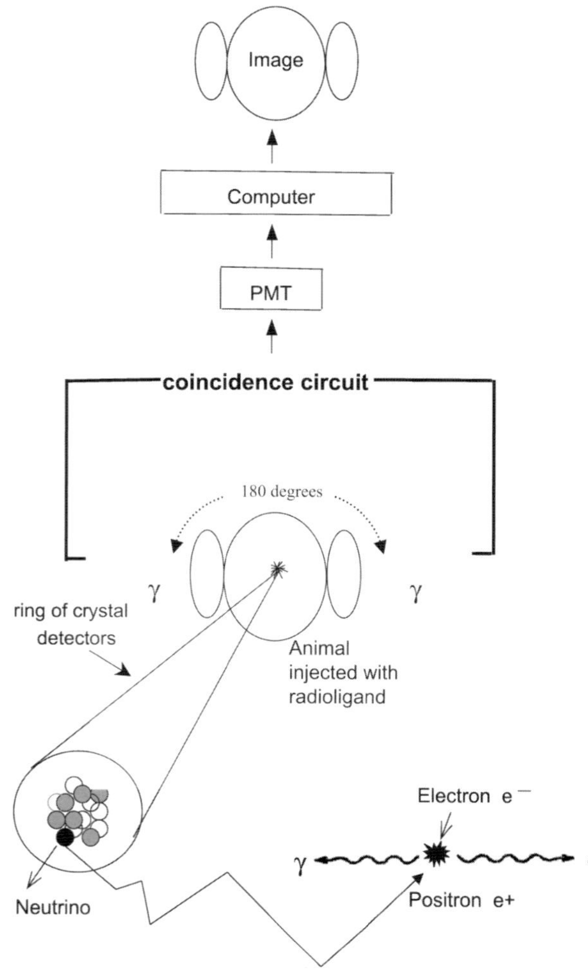

Figure 23.3 Physics of the positron emission tomography measurement. When positron is emitted from a positron emitter, it travels a few mm before colliding with an electron, which results in production of two photons with an energy of 511 keV emitted at almost 180°. Because of the simultaneous emission, the presence of the positron can be detected in temporal coincidence of two gamma-rays. Thus, when two are detected within a time of coincidence window, the compound labeled with positron-emitter is located somewhere in the volume delineated by the two coincidence detectors with a scintillation crystal. These photons are detected to record a line of responses through the scanner's field of view. The raw PET data are essentially a list of these lines of response (known as list-mode data).

Although both SPECT and PET detect radiotracer distribution, it is the chemical versatility of the positron emitters, the ability to measure their concentration quantitatively, with relatively little attenuation by tissue, the greater sensitivity and the superior resolution of PET which differentiates these two methods.

Spatial resolution

The physical characteristic of the imaging systems will directly affect the precision of quantitation in the image. Quantitation is evaluated as a degree of accuracy with which the instrument measures the regional concentration of emitters in a three-dimensional region of interest. Spatial resolution is the ability of the camera to separate adjacent structures and resolve sharp detail.

The spatial resolution of a PET scanner is determined by the size of the scintillator crystals or, in the case of a continuous sheet of scintillator, the thickness of the sheet and the size of the PMTs. The total number of detectors, their size and cross-sectional geometry, the type of crystals used, the number of rings and the packing and shielding of the detectors will have a major influence on spatial resolution of the system. Spatial resolution of PET cameras is characterized by the full width, at half maximum (FWHM) height, of an image of a narrow line source of positron-emitting radioactivity. This will affect quantitation as much as two sources, separated by a distance equivalent of one FWHM, will not resolve a separate source in the image. Spatial resolution is determined by imaging a thin line source of radioactivity, positioned perpendicular to the plan of the scan. A plot of the image intensity (number of counts) along the line shows that this spreading approximates a bell-shaped or gaussian curve. The width of this curve, at one-half of its maximum height (termed the full width at half maximum FWHM), is the measure of resolution. The meaning of the FWHM is that it is the minimum distance by which two points of radioactivity must be separated to be perceived independently in the reconstructed image. Limited resolution, which is visually apparent as blurring of the image, has a major effect on the ability to measure radioactivity accurately, especially in small structures. The true radioactivity in a region can be recovered if the size of the area is at least two times the resolution in all directions. If the size of the organ or brain structure is smaller than the FWHM of instruments, it may lead to misidentification of anatomical structures, blurring of boundaries between closely lying structures, and averaging of activities from surrounding areas outside the area of interest (Mazziotta *et al.,* 1981). This affects, in various ways, the quantitative information from the measurements of the radioactivity concentration within different brain structures. This will depend on their size (Hoffman *et al.,* 1979) and shape (Mazziotta *et al.,* 1981), on their contrast (Kessler *et al.,* 1984), on the statistics of the image, on the relative position of the location of the structure versus the detection system (Bentreim *et al.,* 1986; Miller *et al.,* 1990) and on the methodology used in the data sampling and analysis, such as the region of interest (ROI) strategy (Bohm *et al.,* 1990) and the amount of smoothing introduced by the reconstructed process. As a consequence, any value for regional or tissue radioactivity concentration may include components of "spillover" (contamination from the surrounding tissues or across voxels) and "tissue fraction effects", due to heterogeneity within the voxel (e.g, gray and white matter in brain tissue, and vascular or necrotic region within a tumor). This so-called "partial volume effect" results in errors in the estimation of the true concentration of tissue radioactivity (Hoffman *et al.,* 1979). The correction for partial volume effect requires *a priori* knowledge of the tissue anatomical structure, defined from higher spatial resolution modalities. Appropriate algorithms for partial volume correction are still under investigation (Meltzer *et al.,* 2000; Aston *et al.,* 2002; Frouin *et al.,* 2002).

Since the camera measures concentrations of radioactivity within volumes, spatial resolution is classified in terms of axial, radial and tangential resolutions. Axial resolution is obtained in the direction perpendicular to the plane of the image. This measurement defines the plane thickness for the image. Radial and tangential resolutions reflect the inplane spatial resolution of the image (so-called transaxial resolution). Depending on the design of the camera, the resolution of the system can vary greatly at different positions of the FOV. The best spatial resolution is obtained at the center of FOV (CFOV) and the resolution will deteriorate away from the center.

For very high-resolution applications (detector size <4 mm), the physics of the positron emission and annihilation process can start to become noticeable (Budinger, 1998). Image resolution is then also affected by the positron range (the distance the positron traveled before annihilation) and the noncollinearity of the two 511 keV gamma rays, which in practice are not emitted at exactly 180 degrees apart. The resolution in the final PET image is a convolution of three factors: detector dimensions, positron range and noncollinearity. For very small detector sizes, blurring due to positron range and noncollinearity can become significant and ultimately limiting (Cherry, 2001).

Radiopharmaceuticals

By synthesizing compounds tagged with photon- or positron-emitting radionuclides (Table 23.1) and injecting these into the body, the regional concentration of the labeled compound can be imaged *in vivo* as

TABLE 23.1: List of positron-emitters and their half-lives

Radioisotope	Symbol	Half-life
Bromine-75	^{75}Br	98 min
Bromine-76	^{76}Br	16.1 h
Carbon-11	^{11}C	20.4 min
Cobalt-55	^{55}Co	17.5 h
Cooper-62	^{62}Cu	9.7 min
Cooper-64	^{64}Cu	12.7 h
Fluorine-18	^{18}F	109.8 min
Gallium-68	^{68}Ga	68.1 min
Iodine-124	^{124}I	4.2 days
Iron-52	^{52}Fe	8.3 h
Nitrogen-13	^{13}N	10 min
Oxygen-15	^{15}O	2.03 min
Rubidium-82	^{82}Rb	1.25 min
Technetium-94m	^{94m}Tc	53 min
Yttrium-86	^{86}Y	14.7 h
Zirconium-89	^{89}Zc	78.4 h

The radionuclides in bold are the most commonly used PET radionuclides.

a function of time. Most of the elements that are commonly found in molecules of biological interest, such as oxygen, nitrogen, and carbon, have positron-emitting isotopes. Therefore, theoretically, any biological molecule can be labeled with a positron emitter by direct substitution, resulting in a labeled compound with virtually identical biochemical properties to the native compound.

An important consideration in PET is the half-life of the radionuclide. A radiotracer with a long half-life can be produced at a site different to that of actual administration, though multiple scans in one animal on the same day will not be possible. Alternatively, whilst radiotracers with a short half-life do permit sequential scanning on the same day, there are other limitations. Ideally, a radioligand should be synthesized within 2–3 half-lives of the radionuclide to maintain high radiochemical yield and specific activity. The production of radioligand involves isolation of the target drug, purification, formulation into a sterile, pyrogen-free, isotonic solution and accurate quality control prior to injection. For radiotracers radiolabeled with carbon-11, which has a short half-life of 20 min, this necessitates manufacture at the site of delivery. Another disadvantage occurs if the compound to be studied has a long half-life (several hours) compared to the physical half-life of the label. As a result, a larger amount of radioactivity has to be injected.

Radioligands are administered at "tracer" concentration to the subject, so that the fate of the compound in the body can be studied without perturbing the physiological system under observation. "Tracer" typically means nanomolar or even picomolar concentrations and, in the case of receptor binding studies, the total amount of administered ligand results in receptor occupancy of less than 1–5%. Generally, radiotracers with a very high specific activity (5000–20000 mCi/mmol) are preferred so only small quantities (nmol or pmol) of stable compound are required.

Many positron-labeled compounds have been synthesized, enabling a wide range of biological processes to be measured quantitatively, non-invasively and repeatedly. These processes include metabolism, blood flow and vascular integrity, neurotransmitter properties, blood-brain barrier permeability, pH and enzyme activity, tissue viability and tissue/organ specificity, gene therapy and treatment schedules (Table 23.2). Direct labeling of drugs, both therapeutic and drugs of abuse, is also possible allowing drug pharmacokinetics and pharmacodynamics to be studied *in vivo*. At present, compounds containing N-, S-, or O-methyl (or -ethyl) groups, proteins and antibodies are fairly easily radiolabeled. It is not possible, however, to radiolabel all drugs with sufficient radioactivity to derive useful information from them when used in investigational studies. This may occur if the length of the synthesis and purification of the radiotracer is extensive and leads to substantial decay of radioactivity. Constraints in the availability of potential labeling reagents, including precursors, can also limit the synthesis of radiotracers. The position of the radiolabel should be robust to avoid metabolic degradation, again restricting the number of compounds available. However, this can be fashioned advantageously when demonstrating the metabolic routes of specific compounds *in vivo*.

The flexibility of labeling any molecule that targets a specific biological pathway or marker, and the ability to image the spatial distribution of the labeled molecule *in vivo* over time, opens up many potential applications of PET in basic research.

Analysis of PET data

When the image reconstruction is completed, data can be analyzed by defining a sample region or region of interest (ROI) on the PET image, after correlation with structural imaging and extracting radioactivity versus

TABLE 23.2: Representative PET and SPECT radiotracers

Desired measurement	Examples of PET and SPECT radiotracers
Amino acid transport, protein synthesis	[^{11}C]-Methionine; [^{11}C]-Leucine; [^{11}C]-Tyrosine; [^{123}I]-Iodo-a-methyltyrosine
Cerebral blood flow	$H_2{}^{15}O$; [^{15}O]-Butanol; [^{11}C]-Butanol [^{18}F]-Fluoromethane; [^{123}I]-Iodoamphetamine, [^{99m}Tc]-labeled exametazime
Cerebral blood volume	$C^{15}O$; ^{11}CO
Cerebral energy metabolism	
Oxygen metabolism	$^{15}O_2$
Glucose metabolism	[^{18}F]-Fluorodeoxyglucose; [^{11}C]-Deoxyglucose; [^{11}C]-lucose
Glucose transport	[^{11}C]-3-*O*-Methylglucose
DNA synthesis (proliferation)	[^{18}F]-Deoxyfluorothymidine; [^{11}C]-Thymidine
Enzyme activity	[^{11}C]-Deprenyl; [^{18}F]-Deoxyuracil
Tissue pH	^{11}CO; [^{11}C]-Dimethadione
Tissue drug kinetics	[^{11}C]-Phenytoin; [^{11}C]-Valproate; [^{13}N]-Carmustine (BCN)
Neuroreceptor systems	
Cholinergic muscarinic	
mAChRs-receptor	[^{123}I]-Iododexetimide; [^{123}I]- QNB, [^{11}C]4-NMPB
Cholinergic nicotinic	
nAChRs- receptor	2[^{18}F]F-A-85380; 5[^{123}I]I-A-85380
Opiate	[^{11}C]-Carfentanil; [^{11}C]-Diprenorphine; [^{18}F]-Cyclofoxy
Benzodiazepine	[^{11}C]-Flumazenil; [^{123}I]-Iomazenil
Serotonergic (5-HT)	
Presynaptic serotonin pool	[^{11}C]-α-Methyltriptophan
5-HT_{1A} receptors	[^{11}C]-WAY100,635
5-HT_{2A} receptors	[^{11}C]-MDL100907; [^{18}F]-Altanserin; [^{18}F]-Setoperon; [^{123}I]-R93274
Transporter	[^{11}C]-Fluvoxamine
Biosynthesis	[^{11}C]5-HTP
5-HT reuptake sites	[^{11}C]-McN5652
Dopaminergic	
Biosynthesis	[^{18}F]-Fluoro-L-dopa; [^{18}F]-Fluoro-L-*m*-tyrosine
Dopamine D2 receptors	[^{11}C]-*N*-Methylspiperone; [^{11}C]-Raclopride; [^{18}F]-Fallypride [^{18}F]-Spiperone; [^{18}F]-*N*-Methylspiperone; [^{123}I]-Iodobenzamide; [^{123}I]-Epidepride
Dopamine D1 receptors	[^{11}C]-SCH23390; [^{123}I]-SCH23982
Dopamine reuptake sites	[^{11}C]-Nimiphensine; [^{11}C]-Cocaine; [^{18}F]-GBR; [^{123}I]-β-CIT

time curves from this area. PET images of the brain are usually compared to corresponding anatomical sections in the brain atlases and the ROI manually drawn. For more accurate and precise ROI placement, the MRI scans obtained in the same animals can be registered to the PET images and used as a guide for the ROI placement.

Mathematical kinetic modeling of information, derived from PET data, expands the interpretation of the results, within a preconceived framework of important kinetic behavior, to obtain quantitative parameters of relevance. This involves comparing the concentration of radioactivity in tissue (ROI) with that in arterial plasma and determining plasma-tissue exchange rate constants to construct a pharmacokinetic model (Huang *et al.,* 1986). Modeling is necessary for accurate calculation of all aspects of drug pharmacology and to measure physiological variables, such as tissue perfusion, metabolic rate and receptor binding. A tracer kinetic model is a mathematical description of the fate of the tracer in the living organism, particularly in the organ or in the region under study. The usual method of establishing a mathematical description is to assign the possible distribution of tracer into a limited number of discrete compartments, within which the tracer can be free or specifically or non-specifically bound. The compartments do not necessarily represent actual separate physical compartments but are a simplification of a complex biological interaction. A three compartment (two tissue) model, widely used for *in vivo* receptor binding, is illustrated in Figure 23.4.

When drugs are not metabolized in the body, calculation of drug concentration is relatively straightforward. However, most of the compounds do undergo some metabolism *in vivo* and metabolites are produced. If the radiotracer is metabolized into a radiolabeled metabolite, sometimes it cannot be distinguished, by PET, from the radiolabeled parent compound and, therefore, could be a source of error in data analysis.

Non-human primate PET scanners

The relatively large size of non-human primates, including the size of the brain, allows the performance of imaging with conventional clinical human PET scanners. Clinical PET systems typically produce reconstructed PET images with a spatial resolution of 5 to 8 mm range for the brain and 8 to 15 mm range for the rest of the body. This resolution results in volume resolution elements (which can be interpreted as the volume of tissue from which independent information can be extracted) of 0.1 cc at best, and more typically, 0.5 to 2 cc (Cherry and Gambir, 2001). Even though spatial resolution of human PET scanners allows the visualization of many brain structures in non-human primates, an error in measurement of radioactivity concentration in some of the brain structures may be significant due to "spillover" and partial volume effect. Therefore, spatial resolution of human PET scanners is a concern for accurate quantitative brain imaging in non-human primates, even in macaques and baboons who have a relatively large size of brain (*ca.* 100 cm^3). Lately, dedicated PET systems have been developed specifically for imaging non-human primates, including small monkeys. These systems were designed to obtain higher spatial resolution than clinical PET systems. There are also several advantages of having a tomograph dedicated to the studies in non-human primates, in addition to a

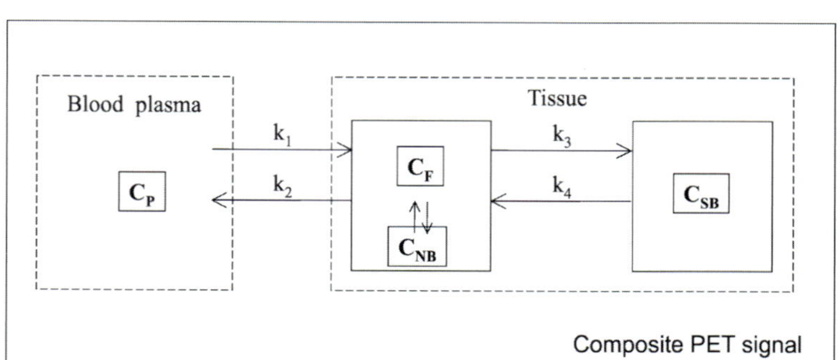

Figure 23.4 General three-compartment model. Radioactivity in the plasma (C_P), free (C_F), specifically bound to the receptor (C_{SB}) and non-specifically bound (C_{NB}) combines to give the radioactivity in the PET image. The model parameters k_1, k_2, k_3, and k_4 reflect the exchange rate for radiolabeled ligand moving between various compartments.

clinical instrument at the PET facility. This will enable access uninterrupted by a clinical programme, it will circumvent any problems associated with ethical and/or legal regulations which prevent the use of a clinical camera for animal studies and, in addition, will allow concomitant PET studies in human subjects and animals making additional use of expensive radiochemical synthesis. Animal PET cameras are more compact and usually less expensive than clinical tomographs and facilitate repeat studies on the same animal.

The first PET tomograph designed for imaging non-human primates was constructed in the early 1990s. One such system, SHR-2000, was developed by Hamamatsu (Hamamatsu, Japan) and was installed in their Central Research Laboratory (Watanabe *et al.*, 1992). Another dedicated PET animal system, ECAT-713, was developed by CTI PET Systems Inc. (Knoxville, TN) and installed at USLA (Cutler *et al.*, 1992). Both these systems offered a relatively large ring diameter and axial FOV. These are 64 cm and 5.4 cm for ECAT-713 and 34.8 and 4.6 cm for the Hamamatsu system, respectively, both of which could accommodate adult vervet monkeys and macaques. The reconstructed spatial resolution of these systems at the CFOV was reported as 3.8 mm transaxial and 4.2 mm axial for the USLA system (Cutler *et al.*, 1992) and 3.0 mm transaxial and 4.4 mm axial for the Hamamatsu system (Watanabe *et al.*, 1992). Hamamatsu has more recently developed a higher resolution PET scanner for imaging non-human primates, the SHR-7700, with axial FOV of 11.4 cm (Watanabe *et al.*, 1997). The axial resolution of the SHR-7700 system is 3.2 mm and the reconstructed transaxial resolution at the COFV is 2.6 mm, yielding a volume resolution element of 22 mm^3 or 0.022 cc. Both Hamamatsu systems have a sophisticated tilt mechanism in the gantry which allows the detector ring to be placed horizontally, thus enabling PET imaging of conscious monkeys trained to sit in a chair (Tsukada *et al.*, 1999a; 1999b).

In recent years, a number of new detector technologies have been proposed for the development of high-resolution animal tomographs for *in vivo* imaging studies in different species, from mice to non-human primates. New scintillators and photodetectors have been explored, as well as other detector technologies that do not employ scintillation detectors. One such small-animal PET system, microPET, was developed at USLA Crump Institute in 1996-1997 (Cherry *et al.*, 1997; Chatziioannou *et al.*, 1999). The scintillator material used for this system was the newly available lutetium oxyorthocilicate (LSO). The LSO scintillator has a higher light output than BGO, very similar stopping power, and a much shorter decay time constant (~40 ns) (Melcher, 2000) than BGO (~300 ns) that results in an increase in counting rate capability and a reduction in random coincident rate. Optical fibers are used to transfer the scintillation light from the individually cut LSO crystals to the multichannel PMTs. The individual fiber-optic read-out of the signals from each crystal allows the miniaturization of the fundamental PET system resolution element, the physical dimensions of the scintillator crystals, to achieve high spatial resolution. The volumetric resolution of the microPET is almost an order of magnitude better than previous generation scanners, allowing the delineation of many more structures in the monkey brain and in rats and mice, while at the same time reducing the overall material costs because of the small size of the scanner. The diameter of the ring of microPET is 17.2 cm with FOV of 11.2 cm in the transverse and 1.8 cm in the axial direction. At the CFOV, the reconstructed resolution is isotropically 1.8 mm FWHM, and produces a volumetric resolution of ~0.006 cc. Based on the same technology as the USLA Crump Institute microPET scanner, Concorde Microsystem Inc. (Knoxville, TN) has recently developed a similar system, microPET® P4, dedicated to studies in non-human primates (Tai *et al.*, 2001). Figure 23.5 shows a photograph of the system. This system has a 27 cm ring diameter with FOV of 8 cm and the same special resolution of ~0.006 cc at the CFOV as USLA microPET. The larger axial field of view, in comparison with the USLA Crump Institute microPET, dramatically improves the absolute detection sensitivity of the P4 at the CFOV from 0.56% for USLA microPET to approximately 2.25% (Chatziioannou *et al.*, 1999; Tai *et al.*, 2001). Currently, microPET® P4 is the highest resolution commercially available PET system for non-human primate PET imaging.

Animal procedures for PET studies

Positioning in the scanner

For PET data analysis, it is critical to keep the animal at the same position during the scanning period to minimize the error to the measurement obtained from the image. Therefore, for most of the PET studies, the monkey is kept anaesthetized over the course of the scanning period. For imaging the brain, the head of the

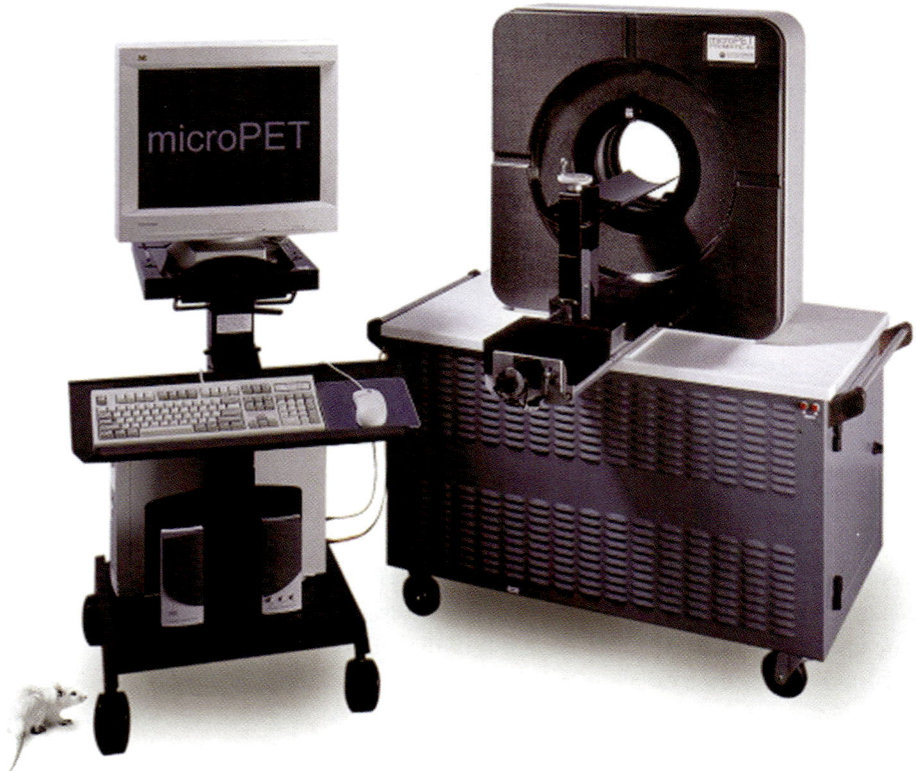

Figure 23.5 Photograph of the microPET®P4 scanner (photograph courtesy of Concorde Microsystems, Knoxville, TN).

anaesthetized monkey is usually fitted with a thermoplastic mask that is worn during the scanning period. The mask is made of the thermoplastic material that is used for PET studies in humans. Strips of material, appropriately sized for use with a monkey head, are pre-cut with openings for the eyes, ears and mouth. The material is heated in a water bath until it becomes pliable (usually around 105°F). It is stretched over the monkey's face and secured in the head-holder board to position the monkey correctly in the PET scanner aperture. An alternative to the thermoplastic mask is a stereotaxic head-holder. If a stereotaxic head-holder is used, the images will be acquired in a coronal view instead of transaxial view when the animal is lying on its back.

Blood sampling

For studies in which the time course of the tracer in the arterial blood needs to be measured, in order to solve the kinetic uptake of tracer into a tissue of interest, a saphena artery is typically cannulated and blood

TABLE 23.3: PET scanners used for imaging studies in non-human primates

Scanner model	Detector ring diameter (cm)	Axial FOV (cm)	Transaxial resolution (CFOV) (mm)	Volumetric resolution (transaxial resolution$^2 \times$ axial resolution) (CFOV) (cm^3)
ECAT HR+	64	15.5	4.6	0.089
SHR-2000	34.8	4.6	3.0	0.040
ECAT-713	64	5.4	3.8	0.061
SHR-7700	50.8	11.4	2.6	0.022
Concordia microPET®P4	27	7.8	1.8	0.006

continuously withdrawn over the course of a PET study. In addition, when using compounds that become metabolized within the body, it is necessary to analyze serial blood samples into their constituent parent compound and metabolite components. From the total blood data collected, the time course of the plasma concentration of the parent compound can be derived. This requires dedicated chromatographic analysis and high sensitivity detectors. If the animal is used for PET studies on a regular basis, an arterial catheter can be implanted chronically and blood samples can be drawn through the vascular access port (Wojnicki *et al.*, 1994).

Quantitative data collection

With the monkey positioned in the gantry of the PET camera, a transmission scan is carried out prior to administering the tracer. Typically, this involves rotating a point or rod source of activity, usually the long-lived positron emitting source, germanium 68/gadolinium 68. When these scan data are ratioed with a blank scan, which is recorded when there is no animal in the camera, factors are produced for the attenuation of the signal that occurs in the animal's head or body, along each coincidence line of response. The net result of this is that the PET image data can be accurately corrected for attenuation. Hence, the camera's voxel-element response can be calibrated against laboratory detectors, used for counting blood samples, and the dose of activity administered. This means that concentrations of tracers measured in tissue are recorded in the same units as those of blood or plasma samples. Data expressed in common units enable kinetic models, using input functions derived from blood samples, to be operated to obtain measures of tracer flux or binding. It also allows a measure of the dose that is present in a unit volume of tissue.

After the transmission scan is acquired, an emission scan will be performed, starting at the same time as the intravenous ligand administration. Depending on the specific goal of the study, the radioligand can be administered as a bolus injection over a few seconds (Villemagne *et al.*, 1999a; Chefer *et al.*, 1999, 2003; Kuge *et al.*, 2001; Howell *et al.*, 2002) or slowly infused over several hours (Carson *et al.*, 1993, 1997; Villemagne *et al.*, 1999b). The cephalic, saphenous or femoral veins are most commonly employed for vascular access to administer the radiotracer. The duration of the scanning period depends on the type of assessment and on the radiopharmaceutical being used and can be varied from 20–90 min up to several hours (Chefer *et al.*, 1999, 2003; Melega *et al.*, 2000; Moore *et al.*, 2000; Kuge *et al.*, 2001; Howell *et al.*, 2002). Blood sample timing has to align with that of the scan times.

When using ligands labeled with short-lived isotopes, co-ordination of timing is of the essence. In particular, the monkey has to be prepared, transported to the PET facility and lying in the PET camera in good time, before the rapidly decaying probe arrives from the radiochemists.

Anaesthesia and immobilization

Since anaesthesia is used to keep the animal still during the scanning period, full consideration must be given to the type and level of anesthetic used, although this is no more true for imaging than for any other type of animal experimentation.

There are several anaesthetic regimes used in PET. The simplest and most consistent regime involves inhalation anaesthetics such as isoflurane or sevoflurane (Carson *et al.*, 1993, 1997; Tsukada *et al.*, 1999a; Kuge *et al.*, 2001). In this procedure, fast acting induction and recovery inhalation regimens afford greater safety as well as stable physiological effects. Another choice, for maintaining anaesthesia over a scanning period, is the use of injectable anaesthetics such as saffan, which is usually given by constant infusion intravenously (Villemagne *et al.*, 1998, 1999a, 1999b; Chefer *et al.*, 1999, 2003). Ketamine, a common agent widely used for chemical restraint of non-human primates, can produce some muscle contraction and, although it has been used to perform short-lasting PET studies in non-human primates, ketamine given alone is insufficient for complete immobilization.

Characterization of the effects of different anaesthetics on biological systems, and careful anaesthetic selection, are required to understand and minimize the effects of anaesthesia on the imaging results. It is necessary to ensure that the anaesthetic agent does not interfere with the tracer biodistribution and kinetics. An underlying assumption is that the process, being measured and modeled, remains constant for the duration of the experiment. Thus, careful monitoring, throughout the study, of anaesthesia and physiological parameters, such as heart rate and blood pressure, is required. This is particularly important when studies are performed with interventions to demonstrate the effect of drug interactions on the system being studied.

Many commonly used anaesthetics and sedatives will affect both physiological and neurochemical processes and often in ways which are poorly characterized. In the brain, GABA and N-methyl-D-aspartate receptor systems are most likely to be directly targeted by the majority of the anaesthetic agents (Franks and Lieb, 1994, 1997). In addition, the contributing factors of the anaesthetic's effects could be: changes in blood-brain barrier permeability (Gjedde and Rasmussen, 1980; Pardridge *et al.,* 1982; LaManna and Harik, 1986); changes in peripheral metabolism, that will affect the ligand metabolism; changes in glucose metabolism (Sokoloff *et al.,* 1977; Crosby *et al.,* 1982; Savaki *et al.,* 1983; Maekawa *et al.,* 1986); changes in cerebral blood flow (CBF) (Hansen *et al.,* 1988; Schulte am Esch and Kochs, 1990; Bjorkman *et al.,* 1992), that can also lead to changes in radioligand delivery rates to the target tissue and affect the coupling between the CBF and metabolism (Ueki *et al.,* 1992); and changes in neurotransmitter concentration in the synaptic cleft that will affect radioligand specific binding. For example, ketamine has regionally selective excitatory and depressant actions on the cerebral metabolic rate of glucose (rCMRGlc), rather than uniform depression as observed with other anaesthetic agents (Nelson *et al.,* 1980; Crosby *et al.,* 1982; Davis *et al.,* 1980; Saija *et al.,* 1989), significantly lowers cerebral metabolic rate of oxygen (Schwedler *et al.,* 1982) and alters CBF (Schwedler *et al.,* 1982; Bjorkman *et al.,* 1992). Different effects of ketamine and isoflurane on receptor binding *in vivo* have been shown for [^{11}C]Raclopride and [^{11}C]methylspiperone, radioligands for dopamine receptors. The effect of isoflurane on raclopride binding has been evaluated by comparison of the results obtained in PET studies in conscious monkeys and that under anaesthesia (Tsukada *et al.,* 1999a). Although the precise mechanisms of the effects of isoflurane on different components of the dopaminergic system still remain unknown, it has been shown to affect both presynaptic and postsynaptic dopaminergic function measured by PET (Tsukada, 1999b).

In some cases, as for example 2-Deoxy-2 [^{18}F]fluoro-D-glucose (FDG), the probes are irreversibly trapped in tissue, therefore enabling the distribution and uptake of the probes to occur while the animal is conscious. After the uptake is completed and the tracer has reached steady state, the animal can be anaesthetized and imaged. This approach permits activation/stimulation type investigations in fully awake animals (Eberling *et al.,* 1995, 1997b; Moore *et al.,* 2000). For some PET studies, the peak of radioactivity at the beginning of the scan represents essential information for mathematical modeling of the kinetic data sets. To eliminate the potential confounding effect of anaesthetics, the monkey can be trained to sit in the non-human primate chair for PET studies in the conscious state (Onoe *et al.,* 1994; Tsukada *et al.,* 1999a, 1999b; Howell *et al.,* 2001). To prevent movement during the PET study, the animal's head can be fixed to the primate chair using an acrylic plate which is surgically implanted on the skull (Onoe *et al.,* 1994). An alternative to an acrylic plate is a customized head-holder attached to the primate chair (Howell *et al.,* 2001), an approach that does not require any surgical implantation. Even though the behavioral observations and objective measures of plasma cortisol levels indicate that the animals tolerate the immobilization procedures well during neuroimaging studies, the potential impact of stress on some physiological measures such as CBF (Fischer *et al.,* 1998) and rCMRGlc (Gur *et al.,* 1987) should be considered.

PET application in non-human primates

In vivo imaging in non-human primates offers a unique resource in research. It provides the opportunity to manipulate appropriate biological and behavioral variables under well-controlled experimental conditions, in an animal model that is closely related to humans, both functionally and anatomically. Another factor that has made PET studies popular in this species is the high expense of primates which precludes extensive terminal experiments.

The major research role of PET in non-human primate research has been in imaging of brain neurotransmitter systems. Some of the early and latest studies in non-human primates in this field were directed primarily to evaluating and validating new PET ligands (Wong *et al.,* 1993; Baldwin *et al.,* 1995; Yousef *et al.,* 1996, Chefer *et al.,* 1999; Mukherjee *et al.,* 1999). The evaluation of novel radiopharmaceuticals generally involves post-mortem tissue sampling and autoradiography using rodents. This is an intensive process in terms of both time and animals. Brain imaging studies of novel compounds, in non-human primates, offer an efficient means of acquiring kinetic data sets from individual animals with a brain similar to human. They also provide the means for repeated studies of that same subject over an arbitrary period of time. PET data, obtained in monkeys, will accurately describe the *in vivo* properties of the tracer and provide information for further application in humans.

The ability to carry out serial studies in a single animal is a clear advantage of *in vivo* imaging. This is of particular significance when the animal is unique by virtue of an experimental procedure as, for example, lesioning or tissue transplantation. With the ability to perform the same studies in human and non-human primates, at the same time, PET imaging facilitates direct comparison and unification of basic and clinical research and provides a connecting link between the animal model and human diseases and pathological states.

Currently, the use of non-human primates for brain imaging is being extended to the testing of new drugs targeted to specific receptor systems (Tedroff *et al.*, 1997; Ginovart *et al.*, 1997; Farde *et al.*, 2000; Mukherjee *et al.*, 2001) and to the studies of neurodegenerative diseases (Wullner *et al.*, 1994; Eberling *et al.*, 1997a, 1997b, 2000; Doudet *et al.*, 2002a, 2002b), stroke (Takamatsu *et al.*, 2000, 2001; Kuge *et al.*, 2001) and drugs of abuse (Melega *et al.*, 1997; Fowler *et al.*, 1998; Villemagne *et al.*, 1998, 1999a, 1999b; Volkow *et al.*, 1995, 1999; Howell *et al.*, 2002; Votaw *et al.*, 2002) using models of these pathological states in non-human primates. These studies include monitoring the effects of neuroprotective and neurotrophic factors (Melega *et al.*, 1998, 2000; Takamatsu *et al.*, 2001) and evaluation of the outcome of graft implantations (Subramanian *et al.*, 1997; Bankiewicz *et al.*, 2000). In imaging studies of tissue transplantation *in vivo*, graft viability can be monitored over time and the PET signal can be correlated with external measures of behavior. At the same time, PET imaging allows the testing of therapeutic interventions prior to clinical or more intensive experimental studies. Such investigations in disease models, where repeat longitudinal measurements can be made, reveal the true worth of *in vivo* imaging over more invasive techniques.

The majority of PET studies, in non-human primates, focus on the dopaminergic system (Onoe *et al.*, 1994; Carson *et al.*, 1997; Tsukada *et al.*, 1999a; Villemagne, 1999a, 1999b; Volkow *et al.*, 1995, 1999; Bankiewicz *et al.*, 2000; Eberling *et al.*, 2000; Melega *et al.*, 1998, 2000) because there are several PET ligands available for different components of this system and the striatum is easily visualized on the PET images.

A glucose analog of ([18F]2-fluoro-2-deoxy-D-glucose (FDG), the most frequently used clinical PET radiotracer, has also been used widely in imaging studies in non-human primates. Uptake of this probe is reflective of regional glucose metabolism and is altered in a wide variety of disorders and disease states. PET studies with FDG have been used to track metabolic development in the monkey brain (Moore *et al.*, 2000), to evaluate the relationship between rCMRglc and age (Eberling *et al.*, 1995; Noda *et al.*, 2002) and performance on a delayed response test of memory in the aged monkey. (Eberling *et al.*, 1997b).

Imaging non-human primates versus rodents

While PET studies of the non-human primate central nervous system will continue to play an important role in select experiments, much of the technology development in imaging is currently being directed at small laboratory animals, such as rats and mice (Cherry, 2001; Cherry and Gambir, 2001; Myers, 2001; Chatziioannou, 2002; Hume and Myers, 2002). As the resolution of imaging systems improves and dedicated small animal PET systems become available, the prospects of *in vivo* imaging of rodents, rather than non-human primates, may look more attractive. Nevertheless, the application of imaging technology in small animals does have some intrinsic problems and constraints which are presented in the following literature (Hume *et al.*, 1998; Hume and Myers, 2002; Myers *et al.*, 2001).

One of the main problems associated with obtaining pharmacokinetic data from PET scans of rodents, rather than non-human primates, is that the physical size of the anatomical structures to be imaged is of the order of the spatial resolution of the existing small animal PET systems. Thus, the problems of partial volume loss and spillover are a real challenge in quantitative imaging studies in rodents. Without the correction factor, spillover would tend to attenuate any effects of those experimental manipulations expected to reduce the specific signal.

Poor sensitivity of small animal PET scanners limits the ability to study the kinetics of the radiotracer and reliably estimate physiological parameters. When using rodents for *in vivo* imaging, the injected radioactive dose must account for scanner performance in that it must fall below the point at which recorded events saturate the electronics and yet be sufficient to produce an image. The low sensitivity of small animal PET cameras can, to some extent, be compensated by a higher injected dose. However, high doses of radioactivity in rodents might be associated with a significant mass of the injected compound which may compromise the specific binding of the radiotracer. It is in the nature of the radioligand labeling process that the final preparation will contain a finite amount of stable, non-radioactive ligand. The lower the specific activity of the radioligand, the greater the mass of non-radiolabeled component it contains (Hume *et al.,* 1998). Therefore, mass limitations can restrict the amount of radioactivity that can be injected in some instances, e.g., very low specific activity compounds and/or the study of easily saturated receptor systems. The applicability of the tracer-kinetic modeling, on which most PET ligand analysis is based, depends on the occupancy of the receptors, under test, not being much greater than the 1–5% required for an application of radioligand with high specific activity. Nonetheless, the use of high affinity radioligands might be problematic in rodents since the permitted injected radioactive dose decreases with increasing affinity of the ligand and the decreasing body weight (Hume *et al.,* 1998). Determination of a reasonable radiotracer dose, based on the values of specific activity and scanner sensitivity, must be considered in order to minimize the degree of receptor occupancy and to achieve count statistics adequate to reconstruct small volumes.

To perform fully quantitative PET studies, information on the time course of tracer delivery to the tissue is required. This "input function" is obtained from direct arterial blood sampling during the PET study. This function becomes extremely challenging in smaller laboratory animals, particularly in longitudinal studies in which blood samples are required from the same animal repeatedly.

The small size of the mouse or other small animals is particularly challenging for maintaining a stable and an appropriate physiological state under conditions of anaesthesia, due to the problems with mechanical ventilation and online measures/adjustments of an appropriate physiological function. In addition, even under anaesthesia, motion caused by breathing will lead to a significant and systematic positional shift that will affect the accuracy of the PET data due to the fact that the higher the resolution of the system, the smaller the movement which can be tolerated.

With certain advantages of using non-human primates for PET imaging over rodents, its application in non-human primates will have a long-term future in the study of receptors, neurotransmitters, drug binding and general pharmacology as well as *biochemistry*/metabolism and physiology *in vivo.* As a scientific tool, PET imaging, in non-human primates, will continue to play a major role in basic and applied science and valuable *in vivo* information will continue to be obtained.

Correspondence

Any correspondence should be directed to Svetlana Chefer, National Institute on Drug Abuse, National Institute of Health, Dept of Health and Human Services, 5500 Nathan Shock Drive, Baltimore, Maryland 21224, USA. Email: schefer@intra.nida.nih.gov

References

Aston, J.A., Cunningham, V.J., Asselin, M.C., Hammers, A., Evans, A.C. and Gunn, R.N. (2002). *J. Cereb. Blood Flow Metab.* 22, 1019–1034.

Baldwin, R.M., Horti, A.G., Bremner, J.D., Stratton, M.D., Dannals, R.F., Ravert, H.T., Zea-Ponce, Y., Ng, C.K., Dey, H.M., Soufer R. Charney, D.S. Mazza, S.M., Sparks, R.B., Stubbs, J.B. and Innis, R.B. (1995). *Nucl. Med. Biol.* 22, 659–665.

Bankiewicz, K.S., Eberling, J.L., Kohutnicka, M., Jagust, W. Pivirotto, P., Bringas, J., Cunningham, J., Budinger, T.F. and Harvey-White (2000). *J. Exp. Neurol.* 164, 2–14.

Bentrelm, B., Soussaline, F., Campagnolo, R., Verrey, B., Wajnberg, P. and Syrota, A. (1986). *J. Comput. Assist. Tomogr.* 10, 287–295.

Bjorkman, S., Akeson, J., Nilsson, F., Messeter, K. and Roth, B. (1992). *Pharmacokinet. Biopharm.* 20, 637–652.

Bohm, C., Greitz, T., Seitz, R. and Eriksson, L. (1990). *J. Cereb. Blood Flow Metab.* 11, A64–A68.

Budinger, T.F. (1998). *Seminars in Nuclear Medicine* 28, 247–267.

Carson, R.E., Channing, M.A., Blasberg, R.G., Dunn, B.B., Cohen, R.M., Rice, K.C. and Herscovitch, P. (1993). *J. Cereb. Blood Flow Metab.* 13, 24–42.

Carson, R.E., Breier, A., de Bartolomeis, A., Saunders, R.C., Su, T.P., Schmall, B., Der, M.G., Pickar, D. and Eckelman, W.C. (1997). *J. Cereb. Blood Flow Metab.* 17, 437–447.

Chatziioannou, A.F. (2002). *Eur. J. Nucl. Med. Mol. Imaging* 29, 98–114.

Chatziioannou, A.F., Cherry, S.R., Shao, Y., Silverman, R.W., Meadors, K., Farquhar, T.H., Pedarsani, M. and Phelps, M.E. (1999). *J. Nucl. Med.* 40, 164–175.

Chefer, S.I., Horti, A.G., Koren, A.O., Gundisch, D., Links, J.M., Kurian, V., Dannals, R.F., Mukhin, A.G. and London, E.D. (1999). *NeuroReport* 10, 2715–2721.

Chefer, S.I., London, E.D., Koren, A.O., Pavlova, O.A., Kurian, V., Kimes, A.S., Horti, A.G. and Mukhin, A.G. (2003). *Synapse* 48, 25–34.

Cherry, S.R. (2001). *J. Clin. Pharmacol.* 41, 482–491.
Cherry, S.R. and Gambir, S.S. (2001). *ILAR J.* 42, 219–232.
Cherry, S.R., Shao, Y., Silverman, R.W., Meadors, K., Siegel, S. and Chatziioannou, A.F. (1997). *IEEE Trans. Nucl. Sci.* 44, 1161–1166.
Crosby, G., Crane, A.M. and Sokoloff, L. (1982). *Anesthesiology* 56, 437–443.
Cutler, P.D., Cherry, S.R., Hoffman, E.J., Digby, W.M. and Phelps, M.E. (1992). *J. Nucl. Med.* 33, 595–604.
Davis, D.W., Mans, A.M., Biebuyck, J.F. and Hawkins, R.A. (1980). *Anesthesiology* 69, 199–205.
De Hevesy, G. and Chiewitz, O. (1935). *Nature* 136, 754–755.
Doudet, D.J., Jivan, S., Ruth, T.J. and Wyatt, R.J. (2002a). *Synapse* 44, 111–115.
Doudet, D.J., Jivan, S., Ruth, T.J. and Holden, J.E. (2002b). *Synapse* 44, 198–202.
Eberling, J.L., Roberts, J.A., De Manincor, D.J., Brennan, K.M., Hanrahan, S.M., Vanbrocklin, H.F., Roos, M.S. and Jagust, W.J. (1995). *Neurobiol. Aging* 16, 825–832.
Eberling, J.L., Bankiewicz, K.S., Jordan, S., VanBrocklin, H.F. and Jagust, W.J. (1997a). *Neuroreport* 8, 2727–2733.
Eberling, J.L., Roberts, J.A., Rapp, P.R., Tuszynski, M.H. and Jagust, W.J. (1997b). *Neurobiol. Aging* 18, 437–443.
Eberling, J.L., Pivirotto, P., Bringas, J. and Bankiewicz, K.S. (2000). *Exp. Neurol.* 165, 342–346.
Farde, L., Andree, B., Ginovart, N., Halldin, C. and Thorberg, S. (2000). *Neuropsychopharmacology* 22, 422–429.
Fischer, H., Andersson, J.L., Furmark, T. and Fredrikson, M. (1998). *Neurosci. Lett.* 251, 137–140.
Fowler, J.S., Volkow, N.D., Logan, J., Gatley, S.J., Pappas, N., King, P., Ding, Y.S. and Wang, G.J. (1998). *Synapse* 28, 111–116.
Franks, N.P. and Lieb, W.R. (1994). *Nature* 367, 607–614.
Franks, N.P. and Lieb, W.R. (1997). *Nature* 389, 334–335.
Frouin, V., Comtat, C., Reilhac, A. and Gregoire, M.C. (2002). *J. Nucl. Med.* 43, 1715–1726.
Gjedde, A. and Rasmussen, M. (1980). *J. Neurochem.* 35, 1382–1387.
Ginovart, N., Farde, L., Halldin, C. and Swahn, C.G. (1997). *Synapse* 25, 321–325.
Gur, R.C., Gur, R.E., Resnick, S.M., Skolnick, B.E., Alavi, A. and Reivich, M. (1987). *J. Cereb. Blood Flow Metab.* 7, 173–177.
Hansen, T.D., Warner, D.S., Todd, M.M., Vust, L.J. and Trawick, D.C. (1988). *Anesthesiology* 69, 332–337.
Hoffman, E.J., Huang, S.C. and Phelps, M.E. (1979). *J. Comput. Assist. Tomogr.* 3, 299–308.
Howell, L.L., Hoffman, J.M., Votaw, J.R., Landrum, A.M. and Jordan, J.F. (2001). *J. Neurosci. Methods* 106, 161–169.
Howell, L.L., Hoffman, J.M., Votaw, J.R., Landrum, A.M., Wilcox, K.M. and Lindsey, K.P. (2002). *Psychopharmacology* 159, 154–160.
Huang, S.C., Barrio, J.R. and Phelps, M.E. (1986). *J. Cereb. Blood Flow Metab.* 6, 515–521.
Hume, S.P. and Myers, R. (2002). *Curr. Pharm. Des.* 8, 1497–1511.
Hume, S.P., Gunn, R.N. and Jones, T. (1998). *Eur. J. Nucl. Med.* 25, 173–176.
Kessler, R.M., Ellis, J.R. Jr. and Eden, M. (1984). *J. Comput. Assist. Tomogr.* 8, 514–522.
Kuge, Y., Yokota, C., Tagaya, M., Hasegawa, Y., Nishimura, A., Kito, G., Tamaki, N., Hashimoto, N., Yamaguchi, T. and Minematsu, K. (2001). *J. Cereb. Blood Flow Metab.* 21, 202–210.
LaManna, J.C. and Harik, S.I. (1986). *J. Cereb. Blood Flow Metab.* 6, 717–723.
Maekawa, T., Tommasino, C., Shapiro, H.M., Keifer-Goodman, J. and Kohlenberger, R.W. (1986). *Anesthesiology* 65, 144–151.
Mazziotta, J.C., Phelps, M.E., Plummer, D. and Kuhl, D.E. (1981). *J. Comput. Assist. Tomogr.* 5, 734–743.
Melcher, C.L. (2000). *J. Nucl. Med.* 41, 1051–1055.
Melega, W.P., Raleigh, M.J., Stout, D.B., Lacan, G., Huang, S.C. and Phelps, M.E. (1997). *Brain Res.* 766, 113–120.
Melega, W.P., Lacan, G., Harvey, D.C., Huang, S.C. and Phelps, M.E. (1998). *Neurosci Lett.* 258, 17–20.
Melega, W.P., Lacan, G, Desalles, A.A. and Phelps, M.E. (2000). *Synapse* 35, 243–249.
Meltzer, C.C., Cantwell, M.N., Greer, P.J., Ben–Eliezer, D., Smith, G., Frank, G., Kaye, W.H., Houck, P.R. and Price, J.C. (2000). *J. Nucl. Med.* 41, 1842–1848.
Miller, T.R., Wallis, J.W. and Grothe, R.A. Jr. (1990). *J. Nucl. Med.* 31, 1732–1739.
Moore, A.H., Hovda, D.A., Cherry, S.R., Villablanca, J.P., Pollack, D.B. and Phelps, M.E. (2000). *Brain Res. Dev.* 120, 141–150.
Mukherjee, J., Yang, Z.Y., Brown, T., Lew, R., Wernick, M., Ouyang, X., Yasillo, N., Chen, C.T., Mintzer, R. and Cooper, M. (1999). *Nucl. Med. Biol.* 26, 519–527.
Mukherjee, J., Christian, B.T., Narayanan, T.K., Shi, B. and Mantil, J. (2001). *Neuropsychopharmacology* 25, 476–488.
Myers, R. (2001). *Nucl. Med. Biol.* 28, 585–593.
Nelson, S.R., Howard, R.B., Cross, R.S. and Samson, F. (1980). *Anesthesiology* 52, 330–334.
Noda, A., Ohba, H., Kakiuchi, T., Futatsubashi, M., Tsukada, H. and Nishimura, S. (2002). *Brain Res.* 936, 76–81.
Onoe, H., Inoue, O., Suzuki, K., Tsukada, H., Itoh, T., Mataga, N. and Watanabe, Y. (1994). *Brain Res.* 663, 191–198.
Pardridge, W.M., Crane, P.D., Mietus, L.J. and Oldendorf, W.H. (1982). *J. Neurochem.* 38, 560–568.
Rousset, O.G., Deep, P., Kuwabara, H., Evans, A.C., Gjedde, A.H. and Cumming, P. (2000). *Synapse* 37, 81–89.
Saija, A., Princi, P., De Pasquale, R. and Costa, G. (1989). *Neuropharmacology* 28, 997–1002.
Savaki, H.E., Desban, M., Glowinski, J. and Besson, M.J. (1983). *J. Comp. Neurol.* 213, 36–45.

Schulte am Esch, J. and Kochs, E. (1990). *Acta Anaesthesiol. Scand.* Suppl. 92, 96–102.

Schwedler, M., Miletich, D.J. and Albrecht, R.F. (1982). *Can. Anaesth. Soc. J.* 29, 222–226.

Sokoloff, L., Reivich, M., Kennedy, C., Des Rosiers, M.H., Patlak, C.S., Pettigrew, K.D., Sakurada, O. and Shinohara, M. (1977). *J. Neurochem.* 28, 897–916.

Subramanian, T., Emerich, D.F., Bakay, R.A., Hoffman, J.M., Goodman, M.M., Shoup, T.M., Miller, G.W., Levey, A.I., Hubert, G.W., Batchelor, S., Winn, S.R., Saydoff, J.A. and Watts, R.L. (1997). *Cell Transplant.* 6, 469–677.

Tai, Y.C., Chatziioannou, A.F., Siegel, S., Yang, Y., Newport, D.F., Goble, R.N., Nutt, R.E. and Cherry, S.R. (2001). *Phys. Med. Biol.* 46, 1845–1862.

Takamatsu, H., Tsukada, H., Kakiuchi, T., Nishiyama, S., Noda, A. and Umemura, K. (2000). *J. Nucl. Med.* 41, 1409–1416.

Takamatsu, H., Tsukada, H., Noda, A., Kakiuchi, T., Nishiyama, S., Nishimura, S. and Umemura, K. (2001). *J. Nucl. Med.* 42, 1833–1840.

Tedroff, J., Torstenson, R., Hartvig, P., Lindner, K.J., Watanabe, Y., Bjurling, P., Westerberg, G. and Langstrom, B. (1997). *Synapse* 25, 56–61.

Tsukada, H. (1999). *Adv. Drug Deliv. Rev.* 37, 175–188.

Tsukada, H., Nishiyama, S., Kakiuchi, T., Ohba, H., Sato, K., Harada, N. and Nakanishi, S. (1999a). *Brain Res.* 849, 85–96.

Tsukada, H., Nishiyama, S., Kakiuchi, T., Ohba, H., Sato, K. and Harada, N. (1999b). *Brain Res.* 841, 160–169.

Ueki, M., Mies, G. and Hossmann, K.A. (1992). *Acta Anaesthesiol. Scand.* 36, 318–322.

Villemagne, V., Yuan, J., Wong, D.F., Dannals, R.F., Hatzidimitriou, G., Mathews, W.B., Ravert, H.T., Musachio, J., McCann, U.D. and Ricaurte, G.A. (1998). *J. Neurosci.* 18, 419–427.

Villemagne, V.L., Rothman, R.B., Yokoi, F., Rice, K.C., Matecka, D., Dannals, R.F. and Wong, D.F. (1999a). *Synapse* 32, 44–50.

Villemagne, V.L., Wong, D.F., Yokoi, F., Stephane, M., Rice, K.C., Matecka, D., Clough, D.J., Dannals, R.F. and Rothman, R.B. (1999b). *Synapse* 33, 268–273.

Volkow, N.D., Fowler, J.S., Logan, J., Gatley, S.J., Dewey, S.L., MacGregor, R.R., Schlyer, D.J., Pappas, N., King, P. and Wang, G.J. (1995). *J. Nucl. Med.* 36, 1289–1297.

Volkow, N.D., Fowler, J.S., Gatley, S.J., Dewey, S.L., Wang, G.J., Logan, J., Ding, Y.S., Franceschi, D., Gifford, A., Morgan, A., Pappas, N. and King, P. (1999). *Synapse* 31, 59–66.

Votaw, J.R., Howell, L.L., Martarello, L., Hoffman, J.M., Kilts, C.D., Lindsey, K.P. and Goodman, M.M. (2002). *Synapse* 44, 203–210.

Watanabe, M., Uchida, H., Okada, H., Shimizu, K., Satoh, N., Yoshikawa, E., Ohmura, T., Yamashita, T. and Tanaka, E. (1992). *IEEE Trans. Med. Imag.* 11, 577–588.

Watanabe, M., Okada, H., Shimizu, K., Ohmura, T., Yoshikawa, E., Kosugi, T., Mori, S. and Yamashita, T. (1997). *IEEE Trans. Nucl. Sci.* 44, 1277–1282.

Wojnicki, F.H., Bacher, J.D. and Glowa, J.R. (1994). *Lab. Anim. Sci.* 44, 491–494.

Wong, D.F., Yung, B., Dannals, R.F., Shaya, E.K., Ravert, H.T., Chen, C.A., Chan, B., Folio, T., Scheffel, U., Ricaurte, G.A., Scheffer, U., Ricaurte, G.A., Neymeyer, J.L., Wagner, H.N. and Kuhar, M. (1993). *Synapse* 15, 130–142.

Wullner, U., Pakzaban, P., Brownell, A.L., Hantraye, P., Burns, L., Shoup, T., Elmaleh, D., Petto, A.J., Spealman, R.D., Brownell, G.L. and Isacson, O. (1994). *Exp. Neurol.* 126, 305–309.

Yousef, K.A., Fowler, J.S., Volkow, N.D., Dewey, S.L., Shea, C., Schlyer, D.J., Gatley, S.J., Logan, J. and Wolf, A.P. (1996). *Nucl. Med. Biol.* 23, 47–52.

PART 4

Current Uses in Biomedical Research

Contents

CHAPTER 24

Use of the Primate Model in Research

William R. Morton, Kelly B. Kyes, Randall C. Kyes, Daris R. Swindler and Kathryn E. Swindler
University of Washington, National Primate Research Center, Seattle, Washington, USA

Introduction

Today, institutions throughout the world maintain numerous collections of non-human primates in research facilities for a myriad of research purposes. Though historically not always the case, various species of non-human primates are currently recognized as the ideal model for many human health-related diseases. Utilization of non-human primate research collections traditionally have centered on health-related issues, including growth and development of the non-human primate, preventive and curative medicine, and psychological studies. The advent of pan-epidemics such as AIDS and SARS, however, and growing world-wide concern regarding both emerging and re-emerging infectious diseases, now dominate the attention of the medical research world in which non-human primates play a vital part. This increased focus has in turn exposed a critical and growing need for increased domestic development and care for this resource and increased efforts at world-wide conservation programs.

T.C. Ruch, in *Diseases of Laboratory Primates* in 1959, prophesized, "In the case of a national emergency, a political upheaval in Asia, *Macaca mulatta* might become totally unavailable, an event which would hamper work of public health and defense significance." How true those words have proven to be. National emergencies that could only be imagined in 1959 have occurred repeatedly (AIDS, bio-terrorism with anthrax, and SARS), as has political turmoil and policy changes in countries where non-human primates originate. India, for example, ceased the export of the Macaque in 1977, requiring many laboratories in the U.S. and elsewhere to establish their own breeding colonies (Swindler, 1998).

This chapter will briefly recapture the relationships of non-human primates to human research endeavors, current use of these animals in biomedical research and specific successes to which they have contributed. It will also highlight the urgency for increased domestic production of this resource as well as improved care, understanding and conservation, and the shift in emphasis to alternative species.

The Laboratory Primate
Copyright 2005 Elsevier
ISBN 0-1208-0261-9

Primatology: An historical overview

Non-human primates have been of significance to even the earliest of civilizations, as indicated in early artifactual findings and paintings from Asia and Africa (Fridman, 2002), and through the Middle Ages to today. Non-human primates have been the subject of observation and experimental study for thousands of years. Egyptian priests dissected baboons for ritualistic purposes as early as 2000 BC (Fridman, 2002). At the ancient Indian city of Mohenjo-Daro, archaeologists uncovered a terracotta figure dated 2500–1500 BC that appears to represent a rhesus monkey (Reynolds, 1967). Hippocrates, who is credited with the theory of the four humors and their effects on the body, dissected monkeys to learn how gall is excreted. In 300 BC, doctors in Alexandria performed anatomical studies of monkeys. There is also evidence of studies done in the Orient. Galen, one of the great early anatomists, wrote 62 treatises on the form and function of the human body, all based on non-human primate anatomy. In that same period, Aristotle (384–322 BC), in *Natural History of Animals,* presented what may be the first classification of non-human primates:

- Monkeys with no tails (Barbary apes).
- Monkeys with tails.
- Dogheaded baboons.

The development of primatology in the centuries that followed is a modern outgrowth of the development of science as a whole. At the end of the Middle Ages, in the sixteenth century, Nicolaus Copernicus and Andreas Vesalius published, within weeks of each other, two books introducing a method of observation relying upon induction, a method which builds upon close observation to propose general principals; that is, a scientific method (Ashley-Montagu, 1943). In the same period, Konrad Gesner published *Historiae Animalium,* five volumes considered to be the first illustrated books on zoology, including descriptions of several kinds of monkeys and apes. Gesner is generally thought of as the founder of the modern science of zoology. His text was followed, in the same year, by Nicholas Tulp's description of the name "orangutan," a Malayan term meaning "man of the woods" or in Latin, *Homo sylvestris*. This term continued to be used popularly into the middle of the nineteenth century for all great apes. It was also the name used by Tyson for the non-human primate he studied in 1699.

Edward Tyson (1650–1708) is credited with the founding of the modern study of primatology. Although not an evolutionist in the sense in which the term is used today, Tyson believed in a "chain of creation" relating various groups of animals, a gradual scheme rather than an evolutionary one. Through his distinguished studies, Tyson is considered not only the founder of primatology but of comparative anatomy as well (Swindler, 1998).

Louis Pasteur, in the nineteenth century continued the dynamic development of science and animal research in his efforts to identify and eradicate disease. Through direct intervention with animals, he originated, and was the first to use, vaccines. A father of animal research, he perfected a technique for vaccinating sheep against anthrax, protected fowl from chicken cholera and, most famously, in 1885, created a successful inoculation against rabies. His work was contemporary with Darwin's Theory of Evolution, establishment by Matthias Schleiden of the cell theory, and Karl Ernst von Baer's recognition of the mammalian egg and the formation of germ layers: the foundation for modern embryology. Great technological advancements in the following centuries, such as development of the electron microscope and ultracentrifuge, advanced the biological research methods of Pasteur's time, from one concerned with organisms and cells to one increasingly equating structure and function at the subcellular and molecular level.

The marriage of scientific advances, comparative anatomy, disease prevention, technological development and field observation comes together in what we recognize as modern primatology. By the early parts of the twentieth century, researchers were actively aware of the possibilities of non-human primates, not only for comparative anatomical studies but also for physiological and psychological investigations, although according to Adolph Schultz (1971), "primatology" as a term was not used in print until Dr. Ruch employed it in 1941 in his famous *Bibliographia Primatologica.*

An excellent study of modern primatology, by Fridman, details not only the discipline's history but the establishment of the national Primate Centers and many of the founding fathers of the science, foremost among them Elie Mechnikov and Robert Yerkes. The first institution focusing on non-human primates was established in 1906, by Madame Rosalia Abrece of Cuba, followed soon by Veronov and by the USSR at Sukhumi in 1927, and Yerkes in the U.S. in 1930. The Yerkes Center, first located in Florida and later at Emory University, Atlanta, Georgia is the oldest U.S. Primate Center and the first to develop a systematic breeding program of captive born chimpanzees. The U.S. Congress, following visitations to the U.S.S.R.

Primate Center at Sukhumi, allotted funding to establish a National Primate Program consisting of seven Regional Primate Research Centers. Histories of the program development have been provided by Goodwin (1972) and Vaitukaitis (1994). Paralleling the development of these centers was the establishment of a number of professional scientific societies centered on the non-human primate as a medical and scientific model, such as the International Primatological Society chartered in 1964.

Anatomy/physiology

From an anatomical perspective, simians are the closest model to humans. These similarities cover a wide range of areas, including skeletal structure, its growth and development, blood, and dentition. In the past, cats were frequently used as a model of locomotion and neural control research in the lab, but because its neural mechanisms for locomotion differ from humans, the findings are not generally as relevant as the studies using non-human primates (Vilensky, 1988). The aging process in the monkey is also similar to humans, providing an excellent animal model for studying bone aging and diseases such as lordosis, kyphosis, osteoporosis, and age osteopenia (Williams and Bowden, 1984; Pope *et al.*, 1989). Additional similarities in many aging systems (Schultz, 1972; Sirianni and Swindler, 1985) and the hematological system have been pointed out.

As early as 1925 it was known that the red blood cells of non-human primates had antigens (substances that stimulate the production of antibodies) similar to those of humans. Then, in 1940, Landsteiner and Winer discovered the Rh factor in rhesus monkeys, one of the most important discoveries in the history of blood group studies in primates. This discovery of a "monkey-type" antigen in human blood was a great impetus to study the blood of other non-human primates using other human antigens. Today we know that there is no clear-cut separation between simian-type blood groups and human-type blood groups (Swindler, 1998).

The similarity between humans and simians in brain biochemistry make non-human primates a preferred model for biomedicine and pharmacology (Lewis *et al.*, 1986), enhanced by the establishment that the circulation and system for supplying blood to the brain is the same for all primates, including humans (Roskosz, 1984).

Olenev (1987) reported that, "Repeated attempts to distinguish nuclear structure or cortical areas which are present only in humans and absent in higher simians produced only modest results. The similarity was distinguished by thousands of characteristics, but the differences by only units."

The similarities in neural structure extend to the aging processes in the brain as well. Rhesus macaques and humans share many likenesses in the aging process of the brain (decreased number and size of neurons, age plaques, etc.), which makes elderly rhesus macaques a model for studying behavioral anomalies and neuropsychology of elderly humans (Price *et al.*, 1994). Additionally, aging chimpanzees are the only animals, besides humans, in which Biondi bodies have been found in the choroid plexus (Oksche *et al.*, 1984).

Essential similarities in growth and development and dental structure between humans and non-human primates has enabled significant studies that provide vital biological information for human medicine and health. Except for timing, the whole sequence of tooth development in all higher primates is similar in terms of formation of the tooth bud, tooth emergence and maturation and post-eruptive phases of tooth wear, allowing non-human primate models for dental disease, dental caries and other environmental and genetic studies. Reports on the growth and development of the skeletal system, specifically bone development and ossification, centers in the Macaque by Watts (1986) and earlier, James Gavan offered comparative information on the effects of nutrition on bone development. Timing was shown to be the significant difference in developmental growth patterns between non-human primates and humans in a longitudinal study of *Macaca nemestrina*, from age 3 months to approximately age 8. This was conducted in the early 1970s by Swindler, at the Washington National Primate Research Center, and provided milestones and predictive data for researchers studying human genetics, disease processes, and the effect of diet on development.

Development of the primate model in research

Earlier studies of the anatomy, growth and development, psychology and disease-response of non-human primates, included the late nineteenth century report of compelling similarities between humans and non-human primates in the nervous system that were not shared by dogs, rats, and other lab animals. These led to the contemporary era of researchers working with non-human primates for medical investigations and experimentation to understand and develop treatments for human diseases.

One of the most successful individuals in this early medical research was Elie Mechnikov (Fridman, 2002). In 1885, Mechnikov created one of the first non-human primate disease models. He induced relapsing fever in rhesus monkeys (*Macaca mulatta*) previously described only in humans. Mechnikov additionally developed a monkey model for typhoid in chimpanzees, leading to testing of previously-developed vaccines. His suggestion to set up a center to breed apes for research was followed by the founding of a laboratory utilizing greater apes for the study of infectious disease in 1903. He is also credited with early development of an animal model of syphilis, which Fridman (2002) contends is still an excellent model of that disease.

In 1928, the rhesus macaque was identified as the ideal animal model for yellow fever. Earlier attempts to use other animals as models had been inconsistent. Stokes *et al.* (1928) found that the symptoms and progression of the disease in the rhesus were virtually identical to the progression in humans. At the time, it was thought that yellow fever was caused by a microorganism called leptospira, but this proved to be incorrect. Rather, the disease was caused by a virus that could be transmitted from monkey to monkey by mosquitoes. The investigators (Stokes *et al.,* 1928) even harvested mosquitoes in their lab and exposed them to the virus to run experiments on infectivity from mosquitoes. They further discovered that the viremia of the disease was so intense that a monkey could be infected simply by rubbing acute phase blood products on intact skin. When Stokes himself became ill with yellow fever, the monkeys were inoculated with his own sera and then exposed to infective mosquitoes to determine the vaccine capabilities of convalescent sera.

Despite anatomical and physiological similarities between simians and humans, the use of non-human primates as medical models for human disease was slow to be accepted. Fridman (2002) believes that this was due to society's failure to comprehend the urgent need for solutions that would eradicate these diseases, lack of financial resources, and negative attitudes toward Darwinism in the early 1900s. These failures contributed to the continued slow pace in recognizing the value of non-human primates as a base part of the medical advances that have been made, could have been made, and have yet to be made.

Research utilization and advances

The creation of a national primate program, currently comprising eight participating national centers in the U.S., has given rise to an extensive body of primate specialists. Scientists engaged in non-human primate-related biomedical research produce an ever-increasing body of literature published as a result of these invaluable resources. In the year 2002, these centers collectively produced nearly 2000 publications and conducted over 800 research projects that covered a variety of areas including (but not limited to) Alzheimer's disease, narcotic addiction, AIDS, Parkinson's disease, and osteoporosis. Over 200 core scientists were employed by the centers and over 150 graduate and post-doctoral students trained at the centers, many from countries such as Japan, Indonesia and Russia. Thus, the contribution made by these centers to the scientific community, both nationally and internationally is quite profound.

The historic significance of monkey models for human health needs have been very briefly highlighted. Current expectations are now greater than ever before, heightened by such early successes as the use of large numbers of non-human primates in the development of a viable vaccine to prevent the devastating consequences of infection with the polio virus. Sibal and Samson (2001) delineated the critical role that non-human primates play in current disease research. Perhaps now, more than any other time in history, the need for development and testing of experimental vaccines in the non-human primate is recognized and appreciated. Diseases such as AIDS, SARS, malaria, Norwalk virus and numerous others still await effective and safe vaccine development.

Acquired Immunodeficiency Syndrome (AIDS)

Acquired Immunodeficiency Syndrome (AIDS) became evident in the early 1980s and now, nearly a quarter of a century later, an effective, economical and deliverable vaccine has yet to be produced. Thousands of non-human primates of multiple species have served the scientific community in delineation of the biology of the AIDS virus (Human Immunodeficiency Virus-Type 1 and Human Immunodeficiency Virus-Type 2, HIV-1, HIV-2) as well as the Simian Immunodeficiency Virus (SIV) and numerous genetically engineered chimeric viruses (SHIV). With a better understanding of the pathogenesis of these lentiviruses in the non-human primate, numerous vaccine strategies have been employed to test experimental vaccines with a number of these vaccines now in human trials.

Malaria

Malaria, a longstanding pandemic, continues to produce devastating effects, not only in many of the developing countries but throughout the tropical and sub-tropical regions of the world. The World Health Organization has estimated that 300 to 500 million new cases of malaria occur yearly while, in Africa alone, nearly 2 million children under the age of 5 die each year from malaria (Sibal, 2001; World Health Organization, 1998). Numerous non-human primate species have been used in malarial research. However the Owl and Squirrel Monkeys are the outstanding non-human primate models as a result of the their susceptibility to infection with *Plasmodium falciparum* and *Plasmodium vivax*, the primary cause of all human malaria. As with AIDS-related research efforts, the scientific community world-wide has focused on the development of varying vaccine strategies for the prevention of malaria which are being tested in these species.

Hepatitis

Hepatitis, in all its forms world-wide, infects over 5.5 billion people based on seroepidemiological estimates. The epidemiologic impact of this is staggering. There now is a rising crisis in infection rates of hepatitis C, with approximately 170 million people now infected (Sibal and Samson, 2001). Feitelson and Larkin (2001) estimated that 350 million people are now chronic carriers of hepatitis B while hepatitis A is believed to have infected more than 5 billion people (Purcell and Emerson, 2001). The chimpanzee, although both endangered in the wild and restricted in its research uses, has served hepatitis-related research needs like no other animal species. It has been invaluable as an animal model for hepatitis C vaccine research and therapeutic development efforts. The importance and contribution of the chimpanzee in hepatitis B research has been reviewed (Prince and Brotman, 2001), while the role of the chimpanzee model for studies of hepatitis C has been presented by Lanford *et al.* (2001).

Tuberculosis

Tuberculosis, an age-old and continuing human health concern, has taken on a new urgency in research efforts to develop new vaccines and therapeutics as new and resistant forms have been identified, most strikingly in association with HIV infection. *Macaca fascicularis* and *mulatta* are now utilized in tuberculosis research and vaccine development (European Commission, 2002).

Neurosciences

Research in the neurosciences, including neuroanatomy, neurobiology and physiology, have used non-human primates extensively. Most importantly, the basis of normal and abnormal brain function and an understanding of brain disease, through experimental non-human primate studies, have achieved major advances over the past 10 years (Judd, 2000). Sibal and Samson (2001) state that studies in the monkey brain have led to advanced understanding of the human brain in areas of sensory perception, memory, movement, learning, emotion, decision-making and communication. Human health conditions such as Parkinson's disease and Alzheimer's disease (see also Chapter 27 by Tigno *et al.* and Chapter 28 by Szabo) have been particularly highlighted in non-human primate-based research. Today, Parkinson's disease is the most common progressive neurodegeneration movement disorder of the central nervous system (Sibal and Samson, 2001; Young, 1999). Although treatment of this new degenerative disorder with levodopa has been well described, its use is known to be associated with serious side effects, thus dictating further research with the monkey models to develop new and safer therapeutics for treatment use.

Alzheimer's disease

In an aging population, Alzheimer's disease is now seen in approximately 10% of persons over 65 and nearly 50% over 85 years of age (NIA, 1999; Sibal and Samson, 2001). Alzheimer's disease affects up to 4 million Americans. The disease causes the formation of plaques and tangles in the brain, affecting memory and other cognitive abilities. On average, those with Alzheimer's die within 8 to 10 years of their diagnosis (National Institute on Aging via the Alzheimer's Disease Education and Referral Center website, June, 2003).

Macaques are an ideal animal model for the study of the progression, treatment, and prevention of Alzheimer's disease and other forms of dementia as they experience a neurological aging processes similar to humans (Sibal and Samson, 2001). Some progress has already been made in the development of strategies to prevent the onset of dementia. Although expert opinions on the benefits of hormone replacement therapy are mixed (Armitage *et al.,* 2003), several studies have demonstrated that estrogen replacement therapy may have protective effects on the brains of postmenopausal women and reduce the risk and/or severity of Alzheimers disease (Tang *et al.,* 1996; Birge, 1997; Inestrosa *et al.,* 1998; McEwen and D'Alves, 1999). The neurologically protective capabilities of Tamoxifen, a selective estrogen receptor modulator (SERM) has been tested in rhesus macaques by Cheng *et al.* (2001). They found that Tamoxifen had an estrogen-like effect on higher brain centers, suggesting that some SERMS may be useful as a neuroprotective agent in menopausal women.

Research programs are also in place to assess the progression and treatment of Alzheimer's disease using non-human primate models. Harder (2000) has studied the neurological progression, treatment, and behavioral assessments of treatment for Alzheimer's disease using the common marmoset. Neurological impairment was induced either pharmacologically or through the use of lesions. The effect of pharmacological agents was studied for their ability to reverse the impairments, using examination of neural tissue and behavioral testing of the marmosets. Bading *et al.* (2002) reported on the vascular capacity of aged squirrel monkeys to remove from the brain, Abeta40, which contributes to cerebrovascular amyloidosis. They found that this capacity diminished with age (comparing aged with middle-aged squirrel monkeys), suggesting that squirrel monkeys are a valuable model for understanding the pathological progression of Alzheimer's.

New emerging infectious disease research

Recent years have seen an emerging emphasis on new infectious disease research. Animal models are needed to study these newly discovered diseases in preparation for any serious epidemics that might arise. In order to accomplish this, we must be able to replicate the disease progression in an animal model if we are to test the effectiveness of a treatment for it (European Commission, 2002).

New frontiers: Xenotransplantation

Researchers and medical professionals have long been interested in the possibility of transplanting organs from animals to humans. There were efforts to perform successful xenotransplants as early as 100 years ago, but they have been largely unsuccessful. Reemtsma (1964) performed a chimpanzee-to-human kidney transplant that was successful for 9 months.

Daar (1999) speculated that there exists a general level of concern with the concept of non-human primate-to-human transplants. This discomfort appears to come from more than one source. First, because non-human primates are morphologically similar to humans, some may fear that the distinction between humans and monkeys will become too blurred if we begin taking organs from these animals and putting them into human bodies. Second, there is the clear risk of zoonotic infection from non-human primate pathogens to humans. Pathogens, benign to the source animal, may not be as benign to humans receiving the transplant under immunosuppressive regimens. The risk of initiating zoonotic epidemics in the human remains a major concern. Third, the cost of breeding animals is becoming more expensive, with time, and an increased demand for this resource, for transplantation purposes, could increase the cost to prohibitive amounts, which would defeat the purpose of the program (Daar, 1999).

Nevertheless, there have been several attempts, and limited successes, in non-human primate-to-human transplants. In 1982, a baboon heart was transplanted into the body of a newborn infant who was born with a terminal congenital heart defect. The transplant succeeded for 3 weeks, after which the infant died (Daar, 1999).

Another partially successful transplant occurred in 1996, when baboon bone marrow was transplanted into a man severely compromised by AIDS, in the hopes that, if the patient's bone marrow could be reconstituted by the baboon marrow, improvement would result from baboon marrow that was resistant to HIV (Kaufman, 1994). While the marrow apparently did not reconstitute, his condition did improve (Getty, 1996) and he continues to survive through the writing of this book.

Because of concerns about the practice of xenotransplantation, the Advisory Group on the Ethics of Transplantation (1997), from the United Kingdom, developed guidelines regarding xenotransplantation, which included the recommendation that non-human primates should not be used for xenotransplantation based primarily on the risk of introduction of pathogens into humans. With proper consideration of the issues involved, the exploration of xenotransplantation between non-human primates and humans continues.

Welfare considerations

The use of non-human primates in biomedical research continues to raise welfare concerns revolving around the more complex nature of their social structure. It is perceived that this complexity may render non-human primates more susceptible to greater stress when placed in a laboratory environment than might be the case with other non-primate research animals (see also Chapter 14, by Reinhardt.) Some question whether non-human primates, with their several affinities to humans, should be used in biomedical research at all. Smith and Boyd (2002) argue that their use may be justified under appropriate circumstances, which include:

- the use of monkeys is the only means of addressing the question at hand;
- the number of animals used is kept to a minimum and the experimental design is sound;
- the treatment of the animals is as sensitive to the animals' needs as possible, including transport, husbandry, environment, and data collection techniques. An example of such innovation is the reduction of stress by training the animal for collection procedures as, for example, training the animal to present a limb for routine blood draws to avoid the need for capture or restraint.

Over the last several decades, the primate programs have developed non-feral breeding programs to develop models for studying specific diseases. Today, for instance, NIH invests millions of dollars in the development of specifically pathogen-free macaque breeding colonies to ensure high-quality, captive-bred, healthy non-human primates for the scientific community.

Additionally, in recent years, the genetic make-up of captive populations has become a topic of discussion. Some, including the European Health Commission (2002), for example, have expressed concerns regarding inbreeding in captive populations which would lead to restricted gene pools. Appropriate breeding mechanisms should be employed to prevent these kinds of concerns.

The Animal Welfare Act

The Animal Welfare Act (AWA) was passed in the U.S. in 1966, to protect animals in laboratories, the entertainment business, trade, and those kept as pets. This legislation is the foundation upon which all U.S. animal protective agencies are based, and all of the policies developed over the years refer back to the AWA as the starting point for the protection of animals in these circumstances.

The AWA outlined appropriate care and treatment of animals in several areas, including (but not limited to): minimization of pain and distress, housing, feeding, and psychological well-being. Each research institute is required to establish and maintain an "Institutional Animal Care and Use Committee" which monitors and enforces the laws and policies designed to protect research animals. These committees are composed of at least 3 members, one of whom must be a veterinarian, and one of whom must be from outside the research institute. If the committee determines that a project is not compliant with Federal regulations and the problem is not corrected immediately, the research project can be closed and the animals removed from the program. Below we examine briefly each of the four major areas of concern for animal welfare.

Minimize pain and distress

This area encompasses a huge range of circumstances. The potential for pain and distress occurs in many circumstances, not just during research procedures. In order to minimize this potential, research institutes must maintain a veterinary care program to monitor the health of the animals. This includes daily observations, consultation with the principal investigators to ensure that the procedures are as painless as possible, and adequate pre- and post-procedural care. All procedures, that involve more than minor or momentary pain, must be conducted with analgesia and/or anesthesia, and the use of paralytics without an accompanying anesthetic is prohibited (AWA, Section 2.31 (d), 1966). Routine medical procedures should also be kept as non-threatening as possible.

Housing

Biomedical research facilities are held to very strict standards for maintaining housing that is appropriate for the animals. Specific requirements are in place for heating, cooling, temperature, and ventilation of the facilities. In addition, facilities are required to allow the animals to become acclimatized to the environment so that they do not suffer unnecessary stress or discomfort. Minimum space requirements must also be met.

One area that has been investigated extensively is the housing of non-human primates in captivity.

Because these animals are social by nature, it is psychologically best for them if they can be housed in a group situation of varying configurations. Several studies have been conducted to determine the optimal means of placing formerly single-caged animals together in pairs or in small groups. Consistently, it has been found that giving the animals several days to become familiar with each other seems to ease the transition and reduce the risk of highly aggressive interactions once the monkeys are housed together.

Psychological well-being programs

Federal law requires the provision of psychological well-being programs for captive non-human primates (National Research Council, Institute for Laboratory Animal Resources, 1996). Programs to address the conditions necessary for optimal psychological well-being of these animals, used in research, are concerned with the humane treatment of the animals, as well as the removal of any effect of stress or trauma from the research design. Distressed animals produce stress hormones that could confound the findings of any biomedical research programs. Thus, a successful Psychological Well-Being program will address both welfare and research concerns for stress-free non-human primates.

An additional, important aspect of psychological well-being is a trusting relationship between the monkeys and their caretakers as a means of enriching the animal's social life. Personnel should routinely devote time to positive interactions with the animals, such as giving fruit and vegetable rewards during observations (Reinhardt and Reinhardt, 2000).

A tangible means of ensuring the psychological well-being of non-human primates is to provide toys and other manipulanda enrichment devices that help prevent boredom. Developmentally normal primates, that are placed in an environmentally restricted, i.e. boring, situation, are more likely to engage in stereotypic and/or self-injurious behavior (Ridley and Baker, 1982; Sackett *et al.,* 1999). The toys must be interesting to the animal, able to be sterilized, and harmless if ingested. Animals grouped in large enough areas, such as outdoor containment facilities, often are provided with climbing structures, swings and "playground" and smaller toys to encourage gross motor activity. Monkeys kept in cages are provided with small toys, such as kong toys, nylabones, puzzle boxes and foraging toys that allow them to exercise their small motor skills and give them activities that provide mental stimulation. They are also provided with healthy treats, such as fruit, often hidden within a foraging toy, to add interest.

Studies have been conducted to determine whether these enrichment strategies are actually of any benefit to the non-human primate. Boinski *et al.* (1999) studied the effect of providing various types of toys on the behavior and cortisol levels in Brown Capuchins (*Cebus appella*). They reported that providing small plastic toys and foraging boxes, with treats hidden in them resulted in the greatest ratio of normal to abnormal behaviors. They also found that plasma cortisol levels decreased when the toys and boxes were made available. These findings provide an objective measure of stress reduction that resulted from the availability of objects that the monkeys found interesting.

Studies have also been conducted to determine the factors that are most likely to predict abnormal behavior, e.g., self-inflicted biting and, stereotypies in captive non-human primates. One of the best predictors of such behavior is social deprivation (Bayne, 2003), particularly if it occurs during the first few years of life (Bellanca and Crockett, 2002). Schapiro (2002) found that housing monkeys socially, i.e., in pairs or groups, not only increased normal, species-typical behavior, it also enhanced the number of immune parameters. This finding suggests that the ideal situation for monkeys is to be housed socially whenever possible. It also provides information that is useful in the study of immunosuppressive disease in humans. Thus, information that was originally sought to help the monkeys themselves has been serendipitous in its contribution to the understanding of human disease.

Conservation and management

For more than a quarter century, the use of non-human primates in biomedical research has been a topic of debate, and even protest, by various conservation organizations. At the heart of the debate has been the concern over the use of wild-caught non-human primates in research and the resulting decline in natural populations. Faced with increasing conservation concerns, a number of habitat countries (those with indigenous non-human primate species) have instituted changes in government policy regarding non-human primate use and exportation. Along with international initiatives and conventions, designed to promote more effective primate conservation and management, these actions are helping to secure the future of naturally-occurring non-human primate populations while ensuring the availability of non-human primate resources for biomedical research.

In 1981, the World Health Organization (WHO) convened a meeting to address the feasibility of establishing, in habitat countries, national programs designed to manage non-human primate populations as sustainable or renewable resources so as to ensure the permanent conservation of the various species and maintain the supply of non-human primates for essential biomedical research (MacKinnon, 1983). That same year, the Ecosystems Conservation Group (of the World Conservation Union, formerly, IUCN) and WHO submitted a joint recommendation that only non-human primates known to be "common" (e.g., longtailed macaques, rhesus macaques, etc.) should be used for biomedical research. They further recommended that "wild caught non-human primates be used primarily for the establishment of self-sustaining breeding colonies, the eventual goal of which should be to captive-breed most or all non-human primates used in research" (p. 1, cited in MacKinnon, 1983). This policy statement has had a significant impact on the way many governments now approach non-human primate management and conservation (Kyes *et al.,* 1998a).

The capture and export of wild-caught non-human primates, particularly during the 1960s and 1970s, contributed to the decline of certain primate species such as the rhesus monkeys of India (Kavanagh *et al.,* 1987). Non-human primate exports, however, decreased considerably after the establishment of the Convention on International Trade in Endangered Species of Wild Flora and Fauna (CITES) in 1973 (Crockett *et al.,* 1996). A number of countries even proceeded to ban all exports of their native non-human primates. India was the first to institute such a ban in 1978. Malaysia followed in 1984. In 1994, Indonesia instituted a policy banning the export of wild-caught non-human primates, allowing only the export of progeny from captive breeding or managed colonies (Kyes *et al.,* 1998a). Most recently (August, 2003), the government of Nepal passed a wildlife farming policy that permits the captive breeding of non-human primates (rhesus macaques) for use in research. The policy is very specific, however, that only the captive-bred progeny may be used in research. Wild-caught monkeys can only be used to establish and replenish the breeding colonies. These government policies have been very effective in reducing pressure on the natural populations and helping to ensure the long-term survival of the species.

There are many more conservation-related initiatives too numerous to mention in this short overview. It is important, however, to note the role of the U.S. National Primate Research Centers in promoting non-human primate management and conservation in habitat countries. The Washington National Primate Research Center, for example, helped in establishing Indonesia's first Primate Research Center at Bogor Agricultural University in 1990 (Kyes *et al.,* 1997) and is currently collaborating with the Nepal Biodiversity Research Society to create the first Primate Research Center in Nepal (Kyes and Chalise, 2002). One of the key objectives of these collaborative international programs is to support conservation-based research and training in habitat countries (Kyes *et al.,* 1998b). The ultimate goal is to help establish a growing body of well-trained, national experts who are capable of implementing the programs needed to ensure the future of their country's important non-human primate resources and the conservation of biodiversity.

Correspondence

Any correspondence should be directed to William Morton, University of Washington, National Primate Research Center, Seattle, Washington, USA.
Email: bmorton@bart.rprc.washington.edu

References

Advisory Group on the Ethics of Transplantation (1997). *Animal Tissues into Humans.* London, Stationery Office.

Agy, M.B., Schmidt, A., Florey, M.J., Kennedy, J., Schaefer, G., Katze, M.G., Corey, L., Morton, W.R. and Bosch, M.L. (1997). *Virology* 238, 336–343.

Armitage, M., Nooney, J. and Evans, S. (2003). *Clinical Endocrinology* 59(2), 145–155.

Ashley-Montague, M.F. (1943). *Edward Tyson, M.D., F.R.S. 1650–1708, and the Rise of Human and Comparative Anatomy in England.* Philadelphia: Am. Philos. Society.

Bader, T.F. (1996). *Am. J. Gastroenterol.* 91(2), 217–222.

Bading, J.R., Yamada, S. and Mackic, J.B. (2002). *J. Drug Target* 10(4), 359–368.

Bauer, J.H. and Hudson, N.P. (1928). *Am. J. Trop. Med.* 8, 371–378.

Bayne, K.A.L. (2003). *Toxicologic Pathology* 31(Supplement), 132–137.

Bellanca, R. and Crockett, C.M. (2002). *Am. J. Primatol.* 58(2), 57–69.

Birge, S.J. (1997). *Neurology* 48, 836–841.

Boinski, S., Swing, S.P., Gross, T.S., and Davis, J.K. (1999). *American Journal of Primatology* 48, 49–68.

Bosch, M.L., Schmidt, A., Agy, M.B., Kimball, L.E. and Morton, W.R. (1997). *AIDS,* 11, 1555–1563.

Charin, G.M. (1979). In Cheboksari (ed.) *Macro-Microstructure of Tissue in Norm, Pathol, and Experim.* 6, pp 30–34 (Russian).

Cheng, C.M., Cohen, M., Wang, J. and Bondy, C.A. (2001). *FASEB Jrnl.* 15, 907–915.

Crockett, C.M., Kyes, R.C. and Sajuthi, D. (1996). *American Jrnl. of Primatology* 40, 343–360.

Daar, A.S. (1999). *Bulletin of the World Health Organization* 77 (1), 54–61.

Darwin, C., (1859). *On the Origin of Species by Means of Natural Selection, or the Preservation of Favoured Races in the Struggle for Life*. London: J. Murray.

Davis, H.L., Suparto, I. and Weeratna, R. (2000). *Vaccine* 18, 1920–1924.

European Commission, Health and Consumer Protection Directorate-General (2002). The need for non-human primates in biomedical research: Statement of the Scientific Steering Committee adopted at its meeting of 4–5 April 2002.

Feitelson, M.A. and Larkin, J.D. (2001). *ILAR J.* 42(2), 127–138.

Fridman, E. (2002). *Medical Primatology: History, Biological Foundations and Applications.* New York: Taylor & Francis.

Gartner, S., Liu, Y., Polonis, V., Lewis, M.G., Elkins, W.R., Hunter, E.A., Miao, J., Corts, K.J. and Eddy, G.A. (1994). *J. Medical Primatology* 23, 155–163.

Getty, J. (1996). The tragic hypocrisy of "animal rights", *Wall St. Journal*, 13 June, 1996.

Goodwin, W.J. (1972). In Goldsmith, E.I. and Moor-Jankowski, J. (eds) *Medical Primatology, Part I,* pp 15–29. Basel: S. Karger.

Gorilla Foundation, The (2003). Available from: http://www.koko.org/ [accessed May 30, 2003].

Harder, J. (2000). Paper presented for: The Common Marmoset Monkey in Biomedical Research: A Model for Human Health and Disease, December 2000, German Primate Center. Goettingen: German Primate Center.

Huxley, T.H. (1863). *Evidence as to Man's Place in Nature.* London: Williams & Norgate.

Inestrosa, N.C., Marzolo, M.P. and Bonnefont, A.B. (1998). *Molecular Neurobiology* 17, 73–86.

Jerome, C.P. and Szostak, L. (1987). *Lab. Animal Science* 37, 508–509.

Joag, S.V. (2000). *Microbes and Infection* 2, 223–229.

Johnston, M.I. (2000). *Molecular Medicine Today* 6, 267–270.

Judd, T. (1999). *Neuropsychotherapy and Community Integration: Brain Illness, Emotions, and Behavior.* Boston: Kluwer Academic/Plenum Publishers.

Kaufman, C.L. (1994). *Blood* 84(8), 2436–2446.

Kavanagh, M., Eudey, A.A. and Mack, D. (1987). In Marsh, C.W. and Mittermeier, R.A. (eds) *Primate Conservation in the Tropical Rain Forest*, pp 147–177. New York: Alan R. Liss, Inc.

Kyes, R.C., Sajuthi, D., Morton, W.R., Smith, O.A., Lelane, A., Pamungkas, J., Iskandriati, D., Iskandar, E. and Crockett, C.D. (1997). *Indonesian J. of Primatology* 1, 1–8.

Kyes, R.C., Daniel, L., Paputungan, U., Bangalino, A., Sondakh, L. and Sajuthi, D. (1998a). *Indonesian J. of Primatology* 2, 43–46.

Kyes, R.C., Sajuthi, D., Iskandar, E., Iskandriati, D., Pamungkas, J. and Crockett, C.M. (1998b). *Tropical Biodiversit.* 5, 39–49.

Kyes, R.C. and Chalise, M.K. (2002). Establishment of a collaborative international program in primatology: The Natural History Society of Nepal and the Washington Regional Primate Research Center. Abstracts of the XIXth Congress of the International Primatological Society, 341–342.

Lanford, R.E., Bigger, C., Bassett, S. and Klimpel, G. (2001). *ILAR J.* 42(2), 117–126.

Lewis, D.A., Campbell, M.J., Foote, S.L. and Morrison, J.H. (1986). *Human Neurobiology* 5(3), 181–188.

MacKinnon, K.S. (1983). Report of a World Health Organization consultancy to Indonesia to determine population estimates of the cynomolgus or long-tailed macaques *Macaca fascicularis* (and other primates) and the feasibility of semi-wild breeding projects of this species. *WHO Primate Resources Programme Feasibility Study: Phase II.* Bogor, Indonesia.

Manson, K., Wyand, M., Miller, C. and Neurath, A.R. (2000). *Antimicrobial Agents and Chemotherapy* 44(11), 3199–3202.

McEwen, B.S. and D'Alkes, S.E. (1999). *Endocrinology Review* 20, 279–307.

Mortimer, P.P. (2001). *Reviews in Medical Virology* 11, 141–148.

NIA (National Institute on Aging) Alzheimer's Disease Education and Referral Center (2003). website http://www.alzheimers.org/index.html [accessed June 5, 2003].

National Research Council, Institute for Laboratory Animal Resources (1996). Guide for the care and use of laboratory animals. Washington, DC: National Academy Press.

Ogata, N., Cote, P.J. and Zanetti, A.R. (1999). *Hepatology* 30(3), 779–786.

Oksche, A., Liesner, R., Tigges, J. and Tigges, M. (1984). *Cell Tissue Res.* 235, 467–469.

Olenev, S.N. (1987). *The Construction of Brain.* Leningrad, Medicine (Russian).

Patton, D.L., Cosgrove Sweeney, Y., Rabe, L.K. and Hillier, S.L. (2002). *Sexually Transmitted Diseases* 29(10), 581–587.

Polakos, N.K., Drane, D. and Cox, J. (2001). *The Journal of Immunology* 166, 3589–3598.

Pope, N.S., Gould, K.G., Anderson, D.C. and Mann, D.R. (1989). *Bone* 10, 109–112.

Price, D.L., Martin, L.J., Sisodia, S.S., Walker, L.C., Voytko, M.L., Wagster, M.V., Cork, L.C. and Koliatsos, V.E. (1994). In Terry, R.D., Katzman, K.L. and Bick, K.L. (eds) *The Aged Nonhuman Primate: A Model for the Behavioral and Brain Abnormalities Occurring in Aged Humans. Alzheimer Disease.* NY: Raven Press, 231–245.

Purcell, R.H. and Emerson, S.U. (2001). *ILAR J.* 42(2), 161–177.

Prince, A.M. and Brotman, B. (2001). *ILAR J.* 42(2), 85–88.

Reemtsma, K. (1964). *Annals of Surgery* 160, 384–410.

Reinhardt, V. and Reinhardt, A. (2000). *Lab. Animal* 29(1), 34–41.

Reynolds, V. (1967). *The Apes, the Gorilla, Chimpanzees, Orangutan and Gibbon: their History and their World.* New York: E.P. Dutton and Co.

Ridley, R.M. and Baker, H.F. (1982). *Psychological Medicine* 12, 61–72.

Roskosz, T. (1984). *Ann. Warsaw Agr. Univ SGGW-AR Vet. Med.* 12, 1–16.

Sackett, G.P., Novak, M.F.S.X. and Kroeker, R. (1999). *Research Rev.* 5, 30–40.

Schapiro, S.J. (2002). *Pharmacology, Biochemistry and Behavior* 73, 271–278.

Schultz, A.H. (1966). *Yerkes Newletter* 3(1), 15–29.

Schultz, A.H. (1971). *The Rise of Primatology in the Twentieth Century*. Proceedings, 3rd International Congress on Primates, Zurich, pp 2–15. Basel: Karger.

Schultz, A.H. (1972). *The Life of Primates.* New York: Universe Books.

Shevtsova, Z.V., Flehmig, B. and Lapin, B.A. (1995). *Zh. Mikrobiol. Epidemiol. Immunobio.* 2, 55–59 [Russian].

Sibal, L.R. and Samson, K.J. (2001). *ILAR J.* 42(2), 74–84.

Sirianni, J.E. and Swindler, D.R. (1985). *Growth and Development of the Pigtailed Macaque.* Boca Raton: CRC Press.

Smith, J.A. and Boyd, K.M. (2002). The Boyd group papers on the use of non-human primates in research and testing. The British Psychological Society, Leicester, UK.

Stokes, A., Bauer, J.H. and Hudson, N.P. (1928). *J. Am. Assoc.* 90, 253–254.

Swindler, D.R. (1998). *Introduction to the Primates.* Seattle: University of Washington Press.

Tang, M.X., Jacobs, D. and Stern, Y. (1996). *Lancet* 348, 429–432.

Tevi-Benissan, C., Makuva, M., Morelli, A., Georges-Courbot, M.C., Matta, M., Georges, A. and Belec, L. (2000). *J. Acquired Immune Deficiency Syndrome* 24(2), 147–153.

Veazey, R.S., Shattock, R.J., Pope, M., Kirijan, J.C., Jones, J., Hu, Q., Ketas, T., Marx, P.A., Klasse, P.J., Burton, D.R. and Moore, J.P. (2003). *Nat. Med.* 9(3), 343–346.

Vaitukaitis, J.L. (1994). *Importance of Primate Models in U.S. Biomedical Research Efforts and the Implications for Involvement in International Collaborative Programs.* Congress of the International Primatological Society.

Vilensky, J.A. (1988). *American Journal of Physiological Anthropology* 75(2), 283–284.

Watts, E.D. (1986). In Dukelow, W.R. and Erwin, J. (eds) *Reproduction and Development: Comparative Primate Biology*, Vol. 3, pp 415–439, New York: Alan R. Liss.

Williams, D.D. and Bowden, D.M. (1984). In Scarpelli, D.G. and Migaki, G. (eds) *Comparative Pathobiology of Major Age-related Diseases: Current Status and Research Frontiers,* pp 207–219. New York: Alan R. Liss.

Young, R. (1999). *Am. Fam. Phys.* 59(8), 2155–2167.

CHAPTER 25

Chronic Diseases

Bert A. 't Hart[1,4,5], Mario Losen[2], Herbert P.M. Brok[1] and Marc H. De Baets[2]

[1]Dept. of Immunobiology, Biomedical Primate Research Centre, Rijswijk, The Netherlands; [2]Institute of Brain and Behavior, University of Maastricht, Maastricht, The Netherlands; [3]Animal Science Department, Biomedical Primate Research Centre Rijswijk, The Netherlands; [4]Dept. Immunology Erasmus Medical Center, Rotterdam, The Netherlands; [5]Multiple Sclerosis Centre ErasMS, The Netherlands

Summary

The phylogenetic proximity between higher primate species and humans is reflected by a high degree of immunological similarity (Bontrop *et al.*, 1999; Bontrop *et al.*, 1995). Non-human primates are therefore important experimental models for immune-based diseases in the human population. Non-human primate disease models are becoming important for the preclinical development and efficacy evaluation of new therapies; in particular those based on biological molecules, which by their high specificity cannot be tested in rodents.

In this chapter we describe several chronic diseases in non-human primates which are caused by an unwanted activity of the immune system. Two diseases are caused by dysregulated activity of autoreactive T-cells, namely collagen-induced arthritis (CIA), a model of rheumatoid arthritis (RA), and experimental autoimmune encephalomyelitis (EAE), modelling aspects of multiple sclerosis (MS). The third model is a classical antibody-based autoimmune disease, namely myasthenia gravis. We shall discuss how these models help us to better understand the etiopathogenesis of the equivalent human diseases, the diagnostic procedures used in preclinical therapy trials and the clinical management during experiments.

Introduction

The conditions under which non-human primates can be used in biomedical research are strictly regulated. The law on animal experimentation prescribes the qualifications of experimenters and the conditions under which experiments in primates can be considered. In general terms, non-human primates can only be used for *in vivo* experimentation where no alternative experimental possibilities exist, such as *in vitro* test systems,

The Laboratory Primate
Copyright 2005 Elsevier
ISBN 0-1208-0261-9

equivalent disease models in rats and mice or the patients themselves. Moreover, higher non-human primate species should not be used when the same information can be obtained in primate species that have a lower ranking on the evolutionary ladder.

An increasingly important area for non-human primate disease models is the safety and efficacy testing of biotechnology-based therapies (Bach *et al.*, 1993; Jonker, 1990; Kennedy *et al.*, 1997; Wierda *et al.*, 2001). The minimal requirement of a valid animal model for safety testing of a new drug is its sensitivity to the drug's specific pharmacological functions. As most biological therapies are only reactive in primates, non-human primates provide the most useful disease models for this purpose. The phylogenetic proximity between humans and non-human primates translates into a high degree of immunological similarity, which is of importance for the research on fundamental pathophysiological processes in immune-based diseases.

In biological terms the species 'man' (*Homo sapiens*) together with the gorilla (*Gorilla gorilla*), chimpanzee (*Pan troglodytes*) and Bonobo (*Pan paniscus*) form the family of Hominidae. The evolutionary linkage of humans and chimpanzees, which is estimated at about 5 million years, is reflected by a high degree of genetic similarity; both species share 98.4% identity of their genome. This also holds true for highly polymorphic genomic regions, such as the major histocompatibility complex (MHC) and T-cell receptor (TCR) gene families (Bontrop *et al.*, 1999; Bontrop *et al.*, 1995). Chimpanzees are therefore a potentially important species as a model of autoimmune diseases in the human population. Indeed, spondylo-arthropathies and neurological disorders in chimpanzees, reminiscent of autoimmune disorders in the human population, have been described (Alford and Satterfield, 1995; Rothschild and Woods, 1992; Rothschild and Woods, 1996). Ethical constraints, however, prohibit the induction of the experimental autoimmune disease models that are current in rodents and lower ranked primate species (see below).

At present, excellent autoimmune disease models exist in Old and New World monkeys. Interestingly, spontaneous RA-like disease can occur in these species when kept in outside enclosures, but the frequency and predictability are usually too low to be of use for therapy tests (Rothschild, 1993; Rothschild *et al.*, 1997). For this reason, experimentally induced disease models have been developed, such as the rhesus monkey (*Macaca mulatta*) and common marmoset (*Callithrix jacchus*) models of collagen-induced arthritis (CIA) and experimental autoimmune encephalomyelitis (EAE), respectively modelling rheumatoid arthritis (RA) and multiple sclerosis (MS) and experimental autoimmune myasthenia gravis (EAMG) in rhesus monkeys.

The rhesus monkey model of collagen-induced arthritis (CIA)

Introduction

Spontaneous cases of spondylo-arthropathy and osteoarthritis have been found in free-ranging colonies of rhesus monkeys, at an incidence of around 20%, but these are rare in animals kept under laboratory conditions.

The most intensely investigated experimental autoimmune arthritis models in laboratory rodent strains are induced by inoculation with inactivated bacteria: adjuvant arthritis (Van Eden, 1990), bacterial cell-walls: Streptococcal cell-wall arthritis; (van den Broek, 1989) or joint-specific antigens such as type II collagen: C-II; collagen-induced arthritis (Holmdahl *et al.*, 1989). By immunisation with C-II extracted from bovine cartilage, but not with the bacterial antigens, a reproducible RA-like autoimmune polyarthritis could be induced in rhesus (Bakker *et al.*, 1990; Rubin *et al.*, 1987; Terato *et al.*, 1989) and Java macaques (Terato *et al.*, 1989). The time of onset and severity of clinical signs appear heterogeneous, likely reflecting the outbred nature of both species.

Whereas susceptibility to RA is associated with the presence of major histocompatibility (MHC) class II region alleles (HLA-DR1/-DR4), no such association has been found in rhesus monkeys (Buckner and Nepom, 2002). However, an unexpected strong influence of the MHC class I region on the susceptibility to CIA was found. More than 95% of monkeys positive for the *Mamu-A26* serotype appeared completely resistant to the disease, although anti-CII IgG antibodies were present (Bakker *et al.*, 1991; Bakker *et al.*, 1992). This resistance seems age-dependent: in *Mamu-A26*+ monkeys older than 20 years a mild arthritis could be induced (Bakker *et al.*, 1992), as well as antigen-specific, as *Mamu-A26* positive and negative monkeys are equally susceptible to autoimmune encephalomyelitis, induced with human myelin basic protein (MBP) (Slierendregt *et al.*, 1995). Whereas, in the human population, females are more susceptible to RA than males, female and male rhesus monkeys are equally susceptible to CIA. However, a gender prevalence to CIA has been reported in Java macaques (*Macaca fascicularis*), which is a closely related species to the rhesus monkey (Terato *et al.*, 1989).

The immunopathogenesis of CIA

In view of the RA-like clinical and patho-morphological presentation of CIA in rhesus monkeys, it is of interest to investigate the sequence of events that gives rise to inflammation and joint-erosion. Many aspects of RA can be investigated in patients, as the joints are relatively easily accessible for tissue collection with needle biopsy. However, as most patients presenting themselves at their doctor already have advanced disease, the very early pathogenic events are usually not seen. In the case of an animal model, where the moment the disease is initiated is exactly known, serial collection of tissue and body fluid samples can be done even before the arthritis is clinically manifest.

The large body of data from rodent models indicate that the induction of CIA depends on a synergy of delayed-type hypersensitivity and immune complex-mediated inflammatory mechanisms. This also seems to be the case in the rhesus monkey model of CIA. In synovial biopsies from knee joints, collected well after the immunisation but before macroscopic signs of clinical arthritis were visible, we found substantial synovial lining cell proliferation and infiltration of CD3-positive and CD68-positive cells, respectively T-cells and macrophages (Kraan *et al.*, 1998). Removal of the hyperplastic synovium at that stage, for instance by suicide gene therapy, ablates the further development of arthritis (Goossens *et al.*, 1999). This strongly suggests that, as in RA, the arthritis in CIA-affected rhesus monkeys starts with (asymptomatic) synovitis. The beneficial effect of T-cell directed therapies, such as cyclosporin A (Bakker *et al.*, 1993) or the humanised anti-CD25 antibody Dacluzimab® (Brok *et al.*, 2001b) indicate that T-cells present in the early inflammatory synovium play an important role in the onset of arthritis. However, it is difficult to prove this directly, with adoptive transfer, as suitable syngeneic recipients are lacking in this outbred species.

In a large panel of disease-susceptible and -resistant monkeys a clear contribution of autoantibodies to the CIA immuno-pathogenesis was found. More specifically, the resistance of a given monkey to CIA appeared mainly to be associated with the failure to produce adequate levels of anti-CII antibodies of the IgM isotype (Bakker *et al.*, 1991; 't Hart *et al.*, 1993). Unpublished data indicate that the anti-collagen IgG antibodies in Mamu-A26+ monkeys are human IgG4-like and may therefore be deficient in complement fixation. As complement factors play an important role in the initiation of CIA, this may explain the disease resistance (Morgan *et al.*, 1981). Thus, although a pathogenic role of IgG autoantibodies has been demonstrated in mouse CIA (Stuart and Dixon, 1983) our results point rather at a protective role of anti-CII IgG autoantibodies as they prevent binding of pathogenic IgM autoantibodies to the cartilage surface.

How can the role of IgM antibodies be explained? Most binding sites of anti-CII antibodies on the surface of intact human articular cartilage are protected by proteinaceous material from the synovial fluid. This layer can be removed by neutrophil elastase digestion (Noyori *et al.*, 1994). We and others have shown that neutrophils indeed play an important role in the initiation of joint inflammation (Schrier *et al.*, 1984; 't Hart *et al.*, 1990). The CII epitope density on the intact cartilage surface is too low for complement fixation by bound anti-CII IgG antibodies (Jasin *et al.*, 1993). However, surface binding with one of the five available antigen binding sites of an anti-CII IgM molecule is already sufficient for complement fixation, so that neutrophil binding via Fc receptor and/or C3 receptor can take place. Erosion of the cartilage surface, under the influence of neutrophil elastase, enhances the exposure of antibody binding sites on collagen and other cartilage antigens. Hence, IgG antibodies can enhance inflammation and degradation of an already affected joint (Jasin *et al.*, 1993).

To date it is unclear how the dominant *Mamu-A26* resistance marker controls expression of arthritis in the rhesus monkey CIA model. Sequencing of MHC class I region genes of *Mamu-A26* positive monkeys revealed that the serotype comprises a group of Mamu-B alleles (unpublished data). The intriguing possibility that similar mechanisms may be involved with both the strong positive association of *HLA-B27* with ankylosing spondylitis (Bowness, 2002), and the negative association of *Mamu-A26* with CIA, justifies detailed investigation. That *Mamu-A26* positive and negative monkeys are equally susceptible to EAE, points at a specific interaction with CII (Slierendregt *et al.*, 1995). However, as we have been unable to detect any difference in the absolute level, or epitope specificity, of anti-CII IgG antibodies between *Mamu-A26*+ or *A26*- monkeys, we think this is a less likely explanation

(Turner *et al.,* 1994). Alternatively, the CIA resistance might be based on MHC-MHC interaction, as, for example, the binding of a motif within *Mamu-A26* molecules by MHC class molecules, as has been shown (Gonzalez-Gay *et al.,* 1995; Zanelli *et al.,* 1995).

Diagnosis of arthritis

A preclinical safety/efficacy experiment of a new therapeutic agent usually involves a placebo group and one or more experimental groups receiving different doses of the drug. On the basis of statistical considerations (power analysis), a group size of 5 animals (male or female, age between 4 and 10 years) is recommended. When monkeys are of south-Indian origin, and preselected for the absence of the Mamu-A26 serotype, a disease incidence of >95% is commonly found. It is pertinent to emphasise here that the rhesus macaque represents an outbred species. Each selection of monkeys is therefore genetically heterogeneous, by definition, and variation in time of disease onset and duration can be expected (Bakker *et al.,* 1990). For this reason we have implemented a set of biomarkers with which, in addition to the scoring of outward clinical signs, the different pathogenic processes can be separately measured. This was done by parallel monitoring of *Mamu-A26*-positive and-negative monkeys after immunisation with CII/CFA ('t Hart *et al.,* 1998a).

Clinical signs of arthritis in the conscious monkeys are scored on a daily basis by trained observers using the scoring Table 25.1. In a standard experiment, monkeys are sedated twice weekly for detailed examination of their joints and for venous blood collection. After an asymptomatic period of variable length, that usually lasts between 4 and 8 weeks after immunisation, the clinical phase of the disease starts with loss of appetite and behavioural signs indicative of painful joints. Usually warmth and swelling of both ankles and wrists can be observed within a few days after disease onset. In a next phase prominent inflammatory swelling of

TABLE 25.1: Integrated clinical and discomfort scoring table for the rhesus monkey CIA model

Disease score	Characteristics	Monitoring	Maximal duration†
0	Asymptomatic	Daily	End of experiment
	No general discomfort signs		
0.5	Fever (>0.5°C)	2× per week	12 weeks
1	Apathy,	Daily	10 weeks
	Less mobility but no pain	Daily	
	Loss of appetite	Daily	
2	Weight-loss,	2× per week	6 weeks
	Warm extremities/joints	2× per week	
	Treatable pain without STS*	Daily	
3	Moderate redness + STS of joints	2× per week	4 weeks
	Normal flexibility of extremities	2× per week	
4	Severe redness + STS of joints with joint stiffness	2× per week	2 weeks
5	Serious lethargy	Daily	18 hours
	Serious untreatable pain	Daily	18 hours
	Serious destruction and/or immobility of joints	2× per week	18 hours ‡
	Body weight loss >25%	2× per week	18 hours ‡

* STS = soft tissue swelling.

† This can only be assessed in the sedated monkey, which cannot be done more than 2× per week for ethical reasons. Any clinical observations that indicate significant changes relevant to the aggravation of arthritis are listed and incorporated in the discomfort score on a daily basis.

‡ The discomfort time combination is used in a cumulative fashion.

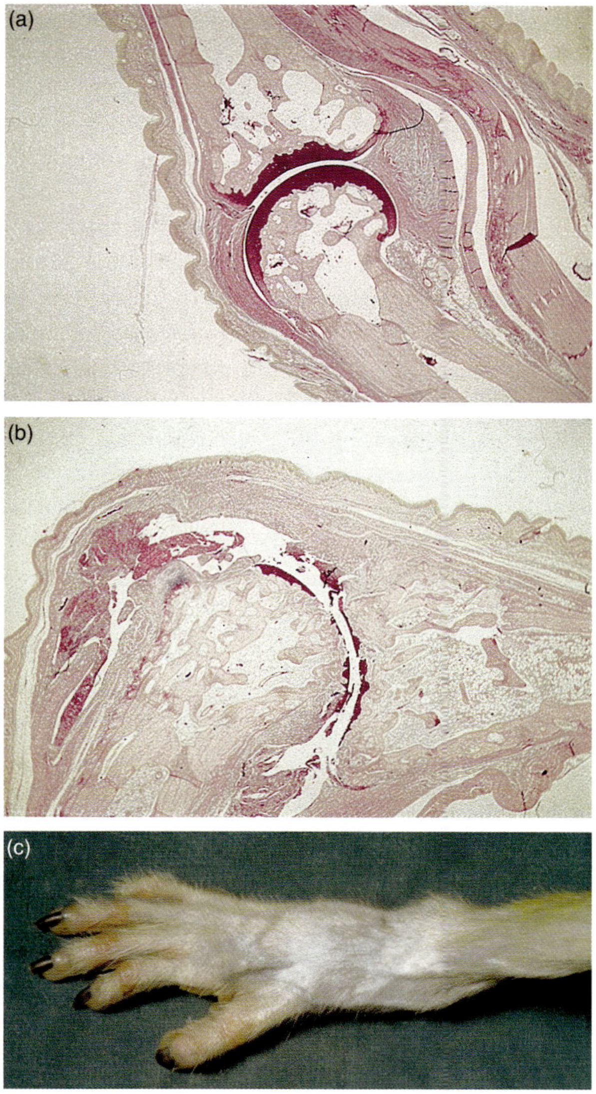

Figure 25.1 Arthritic hand of a CIA-affected rhesus monkey. The bottom figure (C) shows a rhesus monkey hand with prominent inflammatory swelling of the interphalangeal joints and the wrist. Periodic Acid Schiff (PAS) staining of proximal interphalangeal finger joint from a healthy (A) and severely arthritic (B) rhesus monkeys shows the dramatic erosion of joint cartilage in this model. Reprinted with permission from ref. 't Hart *et al.*, *Gene Therapy* (2003), Vol. 10, 890–891.

one or more proximal interphalangeal joints in hands and feet occurs (Figure 25.1). In most monkeys the (meta)carpal/tarsal joints are affected as well. Inflammation of distal interphalangeal joints is only observed in severely arthritic monkeys. The number of arthritic joints is counted and the severity of the joint-swelling is indicated on a graded scale (–, +, ++, +++). At a late stage of the disease the large joints, such as knees, elbows, and hips, become affected and gradually lose flexibility.

Bodyweight

Each time a monkey is sedated for the purpose of blood collection, or test substance administration, the monkey is weighed. As we use young-adult monkeys, which are in the ascending part of their growth curve, a change in the bodyweight curve is a very good objective indicator of disease ('t Hart *et al.*, 1998a).

Serum chemistry

The erythrocyte sedimentation rate (ESR) and serum C-reactive protein (CRP) concentration are routinely used in clinical practice as haematological markers of arthritis (Wollheim, 2000). In CII-immunised rhesus monkeys, the clinical phase of CIA is preceded by an increment of ESR and CRP (Bakker *et al.*, 1990; 't Hart *et al.*, 1998a). During episodes of clinically active arthritis, serum levels of C3, alkaline phosphatase and IL-6 are increased. A meta-analysis of data from more than 100 monkeys revealed that, in high responder monkeys that will develop severe arthritis, a sharp increase of serum CRP level occurs, reaching values above 300 mg/L within days. A more gradual increase of serum CRP levels occurs in monkeys with a less aggressive course and less severe outcome of their disease. Moreover, a relation was found between the absolute serum CRP concentration and the severity of CIA. On the basis of these data we have proposed CRP measurement as an accurate marker of joint inflammation ('t Hart *et al.*, 1998a).

Haematology

During episodes of clinically active disease, leukocytosis, mainly due to an increase of neutrophil counts, as well as thrombocytosis are usually recorded.

Urinary excretion rates of collagen crosslinks

In a standard experiment urine samples are collected once weekly, overnight, into a tray placed under the cage. The trays are covered with a mesh grid to reduce contamination with faeces and spilled food, but spilled drinking water is captured as well. In the morning, the total fluid volume is measured and a sample of 50 ml is centrifuged. The supernatant is stored deep-frozen until analysis. At the end of each experiment the samples are processed for analysis with reversed phase high performance liquid chromatography. The meta-analysis of all data showed that the urinary concentrations, relative to creatinine, of the collagen crosslinks, HP and LP, were increased in all CII-immunised *Mamu-A26–*, but not in *Mamu-A26+* monkeys. Increased levels were always found during episodes of clinical arthritis and the excretion rate curves coincided nicely with other disease markers, such as increased serum CRP concentration and bodyweight loss ('t Hart *et al.*,1998a).

Clinical management

It is inevitable that an experimental model of a very serious disease causes significant discomfort to the animals. Hence, it is of critical importance that the experimental and ethical end-points for each monkey are accurately defined to minimise the suffering. The criteria are listed in an integrated discomfort-scoring table (Table 25.1) which is included in each study protocol. As a rule, monkeys are withdrawn from the experiment when all experimental data have been collected.

Medication

An effective analgesic drug with limited side effects on the model is Temgesic given by i.m. injection at 10 mg/kg. Anti-inflammatory treatment is usually not given as it may affect the outcome of a study.

Multiple sclerosis (MS) and experimental autoimmune encephalomyelitis (EAE)

Concise summary of MS

EAE is not only a valid experimental model of MS, but also one of the most intensely investigated autoimmune models. Concepts developed in this model have contributed significantly to our present understanding of T- and B-cell autoimmunity and the pathogenesis of organ-specific autoimmune diseases. EAE could be induced in all mammalian species tested thus far, including mice, rats, guinea pigs, rabbits and non-human primates (Wekerle, 1993; Wekerle *et al.*, 1994). EAE can be actively induced by immunisation with emulsions, in a strong adjuvant such as complete Freund's adjuvant, of crude central nervous system homogenate, semi-purified myelin or isolated myelin antigens, either as a purified or a recombinant protein or passively transferred with T-cells specific for myelin antigens.

The concept holds that MS is triggered by the activation of myelin-reactive CD4+ T-cells in peripheral lymphoid organs, which can transmigrate the blood-brain-barrier (BBB) using adhesion molecules and specific enzymes to split the tight junctions between endothelial cells. At histological examination, infiltrated mononuclear cells are seen to accumulate in the perivenular Virchow-Rubin space, appearing as characteristic perivascular cuffs. The cognate interaction with resident APC, which present myelin antigens in MHC class II context, together with co-stimulation molecules (CD40, CD80/86), provides the necessary signals to the CD4+ T-cells for further migration into the CNS parenchyma and to trigger a delayed-type hypersensitivity-like reaction. Cells that do not encounter their specific antigen are drained from the CNS or are locally eliminated (Fluegel *et al.*, 2001).

In addition to inflammation, typical histological hallmarks, of the CNS white matter lesions in MS, are demyelination, remyelination, gliosis and axonal pathology. On the basis of the patho-morphological aspect of MS lesions, mirrored against those observed in EAE models, at least four patterns of demyelination have been discerned (Lucchinetti *et al.*, 2000). Two patterns closely resemble the EAE models where lesions are formed by T-cell mediated (pattern I) or T-cell plus antibody mediated (pattern II) anti-myelin autoreactivity. In the other two patterns (III and IV) the likely primary cause of lesion formation is oligodendrocyte dystrophy. Thus, different factors may cause demyelination and only a part of these are reflected in the current EAE models. For a long time the concept has held that axonal structures are initially spared in MS lesions. This concept has been challenged by the observation that axonal suffering and trans-section are consistently found in MS lesions and that the frequency correlates with the degree of inflammation (Trapp *et al.*, 1998). The pathological findings are confirmed with *in vivo* magnetic resonance (MRS) spectroscopy detecting reduced levels of N-acetyl aspartate (a marker of axonal integrity) already at an early disease stage (De Stefano *et al.*, 2001). Neuropathological evidence of axonal damage in EAE, namely intense staining of axons with antibody directed to β-amyloid precursor protein (β-APP), was found in mice (Onuki *et al.*, 2001), rats (Kornek *et al.*, 2000) and non-human primates (Mancardi *et al.*, 2001).

TABLE 25.2: Integrated clinical and discomfort scoring table for the primate EAE models

Discomfort score	Clinical sign	Monitoring	Maximal duration*
0	Asymptomatic No general discomfort signs	Daily	End of experiment
0.5	Reduced alertness, loss of appetite, altered walking pattern without ataxia	Daily	20 weeks
1	Lethargy and/or weight loss less than 15% from start weight	Daily	10 weeks
2	Ataxia (= reduced capacity to keep balance; visual disturbance, including optic neuritis	Daily	6 weeks
2.5	Incomplete paralysis: para- or monoparesis and/or sensory loss and/or brain stem syndrome	Daily	4 weeks
3.0	Complete paralysis hind part of the body one- (hemiplegia) or two-sided (paraplegia)	Daily	1 weeks
4	Complete paralysis all four limbs quadriplegia	Daily	< 18 hours
5	Lethargy (no reaction to external stimuli); incapacity to eat or drink without help, self-mutilation, blindness more than two days, untreatable pain	Daily	< 1 hour

*The discomfort time combination is used in a cumulative fashion.

MS is diagnosed on the guidance of neurological deficit and the presence of abnormalities on magnetic resonance images (MRI) of brain and spinal cord. A detailed discussion of this subject is outside the scope of this chapter, but is reviewed in detail elsewhere (Compston and Coles, 2002). It suffices to state here that non-human primates are more useful for the investigation of lesion formation with MRI than rats and mice as the rodent brain contains only a thin line of white matter (Figure 25.2).

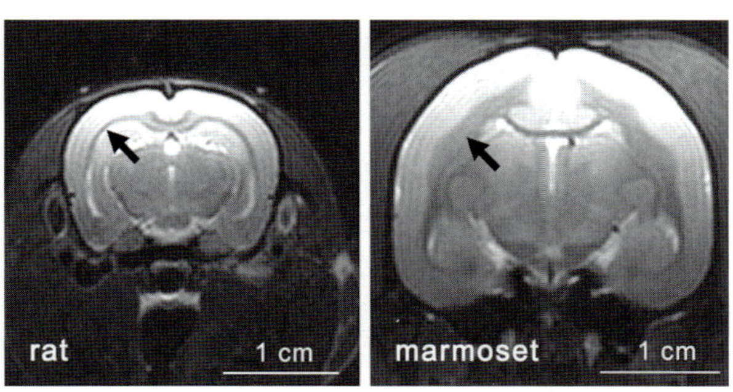

Figure 25.2 T2-weighted brain MR-images of a rat and a common marmoset. In T2-weighted brain MR-images made of naïve animals, the white matter of a rat and a common marmoset appear as dark grey areas (arrow-head). The picture shows the higher proportion of white matter versus grey matter in the common marmoset brain (right) compared to the rat brain (left). Reprinted with permission from 't Hart *et al., Gene Therapy* (2003), Vol. 10, 890–891.

Primate models of EAE

Of the plethora of transgenic, viral and autoimmune models of multiple sclerosis in laboratory strains of mice and rats (Owens and Sriram, 1995; Owens *et al.*, 2001; Wekerle, 1993; Wekerle *et al.*, 1994) only a few have been successfully explored in non-human primates (Brok *et al.*, 2001a). Some attempts have been made to evoke an MS-like disease in chimpanzees, man's closest living relative in nature, by inoculation of cells from MS brain lesions (Lief *et al.*, 1976; Rorke *et al.*, 1979; Wroblewska *et al.*, 1979). However, these experiments have not received a follow-up in more recent years. Of great interest is the finding that spontaneous cases of MS-like disease occur in a free-ranging colony of Japanese macaques that is kept at the Oregon Primate Centre (Gail Marracci and Larry Sherman, personal communication).

The currently most investigated non-human primate models of MS are the autoimmune EAE models in rhesus monkeys and common marmosets. An important aspect of the non-human primate models is that the patterns of neurological deficits are less stereotypical than in most rodent models and resemble, more closely, the variety of aspects seen in MS patients. The clinical and neuropathological presentation of EAE in both species differs fundamentally and may even represent different types of MS (see below). Both models will therefore be discussed separately.

EAE in macaques

The first documented case of EAE was in rhesus macaques, which developed neurological deficit associated with inflammatory demyelination of the CNS after repeated injection with healthy CNS tissue (Rivers and Schwenkter, 1935; Rivers *et al.,* 1933). On the basis of this finding, reproducible EAE models have been established in different macaque species for the study of genetic, immunological pathological and radiological features of MS (including MRI).

Of three tested Macaca species, *M. mulatta* appeared more susceptible to EAE induced with whole brain homogenate, or myelin basic protein, than *M. fascicularis*, while *M. nemestrina* appeared relatively resistant. That MBP-induced EAE in *M. mulatta,* and *M. fascicularis,* respond differently to treatment with combinations of autologous MBP, antibiotics, steroids and copolymer-1, point at a possibly different immunopathogenesis (Alvord *et al.,* 1979). Moreover, some evidence was obtained that these outbred species may respond to different fragments of MBP (Alvord, 1984). Several MBP epitopes have been identified in rhesus monkeys, namely within MBP 29–84 (Slierendregt *et al.,* 1995), MBP 61–82 and MBP 80–105 (Meinl *et al.,* 1997) and MBP 170–186 (Price *et al.,* 1988).

Clinical and neuropathological aspects

Although an MS-like chronic disease pattern has been found, sporadically, EAE in rhesus macaques is typically characterised by an acute clinical course, meaning that serious paralysis develops within 1 to 2 days after the first neurological signs. Large-sized brain white matter lesions, with abundant infiltration of neutrophils and serious destruction of myelin and axon white matter, are found within the brain, but usually not in the spinal cord (Kerlero de Rosbo *et al.,* 2000; Ravkina *et al.,* 1979; Stewart *et al.,* 1991; van Lambalgen and Jonker, 1987). Serial MR images, recorded at 24 hour time intervals, show that large lesions can develop, in most animals, within 4 days after onset and which, on neuropathological examination, appear haemorrhagic/necrotic (Stewart *et al.,* 1991). An example from our own studies is given in Figure 25.3. The destructive aspect of the lesions has triggered the concept that lesions are more likely formed by an acute pathological event that causes severe inflammatory necrosis, than by selective demyelination. EAE in the rhesus monkey therefore resembles more acute disseminated leek-encephalomyelitis, than chronic MS.

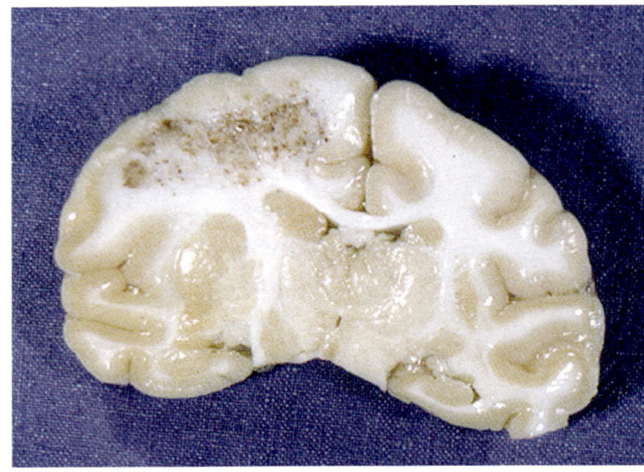

Figure 25.3 Typical lesion in the brain of a rhesus monkey that is affected by myelin-induced EAE. The picture shows a gross anatomical view of a midsagittal coronal brain section in which a large and confluent hemorrhagic/necrotic white matter lesion is evident. Reprinted with permission from Brok *et al., Immunol. Rev.* 183, 173–185; and Poliani *et al., Human Gene Therapy* 12, 905-920.

Immunology

Renewed interest in the rhesus monkey model of EAE has been triggered by a number of recent observations. APC of rhesus monkeys and humans appears to select similar epitopes from the myelin antigens MBP and MOG (Kerlero de Rosbo *et al.,* 2000; Meinl *et al.,* 1995). Moreover, antigens processed by rhesus monkey APC are presented in the correct conformation to specific clones of human T-cells (Geluk *et al.,* 1993; Meinl *et al.,* 1995; 't Hart *et al.,* 1997). This provides the unique possibility to investigate whether APC from MS patients may change the encephalitogenic potential of myelin-specific CD4+ve rhesus monkey T-cells. Also the fact that rhesus monkeys and humans are susceptible to viruses, that have been implicated as possible cause of MS, and that infections follow a similar course, makes the rhesus monkey EAE model especially interesting (Hunt, 1993).

Virus involvement

On the basis of the atypical pathology, the question has been raised as to whether the acute EAE in rhesus monkeys represents an autoimmune or an infectious disease, or a mixture of both. The association of EAE susceptibility with the MHC class II allele *Mamu-DPB*01* (Slierendregt *et al.,* 1995), the protective effect of T-cell depleting therapy with monoclonal antibodies directed to human CD4 or MHC class II (Hu *et al.,* 1997; Jonker *et al.,* 1991; Van Lambalgen and Jonker, 1987) and the induction of EAE by (autologous) transfer of an MBP-specific Th1 cell line (Meinl *et al.,* 1997), all point to a central role of autoreactive T-cells in the disease ethiopathogenesis. However, in contrast to the situation in rodents or common marmosets (see below), immunisation with human myelin, human MBP, or recombinant human MOG induces an identical disease pattern in rhesus monkeys.

The acute disease course appears to depend on genetic factors and on the antigens used to induce the disease. The MHC class II allele (*Mamu-DPB1*01*) appears associated with an increased susceptibility to EAE, induced with bovine MBP, or human myelin in CFA (Slierendregt *et al.,* 1995). However, no MHC effect was found in EAE induced with a recombinant version of the N-terminal extracellular Ig-like domain of human MOG (rhMOG^{Igd}) in CFA (Kerlero de Rosbo *et al.,* 2000). A recently developed third EAE model involves immunisation with a synthetic peptide encompassing amino acids 34–56 of rhMOG^{Igd} (phMOG34–56) in complete adjuvant (Brok *et al.,* 2002). In this model, a possible influence of the *Mamu-DPB1*01* allele was again found (Brok *et al.,* 2000). Whereas *Mamu-DPB1*01* positive monkeys developed chronic EAE, with a relapsing-remitting/secondary progressive course, after several challenges with pMOG34-56 in incomplete adjuvant, monkeys lacking this *Mamu-DP* allele, developed acute fatal EAE. Compared to HLA-DR and -DQ, much less is known about the influence of HLA-DP on MS. A possible role in the epitope spreading, that is thought to determine the disease course after clinical onset, has been proposed (Yu *et al.,* 1998).

The common marmoset

The common marmoset is a small-sized Neotropical primate species weighing 300–500 grams at adult age. Common marmosets breed easily in captivity, giving birth to 1–2 non-identical sets of twin or triplet siblings per year. As fraternal siblings share the placental blood circulation, bone-marrow derived elements from each sibling of a twin or triplet are distributed over the others during ontogeny. This chimerical state creates permanent tolerance towards the allo-antigens among fraternal siblings. Hence, as in inbred rodent strains, it is also possible to elucidate the role of pathogenic cells by adoptive transfer in common marmosets.

A close immunological similarity of humans and common marmosets has been demonstrated at several levels. Highly similar *TCRBV-D-J-C* gene sequences have been found in humans and common marmosets (Uccelli *et al.,* 1997). Moreover, MHC class II region genes of the marmoset were found to encode the evolutionary equivalents of HLA-DR and -DQ molecules (Antunes *et al.,* 1998). Thus far, the common marmoset is the only higher primate species in which the existence of *Caja-DPB1* sequences has not been demonstrated. Each common marmoset, within a sample from four different centres was found to share the monomorphic *Caja-DRB*W1201* allele. The other *Caja-DRB* sequences cluster into two polymorphic lineages, *Caja-DRB1*03,* comprising 7 alleles and DRB*W16, comprising 13 alleles (Bontrop *et al.,* 1999). A Japanese group has also identified five Caja-DRB loci comprising 21 exon 2 alleles (Wu *et al.,* 2000).

The EAE model

In its clinical presentation, and the radiological and pathological aspects of the lesions, EAE in the common marmoset (*Callithrix jacchus*) is an excellent model of chronic MS (Brok *et al.,* 2001a; Genain and Hauser, 2001; 't Hart *et al.,* 2000). The characteristic lesion type resembles closely the pattern II of active MS lesions, which is the most prominent type in chronic MS (Lucchinetti *et al.,* 2000; Raine *et al.,* 1999). The common marmoset is highly susceptible to EAE. Upon a single immunisation with human myelin or recombinant human MOG in complete adjuvant, each monkey tested thus far has developed EAE. The 100% susceptibility appears to have a genetic basis. Although MOG is quantitatively only a minor constituent of CNS myelin (<0.5% of the myelin proteins), the major T- and B-cell reactivity in myelin-immunised marmosets appears directed towards this antigen (Brok *et al.,* 2001a). Data published by McFarland *et al.* seem to confirm this observation, as monkeys immunised with a chimerical MBP/PLP protein develop clinical EAE only when spreading of the autoimmune reactivity to MOG occurs (McFarland *et al.,* 1999). After the finding that all common marmosets share the monomorphic *Caja-DRB*W1201* allele, we could demonstrate that this is a major restriction element for the activation of CD4+ cells specific for the encephalitogenic peptide pMOG14-36 (Brok *et al.,* 2000).

As far as the major myelin antigens are concerned, we only have information on the role of MBP. Compared to the strong encephalitogenic potential of MBP/CFA emulsion in rhesus monkeys (Shaw *et al.,* 1988), MBP is a surprisingly weak antigen in common marmosets (Brok *et al.,* 2000). However, Genain *et al.* (1999) have shown that, when supported with a strong adjuvant like intravenous *Bordetella pertussis* particles, EAE can be induced by immunisation with human MBP in enriched CFA (Massacesi *et al.,* 1995). We prefer not to use such strong adjuvants as these not only mask a variable disease course in individual monkeys, but also change the pathomorphological aspects of lesions. More specifically, administration of *Bordetella pertussis* to monkeys immunised, with human myelin in CFA, causes a synchronised onset of clinical signs with CNS white matter lesions of severe inflammatory/necrotic aspect ('t Hart *et al.,* 1998b).

Neuropathological aspects, histology and immunohistochemistry

An important advantage of non-human primates is that reagents used for the immunohistological typing of tissues from MS patients can also be used for the EAE-models. Thus we have investigated immunological processes within lesions with active inflammation and demyelination. Active lesions contain many CD40+ cells (activated macrophages and B-cells) and fewer CD154+ cells (activated T-cells) (Laman *et al.,* 1998). At the same locations high numbers of cells are also found with an activated appearance and producing a variety of inflammatory mediators, such as cytokines (IL-12) and matrix metallo-proteinases (MMP-9). The observations that antibodies blocking CD40-CD154 ligation or which capture the p40 subunit of the pro-inflammatory cytokines IL-12 and IL-23 abrogating clinical disease and lesion formation (Boon *et al.,* 2001; Brok *et al.,* 2002; Laman *et al.,* 2002), confirm that immunological interactions within the lesions are critical for the EAE pathogenesis.

A particularly interesting aspect of the marmoset model is that the nature of the (auto)antigens, that drive these local immunopathological reactions, can be investigated in an experimental disease that is closely related to human MS. According to a well-supported concept, the CNS white matter lesions in rodent and non-human primate models of EAE, are formed by the synergistic action of a T-cell mediated pathway leading to inflammation, and a B-cell mediated pathway leading to demyelination, anti-MOG antibodies being of particular importance for demyelination (Genain *et al.,* 1995; Linington *et al.,* 1988; Morris-Downes *et al.,* 2002). The finding of MOG-specific autoantibodies in close conjunction with myelin sheaths, in white matter areas where myelin disintegration and lesion formation appears to take place, emphasises that MOG may have a similar important role as target of the autoimmune process in the CNS of common marmosets and MS patients (Genain *et al.,* 1999).

Neuropathological aspects and magnetic resonance imaging (MRI)

MRI is the modality of choice for the visualisation of pathological alterations in soft tissues, such as the brain. MRI is often used to support the diagnosis of MS and, as a surrogate outcome, measure new therapies in clinical trials (McFarland, 2002; McFarland *et al.,* 2002). The tools we have used for the collection of high quality brain MR images of EAE-affected common marmosets are depicted in Figure 25.4. We have started to implement qualitative and quantitative MRI of EAE in the common marmoset model, to investigate the relation between MRI and the neuropathological classification of brain lesions ('t Hart *et al.,* 1998b). Furthermore, we use MRI to monitor the *in vivo* effect of new therapies on the lesions.

The brains of the common marmoset and rat are about the same size but, as shown in Figure 25.2, the marmoset brain contains much more white matter. Lesions can be nicely visualised in T2-weighted (T2w) images (Figure 25.5). Marmosets would seem to be the superior model for unravelling the pathological basis of MRI-detectable lesions ('t Hart *et al.,* 1998b) as well as for the longitudinal monitoring of brain white matter lesion development with MRI (Jordan *et al.,* 1999). In serial brain MR images taken at 2 week intervals, new lesions appear to be continuously formed. This finding illustrates the chronic nature of myelin-induced EAE in this species.

MOG-induced EAE in common marmosets

The conclusion that MOG is a highly relevant antigen for the immuno-pathogenesis of EAE prompted us to investigate the pathogenicity of this quantitatively minor myelin glycoprotein in more detail. We have immunised a random selection of more than 25 monkeys, from the outbred colony at our institute, with a recombinant protein representing the extracellular domain of human MOG. Figure 25.5 shows brain MR images recorded at the height of the disease, illustrating the heterogeneous neuropathological presentation of this model. The most commonly observed pathology, in T2-weighted images, is type C, characterised by many focal lesions (C1) with some contrast enhancement (C4), which are scattered through the brain white matter. The reduced MTR values of the lesions suggest

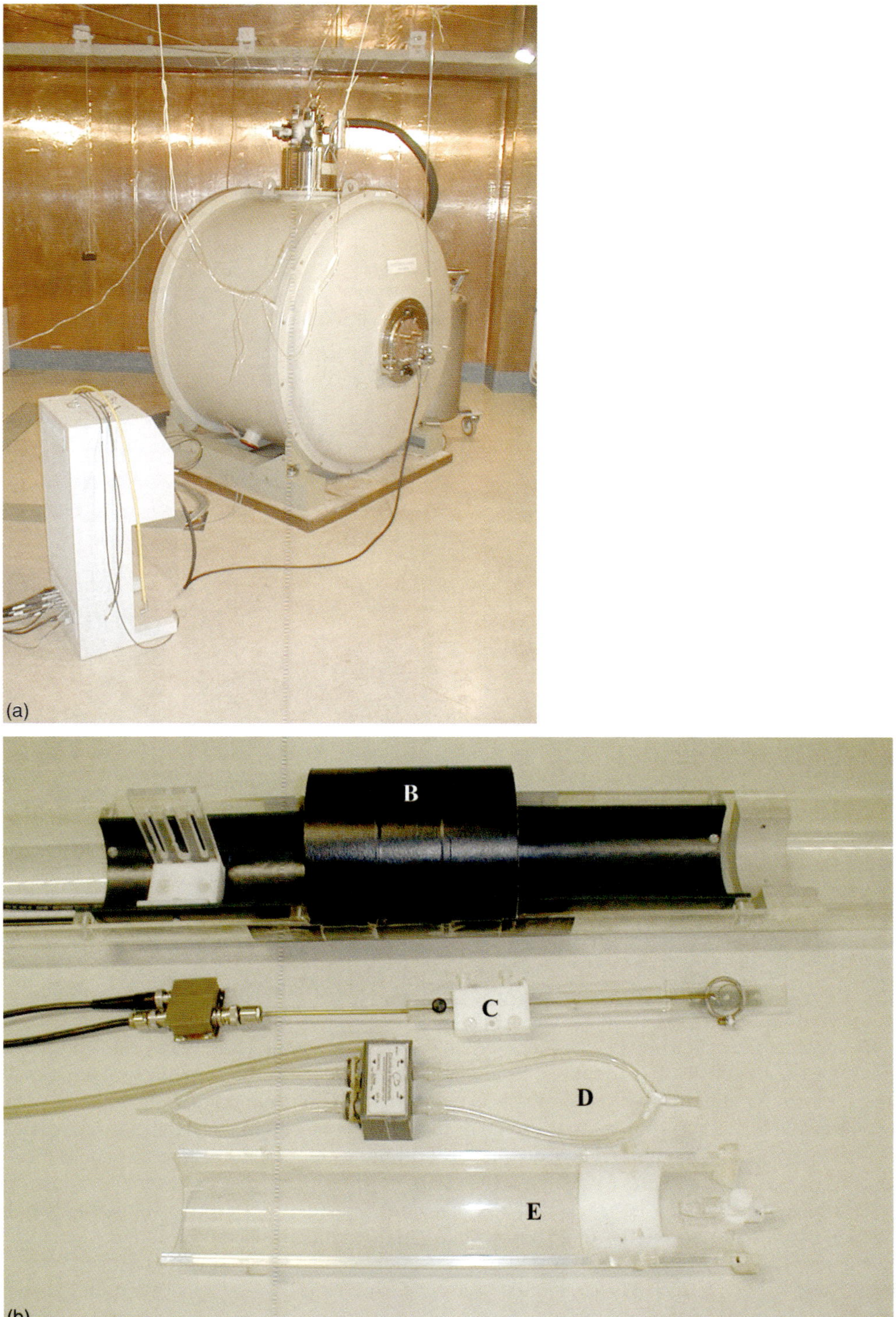

Figure 25.4 Tools for magnetic resonance imaging of the common marmoset brain. High-resolution MRI experiments were performed using a 4.7 T horizontal bore NMR spectrometer (Varian, Palo Alto, California, USA) (**A**), equipped with a high-performance gradient insert (12 cm inner diameter, maximum gradient strength 220 mT/m). A Helmholtz volume coil (Δ 85 mm) (**B**) and an inductively coupled surface coil (Δ 35 mm) (**C**) were used for radio frequency transmission and signal reception, respectively. The animals were immobilised in a specially designed stereotact and placed in an animal cradle (**E**), which was inserted into the NMR spectrometer. During the MRI-experiments the animals were ventilated with isofluorane (1%) in N2O/O2 (70/30) using a remote non-magnetic ventilator valve (Columbus Instruments, Columbus, Ohio, USA) (**D**). Expirated CO2 was monitored, and the body temperature was maintained at 37°C with a heated water pad.

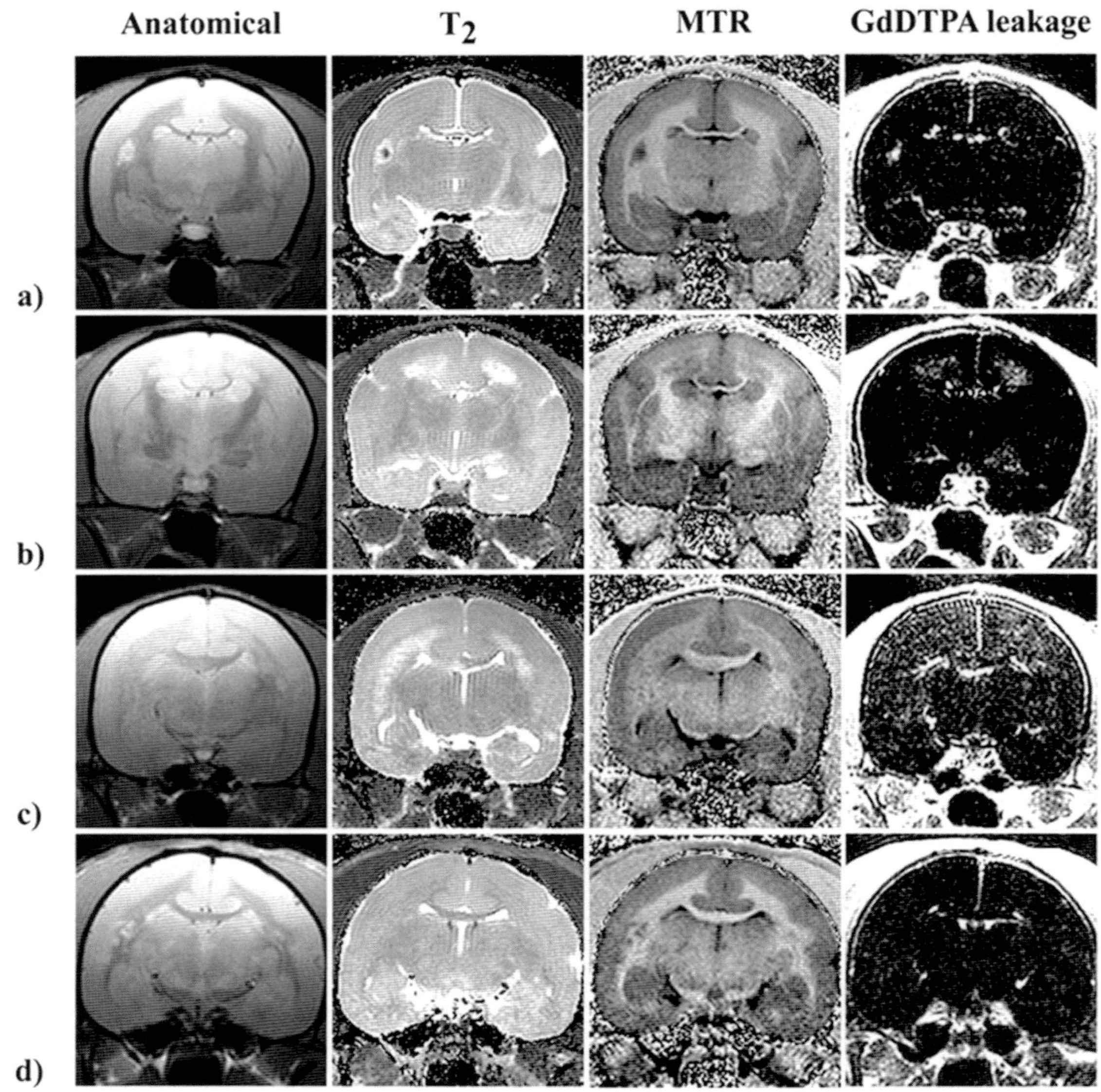

Figure 25.5 MRI-detectable brain white matter alterations in common marmosets affected by rhMOG-induced EAE. EAE was induced in more than 50 randomly collected marmosets from an outbred colony by immunisation with 100 µg of a recombinant protein representing the extracellular domain of human myelin/oligodendrocyte glycoprotein (rhMOG). The typical MRI-detectable changes in the brain of common marmosets at the stage of severe clinical EAE (score 3) fall into four categories (a–d). Patterns a) and d) are the most common cases, where lesions present as focal hyperintense regions on T2-weighted images (row 1; anatomical). Lesions in pattern a) display persistent inflammatory activity of demyelinated lesions probed by signal enhancement on T1-weighted images by intravenously injected gadolinium-diethylenetriamine-penta-acetic acid (DTPA; triple dose), while in pattern d, demyelinated lesions are inactive. The pictures shown in row 4 (GdDTPA leakage) are created by subtraction of the post- and precontrast images.
In T2 maps (row 2) and MTR maps (row 3), are quantitative images in which the actual signal intensities are depicted. The pattern b) pathology is more rarely found and represents early active lesions with strong inflammation, as shown by GdDTPA leakage and T2 hyperintensity and relatively less demyelination. Note the lower MTR reduction compared to patterns a) and d), suggesting that less tissue destruction has taken place. Pattern c) pathology occurred in less than 1% of the examined cases showing ubiguitous oedema that seems to affect the whole white matter, without well defined focal lesions.

oedema and/or loss of tissue mass, likely due to demyelination. The reason that such lesions are hardly visible on the T2 map (C2) is that the T2-weighted signal intensities of lesions and surrounding apparently normal white matter, differ only marginally (Blezer *et al.,* 2004). Pattern A, with large inflammatory lesions (note the focal T2 enhancements in A2 and the intense contrast enhancements in D2) is much less frequently seen. Only after we had developed new quantitative MR strategies, such as T2- and MTR maps which plot the real NMR signal intensities per pixel, were we able to visualise pattern B. This unusual neuropathological pattern possibly represents the ubiquitous vesicularisation of white matter associated with deposition of antibody and complement (Lassmann *et al.,*1988).

Immunopathogenesis of MOG-induced EAE

As was discussed above, the EAE initiating event in each common marmoset is the Caja-DRB*W1201-restricted activation of CD4+ T-cells, specific for an epitope within the encephalitogenic peptide pMOG14–36 (Brok *et al.,* 2000). However, cells and sera from monkeys with chronic EAE, collected at necropsy, display a broad reactivity with a panel of 22-mer peptides covering the cell surface exposed N-terminal MOG domain, which is suggestive of intra-molecular epitope spreading (Brok *et al.,* 2003a). Preliminary data indicate that reactivity with other antigens from the MS patient CNS also emerge, such as PLP or αB-crystalline (data not shown). This phenomenon of epitope spreading is thought to make a major contribution to the chronicity of MS and EAE (Tuohy *et al.,* 1998; Vanderlugt *et al.,* 2000). Recent observations in rhMOG-immunised marmosets support this concept, because peripheral blood T-cells from monkeys, with early-onset EAE, react with a broader panel of MOG peptides than those from monkeys with late-onset EAE. The cervical lymph nodes (CLN) that drain the brain may play a central role in the diversification of the autoimmune reactivity in marmosets (de Vos *et al.,* 2002). Within the CLN of EAE-affected monkeys, but not in unaffected controls, DC-like cells, containing myelin proteins (MBP and PLP), can be found (de Vos *et al.,* 2002). We think that these APC develop from macrophages that have emigrated from the CNS, after having phagocytosed myelin within the lesions. Analogous to the situation in the cryo-lesioned rat model, these APC may locally trigger new T-cell specificities, which migrate to the brain where they modulate ongoing EAE (Lake *et al.,* 1999; Weller *et al.,* 1996). The majority of T-cells, activated in cervical lymph nodes, seem to be anti-inflammatory (Harling-Berg *et al.,* 1999). We hypothesise, however, that the presentation of cryptic epitopes, that have escaped tolerance (Sercarz *et al.,* 1993) by certain disease-associated MHC molecules, may induce T-cells that exacerbate CNS inflammation (Yu *et al.,* 1998).

Clinical management and medication

EAE does not seem to cause serious pain to the monkeys, but the progressive loss of neurological functions can nevertheless be experienced as stressful, causing serious discomfort. As for the CIA model, we use an integrated clinical and discomfort scoring table to monitor the disease course and to minimise the discomfort. However, the different disease courses in common marmosets and rhesus monkeys require different measures.

Clinical management of the rhesus monkey EAE model

Rhesus monkeys normally live in social groups with a strict hierarchy. This may explain why this species is very capable of masking disease symptoms. It is therefore very important that well-trained staff, who are familiar with the normal behaviour of their animals, carry out the daily observations, so that they can see early behavioural changes indicative of neurological problems. After a variable period without apparent disease symptoms, the first clinical signs appear. These are often mild, such as apathy or loss of appetite, but can also be quite aggressive, such as a convulsion. In most monkeys, the clinical condition worsens rapidly and dramatically to a moribund state within 12 hours. It is obvious that such a hyperacute disease course requires very intensive cage-side monitoring to minimise suffering of the monkeys. As remissions do occur, although rarely, monkeys are not sacrificed at the first clinical signs, but only once signs of neurological deficit are manifest (score 2.0 = ataxia).

Clinical management of the marmoset EAE model

The EAE in common marmosets follows a chronic course, but the exact clinical pattern depends on the antigen that is used for disease induction (Brok *et al.,* 2001a). Immunisation with human myelin/CFA typically induces a relapsing-remitting/secondary progressive clinical pattern. Monkeys immunised with rhMOG/CFA, however, display signs of neurological deficit which can lead to significant paralysis (score 3.0) within 1 to 3 weeks without remission. The clinical signs can be described by the categories indicated in the table. Brain lesions develop during the asymptomatic period, and these can be visualised with MRI techniques similar to those used in the diagnosis of MS. Apparently these lesions do not express clinically, a situation that is reminiscent of the clinico-pathological paradox in MS. A complication that is sometimes observed in EAE-affected marmosets is self-mutilation of the genitals, tail or extremities. Monkeys showing this behaviour are immediately sacrificed.

Myasthenia gravis

Myasthenia gravis (MG) is a typical example of an antibody-mediated neurological disease. The understanding of the autoimmune disease, MG, benefits from the corresponding animal model of experimental autoimmune myasthenia gravis (EAMG). Here we will discuss the role of non-human primates in the development of therapies for the treatment of MG.

Myasthenia gravis in humans

MG is a disorder of neuromuscular transmission, in most cases of an autoimmune origin with a few exceptions of genetic causes. MG is characterised by muscle weakness and fatigue that can be restricted to the eye muscles, but generally affects all skeletal muscles. Initial symptoms comprise ocular motor disturbances, ptosis (drooping of the eyelids) or diplopia, oropharyngeal muscle weakness, difficulty with chewing, swallowing, or talking, and limb weakness (Drachman, 1994). The course of the disease is variable but usually progressive. Untreated MG can be lethal, e.g. when respiratory failure occurs. The autoantibodies in 85% of the MG patients are directed against the acetylcholine receptor (AChR), whereas 10% have antibodies against the muscle-specific kinase (MuSK) (Hoch *et al.*, 2001). The cause of MG in a small percentage of patients is yet unknown. Currently available treatments for MG are reduction of the antibody titres in the patients' sera by immunosuppression and, in acute cases, by plasmapheresis or absorption of immunoglobulins using tryptophan-coated columns. However, immunosuppression has serious side effects. Plasmapheresis only improves the patients' condition for a few weeks and requires intensive medical attendance. Hence, plasmapheresis is only indicated in a myasthenic crisis.

Spontaneous myasthenia gravis in animals

MG spontaneously occurs in cats and dogs (Dau *et al.*, 1979; Joseph *et al.*, 1988; Shelton, 2002; Uchida *et al.*, 2002). Similar to the human disease, thymic abnormalities are seen and circulating anti-AChR antibodies can be detected by radioimmunoassay (Dau *et al.*, 1979; Joseph *et al.*, 1988). The immunological mechanisms are probably identical to human MG. To our knowledge, spontaneously occurring MG has not been described in non-human primates, which might be due to a low disease incidence.

Experimental autoimmune myasthenia gravis

The first convincing data suggesting that MG has an autoimmune cause were reported by Patrick and Lindstrom in 1973, who found that immunisation with AChR from the electric eel (*Electrophorus electricus*) or pacific electric ray (*Torpedo californica*) evoked MG-like symptoms in rabbits (Patrick and Lindstrom, 1973). This model was named "experimental autoimmune myasthenia gravis" (EAMG). Since then, MG has been modelled in a range of species, including guinea pigs, rats, mice, pigs and rhesus monkeys (Berman and Patrick, 1980; De Haes *et al.*, 2003; Lennon *et al.*, 1975; Tarrab-Hazdai *et al.*, 1975; Toro-Goyco *et al.*, 1986a). MG-like disease can be induced by antibodies which cross-react with the animals' own AChR. Anti-AChR antibodies damage the neuromuscular synapse by increased internalisation of the AChR (antigenic modulation) and by triggering complement activation. The antibodies alone are both essential and sufficient for the induction of EAMG symptoms. So, two EAMG models are distinguished: (1) chronic EAMG, caused by immunisation with AChR, as described above and (2) passive transfer EAMG, induced by injection of isolated anti-AChR antibodies.

Chronic EAMG

In chronic EAMG, many immunological aspects of MG can be studied, such as AChR processing, APC presentation, the role of MHC, T-cell subsets and cytokines as well as many other issues in human MG that are not directly related to the immune system. The availability of recombinant, mutant and transgenic mice strains offers the possibility to study the involvement of a wide range of proteins, including human immunoglobulins (HuMAb-Mice) (Stassen *et al.*, 2003). Susceptibility to EAMG depends strongly on the mouse strain used (Christadoss *et al.*, 2000; Fuchs *et al.*, 1976). In the most susceptible strains, 50–90 % of the animals develop disease symptoms, whereas others are completely resistant (Christadoss *et al.*, 2000). Also, the severity of disease symptoms is variable. After immunisation of mice with 30 μg *Torpedo* AChR (tAChR), the anti-tAChR serum titers are variable, ranging from 1–5 mg/mL in most cases. The proportion of these antibodies cross-reacting with the mouse AChR was found to be 0.2–2% (Berman and Patrick, 1980).

Chronic EAMG in monkeys

The first reports on EAMG in non-human primates date back to 1975 (Tarrab-Hazdai *et al.*, 1975). Rhesus monkeys (*Macaca mulatta*) immunised with four doses of 100–200 μg tAChR in CFA, administered intradermally with 3 week intervals, were found to develop typical MG signs. Initially, they observed "fatigue, hypo-activity, anorexia and weight loss", and subsequently "extreme flaccid paralysis of the limbs and trunk, sinking of the head and severe difficulties in breathing" (Figure 25.6a).

Electromyography showed a decreased action potential after repetitive nerve stimulation, confirming impaired neuromuscular transmission. This hallmark symptom is used for diagnosis of MG in humans (Drachman, 1994). After injection of 67 μg/kg of the acetylcholine-esterase inhibitor Neostigmine bromide, the symptoms disappeared within 10 min (Figure 25.6b), but this improvement only lasted for about 4 h.

A similar study was performed by Toro-Goyco in 1986, in ten adult rhesus monkeys (Toro-Goyco *et al.*, 1986a). Four monkeys received a total dose of 240 μg tAChR in CFA, divided into three equal doses with intervals of 14 days. Three monkeys developed severe clinical signs of MG and died of respiratory failure. One monkey died earlier without clinical signs, but was found to have anti-tAChR antibodies and muscle inflammation. Five animals, that were immunised with a total dose of 60 μg tAChR, failed to develop EAMG signs, but anti-tAChR antibodies, with titres up to 200 μg/mL, were detected and persisted for more than one year. Antibody titres above 800 μg/mL were found to cause MG symptoms; three animals that died of EAMG had titres between 1 and 2 mg/mL. These anti-tAChR antibody titres are very high compared to human MG patient sera, where maximally 0.24 μg/mL are found (Somnier, 1993; Tindall *et al.*, 1981), the total IgG concentration in serum being between 5 and 16 mg/mL. However, the amount of antibodies cross-reacting with the primate receptor was not determined in that study. Assuming that the degree of cross-reactivity is the same as observed in rodents, (0.2–2%), the anti-primate AChR titre is estimated at 4–40 μg/mL.

The clinical observations demonstrate the similarity of EAMG in rhesus monkeys and MG in humans. Active immunisation of non-human primates could

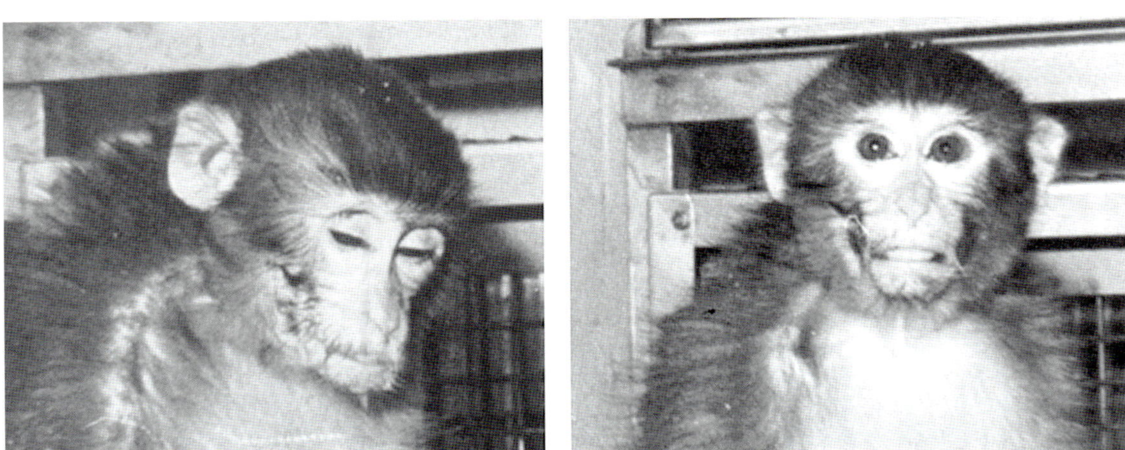

(a) (b)

Figure 25.6 Experimental autoimmune myasthenia gravis in the rhesus monkey. Neostigmine bromide treatment of myasthenic monkey after active immunisation with Torpedo AChR. Left: 5 min after injection of 67 μg/kg neostigmine bromide (note the typical myasthenic drooping of the eyelids). Right: 10 min after injection of neostigmine bromide (Tarrab-Hazdai *et al.*, courtesy of Sara Fuchs).

again become relevant, when new MG therapies, acting upstream of the antibody attack, will become available.

Passive transfer EAMG

Toyka *et al.* have transferred purified IgG from MG patients into mice, which subsequently developed MG symptoms (Toyka *et al.,* 1975). This passive transfer EAMG model is relevant for MG in order to study the effector phase of the disease. Injection of antibodies against the main immunogenic region (MIR), or the acetylcholine (ACh) binding sites of the AChR in experimental animals, induces EAMG within hours or days. The source of antibodies can be serum of MG patients, serum from chronic EAMG animals or monoclonal antibodies (mAbs) produced in cell culture (Richman *et al.,* 1980). Antibodies against the ACh binding sites induce a curare-like effect within 15–30 minutes (Balass *et al.,* 1993). These antibodies block the ligand-induced opening of ion channels.

Antibodies against the MIR of the human AChR are the most prominent cause of MG. In passive transfer EAMG, anti-MIR antibodies induce clinical signs within 8–48 hours depending on the dose and their affinity. The immunopathological mechanisms are antigenic modulation (Tzartos *et al.,* 1985) and complement-mediated focal lysis of the post-synaptic membrane (Fazekas *et al.,* 1986; Lennon *et al.,* 1978).

Passive transfer EAMG in rhesus monkeys

Transfer of IgG, purified from pooled MG patients' sera, into rhesus monkeys (1 g IgG per kg body weight, on three consecutive days) leads to reversible MG symptoms (Heiniger and Toyka, personal communication). We are now in the process of characterising this model in order to test therapeutic proteins in the future.

Two monkeys were injected with 0.8 g IgG of pooled MG serum per kg of body weight on 3 consecutive days. In both monkeys decreased action potential was found after repetitive nerve stimulation on day 8 (Table 25.3). Only one monkey developed clinical signs of MG starting on day 5 and lasting for one week. Serum titres against human AChR ranged from 22.8 μg/ml, after the last IgG injection on day 3, to 7.8 μg/ml on day 9 after the last injection.

Two more monkeys were injected, respectively, with 0.5 mg/kg and 1 mg/kg of the recombinant antibody IgG1-637 on 3 consecutive days. This human antibody binds to the human AChR specifically. It was selected as a Fab from a phage display library made from the thymus of an MG patient (Graus *et al.,* 1997), and subsequently reconstructed to a full IgG1. The antibody is produced in CHO cells as a fully human antibody, with some differences in glycosylation. Due to the IgG1 properties, the antibody is still pathogenic. The concentrations used are based on immunohistochemical staining of neuromuscular junctions in rhesus monkey muscle, using serial dilutions of IgG1-637 and MG patient sera.

None of the animals injected with IgG1-637 developed signs of MG but the monkey injected with the higher dose showed a decrease of the compound muscle action potential after repetitive nerve stimulation on day 8. This suggests that the antibody is actually binding to the *Rhesus* AChR. In the future, higher doses of IgG1-637 will be tested to validate the pathogenic properties of the antibody.

Diagnosis and clinical management of EAMG

Diagnosis of EAMG in rhesus monkeys can be performed in much the same way as in humans, by: (1) Clinical presentation: weakness of muscles after repetitive use, resulting in symptoms such as ptosis, dysphagia, facial muscle weakness, generalised weakness and fatigue, weight loss and reluctance to move (Toro-Goyco *et al.,* 1986b); (2) analysis of anti-AChR and anti-MuSK antibodies by radioimmunoassay; (3) electromyography (EMG) (Pachner and Kantor, 1982)/single fibre EMG (Verschuuren *et al.,* 1990) after repetitive nerve

TABLE 25.3: Effect of passive transfer EAMG in rhesus monkeys

Animal	Antibody	Clinical signs	Decrement
1	Pooled MG IgG 0.8 g/kg on 3 days	+	+
2	Pooled MG IgG 0.8 g/kg on 3 days	–	+
3	IgG1-637 0.5 mg/kg on 3 days	–	–
4	IgG1-637 1.0 mg/kg on 3 days	–	+

stimulation; (4) amelioration of symptoms following administration of acetylcholine esterase inhibitor (Neostigmine-test). In the chronic EAMG models for rhesus monkeys, immunisation with high doses of tAChR can be fatal due to respiratory failure. Neostigmine treatment was found unsuccessful in these cases (Toro-Goyco *et al.,* 1986b) and treatments with other drugs have not been reported.

Unlike the chronic model, the MG symptoms in passively transferred EAMG are reversible. Mild symptoms can be treated with ACh esterase inhibitors. Occurrence of severe symptoms with respiratory problems can be avoided by carefully assessing antibody properties *in vitro*, using the above-mentioned concentrations of polyclonal IgG from MG patient sera and IgG1-637 as a reference.

Myasthenia gravis therapy using the passive transfer EAMG model

Generally, rhesus monkeys are used for efficacy testing of therapeutics that interact specifically with primate proteins. The humans and rhesus monkey AChR structure share a high sequence homology, but the gene encoding the alpha subunit of the human AChR contains an additional exon (MacLennan *et al.,* 1993). Moreover, the character of pathogenic antibodies in MG patients varies widely, ranging from the functional inhibition of ACh binding to anti-MuSK antibodies. Thus, therapeutics tested in rhesus monkeys do not necessarily lead to the same results in (all) human MG patients.

At the level of the antibody attack, two approaches are possible to specifically prevent antibody binding to the AChR: (1) *Absorption of anti-MIR antibodies.* The efficacy of antibody adsorption was demonstrated in a passive transfer model with rhesus monkeys (Ferguson *et al.,* 1995). They injected 10.5 μg/kg radioactive labelled mAb 35, which is a rat anti-MIR antibody. This antibody could subsequently be depleted from the serum using a bi-specific reagent, which consisted of the tAChR crosslinked to an antibody against primate complement receptor (anti-CR1). The mAb 35 bound to the tAChR- anti-CR1 and was depleted from the circulation by the liver. A major risk of this system in a therapeutic application is the possible exacerbation of the autoimmune-response due to an additional response against the auto-antigen part of the bi-specific reagent. So far, this possibility has not been ruled out. (2) *Protection of the MIR by antibody fragments (competition of antibodies for receptor).* The alternative approach, using competitive antibody-fragments does not require AChR fragments and is therefore safer. A competitive recombinant antibody, that protects the MIR, is a promising agent to treat a myasthenic crisis. In a mouse model, Fabs have been demonstrated to protect the AChR against the intact antibodies (Papanastasiou *et al.,* 2000; Toyka *et al.,* 1980). However, these cannot be used in patients because rat antibody fragments will evoke a neutralising immune response. Moreover, Fab antibodies have a short half-life *in vivo*. A potential safety concern is that Fabs are bivalent and still have some pathogenic potential by means of antigenic modulation and increased internalisation of the AChR.

A protective agent should combine the following properties: high affinity for AChR, long half-life, non-immunogenic, no complement activation and no cross-linking of AChRs. In order to obtain a molecule with these properties, we have introduced a mutation in the human antibody IgG1-637 that prevents complement binding to the Fc region (Thommesen *et al.,* 2000). This mutated recombinant antibody (IgG1-637ΔC1q) will be tested *in vivo* for its deficiency to cause EAMG. Since the IgG1-637ΔC1q antibody binds to primate AChR only, the passive transfer model with rhesus monkeys will be used. If this is successful, the next experiment will be the treatment of the disease with IgG1-637ΔC1q after injection of MG patient sera.

Correspondence

Any correspondence should be directed to Bert 't Hart, Biomedical Primate Research Centre, Dept. Immunobiology, Lange Kleiweg 139, 2288 GJ Rijswijk, P.O.Box 3306, 2280 GH Rijswijk, The Netherlands. Tel: (+31)15 284 2691; Fax: (+31)15 284 3999; E-mail: hart@bprc.nl

References

Alford, P.L. and Satterfield, W.C. (1995). *J. Am. Vet. Med. Assoc.* 207, 83–85.

Alvord, E.C., Jr. (1984). *Prog. Clin. Biol. Res.* 146, 523–537.

Alvord, E.C., Jr., Shaw, C.M. and Hruby, S. (1979). *Ann. Neurol.* 6, 469–473.

Antunes, S.G., de Groot, N.G., Brok, H., Doxiadis, G., Menezes, A.A., Otting, N. and Bontrop, R.E. (1998). *Proc. Natl. Acad. Sci. USA* 95, 11745–11750.

Bach, J.F., Fracchia, G.N. and Chatenoud, L. (1993). *Immunol. Today* 14, 421–425.

Bakker, N.P., van Erck, M.G., Zurcher, C., Faaber, P., Lemmens, A., Hazenberg, M., Bontrop, R.E. and Jonker, M. (1990). *Rheumatol. Int.* 10, 21–29.

Bakker, N.P., van Erck, M.G., Botman, C.A., Jonker, M. and 't Hart, B.A. (1991). *Arthritis Rheum.* 34, 616–624.
Bakker, N.P., van Erck, M.G., Otting, N., Lardy, N.M., Noort, R.C., 't Hart, B.A., Jonker, M. and Bontrop, R.E. (1992). *J. Exp. Med.* 175, 933–937.
Bakker, N.P., Van Besouw, N., Groenestein, R., Jonker, M. and 't Hart, L.A. (1993). *Clin. Exp. Immunol.* 93, 318–322.
Balass, M., Heldman, Y., Cabilly, S., Givol, D., Katchalski-Katzir, E. and Fuchs, S. (1993). *Proc. Natl. Acad. Sci. USA.* 90, 10638–10642.
Berman, P.W. and Patrick, J. (1980). *J. Exp. Med.* 151, 204–223.
Blezer, E.L.A., Bauer, J., Brok, H.P.M., Nicolay, K. and 't Hart, B.A. (2004) (submitted).
Bontrop, R.E., Otting, N., Slierendregt, B.L. and Lanchbury, J.S. (1995). *Immunol. Rev.* 143, 33–62.
Bontrop, R.E., Otting, N., de Groot, N.G. and Doxiadis, G.G. (1999). *Immunol. Rev.* 167, 339–350.
Boon, L., Brok, H.P., Bauer, J., Ortiz-Buijsse, A., Schellekens, M.M., Ramdien-Murli, S., Blezer, E., van Meurs, M., Ceuppens, J., de Boer, M., 't Hart, B.A. and Laman, J.D. (2001). *J. Immunol.* 167, 2942–2949.
Bowness, P. (2002). *Rheumatology* (Oxford) 41, 857–868.
Brok, H.P., Uccelli, A., Kerlero De Rosbo, N., Bontrop, R.E., Roccatagliata, L., de Groot, N.G., Capello, E., Laman, J.D., Nicolay, K., Mancardi, G.L., Ben-Nun, A. and 't Hart, B.A. (2000). *J. Immunol.* 165, 1093–1101.
Brok, H.P., Bauer, J., Jonker, M., Blezer, E., Amor, S., Bontrop, R.E., Laman, J.D. and 't Hart, B.A. (2001a*). Immunol. Rev.* 183, 173–185.
Brok, H.P., Tekoppele, J.M., Hakimi, J., Kerwin, J.A., Nijenhuis, E.M., De Groot, C.W., Bontrop, R.E. and 't Hart, B.A. (2001b). *Clin. Exp. Immunol.* 124, 134–141.
Brok, H.P., Van Meurs, M., Blezer, E., Schantz, A., Peritt, D., Treacy, G., Laman, J.D., Bauer, J. and 't Hart, B. (2002). *J. Immunol.* 169, 6554–6563.
Brok, H.P.M., Kerlero de Rosbo, N., van Meurs, M., Bauer, J., Ben-Nun, A., Laman, J.D. and 't Hart, B.A. (2004a) (submitted).
Brok, H.P.M., Laman, J.D., Ouwerling, B., Rus, A., Celebi-Paul, A.G.A., Ben-Nun, A., de Rosbo, K., Bauer, J. and 't Hart, B.A. (2004b) (submitted).
Buckner, J.H. and Nepom, G.T. (2002). *Curr. Opin. Rheumatol.* 14, 254–259.
Christadoss, P., Poussin, M. and Deng, C. (2000). *Clin. Immunol.* 94, 75–87.
Compston, A. and Coles, A. (2002). *Lancet* 359, 1221–1231.
Dau, P.C., Yano, C.S. and Ettinger, S.J. (1979). *Neurology* 29, 1065–1068.
De Haes, A., Proost, J.H., De Baets, M.H., Stassen, M.H., Houwertjes, M.C. and Wierda, J.M. (2003). *Anesthesiology* 98, 133–142.
De Stefano, N., Narayanan, S., Francis, G.S., Arnaoutelis, R., Tartaglia, M.C., Antel, J.P., Matthews, P.M. and Arnold, D.L. (2001). *Arch. Neurol.* 58, 65–70.
de Vos, A.F., van Meurs, M., Brok, H.P., Boven, L.A., Hintzen, R.Q., van der Valk, P., Ravid, R., Rensing, S., Boon, L., 't Hart, B.A. and Laman, J.D. (2002). *J. Immunol.* 169, 5415–5423.
Drachman, D.B. (1994). *N. Engl. J. Med.* 330, 1797–1810.
Fazekas, A., Komoly, S., Bozsik, B. and Szobor, A. (1986). *Clin. Neuropathol.* 5, 78–83.
Ferguson, P.J., Martin, E.N., Greene, K.L., Kuhn, S., Cafiso, D.S., Addona, G. and Taylor, R.P. (1995). *J. Immunol.* 155, 339–347.
Fluegel, A., Berkowicz, T., Ritter, T., Labeur, M., Jenne, D.E., Li, Z., Ellwart, J.W., Willem, M., Lassmann, H. and Wekerle, H. (2001). *Immunity* 14, 547–560.
Fuchs, S., Nevo, D., Tarrab-Hazdai, R. and Yaar, I. (1976). *Nature* 263, 329–330.
Geluk, A., Elferink, D.G., Slierendregt, B.L., van Meijgaarden, K.E., de Vries, R.R., Ottenhoff, T.H. and Bontrop, R.E. (1993). *J. Exp. Med.* 177, 979–987.
Genain, C.P., Cannella, B., Hauser, S.L. and Raine, C.S. (1999). *Nat. Med.* 5, 170–175.
Genain, C.P. and Hauser, S.L. (2001). *Immunol. Rev.* 183, 159–172.
Genain, C.P., Nguyen, M.H., Letvin, N.L., Pearl, R., Davis, R.L., Adelman, M., Lees, M.B., Linington, C. and Hauser, S.L. (1995). *J. Clin. Invest.* 96, 2966–2974.
Gonzalez-Gay, M.A., Zanelli, E., Krco, C.J., Nabozny, G.H., Hanson, J., Griffiths, M.M., Luthra, H.S. and David, C.S. (1995). *Immunogenetics* 42, 35–40.
Goossens, P.H., Schouten, G.J., 't Hart, B.A., Bout, A., Brok, H.P., Kluin, P.M., Breedveld, F.C., Valerio, D. and Huizinga, T.W. (1999). *Hum. Gene Ther.* 10, 1139–1149.
Graus, Y.F., de Baets, M.H., Parren, P.W., Berrih-Aknin, S., Wokke, J., van Breda Vriesman, P.J. and Burton, D.R. (1997). *J. Immunol.* 158, 1919–1929.
Harling-Berg, C.J., Park, T.J. and Knopf, P.M. (1999). *J. Neuroimmunol.* 101, 111–127.
Hoch, W., McConville, J., Helms, S., Newsom-Davis, J., Melms, A. and Vincent, A. (2001). *Nat. Med.* 7, 365–368.
Holmdahl, R., Andersson, M.E., Goldschmidt, T.J., Jansson, L., Karlsson, M., Malmstrom, V. and Mo, J. (1989). *APMIS.* 97, 575–584.
Hu, H., Stavrou, S., Cairns Baker, B., Tornatore, C., Scharff, J., Okunieff, P. and Neville, D.M., Jr. (1997). *Cell Immunol.* 177, 26–34.
Hunt, R.D. (1993). *Herpesviruses of primates.* Springer Verlag, Berlin.
Jasin, H.E., Noyori, K., Takagi, T. and Taurog, J.D. (1993). *Arthritis Rheum.* 36, 651–659.
Jonker, M. (1990). *Semin. Immunol.* 2, 427–436.

Jonker, M., Bakker, K., Slierendregt, B., 't Hart, B. and Bontrop, R. (1991). *Hum. Immunol.* 32, 31–40.
Jordan, E.K., McFarland, H.I., Lewis, B.K., Tresser, N., Gates, M.A., Johnson, M., Lenardo, M., Matis, L.A., McFarland, H.F. and Frank, J.A. (1999). *AJNR Am. J. Neuroradiol.* 20, 965–976.
Joseph, R.J., Carrillo, J.M. and Lennon, V.A. (1988). *J. Vet. Intern. Med.* 2, 75–79.
Kennedy, R.C., Shearer, M.H. and Hildebrand, W. (1997). *Vaccine* 15, 903–908.
Kerlero de Rosbo, N., Brok, H.P., Bauer, J., Kaye, J.F., 't Hart, B.A. and Ben-Nun, A. (2000). *J. Neuroimmunol.* 110, 83–96.
Kornek, B., Storch, M.K., Weissert, R., Wallstroem, E., Stefferl, A., Olsson, T., Linington, C., Schmidbauer, M. and Lassmann, H. (2000). *Am. J. Pathol.* 157, 267–276.
Kraan, M.C., Versendaal, H., Jonker, M., Bresnihan, B., Post, W.J., 't Hart, B.A., Breedveld, F.C. and Tak, P.P. (1998). *Arthritis Rheum.* 41, 1481–1488.
Lake, J., Weller, R.O., Phillips, M.J. and Needham, M. (1999). *J. Pathol.* 187, 259–265.
Laman, J.D., van Meurs, M., Schellekens, M.M., de Boer, M., Melchers, B., Massacesi, L., Lassmann, H., Claassen, E. and 't Hart, B.A. (1998). *J. Neuroimmunol.* 86, 30–45.
Laman, J.D., 't Hart, B.A., Brok, H., Meurs, M., Schellekens, M.M., Kasran, A., Boon, L., Bauer, J., Boer, M. and Ceuppens, J. (2002). *Eur. J. Immunol.* 32, 2218–2228.
Lassmann, H., Brunner, C., Bradl, M. and Linington, C. (1988). *Acta. Neuropathol (Berl).* 75, 566–576.
Lennon, V.A., Lindstrom, J.M. and Seybold, M.E. (1975). *J. Exp. Med.* 141, 1365–1375.
Lennon, V.A., Seybold, M.E., Lindstrom, J.M., Cochrane, C. and Ulevitch, R. (1978). *J. Exp. Med.* 147, 973–983.
Lief, F.S., Rorke, L.B., Kalter, S.S., Hoffman, S.F., Roosa, R.A., Moore, G.T., Cummins, L.B., McCullough, B., Rodriguez, A.R., Eichberg, J. and Koprowski, H. (1976). *J. Neuropathol. Exp. Neurol.* 35, 644–664.
Linington, C., Bradl, M., Lassmann, H., Brunner, C. and Vass, K. (1988). *Am. J. Pathol.* 130, 443–454.
Lucchinetti, C., Bruck, W., Parisi, J., Scheithauer, B., Rodriguez, M. and Lassmann, H. (2000). *Ann. Neurol.* 47, 707–717.
MacLennan, C., Beeson, D., Vincent, A. and Newsom-Davis, J. (1993). *Nucleic Acids Res.* 21, 5463–5467.
Mancardi, G., 't Hart, B., Roccatagliata, L., Brok, H., Giunti, D., Bontrop, R., Massacesi, L., Capello, E. and Uccelli, A. (2001). *J. Neurol. Sci.* 184, 41–49.
Massacesi, L., Genain, C.P., Lee-Parritz, D., Letvin, N.L., Canfield, D. and Hauser, S.L. (1995). *Ann. Neurol.* 37, 519–530.
McFarland, H.F. (2002). *Mult. Scler.* 8, 71–72.
McFarland, H.I., Lobito, A.A., Johnson, M.M., Nyswaner, J.T., Frank, J.A., Palardy, G.R., Tresser, N., Genain, C.P., Mueller, J.P., Matis, L.A. and Lenardo, M.J. (1999). *J. Immunol.* 162, 2384–2390.
McFarland, H.F., Barkhof, F., Antel, J. and Miller, D.H. (2002). *Mult. Scler.* 8, 40–51.
Meinl, E., 't Hart, B.A., Bontrop, R.E., Hoch, R.M., Iglesias, A., de Waal Malefyt, R., Fickenscher, H., Muller-Fleckenstein, I., Fleckenstein, B. and Wekerle, H. (1995). *Int. Immunol.* 7, 1489–1495.
Meinl, E., Hoch, R.M., Dornmair, K., de Waal Malefyt, R., Bontrop, R.E., Jonker, M., Lassmann, H., Hohlfeld, R., Wekerle, H. and 't Hart, B.A. (1997). *Am. J. Pathol.* 150, 445–453.
Morgan, K., Clague, R.B., Shaw, M.J., Firth, S.A., Twose, T.M. and Holt, P.J. (1981). *Arthritis Rheum.* 24, 1356–1362.
Morris-Downes, M.M., Smith, P.A., Rundle, J.L., Piddlesden, S.J., Baker, D., Pham-Dinh, D., Heijmans, N. and Amor, S. (2002). *J. Neuroimmunol.* 125, 114–124.
Noyori, K., Koshino, T., Takagi, T., Okamoto, R. and Jasin, H.E. (1994). *J. Rheumatol.* 21, 293–296.
Onuki, M., Ayers, M.M., Bernard, C.C. and Orian, J.M. (2001). *Microsc. Res. Tech.* 52, 731–739.
Owens, T. and Sriram, S. (1995). *Neurol. Clin.* 13, 51–73.
Owens, T., Wekerle, H. and Antel, J. (2001). *Nat. Med.* 7, 161–166.
Pachner, A.R. and Kantor, F.S. (1982). *Ann. Neurol.* 11, 48–52.
Papanastasiou, D., Poulas, K., Kokla, A. and Tzartos, S.J. (2000). *J. Neuroimmunol.* 104, 124–132.
Patrick, J. and Lindstrom, J. (1973). *Science* 180, 871–872.
Price, W.S., Mendz, G.L. and Martenson, R.E. (1988). *Biochemistry* 27, 8990–8999.
Raine, C.S., Cannella, B., Hauser, S.L. and Genain, C.P. (1999). *Ann. Neurol.* 46, 144–160.
Ravkina, L., Harib, I., Manovitch, Z., Deconenko, E., Letchinskaja, E. and Papilova, E. (1979). *J. Neurol.* 221, 113–125.
Richman, D.P., Gomez, C.M., Berman, P.W., Burres, S.A., Fitch, F.W. and Arnason, B.G. (1980). *Nature* 286, 738–739.
Rivers, T.M. and Schwenkter, F.F. (1935). *J. Exp. Med.* 61, 698–703.
Rivers, T.M., Sprunt, D.H. and Berry, G.P. (1933). *J. Exp. Med.* 58, 39–53.
Rorke, L.B., Iwasaki, Y., Koprowski, H., Wroblewska, Z., Gilden, D.H., Warren, K.G., Lief, F.S., Hoffman, S., Cummins, L.B., Rodriguez, A.R. and Kalter, S.S. (1979). *Ann. Neurol.* 5, 89–94.
Rothschild, B.M. (1993). *J. Med. Primatol.* 22, 313–316.
Rothschild, B.M., Hong, N. and Turnquist, J.E. (1997). *Clin. Exp. Rheumatol.* 15, 45–51.
Rothschild, B.M. and Woods, R.J. (1992). *Semin. Arthritis Rheum.* 21, 306–316.
Rothschild, B.M. and Woods, R.J. (1996). *J. Med. Primatol.* 25, 414–418.

Rubin, A.S., Healy, C.T., Martin, L.N., Baskin, G.B. and Roberts, E.D. (1987). *Lab. Invest.* 57, 524–534.
Schrier, D., Gilbertsen, R.B., Lesch, M. and Fantone, J. (1984). *Am. J. Pathol.* 117, 26–29.
Sercarz, E.E., Lehmann, P.V., Ametani, A., Benichou, G., Miller, A. and Moudgil, K. (1993). *Annu. Rev. Immunol.* 11, 729–766.
Shaw, C.M., Alvord, E.C., Jr. and Hruby, S. (1988). *Ann. Neurol.* 24, 738–748.
Shelton, G.D. (2002). *Vet. Clin. North Am. Small Anim. Pract.* 32, 189–206, vii.
Slierendregt, B.L., Hall, M., 't Hart, B., Otting, N., Anholts, J., Verduin, W., Claas, F., Jonker, M., Lanchbury, J.S. and Bontrop, R.E. (1995). *Int. Immunol.* 7, 1671–1679.
Somnier, F.E. (1993). *J. Neurol. Neurosurg. Psychiatry* 56, 496–504.
Stassen, M.H., Meng, F., Melgert, E., Machiels, B.M., Im, S.H., Fuchs, S., Gerritsen, A.F., van Dijk, M.A., van de Winkel, J.G. and De Baets, M.H. (2003). *J. Neuroimmunol.* 135, 56–61.
Stewart, W.A., Alvord, E.C., Jr., Hruby, S., Hall, L.D. and Paty, D.W. (1991). *Brain* 114 (Pt 2), 1069–1096.
Stuart, J.M. and Dixon, F.J. (1983). *J. Exp. Med.* 158, 378–392.
't Hart, B.A., Simons, J.M., Knaan-Shanzer, S., Bakker, N.P. and Labadie, R.P. (1990). *Free Radic. Biol. Med.* 9, 127–131.
't Hart, B.A., Elferink, D.G., Drijfhout, J.W., Storm, G., van Blooijs, L., Bontrop, R.E. and de Vries, R.R. (1997). *FEBS Lett.* 409, 91–95.
't Hart, B.A., Bakker, N.P., Jonker, M. and Bontrop, R.E. (1993). *Eur. J. Immunol.* 23, 1588–1594.
't Hart, B.A., Bank, R.A., J.A., D.R., Brok, H., Jonker, M., Theuns, H.M., Hakimi, J. and Te Koppele, J.M. (1998a). *Br. J. Rheumatol.* 37, 314–323.
't Hart, B.A., Bauer, J., Muller, H.J., Melchers, B., Nicolay, K., Brok, H., Bontrop, R.E., Lassmann, H. and Massacesi, L. (1998b). *Am. J. Pathol.* 153, 649–663.
't Hart, B.A., van Meurs, M., Brok, H.P., Massacesi, L., Bauer, J., Boon, L., Bontrop, R.E. and Laman, J.D. (2000). *Immunol. Today* 21, 290–297.
Tarrab-Hazdai, R., Aharonov, A., Silman, I., Fuchs, S. and Abramsky, O. (1975). *Nature* 256, 128–130.
Terato, K., Arai, H., Shimozuru, Y., Fukuda, T., Tanaka, H., Watanabe, H., Nagai, Y., Fujimoto, K., Okubo, F., Cho, F. (1989). *Arthritis Rheum.* 32, 748–758.
Thommesen, J.E., Michaelsen, T.E., Loset, G.A., Sandlie, I. and Brekke, O.H. (2000). *Mol. Immunol.* 37, 995–1004.
Tindall, R.S., Kent, M. and Wells, L. (1981). *J. Immunol. Methods* 45, 1–14.
Toro-Goyco, E., Cora, E.M., Kessler, M.J. and Martinez-Carrion, M. (1986a). *P R Health Sci. J.* 5, 13–18.
Toro-Goyco, E., Cora, E.M., Kessler, M.J. and Martinez-Carrion, M. (1986b). *P R Health Sci. J.* 5, 13–18.
Toyka, K.V., Brachman, D.B., Pestronk, A. and Kao, I. (1975). *Science* 190, 397–399.
Toyka, K.V., Lowenadler, B., Heininger, K., Besinger, U.A., Birnberger, K.L., Fateh-Moghadam, A. and Heilbronn, E. (1980). *J. Neurol. Neurosurg. Psychiatry* 43, 836–840.
Trapp, B.D., Peterson, J., Ransohoff, R.M., Rudick, R., Mork, S. and Bo, L. (1998). *N. Engl. J. Med.* 338, 278–285.
Tuohy, V.K., Yu, M., Yin, L., Kawczak, J.A., Johnson, J.M., Mathisen, P.M., Weinstock-Guttman, B. and Kinkel, R.P. (1998). *Immunol. Rev.* 164, 93–100.
Turner, S., Bakker, N.P., 't Hart, B.A., Holt, P.J. and Morgan, K. (1994). *Clin. Exp. Immunol.* 96, 275–280.
Tzartos, S.J., Sophianos, D. and Efthimiadis, A. (1985). *J. Immunol.* 134, 2343–2349.
Uccelli, A., Oksenberg, J.R., Jeong, M.C., Genain, C.P., Rombos, T., Jaeger, E.E., Giunti, D., Lanchbury, J.S. and Hauser, S.L. (1997). *J. Immunol.* 158, 1201–1207.
Uchida, K., Awamura, Y., Nakamura, T., Yamaguchi, R. and Tateyama, S. (2002). *J. Vet. Med. Sci.* 64, 637–640.
van den Broek, M.F. (1989). *APMIS.* 97, 861–878.
Van Eden, W. (1990). *APMIS.* 98, 383–394.
van Lambalgen, R. and Jonker, M. (1987). *Clin. Exp. Immunol.* 68, 100–107.
Van Lambalgen, R. and Jonker, M. (1987). *Clin. Exp. Immunol.* 68, 305–312.
Vanderlugt, C.L., Neville, K.L., Nikcevich, K.M., Eagar, T.N., Bluestone, J.A. and Miller, S.D. (2000). *J. Immunol.* 164, 670–678.
Verschuuren, J.J., Spaans, F. and De Baets, M.H. (1990). *Muscle Nerve* 13, 485–492.
Wekerle, H. (1993). *Curr. Opin. Neurobiol.* 3, 779–784.
Wekerle, H., Kojima, K., Lannes-Vieira, J., Lassmann, H. and Linington, C. (1994). *Ann. Neurol.* 36 Suppl. S47–53.
Weller, R.O., Engelhardt, B. and Phillips, M.J. (1996). *Brain Pathol.* 6, 275–288.
Wierda, D., Smith, H.W. and Zwickl, C.M. (2001). *Toxicology* 158, 71–74.
Wollheim, F.A. (2000). *Curr. Opin. Rheumatol.* 12, 200–204.
Wroblewska, Z., Gilden, D., Devlin, M., Huang, E.S., Rorke, L.B., Hamada, T., Furukawa, T., Cummins, L., Kalter, S. and Koprowski, H. (1979). *Infect. Immun.* 25, 1008–1015.
Wu, M.S., Tani, K., Sugiyama, H., Hibino, H., Izawa, K., Tanabe, T., Nakazaki, Y., Ishii, H., Ohashi, J., Hohjoh, H., Iseki, T., Tojo, A., Nakamura, Y., Tanioka, Y., Tokunaga, K. and Asano, S. (2000). *J. Mol. Evol.* 51, 214–222.
Yu, M., Kinkel, R.P., Weinstock-Guttman, B., Cook, D.J. and Tuohy, V.K. (1998). *Hum. Immunol.* 59, 15–24.
Zanelli, E., Gonzalez-Gay, M.A. and David, C.S. (1995). *Immunol. Today* 16, 274–278.

Chapter 26

Practical Approaches to Pharmacological Studies in Nonhuman Primates

Frank H. Koegler[1,2] and Michael A. Cowley[1]
Oregon National Primate Research Center,
[1] Division of Neuroscience, [2]Division of Reproductive Sciences, Oregon Health & Science University, 505 NW 185th Avenue, Beaverton, OR 97006, USA

Introduction

This chapter is primarily intended to provide information and reference to help with consideration and design of experiments that test pharmacological compounds for efficacy in laboratory nonhuman primates. As such, the material is not geared towards any particular scientific research topic or any particular primate species in the hope that the information can be applicable to nonhuman primate experimentation in general.

Toxicological testing and pharmacokinetic testing are not discussed in detail. Much of the information presented is based on experience in the design and execution of drug efficacy experiments in nonhuman primates, and an appreciation for the types of problems and rewards that nonhuman primate experimentation can present. Finally, two areas of current topical interest with regard to pharmacological testing in nonhuman primates will be addressed.

The nonhuman primate in pharmacological studies

General considerations

In addition to establishing the efficacy of a particular compound to produce a desired effect, animal testing is used to determine how much of a drug is absorbed into

The Laboratory Primate

ISBN 0-1208-0261-9

the blood, how it is metabolized in the body, the toxicity of the drug and its metabolites, and how quickly the drug and its metabolites are excreted. The selection of the experimental animal model for pharmacological experiments is obviously dependent on the type of experiment. In many cases rodent and other animal models suffice and are preferred to the nonhuman primate. In some cases, however, use of the nonhuman primate model is necessary to allow meaningful translation towards understanding human disease and the development of treatments. The obvious primary advantage to using the nonhuman primate as an experimental model, for the testing and development of medicines, is the relative similarity of the whole organism to the human. Mammalian physiology and pathology within the primates, while by no means identical, is grossly similar, whereas comparisons between humans and non-primate experimental models begin to reveal greater differences. In the case of rodents, arguably one of the most popular animals models for basic biomedical science, a host of differences complicate the interpretation of pharmacological testing aimed at developing treatment for human disease. A simple example that may seem benign is the reversed diurnal activity pattern between many primate species and most laboratory rodents. In many cases, these fundamental differences between rodents or other non-primates and humans are not important to the biomedical science performed; however, there are important differences that can confound interpretation of biomedical/pharmacological results obtained in rodent models. A definitive example of a major difference between rodents and nonhuman primates is the nonhuman primate's developed frontal lobes of the brain. Interpretation of rodent experiments evaluating drugs targeting affective disorders, including anxiety and depression, is problematic and thus enhances the value of the primate experiment. Because therapies in the post-genomic era are more likely to be peptide and protein-based, the value of nonhuman primates as models for human physiology and pathology is also likely to increase because of their greater genomic similarity; although the mammalian genome has remarkable similarity across species, minor differences in protein structure and amino acid sequence between species have the potential to render a promising test compound ineffective. In the case of different primate species rather than primate vs. non-primate, these differences are probabilistically minimized.

In addition to technical challenges, that will be addressed later in this chapter, there are two prominent drawbacks to the use of nonhuman primates in pharmacological and other types of biomedical research. The first to be considered is tied directly to the primary advantages of using nonhuman primates for modeling human physiology and pathology. Because of the relative similarity in biology to humans, serious consideration must be given to the ethics of nonhuman primate experimentation (covered in detail elsewhere in this volume). Although it is important to carefully deliberate the ethics of experimentation with animals of all types, non-human primates are thought to be among the most intelligent animals and are most likely to have human-like thought. Given that some, but not all, biomedical experimentation is likely to be adverse to a research animal, the logical extension is that, if an animal is capable of more than superficial emotion, then the animal will experience some sort of negative emotion. The fundamental ethical consideration is whether the emotion experienced by an animal in a pharmacological experiment is less important than the scientific outcome or goals of the experiment. This verbal equation is over-simplified, but nevertheless provides the basis for ethical debate. Ultimately, it must be decided whether the gains towards the reduction of human suffering from disease outweigh the costs of nonhuman primate experimentation.

With regard to the importance of ethical considerations, nonhuman primates are on the opposite end of the spectrum from bacteria, nematodes and insects, with rodents and other non-primate mammals in between. It is important to note that the drawback of using non-human primates does not stem from the actual ethical consideration but it is present when there is a difference of opinion about whether the putative negative emotion, scientific outcome, and importance of research is valued differently between individuals or groups. An extreme hypothetical example is the difference in opinion between an animal rights extremist group, that believes any animal experimentation is unethical, and an investigative group performing experiments in animals without demonstrating the benefit of the experiments to the improvement of human health. In this case, the potential for conflict is high. In more realistic examples, the extremes are moderated but the potential for conflict remains. Ultimately, significant amounts of time and money may be expended in campaigns designed to reduce or eliminate conflict generated by different opinions.

The second major drawback of using nonhuman primates in pharmacological experiments is the actual cost of performing and completing experiments. The supply and demand of nonhuman primates for use in biomedical research has dramatically shifted towards demand in the United States. In 2003, the current lease cost for experimental use of an adult male rhesus at the Oregon National Primate Research Center approached $5,000 per animal for terminal studies. This rate has increased several-fold since 1999 and is a result of decreased supply of animals, increased operating costs, and increased demand for nonhuman primate research. With the advent of the development of SHIV, a genetically engineered version of simian immunodeficiency virus (SIV) that is similar to HIV, the demand for the rhesus macaque specifically, in HIV/AIDS research, has increased (Goodman and Check, 2002). A survey of 6 federally funded National Primate Research Centers in the United States in 2002 revealed that the animal costs of performing a pK study in 15 adult rhesus monkeys for 60 days ranges between $17,754 and $31,535 (personal communication, George Haluska, ONPRC Division of Animal Resources Business Manager/ Research Coord). These costs are based on federal operating subsidies provided to the National Research Centers; therefore, actual costs are in excess of these amounts. Relative to the cost of similar experiments performed in nonprimate species, the cost of performing experiments in nonhuman primates must be considered a disadvantage.

Differential metabolism

When performing experiments with promising test compounds in nonhuman primates, special consideration must be given to interspecies and inter-individual plasma and tissue pharmacokinetics/detoxification. If a compound is being tested with the eventual goal of therapeutic use in humans, then it is especially important to consider the differences between drug metabolism in animal species and human (Wu *et al.*, 1992; Caccia *et al.*, 1995; Sludden *et al.*, 1998; Sharer *et al.*, 1995; for comprehensive bibliography see Caminiti, 1984). The rate of drug metabolism can vary dramatically between different species of nonhuman primate as well as between human and nonhuman primate. For example, the plasma half-life of the anorexic agent dexfenfluramine, after oral administration, is twice as great in cynomolgus macaques as it is in baboons or rhesus monkeys (Caccia *et al.*, 1995). Based on a comparison of the ratio of dexfenfluramine to its active normetabolite between monkey and man, it is estimated that dosing in nonhuman primates should be 10-fold greater to simulate comparable dexfenfluramine and metabolite plasma levels in human. As explained below, by no means should this factor be used as a general guideline but it is helpful to have some knowledge of a particular compound's metabolic fate. Not surprisingly, *in vitro* studies of human and nonhuman primate liver activity demonstrate different levels of drug metabolism depending on the test compound (Stevens *et al.*, 1993). Additionally, hepatic metabolism and conjugation activity of some compounds is higher in the nonhuman primate (Stevens *et al.*, 1993) whereas for others it is higher in human (Sludden *et al.*, 1988). Finally, differences between individuals of the same species can demonstrate marked variation in hepatic metabolism of drugs (Stevens *et al.*, 1993; Sharer *et al.*, 1995). These differences highlight the importance for verifying plasma and tissue levels of test compounds. Samples for drug and metabolite assay can be collected from plasma, urine or tissue via biopsy and extracted for identification by HPLC analysis, RIA or other methods. It is highly recommended that plasma and tissue level measurements of test compounds be made prior to, or during, efficacy studies whenever possible.

Statistical consideration

Because of the unique characteristics of primates as laboratory animals, there are some potential problems with employing standard experimental design for drug testing. Theoretically, all of the rigors of experimental control must be employed to ensure non-confounded results. However, because of the heightened ethical significance of performing procedures in nonhuman primates, and the relatively high actual costs of obtaining and maintaining nonhuman primates as research subjects, special considerations must be made. This is most evident when considering the number of animals for use in a specific experiment. For both ethical and economical reasons, the minimum number of animals needed to generate satisfactory data must be used. Although this concept is not different for experiments using nonprimates, the criteria are different for determining the satisfactory minimum number of animals. Power analyses, generally used to gauge experimental group size, are based on (1) known or predicted variability and (2) assumptions about outcomes or pilot data. To avoid performing experiments that do not generate statistically relevant results, it is preferable to use assumptions that yield subject numbers that err towards larger group size. However, because both the supply of nonhuman primates for use in pharmacological research and the money used for purchasing research subjects is limited, it may not be practical to design an experiment based on standard assumptions about outcomes and significance. Furthermore, because of the phylogenetically advanced nature of nonhuman primates, the ethical considerations of using incorrect numbers of animals for a study cannot be neglected; using too few animals can cause an experiment to be meaningless, whereas using too many causes unnecessary use of animals. Because pilot data are not always available or accurate, assumptions are not fact and the economy of obtaining nonhuman primates can be unfavorable, it is worthwhile to reconsider the standard criteria for determining statistical significance, normally set at a P value of below 0.05. Rather than not reporting results of potentially costly and important nonhuman primate experiments, that fail to achieve the 0.05 level of statistical significance, it may be worth consideration of relaxing the statistical stringency for significance. Obviously, the potential danger of this strategy is the increased likelihood of reporting false positives. If the costs of repeating the experiments are economically, ethically and financially unreasonable, it may be worthwhile to take this risk rather than leave the data unreported.

Behavioral considerations

Not only do the economics of supply and demand and the ethics of using nonhuman primates complicate studies, but the animals themselves can provide challenges for the experimenter. Monkeys and apes are extremely intelligent and require extra attention of the experimenter, animal care taker and other support staff to avoid generating confounded experimental results. Behavioral experiments, such as those measuring effect, activity, or feeding behaviors, are particularly susceptible to confusion because of the emotional nature of nonhuman primates relative to other experimental animals. It is not uncommon to observe transient non-pathogenic anorexia/hypophagia in animals that have undergone relatively mild psychological stress such as a change in animal care staff in a local area (personal observation; Johnson *et al.,* 1996). Transient anorexia may also result from mildly altering a social setting. For example, singly-caged rhesus or cynomolgus monkeys may exhibit transient anorexia for several days after the established social structure of an animal room is disrupted by removing or introducing a single animal (personal observation/personal communication, Judy Cameron). This is extremely confounding to many types of pharmacological experiments, not only those measuring feeding behavior, body weight regulation or general activity. It is difficult to interpret results of behavioral and some physiological studies when animals are experiencing unintended confounding sources of social or other psychological stress. Of particular interest is the difference in sensitivity to stress between individual animals (Streicher *et al.,* 2002). Certain animals are extremely resistant to the effects of social or other stresses. However, the animals that are not resistant are likely to produce confounded experimental data if caution is not taken to ensure social stability and the psychological well- being of animals. Furthermore, unless the sensitivity to stressors is characterized in the individuals of an experimental group, a confounding stressful event (such as a change in animal care staff) will differentially affect the research subjects and create a biased set of outcome measurements.

Prevention of this type of bias, as with most, stems from control over the local environment. This is not fundamentally different from control of experiments using non-primates, but the advanced social nature of primates dictates a higher awareness of social stability and environment. Prevention begins with eliminating or reducing the addition or removal of animals from experimental and living rooms, as well as eliminating rotation of animal care and research staff.

Drug and test compound delivery

Oral delivery

In many cases, the preferred method of drug delivery for pharmacological experiments is voluntary oral consumption. This can be a desirable route of administration in nonhuman primates for testing compounds that are ultimately intended for human oral administration. If the compounds of interest are known to be orally bioavailable and can be disguised and delivered in an appropriate vehicle, such as a small piece of fruit, this method of administration can approximate human oral consumption. Potential degradation in the mouth, stomach and intestine, as well as absorption from the GI tract, is simulated in the nonhuman primate whereas this is not the case with subcutaneous, intramuscular or intravenous injection routes. An advantage of oral delivery is that animals are often eager to consume the fruit or other vehicle, given that the vehicle is palatable and the test compound is sufficiently hidden. Injections, which can be painful, are avoided, thus reducing possible discomfort to the animal. In contrast, a palatable vehicle can be positive reinforcement for accepting a drug treatment. Care must be taken in vehicle selection. Popular vehicles for water soluble compounds include melon (watermelon, cantaloupe, honeydew, banana, etc.) as well as frozen fruit juice mixes prepared with dissolved compounds. These are sweet, yet can be low in calories, thus minimizing the impact of the vehicle on studies that are concerned with caloric intake or whole body metabolism. Using melon or frozen vehicles for water insoluble compounds can be problematic because if the compound is not easily disguised in color, odor or taste, it may easily be separated from the vehicle by the experimental subject. Because nonhuman primates have high dexterity, this can pose problems with delivery of appropriate amounts of compound. Water insoluble compounds are more problematic, yet they can be disguised in banana or other readily available palatable vehicles including jam or peanut butter which is spread onto a cracker. A problem with peanut butter as a vehicle is the potential for protein interactions between the vehicle and the test compound. The impact of caloric content of the vehicle must also be considered when using jam, peanut butter or other energy-rich vehicles.

One potential drawback of oral route of administration is the issue of bioavailability. If test compound integrity cannot withstand the low pH of the gastric environment or cannot be absorbed into the general circulation via the intestine, oral administration has little advantage over other methods. Furthermore, the contents of the stomach may interfere with the absorption rate and yield inconsistent dosing. Mild food deprivation can alleviate this confound. Additionally, with oral administration, there exists the possibility for differential gastrointestinal absorption rate between individuals and also for repeated experimental trials within the same individual. Similarly, there may be inter-individual differences in biodegradation or metabolism of compounds. If possible, it is advisable to perform absorption studies to characterize the blood or tissue levels of test compounds after oral administration (see section later in this chapter). This is not always practical, nor necessary if predicted effects can be are unequivocally and easily obtained.

Another potentially complicating event is the development of a conditioned taste aversion to the vehicle. If an animal experiences drug-induced illness after consumption of a drug paired with a particular vehicle, it may develop a conditioned taste aversion to that vehicle and will be less likely to accept a drug treatment delivered in that vehicle. In the case where the conditioned taste aversion is robust, it can be extremely difficult to overcome. In the case where doses of an orally delivered drug are repeated, it is advisable to use different vehicles to prevent the development of conditioned taste aversions. With careful vehicle selection and counterbalancing the selection between individuals and treatment groups, the effect of different vehicles as an experimental variable can be controlled.

Some experimental design requires oral administration by oral or nasal gastrogavage. This technique is necessary when large volumes must be given orally in a short amount of time or when animals will not voluntarily consume a test compound and it must be enterally absorbed. In the case of the nonhuman primate, gavage can be performed with or without sedation; however, if sedation is not used, the animals must be carefully restrained and trained. Research technicians must also be carefully trained in esophageal/gastric intubation to avoid physical harm to experimental subjects during intubation. Drawbacks of gastrogavage include the potential physical and psychological stresses on the animal and the bioavailability considerations mentioned above for oral administration.

Intravenous infusion

Intravenous catheterization affords excellent access to the internal milieu of nonhuman primates and is a powerful, versatile technique. When drug administration or sample collection must be tightly controlled with regard to amount and time of administration or collection, intravenous access is extremely useful. Drugs or compounds can be infused, at precisely defined rates, directly into the vasculature and are quickly distributed throughout the body. Likewise, repeated blood samples can be taken at precise times and in specific volumes with minimal trauma to vessels or surrounding tissues. Acute catheterization, during which catheters are in place for minutes to hours, is useful for administering controlled infusions of drugs or collecting single or repeat blood samples that must be collected from a specific vessel. Acute intravenous catheterization is especially useful for infusing compounds that have unknown absorption rates or incompatibility with intramuscular or subcutaneous injection. Acute catheterization of peripheral vessels can be easily and reliably accomplished with mild sedation or restraint. However, unless monkeys are sedated or restrained, stability of temporary catheters becomes a problem with increased time of use. Drug formulation solutions for intravenous administration should be carefully evaluated for pH and tonicity and matched to physiologic parameters whenever possible. When pH and tonicity cannot approximate physiologic values, then infusion rate should be minimized to prevent local irritation.

Longer-term catheterization, with catheter-protection systems, greatly increases the viable duration of intravenous catheters. Two methods in use are vest and tether systems and chair restraint systems. It is worthwhile to briefly review the merits and drawbacks of these systems as they are germane to the design and execution of pharmacological experiments. The catheter protection system allows intravenous catheters to be placed chronically in non-restrained, non-sedated nonhuman primates for periods of continuous use ranging from days to many months. The system typically comprises a custom fitted vest or shirt-like garment that is attached to a flexible steel tube or tether through which catheters traverse. The tether is affixed to a swivel that allows animals to turn or otherwise move in home cages without being restricted in range of movement. Catheters are continuously infused with heparinized saline to prevent clotting and can be accessed remotely from an adjacent room without restraining or otherwise disturbing instrumented animals. Drugs can be infused remotely at specific times and at precisely controlled concentrations and infusion rates, an advantage not present with oral, intramuscular or subcutaneous injection. Multiple catheters can be placed simultaneously in different vessels so that blood sampling can be performed while test compounds are infused. This allows precise determination of distribution throughout, and absorption from, the vasculature at the same time that behavioral measurements are being made from awake, nonrestrained, normally behaving animals. The major drawbacks of the chronic catheter system is that it requires specialized technical skill, regular maintenance, and specialized equipment and facilities. From a pharmacology experiment perspective, a potential drawback is the interruption of a behavioral or other study if a problem develops with a catheter protection system. For example, if an animal is able to damage the catheter because of a poorly fitted vest, and a repair requires surgery, the interruption in the experimental protocol can be significant (or permanent) for that particular animal as well as other animals that are housed together with it.

Intracerebroventricular infusion

A particular challenge with regard to pharmacological experiments is delivering drugs or test compounds to the brain. If a target receptor or tissue lies within the blood/brain barrier and the test compound does not permeate, it is undesirable to perform peripheral administration (except perhaps to demonstrate non-target-mediated side effects). Intracerebroventricular (ICV) cannulas with remote catheter systems designed similarly to the intravenous catheter protection system (described above) allow direct administration of compounds inside the blood/brain barrier. Compounds infused to the ventricular system have near immediate access to periventricular structures and tissues and are limited by their diffusion into the neuropil from the cerebrospinal fluid. ICV infusion can be particularly useful for early stage development of compounds that are hypothesized to affect behaviors by affecting brain function at the receptor level. Because experimental compounds need not be engineered to successfully penetrate beyond the blood/brain barrier, proof-of-concept experiments with new compounds can be performed more quickly. The primary advantage of this preparation is that compounds can be infused directly into the ventricular system of awake, non-restrained, non-sedated animals so that behaviors which are otherwise exhibited in non-instrumented animals can be measured. For example, centrally acting compounds thought to affect feeding or other behaviors can be infused remotely and behavior can be videotaped for analysis at a later time.

Although ICV infusion is a powerful technique and affords access by drugs to structures and regions otherwise inaccessible, it is a particularly challenging preparation. Typical targets include one of the lateral ventricles or the third ventricle. Cannulation of the lateral ventricle is preferred because it is a larger, more superficial target. Although targeting either structure will cause incidental brain lesions in superficial structures, cannulation of the third ventricle, although possible, can result in lesions of adjacent hypothalamic structures. Because the predominant flow of native CSF is from the lateral to the third ventricle, and diffusion from the lateral ventricle to the third is rapid (less than one minute for 100 μL contrast media infused over 30 seconds; *personal observation*), the lateral ventricle should be considered even if target regions are hypothalamic. Cannula implantation must be performed with a primate stereotaxic apparatus and cannula placement must be verified with ventricular imaging via contrast media and X-ray during positioning. After cannulas have been implanted, it is important to verify that cannulas are patent and functional on a regular basis. X-ray with contrast media infusion can verify that ventricular integrity remains. Pressure transducers may also be used in guiding cannula placement (Blair-West *et al.*, 1998) Functional evidence of cannula patency is preferred prior to and after completion of experiments and can be achieved by utilizing established neurophysiology. For example, because prolactin secretion is regulated by hypothalamic dopamine neurons and these dopaminergic neurons respond to μ-receptor stimulation, it is possible to verify cannula function by infusing a small dose of opioid receptor agonist (30 μg morphine/ 20 μL aCSF) into the cannula and observing an increase in plasma prolactin (Cameron and Nosbich, 1991). Alternatively, the dipsogenic response to ICV angiotensin-II can be used as a test of cannula integrity (Blair-West *et al.*, 1998, 2001).

There are at least two general types of ICV cannula systems that can be used with nonhuman primates that are not restrained during testing. Both systems utilize a swivel/tether system and the primary difference between the two methods is that one uses a subcutaneously tunneled catheter that is connected to the ICV cannula, whereas the other exits directly from the top of the head and into the catheter/tether. Advantages of the former system are the low profile of the ICV cannula hardware and also the low risk of superficial infection at the scalp because there is no exposed hardware and the catheter exits midscapularly, similar to an intravenous catheter. Another advantage is that after implantation and healing, the central cannula hardware is physically unobtrusive and nearly invisible under the scalp and hair of an animal. Disadvantages, relative to the head-mounted catheter/ tether, include the increased maintenance of the vest/ jacket system which must be inspected every 7 to 14 days after animals have been sedated. If vests are not fitted properly, the peripheral catheter can break and require surgical repair. Another disadvantage of the subcutaneously tunneled catheter is the increased lumen volume due to longer catheters. Because the volume of the ventricular system is small relative to the vascular system, continuous infusion rates must be comparatively small (25–100 μL/hour in a rhesus monkey). Temporary maximum infusion rates of 100 μL in 15 minutes every 30 minutes have been successfully used to deliver artificial cerebrospinal fluid and test compounds into the ventricular system without untoward side effects.

Intranasal delivery

Intranasal delivery (Arora *et al.*, 2002) of test compounds is an alternative method that can be used to quickly deliver test substances to nonhuman primates. Because of the large number of microvilli, the highly vascularized epithelium, and a porous endothelial membrane, the nasal mucosa is a permeable target for drug delivery. Transmucosal delivery can be effective for increasing both systemic and CNS levels of drugs (Von Hogen, 2001), and can be as effective as, or more effective than, intravenous administration of compounds (Rao *et al.*, 1986; Asch *et al.*, 1985; Hussain *et al.*, 1980). Transnasal delivery of compounds in monkeys has also been important for testing viral inoculation (Hurwitz *et al.*, 1997; Weltzin *et al.*, 1996) and has an advantage of bypassing hepatic metabolism. Substances can be delivered manually to monkeys intranasally after sedation or while in restraint chairs. When combined with restraint, local ambient aerosol delivery can also be used to deliver test compounds intranasally and has been used for infection/innoculation with viruses (Weiss *et al.*, 2003). Aerosol delivery is commonly referred to as inhalant delivery and is not restricted to transmucosal transport as the entire respiratory mucosa are exposed. Utility in pharmacological experiments is greater for transnasal delivery than for inhalants because aerosolization, which may be difficult for test compounds, is not necessary for effective delivery to nasal mucosa. Potential drawbacks of intranasal delivery are the limited delivery volume of solution and the prevalence of enzymes/ proteases in primate nasal mucosa (Kaliner *et al.*, 1984).

Intramuscular and subcutaneous injection

Traditionally, the most common form of administering test compounds to nonhuman primates is via hypodermic needle into an intramuscular (IM) or subcutaneous (SC) depot. Convenience and versatility have contributed to the popularity of these methods in monkeys as well as other species. IM/SC injection in monkeys can easily be performed by trained personnel, even in untrained monkeys, given proper restraint. Injection volume can range from less than 1 ml to several ml depending on the injection location. This allows for higher dosing with compounds that have relatively low solubility. Ideally, monkeys can be trained to "present" a hindquarter for intramuscular injection and thus eliminate the need for restraint (Priest, 1991). Whereas intraperitoneal injection is a popular delivery method in rodents, when performed in nonhuman primates a higher degree of restraint is required relative to SC or IM injection. Shortcomings of needle injections into tissue depots include variability of drug absorption rates and drug equilibrium in the body, the limited injection volume, discomfort experienced by the subject and the potential hazard of working with contaminated needles. Caution must be exercised in the case of repeated injections into the same depot to avoid local infection or necrosis from non-physiologically balanced solutions (pH, tonicity).

Transdermal delivery

Transdermal delivery of drug formulations (Panchagnula *et al.,* 1997) has proven to be an effective method in some monkey species (Rupniak *et al.,* 1989a, 1989b; Spilman *et al.,* 1984; Smith *et al.,* 2000). Substances can be applied in paste or solution form directly to cleaned and clipped skin and covered with a dressing. Compounds may also be formulated into release patches or absorbed into pads which are secured to the skin (Gokhale *et al.,* 1992). Advantages of the transdermal delivery methods include the ability to administer compounds continuously over relatively long periods of time without repeated invasive procedures. Disadvantages include difficulty with administering and maintaining accurate doses of drugs/substances. Dose estimates and loss of samples can be calculated by analysis of test compounds remaining in delivery vehicles (pad/patch). However, without measuring blood and tissue levels of compounds it is impossible to know the exact doses administered. Skin irritation can develop at the area of administration and delivery sites must be carefully monitored.

Regardless of the route and method of administration, the importance of verifying blood and/or tissue levels of compounds in pharmacological experiment test subjects cannot be understated. Even in the case where the desired drug effect is manifest and unequivocal, it is important to know the levels of circulating compounds. The most common method is multiple blood sampling before and after treatment. Sample collection is covered elsewhere in this volume; however, it is worth noting the utility of vest-and-tether catheterized monkeys with regard to measuring blood levels of test compounds. With the catheterized monkey, restraint and sedation are not necessary to obtain blood samples and complications of anaesthetic agents and affects of sympathetic arousal on physiology and metabolism are minimized. In some cases tissue levels of drug must be determined, and biopsy or necropsy may be necessary.

Behavior analysis as an aid in pharmacological research

Behavior analysis via video and computer is a powerful tool that must be used to obtain reliable characterization of behaviors resulting from a treatment. In the pharmacology experiment, it is important to determine the effects of test compounds on behavior, regardless of whether the treatment is posited to have a main effect on behavior. If the drug is not intended to affect behavior, it remains important to determine if effective doses have behavioral side effects such as altered activity (reduced or enhanced locomotion), nausea, or depression-like behaviors. Although physical blinds, such as one-way mirrors, allow unobtrusive real-time observations of research subjects, video recording is preferred for subsequent detailed analysis of multiple behaviors. Depending on the resolution of the equipment and media used to record the video, detailed analysis of gross and minor movements and activities is possible, ranging from whole body locomotion, to sexual behavior, penile erection and eye movement. If animals are stationary or caged, video cameras can be positioned with tripods to record animals without having a potentially confounding human observer present. If animals are in group settings and are mobile, it may be necessary to perform focal observations with a camera operator at a distance. Media can be archived or analyzed immediately. Software that allows the synchronization of videotape and a computerized scoring system (Noldus Information Technology, Sterling, VA, USA) enables multiple behaviors to be scored, simultaneously or independently, in real time or in reduced or accelerated relative time. Ideally, a trained scoring technician, blinded to experimental treatment, should be employed for video analysis of behaviors. Disadvantages of video analysis include the time required for proper technician training for use of the scoring systems, inter-observer variance between scorers and actual scoring time needed to perform an analysis. Advantages of video analysis include the ability to score multiple behaviors with repeated analyses of the same observation and relative permanence of media. Because video media with benign content of experimental subjects can serve as powerful propaganda when taken out of the research context, it is worth noting that care should be exercised with regard to storage and distribution of media.

Current pharmacological research in the nonhuman primate model

HIV/AIDS research

Two current areas of nonhuman primate pharmacological research that are particularly timely are the study of HIV-related viruses and the study of the development and prevention of obesity. As stated in the Report on the Global HIV/AIDS Epidemic 2002, (WHO, UNAIDS, Geneva, Switzerland), the World Health Organization estimated that 4.2 million adults and 800,000 children were newly infected with HIV in 2001 worldwide. Although the prevalence and rates of HIV infection vary between geographic regions, HIV/AIDS research is popularly funded by both governmental and private sectors worldwide. Nonhuman primates rarely develop the entire range of AIDS symptoms. However, because the chimpanzee model exhibits similar infection patterns to humans, biomedical experimentation and study of HIV and the related viruses, SIV and SHIV, in monkeys and apes is yielding promising preventive and therapeutic insight (Nath *et al.,* 2000). The majority of pathogenesis studies and vaccine treatment testing for HIV-1 have been performed in the SIV– or SHIV macaque model. Much HIV/SIV/SHIV research with nonhuman primates has focused on vaccination therapy and will not be discussed in this section. However, another area of research is likely to produce pharmacological tools for the study, prevention and treatment of HIV infection. Chemokines are a potential point of pharmacological targeting (Lusso, 2002; Verani and Lusso, 2002) because of their unique ability to bind HIV coreceptors (Strizki *et al.,* 2001; Kazmierski *et al.,* 2003). A potential drawback of using chemokine-based treatments is the inflammatory response of chemokines; however, it is likely that the inflammatory and HIV-recognizing portions of chemokine activity can be separated. (Lusso, 2002).

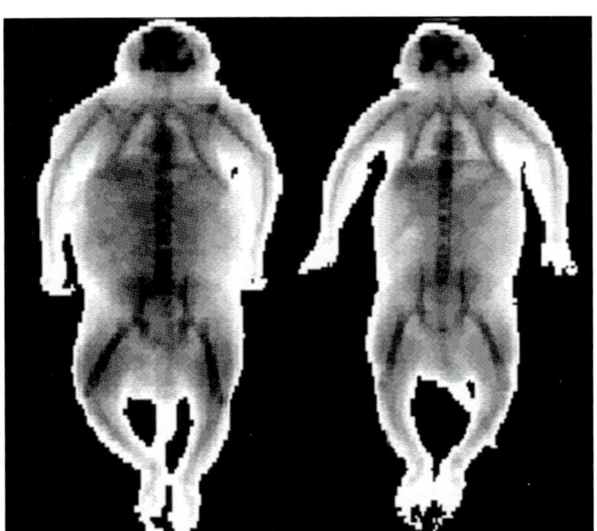

Figure 26.1 DXA scan of obese rhesus before and after chronic daily central melanocortin receptor activation with NDP-MSH.

Obesity research

Because of the increased awareness and interest in the causes, prevention and treatment of human obesity (Mokdad *et al.,* 2003), the spontaneously obese monkey model has great value in experimentation. Because of the prevalence of obesity in the developed world and the subsequent obesity-related health complications that are already manifest (Bray and Macdiarmid, 2000), there is urgency to develop a successful antiobesity treatment, both from a health perspective and a financial perspective. Because of the latter consideration, public disclosure of data generated in nonhuman primate experiments, designed to evaluate anti-obesity treatments, is scarce. An example of a promising target for pharmacological treatment of obesity in primates is the central melanocortin receptor system. Acute stimulation of central melanocortin receptors in the rhesus macaque suppresses eating (Koegler *et al.,* 2001) and chronic stimulation of central melanocortin receptors, with a stable α-melanocyte stimulating hormone analog, causes weight loss in monkeys (Koegler *et al.,* 2000). It is highly likely that further studies targeting the primate central melanocortin system and other likely targets (serotonin, melanin-concentrating hormone, PYY) are underway at pharmaceutical companies (Biftu *et al.,* 2000; Fisher *et al.,* 1998). In addition to current and established rhesus monkey models (Bodkin *et al.,* 1993; Hannah *et al.,* 1991; Jen *et al.,* 1985; Schwartz *et al.,* 1993; Kemnitz *et al.,* 1980,

1989; Kemnitz and Francken, 1986), the development of the baboon as a model for human obesity (Comuzzie *et al.*, 2003) will aid in the ability to test and design new anti-obesity drugs, as well as drugs aimed at treating obesity-related disease including hypertension and diabetes.

Conclusion

The nonhuman primate is currently an irreplaceable model for the development and testing of drugs designed for the treatment of human disease. Although cell culture, invertebrate organisms and non-primate animals may be preferred for many types of biomedical research, pharmacological experiments, that evaluate formulations designed for use in humans, benefit greatly by efficacy testing in the nonhuman primate. The increasing cost of acquiring, maintaining and performing experiments in the nonhuman primate is offset by its value as a model for human physiology and disease. It follows that the nonhuman primate model's greatest advantage can also be seen as one of its disadvantage as an experimental model; the similarity of monkeys and apes to humans. Ultimately, as pharmacological compounds for therapy and prevention of disease are increasingly based on genome-derived protein and peptide structure, the importance of humane, properly conducted pharmacological experiments in nonhuman primates will increase.

Acknowledgements

We wish to thank Diana Takahashi, Marlene (Cooky) Abrams, and Denise Urbanski for their literature research contributions, and Felicity Whelan and Ingrid Louiselle for help with preparation of the manuscript. Special thanks to Judy Cameron, PhD.

Correspondence

Any correspondence should be directed to Michael Cowley, Division of Neuroscience, Oregon National Primate Research Center, Oregon Health and Science University, 505 NW 185th Avenue, Beaverton, OR 97006, USA. Email: cowleym@ohsu.edu

References

Asch, R.H., Balmaceda, J.P., Neves de Castro, M. and Schally A.V. (1985). *Adv. Contracept.* 1(2), 109–117.

Arora, P., Sharma, S. and Garg, S. (2002). *Drug Discov. Today* 7(18), 967–975.

Biftu, T., Feng, D.D., Liang, G.B., Kuo, H., Qian, X., Naylor. E.M., Colandrea, V.J., Candelore, M.R., Cascieri, M.A., Colwell, L.F. Jr., Forrest, M.J., Hom, G.J., MacIntyre, D.E., Stearns, R.A., Strader, C.D., Wyvratt, M.J., Fisher, M.H. and Weber, A.E. (2000). *Bioorg. Med. Chem. Lett.* 10(13), 1431–1434.

Blair-West, J.R., Carey, K.D., Denton, D.A., Weisinger, R.S. and Shade, R.E. (1998). *Am. J. Physiol. Regulatory Integrative Comp. Physiol.* 275, R1639–R1646.

Blair-West, J.R., Carey, K.D., Denton, D.A., Madden, L.J., Weisinger, R.S. and Shade, R.E. (2001) *Am. J. Physiol. Regul. Integr. Comp. Physiol.* 281(5), R1633–R1636.

Bodkin, N.L., Hannah, J.S., Ortmeyer, H.K. and Hansen, B.C. (1993). *Int. J. Obes. Relat. Metab. Disord.* 17(1), 53–61.

Bray, G.A. and Macdiarmid, J. (2000). *West. J. Med.* 172(2), 78–79.

Caccia, S., Bergami, A., Fracasso, C., Garattini, S. and Campbell, B. (1995). *Xenobiotica* 25(10), 1143–1150.

Cameron, J.L. and Nosbisch, C. (1991). *Endocrinology* 128(3), 1532–1540.

Caminiti, B. (1984). In "Animal Models of Toxicant Pharmacokinetics: Species Differences of nonhuman to human primates or other animals. A Bibliography, 1970–1983". Primate Information Center RPRC UW Seattle, WA.

Comuzzie, A.G., Cole, S.A., Martin, L., Carey, K.D., Mahaney, M.C., Blangero, J. and VandeBerg, J.L. (2003). *Obes. Res.* 11(1), 75–80.

Fisher, M.H., Amend, A.M., Bach, T.J., Barker, J.M., Brady, E.J., Candelore, M.R., Carroll, D., Cascieri, M.A., Chiu, S.H., Deng, L., Forrest, M.J., Hegarty-Friscino, B., Guan, X.M., Hom, G.J., Hutchins, J.E., Kelly, L.J., Mathvink, R.J., Metzger, J.M., Miller, R.R., Ok, H.O., Parmee, E.R., Saperstein, R., Strader, C.D., Stearns, R.A., MacIntyre, D.E. *et al.* (1998). *J. Clin. Invest.* 101(11), 2387–2393.

Gokhale, R., Schmidt, C., Alcorn, L., Stolzenbach, J., Schoenhard, G., Farhadieh, B. and Needham, T. (1992). *J. Pharm. Sci.* 81(10), 996–999.

Goodman, S. and Check, E. (2002). *Nature* 417, 684 – 687.

Hannah, J.S., Verdery, R.B., Bodkin, N.L., Hansen, B.C., Le, N.A. and Howard, B.V. (1991). *J. Clin. Endocrinol. Metab.* 72(5), 1067–1072.

Hurwitz, J.L., Soike, K.F., Sangster, M.Y., Portner, A., Sealy, R.E., Dawson, D.H. and Coleclough, C. (1997) *Vaccine* 15(5), 533–540.

Hussain, A., Hirai, S. and Bawarshi, R. (1980). *J. Pharm. Sci.* 69(12), 1411–1413.

Jen, K.L., Hansen, B.C. and Metzger, B.L. (1985). *Int. J. Obes.* 9(3), 213–224.

Johnson, E.O., Kamilaris, T.C., Carter, C.S., Calogero, A.E., Gold, P.W. and Chrousos, G.P. (1996). *Biol. Psychiatry* 40(5), 317–337.

Kaliner, M., Marom, Z., Patow, C. and Shelhamer, J. (1984). *J. Allergy Clin. Immunol.* 73(3), 318–323.

Kazmierski, W., Bifulco, N., Yang, H., Boone, L., DeAnda, F., Watson, C. and Kenakin, T. (2003). *Bioorg. Med. Chem.* 11(13), 2663–2676.

Kemnitz J.W. and Francken G.A. (1986). *Physiol. Behav.* 38(4), 477–483.

Kemnitz, J.W., Goy, R.W., Flitsch, T.J., Lohmiller, J.J. and Robinson, J.A. (1989). *J. Clin. Endocrinol. Metab.* 69(2), 287–293.

Kemnitz, J.W., Kraemer, G.W. and Breese, G.R. (1980). *Pharmacol. Biochem. Behav.* 13(3), 461–465.

Koegler, F.H., Grove, K.L., Schiffmacher, A., Smith, M.S. and Cameron, J.L. (2000). Program No. 371.11. 2000 Abstract Viewer/Itinerary Planner. Washington, DC: Society for Neuroscience, 2000. Online.

Koegler, F.H., Grove, K.L., Schiffmacher, A., Smith, M.S. and Cameron, J.L. (2001). *Endocrinology* 142(6), 2586–2592.

Lusso, P. (2002). *Vaccine* 20(15), 1964–1967.

Mokdad, A.H., Ford, E.S., Bowman, B.A., Dietz, W.H., Vinicor, F., Bales, V.S. and Marks, J.S. (2003). *JAMA* 289(1), 76–79.

Nath, B.M., Schumann, K.E. and Boyer, J.D. (2000). *Trends Microbiol.* 8(9), 426–431.

Panchagnula, R., Stemmer, K. and Ritschel, W.A. (1997). *Methods Find. Exp. Clin. Pharmacol.* 19(5), 335–341.

Priest, G.M. (1991). *Laboratory Primate Newsletter* 30(1), 1–4.

Rao, J.A., Moudal, N.R. and Li, C.H. (1986). *Int. J. Pept. Protein Res.* 28(5), 546–548.

Rupniak, N.M., Tye, S.J., Jennings, C.A., Loper, A.E., Bondi, J.V., Hichens, M., Iversen, S.D. and Stahl, S.M. (1989a). *J. Neurol. Neurosurg. Psychiatry* 52(2), 289–290.

Rupniak, N.M., Tye, S.J., Jennings, C.A., Loper, A.E., Bondi, J.V., Hichens, M., Hand, E., Iversen, S.D. and Stahl, S.M. (1989b). *Neurology* 39(3), 329–335.

Schwartz, S.M., Kemnitz, J.W. and Howard C.F., Jr. (1993). *Int. J. Obes. Relat. Metab. Disord.* 17(1), 1–9.

Sharer, J.E., Shipley, L.A., Vandenbranden, M.R., Binkley, S.N. and Wrighton, S.A. (1995). *Drug. Metab. Dispos.* 23(11), 1231–1241.

Sludden, J., Hardy, S.C., VandenBranden, M.R., Wrighton, S.A. and McLeod, H.L. (1998). *Pharmacology* 56(5), 276–280.

Smith, L.A., Jackson, M.G., Bonhomme, C., Chezaubernard, C., Pearce, R.K. and Jenner, P. (2000). *Clin. Neuropharmacol.* 23(3), 133–142.

Spilman, C.H., Beuving, D.C., Forbes, A.D., Roseman, T.J. and Bennett, R.M. (1984). *J. Pharm. Sci.* 73(2), 282–283.

Stevens, J.C., Shipley, L.A., Cashman, J.R., Vandenbranden, M. and Wrighton, S.A. (1993). *Drug. Metab. Dispos.* 21(5), 753–760.

Streicher, J.M., Lu, N.Z., Cameron, J.L. and Bethea, C.L. (2002). Program No. 75.2. 2002. Abstract Viewer/Itinerary Planner. Washington, DC: Society for Neuroscience, 2002. Online.

Strizki, J.M., Xu, S., Wagner, N.E. *et al.* (2001). *Proc. Natl. Acad. Sci.* 98, 12718–12723.

Verani, A. and Lusso, P. (2002). *Curr. Mol. Med.* 2(8), 691–702.

Von Hoegen, P. (2001). *Adv. Drug Deliv. Rev.* 51, 113–125.

Weiss, W.J., Murphy, T., Lynch, M.E., Frye, J., Buklan, A., Gray, B., Lenoy, E., Mitelman, S., O'Connell, J., Quartuccio S. and Huntley, C. (2003). *J. Med. Primatol.* 32(2), 82–88.

Weltzin, R., Traina-Dorge, V., Soike, K., Zhang, J.Y., Mack, P., Soman, G., Drabik, G. and Monath, T.P. (1996). *J. Infect. Dis.* 174(2), 256–261.

World Health Organization (2002) "Report on the Global HIV/AIDS Epidemic 2002", UNAIDS, Geneva, Switzerland.

Wu, W.N., Pritchard, J.F., Ng, K.T., Hills, J.F., Uetz, J.A., Yorgey, K.A., McKown, L.A. and O'Neill, P.J. (1992). *Xenobiotica* 22(2), 153–169.

Chapter 27

Nonhuman Primate Models of Human Aging

Xenia T. Tigno, Joseph M. Erwin and Barbara C. Hansen
Obesity and Diabetes Research Center, Department of Physiology, University of Maryland School of Medicine, Baltimore, MD, USA

Background

Based on "past changes in birth rate" and progressive increases in life expectancy, global growth of the absolute number and relative proportion of elderly individuals in the human population is reflected in the accelerating incidence of age-related disorders which is precipitating an international public health crisis. The problems associated with human aging must be addressed at multiple levels if disaster is to be avoided, and biomedical and behavioral research, involving nonhuman primates, must play a pivotal role in discovering, devising, and documenting safe and effective means of preventing or treating the debilitating disorders associated with senescence.

Clinical and epidemiological research in humans is important in advancing understanding of healthy and pathological aging. However, many aspects of the biology of aging and age-related disorders can be most productively informed by the use of nonhuman primate models of human aging. Beyond the animal models approach, comparative gerontology is emerging as a field devoted to improving the longevity and well-being of companion animals, and animals in agriculture, zoological collections, and laboratories. The emphasis in this chapter is the study of aging in nonhuman primates, both as models of human aging and to advance fundamental knowledge, promote well-being, and improve clinical care of aging nonhuman primates.

Several styles and avenues of research involving primates are examined here, including controlled experiments, behavioral and health monitoring, some retrospective analyses of databases, post-mortem pathology studies, and systematic documentation of interventions and clinical care. Relevant research is done in laboratories, breeding colonies, zoological gardens and natural settings. The styles of research differ somewhat, depending on the principal purposes for which they are conducted, but it is possible to design studies in ways that address several purposes at once. Data drawn from studies intended to understand evolutionary processes, conserve endangered species, or promote individual animal health, can advance knowledge applicable to human health and aging. At the same time, rigorous and systematic studies aimed at

The Laboratory Primate
Copyright 2005 Elsevier
ISBN 0-1208-0261-9

advancing human health can also promote optimal care for other primates. Some research in aging involves artificial preparations that induce pathology to assess the effectiveness of therapeutic interventions. Some examples would be the chemical elimination of insulin-producing pancreatic beta cells to induce an experimental form of Type 1 diabetes mellitus; chemical destruction of dopamine-producing neurons, in the substantia nigra, to initiate Parkinson's disease and surgical ovariectomy to produce artificial menopause and assess effects of estrogen replacement therapy. Other studies evaluate the consequences of imposing regimens expected to promote health or prolong life as, for example, dietary interventions such as long-term restriction of caloric intake at levels substantially less than normal, feeding diets high in antioxidants, administration of lipid-lowering agents or even use of vaccines intended to protect against development of amyloidosis. Similar studies have assessed the long-term effects of consuming cholesterol-rich diets as well as other interventions expected to have unfavorable health consequences. Still other primate research either identifies individuals with differential risk factors for developing age-related disorders or selects individuals for inclusion that have already presented with spontaneously developed age-related disorders.

Another approach is to maintain laboratory primate colonies, or free-ranging groups, for long periods with careful monitoring of health and behavior. Breeding colonies and zoological specimens can be particularly useful in aging research, especially when longitudinal clinical records have been maintained, along with good records on morbidity, mortality, and post-mortem pathology. Long-term maintenance of laboratory colonies provides special opportunities for longitudinal phenotypic characterization. From the perspective of advancing human health, one of the most important reasons for studying nonhuman primates is that the processes studied are far more likely to be homologous with those operating in humans, rather than just being analogous, as when animals less similar to humans are studied.

Approach

We have combined our collective research experience with an examination of the scientific literature on primate aging. For a literature review, we recommend PrimateLit and PubMed. The PrimateLit database includes a comprehensive searchable on-line bibliography of publications using nonhuman primates from 1940 to the present. Online access to PrimateLit is at http://primatelit.library.wisc.edu/. It includes several especially useful reviews and valuable edited volumes that can serve as sources of additional information on primate aging research. We include detailed summaries of a few relevant areas of research, including brief descriptions of some of our own work. Finally, we suggest some future strategies for the study of aging in primates.

A book by Bowden (1979), entitled *Aging in Nonhuman Primates,* is the most important resource in this field prior to 1980. It stimulated growth in research on primate aging and continues to serve as an inspiring model of integrated multidisciplinary research. After that, the number of research publications on primate aging doubled about every five years into the 1990s, but has recently slowed.

Measurement of cognitive status

An important series of studies of aging in rhesus macaques was conducted by psychologist, Roger Davis, and his collaborators and colleagues. A former student of Harry Harlow at University of Wisconsin, Madison, Davis established the University of South Dakota Primate Laboratory beginning with some two-year-old rhesus monkeys in 1952. He studied the perceptual and cognitive consequences of interventions such as exposure to X-rays, and found evidence of a radiation syndrome that seemed to induce premature aging (Davis and McDonald, 1962). In 1970 he moved his research program to Washington State University and took along his monkeys. His research on aging culminated in the early 1980s with a study of 15 rhesus monkeys, three each of the following ages: 31, 24, 13, 8, and 4. The results of this project are reported in his remarkable book, *Behavior and Pathology of Aging in Rhesus Monkeys* (Davis and Leathers, 1985). Forty investigators from 12 institutions collaborated on a very detailed post-mortem study of these 15 individuals, placed by Davis in the context of behavioral and cognitive histories. It is worth noting that the study was curtailed by a loss of funding to support the monkeys that Davis had worked with for 29 years. The book includes sections on: history, pathology, and cell growth; brain, sense organs, and behavior; endocrine system; heart and lung and bone and muscle. The project stands as a model of collaborative science and testifies to the potential value of detailed life-span cognitive and behavioral monitoring, coordinated with post-mortem tissue analyses.

Another psychologist, Raymond Bartus, with expertise in measuring cognition (learning and memory), focused upon psychopharmacology, and especially the neuropharmacology of aging (Banks *et al.*, 2001; Bartus *et al.*,1979; Bartus and Dean, 1988; Bartus, 1993). He worked productively with Old World monkeys, New World monkeys, and rodents. His influential "catecholamine hypothesis of geriatric memory dysfunction" will be mentioned later in this chapter in the context of current work on the neurobiology of aging.

Diet and cardiovascular health

One of the most remarkable career contributions to comparative primate gerontology is that of Thomas Clarkson, a veterinarian at Wake Forest University, with expertise in cardiovascular pathology. As with a number of other investigators, his work did not begin with a goal of studying aging. His concern with spontaneously occurring atherosclerosis in primates led him to study diet, nutrition, and eventually aging, including the consequences of reproductive endocrinology, social dominance, chemical contraception and both spontaneous and artificial menopause. It is worth noting that his research has involved both New World monkeys (squirrel monkeys, capuchins, woolly monkeys, and spider monkeys), and Old World monkeys (rhesus, stump tailed, and long tailed macaques, vervets) and baboons. The breadth and productivity of Dr. Clarkson's research program reminds us of the importance of paying attention to more than one primate species.

Primate diversity

More than 300 primate species are now recognized by zoological systematists and very few of these species have received any attention at all with regard to aging and its clinical consequences. Remarkably little is known of normal life spans and maximum longevity for most nonhuman primate species (Bowden, 1979; Allman, 1999), although much unpublished information exists in documents, databases, diaries, field notes, and clinical records. The extraordinary diversity of primate species, from gigantic gorillas to tiny mouse lemurs, makes references to aging in "the primate" or "the monkey" seem insufficiently precise. "The primate" or "the monkey" usually refer to the rhesus monkey (*Macaca mulatta*). Of the publications on aging research, far more have involved rhesus macaques than all other primate species combined. There is no doubt that the rhesus macaque is the overwhelming choice of most investigators concerned with modeling human aging and age-related disorders.

The reason most often given for choosing rhesus as an animal model has been that more is known about rhesus than any other primate species. The other often studied species include cynomolgus and Japanese macaques, baboons, chimpanzees, and mouse lemurs. Only ten other species have been represented, including gorillas, bonobos, pigtailed macaques, lion-tailed macaques, Hanuman langurs, patas monkeys, sooty mangabeys, squirrel monkeys, common marmosets, and sifakas. During the past five years, attention to aging has focused on less than 1% of extant primate species, and the vast majority of published research on aging, in the past 60 years, has dealt with less than 1% of the potential primate species diversity. It is encouraging, however, that a few studies have involved comparisons across multiple primate species (Allman, 1999; Nimchinsky, 1999). Indeed, a number of studies of the relatively short-lived mouse lemur (*Microcebus murinus*) confirm the value of expanding research on aging to include more species (Alyard and Perret, 1998; Bons *et al.*, 1994; Dhenain *et al.*, 2003; Gilissen *et al.*, 1998, 1999; Nemoz-Bertholet and Alyard, 2003; Pica and Dhenain, 1998).

Major topics of primate aging research

The recent literature on aspects of primate aging is quite substantial. We have selected a few topics of special interest to us and briefly review each of these below.

Neurobiology of aging

The neurobiology of primate aging is enormously complex, encompassing all aspects of central and peripheral neuronal structure and function in more than 300 species, including humans, along with the changes that occur over the lifespan. Beyond that, the neurobiology of aging is of interest because of the consequences that age-related changes in the nervous system have for the functioning of other systems (e.g., endocrine and cardiovascular), for the health and behavior of the entire organism, for the functioning of individuals in groups (social behavior), and for the function, survival, vulnerability, and evolution occurring in populations. Here we consider two general approaches to the study of aging, each of which has two additional perspectives. The first is the animal model approach which focuses either on artificial models or spontaneous models of human disorders. The other monitors individuals and populations either to characterize what is typical of populations (i.e. to obtain reference values), or to comprehensively document life histories of individuals within populations. These scientific styles need not be mutually exclusive, although some investigators do focus exclusively on one of these aspects.

In the neurobiology of aging, the artificial models produce a condition of concern (e.g., Parkinson's disease or some aspect of Alzheimer's disease) and seek to intervene in ways that may repair the damage, such as transplantation of neurons, gene therapy, or other stimulation of neurons to produce neurotransmitters. The spontaneous models approach includes attention to naturally occurring age-related disorders and to changes that occur, with age, regardless of their clinical consequences. Examples of these would be the apparent changes, with age, in the number of specific neuronal cell types in specific brain regions or to the dendritic arborization or neurotransmitter productivity occurring in brain areas of known function, and the relative abundance of abnormal cell characteristics, such as the beta-amyloid plaques, abnormally phosphorylized *tau* protein, or neurofibrillary tangles associated with Alzheimer's disease.

In addition to the interest in developing "animal models" of human aging, interest is growing in the comparative or veterinary gerontology of primates (Erwin *et al.,* 2002) and in developing a broader understanding of aging throughout the primate order. While the primary goal of basic and clinical studies of primate aging need not have application to humans, such studies clearly attract funding and amplify understanding of the processes that occur in humans. Consequently, it is entirely appropriate that studies intended to promote the health and well-being of nonhuman primates should receive support from programs and agencies concerned mainly with human health. This point is important because representatives of only a few nonhuman primate species are held in human health sciences laboratories where they can serve as animal models of human health and aging. Many more primate species are held in zoological gardens or are accessible through field projects that are orientated mainly toward the conservation of primate populations within their natural ranges. Most zoological gardens have only a few members of each species, but there are now cooperative programs in which animal populations are managed across many institutions.

The experience of one of us (JME), as a zoological curator, led to the development of the Great Ape Aging Project (GAAP) and a "Comparative Neurobiology of Aging Resource" that was developed with support from the National Institute on Aging. The major focus of this project has been on neurobiology, behavior and cognition. However, concerns about adult-onset diabetes, reproductive senescence and estrogen replacement therapy, cardiovascular disorders, and osteoporosis and osteoarthritis, have also been repeatedly raised. By the early 1990s several independent teams had established programs to study various aspects of aging in nonhuman primates, but the efforts lacked coordination and cohesiveness. A conference in 1992 on "The Aging Monkey" drew together many of the scientists who were conducting neurobiological and behavioral research in nonhuman primate models of human aging (Cork, 1993). The emphasis of this conference was directly on "brain science," and neuropathology (Price, 1993), and especially the possible causes of age-related changes in the brain and their consequences for "behavior," which mostly meant individual performance on structured tasks designed to measure memory and cognition (Mishkin, 1993). A number of reviews of topics raised during that landmark conference have been published recently. They include two tomes on the primate nervous system (Bloom, 1997; Bloom *et al.,* 1998) a book entitled *Fundamental Neuroscience,* valuable volumes called *Functional Neurobiology of Aging* (Hof and Duan, 2001), *Functional Endocrinology of Aging* (Mobbs and Hoff, 1998) and *Aging in Nonhuman Primates* (Erwin *et al.,* 2002). The Hof and Mobbs (2001) volume is a superb source for those seeking to understand the ways in which a comparative and evolutionary approach illuminates understanding of the healthy, as well as the dysfunctional, aging brain.

Two of the most common human neurological health concerns, associated with aging, are Alzheimer's and Parkinson's disease (see also Chapter 24, Morton *et al.,* and Chapter 28, Szabo.) Alzheimer's disease currently occurs in more than five million humans in the United States, and Parkinson's in more than 500,000 Americans. It has been possible to create, in nonhuman primates, a realistic and useful model of Parkinson's through chemical lesioning of the substantia nigra using 1-methyl-4-phenyl-1,2,3,6-tetrahydropyridine (MPTP) or 6-hydroxydopamine (6-OHDA). These substances selectively destroy dopamine-producing neurons in the substantia nigra. The resulting dopamine deficiency increases acetylcholine and produces the symptoms of Parkinson's disease. To eliminate symptoms of Parkinsonism, attempts have been made to re-establish appropriately balanced dopamine-acetylcholine concentrations. This was attempted by bolstering dopamine production, either through increasing the number of dopamine-producing cells or by stimulating dopaminergic cells to be more productive. These approaches have used stem cells, fetal-cell transplantation, trophic factors, and gene therapy. The efforts involving nonhuman primates have been effective and these lines of research show promise for human application. Progress has advanced in some cases to human clinical trials that would not have been feasible without demonstration of safety and efficacy studies in nonhuman primates. Detailed reviews regarding experimental therapeutics have been provided by Mufson and Kordower, 2001 and Emborg and Kordower, 2002.

Recent technological developments have enabled advances in the understanding of the neurobiology of aging. The use of unbiased stereological sampling and automated cell counting, for example, promotes precise identification of specific cell types and allows quantification of cell abundance and density (Perl *et al.*, 2000). The counting of neuronal types in brain regions of interest, in post-mortem specimens from individuals of different ages or histories, permits documentation of increases in specific cell types in early life and of cell loss in later life. These methods have dramatically increased the informational yield of CNS tissue banks for humans and nonhuman primates. One of the mysteries revealed by such cell counts has been that rhesus macaques have not been found to lose neurons, with age, in some areas of the brain in which humans typically do. Further, some apparent loss of function is not necessarily accompanied by decline in neuronal number in regions where this was predicted (Morrison and Hof, 1997). This suggested that, for some deficits, the loss of function was at the sub-cellular level, and this hypothesis has been supported by studies that have documented age-related changes in neuronal dendrites and dendritic spines (Nimchinsky *et al.*, 2002; Page *et al.*, 2002; Duan *et al.*, 2003). Some neuronal cell types that are profoundly affected in human Alzheimer's disease (loss of 70% of the cell type in advanced AD), have not been found in macaques. Careful post-mortem examination of the brains of great apes, and many other primate species, has resulted in the discovery, in the anterior cingulate cortex (ACC), of two neuronal cell types that were found to occur exclusively in humans and great apes (Nimchinsky *et al.*, 1999; Hof *et al.*, 2001). This discovery led to recent work on gene expression in the ACC and this underscored the great similarity between humans and chimpanzees, at the molecular genetic level (Uddin *et al.*, 2004). Chimpanzees, however, have not been found to lose neurons with age in the way that humans often do (Hof *et al.*, 2002; Erwin *et al.*, 2001). This suggests that molecular factors probably exist in chimpanzees, and most other nonhuman primates, that protect them against the development of Alzheimer's disease and possibly other neurodegenerative disorders. Identification of these factors could be of interest for the human. Another area of great promise is the use of functional magnetic resonance imaging (fMRI) and other imaging technologies to detect localized brain activity associated with specific tasks, sensory stimulation, motor activity, cognitive processing.

Even though there is no evidence that rhesus macaques develop Alzheimer's disease, some aspects of functional cognitive decline occur in this species. In the macaque neocortex, for example, pyramidal cells that are homologous with some of those that degenerate in Alzheimer's disease, do not degenerate with aging, yet they lose dendritic spines in ways that clearly disrupt synaptic transmission, and similar effects have been found in the hippocampus (Morrison and Hof, 2002). The finding of an association between estrogen insufficiency and the risk of Alzheimer's disease in women stimulated studies in nonhuman primates. Studies confirmed that menopause predicted cognitive decline in rhesus monkeys (Roberts *et al.*,1997) and that estrogen replacement improved cognitive function in aged ovariectomized rhesus (Rapp *et al.*, 2003). Subsequent studies found that estrogen replacement increased dendritic spine density both in hippocampus (Hao *et al.*, 2003) and prefrontal cortex (Tang *et al.*, 2004). Parallel studies in cynomolgus macaques have been conducted (Voytko and Tinkler, 2004), and several other macaque species, as well as patas monkeys (Page *et al.*, 2002; Duan *et al.*, 2003).

Metabolic disorders of obesity and diabetes

Obesity and diabetes develop spontaneously in both nonhuman and human primates. The literature contains reference to these disorders in all of the species shown in Figure 27.1. Such observations have principally been made in primates maintained in zoological gardens. Diabetes in the laboratory primate has been most frequently identified in rhesus, cynomolgus and baboons. The obesity develops in middle age, with its beginnings in rhesus at about 7 to 10 years of age and the maximal body weight commonly reached between 15 and 20 years of age. As shown in Figure 27.2, the laboratory maintained monkey can become very obese, even when fed a healthy diet. This obesity is associated with all of the features of the metabolic syndrome (Hansen, 2001). We have studied the longitudinal progression of these disorders in nonhuman primates (rhesus and cynomolgus) from normal to overtly diabetic and beyond, to the development of severe diabetic complications. This progressive process is diagrammed in Figure 27.3. Monkeys start to become overweight in the period 7 to 15 years of age (phases 2 through 4 here). They then develop progressive hyperinsulinemia and enhanced beta cell response to intravenous glucose (phases 4–6). Meanwhile, glucose tolerance, as defined by the gradually declining glucose disappearance rate in response to intravenous glucose, moves toward impairment. At phase 7 there is "impaired glucose tolerance" immediately preceding the development of overt diabetes (Hansen and Bodkin, 1986). The hyper-responsiveness of the beta cell to glucose has also been documented as a key feature of the progression toward diabetes and long preceding the overt disease (Hansen and Bodkin, 1990).

A number of reviews have appeared on both the nature of diabetes in nonhuman primates (Hansen, 2003) and the development of obesity (Hansen, 2002). We have also discussed more broadly the aging metabolic processes (Hansen, 2002).

As in humans, nonhuman primates develop all of the complications of diabetes that occur after years of the disease. Specifically, diabetic neuropathy develops early after the onset of hyperglycemia and can be measured experimentally by nerve conduction studies, as shown in Figure 27.4. Nephropathy also develops, proceeding from normal to a state of increased glomerular

Macaca mulatta
(rhesus monkey)

Macaca nigra
(crested black macaque)

Macaca radiata
(bonnet macaque)

Macaca nemestrina
(pigtailed macaque)

Macaca fasicularis
(cynomolgus or longtailed macaque)

Macaca cyclopis
(Taiwan or Formosan rock macaque)

Macaca assamensis
(Assamese macaque)

Papio hamadryas
(sacred baboon)

Mandrillus leucophaeus
(drill)

Saguinus fuscicollis
(saddle-backed tamarin)

Saguinus oedipus
(cotton-top tamarin)

Cercopithecus mona
(Mona monkey)

Cercopithecus mitis
(Blue monkey)

Colobus polykomos
(western black-and-white colobus)

Galago crassicaudatus
(African bushbaby)

Cercopithecus aethiops
(African green monkey, vervet, grivet)

Cebus apella
(capuchin)

Cercopithecus cephus
(moustached green guenon)

Cercopithecus diana
(Diana monkey)

Saimiri sciureus
(squirrel monkey)

Pan troglodytes
(chimpanzee)

Pan paniscus
(bonobo)

Figure 27.1 Primate species known to have cases of spontaneous obesity and spontaneous type 2 diabetes mellitus.

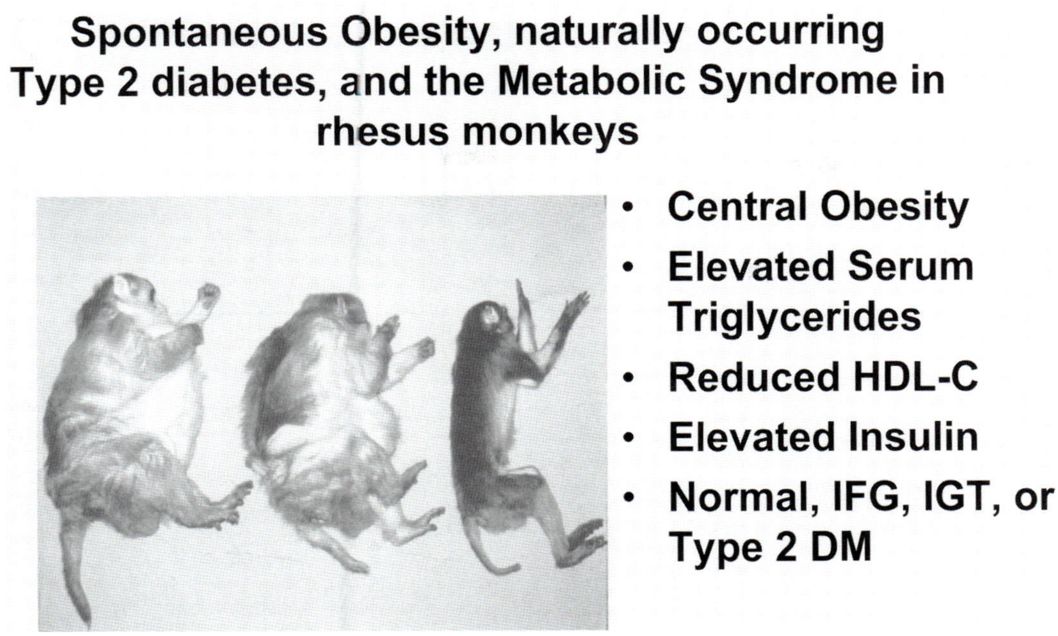

Figure 27.2 Rhesus monkeys, like humans, frequently develop spontaneous obesity at middle age under *ad libitum* fed conditions. Under the same conditions some stay lean all of their lives and some go on to develop overt type 2 diabetes, usually preceded by the features of the Metabolic Syndrome listed.

tuft volume in the prediabetic hyperinsulinemic monkey to severe diabetic glomerular nephropathy as illustrated in Figure 27.5. The glomerular basement membrane width increases with age and is particularly thickened in overt diabetes as shown in non diabetic and diabetic monkeys in Figure 27.6. We and others have also shown the development of retinopathy and changes in the microvascular flow in diabetic monkeys.

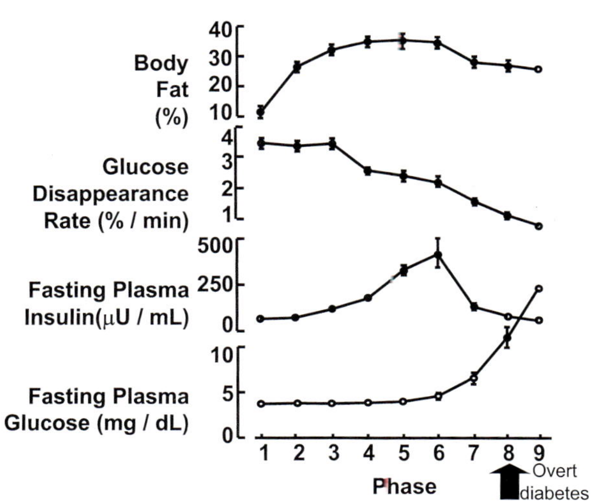

Figure 27.3 The progression from normal lean young adult (phase 1) through obesity to overt diabetes (phase 8) to severe diabetes with complications (phase 9). Monkeys begin to progress through phases 3, 4, and 5 any time between ages 7 and 20, and have been identified to become overtly diabetic as early as age 10 and as late as age 29. Thus, to show the pattern of progression it is necessary to use "phase" on the axis, and to pool across monkeys as they move progressively through these phases. See text.

Reproductive senescence

Some aspects of reproductive biology are unique to human and nonhuman primates. While reproductive cessation occurs in non-primates (Packer *et al.,* 1998), menopause occurs only in humans and other primates. Consequently, nonhuman primates have been especially important in the study of reproductive senescence and its consequences. Naturally occurring menopause has been characterized in macaques (Gilardi *et al.,*1997) and baboons (Martin *et al.,* 1998), and been studied in other species, including chimpanzees. It has become a topic of interest for the development of primate models of human health (Bellino and Wise, 2003), principally with regard to the effects of age-associated and menopause-related changes in endocrine profiles on metabolism, bone density, cardiovascular disorders, and cognitive function. Features in nonhuman primates that resemble human menopause include similarity of hormone profiles across the menopausal transition, progression through irregular cycles to cycle cessation, declining fertility with age, weight gain and increased body fat content percentage, increased insulin resistance and

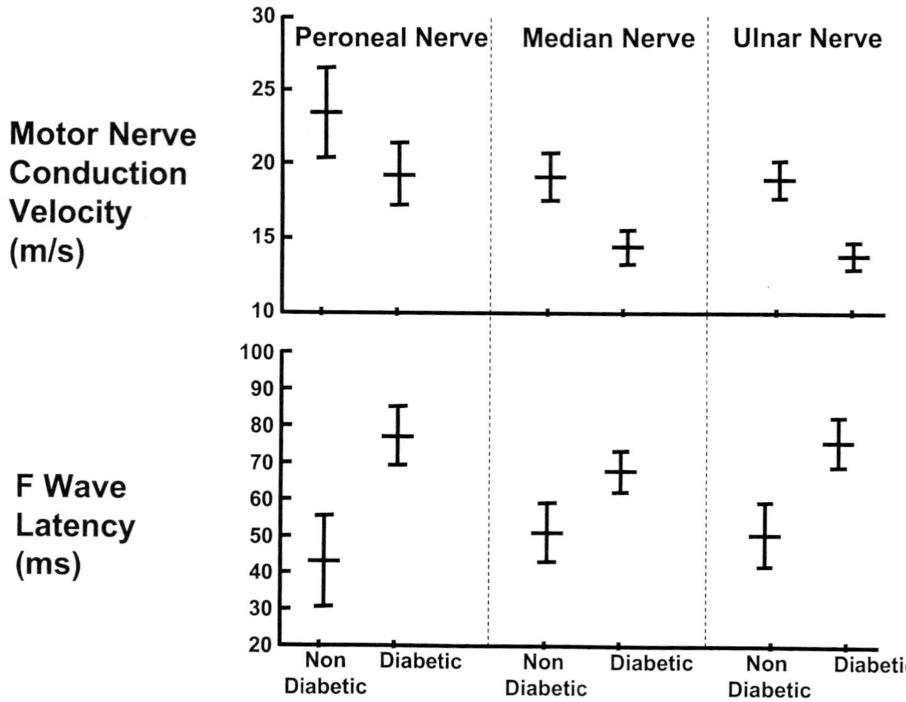

Figure 27.4 The functioning of peripheral nerves (peroneal, median and ulnar) in non diabetic and in diabetic monkeys as measured by motor nerve conduction velocity and F wave latency evaluation.

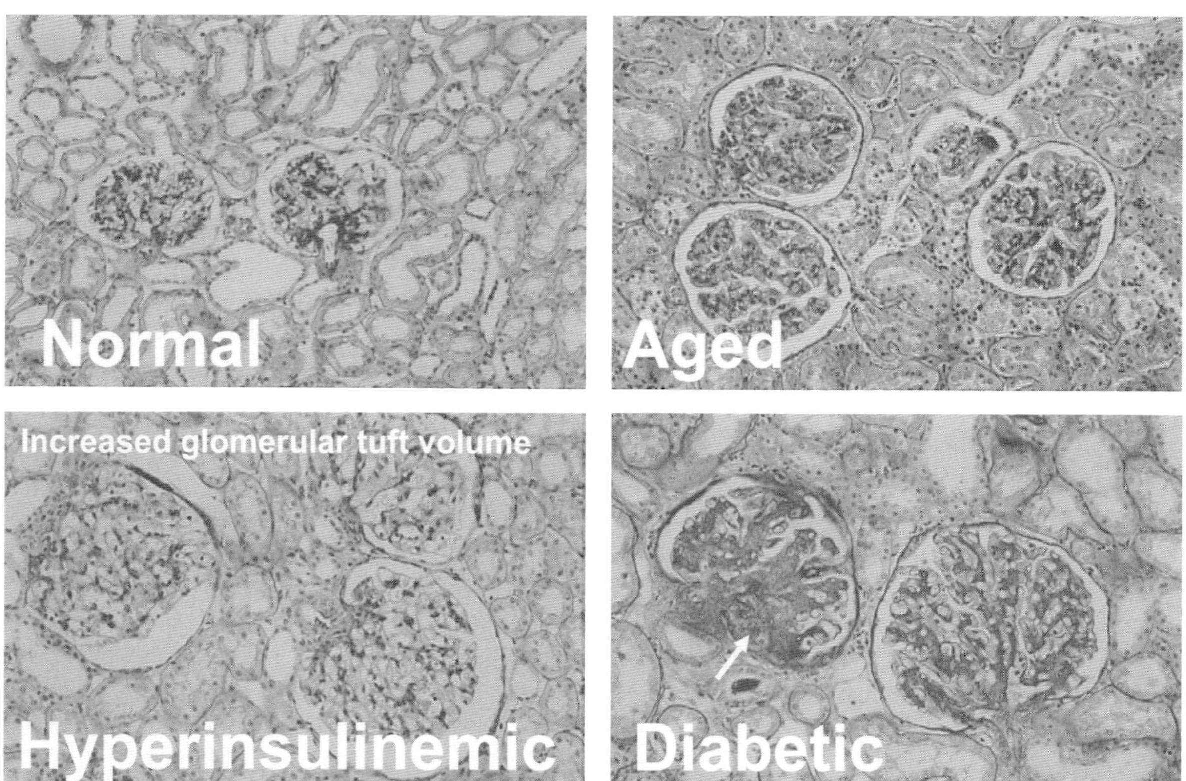

Figure 27.5 Kidney glomeruli in normal, aged (non diabetic), hyperinsulinemic (prediabetic) and overtly diabetic monkeys.

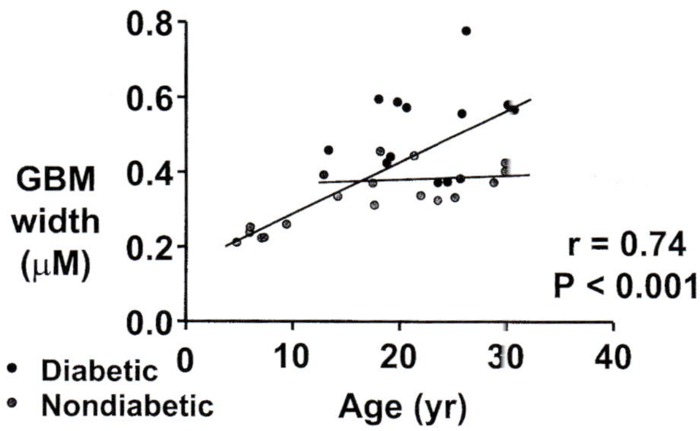

Figure 27.6 The increasing glomerular basement membrane width (GBM) with age and with diabetes in rhesus monkeys.

glucose intolerance, increased levels of LDL ("bad") cholesterol and decreased HDL ("good") cholesterol, declines in levels of serum dehydroepiandrosterone (DHEAS), and similar reactions to estrogen replacement such as temperature-regulation, bone metabolism, lipid profiles, and cognitive changes. The results of studies of nonhuman primates have revealed only a few differences from humans, including short post-reproductive lifespans relative to humans, the timing of peri-menopausal endocrine changes, and the seasonality of menstrual cycles. Rhesus macaques, for example, are seasonal breeders. The females cease cycling for a portion of the year, even if not pregnant, and sperm production by males concurrently subsides. Closely related macaques, such as the cynomolgus monkey, however, are not seasonal breeders, offering some advantages over rhesus for studies of reproductive biology. As previously mentioned, the enormous body of background information already gained on rhesus is often a reason for their selection for research.

Biomarkers of aging

Senescence occurs at variable rates, even within the same population. Because aging is accompanied by an increased incidence of diseases such as stroke, heart disease, cancer and type 2 diabetes, there are needs for indices for "biologic aging" which would be representative of changes in physiologic function attributable to chronologic age alone. The close phylogenetic relationship between non-human primates and humans, and the fact that primate lifespans are but a fraction of that of humans, make biomarkers of primate aging even more relevant. The issue on whether a candidate biomarker is a valid and reliable measure of aging has been previously reviewed (Ingram *et al.*, 2001). For species with long lifespans, such as primates, the National Institute of Aging has suggested the following criteria for establishing validity: significant cross-sectional correlation with age; significant longitudinal change in the same direction as the cross-sectional correlation; significant stability of individual differences over time and, finally, the rate of age-related change should be proportional to differences in lifespan among related species. The rate of change of a marker, for example, should be twice that of a human if the species had a lifespan half of that of human. Aside from being able to establish normative values across different age groups, such biomarkers would also be essential in assessing whether particular interventions, be they preventive or curative, are indeed effective in extending longevity and improving the quality of life.

Hematologic variables

A number of primate hematologic variables have been examined in an effort to identify which of them demonstrate alterations attributable to chronologic aging. In captive rhesus macaques, 15 of these variables were shown to be significantly altered in relation to age (Smucny *et al.*, 2001). The data of the 345 rhesus monkeys, maintained in three different primate research facilities in Maryland, Wisconsin and Oregon, were obtained from the Primate Aging Database. Based on the assumption that rhesus monkeys have attained puberty and completed many developmental processes by 7 years, mean values for each age group were compared to the mean values at 7 years of age. Monkeys aged 20–30 were considered "old adults." Regression analyses were performed and 15 variables demonstrated age-related changes which were significant. These were body weight, albumin, albumin/globulin ratio, alkaline phosphatase, blood urea nitrogen (BUN), blood urea nitrogen/creatinine ratio, creatinine, % eosinophils, globulin, hematocrit (Hct), hemoglobin (Hgb), mean corpuscular hemoglobin (MCH), mean corpuscular volume (MCV), red blood cell count (RBC) and triglycerides. Among these variables, albumin, albumin/globulin (A/G) ratio, creatinine, MCH and MCV decreased by age 30. On the other hand, BUN, BUN/creatinine ratio, globulin, hemoglobin, RBC count and triglycerides increased with age. Body weight, % eosinophils and hematocrit all demonstrated a U-shaped pattern, increasing with adulthood, then declining by age 30.

A similar analysis was performed on hematologic and biochemical parameters of captive chimpanzees in the Yerkes Regional Primate Research Center

(Herndon and Tigges, 2001). In this study, data from 146 females and 106 males were examined cross-sectionally as well as longitudinally over a 9-year period. Several hematologic variables were identified to have significant sex-related differences. RBC, Hct, Hgb and mean corpuscular hemoglobin concentration (MCHC) were higher in males than females, whereas white blood cell count (WBC), lymphocyte count (Lymph), and eosinophil count (Eos) were higher in females. Most of the variables, such as Hct, Hgb, MCVand MCH, demonstrated a curvilinear relationship with age, all increasing in early maturation and reaching a plateau at age 10, while platelet counts decreased rapidly in the early years. Segmented neutrophils (SEG) also demonstrated a curvilinear pattern, decreasing after age 20, whereas RBC, WBC and basophil count (BASO) did not change significantly with age. MCHC and band neutrophil count (BAND) all decreased monotonically with age, whereas monocyte count (MONO) increased significantly.

To compare values obtained from these primates held in captivity, with those in the wild, hematologic values were also examined of the three wild groups of toque macaques (*Macaca sinica*), in the Polonnaruwa Sanctuary in Sri Lanka (Ekanayake *et al.,* 2003). Thirty-five males and thirty-seven females were included in the study, 16 of which were infants. The age range of the study population was from 0.33 to 24.5 years, and age classes were defined as: Infants (<1 year), Young juveniles (1–3 years), Old juveniles (3–6.5 years) and Adults (>6.5 years). Values for RBC, Hgb, MCHC and erythrocyte sedimentation rates (ESR) were found to be lower in infants than other ages, whereas MCV was higher. Gender differences were noted in plasma protein concentrations, total white blood cell (WBC) count, absolute neutrophil (Neutro) counts and ESR, where females had higher values. Similar to the findings in monkeys in captivity, most of the age-related changes were related to the maturation process. Haemoglobin, for example, increased with age up to 6.5 years, but was lower in adults. Red blood cell count appeared to increase with age, but mainly because infancy levels were very low. White blood cell count, however, did not show age differences. Differential and absolute counts of neutrophils increased significantly with age, but leveled off beyond 3 years of age. Eosinophil differential counts did not vary across age classes, whereas lymphocyte differential counts decreased significantly, over-all, with aging. In contrast, plasma protein concentrations and ESR significantly increased as aging progressed.

Immune function

Changes attributable to aging in the immune system are of primary interest because of the increasing tendency towards infection and the development of cancers among the elderly. In a number of mammalian species, both humoral and cellular responses appear to decline with aging. Because of the influence of helper T cells on B cell function, however, much of the age-related dysregulation of the immune system may be due largely to senescence of T cells (Pawelec *et al.,* 2002). Thymic hormone production is thought to decline with age, and attempts have been made to enhance immune function by administration of thymic preparations (Ershler *et al.,* 1988). In a study involving 60 pig-tailed macaques (*Macaca nemestrina*), aged 2 to 32 years, older animals were seen to demonstrate diminished responses to mitogens (Bowden *et al.,* 1994). In this study, CD8+T-cells (class I restricted), of the memory subtype, increased while naïve CD8+T cells decreased with age in females. CD4+T cells (class II restricted), which included memory T cells, also increased with age among females. IgA also increased, and Bowden *et al.* (1994) suggested the use of memory and naïve CD8+T lymphocytes, CD4+T lymphocytes and IgA as biomarkers of aging. In comparison, when lymphocyte subsets were examined in cynomolgus monkeys (n = 195, age range between 1 month and 31 years), CD4+T cells were found to decrease, and CD8+T cells increased during the first ten years of life. CD20+B lymphocytes also decreased with age until the age of 5, after which no significant changes occurred. CD16+ Natural killer cells increased up to age 5, peaking between 4 and 10 years (Nam *et al.,* 1998). More recently it was demonstrated that, parallel to thymic involution, which occurs at around 11 years of age among cynomolgus monkeys, CD4+CD8+double-positive (DP) T cells begin to increase abruptly with age (Lee *et al.,* 2003a,b) and correlated with an increase in percentage of memory T cells. Using signal joint TCR rearrangement excision circle (sjTREC) technology to assess thymic function, it is reported that this index declines with age both in rhesus macaques and in sooty mangabeys (Sodora *et al.,* 2000), suggesting a diminution in T cell population.

In rhesus macaques, the ratio of CD4+/CD8+T cells declines with age, and this decline is predominantly due to the decreasing number of CD4+ lymphocytes (Dykhuizen *et al.,* 2000). Using flow cytometric assays to identify CD4+ cells in rhesus monkeys, Grossman also reported a decline, with aging, in the response of CD4+ lymphocytes to stimulation with anti-CD3 mAb and this was partly due to a decrease in the number of cells (Grossman *et al.*, 1995). In a model of immunosenescence in healthy baboons (*Papio cynocephalus anubis*), ranging in age from 6 months to 26 years, the B cell population decreased and the T-cell population, both CD4+ and CD8+, increased with age. In a further attempt to characterize immunosenescence, Macarucci and colleagues have examined age-related changes in T-lymphocyte subsets using leukocytes derived from 37 male rhesus, ages 6–28 years. Expression of mRNA and protein for IL-10, IL-6, interferon gamma (IFNγ), IL-1β and TNFα were measured following stimulation with lipopolysaccharide (LPS) or phytohemaglutinin (PHA). The Th1 subset consists of CD4+T-helper lymphocytes that produce IL-2 and IFNγ and are important for defense against intracellular pathogens. The Th2 subset, on the other hand, consists of CD4+T helper lymphocytes that are involved in humoral immunity, protect against extra cellular pathogens, and elaborate IL-4, IL-5, IL-10, and IL-13 cytokines. The study did not show any change in the total number of circulating peripheral blood mononuclear cells (PBMCs) or neutrophils with age in the monkeys. However, an increase in both protein and mRNA expression of IL-10 was noted in both whole blood and stimulated and unstimulated PBMCs with age. Protein levels of IFNγ measured from the supernatant of PHA-stimulated PBMCs, on the other hand, showed a decline with aging, although mRNA levels appeared to be unchanged. There did not appear to be any age-associated changes in the IL-1β or TNFα protein levels in either whole blood or PBMC, while mRNA expression of IL-6 (an acute phase inflammatory marker) by PBMCs, following *in vitro* stimulation with LPS, appeared to show an age-related increase. The study suggested that some of the circulating T cells "switched" from a Th0/Th1 profile to a Th2 cytokine profile, which may partly explain the susceptibility of the elderly towards neoplastic disease and infections. Autoantibody production has also been shown to be age-related. In a study performed on 188 healthy male and female baboons from 1–24 years of age, production of antinuclear antibodies, anticell extract antibodies and natural autoantibodies were seen to increase with age in the absence of the disease and even if serum immunoglobulin concentrations did not parallel the changes (Attanasio *et al.*, 2001). This is thought to represent an increase in B1 lymphocyte activity.

Considering that the intestines are directly exposed to various pathogens and constitute a major effector organ of the immune response, the influence of aging on mucosal immunity has also been studied. Gastrointestinal immunity does appear to be compromised with aging as evidenced by lower antigen-specific immunoglobulin A (IgA) antibody levels in the intestinal lavage samples of old compared to young rhesus monkeys (Taylor *et al.*, 1992). While neither the local production nor secretion of IgA antibodies in the intestinal mucosa appears to be reduced, it has been suggested that the migration of IgA positive immunoblasts from the site of induction (Peyer's patches) to the effector site, impairs the mucosa. (Schmucker *et al.*, 2001).

Biochemical variables

Changes in many biochemical and physiologic variables appear to correlate with aging in both non-human primates and humans. For instance, deterioration of cognitive function as well as visual and neurodegenerative decline closely parallel the observed changes in humans (Lane, 2000). Beta amyloid depositions and plaques have been demonstrated, particularly after the age of 20 in rhesus monkeys, as noted above. In female monkeys, estrogen and Follicle Stimulating Hormone (FSH) levels decrease with aging, while progesterone and Luteinizing Hormone (LH) levels show little change around menopause. Bone mass declines with age, in both male and female monkeys, and osteocalcin, a vitamin-dependent calcium binding protein, was found to decrease with age in males but increased in postmenopausal females. However, although osteoarthritis was common in both genders with aging, vitamin D and calcium metabolism appeared unchanged.

The clinical chemistry parameters, glucose, BUN, creatinine, phosphate, aspartate transaminase (AST), alkaline phosphatase and triglycerides levels changed significantly with age in captive chimpanzees, but only during the first decade of life. However, bilirubin, protein, globulin and the A/G ratio demonstrated age-related changes throughout (Herndon and Tigges, 2001). Alkaline phosphatase levels were also found to be significantly higher in juvenile cynomolgus monkeys than adults whereas BUN levels appeared to increase with age (Kapeghian *et al.*, 1984). In rhesus monkeys, the greatest age-related alterations were in the albumin/globulin ratio, BUN, BUN/creatinine ratio and triglycerides (Smucny *et al.*, 2001). From a 5 year longitudinal study of 29 male rhesus monkeys on dietary restriction, serum glutamic oxalacetic transaminase, alkaline phosphatase, total protein, globulin, BUN, Creatinine and phosphates were suggested as potential biomarkers for aging (Nakamura, 1994). A follow-up study involving 33 male rhesus calorie restricted monkeys, aged 4–27 years, revealed that the percentage of lymphocytes, circulating levels of alkaline phosphatase, albumin, creatinine and calcium appeared to serve as suitable markers of the aging process.

Because senescence is accompanied by a decline in reproductive, metabolic and neuroendocrine functions, several hormones and metabolic markers have been suggested to serve as biomarkers of the aging process. In a study involving old (~25.7 year old) and young (~5.4 year old) female rhesus monkeys, mean growth hormone (GH), estrogen and Insulin-like Growth Factor (IGF-I) levels were found to be significantly lower in the older monkeys than in the young, whereas mean Luteinizing Hormone (LH) levels were significantly higher (Woller *et al.,* 2002).

Another candidate hormone that has been suggested is plasma melatonin concentration. Melatonin, which is primarily secreted by the pineal gland, exhibits a circadian pattern of secretion with peak levels occurring at around 2:00 a.m. in monkeys (Roth *et al.,* 2001). The peak nocturnal levels of melatonin have been shown to decrease, with aging both in humans and monkeys and this decline has been shown to be abolished by long-term calorie restriction.

Increased production of superoxide radicals is believed to be closely involved in the aging process, and activity levels of the enzyme, superoxide dismutase (SOD), were measured in the liver, brain and heart of rodents and 12 primate species. While no general correlation could be found between SOD and longevity of the species, the ratio of SOD activity to specific metabolic rate of the tissue or the whole organism, was found to increase with increasing maximum lifespan, suggesting a protective effect of SOD against by-products of oxygen metabolism in species with longer lifespans (Tolmasoff, 1980).

Recently, much attention has been given to leptin because of its association with obesity and type 2 diabetes. Studies in rodents have previously indicated that leptin inhibits gene expression of lipogenic enzymes and food intake, giving it a major role in energy balance, total body fat and visceral fat distribution (Ma *et al.,* 2002; Nogalska *et al.,* 2003). Aging rats show increased fat mass, abdominal adiposity, hyperleptinemia and insulin resistance, and it has been suggested that leptin-resistance could partly explain the metabolic decline in aging (Gabriely *et al.,* 2002). This may be due, in part to a decreased number of hypothalamic receptors for leptin, despite elevated levels of circulating hormone (Fernandez-Galaz *et al.,* 2001). It has been suggested that the anti-aging effect of calorie restriction may also be due, in part to its ability to reduce concentrations of plasma leptin (Shimokawa and Higami, 2001). However, studies in male rhesus monkeys, with ages ranging from 10 to 80 months, demonstrated a decline in leptin levels through the juvenile period until the onset of puberty, and leptin levels varying seasonally after achievement of sexual maturity (Mann *et al.,* 2000). In wild baboons (*Papio anubis, P. hamadryas,* and *anubis/hamadryas* hybrids), serum leptin levels were highest in the young and tended to decrease with dental age. Furthermore, no association between serum leptin levels and age could be found in human studies (Bruunsgaard *et al.,* 2000; Maugeri *et al.,* 2002; Rzepka *et al.,* 2002). The exact role of leptin in general metabolism and aging remains equivocal at this point.

Among the reproductive hormones, serum dehydroepiandrosterone sulfate (DHEAS) appears to be one of the more consistent markers of longevity. The secretion of this hormone from the adrenal cortex, drops with age both in human and non-human primates (Sapolsky *et al.,* 1993; Vermeulen, 1994; Lane *et al.,* 1997; Goncharova *et al.,* 2000; Kemnitz *et al.,* 2000; Goncharova and Lapin, 2002; Roberts, 1999). This hormone is the most abundant steroid hormone in circulation and is a precursor of both androgens and estrogens (Sapolsky *et al.,* 1993). Apart from its reproductive functions, it is also believed to be involved in memory and cognition, metabolism, vascular biology and immunity (Buvat, 2003). In a colony of wild yellow baboons living in a national park in East Africa, DHEAS concentrations showed a remarkable decline with age from youth to old age (Sapolsky *et al.,* 1993). A reduction with age has been reported in both male and female rhesus monkeys, with a rate of decline twice that observed in humans. Furthermore, calorie restriction reportedly slowed the rate of decline in DHEAS, after maturity was attained, and DHEAS was proposed as a marker of longevity (Lane *et al.,* 1997). Measurement of DHEAS in 793 rhesus monkeys of both sexes showed that the most rapid decline occurs from infancy until 5 years of age, after which a gradual decline continues into old age (Kemnitz *et al.,* 2000). The age-related reduction in DHEAS was again observed in a study involving female rhesus monkeys where levels of both cortisol and DHEAS were compared in older (20–26 year old) and young (6–9 year old) subjects, and where DHEAS concentrations were found to be much lower in the older group while cortisol levels did not show any change (Goncharova *et al.,* 2000). The decline in the hormone may be reflective of senescence of the adrenal gland itself (Vermeulen, 1994), or a consequence of dysregulation of the hypothalamo-pituitary-adrenal gland axis (Goncharova and Lapin, 2002).

The fall in DHEAS may be due to a decrease in the number of DHEA-synthesizing cells in the adrenals because of age-related decreased blood flow and oxygen deficit to the adrenal glands (Roberts, 1999). In addition to its metabolic and reproductive actions, DHEA may also have antiatherogeneic effects (Khalil *et al.,* 2000). A study on 403 elderly men found DHEAS to be positively associated with bone mineral density (van den Beld and Lamberts, 2000). Since DHEA appears to have beneficial effects, DHEA replacement therapy was attempted in a group of 280 elderly men and women (60–79 years old), who received the hormone (50 mg) or placebo, orally for a year Results of the study suggest that bone turnover improved and osteoclastic activity was reduced in women who were older than 70 years, in addition to improvement of skin status and parameters of libido, with no apparent untoward effects (Baulieu *et al.,* 2000). However, at this stage, data are not sufficient to warrant recommendation of DHEA supplementation to the elderly (Legrain and Girard, 2003). Further studies in non-human primates are warranted.

A recent study in our laboratory examined longitudinal profiles of DHEA and DHEAS in individual rhesus males (Erwin *et al.,* in press). No consistent age-related decline was detected in individuals even though a modest decline occurred for the population studied. This suggests that DHEAS may not be a very reliable biomarker of aging after all.

Cardiovascular changes

The aging process is commonly associated with changes in cardiovascular function. In normal human subjects, it is known that the heart enlarges with age, with an overall increase in mean cardiothoracic ratio and an increase in the aortic knob diameter (Ensor *et al.,* 1983). Echocardiographic measurements, taken of 58 nonagenarians, revealed that although the majority of the subjects had normal sized ventricles, over half had enlarged left atria (Tunick *et al.,* 1990). In Japanese monkeys (*Macaca fuscata*), accumulation of cardiac lipofuscin appears to correlate with the aging process (Nakano *et al.,* 1989, 1990). Deposition is apparently initiated upon reaching sexual maturity in different primate species (Nakano *et al.,* 1993). Physiologic function in the resting state, however, seems to be preserved as shown by the absence of any significant differences in left ventricular function between older adult (17 ± 1 year old) healthy monkeys compared to younger adult (3 ± 11 year old) monkeys (Sato *et al.,* 1995). This study demonstrated that both left ventricular contractility (dP/dt) and the isovolumic relaxation period (tau) were similar in the old and young subjects. However, stimulation with adrenergic drugs such as isoproterenol, norepinephrine and forskolin, resulted in significantly reduced inotropic and chronotropic actions, suggesting decreased responsiveness to adrenergic stimulation among the elderly monkeys. In addition, stimulation with phenylephrine also resulted in diminished vascular responsiveness in the older subjects. More recently, measurements were performed on chronically instrumented conscious monkeys (*Macaca fascicularis*) demonstrating again, no differences in left ventricular pressure, left ventricular contractility (dP/dT) cardiac index, mean arterial pressure, total peripheral resistance or heart rate between young (*circa* 6-year-old) and old (*circa* 20-year-old) monkeys (Takagi *et al.,* 2003). However, stimulation of beta-adrenergic receptors with isoproterenol, norepinephrine or forskolin resulted in diminished inotropic responses in the old male monkeys, compared to the young males, as well as reduced effect on vascular resistance in the older male subjects. Among the female monkeys, however, neither cardiac nor vascular responses to beta adrenergic receptor stimulation appeared to be attenuated with aging. The age-related reactivity to beta-adrenergic stimulation may explain the reduced cardiac reserve in the elderly. It has been suggested that this may be due to an increase in the L-type calcium channel current (ICa, L) with cardiac hypertrophy, resulting in prolongation of the cardiac action potential and a propensity for arrhythmias with senescence. Aging is also associated with chronic activation of the sympathetic nervous system (Seals and Bell, 2004). It has been proposed that this is driven by the accumulation of body fat as one ages and is intended to result in beta-adrenergic thermogenic action. It is unclear whether the impaired response of the heart and the vasculature to beta-adrenergic stimulation is a cause or a corollary to this age-associated chronic elevation of sympathetic activity.

A study involving autopsies performed on 186 rhesus monkeys between 20 and 36 years of age places the age of biosenescence, in captivity, at around 25 years (Uno and Walker, 1993). Both cerebral beta-amyloidosis and cerebral angiopathy were commonly observed in the basal pre-frontal gyrus, followed by the amygdala. The ability of circulating amyloid-beta peptide to be transported across the blood-brain barrier may contribute to accumulation of the substance in the aging monkey brain (Mackic *et al.,* 2002). Aging is also associated with thinning of capillary walls and a decrease in the number of endothelial mitochondria in the cortex of *Macaca nemestrina* monkeys (Burns *et al.,* 1979).

The aging arteries, on the other hand, demonstrate a dramatic increase in calcium content, as evidenced by a study performed on various arteries of Japanese monkeys (Tohno *et al.*, 2001). Furthermore, even in healthy, old (around 20 years of age) non-human primates, with no evidence of atherosclerosis, the intimal layer of the aorta thickens, angiotensin II (Ang II) levels rise, angiotensin converting enzyme (ACE) activity increases and arterial remodeling, similar to that seen in the development of atherosclerosis, is evident. These findings were based on measurements of aortic matrix metalloproteinase-2 (MMP-2) and its regulators, membrane type-1 matrix metalloproteinase (MT1-MMP) and tissue inhibitor of matrix metalloproteinase-2 (TIMP-2), as well as expression of Ang II and ACE in young (6.4 year old) and old (20 year old) subjects (Wang *et al.*, 2003). Despite these changes, there are no significant differences in baseline aortic pressure and total peripheral resistance between old and young monkeys. This was shown in a study using chronically instrumented monkeys consisting of two groups, one of 9 old male *Macaca fascicularis* (*ca.* 20 years of age) and the other of 13 young males (*ca.* 7 years) where cardiac output, left ventricular pressures and aortic pressures were measured (Asai *et al.*, 2000). However, in response to acetylcholine, the drop in total peripheral resistance was significantly greater in the young monkeys. The authors believe that this diminished reactivity, in the old monkeys, is partly explained by the reduced endothelial density and enhanced endothelial apoptosis which then lead to impairment of endothelial function, despite the absence of atherosclerotic disease.

Metabolic changes and calorie restriction

Human studies based on fasting plasma glucose (FPG), glucose (OGTT-G) and insulin (OGTT-I) levels, during an oral glucose tolerance test, as well as glucose uptake during euglycemic hyperinsulinemic clamps, suggest that glucose homeostasis deteriorates with age (Andres, 1971; Davidson, 1979; Reaven *et al.*, 1989; Elahi *et al.*, 1993; Muller *et al.*, 1996; Elahi *et al.*, 2002). However, studies in humans are difficult to interpret, partly because of the lack of standardization of procedures across laboratories and partly because of the presence of confounders such as lifestyle, diet, ethnicity and other significant variables. In the study by Asai *et al.* (2000) on chronically instrumented monkeys, total cholesterol, triglyceride (Bodkin *et al.*, 2003) and fasting blood sugar levels did not differ between the older and younger age groups. In a study comparing control *ad libitum-fed* monkeys with dietary-restricted monkeys, the glucose disappearance rate, following an intravenous glucose tolerance test, was found to be lower in the control group at around 18 years of age compared to the values obtained from the same monkeys 8 years earlier (Kemnitz *et al.*, 2000; Gresl *et al.*, 2001). Fasting plasma insulin increased, suggesting a decrease in insulin sensitivity, but calorie restriction prevented this increase (Bodkin *et al.*, 2003). A gradual decrease in caloric intake, in monkeys over a 3-month period, to 70% of their *ad libitum* intake, appeared to retard aging as well as improve metabolic indices. In our laboratory, we have found that when calories are restricted and titrated on a weekly basis so that the monkeys maintain their stable optimal adult body weight of around 10 kilograms (Bodkin *et al.*, 1995, 2003; Hansen *et al.*, 1995) we have shown that dietary restriction leads to increased average lifespan and prevention of hyperinsulinemia. Thus, in both types of calorie restriction (CR), many of the physiologic changes associated with aging are prevented. Apart from indices of glucose metabolism, calorie restriction also appears to reduce body fat and serum triglycerides (Lane *et al.*, 1995, 1999) and lead to higher levels of the HDL2B sub fraction which is indicative of a reduction in cardiovascular risk factors (Verdery *et al.*, 1997). Another study, in cynomolgus monkeys, demonstrated that caloric restriction not only increased insulin sensitivity, it also decreased body weight and intra-abdominal fat, which are other known risk factors for cardiovascular disease (Cefalu *et al.*, 1997). Furthermore, although concentrations of plasma LDL cholesterol were similar in control monkeys and monkeys that were restricted for over 5 years, LDL from the CR monkeys had a lower molecular weight and the monkeys were depleted in triglyceride and phospholipid levels compared to those in the control group (Edwards *et al.*, 1998).

Energy efficiency appears to be increased in response to calorie restriction of aging monkeys. Thus, when adjusted on the basis of either surface area, or lean body mass, long-term calorie restricted monkeys require fewer calories to maintain stable body weight *at a reduced level* (compared to *ad libitum* fed monkeys of similar age) (Delany *et al.*, 1999), as shown in Figure 27.7. Food intake is substantially lower in the restricted monkeys (Hansen *et al.*, 1995), although by all measures they are healthier.

Since caloric restriction has been shown to delay aging and increase lifespan, immunologic function was studied in rhesus monkeys after two to four years of calorie restriction. Surprisingly, mitogen-induced

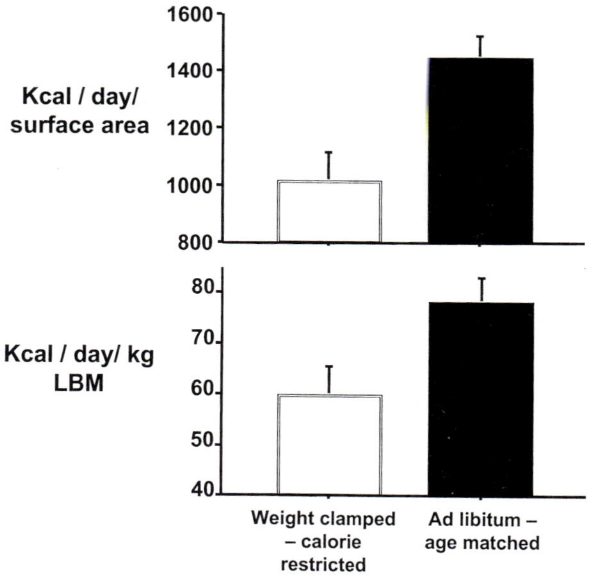

Figure 27.7 Six rhesus monkeys, calorie restricted to maintain normal adult lean body mass for over 20 years have reduced calorie utilization compared to *ad libitum* monkeys, even when adjusted for surface area or lean body mass (LBM).

proliferation of peripheral blood mononuclear cells (PBMC) was reduced in restricted monkeys compared to controls, as were natural killer cell and antibody responses (Roecker *et al.*, 1996). Weindruch, however, reported that the reduced PBMC response occurs only in monkeys who were restricted in very early life (< 1 year), but not in those restricted from young adulthood (Weindruch *et al.*, 1997). Caloric restriction also resulted in lymphopenia.

No significant differences between IL-10 and IL-6 levels were observed between CR and control monkeys, but both protein and gene expression of IFNγ were found to be significantly higher among the CR monkeys. (Mascarucci *et al.*, 2002). In addition to the already discussed effects of calorie restriction, lower basal insulin secretion and increased levels of DHEAS in monkeys, and reduction of markers of inflammation, such as C-reactive Protein (CRP), interleukin 6 (IL-6) and plasminogen activator inhibitor type 1 in humans (PAI-1) (Heilbronn and Ravussin, 2003) have been reported.

Of greatest interest to those interested in the study of aging is the convincing evidence of the improvement of health, reduction of morbidity and extension of average lifespan in the long-term calorie restricted monkey (Hansen *et al.*, 1999). This 25 year study of 117 laboratory maintained rhesus monkeys, which included a large group of *ad libitum* fed monkeys and a small group of calorie restricted monkeys, showed that the *ad libitum* fed monkeys have an average lifespan of ~25 years, while the average lifespan of the calorie restricted monkeys was 32 years. Whether ultimate maximal lifespan is extended by long-term sustained calorie restriction will not be known for about 10 more years, but evidence to date suggests that this effect if any will be minor compared to the increase in average lifespan. The latter reflects the generally healthier state of the calorie restricted monkeys and their lack of many of the diseases of aging.

Correspondence

Any correspondence should be directed to Barbara C. Hansen, Ph.D., Obesity and Diabetes Research Center, Department of Physiology, University of Maryland School of Medicine, 10 South Pine Street, 6-00 MSTF, Baltimore, Maryland 21201, USA. Tel: 410 706-3168; Fax: 703 356-4143; Email: bchansen@aol.com

References

Alyard, F. and Perret, M. (1998). *Physiol. Behav.* 64, 513–519.

Allman, J. (1999). In Allman, J. (ed.) *Evolving Brains*. W.H. Freeman, New York.

Andres, R. (1971). *Med. Clin. North Am.* 55, 835–846.

Asai, K., Kudej, R.K. and Shen, Y.T. (2000). *Arterioscler. Thromb. Vasc. Biol.* 20, 1493–1499.

Attanasio, R., Brasky, K.M. and Robbins, S.H. (2001). *Clin. Exp. Immunol.* 123, 361–365.

Banks, W.A., Phillips-Conroy, J.E., Jolly, C.J. and Morley, J.E. (2001). *J. Clin. Endocrinol. Metab.* 86, 4315–4320.

Bartus, R.T. (1993). *Neurobiol. Aging* 14, 711–714.

Bartus, R.T. and Dean, R.L. (1988). *Neurobiol. Aging* 9, 351–356.

Bartus, R.T., Fleming, D. and Johnson, H.R. (1978). *J. Gerontol.* 33, 858–871.

Bartus, R.T., Dean, R.L. 3rd, Fleming, D.L. (1979). *J. Gerontol.* 34, 209–219.

Bartus, R.T., Dean, R.L. and Beer, B. (1982). *Neurobiol. Aging* 3, 61–68.

Baulieu, E.E., Thomas, G. and Legrain, S. (2000). *Proc. Natl. Acad. Sci. U S A* 97, 4279–4284.

Bellino, F.L. and Wise, P.M. (2003). *Biol. Reprod.* 68, 10–18.

Bloom, F. (1997). In Bloom, F., Bjorkland, A., Hokfelt, T. (eds), *The Primate Nervous System, Part I,* Vol. 13. Elsevier, Amsterdam.

Bloom, F., Bjorkland, A. and Hokfelt, T. (1998). In Bloom, F., Bjorkland, A. and Hokfelt, T. (eds)

The Primate Nervous System, Part II, Vol. 14. Elsevier, Amsterdam.
Bodkin, N.L., Ortmeyer, H.K. and Hansen, B.C. (1995). *J. Gerontol. A Biol. Sci. Med. Sci.* 50, B142–B147.
Bodkin, N.L., Alexander, T.M. and Ortmeyer, H.K. (2003). *J. Gerontol. A Biol. Sci. Med. Sci.* 58A, 212–219.
Bons, N., Mestre, N. and Ritchie, K. (1994). *Neurobiol. Aging* 15, 215–220.
Bowden, D.M. (1979). In Bowden, D.M. (ed.) *Aging in Nonhuman Primates.* Van Nostrand Reinhold, New York.
Bowden, D.M., Short, R., Williams, D.D. and Clark, E.A. (1994). *J. Gerontol.* 49, B93–B103.
Bruunsgaard, H., Pedersen, A.N. and Schroll, M. (2000). *Life Sci.* 67, 2721–2731.
Burns, E.M., Kruckeberg, T.W., Comerford, L.E. and Buschmann, M.T. (1979). *J. Gerontol.* 34, 642–650.
Buvat, J. (2003). *World J. Urol.* 21, 346–355.
Cefalu, W.T., Wagner, J.D. and Wang, Z.Q. (1997). *J. Gerontol. A Biol. Sci. Med. Sci.* 52, B10–B19.
Clarkson, T. (1971). *North Carolina Medical Journal* 32, 88–96.
Clarkson, T. and Appt, S. (2003). *Steroids* 68, 941–951.
Clarkson, T., Bullock, B. and Lehner, D. (1968). *Progress in Biochemical Pharmacology* 4, 420–428.
Clarkson, T., Koritnik, D., Weingand, K. and Miller, C. (1985). *Metabolism: Clinical and Experimental* 34, 51–59.
Clarkson, T., Adams, M. and Kaplan, J. (1989). *American Journal of Obstetrics and Gynecology* 160, 1280–1286.
Cork, L.C. (1993). *Neurobiol. Aging* 14, 613.
Cornblath, D.R., Hillman, M.A. and Striffler, J.S. (1989). *Diabetes* 38, 1365–1370.
Cusumano, A.M., Bodkin, N.L. and Hansen, B.C. (2002). *Am. J. Kidney Dis.* 40, 1075–1085.
Davidson, M.B. (1979). *Metabolism* 28, 688–705.
Davis, R. and Leathers, C. (1985). *Behavior and Pathology of Aging in Rhesus Monkeys.* Alan R. Liss, New York.
Davis, R. and McDonald, A. (1962). *Psychological Reports* 11, 383–386.
Delany, J.P., Hansen, B.C. and Bodkin, N.L. (1999). *J. Gerontol. A Biol. Sci. Med. Sci.* 54A, B5–11; discussion B2–B3.
Dhenain, M., Chenu, E. and Hisley, C. (2003). *Magnetic Resonance in Medicine* 50, 984–992.
Duan, H., Wearne, S.L. and Rocher, A.B. (2003). *Cereb. Cortex* 13, 950–961.
Dykhuizen, M., Ceman, J. and Mitchen, J. (2000). *Cytometry* 40, 69–75.
Edwards, I.J., Rudel, L.L. and Terry, J.G. (1998). *J. Gerontol. A Biol. Sci. Med. Sci.* 53, B443–B448.
Ekanayake, D.K., Horadagoda, N.U. and Sanjeevani, G.K. (2003). *Am. J. Primatol.* 61, 13–28.
Elahi, D., Muller, D.C. and Egan, J.M. (2002). *Novartis Found Symp.* 242, 222-42; discussion 42–46.
Elahi, D., Muller, D.C. and McAloon-Dyke, M. (1993). *Exp. Gerontol.* 28, 393–409.
Emborg, M. and Kordower, J. (2002). In Erwin, J. and Hof, P. (eds) *Aging in Nonhuman Primates,* pp 102–117. Karger, Basel.
Ensor, R.E., Fleg, J.L. and Kim, Y.C. (1983). *J. Gerontol.* 38, 307–314.
Ershler, W.B., Coe, C.L. and Laughlin, N. (1988). *J. Gerontol.* 43, B142–B146.
Erwin, J., Hof, P., Ely, J. and Perl, D. (2002). In Erwin, J. and Hof, P. (eds) *Aging in Nonhuman Primates,* Vol. 31, pp 1–21. Karger, Basel.
Erwin, J., Nimchinsky, E. and Gannon, P. (2001). In Hof, P. and Mobbs, C. (eds), *Functional Neurobiology of Aging,* pp 447–456. Academic Press, San Diego.
Fernandez-Galaz, C., Fernandez-Agullo, T. and Campoy, F. (2001). *J. Endocrinol.* 171, 23–32.
Gabriely, I., Ma, X.H. and Yang, X.M. (2002). *Diabetes* 51, 1016–1021.
Gilardi, K.V., Shideler, S.E. and Valverde, C.R. (1997). *Biol. Reprod.* 57, 335–340.
Gilissen, E., Ghosh, P., Jacobs, R. and Allman, J. (1998). *American Journal of Primatology* 45, 291–299.
Gilissen, E.P., Jacobs, R.E. and Allman, J.M. (1999). *J. Neurol. Sci.* 168, 21–27.
Goncharova, N.D., Oganyan, T.E., Taranov, A.G. (2000). *Neurosci. Behav. Physiol.* 30, 717–721.
Goncharova, N.D. and Lapin, B.A. (2002). *Mech. Ageing Dev.* 123, 1191–1201.
Gresl, T.A., Colman, R.J. and Roecker, E.B. (2001). *Am. J. Physiol. Endocrinol. Metab.* 281, E757–E765.
Grossman, A., Rabinovitch, P.S. and Lane, M.A (1995). *J. Cell Physiol.* 162, 298–303.
Hansen, B.C. (2001). In Borntrop, P. (ed.) *International Textbook of Obesity*, pp 181–201. John Wiley & Sons, Ltd., Sussex.
Hansen, B.C. (2002a). In Bouchard, G.Ba.C (ed.) *Handbook of Obesity*. Marcel Dekker, Inc., New York.
Hansen, B.C. (2002b). In Erwin, J.M. and Hoff, P.R. (eds) *Interdisciplinary Topics,* Vol. 32. Karger, Basel.
Hansen, B.C. (2003, in press). In LeRoith, D., Olefsky, J.M. and Taylor, S. (eds) *Diabetes Mellitus: A Fundamental and Clinical Text.* J.B. Lippincott Co., Philadelphia.
Hansen, B.C. and Bodkin, N.L. (1986). *Diabetologia* 29, 713–719.
Hansen, B.C. and Bodkin, N.L. (1990). *Am. J. Physiol.* 259 (Regulatory Integrative Comp Physiol 28), R612–R617.
Hansen, B.C., Ortmeyer, H.K. and Bodkin, N.L. (1995). *Obes. Res.* 3 Suppl 2, 199s–204s.
Hansen, B.C., Bodkin, N.L. and Ortmeyer, H.K. (1999). *Toxicol. Sci.* 52, 56–60.
Hao, J., Janssen, W.G. and Tang, Y. (2003). *J. Comp. Neurol.* 465, 540–550.
Heilbronn, L.K. and Ravussin, E. (2003). *Am. J. Clin. Nutr.* 78, 361–369.

Herndon, J.G. and Tigges, J. (2001). *Comp. Med.* 51, 60–69.
Hof, P. and Duan, H. (2001). In Hof, P. and Mobbs, C. (eds) *Functional Neurobiology of Aging*, pp 435–445. Academic Press, San Diego.
Hof, P.R., Duan, H. and Page, T.L. (2002). *Brain Res.* 928, 175–186.
Hof, P.R., Nimchinsky, E.A., Perl, D.P. and Erwin, J.M. (2001). *Neurosci. Lett.* 307, 139–142.
Ingram, D.K., Nakamura, E. and Smucny, D. (2001). *Exp. Gerontol.* 36, 1025–1034.
Jayashankar, L., Brasky, K.M., Ward, J.A. and Attanasio, R. (2003). *Clin. Diagn. Lab. Immunol.* 10, 870–875.
Kapeghian, J.C., Bush, M.J. and Verlangieri, A.J. (1984). *J. Med. Primatol.* 13, 283–288.
Kemnitz, J.W., Roecker, E.B. and Haffa, A.L. (2000). *J. Med. Primatol.* 29, 330–337.
Khalil, A., Fortin, J.P., LeHoux, J.G. and Fulop, T. (2000). *J. Lipid Res.* 41, 1552–1561.
Lane, M.A. (2000). *Exp. Gerontol.* 35, 533–541.
Lane, M.A., Ball, S.S. and Ingram, D.K. (1995). *Am. J. Physiol.* 268, E941–E948.
Lane, M.A., Ingram, D.K. and Roth, G.S. (1999). *Toxicol. Sci.* 52, 41–48.
Lane, M.A., Ingram, D.K., Ball, S.S. and Roth, G.S. (1997). *J. Clin. Endocrinol. Metab.* 82, 2093–2096.
Lee, W.W., Nam, K.H. and Terao, K. (2003a). *Immunology* 109, 217–225.
Lee, W.W., Nam, K.H., Terao, K. and Yoshikawa, Y. (2003b). *Exp. Anim.* 52, 309–316.
Legrain, S. and Girard, L. (2003). *Drugs Aging* 20, 949–967.
Ma, X.H., Muzumdar, R. and Yang, X.M. (2002). *J. Gerontol. A Biol. Sci. Med. Sci.* 57, B225–B231.
Mackic, J.B., Bading, J. and Ghiso, J. (2002). *Vascul. Pharmacol.* 38, 303–313.
Mann, D.R., Akinbami, M.A., Gould, K.G. and Castracane, V.D. (2000). *Biol. Reprod.* 62, 285–291.
Martin, L.J., Carey, K.D. and Comuzzie, A.G. (1998). *Mech. Ageing Dev.* 124, 865–871.
Mascarucci, P., Taub, D. and Saccani, S. (2002). *J. Interferon Cytokine Res.* 22, 565–571.
Maugeri, D., Bonanno, M.R. and Speciale, S. (2002). *Arch. Gerontol. Geriatr.* 34, 47–54.
Mishkin, M. (1993). *Neurobiol. Aging* 14, 615–617.
Mobbs, C. and Hof, P. (1998). *Functional Endocrinology of Aging.* Karger, Basel.
Morrison, J.H. and Hof, P. (1997). *Science* 412–419.
Morrison, J.H. and Hof, P.R. (2002). *Prog. Brain Res.* 136, 467–486.
Mufson, E. and Kordower, J. (2001). In Hof, P. and Mobbs C. (eds), *Functional Neurobiology of Aging,* pp 243–281. Academic Press, San Diego.
Muller, D.C., Elahi, D., Tobin, J.D. and Andres, R. (1996). *Semin. Nephrol.* 16, 289–298.
Nakamura, E., Lane, M.A., Roth, G.S. (1994). *Exp. Gerontol.* 29, 151–177.
Nakano, M., Mizuno, T., Katoh, H. and Gotoh, S. (1989). *Mech. Ageing Dev.* 49, 41–48.
Nakano, M., Mizuno, T. and Gotoh, S. (1990). *Mech. Ageing Dev.* 52, 93–106.
Nakano, M., Mizuno, T. and Gotoh, S. (1993). *Mech. Ageing Dev.* 66, 243–248.
Nam, K.H., Akari, H. and Terao, K. (1998). *Exp. Anim.* 47, 159–166.
Nemoz–Bertholet, F. and Alyard, F. (2003). *Exp. Gerontol.* 38, 407–414.
Nimchinsky, E.A., Gilissen, E. and Allman, J.M. (1999). *Proc. Natl. Acad. Sci. U S A* 96, 5268–5273.
Nimchinsky, E.A., Sabatini, B.L. and Svoboda, K. (2002). *Annu. Rev. Physiol.* 64, 313–353.
Nogalska, A., Pankiewicz, A., Goyke, E. and Swierczynski, J. (2003). *Exp. Gerontol.* 38, 415–422.
Packer, C., Tatar, M. and Collins, A. (1998). *Nature* 392, 807–811.
Page, T.L., Einstein, M. and Duan, H. (2002). *Neurosci. Lett.* 317, 37–41.
Pawelec, G., Barnett, Y. and Forsey, R. (2002). *Front. Biosci.* 7, d1056–d1183.
Perl, D.P., Good, P.F. and Bussiere, T. (2000). *J. Chem. Neuroanat.* 20, 7–19.
Pica, J. and Dhenain, M. (1998). *Quarterly Journal of Experimental Psychology* B51, 337–348.
Price, D.L. (1993). *Neurobiol. Aging* 14, 685–686.
Rapp, P.R., Morrison, J.H. and Roberts, J.A. (2003). *J. Neurosci.* 23, 5708–5714.
Reaven, G.M., Chen, N., Hollenbeck, C. and Chen, Y.D. (1989). *J. Am. Geriatr. Soc.* 37, 735–740.
Roberts, E. (1999). *Biochem. Pharmacol.* 57, 329–346.
Roberts, J.A., Gilardi, K.V., Lasley, B. and Rapp, P.R. (1997). *Neuroreport* 8, 2047–2051.
Roecker, E.B., Kemnitz, J.W., Ershler, W.B. and Weindruch, R. (1996). *J. Gerontol. A Biol. Sci. Med. Sci.* 51, B276–B279.
Roth, G.S., Lesnikov, V. and Lesnikov, M. (2001). *J. Clin. Endocrinol. Metab.* 86, 3292–3295.
Rzepka, E., Adamczak, M., Kokot, F. and Wiecek, A. (2002). *Pol. Arch. Med. Wewn.* 107, 125–133.
Sapolsky, R.M., Vogelman, J.H., Orentreich, N. and Altmann, J. (1993). *J. Gerontol.* 48, B196–B200.
Sato, N., Kiuchi, K. and Shen, Y.T. (1995). *Am. J. Physiol.* 269, H1664–H1671.
Schmucker, D.L., Thoreux, K. and Owen, R.L. (2001). *Mech. Ageing. Dev.* 122, 1397–1411.
Seals, D.R. and Bell, C. (2004). *Diabetes* 53, 276–284.
Shimokawa, I. and Higami, Y. (2001). *Mech. Ageing Dev.* 122, 1511–1519.
Silhol, S., Calenda, A. and Jallageas, V. (1996). *Neurobiol. Dis.* 3, 169–182.
Smucny, D.A., Allison, D.B. and Ingram, D.K. (2001). *J. Med. Primatol.* 30, 161–173.
Sodora, D.L., Douek, D.C. and Silvestri, G. (2000). *Eur. J. Immunol.* 30, 1145–1153.

Takagi, G., Asai, K. and Vatner, S.F. (2003). *Am. J. Physiol. Heart Circ. Physiol.* 285, H527–H534.

Tang, Y., Janssen, W.G. and Hao, J. (2004). *Cereb. Cortex* 14, 215–223.

Taylor, L.D., Daniels, C.K. and Schmucker, D.L. (1992). *Immunology* 75, 614–618.

Tohno, S., Tohno, Y. and Hayashi, M. (2001). *Biol. Trace Elem. Res.* 82, 77–86.

Tolmasoff, J.M., Ono, T. and Cutler, R.G. (1980). *Proc. Natl. Acad. Sci. U S A* 77, 2777–2781.

Tunick, P.A., Freedberg, R.S. and Kronzon, I. (1990). *Gerontology* 36, 206–211.

Uddin, M., Wildman, D.E. and Liu, G. (2004). *Proc. Natl. Acad. Sci. U S A* 101, 2957–2962.

Uno, H. and Walker, L.C. (1993). *Ann. N.Y. Acad. Sci.* 695, 232–235.

van den Beld, A.W. and Lamberts, S.W. (2000). *Prostate Suppl.* 10, 2–8.

Verdery, R.B., Ingram, D.K., Roth, G.S. and Lane, M.A. (1997). *Am. J. Physiol.* 273, E714–E719.

Vermeulen, A. (1994). *Verh. K. Acad. Geneeskd. Belg.* 56, 267–280.

Voytko, M.L. and Tinkler, G.P. (2004). *Front. Biosci.* 9, 1899–1914.

Wang, M., Takagi, G. and Asai, K. (2003). *Hypertension* 41, 1308–1316.

Weindruch, R., Lane, M.A. and Ingram, D.K. (1997). *Aging (Milano)* 9, 304–308.

Woller, M.J., Everson-Binotto, G. and Nichols, E. (2002). *J. Clin. Endocrinol. Metab.* 87, 5160–5167.

CHAPTER 28

Primate Models of Neurological Disease

Charles Akos Szabo
Department of Medicine/Neurology,
University of Texas Health Science Center at
San Antonio, 7703 Floyd Curl Drive,
San Antonio, TX 78284-7883, USA

Introduction

Due to their phylogenetic proximity to humans, primate models bear particular significance for understanding neurological diseases in humans (Rapoport, 1991; Sibal and Samson, 2001). The cerebral organization of primates approximates that of humans more closely than does any other animal and, hence, offers a unique opportunity to study cortical processing and functional connectivity. Furthermore, a few neurological diseases found in humans, such as epilepsy, occur naturally in some primates (Naquet and Valin, 1998). Cerebral amyloidosis and memory changes due to natural aging are also found in primates (Erickson and Barnes, 2003). Experimental models for Parkinson's disease, multiple sclerosis, and epilepsy have been extensively studied and show great promise in the development and testing of medical or surgical therapies. In this chapter, the most important primate models of human neurological diseases will be presented.

Amnestic syndromes

Memory disturbances accompany several neurological illnesses in humans, including normal aging, Korsakoff's syndrome, temporal lobe epilepsy and Alzheimer's disease. The neurological substrate of memory is embodied by the limbic system, including the amygdala, hippocampi, rhinal cortex, thalamus, and mammillary bodies, and its cortical projections to the temporal, frontal and parietal lobe cortices (Gaffan and Parker, 2000). Two approaches were employed to study the pathophysiological mechanisms underlying memory dysfunction in humans. The first approach looked at effects of surgical or accidental lesions of the central nervous system, such as unilateral or bilateral temporal lobectomy in patients with temporal lobe epilepsy (Scoville and Milner, 1957; Squire and Zola-Morgan, 1991), or Wernicke-Korsakoff syndrome

The Laboratory Primate
Copyright 2005 Elsevier
ISBN 0-1208-0261-9

(Victor *et al.,* 1971). The second approach entailed the assessment of memory functioning in animals to identify similarities shared across species. Once visuospatial discrimination and learning paradigms for testing similar memory functions in humans and animals were developed, selective and combined lesions of specific structures in animals were employed to better our understanding of their roles in memory.

Primate models for amnesia demonstrated several advantages over rodent models. First, the neurological substrate of memory is essentially the same in the human as in the primate (Rapoport, 1991; Erickson and Barnes, 2003). Second, nonhuman primates can be trained to perform more complex discrimination tasks that help to study different aspects of memory. Furthermore, due to their longevity, memory testing can be performed over longer intervals. Most of the paradigms testing visual discrimination were developed for the macaques, including rhesus (*Macaca mulatta*) and cynomolgus (*Macaca fascicularis*) monkeys. Macaques were favored over other nonhuman primates because of their availability, ease of handling, and ability to cooperate with neuropsychological testing (Weinstein, 1941).

Memory testing in the nonhuman primate

The visuospatial memory and discriminations tasks used in nonhuman primates have been adapted in large part from human research. The manually controlled tasks were routinely performed in modification of the Wisconsin General Testing Apparatus (Weinstein, 1941; Harlow, 1944). The apparatus was accessed by the macaque from its cage, and consisted of a tray with wells that could be baited with a food reward and covered by objects or cardboard plaques. A sliding door could be lowered between the animal and the tray in order to introduce progressive delays before it could respond to a stimulus. Computer-controlled tasks were performed with the sample and choices displayed on a large touch-sensitive screen that allowed animals to manually indicate their choices (Gaffan, 1994).

The earliest paradigm to assess visual recognition memory in the Wisconsin General Testing Apparatus was the "matching-from-sample" paradigm. This procedure required the recognition of the similarity between the sample shape of an object and an identical shape presented in an array. This response was reinforced by a reward concealed under the matching shape. In the computer-generated adaptation of this test, enabling the generation of a large number of stimuli, a sample image is presented before it is compared to a novel choice object. The ability of the animal to remember the sample over time was then tested by delaying presentation of the sample and choice items for seconds to minutes. This paradigm was also able to test recognition of a list of up to 10 sample items presented sequentially (Eacott *et al.,* 1994). Other visuospatial memory tasks used to test working memory included the delayed spatial memory task, for which the animal had to remember which of two trays was baited, delayed spatial alternation tasks requiring the animal to alternate between sides on repeated trials, or delayed object alternation tasks rewarding the monkey when it learned to alternate between objects from one trial to the next (Friedman and Goldman-Rakic, 1988) (Figure 28.1). The "non-matching-from-sample" paradigm, on the other hand, required that the animal learn a new condition or rule to be rewarded. This entailed the macaques identifying the novel choice object when presented with the "sample" or an object that it was already familiar with from a previous trial. More recently, computer-generated discrimination tasks were employed to study "object in place" and "object-reward association learning", which have been shown to assess episodic memory in humans (Gaffan, 1994). The "object in place" paradigm requires the animal to learn to associate figures with background images in order to obtain a reward. The animals learn to choose between object pairs that were presented previously, but recognizing the correct choice in the particular combination of object and background. Object-reward association requires learning and retaining rewarded objects from a pair from one trial to the next, often repeated after progressive delays (Aggleton and Mishkin, 1983). Learning was quantified by measuring the number of trials to achieve an 85–90% correct response rate (Figure 28.1).

The use of autoradiography, to measure metabolic changes associated with visuospatial processing and memory tasks, was implemented to elucidate the cortical-subcortical interactions related to visuospatial processing and working memory (Friedman and Goldman-Rakic, 1988; Friedman and Goldman-Rakic, 1994). Rhesus monkeys were trained to consistently achieve an acceptable correct response rate, using short to medium delays and short intertrial intervals. The working memory tasks included the spatial delay, spatial delayed alternation, and the object alternation tasks. A control group was trained to perform simpler tasks requiring similar motor activation without memory processing. These included the visual pattern discrimination task, in which the animals had to choose the same pattern with each trial, or the sensory motor task, in which the animal was rewarded with either choice. This procedure was developed in order to perform continuous testing over the duration of the 45–60 minute uptake period needed for intravenously administered radioactive tracers, such as 2-deoxyglucose. Glucose uptake and metabolism was increased in hippocampus and dentate gyrus, dorsomedial thalamic regions and in the dorsolateral prefrontal cortex and inferior parietal cortex (Friedman and Goldman-Rakic, 1988; Friedman *et al.,* 1990; Friedman and Goldman-Rakic, 1994). This supported the lesional studies and electrophysiological studies in the observation that working memory relied on a network of cortical and subcortical structures. As autoradiography require sacrifice of the animals, these studies highlight the need, and proposed methodology, to study working memory *in vivo* with noninvasive functional neuroimaging.

Applications of paradigms

Cognitive deficits, associated with natural aging, have been demonstrated in macaques. The lifespan of macaques can exceed 40 years, with menopause occurring in the mid-twenties (Tigges *et al.,* 1988; Walker, 1995). An age dependent decline has been demonstrated in the delayed spatial responses and delayed nonmatching-to-sample tests resembling deficits in humans. However, decline in visuospatial memory and learning may not be a measure of selective dysfunction in the medial temporal lobe, but rather a reflection of more diffuse cerebral dysfunction. Aging

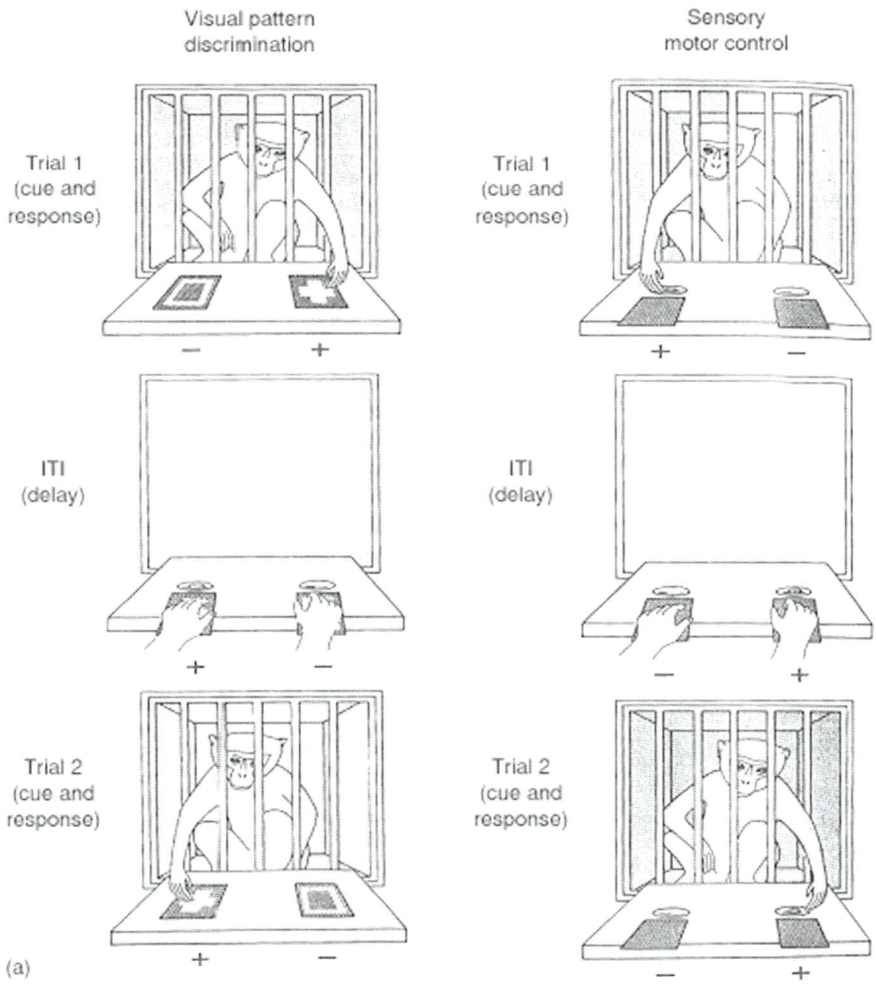

Figure 28.1 Working memory and control tasks. (A) Working memory (delayed spatial response, delayed spatial alternation, delayed object alternation) tasks and (B) control (visual pattern discrimination, sensory motor control) tasks were performed by rhesus monkeys in a modified Wisconsin General Test Apparatus. In the delayed spatial response test the wells were baited under the monkey's supervision, while in the remaining tasks the wells were baited while the screen was lowered. The screen was also used to introduce a delay between the responses. See the text for further description of the tests. From Friedman and Goldman-Rakic (1988). Reprinted by permission from The Society of Neuroscience.

(Continued)

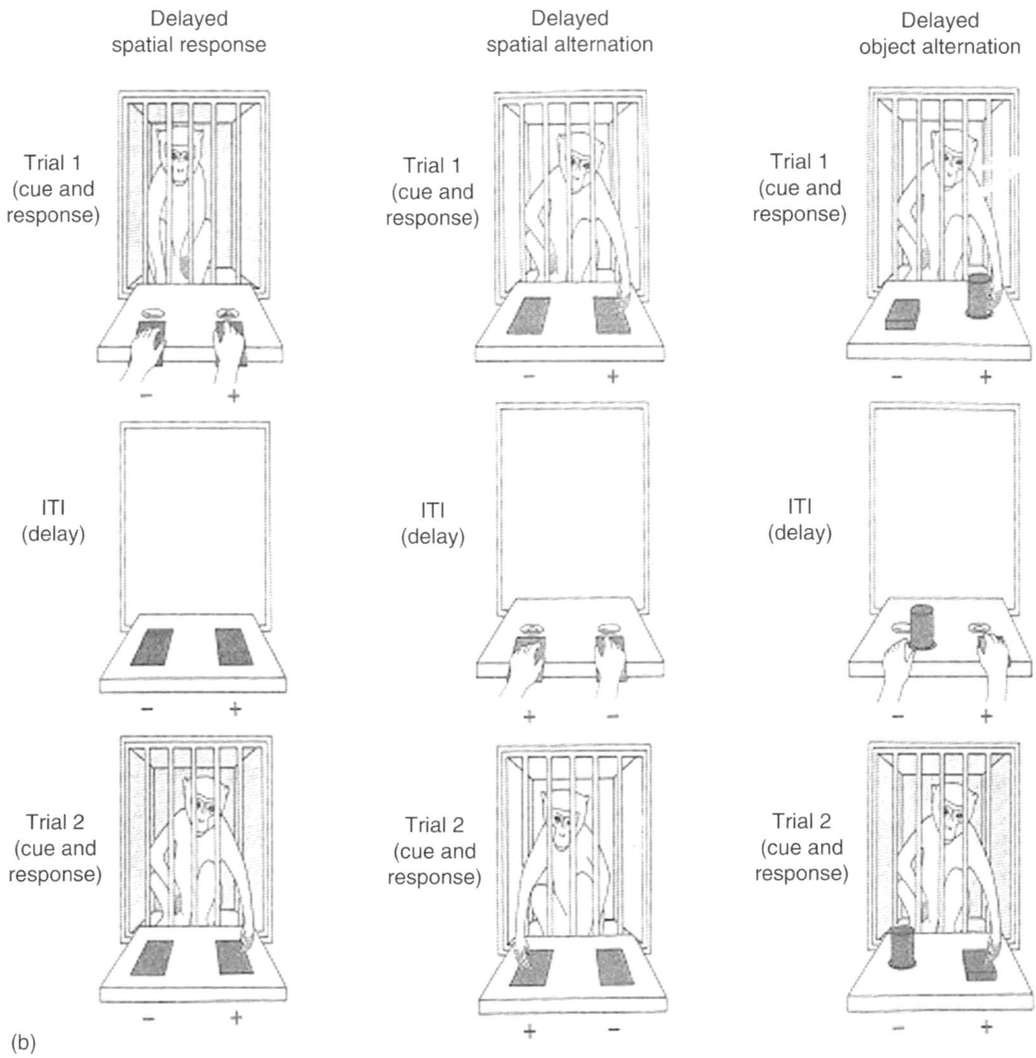

Figure 28.1 *(Continued).*

macaques demonstrated declines in visuospatial orientation and visually guided reaching. These reflect dysfunction in the posterior parietal lobe, in standard and concurrent object discrimination tasks, requiring integrity of the occipitotemporal association areas, and classical response delay, which relies on the dorsolateral prefrontal cortex and rostral striatum (Bachevalier *et al.,* 1991). Variability of the deficits were apparent among aging macaques, some performing as well as the younger controls on specific tests. The age-related decline was not uniform over all cognitive domains, appearing earlier in delayed spatial responses, reflecting frontal lobe dysfunction, before other visuospatial or memory abilities were affected. Acquisition of motor skills was not affected by age, though slowing reaction time was demonstrated in another study (Davis and Ruggiero, 1973). Although the age-related decline in recognition memory functioning resembles medial temporal lesions (Rapp and Amaral, 1991), age-related decline in memory is not due to cell loss in the hippocampus, dentate gyrus, or other cerebral structures (Merrill *et al.,* 2000; Keuker *et al.,* 2003). The substrate of aging is still unclear, though age-related changes in myelination or degenerative changes involving ascending cholinergic or adrenergic pathways may play an important role.

The characterization of a nonhuman primate model for human amnesic syndromes, such as bilateral temporal lobe damage and Wernicke-Korsakoff syndrome, stimulated research into structures and pathways related to memory functioning. The general approach of the studies was to compare recognition memory before and after chemical (Brozoski *et al.,* 1979; Murray and Mishkin, 1998) or surgical (Mishkin, 1978; Aggleton and Mishkin, 1983; Squire and Zola-Morgan, 1983) lesioning of selected cerebral structures or to compare performance on cognitive testing between surgical and unoperated controls. The retention of preoperatively learned skills and the postoperative acquisition of new skills were also taken into consideration. Despite extensive research into memory functioning in the primate, there is still controversy regarding the neurological substrate of amnesic syndromes.

Memory impairments were demonstrated after medial temporal (Mishkin, 1978; Mahut *et al.,* 1981), thalamic (Aggleton and Mishkin, 1983; Zola-Morgan and Squire, 1985) and frontal lobe lesions (Brozoski *et al.,* 1979). The delayed nonmatching-to-sample tasks demonstrated differences between lesions affecting various structures in the medial temporal lobes. These included selective, bilateral lesions of the amygdala, hippocampal/parahippocampal regions, amygdalohippocampal structures, amygdalohippocampal/parahippocampal regions, and rhinal cortex with the poorest performance on this task occurring after hippocampal/entorhinal and perirhinal cortex ablations (Squire and Zola-Morgan, 1991). Bilateral rhinal cortex ablations posed additional difficulties in recognition of objects presented in large sets and impairment of retention of preoperatively learned tasks, reflecting a general impairment in the processing of visual stimuli (Eacott *et al.,* 1994). The absence of deficits on delayed nonmatching-to-sample task for objects and position, with selective amygdalohippocampal damage with ibotenic acid, further supported the observation that the rhinal cortex was both necessary and sufficient for visual recognition and learning (Murray and Mishkin, 1998). However, in spite of the important role of the medial temporal lobe structures in memory and learning, deficits on delayed spatial responses and nonmatching-to-sample tests were by no means specific to lesions in these structures.

Neuropsychological research in nonhuman primates was driven primarily by the interest in the amnesic state associated with Wernicke-Korsakoff syndrome (Victor *et al.,* 1971). Wernicke's encephalopathy presents predominantly in alcoholic patients suffering from malnutrition and hypovitaminosis. Wernicke's encephalopathy is associated with mental status changes, ataxia and conjugate gaze palsies due to necrotic lesions in the paraventricular regions of the thalamus and hypothalamus, mammillary bodies, periaqueductal region of the midbrain, and in the floor of the fourth ventricle. Korsakoff psychosis is the residual cognitive and behavioral correlate of Wernicke's encephalopathy, but has been associated with stroke and encephalitis. Korsakoff psychosis was characterized as a subacutely evolving amnesic syndrome, in which newly presented material is registered but not retained for more than a few minutes. In addition, patients develop a retrograde amnesia, or an inability to recall events predating the onset of their illness. In one series reporting patients with Korsakoff's psychosis due to a variety of organic causes, the thalamus appeared to be involved in all patients with profound amnesia (Victor *et al.,* 1971). This finding was controversial as other studies encountered severe amnesia in a few patients without thalamic involvement (Mair *et al.,* 1979). This controversy stimulated interest in the role of the thalamus in memory functioning. Delayed nonmatching-to-sample tests detected cognitive deficits in macaques with thalamic lesions, specifically when the lesions affected the magnocellular division of thalamic mediodorsal nucleus (Aggleton and Mishkin, 1983). Deficits on the delayed nonmatching-to-sample, delayed object-reward association, and scene memory tests were not as severe as after bilateral rhinal cortex ablations (Eacott *et al.,* 1994; Parker *et al.,* 1997). The pattern of deficits was similar to those observed after bilateral frontal ablations in the macaque (Parker and Gaffan, 1998), which may not be surprising as the magnocellular division of the mediodorsal thalamus is extensively connected to the frontal lobes (Goldman-Rakic and Porrino, 1985; Giguere and Goldman-Rakic, 1988). Furthermore, bilateral thalamic, as well as frontal, lesions appeared to affect a broader range of memory tasks, but none as severely as those due to combined lesions of the amygdala, fornix and temporal lobe stem (Gaffan, 1994; Parker and Gaffan, 1998). Hence, Gaffan and Parker proposed that Korsakoff psychosis results from multiple lesions in the limbic system, which in some cases may not involve the thalamus (Gaffan and Parker, 2000). High-resolution magnetic resonance and positron emission tomography in amnesic patients and macaques, with varying lesions of the limbic system, may succeed in demonstrating chronic structural or functional changes in the thalamus, even in the absence of acute lesions.

The delayed spatial response task also demonstrated cognitive dysfunction in an experimental form of parkinsonism in rhesus monkeys (Fernandez-Ruiz *et al.,* 1999). Memory functioning was compared in rhesus monkeys before and after a single dose of levodopa (L-DOPA) ranging between 100 and 500 mg. While L-DOPA produced a striking improvement on spatial delayed tasks, high doses of L-DOPA led to hyperactivity and inability to cooperate with testing. This adverse effect was also noted in healthy rhesus monkeys receiving levodopa.

Parkinson's disease

Parkinson's disease (see also Chapter 24, Morton *et al.*, and Chapter 27, Tigno *et al.*) is one of the most common neurodegenerative illnesses in humans, occurring in 1% of the population over 65 years old (Koller, 1992). Parkinson's disease is associated with the gradual development of bradykinesia, rigidity, resting and postural tremors, postural instability and a fixed posture. Some patients develop freezing or an arrest of movement. The symptoms are due to the loss of dopamine-producing neurons in the substantia nigra, pars compacta, and, to a lesser extent, in the ventral tegmental area (Forno, 1996), which project to the basal ganglia and the limbic and cortical regions, respectively. Clinical signs emerge when the striatal content of dopamine is reduced by about 80%. The hallmark of the disease is the formation of intracytoplasmic, round, eosinophilic inclusions, referred to as Lewy bodies, within dopaminergic neurons of the substantia nigra.

The treatment of Parkinson's disease has focused on replacement of the missing dopamine (Fahn, 1988). The most efficacious therapy has been the oral administration of the natural precursor of dopamine, levodopa, which is converted to dopamine by the enzyme aromatic amino acid decarboxylase in dopaminergic neurons. The main limitations of chronic replacement therapy with levodopa are the development of motor complications, in particular motor fluctuations and dyskinesia, and of psychosis with visual hallucinations and delusions. Two-thirds of patients with Parkinson's disease, given levodopa, chronically develop abnormal extra movements or dyskinesia in the form of chorea or dystonia. Dyskinesia is thought to represent an over-sensitization of dopamine receptors in the striatum due to fluctuations in dopamine levels resulting from levodopa therapy. Chronic treatment with dopamine receptor agonists is less likely to produce dyskinesias, but is also less efficacious than treatment with levodopa. There is a need for further research into the mechanisms of dyskinesia to improve replacement therapies. Furthermore, as dopamine replacement therapies and dopamimetic agents do not affect disease progression, there is great interest in novel surgical therapies and therapies that can potentially arrest or reverse the disease process.

Animal models have played significant roles in both instances. The primate model has become one of the most important models, in part because of the similarity of the organization of the motor system and their neurotransmitter systems (Haber, 1986; Rapoport, 1991), and in part due to the similarities of the motor syndrome of parkinsonism and levodopa-related dyskinesia (Jenner, 2000). While circling behavior or other stereotypic movements, such as grooming or orofacial dyskinesia, were described in rodents, choreoathetotic movements characteristic of human dyskinesia have been observed only in primates (Langston, 2000).

The MPTP model in the nonhuman primate

In the early 1980s several people developed parkinsonism following intravenous self-administration of a "synthetic heroin" powder, which they had bought on the street and which contained a noxious byproduct, N-methyl-4-phenyl-1,2,3,6-tetrahydropyridine (MPTP). Their symptoms included akinesia, rigidity and a postural tremor. It is now known that MPTP readily crosses the blood-brain barrier and is converted to the 1-methyl-4-phenyl-2,3-dihydropyridinium ion (MPP+) by the enzyme monoamine oxidase-B (MAO-B) in glia and serotonergic neurons. MPP+ is selectively taken up into dopaminergic neurons of the substantia nigra, via the dopamine transporter, and its toxicity is mitigated by inhibiting the production of ATP in the mitochondria and by augmenting the production of free radicals (Tolwani *et al.*, 1999). MPTP was administered to adult rhesus monkeys (*Macaca mulatta*) with similar effect on motor activity (Burns *et al.*, 1983). After the initial doses of MPTP, the motor symptoms were transient and included intermittent eyelid closure, decrease in spontaneous movements, loss of facial expression and a postural tremor. After a cumulative dose of 1.7 mg/kg of MPTP over one week, the symptoms persisted, characterized by changes in posture, brady- or akinesia and postural tremor. The motor symptoms included slowness of movement, a flexed posture of the neck and thoracic spine, flexion of the upper and lower extremities, balance impairment, tremor, a failure to initiate or complete movement or "freezing" and impairment of balance. Cerebrospinal fluid studies showed an acute depletion of dopamine metabolites, while examination of postmortem brain demonstrated degeneration and loss of pigmented neurons in the substantia nigra, with relative sparing of neurons in the ventral tegmental area (Figure 28.2).

The MPTP-treated model has been applied to several nonhuman primate species, including the rhesus monkeys (*Macaca mulatta*; Burns *et al.*, 1983), cynomolgus monkeys (*Macaca fascicularis*; Goulet and Madras, 2000), marmosets (*Callithrix jacchus*; Albanese

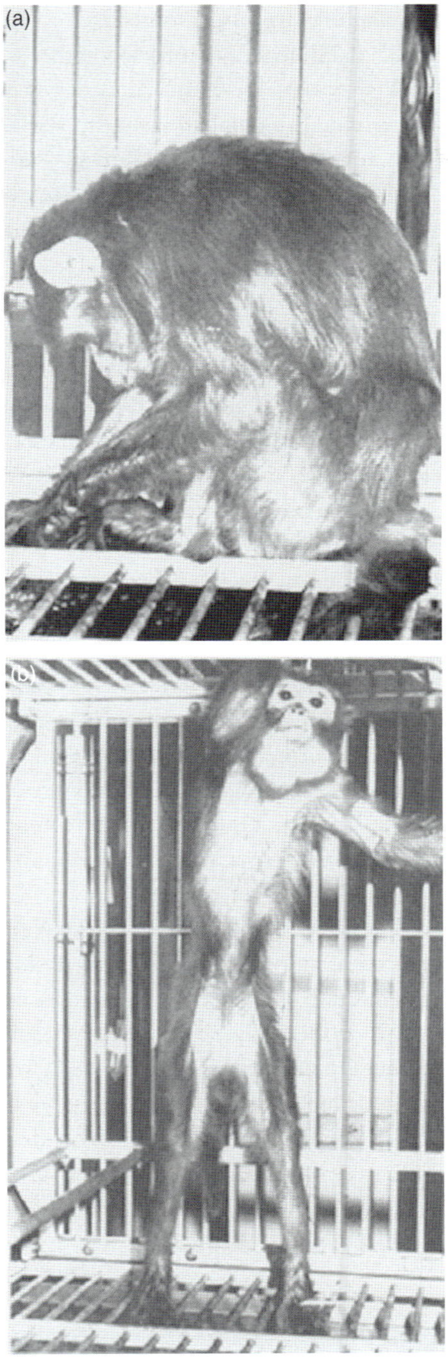

Figure 28.2 The MPTP-treated monkey. Photographs of the (A) the flexed posture associated with akinesia in an MPTP-treated monkey before and (B) with symptom reversal after administration of oral levodopa. The two photographs were taken two hours apart. From Burns *et al.* (1983). Reprinted by permission from the author.

et al., 1993), squirrel monkeys (*Saimiri scuireus*; Boyce *et al.,* 1990) and St. Kitts green monkeys (*Cercopithecus aethiops sebaeus*; Elsworth *et al.,* 1987). Sensitivity to MPTP varies among primate species, being greater in the rhesus than squirrel monkey, for example (Tolwani *et al.,* 1999). Nonetheless, the dosage of MPTP used typically falls within a range of 2 to 10 mg/kg given over 1 to 10 days in the subacute model (Burns *et al.,* 1983; Bedard *et al.,* 1986; Pearce *et al.,* 1998; Goulet and Madras, 2000; Langston, 2000), and 15–45 mg/kg in the chronic models (Albanese *et al.,* 1993; Boyce *et al.,* 1990). The motor symptoms gradually worsen over the course of weeks, after MPTP administration, but in chronic models would improve if administration of MPTP were discontinued. In fact, not only the motor symptoms, but also the neurotransmitter levels normalized and the cytoarchitectural changes of the dopaminergic neurons regressed over several months in some chronic models (Albanese *et al.,* 1993). The route of administration of MPTP varied among these primate models and included intravenous (Burns *et al.,* 1983), intraperitoneal (Boyce *et al.,* 1990), or intra-arterial administration, the latter exemplified by the hemi-lesioned MPTP primate model (Bankiewicz *et al.,* 1986). In the hemi-lesioned model, MPTP is injected unilaterally into one internal carotid artery, leading to ipsilateral damage to the nigrostriatal system and contralateral parkinsonism. The advantages of this model include the reduced severity of the motor impairment, thereby reducing morbidity and mortality, and the ability to compare the motor changes with the unaffected side.

Although the MPTP-treated primate model resembles Parkinson's disease, because of the selective involvement of the nigrostriatal system and similarities of the motor symptoms, it is important to remember that the MPTP model does not reflect chronic degenerative changes that occur in Parkinson's disease. In some models, the parkinsonian signs subside, even if not completely, once MPTP administration is discontinued. Secondly, the pathological hallmark of Parkinson's disease, namely the formation of Lewy body inclusions, is not associated with MPTP-induced damage. Finally, the dopaminergic depletion, in humans, primarily involves the putamen, whereas, in the MPTP model, the caudate is equally or more affected than the putamen (Albanese *et al.,* 1993).

Applications of the MPTP model

Dyskinesia

Even before the development of the MPTP-treated model of parkinsonism in nonhuman primates, dyskinesia had been induced in rhesus and squirrel monkeys by the intraperitoneal injection or oral administration of supratherapeutic doses of levodopa, respectively (Mones, 1972; Boyce *et al.,* 1990). The monkeys became hyperactive within 30 minutes of injection, demonstrating stereotyped movements of the shoulders, hips, tongue and mouth. One to two hours after injection, animals exhibited choreoathetotic movements involving the trunk, face and extremities that resolved within six hours. Dystonic posturing or repeated brief periods of sustained contraction of agonist and antagonist muscles occurred less predictably after levodopa administration, although its severity was dose dependent. When it did occur, the episodes of dystonia lasted less than five seconds and predominantly involved the lower extremities, resulting in either a brief flexion of the hip and knee with plantar flexion and eversion of the foot or extension of the hip and knee with plantar flexion of the ankle. At high doses of levodopa, the animals froze in a climbing position at the top of the cage, suggesting the possibility of behavioral abnormalities resembling a psychotic trance (Boyce *et al.,* 1990).

Levodopa-induced dyskinesia is seen in 60–80% of patients with Parkinson's disease within six years of therapy (Nutt, 1990). Dyskinesia can be disabling and limit the use of levodopa, which is still the most effective symptomatic therapy. It also complicates drug therapy since once a patient develops dyskinesia, it can reoccur with the use of other antiparkinsonian therapies, such as dopamine agonists. The mechanisms underlying the development of dyskinesia are unknown. The MPTP-treated primate model offers an inexpensive and reliable way to study the biochemical and molecular changes induced by chronic levodopa therapy and the effects of potential therapies on the dyskinesia.

Difficulties exist in comparing the effects of different therapies directed toward reducing dyskinesia in the nonhuman primate model. These include differences in the duration of the effects of dopamine agonists and the reversibility of the levodopa-induced dyskinesia, failure to blind observers to previous treatments and differences among laboratories characterizing the symptoms and their severity (Langston, 2000). As with the Primate Parkinsonism Rating Scale, which was adapted from the Unified Parkinson's Disease Rating Scale and modified for use in the squirrel monkey (Langston, 2000), the Global Primate Dyskinesia Rating Scale was derived from clinical research tools (Langston, 2000). Like the other rating scales, these demonstrated excellent intra- and inter-rater reliability for the measurement of abnormal limb and truncal movement (Langston, 2000). In contrast to other scales, the Global Primate Dyskinesia Rating Scale tolerates some uncertainty with regards to the classification of excessive movements in primates, which may be either normal for age or a nonspecific reaction to levodopa, such as hyperactivity (Tables 28.1 and 28.2).

MPTP-treated primates can be "primed" to develop dyskinesia by previous exposure to levodopa. The appearance of dyskinesia can be provoked more rapidly in MPTP-treated primates and in Parkinson's disease. The threshold for levodopa-induced dyskinesia is lowered by MPTP, as it is observed even in animals with mild parkinsonism after a single injection of MPTP (Langston, 2000).

There has been a controversy over which receptor type is responsible for the emergence of dyskinesia in the MPTP-lesioned primates. Initial studies proposed that levodopa-induced dyskinesia was a result of its effect on the D_1 dopamine receptor (Bedard *et al.,* 1986). Later studies compared the effects of selective D_1 and D_2 dopamine receptor agonists in drug-naïve cynomolgus monkeys, and found that dyskinesia could be produced by some drugs in both categories (Blanchet *et al.,* 1993, Bedard *et al.,* 1993). Short- or

TABLE 28.1: Global primate dyskinesia rating scale (GPDRS)

Score	
0	No evidence of dyskinesia
1	Subtle movements suggestive of dyskinesia, but could be normal
2	Mild dyskinesia: definitely present, but mild and of low amplitude
3	Moderate dyskinesia: intermediate or even higher amplitude, but not violent or extreme
4	Severe dyskinesia: extreme movements, including flinging of the arms, and/or violent jerks or thrusts of the extremities or trunk. May have marked gyrations of the hips (hoola-hooping). May be incapacitating.

Ratings based on the most severe dyskinesia during the observation period. From Langston (2000). Reprinted by permission of John Wiley & Sons, Inc.

TABLE 28.2: Parkinsonian scale modified for the squirrel monkey

Spatial hypokinesia (Movement around cage)	
0	Normal (use entire cage space)
1	Utilizes most of the cage (>75% of cage space), but may be slow
2	Definitely slowed, but uses more than 50% of cage space
3	Definitely slowed, but uses less than 50% of cage space
4	Little or no movement; stays in a confined area of the cage
Body bradykinesia	
0	Normal body movements, actively using the cage or bars
1	Slow or deliberate body movements, may be normal for age
2	Moderately slow, intermittent limb dragging, but still moves extremities without provocation
3	Marked slowness, requires provocation to move arms or legs
4	Frozen, little or no body movements regardless of provocation
Manual dexterity (right arm/left arm)	
0	Normal
1	Mildly slow and/or some loss of maneuverability of food items; could be normal for age
2	Moderate slowness, noticeable effort needed to grab or maneuver food
3	Marked slowness, with multiple attempts needed to grab food, may use both hands, may drop food
4	Severe slowness, with inability to grab or maneuver food; may need to be hand fed
Balance	
0	Normal
1	Slight tendency to hold onto cage, may be normal for age; no falls
2	Uses both hands intermittingly for support; rare/occasional falls
3	Uses both hands for support at all times; frequent falls
4	Continually hanging on for support; falls with no attempt to move
Freezing (observation over 4 minutes)	
0	None (no freezing observed)
1	Occasional mild (<5 sec duration) freezing episodes
2	Frequent mild freezing episodes (<5 sec in duration), or rare severe episodes (>5 sec in duration)
3	Frequent severe freezing observed (>5 sec duration)
4	Frozen most of the time

From Langston (2000). Reprinted by permission from John Wiley & Sons, Inc.

long-acting D_2 receptor agonists were more likely to be associated with persistent dyskinesia in levodopa-primed animals, whereas the use of long-acting D_1 receptor agonists was associated with a gradual resolution of the dyskinesia (Pearce *et al.,* 1998). On the other hand, administration of the D_2 receptor agonists, ropirinole and bromocriptine, was associated with milder dyskinesia that did not debilitate the animals. These studies supported the early use of long-acting dopamine agonists as the initial therapy for Parkinson's disease.

Based upon the studies in primates, it appears that levodopa's causative role in producing dyskinesia may not be mitigated via dopamine receptors alone. Levodopa can be converted into noradrenaline in the locus coeruleus, displace 5-hydroxytryptamine from serotonergic neurons and enhance glutamate secretion. N-methyl-D-aspartate (NMDA) antagonists can improve parkinsonian symptoms and dyskinesia by reducing glutaminergic transmission (Papa and Chase, 1996; Mitchell and Carroll, 1997). Other agents, such as the selective α-adrenergic antagonist, indoxan, and adenosine A_{2A} receptor antagonists, have also been shown to reduce dyskinesia in MPTP-lesioned marmosets (Henry *et al.,* 1999; Kanda *et al.,* 1998). Although these findings suggest that several neurotransmitter systems may be involved in the production of levodopa-induced dyskinesia, more attention needs to be directed toward levodopa's effects on the second messengers or metabolic changes in striatal cells (Jenner, 2000).

Implantation therapies

The currently available medical therapies aim to reduce the motor symptoms of Parkinson's disease. Two approaches to alter the course of the disease would be either to prevent cell death or replenish the dopamine-producing capability of the central nervous system. The intracerebral implantation of dopamine secreting cells offers one opportunity to alter the course of Parkinson's disease (Widner *et al.,* 1992). Another approach involves the insertion, into striatal cells, of a gene that codes for tyrosine hydroxylase, the rate-limiting enzyme that converts tyrosine into levodopa. These types of therapies aim to achieve a continuous, site-specific delivery of dopamine or L-DOPA.

The advantage of the neurotoxin MPTP-treated model over the models using the neurotoxin 6-hydroxydopamine (6-OHDA) in rodents and primates has been the prolonged effects of MPTP, allowing observation of the efficacy and viability of transplants in long-term studies. Unilateral MPTP models have been employed by some centers to reduce long-term morbidity and mortality.

Allogenic transplantation as a treatment for MPTP- or 6-OHDA-induced parkinsonism in marmosets, St. Kitts green monkeys or rhesus monkeys, was initially studied with the grafting of adrenal medullary or fetal mesencephalic cells into the substantia nigra or striatum (Sladek *et al.,* 1986; Bakay *et al.,* 1987; Bankiewicz *et al.,* 1991; Taylor *et al.,* 1991; Annett *et al.,* 1993). The goals of this approach include provision of dopamine or of trophic factors to stimulate sprouting in residual host dopaminergic neurons, activation of the host parenchyma to release trophic factors, and disruption of the blood brain barrier to increase the concentration of catecholamines in the central nervous system.

The availability of the primate model also gave several researchers an opportunity to study cell viability and the pathological changes associated with different types of grafts. The efficacy of cell transplantation could be compared to sham surgeries limited to mechanical injury to the host central nervous system, hence controlling for the desired effects of transplantation (Taylor *et al.,* 1995). Immune responses to grafts could be monitored by direct examination of the brain post-transplant (Bakay *et al.,* 1998). Neuroimaging with ^{18}Fluoro-DOPA PET was shown to be a noninvasive technique that could monitor the viability of the grafts *in vivo* (Subramanian *et al.,* 1997).

In contrast with adrenal medullary and mesencephalic cells, PC12 cells, derived from a pheochromocytoma cell line and producing primarily dopamine and L-DOPA and not norepinephrine, can be generated in tissue culture and stored, cloned or purified. Adrenal medullary and mesencephalic cells have to be harvested from human donors or other animal species. With the development of a capsule, consisting of a semipermeable polymer membrane to encapsulate graft cells, acute tissue rejection could be averted and the need for immunosuppression avoided (Aebischer *et al.,* 1994; Lindner and Emerich, 1998). Despite the significant and long-lasting improvement of motor function, efficacy of this approach varies depending on differences in dopamine output.

Other strategies to enhance dopamine production in the central nervous system include the insertion of therapeutic genes into cells which are then implanted into the brain, or direct insertion of the genes into nerve cells in symptomatic brain nuclei. These therapeutic strategies have primarily been developed in rodent models. Because of the difficulty of delivering genes or gene products to nondividing cells, herpes simplex virus type I and adenoviral vectors have been employed to deliver the human tyrosine hydroxylase gene into striatal neurons. A similar approach was developed in the MPTP-treated St. Kitts green monkeys (*Cercopithecus aethiopis sabaeus*). Not only was the gene coding for tyrosine hydroxylase, the enzyme responsible for the conversion of tyrosine into L-DOPA, expressed, but, by using the same adenovirus-associated vector, the gene coding for aromatic acid decarboxylase, the enzyme necessary for the conversion of L-DOPA to dopamine, was also delivered to the caudate nucleus (During *et al.,* 1998). Postmortem examinations of animals that exhibited neurological improvement showed increased dopamine in the tissue surrounding the injection site and recovery of dopaminergic terminals, despite the vector DNA being limited to a small number of transfected cells (During *et al.,* 1998). The use of lentiviruses to transfer genes for nerve growth factors represents another potential therapy for Parkinson's disease, or other neurodegenerative diseases, such as Alzheimer's disease (Kordower *et al.,* 1999).

Alzheimer's disease and amyloid angiopathy

Alzheimer's disease is the most common type of dementia. It is one of the most common neurodegenerative diseases (Adams and Victor, 1993). The cause of Alzheimer's disease is unknown, but genetic factors play an important role in early onset cases (Sandbrink and Bayreuther, 1996). The onset of the cognitive decline is insidious, presenting initially with memory disturbances. As the disease progresses, visuospatial and verbal skills deteriorate, and patients become apraxic, unable to care for themselves. Personality changes may predate the onset of the amnesia, and paranoia and aggressiveness are the most problematic to manage. The underlying pathology includes senile or "neuritic" plaques, intracytoplasmic neurofibrillary tangles and granulovacuolar degeneration of neurons, most commonly in the hippocampus. Cell loss is encountered diffusely in the cholinergic nucleus basalis of Meynert and cortical association cortices, especially of the glutaminergic corticocortical and hippocampal pyramidal cell pathways (Braak and Braak, 1991). MRI demonstrates diffuse cortical atrophy, whereas PET findings show reductions of temporoparietal association cortex initially, with secondary involvement of the prefrontal association cortex (Benson *et al.,* 1983; Cutler *et al.,* 1985).

Impairment of cholinergic function has been studied in several primate models. These studies included lesional studies damaging the nucleus basalis of Meynert or chemically induced cholinergic dysfunction. Lesions of the nucleus induced by electrocoagulation or injection of ibotenic acid, resulted in hypometabolism of the frontotemporal cortices and learning impairments (Kiyosawa *et al.,* 1987; Ridley *et al.,* 1986). Treatment with scopolamine, a muscarinic receptor blocker, also causes cholinergic dysfunction, but its effects are mild and transient. Scopolamine-induced attention deficits on continuous performance tasks in rhesus monkeys could be reversed by combinations of tacrine and milameline, medications used for the treatment of Alzheimer's disease (Callahan, 1999). Dizocilpine, a NMDA receptor antagonist, also causes impairment of conditional visuospatial memory and visual discrimination tasks in marmosets, possibly due to disruption of glutaminergic association pathways (Harder *et al.,* 1998). This model simulates reduced glutaminergic function due to acetylcholine depletion, which, along with serotonin, modulates glutaminergic cells in the hippocampus and rhinal cortex. The effects of the glutaminergic blockade could be reversed by selective serotonin antagonists, which restore cholinergic function and could be a potential treatment of cognitive deficits experienced in early stages of Alzheimer's disease (Harder and Ridley, 2000).

Although no chronic degenerative central nervous system diseases have been described in nonhuman primates, they appear to age similarly to humans (Erickson and Barnes, 2003). Normal aging in monkeys is associated with slowly progressive cerebral dysfunction (Bachevalier *et al.,* 1991). Cell loss or apoptosis does not appear to be the cause underlying the dysfunction. The onset of the dysfunction may coincide with cerebral amyloid deposition in aging monkeys. Cerebral amyloid deposition, in the form of senile or "neuritic" plaques, occurs as a part of normal aging in humans and other primates. The amyloid-β is primarily deposited in rhesus monkeys from 25 years of age (Uno and Walker, 1993) and from 15 years of age in squirrel monkeys (Walker, 1990).

There is a considerable difference, between primates, in the type and location of amyloid deposition. Amyloid deposition in most primates occurs initially in the blood vessel walls. Cerebrovascular amyloid is also found in humans, and it is responsible for 15–20% of hemorrhagic stroke in the elderly, but not for the cognitive decline associated with Alzheimer's disease. Cerebral amyloid angiopathy (CAA) has been described in aging dogs and primates, ranging from lemurs, such as *Lemur fulvus* (Strittmatter *et al.,* 1993) and *Microcebus murinus* (Bons *et al.,* 1994), to chimpanzees (*Pan troglodytes*) (Gearing *et al.,* 1993). Models of natural CAA found in rhesus (*Macaca mullata*) and squirrel monkeys (*Saimiri sciureaus*) have been extensively studied, while marmosets (*Callithrix jacchus*) (Baker *et al.,* 1993) and cynomolgus monkeys (*Macaca fascicularis*) (Nakamura *et al.,* 1995) represent less studied models.

Parenchymal amyloid is more common than vascular amyloid in rhesus monkeys, and is deposited diffusely in the neocortex, while the basal ganglia and cerebellum are relatively unaffected (Walker, 1997). In squirrel monkeys, vascular amyloid is more common. The distribution is similar to that of rhesus monkeys, but also includes the amygdala and hippocampus. Leptomeningeal vessels are less affected in squirrel monkeys and, among vessel types, the capillaries are more affected than those of the rhesus monkey. The frontal and temporal lobes are more affected than the occipital lobes in both animals, while both are similarly affected in humans with CAA and Alzheimer's disease. There is no known genetic model for cerebral amyloidosis in nonhuman primates. Although most nonhuman primates are homozygous for the apolipoprotein E–ϵ4 allele, which is associated with increased amounts of amyloid deposition in humans, with and without dementia, differences in particular amino acids in the APOE allele may render monkeys less susceptible to neurodegenerative disorders, such as Alzheimer's disease (Morelli *et al.,* 1996).

Amyloid-β peptides may be derived from several proteins, including the extracellular matrix proteins α1-antichymotrypsin, apolipoprotein E, ubiquitin, serum amyloid P, acetylcholinesterase, cystatin C, and complement components (Walker, 1997). Cerebral parenchymal amyloid accumulation has been centrally implicated in the pathogenesis of Alzheimer's disease for the following reasons: the compact, neuritic plaques, formed by amyloid-β deposits, are a pathological hallmark of Alzheimer's disease, mutations of the amyloid precursor protein have been associated with the disease, and fibrillary amyloid exhibits neuronal toxicity (Sani *et al.,* 2003). Conversely, selective lesions of the basal forebrain cholinergic structures have been associated with increases in amyloid precursor protein-like deposition in the hippocampus and neocortex (Ramirez *et al.,* 2001). Once fibrillar amyloid-β is deposited in one area of the brain, it begins to accumulate at remote sites as well (Sani *et al.,* 2003). The increase of amyloid precursor protein in the central nervous system may lead to increased Aβ production and plaque formation, which in turn could have detrimental effects on cholinergic function (Galdiczki *et al.,* 1994; Kar *et al.,* 1998). The event that may mark the onset of this cycle is unclear.

There has been a growing interest in developing radioactively labeled ligands that can cross the blood-brain barrier and bind to amyloid (Majocha *et al.,* 1992). This technique would allow the visualization of amyloid deposits, which at this time can only be recognized pathologically. It would allow correlation of cognitive impairment with the location and load of amyloid deposits. The development of ligands in the form of monoclonal antibodies, directed toward the amyloid components, could also be used to deliver therapeutic agents to reduce the amyloid load.

Multiple sclerosis

Multiple sclerosis is an inflammatory disease of the central nervous system (Adams and Victor, 1993). The disease process is characterized by a T-lymphocyte mediated destruction of myelin, which is the lipoprotein sheath that insulates and supports axons. The inflammatory process leads to demyelination, oligodendrocyte death, and secondary gliosis. The disease can be remitting-relapsing or chronic progressive, though in both cases it is associated with increasing damage to the white matter and neural pathways over time.

Experimental models of demyelinating disease

The first experimental animal model was developed, in macaque monkeys, by intramuscular injections of aqueous emulsions and alcohol-ether extracts of normal rabbit brains (Rivers and Schwentker, 1933). After 40–80 injections of these mixtures, animals developed neurological deficits characteristic of demyelinating disease. Pathological changes in the central nervous system demonstrated hemorrhagic and necrotic lesions, commonly associated with acute disseminating encephalomyelitis. Perivascular cuffing by mononuclear cells was associated with destruction of myelin in the vicinity of the blood vessels. A model of chronic experimental allergic encephalomyelitis (EAE) was developed subsequently with the addition of adjuvants, including aquafor, paraffin oil, and heat-killed tubercle bacilli, producing encephalomyelitis with a reduced number of injections (Freund *et al.,* 1947, Ferraro and Cazzullo, 1948). By this time, histopathological studies were able to identify lymphocytes, mononuclear cells, and microglia as the mediators of demyelination. Different clinical types of encephalomyelitis could be produced, including one that affected the optic nerves, another primarily producing brainstem and cerebellar symptoms, and a third with mild generalized symptoms that resolved without any residuals. A more recent model involved the marmoset (*Callithrix jacchus*), injected with human white matter to produce a remitting and relapsing form of EAE ('t Hart *et al.,* 1998). This model produced neurological symptoms and histopathological changes typical for human multiple sclerosis and demonstrated a central role of T-lymphocytes in this model. As the placentas of marmosets, arising from separate ovas, fuse, resulting in a cross-circulation of bone marrow derived elements between the fetuses, these animals are born as bone marrow chimeras. Hence, there is no crossreactivity of T-cells between siblings, allowing the adoptive transfer of immunocompetent lymphocytes (Genain and Hauser, 1997). The transfer of myelin basic protein reactive T-cell clones was successfully achieved in the marmoset. The resulting illness developed in 14–21 days and was associated with clinical signs of EAE, pleocytosis in the cerebrospinal fluid and foci of gadolinium enhancement on MRI (Genain *et al.,* 1994), demonstrating that T-cells alone could mediate EAE in a nonimmunized healthy animal. Nonetheless, the pathological changes were not characterized by demyelinating lesions but rather meningeal and subpial inflammation, perivascular parenchymal cuffing with adjacent necrotic lesions. This finding suggested that other antigens or immunocompetent cells were required to activate relapsing and remitting demyelinating disease.

In a study performed to correlate neuroimaging findings and histopathological findings of disease or plaque activity, Bordetella pertussis was eliminated from the vaccination, resulting in a chronic disease of moderate severity ('t Hart *et al.,* 2000) In this chronic form of EAE, actively demyelinating, inactive demyelinated, and remyelinated plaques could be identified by MRI and brain pathology. T2-weighted images identified active and inactive remyelinating plaques and were able to reliably correlate assessment of total plaque load on pathological examination. Gadolinium enhancement was only identified in active lesions. Hence, nonhuman primates can provide a neuroimaging model for chronic relapsing and remitting or chronic progressive MS in humans. Such a model may be used to test the efficacy and safety of immunosuppressive and immunomodulatory therapies.

Epilepsy

Epilepsy is a condition of repetitive unprovoked seizures (Adams and Victor, 1993). Seizures are episodic changes in behavior associated with an electrical discharge synchronizing populations of neurons in the cerebral cortex. To classify human epilepsies as focal or generalized, clinicians rely on a seizure description combined with electroencephalography (EEG). As seizures are rarely recorded in brief EEG samples, clinicians rely on detection of interictal (between seizures) discharges, which serve as markers for a type of epilepsy. Focal epilepsies generally begin with focal symptomatology and are associated with interictal discharges that are focal or lateralized to one hemisphere. Generalized epilepsies are associated with bilateral motor symptoms at onset and interictal discharges that involve both hemispheres simultaneously. Focal epilepsies are generally symptomatic, or related to a localized structural lesion, while most generalized epilepsies are idiopathic, and hence are probably inherited. Photosensitivity describes an enhanced response of the brain to a photic stimulus, such as flickering lights. Photosensitivity is rare, but more frequently encountered in idiopathic generalized epilepsies.

Natural and experimental models of epilepsy have been extensively studied in nonhuman primates. The red baboon (*Papio hamadryas papio*) is a natural model of generalized epilepsy associated with photosensitivity. The discovery of photosensitivity in these animals led to extensive research into the origins and propagation of epileptic seizures. Both generalized and focal epilepsy models were generated experimentally. These models became the vehicle for the electroclinical correlation of ictal motor behaviors, the substrate to study the mechanisms underlying focal and generalized discharges, and for the testing of seizure medications.

Experimental models of generalized epilepsy

The first experimental primate models of absence epilepsy were adapted from the cat. The epileptic discharges were induced by a 1% solution of conjugated estrogens applied bilaterally to the cortex of rhesus monkeys (Marcus *et al.,* 1968). The resulting 2.5–3 Hz generalized spike- and-wave discharges resembled, morphologically, the discharges recorded in humans more closely than the cat model. Seizures and discharges were increased in severity and frequency initially, including myoclonic and generalized tonic-clonic seizures, in addition to absence seizures, but occurred at a more stable rate over time. These discharges became more prolonged with hyperventilation induced by increasing resistance to an open end of a T tube attached to a tracheostomy tube. A chronic model of absence epilepsy was developed using cobalt powder applied to the pial surface bilaterally (Marcus *et al.,* 1972). The advantage of this preparation was that the applications were made through burr holes and that the animals could be monitored chronically.

Experimental models of focal epilepsy

The focal cortical application of aluminum hydroxide gel, in the rhesus monkey, was the basis of an experimental model of focal epilepsy (Kopeloff *et al.,* 1942; Ward, 1969). This model was later adapted by Joan Lockard and her laboratory, who aimed to bridge the gap between rodent models and human epilepsies (Lockard, 1980). They established methods for continuous polygraphic monitoring of seizures, EEG sampling of interictal discharges and monitoring of drug pharmacokinetics. Animals began to exhibit interictal discharges and focal seizures 6–10 weeks after subpial injection of 0.2 cc of aluminum hydroxide in the left pre- and postcentral gyrus (Lockard, 1980). Thereafter, the seizures became more frequent and severe, evolving into generalized tonic-clonic seizures. Eventually, the seizure frequencies stabilized. Each monkey was implanted with an EEG plug comprising an 8-channel, 9 skull-screw electrodes. The animals could be maintained for prolonged periods, extending beyond 12 months in one study (Lockard and Levy, 1976). Removal of the alumina focus and local gliotic tissue rendered the animals seizure free, even in the setting of spike and seizure propagation to the contralateral hemisphere (Harris and Lockard, 1981).

One application of this model was to compare the efficacy of antiepileptic medication for the treatment of partial and secondarily generalized tonic-clonic seizures. The efficacy of controlling these two seizure types was related to the dose of the medication. Valproic acid and clonazepam controlled secondarily generalized tonic-clonic seizures at low therapeutic doses, while partial or focal motor seizures were controlled only at higher therapeutic doses (Lockard *et al.,* 1977; Lockard *et al.,* 1979). The use of phenytoin and phenobarbital as prophylactic agents to prevent the evolution of seizures was not successful (Lockard and Levy, 1976). However, treated animals developed less severe seizures than their untreated counterparts, exhibiting only focal motor symptoms. The greatest reduction of seizure frequency was noted in animals receiving high-dose therapies, even after the medications were discontinued. These results supported the use of antiepileptic medication for prophylaxis, but their efficacy at preventing or modifying the risk of epilepsy may require high therapeutic doses. The prolonged effect of valproic acid, lasting for weeks after cessation of the medication was first demonstrated in this model (Lockard and Levy, 1976).

This monkey model was also used to study potential mechanisms underlying focal epilepsy (Bakay and Harris, 1981; Ribak *et al.,* 1985; Ribak *et al.,* 1989; Houser *et al.,* 1986). These authors evaluated the role of the inhibitory neurotransmitter, γ-aminobutyric acid (GABA), in the evolution of the epileptic focus. These studies employed immunocytochemistry for glutamate decarboxylase (GAD), the enzyme responsible for formation of GABA, which was colocalized in basket and chandelier cells in the monkey cortex. GAD-positive cells were reduced by 25–50% in the epileptic foci, but their reduction was less severe in adjacent cortices that demonstrated spiking on electrocorticography. The loss of GAD-positive cells did not only progress, but was evident even before the evolution of seizures, suggesting that the loss of inhibitory cells predisposes to the expression of spikes and seizures.

The alumina gel model was also modified to develop a model for temporal lobe epilepsy in rhesus monkeys (Ribak *et al.,* 1999). Stereotactically guided injections of the toxin were delivered into the hippocampus, amygdala, rhinal cortices, and the lateral temporal neocortex (Ribak *et al.,* 1998). Complex partial seizures, characterized by an unresponsive stare, head turning, oral and manual automatisms, were noted in all animals, except for those receiving the neocortical injections. Seizures frequently generalized secondarily and rapidly progressed into status epilepticus. Clinical seizure activity evolved within 12 to 14 days after injections into the hippocampus and dentate gyrus. Complex partial seizures appeared within 2 to 3 weeks of rhinal injections, but only 3 to 6 weeks after amygdalar injections. Light microscope examination of the temporal lobe structures, after the animals were sacrificed, demonstrated variability in cell loss and plasticity. While the hippocampal injections produced injury predominantly within the vicinity of the toxin, injections in the rhinal cortices caused mossy fiber sprouting without severe neuronal loss or gliosis in the hippocampus. Amygdalar injections produced changes most closely resembling the pathology underlying with chronic temporal lobe epilepsy in humans, including cell loss in the CA1 region of the hippocampus, dentate gyrus, and layer III of the entorhinal cortex, dendritic swelling and degeneration in CA3, as well as gliosis and mossy fiber sprouting in the dentate gyrus. Similar models of temporal lobe epilepsy were developed by induction of status epilepticus in pig-tailed monkeys (*Macaca nemestrina*) (Gunderson *et al.,* 1999; Wenzel *et al.,* 2000) with unilateral entorhinal infusions of bicuculline. These models may not only add to our knowledge about pathophysiological changes leading to temporal lobe epilepsy in humans, but could help to test new medical or surgical therapies.

Natural model of generalized, photosensitive epilepsy

In contrast to the experimental models of Parkinson's disease, multiple sclerosis, and epilepsy, the baboon is unique in that it represents a natural model for a generalized photosensitive epilepsy in humans (Naquet and Valin, 1998). The epilepsy of the red baboon (*Papio hamadryas papio*), which originates from the West African Coast, has been studied over four decades. *P. h. papio* has rare generalized myoclonic and tonic-clonic seizures (Naquet and Meldrum, 1972; Killam, 1979; Menini and Silva-Barrat, 1998). Generalized interictal epileptic discharges have been reported in EEG studies, and were maximal over the frontocentral regions. They tended to occur in relaxed wakefulness or light sleep, and were diminished when the baboon was aroused or attentive. Photosensitivity is prevalent in 10–60% of the baboons, and varies according to the region of origin. Photoepileptic responses were most reliably identified in baboons older than 2 years of age. Clinical symptoms induced by photic stimulation appear within seconds of the epileptic discharges, and are characterized by bilateral clonic activity of the eyelids and periorbital musculature followed by the lower face and neck, and occasionally of the entire body (Naquet and Valin, 1998; Killam, 1979; Menini and Silva-Barrat, 1998). Hyperventilation, overexertion, heat, and restraint can also exacerbate seizures in this species (Menini and Silva-Barrat, 1998; Serbanescu *et al.,* 1973). Circadian rhythms also affect seizure threshold with photosensitivity being maximally exhibited in the morning (Naquet and Valin, 1998; Killam, 1979; Serbanescu *et al.,* 1973; Ehlers and Killam, 1980). The role of sex hormones in the predisposition of seizures or photosensitivity in female baboons has not been studied, but may play an important role in the observed fluctuations in the response to photic stimulation (Wada *et al.,* 1972). Antiepileptic medications, such as benzodiazepines or barbiturates, suppress interictal discharges and photosensitivity, whereas proconvulsants, such as allyl-glycine, which inhibits GABA synthesis, and bicuculline, a GABA receptor blocker, exacerbated spontaneous seizures and aggravated photosensitivity (Meldrum *et al.,* 1975; Menini and Silva-Barrat, 1998).

The electroclinical features of the generalized epilepsy and photosensitivity of *P.h. papio* are similar to those seen in humans, specifically to juvenile myoclonic epilepsy (Menini and Silva-Barrat, 1998). The similarities include the presence of generalized interictal discharges and rare spontaneous seizures that are brought on by awakening, stress or hyperventilation, and photosensitivity and the expression of these varies with age and gender. One difference between baboon and human epilepsy is the occipital predominance of the photic activation in humans compared to the frontocentral predominance in baboons.

The baboon has also been studied as an electrophysiological model of photosensitive epilepsy. As implantation of intracerebral electrodes and lesion studies are not clinically justified in humans with generalized epilepsies, the baboon provided an opportunity to investigate the generation and propagation of afterdischarges related to photic stimulation. Frontocentral afterdischarges appear within 40 milliseconds of the stimulus, suggesting a subcortically mediated activity. Nonetheless, bilateral occipital lobe ablation completely suppresses activation of frontocentral afterdischarges while destruction of the superior colliculus or pulvinar unilaterally only transiently affects the bilateral appearance of frontocentral discharges, suggesting the importance of cortical generators (Menini and Silva-Barrat, 1998). Unilateral stimulation of the occipital lobe, after sectioning of the corpus callosum, results in frontocentral afterdischarges and seizures that remain lateralized to the same hemisphere (Fukuda *et al.,* 1988). Similar results were achieved with monocular stimulation following destruction of the temporal retinal hemifield (Fukuda *et al.,* 1989). Asynchronous, but bilateral, discharges were observed when stimulating the healthy eye in the same animal. Although frontocentral discharges only propagate to subcortical structures secondarily, the centrum medianum of the thalamus (Arfel *et al.*, 1972) and the hypothalamus (Riche, 1973) may be important in the manifestation of seizures.

Other baboon subspecies also demonstrate a photosensitive generalized epilepsy (Killam *et al.,* 1967; Corcoran *et al.,* 1979; Ticku *et al.,* 1991; Szabo *et al.,* 2004). The prevalence of photosensitivity appears to be lower in the olive (*P. h. anubis*) and yellow baboons (*P. h. cynocephalus*) than in the red baboon (Killam *et al.,* 1967). Investigation of other cercopithecanae, including *Erythrocebus patas, Cercopithecus aethiops sebaeus,* and *Macaca mulatta,* revealed that photoepileptic responses were rare (Killam *et al.,* 1966). Chimpanzees were not found to be photosensitive (Naquet *et al.,* 1967).

Due to its phylogenetic proximity to humans, the baboon may represent one of the best models for idiopathic human epilepsies (Naquet and Valin, 1998). Unfortunately, no genetic studies have been attempted to define the genes underlying its epilepsy. This is in a large part due to lower birth rates, longer pregnancies and slower maturation of primates as compared to rodents. In order to perform genetic linkage studies, large populations of pedigreed animals are required and few colonies have maintained pedigrees for this purpose. The Southwest Foundation for Biomedical Research in San Antonio maintains such a colony, containing over 1,500 baboons of *P. h. anubis* and its hybrid species with *P. h. cynocephalus, Hamadryas anubis* and *P. h. papio,* which have been pedigreed and genotyped (Vandeberg and Williams-Blangero, 1997; Rogers and Hixson, 1997). Currently, an epidemiological study is underway to characterize the epileptic and subclinical electroencephalographic phenotypes and their heritability exhibited in seizure pedigrees (Szabo *et al.,* 2004).

Summary

Nonhuman primate models have been instrumental in the evaluation of underlying pathophysiological mechanisms and testing medical and surgical therapies for neurological diseases in humans. Because of technological improvements and the increasing role of genetics research, rodent models have reduced the need for research in nonhuman primates. Nonetheless, due to the similarities of cerebral organization and neurotransmitter pathways between humans and nonhuman primates, some primate models are irreplaceable. These include the MPTP model for Parkinson's disease, the EAE model for multiple sclerosis, and the experimental models for focal epilepsy. The natural models for aging and cerebral amyloidosis will gain increasing importance in dementia research. Future research may apply current neuropsychological testing batteries to explore the cognitive and behavioral correlates of cerebral amyloidosis, EAE and epilepsy models. Recent developments in neuroimaging will allow *in vivo* studies in nonhuman primates, especially in larger primates such as baboons or macaques (Genain, 1999; Black *et al.,* 2001a; Black *et al.,* 2001b; Greer *et al.,* 2002). Functional magnetic resonance imaging and radioligand positron emission tomography will contribute to our knowledge of functional connectivity of cortical and subcortical regions, the pathophysiological changes underlying neurologic diseases, disease progression and the efficacy of therapeutic interventions in these diseases (Subramanian *et al.,* 1997; Doudet *et al.,* 1998; Stefanacci *et al.,* 1998; Disbrow *et al.,* 1999; Newsome and Stein-Aviles, 1999).

Acknowledgements

I would like to thank Jeff T. Williams and R. Stanley Burns for reviewing this chapter, and Koyle Knape for his help with the literature search, tables and figures.

Correspondence

Any correspondence should be directed to Charles Szabo, Dept of Medicine/Neurology, University of Texas Health Science Center at San Antonio, 7703 Floyd Curl Drive, San Antonio, TX 78284-7883, USA. Email: caszabo@university-health-sys.com

References

Adams, R.D. and Victor, M. (1993) *Principles of Neurology, Fifth Edition.* McGraw-Hill, New York.

Aebischer, P., Goddard, M., Signore, A.P. and Timpson, R.L. (1994). *Exp. Neurol.* 126, 151–158.

Aggleton, J.P. and Mishkin, M. (1983). *Neuropsychologia* 21, 189–197.

Albanese, A., Granata, R., Gregori, B., Piccardi, M.P., Colosimo, C. and Tonali, P. (1993). *Neuroscience* 55, 823–832.

Andringa, G., Stoof, J.C. and Cools, A.R. (1999). *Psychopharmacology* 146, 328–334.

Annett, L.E., Maartel F.L., Rogers, D.C., Ridley R.M., Baker, H.F. and Dunnett, S.B. (1993). *Exp. Neurol.* 125, 228–246.

Arfel, G., Walter, S. and Christolomme, A. (1972). *Rev. EEG Neurophysiol. Clin.* 2, 208–212.

Bachevalier, J., Landis, L.S., Walker, L.C., Brickson, M., Mishkin, M., Price, D.L. and Cork, L.C. (1991). *Neurobiol. Aging* 12, 99–111.

Bakay, R.A., Barrow D.L., Fiandaca M.S., Iuvone P.M., Schiff, A. and Collins, D.C. (1987). *Ann. N.Y. Acad. Sci.*, 495, pp 623–640.

Bakay, R.A. and Harris, A.B. (1981). *Brain Res* 206, 387–404.

Bakay, R.A., Boyer, K.L., Freed, C.R. and Ansari, A.A. (1998). *Cell Transplant.* 7, 109–120.

Baker, H.F., Ridley, R.M., Duchen, L.W., Crow, T.J. and Bruton, C.J. (1993). *Int. J. Exp. Path.* 74, 441–454.

Bankiewicz, K.S., Oldfield, E.H., Chiueh, C.C., Doppman, J.L., Jacobowitz, D.M. and Kopin, I.J. (1986). *Life Sci.* 39, 7–16.

Bankiewicz, K.S., Plunkett, R.J., Jacobowitz, D.M., Kopin I.J. and Oldfield, E.H. (1991). *J. Neurosurg.* 74, 97–104.

Bedard, P.J., Di Paolo, T., Falardeau, P. and Boucher, R. (1986). *Brain Res.* 379, 294–299.

Bedard, P.J., Gomez-Mancilla, B., Blanchette, P., Gagnon, C., Falardeau, P. and DiPaolo, T. (1993). *Adv. Neurol.* 60, 113–118.

Benson, D.F., Kuhl, D.E., Hawkins R.A., Phelps, M.E., Cummings J.L. and Tsai, S.Y. (1983). *Arch. Neurol.* 40, 711–714.

Black, K.J., Snyder, A.Z., Koller, J.M., Gado, M.H. and Perlmutter, J.S. (2001a). *NeuroImage* 14, 736–743.

Black, K.J., Koller, J.M., Snyder, A.Z. and Perlmutter, J.S. (2001b). *NeuroImage* 14, 744–748.

Blanchet, P., Bedard, P.J., Britton, D.R. and Kebabian, J.W. (1993). *J. Pharm. & Exp. Ther.* 267, 275–259.

Bons, N., Mestre, N., Ritchie, K., Peters, A., Podlisny, M. and Selkoe, D. (1994). *Neurobiol. Aging* 15, 215–220.

Boyce, S., Rupniak, N.M., Steventon, M.J. and Iversen, S.D. (1990). *Psychopharmacology* 102, 21–27.

Braak, H. and Braak, E. (1991). *Acta Neuropathologica* 82, 239–259.

Brozoski, T.J., Brown R.M., Rosvold H.E. and Goldman P.S. (1979). *Science* 205, 929–932.

Burns, R.S., Chiueh, C.C., Markey, S.P. Ebert, M.H., Jacobowitz, D.M. and Kopin, I.J. (1983). *Proc. Nat. Acad. Sci. USA.* 80, 4546–4550.

Callahan, M.J. (1999). *Psychopharmacology* 144, 234–238.

Corcoran, M.E., Cain, D.P. and Wada, J.A. (1979). *Le J. Can. Sciences Neurologiques* 6, 129–131.

Crossman, A.R., Peggs, D., Boyce, S., Luquin, M.R. and Sambrook, M.A. (1989). *Neuropharmacology* 28, 1271–1273.

Cutler, N.R., Haxby J.V., Duara, R., Grady, C.L., Kay, A.D., Kessler R.M., Sudaram, M. and Rapoport, S.I. (1985). *Ann. Neurol.* 18, 298–309.

Davis, R.T. and Ruggiero, F.T. (1973). *Am. J. Physiol. Anthrop.* 38, 573–580.

Disbrow, E., Roberts, T.P.L., Slutsky, D. and Krubitzer, L. (1999). *Brain Res.* 829, 167–173.

Doudet, D.J., Chan, G.L.Y., Holden, J.E., McGeer, E.G., Aigner T.A., Wyatt, R.I. and Ruth T.J. (1998). *Synapse* 29, 225–232.

During, M.J., Samulski, R.J., Elsworth, J.D., Kaplitt, M.G., Leone, P., Xiao, X., Li, J., Freese, A., Taylor, J.R., Roth, R.H., Sladek, J.R., O'Malley, K.L. and Redmond, D.E., Jr. (1998). *Gene Ther.* 5, 820–827.

Eacott, M.J., Gaffan, D. and Murray, E.A. (1994). *Eur. J. Neurosci.* 6, 1466–1478.

Ehlers, C.L. and Killam, E.K. (1980). *Am. J. Phys.* 38 (Regulatory Integrative Comp. Physiol. 8), R35–41.

Ellis, J.E., Byrd, L.D. and Bakay, R.A. (1992). *Exp. Neurol.* 115, 376–387.

Elsworth, J.D., Deutch, A.Y., Redmond, D.E., Sladek, J.R., Jr. and Roth, R.H. (1987). *Life Sci.* 40, 193–202.

Erickson, C.A. and Barnes, C.A. (2003). *Exp. Gerontology* 38, 61–69.

Fahn, S. (1988). In Byngaarden, J.B. and Smith, L.H. (eds) *Cecil's Textbook of Medicine.* W.B. Saunders, Philadelphia.

Fernandez-Ruiz, J., Doudet, D. and Aigner, T.G. (1999). *Psychopharmacology* 147, 104–107.

Ferraro, A. and Cazzullo, C.L. (1948). *J. Neuropath. and Exp. Neurol.* 7, 235–260.

Forno, L.S. (1996). *J. Neuropath. Exp. Neurol.* 55, 259–272.

Freund, J., Stern, E.R. and Pisani T.M. (1947). *J. Immunology* 57, 179–194.

Friedman, H.R., Janess, J. and Goldman-Rakic, P.S. (1990). *J. Cognit. Neurosci.* 2, 18–31.

Friedman, H.R. and Goldman-Rakic, P.S. (1988). *J. Neurosci.* 8, 4693–4706.

Friedman, H.R. and Goldman-Rakic, P.S. (1994). *J. Neurosci.* 14, 2775–2788.

Fukuda, H., Valin, A., Bryere, P., Riche, D., Wada J.A. and Naquet, R. (1988). *Electroencephalogr. Clin. Neurophysiol.* 69, 363–370.

Fukuda, H., Valin, A., Menini, C., Boscher, C., de la Sayette, V., Riche, D., Kunimoto, M., Wada, J.A. and Naquet, R. (1989). *Epilepsia* 30, 623–630.

Gaffan, D. (1994). *J. Cogn. Neurosci.* 6, 305–320.

Gaffan, D. and Parker, A. (2000). *Brain* 123, 816–827.

Gagnon, C., Bedard, P.J. and Di Paolo, T. (1990). *Eur. J. Pharmacology* 78, 115–120.

Galdiczki, Z., Fukuyama, R, Wadhwani, K.C., Rapoport, S.I. and Ehrenstein, G. (1994). *Brain Res.* 646, 332–336.

Gearing, M., Rebeck G.W., Hyman, B.T., Tigges, J. and Mirra, S.S. (1993). *Proc. Natl. Acad. Sci. USA* 91, 9382–9386.

Genain, C.P., Lee-Paritz, D., Nguyen, M.H., Massacesi, L., Joshi, N., Ferrante, R., Hoffman, K., Moseley, M., Letvin, N.L. and Hauser, S.L. (1994). *J. Clin. Invest.* 94, 1339–1345.

Genain, C.P. and Hauser, S.L. (1997). *J. Mol. Med.* 75, 187–197.

Genain, C.P. (1999). *AJNR.* 20, 955–956.

Giguere, M. and Goldman-Rakic, P.S. (1988). *J. Comp. Neurol.* 277, 195–213.

Goetz, C.G., Stebbins, G.T., Shale, H.M., Lang, A.E., Chernik, D.A., Chmura, T.A., Ahlskog, J.E. and Dorflinger, E.E. (1994). *Movement Disorders* 9, 390–394.

Goldman-Rakic, P.S. and Porrino, L.J. (1985). *J. Comp. Neurol.* 242, 535–560.

Goldman-Rakic, P.S. (1996). *Proc. Natl. Acad. Sci. USA.* 93, 13473–13480.

Goulet, M. and Madras, B.K. (2000). *J. Pharmacology & Exp. Ther.* 292, 714–724.

Greer, P.J., Villemagne, V.L., Ruszkiewicz, J., Graves, A.K., Meltzer, C.C., Mathis, C.A. and Price, J.C. (2002). *Brain Res. Bull.* 58, 429–438.

Gunderson, V.M., Dubach, M., Szot, P., Born, D.E., Wenzel, H.J., Maravilla, K.R., Zierath, D.K. and Robbins, C.A. (1999). *Dev. Neurosci.* 21 352–364.

Haber, J.S. (1986). *Human. Neurobiol.* 5, 159–168.

Harder, J.A., Aboobaker, A.A., Hodgetts, T.C., Ridley, R.M. (1998). *Br. J. Pharmacology* 125, 1013–1018.

Harder, J.A. and Ridley, R.M. (2000). *Neuropharmacology* 39, 547–552.

Harlow, H.F. (1944). *J. Gen. Psychology* 30, 3–12.

Harris, A.B. and Lockard, J.S. (1981). *Epilepsia* 22, 107–122.

Houser, C.R., Harris, A.B. and Vaughn, J.E. (1986). *Brain Res.* 383, 129–145.

Henry, B., Fox, S.H., Peggs, D., Crossman, A.R. and Brotchie, J.M. (1999). *Movement Disorders* 14, 744–753.

Jenner, P. (2000). *Ann. Neurol.* 47 (4 Suppl. 1), S90-9; discussion S99–104.

Kanda, T., Jackson, M.J., Smith, L.A., Pearce, R.K.B., Nakamura, J., Kase, H., Kuwara, Y. and Jenner, P. (1998). *Ann. Neurol.* 43, 507–513.

Kar, S., Iss, A.M., Seto, D., Auld, D.S., Collier, B. and Quirion, R. (1998). *J. Neurochem.* 70, 2179–2187.

Kemper, T.L., Blatt, G.J., Killiany R.J. and Moss, M.B. (2001). *Acta Neuropathologica* 101, 145–153.

Keuker, J.I., Luiten, P.G. and Fuchs, E. (2003). *Neurobiol. Aging* 24, 157–165.

Killam, E.K. (1979). *Federation Proc.* 38, 2429–2433.

Killam, K.F., Killam, E.K. and Naquet, R. (1966) *C. R. Acad. Sci.* 262, 1010–1012.

Killam, E.K., Stark, L.G. and Killam, K.F. (1967). *Life Sci.* 6, 1569–1574.

Kiyosawa, M., Pappata, S., Duverger, D., Riche, D., Cambon, H., Mazoyer, B., Samson, Y., Crouzel, C., Naquet, R., MacKenzie, E.T. and Baron, J-C. (1987). *J. Cer. Blood Flow & Metab.* 7, 812–817.

Koller, W.C. (1992). *Neurology* 42, 27–31.

Kopeloff, L.M., Barrera, S.E. and Kopeloff, N. (1942). *Am. J. Psychiatry* 98, 881–902.

Kordower, J.H., Bloch, J., Ma, S.Y., Chu, S.Y., Palfi, S., Roitberg, B.Z., Emborg, M., Hantraye, P., Deglon, N. and Aebischer, P. (1999). *Exp. Neurol.* 160, 1–16.

Langston, J.W., Quik, M., Petzinger, G., Jakowec, M. and Di Monte, D.A. (2000). *Ann. Neurol.* 47 (Suppl. 1), S79–S89.

Lindner, M.D. and Emerich, D.F. (1998). *Cell Transplant.* 7, 165–174.

Lockard, J.S. (1980). In Lockard, J.S. and Ward Jr., A.A. (eds.) *Epilepsy: A Window to Brain Mechanisms,* pp 11–49, Raven Press, New York.

Lockard, J.S. and Levy, R.H. (1976). *Epilepsia* 17, 477–479.

Lockard, J.S., Levy, R.H., Congdon, W.C., DuCharme, L.L. and Patel, I.H. (1977). *Epilepsia* 18, 205–224.

Lockard, J.S., Levy, R.H., Congdon, W.C., DuCharme, L.L. and Salonen, L.D. (1979). *Epilepsia* 20, 683–695.

Mahut, H., Moss, M. and Zola-Morgan, S. (1981). *Neuropsychologia* 19, 201–225.

Mair, W.G.P., Warrington, E.K. and Weiskrantz, L. (1979). *Brain* 102, 749–783.

Majocha, R.E., Reno, J.M., Friedland, R.P., VanHaight, C., Lyle, L.R. and Marotta, C.A. (1992). *J. Nucl. Med.* 33, 2184–2189.

Malamut, B.L., Saunders, R.C. and Mishkin, M. (1984). *Behavioral Neurosci.* 98, 759–769.

Marcus, E.M., Watson, C.W. and Simon, S.A. (1968). *Epilepsia* 9, 233–248.

Marcus, E.M., Fullerton, A., Losh, E. and Bowker, R. (1972). *Epilepsia* 13, 343–344.

Meguro, K., Blaizot, X, Kondoh, Y., Le Mestric, C., Baron, J-C., Chavoix and C. (1999). *Brain* 122, 1519–1531.

Meldrum, B.S., Anlezark, G., Balzamo, E., Horton, R. and Trimble M. (1975). *Adv. Neurol.* 10, 117–128.

Menini, C. and Silva-Barrat, C. (1998). In Zifkin, B.J., Andermann, F.A., Beaumanoir, A. and Rowan, A.J. (eds) *Reflex Epilepsies and Reflex Seizures: Advances in Neurology, Vol. 75,* pp 29–47. Lippincott-Raven Publishers, Philadelphia.

Merrill, D.A., Roberts, J.A. and Tuszynski, M.H. (2000). *J Comp. Neurol.* 342, 396–401.

Mishkin, M. (1978). *Nature* 273, 297–298.

Mitchell, I.J. and Carroll, C.B. (1997). *Neurosci. & Biobehavioral Rev.* 21, 469–475.

Mones, R.J. (1972). *Mt. Sinai. J. Med.* 39, 197–201.

Morelli, L., Wei, L., Amorim, A., McDermid, J., Abee, C.R., Frangione, B., Walker, L.C. and Levy, E. (1996). *FEBS. Lett.* 379, 132–134.

Murray, E.A. and Mishkin, M. (1998). *J. Neurosci.* 18, 6568–6582.

Nakamura, S., Tamaoka, A., Sawamura, N., Shoji, S., Nakayama, H., Ono, F., Sakakibara, I., Yoshikawa, Y., Mori, H., Goto, N. and Doi, K. (1995). *Neurosci. Lett.* 201, 151–154.

Naquet, R.G., Killam, K.F. and Rhodes, J.M. (1967). *Life Sci.* 6, 1575–1578.

Naquet, R.G., Cartier, J. and Menini, C. (1975). *Adv. Neurol.* 10, 107–118.

Naquet, R., and Meldrum, B.S. (1972). In Purpura, D.P., Penry, J.K., Tower, D.B., Woodbury, D.M. and Walter, R.D. (eds) *Experimental Models of Epilepsy,* pp 373–406, Raven Press, New York.

Naquet, R.G. and Valin, A. (1998). In Zifkin, B.J., Andermann, F., Beaumanoir, A. and Rowan, J. (eds) *Reflex Epilepsies and Reflex Seizures: Advances in*

Neurology, Vol. 75, pp 15–28, Lippincott-Raven Publishers, Philadelphia.
Newsome, W.T. and Stein-Aviles, J.A. (1999). *I.L.A.R. Journal* 40, 78–91.
Nutt, J.G. (1990). *Neurology* 40, 340–345.
Papa, S.M. and Chase, T.N. (1996). *Ann. Neurol.* 39, 574–578.
Parker, A., Eacott, M.J. and Gaffan, D. (1997). *Eur. J. Neurosci.* 9, 2423–2431.
Parker, A. and Gaffan, D. (1998). *Eur. J. Neurosci.* 10, 3044–3057.
Pawlik, M., Fuchs, E., Walker, L.C. and Levy, E. (1999). *Neurobiol. Aging* 20, 47–51.
Pearce, R.K., Banerji, T., Jenner, P. and Marsden, C.D. (1998). *Movement Disorders* 13, 234–241.
Pearce, R.K., Jackson, M., Britton, D.R., Shiosaki, K., Jenner, P. and Marsden, C.D. (1999). *Psychopharm.* 142, 51–60.
Pearce, R.K., Jackson, M., Smith, L., Jenner, P. and Marsden, C.D. (1995). *Movement Disorders* 10, 731–740.
Ramirez, M.J., Ridley, R.M., Baker, H.F., Maclean, C.J., Honer, W.G. and Francis, P.T. (2001). *J. Neural Transm.* 108, 809–826.
Rapoport, S.I. (1991). *Brain Res. Rev.* 15, 267–294.
Rapp, P.R. and Amaral, D.G. (1991). *Neurobiol. Aging* 12, 481–486.
Ribak, C.E., Hunt, C.A., Bakay, R.A.E. and Oertel, W.H. (1985). *Brain Res.* 363, 78–90.
Ribak, C.E., Joubran C., Kesslak, J.P. and Bakay, R.A.E. (1989). *Epilepsy Res.* 4, 126–138.
Ribak, C.E., Seress, L., Weber, P., Epstein, C.M., Henry, T.R. and Bakay, R.A.E. (1998). *J. Comp. Neurol.* 401, 266–290.
Ribak, C.E. and Bakay, R.A.E. (1999). In Delgado-Escueta, A.V., Wilson, W.A., Olsen, R.W. and Porter, R.J., *Jasper's Basic Mechanisms of the Epilepsies: Advances of Neurology, Volume 79*, pp 737–741. Lippincott Williams & Wilkins, Philadelphia.
Riche, D. (1973). *J. Hirnforschung* 14, 527–547.
Ridley, R.M., Murray, T.K., Johnson, J.A. and Baker, H.F. (1986). *Brain Res.* 376, 108–116.
Rivers, T.M. and Schwenker, F.F. (1933). *J. Exp. Med.* 61, 689–702.
Rogers, J. and Hixson, J.E. (1997). *Am. J. Hum. Genet.* 61, 489–493.
Sandbrink, R. and Beyreuther, K. (1996). *Mol. Psychiatry* 1, 438–444.
Sani, S., Traul, D., Klink, A., Niaraki, N, Gonzabo-Ruiz, A., Wu, C.K. and Geula, C. (2003). *Acta Neuropathologica* 105, 145–156.
Scoville, W.B. and Milner, B. (1957). *J. Neurol. Neurosurg. Psychiatry* 20, 11–21.
Serbanescu, T., Naquet, R. and Menini C. (1973). *Brain Res.* 52, 145–158.
Sibal, L.R. and Samson, K.J. (2001). *I.L.A.R. Journal* 42, 74–84.
Sladek, J.R., Jr., Collier, T.J., Haber, S.N., Roth R.H. and Redmond, D.E., Jr. (1986). *Brain Res. Bull.* 17, 809–818.
Squire, L.R. and Zola-Morgan, S. (1983). In Deutsch, J.A. (ed.) *The Physiological Basis of Memory: Second Edition*, pp 200–268. New York Academy of Sciences, New York.
Squire, L.R. and Zola-Morgan, S. (1991). *Science* 253, 1380–1386.
Stefanacci, L., Reber, P., Costanza, J., Wong, E., Buxton, R., Zola, S., Squire, L. and Albright, T. (1998). *Neuron* 20, 1051–1057.
Strittmatter, W.J., Saunders A.M. and Schmechel, D. (1993). *Proc. Natl. Acad. Sci.* 90, 1977–1981.
Subramanian, T., Emerich, D.F., Bakay, R.A.E., Hoffman, J.M., Goodman, M.M., Shoup, T.M., Miller, G.W., Levey, A.I., Hubert, G.W., Batchelor, S., Winn, S.R., Saydoff, J.A. and Watts, R.L. (1997). *Cell Transplant.* 6, 469–477.
Szabo, C.A., Leland, M.M., Sztonak, L., Restrepo, S., Haines, R.J., Mahaney, M.C. and Williams, J.T. (2004). *Am. J. Primat.* 62, 95–106.
Taylor, J.R., Elsworth, J.D., Roth, R.H., Sladek, J.R., Collier, T.J. and Redmond, D.E., Jr. (1991). *Exp. Brain. Res.* 85, 335–348.
Taylor, J.R., Elsworth, J.D., Sladek, J.R., Jr., Collier, T.J., Roth, H.R. and Redmond, D.E., Jr. (1995). *Cell Transplant.* 4, 13–26.
t' Hart, B.A., Bauer, J., Muller, H.-J., Melchers, B., Nicolay, K., Bontrop, R.E., Lassmann, H. and Massacesi, L. (1998). *Am. J. Path.* 153, 649–663.
t' Hart, B.A., van Meurs, M., Brok, H.P.M., Massacesi, L., Bauer, J., Boon, L., Bontrop, R.E. and Laman, J.D. (2000). *Immunology Today* 21, 290–297.
Ticku, M.K., Lee, J.C., Murk, S., Mhatre, M.C., Story, J.L., Kagan-Hallett, K., Luther, J.S., MacCluer, J.W., Leland, M.M. and Eidelberg, E. (1991). In Engel. J., Jr. (ed.) *Molecular Neurobiology of Epilepsy, Epilepsy Res. Suppl.* 9, 141–149.
Tigges, J., Gordon, T.P., McClure, H.M., Hall, E.C. and Peters, A. (1998). *Am. J. Primat.* 15, 263–273.
Tolwani, R.J., Jakowec, M.W., Petzinger, G.M., Green, S. and Waggie, K. (1999). *Lab. Anim. Sci.* 49, 363–371.
Uno, H. and Walker, L.C. (1993). *Ann. N.Y. Acad. Sci.* 695, 232–235.
Vandeberg, J.L. and Williams-Blangero, S. (1997). *J. Med. Primatol.* 26, 113–119.
Victor, M., Adams, R.D. and Collins, G.H. (1971). *The Wernicke-Korsakoff Syndrome.* Blackwell, Oxford.
Wada, J.A., Teraro, A. and Booker, H.E. (1972). *Neurology* 22, 1272–1285.
Walker, M.L. (1995). *Am. J. Primat.* 35, 59–71.
Walker, L.C. (1997). *Brain Research – Brain Res. Rev.* 25, 70–84.
Ward, A.A., Jr. (1969). In Jaspers, H.A., Ward, A.A., Jr. and Pope, A. (eds) *Basic Mechanisms of the Epilepsies*, pp 13–35, Brown and Company, Boston.

Watts, R.L., Mandir, A.S. and Bakay, R.A. (1995). *Cell Transplant.* 4, 27–38.

Watts, R.L., Subramanian, T., Freeman, A., Goetz, C., Penn, R.D., Stebbins, G.T., Kordower, J.H. and Bakay, R.A.E. (1997). *Exp. Neurol.* 147, 510–517.

Weinstein, B. (1941). *J. Comp. Psychol.* 31,195–213.

Wenzel, H.J., Born, D.E., Dubach, M.F., Gunderson, V.M., Maravilla, K.R., Robbin, C.A., Szot, P., Zierath, D. and Schwartzkroin, P.A. (2000). *Epilepsia* 41 (Suppl. 6): S70–S75.

Widner, H., Tetrud, J., Rehncrona, S., Snow, B., Bruadin, P., Gustavii, B., Bjorklund, A., Lindvall, O. and Langston, J.W. (1992). *N.E.J.M.* 327, 1556–1563.

Zola-Morgan, S., Squire, L.R., Amaral, D.G. and Suzuki, W.A. (1989). *J. Neurosci.* 9, 4355–4370.

Zola-Morgan, S. and Squire, L.R. (1985). *Ann. Neurol.* 17, 558–564.

Zola-Morgan, S. and Squire, L.R. (1984). *J. Neurosci.* 4, 1072–1085.

CHAPTER 29

Genetics: A Survey of Nonhuman Primate Genetics, Genetic Management and Applications to Biomedical Research

Jeffrey Rogers
Department of Genetics, Southwest Foundation for Biomedical Research and Southwest National Primate Research Center, San Antonio, Texas 78227, USA

The goal of this chapter is to summarize current information concerning specific aspects of the genetics of nonhuman primates. Focus is placed on aspects of genetics that have direct impact on studies of laboratory primates. The review begins with a brief summary of information and resources related to the content and function of the primate genome. Next, studies related to genetic variation within species and genetic differences between species are discussed, and current ideas about the genetic management of captive colonies are reviewed. The final sections of the chapter discuss current trends and future opportunities for the genetic analysis of captive primates.

The analysis of primate genomes

Genome structure and content

During the past few years, tremendous advances have been made in our understanding of human genetics and genomics. Progress has also been rapid in studies of other mammalian genomes. The complete sequencing of the human genome (Intern. Human Genome Sequencing Consortium 2001; Venter *et al.,* 2001,

The Laboratory Primate
Copyright 2005 Elsevier
ISBN 0-1208-0261-9

http://www.ncbi.nlm.nih.gov/genome/guide/human), and the subsequent completion of the sequencing of the mouse (http://www.ncbi.nlm.nih.gov/genome/guide/mouse) and rat (http://www.ncbi.nlm.nih.gov/genome/guide/rat) genomes have generated remarkable new information about these species, and simultaneously opened new opportunities for future advances. Such data are not yet available on the same scale for nonhuman primates. Whole genome DNA sequence data will be published in the coming months for chimpanzees (*Pan troglodytes*), and the National Human Genome Research Institute of the U.S. National Institutes of Health has also given the rhesus macaque (*Macaca mulatta*) high priority for sequencing (http://www.genome.gov/10002154). Efforts already underway (http://www.hgsc.bcm.tmc.edu/) will likely produce whole genome sequence data for rhesus monkeys in two to three years. Both the chimpanzee and rhesus projects will yield DNA sequence data across the entire genome, but with less depth and redundancy than the human DNA sequence. Therefore, these projects will generate outstanding new information, but with somewhat reduced accuracy and completeness as compared to the human and mouse sequences.

As a result, in a short time, it will be possible to compare genome content and structure among three primate species, including humans. In addition, smaller but still substantial amounts of genomic information is being generated for a number of other primates (see for example the Genbank database at www.ncbi.nlm.nih.gov or other databases such as www.nisc.nih.gov/index). These results will undoubtedly transform the field of nonhuman primate genetics over the coming years. At present, generalizations about primate genomics are dominated by information from the human genome, and any general conclusions beyond that are premature and subject to major revision as new data rapidly become available.

Nevertheless, some basic information is appropriate here. It is clear that the human (and other mammalian) genomes consist of 30–40,000 genes interspersed with a complex array of sequences that do not code for protein. The complement of functional protein coding genes is likely to be very similar across primates, but important differences between species in the catalog of functional genes are already known (see for example Johnson *et al.*, 2001; Johnson *et al.*, 2002; Chou *et al.*, 2002). A currently unknown proportion of the rest of the genome, i.e. segments that do not code for functional proteins, do code for functional RNA molecules, act as regulatory sequences or serve other functions in the cell. Much of this non-protein-coding DNA is single copy, but about one-third of the human genome consists of repetitive elements of one type or another. Repetitive sequences are subject to different evolutionary forces than are coding sequences of genes, and different primate species have evolved different complements of repetitive elements. Though they are often assumed to have no effect on phenotype, some repetitive elements can affect function. One hypothesis is that insertions of Alu repeats have had significant effects on gene function through primate evolution (Stewart, 2000).

It is beyond the scope of this review to summarize the recent advances in understanding of the content or structure of the primate genome that have come from the completion of the human sequence. The reader is referred to the review papers cited above, and warned that this field of biology is developing at a rapid pace. Examples of genes that have been silenced in one species or another with possible evolutionary consequences (e.g. Chou *et al.,* 2002), or gene families that differ in gene number across species (e.g. Johnson *et al.,* 2001), are being identified regularly. Once large-scale analysis of the chimpanzee and rhesus genomes is possible, along with comparative information from other species, we should expect significant new insights in the genetic similarities and differences that occur across the order Primates.

Gene expression

One area of intensive research is the study of gene expression at the genomic (i.e. global) level. With various technologies, it is now possible to quantify the level of expression of thousands of genes in a single experimental sample of RNA. Looking forward, one of the exciting areas within primate genetics will certainly be the comparative analysis of gene expression. Closely related primate species show remarkably different phenotypes, either anatomical or physiological, while maintaining a basic complement of genes that seem to change only slowly. Evolutionary changes in patterns of gene expression are likely to explain many (though certainly not all) morphological and physiological differences among primates. Simultaneous analysis of quantitative expression of many genes has been performed in chimpanzees (Bigger *et al.,* 2001) and rhesus monkeys (e.g. Zou *et al.,* 2002; Mirnics *et al.,* 2000). This aspect of primate genetics is likely to grow substantially over the coming years.

Intra-specific variation

Studies too numerous to catalog here have demonstrated that nonhuman primate species exhibit substantial amounts of intra-specific genetic variation. The first genetic polymorphisms identified and analyzed in primates were blood group and protein polymorphisms. During the 1960s and 1970s researchers employed immunological methods to detect differences among animals, and gel-based electrophoresis was used to survey individual variation in the mobility of many serum proteins. Classic analyses of electrophoretic variation in primates were performed by Barnicot, Harris and others (see Barnicot *et al.,* 1965; Meera Khan, 1987; Stone *et al.,* 1993). These were the first empirical demonstrations that genetic variation is common among individuals within nonhuman primate species. The study of protein polymorphisms decreased dramatically once methods were developed to examine the more polymorphic, and thus more informative, differences in DNA sequences. But interest in proteins and protein evolution has increased again. Molecular biologists are now integrating data from all organizational levels from DNA through proteins and cells up to the level of whole organisms. Significantly expanded studies of variation in protein expression are possible because technologies now allow investigation of the presence, absence or variability of hundreds or thousands of proteins simultaneously (Rapsilber *et al.,* 2002; Zhu *et al.,* 2003). This work on proteins at a greatly expanded scale is referred to as proteomics, and will eventually be applied to large-scale analyses of protein-expression in nonhuman primates.

In the late 1980s, a number of laboratories began using multilocus DNA fingerprinting to investigate molecular differences among individuals within primate species (Inoue *et al.,* 1990; Ely *et al.,* 1991; Wickings, 1993). These methods can detect a substantial amount of genetic variability, but do not distinguish individual loci. Short segments of DNA that complement repetitive sequences in the primate genome are radioactively labeled, then hybridized to genomic DNA that has been digested with endonucleases and size-fractionated in agarose gels. The presence and relative sizes of the complementary fragments are revealed through autoradiography. This approach is effective because it detects variation at multiple loci simultaneously. By comparing the complex patterns of digested fragments across individual animals, one can readily identify individual differences. These methods were often used for paternity testing and can be very effective for that purpose, since a large amount of variation is identified by each experiment (i.e. by each radiolabeled probe). But the simultaneous detection of an unquantified number of independent and unidentified chromosomal segments makes these DNA fingerprints inappropriate for analyses that require specific genotypes at specific loci to be assigned to individual animals.

At the time of this writing, the most widely used and informative genetic markers (i.e. genetic polymorphisms) in primate research are microsatellites. These loci (which are sometimes called short tandem repeats, STRs, or simple sequence repeats, SSRs) are short segments of DNA that contain a series of tandem repeats. Most microsatellites have either dinucleotide or tetranucleotide repeats, in which the repeat units are two or four base pairs respectively. A number of trinucleotide and pentanucleotide repeat loci are also known. Several thousand microsatellites occur in the human genome (Kong *et al.,* 2002), and many have also been described in nonhuman primates (Di Fiore, 2003). These STR loci are highly informative in terms of genetic differences among individual animals because they have high mutation rates and can sometimes accumulate large numbers of alleles per locus. New mutations add or delete repeat units so that, over time, a microsatellite locus generates new alleles that differ in the number of repeat units.

Microsatellites are valuable for genetic studies because they occur frequently in the genome, and because many are highly polymorphic, exhibiting 10–15 alleles or more in a single primate population. Most of these loci (but not all) appear to have no functional effect, so the variation develops as the result of mutation, and the fate of newly generated alleles is governed by genetic drift. In captive colonies of primates, these loci are useful for pedigree analyses, especially for identifying the sires or confirming the dams of newborn infants. The standard approach for assaying variation among individuals for a given microsatellite locus is to identify single-copy DNA sequences on both sides of the variable stretch of repeats, and to design 20-basepair oligonucleotides that complement those single-copy sequences. These oligonucleotides can be used as primers in polymerase chain reaction (PCR) amplification reactions. Those primers are generally designed to amplify a segment of DNA 100–250 basepairs in length that contains the variable microsatellite repeats. Different alleles at the microsatellite will produce different length PCR products because the primers bind to specific single-copy sites, and the number of repeat units between those two sites varies from individual to individual. These different length PCR products are easily distinguished from one another in polyacrylamide gels or in polymer capillary tubes.

Microsatellites were first recognized in humans as a result of comparing DNA sequences across individuals (Weber and May, 1989; Litt and Luty, 1989). Subsequent cloning efforts identified a large number of polymorphic $(CA)_n$, $(GATA)_n$ and other repeat motifs

(Kong *et al.,* 2002). Since the publication of the complete human genome sequence, one can now search any particular chromosomal region for extended runs of di-, tri- or tetranucleotides and develop PCR primers to amplify these potentially variable loci.

There are three different approaches to the identification and characterization of microsatellite polymorphisms in nonhuman primate genomes. A number of investigators have cloned microsatellites directly from the species of interest (Inoue and Takenaka, 1993; Witte and Rogers, 1999; Lawlor *et al.,* 2001; see also Di Fiore, 2003). This approach is straightforward, but can be time consuming. It is much simpler to make use of available information about known microsatellites in other species. Given that thousands of microsatellites have been described in the human genome, it is simple to test PCR primers that amplify human microsatellites to determine whether they will also amplify informative microsatellites in other species. This approach was first used to identify polymorphisms in chimpanzees (Morin *et al.,* 1994; Deka *et al.,* 1994) and then in other species (Rogers *et al.,* 1995; Blanquer-Maumont and Crouau-Roy, 1995). Subsequently human PCR primers have proven useful for rhesus monkeys (Smith *et al.,* 2000; Hadfield *et al.,* 2001), vervets (Newman *et al.,* 2002), and other species (Witte and Rogers, 1999; Nair *et al.,* 2000). The more closely related a given species is to humans, the more likely human microsatellites and PCR primers are to be useful in that species. Most human STRs can be amplified in apes, and about 25% of human microsatellites can be amplified and are polymorphic in Old World monkeys (Morin *et al.,* 1998). The probability of success when using human PCR primers to amplify polymorphisms in New World monkeys or strepsirhines is significantly less than it is in Old World monkeys (Witte and Rogers, 1999).

New technologies make it practical to rapidly assay a class of DNA polymorphisms even more common in the genome than microsatellites. Single nucleotide polymorphisms (SNPs) are individual basepairs that have undergone mutation and thus have two (or in rare cases 3 or 4) alternative nucleotides present in a given population. This is the simplest form of genetic mutation, since it is nothing more than the substitution of one nucleotide for another in a given DNA sequence, thus creating a new allele. The presence and frequency of SNPs in the human genome has been known and appreciated for some time but, until recently, the methods for genotyping were relatively slow and thus expensive. The most common approach to studying human basepair polymorphisms during the 1980s and early 1990s was the use of restriction fragment length polymorphism or RFLP methods, which are no longer widely used because they are slow and tedious.

Researchers have now identified millions of SNPs in the human genome, and information about these polymorphisms is available in online databases (e.g. http://www.snp.cshl.org). Little is known about SNPs in nonhuman primates, although it is reasonable to expect that the genomes of commonly used laboratory primates will exhibit as many SNPs as does the human genome. One question worthy of study over the coming years is the proportion of human SNPs that will also be polymorphic and hence informative in nonhuman primates. If SNPs occur at specific nucleotide sites because those sites have unusually high mutation rates, and if that high mutation rate is shared across species, then there may be many SNPs which are shared across primate species as a result of recurrent mutation. But because there are so many fixed genetic differences between even closely related primates such as chimpanzees and humans, it is unlikely that many SNPs could be shared as a result of long-term retention of a single ancestral polymorphism in two or more independent evolutionary lineages.

In addition to microsatellites, SNPs and proteins, other types of genetic polymorphism have been described within nonhuman primate species. Given the proportion of the mammalian genome that consists of repetitive elements, it is not a surprise that the presence or absence of some repetitive elements in specific chromosomal sites can be variable. The large number of Alu repeats, LINE repeats and others is the result of continuous insertion of new copies into primate genomes. Many polymorphic Alu repeats are known in the human genome (Salem *et al.,* 2003a). Cole *et al.* (1997) found an Alu insertion that is polymorphic in baboons, and this polymorphism has been studied in wild baboon populations (Szmulewicz *et al.,* 1999). The presence or absence of particular LINE element insertions also differs across primate species (Ovchinnikov *et al.,* 2002; Mathews *et al.,* 2003). The presence of Alu, LINE or other insertions can be a useful phylogenetic character, since these insertions are relatively rare events, and unlikely to occur separately in different evolutionary lineages in exactly the same location (Salem *et al.,* 2003b).

Resources for the genetic analysis of primates

Genetic studies of primates require a variety of tools, resources and information. Accurate molecular genetic information about the species under study, especially

DNA sequence data, is obviously important. Given the wide use of PCR to amplify genomic DNA or cDNA sequences, there is clearly a need to have the relevant sequence information for a given gene or DNA segment from the species one intends to study. However, this is not always available and the relevant information from a closely related species is often a useful alternative because sequences are generally conserved among closely related species. One of the most widely used sources of DNA sequence information is the integrated set of online databases maintained by the National Center for Biotechnology Information, NCBI (http://www.ncbi.nlm.nih.gov). The NCBI site, which includes access to the very large Genbank database of sequences, archives a tremendous amount of genomic and cDNA sequence information for multiple primates. Those data can be used to design experimental protocols in the same or related species, and are especially valuable in designing PCR primers to amplify a given DNA sequence. Additional information is available through the NIH Intramural Sequencing Center at http://www.nisc.nih.gov/index. This site distributes data for a number of genes that are being sequenced from a wide variety of mammals and nonmammalian model organisms. The list of species under study by this project includes three hominoids, three Old World monkeys, three New World monkeys and three strepsirhines, making this an important resource for many primatologists.

Significant comparative genetic information is also archived and distributed through other websites. The Genome Informatics website from the University of Santa Cruz (http://www.genome.ucsc.edu) provides easy access to primary sequence data and comparative alignments for several species. This site will maintain comparative data related to human vs. chimpanzee sequence alignments. The Jackson Laboratory in Bar Harbor, Maine is the center of the world of mouse genetics, and its website (http://www.informatics.jax.org) is focused on the mouse genome, but includes helpful tools and data for comparative genetics beyond the rodents. Finally, although several other resources could also be mentioned, the Southwest National Primate Research Center is maintaining a database related to linkage maps for nonhuman primates. Unpublished information about the baboon genetic linkage map is available at http://www.snprc.org/baboon/genome/index.

A second critical resource is biological material suitable for laboratory analysis. Purified genomic DNA (or tissue samples that can serve as a source of DNA or RNA) from known individuals of the species under study are the most widely needed materials. The Integrated Primate Biomaterials and Information Repository (IPBIR) is a repository and distribution service funded by the U.S. National Science Foundation to provide investigators with biomaterials from a wide range of primate species. The website maintained by IPBIR (http://www.ipbir.org) and its member partners (the Coriell Institute, the Center for Reproduction of Endangered Species at the San Diego Zoo, Princeton University and the International Species Identification Service, ISIS) lists the materials that are available and the procedures for obtaining them. This is an important resource, since it is not always possible for interested researchers to obtain DNA or tissue from primates rarely held in captivity, such as lemurs or endangered apes, and it is sometimes valuable to have very large amounts of DNA available from a small set of individuals representing a given species. This allows many laboratories to use the same DNA sample for comparative analyses. IPBIR is creating cell lines for some of their samples, and thus can distribute modest quantities of DNA to a large number of investigators. In addition, the National Institutes of Health support eight national primate research centers (see http://www.ncrr.nih.gov/compmed/cm_nprc.asp). These centers are charged with the responsibility of assisting researchers who wish to obtain information or materials for research studies involving primates. The NIH primate centers can also provide access to captive colonies of animals for direct study. The primate center colonies consist primarily of macaques and baboons, but do include substantial numbers of chimpanzees, squirrel monkeys, marmosets and other species. For example, the Southwest National Primate Research Center (www.snprc.org) maintains colonies of nine different primate species, and has an active program sharing tissues, DNA or other material from each of them.

Genetic relationships among primates

The basics of primate phylogeny

One of the major successes of primate genetics over the past twenty years has been establishment of the basic phylogeny of nonhuman primates. Much of what is known about the evolutionary relationships among extant primates was determined through comparative anatomy, without significant input from genetic data

(Le Gros Clark, 1971; Szalay and Delson, 1979; Martin, 1990). But molecular genetics has contributed a great deal to the study of primate phylogeny, and current practice requires integration of genetic data with morphological comparisons. Unresolved issues remain, such as the relationships among the four major lineages of platyrrhine (New World) primates (Canavez *et al.,* 1999; Singer *et al.,* 2003) and among gibbon species (Mueller *et al.,* 2003). However, the broad outline and much of the fine detail of primate evolutionary relationships is now known with substantial confidence. For more thorough discussion of primate evolutionary relationships and the history of diversification, readers should consult works by Goodman and his colleagues (e.g. Goodman *et al.,* 1998; Meireles *et al.,* 1999; Page *et al.,* 1999; Page and Goodman, 2001). Other important studies have been published in the last few years, and have clarified the relationships among most primate genera (e.g. Tosi *et al.,* 2003a, 2003b; Pastorini *et al.,* 2002; Morales *et al.,* 1999; Stewart and Disotell, 1998; and Yoder *et al.,* 1996, 2000).

It is now clear that the genus *Pan* (consisting of chimpanzees, *P. troglodytes* and bonobos, *P. paniscus*) is the closest extant group to humans, with *Gorilla* somewhat more distant and *Pongo* (the orangutan) the outgroup to the African hominoid clade. The living hominoid primates (superfamily Hominoidea) form a monophyletic group that includes humans, the three genera mentioned above, gibbons and siamangs (genus *Hylobates*). The Old World monkeys (superfamily Cercopithecoidea) include most of the widely used laboratory primates (macaques, baboons, African green monkeys) and share a most recent common ancestor with the hominoids about 24–26 million years ago (Stewart and Disotell, 1998). The hominoids and cercopithecoids together make up the infraorder Catarrhini. The catarrhines diverged from the platyrrhines or New World monkeys about 36–45 million years ago. Together the platyrrhines and catarrhines are the sister group to tarsiers (genus *Tarsius*), and the three lineages are placed within the suborder Haplorrhini. The last major group of extant primates is the suborder Strepsirhini, which includes the lemurs, lorises, and galagos. The strepsirhines are a diverse and extraordinarily interesting radiation of primates, but are used as laboratory research subjects less often than are Old World monkeys and even apes, so strepsirhines are not discussed extensively in this review. One institution that has important colonies of lemurs and other strepsirhines is the Duke University Primate Center (http://www.duke.edu/web/primate/home) and studies of these captive animals are possible through this facility.

The importance of further phylogenetic and taxonomic study

While the major outline of primate phylogeny is well established, many details are still not fully resolved. The problem of defining species is a recurrent source of discussion and debate, as it is regarding many groups of organisms. The number of extant species of gorillas is one subject of current active debate (see Jensen-Seaman, 2003), as is the number of species of mangabeys, galagos and Asian leaf monkeys. It may seem paradoxical that the questions of evolutionary relationships and the sequence of divergence events that occurred millions of years in the past are more readily answered than the issue of how many species exist today. The problem is not quantifying the genetic and phenotypic variation present today, but finding agreement concerning the translation of complex patterns of variation into a set of clearly defined and mutually exclusive taxa. Phenotypic and genetic variation across primate groups is often, but not always, organized into easily recognized and diagnosed species. For discussion of these problems, see Jolly (1993) and Groves (2001).

This problem of defining and diagnosing species is important to laboratory studies in a biomedical context, just as it is to analyses of evolutionary process or efforts to conserve primate diversity. Some of the most commonly studied laboratory primates are plagued by controversial or contentious taxonomy, and disagreements or confusion about taxonomy or nomenclature can make it difficult to communicate effectively and unambiguously about the animal subjects used in a research study. Baboons (genus *Papio*) are a good example. This genus consists of several geographically and morphologically distinguishable populations (Jolly, 1993; Groves, 2001). Many investigators are inclined to consider these populations as members of a single species, because hybridization between morphologically different populations is often observed in the wild. This suggests that gene flow can occur between these populations, and, therefore, the gene pools of the two "types" are not isolated. On the other hand, a number of biologists prefer to identify recognizably distinct populations as different formal species, even when occasional interbreeding at the margins is observed or cannot be excluded (Cracraft, 1983; Nixon and Wheeler, 1990). The result is that some investigators have used five different species names for the five widely recognized types of *Papio* baboons (Hill, 1970; Groves, 2001) while others refer to all five as subspecies with the single species of

Papio hamadryas (e.g. Phillips-Conroy and Jolly, 1986; Williams-Blangero *et al.,* 1990; Rogers, 2000). Similar disagreements have occurred regarding the number of species of squirrel monkeys, owl monkeys and other species.

This is not mere academic trivia because the various populations of baboons and squirrel monkeys have different physiological and behavioral phenotypes, and differences among research studies, that appear to be contradictions of results, may in some instances be due to unrecognized differences in the genetic background of the animals. Taxonomic confusion can thus lead to inappropriate or inadvertent comparisons between physiologically or morphologically different sets of animals. Effective communication concerning research results must include reliable and broadly understood information concerning the type of animal subjects used.

Similarity to humans

The overall genetic similarity among primate species tracks the phylogeny presented above, and the evolutionary history it represents. In broad terms, all New World monkeys are more similar genetically to each other than any of them is to any Old World monkey or ape. This means that, in terms of overall genetic constitution, phylogeny is a useful guide to genetic similarity. But this does not hold true for every characteristic in every possible comparison. The total estimated divergence across the entire DNA sequence is a reasonable predictor of sequence similarity, or difference between two species at any given gene. However, some pairs of species may be more different at a given locus than is predicted by phylogenetic position, simply due to divergent selection on that gene, unusually high mutation rates or other mechanisms. Not every gene shows the average divergence (e.g. Stewart *et al.,* 1987; Messier and Stewart, 1997).

Calculating DNA sequence similarity is not a simple matter. Once again, the definitions and assumptions that underlie research methods are critical. Statements concerning the average genetic difference (or alternatively the average sequence identity) between any two species depend on which types of DNA sequences are included in the calculation. It is widely (and accurately) repeated that humans and chimpanzees are about 1.1–1.4% different at the level of whole genome DNA sequence (Chen and Li, 2001; Ebersberger *et al.,* 2002). However, this calculation is based on aligning overall DNA sequences between the two species, then removing the segments that are present in one and not the other, and finally counting differences among the remaining homologous nucleotides. This is obviously a sensible approach, but Britten (2002) has argued that removing the segments that are not shared distorts the calculation. He performed the calculation differently and found that humans and chimpanzees should be described as sharing only 95% of their sequence. The difference between the two values is not a biological disagreement but is due to methodological differences regarding the appropriate comparison. Wildman *et al.* (2002) made the calculation a third way, and found that there is only 0.6% difference in the functional portion of the genome they consider to be the most valid basis for analysis. It is clear that the genome is a complex entity made up of a wide variety of coding, non-coding, repetitive and single-copy sequences, with insertions and deletions common among even very closely related species (Frazer *et al.,* 2003). Calculations of "genetic similarity" must include precise definitions of what types of sequences were included in order to provide a meaningful basis for comparison.

Genetic management of primates

Ideas about the genetic management of captive nonhuman primates have developed over a number of years. Early efforts to formulate general principles and recommendations (Curie-Cohen *et al.,* 1983; Dyke *et al.,* 1990; Smith, 1982; Stone *et al.,* 1993; Williams-Blangero, 1991, 1993) were very important and ultimately quite influential. Nevertheless, it was several years before these fundamental notions of genetic management achieved general acceptance and were put into wide practice. More recent treatments (VandeBerg, 1995; VandeBerg and Williams-Blangero, 1996; Williams-Blangero *et al.,* 2002) have continued to develop those ideas and principles based on new experience and novel genetic and/or demographic information. At the time of this writing, the basic fundamentals of primate genetic management are now broadly agreed.

Any research program, regardless of specific research goals, must use animals of known genetic background. While this is especially true for programs in which genetic analysis is a major goal, no study, regardless of focus, should use animals of unknown genetic background. Phenotypic and genetic variation within primate species is significant enough (see above) that research results should be associated with some degree of genetic or taxonomic information.

For most primates, subspecies identity, or region of geographic origin, is a reasonable surrogate for detailed genetic characterization. For example, the genetic differences among subspecies of chimpanzees (Morin *et al.,* 1994; Gagneux *et al.,* 2001); gorillas (Jensen-Seaman *et al.,* 2001, 2003); baboons (Williams-Blangero *et al.,* 1991; Newman *et al.* (in press); rhesus macaques (Champoux *et al.,* 1997; Melnick *et al.,* 1993; Tosi *et al.,* 2003); other macaques (Ashley *et al.,* 1989); squirrel monkeys (VandeBerg *et al.,* 1990) and other primate species can be substantial.

In many of these cases, there are clear phenotypic differences among subspecies, and these differences can have consequences for research results (e.g. Abee, 1989; Williams- Blangero *et al.,* 1990; Trichel *et al.,* 2002).

The high level of molecular divergence between animals originating in different geographic regions, or assigned to different subspecies, (especially in mitochondrial DNA) makes it possible in some cases to classify individual primates by subspecies based on genetic analysis. This approach depends, of course, on the assumption that the animals being tested are not admixed across subspecies or localities. If one can assume that the individuals in question are pure bred or wild caught, then it is often possible to assign them to a taxonomic group with reasonable confidence (of course, morphological comparisons remain critically important in the assignment of individuals to taxa.) Mitochondrial sequences, including the D-loop, along with other rapidly evolving parts of the genome, are useful in these contexts.

Breeding programs for the production of research primates should be based on three principles. First, breeding should be structured to maintain appropriate levels of genetic diversity within the population. Substantial genetic variation exists in all primate species studied to date, and loss of genetic variation can influence the value of a population for some types of research studies. Loss or reduction of variation can bias the results of experiments. If study subjects are derived from a captive colony that has a different genetic composition than is normally found in that species, then conclusions based on that unusual population may not generalize across the species (Williams-Blangero *et al.,* 2002). Extreme loss of variation may place the future health and fertility of the animals at risk, if inbreeding or loss of variability proceed far enough (Ralls *et al.,* 1979).

The second principle of captive breeding is that there should be an explicit plan for the genetic composition of the colony. Explicit decisions and plans should be made concerning the handling of animals from different geographic or taxonomic origins. As discussed above for baboons, squirrel monkeys and other species, biomedically important variation exists among populations within many individual primate species. In most circumstances, it is prudent to maintain genetically distinct populations as separate and isolated breeding stocks. But in special cases, it may be valuable to create crosses between genetically distinct populations, in order to produce genetically mixed progeny. For example, crossing animals from two subspecies that differ in a genetically determined trait will produce a mixed pedigree in which the genes causing that difference could be mapped by linkage (Rogers *et al.,* 1999). This type of circumstance will occur only rarely, but in all situations the decision about admixture should not be left to chance or accident.

Third, carefully recorded and documented pedigrees should be maintained for all offspring born in the colony. This means that the sire and dam for each infant should be a permanent part of that infant's colony record. By simply recording sire and dam for every infant born, a colony manager will eventually build the database needed to calculate the kinship (i.e. genealogical relationship) between any two animals in the colony. Wild-caught founders can be assumed to be unrelated or, where more specific information is available, it may be possible to determine relationships among them. In either case, once parentage is known for all infants born in captivity, the full pedigree can be constructed and maintained for any number of generations. For the baboon colony at the Southwest National Primate Research Center, basic sire-dam information for hundreds of infants born over the last twenty years allows the colony managers to construct a pedigree that spans up to six generations and links over 2000 individuals into multi-generation families. This background information is invaluable in selecting animals for studies that focus on issues of genetics. But the pedigree data are just as valuable when selecting animals for studies in which genetics is not a component of the study design. In those cases unrelated animals should be used. Inadvertently assigning closely related animals to research studies can result in spurious results because genetically related animals may tend to be more similar to each other than randomly selected animals, and thus do not produce independent data points. Genealogically related animals may also respond more similarly than unrelated animals to an experimental challenge. Unknowingly assigning siblings or half-siblings to control or treatment groups in an experiment can result in false positive or false negative results depending on the circumstances.

Experience shows that the use of single-male breeding groups is a simple approach to producing animals of known parentage. Maternity is generally easy to assign in primates, given the close bond between mother and offspring. Thus it is the identity of the sire that is more often the open question. But studies at SNPRC have shown that even when infants are produced in single-male breeding groups, and attention is given to identifying maternity and paternity through caging records, as many as 1% of maternity assignments and 5% of paternity assignments can be erroneous (VandeBerg, 1992). This error rate is probably due to a combination of human error in record keeping, mistakes in caging records, mistakes in the dates that animals are moved in or out of specific cages, exchange or kidnapping of infants among breeding females in the same cage, and in some cases the siring of infants by males in adjacent but not sufficiently separated cages. The conclusion must be that whenever paternity or maternity is unknown, but even when parentage is inferred through caging records or observations, the confirmation of pedigree relationships through genetic testing is important.

As discussed above, there are a number of different types of genetic markers that can be used for pedigree testing in primates. DNA fingerprinting has been used effectively, but many practitioners prefer the use of panels of polymorphic microsatellite loci. The microsatellites provide genetic data that are attributable to specific chromosomal locations, and can be tracked through multi-generation pedigrees. Techniques for genotyping microsatellites have improved to the point that they are much faster than fingerprinting methods. Finally, single locus markers, such as STRs, can be chosen to track specific chromosomal regions that contain functional genes of interest. The MHC gene cluster is a chromosomal segment of particular interest to many primate researchers. Microsatellites near to the MHC gene cluster can be used to conduct paternity tests, and simultaneously to monitor the inheritance of MHC haplotypes that may be relevant to the selection of animals for research studies (see Penedo *et al.*, 2003).

Current applications to biomedical research

While phylogenetics and the analysis of genetic variability in wild populations are active areas of research, much of primate genetics is done in the context of biomedical research. It is not possible here to review the entire field, but there are several aspects of primate genetics that warrant specific discussion. As more information becomes available concerning the genomic content and structure of selected primate species, and this type of knowledge is extended to other lesser known species, more opportunities will develop to pursue work in the following areas.

Transmission genetics and estimation of heritability

An increasing number of colony managers are developing multigeneration pedigree information for the nonhuman primates under their care. As more pedigreed colonies become available, the use of nonhuman primates for analyses of the genetic basis of phenotypic variation can increase. Once a set of 200–250 animals can be linked into multigeneration pedigrees, various methods can be employed to determine the proportion of variation in any given phenotype that is attributable to genetic differences among the animals (Almasy and Blangero, 1998). Large extended pedigrees of macaques, baboons, vervet monkeys, marmosets and other species have been used to determine the heritability of various traits, i.e. the proportion of the phenotypic variance that is controlled by genetic variation (Falconer, 1999). This is an important step in development of primate models of disease, since genetic factors are known to influence the risk that a given person will develop specific diseases, and laboratory primates can be used to investigate the genetic mechanisms that underlie that process (see below). Heritability studies can also provide basic information regarding primate biology. For example, Williams-Blangero *et al.* (1995) showed that the age at first birth is heritable in a population of baboons (*Papio hamadryas*). Ha *et al.* (2002) found a strong genetic effect on variation in birth weight among pig-tailed macaques. Studies like these are relevant to our understanding of growth, development and reproductive maturation in primates. Lifespan is another basic characteristic that is heritable in nonhuman primates (Martin *et al.*, 2002). Traits related to body size and morphology have also been shown to be influenced by genetic variation. Cheverud *et al.* (1990) demonstrated the heritability of brain size and other features in macaques, while Mahaney *et al.* (1993) showed that individual variation in the weights of specific internal organs is heritable in baboons. Genetic variation has also been shown to affect handedness (Hopkins *et al.*, 2001), skeletal nonmetric traits and the volume of specific brain regions (Lyons *et al.*, 2001).

Most quantitative genetic analyses of heritability in nonhuman primates have focused on phenotypes related to risk factors for disease. Bone density, a risk factor for osteoporotic fracture, is heritable in baboons (Mahaney *et al.,* 1997) and macaques (Lipkin *et al.,* 2001). Risk factors for cardiovascular disease, such as levels of high-density lipoprotein cholesterol (Rainwater *et al.,* 2002) or low-density lipoprotein cholesterol (Kammerer *et al.,* 2002) are heritable in baboons, as are other biological parameters known to be related to obesity (Comuzzie *et al.,* 2001) and hypertension (Kammerer *et al.,* 2001). Risk of cancer is heritable among tamarins (Cheverud *et al.,* 1993).

Nonhuman primates provide opportunities to examine the genetic basis of variation in a wide range of phenotypes, and this can contribute both to our knowledge of the primates themselves and to our understanding of human biology.

Gene-environment interaction

One of the most challenging but also most important aspects of genetics is the analysis of gene by environment interaction. It is clear that a large number of phenotypes, including aspects of growth, reproduction, aging, onset of many diseases, behavior and others are influenced by a complex interaction of the genetic factors inherited by a person or a monkey, and the environmental conditions experienced during their lifetime. Such interactions can take several forms, but many result in nonadditive effects of the causative variables. Identification of these multiple variables, and investigation of the mechanisms by which interaction effects occur, requires animal models in which both the genetics and the environment can be controlled to a greater degree than is possible in human studies. Nonhuman primates provide an outstanding opportunity to examine the impact of gene by environment interaction. In one example, Bennett *et al.* (2002) showed that genotype at the serotonin transporter locus interacts with early rearing experience to influence serotonin function in macaques. Primates with shorter generation times, such as marmosets or squirrel monkeys, provide the investigator with the chance to observe the animal's entire lifespan in a reasonable number of years, making longitudinal analyses of environmental and genetic interactions practical. Slightly longer studies of baboons and macaques are also very informative and, for some phenotypes, these species are preferred.

Genetic linkage analysis

Modern methods of genetic linkage analysis make it possible to use multigeneration pedigrees to locate within the genome the specific genes that influence inter-individual variation in a given phenotype (see Rogers *et al.,* 1999 and references therein). Three types of information are required for such analyses: (a) genotype data for a series of genetic polymorphisms scored in a population of animals, (b) phenotype data for the same animals, and (c) knowledge of the pedigree relationships among those animals. This information must be available for at least 500–600 individuals, and the details of pedigree structure will influence statistical power to detect genetic effects. The genetic polymorphisms must be highly variable and numerous enough to cover the entire genome at intervals of about 10 centiMorgans, or 10% recombination between loci. When such information is available, it is possible to use linkage analysis to search for a chromosomal region, or regions within the genome, that contain genes that influence variation in that phenotype (Blangero and Almasy, 1997). Identification of the functional gene requires a substantial amount of additional study of the region of interest. Nevertheless, the whole genome linkage approach does allow the identification of chromosomal segments that contain genes that influence variation in any number of different biological processes, from risk factors to disease, to normal variation in metabolism or normal variation in growth, development, anatomy or behavior.

Genetic linkage maps of the entire human genome were first developed almost 20 years ago. Similar maps have been constructed for other mammalian species, including the mouse, rat, pig, horse and dog. The first genetic linkage map for a nonhuman primate was developed for baboons, *Papio hamadryas* (Rogers *et al.,* 2000). The initial baboon linkage map was produced through collaboration between researchers at the Southwest Foundation for Biomedical Research and Sequana Therapeutics, Inc. (La Jolla, CA). However, the map has undergone further development. Additional information is available on the website of the Southwest National Primate Research Center (www.snprc.org). Genetic linkage maps are also under development for the rhesus macaque (Rogers, unpublished data) and the vervet monkey, *Chlorocebus aethiops* (N. Freimer, pers. comm.).

Complex phenotypes that are influenced by multiple genes and multiple environmental factors can be investigated through linkage screening. The chromosomal regions found to harbor causative genes are generally referred to as quantitative trait loci (QTLs). Several QTLs have been localized in baboons using the available linkage map (e.g. Kammerer *et al.,* 2001; Rainwater *et al.,* 2002). Work is underway to identify the specific functional mutations in some of these QTLs, but none have yet been published. The task of identifying these causative mutations remains a major challenge.

Genetic response to challenges

The types of studies discussed above are designed to identify genetic causes of variation among individuals. It is also possible to use genetic information to monitor changes within an individual in response to various challenges. Researchers at the Washington National Primate Research Center are challenging rhesus macaques with infectious viruses, and using gene arrays to quantify changes in the expression of specific genes (Agy *et al.*, 2003). Gene expression can change in response to a number of environmental factors. Another example is work by Cox *et al.* (2002) investigating the response, at the level of mRNA expression, of baboons fed a series of controlled diets. Genetic tools can be used either to hunt for genetic causes, or to monitor the consequences of environmental treatments at the level of gene function.

Gene therapy

The potential to deliver specific genes to a human patient as a treatment for disease opens extraordinary possibilities for advancing human health. As a result, gene therapy is an exciting aspect of biomedical research that is being pursued with great energy. But delivery of the therapeutic genes safely, without adverse effects, is critical and challenging. Human trials of experimental procedures have been promising, but have also had failures. The use of animal models to test gene therapy vectors (the delivery systems used to introduce gene constructs into cells), and to evaluate the effects of introducing such genetic construct into an individual, is critically important. Nonhuman primates have been used in a variety of ways (e.g. Gao *et al.*, 2002; Song *et al.*, 2002; t'Hart *et al.*, 2003) and will continue to play an essential role in the development of gene therapy systems for human medicine.

Future directions in primate genetics

Development of whole genome sequence for nonhuman primate species

The sequencing of the human genome is one of the major scientific achievements of recent history. The genomes of the laboratory mouse and the laboratory rat have also been sequenced to a high level of accuracy and completeness. The National Human Genome Research Institute has established a large program to sequence the complete genomes of other model organisms, and species such as the fruitfly (*Drosophila*), cow and honeybee have been selected for analysis for a variety of reasons (www.nhgri.nih.gov). Two nonhuman primates have been approved for large-scale sequencing, the chimpanzee and the rhesus macaque.

Whole genome sequencing of any species will make that organism much more useful for a number of types of research. Immediate and direct access to this wealth of information about the fundamental molecular biology of an organism allows researchers, with interest in physiology, developmental biology, behavior, pathology or other fields, to perform a wide range of analyses that would not be possible without the sequence information. Techniques for whole genome sequencing have evolved over the past ten years (Green, 2001), but most of the current sequencing programs incorporate a combination of whole genome shotgun sequencing of randomly primed sequences, with a structured approach in which selected large insert BAC clones are sequenced to a high resolution. Different genome projects differ in their coverage of the genome, meaning that some genomes (such as the human and mouse) have been sequenced in a highly redundant manner to ensure complete coverage of as much of the genome as is possible, whereas other projects produce fewer total sequence reads and thus cover areas less redundantly. By reducing the coverage across the genome, the cost of sequencing is significantly reduced, but the reliability of the final assembled sequence is also reduced to some degree.

Current plans call for moderate coverage of the chimpanzee and rhesus monkey genomes (www.nhgri.nih.gob). The sequence produced for each of these species will be highly reliable, but will not be as comprehensive as the mouse or human sequences. The data for these two nonhuman primates cannot fail to create many new research opportunities. The chimpanzee sequence will of course be of great value to anyone interested in the evolution of the human genome or evolution of the human phenotype. The sequence of the rhesus macaque will be extraordinarily valuable to anyone using this or any closely related species (e.g. cynomolgus macaques or baboons) for studies of genetic influences on phenotypic variation or genetic (gene expression) consequences of environmental treatments. It is possible that other nonhuman primates will also be completely sequenced in the future, especially since the cost of sequencing may continue to fall.

Genetic analysis of normal variation

With the development of genetic linkage maps for additional primate species beyond baboons and rhesus macaques, and with the whole genome sequencing of at least two nonhuman primates, it will be possible for researchers to investigate the genetic basis of individual variation in any number of physiological, anatomical or behavioral traits. While genomic analysis of nonhuman primates has, to date, focused primarily on models of human diseases, there will be growing opportunities to study variation beyond that associated with risk of disease. The field of primate genomics could conceivably expand to much broader efforts aimed at understanding a variety of aspects of the basic biology of nonhuman primates.

Pharmacogenomics

Individual variation in response to drug treatment is another important aspect of biomedical research. Primates are often used to study the efficacy or safety of new pharmaceuticals. Pharmacogenetics is the study of the genetic basis of individual variation in response to drug treatment, either variation in effectiveness of treatment or variation in side effects (Mancama *et al.,* 2003; Terra and Johnson, 2002; Ulrich *et al.,* 2003). Whole genome expression analysis may soon be used to monitor, in great detail, the cellular and organ-level responses to currently used, or newly developed, drugs and to measure those responses at the level of gene expression. This type of research could conceivably be performed on well-chosen pedigrees of nonhuman primates in order to combine the power of primate model systems with the tools of quantitative genetics, expression array technology, gene mapping strategies and other genomic approaches.

Transgenic primates

Transgenic mice have been a powerful tool for understanding a wide range of biological processes at the molecular level. At present, it is not possible to create transgenic primates. Researchers have successfully introduced a foreign gene into a primate embryo, and have obtained detectable expression in the newborn infant (Chan *et al.,* 2001). However, the level of expression was limited, and not what is routinely possible in rodents. Prospects for the production of transgenic primates are difficult to judge, and there may be fundamental differences that make this process much more difficult in primates as compared to other mammals (Simerly *et al.,* 2003). Nevertheless, several groups of researchers continue to pursue this goal, and if successful, these methods would have an extraordinary impact on the study of primate genetics.

Correspondence

Any correspondence should be directed to Jeffrey Rogers, Dept. of Genetics, Southwest Foundation for Biomedical Research, 7620 N.W. Loop 410, San Antonio, Texas 78227. Phone: 210-258-9532; Fax: 210-670-3344; Email: jrogers@darwin.sfbr.org

References

Abee, C.R. (1989). *ILAR News.* 31, 11–20.

Agy, M.B., Li, Y., Thomas, M.J., Korth, M.J., Geiss, G.K. Kwieciszewki, B.K., Bumgarner, R.E. and Katze, M.G. (2003). *J. Med. Primatol* 32, 305.

Almasy, L. and Blangero, J. (1998). *Am. J. Hum. Genet.* 62, 1198–1211.

Ashley, M.V., Williams, A., Tenaza, R. and Melnick, D.J. (1989). *Am. J. Phys. Anthropol.* 78, 185.

Barnicot, N.A., Jolly, C.J., Huehns, E.R. and Dance, N. (1965). In Vagtborg, H. (ed.) *The Baboon in Medical Research. Proceedings of the First International Symposium on the Baboon and Its Use as an Experimental Animal,* pp 323–338. Austin, University of Texas Press.

Bennett, A.J., Lesch, K.P., Heils, A., Long, J.C., Lorenz, J.G., Shoaf, S.E., Champoux, M., Suomi, S.J., Linnoila, M.V. and Higley, J.D. (2002). *Mol. Psychiatry.* 7, 118–122.

Bigger, C.B., Brasky, K.M. and Lanford, R.E. (2001). *J. Virol.* 75, 7059–7066.

Blangero, J. and Almasy, L. (1997). *Genet. Epidemiol.* 14, 959–964.

Blanquer-Maumont, A. and Crouau-Roy, B. (1995). *J. Mol. Evol.* 41, 492–497.

Britten, R.J. (2002). *Proc. Natl. Acad. Sci. USA* 99, 13633–13635.

Canavez, F.C., Moreira, M.A.M., Ladasky, J.J., Pissinatti, A., Parham, P. and Seuanez, H.N. (1999). *Mol. Phylogenet. Evol.* 12, 74–82.

Chan, A.W., Chong, K.Y., Martinovich, C., Simerly, C. and Schatten, G. (2001). *Science* 291, 309–312.

Champoux, M., Higley, J.D. and Suomi, S.J. (1997). *Dev. Psychobiol.* 31, 49–63.

Chen, F.-C. and Li, W.-H. (2001). *Amer. J. Hum. Genet.* 68, 444–456.

Cheverud, J.M., Falk, D., Vannier, M., Konigsberg, L., Helmkamp, R.C. and Hildebolt, C. (1990). *J. Hered.* 81, 51–57.

Cheverud, J.M., Tardif, S., Henke, M.A. and Clapp, N.K. (1993). *Hum. Biol.* 65, 1005–1012.

Chou, H.H., Hayakawa, T., Diaz, S., Krings, M., Indriati, E., Leakey, M., Paabo, S., Satta, Y., Takahata, N. and Varki, A. (2002). *Proc. Natl. Acad. Sci. U.S.A.* 99, 11736–11741.

Clarke, A.S., Kammerer, C.M., George, K.P., Kupfer, D.J., McKinney, W.T., Spence, M.A. and Kraemer, G.W. (1995). *Biol. Psychiatry* 38, 572–577.

Cole, S.A., Birnbaum, S. and Hixson, J.E. (1997). *Gene* 193, 197–201.

Comuzzie, A.G., Martin, L.J., Cole, S.A., Rogers, J., Mahaney, M.C., Blangero, J. and VandeBerg, J.L. (2001). *Obes. Res.* 9(Suppl.), 71S.

Cox, L.A., Birnbaum, S. and VandeBerg, J.L. (2002). *Genome Res.* 12, 1693–1702.

Cracraft, J. (1983). In Johnson, R.F. (ed.) *Current Ornithology*, Volume 1, pp 159–187. New York: Plenum Press.

Curie-Cohen, M., VandeBerg, J.L. and Stone, W.H. (1983). *J. Hum. Evol.* 12, 573–585.

Deka, R., Shriver, M.D., Yu, L.M., Jin, L., Aston, C.E., Chakraborty, R. and Ferrell, R.E. (1994). *Genomics* 22, 226–230.

Di Fiore, A. (2003). *Yearb. Phys. Anthropol.* 46, 62–99.

Disotell, T.R. (1994). *Am. J. Phys. Anthropol.* 94, 47–57.

Dyke, B., Williams-Blangero, S., Dyer, T.D. and VandeBerg, J.L. (1990). *Prog. Clin. Biol. Res.* 344, 563–574.

Ebersberger, I., Metzler, D., Schwarz, C. and Paabo, S. (2002). *Am. J. Hum. Genet.* 70, 1490–1497.

Ely, J., Alford, P. and Ferrell, R.E. (1991). *Am. J. Primatol.* 39, 54.

Falconer, D.S. and MacKay, T. (1999). *Introduction to Quantitative Genetics*, 4th edition. Essex, England, Longman.

Frazer, K.A., Chen, X., Hinds, D.A., Pant, P.V., Patil, N. and Cox, D.R. (2003). *Genome Res.* 13, 341–346.

Gagneux, P., Wills, C., Gerloff, U., Tautz, D., Morin, P.A., Boesch, C., Fruth, B., Hohmann, G., Ryder, O.A. and Woodruff, D.S. (1999). *Proc. Natl. Acad. Sci. U.S.A.* 96, 5077–5082.

Gao, G.P., Alvira, M.R., Wang, L.L., Calcedo, R., Johnston, J. and Wilson, J.M. (2002). *Proc. Nat. Acad. Sci. USA* 99, 11854–11859.

Goodman, M. (1999). *Am. J. Hum. Genet.* 64, 31–39.

Goodman, M., Porter, C.A., Czelusniak, J., Page, S.L., Schneider, H., Shoshani, J., Gunnell, G. and Groves, C.P. (1998). *Mol. Phylogenet. Evol.* 9, 585–598.

Green, E.D. (2001). *Nat. Rev. Genet.* 2, 573–583.

Groves, C. (2001) *Primate Taxonomy*. Washington, D.C., Smithsonian Institution Press.

Ha, J.C., Ha, R.R., Almasy, L. and Dyke, B. (2002). *Am. J. Primatol.* 56, 207–213.

Hadfield, R.M., Pullen, J.G., Davies, K.F., Wolfensohn, S.E., Kemnitz, J.W., Weeks, D.E., Bennett, S.T. and Kennedy, S.H. (2001). *Am. J. Primatol.* 54, 223–231.

Harris, E.E. and Disotell, T.R. (1998). *Mol. Biol. Evol.* 15, 892–900.

Hill, W.C.O. (1970). *Primates (Volume VIII): Cynopithecinae*. New York, Wiley-Interscience.

Hopkins, W.D., Dahl, J.F. and Pilcher, D. (2001). *Psychological Science* 12, 299–303.

Inoue, M. and Takenaka, O. (1993). *Primates* 34, 37–45.

Inoue, M., Takenaka, A., Tanaka, S., Kominami, R. and Takenaka, O. (1990). *Journal Primates* 31, 563–570.

Jensen-Seaman, M.I., Deinard, A.S. and Kidd, K.K. (2001). *J. Hered.* 92, 475–480.

Jensen-Seaman, M.I., Deinard, A.S. and Kidd, K.K. (2003). In Taylor, A.B. and Goldsmith, M.L. (eds) *Gorilla Biology: A Multidisciplinary Perspective*, pp 247–268. New York, Cambridge University Press.

Jolly, C.J. (1993). In Kimbel, W.H. and Martin, L.B. (eds) *Species, Species Concepts and Primate Evolution*, pp 67–107. New York, Plenum Press.

Johnson, M.E., Viggiano, L., Bailey, J.A., Abdul-Rauf, M., Goodwin, G., Rocchi, M., and Eichler, E.E. (2001). *Nature* 413, 514–519.

Johnson, R.M., Gumucio, D. and Goodman, M. (2002). *Comp. Biochem. Physiol.* A133(3), 877–883.

Kammerer, C.M., Cox, L.A., Mahaney, M.C., Rogers, J. and Shade, R.E. (2001). *Hypertension* 37, 398–402.

Kammerer, C.M., Rainwater, D.L., Cox, L.A., Schneider, J.L., Mahaney, M.C., Rogers, J. and VandeBerg, J.L. (2002). *Arterioscler. Thromb. Vasc. Biol.* 22, 1720–1725.

Kong, A., Gudbjartsson, D.F., Sainz, J., Jonsdottir, G.M., Gudjonsson, S.A., Richardsson, B., Sigurdardottir, S., Barnard, J., Hallbeck, B., Masson, G., Shlien, A., Palsson, S.T., Frigge, M.L., Thorgeirsson, T.E., Gulcher, J.R. and Stefansson, K. (2002). *Nat. Genet.* 31, 241–247.

Lawlor, R.R., Richard, A.F. and Riley, M.A. (2001). *Am. J. Primatol.* 55, 253–259.

Le Gros Clark, W.E. (1978). *The Antecedents of Man*. New York, Quadrangle Books.

Lipkin, E.W., Aumann, C.A. and Newell-Morris, L.L. (2001). *Bone* 29, 249–257.

Litt, M. and Luty, J.A. (1989). *Am. J. Hum. Genet.* 44, 397–401.

Lyons, D.M., Yang, C., Sawyer-Glover, A.M., Moseley, M.E. and Schatzberg, A.F. (2001). *Arch. Gen. Psychiatry* 58, 1145–1151.

Mahaney, M.C., Williams-Blangero, S., Blangero, J. and Leland, M.M. (1993). *Hum. Biol.* 65, 991–1003.

Mahaney, M.C., Morin, P., Rodriguez, L.A., Newman, D.E. *et al.* (1997). *J. Bone Min. Res.* 12(Suppl. 1), S118.

Mancama, D., Arranz, M.J. and Kerwin, R.W. (2003) *Curr. Opin. Mol. Ther.* 5, 642–649.

Martin, R.D. (1990). *Primate Origins and Evolution*. Princeton, Princeton University Press.

Martin, L.J., Mahaney, M.C., Bronikowski, A.M., Dee Carey, K., Dyke, B. and Comuzzie, A.G. (2002). *Mech. Ageing Dev.* 123, 1461–1467.

Mathews, L.M., Chi, S.Y., Greenberg, N., Ovchinnikov, I. and Swergold, G.D. (2003). *Am. J. Hum. Genet.* 72, 739–748.

Meera Khan, P. (1987). *Genetica* 73, 25–36.

Meireles, C.M., Czelusniak, J., Schneider, M.P., Muniz, J.A., Brigido, M.C., Ferreira, H.S. and Goodman, M. (1999). *Mol. Phylogenet. Evol.* 12, 10–30.

Melnick, D.J., Hoelzer, G.A., Absher, R. and Ashley, M.V. (1993). *Mol. Biol. Evol.* 10, 282–295.

Messier, W. and Stewart, C.-B. (1997). *Nature* 385, 151–154.

Mirnics, K., Middleton, F.A., Marquez, A., Lewis, D.A. and Levitt, P. (2000). *Neuron* 28, 53–67.

Morales, J.C., Disotell, T.R. and Melnick, D.J. (1999). In Dolhinow P. and Fuentes, A. (eds) *The Nonhuman Primates*, pp 18–28. Mountain View, Mayfield Publications.

Morin, P.A., Moore, J.J., Chakraborty, R., Jin, L., Goodall, J. and Woodruff, D.S. (1994). *Science* 265, 1193–1201.

Morin, P.A., Mahboudi, P., Wedel, S. and Rogers, J. (1998). *Genomics* 53, 12–20.

Mueller, S., Hollatz, M. and Wienberg, J. (2003). *Hum. Genet.* 113, 493–501.

Nair, S., Ha, J. and Rogers, J. (2000). *Primates* 41, 343–350.

Newman, T.K., Fairbanks, L.A., Pollack, D. and Rogers, J. (2002). *Am. J. Primatol.* 56, 237–243.

Nixon, K.C. and Wheeler, Q.D. (1990). *Cladistics* 6, 211–223.

Ovchinnikov, I., Rubin, A. and Swergold, G.D. (2002). *Proc. Natl. Acad. Sci. USA* 99, 10522–10527.

Page, S.L. and Goodman, M. (2001). *Mol. Phylogenet. Evol.* 18, 14–25.

Page, S.L., Chiu, Ch. and Goodman, M. (1999). *Mol. Phylogenet. Evol.* 13, 348–359.

Pastorini, J., Forstner, M.R. and Martin, R.D. (2002). *J. Hum. Evol.* 43, 463–478.

Penedo, M.C.T., Ward, T., Philbrick, E., Marthas, M. and Doxiadis, G.G.M. (2003). *J. Med. Primatol.* 33, 293.

Phillips-Conroy, J.E. and Jolly, C.J. (1986). *Am. J. Phys. Anthropol.* 71, 337–350.

Porter, C.A., Czelusniak, J., Schneider, H., Schneider, M.P.C., Sampaio, I. and Goodman, M. (1999). *Am. J. Primatol.* 48, 69–75.

Ralls, K., Brugger, K. and Ballou, J. (1979). *Science* 206, 1101–1103.

Rainwater, D.L., Kammerer, C.M., Cox, L.A., Rogers, J., Carey, K.D., Dyke, B., Mahaney, M.C., McGill, H.C., Jr. and VandeBerg, J.L. (2002). *Atherosclerosis* 163, 241–248.

Rapsilber, J., Ryder, U., Lamond, A.I. and Mann, M. (2002). *Genome Res.* 12, 1231–1245.

Rogers, J. (2000). In Whitehead, P. and Jolly, C.J. (eds) *Old World Monkeys,* pp 57–76. Cambridge, MA, Cambridge University Press.

Rogers, J., Witte, S.M., Kammerer, C.M., Hixson, J.E. and MacCluer, J.W. (1995). *Genomics* 28, 251–254.

Rogers, J., Mahaney, M.C., Almasy, L., Comuzzie, A.G. and Blangero, J. (1999). *Yearb. Phys. Anthropol.* 42, 127–151.

Rogers, J., Mahaney, M.C., Witte, S.M., Nair, S., Newman, D., Wedel, S., Rodriguez, L.A., Rice, K.S., Slifer, S.H., Perelygin, A., Slifer, M., Palladino-Negro, P., Newman, T., Chambers, K., Joslyn, G., Parry, P. and Morin, P.A. (2000). *Genomics* 67, 237–247.

Salem, A.H., Kilroy, G.E., Watkins, W.S., Jorde, L.B. and Batzer, M.A. (2003a). *Mol. Biol. Evol.* 20, 1349–1361.

Salem, A.H., Ray, D.A., Xing, J., Callinan, P.A., Myers, J.S., Hedges, D.J., Garber, R.K., Witherspoon, D.J., Jorde, L.B. and Batzer, M.A. (2003b). *Proc. Natl. Acad. Sci. U.S.A.* 100, 12787–12791.

Simerly, C., Dominko, T., Navara, C., Payne, C., Capuano, S., Gosman, G., Chong, K.Y., Takahashi, D., Chace, C., Compton, D., Hewitson, L. and Schatten, G. (2003). *Science* 300, 297.

Singer, S.S., Schmitz, J., Schwiegk, C. and Zischler, H. (2003). *Mol. Phylogenet. Evol.* 26, 490–501.

Smith, D.G. (1982). *Lab. Anim. Sci.* 32, 540–546.

Smith, D.G., Kanthaswamy, S., Viray, J. and Cody, L. (2000). *Am. J. Primatol.* 50, 1–7.

Song, S.H., Scott-Jorgensen, M., Wang, J.M., Poirier, A., Crawford, J., Campbell-Thompson, M. and Flotte, T.R. (2002). *Molec. Therapy* 6, 329–335.

Stewart, C.B. (2000). *Am. J. Phys. Anthropol.* Suppl. 30, 292.

Stewart, C.B. and Disotell, T.R. (1998). *Curr. Biol.* 8, R582–R588.

Stewart, C.B., Schilling, J.W. and Wilson, A.C. (1987). *Nature* 330, 401–404.

Stone, W.H., Ely, J.J., Manis, G.S. and VandeBerg, J.L. (1993). *Primates* 34, 365–376.

Szalay, F.S. and Delson, E. (1979). *Evolutionary History of the Primates.* New York, Academic Press.

Szmulewicz, M.N., Andino, L.M., Reategui, E.P., Woolley-Barker, T., Jolly, C.J., Disotell, T.R. and Herrera, R.J. (1999). *Am. J. Phys. Anthropol.* 109, 1–8.

Terra, S.G. and Johnson, J.A. (2002). *Amer. J. Cardiovasc Drugs* 2, 287–296.

Tosi, A.J., Morales, J.C. and Melnick, D.J. (2000). *Mol. Phylogenet. Evol.* 17, 133–144.

Tosi, A.J., Disotell, T.R., Morales, J.C. and Melnick, D.J. (2003a). *Mol. Phylogenet. Evol.* 27, 510–521.

Tosi, A.J., Morales, J.C. and Melnick, D.J. (2003b). *Evolution* 57, 1419–1435.

Trichel, A.M., Rajakumar, P.A. and Murphey-Corb, M. (2002). *J. Med. Primatol.* 31, 171–178.

Ulrich, C.M., Robien, K. and McLeod, H.L. (2003). *Nat. Rev. Cancer* 3, 912–920.

VandeBerg, J.L. (1992). In Martin, R.D., Dixson, A.F. and Wickings, E.J. (eds) *Paternity in Primates: Genetic Tests and Theories*, pp 18–31. Basel, S. Karger.

VandeBerg, J.L. (1995). In Bennett, B.T., Abee, C.R. and Henrickson, R. (eds) *Nonhuman Primates in Biomedical Research: Biology and Management*, pp. 129–146. San Diego, Academic Press.

VandeBerg, J.L., Aivaliotis, M.J., Williams, L.E. and Abee, C.R. (1990). *Biochem. Genet.* 28, 41–56.

VandeBerg, J.L. and Williams-Blangero, S. (1996). *Lab. Anim. Sci.* 46, 146–151.

VandeBerg, J.L. and Williams-Blangero, S. (1997). *J. Med. Primatol.* 26, 113–119.

Venter, J.C., Adams, M.D., Myers, E.W. and Li, P.W. (2001). *Science* 291, 1304–1351.

Weber, J.L. and May, P.E. (1989). *Am. J. Hum. Genet.* 44, 388–396.

Wickins, E.J. (1993). *Primates* 34, 323–331.

Wildman, D.E., Uddin, M., Liu, G., Grossman, L.I. and Goodman, M. (2002). *Proc. Natl. Acad. Sci. U.S.A.* 100, 7181–7188.

Williams-Blangero, S. (1991). *Yearb. Phys. Anthropol.* 34, 69–96.

Williams-Blangero, S. (1993). *Lab. Anim. Sci.* 43, 535–540.

Williams-Blangero, S. and Blangero, J. (1995). *Am. J. Primatol.* 37, 233–239.

Williams-Blangero, S., VandeBerg, J.L., Blangero, J., Konigsberg, L. and Dyke, B. (1990). *Am. J. Primatol.* 20, 67–81.

Williams-Blangero, S., VandeBerg, J.L. and Dyke, B. (2002). *J. Med. Primatol.* 31, 1–7.

Witte, S.M. and Rogers, J. (1999). *Am. J. Primatol.* 47, 75–84.

Yoder, A.D., Cartmill, M., Ruvolo, M., Smith, K. and Vilgalys, R. (1996). *Proc. Natl. Acad. Sci. USA* 93, 5122–5126.

Yoder, A.D., Rasoloarison, R.M., Goodman, S.M., Irwin, J.A., Atsalis, S., Ravosa, M.J. and Ganzhorn, J.U. (2000). *Proc. Natl. Acad. Sci. USA* 97, 11325–11330.

Zou, J., Young, S., Zhu, F., Gheyas, F., Skeans, S., Wan, Y., Wang, L., Ding, W., Billah, M., McClanahan, T., Coffman, R.L., Egan, R. and Umland, S. (2002). *Genome Biol.* 3(5), 10–20.

Zhu, H., Bilgin, M. and Snyder, M. (2003). *Ann. Rev. Biochem.* 72, 783–812.

CHAPTER 30

The Respiratory System and its Use in Research

Charles G. Plopper
California National Primate Research Center and Department of Anatomy, Physiology and Cell Biology, School of Veterinary Medicine, University of California, Davis, CA, USA

Jack R. Harkema
Department of Pathology, College of Veterinary Medicine, Michigan State University, East Lansing, MI, USA

Introduction

The respiratory system of mammals is organized into two compartments based on function: the conducting airways and the gas exchange area. The basic function of the conducting airways is to direct air from the outside environment to the gas exchange area during inspiration and back out during expiration. The three dimensional architectural organization of the conducting airways, and the cell populations that compose the walls of the airways, are also organized to serve a protective function for the much more fragile gas exchange area. This protective function includes warming and humidifying the inspired air, filtering and detoxifying the air as it passes in, clearing the debris and dead cells accumulating on the surfaces of the passages, as a result of the deposition of the inhaled material, and the cellular reaction to it, and recapturing of heat and water as air passes out of the system during expiration. The conducting airways include the nasal cavity, which begins with the external nares and ends in the nasopharynx; the larynx, which opens into the pharynx and joins it to the trachea; the trachea, which carries air from the larynx along the neck and into the thoracic cavity; the extrapulmonary bronchi which branch from the trachea in the mediastinum and carry air to the left and right lungs; and the intrapulmonary bronchi and bronchioles, which form a branching tree within the lungs to carry air to the gas exchange area. The gas exchange area itself makes up the majority of the lung volume and is organized into a series of branched passages consisting of individual air pockets, the alveoli, whose walls are the gas exchange membrane, the interalveolar septum.

The Laboratory Primate
Copyright 2005 Elsevier
ISBN 0-1208-0261-9

While the general organization of these two compartments is fundamentally the same in all mammals, there is sufficient variation in the architecture of the air passages and their cellular organization, between species, to produce considerable differences in the response to a given stimulus or biological event. This chapter will define the organization of these compartments in the rhesus macaque (*Macaca mulatta*) and compare cellular composition and architecture in the rhesus to other common laboratory species and to humans.

Nasal cavity

The nose is the portal of entry for inhaled air in the respiratory system of laboratory monkeys and most other mammalian species. It is a structurally and functionally complex organ in the upper respiratory tract with many important and diverse functions. Besides being the principal organ for the sense of smell (olfaction), the nose also serves as an efficient filter, heater, and humidifier of inhaled air. This nasal air conditioning system is an essential part of the upper respiratory defense mechanism that protects the distal trancheobronchial airways and the more delicate alveolar gas exchange tissues in the lung. The nose has been described as a "scrubbing tower" that removes various airborne infectious and toxic agents that could potentially be harmful to the lower airways and the pulmonary parenchyma (Brain, 1970). The mucous membranes (nasal mucosa) lining the nasal airways also metabolize and detoxify many commonly inhaled toxicants (Dahl and Hadley, 1991). The nose and associated paranasal sinuses (e.g., maxillary sinus) may be afflicted by many diseases. The majority of these conditions are a consequence of viral or bacterial infections, allergic reactions, or aging. However, exposure to toxic agents may also cause or exacerbate certain nasal diseases.

Architecture

The nasal cavity of the laboratory monkey, like other mammalian species, is divided into two airway passages by the nasal septum. Each nasal passage extends from the nostrils to the nasopharynx. The nasopharynx is defined as the airway posterior to the termination of the nasal septum and proximal to the termination of the soft palate. Inhaled air flows through the nostril openings, or nares, into the vestibule, which is a slight dilatation just inside the nares and before the main chamber of the nose. Unlike the more distal main nasal chamber, that is surrounded by bone, the nasal vestibule is surrounded primarily by cartilage. The luminal surface is lined by a squamous epithelium similar to that of external skin. After passing through the nasal vestibule, inhaled air courses through the narrowest part of the entire respiratory tract, the nasal valve (ostium internum), into the main nasal chamber. Each nasal passage of the main chamber is defined by a lateral wall, a septal wall, a roof, and a floor. The lumen of the main chamber is lined by well-vascularized and innervated mucous membranes that are covered by a continuous layer of mucus. The nasal mucous layer is moved distally, by underlying cilia to the oropharynx where it is swallowed into the esophagus.

Turbinates, bony structures lined by the well-vascularized mucosal tissue, project into the airway lumen from the lateral walls into the main chamber of the nose. Nasal turbinates increase the inner surface area of the nose, which is important in the filtering, humidification, and warming of the inspired air. Although there are some general similarities in the nasal passages of most mammalian species, there are also striking interspecies differences in nasal architecture (Figure 30.1). From a comparative viewpoint, monkeys and humans have relatively simple noses with breathing as the primary function (microsmatic species). Other mammals (e.g., dogs, cats, rodents) have more complex noses with olfaction as the primary function (macrosmatic species). In addition, the nasal and oral cavities of primates are arranged in a manner to allow for both nasal and oronasal breathing. In contrast, most laboratory rodents (e.g., rats, mice, hamsters, guinea pigs) are obligate nose breathers, due to the close apposition of the epiglottis to the soft palate.

Interspecies variability in nasal gross anatomy has been emphasized in previous reviews (Negus, 1958; Gross and Morgan, 1992; Harkema, 1992) and demonstrated in studies using various methods, including silicone rubber casts, tissue morphometric techniques, computerized tomography and magnetic resonance imaging of the nasal airways. Selected data from the nasal cavities of humans and common laboratory animals, including the rhesus monkey, is presented in Table 30.1 (Schreider, 1983). Gross *et al.* (1987) reported the total surface areas and volumes of the nasal cavity and maxillary sinus of three rhesus monkey noses (Table 30.2). Interestingly macaque monkeys (Macaca rhesus, M. radiata, and M. cynomologus) only have one pair of sinuses, the maxillary sinuses (Rae *et al.*, 2002).

Marked differences in airflow patterns among mammalian species are primarily due to variation in

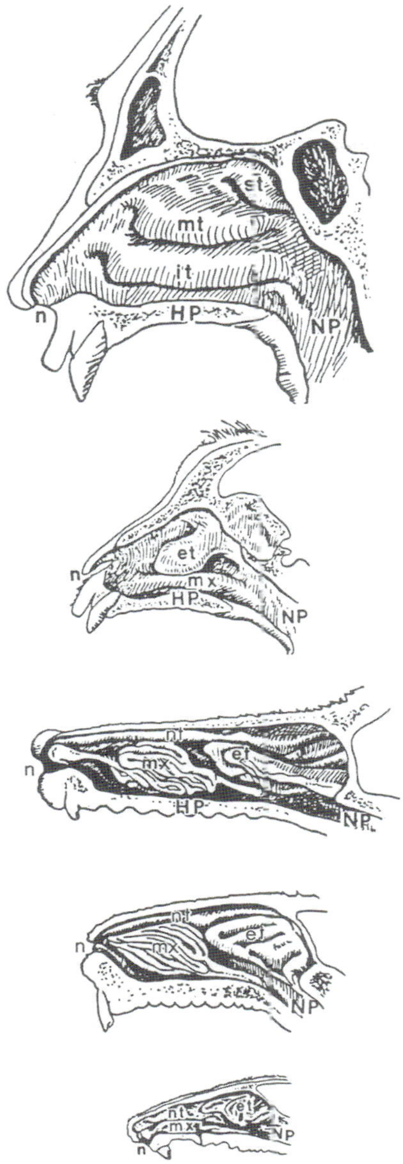

Figure 30.1 Representation of the exposed mucosal surface of the nasal lateral wall of the human, monkey, dog, rabbit and rat. HP = hard palate; N = naris; NP = nasopharynx; et = ethmoturbinate; nt = nasoturbinate; mx = maxilloturbinate; mt = middle turbinate; it = inferior turbinate; st = superior turbinate. (From Harkema *et al.*, *The Airway Epithelium: Physiology, Pathophysiology, and Pharmacology*, Vol. 55, Farmer, F.G. and Hay, D.W.T., eds., Marcel Dekker, New York, 1991. With permission).

the shape of nasal turbinates. Differences in the complexity of turbinate structure among humans, monkeys and other mammalian species are illustrated in Figure 30.1. Each nasal passage of the macaque monkeys contains a dorsally located ethmoturbinate and a more ventral maxilloturbinate. The human nose has three turbinates – the superior, middle, and inferior. These intranasal structures of primates are relatively simple in shape compared to the turbinates of most nonprimate laboratory species (e.g., dog, rat, mouse, rabbit) that have complex folding and branching patterns. In laboratory rodents (e.g., rat, mouse, hamster, guinea pig), evolutionary pressures, concerned chiefly with olfactory function and dentition, have defined the shape of the turbinates and the type and distribution of the cells lining the turbinates. The complex maxilloturbinates of small laboratory rodents and rabbits may provide far better protection of the lower respiratory tract, by better filtration, absorption, and disposal of airborne particles and gases, than do the simple turbinates of the primate nose. The highly complex shape of the predominantly olfactory neuroepithelium lined ethmoturbinates, in the distal half of the nasal cavity of small laboratory animals, is suitably designed for acute olfaction.

Cellular composition

Besides the differences in the gross architecture of the nose between different laboratory animal species, there are also species differences in the surface epithelial populations of the mucosal tissue lining the nasal passages. These differences between species are found in the distribution of nasal epithelial populations and in the types of nasal cells within these populations. Three distinct surface epithelia are located in specific positions within the main nasal chamber of most mammalian species, including the nonhuman primate (Harkema *et al.*, 1991). These include the squamous epithelium, primarily restricted to the nasal vestibule, ciliated respiratory epithelium in the main chamber and nasopharynx, nonciliated transitional epithelium lying between squamous epithelium and the respiratory epithelium, in the proximal or anterior aspect of the main chamber, and olfactory epithelium located in the dorsal or dorsoposterior aspect of the nasal cavity. Figure 30.2 illustrates the distribution of these distinct epithelial cell populations in the nasal cavity of the macaque monkey and laboratory rat.

The major difference in nasal epithelium between animal species is the percentage of the nasal airway that is covered by olfactory epithelium. For example, the olfactory epithelium covers a much greater percentage of nasal cavity in rats, which have a more acute sense of smell, compared to that in monkeys or humans. Approximately 50% of the nasal cavity surface area in the laboratory rat is lined by this sensory neuroepithelium (Gross *et al.*, 1982). Olfactory epithelium covers only 14% of the nasal cavity of the adult rhesus monkey (Gross *et al.*, 1987). In humans, this nasal epithelium is limited to an area of about 500 mm^2, which is

TABLE 30.1: Interspecies comparison of nasal cavity characteristics

	Sprague-Dawley Rat	Guinea Pig	Beagle Dog	Rhesus Monkey	Man
Body weight	250 g	600 g	10 kg	7 kg	~ 70 kg
Naris cross-section	0.7 mm^2	2.5 mm^2	16.7 mm^2	22.9 mm^2	140 mm^2
Bend in naris	40°	40°	30°	30°	
Length	23 cm	3.4 cm	10 cm	5.3 cm	7–8 cm
Greatest vertical diameter	9.6 mm	12.8 mm	23 mm	27 mm	40–45 mm
Surface area (both sides of nasal cavity)	10.4 cm^2	27.4 cm^2	220.7 cm^2	61.6 cm^2	181 cm^2
Volume (both sides)	0.4 cm^3	0.9 cm^3	20 cm^3	8 cm^3	16–19 cm^3 (does not include sinuses)
Bend in nasopharynx	15°	30°	30°	80°	~90°
Turbinate complexity	Complex scroll	Complex scroll	Very complex membranous	Simple scroll	Simple scroll

Reprinted with permission from: Nasal Airway Anatomy and Inhalation Deposition in Experimental Animals and People. In *Nasal Tumors in Animals and Man. Vol. III. Experimental Nasal Carcinogenesis,* G. Reznik and S.F. Stinson, eds., Copyright CRC Press, Inc., Boca Raton, FL, 1–26 (1983).
Source: Schreider, 1983.

only 3% of the total surface area of the nasal cavity (Sorokin, 1988). Mice, rabbits, and dogs are much closer to rats than to humans or monkeys with respect to the relative amount of olfactory epithelium within their nasal passages.

Olfactory epithelium

The olfactory epithelium is composed of three basic cell types (sensory, sustentacular, and basal cells)

TABLE 30.2: Morphometric analysis of three rhesus monkey noses

	#1	#2	#3
Weight	7.27 kg	6.99 kg	7.26 kg
Volume			
Nasal cavity	4.44 cm^2	5.94 cm^2	8.17 cm^2
Maxillary sinus	1.00 cm^2	3.75 cm^2	2.00 cm^2
Surface area			
Nasal cavity	48.70 cm^2	68.65 cm^2	75.55 cm^2
Maxillary sinus	7.48 cm^2	15.30 cm^2	10.4 cm^2

Source: Gross *et al.,* 1987.

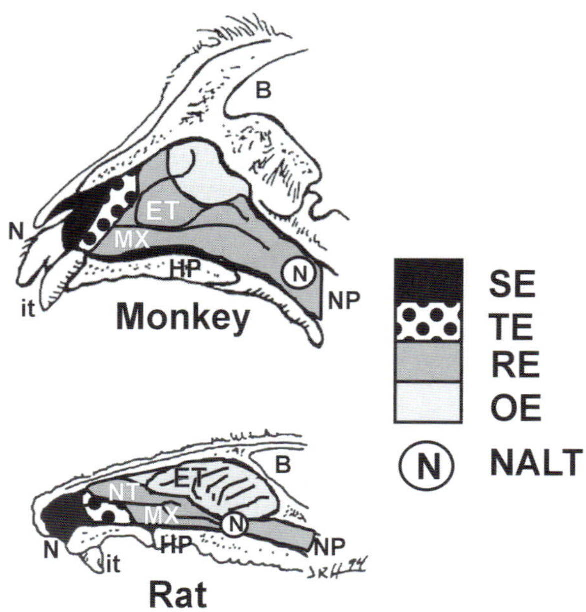

Figure 30.2 Distribution of nasal surface epithelia covering the nasal lateral wall and turbinates of the monkey and rat. SE = squamous epithelium, TE = nasal transitional (nonciliated cuboidal) epithelium; RE = Respiratory epithelium; OE = olfactory epithelium; NALT = nasal associated lymphoid tissue. (From Harkema, J.R., *Comparative Biology of the Normal Lung*, Parent, R.A., ed., Boca Raton, CRC Press Inc., 1992. With permission).

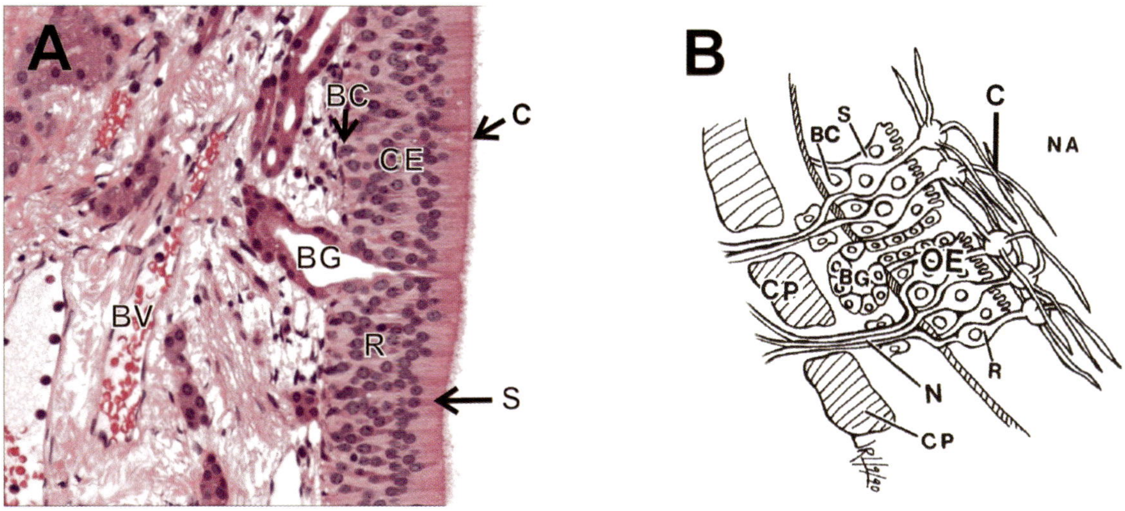

Figure 30.3 (A) Light photomicrograph of nasal olfactory epithelium (OE) from the nasal septum of a rhesus monkey. (B) Cartoon of the OE and underlying lamina propria. BC = basal cell; BG = Bowman's gland; BV = blood vessel; C = immotile cilia of olfactory sensory receptor cells (R); CP = cribiform plate between nasal and cranial cavities; S = sustentacular (support) cell; N = olfactory nerve in lamina propria; NA = Nasal airway.

(Figure 30.3). The olfactory sensory cells (or receptor cells) are bipolar neurons interposed between the sustentacular cells. The dendritic portions of these neurons extend above the epithelial surface and terminate into a bulbous olfactory knob from which protrude 12 or more immotile cilia. These cilia are enmeshed with each other and with microvilli in the surface fluid and provide an extensive surface area for reception of odorants. The axon of the olfactory sensory cell originates from the base of the cell and passes through the basal lamina to join axons from other sensory cells forming nonmyelinated nerves in the lamina propria. These axons perforate the cribiform plate to synapse with neurons in the olfactory bulb of the brain. Unlike other neurons in the body, the olfactory sensory cell can regenerate, having a 28- to 30-day turnover rate in the rat (Graziadei, 1977). Basal cells are generally considered the progenitor or stem cells for the regenerating olfactory epithelium. Regeneration of olfactory epithelium, after experimental injury, has been found to be an excellent model for the study of neurogenesis and axon regeneration in mammals.

Sustentacular cells in olfactory epithelium have been considered as support cells for the sensory cells. These cells have abundant smooth endoplasmic reticulum (SER) and xenobiotic-metabolizing enzymes (e.g., esterases, cytochrome P-450). The metabolism in these cells may be important in detoxification of inhaled xenobiotics and in the function of smell.

Other important sites of xenobiotic metabolism, associated with olfactory epithelium, are the Bowman's glands. These structures, located in the underlying lamina propria and interspersed among the olfactory nerve bundles, are simple tubular-type glands composed of small compact acini. Ducts from these glands transverse the basal lamina at regular intervals and extend through the olfactory epithelium to the luminal surface. Bowman's glands contain copious amounts of neutral and acidic mucosubstances that contribute to the mucous layer covering the luminal surface of the olfactory epithelium.

With few exceptions, the olfactory epithelium has greater xenobiotic metabolizing activity than does the respiratory epithelium in most animal species (Dahl and Hadley, 1991). Immunohistochemical analyses suggest that sustentacular cells, in the olfactory epithelium, and Bowman's glands, in the underlying lamina propria, tend to have especially high concentrations of xenobiotic-metabolizing enzymes. The presence of these enzymes in the olfactory epithelium readily explains numerous observations of nasal toxicity in animals exposed to certain inhaled toxicants. With its generally higher enzyme activity, the olfactory mucosa is more sensitive to the toxic effects of many metabolized materials.

Squamous epithelium

The nasal vestibule is completely lined by squamous epithelium. It is a stratified epithelium composed of basal cells along the basal lamina and several layers of squamous cells, which become progressively flatter toward the luminal surface. It has been estimated that this epithelium covers only 12% of the total nasal surface area in the rhesus monkey (Gross *et al.,* 1987). This nasal epithelium, covering the most proximal aspect of the nasal airway, probably functions like the epidermis in the skin, to protect the underlying tissues from potentially harmful atmospheric agents (Figure 30.4).

Transitional epithelium

Distal to the stratified squamous epithelium, and proximal to the ciliated respiratory epithelium, is a narrow zone of nonciliated, microvilli-covered surface epithelium, which has been referred to as nasal, nonciliated, respiratory epithelium or nasal transitional epithelium (Figure 30.4). In the rhesus monkey, the transitional epithelium makes up less than 0.5% of the total nasal epithelium (Gross *et al.,* 1987). Common, distinctive features of this nasal epithelium in all laboratory animal species and humans include: (1) anatomical location in the proximal aspect of the nasal cavity between the squamous epithelium and the respiratory epithelium, (2) the presence of nonciliated cuboidal or columnar surface cells and basal cells, (3) a scarcity of mucous (goblet) cells and a paucity of intraepithelial mucosubstances, and (4) an abrupt morphological border with squamous epithelium, but a less abrupt border with respiratory epithelium.

In rodents, this surface epithelium is thin (i.e., one to two cells thick), pseudostratified, and composed of three distinct cell types (basal, cuboidal, and columnar) (Monteiro-Riviere and Popp, 1984). In contrast, transitional epithelium in monkeys is thick (i.e., four to five cells), stratified, and composed of at least five different cell types (Figure 30.5). The luminal surfaces of transitional epithelial cells, lining the nasal airway, possess numerous microvilli. Luminal, nonciliated cells in the transitional epithelium of rodents have no secretory granules but do have abundant SER in their apices (Harkema and Plopper, 1987a) (Figure 30.5). SER is an important intracellular site for xenobiotic metabolizing-enzymes, including cytochromes P-450. Though the transitional epithelium normally contains few or no secretory cells in monkeys and rodents, exposure to high ambient concentrations of irritating pollutants, like ozone, can cause a rapid appearance of numerous mucous-secreting cells (i.e., mucous cell metaplasia) (Harkema and Plopper, 1987a; Harkema and Hotchkiss, 1999).

Respiratory epithelium

Most of the non olfactory nasal epithelia of laboratory animals and humans are ciliated respiratory epithelium (Figure 30.6). Approximately 75% of the nasal cavity in the adult rhesus monkey is lined by respiratory epithelium (Gross *et al.,* 1987), compared to only 46% in the laboratory rat (Gross *et al.,* 1982). Though this pseudostratified nasal epithelium is similar to ciliated epithelium lining other proximal airways

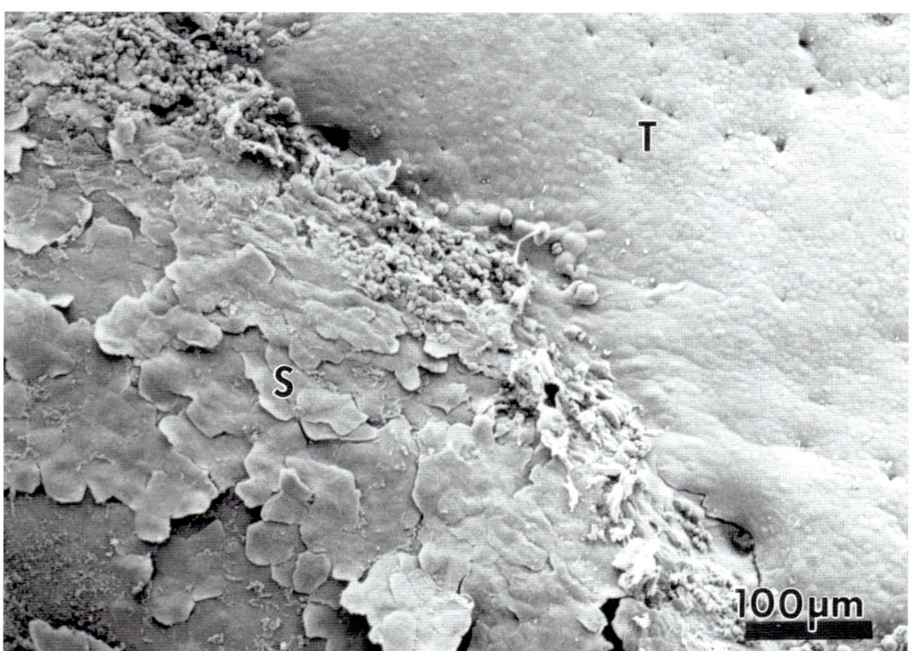

Figure 30.4 Scanning electron photomicrograph of the junction between nasal squamous epithelium (SE) and nonciliated cuboidal (transitional) epithelium in the bonnet monkey (*Macaca radiata*). (From Harkema, J.R. *et al., Am. J. Anat.* 180: 266–79, 1987. With permission).

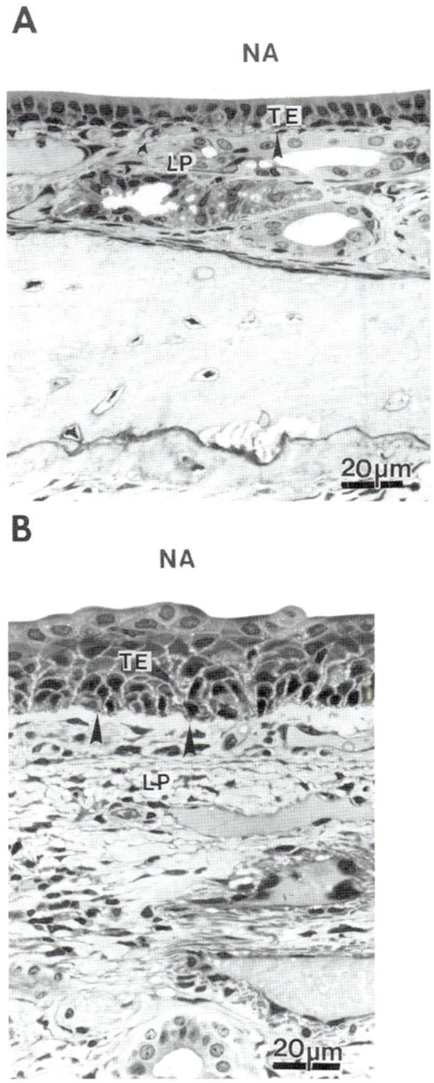

Figure 30.5 Light photomicrograph of nasal nonciliated cuboidal epithelium in a F344 rat (A) and in a bonnet monkey (B). E = nasal epithelium; L = airway lumen; LP = underlying lamina propria. The rat epithelium is thin, one or two cells in thickness, and pseudostratified, while the monkey epithelium is thick, four to six cells in thickness, and stratified.

(i.e., tracheobronchial airways), it also has unique features. Harkema and Plopper (1987b) reported that, like the respiratory epithelium of other mammals, the nasal respiratory epithelium of bonnet monkeys is composed primarily of ciliated cells, mucous goblet cells, and basal cells. However, the nasal respiratory epithelium of this macaque monkey, unlike that of the rat, also contains small mucous granule cells and cells with intracytoplasmic lumina (Figures 30.6 and 30.7). Brush cells, that are present in the nasal respiratory epithelium of laboratory rodents, are not found in the nasal epithelium of monkeys. A comparison of the abundance and percentage of epithelial cell types in the nasal respiratory and transitional epithelium, lining the anterior nasal septum of the bonnet monkey, is presented in Table 30.3

Vomeronasal organ

The vomeronasal organ (or Jacobson's organ) is a paired tubular diverticulum located in the vomer bone in the ventral portion of the proximal nasal septum of most mammals. Like the olfactory epithelium, it is a chemosensory structure that contributes to the sense of smell, in macrosmotic species (e.g., laboratory rodents, dogs, rabbits). In laboratory rodents, the lateral wall of this organ is lined with tall columnar, respiratory-like, epithelium (nonchemosensory), while the medial wall is lined with a sensory neuroepithelium (chemosensory) similar in morphology to the olfactory epithelium lining the main nasal chamber. Vomeronasal sensory neurons project from the vomeronasal organ to the accessory olfactory bulb of the brain. The lumen of the vomeronasal organ communicates, anteriorly, with the nasopalatine duct. Therefore, the vomeronasal chemosensory system may detect pheromones and other chemicals in both the oral or nasal cavities.

The presence and functionality of the vomeronasal organ in primate species is variable (Smith and Siegel, 2001). The vomeronasal organ has been identified in New World monkeys, prosimians, chimpanzees and even humans. New world monkeys and prosimians have well developed vomeronasal organs with a sensory epithelium. However, the vomeronasal organs of chimpanzees and humans are nonchemosensory homologues consisting of bilateral septal tubes lined only by nonsensory ciliated epithelium. Macaques have no structures that resemble the vomeronasal organs of either prosimians or humans.

Immune tissues

Scattered throughout the lamina propria of the nasal mucosa are lymphocytes, plasma cells and mast cells that are important participants in the immune response. Antibodies produced by plasma cells protect the nasal airways against inhaled antigens and invasion of infectious agents. In addition to these widely scattered individual or small aggregates of lymphoid cells, there are larger focal sites of discrete lymphoid tissue, designated as nasal-associated lymphoid tissue (NALT). They are located in the nasopharyngeal mucosa of humans and laboratory animals, including monkeys. NALT is more abundant in macaque monkeys than in laboratory rodents and is located on both the lateral and septal walls of the monkey's proximal nasopharynx

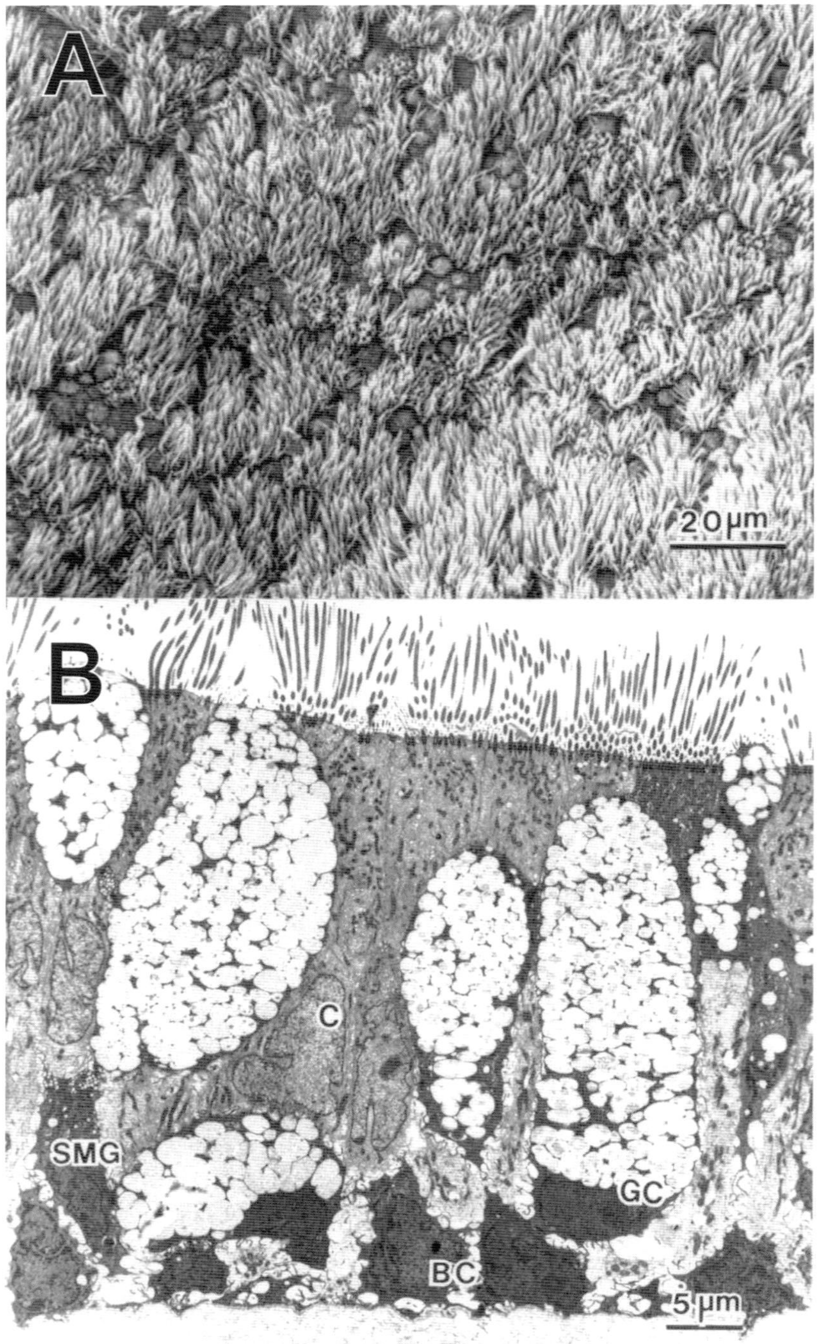

Figure 30.6 (A) Scanning and transmission (B) electron photomicrographs of the luminal surface of nasal ciliated respiratory epithelium from the anterior nasal septum of a bonnet monkey. C = ciliated cell; GC = goblet cell; SMG = small mucous granule cell; BC = basal cell. (From Harkema, J.R. *et al., Am. J. Anat.* 180: 266–79, 1987. With permission).

(Harkema and Plopper, 1987b). The correlate of NALT in humans is Waldeyer's ring, the oropharyngeal lymphoid tissues composed of the adenoid and the bilateral tubular, palatine, and lingual tonsils (Brandtzaeg, 1984). A lymphoepithelium covers the luminal side of NALT and is composed of lymphoid cells and nonciliated, cuboidal cells with luminal micovilli similar to membranous cells (i.e., M cells) found in the gut- and bronchus-associated lymphoid tissues (GALT and BALT, respectively). There are few, if any, mucous cells or ciliated cells in this specialized airway epithelium. The nonciliated cuboidal cells are thought to be involved in the uptake and translocation of inhaled antigen from the nasal lumen to the underlying

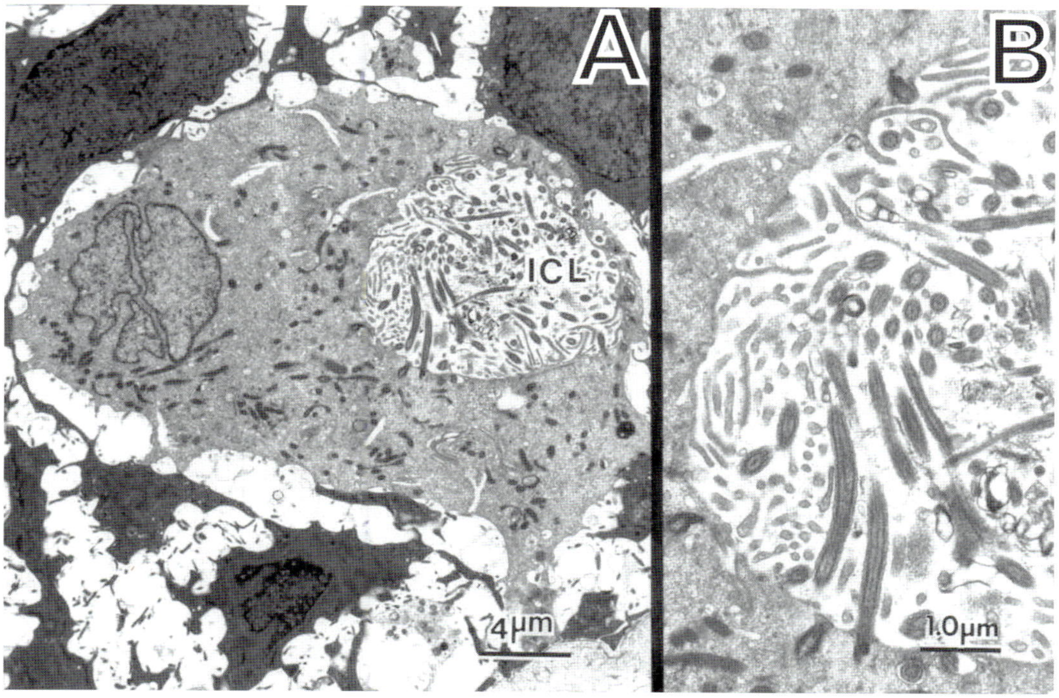

Figure 30.7 (A) Transmission electron photomicrographs of an epithelial cell with an intracytoplamic lumen (arrow). Bar = 3 microns. (B) The surface of the intracytoplasmic lumina is lined by cilia and long microvilli. Bar = 1 micron. (From Harkema, J.R. *et al., Am. J. Anat.* 180: 266–79, 1987. With permission).

lymphoid structures. The location of NALT, at the entrance of the nasopharyngeal duct, is a very strategic position as most of the nasal secretions and inhaled air, both presumably laden with antigenic material, pass over this area. Though the function of NALT, and its place in the general mucosal-associated lymphoid system, are not fully understood, these mucosal lymphoid tissues presumably have an important function in regional immune defense of the upper airways.

Pharynx

The pharynx connects the nasal and oral airways with the laryngeal airway. In human and nonhuman primates, like the rhesus monkey, the pharynx is situated posterior to the nasal cavity, mouth, and larynx. In many other laboratory mammals (e.g., rodent and dog), apart from the anterior portion of the nasopharynx, that lies

TABLE 30.3: Abundance and percentage (%) of epithelial cell types in transitional and respiratory epithelium of bonnet monkey anterior nasal septum (mean ± standard error of the mean)

Epithelium	N	Total no. of nuclei[a]	% Basal cells	Small mucous granule cells	Nonciliated cells without secretory granules	Nonciliated cells with few secretory granules	Goblet cells	Ciliated cells	Cells with intracytoplasmic lumina
Transitional	4	436 ± 26	35.5 ± 1.5	22.2[b] ± 4.0	31.5 ± 5.0	7.6 ± 2.2	3.2[b] ± 0.6	–	–
Respiratory	6	358 ± 32	39.5 ± 1.7	4.9 ± 1.2	–	–	19.1 ± 2.0	35.2 ± 2.1	1.3 ± 0.9

[a]Number of nuclei per mm of basal lamina.
[b]Significantly different (p <0.05) from respiratory epithelium.
(From Harkema, J.R. *et al., Am. J. Anat.* 180: 266–79, 1987b. With permission.)

ventral to the distal aspect of the main nasal cavity, the pharynx is distal to most of the nasal airway and dorsal to the oral cavity and larynx. The pharynx is a musculomembranous tube which is approximately 130 mm in the adult human, 115 mm in the adult male beagle dog, 35 mm in the adult male rhesus monkey and 22 mm in the adult Sprague-Dawley rat (Schreider and Raabe, 1981). It can be anatomically divided into nasal, oral, and laryngeal regions. The nasopharynx is lined with ciliated respiratory epithelium with mucous goblet cells, while the oropharynx and laryngopharynx are lined by non-keratinized, squamous epithelium.

Larynx

The larynx of the rhesus macaque is organized like that of most mammals. There are five primary cartilages: the epiglottis, the u-shaped thyroid cartilage, paired arytenoid cartilages with muscular, vocal, cornual processes and the ring-shaped cricoid cartilage. As with humans, the cuneiform cartilage is in the aryepiglottic fold interspersed between arytenoids and epiglottic cartilages. Principle paired muscles include the lateral and posterior cricoarytenoideus, cricothyroideus and vocalis, along with the unpaired arytenoideus transversus. Stratified squamous epithelium lines the epiglottis, false vocal cords, and arytenoid cartilages (Sutton *et al.,* 1977; Stearns and Cummings, 1982). The respiratory epithelium, with mucous goblet cells, basal cells and ciliated cells, begins distal to the free margin of the vocal cord. This boundary with stratified squamous epithelium varies with age (Stearns and Cummings, 1982). One of the major differences between the rhesus larynx and the human larynx is the presence of the laryngeal sacs which are located in the hyoid apparatus and open into the lateral aspect of the anterior larynx through the ostium (Hilloowala and Lass, 1978). Other differences include the size of the overall larynx as well as the internal air passage (Fitch, 1997; Morgan *et al.,* 1991; Patra *et al.,* 1986; Schreider and Raabe, 1981; Taylor *et al.,* 1976), and the position of the larynx in the cervical region (Flugel and Rohen, 1991).

Lung organization

As illustrated in Figure 30.8, once the trachea enters the thoracic cavity it divides in the mediastinum into two primary bronchi which supply air to the left and right lungs. The bronchi traverse the thoracic cavity for a short distance before dividing into the lobar bronchi which supply each lung. Table 30.4 summarizes the general organizational characteristics of the respiratory system in the rhesus monkey and compares it to the Swiss Webster mouse, the Sprague-Daley rat, and the adult male human. The lung volume of the rhesus monkey is approximately 8% of body weight. This is approximately twice what it is in mice and rats and slightly larger than it is in humans. The lungs of the rhesus monkey are generally separated into six lobes (Table 30.4). The left lung has two lobes, cranial and caudal, of which the cranial is further segmented into a cranial and a caudal portion. In most cases, the cranial and caudal lobes are completely separated from each other and supplied by independent lobar bronchi and pulmonary vessels. The cranial and caudal segments of the cranial are fused to each other near the lobar bronchus lobe. This is in marked contrast to mice and rats where the entire left lung is fused into a single lobe, with one large axial airway and many smaller minor daughter side branches. Humans have a similar division with a proximal portion fused to varying degrees.

The right lung has four distinct lobes in the rhesus monkey. In approximately 20% of cases there will be a small amount of fusion in the most proximal portions, especially between the caudal and middle lobes and the caudal and accessory lobes. This is in contrast to mice and rats where each lobe is completely separate. Humans have three lobes in the right lung and these are fused to variable degrees in the most proximal portion.

The lungs are lined by a connective tissue band with a mesothelial surface facing the plural space. This connective tissue band is remarkably thinner in monkeys than it is in humans but still much thicker than it is in mice and rats. As with the lungs of many larger species such as pigs, cows, and horses, the human lung has extensive interlobular and intralobular connective tissue which joins major vessels and the bronchi to the plural surface. In smaller species, such as the rhesus monkey, there is little interlobular connective tissue.

Tracheobronchial airways

Architecture

As illustrated in Figures 30.8 and 30.9, the conducting airways from the trachea, and its division into primary bronchi, form a complex series of branching tubes

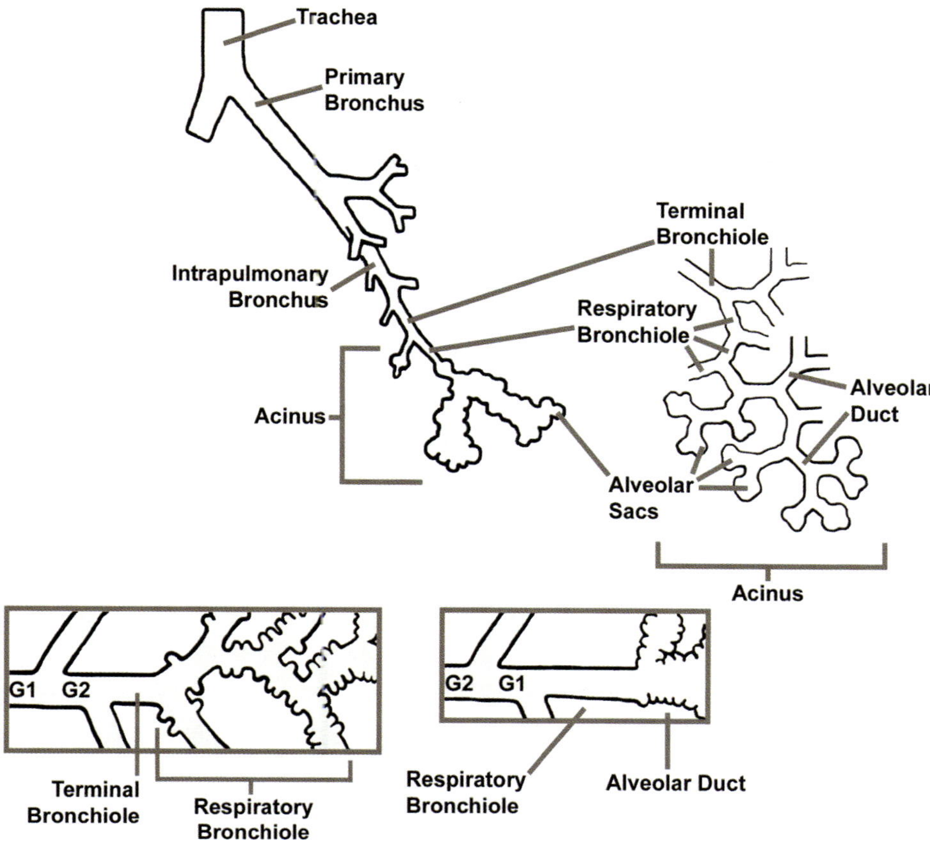

Figure 30.8 Diagrammatic representation of the organization of airspaces in the mammalian respiratory system, including trachea, primary bronchi, intrapulmonary bronchi and the acinus. Primates, including rhesus monkeys and humans, have an extensive area of transition between alveolar ducts and the most distal conducting airway, the terminal bronchiole, in the central portions of the acinus. This zone of transition is called respiratory bronchioles because the majority of the wall is structured like a bronchiole, and bronchiolar epithelium is interrupted by alveolar outpockets organized for gas exchange with epithelial and endothelial cell composition of interalveolar septa in the alveolar ducts. Compare with casts of acute airways in Figure 30.9.

TABLE 30.4: Comparison of species differences in gross anatomy of the lungs

Parameter	Monkey	Mouse	Rat	Human
Body Weight (kg)	3.71	0.023 ± 0.002	0.360 ± 0.004	74 ± 4
Lung Volume (Fixed, ml)	2393 ± 100	1.1 ± 0.05	11.4 ± 1.2	4,341 ± 285
Lung Volume/Body Weight	0.0825	0.048	0.032	0.059
Location:				
Number	6	5	5	5
Left Lung	Cranial	Single	Single	Superior
	Cranial Segment			
	Caudal Segment			
	Caudal			Inferior
Right Lung	Cranial	Cranial	Cranial	Superior
	Middle	Middle	Middle	Middle
	Caudal	Caudal	Caudal	Inferior
	Accessory	Accessory	Accessory	
Pleura	Thin	Thin	Thin	Thick
Interlobular Connective Tissue	Little	Little, if any	Little, if any	Extensive

Compiled from: Harkema *et al.*, 1991; McBride, 1992; Mercer *et al.*, 1987; Tyler and Julian, 1992; Weibel, 1989.

which extend to the gas exchange area. The more proximal of these branches are termed bronchi and are usually characterized by their histological composition, including the presence of mucus and basal cells in the epithelium, some mucosal glands in the interstitium and a significant amount of cartilage in the interstitial spaces. More distally, the airway branches are called bronchioles because the wall is thinner, there is a reduction in the complexity of the airway epithelial population, the preponderance of the wall is composed of smooth muscle and there is little to no cartilage. Figure 30.9 compares the branching pattern in weaning-age animals: a Sprague-Daley rat (28 days postnatal age) and a rhesus monkey (6 months postnatal age). Aside from the substantial differences in airway size, there are clear disparities in how the airways themselves are organized. In monkeys, airways branch at a greater degree of angle in the more proximal airways and the two-paired daughter branches are at nearly the same angle. However, the overall pattern of distribution of airways is quite similar between spaces. The tracheobronchial airways occupy approximately 2% of the lung volume in the rhesus monkey, in contrast to that in mice and rats (Table 30.5). In primates, cartilage is found in the walls of the tracheobronchial airways, from the trachea distally to the smallest bronchioles (Table 30.5). In the distal bronchioles of humans and rhesus monkeys, cartilage is restricted to a small zone in the bifurcation area. By contrast, cartilage ends at the lobar bronchus in mice and rats. All species have a significant number of generations of intrapulmonary airways that are very thin walled and with minimal cartilage, i.e. non-respiratory bronchioles. The organization of the zone of transition, between conducting airways and the gas exchange area, separates the lungs of primates

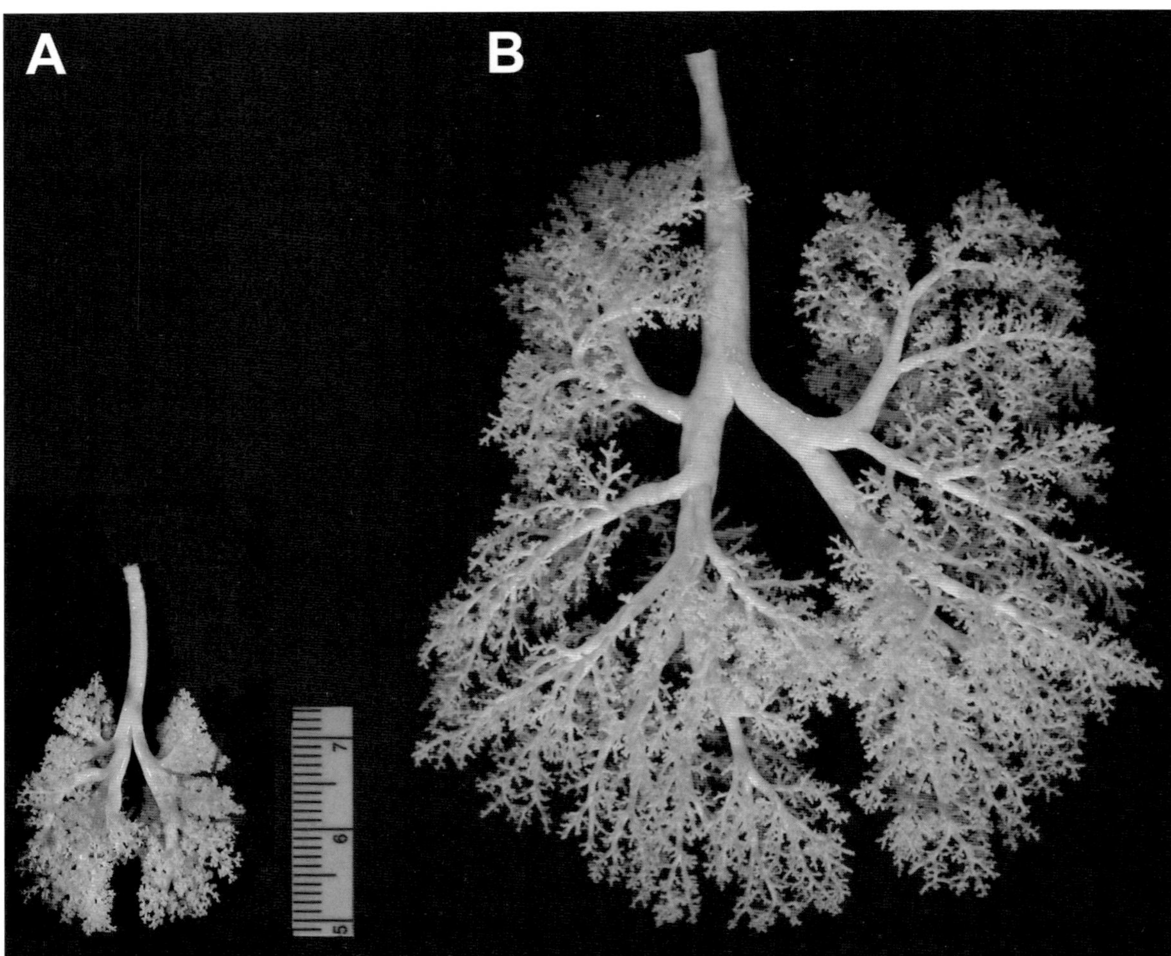

Figure 30.9 Comparison of the organization of the airspaces of the tracheobronchial airways represented by silicon casts. Both the rat (A) and rhesus monkey (B) were at approximately weaning age when the lungs were fixed and the casting material introduced by negative pressure to fill down to the respiratory bronchioles. The trachea of each animal is at the top. Compare to Figure 30. 8.

TABLE 30.5: Comparison of species differences in tracheobronchial airway organization

Parameter	Monkey	Mouse	Rat	Human
% of lung	1.8	11	5.7	
Cartilage in wall	Trachea to Distal Bronchiole	Trachea Lobar Bronchi	Trachea Lobar Bronchi	Trachea to Distal Bronchiole
Nonrespiratory Bronchioles	Several Generations	Several Generations	Several Generations	Several Generations
Respiratory Bronchioles	Several Generations	None or one	None or one	Several Generations
Generations to Alveolarized Bronchiole (Axial path)	13 – 17	13 – 17	13 – 17	17 – 21
Branching Pattern	Dichotomous/ Tricotomous	Monopodial	Monopodial	Dichotomous

Compiled from: Mariassy, 1992; Mercer and Crapo, 1981; McBride, 1992; Miller and Mercer, 1993; Plopper and Hyde, 1992; Tyler and Julian, 1992.

and carnivores from other mammalian species. In primates and carnivores there is an extensive transition zone, with the walls of the distal airways having a mixture of bronchiolar epithelial sub-populations mixed with the alveolar gas exchange area. The average number of branches from the trachea to the bronchioles is approximately the same for most mammalian species (Table 30.5). Branching itself is relatively unique to primates, with branches separating from the parent airway at an approximately 45 degree angle and being almost uniformly equal in size and in diameter (i.e. dichotomous branching). This is also true for the human lung.

Cellular composition

As summarized in Figure 30.10, the walls of tracheobronchial airways are highly complex cellular structures. All of the compartments making up the wall interact closely with each other in determining function and are present to varying degrees in all species. Figure 30.11 compares the histological composition of conducting airways in the Swiss Webster mouse and the adult male rhesus monkey. The full extent of the walls of more proximal airways is not illustrated. All the images are at the same magnification. In all species, the interstitium of the tracheal wall contains C-shaped cartilages and there is a band of smooth muscle which joins the open end of the cartilages (Table 30.6). The trachea and proximal airways have extensive sub-mucosal glands beneath the epithelium in rhesus monkeys and humans. These glands are present to a variable extent in smaller laboratory species.

As emphasized in Figure 30.11, there is a substantial difference in the amount of epithelium that lines the luminal surface in the trachea. The thickness of the epithelium in the trachea of rhesus monkeys is approximately twice that of mice and rats and one half to one-third that in humans (Table 30.6). Other major differences between species are the composition of the epithelium, the density of cells lining the surface, and the proportion of cell phenotypes in the epithelium (Table 30.6). A substantial percentage of the airway in primates is occupied by mucus cells, but they are not generally found to the same extent in the trachea of healthy, pathogen-free mice and rats. The proportion of cilated cells in the epithelium is relatively similar in all species, but the proportion of basal cells varies with species. As would be expected with the differences in cell populations that line the trachea in different species, there is considerable variation in the carbohydrate content of the secretory product (Table 30.7). Primates, in general, have a more heavily sulfated secretory product that is not usually found in laboratory mammals (Table 30.7). The difference in secretory product composition is also reflected, to some degree, in the composition of the carbohydrates in tracheal sub-mucosal glands (Table 30.8).

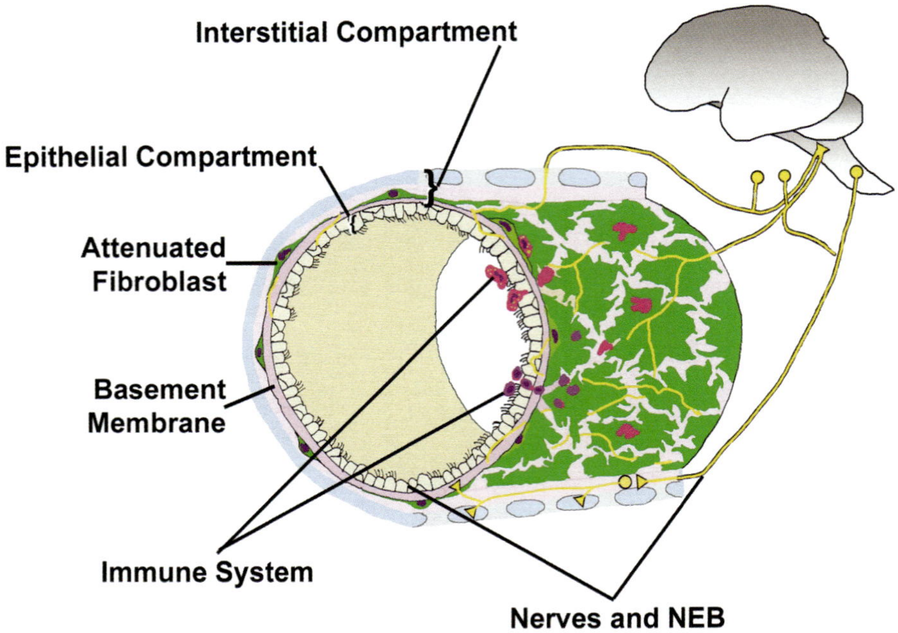

Figure 30.10 Diagrammatic representation of the wall of a tracheobronchial airway with the compartments which vary by species: epithelium, interstitium, nerves and immune/inflammatory cells. The interstitial compartment includes smooth muscle, fibroblasts, cartilage, and an extensive basement membrane zone, found in primates.

As is summarized in Figure 30.11, the organization of the tracheobronchial airways varies by airway generation within the airway tree. The largest, most proximal intrapulmonary bronchi exhibit some of the most organizational variation between species (Table 30.9). While smooth muscle is present in the walls of all mammalian species, there is a substantial difference in the amount of cartilage found in the lobar bronchus in laboratory mammals and in the distribution of sub-mucosal glands. Epithelium is reduced in thickness in more distal airways compared to the trachea (See Tables 30.10 and 30.11). There also are major differences in the organization of the surface epithelial population, with mucus cell and basal cells predominating in primates and Clara cells being the principal nonciliated cell population in other laboratory animals (Table 30.9). These differences are even more marked in more distal conducting airways such as the mid-level bronchi (Table 30.10). While much of the data regarding the organization of mid-level airways is lacking, what is available suggests that the secretory product, as well as the extent of the attachment, i.e. basal cell population, is substantially different.

In the most distal conducting airways, the bronchioles, the major differences between species are related to the epithelial surface lining (Table 30.11). In laboratory mammals, the Clara cell is the primary secretory cell phenotype, there are no mucous cells, and the numbers of basal cells in the epithelium are related to the extent of alveolorization. The bronchioles of rhesus monkeys have an extensive smooth muscle portion which is arranged in large bundles and is interspersed with extensive connective tissue not generally observed in laboratory mammals (Figure 30.11).

Gas exchange area

Architecture

In all species, the majority of the lung consists of the alveolar gas exchange area (Table 30.12). In primates, this portion of the lung makes up more than 90% of lung volume with alveolar air space taking up almost 80% of that volume. Alveolar surface area is closely related to metabolic body weight. The capillary bed, lining individual alveoli in the septa, has a surface area roughly comparable to the alveolar surface area for adequate transport of gas to and from exchange surface. Capillary blood volume also varies in relationship to the extent of the gas exchange area. The size of individual alveoli varies in proportion to body size, being larger in bigger species (Table 30.12).

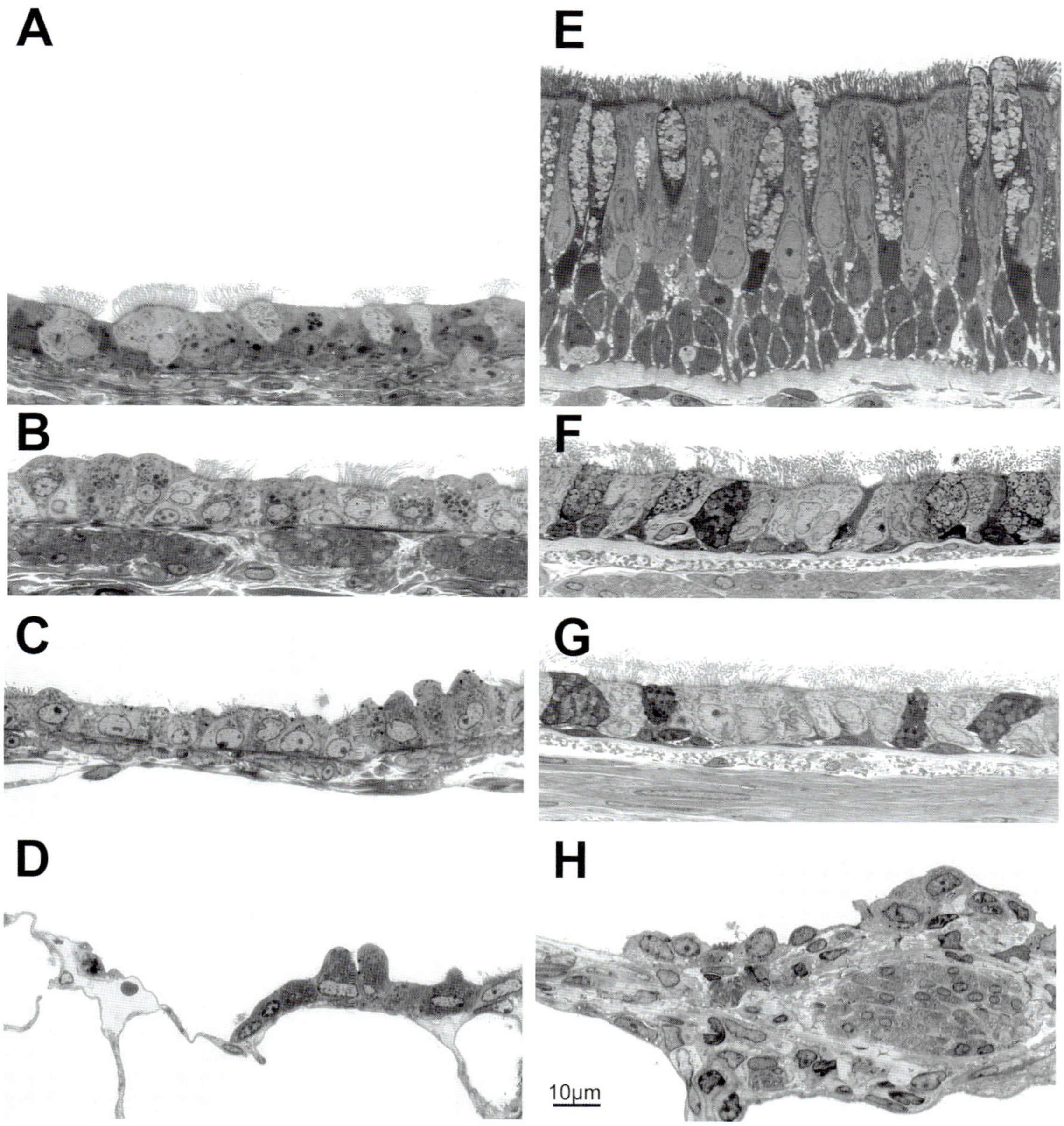

Figure 30.11 Histological comparison of the airspace walls in the same airway generations of an adult male rhesus monkey (E,F,G,H) and an adult male Swiss Webster mouse (A,B,C,D). The luminal airspace in the upper part of each field the epithelium is set so the basal lamina matches for both species. The epithelium is more complex and taller in the proximal airways of monkeys than at the same site in mice. Smooth muscle occupies a larger portion of the interstitium in monkeys than nice at all airway levels. All micrographs are at the same magnification.

Cellular composition

The cellular and acellular parts of the tissue surrounding individual alveolar air pockets makes up around 10% of all the tissue in the lungs (Table 30.13), the rest being walls of blood vessels and conducting airways. The alveolar wall is composed of capillary epithelium, connective tissue and alveolar epithelium on the airside. The cellular and acellular compartment for the blood-air barrier is thicker in primates than in small laboratory mammals (Table 30.13). The epithelium of the alveolar blood air barrier represents about a quarter of this thickness in primates and nearly a third in small laboratory mammals (Table 30.13). About 25% of the blood-air barrier thickness is made up of capillary endothelial cells, but this is much less in humans. The acellular interstitial compartment makes up a larger proportion of the barrier, in larger species, and ranges from 40% in mice to almost 60% in humans.

TABLE 30.6: Comparison of species differences in cells of the trachea

Parameter	Monkey	Mouse	Rat	Human
Wall:				
Smooth Muscle	Present	Present	Present	Present
Cartilage	Present	Present	Present	Present
Submucosal Glands	Present	Present (proximal 1/3)	Present	Present
Epithelium:				
Thickness (μm)	20–30	11–14	7–10	50–100
Cells/mm b.m.	181 ± 51	215	148 ± 3	303 ± 20
Mucous Goble cells (%)	17	<1	<1	9
Serous Cells (%)	<1	<1	} 44	} 9
Clara cells (%)	<1	49		
Ciliated Cells (%)	33	39	27	49
Basal Cells (%)	42	10	20	33
Other Cells (%)	8	1	4	—

Compiled from: Harkema *et al.*, 1991; Hyde *et al.*, 1992; Mariassy 1992; Mercer *et al.*, 1994; Plopper *et al.*, 1998; St. George *et al.*, 1988.

Overview of research uses

Nasal cavity

In recent years there has been a marked increase in nasal toxicology studies using laboratory animals, including nonhuman primates, to ensure the safety of intranasal delivery of pharmaceutical agents and to assess the potential human risk of nasal injury from the inhalation of ambient levels of environmental or occupational toxicants (Miller, 1995). For example, the nasal toxicology and pathology of inhaled ozone and formaldehyde has been extensively investigated using the laboratory monkey (Harkema and Plopper 1987a,b; Monticello and Morgan, 1989). Many inhaled nasal toxicants, like ozone and formaldehyde, cause injury to specific regions within the nasal passages and these site-specific nasal lesions are dependent on both the regional uptake of the inhaled toxicant and the regional sensitivity of the nasal mucosa. To better estimate regional uptake of inhaled toxicants in the nose, 3-dimensional, anatomically accurate, computational fluid dynamic models of the adult monkey, rat, and human nasal passages have been developed (Kepler and Richardson, 1998; Kimbell *et al.,* 2001). These sophisticated computational models, along with detailed mapping of toxicant-induced nasal histopathology in laboratory animals, have provided valuable insight into dose-response relationships, interspecies variability and mechanisms of nasal toxicity which all serve to reduce the uncertainty in human risk estimates from toxicant exposure.

Larynx

The larynx apparatus of macaque monkeys has been used as a model for defining factors controlling vocalization. The impact, on vocal sound quality, of the size of the larynx (Fitch, 1997), three-dimensional architecture (Rendall *et al.,* 1998), position in the neck (Flugel and Rohen, 1991) and the presence of a laryngeal sac (Hilloowala and Lass, 1978) has been characterized. The organization of neural structures, including chemoreceptors (Ide and Munger, 1980), muscle spindles (Larson *et al.,* 1974, Raman and Devanandan, 1989), and neural fiber networks (Murano *et al.,* 1993) have been defined. The functional characteristics of the extrinsic and intrinsic muscles, and their response to denervation, has been evaluated (Sahgal and Hast, 1974, 1986). The macaque larynx has also been used to develop and evaluate surgical approaches for reconstitution of laryngeal deficiencies (Crumley, 1991; Fearon and McMillin, 1985) and the role of central nervous system connections (Simonyan and Jurgens, 2002; Vanderhorst *et al.,* 2001; Sutton *et al.,* 1978).

TABLE 30.7: Comparison of species differences in carbohydrate content of clara (C), serous (S) and mucous (M) cells in tracheal epithelium

Parameter	Monkey	Mouse	Rat	Human
Abundance	+ +	+/–	+/–	+ + +
Periodic Acid Schiff	+ (M)	+ (M)	+ (S)	+ (M)
		+/– (C)	+ (M)	
Alcian Blue	+ (M)	+ (M)	– (S)	+
		– (C)	+ (M)	
High Iron Diamine	+ and – (M)	– (M)	– (S)	+ and – (M)
	and – (M)	– (C)	– (M)	

Compiled from: St. George and Wang, 1992; St. George *et al.*, 1988; Harkema *et al.*, 1991.

TABLE 30.8: Comparison of species differences in carbohydrate content of mucous (M) and serous (S) cells in tracheal submucosal glands

Parameter	Monkey	Mouse	Rat	Human
Periodic Acid Schiff	+ (S)	+ (S)	+ (S)	+ (S)
	+ (M)	+ (M)	+ (M)	+ (M)
Alcian Blue	– (S)	– (S)	– (S)	– (S)
	+ (M)	+ (M)	+ (M)	+ (M)
High Iron	– (S)	– (S)	– (S)	– (M)
Diamine	+ (M)	+/– (M)	+/– (M)	+ (M)

Compiled from St. George and Wang, 1992; St. George *et al.,* 1988; Harkema *et al.,* 1991.

TABLE 30.9: Comparison of species differences in proximal intrapulmonary airways

Parameter	Monkey	Mouse	Rat	Human
Wall:				
Smooth Muscle	Present	Present	Present	Present
Cartilage	Present	Absent	Absent	Present
Submucosal Glands	Present	Absent	Absent	Present
Epithelium:				
Thichkness (μm)	27	8–16	13	40–50
Cells/mm b.m.	175	109	116	?
Mucous Goblet Cells (%)	15	<1	<1	10
Serous Cells (%)	} 5	<1	20	} 3
Clara Cells (%)		61	<1	
Cillated Cells (%)	47	36	53	37
Basal Cells (%)	32	<1	14	32
Other (%)	2	2	12	18

Compiled from: Harkema *et al.,* 1991, 2000; Hyde *et al.,* 1992; Mariassy, 1992; Plopper *et al.,* 1994, 1998; St. George *et al.,* 1988.

TABLE 30.10: Comparison of species differences in midlevel intrapulmonary airways

Parameter	Monkey	Mouse	Rat	Human
Wall:				
Cartilage	Present	Absent	Absent	Present
Smooth Muscle	Present	Present	Present	Present
Submucosal Glands	Present/Absent	Absent	Absent	Present/Absent
Epithelium:				
Thickness	15	10–12	7.8	?
Cells/mm b.m.	?	?	?	?
Mucous Goblet Cells (%)	14	<1	} 36	9
Serous Cells (%)	<1	<1		} 2
Clara Cells (%)	<1	60		
Ciliated Cells (%)	49	40	64	37
Basal Cells (%)	29	<1	<1	29
Other Cells (%)	5	?	?	16

Compiled from: Harkema *et al.*, 1991, 2000; Hyde *et al.*, 1992; Mariassy, 1992; Plopper *et al.*, 1994, 1998; St. George *et al.*, 1988.

Trachea and lungs

The rhesus monkey has been used extensively as a model for diseases of the lower respiratory tract. These studies have included evaluation of therapeutic approaches to disease, definition of mechanisms of infection and of acute and chronic inflammatory diseases, and regulation of differentiated epithelial function.

A major emphasis in recent years has been testing strategies for gene therapy. A wide variety of vectors has been tested for safety and efficacy in adult animals. These vectors include adenoviruses (Beck *et al.*, 2002; Chirmule *et al.*, 2000; McDonald *et al.*, 1997; Scaria *et al.*, 2000; Shean *et al.*, 2002; St. George *et al.*, 1986; Afione *et al.*, 1996; Brody *et al.*, 1994); Sendai virus (Li *et al.*, 2000) and lentivirus (Tarantal *et al.*, 2001;

TABLE 30.11: Comparison of species differences in terminal bronchioles

Parameter	Monkey	Mouse	Rat	Human
Wall:				
Cartilage	Absent (Bifurcation)	Absent	Absent	Absent (Bifurcation)
Smooth Muscle	Present	Present	Present	Present
Submucosal Glands	Absent	Absent	Absent	Absent
Epithelium:				
Thickness	?	7–8	5–8	?
Cells/mm	?	?	?	?
Mucous Goblet (%)	} ~20	-0-	-0-	} 35
Serous Cells (%)		-0-	-0-	
Clara Cells (%)		60–80	35–60	
Ciliated Cells (%)	~50	40–20	65–40	52
Basal Cells (%)	~10	<1	<1	<1
Other (%)	~ 5	-0-	-0-	13

Compiled from: Harkema *et al.*, 1992, 2000; Mercer *et al.*, 1994; Miller *et al.*, 1993; Plopper *et al.*, 1994, 1998; Plopper and Hyde, 1992; Tyler and Julian, 1992.

TABLE 30.12: Comparison of species differences in the gas exchange area (parenchyma) of the lungs

Parameter	Monkey	Mouse	Rat	Human
Parenchyma (% of Lung Volume)	93.0 ± 0.7	82.9 ± 0.02	85.6 ± 2.3	90.0 ± 5.9
Alveolar Airspace (% of Lung Volume)	79.7 ± 0.2	65 ± 2	61.4 ± 3	78.02 ± 5.6
Parenchymal Tissue (% Lung Volume)	12.3 ± 0.2	17.9 ± 0.7	24.2 ± 0.02	11.77 ± 1.54
Alveolar Surface Area (cm^2)	133,000 ± 12,700	680 ± 85	7,006 ± 1,177	1,430,000 ± 120,000
Alveolar Surface/Body Weight	36,000	27,200	13,333	19,300
Capillary Surface Area (cm^2)	116,000 ± 15,400	590 ± 60	6,153 ± 1,165	1,260,000 ± 120,000
Capillary Blood Volume (ml)	15.5 ± 2.7	0.084 ± 0.009	0.63 ± 0.07	213 ± 31
Alveolar Size (MLI) µm	196	80	100	210

Compiled from: Crapo *et al.*, 1982; Geelhaar and Weibel, 1971; Gehr *et al.*, 1978; Gehr and Erni, 1980, 1981; Hugonnaud *et al.*, 1977; Kapanci *et al.*, 1969; Mercer and Crapo, 1987, 1992; Mercer *et al.*, 1987; Pinkerton *et al.*, 1992; Plopper *et al.*, 1993; Tyler *et al.*, 1988; Weibel *et al.*, 1981.

Larson *et al.*, 2000). Liposomes (McDonald *et al.*, 1997; Fortunati *et al.*, 1996) and perflubron (Weiss *et al.*, 2002) also have been tested as delivery systems using macaques.

The rhesus monkey has received increasing interest since it was established that simian immunodeficiency virus (SIV) produces many of the same clinical manifestations in macaques that the human version (HIV) does in man. The pathogenesis of infection (Durand-Joly *et al.*, 2000) has been defined and compared to that occurring in SIV-infected monkeys (Croix *et al.*, 2002; Board *et al.*, 2002; Mankowski *et al.*, 1998), as well as the impact of SIV-infection itself on other aspects of pulmonary immunology (Israel *et al.*, 1999; Soppor *et al.*, 2003) and viral infection (Fuller *et al.*, 2001). Another pathogen recently studied in SIV-infected macaques is *Mycobacterium tuberculosis* (Safi *et al.*, 2003; Lai *et al.*, 2003). Macaques also may serve as a good model for Q-fever (Waag *et al.*, 1999), sepsis (Todoroki *et al.*, 2000) and Legionnaire's disease (Baskerville *et al.*, 1983). Interest in macaques as models for respiratory infections has stimulated the use of the model for antimicrobial therapies and agents. Antimicrobial peptides in the respiratory tract (Bals *et al.*, 2001) have been characterized, as well as the response to physotigmine (Jeevaratham *et al.*, 1998), a thromboxane prostaglandin

TABLE 30.13: Comparison of species differences in composition of interalveolar septa

Parameter	Monkey	Mouse	Rat	Human
% of Lung Tissue	8.3 ± .03		12.6 ± 0.5	11.1 ± 0.9
Blood-Air-Barrier Thickness				
(µm) (Harmonic Mean):	0.67 ± 0.06	0.32 ± 0.01	0.38 ± 0.03	0.62 ± 0.04
% Alveolar Epithelium	24.4	36	31	24.9
Type I Cell			22.3	12.5
Type II Cell			8.2	12.4
% Capillary Endothelium	25.2	24	24.2	16.4
% Septal Interstitum	49.6	40	45.4	58.7

Compiled from: Crapo *et al.*, 1982; Geelhaar and Weibel, 1971; Gehr *et al.*, 1978; Gehr and Erni, 1980, 1981; Hugonnaud *et al.*, 1977; Kapanci *et al.*, 1969; Mercer and Crapo, 1987, 1992; Mercer *et al.*, 1987; Pinkerton *et al.*, 1992; Plopper *et al.*, 1993; Tyler *et al.*, 1988; Weibel *et al.*, 1981.

endoperoxide receptor antagonist (Ford-Hutchinson *et al.*, 1989); cyclosporine A (Komisar *et al.*, 2001); and interferon beta (Martin *et al.*, 2002). Experimental approaches to enhancement of immunity, by serum transfer, also has been evaluated (Dykewicz *et al.*, 1988; Patterson *et al.*, 1977 and 1978; Saban *et al.*, 1994; McDonald *et al.*, 1993a). Strategies also have been evaluated in rhesus monkeys for modifying the respiratory response to enterotoxin B (Lowell *et al.*, 1996; Tseng *et al.*, 1995).

The tracheobronchial airways of rhesus monkeys have been used to define the biology of mucociliary epithelium. The capability of the airways to metabolically activate xenobiotics has been evaluated for both the cytochrome P-450 monooxygenases (Daniel *et al.*, 1983; Lee *et al.*, 1998) and flavin-containing monooxygenases (Yueh *et al.*, 1997; Krueger *et al.*, 2001). The composition of secretory products has been defined for complex carbohydrates in surface epithelium and glands throughout the airway tree in adults (Choi *et al.*, 2000; Heidsiek *et al.*, 1987; Plopper *et al.*, 1989; St. George *et al.*, 1984, 1985, 1986) and mucin antigens (St. George *et al.*, 1985). The expression and distribution of the cysticfibrosis transmembrane receptor (Dupuit *et al.*, 1995) and keratins (Huang, *et al*, 1989) has been characterized in airway epithelium. The pattern for differentiation of Tracheobronchial epithelium (Plopper *et al.*, 1986a; Tyler *et al.*, 1988, 1989) and submucosal glands has been defined (Plopper *et al.*, 1986b). While regulatory control of epithelial differentiation has been studied *in vivo* (St. George *et al.*, 1991), the majority of the work has been done using biphasic epithelial cultures. The role of vitamin A in regulation of epithelial differentiation has been established for mucin genes (An *et al.*, 1994); cytokeratin markers of squamous cell differentiation (Huang *et al.*, 1994); small proline-rich protein (SPR) (An *et al.*, 1992, 1993a, 1993b; Patterson *et al.*, 2001); an antioxidant, thioredoxin (An and Wu, 1993); the cytokine, IL8 (Chang *et al.*, 2000); the growth factor EGF (Miller *et al.*, 1993); and RNPAI (An and Wu, 1993). Extra-cellular calcium also plays a role in epithelial differentiation in vitro (Martin *et al.*, 1991).

The rhesus monkey has been used as a model for oxidant-induced inflammation and lung injury, relying on the air pollutant ozone (Plopper *et al.*, 1998; Sterner-Kock *et al.*, 2000). The role which neutrophils play in the acute injury (Hyde *et al.*, 1992), the repair of epithelial injury (Hyde *et al.*, 1999) and the mechanism by which neutrophils identify migration sites (Miller *et al.*, 2001) and adhere to injured areas (McDonald *et al.*, 1993) have been defined. The induction of protective mechanisms by ozone exposure has been defined for heat shock proteins (Wu *et al.*, 1999) and glutathione pools (Duan *et al.*, 1996; Plopper *et al*, 1998). A model of experimental allergic asthma, using a known human allergen, house dust mite, has been validated for rhesus monkeys (Schelegle *et al.*, 2001). Ozone exposure exacerbates the impact of allergen exposure on this model when applied to infant monkeys (Schelegle *et al.*, 2003). This includes modulating hypercontractility of airway smooth muscle tested *in vitro* (Kott *et al.*, 2002) and modulated by 5-lipoxygenase (Johnson *et al.*, 1988) and a variety of other mediators (Foster *et al.*, 1996; Chand *et al.*, 1980). This exposure disrupts airway growth (Powell, 2003) and the development and function of a key element in the airway wall, the basement membrane zone (Evans *et al.*, 2002a, 2002b, 2003).

In addition to challenges by oxidants and allergens, rhesus monkeys also have been used to test other respiratory toxicants (Martonen *et al.*, 2001) including environmental tobacco smoke (Slotkin *et al.*, 2000), nicotine (Sekhon *et al.*, 2001, 2002), marijuana (Fligiel *et al.*, 1991) and formaldehyde (Casanova *et al.*, 1991; Monticello *et al.*, 1989).

The rhesus monkey has long been used as a model for fetal lung development, especially in relation to the gas exchange area. Example studies include Gilbert *et al.* (2002); Zhang *et al.* (1997); Cochrane *et al.* (1998) and Socol *et al.* (1981).

Correspondence

Any correspondence should be directed to Charles Plopper, California National Primate Research Center and Department of Anatomy, Physiology and Cell Biology, School of Veterinary Medicine, University of California, Davis, CA, USA. Email: cgplopper@ucdavis.edu

References

Afione, S.A., Conrad, C.K., Kearns, W.G., Chunduru, S., Adams, R., Reynolds, T.C., Guggino, W.B., Cutting, G.R., Carter, B.J. and Flotte, T.R. (1996). *J. Virol.* 70, 3235–3241.

An, G., Huang, T.H., Tesfaigzi, J., Garcia-Heras, J., Ledbetter, D.H., Carlson, D.M. and Wu, R. (1992). *Am. J. Respir. Cell. Mol Biol.* 7, 104–111.

An, G., Tesfaigzi, J., Carlson, D.M. and Wu, R. (1993a). *J. Cell. Physiol.* 157, 562–568.

An, G., Luo, G. and Wu, R. (1994). *Am. J. Respir Cell. Mol. Biol.* 10, 546–551.

An, G., Tesfaigzi, J., Chuu, Y.J. and Wu, R. (1993b). *J. Biol. Chem.* 268, 10977–10982.

An, G. and Wu, R. (1992). *Biochem. Biophys. Res. Commun.* 183, 170–175.

An, G. and Wu, R. (1993). *Biochem. Biophys. Acta.* 1172, 292–300.

Brain, J.D. (1970). *Ann. Otol. Rhinol Laryngol.* 79, 529–539.

Bals, R., Lang, C., Weiner, D.J., Vogelmeier, C., Welsch, U. and Wilson, J.M. (2001). *Clin. Diagn. Lab. Immunol.* 8, 370–375.

Baskerville, A., Fitzgeorge, R.B., Broster, M. and Hambleton, P. (1983). *J. Pathol.* 139, 349–362.

Bayle, J.Y., Johnson, L.G., St George, J.A., Boucher, R.C. and Olsen, J.C. (1993). *Hum. Gene. Ther.* 4, 161–170.

Beck, S.E., Laube, B.L., Barberena, C.I., Fischer, A.C., Adams, R.J., Chesnut, K., Flotte, T.R. and Guggino, W.B. (2002). *Mol. Ther.* 6, 546–554.

Board, K.F., Patil, S., Lebedeva, I., Capuano, S., 3rd, Trichel, A.M., Murphey-Corb, M., Rajakumar, P.A., Flynn, J.L., Haidaris, C.G. and Norris, K.A. (2002). *J. Infect. Dis.* 187, 576–588.

Bout, A., Perricaudet, M., Baskin, G., Imler, J.L., Scholte, B.J., Pavirani, A. and Valerio, D. (1994). *Hum. Gene. Ther.* 5, 3–10.

Brandtzaeg, P. (1984). *Immunology of the Lung and Upper Respiratory Tract.* J. Bienenstock. New York, McGraw-Hill, pp 28–95.

Brody, S.L., Metzger, M., Danel, C., Rosenfeld, M.A. and Crystal, R.G. (1994). *Hum. Gene. Ther.* 5, 821–836.

Caldwell, R.A., Grubb, B.R., Tarran, R., Boucher, R.C., Knowles, M.R. and Barker, P.M. (2002). *J. Gen. Physiol.* 119, 3–14.

Casanova, M., Morgan, K.T., Steinhagen, W.H., Everitt, J.I., Popp, J.A. and Heck, H.D. (1991). *Fundam. Appl. Toxicol.* 17, 409–428.

Chand, N., Dhawan, B.N., Srimal, R.C., Rahmani, N.H., Shukla, R.K. and Altura, B.M. (1980). *J. Appl. Physiol.* 49, 729–734.

Chang, M.M., Wu, R., Plopper, C.G. and Hyde, D.M. (1998). *Am. J. Physiol.* 275, L524–L532.

Chang, M.M., Harper, R., Hyde, D.M. and Wu, R. (2000). *Am. J. Respir. Cell. Mol. Biol.* 22, 502–510.

Chirmule, N., Raper, S.E., Burkly, L., Thomas, D., Tazelaar, J., Hughes, J.V. and Wilson, J.M. (2000). *J. Virol.* 4, 3345–3352.

Choi, H.K., Finkbeiner, W.E. and Widdicombe, J.H. (2000). *J. Anat.* 197 Pt 3, 361–372.

Cochrane, C.G., Revak, S.D., Merritt, T.A., Schraufstatter, I.U., Hoch, R.C., Henderson, C., Andersson, S., Takamori, H. and Oades, Z.G. (1998). *Pediatr. Res.* 44, 705–715.

Conrad, C.K., Allen, S.S., Afione, S.A., Reynolds, T.C., Beck, S.E., Fee-Maki, M., Barrazza-Ortiz, X., Adams, R., Askin, F.B., Carter, B.J., Guggino, W.B. and Flotte, T.R. (1996). *Gene Ther.* 3, 658–668.

Crapo, J.D. and Barry, B.E. (1982) *American Review of Respiratory Diseases*, 125, 332–337.

Croix, D.A., Board, K., Capuano, S., 3rd, Murphey-Corb, M., Haidaris, C.G., Flynn, J.L., Reinhart, T. and Norris, K.A. (2002). *AIDS Res. Hum. Retroviruses* 18, 391–401.

Crumley, R.L. (1991). *Arch. Otolaryngol Head Neck Surg.* 117, 1113–1117.

Dahl, A.R. and W.M. Hadley (1991). *Crit. Rev. Toxicol.* 21, 345–372.

Daniel, F.B., Schut, H.A., Sandwisch, D.W., Schenck, K.M., Hoffmann, C.O., Patrick, J.R. and Stoner, G.D. (1983). *Cancer Res.* 43, 4723–4729.

Duan, X., Buckpitt, A.R., Pinkerton, K.E., Ji, C. and Plopper, C.G. (1996). *Am. J. Respir. Cell. Mol. Biol.* 14, 70–75.

Dupuit, F., Bout, A., Hinnrasky, J., Fuchey, C., Zahm, J.M., Imler, J.L., Pavirani, A., Valerio, D. and Puchelle, E. (1995). *Gene Ther.* 2, 156–163.

Durand-Joly, I., Wakefield, A.E., Palmer, R.J., Denis, C.M., Creusy, C., Fleurisse, L., Ricard, I., Gut, J.P. and Dei-Cas, E. (2000). *Med. Mycol.* 38, 61–72.

Dykewicz, M.S., Patterson, R. and Harris, K.E. (1988). *J. Lab. Clin. Med.* 111, 459–465.

Engle, M.J., Kemnitz, J.W., Rao, T.J., Perelman, R.H. and Farrell, P.M. (1996). *Am. J. Perinatol.* 13, 399–407.

Evans, M.J., Fanucchi, M.V., Van Winkle, L.S., Baker, G.L., Murphy, A.E., Nishio, S.J., Sannes, P.L. and Plopper, C.G. (2002a). *Am. J. Physiol. Lung Cell. Mol. Physiol.* 283, L1263–L1270.

Evans, M.J., Van Winkle, L.S., Fanucchi, M.V., Baker, G.L., Murphy, A.E., Nishio, S.J., Schelegle, E.S., Gershwin, L.J., Sannes, P.L. and Plopper, C.G. (2002b). *Lab. Invest.* 82, 1747–1754.

Evans, M.J., Fanucchi, M.V., Baker, G.L., Van Winkle, L.S., Pantle, L.M., Nishio, S.J., Schelegle, E.S., Gershwin, L.J., Miller, L.A., Hyde, D.M., Sannes, P.L. and Plopper, C.G. (2003). *Am. J. Physiol. Lung Cell. Mol. Physiol.* 285, L931–L939.

Fearon, B. and McMillin, B.D. (1985). *Ann. Otol. Rhinol. Laryngol.* 94, 631–633.

Fitch, W.T. (1997). *J. Acoust. Soc. Am.* 102, 1213–1222.

Fligiel, S.E., Beals, T.F., Tashkin, D.P., Paule, M.G., Scallet, A.C., Ali, S.F., Bailey, J.R. and Slikker, W., Jr. (1991). *Pharmacol. Biochem. Behav.* 40, 637–642.

Flugel, C. and Rohen, J.W. (1991). *Mech. Ageing Dev.* 61, 65–83.

Ford-Hutchinson, A.W., Girard, Y., Lord, A., Jones, T.R., Cirino, M., Evans, J.F., Gillard, J., Hamel, P., Leveille, C., Masson, P. *et al.* (1989). *Can. J. Physiol. Pharmacol.* 67, 989–993.

Fortunati, E., Bout, A., Zanta, M.A., Valerio, D. and Scarpa, M. (1996). *Biochem. Biophys. Acta* 1306, 55–62.

Foster, A., Zeller, W. and Pfannkuche, H.J. (1996). *Lab. Anim. Sci.* 46, 327–334.

Fuller, C.L., Choi, Y.K., Fallert, B.A., Capuano, S., 3rd, Rajakumar, P., Murphey-Corb, M. and Reinhart, T.A. (2002). *Am. J .Pathol.* 161, 969–978.

Geelhaar, A. and Weibel, E.R. (1971). *Respiration Physiology*, 11, 354–366.

Gehr, P., Bachofen, M. *et al.* (1978). *Respiration Physiology*, 32, 121–140.

Gehr, P. and Erni, H. (1980). *Respiration Physiology*, 41, 199–210.

Gehr, P., Mwangi, D.K. *et al.* (1981). *Respiratory Physiology*, 44, 61–86.

Gerard, C.J., Dell'Aringa, J., Hale, K.A. and Klump, W.M. (2003). *Gene. Ther.* 10, 1744–1753.

Gilbert, W.M., Eby-Wilkens, E., Plopper, C., Whitsett, J.A. and Tarantal, A.F. (2002). *Obstet. Gynecol.* 98, 466–470.

Graziadei, P.P.C. (1977). *Functional Anatomy of the Mammalian Chemoreceptor system. Chemical Signals in Vertebrates*. Edited by Muller-Schwarze, D. and Mozell, M.M. Plenum Press, New York, pp, 435–454.

Gross, E.A. and K.T. Morgan (1992). *Comparative Biology of the Normal Lung*. Boca Raton, CRC Press, Inc.

Gross, E.A., Swenberg, J.A. *et al.* (1982). *J. Anat.* 135, 83–88.

Gross, E.A., Starr, T.B. *et al.* (1987). *Toxicologist* 7, 193.

Harkema, J.R., Mariassy, A.T. *et al.* (1991). *The Airway Epithelium: Physiology, Pathophysiology and Pharmacology*, 3–39.

Harkema, J.R. (1992). *Comparative Biology of the Normal Lung*. Boca Raton, CRC Press, Inc.

Harkema, J.R. and Hotchkiss, J.A. (1999). *Am. J. Respir. Cell. Mol. Biol.* 20, 517–529.

Harkema, J.R. and Mariassy, A. (1991). *The Airway Epithelium: Physiology, Pathophysiology, and Pharmacology*. New York, Marcel Dekker, Inc.

Harkema, J.R., and Plopper, C.G. (1987a). *Am. J. Pathol.* 128, 29–44.

Harkema, J.R. and Plopper, C.G. (1987b). *Am. J. Anat.* 180, 266–279.

Harkema, J.R. and Plopper, C.G. (2000). *Pulmonary Immunotoxicology*, 1–59.

Heidsiek, J.G., Hyde, D.M., Plopper, C.G. and St George, J.A. (1987). *J Histochem. Cytochem.* 35, 435–442.

Hilloowala, R.A. and Lass, N.J. (1978). *Am. J. Phys. Anthropol.* 49, 129–131.

Huang, T.H., Ann, D.K., Zhang, Y.J., Chang, A.T., Crabb, J.W. and Wu, R. (1994). *Am. J. Respir. Cell. Mol. Biol*, 10, 192–201.

Huang, T.H., St George, J.A., Plopper, C.G. and Wu, R. (1989). *Differentiation* 41, 78–86.

Hugonnaud, C., Gehr, P. *et al.* (1977). *Respiration Physiology* 29, 1–10.

Hyde, D.M., Hubbard, W.C. *et al.* (1992). *American Journal of Respiratory Cell and Molecular Biology* 6, 481–497.

Hyde, D.M., Hubbard, W.C., Wong, V., Wu, R., Pinkerton, K. and Plopper, C.G. (1992). *Am. J. Respir. Cell Mol. Biol.* 6, 481–497.

Hyde, D.M., Miller, L.A., McDonald, R.J., Stovall, M.Y., Wong, V., Pinkerton, K.E., Wegner, C.D., Rothlein, R. and Plopper, C.G. (1999). *Am. J. Physiol.* 277, L1190–L1198.

Ide, C. and Munger, B.L. (1980). *Am. J. Anat.* 158, 193–209.

Ilievski, V. and Fleischman, R.W. (1981). *Lab. Anim. Sci.* 31, 524–525.

Israel, Z.R., Gettie, A., Ishizaka, S.T., Mishkin, E.M., Staas, J., Gilley, R., Montefiori, D., Marx, P.A. and Eldridge, J.H. (1999). *AIDS Res. Hum. Retroviruses* 15, 1121–1136.

Jeevaratham, K., Dasgupta, S., Pravinkumar, Pant, S.C., Sachan, A.S., Selvamurthy, W., Ray, U.S., Mukhopadhyay, S. and Purkayastha, S.S. (1998). *Indian J. Physiol. Pharmacol.* 42, 25–38.

Johnson, H.G., Stout, B.K. and Ruppel, P.L. (1988). *Prostaglandins* 35, 459–466.

Kapanci, Y. and Weibel, E.R. (1969). *Lab. Invest.* 20, 101–118.

Kepler, G.M., and Richardson, R.B. (1998). *Toxicol. Appl. Pharmacol.* 150, 1–11.

Kimbell, J.S. and Subramaniam, R.P. (2001). *Toxicol. Sci.* 64, 100–110.

Komisar, J.L., Weng, C.F., Oyejide, A., Hunt, R.E., Briscoe, C. and Tseng, J. (2001). *Toxicol. Pathol.* 29, 369–378.

Kott, K.S., Pinkerton, K.E., Bric, J.M., Plopper, C.G., Avadhanam, K.P. and Joad, J.P. (2002). *J. Appl. Physiol.* 92, 989–996.

Krueger, S.K., Martin, S.R., Yueh, M.F., Pereira, C.B. and Williams, D.E. (2001). *Drug Metab. Dispos.* 30, 34–41.

Lai, X., Shen, Y., Zhou, D., Sehgal, P., Shen, L., Simon, M., Qiu, L., Letvin, N.L. and Chen, Z.W. (2003). *Clin. Exp. Immunol.* 133, 182–192.

Larson, C., Sutton, D. and Lindeman, R.C. (1974). *Folia Primatol. (Basel)* 22, 315–323.

Larson, J.E., Morrow, S.L., Delcarpio, J.B., Bohm, R.P., Ratterree, M.S., Blanchard, J.L. and Cohen, J.C. (2000). *Mol. Ther.* 2, 631–639.

Lee, C., Watt, K.C., Chang, A.M., Plopper, C.G., Buckpitt, A.R. and Pinkerton, K.E. (1998). *Drug Metab. Dispos.* 26, 396–400.

Li, H.O., Zhu, Y.F., Asakawa, M., Kuma, H., Hirata, T., Ueda, Y., Lee, Y.S., Fukumura, M., Iida, A., Kato, A., Nagai, Y. and Hasegawa, M. (2000). *J. Virol.* 74, 6564–6569.

Lowell, G.H., Colleton, C., Frost, D., Kaminski, R.W., Hughes, M., Hatch, J., Hooper, C., Estep, J., Pitt, L., Topper, M., Hunt, R.E., Baker, W. and Baze, W.B. (1996). *Infect. Immun.* 64, 4686–4693.

Lund, W.S. and Palmer, J.F. (1969). *J. Laryngol. Otol.* 83, 69–78.

Mankowski, J.L., Carter, D.L., Spelman, J.P., Nealen, M.L., Maughan, K.R., Kirstein, L.M., Didier, P.J., Adams, R.J., Murphey-Corb, M. and Zink, M.C. (1998). *Am. J. Pathol.* 153, 1123–1130.

Mariassy, A.T. (1992). *Treatise on Pulmonary Toxicology: Comparative Biology of the Normal Lung*, 63–76.

Martin, P.L., Vaidyanathan, S., Lane, J., Rogge, M., Gillette, N., Niggemann, B. and Green, J. (2002). *J. Interferon Cytokine Res.* 22, 709–717.

Martin, W.R., Brown, C., Zhang, Y.J. and Wu, R. (1991). *J. Cell. Physiol.* 147, 138–148.

Martonen, T.B., Katz, I.M. and Musante, C.J. (2001). *Inhal. Toxicol.* 13, 307–324.

Massion, P.P., Funari, C.C., Ueki, I., Ikeda, S., McDonald, D.M. and Nadel, J.A. (1993). *Am. J. Respir. Cell. Mol. Biol.* 9, 361–370.
McBride, J.T. (1992). *Treatise on Pulmonary Toxicology: Comparative Biology of the Normal Lung*, 49–61.
McDonald, R.J., Lukason, M.J., Raabe, O.G., Canfield, D.R., Burr, E.A., Kaplan, J.M., Wadsworth, S.C. and St George, J.A. (1997). *Hum. Gene. Ther.* 8, 411–422.
McDonald, R.J., Liggitt, H.D., Roche, L., Nguyen, H.T., Pearlman, R., Raabe, O.G., Bussey, L.B. and Gorman, C.M. (1998). *Pharm. Res.* 15, 671–679.
McDonald, R.J., Pan, L.C., St George, J.A., Hyde, D.M. and Ducore, J.M. (1993a). *Inflammation* 17, 715–722.
McDonald, R.J., St George, J.A., Pan, L.C. and Hyde, D.M. (1993b). *Inflammation* 17, 145–151.
Mercer, R.R. and Crapo, J.D. (1981). *Journal of Applied Physiology* 63(2), 785–794.
Mercer, R.R. and Crapo, J.D. (1992). *Treatise on Pulmonary Toxicology: Comparative Biology of the Normal Lung*, 109–119.
Mercer, R.R. and Laco, J.M. (1987). *Journal of Applied Physiology* 62(4), 1480–1487.
Mercer, R.R. and Russell M.L. (1994). *Am. J. Respir. Cell. Mol. Biol.* 10(6), 613–624.
Miller, F.J. and Mercer, R.R. (1993). *Aerosol Science and Technology* 18, 257–271.
Michalewski, K. and Szostakiewicz-Sawicka, H. (1990). *Folia Morphol (Warsz).* 49, 53–60.
Miller, F.J. (1995). *Nasal Toxicology and Dosimetry of Inhaled Xenobiotics: Implications for Human Health.* Washington, Taylor & Francis.
Miller, L.A., Barnett, N.L., Sheppard, D. and Hyde, D.M. (2001). *J Histochem. Cytochem.* 49, 41–48.
Miller, L.A., Cheng, L.Z. and Wu, R. (1993). *Cancer Res.* 53, 2527–2533.
Monteiro-Riviere, N.A. and Popp, J.A. (1984). *Am. J. Anat.* 169, 31–43.
Monticello, T.M. and Morgan, K.T. (1989). *Am. J. Pathol.* 134, 515–527.
Morgan, K.T., Kimbell, J.S., Monticello, T.M., Patra, A.L. and Fleishman, A. (1991). *Toxicol. Appl. Pharmacol.* 110, 223–240.
Murano, K., Ngaotepprutaram, P., Chung, S.K. and Furuta, S. (1993). *Acta Otolaryngol. Suppl.* 506, 75–79.
Negus, V.E. (1958). *The Comparative Anatomy and Physiology of the Nose and Paranasal Sinuses.* Edinburgh, E&S Livingstone.
Nouwen, E.J., Dauwe, S. and De Broe, M.E. (1990). *Differentiation* 45, 192–198.
Patra, A.L., Gooya, A. and Menache, M.G. (1986). *Anat. Rec.* 215, 42–50.
Patterson, R., Harris, K.E. and Suszko, I.M. (1977). *Am. Rev. Respir. Dis.* 115, 929–936.
Patterson, R., Suszko, I.M. and Harris, K.E. (1978). *Trans. Assoc. Am. Physicians* 91, 180–185.
Patterson, T., Vuong, H., Liaw, Y.S., Wu, R., Kalvakolanu, D.V. and Reddy, S.P. (2001). *Oncogene* 20, 634–644.
Pearl, A.W., Gannon, P.J. and Urken, M.L. (1998). *Laryngoscope* 108, 1062–1065.
Phalen, R.F. and Oldham, M.J. (1983). *Am. Rev. Respir. Dis.* 128, S1–S4.
Pinkerton, K.E., Gehr, P. *et al.* (1992). *Treatise on Pulmonary Toxicology: Comparative Biology of the Normal Lung*, 121–128.
Plopper, C.G. and Chu, F.P. (1994). *American Journal of Pathology* 144(2), 404–420.
Plopper, C.G. and Dungworth, D.L. (1973). *American Journal of Pathology* 71, 395–408.
Plopper, C.G. and Hatch, G.E. (1998). *Am. J. Respir. Cell. Mol. Biol.* 19(3), 387–399.
Plopper, C.G. and Hyde, D.M. (1992). *Treatise on Pulmonary Toxicology*, 85–92.
Plopper, C.G., Alley, J.L. and Weir, A.J. (1986a). *Am. J. Anat.* 175, 59–71.
Plopper, C.G., Weir, A.J., Nishio, S.J., Cranz, D.L. and St George, J.A. (1986b). *Anat. Embryol. (Berl).* 174, 167–178.
Plopper, C.G., Hatch, G.E., Wong, V., Duan, X., Weir, A.J., Tarkington, B.K., Devlin, R.B., Becker, S. and Buckpitt, A.R. (1998). *Am. J. Respir. Cell. Mol. Biol.* 19, 387–399.
Plopper, C.G., Heidsiek, J.G., Weir, A.J., George, J.A. and Hyde, D.M. (1989). *Am. J. Anat.* 184, 31–40.
Powell, K. (2003). *Nat. Med.* 9, 490.
Rae, T.C. and Koppe, T. (2002). *Am. J. Phys. Anthropol.* 117, 293–296.
Raman, R. and Devanandan, M.S. (1989). *Anat. Rec.* 223, 433–436.
Rendall, D., Owren, M.J. and Rodman, P.S. (1998). *J. Acoust. Soc. Am.* 103, 602–614.
Revak, S.D., Merritt, T.A., Cochrane, C.G., Heldt, G.P., Alberts, M.S., Anderson, D.W. and Kheiter, A. (1996). *Pediatr. Res.* 39, 715–724.
Saban, R., Haak-Frendscho, M., Zine, M., Ridgway, J., Gorman, C., Presta, L.G., Bjorling, D., Saban, M. and Jardieu, P. (1994). *J. Allergy Clin. Immunol.* 94, 836–843.
Safi, H., Gormus, B.J., Didier, P.J., Blanchard, J.L., Lakey, D.L., Martin, L.N., Murphey-Corb, M., Vankayalapati, R. and Barnes, P.F. (2003). *AIDS Res. Hum. Retroviruses* 19, 585–595.
Sahgal, V. and Hast, M.H. (1974). *Acta. Otolaryngol.* 78, 277–281.
Sahgal, V. and Hast, M.H. (1986). *J. Laryngol. Otol.* 100, 553–560.
Sapir, S., Campbell, C. and Larson, C. (1981). *Laryngoscope* 91, 457–468.
Scaria, A., Sullivan, J.A., St George, J.A., Kaplan, J.M., Lukason, M.J., Morris, J.E., Plog, M., Nicolette, C., Gregory, R.J. and Wadsworth, S.C. (2000). *Mol. Ther.* 2, 505–514.
Schelegle, E.S., Gershwin, L.J., Miller, L.A., Fanucchi, M.V., Van Winkle, L.S., Gerriets, J.P., Walby, W.F., Omlor, A.M., Buckpitt, A.R., Tarkington, B.K., Wong, V.J.,

Joad, J.P., Pinkerton, K.B., Wu, R., Evans, M.J., Hyde, D.M. and Plopper, C.G. (2001). *Am. J. Pathol.* 158, 333–341.
Schreider, J.P. and Raabe, O.G. (1981). *Anat. Rec.* 200, 195–205.
Schroeder, T.H., Reiniger, N., Meluleni, G., Grout, M., Coleman, F.T. and Pier, G.B. (2001). *J. Immunol.* 166, 7410–7418.
Sekhon, H.S., Keller, J.A., Benowitz, N.L. and Spindel, E.R. (2001). *Am. J. Respir. Crit. Care. Med.* 164, 989–994.
Sekhon, H.S., Keller, J.A., Proskocil, B.J., Martin, E.L. and Spindel, E.R. (2002). *Am. J. Respir. Cell. Mol. Biol.* 26, 31–41.
Shean, M.K., Baskin, G., Sullivan, D., Schurr, J., Cavender, D.E., Shellito, J.E., Schwarzenberger, P.O. and Kolls, J.K. (2002) *Hum. Gene Ther.* 11, 1047–1055.
Schreider, J.P. (1983). *Nasal Tumors in Animals and Man. Vol. III. Experimental Nasal Carcinogenesis.* Boca Raton, CRC Press Inc.
Schreider, J.P. and Raabe, O.G. (1981). *Anat. Rec.* 200, 195–205.
Simonyan, K. and Jurgens, U. (2002). *Brain. Res.* 949, 23–31.
Slotkin, T.A., Pinkerton, K.E. and Seidler, F.J. (2000). *Brain Res. Dev. Brain Res.* 124, 53–58.
Smiley-Jewell, S.M., Tran, M.U., Weir, A.J., Johnson, Z.A., Van Winkle, L.S. and Plopper, C.G. (2002). *J. Appl. Physiol.* 93, 1506–1514.
Smith, T.D. and Siegel, M.I. (2001). *J. Anat.* 198, 77–82.
Socol, M.L., Druzin, M.L., Murata, Y., Strassner, H.T., Whittle, M.J., Manning, F.A. and Martin, C.B., Jr. (1981). *Am. J. Obstet. Gynecol.* 140, 708–709.
Soppor, S., Nierwetberg, D., Halbach, A., Sauer, U., Scheller, C., Stahl-Hennig, C., Matz-Rensing, K., Schafer, F., Schneider, T., ter Meulen, V. and Muller, J.G. (2003). *Blood* 101, 1213–1219.
Sorokin, S.P. (1988). *Cell and Tissue Biology: A Textbook of Histology*. L. Weiss. Baltimore, Urban & Schwarzenberg 751–814.
St George, J.A., Harkema, J.R. *et al.* (1988). *Toxicology of the Lung.*
St George, J.A. and Wang, S. (1992). *Treatise on Pulmonary Toxicology: Comparative Biology of the Normal Lung*, 77–83.
St George, J.A., Cranz, D.L., Zicker, S.C., Etchison, J.R., Dungworth, D.L. and Plopper, C.G. (1985). *Am. Rev. Respir. Dis.* 132, 556–563.
St George, J.A., Nishio, S.J., Cranz, D.L. and Plopper, C.G. (1986). *Anat. Rec.* 216, 60–67.
St George, J.A., Nishio, S.J. and Plopper, C.G. (1984). *Anat. Rec.* 210, 293–302.
St George, J.A., Pennington, S.E., Kaplan, J.M., Peterson, P.A., Kleine, L.J., Smith, A.E. and Wadsworth, S.C. (1996). *Gene Ther.* 3, 103–116.
St George, J.A., Read, L.C., Cranz, D.L., Tarantal, A.F., George-Nascimento, C. and Plopper, C.G. (1991). *Am. J. Respir. Cell. Mol. Biol.* 4, 95–101.
Stearns, M.P. and Cummings, C.W. (1982). *Ann. Otol. Rhinol. Laryngol.* 91, 370–371.
Sterner-Kock, A., Kock, M., Braun, R. and Hyde, D.M. (2000). *Am. J. Respir. Crit. Care Med.* 162, 1152–1156.
Sutton, D., Taylor, E.M. and Lindeman, R.C. (1977). *Acta. Anat. (Basel)* 97, 57–67.
Sutton, D., Taylor, E.M. and Lindeman, R.C. (1978). *Pediatrics* 61, 519–527.
Tarantal, A.F., Lee, C.I., Ekert, J.E., McDonald, R., Kohn, D.B., Plopper, C.G., Case, S.S. and Bunnell, B.A. (2001). *Mol. Ther.* 4, 614–621.
Tashiro, M., Beppu, Y., Sakai, K. and Kido, H. (1996). *Arch. Virol.* 141, 1571–1577.
Taylor, E.M., Sutton, D. and Lindeman, R.C. (1976). *Growth* 40, 69–81.
Todoroki, H., Nakamura, S., Higure, A., Okamoto, K., Takeda, S., Nagata, N., Itoh, H. and Ohsato, K. (2000). *Surgery* 127, 209–216.
Tseng, J., Komisar, J., Chen, J.Y., Hunt, R., Johnson, A.J., Pitt, L. and Ruble, D. (1995). *Adv. Exp. Med. Biol.* 371B, 1615–1619.
Tyler, W.S. and Julian, M.D. (1992). *Treatise on Pulmonary Toxicology: Comparative Biology of the Normal Lung*, 37–48.
Tyler, W.S. and Tyler, N.K. (1988). *Toxicology* 50, 131–144.
Tyler, N.K., Hyde, D.M., Hendrickx, A.G. and Plopper, C.G. (1988). *Am. J. Anat.* 182, 215–223.
Tyler, N.K., Hyde, D.M., Hendrickx, A.G. and Plopper, C.G. (1989). *Anat. Rec.* 225, 297–309.
Vanderhorst, V.G., Terasawa, E. and Ralston, H.J., 3rd. (2001). *Neuroscience* 107, 117–125.
Waag, D.M., Byrne, W.R., Estep, J., Gibbs, P., Pitt, M.L. and Banfield, C.M. (1999). *Lab. Anim. Sci.* 49, 634–638.
Weibel, E.R. (1973). *Physiological Reviews* 53(2), 419–495.
Weibel, E.R. (1989). *Respiratory Physiology: An Analytical Approach*, 1–56. Weibel, E.R., and Gehr, P. (1981). *Respiratory Physiology* 44, 39–60.
Weichman, B.M., Muccitelli, R.M., Tucker, S.S. and DeVan, J.F. (1985). *J. Pharmacol. Exp. Ther.* 233, 345–351.
Weiss, D.J., Baskin, G.B., Shean, M.K., Blanchard, J.L. and Kolls, J.K. (2002). *Mol. Ther.* 5, 8–15.
Wilson, J.M., Engelhardt, J.F., Grossman, M., Simon, R.H. and Yang, Y. (1994). *Hum. Gene Ther.* 5, 501–519.
Wu, R., Zhao, Y.H., Plopper, C.G., Chang, M.M., Chmiel, K., Cross, J.J., Weir, A., Last, J.A. and Tarkington, B. (1999). *Am. J. Physiol.* 277, L511–L522.
Yueh, M.F., Krueger, S.K. and Williams, D.E. (1997). *Biochem. Biophys. Acta.* 1350, 267–271.
Zhang, Q., Wu, W.X., Hong Ma, X., Nathanielsz, P.W. and Brenna, J.T. (1997). *Prostaglandins Leukot. Essent. Fatty Acids* 57, 311–321.
Zoetelief, J., Wagemaker, G. and Broerse, J.J. (1998). *Int. J. Radiat. Biol.* 74, 265–272.

CHAPTER 31

Reproduction: Male

Gerhard van der Horst
Department of Medical Biosciences, University of the Western Cape, Bellville, 7530, South Africa

Introduction

Non-human primates have been used extensively for research in the field of reproduction. Captive breeding of several of these species has been necessary because of declining numbers in the wild, and also for the purpose of finding a good model for human reproductive research and reproductive toxicology. An important question is which non-human primate represents the best model for human reproduction? Table 31.1 shows a comparison of the metabolism of various compounds in different mammals, including humans. Firstly, it is clear that, on this basis, non-human primates (NHP) are better models than other laboratory animals such as mice and rats. Secondly, it is apparent that among the different NHP, the Old World Primates generally represent a more relevant/applicable human model than the New World Primates. Therefore, although three New World Primate monkey species, viz. marmoset, capuchin and squirrel monkeys, have been used extensively (Table 31.2), more emphasis will be placed on the Old World Primates than the New World Primates in this chapter.

Which non-human primate models are used/or should be used?

In the search for a relevant research model for human male reproduction, the following important aspects should be taken into consideration. How abundant is the species, are they disease free, is there disease control, have they been bred for several generations under controlled laboratory conditions, and do research results show that comparisons with the human, in the field of male reproduction, are relevant?

Many NHP species, particularly among the Old World Primates, would meet many of these criteria but, in reality, relatively few research centres can claim good control over all of the above aspects. Table 31.2 represents those NHP that have been most often successfully used in male reproductive studies with applications in human reproduction. The big apes such

The Laboratory Primate

ISBN 0-1208-0261-9

TABLE 31.1: Comparative metabolism of some drugs in different laboratory animals and primates

	Glucoronic acid conjugation	Amino acid conjugation	O- methylation	Metabolic model
	Phenol excretion in urine (conjugation ratio)	Indolacetic acid excretion (%)	DMBA* excretion units	Invalid as model for humans when testing 47 drugs (%)
Humans	7	15	19	–
Old World Primates	7	44	12	4
New World Primates	0.5	30	54	–
Rat	1	0	0	68
Mouse	1	0	–	–
Cat	93	0	–	–
Dog	–	0	–	52
Pig	0.1	–	–	–

*DMBA = Diiodomethoxybenzoic acid.
Compiled from Smith and Williams, 1974; Muntzing *et al.*, 1975; Caldwell, 1985.

as chimpanzees and gorillas are excluded for ethical reasons. It is the author's opinion that the big apes should only be used in experiments to propagate the species and/or, in exceptional cases, when all other NHP models fail. This will be indicated in relevant sections.

Both rhesus monkeys and cynomolgus monkeys (macaques) have been used extensively in male reproduction studies in many centres across the world and are well-established models. African green (vervet) monkeys are the most widely distributed of all African

TABLE 31.2: Some of the most frequently used non-human primate species and examples of diverse applications in humans

Non-human primate species	New or Old World primate	Type of application	Reference
Rhesus monkey, *Macaca mulatta*	OWP	Vas occlusion	Kuckuck *et al.* 1975
		Pathology, male reproductive system	Baskerville, 1992
		In vitro fertilization	Wolf *et al.* 1989
		Penile erection/impotence	Giraldi *et al.* 1990
Cynomolgus monkey, *Macaca fascicularis*	OWP	Micro-insemination	Ogura *et al.* 2000
		Germ cell transplantation	Schlatt *et al.* 2002
African green or vervet monkey, *Chlorocebus aethiops*, previously known as *Cercopithecus aethiops*	OWP	Sperm morphology	Seier *et al.* 1996
		Cryopreservation	Seier *et al.* 1993
		Quantitative sperm motility	Van der Horst *et al.* 1999
Chacma baboon, *Papio ursinus*	OWP	Semen quality	Bornman *et al.* 1988
Olive baboon, *Papio anubis*	OWP	Male contraception	Gichuhi *et al.* 1999
Common marmoset, *Callithrix jacchus*	NWP	Electro-ejaculation	Kuederling *et al.* 2000
		Reproductive toxicology	Rune *et al.* 1991
Capuchin monkey, *Cebus apella*	NWP	Ultrasound contraception	Fahim *et al.* 1977
Squirrel monkeys, *Saimiri sciureus*	NWP	Electro-ejaculation	Kuehl and Dukelow, 1974

OWP = Old World Primate; NWP = New World Primate.

monkeys (Eley, 1992). The Primate Unit of the Medical Research Council of South Africa has a very successful captive breeding program of African green monkeys (*Chlorocebus aethiops* previously known as *Cercopithecus aethiops*). Currently about 250 monkeys, who are second/third generation bred and born under captive conditions (disease free) in the Unit, are used for research purposes, including male reproduction (Seier, 1986). This species has become increasingly popular as a good model for human reproduction.

The Institute of Primate Research in Nairobi, Kenya, uses the olive baboon, *Papio anubis,* and one *Cercocebus* and one *Cercopithecus* monkey species (Grey Mangabey and Brazza's monkeys) for male reproductive studies applicable to humans (Table 31.2).

Main applications in male reproduction: models for biomedical research

General aspects

The main current applications of NHP, for biomedical research in reproduction, are in the field of fertility/infertility, *in vitro* fertilization/assisted reproductive technologies, male contraception and reproductive toxicology. Before discussing these it is important to establish that baseline information, on the reproductive system of a NHP model, is available and applicable to the human. Some basic features of the male reproductive system will first be discussed.

Structural features of the male reproductive system and spermatogenesis

Generally, the macro anatomy and histology of the male reproductive systems of NHP are similar to those of humans. The gonad somatic index (gonads as a fraction of body mass) of vervet monkeys, for example, is 0.36 (van der Horst, 1995) and resembles that of other primates (Short, 1980). In humans, the average testis weight is about 16 g or about 32 g in total. For a 70 kg man the gonad somatic index will only be 0.04 which is about ten times less than in the NHP. The external morphology and organization of the seminiferous tubules have been elucidated in several NHPs and there is a great similarity to humans, as is evident in Figure 31.1a and b.

Spermatogenesis in the seminiferous epithelium of humans is in the form of a helical wave. In both New World Primates and Old World Primates, spermatogenesis shows a similar type of helical wave subject to slight modifications in various primate species (Dietrich *et al.,* 1986; Kerr, 1992; Millar *et al.,* 2000). This can also be seen in Figure 31.2, which shows that there is more than one spermatogenic stage represented in a transverse

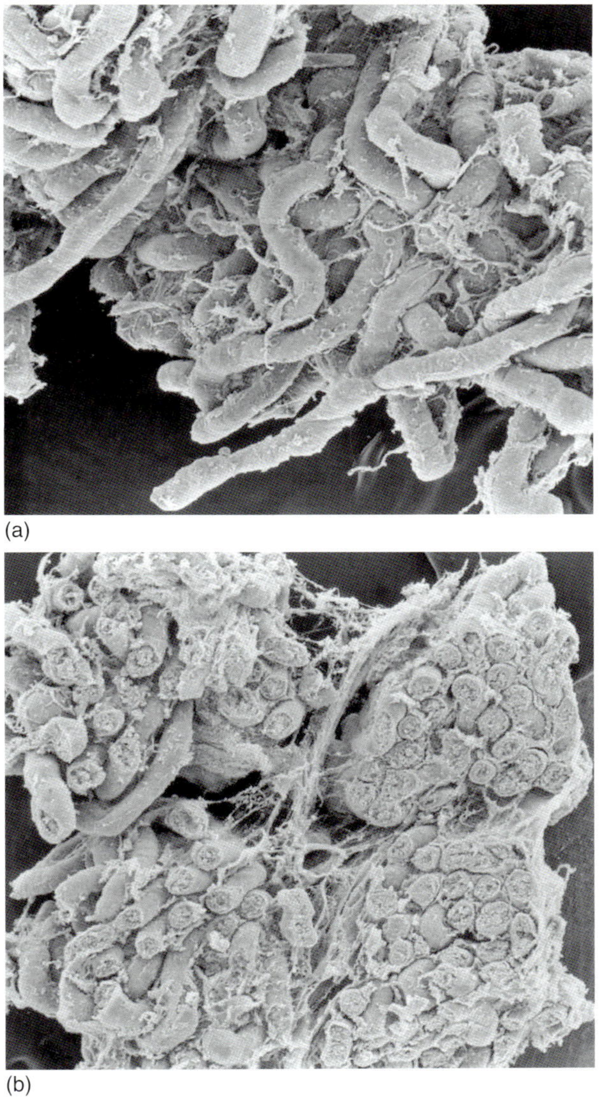

(a)

(b)

Figure 31.1 Scanning electron micrographs of the seminiferous epithelium of a typical NHP. (a) External morphology of the seminiferous tubules of the vervet monkey, *Chlorocebus aethiops;* (b) Scanning electron micrograph showing the seminiferous tubules of *Chlorocebus aethiops* in a few testicular septa in transverse section.

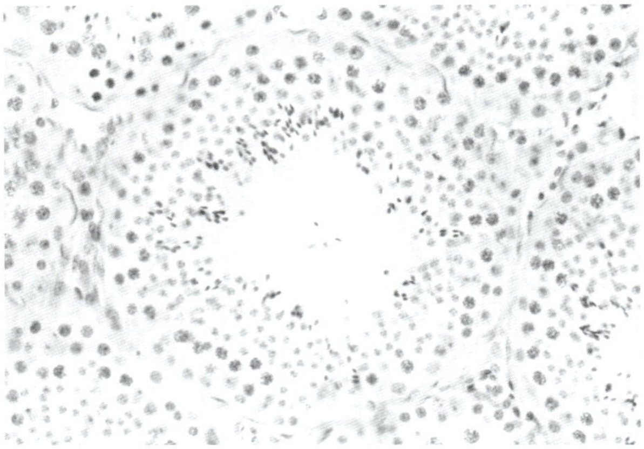

Figure 31.2 Bright field microscopy of a transverse section through a seminiferous tubule of *Chlorocebus aethiops* showing different stages in the same transverse section. This is also typical for humans.

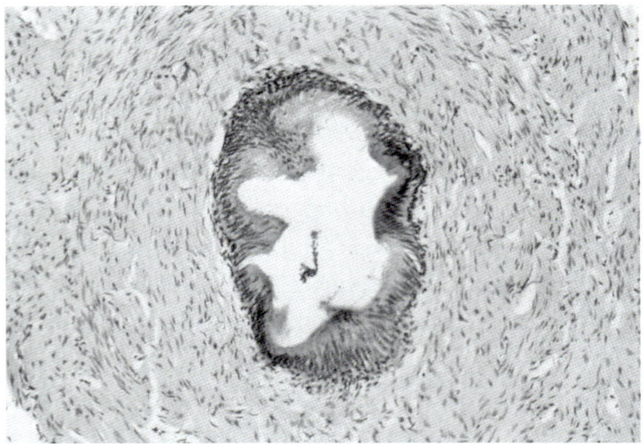

Figure 31.3 Transverse section through the vas deferens of *Chlorocebus aethiops* clearly showing the three smooth muscle coats.

section of a seminiferous tubule of *Chlorocepus aethiops*. According to Weinbauer *et al.* (2001) common marmosets could provide a new animal model for experimental studies of human spermatogenesis based on the organizational similarities in the testis. In this example a New World Primate seems to be a relevant model for the human.

Sperm maturation in the epididymis appears to be quite uniform in NHP and humans. Both histological evidence and data, on the development of sperm motility along the epididymis, indicates similarities in sperm motility and maturation in both NHPs and humans (Yeung *et al.,* 1993; Yeung *et al.,* 1996; van der Horst *et al.,* 1999). Here sperm from the caput show negligible motility and no forward progression. In contrast, sperm from the corpus show a high degree of motility and this continues to improve, together with progressive motility, as sperm pass through the cauda epididymis.

Figures 31.3 and 31.4, respectively, show micrographs of the vas deferens and seminal vesicle of the vervet monkey, *Chlorocebus aethiops*. When histological comparisons of these structures are made with those of humans (Ross and Romrell, 1989), there are striking similarities. In the case of the vas deferens, it is particularly the folded columnar epithelium, with the arrangement of smooth muscle coats, that appear similar in NHP and humans.

In summary it appears that macro anatomically, as well as histologically, the reproductive system of male NHPs shows great similarities to that of humans. The main difference is that daily sperm output is considerably greater and more effective in a NHP, such as the rhesus monkey, than in humans (Amann and Howards, 1980).

Fertility/infertility, *in vitro* fertilization and reproductive technologies

An important basis for this research is a good understanding of semen quality and definition of the fertility status of a semen sample. In applied biomedical research, in the field of fertility/infertility, it is important to use methods that will produce repeatable results when collecting semen routinely from NHPs. Standard semen analysis should be performed using similar criteria as those used for human semen analysis and, preferably, the same internationally accepted methodology (e.g. WHO criteria). The constructed baseline information can then be related to fertility status. In this regard, the best way of defining fertility status of a particular NHP is its record as a proven father with

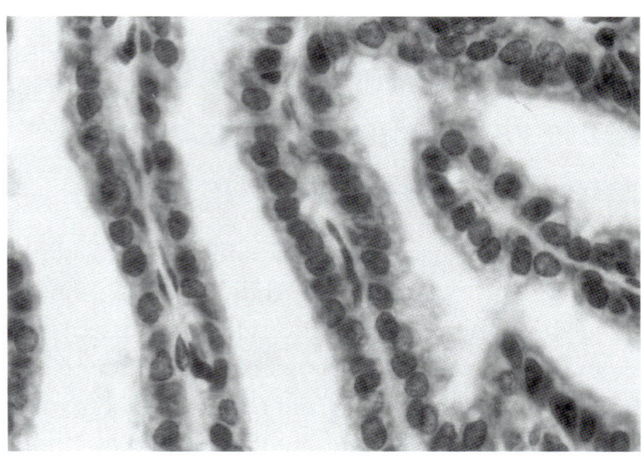

Figure 31.4 Section through the seminal vesicle of *Chlorocebus aethiops*. The structure and details of the epithelial lining is almost identical to that of humans.

several different females in a breeding colony. Some NHP semen parameters can then be compared to humans to establish the relevance of the model.

In most instances, human semen is obtained by masturbation and, in exceptional cases such as spinal injuries, it is obtained by rectal probe electro-stimulation. Rectal probe electro-stimulation and penile vibro stimulation are the two main methods used to obtain ejaculates from NHP. Several papers comment on the advantages/disadvantages of penile vibro stimulation *versus* rectal probe stimulation. While penile vibro stimulation represents a more humane method of collecting semen in NHP (Lanzendorf *et al.,* 1990) and may produce better quality semen (Matsubayashi, 1982), the success of these techniques is often dependent on the operator. Semen samples of high quality are routinely obtained from vervet monkeys using rectal probe electro-stimulation under Ketamine anaesthesia (Seier *et al.,* 1989, 1996; van der Horst *et al.,* 1999; Abrahams *et al.,* 2002, Mdhluli, 2003).

The ejaculates of NHPs typically consist of two main fractions, a liquid fraction and a coagulum. The coagulum very slowly undergoes liquefaction, but it never fully liquefies *in vitro*. Trypsin (Protease) has often been used to liquefy the coagulum component, fully, for accurate sperm concentration determinations (Seier *et al.,* 1989). Furthermore, 1% pronase in Eagle's medium has been employed successfully to liquefy capuchin monkey semen coagulum and the sperm was apparently little affected by this procedure when compared to sperm found in the vagina after normal copulation (Bush *et al.,* 1975).

Despite the differences in the consistency of the ejaculates of human and NHP semen, there are many similarities. Several reports are available on semen characteristics of non-human primates. Harrison (1980) investigated semen parameters in the cynomolgus monkey, *Macaca mulatto,* and Kraemer and Vera Cruz (1969) and Bornman *et al.* (1988) made similar investigations in the baboon (*Papio sp*). In *Papio ursinus,* semen quality evaluations, by usual human spermiogram methods, were applicable with only minor modifications to the procedures. Valerio and Dalgard (1975) were the first to report on semen characteristics in the vervet monkey. Seier *et al.* (1989) performed an in-depth study on semen characteristics of 91 ejaculates (obtained by electro-ejaculation) of 47 vervet monkeys and established the spermiogram of 10 breeding males (proven "fathers") compared to vervet monkeys of unknown fertility. Hiyaoka and Cho (1990) developed a manual penis manipulation method to obtain ejaculates in vervet monkeys and the results on semen volume, sperm density and percent motile sperm, were within the range proposed by Seier *et al.* (1989) who used rectal probe electro-stimulation.

Some of these results are quantified in Table 31.3, showing typical semen parameters for some representative NHP and humans. The species in Table 31.3 are

TABLE 31.3: Semen characteristics of some NHP species compared to minimum acceptable values for humans according to WHO criteria

Semen parameters	Squirrel monkey[a]	Vervet monkey[b]	Chacma baboon[c]	Chimpanzee[d]	Human[e]
Volume (ml)	0.2–1.5	0.86	Few drops	1.1	>2.5
pH	–	7.6	–	–	>7.0
Normal morphology (%)	–	98 (Head) 92 (MP) 70 (PP)	50		>30%
Sperm concentration (x10^6/ml)	205.9	280.5	204.8	548	>20 (range about 8–200)
Motility (%)	52	55	60.9	30	60
Progressive motility (%)	–	52	4.2 on a scale of 0–6		>More than 25% a and b motility
Vitality (%)	–	78	65		>60

a and d = Hendrickx and Dukelow, 1998; b = Seier *et al.,* 1989; c = Bornman *et al.,* 1988; e = WHO, 2000.
MP = Mid piece; PP = Principal piece.

representative of all of the main NHP Taxa. Both New and Old World Primates are represented in the table, with a monkey, a baboon and one of the big apes representing the Old World Primates. In many instances the semen parameters show similarities in volume, percentage motile sperm, percentage progressively motile sperm, pH, and vitality. Sperm concentration in the NHP is generally higher than in humans as is the percentage of morphologically normal sperm. The averages for sperm concentration in Table 31.3 may be misleading since, in each instance, including humans, there are very large ranges. These ranges often overlap when the various NHP's and humans are compared.

The most common morphological abnormality in humans is amorphous sperm with an uneven shape of the head. In contrast, sperm head abnormalities are rare in most NHPs and midpiece and principal piece abnormalities are more common (Seier *et al.,* 1996; Cui *et al.,* 1991). These differences should be remembered in applied biomedical research studies. In contrast, the average percentage sperm motility of the various NHPs and humans falls within a remarkably narrow range.

There has been good progress in quantifying sperm motility in a broad range of mammalian species, particularly humans (Mortimer, 1994). Few studies report on computer aided sperm motility analysis (CASMA) in non-human primates (Yeung *et al.,* 1993, 1996 – cynomolgus monkeys; Gould *et al.,* 1993 – chimpanzees; van der Horst *et al.,* 1999 – vervet monkey). In all other studies, use was made of a subjective rating of sperm motility and this may vary greatly between laboratories. The usefulness of a definition of the normal quantitative motile status of NHP sperm is apparent. This information may not only provide exact baseline data for sperm motility, for a particular species, but, together with quantitative physical and biochemical semen parameters, it assists in establishing cut-off points of fertility assessment. The applications of this are particularly important in sperm capacitation, male fertility assessment, contraceptive research and reproductive toxicology. Table 31.4 shows some comparative parameters of CASMA of humans and a non-human primate (vervet monkey). The results suggest close similarities between NHP and humans. The CASMA data for cynomolgus monkeys and the chimpanzee are similar to those of vervet monkeys.

Table 31.4 shows that sperm velocity increases when sperm are incubated for a short period of time in culture medium. In humans, there is a concomitant decrease in linearity and an increase in amplitude of lateral head displacement. Normally, in a potentially fertile semen sample, more than 20% of human sperm become hyperactivated (erratically and very fast swimming sperm) within one to three hours after incubation. Hyperactivation is an important parameter indicating that sperm capacitation is taking place. Sperm cannot undergo the acrosome reaction (pre-requisite for fertilization) unless they have become capacitated. Hyperactivation has been described in macaque sperm (Vandevoort and Overstreet, 1995; Gwathmey *et al.,* 2000). That vervet monkey sperm will become hyperactivated in a capacitation medium, such as human oviduct fluid (HTF), has proved useful for *in vitro* studies and human clinical applications including IVF, GIFT and IUI (Mortimer, 1994).

TABLE 31.4: Quantitative sperm motility parameters of ejaculated sperm as measured in seminal plasma and the culture medium, Ham's F10, after about 10 minutes immersion at 36°C. Data ± SEM

Sperm motility	Vervet monkey		Human	
Parameter	In semen	In Ham's F10	In semen	In Ham's F10
VCL (µm/s)	109.8 (±7.3)	164.4 (±4.4)	92.3 (±3.3)	145 (±13.2)
LIN (%)	61.4 (±4.5)	76 (±1.95)	59.4 (±2.9)	24 (3.5)
ALH (µm)	4.6 (±0.21)	4.66 (0.15)	4.7 (0.21)	8.5 (0.8)

VCL = curvilinear velocity; LIN = Linearity of swimming track; ALH = Amplitude of lateral head displacement.
(van der Horst, 1995; van der Horst *et al.,* 1999).

Sperm vitality, functional and fertilizing ability testing, such as the hypo-osmotic swelling test (HOS), cervical mucous penetration assay, the assessment of the acrosome reaction and the hemizona assay, are important adjuncts to human semen analysis (Mortimer, 1994). Very few of these approaches have been tested in the NHP. However, Lohiya *et al.* (2002) used the HOS test successfully in fertility studies on langur sperm. Mdhluli (2003), furthermore, showed that the four classes of peanut agglutinin labelling (PNA), demonstrating various phases of the human acrosome reaction, were identical in the vervet monkey.

Using rhesus monkey gametes, Hewitson and Schatten (2002) have shown that the cytoskeletal events during fertilization, by IVF and intracytoplasmic sperm injection (ICSI), are very similar to those in human fertilization. Indeed, manipulations of non-human primate gametes may help in testing the safety of, and improve current strategies for, reproduction, as well as developing new techniques. Furthermore, Sutovsky *et al.* (1996), suggested that the similarities between fertilization in rhesus monkeys and humans provide a model system in which to investigate the cellular and molecular biological basis of human reproduction.

Results from NHP models have revealed the necessity for caution in clinical procedures such as intracytoplasmic sperm injection (ICSI). It has been shown that rhesus monkey spermatozoa, with exogenously bound DNA, retain their full reproductive capacity in ICSI. However, there is concern that ICSI could, theoretically, also transmit infectious material (Chan *et al.*, 2000). Furthermore, fertilization of rhesus monkey zygotes, by ICSI, resulted in abnormal nuclear remodelling, leading to asynchronous chromatin decondensation of the sperm, causing a delay in the onset of DNA synthesis. It raises concerns that the ICSI procedure may result in chromatin damage during DNA decondensation. This highlights the need for devising improved pre-clinical assessment prior to global acceptance of this, and other, novel methods of assisted reproduction (Hewitson *et al.*, 2000).

Cryopreservation

Cryopreservation of human sperm is important for many reasons. Sperm banks are an important resource for male genetic material to assist couples where the male is sterile. Sperm banks provide genetically safe material and can provide HIV free semen. Sperm banks ensure the preservation of sperm from fathers who may subsequently undergo a vasectomy or develop a pathological condition that will eventually render them sterile.

Studies on optimal cryopreservation of NHP sperm are therefore not only important to preserve genetic material, especially of endangered NHP, but to assist in the improvement of preservation techniques for human sperm. In this regard, NHP sperm have been found to be very good models for cryopreservation of ejaculate semen and epididymal sperm. Seier *et al.* (1993) developed a cryopreservation protocol for vervet monkeys that yielded 68% post-thaw sperm motility. The incidence of post-thaw intact acrosomes in a cryopreservation method for vervet monkey semen was found to be similar to that reported for man and cynomolgus monkey semen (Conradie *et al.*, 1994). Sankai *et al.* (1994) used a similar sperm freezing protocol to that of Seier *et al.* (1993) for cynomolgus monkeys and reported successful fertilization and embryonic development to hatched blastocysts.

The effective combination of cryopreservation of epididymal spermatozoa and ICSI is possible in reproduction of NHPs and has potential in the conservation of highly endangered non-human primate species. The cynomolgus monkey is a reliable biomedical research model to study the potential risks and benefits associated with assisted reproductive techniques prior to approval for clinical trials on humans (Ng *et al.*, 2002).

Contraception

Hormonal, immunological, mechanical and other diverse approaches have been used in NHP to test their suitability as models for human male contraception. Table 31.5 lists some of the non-human primate species used and the contraceptive approach utilized.

Hormonal studies represent the major approach to male contraception. Reproductive steroids (Adams *et al.*, 1987), synthetic steroids (Ramachandra *et al.*, 2002) and GnRH analogues (Fraser *et al.*, 1994) have been used, with various levels of success, as male contraceptives in both New World and Old World Primates. MENT, chemically defined as 7alpha-methyl-nortestosterone, has the advantage that no additional testosterone therapy is required as has been demonstrated in the marmoset (Ramachandra *et al.*, 2002). Hormonal oral contraceptive intake in NHPs suggest similarities with humans. However, it is not clear what the long-term effects of exogenous androgens may be, particularly in view of the risk of prostate cancer.

An improvement in the development of immunological approaches to male contraception has been due to new technologies of molecular biology. This approach allows the investigator to be very specific in targeting a male reproductive enzyme, such as LDH-4(C),

TABLE 31.5: Some non-human primate species used for various contraceptive approaches

Contraceptive approach	Non-human primates used	References
Hormonal		
17-beta estradiol and Levonorgestrel	Cynomolgus monkey	Adams *et al.*, 1987
7-alpha-methyl 19 nortestosterone (MENT)	Bonnet monkey	Ramachandra *et al.*, 2002
GnRH	Marmoset	Fraser *et al.*, 1994
Immunological		
Immunogen to CD 52 Protein	Chimpanzee	McCauley *et al.*, 2002
Immunogen to LDH-4(C)	Baboon	Goldberg *et al.*, 2001
Riboflavin Carrier Protein (RCP) - Vaccine	Bonnet monkey	Adiga *et al.*, 1997
Various immunogens to acrosome	Rhesus monkey	Archibong *et al.*, 1995
Immunogen to PH-20 Protein	Cynomolgus and Macaque	Deng *et al.*, 2002
Immunogen to ZP3	Bonnet monkey	Afzulpurkar *et al.*, 1997
Mechanical		
Vasectomy	Chimpanzee	Hoffman *et al.*, 1997
Vas occlusion with SMA	Langurs	Lohiya *et al.*, 1998
Other/Natural plant products isolated		
Penta cyclic Triterpene, Oleanolic acid	African green monkey	Mdhluli, 2003

Modified from van der Horst *et al.*, 2004.

for example. According to McCauley *et al.* (2002), sperm agglutination antigen-1 (SAGA-1) is an attractive candidate contraceptive immunogen. However, the immunogen developed specifically compromises fertility in the chimpanzee and human but not in bonnet monkeys, macaques and the baboon (McCauley *et al.*, 2002). In this example of immunological contraception, most NHP do not represent good models for human male contraception. Furthermore, in some instances, cross-reaction of the immunogen with other of the body's cells or tissues, presents problems (Gupta and Koothan, 1990). Despite some of the problems outlined above, both hormonal as well as immunological approaches show potential in male contraception and, in many instances, several non-human primate species appear to be appropriate models to study.

Vasectomy is not reversible in all humans but vas occlusion with styrene maleic acid, in langurs, appears to be fully reversible (Lohiya *et al.*, 1998). Experimental vasectomy, in rhesus monkeys, showed that antisperm antibodies might, in some cases, impair the restoration of fertility after vasovasostomy (Alexander, 1977).

Preliminary studies show that several plant products, such as the pentacyclic triterpene, oleanolic acid, show potential for use as a human male contraceptive (Mdhluli and van der Horst, 2002) since OA does not appear to be toxic or have any side effects in African green monkeys (Mdhluli, 2003).

Alkyloxynol-741 ("agent 741") is used, in the People's Republic of China, as a substitute for nonoxynol-9 in vaginal contraceptive formulations (Diao *et al.*, 1990). Postcoital vaginal spermicidal studies were performed comparing the spermicidal activity of the above two substances in the stumptailed macaque (*Macaca arctoides*). The spermicidal potency of alkyloxynol-741 was found to be greater than nonoxynol-9 (Diao *et al.*, 1990).

To summarize, it appears that, in most instances, NHP would appear to be good models for almost all male contraception studies in humans, the exception being for some immunological approaches.

Reproductive toxicology

Potentially toxic substances, which may directly or indirectly affect humans, have been studied in many NHPs. Among these are organochlorines. Arnold *et al.* (2001) found no statistically significant changes, related to treatment with organochloride toxaphane, in body weight, food and water consumption or haematological parameters during the 75-week pre-mating phase in cynomolgus monkeys. In a further study of organochlorines (Bryce *et al.,* 2001) two major clinical findings were inflammation and/or enlargement of the tarsal gland and impacted diverticulae in the upper and lower eyelids. Toxaphene (organochloride) administration resulted in an increase in metabolism of aminopyrene, methoxyresorufin and ethoxyresorufin. These three substrates are altered specifically by cytochrome P450-based hepatic monooxygenase enzymes. Histopathological examination of tissues by light microscopy was unremarkable by light microscopy in cynomolgus monkeys.

Cooke *et al.* (2001) established that organochlorine residues in the testis were lower than in most of the other reproductive tract and non-reproductive tract tissues examined. For example, testicular aldrin and dieldrin levels were <5% of the epididymal content; testicular cis- and trans-nonachlor <25% and testicular toxaphene <15% of the epididymal content. The reasons for the low degree of accumulation by the testis, in comparison with other tissues, are unknown according to Cooke *et al.* (2001). However, the lower testicular content in cynomolgus monkeys may afford germ cells some protection from the potentially toxic effects of these chemicals.

Baldwin *et al.* (1983) tested Zeranol, an anabolic agent produced commercially for use in cattle and sheep that are intended for human consumption. Toxicity testing (acute, subacute, and chronic) in several species, including rhesus monkeys, by various routes of administration, revealed extremely low toxicity.

It is difficult to make direct comparisons between humans and NHPs in terms of toxicology. However, some extrapolations can be made, in this contex, from experiments on NHP as the closest living relatives of humans.

Correspondence

Any correspondence should be directed to Gerhard van der Horst, Department of Medical Biosciences, University of the Western Cape, Bellville, 7530, South Africa. Email: gvdhorst@uwc.ac.za

References

Abrahams, H., Seier, J., Johnson, Q. and van der Horst, G. (2002). In van der Horst, G., Franken, D., Bornman, R., de Jager, T. and Dyer, S. (eds) *Proceedings of the 9th International Symposium on Spermatology*, pp 121–126. Monduzzi Editore.

Adams, M.R., Clarkson, T.B., Koritnik, D.R. and Nash, H.A. (1987). *Fertil. Steril.* 47, 1010–1018.

Adiga, P.R., Subramanian, S., Rao, J. and Kumar, M. (1997). *Hum. Reprod. Update* 3, 325–334.

Afzalpurkar, A., Shibahara, H., Hasegawa, A., Koyama, K. and Gupta, S.K. (1997). *Hum. Reprod.* 12, 2664–2670.

Alexander, N.J. (1977). *Fertil. Steril.* 28, 562–569.

Amann, R.P. and Howards, S.S. (1980). *J. Urol.* 124, 211–215.

Archibong, A.E., Lee, C.Y. and Wolf, D.P. (1995). *J. Androl.* 16, 318–326.

Arnold, D.L., Bryce, F., Baccanale, C., Hayward, S., Tanner, J.R., MacLellan, E., Dearden, T. and Fernie, S. (2001). *Food Chem. Toxicol.* 39, 467–476.

Baldwin, R.S., Williams, R.D. and Terry, M.K. (1983). *Regul Toxicol Pharmacol.* 3, 9–25.

Baskerville, A., Cook, R.W., Dennis, M.J., Cranage, M.P. and Greenaway, P.J. (1992). *J. Comp. Pathol.* 107, 49–57.

Bornman, M.S., van Vuuren, M., Meltzer, D.G., van der Merwe, C.A. and Rensburg, S.J. (1988). *J. Med. Primatol.* 17, 57–61.

Bryce, F., Iverson, F., Andrews, P., Barker, M., Cherry, W., Mueller, R., Pulido, O., Hayward, S., Fernie, S. and Arnold, D.L. (2001). *Food Chem. Toxicol.* 39, 243–251.

Bush, D.E., Russell, L.H. Jr, Flowers, A.I. and Sorensen, A.M. Jr. (1975). *Lab. Anim. Sci.* 25, 588–593.

Caldwell, J. (1985). *Hazelton Laboratories* 1, 1–18.

Chan, A.W., Luetjens, C.M., Dominko, T., Ramalho-Santos, J., Simerly, C.R., Hewitson, L. and Schatten, G. (2000). *Mol. Hum. Reprod.* 6, 26–33.

Conradie, E., Oettle, E.E. and Seier, J.V. (1994). *J. Med. Primatol.* 23, 315–316.

Cooke, G.M., Newsome, W.H., Bondy, G.S., Arnold, D.L., Tanner, J.R., Robertson, P., Whalen, C.M., Angers, G. and Masse, A. (2001). *Reprod. Toxicol.* 15, 333–338.

Cui, K.H., Flaherty, S.P., Newble, C.D., Guerin, M.V., Napier, A.J. and Matthews, C.D. (1991). *J. Androl.* 12, 214–220.

Deng, X., Meyers, S.A., Tollner, T.L., Yudin, A.I., Primakoff, P.D., He, D.N. and Overstreet, J.W. (2002). *J. Reprod. Immunol.* 54, 93–115.

Diao, X.H., Zou, S., Quigg, J., Kaminski, J. and Zaneveld, L.J. (1990). *Contraception* 42, 677–682.

Dietrich, T., Schulze, W. and Riemer, M. (1986). *Urologe* A, 25, 179–186.

Eley, R.M. (1992). *Utafiti.* 4, 1–33.

Fahim, M.S., Fahim, Z., Harman, J., Thompson, I., Montie, J. and Hall, D.G. (1977). *Fertil. Steril.* 28, 823–831.

Fraser, H.M., Recio, R., Conn, P.M. and Lunn, S.F. (1994). *J. Clin. Endocrinol. Metab.* 78, 121–125.

Gichuhi, P.M., Makokha, A. and Chai, D. (1999). *Contraception* 59, 131–135.

Giraldi, A. and Wagner, G. (1990). *Pharmacol. Toxicol.* 67, 235–238.

Goldberg, E., VandeBerg, J.L., Mahony, M.C. and Doncel, G.F. (2001). *Contraception* 64, 93–98.

Gould, K.G., Young, L.G., Smithwick, E.B. and Pythyon, S.R. (1993). *Am. J. Primatol.* 29, 221–232.

Gupta, S.K. and Koothan, P.T. (1990). *Arch. Immunol. Ther. Exp.* (Warsz). 38, 47–60.

Gwathmey, T., Blackmore, P.F. and Mahony, M.C. (2000). *J. Androl.* 21, 534–540.

Harrison, R.M. (1980). *J. Med. Primatol.* 9, 265–273.

Hendrickx, A.G. and Dukelow, W.R. (1998). In Bennett, B.T., Abee, C.R. and Henrickson, R. (eds). *Nonhuman Primates in Biomedical Research,* p 180. Academic Press.

Hewitson, L. and Schatten, G. (2002). *Reprod. Biomed. Online.* 5, 50–55.

Hewitson, L., Simerly, C., Dominko, T. and Schatten, G. (2000). *Theriogenology* 53, 95–104.

Hiyaoka, A. and Cho, F. (1990). *Jikken Dobutsu* 39, 121–124.

Hoffman, K., Howell, S., Schwandt, M., and Fritz, J. (2002). *Lab. Anim. (NY)* 31, 45–48.

Kerr, J.B. (1992). *Baillieres Clin. Endocrinol. Metab.* 6, 235–250.

Kraemer, D.C., and Vera Cruz, N.C. (1969). *J. Reprod. Fertil.* 20, 345–348.

Kuckuck, L., Chhina, G.S. and Manchanda, S.K. (1975). *Indian J. Physiol. Pharmacol.* 19, 20–27.

Kuederling, I., Schneiders, A., Sonksen, J., Nayudu, P.L. and Hodges, J.K. (2000). *Am. J. Primatol.* 52, 149–154.

Kuehl, T.J. and Dukelow, W.R. (1974). *Lab. Anim. Sci.* 24, 364–366.

Lanzendorf, S.E., Gliessman, P.M., Archibong, A.E., Alexander, M. and Wolf, D.P. (1990). *Mol. Reprod. Dev.* 25, 61–66.

Lohiya, N.K., Manivannan, B., Mishra, P.K., Pathak, N. and Balasubramanian, S.P. (1998). *Contraception* 58, 119–128.

Lohiya, N.K., Manivannan, B., Mishra, P.K., Pathak, N., Sriram, S., Bhande, S.S. and Panneerdoss, S. (2002). *Asian J. Androl.* 4, 17–26.

Matsubayashi, K. (1982). *Jikken Dobutsu* 31, 1–6.

McCauley, T.C., Kurth, B.E., Norton, E.J., Klotz, K.L., Westbrook, V.A., Rao, A.J., Herr, J.C. and Diekman, A.B. (2002). *Biol. Reprod.* 66, 1681–1688.

Mdhluli, M.C. (2003). *Toxicological and antifertility investigations of oleanolic acid in male vervet monkeys (Chlorocebus aethiops).* Unpublished PhD, University of the Western Cape.

Mdhluli, M.C. and van der Horst, G. (2002). *Laboratory Animals* 36, 432–437.

Millar, M.R., Sharpe, R.M., Weinbauer, G.F., Fraser, H.M. and Saunders, P.T. (2000). *Int. J. Androl.* 23, 266–277.

Mortimer, D. (1994). *Practical Laboratory Andrology,* pp 199–213. Oxford University Press.

Muntzing, J., Varkarakis, M.J., Saroff, J. and Murphy, G.P. (1975). *J. Med. Primatol.* 4, 245–251.

Ng, S.C., Martelli, P., Liow, S.L., Herbert, S. and Oh, S.H. (2002). *Theriogenology* 58, 1385–1397.

O'Hern, P.A., Bambra, C.S., Isahakia, M. and Goldberg, E. (1995). *Biol. Reprod.* 52, 331–339.

Ogura, A., Inoue, K., Ogonuki, N., Suzuki, O., Mochida, K., Matsuda, J. and Sankai, T. (2000). *Int. J. Androl. 23 Suppl.* 2, 60–62.

Ramachandra, S.G., Ramesh, V., Krishnamurthy, H.N., Kumar, N., Sundaram, K., Hardy, M.P. and Rao, A.J. (2002). *Reproduction* 124, 301–309.

Ross, M.H. and Romrell, L.J. (1989). In Williams and Wilkins (eds) *Histology. A Text and Atlas,* pp 629–647. Williams & Wilkins.

Rune, G.M., deSouza, P., Krowke, R., Merker, H.J. and Neubert, D. (1991). *Arch. Androl.* 26, 143–154.

Sankai, T., Terao, K., Yanagimachi, R., Cho, F. and Yoshikawa, Y. (1994). *J. Reprod. Fertil.* 101, 273–278.

Schlatt, S., Foppiani, L., Rolf, C., Weinbauer, G.F., and Nieschlag, E. (2002). *Hum. Reprod.* 17, 55–62.

Seier, J.V. (1986). *J. Med. Primatol.* 15, 339–349.

Seier, J.V., Conradie, E., Oettle, E.E. and Fincham, J.E. (1993). *J. Med. Primatol.* 22, 355–359.

Seier, J.V., Fincham, J.E., Menkveld, R. and Venter, F.S. (1989). *Lab. Anim.* 23, 43–47.

Seier, J.V., van der Horst, G. and Laubscher, R. (1996). *J. Med. Primatol.* 25, 397–403.

Short, R.V. (1980). In Austin, C. and Short, R.V. (eds) *Reproduction in Mammals,* Volume 8, Chapter 1, pp 1–33. Cambridge University Press.

Smith, R.L. and Williams, R.T. (1974). *J. Med. Primatol.* 3, 138–152.

Sutovsky, P., Hewitson, L., Simerly, C.R., Tengowski, M.W., Navara, C.S., Haavisto, A. and Schatten, G. (1996). *Hum. Reprod.* 11, 1703–1712.

Valerio, A. and Dalgard, W.D. (1975). In Perkins, F.T. and Donoghue, O. (eds) *Breeding Simians for Developmental Biology,* p 49–62. London Laboratory Animals Ltd.

van der Horst, G., Seier, J. and Mdhlui, M. (2004). Subhuman primates as models for the development of male contraceptives. *Gynecologic and Obstetric Investigation* 57(1), 15–17.

van der Horst, G. (1995). *Computer aided sperm motility analysis of selected mammalian species.* Unpublished PhD thesis, University of Stellenbosch.

van der Horst, G., Seier, J.V., Spinks, A.C. and Hendricks, S. (1999). *Int. J. Androl.* 22, 197–207.

Vandevoort, C.A. and Overstreet, J.W. (1995). *J. Androl.* 16, 327–333.

Weinbauer, G.F., Schlatt, S., Walter, V. and Nieschlag, E. (2001). *J. Endocrinol.* 168, 25–38.

Wolf, D.P., Vandevoort, C.A., Meyer-Haas, G.R., Zelinski-Wooten, M.B., Hess, D.L., Baughman, W.L. and Stouffer, R.L. (1989). *Biol. Reprod.* 41, 335–346.

World Health Organization. (2000). *WHO Manual for the Examination of Human Semen and Sperm-cervical Mucus Interaction* (Fourth edition). Cambridge University Press.

Yeung, C.H., Cooper, T.G., Oberpenning, F., Schulze, H. and Nieschlag, E. (1993). *Biol. Reprod.* 49, 274–280.

Yeung, C.H., Morrell, J.M., Cooper, T.G., Weinbauer, G.F., Hodges, J.K., and Nieschlag, E. (1996). *Int. J. Androl,* 89, 113–121.

CHAPTER 32

Reproduction: Female

W. Richard Dukelow
Professor Emeritus, Michigan State University, USA

Historical perspective

The function of the sperm was discovered in 1788 and description of ova followed 50 years later. From that point on, efforts were made to better understand the fertilization process. Attempts were even made to fertilize oocytes *in vitro* to produce embryos but these were unsuccessful. It wasn't until the 1890s that true scientific methods began to be applied to reproductive biology.

Walter Heape, of Cambridge University, is credited with being the father of modern reproductive biology. His efforts, primarily with the rabbit model, demonstrated ovum and embryo recovery and even embryo transplantation to yield living offspring. Heape was independently wealthy and traveled the world. In India he became fascinated by rhesus monkeys and transported five of these animals back to his home in England. There he studied their menstrual cycles and compared the rhesus to what was known of human reproduction. His subsequent publication (Heape, 1897) represented the first published scientific report in this research area.

Hartman (1933) demonstrated that the rhesus monkey ovulates on approximately day 13 of the 28-day menstrual cycle. Yet the optimal time for mating is on day 11 or 12 (Van Wagenen, 1945). These studies suggested a time interval that would eventually be described as "capacitation time," a phenomenon that remained unknown until its discovery in 1951.

Up until about 30 years ago the vast majority of research was conducted on the rhesus monkey. Despite a distinct seasonality of ovulation, and other problems, the background information accumulated on the rhesus provided a sound basis for the use of this species as a model for fertilization and early embryogenesis in man. In the late 1970s exportation of rhesus monkeys from India was banned and this resulted in an increased effort to evaluate other nonhuman primates as research models for man. Today the use of the rhesus has been partially replaced by the cynomolgus or long-tailed macaque.

While the total number of nonhuman primate species used in research has declined, there are still large numbers used for experimental purposes. Included are several prosimians (lemurs and bushbabies), New World species (squirrel monkeys, marmosets, tamarins), Old World species (rhesus, cynomolgus, pig-tailed macaques, baboons, African green monkey), and apes

The Laboratory Primate

ISBN 0-1208-0261-9

(chimpanzees). Endangered species should not be used as animal models for man and are precluded from consideration in this chapter (Hendrickx and Dukelow, 1995a).

Over the years breeding programs, especially those employing timed-mating, have contributed basic information on the estrus and menstrual cycles of primates and these are reviewed in other chapters of this publication.

Follicular growth and ovulation

The selection of the dominant follicle and endocrinological events leading to ovulation have been defined by Hodgson (1982) and diZerega and Hodgson (1981). The macaque monkey has been extensively used to define the development of the follicle immediately prior to the time of ovulation, the ovulation process itself and the formation and initiation of progesterone secretion by the corpus luteum. The nonhuman primate model is particularly valuable for the study of the human process. This has been greatly aided by the rediscovery of abdominal laparoscopy in the 1960s (Jewett and Dukelow, 1973; Harrison, 1980).

Ovum and follicular morphology

As ovulation approaches, the dominant follicle protrudes from the ovarian surface and shows increased vascularity. The major volume of the follicle is a fluid-filled space, the antrum, which is lined by granulosa cells. These cells will eventually transform into the cells of the corpus luteum after ovulation. At the base of the follicle is a small pedestal of cumulus cells, the cumulus oophorus, that projects from the follicular wall. The cumulus cells also surround the ovum. The basic vestments of the ovum are: (a) a loosely packed layer of large granulosa cells, (b) an inner layer of small compact cells that constitute the corona radiata, and (c) the mucopolysaccharide layer that immediately surround the ovum itself. The latter is termed the zona pellucida. At the time of ovulation, the ovum and surrounding vestments are expelled from the follicle and the cumulus cells aid in facilitating the pickup of the ovum by the wavelike action of the cilia located on the fimbriated end of the oviduct. All of these events, including the growth of the follicle, the capacitation of the sperm, the growth of the cilia to influence ovum transport and the muscular action of the female reproductive tract are under the control of the endocrine system.

Preovulatory events on the ovarian surface

Laparoscopy can be utilized to accurately define changes on the follicular surface (Jewett and Dukelow, 1972; Dukelow, 1975). Using cynomolgus macaques and squirrel monkeys (Harrison and Dukelow, 1974) these investigators characterized follicular morphological changes in the 24 to 36 hour period prior to ovulation. These changes occur in a fixed sequence allowing one to predict the time of natural ovulation with reasonable accuracy. The endocrinological changes associated with ovulation in macaques have been characterized by Weick *et al.* (1973). There is a continuous, symmetrical luteinizing hormone (LH) peak with maximal plasma concentrations ranging from 42 to 81 ng/ml. The follicle stimulating hormone (FSH) follows a similar time course with peak levels ranging from 67 to 230 ng/ml, occurring either simultaneously with the LH peak or three hours later. The mean duration of these surges is approximately 50 hours. The first unambiguous increase in plasma LH concentration is accompanied by an increase in estradiol concentration that peaks 9 to 15 hours before the LH peak. Plasma progesterone levels are low (less than 0.1 ng/ml) until at least 15 hours after the initial rise in LH, becoming detectable 6 to 9 hours after the estradiol peak. After a small progesterone peak, which occurs within 6 hours of the LH peak, the progesterone concentration declines for about 18 hours before starting the major rise associated with corpus luteum secretion. The earliest time of ovulation is about 28 hours after the LH peak.

Species differences in reproductive parameters are important in selecting appropriate animal models for man. It has become fashionable in recent years to use nonhuman primates as human models because of their menstrual cyclicity and their evolutionary relationship to man. Such a rationale has merit in many research areas but there are some notable exceptions that must be considered. Implantation, for example, is superficial in nonhuman primates compared to humans. A researcher looking for an appropriate model may wish to select the guinea pig to study implantation. Not all nonhuman primates menstruate (many New World primates do not) and some species have a questionably active corpus luteum. While many nonhuman primates have cycle lengths and characteristics approximating those of the human, most have a shorter gestation length and many show seasonal responses, reflecting their wild condition, that are not found in humans. These differences must be recognized, not only when making nonhuman vs. human primate comparisons but also between species of nonhuman primates as well. Traditionally the rhesus has served as the most common nonhuman primate model of man in reproductive studies. However, this species experiences an anovulatory period during the summer months (sometimes referred to as "summer sterility") that makes it a less desirable model for ovulation studies in contrast to macaques such as the cynomolgus, stump-tailed or pig-tailed macaques, that ovulate and can conceive throughout the year.

The site of the developing follicle can be identified on the ovarian surface, two days before ovulation, by a generalized swelling and darkening of the site and an increase of about 35% in the total size of the ovary. Within 30 hours of ovulation, a stellate pattern of blood vessels occurs on the follicular surface and, by 8 to 10 hours before ovulation, the blood vessels have become more pronounced and the follicular cone or stigma is established. At this time the oviductal fimbria moves to envelop the developing follicle and to prepare for the pickup of the ovum from the ovarian surface. Two to three hours before ovulation a lightening at the base of the follicle occurs, probably indicating the start of luteinization and the source of preovulatory progesterone. Observations were made of 78 actual ovulations in cynomolgus macaques. In 19 of these the time of ovulation could be determined within a 24 hour period (Rawson and Dukelow, 1973).

In the squirrel monkey, similar vascular patterns are observed but ovulation is preceded by much more extensive hemorrhaging at the base of the follicle. Accordingly, discrete blood vessel patterns are difficult to observe. In the galago (bushbaby) a very pronounced protrusion of the follicle occurs from the ovarian surface and the follicular vessels can be observed at the base of the follicle and occasionally near the apex. The formation of clear areas (the stigmata) is not as evident in either the squirrel monkey or the galago as it is in the cynomolgus macaque. Basically similar patterns of periovulatory follicular development have been described in the Japanese macaque, the baboon and the chimpanzee. In the capuchin monkey the stellate pattern of blood vessels becomes pronounced about 24 hours before ovulation and concominant with the highest level of estradiol and at the time of a slight increase in the level of progesterone (Nagle *et al.,* 1980). These workers reported that ten hours before ovulation, and with a decreasing pattern of estradiol and an increasing level of progesterone, the follicle shows a conical aspect with hemorrhages occurring in the follicular wall. Ovulation time in this species was characterized by a tight adherence of the fimbria of the oviduct to the follicular surface and this was maintained for at least two hours.

In the cynomolgus macaque the movement of the fimbria to the dominant follicle is similar to the pattern seen in humans. In the rhesus monkey the fimbria is normally found more closely surrounding one pole of the ovary in a cap-like fashion. In the pig-tailed macaque the fimbrial position is intermediate between that of the rhesus and cynomolgus macaques.

Preovulatory follicle diameter varies according to body size in primates. In humans this is about 10 mm. Corresponding values for other primates are: gorilla, 7 mm; rhesus, 6.3–6.6 mm; baboon, 6 mm; gibbon, 5–6 mm, New World monkeys, 5–6 mm; tupia and tarsiers, 1 mm. The quantity of granulosa cells is similar in all of these species with the follicular size difference primarily due to the secretion of fluid into the anthrum. The theca interna layer of the follicle is usually 2–4 cells in depth. The size of the total ovary is generally proportional in body size except in the platyrrhine species in which the ovaries are disproportional large. The bonnet macaque shows unusual follicular development that is characterized by invaginations or foldings of the follicular wall into the follicular anthrum.

Exogenous administration of steroids is normally used to stimulate and mimic the events of a normal reproductive cycle. In most macaque species, exogenous estradiol initiates an LH rise and an increased level of serum progesterone occurs several hours before the LH peak but after the initiation of the LH surge. In most macaques a secondary FSH surge occurs 2–3 days after the LH peak. The significance of this second FSH surge is not certain, but it may enhance postovulatory progesterone and estrogen synthesis.

Induced ovulation

Numerous regimens have been used over many years to induce ovulation. Steroids are used to alter the cycle and these are normally followed by gonadotropin (LH, FSH) or related compounds (Pregnant Mares Serum-PMS, Human Menopausal Gonadotropin-HMG, Human Chorionic Gonadotropin-HCG, and various releasing factors or synthesized compounds) to induce ovum and follicle maturation and ovulation itself. Administration of 0.5 mg of progesterone daily to the rhesus will block ovulation and effectively blocks the LH surge and ovulation. This is the basic principle used in the synchronization of the reproductive cycles of a number of laboratory and domestic species and, of course, was crucial in the development of the contraceptive "pill" for humans. When the progesterone is stopped, and estradiol administered, the animals experience a rise in LH and ovulation.

Seasonal responses to ovulation induction regimens occur in several primate species. The rhesus macaque experiences a distinct seasonal effect on ovulation and this must be considered in breeding programs or studies of basic reproductive phenomena. Seasonal responses in the squirrel monkey are even more pronounced (Harrison and Dukelow, 1973). Natural seasonality occurs when the animals receive an ovulatory regimen of five days of 5 mg of progesterone followed by four days of FSH (one mg) and a single injection of HCG. In the northern hemisphere such a regimen does not work during July, August and September. Various techniques have been used to overcome this seasonality (Kuehl and Dukelow, 1975a). To more closely mimic the natural cycle, the administration of sequential estradiol and progesterone pretreatments, before the FSH-HCG treatments, increases the ovulation rate. Alternatively, merely increasing the level of FSH from 1 to 2 mg per day or from a 4-day to a 5-day regimen results in much higher levels of ovulation.

Protein hormone and ovulation induction

Gonadotropins from a wide variety of domestic animal sources have been used in primates to induce ovulation with varying degrees of success. Today we recognize that while a variety of FSH sources can be used to stimulate primate follicular growth, there appears to be a species-specific requirement for the LH to induce the actual ovulation. Accordingly, many of the hormones used with nonhuman primates (and in humans) derive from either primate sources or are produced by modern synthetic procedures. For primates, the most common sources of FSH are human menopausal gonadotropin (HMG) or pituitary compounds from human or nonhuman primate sources. HCG is most commonly used for the induction of ovulation. In addition, clomiphene citrate, a nonsteroidal estrogenic compound that augments endogenous FSH release, is used in both human and nonhuman primates, although the ease of controlling the exact time of ovulation is more difficult with this compound. In rhesus, a "staircase" regimen of increasing levels of HMG can be used with close monitoring of the urinary estrogen levels. Once estrogen excretion reaches the normal preovulatory level of about 7 mg for 24 hours, the HMG treatment can be stopped and the estrogen level is allowed to fall. Then a single injection of 500 IU of HCG is used to induce ovulation. The dose of HMG required to induce this follicular development can vary markedly from animal to animal. In baboons the HMG-HCG regimen normally gives single ovulations but some reports suggest that increased levels result in multiple ovulations. These hormones have also been implicated in some changes in ovum maturation and possible defects of the zona pellucida.

Excessive use of gonadotropins

Excessive use of gonadotropins and related products can result in antibody reactions that diminish the value of the product or, in some cases, may render the animal nonresponsive in later courses of treatment. This is most evident in the use of pregnant mares serum (PMS). This compound has been used in many species for "superovulation" (production of more than the normal number of ova/embryos) but is only effective for a limited number of treatments. It should not be used unless sacrifice of the animal is anticipated after a single trial. With other compounds, preliminary experiments should be undertaken to determine the "minimum effective dose" (MED) to achieve follicular growth and ovulation. For example, in the squirrel monkey, the FSH dose used in season appears to be about l mg per day for four days. For HCG in the squirrel monkey, the MED is between 100 and 250 IU (Dukelow, 1979). In sheep, where 1500 to 2000 IU is normally used clinically, the MED is approximately 100 IU. Interestingly, in human IVF procedures, 5,000 to 10,000 IU of HCG are routinely given, approximately seven times the amount necessary to ovulate a cow! Administration of excessive HCG to rabbits, to induce ovulation, resulted in 9.7% of 6-day blastocysts identified as chromosomally defective. (Fujimoto *et al.*, 1974). Excessive administration of gonadotropins can be detrimental.

LH-releasing hormone (LHRH) or GnRH (gonadotropin releasing hormone), has been used to elicit an LH response in a number of nonhuman primates and in humans. Norman and Spies (1979) found that female fetuses, infants and prepubertal animals can release LH in response to LHRH. Baboons and ring-tailed lemurs also give this response. In the chimpanzee, LHRH antagonists can block gonadotropin release (Gosselin *et al.*, 1979) as in humans.

Ovum and embryo recovery techniques

Nonhuman primate ovum and embryo recovery techniques have primarily been studied in the baboon, rhesus monkey and the squirrel monkey. In their classic studies in the baboon, Hendrickx and Kraemer (1968) used an *in situ* flushing procedure where a glass speculum was inserted into the cleaned vagina of an anesthetized female. The reproductive tract was then exposed, by midventral incision, and plastic catheters placed about 3 mm into the infundibular end of the oviduct. An 18-gauge catheter was then inserted through the uterine fundus into the uterine lumen. The cervix was clamped and the collection medium was forced through the uterus and out the oviduct. After this procedure, the cervical clamp was removed and additional fluid forced through the uterus and cervix through the vaginal speculum.

Laparoscopic techniques have been used to recover follicular oocytes and, more recently, this procedure has been supplanted by ultrasonic observation of the developing follicle. These procedures have been adapted to many uses including intrafollicular injections and topical application of compounds to the ovarian surface (Dukelow, 1978) as well as for direct deposition of sperm suspensions or fertilized ova into the oviductal fimbria (Figure 32.1).

Laparoscopic techniques have also been used to recover uterine blastocysts in the squirrel monkey. The animal is first subjected to an ovulation induction regimen and then exposed to the male for a period of 12 to 24 hours. A preliminary laparoscopic or an ultrasound examination is performed about 36 hours after the HCG injection to determine if ovulation has occurred. If positive, the uterine flushing procedure illustrated in Figure 32.2 is then carried out at 4, 5, or 6 days depending on the stage of embryo desired. With this technique the uterus is manipulated with a Verres cannula and grasped with the alligator forceps to steady the uterus. A 25-gauge, 5/8 inch needle is inserted through the lower abdominal wall and the uterine fundus. Prewarmed media (2–3 ml) is then flushed through

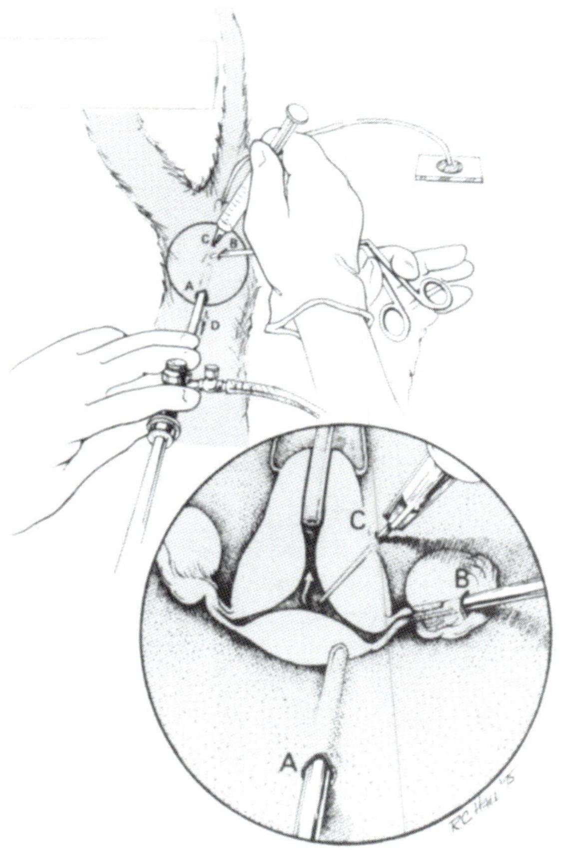

Figure 32.1 Laparoscopic embryo transfer procedure used with the squirrel monkey.

the polyethylene catheter previously inserted into the external os of the vagina.

Production of precisely aged embryos

The production of precisely timed aged embryos can be accomplished by controlled breeding with a short period of exposure to the male about the time of ovulation. While sperm will survive in the female reproductive tract of most primates for up to five days, the fertilizable life of the ovum is much shorter, ranging from 12 to 24 hours. Without supplemental ultrasonic observation or serum/urinary endocrine monitoring, the time of initial fertilization can be estimated to within two days. A thorough discussion of such breeding techniques has been published (Hendrickx and Dukelow, 1995b). More precise estimates of fertilization time can

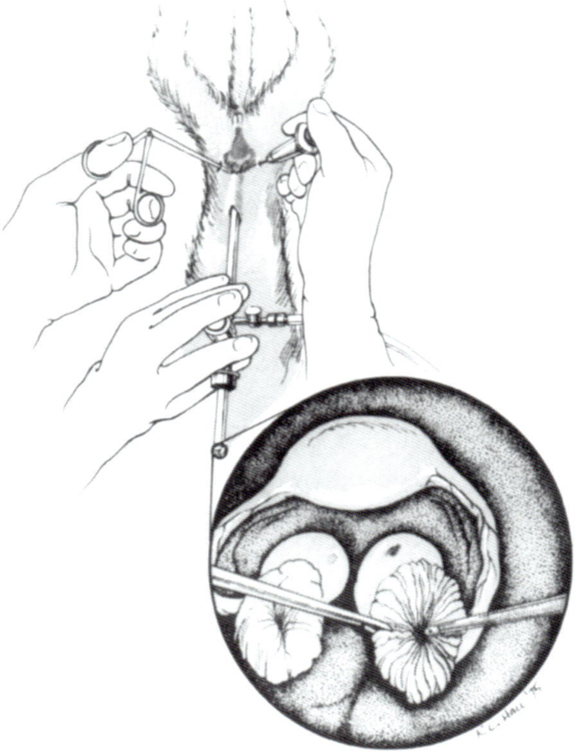

Figure 32.2 Diagrammatic representation of the procedure for laparoscopic flushing of the squirrel monkey uterus.

be obtained by the *in vitro* fertilization technique or related new technologies. In the terminology of human reproduction these procedures are termed ART or "Assisted Reproductive Techniques."

Basically three types of criteria can be used to measure the normality of *in vivo* or *in vitro* fertilized oocytes cultured *in vitro* prior to transfer to a recipient female. These criteria are developmental, chromosomal and biochemical. Table 32.1 illustrates comparison of both *in vivo* and *in vitro* fertilization and development of several primate species including the human. Since differences between the two methods of fertilization were not significant, the values were pooled. The similarity of the rates of development of the nonhuman species to the human are clear. This emphasizes the usefulness of nonhuman primate models, in general, and precludes the concept of a single primate species as the only acceptable model for human reproduction.

Extensive biochemical or chromosomal analysis of fertilized embryos is not ethically possible with humans and offers one of the major advantages of the application of such studies with nonhuman primate species. Studies carried out on squirrel monkey IVF oocytes demonstrate that, by the metaphase II stage, the incidence of chromosomal abnormalities was 7.4 to 14%, a value comparable to that found with other species and with normal oocytes (Asakawa *et al.,* 1982; Dukelow, 1993). There is no indication of an effect of IVF on increasing the level of chromosomal abnormalities. The incidence of triploidy of IVF squirrel monkey oocytes was 16.7%. Triploidy is commonly encountered with IVF in all laboratory species (Pierce *et al.,* 1993) and has been reported in man as well (Lopata *et al.,* 1978). The cause of the triploidy is not known. There does not seem to be a relationship between sperm concentration in the culture and the level of triploidy or percent fertilization.

While fertilization rate *in vitro* in man (Soupart and Strong, 1974) is high (70–90%) the number resulting in the delivery of living young is low (10–25%). This greater embryonic and fetal death loss occurs after fertilization and could reflect the incidence of triploidy

TABLE 32.1: Comparative rates of primate preimplantation development

Species	Hours after Fertilization						
	Two Polar Bodies	2-cell	4-cell	8-cell	16-cell	M	B
Squirrel M.	6–22	16–40	46–52	57–72	96		
Marmoset			24–48	48–72	96–144	120–192	168–240
Rhesus		24–36	36–48	48–72	72–96	88–111	144–192
Cynomolgus		16–36	37–48	49–72	61–108	97–120	
Baboon	6–24	24	48	48–72	96–120	120–148	96–144
Human	12	30–38	38–52	51–72	85–96	96–135	123–147

From: Dukelow (1993).
M = Morula; B = Blastocyst.

that occurs. Greater emphasis of research is needed to reduce this incidence.

There have been few biochemical or metabolic aspects of development that have been examined. In the squirrel monkey there have been studies on protein synthesis, uptake of steroid hormones, oxygen consumption and overall viability of IVF embryos (Hutz *et al.*, 1983). Incorporation of labeled leucine, as an indicator of protein synthesis, declined with oocyte maturation and remained constant after IVF, as assessed by autoradiography. There was a nonsignificant increase at first cleavage. Uptake of estradiol and progesterone increased after IVF in these embryos, but there was no further change in the uptake of either steroid after the first cleavage. Uridine incorporation and uptake, as a measure of RNA synthesis, decreased in oocytes recovered after 36 hours after HCG administration compared to oocytes recovered after 16 hours. A doubling of uridine incorporation occurred after fertilization, with further increase as development progressed after the first cleavage division.

Contraceptive effects

Since a major action of contraceptive steroids has been the blocking of ovulation (with a minor effect on spermatozoal capacitation), it is understandable that normal and induced ovulation regimens should be studied in nonhuman primates using various antiprogestational agents. Much of this work is beyond the scope of this review, but it should be mentioned that ovulation in rhesus monkeys has been suppressed by intranasal administration of progesterone or norethisterone (Anand Kumar *et al.*, 1977). Additionally, the ovulation induction regimen (FSH-HCG), previously described for use in the squirrel monkey, has been used to test the effect of megestrol acetate with a dose-response blockage of ovulation (Harrison and Dukelow, 1971). Similarly, synthetic TPAL (threonyl-prolyl-arginyl-lysine), a polypeptide reported to occur naturally in the hamster, has been tested at two doses for antiovulatory activity in the squirrel monkey. This treatment did not alter the ovulatory response in the primates but emphasizes the usefulness of the induced ovulatory nonhuman primate as a test system (Kuehl and Dukelow, 1978).

Intrauterine devices (IUD) are widely used for human contraception around the world. These devices are lodged in the uterus with either the device or attached material projecting through the cervix. These effective contraceptives can affect embryo implantation and also alter the biochemical composition of the cervical mucus that could have anti-sperm activity. Nasir-Ud-Din *et al.* (1979) have described changes in the cervical mucus in the bonnet monkey. The anatomy of the cervical canal differs widely in various primates and for IUD research an animal model is needed with a relatively straight canal such as the patas monkey. The macaques are not ideal for such studies because of the tortuous nature of their cervices.

Embryo transfer

The first successful surgical transfer of nonhuman primates (natural mating) was reported by Kraemer *et al.* (1976) in the baboon. A 5-day old embryo was transferred to a synchronized female and 174 days later an infant was delivered by cesarian section. In the rhesus monkey a simple surgical flushing procedure was developed by Hurst *et al.* (1976). Of 22 flushes, nine embryos and two unfertilized ova were recovered. By a similar technique Marston *et al.* (1977) recovered fertilized rhesus embryos and these were transferred to the opposite oviduct or the uterus of the same animal. Of eight such transfers to the oviduct, all resulted in pregnancy. Additionally these workers transferred one 5-cell embryo and two 8-cell embryos to the uterus without a resulting pregnancy. Later, two 6-cell embryos were transferred to the uterus and pregnancy resulted. It was these experiments, showing that early embryos could be transferred directly to the uterus, that led other English workers, using the same technique, to transfer human *in vitro* fertilized embryos to produce a living offspring.

In 1980, uterine embryos were successfully recovered by a nonsurgical technique from a baboon (Pope *et al.*, 1980). Of 80 flushes, 37 embryos were recovered from 33 baboons. The procedure was subsequently used to recover a 4-cell embryo that was then transferred, and this resulted in the birth of the first nonsurgical, *in vivo* fertilized nonhuman primate (Pope *et al.*, 1983).

The final major event in this series of pioneering events occurred on July 25, 1983, exactly five years to the day from the birth of the first IVF, nonsurgically transferred human embryo. On that date a baboon infant was delivered after a normal gestation from a prima gravida baboon (Figure 32.3). Earlier this female received four embryos that had been collected from follicles on the ovary of a female that had been autopsied.

Figure 32.3 The first nonhuman primate born as a result of *in vitro* fertilization, July 25, 1983 (Clayton and Kuehl, 1984).

After 24 hours of culture, sperm were added. At 24 hours, a total of 19 of 22 mature ova were fertilized. From this trial, transfers were made (Clayton and Kuehl, 1984). Baboons have also been produced following natural mating procedures, freezing of the embryos, thawing and transfer to recipient females. The role of embryo technologies in genetic management and conservation of wildlife has recently been reviewed (Loskutoff, 2003).

In vitro fertilization

The first successful IVF with laboratory animals (rabbits) was in 1951. Following efforts with other laboratory and domestic animals, and success with humans (Edwards *et al.,* 1969; Steptoe and Edwards, 1978), several laboratories attempted IVF in the nonhuman primate species. Much of this work has been published by Wolf and Stouffer (1993).

Probably the first successful nonhuman primate IVF was with the baboon and reported by Kraemer in 1972 at the International Congress of Animal Reproduction and Artificial Insemination in Munich. These studies were published in a much later report (Kraemer *et al.,* 1979). The first reported primate birth following IVF-embryo transfer was by Clayton and Kuehl (1984) and was also in the baboon. IVF has since been achieved in five nonhuman primate species, the squirrel monkey, the marmoset, the baboon, the rhesus macaque and the cynomolgus macaque.

Two reports appeared in 1972 announcing successful IVF of squirrel monkey oocytes (Cline *et al.,* 1972; Johnson *et al.,* 1972). The first full publication of the former studies appeared in 1973 (Gould *et al.,* 1973). In this report 22 mature oocytes were recovered and of these, 11 had sperm in the perivitelline space, extrusion of the second polar body or pronuclear formation. Six of these zygotes cleaved to the 2-cell stage. Expanded reports of IVF from the latter studies appeared in 1975 (Kuehl and Dukelow, 1975b) with 32 of 79 oocytes fertilized *in vitro*. IVF studies with the squirrel monkey had the advantage of a wide variety of background techniques for ovulation induction and semen collection. Over a five year period, 745 oocytes were aspirated from 2,168 follicles. Of these, 18.4% were atretic and of the remaining 608 oocytes, 38% matured to the metaphase II stage. Of these, 33.5% were fertilized *in vitro*. More recent studies have achieved fertilization rates approximating 70 to 80%. The quantity of cumulus cells on the oocyte at the time of IVF can alter the fertilization rate (Chan *et al.,* 1982). If cumulus cells are absent, maturation is reduced but even as little as one quarter of the oocyte covered with cumulus cells results in a 70% fertilization rate *in vitro*. These workers also found that if oocytes were collected 15 to 16 hours after HCG and allowed to incubate in TC199 culture medium for an additional 21 hours before sperm addition (i.e. fertilization about 37 hours after HCG), a higher level of fertilization was achieved. Furthermore, if 1 to 10 µm of dibutyryl cAMP is added to the culture medium, IVF increased from 60 to 90%. The latter is probably a result of stimulation of sperm capacitation, the acrosome reaction and whiplash motility, all of which are required for penetration of the oocyte.

In the marmoset, oocytes were recovered surgically 24 hours after HCG and incubated in minimum essential medium (MEM) supplemented with 10% heat-inactivated human cord serum. Incubation of oocytes for 2–5, 9–11 and 21–29 hours before insemination resulted in IVF rates of 50%, 86% and 90% respectively (Lopata *et al.,* 1988).

Cynomolgus monkey oocytes have been successfully fertilized after incubation with homologous sperm with some development to the morula stage (Kreitmann *et al.,* 1982). In the rhesus monkey, follicles have been aspirated by laparoscopy 30 hours after HCG (with pretreatment with PMS) and 43% of the oocytes showed signs of fertilization (Bavister *et al.,* 1983). The times of cleavage were comparable to earlier reports with *in vivo* fertilized rhesus oocytes but were somewhat faster than cleavage rates reported by others (Kreitmann *et al.,* 1982). The Oregon National Primate Research Center has, in recent years, produced more IVF rhesus offspring and their procedures have been well described (Wolf and Stouffer *et al.,* 1993).

At times, developments in human IVF have served as models for application to other primates such as the gorilla (Loskutoff *et al.,* 1991; Pope *et al.,* 1997).

Other manipulative techniques and future clinical application

Development of embryo technologies often occur with common laboratory or domestic species, where databases exist, and are then are applied to primates, humans or wild species (Wildt *et al.,* 1986).

Xenogenous fertilization

In the 1950s, before the development of effective embryo freezing techniques, pseudopregnant rabbits were used to transport embryos of other species long distances. The female rabbit was ovulated and for a period of 18 days she was in "pseudopregnancy." Embryos to be transferred were surgically placed in the oviduct and the rabbit shipped between continents! The fertilization of ova in the oviduct of a foreign species, xenogenous fertilization (XF), (DeMayo *et al.,* 1980; Hirst *et al.,* 1981), has been achieved with hamsters, mice, squirrel monkeys, bovine and even human ova (Edwards *et al.,* 1966). Fertilization rates ranged up to 60% (hamster oocytes) and the fertilization rate for squirrel monkeys was 35.3%. The procedure also has been successful with frozen squirrel monkey oocytes (DeMayo *et al.,* 1985). This procedure is of doubtful clinical value with humans, especially considering the success of IVF programs, but it does have strong implications in reproductive studies to preserve endangered animal species that manifest poor reproduction in captivity-stress conditions.

GIFT, ICSI and SUZI procedures

Two ART procedures have been adapted clinically to augment IVF procedures. The first of these is termed "GIFT" (gamete intrafallopian tube transfer) which is essentially xenogenous fertilization but within the human reproductive tract. Basically a recovered oocyte and portions of a semen sample are transferred into the fallopian tube (oviduct). Fertilization occurs in the oviduct and pregnancy continues normally under these conditions. This procedure is frequently used in cases of male-factor diminished fertility.

A second procedure, also used with male-factor infertility, is termed "ICSI" (intracytoplasmic sperm injection). In this procedure the oocyte is held by suction to a small pipette under a dissecting microscope and a single sperm is injected directly into the cellular cytoplasm. There have been reports of reduced blastocyst development with human ICSI compared with IVF (Griffiths *et al.,* 2000; Aytoz *et al.,* 1999). Hewitson *et al.* (2000) used a primate model to demonstrate that ICSI yielded embryos with abnormal nuclear development. A related technique involves placing of the sperm just under the zona pellucida and is termed subzonal sperm injection (SUZI) (Mann, 1988). Both of these procedures are clinically routine in human IVF laboratories with fertilization rates comparable, or even higher, than that found with IVF.

In zoo species and humans, immotile, nonviable sperm (Iritani *et al.,* 1998), testicular sperm (Silber *et al.,* 1995; Wright *et al.,* 1998) and even spermatids (Tesarik *et al.,* 1996) have been used to produce embryos by microinjection techniques.

Embryo biopsy, splitting and cloning (SCNT)

The separation of individual or groups of blastomeres from embryos, with subsequent fusion and transfer of the resultant embryos, has been studied in mice for many years and was recently summarized (Critser *et al.,* 2003). This allows production of multiple, genetically identical offspring (Loskutoff *et al.,* 1993). The pioneering studies of Willadsen (1982) demonstrated the practical application of blastomere separation (embryo splitting) to produce monozygotic twins, triplets and "multiplets" in domestic animal species. The procedure is now routinely used in the commercial cattle industry (Seidel, 1983). For primates, human and nonhuman, blastomere separation coupled with IVF offers a diagnostic tool that could have tremendous implications relative to identifying abnormal embryos and those with potential congenital defects. The isolation and culture of a few cells from a developing embryo fertilized *in vitro,* while the rest of the nondifferentiated embryo continues normal growth and development preparatory to transfer, would allow cytogenetic and biochemical assays to be carried out on the isolated segment for diagnostic purposes. Additionally, in nonhuman primates, the procedure would provide identical twin embryos at a very early stage of development for basic biochemical and cytogenetic studies.

In recent years the scientific world has become enamoured with the possibility of "cloning" and of stem cell reproduction (Critser *et al.,* 2003). Unfortunately there have been innumerable definitions of the term "cloning" ranging from replication of specific DNA sequences to true transfer of genetic material between embryos (Somatic cell nuclear transplantation (SCNT) from adult cells) to the mere production of twins (Campbell *et al.,* 1996; Wakayama and Yanagimachi 1999a,b; Wakayama *et al.,* 1998, 2000; Colman, 2000). This has also led to misinterpretations of the procedure on the part of the public.

By strictest definition, cloning involves the removal, under a dissecting microscope with micromanipulators, of the nuclear DNA from an embryo. Then similar genetic material from another tissue (such as the mammary gland) is transferred into the evacuated embryo. Growth continues in the stem cells. Such studies have tremendous potential for identifying cell development and have possible application to solving the problems of many devastating diseases. It does not imply exact duplication of the embryo. To date, six species have been "cloned" including one sheep, one domestic cat, mice, goats, cows and several pigs. The success rate is low. In sheep, where telemere length is abnormally shortened after SCNT, there may be premature aging (Shiels *et al.,* 1999). There have been reports of cloning in humans but the procedures used have not been published and the results are extremely questionable. At the date of this writing no such studies have been successful in nonhuman primates.

The ethics and legal issues involved with SCNT (and other assisted reproductive technology in general) have been widely discussed (Meslin, 2000; Solter, 2000; Varmus, 2000; Chougule, 2001; Craft, 2001; Jaenisch and Wilmut, 2001).

Conclusion

Progress in the manipulation of the reproductive process to reduce human fertility, enhance fertility and increase production of food-producing animals and endangered species has been significant. Such studies have been carefully conducted with full recognition of the scientific approaches and ethical considerations of the results. Nonhuman primates have made significant contributions to such studies, serving as models for man. Future studies will continue to utilize this important animal model.

Correspondence

Any correspondence should be directed to W. Richard Dukelow, High Meadows Enterprises, 325 Spring Creek Road, Somers, MT 59932, USA.

References

Anand Kumar, T.C., David, G.F.X. and Puri, V. (1977). *Nature* 270, 532–534.

Asakawa, T., Chan, P.J. and Dukelow, W.R. (1982). *Biol. Reprod.* 27, 118–124.

Aytoz, A., Van den Abbeel, E., Bonduelle, M., Camus, M., Joris, H., Van Steirteghem, A. and Devroey, P. (1999). *Human Reprod.* 14, 2619–2624.

Bavister, B., Boatman, D., Liebfried, L., Loose, M. and Vernon, M. (1983). *Biol. Reprod.*, 983–999.

Campbell, K.H., McWhir, J., Ritchie, W.A. and Wilmut, I. (1996). *Nature* 380, 64–66.

Chan, P.J., Hutz, R.J. and Dukelow, W.R. (1982). *Fertil. Steril.* 38, 609–615.

Chougule, G. (2001). *Science* 292, 639.

Clayton, O. and Kuehl, T.J. (1984). *Theriogenology* 21, 228.

Cline, E.M., Gould, K.G. and Foley, D.W. (1972). *Fed. Proc.* 31, 277.

Colman, A. (2000). *Cloning* 1, 185–200.

Craft, I. (2001). *Lancet* 357, 1368.

Critser, J.K., Riley, L.K. and Prather, R.S. (2003). In Holt, W.V., Pickard, A.R., Rodger, J.C. and Wildt, D.E. (eds) *Reproductive Science and Integrated Conservation,* pp 195–208. Cambridge University Press, U.K.

DeMayo, F.J., Mizoguchi, H. and Dukelow, W.R. (1980). *Science* 208, 1468.

DeMayo, F.J., Rawlins, R.G. and Dukelow, W.R. (1985). *Fertil. Steril.* 43, 295–300.

diZerega, G.S. and Hodgson, G.D. (1981). *Endocrine Reviews* 2, 27–49.

Dukelow, W.R. (1975). *J. Reprod. Fertil. Suppl.* 22, 23–51.

Dukelow, W.R. (1978). In Daniel, J.C. (eds) *Methods of Mammalian Reproduction,* pp 438. Academic Press, New York.

Dukelow, W.R. (1979). *Science* 206, 234–235.

Dukelow, W.R. (1993). In Wolf, D.P, Stouffer, R.L. and Brenner, R.M. (eds) *In Vitro Fertilization and Embryo Transfer in Primates,* pp 73–84. Springer-Verlag, New York.

Dukelow, W.R., Fan, Bi-qin and Sacco, A.G. (1986). In Liss, A.R. *Comparative Primate Biology. Volume 3. Reproduction and Development,* pp 263–275. Springer-Verlag, New York.

Edwards, R.G., Donahue, R.P., Baramki, T.A. and Jones, H.W. (1966). *Amer. J. Obstet. Gynecol.* 96, 192–200.

Edwards, R.G., Bavister, B.D. and Steptoe, P.C. (1969). *Nature* 221, 632–635.

Fujimoto, S., Pahlavan, N. and Dukelow, W.R. (1974). *J. Reprod. Fertil.* 40, 177–181.

Gosselin, R.E., Fuller, G.B., Coy, D.H., Schally, A.V. and Hobson, W.C. (1979). *Proc. Soc. Exptl. Biol. Med.* 161, 21–24.

Gould, K.G., Cline, E.M. and Williams, W.L. (1973). *Fertil. Steril.* 24, 260–268.

Griffiths, T.A., Murdock, A.P. and Herbert, M. (2000). *Human Reprod.* 15, 1592–1596.

Harrison, R.M. (1980). In Harrison, R.M. and Wildt, D.E. (eds) *Animal Laparoscopy,* pp 73–93. Williams and Wilkins, Baltimore.

Harrison, R.M. and Dukelow, W.R. (1971). *J. Reprod. Fertil.* 25, 99–101.

Harrison, R.M. and Dukelow, W.R. (1973). *J. Medical Primatol.* 2, 277–283.

Harrison, R.M. and Dukelow, W.R. (1974). *Primates* 15, 305–309.

Hartman, C. (1933). *Amer. J. Obstet. Gyn.* 26, 600–608.

Heape, W. (1897). *Phil. Trans. Royal Soc. (London)* 188(B), 135–166.

Hendrickx, A.G. and Kraemer, D.C. (1968). *Anat. Rec.* 162, 111–121.

Hendrickx, A.G. and Dukelow, W.R. (1995a). In Bennet, B.T., Abee, C.R. and Hendrickson, R. (eds) *Nonhuman Primates in Biomedical Research: Biology and Management,* pp 147–191. Academic Press, New York.

Hendrickx, A.G. and Dukelow, W.R. (1995b). In Bennet, B.T., Abee, C.R. and Hendrickson, R. (eds) *Nonhuman Primates in Biomedical Research: Biology and Management,* pp 335–374, Academic Press, New York.

Hewitson, I., Simerly, C., Dominko, T. and Schatten, G. (2000). *Theriogenology* 53, 95–104.

Hirst, P.J., DeMayo, F.J. and Dukelow, W.R. (1981). *Theriogenology* 15, 67–75.

Hodgson, G.D. (1982). *Fertil. Steril.* 38, 281–300.

Hurst, P.R., Jefferies, K., Eckstein, P. and Wheeler, A.G. (1976). *Biol. Reprod.* 15, 429–434.

Hutz, R.J., Chan, P.J. and Dukelow, W.R. (1983). *Fertil. Steril.* 40, 521–524.

Iritani, A., Hosoi, Y. and Torii, R. (1998). In Lauria, F., Gandolfi, F., Enne, G. and Gianoroli, J. (eds) *Gametes: Development and Function,* pp 393–404. Serono Symposium, Rome.

Jaenisch, R. and Wilmut, I. (2001). *Science* 291, 2552.

Jewett, D.A. and Dukelow, W.R. (1972). *J. Reprod. Fertil.* 31, 287–290.

Jewett, D.A. and Dukelow, W.R. (1973). *J. Medical Primatol.* 2, 108–113.

Johnson, M.P., Harrison, R.M. and Dukelow, W.R. (1972). *Fed. Proc.* 31, 278.

Kraemer, D.C., Moore, G.T. and Kramen, M.A. (1976). *Science* 192, 1246–1247.

Kraemer, D.C., Flow, B.L., Schriver, M.D., Kinney, G.M. and Pennycook, J.W. (1979). *Theriogenology* 11, 51–62.

Kreitmann, O., Lynch, A., Nixon, W. and Hodgen, G. (1982). In Hafez, E. and Semm, K. (eds) *In Vitro Fertilization and Embryo Transfer,* pp 303–324. MTP Press, Lancaster, U.K.

Kuehl, T.J. and Dukelow, W.R. (1975a). *J. Medical Primatol.* 4, 23–31.

Kuehl, T.J. and Dukelow, W.R. (1975b). *J. Medical Primatol.* 4, 209–216.

Kuehl, T.J. and Dukelow, W.R. (1978). *J. Reprod. Fertil.* 52, 25–28.

Kuehl, T.J. and Dukelow, W.R. (1979). *Biol. Reprod.* 21, 545–556.

Lopata, A., Brown, J., Leaton, J., McTalbot, J. and Wood, C. (1978). *Fertil. Steril.* 30, 27–35.

Lopata, A., Summers, P.M. and Hearn, J.P. (1988). *Fertil. Steril.* 50, 503–509.

Loskutoff, N.M. (2003). In Holt, W.V., Pickard, A.R., Rodger, J.C. and Wildt, D.E. (eds) *Reproductive Science and Integrated Conservation,* pp 183–194. Cambridge University Press, U.K.

Loskutoff, N.M., Kraemer, D.C., Raphael, B.L., Huntress, S.L. and Wildt, D.E. (1991). *Amer. J. Primatol.* 24, 151–166.

Loskutoff, N.M., Johnson, W.H. and Betteridge, K.J. (1993). *Theriogenology* 39, 95–108.

Mann, J.R. (1988). *Biol. Reprod.* 38, 1077–1083.

Marston, J.H., Penn, R. and Sivelle, P.C. (1977). *J. Reprod. Fertil.* 49, 175–176.

Meslin, E.M. (2000). *J. Women's Health & Gender-Based Medicine* 9, 831–841.

Nagle, C.A., Riarte, A., Quiroga, S., Azorero, R.M., Carril, M., DeNari, J.H. and Rosner, J.M. (1980). *Biol. Reprod.* 23, 629–635.

Nasir-Ud-Din, McArthur, J.W. and Jeanloz, R.W. (1979). In Alexander, N.J. (ed.) *Animal Models for Research on Contraception and Fertility,* pp 396–403.

Norman, R.L. and Spies, H.G. (1979). *Endocrinology* 105, 655–659.

Pierce, D.L., Johnson, M.P., Kaneene, J.B. and Dukelow, W.R. (1993). *Amer. J. Primatol.* 29, 37–48.

Pope, C.E., Dresser, B.L., Chin, N.W., Liu, J.H., Loskutoff, N.M., Behnke, E.J., Brown, C., McRae, M.A.,

Sinoway, C.E., Campbell, M.K., Cameron, K.N., Owens, O.M., Johnson, C.A., Evans, R.R. and Cedars, M.I. (1997). *Amer. J. Primatol.* 41, 247–260.

Pope, C.E., Pope, V.Z. and Beck, L.R.(1980). *Biol. Reprod.* 23, 657–662.

Pope, C.E., Pope, V.Z. and Beck, L.R. (1983). *Teratol.* 19, 144.

Rawson, J.M.R. and Dukelow, W.R.(1973). *J. Reprod. Fertil.* 34, 187–190.

Seidel, G. (1983). *J. Exp. Zool.* 228, 347–355.

Shiels, P.G., Kind, A.J., Campbell, K.H.S., Waddington, D., Wilmut, I., Colman, A. and Schnieke, A.E. (1999). *Nature* 399, 316–317.

Silber, S.J., Van Steirteghem, A.C., Lui, J., Nay, Z., Tournaye, G. and Devroey, P. (1995). *Human Reprod.* 10, 148–152.

Solter, D. (2000). *Nature Reviews Genetics* 1, 199–207.

Soupart, P. and Strong, P. (1974). *Fertil. Steril.* 25, 11–44.

Steptoe, P.C. and Edwards, R.G. (1978). *Lancet* 2, 366.

Tesarik, J., Rolet, F., Brami, C., Sedbon, E., Thorel, J., Tibi, C. and Thebault, A. (1996). *Human Reprod.* 11, 780–781.

Van Wagenen, G. (1945). *Anat. Rec.* 91, 304.

Varmus, H.E. (2000). *J. Law, Medicine & Ethics* 28 (Suppl. 4), 46–53.

Wakayama, T. and Yanagimachi, R. (1999a). *Nature Genetics* 22, 127–128.

Wakayama, T. and Yanagimachi, R. (1999b). *Seminars in Cell & Devel. Biology* 10, 253–258.

Wakayama, T., Perry, A.C., Zuccotti, M., Johnson, K.R. and Yanagimachi, R. (1998). *Nature* 394, 369–374.

Wakayama, T., Tateno, H., Mombaerts, P. and Yanagimachi, R. (2000). *Nature Genetics* 24, 108–109.

Weick, R.E., Dierschke, D.J., Karsch, F.J., Butler, W.R., Hotchkiss, J. and Knobil, E. (1973). *Endocrinology* 93, 1140–1147.

Wildt, D.E., Schiewe, M.C., Schmidt, P.M., Goodrowe, K.L., Howard, J.G., Phillips, L.G., O'Brien, S.J. and Bush, M. (1986). *Thereogenology* 25, 33–51.

Willadsen, S.M. (1982). In Adams, C.E. (ed.) *Mammalian Egg Transfer,* pp 185–210.

Wolf, D.P. and Stouffer, R.L. (1993). In Wolfe, D.P., Stouffer, R.L. and Brenner, R.M. (eds) *In Vitro Fertilization and Embryo Transfer in Primates,* pp 85–99. Springer-Verlag, New York.

Wolf, D.P., Stouffer, R.L. and Brenner, R.M. (1993). In Wolf, D.P., Stouffer, R.L. and Brenner, R.M. (eds) *In Vitro Fertilization and Embryo Transfer in Primates,* pp 403. Springer-Verlag, New York.

Wright, G., Tucker, M.J., Morton, P.C., Swetzer-Yoder, C.L. and Smith, S.E. (1998). *Fertility* 10, 221–226.

CHAPTER 33

The Baboon as an Appropriate Model for the Study of Multifactoral Aspects of Human Endometriosis

Jason M. Mwenda[1], Cleophas M. Kyama[1,2], Daniel C. Chai[1], Sophie Debrock[2] and Thomas M. D'Hooghe[2]
[1]Division of Reproductive Biology, Institute of Primate Research, Karen, Nairobi, Kenya;
[2]Leuven University Fertility Centre, Department of Obstetrics & Gynaecology, University Hospital Gasthuisberg, Leuven, Belgium

Introduction

Endometriosis is defined as a progressive disease in which functional endometrial tissue fragments are deposited outside the uterine cavity. It is found in visceral and peritoneal surfaces within the pelvis and is associated with pelvic pain, adhesion formation and infertility. Endometriosis occurs in 30–40% of women with infertility and is a progressive disease in 40–50% of women (D'Hooghe and Hill, 2002). Women with early menarche, short cycle, long duration of menstrual flow or dysmenorrhoea are perceived to be at a higher risk of disease development (Cramer and Missmer, 2002). Some clinical investigations have shown that the prevalence of endometriosis may be lower in black women and higher in oriental women than in the Caucasian population (Miyazawa, 1976; Arumugam *et al.*, 1992). Among Caucasian women with infertility, the prevalence of endometriosis has been reported to vary between 13% in the period 1970–1987 and 33% in the period 1988–2000 (D'Hooghe *et al.*, 2003). The prevalence of endometriosis in adolescent African-American women presenting with pain or abnormal vaginal bleeding, was 65% including mild (50%), moderate (39%) or severe (11%) endometriosis according

The Laboratory Primate
Copyright 2005 Elsevier
ISBN 0-1208-0261-9

to the Acosta classification (Chatman and Ward, 1982; Acosta *et al.,* 1961). Among indigenous-African women, the prevalence of endometriosis was lower in black women (2%) than in women of mixed race (4%) or white patients (6%) (Wiswedel and Allen, 1989). There is a general perception that endometriosis rarely occurs among black women. This is based on the hypothesis that pregnancy at early age, early marriage and high incidence of pelvic inflammatory disease tend to preclude the development of endometriosis. Endometriosis contributes to more than 100,000 hysterectomies each year (Carlson *et al.,* 1994). The estimated total hospitalisation costs, represented by hospital charges for women with endometriosis as the primary diagnosis, in the USA, was $540 million for 1991 and $579 million for 1992. These cost-estimates suggest that endometriosis-related hospitalisation is a major burden on healthcare systems (Zhao *et al.,* 1998) worldwide. Studies focusing on endometriosis should be geared towards intervention strategies to block early establishment of the disease. This chapter provides an overview of the baboon as an appropriate model for elucidating the pathophysiology, immunological mechanisms and spontaneous evolution of endometriosis.

Animal models for endometriosis research

Most of the current knowledge about endometriosis has been acquired from animal experimentation. Appropriate animal models are important in biomedical research because disease presentation, in these models, may mimic the pathophysiological features and immunological responses observed in humans. An animal model for endometriosis could provide an excellent *in vivo* model system to investigate the role of peritoneal adhesion, peritoneal angiogenesis/phagocytosis and dioxins in the establishment of the disease.

Animal models play a critical role in studies of endometriosis

Chronic pelvic pain, emotional distress and associated infertility are hallmarks of endometriosis among women of reproductive age (Lebovic *et al.,* 2001; Taylor *et al.,* 2002). Although the disease has been well described and known for over a century, the current knowledge on mechanisms that cause the disease remain elusive. Adequate knowledge regarding pathogenesis, pathophysiology of associated infertility and spontaneous evolution has been hampered by the following factors: (1) Late presentation and diagnosis of the disease. This frustrates efforts to initiate clinical experiments with the aim of understanding aetiology or disease progression. (2) It is difficult to perform randomised multicenter oriented research, due to the lack of adequate patients and controls for such studies (Hill, 1992). (3) Spontaneous endometriosis occurs only in human and non-human primates (Mackenzie and Casey, 1975; D'Hooghe *et al.,* 1996c). This is a factor which limits valuable studies of endometriosis because they have to be carried out exclusively in humans and non-human primates. (4) Due to ethical and moral reasons, properly controlled invasive studies cannot be carried out in humans. Thus, an appropriate animal model is of paramount importance in understanding the pathophysiology of endometriosis.

Rodent models for endometriosis

The rat, rabbit and the hamster (Vernon and Wilson, 1985; Mann *et al.,* 1986; Werlin and Hodgen, 1983) have been extensively used in the experimental study of endometriosis. They have been used on studies to evaluate the effects of the exposure of dioxins in relation to onset of endometriosis (Rier, 2002). Rodent (rat and rabbit) models offer some advantages because of low cost and ease of manipulation. However, data derived from rodent models for endometriosis studies have numerous drawbacks: rodents lack a menstrual cycle, they do not develop spontaneous endometriosis and they exhibit a wider phylogenetic gap with humans. The rat has spontaneous ovulation but its luteal phase is shorter than in humans, while the rabbit lacks a luteal phase. Endometriosis may be induced surgically through autotransplantation of endometrial fragments in rodents (Vernon *et al.,* 1985). However it is not physiological and leads to severed uterus and additional complications associated with adhesions, which interfere with fertility. The endometriotic lesions consist of cysts containing clear serous fluid in the rat, whilst vascularised hemorrhagic solid masses, observed in the rabbit, are quite different from the variety of pigmented and non-pigmented lesions exhibited in human endometriosis (Jansen and Russell, 1986, Martin *et al.,* 1989). It is apparent that surgical induction of endometriosis, in these rodent models, does not address the basic cellular mechanisms of endometrial attachment and invasion that are critical for the early establishment of ectopic lesions observed in women with endometriosis. These disadvantages, therefore, have led to exploring possibilities of using some immunocompromised animals, such as athymic (nude) mouse and severe combined immunodeficient mouse (SCID), as models for endometriosis (Zamah *et al.*, 1984; Bergqvist *et al.*, 1985). The advantage of these models is that they do not reject xenographic human endometrial tissue (Bruner *et al.*, 2002), which can be introduced either subcutaneously or into the peritoneal cavity. The nude mouse model has been utilized, as a human endometrial tissue recipient, to examine the role of steroid-regulated endometrial matrix metalloproteinases (MMPs) in the establishment of experimental endometriosis (Bruner *et al.*, 1997). In the nude mouse model, ectopic adhesion of endometrial cells to the peritoneum, and the formation of endometriotic-like lesions, involve cooperation of epithelial and stromal cells through secretion of growth factors, cytokines and/or chemokines (Beliard *et al.*, 2003). It is still controversial whether the data generated from this rodent model can be extrapolated to the human disease situation, given the wide species difference between mice and humans.

Non-human primate models for endometriosis

Non-human primates are valuable models for biomedical research because of close phylogenetic characteristics and other similarities including anatomy, immunology, reproduction and physiology to humans (Stevens, 1997). The use of non-human primates is limited due to unavailability in developed countries and the cost of primate experimentation. Hence, primate studies have been restricted to unravelling critical questions that cannot be answered using the other experimental animals. Thus primate facilities, in source developing countries, provide an important resource that can be utilized in a sustainable manner. Non-human primate studies provide a critical step in preclinical product development, tested initially on other laboratory animals. Endometriosis occurs only in menstruating species including non-human primates and humans. Non-human primates, although expensive to maintain in captivity, offer unique advantages in endometriosis research when compared to rodents. They are close, phylogenetically, to humans; they have comparable menstrual cycle features; they are known to develop spontaneous endometriosis (Merrill, 1968) and experimental induction of endometriosis results in macroscopic lesions, which mimic human disease (Jacobson, 1926; Mann *et al.*, 1986; D'Hooghe *et al.*, 1995a). Finally, the similar immunological parameters and cross-reactivity of most anti-human monoclonal antibodies to non-human primate cell surface antigens may lead to a better understanding of the dynamics of the establishment and progression of endometriosis disease (D'Hooghe *et al.*, 1996a; D'Hooghe *et al.*, 2001a).

The genomes of Great Apes and humans differ by only 1.5% in nucleotide sequence (Taylor *et al.*, 2002). Although the chimpanzee and gorilla are close to humans in terms of anatomy and physiology of reproduction, they cannot be used as experimental models because they are endangered species (D'Hooghe, 1997a). Limited studies on endometriosis have been done using the De Brazza monkey (Binhazim *et al.*, 1989), also an endangered species. Until 12 years ago, most endometriosis research had been done on rhesus and cynomolgus monkeys (Lindberg and Busch, 1984; Fanton and Hubbard, 1983).

Rhesus monkeys have been extensively studied in reproductive medicine. The occurrence of spontaneous endometriosis in rhesus monkeys has enabled scientists to investigate the genetic epidemiology of this disease (Zondervan *et al.*, 2002). The exposure of rhesus monkeys to 0.5 or 25 PPT TCDD (parts per trillion 2,3,7,8-tetrachlorodibenzo-*p*-dioxin) via their food, for approximately 4 years, demonstrated a dose dependent increase in incidence and severity of endometriosis when they were assessed for spontaneous endometriosis 10 years after withdrawal of treatment (Rier, 2002). Spontaneous endometriosis in rhesus monkeys has been associated with irradiation, but only after at least 6 years of exposure (Wood, 1991). Experimental endometriosis in rhesus monkeys has been described as presenting as chocolate cysts, blue-green cysts, reddish nodules and blue-red nodules (Mann *et al.*, 1986; DiZerega *et al.*, 1980; Schenken *et al.*, 1984; Jacobson, 1926).

Natural progression of experimental endometriosis in cynomolgus monkeys has been reported, but complete regression of macroscopic disease, after pregnancy, occurred in monkeys with minimal and mild disease (Schenken *et al.*, 1987). Extensive studies carried out at the Institute of Primate Research (IPR) Kenya have demonstrated that the baboon is a better model for studies on human endometriosis than the rhesus and cynomolgus monkeys (D'Hooghe, 1997a), as reviewed in the following part of this chapter.

The role of the baboon model for study of human endometriosis

The baboon is one of the Old World primate species that is widely used in biomedical and behavioral research. Baboons are abundant in the African savannah land and breed fairly well in captivity (Birrell *et al.*, 1996) as opposed to rhesus monkey, whose seasonal breeding may be disrupted when housed indoors and maintained under constant light/dark and temperature/ humidity environment (Zondervan *et al.*, 2002). Baboons adapt well in captivity and they have an interesting social life (Smith *et al.*, 2003).

Advantages of the baboon model for studies focused on endometriosis

Baboons offer a clear advantage for both basic and clinical studies on endometriosis, when compared to rhesus and cynomolgus monkeys, because of the following reasons: (a) baboons have 42 chromosomes, a very close phylogenetic homology to the 46 chromosomes in humans, and share many genetic characteristics (Marks, 1982). (b) The reproductive anatomy and physiology of baboons is well documented and include similar menstrual cycle characteristics to that of women (Stevens, 1997). Baboons have a menstrual cycle of 33 days, which is the same as in humans. Perineal skin inflation and deflation, in the baboon, with relative precision, correspond to the follicular and luteal phases respectively, enabling non-invasive follow-up of menstruation and follicular and luteal phases. (c) The baboon has been demonstrated as a model for studies in cardiovascular and endoscopic surgery, endocrinology, teratology and toxicology (D'Hooghe *et al.*, 1993). (d) The baboon menstrual cycle continues throughout the year, even in captivity, as opposed to the rhesus monkey whose breeding season can be interrupted in captivity. (e) Baboons are large primates and versatile, allowing repeated blood sampling, humane manipulation and performance of even complicated surgical procedures. (f) Accessibility of the baboon uterine cavity, through the cervix, permits sampling of endometrium without hysterotomy (D'Hooghe *et al.*, 1996b). (g) Endometriosis in the baboon is associated with lesions which undergo active remodelling with some disappearing while other new lesions are formed (D'Hooghe *et al.*, 1992).

Some lesions develop more aggressively and progress to the typical or coloured lesions found in women (Martin *et al.*, 1989). (h) Baboons experience a high prevalence of spontaneous endometriosis. (i) Laparoscopic appearances, pathological aspects and pelvic localization of the implants are similar to those of the human disease. Hence, the baboon model can be utilized to elucidate the mechanisms involved in the pathogenesis of endometriosis in women.

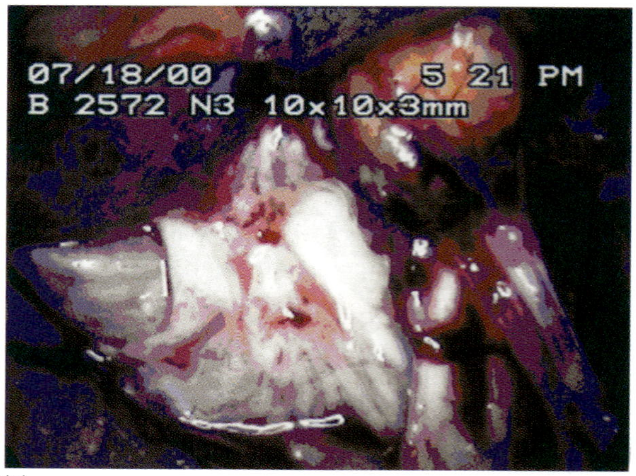

(a)

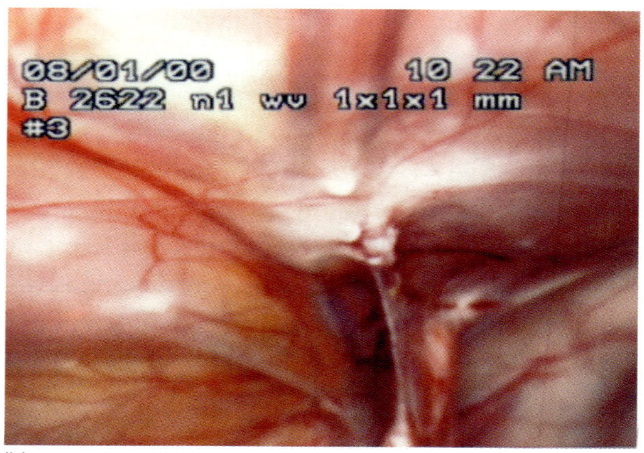

(b)

Figure 33.1 **(a)** Red polipoid endometriotic lesions surrounded by fibrosis and adhesions in the pouch of Douglas from a baboon with induced edometriosis after intrapelvic injection of menstrual endometrium. **(b)** Peritoneal endometriosis: white vesicles and endometriosis-related adhesion in a baboon with induced endometriosis after intrapelvic injection of menstrual endometrium.

Prevalence of endometriosis in baboons

In our previous studies, we have shown that baboons of proven fertility have a prevalence of 25% spontaneous endometriosis with laparoscopic appearance and pelvic localization similar to that found in human disease (D'Hooghe *et al.*, 1991). The prevalence of endometriosis in baboons without previous hysterotomy (8%) was comparable to the reported 7.5% prevalence of endometriosis in asymptomatic women undergoing sterilisation (Kirshon *et al.*, 1989). An increased prevalence of endometriosis (27%) has been observed in baboons that had been maintained in captivity for more than 2 years (D'Hooghe *et al.*, 1995b). This could be caused by the high number of menstrual cycles in captivity without interruption by pregnancy, which does not occur in wild baboons that are mostly pregnant or breast-feeding. It is also possible that the increased prevalence could be due to captivity-associated stress. Due to limited social interactions and constrained physical exercise, baboons in captivity may lead a stressful lifestyle (D'Hooghe *et al.*, 1995b), which is an exposing risk factor for development of endometriosis. Recent studies suggest that immune dysfunction may be associated with a stressful lifestyle and thus sub-optimal immune functions may be associated with the pathogenesis of endometriosis (Lebovic *et al.*, 2001, D'Hooghe *et al.*, 1995c).

Spontaneous evolution of endometriosis in baboons

According to current knowledge, the natural history of endometriosis seems to have been poorly studied. Endometriosis appears to be a dynamic and progressive disease (Chatman and Ward, 1982). In baboons, spontaneous endometriosis has been reported as a progressive disease. An increase in the number of endometriotic lesions was observed in baboons after 10 months with spontaneous endometriosis compared to the number of lesions scored during the first laparoscopy. The remodelling, defined by transition of lesions between typical, subtle, and suspicious implants, was observed in 23% of lesions. Studies on baboons indicate that endometriosis is a dynamic and moderately progressive disease with periods of development and regression, and active remodelling between different types of lesions (D'Hooghe *et al.*, 1996e), as seen in women with endometriosis (Weigerinck *et al.*, 1993). A study of 24 baboons, with an initially normal pelvis, demonstrated a cumulative endometriosis incidence of 29% and 64% within 12 and 32 months respectively (D'Hooghe *et al.*, 1996c). The effect of pregnancy on endometriosis is not well understood. Endometriosis has been said to undergo regression during pregnancy. Thus, it is not surprising that during early clinical practice, marriage and childbearing at young age were the method of choice for prophylaxis against endometriosis (Meigs, 1953). Conversely, other investigators reported enlargement of endometriotic lesions during the first trimester and regression during the remaining part of pregnancy (McArthur and Ulfeelder, 1965). However, more recent studies in the baboons have shown that pregnancy did not have a significant effect on endometriosis during the first and second trimester (D'Hooghe *et al.*, 1997b).

Induction of endometriosis in the baboon

Ideally, experimental induction of endometriosis in the baboon should lead to clinical symptoms of the disease for a model system to be relevant to the human disease (Vernon and Hodgen, 1987). The induction of endometriosis is based on the principle and hypothesis of retrograde menstruation. Retrograde menstruation is the reverse flow of menses through the fallopian tubes during menstruation. Ectopic implantation of endometrial fragments is important in the pathogenesis of endometriosis (Sampson, 1927). There is experimental evidence that intrapelvic injection of menstrual endometrium in baboons causes endometriosis (D'Hooghe *et al.*, 1995a). These concur with reports of increased prevalence of endometriosis in women with outflow obstruction of menses, a reflection of increased retrograde menstruation (Olive and Henderson, 1987; Baggish and Baltoyannis, 1987). Menstrual endometrium is said to induce endometriosis more successfully than luteal endometrium. Furthermore, a higher success rate of inducing the disease is achieved through injection intraperitoneally than by the subperitoneal route (D'Hooghe *et al.*, 1995a).

Induction method

The peritoneal cavity and reproductive tract of animals, to be used for induction, must be documented to be free of endometriotic lesions or adhesions prior to the deposition of endometrial tissues under minimal invasive laparoscopy. Usually the baboons are screened by videolaparoscopy during the mid-luteal phase (approximately 25th day of the cycle) and animals are then allowed to recover for one menstrual cycle as previously described (D'Hooghe *et al.*, 1999). On either the first or second day after the onset of the next menses, menstrual endometrium is extracted from each animal, by uterine curettage, and minced through an 18 gauge needle. A standardized amount (D'Hooghe *et al.*, 1995a; Fazleabas *et al.*, 2002) is autologously seeded onto ectopic sites (uterosacral ligaments, uterovesical fold, pouch of Douglas, ovaries). Subsequently, a videolaparoscopy is carried out after at least one month to assess the success of induction.

The number, surface and volume of the endometriotic lesions and adhesions are measured to allow calculation of the rAFS score and stage according to the revised classification system of the American Society for Reproductive Medicine (Revised American Society for Reproductive Medicine: 1996, 1997). Endometriosis is histologically confirmed by the presence

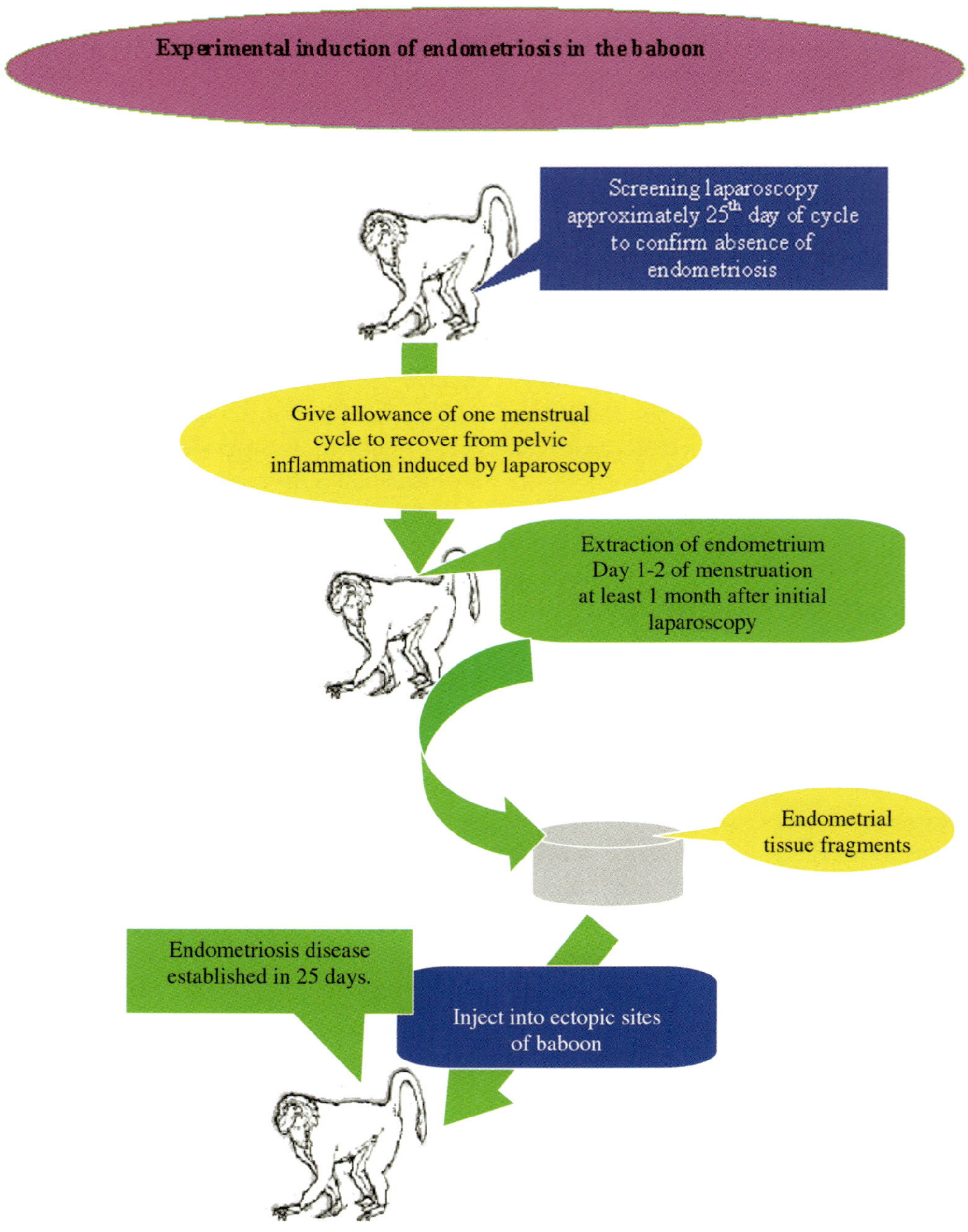

Figure 33.2 IPR Protocol for induction of endometriosis in the baboon.

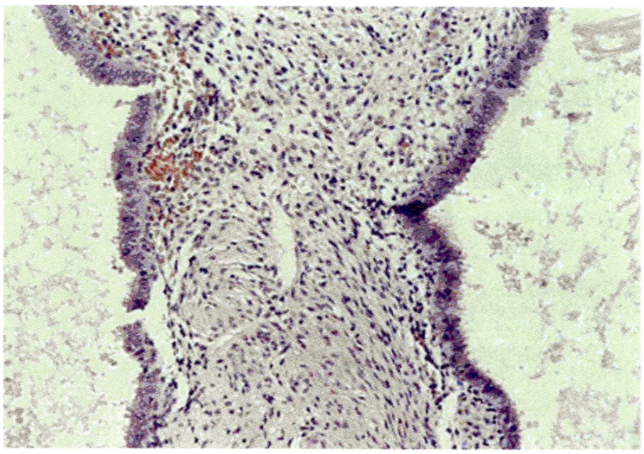

(a)

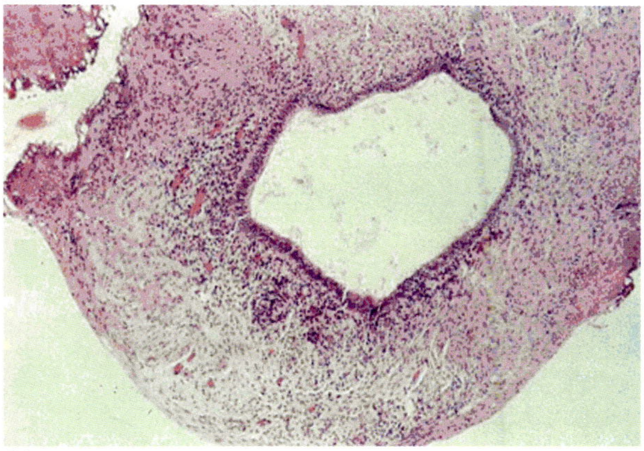

(b)

Figure 33.3 Histological appearance of endometriosis in the baboon. (a) Suspected endometriosis lesion showing endometrial glandular tissue with subepithelial vascularization. HE ×150. (b) Histological appearance of endometriosis in baboons (following intrapelvic injection of menstrual endometrium), characterized by endometrial gland and stromal cells with extensive vascularization. HE ×150.

of endometrial gland and stroma at ectopic pelvic sites in biopsies taken from the endometriotic implants (D'Hooghe *et al.,* 1991).

Endometriosis and infertility

More recent studies have shown that women who have endometriosis are much more likely to be infertile, even among those who ovulate and have anatomically patent fallopian tubes (Harada *et al.,* 2001). Further studies by Taylor *et al.* (2002) reported that endometriosis occurs in 30–40% of women with infertility.

The severe form of this condition has been associated with scarring and adhesions in the pelvis leading to mechanical blockage, prevention of the fusion of sperm and egg and disruption of normal pelvic anatomy. However, endometriosis has been associated with infertility even in the absence of adhesions or anatomic distortion of pelvis. Infertility of women with minimal endometriosis has been suggested to be caused by ovulatory dysfunction (Tummon *et al.,* 1988). Koninckx *et al.* (1980) suggested that minimal endometriosis might be due to lutenized unruptured follicle syndrome. D'Hooghe *et al.* (1996d) reported a higher incidence, and recurrence of corpus luteum without fresh ovulation stigma, in baboons with mild endometriosis than in controls, and this was associated with a reduced egg recovery. This study suggested that the existence of the luteinized unruptured follicle (LUF) syndrome could be a contributing factor to mild endometriosis-associated subfertility. However, minimal endometriosis, in baboons, appears not to be associated with infertility. Baboons with minimal endometriosis were found to have a cycle fecundity rate of 18%, while baboons with a normal pelvis had a cycle fecundity rate of 24% (D'Hooghe *et al.,* 1994). This corresponds relatively to the normal monthly fecundity rate (MFR) of 20%, observed in fertile women and in patients with minimal endometriosis (Rodrigez-Escudero *et al.,* 1988). However, mild, moderate or severe endometriosis in baboons is associated with infertility, even in the absence of ovarian endometriotic cysts (D'Hooghe *et al.,* 1994).

The presence of endometriosis has been postulated to stimulate the production of toxins in the pelvic environment. Cytokines, which impair fertilization, are present in the peritoneal secretions of most patients with endometriosis. Tumor Necrosis Factor-alpha (TNF-α) is cytotoxic to gametes (Halme, 1991) and sperm motility is inhibited in proportion to TNF-α concentration (Liang, 1994). Peritoneal macrophages can phagocytise sperm *in vitro* and these macrophages are more activated in women with endometriosis than in those without the disease (Muscato *et al.,* 1982). Sperm function can be impaired after exposure to peritoneal fluid of patients with endometriosis (Aeby *et al.,* 1996). In baboons, endometriosis-associated infertility may also be due to up regulation of some of these cytokines found in women with endometriosis. Further investigations are required to study the expression of these cytokines in baboons with and without endometriosis.

Pathogenesis of endometriosis

Retrograde menstruation has been reported in 83% of baboons (D'Hooghe *et al.,* 1996b), and in 70–90% of women with spontaneous endometriosis (Blumenkrantz *et al.,* 1981). The existence of endometrial cells in the peritoneal fluid has been reported in 59–79% of women during menses or during the early follicular phase (Koninckx *et al.,* 1980; Kruitwagen *et al.,* 1991).

According to Sampson's hypothesis (1927), menstrual debris, refluxed into the peritoneal cavity, contains viable endometrial cells that can implant and develop into endometriotic lesions. However, the survival and growth of implanted endometrial tissues are associated with angiogenesis and the potentially decreased natural killer (NK) cell cytotoxicity (Healy *et al.,* 1998). D'Hooghe and colleagues (1995d) reported no difference in lymphocyte-mediated cytotoxicity and NK cell activity between baboons with and without endometriosis. The experimental induction of endometriosis in baboons, using menstrual endometrium, supports the Sampson theory of retrograde menstruation. In humans, endometrial expression of matrix metalloproteinase (MMP-3, MMP-7 and MMP-11) occurs during menstrual breakdown and subsequent estrogen-mediated growth, but not during the secretory phase (Osteen *et al.,* 1999). Although malignant transformation of ectopic endometrial lesions occurs only rarely, the cellular mechanisms required for establishment of ectopic endometrial growth represent invasive events similar to those observed in cancer metastasis. This involves extensive degradation of the extracellular matrix (Spuijbroek *et al.,* 1992), suggesting the role of MMPs. In baboons, studies have shown that MMP-7 is a dominant metalloproteinase during establishment of endometriosis and may regulate the invasive events of the endometriotic tissue (Fazleabas *et al.,* 2002). Detailed studies in the baboon model may help understand the role of MMPs in the establishment of endometriosis and probably elucidate the mechanisms that lead to the aberrant expression of these proteases.

Endometrial fragments may survive immune attack due to diverse and somewhat complicated mechanisms for subversion and evasion of the immune system defence strategies. It has been hypothesed that endometrial cells, in the peritoneal fluid, avoid immunosurveillance and implant into the peritoneum (Somigliana *et al.,* 1996). Lymphocytes adhere to endometrial cells through Lymphocyte Function-Associated Antigen – 1/ Intercellular Adhesion Molecule-1 (LFA-1/ICAM-1) dependent pathway and present them as a target to Natural Killer (NK) cells (Vigano *et al.,* 1998). However, the endometrial cells are speculated to secrete a soluble form of ICAM (Intercellular adhesion molecule)-1, which can also bind LFA-1 and thus disrupt the recognition of endometrial implant by immune cells through NK cell mediated cytotoxicity (Vigano *et al.,* 1998). This occurs through the binding of sICAM-1 to leukocyte-associated ligands causing a competitive inhibition to the surface bound ICAM-1.

The Fas-FasL (FasL) system is a major pathway in programmed cell death (Nothnick, 2001). It has been speculated that expression of FasL by endometrial cells induces apoptosis of T cells through ligation of Fas. This mechanism, though not well defined, allows the endometrial fragments to escape cell death and to implant and develop to endometriotic lesions. Garcia-Velasco *et al.* (1999) suggested that peritoneal macrophages stimulate Fas-Fas ligand (FasL) mediated apoptosis of immune cells in endometriosis, allowing endometrial tissues to avoid immune surveillance. The failure of immune cells to transmit death signals, or the ability of endometrial fragments to avoid apoptosis, may be associated with the pathogenesis and severity of the disease.

The baboon as a model to evaluate the safety and efficacy of new products for treatment of endometriosis and associated infertility

Endometriosis is characterized by painful periods (dysmenorrhoea), pain during sexual intercourse (dyspareunia), chronic pelvic pain and infertility. Current treatment is limited because the causes of the disease are not well understood. These treatment regimens include hormonal drugs to suppress the menstrual cycle, surgical ablation of the endometriotic lesions and hysterectomy, with removal of ovaries and fallopian tubes. Future potential targets in the treatment or management of endometriosis may include inflammatory cytokines, MMPs, adhesion and growth factors. In baboons, spontaneous retrograde menstruation and experimental intrapelvic injection of endometrium is associated with intrapelvic inflammation (D'Hooghe *et al.,* 2001). Subsequently, inflammatory cytokines and growth factors are secreted and recruited to the peritoneal cavity. In women with endometriosis, activated macrophages, endometriotic lesions and/or mesothelial cells of the peritoneum, secrete cytokines such as Tumor Necrosis Factor-alpha (TNFα) and Interleukin-1 (IL-1), which in turn modulate other cytokines and chemokines such as Interleukin-8 (IL-8). Red lesions are associated with an active form of endometriosis with increased cell division and vascularisation (Bullimore, 2003). TNF-α and IL-8 concentrations in peritoneal fluid correlate with the size and number of active lesions (Harada *et al.,* 1994). Endometriosis is a multifactoral disease and current

Figure 33.4 A female baboon (*Papio anubis*) feeding her infant in the wild.

conventional medical therapy has many limitations (Nothnick, 2001). Therefore, the use of baboons to evaluate novel products to prevent or treat endometriosis more effectively has been accorded high priority by pharmaceutical companies.

Baboons may be used to evaluate candidate drugs for endometriosis-associated subfertility (D'Hooghe *et al.*, 2003). We have recently demonstrated neutralization of TNFα activity with recombinant human Tumor Necrosis Factor-alpha Binding Protein-1 (r-hTBP-1) which may reverse the chronic inflammatory state associated with endometriosis in baboons (D'Hooghe *et al.*, 2001). In this preliminary study, endometriosis was surgically induced in baboons (Figure 32.2) and treatment with r-hTBP-1 partially inhibited the development of endometriotic lesions and fully prevented the establishment of endometriosis-related adhesions (D'Hooghe *et al.*, 2001). It is important to note that the treatment of endometriosis did not result in inhibition of the follicular phase, ovulation and/or menstruation. Thus, it would appear that r-hTBP-1 treatment in baboons prevented the development of endometriotic lesion without any effect on estrogen levels. These findings demonstrate that the baboon is an appropriate model for testing novel products or drug regimens for treatment and prevention of human endometriosis.

Conclusion

Endometriosis is a debilitating condition in women of reproductive age and a major disease burden on healthcare systems worldwide. Several animal models have been used to study the pathophysiology of endometriosis. However, the extensive experimental studies, using the baboon as a model of human endometriosis, support and confirm the critical role of the baboon as a superior model for understanding human endometriosis.

Future research in endometriosis should focus on using baboons to unravel the mechanism leading to early establishment of the disease. It is important to understand the interaction between peritoneal and ectopic endometrial cells in the pelvic cavity during menstruation. Further research is needed to understand the role of inflammatory cytokines, MMPs and the establishment of a receptive peritoneal environment, which allows the development and progression of endometriosis. Accurate diagnosis of endometriosis is critical but is a complex process that involves surgical procedures and expensive instrumentation that is not universally available. A non-invasive diagnostic method will be useful in epidemiological studies on endometriosis. The current treatment options, both medical and surgical, are insufficient. Preliminary evidence, suggesting that r-hTBP-1 may prevent the development of endometriosis, raises the hope of developing novel and effective drug regimens that may inhibit the establishment and progression of endometriosis.

Correspondence

Any correspondence should be directed to Dr. Jason M. Mwenda, Institute of Primate Research, P.O. Box 24481, Karen, Nairobi, Kenya. Tel: +254-20-882571/4. Fax: +254-20-882546. E-mail: jmmwenda@arcc.or.ke or jmmathiu@yahoo.co.uk

References

Acosta, A.A., Buttram, V.C. Jr. and Besch, P.K. (1961). *Obstet. Gynecol.* 42, 19.

Aeby, T.C., Huang, T. and Nakayama, R.T. (1996). *Am. J. Obstet. Gynecol.* 174, 1779–1785.

Arumugam, K. and Templeton, A.A. (1992). *Aust. NZ. J. Obstet. Gynecol.* 32, 164–165.

Baggish, M.S. and Baltoyannis, P. (1987). *Fertil. Steril.* 48, 24–28.

Beliard, A., Noel, A., Goffin, F., Frankenne, F. and Foidart, J.M. (2003). *Fertil. Steril.* 79, 724–729.

Bergvist, A., Jeppsson, S., Kullander, S. and Ljungberg, O. (1985). *Am. J. Pathol.* 121, 337–341.

Binhazim, A.A., Tarara, R.P. and Suleman, M.A. (1989). *J. Comp. Pathol.* 101, 471–474.

Birrell, A.M., Hennessy, A., Gillin, A., Horvath, J., Tiller. (1996). *J. Med. Primatol.* 25, 287–293.

Blumenkrantz, M.J., Gallaagher, N., Bashore, R.A. and Tenckhoff, H. (1981). *Obstet. Gynecol.* 142, 890–895.

Bruner, K.L., Matrisian, L.M., Rodgers, W.H., Gorstein, F. and Osteen, K.G. (1997). *J. Clin. Invest.* 99, 2851–2857.

Bruner-Tran, K.L., Webster-Clair, D. and Osteen, K.G. (2002). *Ann. N.Y. Acad. Sci.* 955, 328–339.

Bullimore, D.W. (2003). *Med. Hypotheses* 60, 84–88.

Carlson, K., Miller B. and Fowler F.J. (1994). *Obstet. Gynecol.* 83, 556–565.

Chatman, D.L. and Ward, A.B. (1982). *J. Reprod. Med.* 27, 156–160.

Cramer, D.W. and Missmer, S.A. (2002). *Ann. N. Acad. Sci.* 955, 11–22.

D'Hooghe, T.M., Bambra, C.S., Raeymaekers, B.M., De Jonge, I., Lauweryns, J.M. and Koninckx, P.R. (1995a). *Am. J. Obstet. Gynecol.* 173, 125–134.

D'Hooghe, T.M., Bambra, C.S., De Jonge, I., Lauweryns, J.M. and Koninckx, P.R. (1995b). *Acta. Obstet. Gynecol. Scand.* 74, 1–4.

D'Hooghe, T.M., Bambra, C.S., Raeymaekers, B.M., De Jonge, I., Hill, J.A. and Koninckx, P.R. (1995c). *Fertil. Steril.* 64, 172–178.

D'Hooghe, T.M., Scheerlinck, J.P., Koninckx, P.R., Hill, J.A. and Bambra, C.S. (1995d). *Hum. Reprod.* 10, 558–562.

D'Hooghe, T.M., Bambra, C.S., Hill, J.A. and Koninckx, P.R. (1996a). *Hum. Reprod.* 11, 1736–1740.

D'Hooghe, T.M., Bambra, C.S., Raeymaekers, B.M. and Koninckx, P.R. (1996b). *Hum. Reprod.* 11, 2022–2025.

D'Hooghe, T.M., Bambra, C.S., Raeymaekers, B.M. and Koninckx, P.R. (1996c). *Obstet. Gynecol.* 88, 462–466.

D'Hooghe, T.M., Bambra, C.S., Raeymaekers, B.M. and Koninckx, P.R. (1996d). *Soc. Gynecol. Investig.* 3, 140–144.

D'Hooghe, T.M., Bambra, C.S., Raeymaekers, B.M. and Koninckx, P.R. (1996e). *Fertil. Steril.* 65, 645–649.

D'Hooghe, T.M. (1997a). *Fertil. Steril.* 68, 613–625.

D'Hooghe, T.M., Bambra, C.S., De Jonge, I., Lauweryns, J.M., Raeymaekers, B.M. and Koninckx, P.R. (1997b). *Arch. Gynecol. Obstet.* 261, 15–19.

D'Hooghe, T.M., Bambra, C.S., Raeymaekers, B.M. and Hill, J.A. (1999). *Fertil. Steril.* 72, 1134–1141.

D'Hooghe, T.M., Bambra, C.S., Cornillie, F.J., Isahakia, M. and Koninckx, P.R. (1991). *Biol. Reprod.* 45, 411–416.

D'Hooghe, T.M., Bambra, C.S., Isahakia, M. and Koninckx, P.R. (1992). *Fertil. Steril.* 58, 409–412.

D'Hooghe, T.M., Bambra, C.S., Farah, I.O., Raeymaekers, B.M. and Koninckx, P.R. (1993). *Am. J. Obstet. Gynecol.* 169, 1352–1356.

D'Hooghe, T.M., Bambra, C.S. and Koninckx, P.R. (1994). *Gynecol. Obstet. Invest.* 37, 63–65.

D'Hooghe, T.M. and Hill, J.A. (2002). In *Novak's Gynecology*. Williams and Wilkins, 13th edition, pp 931–972.

D'Hooghe, T.M., Debrock, S., Hill, J.A. and Meuleman, C. (2003a). *Sem. Reprod. Med.* 21, 243–254.

D'Hooghe, T.M., Debrock, S., Meuleman, C., Hill, J.A. and Mwenda, J.M. (2003b). *Obstet. Gynecol. Clin. North. Am.* 30, 221–244.

D'Hooghe, T.M., Nugent, N., Chai, D., Deer, F., Debrock, S. and Mwenda, J. (2001a). Oral presentation at the Annual Meeting of the American Society for Reproductive Medicine, Orlando, US, October 22nd–24th 2001.

D'Hooghe, T.M., Pudney, J. and Hill, J.A. (2001b). *Am. J. Primatol.* 53, 47–54.

DiZerega, G.S., Barber, D.L. and Hodgen, G.D. (1980). *Fertil. Steril.* 33, 649–653.

Fanton, J.W. and Hubbard, G.B. (1983). *Lab. Anim. Sci.* 33, 597–601.

Fazleabas, A.T., Brudney, A., Gurates, B., Chai, D. and Bulun, S.A. (2002). *Ann. N.Y. Acad. Sci.* 955, 308–317.

Garcia–Velasco, J.A., Arici, A., Zreik, T., Noftolin, F. and Mor, G. (1999). *Mol. Hum. Reprod.* 5, 642–650.

Halme, J. (1991). *Ann. N.Y. Acad. Sci.* 622, 266–274.

Harada, T., Enatsu, A., Mitsunari, M., Nagano, Y., Ito, M., Tsudo, T., Tanaguchi, F., Iwabe, T., Takinawa, M. and Terakawa, N. (1994). *Gynecol. Obstet. Invest.* 47, 34–39.

Harada, T., Iwabe, T. and Tarakawa, N. (2001). *Fertil. Steril.* 76, 1–10.

Healy, D.L., Rogers, P.A., Hii, L. and Winfield, M. (1998). *Hum. Reprod. Update* 4, 736–740.

Hill, J.A. (1992). *Infertility and Reproductive Medicine Clinics of North America* 3, 583–596.

Jacobson, V.C. (1926). *Arch. Pathol. Lab. Med.* 1, 169–174.

Jansen, R.P.S. and Russell, P. (1986). *Am. J. Obstet. Gynecol.* 155, 1160–1163.

Kirshon, B., Poindexter, A.N. 3rd and Fast, J. (1989). *J. Reprod. Med.* 34, 215–217.

Koninckx, P.R., De Moor, P. and Brosens, I.A. (1980). *Br. J. Obstet. Gynaecol.* 87, 929–934.

Kruitwagen, R.F., Poels, L.G., Willemsen, W.N., de Ronde, I.J., Jap, P.H. and Rolland, R. (1991). *Fertil. Steril.* 55, 297–303.

Lebovic, D.I., Mueller, M., Taylor, R. (2001). *Fetil. Steril.* 75, 1–10.

Liang, X.S., Hu, L.N. and Wu, P. (1994). *Zhonghua Fu Chan Ke Za Zhi.* 29, 524–526.

Lindberg, B.S. and Busch, C. (1984). *Ups. J. Med. Sci.* 89, 129–134.

Mackenzie, W.F. and Casey, H.W. (1975). *Am. J. Pathol.* 80, 341–344.

Mann, D.R., Collins, D.C., Smith, M.M., Kessler, M.J. and Gould, K.G. (1986). *J. Clin. Endocrin. Metab.* 63, 277–283.

Marks, J. (1982). *Cytogenet. Cell Genet.* 34, 261–264.

Martin, D.C., Hubert, G.D., Vander Zwaag, R., El-Ieky, F.A. (1989). *Fertil. Steril.* 51, 63–67.

McArthur, J.W. and Ulfeelder, H. (1965). *Obstet. Gynecol. Sur.* 20, 709.

Meigs, J.V. (1953). *Obstet. Gynecol.* 46, 2–13.

Merrill, J.A. (1968). *Am. J. Obstet. Gynecol.* 101, 569–570.

Miyazawa, K. (1976). *Obstet. Gynecol.* 48, 407–409.

Muscato, J.J., Haney, A.F. and Weinberg, J.B. (1982). *Am. J. Obstet. Gynecol.* 70, 115–122.

Nothnick, W.B. (2001). *Fertil. Steril.* 76, 223–231.

Olive, D.L. and Henderson, D.Y. (1987). *Obstet. Gynecol.* 69, 412–415.

Osteen, K.G., Keller, N.R., Feltus, F.A and Melner, M.H. (1999). *Gynecol. Obstet. Invest.* 48, 2–13.

Revised American Society for Reproductive Medicine classification of endometriosis (1996, 1997). *Fertil. Steril.* 67, 817–821.

Rier, S.E. (2002). *Ann. N.Y. Acad. Sci.* 955, 201–212.

Rodriguez-Escudero, F.J., Neyro, J.L., Corcostegui, B. and Benito, J.A. (1988). *Fertil. Steril.* 50, 522–524.

Sampson, J.A. (1927). *Am. J. Obstet. Gynecol.* 14, 422–469.

Schenken, R.S., Asch, R.H., Williams, R.F. and Hodgen, G.D. (1984). *Fertile. Steril.* 41, 122–130.

Schenken, R.S., Williams R.F. and Hodgen, G. (1987). *Am. J. Obstet. Gynecol.* 157, 1392–1396.

Smith, K, Alberts, S.C. and Altmann, J. (2003). *Proc. R. Soc. Lond. B. Biol. Sci.* 270, 503–510.

Somigliana, E., Vigano, P., Gaffuri, B., Guarneri, D., Busacca, M. and Vignali, M. (1996). *Hum. Reprod.* 11, 1190–1194.

Spuijbroek, M.D., Dunselman, G.A., Menheere, P.P. and Evers, J.L. (1992). *Fertil. Steril.* 58, 929–933.

Stevens, V.C. (1997). *Hum. Reprod. Update* 3, 533–540.

Taylor, R.N., Lundeeni, S.G. and Giudice, L.C. (2002). *Fertil. Steril.* 78, 694–698.

Tummon, I.S., Maclin, V.M., Radwanksa, E., Binor, Z. and Dmowski, W.P. (1988). *Fertil. Steril.* 50, 716–720.

Vernon, M.W. and Wilson, E.A. (1985). *Fertil. Steril.* 44, 684–694.

Vernon, M.W., Hodgen, G.D. (1987). New York, Alan R. Liss, Inc, 207–223.

Vigano, P., Gaffuri, B., Somigliana, E., Busacca, M., Di Blasio, A.M., Vignali, M. (1998). *Mol. Hum. Reprod.* 4, 1150.

Werlin, L.B. and Hodgen, G.D. (1983). *J. Clin. Endocrinol. Met.* 56, 844–848.

Wiegerinck, M.A., Van Dop, P.A. and Brosens, I.A. (1993). *Fertil. Steril.* 60, 461–464.

Wiswedel, K. and Allen, D.A. (1989). *S. Afric. Med. J.* 76, 65–66.

Wood, D.H. (1991). *Rad. Research* 126, 132–140.

Zamah, N.M., Dodson, M.G., Stephens, L.C, Buttram, V.C Jr., Besch, P.K. and Kaufman, R.H. (1984). *Am. J. Obstet. Gynecol.* 149, 591–597.

Zhao, S.Z., Wong, J.M., Davis, M.B., Gersh, G.E. and Johnson, K.E. (1998). *Am. J. Manag. Care* 8, 1127–1134.

Zondervan, K., Cardon, L., Desrosiers, R., Hyde, D., Kemnitz, J., Mansfield, K., Roberts, J., Scheffler, J., Weeks, D.E. and Kennedy, S. (2002). *Ann. N.Y. Acad. Sci.* 955, 233–238.

CHAPTER 34

Virology Research

Peter Barry[1,2], Marta Marthas[1], Nicholas Lerche[1], Michael B. McChesney[1] and Christopher J. Miller[1,2]

[1]California National Primate Research Center and [2]Center for Comparative Medicine, University of California, Davis, California 95616, USA

Introduction and scope

Viral infections are responsible for significant morbidity and mortality in humans. In 1918, the influenza A pandemic was responsible for over 40 million deaths worldwide. As of December 2002, 60 million people had been infected with HIV and one-third of those infected were dead. Further, emerging infectious diseases often have a viral etiology as evidenced by the recent outbreaks of a newly discovered coronavirus that produced Severe Acute Respiratory Syndrome (SARS) epidemics in Asia and North America and of monkeypox in the U.S. Chronic infections with hepatitis B and C viruses and herpes viruses are also significant public health threats and will remain so for the forseeable future. Clearly viral infections remain a significant threat to human health. Thus, considerable efforts are expended to develop vaccines and therapeutics to control both acute and persistent viral infections.

The outcome of human viral infections depends on a complex interaction between the virus and the host. Genetic polymorphisms in the human population and the dose, strain and route of virus inoculation all contribute to the wide range of clinical outcomes in viral infections. The study of viral pathogenesis, and the preclinical development of viral vaccines and therapeutics, requires the use of animal models. The exquisite specificity of viruses for their cellular receptors and the absolute requirement of viruses for appropriate intracellular machinery defines the host cell tropism of viruses and explains why viruses tend be adapted for infection of particular host species. The relative genetic homology of humans and nonhuman primates (NHPs) means that there is considerable conservation of the critical molecules used by viruses in their life cycle.

The Laboratory Primate
Copyright 2005 Elsevier
ISBN 0-1208-0261-9

Thus NHPs often are the only species that can be infected with human pathogens. Alternatively, a number of viruses endemic to NHPs are distinct but closely related to human pathogens and are good models for the related human infection. Further, the development, genetics, function and anatomy of the immune system of nonhuman primates and humans are very similar which makes nonhuman primates very useful for understanding human immunity.

Excellent reviews document the historical role of NHPs in viral research (Renegar, 1992; Soike *et al.*, 1984), including the critical role that monkeys played in elucidating the pathogenesis of poliovirus and in the development of the poliovirus vaccine (Sabin, 1965). This review will focus on contemporary studies employing NHPs to model viral diseases currently of greatest public health concern.

Acute viral diseases

Systemic infections

Hantavirus

Hantavirus (genus *Hantavirus*, family *Bunyaviridae*) infection in humans is associated with two severe and potentially fatal diseases. Hemorrhagic fever with renal syndrome (HFRS) is associated primarily with hantaviruses found in Europe and Asia, and is characterized by fever, thrombocytopenia, renal failure and, in severe cases, hemorrhage due to increased capillary permeability (Lee *et al.*, 1982). Hantavirus pulmonary syndrome (HPS) is associated with infection by hantaviruses indigenous to the Americas (e.g. Andes, Muerto Canyon, and Sin Nombre viruses) and is characterized by fever, myalgia, and rapid onset of severe respiratory distress (Nichol *et al.*, 1993). The mortality associated with HPS is 40–50% (Lednicky, 2003). The reservoirs of hantaviruses are specific rodent hosts, in which persistent, nonpathogenic infections are common (Lednicky, 2003; Schmaljohn and Hjelle, 1997). Transmission to humans is by inhalation of aerosolized rodent urine or feces (McElroy *et al.*, 2002). Human to human transmission of hantavirus has been documented only in cases of HPS caused by Andes virus (Padula *et al.*, 1998). No effective hantavirus vaccines have been developed and, while the viruses causing HFRS respond to antiviral therapy with ribivarin, there is currently no effective antiviral drug therapy for HPS (Chapman *et al.*, 1999).

Several species of NHPs are susceptible to experimental infection with hantaviruses (Groen *et al.*, 1995; Yanagihara *et al.*, 1988), and NHP models are emerging as valuable tools for investigating hantavirus pathogenesis and for evaluating antiviral therapies and candidate vaccines. Experimental infection of cynomolgus macaques with Puumala virus produced typical signs of HFRS, including lethargy, anorexia, proteinuria, and/or hematuria, along with cytokine, creatinine, nitric oxide and C-reactive protein responses similar to those of human HFRS (Klingstrom *et al.*, 2002). NHP models have also been used to evaluate candidate hantavirus vaccines. Prototype DNA vaccines, against hantaviruses causing HFRS, have elicited high titer neutralizing antibody in rhesus macaques (Hooper *et al.*, 2001).

In a first attempt to develop a model of HPS, cynomolgus macaques were experimentally infected with Andes virus by the intravenous or aerosol route. Viremia was detected by RT-PCR in 4 of 6 exposed animals, but none developed clinical signs of HPS. One animal developed a 3-fold decrease in baseline platelet count coincident with high viremia and consistent with thrombocytopenia seen in human HFRS. All animals, however, showed significantly reduced lymphocyte counts, and all developed IgM and IgG antibodies against viral nucleocapsid protein and neutralizing antibody titers (McElroy *et al.*, 2002). Future refinements of primate models should greatly facilitate progress in the treatment and prevention of hantavirus-related diseases.

Ebola virus

Ebola virus (family *Filoviridae*) has been responsible for outbreaks of hemorrhagic fever with high mortality in both humans and NHPs (Evans *et al.*, 2003; Walsh *et al.*, 2003). Many species of NHPs are susceptible to experimental infection with Ebola virus, the outcome of which is almost uniformly fatal. The lethality of Ebola virus infection in primates has been used as support for the conclusion that NHPs are not the natural reservoir host of Ebola in nature. The course of Ebola virus infection in NHPs closely follows that observed in infected humans (Baskerville *et al.*, 1978; Johnson *et al.*, 1995; Jaax *et al.*, 1996). As a result, nonhuman primates have been used extensively in studies of Ebola virus pathogenesis (Baskerville, 1978; Johnson *et al.*, 1995; Jaax *et al.*, 1996; Riyabchikova *et al.*, 1999; Ignatiev *et al.*, 2000). Studies in rhesus and cynomolgus macaques have demonstrated that lethal Ebola virus infections can occur by oral, conjunctival or aerosol routes of exposure (Jaax *et al.*, 1996; Jahrling *et al.*, 1996; Johnson, E. *et al.*, 1995). Experimental infections, in African green monkeys and baboons, have confirmed that macrophages and monocytes are the first cells to be infected and play a major role in subsequent systemic distribution of virus (Riyabchikova *et al.*, 1999). Experimental studies have also documented the absence of an inflammatory response in Ebola virus-infected NHPs (Baskerville *et al.*, 1978; Jaax *et al.*, 1996), suggesting virus-induced impairment of immune function. Studies in baboons have shown that the severity of Ebola virus-induced coagulopathies is correlated with increases in serum levels of interferon and tumor necrosis factor-α (Ignatiev *et al.*, 2000). Interestingly, differences have been observed in Ebola virus pathogenesis in different species of nonhuman primates. Following infection with Ebola-Zaire, baboons developed signs of hemorrhagic disease, while African green monkeys, given the same dose, developed generalized fibrin thrombosis in the absence of overt hemorrhage (Riyabchikova *et al.*, 1999). These findings indicate that species-specific differences need to be considered when selecting a NHP model of Ebola virus infection, and that use of different model systems may be necessary to address the entire spectrum of human disease associated with Ebola virus infection.

NHP models have also figured prominently in efforts to develop an Ebola virus vaccine (Sullivan *et al.*, 2000; Sullivan *et al.*, 2003). Passive immunization of cynomolgus macaques with hyperimmune globulin alone, or in combination with recombinant interferon-α, was not protective (Geisbert *et al.*, 2002; Jaax *et al.*, 1996; Jahrling *et al.*, 1996; Jahrling *et al.*, 1999). Similarly, cynomolgus macaques, immunized with liposome-encapsulated irradiated Ebola virus, produced virus-neutralizing antibodies but were not protected against lethal challenge, suggesting that both humoral and cell-mediated immunity are required for protection (Rao *et al.*, 2002). Newer approaches to vaccine development have focused on the induction of both cellular and humoral immune responses, incorporating the use of DNA vaccines. A "prime-boost" protocol combining DNA immunization, followed by a boost with a recombinant adenoviral vector expressing Ebola viral proteins, induced both cellular and humoral immune responses in cynomolgus macaques. Immunized animals were completely protected from lethal challenge with Ebola-Zaire, indicating that a preventive vaccine against Ebola virus is feasible (Sullivan *et al.*, 2000). This prime-boost protocol, while effective as a preventive vaccine regimen, required more than 6 months to complete, precluding its use in controlling an acute epidemic (Sullivan *et al.*, 2003). Recent studies have demonstrated that immunization of cynomolgus macaques, with recombinant adenovirus vectors encoding Ebola virus glycoprotein and nucleoprotein, in the absence of DNA priming, resulted in earlier antibody production and induction of Ebola virus-specific CD8+ T-cell responses. Monkeys immunized with only a single dose of this construct and challenged 28 days later were completely protected against both high and low-dose viral challenge (Sullivan *et al.*, 2003).

Respiratory virus infections

The disease potential and pathology of respiratory pathogens differs in primates and rodent animal models for two major reasons. First, the anatomy and development of the respiratory system in rodents is significantly different from that in the human or nonhuman primate (Plopper *et al.*, 1992). Second, many viruses that are pathogenic in humans are also pathogenic in another primate species, but not in rodents, mainly due to the need for specific host cell factors for viral replication. Respiratory virus infections of humans have been modeled in a wide range of primates including chimpanzees (Belshe *et al.*, 1977; Hancock *et al.*, 2000; Teng *et al.*, 2000), macaques (Blake and Trask, 1921; Bukreyev *et al.*, 2002; Fouchier *et al.*, 2003; Ponnuraj *et al.*, 2001; Rimmelzwaan *et al.*, 2001; van den Hoogen *et al.*, 2001; Zaucha *et al.*, 2001) and African green monkeys (Durbin *et al.*, 2000; Kakuk *et al.*, 1993). Recently two new viral pathogens, a pneumovirus and a coronavirus, responsible for Severe Acute Respiratory Syndrome, were successfully transmitted to cynomolgus monkeys and this helped to establish the etiology of these newly discovered respiratory tract infections (Fouchier *et al.*, 2003; van den Hoogen *et al.*, 2001).

Primate models have been used for many years to study respiratory tract infection and immunity to measles virus (Figure 34.1) (Blake and Trask, 1921b; Putz *et al.*, 2003), influenza (Renegar, 1992; Rimmelzwaan *et al.*, 2001; Soike *et al.*, 1984), parainfluenza (Durbin *et al.*, 2000; Hall *et al.*, 1993; Schmidt *et al.*, 2002), respiratory syncytial virus (Belshe *et al.*, 1977; Cheng *et al.*, 2001; Hancock *et al.*, 2000; Kakuk *et al.*, 1993; Leaman *et al.*, 2002; Teng *et al.*, 2000) and, more recently, monkeypox (Zaucha *et al.*, 2001). The majority of published work is focused on vaccine development. Possibly the best studied example of differing host species pathogenicity, and hence the use of an animal model, is the case of measles virus. Both rodent and primate animal models have been used for measles vaccine research, but wildtype strains of measles virus

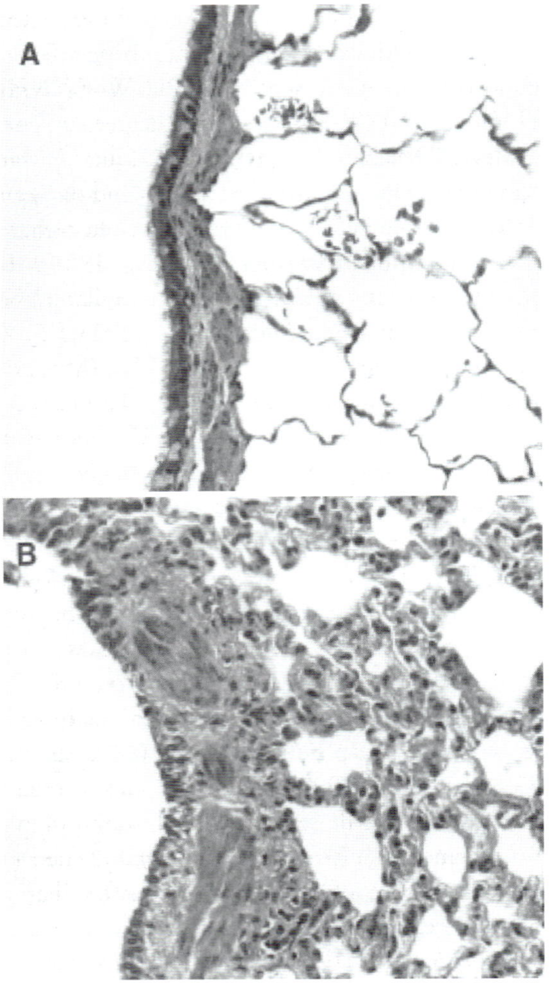

Figure 34.1 Histologic sections of lung tissue from rhesus monkeys 7 days after measles virus challenge. (A) Lung of a monkey that had received the live, attenuated measles vaccine 1 year prior to challenge. This is essentially normal lung. (B) Lung of an unvaccinated monkey. The wall of the respiratory bronchiole (left) is thickened with mononuclear cell infiltration and the alveolar septae are also thickened with inflammation and fibrin. These are features of viral bronchiolitis and interstitial pneumonia (Reprinted from McChesney, M.B. *et al.*, 1997, *Virology* 233, pp 74–84, with permission from Elsevier).

do not replicate in rodent species, with the exception of cotton rats, so that vaccine-induced protective immunity can be tested only in humans or susceptible nonhuman primates (Putz *et al.*, 2003). Similar restrictions are a concern for animal models of influenza virus (Renegar, 1992; Soike *et al.*, 1984).

Primate models have played a critical role in understanding the problem of enhanced or atypical lung disease that occurred in children vaccinated with whole-inactivated viral vaccines against measles or respiratory syncytial virus, when these children were exposed to wildtype virus. This vaccine-induced immunopathology has been extensively modeled in mice and a Type 2 CD4+ T cell has been found to play a pivotal role (Graham *et al.*, 2002). However, the laboratory mouse is not permissive for measles or respiratory syncytial virus replication and the question of protective versus immunopathologic vaccine immunity can only be addressed in a primate model. Using nonhuman primate models, it has now been shown that hypersensitivity reactions are largely responsible for the enhanced pathology associated with respiratory syncytial virus infection after immunization with formalin-inactivated RSV vaccines (De Swart *et al.*, 2002; Kakuk *et al.*, 1993; Ponnuraj *et al.*, 2001). Studies in macaques have demonstrated a role for immune complexes and eosinophils in the pathogenesis of atypical measles (Polack *et al.*, 1999).

Chronic viral diseases

Chronic viral infections are characterized by the ability of the pathogen to evade host antiviral immunity over the course of years. This reflects a highly evolved relationship between host and virus that involves the interactions of numerous viral proteins and the host immune system. This highly evolved relationship also means that human chronic viral pathogens often do not infect NHPs and thus, homologous viral pathogens of primates are used as models of human viral diseases. Further, for models of chronic viral diseases, the animal species must be long-lived and possess an immune system that is highly homologous to humans.

Primate T-lymphotropic viruses

Primate T-lymphotropic viruses (PTLV) comprise a closely related group of retroviruses in the Oncovirus subfamily of *Retroviridae* that includes viruses of humans and NHPs. The first member of this group to be identified was human T-lymphotropic virus type-I (HTLV-I), isolated from the blood of patients with Adult T-cell leukemia/lymphoma (ATLL) (Poiesz *et al.*, 1980). HTLV-I is the only human retrovirus known to be oncogenic and the etiologic link of this virus to ATLL is now well established (Johnson, J.M. *et al.*, 2001). HTLV-1 infection is also etiologically associated with a chronic, progressive myelopathy known as tropical spastic paraparesis/HTLV-1 associated myelopathy (TSP/HAM) (Levin and Jacobson, 1997). The specific mechanisms underlying HTLV-1 pathogenesis and host immune response are not completely understood and there is currently no vaccine for prevention of HTLV-1 infection, or effective treatment for HTLV-1-related malignancies (Dekaban *et al.*, 1995; Franchini, 1995; Franchini and Streicher, 1995).

Other members of the PTLV group include HTLV-2 (Kalyanaraman *et al.*, 1982), a human virus not conclusively linked to cancer, and the cognate simian T-lymphotropic viruses types 1, 2, and 3 (STLV-1, -2, -3) (Kalyanaraman *et al.*, 1982; Meertens and Gessain, 2003; Miyoshi *et al.*, 1982). In NHPs with naturally acquired infections, STLV-I induces leukemia or lymphoma, predominantly in African species, with many clinical and pathological similarities to ATLL (Tsujimoto *et al.*, 1987). Both human and simian lymphomas occur in only a small proportion of infected individuals, usually after a prolonged period of latency (Blattner, 1989; Tsujimoto *et al.*, 1987). To date, no counterpart to TSP/HAM has been recognized in STLV-1-infected NHPs.

Two approaches have been used to develop and utilize NHP models of human HTLV-I infection and disease. One approach takes advantage of the fact that several species of NHPs are susceptible to experimental infection with HTLV-I, including various species of macaque (Beilke *et al.*, 1996; Ibuki *et al.*, 1997; Murata *et al.*, 1996), the squirrel monkey (Kazanji, 2000; Kazanji *et al.*, 2000) and the common marmoset (Yamanouchi *et al.*, 1985). Experimentally infected Japanese macaques (*M. fuscata*) have been used to demonstrate the efficacy of passive immunization with hypreimmune serum in preventing HTLV-I infection (Murata *et al.*, 1996). A cynomolgus macaque model was used to demonstrate the long-term persistence of protective immunity following immunization with recombinant vaccinia virus expressing the HTLV-I envelope gene (Ibuki *et al.*, 1997). Although NHPs infected with HTLV-I do not develop tumors, other HTLV-I-related pathological conditions, including polymyositis, uveitis and arthritis, were observed in a rhesus macaque following experimental infection with HTLV-I (Beilke *et al.*, 1996). The squirrel monkey model has been used to investigate the distribution of virus to various tissues, in early HTLV-I infection, and the relationship between viral gene expression and the host humoral and cellular immune response to infection (Kazanji, 2000; Kazanji *et al.*, 2000). Infection of common marmosets with HTLV-I by the oral route has provided support for the concept of milk-borne infection from HTLV-I-infected mothers to nursing infants (Yamanouchi *et al.*, 1985).

Another approach to HTLV-1 primate model development has exploited the naturally occurring infections with STLV-1 in many species of Old World NHPs. HTLV-1 and STLV-1 share a high degree of genetic relatedness (Dekaban *et al.*, 1995). The routes of transmission are the same (Lazo *et al.*, 1994), and the course of infection of STLV-1 in primates parallels that of HTLV-1 in humans (Tsujimoto *et al.*, 1987). The respective host-virus systems give rise to similar proportions of malignancies, and tumors induced by STLV-1 are indistinguishable from human ATLL (McCarthy *et al.*, 1990; Tsujimoto *et al.*, 1985). The method of replication of both STLV-1 and HTLV-1, namely clonal expansion of virus infected T-cells, is virtually identical (Gabet *et al.*, 2003; Morteaux *et al.*, 2003). The exact mechanisms of PTLV oncogenesis have not been fully elucidated. Results of recent studies suggest that the malignant transformation in ATLL is a "multi-hit" phenomenon, marked by discreet genetic events (Arima and Tei, 2001). The Tax gene, present in both STLV-1 and HTLV-1, is thought to play a major role in leukemogenesis, due to its pleiotropic actions (Dekaban *et al.*, 1995; Franchini, 1995; Franchini and Streicher, 1995). These include transcriptional suppression of DNA polymerase β (Jeang *et al.*, 1990) and the functional suppression of p16 and p53, genes that are important regulators of the cell cycle (Matsuoka, 2003). STLV-1 infection of nonhuman primates provides an excellent model system to investigate mechanisms of PTLV oncogenesis *in vivo*.

HIV infection and AIDS

For two decades, NHP models for AIDS, especially Simian immunodeficiency virus (SIV) infection of macaques, have provided important insights into the transmission and pathogenesis of AIDS as well as virus-specific immune responses to anti-SIV vaccine regimens. Although an effective vaccine against human AIDS is not yet available, significant progress is being made towards this goal (reviewed in Robinson, 2002).

Studies of primate lentivirus transmission, in NHP models, has permitted dissection of the earliest events that occur in an infected host, after exposure to virus under defined conditions. These include the effects of a viral inoculum dose, viral virulence and the route of exposure (reviewed in Pope and Haase, 2003). Experiments in NHP have also determined the temporal relationship between lentiviral replication, after known virus exposure/infection, systemic infection, the development of antiviral immune responses and disease progression. Perhaps most important for vaccine development,

NHP studies have compared lentivirus-specific immune responses early after virus exposure, in both vaccinated and unvaccinated animals, to assess potential correlates of vaccine-mediated protection (reviewed in McMichael and Hanke, 2003). The results of these experiments have begun to identify key strategies that could be used for reducing or eliminating HIV transmission from one infected person to another. Understanding each of these areas is essential to developing a vaccine to prevent human HIV infection or AIDS and these studies are neither feasible nor ethical in humans.

Primate species/virus systems

Table 34.1 compares the key similarities and differences between the major NHP models of HIV/AIDS and human HIV-infection. The chimpanzee/HIV-1 model uses HIV-1 but HIV infection in chimps is generally non-pathogenic and thus does not model AIDS in humans. This limitation, coupled with the very small number of chimps, and their endangered species status, make the chimp/HIV model for AIDS largely impractical for HIV vaccine research. A variety of NHP species can be infected with isolates of HIV-2, which is genetically similar to SIV.

Many African NHP species are naturally infected with species-specific SIV variants, but these endemic SIV infections apparently do not result in AIDS. Asian macaques are not naturally infected with SIV but experimental inoculation of macaques, with SIV variants isolated from African monkeys, results in AIDS (reviewed in Gardner, 2003). A number of macaque and baboon species have been used for transmission, pathogenesis and vaccine studies with HIV-2 (Locher *et al.,* 2001). Infection of macaques with SIV or hybrid SIV/HIV viruses, containing the HIV-1 envelope (SHIV), are generally accepted as the most practical NHP models for human HIV-infection and AIDS and they are widely used to evaluate vaccine strategies against AIDS (reviewed in Letvin *et al.,* 2002; McChesney *et al.,* 1999; McMichael and Hanke, 2003; Robinson, 2002).

Virus challenge: viral properties and mucosal transmission

It is well-established that the capacity of a primate lentivirus to cause immunodeficiency disease, in Asian macaque species, is associated with its ability to replicate to high levels after establishing systemic infection (McChesney *et al.,* 1999). Although several biological properties, including co-receptor use and growth in transformed T cell lines and primary macaque blood cells, were proposed to account for virulence of SIV and SHIV isolates, it has been demonstrated that none of these reliably predict the ability to induce AIDS in infected macaques (Table 34.2). Persistently high viral replication, as measured by high levels of viral RNA or core antigen in plasma, is also the best predictor of the reliability of transmission by mucosal exposure for SIV or SHIV isolates (Table 34.2).

NHP models for HIV breast milk transmission

In developed countries, where infant formula is readily available and breastfeeding is not essential to infant survival, vertical HIV transmission can be prevented effectively by antiretroviral treatment during pregnancy and delivery combined with a short course of antiretroviral drugs given to the newborn. However, in most developing nations, even with some prenatal care or limited access to antiretroviral drugs, there are few alternatives to breast feeding because formula and clean water are rarely available or affordable (Mbori-Ngacha *et al.,* 2001). In these settings, exposure to breast milk remains a major risk for HIV infection and a neonatal vaccine, that can protect against HIV breast milk transmission, is urgently needed.

To evaluate strategies that may prevent HIV infection by breast feeding, HIV breast milk transmission has been modeled in infant rhesus macaques. The most direct model is SIV-infection of pregnant macaques who are allowed to suckle their infants. Early studies using this model found low or unreliable SIV transmission to infants (McClure *et al.,* 1991). Recently Amedee and colleagues (Amedee *et al.,* 2003) reported efficient breast milk transmission of SIV to infants of SIV-infected dams. A limitation of this model is that the duration of breast feeding, until SIV infection of infants, varied from a few weeks to months and was not associated with levels of viral RNA in milk (Amedee *et al.,* 2003).

To conserve adult female macaques needed for breeding and to better control viral and host variables, an alternative macaque model of HIV breast milk transmission is oral SIV or SHIV inoculation of newborn and infant macaques that may be nursery-reared. Advantages of experimental oral virus inoculation include the ability to use a specific dose of a well-characterized virus inoculum, to vary the total number of virus exposures and the intervals between virus exposures, and to assess how infant development may modulate outcome of virus exposure. Disadvantages are the need for labor-intensive nursery rearing of SIV-exposed infant macaques in specialized BSL-2

TABLE 34.1: Comparison of available nonhuman primate models to human HIV infection

NHP host species /virus	Similarities	Differences
Chimpanzee/HIV-1	NHP species most like humans, and only NHP species in which HIV-1 establishes persistent infection Genital mucosal exposure to high doses of some HIV-1 isolates has resulted in infected chimpanzees	Virus replicates only after adaptation by passage in PBMC of this species. Viral levels in blood are low in most, and transiently detected in many, HIV-1 infected animals Infected animals rarely develop disease; only after serial animal passage of virus
Macaque/HIV-2* or Baboon/HIV-2*	HIV-2 establishes persistent infection in multiple macaque species and in baboons Some macaque species develop AIDS after long-term infection with an HIV-2 adapted to that species	Viral levels in blood are low in many HIV-2 infected animals, except after virus is adapted to a host species
Macaque/SHIV*	SHIVs are chimeric viruses made by molecular biology techniques which encode some HIV-1 or HIV-2 genes (e.g. envelope) in an otherwise SIV genome	Virus replicates to high levels only after adaptation by passage in a species; viremia controlled spontaneously in many animals Biological properties of SHIVs in NHP differ substantially from HIV in humans and SIV in macaques. Pathogenic SHIVs induce rapid, severe, though often transient, reduction in CD4+ T cells in peripheral blood; AIDS results from sustained CD4+ T cell loss (observed in 25–40% of animals infected with pathogenic SHIVs) Mucosal transmission of most SHIVs is inefficient and unreliable
Macaque/SIV*	SIVs that infect macaques are most similar genetically to HIV-2 (SIV from wild chimpanzees most similar to HIV-1) Efficiency of pathogenic SIV transmission by mucosal exposure (genital, rectal, oral) is route and dose dependent Pathogenic SIV isolates reliably cause a spectrum of immunodeficiency disease remarkably similar to AIDS in HIV-1 infected humans SIV isolates with deletions in accessory genes are attenuated and induce slow progression to AIDS as seen in humans infected with HIV-1 containing nef gene deletions	Most pathogenic SIV isolates cause simian AIDS within 6 to 12 months after infection by parenteral or mucosal exposure to high (100,000 TCID50%) doses of virus

*Multiple virus isolates with distinct biological properties.

TABLE 34.2: Biological properties of SIV and SHIV isolates in rhesus macaques							
Virus isolate	Growth in macrophage	Co-receptor use		Replication in macaques (after IV inoc.)	Induction of AIDS	Reliability of mucosal virus transmission	
		CXCR4	CCR5			Vaginal	Oral
SIVmac251	+	–	+	High	+	High	High
SIVmac239	–	–	+	High	+	High	High
SHIV89.6PD	+	+	+	High	+	High	Not done
SHIV33A	+	+	–	High	+	High	High
SHIV89.6	+	+	+	Moderate	–	Moderate	High
SHIV33	–	+	–	Moderate	–	Low	None
SIVmac1A11	+	–	+	Low	–	Low	Low
SHIVHxb2	+	+	–	Low	–	Low	Not done

containment facilities, increased clinical care, by veterinary staff, of SIV-infected infants and the absence of anti-microbial factors normally present in macaque breast milk, which may more closely mimic conditions found in breast milk of HIV-infected women (Farquhar *et al.,* 2002). Studies using oral inoculation of infant macaques, with SIV or SHIV, have demonstrated that antiretroviral drugs (Van Rompay *et al.,* 2001), HIV neutralizing monoclonal antibodies (Baba *et al.,* 2000; Hofmann-Lehmann *et al.,* 2001), and passively acquired SIV-specific, non-neutralizing antibodies (Van Rompay *et al.,* 1998) can prevent infection if given before or, in some cases, soon after virus exposure. Recently, Van Rompay and colleagues (Van Rompay *et al.,* 2003) have shown that active immunization of infant rhesus macaques, with live-attenuated SIV and recombinant poxvirus vectors expressing SIV structural proteins, can protect against high viremia and delay onset of AIDS after oral inoculation with virulent SIV. Collectively, the results of these studies, using the SIV/infant macaque model, suggest that both passive and active immunization strategies may be successful in reducing breast milk transmission of HIV.

Influence of immunogenetics on NHP models for AIDS and HIV vaccine development

The major histocompatibility complex (MHC) plays an essential role in immune responses to viral infections and the enormous variation observed, for human MHC genes, is thought to reflect selective pressure by microbial, especially viral, pathogens. There is consensus that some MHC alleles (Carrington and Bontrop, 2002) and haplotypes can significantly modulate the rate of progress to AIDS in HIV-infected people (Carrington, 2003; Carrington and O'Brien, 2003; Cullen *et al.,* 2002; Cullen *et al.,* 2003; Trachtenberg *et al.,* 2003). In addition, concordance between MHC class I alleles of HIV-infected women and their infants is associated with increased risk of vertical HIV transmission (Polycarpou *et al.,* 2002). Further, there is evidence that specific human MHC alleles influence both the response to HIV-infection and immunization with HIV vaccines (Kaslow *et al.,* 2001). It has even been proposed that the relatively low genetic variation at MHC class I in chimpanzees is a direct result of long-term selection by an ancient SIV during a devastating pandemic in wild chimpanzees (de Groot *et al.,* 2002).

In contrast to the chimpanzee, recent studies of the MHC of Asian macaques, indicate that the allelic and haplotype variation among individual animals is similar to, or greater than, MHC variation observed for humans (Doxiadis *et al.,* 2000). Also, as shown for HIV-infected humans, some rhesus MHC alleles and haplotypes have been reported to influence the rate of disease progression in SIV-infected animals (Carrington and Bontrop, 2002; O'Connor *et al.,* 2003; Pal *et al.,* 2002; Sauermann *et al.,* 2000) and may modulate vaccine protection against SIV challenge (O'Connor *et al.,* 2003; Pal *et al.,* 2002). These observations underscore the need to identify more MHC alleles and to define complete MHC halplotypes in NHP species used for AIDS vaccine development (Friedrich and Watkins, 2003; Sauermann, 2001).

Ability of NHP studies to predict outcomes of human clinical HIV vaccine trials

The ultimate test of the relevance and utility of NHP models of AIDS, for HIV vaccine development, is their accuracy in predicting efficacy of specific vaccine strategies in human clinical trials. To date, only results of a single clinical trial of HIV vaccine efficacy are available for comparison with efficacy observed in NHP models. This double-blind, placebo controlled human trial, jointly sponsored by the NIH and Vaxgen, found no overall efficacy of a clade B HIV recombinant gp120 protein vaccine in groups at high-risk for HIV exposure in the USA; i.e. there were no statistically significant differences between individuals, receiving placebo or vaccine, in either the proportion of individuals who become HIV-infected, or in viral load after infection (Cohen, 2003a; Cohen, 2003b). This outcome was mirrored in prior NHP efficacy trials of recombinant HIV envelope (gp120 or 160), for chimpanzees challenged with non-homologous laboratory isolates of HIV-1 (Girard *et al.*, 1996), and for macaques immunized with recombinant HIV envelope (gp120 or gp160). Both examples showed little or no protection against infection by parenteral or mucosal challenge with pathogenic SIV or SHIV isolates (reviewed in Robinson, 2002).

Results are expected to be available, by the end of 2003, of a second large human clinical trial to assess the efficacy of recombinant canarypox vector, expressing HIV envelope, and other structural and regulatory proteins given as primary immunization followed by recombinant HIV envelope protein as a booster immunization. Preclinical studies in macaques immunized with recombinant canarypox expressing SIV and/or HIV antigens found that, despite only modest levels of virus-specific immune responses, vaccinated animals had reduced viral load and slower disease progression compared to unvaccinated controls after challenge (reviewed in Robinson, 2002). If the results of NHP efficacy studies consistently reflect those for human trials of similar HIV vaccine strategies, this will reinforce the value and relevance of NHP models of AIDS for HIV vaccine development, especially to identify immune correlates and perhaps, to understand mechanisms of HIV vaccine efficacy.

Simian cytomegalovirus

Use of NHPs as models for human cytomegalovirus (HCMV) has expanded dramatically in the last few years because of the increasing appreciation of the impact of HCMV infection on human health. HCMV is a member of the *Herpesviridae* family of viruses (Mocarski, 1993). It has been recognized for over 30 years as a serious threat to the developing fetus where it can cause a wide spectrum of pathological outcomes (Weller, 1971). HCMV is also a major cause of morbidity and mortality in those with either AIDS or an immunosuppressed immune system (Alford and Britt, 1993). There is no licensed vaccine for HCMV, although there has been a long-standing recognition of the medical need for one (Committee to Study Priorities for Vaccine Development and Medicine, 1999).

HCMV is species-specific and will not grow in the NHPs available for study. This limits experimental investigations to tissue culture settings and natural history studies in humans, leaving many experimental questions unanswered. However, the NHP models are ideally suited to fill the experimental void concerning many critical aspects of HCMV natural history and to develop novel immunological and chemotherapeutic strategies that can either prevent infection or limit disease.

The vast majority of studies have involved infection of rhesus macaques with rhesus CMV (RhCMV). However, important observations have also been made with CMV of other macaque species, the African green monkey, baboon and chimpanzee. All of these simian systems are also relevant surrogates for HCMV. For this review, a prototypical simian CMV phenotype is presented, based on studies with the different NHP CMV. Comparative studies between different NHP CMV are extremely limited. It should be stressed that each CMV representative has co-evolved with its host species. Accordingly, the evolution of virus-host relationships may have led to important and, as yet, undiscovered distinctions between the different CMV isolates.

CMV has been isolated from multiple genera and species of old and new world NHP hosts (Asher *et al.*, 1974; Black *et al.*, 1963; Eizuru *et al.*, 1989). It is probable that every NHP species has an associated species-specific CMV. Analyses of the genomes, protein expression and serological cross-relatedness demonstrate that CMV isolates from different NHP species are unique to each host (Davison, A.J. *et al.*, 2003; Eizuru *et al.*, 1989; Gibson, 1983; Hansen *et al.*, 2003; Kilpatrick *et al.*, 1976; Minamishima *et al.*, 1971; Tinghitella *et al.*, 1982). There is no evidence to date that the CMV of one host species can grow in another host species following natural exposure. However, simian CMV replication *in vitro* is not restricted to cells of the host species. The CMV of African green monkeys (Black *et al.*, 1963) and Rh CMV (Alcendor *et al.*, 1993) can productively infect human and NHP cells in culture. Further, AGMCMV can productively infect rhesus macaques and replicate for years (Swack *et al.*, 1971). With the expansion of NHP CMV sequences on GenBank, it is now possible to confirm the species identity of a primary isolate by limited sequence analysis.

Like HCMV, simian CMV is a common infectious agent in NHP populations (Andrade *et al.*, 2003; Black *et al.*, 1963; Eizuru *et al.*, 1989; Kessler *et al.*, 1989; Minamishima *et al.*, 1971; Swack and Hsiung, 1982; Swack *et al.*, 1971; Vogel *et al.*, 1994). Virtually 100% of monkeys in breeding facilities are seropositive by one year of age and 50% of infants are seropositive by 6 months of age (Vogel *et al.*, 1994). It is likely that multiple strains of CMV are present within each breeding facility (Alcendor *et al.*, 1993). Comparable rates of seroprevalence have been reported in monkeys trapped in the wild (Eizuru *et al.*, 1989; Minamishima *et al.*, 1971; Ohtaki *et al.*, 1986; Swack *et al.*, 1971). The routes of transmission are not known, but it is likely that virus is horizontally transmitted from mother to infant via breast milk and saliva, similar to identified modes in humans. Virus is also normally excreted in urine, adding another route of virus spread. There is no evidence consistent with transmission in utero, although low rates cannot be excluded (Vogel *et al.*, 1994). Maternal-fetal transmission is a critical component of HCMV congenital transmission, and this is one aspect of HCMV natural history that cannot be modeled in NHP at this time. It is most likely that absence of congenital infection, in NHP, has to do with the high seroprevalence of HCMV in breeding age females.

CMV seroprevalence changes dramatically if animals are reared in smaller cohorts from birth. When infants are separated from the dam at, or soon after, birth, and hand reared in a nursery, the animals remain essentially CMV-free well past the age of sexual maturity (Barry, unpublished; Minamishima *et al.*, 1971).

Primary infection

Simian CMV infection in immunocompetent hosts, either experimental or following natural exposure, does not result in any clinical signs of disease. This is similar to the vast majority of HCMV infections. Mononucleosis, a relatively rare outcome of primary HCMV infection, has never been associated with CMV infection in the NHP. In addition, there have been no published reports, or anecdotal observations, of CMV disease in monkeys that were culled or died of reasons other than acquired immunodeficiency or immunosuppression. It can be concluded, from the absence of disease, that host immune responses to primary CMV infection are highly protective.

The general course of primary infection in a seronegative host involves rapid dissemination from the site of infection to distal sites throughout the body (Lockridge *et al.*, 1999). In monkeys naturally exposed to CMV, virus is probably transmitted across the oral or genital mucosa. Experimental inoculations have been done by the oral, intravenous and subcutaneous (sc) routes. Viral DNA can generally be detected in the blood within 7 days of infection and in multiple tissues within two weeks (Lockridge *et al.*, 1999). Antiviral immune responses are rapid and increase in intensity as viral plasma DNA loads decrease. Cellular antiviral responses follow the same course of development as the humoral responses. A variety of hematological changes follow experimental IV inoculation, although no consistent pattern is observed (Lockridge *et al.*, 1999).

Persistent infection

There are two prominent hallmarks of the persistent phase of RCMV infection: chronic viral shedding and stability of the antiviral immune responses. These characteristics are identical to HCMV. Infected monkeys can remain viuric for years following primary infection, probably for the life of the host (Asher *et al.*, 1974; Swack and Hsiung, 1982). The frequency of CMV shedding is variable between animals, although some monkeys appear to be constantly shedding infectious virus at the oral and genital mucosa. Historically, shedding has been assayed by culturing virus. More recently, sensitive molecular techniques, such as real-time PCR, have been used to detect and quantify CMV DNA purified from mucosal swabs (Huff *et al.*, 2003). Approximately 50% of seropositive monkeys are DNA-positive in mucosal fluids at any one time. Thus, there is active and ongoing virus replication at mucosal surfaces within a persistently infected host. Occasionally, antigen-positive cells can be detected by immunohistochemistry in other tissues, although the tissues are usually histologically normal without an accompanying inflammatory response.

The relative constant exposure to CMV antigens probably explains the pattern of antiviral immune responses observed in long-term infected monkeys. Both antibody titers and cellular responses stay relatively stable over time. End-point antibody titers, to total viral antigen preparations, hover around the plateau level achieved at the end of the primary infection. Cellular responses to RhCMV antigens also exhibit little fluctuation (Kaur *et al.*, 2002).

The maintenance of a stable virus-host relationship (i.e., no disease) requires a considerable expenditure of the immunological repertoire of the infected host. Up to 5.8% and 5.3% of memory $CD4^+$ and $CD8^+$ T-lymphocytes, respectively, have been shown to be CMV-specific in healthy, CMV-positive rhesus macaques (Kaur *et al.,* 2002; Pitcher *et al.,* 2002). Comparably high frequencies of CMV-specific, memory T cells have been observed in persistently infected humans (Waldrop *et al.,* 1998; Waldrop *et al.,* 1997).

The three conditions in which HCMV is a serious pathogen (immunodeficiency, immunosuppression, and intrauterine infection) have strong parallels in NHP. These conditions are summarized below.

Immunodeficiency

The first published descriptions of CMV disease in NHP occurred in the context of rhesus macaques coinfected with the Type D simian retrovirus (SRV) and a lentivirus, the simian immunodeficiency virus (SIV) (Henrickson *et al.,* 1983; King *et al.,* 1983; Lerche *et al.,* 1984; Letvin *et al.,* 1983a; Letvin *et al.,* 1983b; Osborn *et al.,* 1984). The acquired immunodeficiency disease, caused by both SRV and SIV, was characterized by persistent lymphadenopathy, severe wasting, chronic diarrhea, high morbidity and mortality and multiple opportunistic infections, including activated CMV. Activated CMV disease is generally characterized by an abundance of cytomegalic cells containing cytoplasmic and/or intranuclear inclusions and often associated with tissue necrosis and neutrophilic infiltration in tissues. This pattern of CMV infection in immunodeficient monkeys is distinct from the paucity of viral inclusions in healthy macaques and is strikingly similar to CMV disease in human AIDS patients. The incidence of CMV disease in monkeys with SAIDS caused by SIV or SRV is variable. Some studies have reported that up to one-third to one-half of CMV seropositive animals have evidence of CMV disease at necropsy (Baskin *et al.,* 1988; Kaur *et al.,* 2003; King *et al.,* 1983; Kuhn *et al.,* 1999; Osborn *et al.,* 1984). Similar to HCMV, activated simian CMV can be observed in multiple tissues, (Baskin, 1987). However, CMV disease is not always systemic and can occur in only a single tissue (Sequar *et al.,* 2002). CMV retinitis, an important clinical problem in some human AIDS patients, has not been described in a SAIDS monkey.

There are multiple distinctions between the course of RhCMV infection in an immunodeficient and an immunocompetent host. Notable changes include an increased frequency of detectable RhCMV DNA in blood, elevated genome copy numbers in tissues and declining measures of anti-CMV immune functions, such as CTL activity, cytokine secretion, and neutralizing antibody titers (Kaur *et al.,* 2003; Sequar *et al.,* 2002). The kinetics of these changes are variable. If SIV infection occurs during the primary phase of RhCMV infection, the onset of RhCMV disease and SAIDS can be rapid (10–27 weeks post SIV). If the monkeys are inoculated with SIV during the persistent phase of RhCMV infection, the time of death can be greatly extended (10 weeks – >1 year). One study observed a statistically significant decrease in the time of death in animals with RhCMV end-organ disease compared to animals without histological evidence of activated RhCMV (Kaur *et al.,* 2003).

Transplantation

CMV sequelae have also been observed in immunosuppressed NHP receiving either allografts or xenografts. Although there are not many references in the literature, the occurrence of activated CMV in transplant-recipients appears to be related to the use of intense immunosuppression regimes, such as cyclophosphamide, corticosteroids and/or anti-thymocytic globulin, designed to prevent graft rejection. A transplant recipient can develop CMV histopathology, similar to CMV disease in human allograft recipients, including pneumonitis and vasculopathies (Ghanekar *et al.,* 2002; Mueller *et al.,* 2002; Ohtaki *et al.,* 1986; Ohtaki *et al.,* 1988; Teranishi *et al.,* 2003). Further studies are required to determine the utility of NHP as an experimental model to study transplantation-associated CMV disease, primarily because the frequency of CMV reactivation in this setting is not known.

Fetal infection

RhCMV can cause a range of developmental defects in experimentally inoculated rhesus macaque fetuses that are almost identical to those observed in human infants congenitally infected with HCMV. It should be stressed that fetal infection in this NHP is a model of intrauterine pathogenesis and not transplacental transmission. No studies have been reported documenting either natural or experimental maternal-fetal CMV transmission in NHP. The published studies of CMV-induced fetal disease required direct *in utero* inoculation of fetuses with virus (Chang *et al.,* 2002; London *et al.,* 1986; Tarantal *et al.,* 1998). Using ultrasound guidance, needles can be directed through the abdominal wall of the dam to deliver virus to precise locations within the developing fetus at defined stages of gestation. Growth and developmental outcomes can be prospectively monitored, by ultrasound, and fetal samples (blood, amniotic fluid, and tissue) can be obtained by needle biopsy.

Rhesus macaque fetuses have been inoculated with RhCMV by the intramniotic (IA), intracranial (IC), and intraperitoneal (IP) routes from late in the first trimester (gestation day 50) through mid-gestation (day 80) (Chang *et al.*, 2002; London *et al.*, 1986; Tarantal *et al.*, 1998). Mild to severe fetal outcomes have been observed in approximately 50% of inoculated fetuses. No RhCMV sequelae have ever been observed in the dams, although there has been suggestive evidence of retrograde transmission of the virus across the placenta (London *et al.*, 1986).

The developing brain is highly sensitive to CMV disease with a spectrum of developmental abnormalities. These include microcephaly, lissencephaly, ventricular dilatation, leptomeningitis, encephalitis and periventricular calcifications. All of these are hallmarks of congenital HCMV infection (Alford and Britt, 1993). RhCMV histopathology is not limited to the brain and systemic effects such as intrauterine growth restriction, disseminated RhCMV disease and isolation of virus in blood and tissues have been seen (Chang *et al.*, 2002; Tarantal *et al.*, 1998). Placental abnormalities (deciduitis, infarction, calcification, and lymphocytic infiltration) have been seen in some of the inoculated fetuses (London *et al.*, 1986).

Cercopithecine herpesvirus 1 (Herpes B virus)

Cercopithecine herpesvirus 1, generally referred to as herpes B virus (BV), is an alphaherpesvirus of the rhesus macaque with strong genetic, virological and immunological relatedness to herpes simplex virus (HSV) of humans. BV infection of the rhesus monkey seems to accurately model HSV infection of humans, but the deadly zoonotic potential of BV has prevented its use in an animal model. However, it was recently listed as a categorical pathogen by the U.S. Public Health Service and could be an agent of bioterror. Thus a brief discussion of BV is included here.

BV is endemic in Asian macaque populations (Weigler, 1992), and closely related alphaherpesviral variants are present in African NHP (Eberle and Hilliard, 1995). BV deservedly bears the most attention of the NHP alphaherpesviruses because of its pathogenic potential in humans. Zoonotic infection of humans with B virus is almost invariably fatal (>70%) in the absence of antiviral chemotherapies, and severe, non-fatal infections can result in encephalomyelitis or severe neurological impairment (Huff and Barry, 2003). BV is the only simian herpesvirus that is known to cause disease in humans. The most salient feature of BV natural history, in terms of its occupational risk, is that an overwhelming majority of macaques shed B virus without overt signs of disease (Huff, J.E. *et al.*, 2003; Weigler *et al.*, 1993; Zwartouw and Boulter, 1984). Thus, every BV-seropositive macaque must be considered as a potential source of infectious BV, whether through its bodily fluids or its tissues.

Despite the preceding risk assessment, there have only been approximately 40 documented cases of human infection since the first reported transmission to humans in 1932 (Huff and Barry, 2003). The disproportionality between the number of zoonotic infections and the high seroprevalence of BV in breeding age animals is a function of the BV life cycle. The biology of BV infection in macaques is characterized by a lifelong persistence with infrequent, and usually subclinical, shedding at mucosal surfaces. Although the natural history of BV has not been described in detail, the replication cycle of HSV serves as a precedent (Roizman and Sears, 1993). Virus is transmitted across a mucosal surface, such as the oral or genital mucosa. Following localized replication in mucosal epithelial cells, the virus is transmitted directly to sensory nerve endings. The prevailing thought is that there is no associated bloodborne stage of infection, except in rare systemic infections (Simon *et al.*, 1993). The virus particle is carried, by axonal transport, to the dorsal root ganglia where the virus establishes a true latent infection in neurons.

Latency is noted for a lack of viral replication and an extremely limited pattern of viral transcription. Periodic reactivation from latency results in the production of progeny virions which transport back down the axon to mucosal epithelial cells, where they replicate and the infectious virus is released from the mucosal epithelium. For BV, most episodes of recurrent viral shedding are asymptomatic (Huff *et al.*, 2003; Weigler *et al.*, 1993; Zwartouw and Boulter, 1984), and represent a constant risk of zoonoses. Clinical signs of either primary or recurrent infection (oral herpetic lesions such as gingivostomatitis, oral and lingual ulcers, and conjunctivitis) are the exception (Carlson *et al.*, 1997; Keeble *et al.*, 1958; Weigler, 1992), and usually require immediate euthanasia of the animal. Virus isolation and molecular detection of BV DNA indicate that the frequency of shedding in a population is low (1–5%), although more studies are needed to establish a true rate. There is evidence to suggest that shedding frequency may go up during breeding season (Huff *et al.,* 2003; Weigler *et al.,* 1993).

There is a strong correlation between the seroconversion to BV and the age of the animal. Seroconversion rates increase sharply when monkeys reach the age of sexual maturity, with prevalence rates of 80–100% in adult populations (Weigler, 1992; Weigler *et al.,* 1993). There is no evidence for vertical transmission of BV.

Human B virus infections have generally involved direct contact with macaques or their tissues or fluids (Huff and Barry, 2003). Methods of contact have included a bite, scratch, contact of a mucosal surface with a macaque body fluid or tissue, or a contaminated needle puncture or cage scratch (Huff and Barry, 2003). There has only been one documented case of human-to-human transmission, although the potential for secondary transmission is probably low (Holmes *et al.,* 1990). BV disease can begin within a few days to a month, although disease progression is variable in terms of the kinetics and clinical signs. Guidelines for reducing potential exposure and treatment of suspected BV infections in humans have recently been published (Holmes *et al.,* 1995).

Hepatitis viruses

Hepatitis B virus

Hepatitis B virus (HBV), a small double-shelled hepadenavirus virus that contains a partially double-stranded DNA genome of approximately 3200 bases, is found in several species, including woodchuck, ground squirrel, a range of bird species such as duck, goose and grey heron (Marion *et al.,* 1980; Mason *et al.,* 1980; Summers *et al.,* 1978), and the NHPs chimpanzee (*Pan troglodytes*), woolly monkey (*Lagothrix lagothrica*), orangutan (*Pongo pygmaeus*), gibbon (*Hylobates sp.*) and gorilla (*Gorilla gorilla*) (Grethe *et al.,* 2000; Lanford *et al.,* 1998; Mimms *et al.,* 1993; Vaudin *et al.,* 1988; Warren *et al.,* 1999). HBV, isolated from gibbons and chimpanzees, share an early phylogenetic lineage, indicating that these viruses were indigenous to their respective hosts (Norder *et al.,* 1996). Despite the species specificity of natural HBV isolates, experimental infection of chimpanzees with human and gibbon HBV can be accomplished (Gallagher, 1991). Gibbons can be infected through experimental exposure to human saliva containing HBV (Bancroft *et al.,* 1977; Scott *et al.,* 1980). Replication of human HBV in a number of primate species suggests that natural HBV cross-transmission can occur. HBV can be present in the blood and other body fluids, including saliva/nasopharyngeal fluids, semen, cervical secretions and leukocytes (Alter *et al.,* 1977; Davison, F. *et al.,* 1987). However, HBV transmission from gibbon or chimpanzee to human has never been documented. CD8+ T cells mediate viral clearance during acute HBV infection in chimpanzees (Thimme *et al.,* 2003).

Hepatitis C virus

Hepatitis C virus (HCV) is a member of the Flaviviridae family and has a single-stranded positive sense RNA genome. The chimpanzee is the only experimental animal susceptible to infection with hepatitis C virus (HCV). The chimpanzee model of HCV infection was instrumental in the initial studies on non-A, non-B hepatitis, including observations on the clinical course of infection, determination of the physical properties of the virus and eventual cloning of the HCV nucleic acid (reviewed in Lanford and Bigger, 2002). Other NHP models of HCV have been developed using surrogate viruses such as GB virus-B (reviewed in Beames *et al.,* 2001). GB virus-B virus is closely related to HCV and it is hepatotropic. In addition, the level of GBV-B viremia observed in infected tamarins, the animal model for GBV-B, is greater than 1000-fold higher than for HCV. Tamarins are much easier to house than chimpanzees and a tissue culture system for GBV-B, using primary tamarin hepatocytes, is available (Beames *et al.,* 2000). Thus, primates are likely to play an important role in developing effective HCV vaccines.

Conclusion

Early success in developing effective antibiotics and vaccines provided some hope that modern medical advances would minimize the impact of infectious diseases on human health and that research effort could shift to meet the challenges of chronic degenerative diseases of humans. However, the AIDS epidemic, the emergence of multi-drug resistant tuberculosis and the numerous emerging diseases of viral etiology, including SARS, hantavirus, West Nile virus, etc., have made it very clear that infectious diseases still have the capacity to alter human society dramatically. Thus, infectious disease research remains a very high priority and NHP models of human viral diseases will remain a critical tool in understanding pathogenesis and in developing vaccines and therapies to viral agents that are a major public health challenge.

Correspondence

Any correspondence should be directed to Dr. Christopher J. Miller, California National Primate Research Center, University of California, Davis, CA 95616. Telephone: (530) 752-0447. Fax: (530) 752-2880. E-mail: cjmiller@ucdavis.edu

References

Alcendor, D.J., Barry, P.A., Pratt-Lowe, E. and Luciw, P.A. (1993). *Virol.* 194, 815–821.

Alford, C.A. and Britt, W.J. (1993). In Lopez, C. (ed.) *The Human Herpesviruses*, pp 227–255. Raven Press Ltd., New York.

Alter, H.J., Purcell, R.H., Gerin, J.L., London, W.T., Kaplan, P.M., McAuliffe, V.J., Wagner, J. and Holland, P.V. (1977). *Infect. Immun.* 16, 928–933.

Amedee, A.M., Lacour, N. and Ratteree, M. (2003), *J. Med. Primatol.* 32, 187–193.

Andrade, M.R., Yee, J., Barry, P.A., Spinner, A., Roberts, J.A., Cabello, H., Leite, J.P. and Lerche, N.W. (2003). *Amer. J. Primatol.* 59, 123–128.

Arima, N. and Tei, C. (2001). *Leukemia and Lymphoma* 40(3-4), 267–268.

Asher, D.M., Gibbs, J.C.J., Lang, D.J., Gadjusek, D.C. and Chanock, R.M. (1974). *Proc. Soc. Exp. Biol. Med.* 145, 794–801.

Baba, T.W., Liska, V., Hofmann-Lehmann, R., Vlasak, J., Xu, W., Ayehunie, S., Cavacini, L.A., Posner, M.R., Katinger, H., Stiegler, G., Bernacky, B.J., Rizvi, T.A., Schmidt, R., Hill, L.R., Keeling, M.E., Lu, Y., Wright, J.E., Chou, T.C. and Ruprecht, R.M. (2000). *Nat. Med.* 6(2), 200–206.

Bancroft, W.H., Snitbhan, R., Scott, R.M., Tingpalapong, M., Watson, W.T., Tanticharoenyos, P., Karwacki, J.J. and Srimarut, S. (1977). *J. Infect. Dis.* 135, 79–85.

Barnard, J.A., Beauchamp, R.D., Coffey, R.J. and Moses, H.L. (1989). *Proc. Natl. Acad. Sci.* 86, 1578–1582.

Baskerville, A., Bowen, A.E.T., Platt, G.S., McArdell, L.B. and Simpson, D.I.H. (1978). *Journal of Pathology* 125, 131–138.

Baskin, G.B. (1987). *Amer. J. Path.* 129, 345–352.

Baskin, G.B., Murphey-Corb, M., Watson, E.A. and Martin, L.N. (1988). *Vet. Pathol.* 25(6), 456–467.

Beames, B., Chavez, D., Guerra, B., Notvall, L., Brasky, K.M. and Lanford, R.E. (2000). *J. Virol.* 74, 11764–11772.

Beames, B., Chavez, D. and Lanford, R.E. (2001). *Ilar J.* 42, 152–160.

Beilke, M.A., Traina-Dorge, V., England, J.D. and Blanchard, J.L. (1996). *Arthritis and Rheumatology* 39(4), 610–615.

Belshe, R.B., Richardson, L.S., London, W.T., Sly, D.L., Lorfeld, J.H., Camargo, E., Prevar, D.A. and Chanock, R.M. (1977). *J. Med. Virol.* 1(3), 157–162.

Black, P.H., Hartley, J.W. and Rowe, W.P. (1963). *Proc. Soc. Exp. Biol. Med.* 112, 601–605.

Blake, F.G. and Trask, J.D. (1921). *J. Exp. Med.* 33, 413–422.

Blattner, W. (1989). *Annals of Internal Medicine* 111, 4–6.

Bukreyev, A., Skiadopoulos, M.H., McAuliffe, J., Murphy, B.R., Collins, P.L. and Schmidt, A.C. (2002). *Proc. Natl. Acad. Sci. U S A* 99(26), 16987–16991.

Carlson, C.S., O'Sullivan, M.G., Jayo, M.J., Anderson, D.K., Harber, E.S., Jerome, W.G., Bullock, B.C. and Heberling, R.L. (1997). *Vet. Pathol.* 34(5), 405–414.

Carrington, M. (2003). *Methods Mol. Biol.* 210, 325–332.

Carrington, M. and Bontrop, R.E. (2002). *Aids* 16 Suppl 4, S105–114.

Carrington, M. and O'Brien, S.J. (2003). *Annu. Rev. Med.* 54, 535–551.

Chang, W.L., Tarantal, A.F., Zhou, S.S., Borowsky, A.D. and Barry, P.A. (2002). *J. Virol.* 76(18), 9493–9504.

Chapman, L.E., Mertz, G.J., Peters, C.J. and Group, R.S. (1999). *Antiviral Therapy* 4, 211–219.

Cheng, X., Zhou, H., Tang, R.S., Munoz, M.G. and Jin, H. (2001). *Virology* 283(1), 59–68.

Cohen, J. (2003a). *Science* 299(5612), 1495.

Cohen, J. (2003b). *Science* 299(5611), 1290–1291.

Committee to Study Priorities for Vaccine Development and I. o. Medicine (1999). In Stratton, K.R., Durch, J.S. and Lawrence, R.S. (eds) *Vaccines for the 21st Century: A Tool for Decision Making*. National Academy Press, Washington, D.C.

Cullen, M., Perfetto, S.P., Klitz, W., Nelson, G. and Carrington, M. (2002). *Am. J. Hum. Genet.* 71(4), 759–776.

Cullen, M., Malasky, M., Harding, A. and Carrington, M. (2003). *Immunogenetics* 54(12), 900–910.

Davison, A.J., Dolan, A., Akter, P., Addison, C., Dargan, D.J., Alcendor, D.J., McGeoch, D.J. and Hayward, G.S. (2003). *J Gen. Virol.* 84(Pt 1), 17–28.

Davison, F., Alexander, G.J., Trowbridge, R., Fagan, E.A. and Williams, R. (1987). *J. Hepatol.* 4, 37–44.

de Groot, N.G., Otting, N., Doxiadis, G.G., Balla-Jhagjhoorsingh, S.S., Heeney, J.L., van Rood, J.J., Gagneux, P. and Bontrop, R.E. (2002). *Proc. Natl. Acad. Sci. U S A* 99(18), 11748–11753.

De Swart, R.L., Kuiken, T., Timmerman, H.H., Amerongen Gv, G., Van Den Hoogen, B.G., Vos, H.W., Neijens, H.J., Andeweg, A.C. and Osterhaus, A.D. (2002). *J. Virol.* 76(22), 11561–11569.

Dekaban, G.A., Digilio, L. and Franchini, G. (1995). *Current Opinion in Genetics and Development* 5, 807–813.

Doxiadis, G.G., Otting, N., de Groot, N.G., Noort, R. and Bontrop, R.E. (2000). *J. Immunol.* 164(6), 3193–3199.

Durbin, A.P., Elkins, W.R. and Murphy, B.R. (2000). *Vaccine* 18(22), 2462–2469.
Eberle, R. and Hilliard, J.K. (1995). *Infect. Agents Dis.* 4, 55–70.
Eizuru, Y., Tsuchiya, K., Mori, R. and Minamishima, Y. (1989). *Arch. Virol.* 107, 65–75.
Evans, D.T., Chen, L.M., Gillis, J., Lin, K.C., Harty, B., Mazzara, G.P., Donis, R.O., Mansfield, K.G., Lifson, J.D., Desrosiers, R.C., Galan, J.E. and Johnson, R.P. (2003). *J. Virol.* 77, 2400–2409.
Farquhar, C., VanCott, T.C., Mbori-Ngacha, D.A., Horani, L., Bosire, R.K., Kreiss, J.K., Richardson, B.A. and John-Stewart, G.C. (2002). *J. Infect. Dis.* 186(8), 1173–1176.
Fouchier, R.A., Kuiken, T., Schutten, M., van Amerongen, G., van Doornum, G.J., van den Hoogen, B.G., Peiris, M., Lim, W., Stohr, K. and Osterhaus, A.D. (2003). *Nature* 423(6937), 240.
Franchini, G. (1995). *Blood* 86(10), 3619–3639.
Franchini, G. and Streicher, H. (1995). Human T-cell leukemia virus. *Clinical Haematology* 8(1), 131–148.
Friedrich, T.C. and Watkins, D.I. (2003). *International Perspectives: the future of nonhuman primate resources*, pp 122–127. The National Academies Press, Washington, DC.
Gabet, A.S., Gessain, A. and Wattel, E. (2003). *International Journal of Cancer* 107, 74–83.
Gallagher, M., Fields, H.A., De la Torre, N., Ebert, J.W. and Krawczynski, K.Z. (1991). In Hollinger, E.B., Lemon, S.M. and Margolis, H.S. (eds) *Characterization of HBV2 by Experimental Infection of Primates*. Williams & Wilkins, Baltimore.
Gardner, M.B. (2003). *J. Med. Primatol.* 32, 180–186.
Geisbert, T.W., Pushko, P., Anderson, K., Smith, J., Davis, K.J. and Jahrling, P.B. (2002). *Emerg. Infect. Dis.* 8, 503–507.
Ghanekar, A., Lajoie, G., Luo, Y., Yang, H., Choi, J., Garcia, B., Cole, E.H., Greig, P.D., Cattral, M.S., Phillips, M.J., Cardella, C.J., Levy, G.A , Zhong, R. and Grant, D.R. (2002). *Transplantation* 74(1), 28–35.
Gibson, W. (1983). *Virol.* 128, 391–406.
Girard, M., Yue, L., Barre-Sinoussi, F., van der Ryst, E., Meignier, B., Muchmore, E. and Fultz, P.N. (1996). *J. Virol.* 70(11), 8229–8233.
Graham, B.S., Rutigliano, J.A. and Johnson, T.R. (2002). *Virology* 297(1), 1–7.
Grethe, S., Heckel, J.O., Rietschel, W. and Hufert, F.T. (2000). *J. Virol.* 74, 5377–5381.
Groen, J., Gerding, M., Koeman, J.P., Roholl, P.G., van Amerongen, G., Jordans, H.G., Niesters, H.G. and Osterhaus, A.D. (1995). *Journal of Infectious Diseases* 172(1), 38–44.
Hall, S.L., Sarris, C.M., Tierney, E.L., London, W.T. and Murphy, B.R. (1993). *J. Infect. Dis.* 167(4), 958–962.
Hancock, G.E., Smith, J.D. and Heers, K.M. (2000). *J. Infect. Dis.* 181(5), 1768–1771.
Hansen, S.G., Strelow, L.I., Franchi, D.C., Anders, D.G. and Wong, S.W. (2003). *J. Virol.* 77, 6620–6636.
Henrickson, R.V., Maul, D.H., Osborn, K.G., Sever, J.L., Madden, D.L., Ellingsworth, L.R., Anderson, J.H., Lowenstine, L.J. and Gardner, M.B. (1983). *Lancet* 1(8321), 388–390.
Hofmann-Lehmann, R., Rasmussen, R.A., Vlasak, J., Smith, B.A., Baba, T.W., Liska, V., Montefiori, D.C., McClure, H.M., Anderson, D.C., Bernacky, B.J., Rizvi, T.A., Schmidt, R., Hill, L.R., Keeling, M.E., Katinger, H., Stiegler, G., Posner, M.R., Cavacini, L.A., Chou, T.C. and Ruprecht, R.M. (2001). *J. Med. Primatol.* 30(4), 190–196.
Holmes, G.P., Hilliard, J.K., Klontz, K.C., Rupert, A.H., Schindler, C.M., Parrish, E., Griffin, D.G., Ward, G.S., Bernstein, N.D., Bean, T.W. (1990). *Ann. Intern. Med.* 112(11), 833–839.
Holmes, G.P., Chapman, L.E., Stewart, J.A., Straus, S.E., Hilliard, J.K. and Davenport, D.S. (1995). *Clin. Infect. Dis.* 20(2), 421–439.
Hooper, J.W., Custer, D.M., Thompson, E. and Schmaljohn, C.S. (2001). *Journal of Virology* 75(18), 8469–8477.
Huff, J.E., Eberle, R., Capitanio, J., Zhou, S.S. and Barry, P.A. (2003). *J. Genl. Virol.* 84, 83–92.
Huff, J.L. and Barry, P.A. (2003). *Emerg. Infect. Dis.* 9(2), 246–250.
Ibuki, K., Funahashi, S.I., Yamamoto, H., Nakamura, M., Igarashi, T., Miura, T., Ido, E., Hayami, M. and Shida, H. (1997). *Journal of General Virology* 78(1), 147–152.
Ignatiev, G.M., Dadaeva, A.A., Luchko, S.V. and Chepurnov, A.A. (2000). *Immunology Letters* 71, 131–140.
Jaax, N.K., Davis, K.J., Geisbert, T.J., Vogel, P., Jaax, G.P., Topper, M. and Jahrling, P.B. (1996). *Archives of Pathology and Laboratory Medicine* 120(2), 140–155.
Jahrling, P.B., Geisbert, J., Swearengen, J.R., Jaax, G.P., Lewis, T., Huggins, J.W., Schmidt, J.J., LeDuc, J.W. and Peters, C.J. (1996). *Archives of Virology* Supplement(11), 135–140.
Jahrling, P.B., Geisbert, T.W., Geisbert, J.B., Swearengen, J.R., Bray, M., Jaax, N.K., Huggins, J.W., LeDuc, J.W. and Peters, C.J. (1999). *Journal of Infectious Diseases* 179 (Supplement 1), S224–234.
Jeang, K.T., Widen, S.G., Semmes, O.G.T. and Wilson, S.H. (1990). *Science* 247(4946), 1082–1084.
Johnson, E., Jaax, N., White, J. and Jahrling, P. (1995). *International Journal of Experimental Pathology* 76(4), 227–236.
Johnson, J.M., Harrod, R. and Franchini, G. (2001). *Journal of Experimental Pathology* 82, 135–147.
Kakuk, T.J., Soike, K., Brideau, R.J., Zaya, R.M., Cole, S.L., Zhang, J.Y., Roberts, E.D., Wells, P.A. and Wathen, M.W. (1993). *J. Infect. Dis.* 167(3), 553–561.

Kalyanaraman, V.S., Sarngadharan, M.G., Robert-Guroff, M., Miyoshi, I., Golde, D. and Gallo, R.C. (1982). *Science* 218, 571–573.

Kaslow, R.A., Rivers, C., Tang, J., Bender, T.J., Goepfert, P.A., El Habib, R., Weinhold, K. and Mulligan, M.J. (2001). *J. Virol* 75(18), 8681–8689.

Kaur, A., Hale, C.L., Noren, B., Kassis, N., Simon, M.A. and Johnson, R.P. (2002). *J. Virol.* 76(8), 3646–3658.

Kaur, A., Kassis, N., Hale, C.L., Simon, M., Elliott, M., Gomez-Yafal, A., Lifson, J.D., Desrosiers, R.C., Wang, F., Barry, P., Mach, M. and Johnson, R.P. (2003). *J. Virol.* 77(10), 5749–5758.

Kazanji, M. (2000). *AIDS Rearch and Human Retroviruses* 16(16), 1741–1746.

Kazanji, M., Ureta-Vidal, A., Ozden, S., Tangy, F., De Thoisy, B., Fiette, L., Talarmin, A., Gessain, A. and De The, G. (2000). *Journal of Virology* 74(10), 4860–4867.

Keeble, S.A., Christofinis, G.J. and Wood, W. (1958). *J. Path. Bact.* 76, 189–199.

Kessler, M.J., London, W.T., Madden, D.L., Dambrosia, J.M., Hilliard, J.K., Soike, K.F. and Rawlins, R.G. (1989). *Puerto Rican Health Sci. J.* 8, 95–97.

Kilpatrick, B.A., Huang, E.S. and Pagano, J.S. (1976). *J. Virol.* 18(3), 1095–1105.

King, N.W., Hunt, R.D. and Letvin, N.L. (1983). *Am. J. Pathol.* 113(3), 382–388.

Klingstrom, J., Plyusnin, A., Vaheri, A. and Lundkvist, A. (2002). *Journal of Virology* 76(1), 444–449.

Kuhn, E.M., Stolte, N., Matz-Rensing, K., Mach, M., Stahl-Henning, C., Hunsmann, G. and Kaup, F.J. (1999). *Vet. Pathol.* 36(1), 51–56.

Lanford, R.E. and Bigger, C. (2002). *Virology* 293, 1–9.

Lanford, R.E., Chavez, D., Brasky, K.M., Burns, R.B. 3rd, and Rico-Hesse, R. (1998). *Proc. Natl. Acad. Sci. USA* 95, 5757–5761.

Lazo, A., Bailer, R., Lairmore, M.D., Yee, J., Andrews, J., Stevens, V.C. and Blakeslee, J.R. (1994). *Leukemia* 8 (Supplement 1), S222–S226.

Leaman, D.W., Longano, F.J., Okicki, J.R., Soike, K.F., Torrence, P.F., Silverman, R.H. and Cramer, H. (2002). *Virology* 292(1), 70–77.

Lednicky, J.A. (2003). *Archives of Pathology and Laboratory Medicine* 127, 30–35.

Lee, H.W., Baek, L.J. and Johnson, K.M. (1982). *Journal of Infectious Diseases* 146, 638–644.

Lerche, N.W., Henrickson, R.V., Maul, D.H. and Gardner, M.B. (1984). *Lab. Anim. Sci.* 34(2), 146–150.

Letvin, N.L., Aldrich, W.R., King, N.W., Blake, B.J., Daniel, M.D. and Hunt, R.D. (1983a). *Lancet* 2(8350), 599–602.

Letvin, N.L., Eaton, K.A., Aldrich, W.R., Sehgal, P.K., Blake, B.J., Schlossman, S.F., King, N.W. and Hunt, R.D. (1983b). *Proc. Natl. Acad. Sci. USA* 80(9), 2718–2722.

Letvin, N.L., Barouch, D.H. and Montefiori, D.C. (2002). *Annu. Rev. Immunol.* 20, 73–99.

Levin, M.C. and Jacobson, S. (1997). *Journal of Neurovirology* 3, 126–140.

Locher, C.P., Witt, S.A., Herndier, B.G., Tenner-Racz, K., Racz, P. and Levy, J.A. (2001). *Immunol. Rev.* 183, 127–140.

Lockridge, K.M., Sequar, G., Zhou, S.S., Yue, Y., Mandell, C.M. and Barry, P.A. (1999). *J. Virol.* 73, 9576–9583.

London, W.T., Martinez, A.J., Houff, S.A., Wallen, W.C., Curfman, B.L., Traub, R.G. and Sever, J.L. (1986). *Teratol.* 33, 323-331.

Marion, P.L., Oshiro, L.S., Regnery, D.C., Scullard, G.H. and Robinson, W.S. (1980). *Proc. Natl. Acad. Sci. USA* 77, 2941–2945.

Mason, W.S., Seal, G. and Summers, J. (1980). *J. Virol.* 36, 829–836.

Matsuoka, M. (2003). *Oncogene* 22, 5131–5140.

Mbori-Ngacha, D., Nduati, R., John, G., Reilly, M., Richardson, B., Mwatha, A., Ndinya-Achola, J., Bwayo, J. and Kreiss, J. (2001). *Jama* 286(19), 2413–2420.

McCarthy, T.J., Kennedy, J.L., Blakeslee, J.R. and Bennett, B.T. (1990). *Laboratory Animal Science* 40, 79–81.

McChesney, M.B. *et al.* (1997). *Virology* 233, 77–84.

McChesney, M.B., Sawai, E.T. and Miller, C.J. (1999). In Chen, I. and Ahmed, R. (eds) *Persistent Viral infections,* pp 321–345. John Wiley and Sons Ltd., Chichester, England.

McClure, H.M., Anderson, D.C., Fultz, P.N., Ansari, A.A., Jehuda-Cohen, T., Villinger, F., Klumpp, S.A., Switzer, W., Lockwood, E. and Brodie, A. (1991). *J. Med. Primatol.* 20(4), 182–187.

McElroy, A.K., Bray, M., Reed, D.S. and Schmaljohn, C.S (2002). *Journal of Infectious Diseases* 186, 1706–1712.

McMichael, A.J. and Hanke, T. (2003). *Nat. Med.* 9(7), 874–880.

Meertens, L. and Gessain, A. (2003). *Journal of Virology* 77(3), 782–789.

Mimms, L.T., Solomon, L.R., Ebert, J.W. and Fields, H. (1993). *Biochem. Biophys. Res. Commun.* 195, 186–191.

Minamishima, Y., Graham, B.J. and Benyesh-Melnick, M. (1971). *Infection and Immunity* 4(4), 368–373.

Miyoshi, I., Ohtsuki, Y., Fujishita, M., Yoshimoto, S., Kubonishi, I. and Minezawa, M. (1982). *Gann.* 73, 848–849.

Mocarski, E.S.J. (1993). In Lopez, C (ed.) *The Human Herpesviruses,* pp 173–226. Raven Press, Ltd., New York.

Morteaux, F., Gabet, A.S. and Wattel, E. (2003). *Leukemia* 17, 26–38.

Mueller, N.J., Barth, R.N., Yamamoto, S., Kitamura, H., Patience, C., Yamada, K., Cooper, D.K., Sachs, D.H., Kaur, A. and Fishman, J.A. (2002). *J. Virol.* 76(10), 4734–4740.

Murata, N., Hakoda, E., Machida, H., Ikezoe, T., Sawada, T., Hoshino, H. and Miyoshi, I. (1996). *Leukemia* 10(12), 1971–1974.

Nichol, S.T., Spiropoulou, C.F., Morzunov, G., Rollin, P.E., Ksiazek, T.G., Feldman, H., Sanchez, A., Childs, J., Zaki, S. and Peters, C.J. (1993). *Science* 262, 914–917.

Norder, H., Ebert, J.W., Fields, H.A., Mushahwar, I.K. and Magnius, L.O. (1996). *Virology* 218, 214–223.

O'Connor, D.H., Mothe, B.R., Weinfurter, J.T., Fuenger, S., Rehrauer, W.M., Jing, P., Rudersdorf, R.R., Liebl, M.E., Krebs, K., Vasquez, J., Dodds, E., Loffredo, J. Martin, S., McDermott, A.B., Allen, T.M., Wang, C., Doxiadis, G.G., Montefiori, D.C., Hughes, A., Burton, D.R., Allison, D.B., Wolinsky, S.M., Bontrop, R., Picker, L.J. and Watkins, D.I. (2003). *J. Virol.* 77(16), 9029–9040.

Ohtaki, S., Kodama, H., Hondo, R. and Kurata, T. (1986). *Acta. Patholog. Jpn.* 36, 1553–1563.

Ohtaki, S., Hondo, R., Kodama, H. and Kurata, T. (1988). *Acta. Patholog. Jpn.* 38, 967–978.

Osborn, K.G., Prahalada, S., Lowenstine, L.J., Gardner, M.B., Maul, D.H. and Henrickson, R.V. (1984). *Amer. J. Pathol.* 114, 94–103.

Padula, P.J., Edelstein, A., Miguel, S.D., Lopez, N.M., Rossi, C.M. and Rabinovich, R.D. (1998). *Virology* 241, 323–330.

Pal, R., Venzon, D., Letvin, N.L., Santra, S., Montefiori, D.C., Miller, N.R., Tryniszewska, E., Lewis, M.G., VanCott, T.C., Hirsch, V., Woodward, R., Gibson, A., Grace, M., Dobratz, E., Markham, P.D., Hel, Z., Nacsa, J., Klein, M., Tartaglia, J. and Franchini, G. (2002). *Virol.* 76(1), 292–302.

Pitcher, C.J., Hagen, S.I., Walker, J.M., Lum, R., Mitchell, B.L., Maino, V.C., Axthelm, M.K. and Picker, L.J. (2002). *J. Immunol.* 168(1), 29–43.

Plopper, C., St George, J., Cardoso, W., Wu, R., Pinkerton, K. and Buckpitt, A. (1992). *Chest* 101(3 Suppl), 2S–5S.

Poiesz, G.J., Ruscetti, F.W., Gazdar, A.F., Bunn, P.A., Minna, J.S. and Gallo, R.C. (1980). *Proceedings of the National Academy of Sciences USA* 77, 7415–7421.

Polack, F.P., Auwaerter, P.G., Lee, S.H., Nousari, H.C., Valsamakis, A., Leiferman, K.M., Diwan, A., Adams, R.J. and Griffin, D.E. (1999). *Nat. Med.* 5(6), 629–634.

Polycarpou, A., Ntais, C., Korber, B.T., Elrich, H.A., Winchester, R., Krogstad, P., Wolinsky, S., Rostron, T., Rowland-Jones, S.L., Ammann, A.J. and Ioannidis, J.P. (2002). *AIDS Res. Hum. Retroviruses* 18(11), 741–746.

Ponnuraj, E.M., Hayward, A.R., Raj, A., Wilson, H. and Simoes, E.A. (2001). *J. Gen. Virol.* 82(Pt 11), 2663–2674.

Pope, M. and Haase, A.T. (2003). *Nat. Med.* 9(7), 847–852.

Putz, M.M., Bouche, F.B., de Swart, R.L. and Muller, C.P. (2003). *Int. J. Parasitol.* 33(5-6), 525–545.

Rao, M., Bray, M. Alving, C.R., Jahrling, P.B. and Matyas, G.R. (2002). *Journal of Virology* 76(18), 9176–9185.

Renegar, K.B. (1992). Lab. *Anim. Sci.* 42(3), 222–232.

Rimmelzwaan, G.F., Kuiken, T., van Amerongen, G., Bestebroer, T.M., Fouchier, R.A. and Osterhaus, A.D. (2001). *J. Virol.* 75(14), 6687–6691

Riyabchikova, E.I., Kolesnikova, L.V. and Luchko, S.V. (1999). *Journal of Infectious Diseases* 179(Supplement 1), S199–S202.

Robinson, H.L. (2002). *Nat. Rev. Immunol.* 2(4), 239–250.

Roizman, B. and Sears, A.E. (1993). In Lopez, C (ed.) *The Human Herpesviruses,* pp. 11–68. Raven Press, New York.

Sabin, A.B. (1965). *J. Amer. Med. Assoc.* 194, 872–876.

Sauermann, U., Stahl-Hennig, C., Stolte, N., Muhl, T., Krawczak, M., Spring, M., Fuchs, D., Kaup, F.J., Hunsmann, G. and Sopper, S. (2000). *J. Infect. Dis.* 182(3), 716–724.

Sauermann, U. (2001). *Curr. Mol. Med.* 1(4), 515–522.

Schmaljohn, C. and Hjelle, H. (1997). *Emerging Infectious Diseases* 3, 95–104.

Schmidt, A.C., Wenzke, D.R., McAuliffe, J.M., St Claire, M., Elkins, W.R., Murphy, B.R. and Collins, P.L. (2002). *J. Virol.* 76(3), 1089–1099.

Scott, R.M., Snitbhan, R., Bancroft, W.H., Alter, H.J. and Tingpalapong, M. (1980). *J. Infect.* Dis. 142, 67–71.

Sequar, G., Britt, W.J., Lakeman, F.D., Lockridge, K.M., Tarara, R.P., Canfield, D.R., Zhou, S.S., Gardner, M.B. and Barry, P.A. (2002). *J. Virol.* 76(15), 7661–7671.

Simon, M.A., Daniel, M.D., Lee-Parrotz, D., King, N.W. and Ringler, D.J. (1993). *Lab. Animal Sci.* 43, 545–550.

Soike, K.F., Rangan, S.R. and Gerone, P.J. (1984). *Adv. Vet. Sci. Comp. Med.* 28, 151–199.

Stornello, C. (1991). *Lancet* 338, 1024–1025.

Sullivan, N.J., Sanchez, A., Rollin, P.E., Zang, Z.Y. and Nabel, G.J. (2000). *Nature* 408, 605–609.

Sullivan, N.J., Geisbert, T.W., Geisbert, J.B., Xu, L., Yang, Z., Roederer, M., Koup, R.A., Jahrling, P.B. and Nabel, G.B. (2003). *Nature* 424, 681–684.

Summers, J., Smolec, J.M. and Snyder, R. (1978). *Proc. Natl. Acad. Sci. U S A* 75, 4533–4537.

Swack, N.S., Liu, O.C. and Hsiung, G.D. (1971). *Am. J. Epidem.* 94, 397–402.

Swack, N.S. and Hsiung, G.D. (1982). *J. Med. Primatol.* 11, 169–177.

Tarantal, A.F., Salamat, S., Britt, W.J., Luciw, P.A., Hendrickx, A.G. and Barry, P.A. (1998). *J. Infect. Dis.* 177, 446–450.

Teng, M.N., Whitehead, S.S., Bermingham, A., St Claire, M., Elkins, W.R., Murphy, B.R. and Collins, P.L. (2000). *J. Virol.* 74(19), 9317–9321.

Teranishi, K., Alwayn, I.P., Buhler, L., Gollackner, B., Knosalla, C., Huck, J., Duthaler, R., Katopodis, A., Sachs, D.H., Schuurman, H.J., Awwad, M. and Cooper, D.K. (2003). *Xenotransplantation* 10(4), 357–367.

Thimme, R., Wieland, S., Steiger, C., Ghrayeb, J., Reimann, K.A., Purcell, R.H. and Chisari, F.V. (2003). *J. Virol.* 77, 68–76.

Tinghitella, T.J., Swack, N., Baumgarten, A. and Hsiung, G.D. (1982). *Infect. Immun.* 37(2), 823–825.

Trachtenberg, E., Korber, B., Sollars, C., Kepler, T.B., Hraber, P.T., Hayes, E., Funkhouser, R., Fugate, M., Theiler, J., Hsu, Y.S., Kunstman, K., Wu, S., Phair, J., Erlich, H. and Wolinsky, S. (2003). *Nat. Med.* 9(7), 928–935.

Tsujimoto, H., Seiki, M., Nakamura, H., Watanabe, T., Sakakibara, I., Sasagawa, A., Honjo, S., Hayami, M. and Yoshida, M. (1985). *Japanese Journal of Cancer Research* 76, 911–914.

Tsujimoto, H., Noda, Y., Ishikawa, K., Nakamura, H., Fukasawa, M., Sakakibara, I., Sasagawa, A., Honjo, S. and Hayami, M. (1987). *Cancer Research* 47(1), 269–274.

van den Hoogen, B.G., de Jong, J.C., Groen, J., Kuiken, T., de Groot, R., Fouchier, R.A. and Osterhaus, A.D. (2001). *Nat. Med.* 7(6), 719–724.

Van Rompay, K.K.A., Berardi, C.J., Dillard-Telm, S., Tarara, R.P., Canfield, D.R., Valverde, C.R., Montefiori, D.C., Stefano Cole, K., Montelaro, R.C., Miller, C.J. and Marthas, M.L. (1998). *Journal of Infectious Diseases* 177(5), 1247–1259.

Van Rompay, K.K., McChesney, M.B., Aguirre, N.L., Schmidt, K.A., Bischofberger, N. and Marthas, M.L. (2001). *J. Infect. Dis.* 184(4), 429–438.

Van Rompay, K.K., Greenier, J.L., Cole, K.S., Earl, P., Moss, B., Steckbeck, J.D., Pahar, B., Rourke, T., Montelaro, R.C., Canfield, D.R., Tarara, R.P., Miller, C., McChesney, M.B. and Marthas, M.L. (2003). *J. Virol.* 77(1), 179–190.

Vaudin, M., Wolstenholme, A.J., Tsiquaye, K.N., Zuckerman, A.J. and Harrison, T.J. (1988). *J. Gen. Virol.* 69 (Pt 6), 1383–1389.

Vogel, P., Weigler, B.J., Kerr, H., Hendrickx, A. and Barry, P.A. (1994). *Lab. Anim. Sci.* 44, 25–30.

Waldrop, S.L., Pitcher, C.J., Peterson, D.M., Maino, V.C. and Picker, L.J. (1997). *J. Clin. Invest.* 99(7), 1739–1750.

Waldrop, S.L., Davis, K.A., Maino, V.C. and Picker, L.J. (1998). *J. Immunol.* 161(10), 5284–5295.

Walsh, P.D., Abernethy, K.A., Bermejo, M., Beyers, R., De Wachter, P., Akou, M.E., Huljbregts, B., Mambounga, D.I., Toham, A.K., Kilbourn, A.M., Lahm, S.A., Latour, S., Maiseis, F., Mbina, C., Mihindou, Y., Oblang, S.N., Effa, E.N., Starkey, M.P., Telfer, P., Thibault, M., Tutin, C.E.G., White, L.J.T. and Wilkie, D.S. (2003). *Nature* 422, 611–614.

Warren, K.S., Heeney, J.L., Swan, R.A., Heriyanto and Verschoor, E.J. (1999). *J. Virol.* 73, 7860–7865.

Weigler, B.J. (1992). *Clin. Infec. Dis.* 14, 555–567.

Weigler, B.J., Hird, D.W., Hilliard, J.K., Lerche, N.W., Roberts, J.A. and Scott, L.M. (1993). *J. Infect. Dis.* 167, 257–263.

Weller, T.H. (1971). *N.E.J. Med.* 285, 203–214.

Yamanouchi, K., Kinoshita, K., Moriuchi, R., Katamine, S., Amagasake, T., Ikeda, S., Ichimaru, M., Miyamoto, T. and Hino, S. (1985). *Japanese Journal of Cancer Research* 76(6), 481–487.

Yanagihara, R., Amyz, H.L., Lee, P.W., Asher, D.M., Gibbs, C.J.J. and Gajdusek, D.C. (1988). *Archives of Virology* 101(1-2), 125–130.

Zaucha, G.M., Jahrling, P.B., Geisbert, T.W., Swearengen, J.R. and Hensley, L. (2001). *Lab. Invest.* 81(12), 1581–1600.

Zwartouw, H.T. and Boulter, E.A. (1984). *Lab. Anim.* 18, 65–70.

CHAPTER 35

Parasitic Diseases of Nonhuman Primates

Jeanette E. Purcell[1] and Mario T. Philipp[2]
Divisions of Veterinary Medicine[1], and Bacteriology and Parasitology[2], Tulane National Primate Research Center, Tulane University Health Sciences Center, Covington, Louisiana, USA

Introduction

A wide array of parasite species may infect nonhuman primates that are used in biomedical research. Both helminth and protozoal parasites are common, as indicated by several surveys of nonhuman primates imported for research purposes (Wong and Conrad, 1978; Karr *et al.*, 1980; Sano *et al.*, 1980; Muriuki *et al.*, 1998; Munene *et al.*, 1998). Animals maintained in outdoor facilities in permissive environments in the United States also have been shown to be infected with helminths and protozoa (Sestak *et al.*, 2003b). Parasitic infections may affect animal health and ultimately the studies for which these animals are used. Moreover, many of these parasites are zoonotic pathogens and therefore pose a risk for animal personnel.

In this chapter we first review parasites that may cause disease in immune-competent animals. Next we describe parasites that cause disease in immune deficient animals. We focus on the simian immunodeficiency virus (SIV)-macaque model of human immunodeficiency virus. SIV-infected macaques are widely used in primate research facilities. It is imperative that veterinarians who work with such animals be knowledgeable of parasites that may cause disease, when SIV-induced immunodeficiency occurs. Finally, we discuss those parasitic infections that occur commonly but are clinically benign. Our primary goal is to alert the clinical veterinarian to the potential clinical and pathologic manifestations of these parasitic diseases. More general, all-encompassing reviews of parasites that may be found in nonhuman primates are available (Toft *et al.*, 1998; Bernacky *et al.*, 2002).

Parasitic diseases of immune-competent nonhuman primates

Strongyloidiasis

Strongyloidiasis is caused by *Strongyloides cebus* in New World monkeys, and *Strongyloides fulleborni* in Old World monkeys and great apes. *Strongyloides* is most prevalent in tropical and subtropical regions of the world.

The Laboratory Primate

ISBN 0-1208-0261-9

Infection with this parasite is of particular clinical significance in animals maintained outdoors in these environments. Animals maintained indoors also may harbor *Strongyloides* if not properly screened and treated during quarantine.

Members of the genus *Strongyloides* have a unique lifecycle with two separate modes of replication. In the parasitic sexual cycle, only the adult filariform female is parasitic within the small intestine of the host. Eggs produced by the female via mitotic parthenogenesis are passed in the stool. The eggs then hatch to release rhabditiform larvae. These larvae undergo two molts to become filariform larvae, or four molts to become minute free-living males and females. These free-living parasites produce rhabditiform larvae which then molt into infective filariform larvae. Transmission to a new host most commonly occurs when the infective (filariform) larvae penetrate the skin. The larvae reach the blood and travel to the lung, whereupon they are coughed up and swallowed. Once larvae reach the intestines they develop into adult females, completing the lifecycle. In the free-living cycle, rhabditiform larvae develop through four molts to free-living male and female parasites.

Nonhuman primates with *Strongyloides* infection are most often asymptomatic. However, some animals may present with moderate to severe diarrhea (Sestak *et al.,* 2003b). Cough may occur when larvae migrate through the lung. At necropsy, multifocal erosive and ulcerative enteritis lesions, caused by parasitic females and larvae, are generally limited to the small intestine. In severe cases, lesions may be found in the colon, and penetration of the bowel may occur with resultant peritonitis.

S. fulleborni is a recognized pathogen of humans (Pampiglione and Ricciardi, 1972). This zoonotic infection (Evans *et al.,* 1991; Vince *et al.,* 1979; Hira and Patel, 1980; Crouch and Shield, 1982) occurs in Africa and Papua New Guinea. Transmission is reported to occur from mother to infant through maternal milk. Experimental human to human (Pampiglione and Ricciardi, 1972) and nonhuman primate to human infections (Sandground, 1925) have been performed. It is prudent to inform nonhuman primate handlers of this zoonotic agent. Handlers should wear appropriate protective clothing and be diligent in handwashing when working with nonhuman primates.

The most important diagnostic stages are eggs, rhabditiform larva, and infective filariform larva (Georgi and Georgi, 1990). Larvae may be identified at necropsy or during histologic examination of the small intestine. Given the short lifecycle and environmental persistence of *Strongyloides,* this parasite is notoriously difficult to eradicate. The development of *Strongyloides* larval stages is favored by a moist environment. Thus, effective eradication requires diligent daily sanitization and drying of enclosures. The difficulty of achieving this may be fully appreciated when nonhuman primates are maintained outdoors in tropical or subtropical climates. Drugs for *Strongyloides* include ivermectin (Battles *et al.,* 1988) and benzimidazoles. In our experience, treatment of *S. fulleborni* with a combination of ivermectin and fenbendazole is successful in rhesus macaques (J.E. Purcell, unpublished observations).

It is important for veterinary clinicians who care for nonhuman primates to be aware that *S. fulleborni* and other species of *Strongyloides* differ in their life cycle and signs of infection from that of *S. stercoralis* of humans. Eggs produced by *S. stercoralis* hatch almost immediately within the small intestine of the host to become rhabditiform larvae. Autoinfection of the human host occurs when the rhabditiform larvae penetrate the colonic or perianal mucosa. In humans that are immune compromised, autoinfection may lead to "hyperinfection syndrome" or disseminated strongyloidiasis, which may be fatal (Haque *et al.,* 1994; Gotuzzo *et al.,* 1999). There is currently no acceptable naturally-occuring animal model of the *Strongyloides* hyperinfection syndrome that occurs in humans. Consequently, most of the information on the lifecycle of *S. stercoralis* was obtained by experimental infections of humans with the parasite (Freedman, 1991). Interestingly, there is an increased incidence of hyperinfection syndrome in patients co-infected with *S. stercoralis* and human T-cell lymphotrophic virus-1 (Satoh *et al.,* 2003; Adedayo *et al.,* 2001). To our knowledge, hyperinfection syndrome due to *S. stercoralis* is not associated with HIV infection. Similarly, nonhuman primates co-infected with SIV and *S. fulleborni* have not been reported to experience more severe disease.

It seems that, in the human host, there is a unique relationship between *S. stercoralis* and HTLV-1 that increases the incidence of the hyperinfection syndrome. HTLV-1 proviral load directly influences the *S. stercoralis* load via suppression of the host immunity (Satoh *et al.,* 2003). It is thought that HTLV-1 proviral load may be a useful predictive marker of the development of hyperinfection syndrome in patients infected with both *S. stercoralis* and HTLV-1 (Satoh *et al.,* 2003). Features of the human disease, including hyperinfection syndrome, have been experimentally produced in immune compromised Patas monkeys (*Erythrocebus patas*) infected with *S. stercoralis* (Neva, 1986; Barrett *et al.,* 1988). Although the incidence of natural infection with simian T-cell lymphotrophic virus-1 (STLV-1) in macaque sp. has not been reported, STLV-1 is enzootic in many nonhuman primate species (Mwenda *et al.*, 1999). It may be worthwhile to study the course of disease in nonhuman primates co-infected with *S. stercoralis* and STLV-1 as a possible model to elucidate the pathogenesis of human hyperinfection syndrome.

Amoebiasis

Amoebiasis refers to infection with the protozoan *Entamoeba histolytica. E. histolytica* is the cause of amebic colitis and liver abscessation in humans (http://homepages.lshtm.ac.uk/entamoeba/; Haque *et al.*, 2003). Amebiasis is the third leading cause of death in humans from parasitic disease worldwide (http://homepages.lshtm.ac.uk/entamoeba/). *E. histolytica* is found worldwide and is endemic in tropical regions and sporadic in temperate regions. Several nonhuman primate species are hosts for *E. histolytica* (Amyx *et al.*, 1978; Loomis *et al.*, 1983; Palmieri *et al.*, 1984; Marquez-Monter, 1991; Verweij *et al.*, 2003). Thus, nonhuman primates infected with *E. histolytica* may serve as a source of infection for humans.

A substantial amount of work is currently devoted to understanding the epidemiology and pathogenesis of *E. histolytica* in humans, with the ultimate goal of developing a vaccine and improving chemotherapy (Stanley, 1997; Stauffer and Ravdin, 2003). Prophylaxis and therapy of nonhuman primate amebiasis should directly benefit from such efforts.

It was once thought that *E. histolytica* existed in both pathogenic and non-pathogenic forms. However, it is now known that the nonpathogenic form is actually the genetically distinct *Entamoeba dispar* (Diamond and Clark, 1993). It appears that *E. dispar* more commonly infects nonhuman primates than *E. histolytica* (Smith and Meerovitch, 1985; Jackson *et al.*, 1990; Tachibana *et al.*, 2000, 2001; Rivera and Kanbara, 1999).

E. histolytica is acquired by ingestion of viable cysts from contaminated water or other environmental sources. Cockroaches may serve as mechanical vectors of *E. histolytica* (Hsiu-Hua *et al.*, 2003). Motile trophozoites are released from cysts in the small intestine and may remain in the large bowel without untoward clinical manifestations (Verweij *et al.*, 2003). Both trophozoites and cysts are found in the intestinal lumen, but only trophozoites invade tissue. In humans, a number of virulence factors have been linked to the ability of *E. histolytica* to invade through the interglandular epithelium (Nok and Rivera, 2003; Stanley, 2003).

Lesions and clinical signs are due to tissue invasion by trophozoites. Nonhuman primates infected with *E. histolytica* may be either asymptomatic or have severe amebic dysentery. When trophozoites invade the colonic mucosa, microulcerations occur. Submucosal extension of ulcerations under viably-appearing surface mucosa causes the classic "flask-shaped" ulcers, which contain trophozoites. In severe cases, colonic ulceration may lead to perforation and peritonitis. A non-pathogenic ameba, *E. chattoni*, has been shown to invade the cecal mucosa of humans and macaques shortly after death of the host (Sargeaunt *et al.*, 1992; Vogel *et al.*, 1996). The presence of this parasite in histologic sections should not be confused with *E. histolytica.*

In leaf-eating monkeys, infection with *E. histolytica* may result in two clinical outcomes, amebic gastritis and hepatic abscessation. In these animals the stomach is the preferred niche for *E. histolytica*, and infection may result in gastritis (Loomis *et al.*, 1983; Palmieri *et al.*, 1984). This is probably due to the higher gastric pH found in these species. Clinicians should consider *E. histolytica* as a cause of gastritis when working with these nonhuman primate species. *E. histolytica* may not be isolated from the feces of animals with amebic gastritis. Hepatic abscessation due to *E. histolytica* also has been described in leaf-eating nonhuman primates (Amyx *et al.*, 1978; Loomis *et al.*, 1983; Palmieri *et al.*, 1984; Marquez-Monter *et al.*, 1991). This apparently occurs when *E. histolytica* gains access to the liver via the portal circulation. The initial destruction of the liver results secondarily in bacterial infection and liver abscessation. Extraintestinal amebiasis due to *E. histolytica*, observed in leaf-eating monkeys, is apparently identical to that which occurs in humans (Marquez-Monter *et al.*, 1991).

E. histolytica can be readily distinguished from other amebae by identification of cysts or trophozoites in feces or of trophozoites in histologic sections. However, differentiation of *E. histolytica* from *E. dispar* by fecal microscopy is unreliable and should not be used alone. While stool culture followed by isoenzyme analysis may be used to differentiate *E. histolytica* from *E. dispar*, this technique takes several weeks to perform (Haque *et al.*, 1998). Enzyme-linked immunosorbent assays (ELISAs), that detect antigens in stool samples, and polymerase chain reaction (PCR) may allow clinicians to differentiate *E. histolytica* from *E. dispar* without the use of fecal microscopy or isoenzyme analysis (Haque *et al.*, 1998). Based on studies using PCR to diagnose amebiasis in nonhuman primates, it appears that *E. dispar* is more prevalent than *E. histolytica*. In a survey of 107 chimpanzees, 56% were positive for *E. dispar* DNA and no samples contained *E. histolytica* DNA (Tachibana *et al.*, 2000). In another survey of 268 nonhuman primates belonging to 20 species, 53% of animals were positive for *E. histolytica/E. dispar* based on fecal microscopy. All of the *E. histolytica/E. dispar* positive animals were macaques. *E. dispar* DNA was isolated from 137 of the 141 positive samples. No *E. histolytica* DNA was isolated from the samples. These results indicate that *E. dispar* is the predominant amoeba in macaques (Tachibana *et al.,* 2001). In addition, it appears from this study that a combination of fecal microscopy and PCR is optimal for diagnosis of amoebiasis. Similarly, in a survey of *Macaca fuscata,* 43% of stool samples were positive for *E. dispar* DNA. No samples were positive for *E. histolytica* DNA (Rivera and Kanbara, 1999). *E. histolytica* DNA was isolated from 2 species of New World and 3 species of Old World nonhuman primates at a facility where a spider monkey (*Ateles belzebuth hybridus*) was diagnosed with *E. histolytica* dysentery (Verweij *et al.,* 2003). In this report, using real-time PCR, the serine-rich antigen gene (*shrep*) of *E. histolytica* was amplified and digested with the restriction endonuclease *Alu*I, providing epidemiological evidence of a common source of infection (Verweij *et al.,* 2003).

E. histolytica cysts are shed in the feces and may remain infectious for several weeks in a moist environment (Flynn, 1973). Although trophozoites may be present in stool, particularly in animals with diarrhea, they are sensitive to air and gastric acid. Therefore, trophozoites are considered to be non-infectious. Trophozoites of *E. histolytica* are easily destroyed by commonly used disinfectants in nonhuman primate facilities. However, cysts are resistant to acidification, chlorination, and desiccation and require steam to be destroyed.

Successful treatment of amoebiasis is not documented in nonhuman primates. Benzimidazole derivatives have been shown to be ineffective *in vitro* against *E. histolytica* (Katiyar *et al.,* 1994). In humans, agents used to treat amoebiasis include Iodoquinol, Paromomycin, and Metronidazole (Petri and Singh, 1999).

Giardiasis

Giardiasis in humans and nonhuman primates is caused by the flagellated protozoan *Giardia lamblia* (*intestinalis, duodenalis*) (Meyer and Jarrol, 1980; Adam, 1991; Georgi and Georgi, 1990).

Giardia has a direct lifecycle and is transmitted by the fecal-oral route (Georgi and Georgi, 1990). Contaminated water has been implicated as the source of *Giardia* infection in humans (Istre *et al.,* 1984). *Giardia* is a common parasite, inhabiting the small intestine of numerous nonhuman primate species (Hamlen and Lawrence, 1994; Ghandour *et al.,* 1995b; Kalishman *et al.,* 1996; Graczyk *et al.,* 2002; Sestak, 2003b). As with humans, nonhuman primates infected with *Giardia* may have clinical signs associated with severe diarrhea (Hamlen and Lawrence, 1994). However, parasitologic surveys of nonhuman primate colonies have revealed that animals may harbor *Giardia* asymptomatically (Hamlen and Lawrence, 1994; Kalishman *et al.,* 1996; Sestak, 2003b).

At necropsy, *Giardia* is present on the epithelial surface of the small intestine. The mucosa may appear normal or there may be villus atrophy with mild inflammation of the lamina propria (Baskin, 1996). Given the findings of asymptomatic carriers and mild pathologic lesions, it is conceivable that *Giardia* is present in nonhuman primate colonies without overt clinical consequences. To our knowledge, giardiasis is not reported to occur more commonly in SIV-infected macaques.

Clinicians, and animal technicians, should be notified that nonhuman primates infected with *Giardia* pose a zoonotic risk for humans. Personnel should follow accepted personal hygiene practices when working with nonhuman primates. Diagnosis of giardiasis may be made by finding trophozoites and/or cysts in the feces or intestine of the affected nonhuman primate. A fecal antigen detection assay and fluorescent antibody test have been developed and used to detect *G. lamblia* cysts in fecal smears (Stibbs and Ongerth, 1986; Stibbs, 1989; Johnston *et al.,* 2003).

Effective control of *Giardia* requires the destruction of cysts in the environment with phenolic compounds, heat, or dessication. Reported successful treatment of giardiasis includes metronidazole (Hamlen and Lawrence, 1994). Benzimidazoles derivatives have been shown to have high activities against *G. lamblia* (Katiyar *et al.,* 1994). In our experience, one of these derivatives, fenbendazole, is effective in treating *G. lamblia* in rhesus macaques (J. Dufour, personal communication).

Toxoplasmosis and encephalitozoonosis

Toxoplasmosis is caused by *Toxoplasma gondii,* an obligate intracellular coccidian that infects both birds and mammals. The overwhelming majority of naturally occurring infections of significance are reported in New World nonhuman primates, especially squirrel monkeys (*Saimiri sp.*) (Inoue, 1997; Dietz *et al.,* 1997; Pertz *et al.,* 1997; Juan-Salles *et al.,* 1998; Frenkel and Escajadillo, 1987). Most reports describe spontaneous infections in a few animals (Inoue 1997; Dietz *et al.,* 1997). However, there are also reports of epizootics in New World species (Cunningham *et al.,* 1992).

A detailed description of the lifecycle of *T. gondii* may be found in Georgi and Georgi (1990). Briefly, the lifecycle of this parasite involves two distinct stages, the feline and non-feline stages. In the non-feline stage, tissue cysts that contain tissue bradyzoites, or sporulated oocysts that are present in the feces of the definitive host (the feline), are ingested by an intermediate host (here the nonhuman primate). Sporozoites are released and transform within the small intestinal epithelium to form tachyzoites. Tachyzoites are able to infect and replicate in all mammalian cells except red blood cells. Tachyzoites develop into bradyzoites 7-to-10 days after systemic infection. Bradyzoites are contained within cysts in a variety of host organs but persist primarily within the central nervous system and muscle. Aerosol transmission of *T. gondii* occurred when experimentally-inoculated squirrel monkeys were socially housed (Furuta *et al.,* 2001). Horizontal transmission of *T. gondii* should therefore be considered possible.

New World monkeys are among the most vulnerable species to *T. gondii,* and infection is often rapidly fatal (Cunningham *et al.,* 1992; Dietz *et al.,* 1997). Animals may present with lethargy or depression, but more commonly are found dead. When squirrel monkeys are experimentally infected with *T. gondii,* death occurs within 7–9 days (Harper *et al.,* 1985). Pathologic findings associated with *T. gondii* are attributed to cellular damage due to both tachyzoite and bradyzoite development and rupture. Lesions consist of multifocal necrosis and are frequently reported in liver, spleen, mesenteric lymph nodes, lungs, heart, and adrenals. The liver is the most frequently affected organ, and Kupffer cells, hepatocytes, as well as necrotic lesions, frequently contain both tachyzoites and bradyzoites. In one report of acute toxoplasmosis in 5 squirrel monkeys, multifocal hepatosplenic necrosis and diffuse interstitial pneumonia were present. Tachyzoites were identified immunohistochemically in both of these organs (Inoue, 1997). In another report, 9 New World monkeys (marmosets, tamarins and one saki) died of acute toxoplasmosis. *T. gondii* was identified histologically in necrotic lesions of lung, intestine and liver. In another epizootic of captive squirrel monkeys (Cunningham *et al.,* 1992), 100% morbidity and 30% mortality was due to pulmonary edema. One monkey died of both heart and liver failure.

Lesions of the central nervous system include focal hemorrhage, infarcts, necrosis, and gliosis (Dubey *et al.,* 1985). Cysts of *T. gondii* have been observed in the brains of infected animals associated with perivascular cuffing and cellular necrosis. It has been proposed that the presence and subsequent disintegration of cysts may result in encephalitis clinically (Frenkel and Escajadillo, 1987). Toxoplasmosis, affecting the central nervous system and eye, has been described in patients infected with HIV (Grigg *et al.,* 2001; Ghosn *et al.,* 2003; Gray *et al.,* 2003; Nissapatorn *et al.,* 2003). A report of central nervous system toxoplasmosis in a SIV-infected Barbary macaque (*Macaca sylvana*), was reported (Sasseville *et al.,* 1995). Clinicians should be alerted to consider toxoplasmosis when signs of central nervous system disease are observed in SIV-infected monkeys.

Toxoplasmosis must be distinguished from encephalitozoonosis, which is due to the microsporidian *Encephalitozoon cuniculi.* The biology of *E. cuniculi* has been described extensively (Wilson, 1979, see section on microsporidiosis in this chapter). The squirrel monkey is the only nonhuman primate reported to be infected with *E. cuniculi.* While reports of encephalitozoonosis in squirrel monkeys are few and not of recent date, the rate of seroprevalence in one report was high (179/250) (Shadduck and Baskin, 1989). In another report, 22 cases of encephalitozoonosis were described in a colony of squirrel monkeys (Zeman and Baskin, 1985). Infection with *E. cuniculi* in squirrel monkeys is asymptomatic and lesions are not visible by gross examination. Therefore, it is conceivable that *E. cuniculi* infections occur in squirrel monkey colonies but are not reported. *E. cuniculi* infection in the squirrel monkey results in small (100–200 μm) cerebral granulomas that may be identified individually or contained in pseudocysts within lesions (Baskin, 1996). Granulomas of liver, adrenals and chorionic villi also have been reported (Baskin, 1996). Differentiation of *T. gondii* from *E. cuniculi* has been described extensively (Baskin, 1996).

Diagnosis of toxoplasmosis in New World monkeys is made by identification of tachyzoites or cysts containing bradyzoites associated with the characteristic lesion. Immunohistochemical analysis also may be performed. Seroconversion may prove helpful in assessing outbreaks for epidemiological purposes.

Toxoplasmosis in New World monkeys may be prevented by daily sanitization of enclosures to remove oocysts prior to sporulation. It is also recommended that any meat fed to nonhuman primates be thoroughly cooked. Finally, it is imperative to control vermin, which may contain infective tissue bradyzoites. Successful treatment of naturally occurring toxoplasmosis in nonhuman primates has not been described. In a report of squirrel monkeys experimentally infected with *T. gondii,* sulfonamides alone or in combination with pyrimethamine or trimethoprim were significantly more effective than spiramycin in treating toxoplasmosis (Harper *et al.*, 1985).

American trypanosomiasis

Trypanosoma cruzi is the etiologic agent of American Trypanosomiasis, or Chagas' disease, in humans and animals. *T. cruzi* is found only in the New World. Wild and domestic animals harboring *T. cruzi* are found with spotty distribution throughout South and Central America and select areas of southern United States (http://www.who.int/ctd/chagas). Nonhuman primates imported from these areas may harbor the parasite and serve as a source for human infection (Sullivan *et al.*, 1993; Ziccardi and Lourenco-de-Oliveira, 1997; Pung *et al.*, 1998; Ndao *et al.*, 2000; Ward *et al.*, 2001). *T. cruzi* has been identified in captive-bred baboon and macaque colonies in the southern United States (Zabalgoitia *et al.*, 2003; Schielke *et al.*, 2002; Arganaraz *et al.*, 2001; Pung *et al.*, 1998).

Trypanosomes, including *T. cruzi,* are hemoflagellates. They are elongated, spindle-shaped cells with a single nucleus lying near the middle of their length. A single flagellum arises near a small granule of extranuclear DNA, called a kinetoplast, and passes out of the anterior end of the cell. *T. cruzi* is vectorially transmitted among its mammalian hosts by hematophagous triatomine insects, often called reduviid bugs. The bug becomes infected by taking a blood meal from infected animals or humans. In the insect, *T. cruzi* multiplies and undergoes morphologic change. Transmission to the mammalian host occurs upon a subsequent blood meal by the vector. Infective forms of *T. cruzi,* present in the feces of the insect are deposited on to the abraded skin of the mammalian host (Georgi and Georgi, 1990). Blood-sucking lice, collected from the hair and skin of *T. cruzi*-infected baboons, were found to contain *T. cruzi* nuclear and kinetoplast DNA (Arganaraz *et al.*, 2001). This finding suggests that the louse may have contributed to the dissemination of *T. cruzi* in the baboon colony.

Chagasic heart disease, as a result of *T. cruzi* infection, is reported to occur in nonhuman primates. In one study, 24% of naturally-infected baboons had electrocardiographic evidence of chagasic heart disease (Zabalgoitia *et al.*, 2003). This was characterized by right atrial enlargement, and biventricular systolic and diastolic abnormalities. Rhesus macaques also are susceptible to infection, as shown experimentally, and reproduce signs of the acute (chagoma), indeterminate (myocarditis and myositis) and chronic (electrocardiographic abnormalities) disease phases (Carvalho *et al.*, 2003; Bonecini-Almeida *et al.*, 1990). Histologic evidence of chagasic myocarditis was detected in an SIV-infected rhesus macaque latently infected with *T. cruzi* (Kunz *et al.*, 2002).

Diagnosis of *T. cruzi* is made by identification of parasites in blood smears, artificial xenodiagnosis, serology, and PCR for kinetoplast DNA and nuclear DNA (Schielke *et al.*, 2002; Carvalho *et al.*, 2003). PCR analysis appears to be advantageous in the diagnosis of *T. cruzi* (Ndao *et al.*, 2000; Sheilke *et al.*, 2002). In one study, *T. cruzi* was diagnosed in squirrel monkeys by blood smear (6% positive), by ELISA 10.4%, and by PCR 26.5% (Ndao *et al.*, 2000).

To our knowledge, there are no reports of effective treatment of *T. cruzi* infection in nonhuman primates.

Parasitic diseases of immune-compromised nonhuman primates

Cryptosporidiosis

Cryptosporidiosis is caused by infection with *Cryptosporidium* spp. Cryptosporidia are tiny (4–8 µm in diameter) coccidians that cause clinically significant disease in two populations of nonhuman primates, immunologically intact infants and immunodeficient animals of any age (Sestak *et al.*, 2003a). Mammals, including nonhuman primates, become infected with *Cryptosporidium* spp. through food or water contaminated with oocysts (Fayer *et al.*, 2000). Fecal contamination of water supply has been linked to human outbreaks of cryptosporidiosis (MacKenzie *et al.*, 1995) and should be considered as a possible source in outbreaks in nonhuman primates.

Oocysts excyst in the small intestine and liberate sporozoites, which infect intestinal epithelial cells. Subsequent sexual and asexual phases of the lifecycle produce thin-walled oocysts that release sporozoites to "autoinfect" the host by re-entry of intestinal epithelial cells. Thick-walled oocysts are shed in the feces and are immediately infective to other animals or humans, and are extremely resilient in the environment.

In immune-competent animals, infection with *Cryptosporidium* spp. is either asymptomatic or results in mild self-limiting diarrhea. In a survey of common marmosets (*Callithrix jacchus*), cryptosporidia were isolated from 25% of animals <1 year of age, and 4% of animals >1 year of age. None of the animals, positive for *Cryptosporidium* spp., had diarrhea (Kalishman *et al.,* 1996). In a survey of nonhuman primates in Kenya, 40/51 vervet monkeys (*Cercopithecus aethiops*), and 19/63 olive baboons (*Papio anubis*) were positive for *Cryptosporidium* fecal oocysts. Again, none of the animals positive for *Cryptosporidium* spp. in this study had diarrhea (Muriuki *et al.,* 1997). The findings in these surveys provide evidence that nonhuman primates infected with *Cryptosporidium* spp. may be asymptomatic and provide a reservoir of infection for humans. In infant pig-tail macaques (*Macaca nemestrina*) infected with *Cryptosporidium* spp., severe watery diarrhea and subsequent dehydration often occurs (Miller *et al.,* 1990a). *Cryptosporidium* was reported to be the second most-common enteric pathogen detected in a population of infant *M. nemestrina* at a U.S. National Primate Center (Miller *et al.,* 1990a). In this outbreak, 52% of infants were positive for *Cryptosporidium* spp. oocysts. At the same Center, investigators demonstrated the highly infectious nature of cryptosporidia by producing significant enteritis and oocyst shedding when animals were inoculated with a very low number of parasites. For example, in this study either 2×10^5 or as few as 10 oocysts, administered to nonhuman primates by nasogastric tube, resulted in clinical signs that were indistinguishable from those observed in human children infected with *Cryptosporidium* spp. (Miller *et al.,* 1990b). Lesions of the small intestine, due to cryptosporidiosis, include mild to moderate blunting and fusion of villi. Organisms appear on the epithelial surface of the small intestine as round basophilic structures (Georgi and Georgi, 1990).

Cryptosporidiosis in the SIV-infected macaque illustrates the potentially disseminated nature of this parasitic infection within the host. In these animals, cryptosporidiosis is a common opportunistic infection and may be found in many sites. Replication of the parasite may occur in enteric, biliary, and/or pancreatic epithelium (Kaup *et al.,* 1998; Yanai *et al.,* 2000) and hyperplasia and fibrosis of biliary and pancreatic ducts may be observed (Baskerville *et al.,* 1991). In some cases, evidence of chronic pancreatitis has been described (Kaup *et al.,* 1994). Cryptosporidia also were found in the conjunctiva of 6 SIV-infected macaques with clinical conjunctivitis (Baskin, 1996). Disseminated cryptosporidiosis in 2 SIV-infected macaques was associated with moderate to severe bronchopneumonia (Yanai *et al.,* 2000). In a recent report, protozoa resembling *Cryptosporidium muris* were isolated from the stomachs of 15 cynomolgus macaques (*Macaca fascicularis*) that were on experimental immunosuppressive agents (Dubey *et al.,* 2002). To our knowledge, this is the first report of *C. muris*-like organisms in the stomachs of monkeys. Clinicians should be alerted to this potential opportunistic infection in immune-compromised nonhuman primates.

Because cryptosporidia lack host specificity, these organisms should be regarded as zoonotic. The zoonotic potential was exemplified in a report of inadvertent transmission of cryptosporidia from infant macaques to humans during an outbreak (Miller *et al.,* 1990a). This, together with reports of asymptomatic infections (Kalishman *et al.,* 1996; Muriuki, 1997), reinforces the need for diligence among nonhuman primate handlers in using proper personal protective equipment (PPE) and adhering to personal hygiene practices.

Diagnosis of cryptosporidiosis is made by demonstration of characteristic small oocysts (4–5 μm) in the feces (Georgi and Georgi, 1990). Cryptosporidia are smaller than other coccidians and may be mistakenly identified as yeasts. Visualization and differentiation from yeasts may be enhanced by the use of special stains with phase contrast microscopy. Furthermore, *Cryptosporidium* is acid-fast and, unlike yeasts, stains with iodine (Georgi and Georgi, 1990). Finally, several immunofluorescent antibody assays and enzyme immunoassays have become available for use in humans (Ungar, 1990; Fayer *et al.,* 2000; Johnston *et al.,* 2003) and, as mentioned before, post-mortem diagnosis may be made by identification of *Cryptosporidium* spp. associated with the brush border of enterocytes of the small intestine, using light or electron microscopy.

The ability of *Cryptosporidium* spp. to autoinfect the host, and its resilience to environmental factors, make it difficult to eradicate. Cryptosporidia may survive for months, unless exposed to extreme temperatures (below 0°C or above 65°C), dessicated, or immersed in 5% ammonia. Furthermore, oocysts are resistant to highly concentrated bleach solutions (Georgi and Georgi, 1990).

There is currently no effective treatment for cryptosporidiosis (Leav *et al.,* 2003). Therapy for nonhuman primates is supportive in nature, i.e. fluid therapy.

Microsporidiosis

Members of the phylum *Microspora* that infect humans and animals, were once thought to be the most primitive of eukaryotes. It is now known that *Microspora* are highly specialized obligate intracellular, spore-forming fungi (Roger and Silberman, 2002; Keeling and Fast, 2002). In nonhuman primates, two microsporidian species are of clinical significance, *Enterocytozoon bieneusi*, which affects immunocompromised macaques (Mansfield *et al.*, 1997; Chalifoux *et al.*, 1998, 2000; Schwartz *et al.*, 1998; Sestak *et al.*, 2003a) and *Encephalitozoon cuniculi*, which is found in New World monkeys (Zeman and Baskin, 1985). *E. cuniculi* were described earlier with *Toxoplasma gondii*.

Enterocytozoonosis

Enterocytozoonosis in humans and animals is caused by *Enterocytozoon bieneusi*. Little is known of the reservoir, mode of transmission, or basic biology of this organism. *E. bieneusi* is a common microsporidian parasite of humans infected with HIV (Beaugerie *et al.*, 1992; French *et al.*, 1995; Wanachiwanawin *et al.*, 2002). Similarly, *E. bieneusi* causes significant disease in rhesus macaques infected with SIV (Mansfield *et al.*, 1997; Chalifoux *et al.*, 1998, 2000; Schwartz *et al.*, 1998; Sestak *et al.*, 2003a). While *E. bieneusi* may be isolated from immunologically-intact rhesus macaques (Mansfield *et al.*, 1998; Sestak *et al.*, 2003a), the rate of infection appears to be higher in SIV-infected animals. PCR performed on DNA isolated from feces, detected *E. bieneusi* infection in 22 (16.7%) of 131 SIV-negative and 19 (33.8%) of 53 SIV-positive rhesus macaques (Mansfield *et al.*, 1998). In another study, 12 animals were assessed prior to inoculation with SIV. All animals had low (n = 5) to undetectable (n = 7) quantities of *E. bieneusi* in the feces. Following SIV inoculation, the number of animals shedding *E. bieneusi* increased (n = 10) and the quantity of *E. bieneusi* in the feces also increased. The presence of the organism in the feces correlated with viral load and decreasing $CD4^+$ T-lymphocyte counts (Sestak *et al.*, 2003a). It appears from this study that immunosuppression facilitiates the shedding of *E. bieneusi* in feces.

In SIV-infected macaques, *E. bieneusi* causes enteropathy and cholangiohepatitis (Mansfield *et al.*, 1997). *E. bieneusi* infection of the hepatobiliary system and small intestine was identified retrospectively in 18 SIV-infected macaques using histochemical techniques, *in situ* hybridization, PCR, and ultrastructural examination. Organisms were identified in the bile ducts and gall bladder and were associated with nonsuppurative and proliferative cholecystitis (Mansfield *et al.*, 1997). Hepatic involvement was characterized by portal fibrosis and nodular hepatocellular regeneration and hyperplasia. Sequencing of a segment of the small subunit ribosomal RNA revealed, respectively, 97 and 100% identity to two clones of this gene that were derived from *E. bieneusi* of human origin (Mansfield *et al.*, 1997). In another retrospective study, *E. bieneusi* was identified as the cause of proliferative serositis of previously-unknown etiology in SIV-infected rhesus macaques (Chalifoux *et al.*, 2000).

Transmission of duodenal *E. bieneusi* spores from an AIDS patient to SIV-infected rhesus macaques, resulted in shedding of spores within one week of inoculation (Tzipori *et al.*, 1997). This, along with extensive morphologic and genetic similarities between simian and human isolates of *E. bieneusi* (Mansfield *et al.*, 1997), provide evidence that experimentally infected macaques may serve as a useful model of enterocytozoonosis in human AIDS. Personnel working with nonhuman primates should be aware of the zoonotic potential of *E. bieneusi*-infected nonhuman primates.

Diagnosis of microsporidiosis in nonhuman primates is commonly performed by calcoflour staining of fecal samples (García, 2002; Sestak *et al.*, 2003a). PCR for *E. bieneusi* may be performed on fecal samples or percutaneously collected bile (Mansfield *et al.*, 1998; Sestak *et al.*, 2003a) and immunohistochemical techniques may be used on intestinal specimens (Mansfield *et al.*, 1998).

To our knowledge, there is no effective treatment for enterocytozoonosis in nonhuman primates.

Trichomoniasis

Trichomonas spp. are mucosoflagellates that are isolated frequently from normal healthy nonhuman primates and are generally of no clinical significance. However, they may be pathogenic in immunodeficient hosts. *Trichomonas* spp. inhabit the crypts of the large intestine and are rarely associated with clinical signs or pathology. Even when trichomonas invade the colonic mucosa, they are rarely associated with pathology. In the SIV-infected macaque, *Trichomonas* spp. are found outside their normal ecological niche of the crypts of the large intestine, in the stomach. When organisms are present in low numbers, no gross lesions are evident (Blanchard and Baskin, 1988). However, as the infection burden increases, severe gastritis occurs. The gastric mucosa becomes erythematous and thickened, with a granular, corrugated appearance (Blanchard and Baskin, 1988). Histologically, protozoa, along with sloughed cells and inflammatory cells, are observed in gastric glands. Inflammatory cells are found in the lamina propria. In some cases, *Trichomonas* organisms may penetrate to the submucosa to induce multifocal pyogranulomatous inflammation. The morphology of *Trichomonas* spp. identified in SIV-infected macaques has been described (Blanchard and Baskin, 1988).

Diagnosis of gastritis due to *Trichomonas* infection may be made antemortem, by obtaining gastric biopsies, or postmortem with tissue sections. *Trichomonas* organisms do not have a characteristic appearance histologically. However, Giemsa-stained smears of gastric mucosa aid in differentiation of *Trichomonas* from the unflagellated amebic protozoa and from flagellated protozoa with characteristic cyst forms such as *Giardia* and *Chilomastix*. Special techniques are required for the speciation of *Trichomonas* spp. on the basis of morphologic characteristics. As for treatment, metronidazole was used to successfully treat gastrointestinal trichomoniasis in a colony of squirrel monkeys (Brady *et al.*, 1988).

Commonly occurring benign parasitic infections of nonhuman primates

Balantidiasis

Balantidiasis is caused by the ciliated ameba *Balantidium coli*. *B. coli* is thought to be part of the normal flora of the pig and rat (Georgi and Georgi, 1990). In general, *B. coli* is not a frequent cause of diarrhea in humans or nonhuman primates. However, epidemics of colonic balantidiasis in tropical countries have been reported, and *B. coli* may be a primary pathogen causing typhlitis in great apes (Lee *et al.*, 1990).

B. coli is considered a nonpathogenic inhabitant of the cecum of immune-competent monkeys and is frequently isolated during routine fecal examinations of nonhuman primates with or without enterocolitis (Sestak *et al.*, 2003b). Diarrhea and severe enterocolitis occurred when hydrocortisone-treated monkeys were experimentally inoculated with *B. coli* cysts of human origin (Yang *et al.*, 1995). Interestingly, *B. coli* has been found in association with colitis lesions in apparently immune-competent *Macaca nemestrina* (R. Veazey, personal communication). The possibility that *B. coli* may be a primary pathogen causing colitis in *Macaca nemestrina* is currently being investigated at the Tulane National Primate Research Center.

Experimental infection of rhesus macaques with SIV has not been reported to cause an increase in clinical disease due to *B. coli*. Intestinal lesions were studied in 32 rhesus macaques infected with SIV. *B. coli* was identified to be the most common opportunistic infection in these animals (18/32 animals) and organisms were usually identified in the intestinal lumen and occasionally within mucosal folds. Typhlocolitis was seen in only one animal. *B. coli* was not correlated with diarrhea or intestinal findings in this study (Kuhn *et al.*, 1997). To our knowledge, there are no other reports of *B. coli* infection in SIV-infected macaques. The relatively nonpathogenic nature of *B. coli* is also illustrated in HIV-infected humans. Only recently were there separate case reports that identified *B. coli* as the etiologic agent of colitis in HIV-infected humans. The first report was on an HIV-infected patient with concurrent disseminated histoplasmosis (Clyti *et al.*, 1998). In the second case, *B. coli* was isolated from an HIV-infected patient that had intermittent diarrhea and later died (Cermeno *et al.*, 2003). However, *B. coli* was not confirmed to be the cause of diarrhea or death in this patient. Based on this information, *B. coli* does not appear to be an opportunistic pathogen for HIV-infected humans or SIV-infected monkeys.

Given the infrequent reports of infection and the lack of significant physiopathologic findings when isolated from nonhuman primates *B. coli* should be considered non-pathogenic in most nonhuman primates species.

Although *B. coli* may be identified as cysts, on fecal flotation, or trophozoites, in direct fecal smear examination, isolation may be incidental in most nonhuman primates and does not incriminate *B. coli* as the etiologic agent of diarrhea.

Treatment of balantidiasis in nonhuman primates is not well-documented, but metronidizole has been effective in treating infected macaques (J. E. Purcell, unpublished observations). In humans, balantidiasis is treated with tetracycline (Steiner *et al.*, 1997).

Trichuriasis

Trichuriasis is caused by the nematode *Trichuris trichiura*. Trichurid parasites, commonly known as "whipworms" are found worldwide, but at a higher frequency in tropical and subtropical environments.

Trichuris trichiura is a common inhabitant of the cecum of New World and Old World nonhuman primates (Orihel and Seibold, 1972). Like *B. coli, Trichuris* is frequently identified during routine fecal examination but is rarely of clinical significance. *T. trichiura* of nonhuman primates has previously been described as morphologically indistinguishable from the human parasite (Flynn, 1973). Subtle morphologic differences between *T. trichiura* of monkeys and humans have been identified by scanning electron microscopy (Ooi *et al.*, 1993), but these subtle differences were not deemed sufficient to create a new species for *T. trichiura* of nonhuman primates. This, and the fact that *Trichuris* ova have been experimentally transmitted from nonhuman primates to humans (Horii and Usui, 1985), provide evidence for the zoonotic potential of *T. trichiura* of nonhuman primates.

Trichuriasis is a well-recognized cause of intestinal intussusception in human infants (Stephenson *et al.*, 2000). A report of intestinal intussusception in two juvenile baboons infected with *Trichuris* demonstrates the pathogenic potential of this parasite in nonhuman primates (Hennessey *et al.*, 1994). SIV-infected macaques are not commonly reported to experience more severe disease from *Trichuris*. However, in one study, 25% of SIV-infected macaques were shown to be infected with *Trichuris* at postmortem examination (Kuhn *et al.*, 1997). The incidence of diarrhea in the infected animals was 50%. Transmission electron microscopy showed the anterior ends of the parasites invading the cecal mucosa and associated with mild focal inflammation.

Diagnosis of trichuriasis in nonhuman primates is made by identification of either characteristic bipolar-plugged eggs in fecal samples, or adults in the cecum.

Trichuris eggs are highly resistant and if treatment of nonhuman primates is deemed to be necessary, animals should be removed from contaminated enclosures and caging should be sanitized regularly (Flynn, 1973).

Butamisole, mebendazole, flubendazole (Kumar *et al.*, 1978) and levamisole have been determined to be successful in treating nonhuman primates with trichuriasis.

Cestodiasis

Most recently a number of reports of cestodiasis in nonhuman primates refer to infection with larval-stage cestodes of the Families Taeniidae (genus *Echinococcus*) and Mesocestoididae (genus *Mesocestoides*).

Hydatidosis (echinococcosis)

Larval echinococcosis may be present in nonhuman primates that once lived in, or were imported from, regions where these parasites are endemic. Hydatidosis or echinococcosis is the result of infection with the larval stage of cestodes of the genus *Echinococcus*. Hydatid cysts of *Echinococcus granulosus* and alveolar hydatids of *E. multilocularis* represent the two types of hydatids. *E. granulosus* is found most prevalently in Australia, Argentina, Chile, Africa, E. Europe, the Middle East, New Zealand, and the Mediterranean countries, while *E. multilocularis* is found in Canada, the United States, Central and Northern Europe, and Asia. The prevalence rates of *E. multilocularis* in foxes, in some areas of Europe, are very high, with average rates of more than 40% (Eckert, 1997). From the years 1987 to 2000, investigators in Zurich diagnosed alveolar echinococcosis in 15 captive primates (Deplazes and Eckert, 2001).

Nonhuman primates become infected with *E. granulosus* by ingesting the eggs from the feces of the canine definitive host. Embryos are liberated in the gastrointestinal tract, enter the portal circulation and travel to many organs, but mainly targeting liver and lungs where they develop into fluid-filled, unilocular hydatid cysts. Hydatid cysts contain a thick inner membrane from which develop brood capsules that, in turn, produce multiple scolices. When brood capsules rupture, scolices are released together with a sediment, often referred to as "hydatid sand". These cysts expand slowly within the intermediate host. The nonhuman primate is somewhat of an "aberrant" intermediate host and cysts develop in some cases for many years and become quite large.

Infection with *E. multilocularis* occurs following ingestion of eggs from the feces of the canine or feline definitive hosts or by eating the rodent intermediate host. Larvae of *E. multilocularis* remain in a proliferative stage and the cyst is always multilocular. Clinical signs and pathology associated with hydatid cysts of *E. granulosus* are dependent upon the type of infected tissue, as may be inferred from the following examples. Hydatidosis was reported in free-ranging baboons in Saudi Arabia (Ghandour *et al.*, 1995a). A baboon with hydatidosis presented with tachypnea and an elevated white blood-cell count. Radiographs showed three radiopaque masses in lung fields and hepatosplenomegally. At necropsy, intact hydatid cysts were identified in the lungs and liver and a ruptured hydatid cyst was present, with adhesions to the thoracic wall (García *et al.*, 2002). In another report, *E. granulosus* was diagnosed in a colony of pig-tailed macaques (*Macaca nemestrina*) (Plesker *et al.*, 2001). In this report, cysts were identified at necropsy in one animal and ultrasonographically in 4 serologically positive animals. Clinical signs included anorexia and icterus. Other findings of hydatidosis in baboons have been reported (Boever and Britt, 1975; Goldberg *et al.*, 1991; Markovics *et al.*, 1992). As with humans, rupture of hydatid cysts may cause fever, pruritis and anaphylaxis.

Alveolar hydatidosis usually presents as a slow growing infiltrative malignant tumor of the liver with possible metastatic lesions in lung and brain. The first report of alveolar echinococcosis in an Old World monkey was described in an 18-month-old cynomolgus macaque (*Macaca fascicularis*) that was housed outdoors in an *E. multilocularis* endemic area. This animal had typical cystic structures in a massively enlarged liver. Alveolar hydatidosis, due to *E. multilocularis*, was recently described in an 11-year-old rhesus monkey (*Macaca mulatta*) (Brack *et al.*, 1997). In addition, disseminated alveolar echinococcosis has been described in a 22-year-old gorilla (*Gorilla gorilla*), and a lemur (*Lemur catta*) (Kondo *et al.*, 1996).

Larval cestodiasis in nonhuman primates is generally not of zoonotic concern. Humans become infected with *Echinococcus* by the same modes as nonhuman primates. Pathologists that perform postmortem examinations of nonhuman primates should therefore wear proper personal protective equipment to prevent inadvertent ingestion of larval stages of cestodes.

Immunoblot and ELISA are currently available for detection of hydatidosis in humans. These tests were used to diagnose hydatidosis in a baboon (García *et al.*, 2002a) and may prove useful for diagnosis of hydatidosis in other nonhuman primates. An *E. multilocularis*-specific ELISA may help to differentiate *E. multilocularis* from *E. granulosis* (Brack *et al.*, 1997) and PCR has been used successfully to confirm diagnoses of alveolar hydatids in a gorilla and a lemur (Kondo *et al.*, 1996).

A potential for nonhuman primates to be infected with the larval stages of *Taenia, Spirometra, Mesocestoides*, and *Multicepts* has been reported (Maravilla *et al.*, 1998). Adult cestodes of the families Anaplocephalidae, Davaineidae, Hymenolepiididae, and Dilepididae also may be found in the intestine (Galan-Puchades *et al.*, 2000). Infection with the latter may result in asymptomatic infection, failure to gain or maintain body weight and generalized poor condition of the animal.

There is currently no effective therapeutic treatment for larval cestodiasis. Surgical resection is generally impractical and not recommended due to the potentially invasive nature of the cysts and high probability of cyst rupture with dissemination of cyst contents. The drug of choice for treatment of the adult stages is praziquantel (Tanowitz *et al.*, 1993).

Pulmonary acariasis

Pulmonary acariasis is caused by the lung mite, *Pneumonyssus simicola,* and to a lesser extent by *Pneumonyssoides* sp. Pulmonary acariasis was described as tick invasion of the lungs, at least as early as 1929, and was reported to be found in 100% of animals in some macaque colonies. Today, *P. siminicola* is still identified in macaques, and to a lesser extent in baboons, particularly those imported from endemic areas. The apparent decrease in the incidence and severity of *P. siminicola* infections may be due to the widespread use of medications, such as ivermectin, during quarantine and as part of routine colony health programs. The most recent published report of pulmonary acariasis in a nonhuman primates was in 1987 (Kim and Cole, 1987). It can be concluded that *P. siminicola* does not generally result in clinically significant disease. Furthermore, veterinary pathologists recognize the characteristic pulmonary lesion due to this parasite and no longer report on this. The parasite is included in this chapter because the lesions of pulmonary acariasis may resemble tuberculosis radiographically and grossly. It is important to include pulmonary acariasis on the list of differential diagnoses when characteristic radiographic lesions are identified.

Most nonhuman primate species infected with the etiologic agent of pulmonary acariasis are asymptomatic. Infection with *P. siminicola,* however, predisposes monkeys to other pulmonary diseases, due to bronchiolar epithelial changes (Kim and Kalter, 1975). Grossly, lesions are bulla that are 1–10 mm, yellow to grey, and may contain one or more mites, often referred to as "mite houses". Lesions are found throughout the pulmonary parenchyma but tend to localize immediately under the pulmonary pleura. Fibrous adhesions may be found between pulmonary and parietal pleura. Microscopically, chronic bronchitis and bronchiectiasis with eosinophilic granulomatous inflammation is observed (Baskin, 1996). Cross-sections of lesions containing mites often reveal, in surrounding macrophages, a golden brown to black refractile pigment thought to be a product of the mite.

Diagnosis is readily made by microscopic examination. It may be possible to diagnose pulmonary acariasis antemortem by tracheobronchial lavage. Treatment of pulmonary acariasis is usually performed with ivermectin.

Concluding remarks

Nonhuman primates, that are used in biomedical research, are exposed to parasitic infections that may compromise their well-being, distort experimental results, and endanger the professional in touch with the animals. It is our hope that the succinct review we have made of both disease-causing and benign, but potentially confounding, parasites may aid the professional in differentially diagnosing, curing, or altogether avoiding parasitic diseases in the animals in their care.

Acknowledgments

This work was supported by grant RR00164 from the National Center for Research Resources, National Institutes of Health. Secretarial help from Avery Labat is gratefully acknowledged.

Correspondence

Any correspondence should be directed to Jeanette Purcell, Biologic Research Laboratory, Chicago, Illinois, USA.

References

Adam, R.D. (1991). *Microbiol. Rev.* 55, 706–732.

Adedayo, A.O., Grell, G.A. and Bellot, P., (2001). *Am. J. Trop. Med. Hyg.* 65, 650–651.

Amyx, H.L., Asher, D.M., Nash, T.E., Gibbs, C.J., Jr. and Gajdusek, D.C. (1978). *Am. J. Trop. Med. Hyg.* 27, 888–891.

Arganaraz, E.R., Hubbard, G.B., Ramos, L.A., Ford, A.L., Nitz, N., Leland, M.M., Vandeberg, J.L. and Teixeira, A.R. (2001). *Rev. Inst. Med. Trop. Sao Paulo.* 43, 271–276.

Barrett, K.E., Neva, F.A., Gam, A.A., Cicmanec, J., London, W.T., Phillips, J.M. and Metcalfe, D.D. (1988). *Am. J. Trop. Med. Hyg.* 38, 574–581.

Baskerville, A., Ramsay, A.D., Millward-Sadler, G.H., Cook, R.W., Cranage, M.P. and Greenaway, P.J. (1991). *J. Com. Pathol.* 105, 415–421.

Baskin, G.B. (1996). *J. Parasitol.* 82, 630–632.

Battles, A.H., Greiner, E.C. and Collins, B.R. (1988). *Lab. Anim. Sci.* 38, 474–476.

Beaugerie, L., Teilhac, M.F., Deluol, A.M., Fritsch, J., Girard, P.M., Rozenbaum, W., Le Quintrec, Y. and Chatelet, F.P. (1992). *Ann. Intern. Med.* 117, 401–402.

Bernacky, B.J., Gibson, S.V., Keeling, M.E. and Abee, C.R. (2001). In Fox, J.G., Anderson, L.C., Loew, F.M. and Quimby, F.W. (eds) *Laboratory Animal Medicine.* Second Edition, pp 757–772. San Diego: Academic Press.

Blanchard, J.L. and Baskin, G.B. (1988). *J. Infect. Dis.* 157, 1092–1093.

Boever, W.J. and Britt, J. (1975). *J. Am. Vet. Med. Assoc.* 167, 619–621.

Bonecini-Almeida Mda, G., Galvao-Castro, B., Pessoa, M. H., Pirmez, C. and Laranja, F. (1990). *Mem. Inst. Oswaldo Cruz.* 85, 163–171.

Brack, M., Tackmann, K., Conraths, F.J. and Rensing, S. (1997). *Trop. Med. Int. Health.* 2, 754–759.

Brady, A.G., Pindak, F.F., and Abee, C.R. (1988). *Am. J. Primatol.* 14, 65–71.

Carvalho, C.M., Andrade, M.C., Xavier, S.S., Mangia, R.H., Britto, C.C., Jansen, A.M., Fernandes, O., Lannes-Vieira, J. and Bonecini-Almeida, M.G. (2003). *Am. J. Trop. Med. Hyg.* 68, 683–691.

Cermeno, J.R., Hernandez De Cuesta, I., Uzcategui, O., Paez, J., Rivera, M. and Baliachi, N. (2003). *Aids* 17, 941–942.

Chalifoux, L.V., MacKey, J., Carville, A., Shvetz, D., Lin, K.C., Lackner, A. and Mansfield, K.G. (1998). *Vet. Pathol.* 35, 292–296.

Chalifoux, L.V., Carville, A., Pauley, D., Thompson, B., Lackner, A.A. and Mansfield, K.G. (2000). *Arch. Pathol. Lab. Med.* 124, 1480–1484.

Clyti, E., Aznar, C., Couppie, P., el Guedj, M., Carme, B. and Pradinaud, R. (1998). *Bull. Soc. Pathol. Exot.* 91, 309–311.

Crouch, P.R. and Shield, J.M. (1982). *PNG Med. J.* 25, 164–165.

Cunningham, A.A., Buxton, D. and Thomson, K.M. (1992). *J. Comp. Pathol.* 107, 207–219.

Deplazes, P. and Eckert, J. (2001). *Vet. Parasitol.* 98, 65–87.

Diamond, L.S. and Clark, C.G. (1993). *J. Eukaryot. Microbiol.* 40, 340–344.

Dietz, H.H., Henriksen, P., Bille-Hansen, V. and Henriksen, S.A. (1997). *Vet. Parasitol.* 68, 299–304.

Dubey, J.P., Kramer, L.W. and Weisbrode, S.E. (1985). *J. Am. Vet. Med. Assoc.* 187, 1272–1273.

Dubey, J.P., Markovits, J.E. and Killary, K.A. (2002). *Vet. Pathol.* 39, 363–371.

Eckert, J. (1997). *Parasitologia.* 39, 337–344.

Evans, A.C., Markus, M.B., Joubert, J.J. and Gunders, A.E. (1991). *S. Afr. Med. J.* 80, 410–411.

Fayer, R., Morgan, U. and Upton, S.J. (2000). *Int. J. Parasitol.* 30, 1305–1312.

Flynn, R. (1973). *Parasites of Laboratory Animals,* Iowa State University Press, Ames.

Freedman, D.O. (1991). *Rev. Infect. Dis.* 13, 1221–1226.

French, A.L., Beaudet, L.M., Benator, D.A., Levy, C.S., Kass, M. and Orenstein, J.M. (1995). *Clin. Infect. Dis.* 21, 852–858.

Frenkel, J.K. and Escajadillo, A. (1987). *Am. J. Trop. Med. Hyg.* 36, 517–522.

Furuta, T., Une, Y., Omura, M., Matsutani, N., Nomura, Y., Kikuchi, T., Hattori, S. and Yoshikawa, Y. (2001). *Exp. Anim.* 50, 299–306.

Galan-Puchades, M.T., Fuentes, M.V. and Mas-Coma, S. (2000). *Folia Parasitol. (Praha).* 47, 23–28.

Garcia, K.D., Hewett, T.A., Bunte, R. and Fortman, J.D. (2002). *Contemp. Top. Lab. Anim. Sci.* 41, 61–64.

Garcia, L.S. (2002). *J. Clin. Microbiol* 40, 1892–1901.

Georgi, J.G. and Georgi, M.E. (1990). In *Parasitology for Veterinarians.* Fifth edition. Dyson, Kilmer, Dick (eds). W.B. Saunders.

Ghandour, A.M., Zahid, N.Z., Banaja, A.A., Kamal, K.B. and Boug, A.I. (1995a). *Ann. Trop. Med. Parasitol.* 89, 313–316.

Ghandour, A.M., Zahid, N.Z., Banaja, A.A., Kamal, K.B. and Bouq, A.I. (1995b). *J. Trop. Med. Hyg.* 98, 431–439.

Ghosn, J., Paris, L., Ajzenberg, D., Carcelain, G., Darde, M.L., Tubiana, R., Boss., P., Bricaire, F. and Katlama, C. (2003). *Clin. Infect. Dis.* 37, e112–e114.

Goldberg, G.P., Fortman, J.D., Beluhan, F.Z. and Bennett, B.T. (1991). *Lab. Anim. Sci.* 41, 177–180.

Gotuzzo, E., Terashima, A., Alvarez, H., Tello, R., Infante, R., Watts, D.M. and Freedman, D.O. (1999). *Am. J. Trop. Med. Hyg.* 60, 146–149.

Graczyk, T.K., Bosco-Nizeyi, J., Ssebide, B., Thompson, R.C., Read, C. and Cranfield, M.R. (2002). *J. Parasitol.* 88, 905–909.

Gray, F., Chretien, F., Vallat-Decouvelaere, A.V. and Scaravilli, F. (2003). *J. Neuropathol. Exp. Neurol.* 62, 429–440.

Grigg, M.E., Ganatra, J., Boothroyd, J.C. and Margolis, T.P. (2001). *J. Infect. Dis.* 184, 633–639.

Hamlen, H.J. and Lawrence, J.M. (1994). *Lab. Anim. Sci.* 44, 235–239.

Haque, A.K., Schnadig, V., Rubin, S.A. and Smith, J.H. (1994). *Mod. Pathol.* 7, 276–288.

Haque, R., Ali, I.K., Akther, S. and Petri, W.A., Jr. (1998). *J. Clin. Microbiol.* 36, 449–452.

Haque, R., Huston, C.D., Hughes, M., Houpt, E. and Petri, W.A. Jr., (2003). *N. Engl. J. Med.* 17, 1565–1573.

Harper, J.S., 3rd, London, W.T. and Sever, J.L. (1985). *Am. J. Trop Med. Hyg.* 34, 50–57.

Hennessey, A., Phippard, A.F., Harewood, W.J., Horam, C.J. and Horvath, J.S. (1994). *Lab. Anim.* 28, 270–273.

Hira, P.R. and Patel, B.G. (1980). *Trop. Geogr. Med.* 32, 23–29.

Horii, Y. and Usui, M. (1985). *Trans. R. Soc. Trop. Med. Hyg.* 79, 423.

Hsiu-Hua, Pai, Ko, Y.C., Chen E.R. (2003). *Acta Tropica.* 87, 355–359.

http://www.who.int/ctd/chagas

http://homepages.lshtm.ac.uk/entamoeba

Inoue, M. (1997). *J. Vet. Med. Sci.* 59, 593–595.

Istre, G.R., Dunlop, T.S., Gaspard, G.B. and Hopkins, R.S. (1984). *Am. J. Public. Health* 74, 602–604.

Jackson, T.F., Sargeaunt, P.G., Visser, P.S., Gathiram, V., Suparsad, S. and Anderson, C.B. (1990). *Arch. Invest. Med. (Mex).* 21 Suppl 1, 153–156.

Johnston, S.P., Ballard, M.M., Beach, M.J., Causer, L. and Wilkins, P.P. (2003). *J. Clin. Microbiol.* 41, 623–626.

Juan-Salles, C., Prats, N., Marco, A.J., Ramos-Vara, J.A., Borras, D. and Fernandez, J. (1998). *J. Zoo. Wildl. Med.* 29, 55–60.

Kalishman, J., Paul-Murphy, J., Scheffler, J. and Thomson, J.A. (1996). *Lab. Anim. Sci.* 46, 116–119.

Karr, S.L., Jr., Henrickson, R.V. and Else, J.G. (1980). *J. Med. Primatol.* 9, 200–204.

Katiyar, S.K., Gordon, V.R., McLaughlin, G.L. and Edlind, T.D. (1994). *Antimicrob. Agents Chemother.* 38, 2086–2090.

Kaup, F.J., Kuhn, E.M., Makoschey, B. and Hunsmann, G. (1994). *J. Med. Primatol.* 23, 304–308.

Kaup, F., Matz-Rensing, K., Kuhn, E., Hunerbein, P., Stahl-Hennig, C. and Hunsmann, G. (1998). *Pathobiology* 66, 159–164.

Keeling, P.J. and Fast, N.M. (2002). *Annu. Rev. Microbiol.* 56, 93–116.

Kim, J.C. and Cole, R. (1987). *Am. J. Vet. Res.* 48, 511–514.

Kim, J.C. and Kalter, S.S. (1975). *J. Med. Primatol.* 4, 70–82.

Kondo, H., Wada, Y., Bando, G., Kosuge, M., Yagi, K. and Oku, Y. (1996). *J. Vet. Med. Sci.* 58, 447–449.

Kuhn, E.M., Matz-Rensing, K., Stahl-Hennig, C., Makoschey, B., Hunsmann, G. and Kaup, F.J. (1997). *Zentralbl Veterinarmed B.* 44, 501–512.

Kumar, V., Ceulemans, F. and De Meurichy, W. (1978). *Acta Zool. Pathol. Antverp.* 3–9.

Kunz, E., Matz-Rensing, K., Stolte, N., Hamilton, P.B. and Kaup, F.J. (2002). *Vet. Pathol.* 39, 721–725.

Leav, B.A., Mackay, M. and Ward, H.D. (2003). *Clin. Infect. Dis.* 36, 903–908.

Lee, R.V., Prowten, A.W., Anthone, S., Satchidanand, S.K., Fisher, J.E. and Anthone, R. (1990). *Rev. Infect. Dis.* 12, 1052–1059.

Loomis, M.R., Britt, J.O., Jr., Gendron, A.P., Holshuh, H.J. and Howard, E.B. (1983). *J. Am. Vet. Med. Assoc.* 183, 1188–1191.

MacKenzie, W.R., Kazmierczak, J.J. and Davis, J.P., (1995). *Epidemiol. Infect.* 115, 545–553.

Mansfield, K.G., Carville, A., Shvetz, D., MacKey, J., Tzipori, S. and Lackner, A.A. (1997). *Am. J. Pathol.* 150, 1395–1405.

Mansfield, K.G., Carville, A., Hebert., D., Chalifoux, L., Shvetz, D., Lin, K.C., Tzipori, S., and Lackner, A.A. (1998). *J. Clin. Microbiol.* 36, 2336–2338.

Maravilla, P., Avila, G., Cabrera, V., Aguilar, L. and Flisser, A. (1998). *J. Parasitol.* 84, 882–886.

Markovics, A., Pipano, E., Harmelin, A., Orgad, U., Yakobson, B. and Nyska, A. (1992). *J. Comp. Pathol.* 106, 435–438.

Marquez-Monter, H., Fuentes-Orozco, R., Correa-Lemus, I. and Becker, I. (1991). *Arch Invest. Med. (Mex).* 22, 75–78.

Meyer, E.A. and Jarroll, E.L. (1980). *Am. J. Epidemiol.* 111, 1–12.

Miller, R.A., Bronsdon, M.A., Kuller, L. and Morton, W.R. (1990a). *Lab. Anim. Sci.* 40, 42–46.

Miller, R.A., Bronsdon, M.A. and Morton, W.R. (1990b). *J. Infect. Dis.* 161, 312–315.

Munene, E., Otsyula, M., Mbaabu, D.A., Mutahi, W.T., Muriuki, S.M. and Muchemi, G.M. (1998). *Vet. Parasitol.* 78, 195–201.

Muriuki, S.M., Farah, I.O., Kagwiria, R.M., Chai, D.C., Njamunge, G., Suleman, M. and Olobo, J.O. (1997). *Vet. Parasitol.* 72, 141–147.

Muriuki, S.M., Murugu, R.K., Munene, E., Karere, G.M. and Chai, D.C. (1998). *Acta Trop.* 71, 73–82.

Mwenda, J.M., Sichangi, M.W., Isahakia, M., van Rensburg, E.J. and Langat, D.K. (1999). *Ann. Trop. Med. Parasitol.* 93, 289–297.

Ndao, M., Kelly, N., Normandin, D., Maclean, J.D., Whiteman, A., Kokoskin, E., Arevalo, I. and Ward, B.J. (2000). *Comp. Med.* 50, 658–665.

Neva, F.A. (1986). *J. Infect. Dis.* 153, 397–406.

Nissapatorn, V., Lee, C.K. and Khairul, A.A. (2003). *Singapore Med. J.* 44, 194–196.

Nok, A. and Rivera, W. (2003). *Parasitol. Res.* 89, 302–307.

Ooi, H.K., Tenora, F., Itoh, K. and Kamiya, M. (1993). *J. Vet. Med. Sci.* 55, 363–366.

Orihel, T.C. and Seibold, H.R. (1972). In Fiennes, R.N.T.W. (ed.) *Pathology of Simian Primates,* Part 2, pp 76–103. Karger, Basel.

Pai, H.H., Ko, Y.C. and Chen, E.R. (2003). *Acta Trop.* 87, 355–359.

Palmieri, J.R., Dalgard, D.W. and Connor, D.H. (1984). *J. Am. Vet. Med. Assoc.* 185, 1374–1375.

Pampiglione, S. and Ricciardi, M.L. (1972). *Lancet* 1, 663–665.

Pertz, C., Dubielzig, R.R. and Lindsay, D.S. (1997). *J. Zoo. Wildl. Med.* 28, 491–493.

Petri, W.A., Jr. and Singh, U. (1999). *Clin. Infect. Dis.* 29, 1117–1125.

Plesker, R., Bauer, C., Tackmann, K. and Dinkel, A. (2001). *J. Vet. Med. B. Infect. Dis. Vet. Public Health* 48, 367–372.

Pung, O.J., Spratt, J., Clark, C.G., Norton, T.M. and Carter, J. (1998). *J. Zoo. Wildl. Med.* 29, 25–30.

Rivera, W.L. and Kanbara, H. (1999). *Parasitol Res.* 85, 493–495.

Roger, A.J. and Silberman, J.D. (2002). *Nature* 418, 827–829.

Sandground, J.H. (1925). *J. Parasitol.* 12 59–82.

Sano, M., Kino, H., de Guzman, T.S., Ishii, A.I., Kino, J., Tanaka, T. and Tsuruta, M. (1980). *Int. J. Zoonoses* 7, 34–39.

Sargeaunt, P.G., Patrick, S. and O'Keeffe, D. (1992). *Trans. R. Soc. Trop. Med. Hyg.* 86, 633–634.

Sasseville, V.G., Pauley, D.R., MacKey, J.J. and Simon, M.A. (1995). *Vet. Pathol.* 32, 81–83.

Satoh, M., Kiyuna, S., Shiroma, Y., Toma, H., Kokaze, A. and Sato, Y. (2003). *Clin. Exp. Immunol.* 133, 391–396.

Schielke, J.E., Selvarangan, R., Kyes, K.B. and Fritsche, T.R. (2002). *Contemp. Top. Lab. Anim. Sci.* 41, 42–45.

Schwartz, D.A., Anderson, D.C., Klumpp, S.A. and McClure, H.M. (1998). *Arch. Pathol. Lab. Med.* 122, 423–429.

Sestak, K., Aye, P.P., Buckholt, M., Mansfield, K.G., Lackner, A.A. and Tzipori, S. (2003a). *J. Med. Primatol.* 32, 74–81.

Sestak, K., Merritt, C.K., Borda, J., Saylor, E., Schwamberger, S.R., Cogswell, F., Didier, E.S., Didier, P.J., Plauche, G., Bohm, R.P., Aye, P.P., Alexa, P., Ward, R.L. and Lackner, A.A. (2003b). *Infect. Immun.* 71, 4079–4086.

Shadduck, J.A. and Baskin, G. (1989). *Lab. Anim. Sci.* 39, 328–330.

Smith, J.M. and Meerovitch, E. (1985). *J. Parasitol.* 71, 751–756.

Stanley, S.L., Jr. (1997). *Clin. Microbiol. Rev.* 10, 637–649.

Stanley, S.L., Jr. (2003). *Lancet* 361, 1025–1034.

Stauffer, W. and Ravdin, J.I. (2003). *Curr. Opin. Infect. Dis.* 16, 479–485.

Steiner, T.S., Thielman, N.M. and Guerrant, R.L. (1997). *Annu. Rev. Med.* 48, 329–340.

Stephenson, L.S., Holland, C.V. and Cooper, E.S. (2000). *Parasitology* 121 Suppl, S73–S95.

Stibbs, H.H. (1989). *J. Clin. Microbiol.* 27, 2582–2588.

Stibbs, H.H. and Ongerth, J.E. (1986). *J. Clin Microbiol.* 24, 517–521.

Sullivan, J.J., Steurer, F., Benavides, G., Tarleton, R.L., Eberhard, M.L. and Landry, S. (1993). *Am. J. Trop. Med. Hyg.* 49, 254–259.

Tachibana, H., Cheng, X.J., Kobayashi, S., Fujita, Y. and Udono, T. (2000). *Parasitol. Res.* 86, 537–541.

Tachibana, H., Cheng, X.J., Kobayashi, S., Matsubayashi, N., Gotoh, S. and Matsubayashi, K. (2001). *Parasitol. Res.* 87, 14–17.

Tanowitz, H.B., Weiss, L.M. and Wittner, M. (1993). *Gastroenterologist* 1, 265–273.

Teare, J.A. and Loomis, M.R. (1982). *J. Am. Vet. Med. Assoc.* 181, 1345–1347.

Toft, J.D. and Eberhard, J.D. (1998.) In Bennett, B.T., Abee, C.R., and Henrickson, R. (eds) *Nonhuman Primates In Biomedical Research,* pp 111–179. San Diego: Academic Press.

Tzipori, S., Carville, A., Widmer, G., Kotler, D., Mansfield, K. and Lackner, A. (1997). *J. Infect. Dis.* 175, 1016–1020.

Ungar, B.L. (1990). *J. Clin. Microbiol.* 28, 2491–2495.

Verweij, J.J., Vermeer, J., Brienen, E.A., Blotkamp, C., Laeijendecker, D., van Lieshout, L. and Polderman, A.M. (2003). *Parasitol. Res.* 90, 100–103.

Vince, J.D., Ashford, R.W., Gratten, M.J. and Bana-Koiri, J. (1979). *P.N.G. Med. J.* 22, 120–127.

Vogel, P., Zaucha, G., Goodwin, S.D., Kuehl, K. and Fritz, D. (1996). *Am. J. Trop. Med. Hyg.* 55, 595–602.

Wanachiwanawin, D., Chokephaibulkit, K., Lertlaituan, P., Ongrotchanakun, J., Chinabut, P. and Thakerngpol, K. (2002). *Southeast Asian J. Trop. Med. Public Health* 33, 241–245.

Ward, B.J., Kelly, N. and Ndao, M. (2001). *J. Med. Primatol.* 30, 188.

Wilson, J.M. (1979). *Med. Biol.* 57, 84–101.

Wong, M.M. and Conrad, H.D. (1978). *Lab. Anim. Sci.* 28, 412–416.

Yanai, T., Chalifoux, L.V., Mansfield, K.G., Lackner, A.A. and Simon, M.A. (2000). *Vet. Pathol.* 37, 472–475.

Yang, Y., Zeng, L., Li, M. and Zhou, J. (1995). *J. Trop. Med. Hyg.* 98, 69–72.

Zabalgoitia, M., Ventura, J., Anderson, L., Carey, K.D., Williams, J.T. and Vandeberg, J.L. (2003). *Am. J. Tro. Med. Hyg.* 68, 248–252.

Zeman, D.H. and Baskin, G.B. (1985). *Vet. Pathol.* 22, 24–31.

Ziccardi, M. and Lourenco-de-Oliveira, R. (1997). *Mem. Inst. Oswaldo Cruz.* 92, 465–470.

Glossary

acclimatory adjustments	Reversible physiological adjustments to stressful environments.
acrosome reaction	An anterior prolongation of a spermatozoon that releases egg-penetrating enzymes.
adaptation	Changes in gene frequencies resulting from selective pressures being placed upon a population by environmental factors; results in a greater fitness of the population to its ecological niche.
adenocarcinoma	A malignant tumor originating in glandular epithelium.
adenolymphitis	In human embryology, the cellular, outermost extraembryonic membrane, composed of trophoblast lined with mesoderm; it develops villi about 2 weeks after fertilization, is vascularized by allantoic vessels a week later, gives rise to the placenta, and persists until birth.
adnexal	Conjoined, subordinate, or associated anatomic parts (the uterine *adnexa* include the ovaries and fallopian tubes).
affiliative behavior	Close-proximity behavior that includes touching, grooming, and hugging.
agonistic behavior	Behavior that involves fighting, threats, and fleeing.
akinesia	Loss or impairment of voluntary activity (as of a muscle).
algorithm	1. a step-by-step method of solving a problem or making decisions, as in making a diagnosis. 2. an established mechanical procedure for solving certain mathematical problems.
Alkaline Phosphatase (ALP)	Alkaline phosphatases are a group of nonspecific enzymes that hydrolyze many types of monophosphate esters in most cells.
allantois	A sack within the amniote egg in which waste products produced by the embryo are deposited.
allele	Any alternate form of a gene that may occur at a given locus.
Allen's rule	A rule which states that among endotherms, populations of the same species living near the equator tend to have more protruding body parts and longer limbs than do populations farther away from the equator.
allografts	A homograft between allogeneic individuals.
allogrooming	Grooming another animal.
alloparenting	Animals other than mother looking after infants or juveniles.
allopatric species	Species occupying mutually exclusive geographical areas.
altricial	Born helpless, long period of dependency.
amastigotes	Any of the bodies representing the morphologic (leishmanial) stage in the life cycle of all trypanosomatid protozoa.
ameloblastic	Any of a group of columnar cells that produce and deposit enamel on the surface of a developing vertebrate tooth.
amenorrhoea	Abnormal absence or suppression of menstruation.
aminotransfinase enzyme	A sub-subclass of enzymes of the transferase class that catalyze the transfer of an amino group from a donor (generally an amino acid) to an acceptor (generally a 2-keto acid). Most are pyridoxal phosphate proteins. Called also aminotransferase.

The Laboratory Primate
Copyright 2005 Elsevier
ISBN 0-1208-0261-9

amnestic syndrome Memory disturbance.

amniocentesis A medical technique in which amniotic fluid is removed for study of the fetus.

amnion A fluid-filled sack, formed from embryonic tissue, that contains the embryo in the amniote egg.

amygdala One of the four basal ganglia in each cerebral hemisphere that is part of the limbic system and consists of an almond-shaped mass of gray matter in the roof of the lateral ventricle.

amyloidosis A disorder characterized by the deposition of amyloid in organs or tissues of the animal body.

aneurysm An abnormal blood-filled dilatation of a blood vessel, especially an artery, resulting from disease of the vessel wall.

angiogenesis The formation and differentiation of blood vessels.

angiopathy A disease of the blood or lymph vessels.

anion The ion in an electrolyzed solution that migrates to the anode, a negatively charged ion.

antrum The cavity of a hollow organ or a sinus.

anxiolytic A drug that relieves anxiety.

apraxia Loss or impairment of the ability to execute complex coordinated movements without impairment of the muscles or senses.

arytenoid cartilage Relating to or being either of two small cartilages to which the vocal cords are attached and which are situated at the upper back part of the larynx.

astrocytosis The proliferation of astrocytes owing to the destruction of nearby neurons during a hypoxic or hypoglycemic episode.

atony Lack of physiological tone especially of a contractile organ.

atresia 1. absence or closure of a natural passage of the body (atresia of the small intestine); 2. absence or disappearance of an anatomical part (as an ovarian follicle) by degeneration.

atretic Of, relating to, or marked by atresia (*atretic* follicles).

attenuation A decrease in the pathogenicity or vitality of a microorganism or in the severity of a disease.

autopsy A surgical procedure after death which involves the examination of body tissues, often to determine cause of death.

balantidiasis Infection with or disease caused by protozoans of the genus *Balantidium*.

blastomeres A cell produced during cleavage of a fertilized egg.

bolus injection A concentrated mass of pharmaceutical preparation given intravenously for diagnostic purposes.

bradycardia Relatively slow heart action whether physiological or pathological.

bradykinesia Extreme slowness of movements and reflexes.

bradyzoites A small, comma-shaped form of Toxoplasma gondii, found in clusters enclosed by an irregular wall (pseudocyst) in the tissues, chiefly muscles and the brain, in chronic (latent) toxoplasmosis; considered to be the slow-growing form.

capacitation The change undergone by sperm in the female reproductive tract that enables them to penetrate and fertilize an egg.

capnography Monitoring of the concentration of exhaled carbon dioxide in order to assess the physiologic status of patients with acute respiratory problems or who are receiving mechanical ventilation and to determine the adequacy of ventilation in anesthetized patients.

carcinoma A malignant tumor that begins in the skin or in tissues that line or cover internal organs.

cation The ion in an electrolyzed solution that migrates to the cathode; a positively charged ion.

cell-mediated immunity (CMI) The part of the immune system which reacts, to foreign material, with specific white blood cells including killer cells, lymphocytes and macrophage.

chagoma A swelling resembling a tumor that appears at the site of infection in Chagas' disease.

cholangiohepatitis Severe inflammation of the bile passages, often associated with liver fluke infestation that causes obstruction of the bile ducts.

chorea Any of various nervous disorders, of infectious or organic origin, marked by spasmodic movements of the limbs and facial muscles and by incoordination.

choreoathetotic Resembling or characteristic of a condition marked by choreic and athetoid movements.

choriocarcinoma A malignant tumor developing in the uterus from trophoblast and, rarely, in the testes from a neoplasm.

chorionic plate In human embryology, the cellular, outermost extraembryonic membrane, composed of trophoblast lined with mesoderm; it develops villi about 2 weeks after fertilization, is vascularized by allantoic vessels a week later, gives rise to the placenta, and persists until birth.

coagulopathies Diseases or conditions affecting the blood's ability to coagulate.

collimation (mechanical) To make parallel (e.g. rays of light).

congophylic Staining with Congo red.

consort behaviour Male and female spending time together grooming, sleeping, eating, mating.

cornu A horn-shaped anatomical structure, as in either of the lateral divisions of a bicornuate uterus, one of the lateral processes of the hyoid bone, or one of the gray columns of the spinal cord.

corona radiata The radiating crown of projection fibers which pass from the internal capsule to every part of the cerebral cortex.

coronal Of, or relating to, the frontal plane that passes through the long axis of the body.

corpus luteum A yellowish mass of progesterone-secreting endocrine tissue that consists of pale secretory cells derived from granulosa cells, that forms immediately after ovulation from the ruptured graafian follicle in the mammalian ovary. It regresses quickly if the ovum is not fertilized but persists throughout the ensuing pregnancy if it is fertilized.

costodiaphragmatic angle Relating to or involving the ribs and diaphragm.

cribriform Perforated with small apertures like a sieve.

cricoid cartilage A cartilage of the larynx which articulates with the lower cornua of the thyroid cartilage and with which the arytenoid cartilages articulate.

cryptorchidism A condition in which one or both testes fail to descend normally.

cryptosporidia A genus of protozoans of the order Coccidia that are parasitic in the gut of many vertebrates including humans and that sometimes cause diarrhea.

cumulus cells The projecting mass of granulosa cells that bears the developing ovum in a graafian follicle – called also *discus proligerus.*

cumulus eophorus The projecting mass of granulosa cells that bears the developing ovum in a graafian follicle – called also *discus proligerus.*

cytokines Proteins, secreted by cells, involved in regulating cellular activity particularly within the immune system.

cytotrophoblast A layer of extraembryonic ectodermal tissue on the outside of the blastocyst. It attaches the blastocyst to the endometrium of the uterine wall and supplies nutrition to the embryo.

dermatitis Inflammation of the skin – called also *dermitis.*

dexamethasone A synthetic glucocorticoid, $C_{22}H_{29}FO_5$, used especially as an anti-inflammatory and antiallergic agent.

differential diagnosis	The determination of which two or more diseases, with similar symptoms, is the one from which a patient is suffering, based on an analysis of the clinical data.
dizygotic	Twins.
DNA polymorphisms	The ability of DNA to exist in several different forms.
double blind placebo	Pertaining to a clinical trial or other experiment in which neither the subject nor the person administering treatment knows which treatment the subject is receiving. A dummy treatment administered to the control group in a controlled clinical trial in order that the specific and nonspecific effects of the experimental treatment can be distinguished.
dyserythropoiesis	Defective development of erythrocytes.
dyskinesia	Difficulty of moving, distortion or impairment of voluntary movement, as in tic, spasm, or myoclonus.
dysmenorrhoea	Painful menstruation.
dyspnoea	Difficult or labored respiration.
dystonia	A state of disordered tonicity of tissues (as of muscle).
echogenic	Reflecting ultrasound waves.
ectatic veins	Of, relating to, or involving dilatation, expansion or distention of the veins.
ectopic lesions	Occurring in an abnormal sites.
electrophoresis	The movement of suspended particles through a fluid or gel under the action of an electromotive force applied to electrodes in contact with the suspension.
ELISA assay	A quantitative *in vitro* test for an antibody or antigen.
encephalitozoonosis	Infection with protozoa.
endemic	Restricted or peculiar to a locality or region (*endemic* diseases) (an *endemic* species).
endometriosis	The presence and growth of functioning endometrial tissue in places other than the uterus. Often results in severe pain and infertility.
endotracheal	Placed within the trachea.
enteropathy	A disease of the intestinal tract.
enzootic viral infections	Present in an animal community at all times, but occurring in only small numbers of cases.
ependymoma (tumour)	A glioma arising in or near the ependyma.
epidemiology	The statistical study of the distribution and determinants of disease in populations.
epitope	A small part of a protein which is an antigen.
erythema	Abnormal redness of the skin due to capillary congestion (as in inflammation).
erythropoiesis	The production of red blood cells (as from the bone marrow).
estradiol	A natural estrogenic hormone that is a phenolic alcohol $C_{18}H_{24}O_2$ secreted chiefly by the ovaries, is the most potent of the naturally occurring estrogens, and is administered in its natural or semisynthetic esterified form especially to treat menopausal symptoms.
ethmoturbinate	Pertaining to the superior and middle nasal conchae.
etiology	The cause or causes of a disease or abnormal condition.
etiopathogenesis	The development of morbid conditions or of disease; more specifically the cellular events and reactions and other pathologic mechanisms occurring in the development of disease.
euglycaemic	Pertaining to, characterized by or conducive to a blood glucose level that is within the normal range.
extravasation	A discharge or escape of blood, or some other fluid normally found in a vessel or tube, into the surrounding tissues.
extubation	The removal of a tube, especially from the larynx after intubation.

feral Savage; wild; living in the wild state, especially after having been domesticated.

filariform larvae *Of a larval nematode*: resembling a filaria especially in having a slender elongated form and in possessing a delicate capillary esophagus.

fMRI Functional magnetic resonance imaging: noninvasive technology to understand the organization of functional systems within and throughout the entire brain, in response to specific stimuli.

foliovore Eats leaves, fruit that's too hard for smaller primates to eat – lower metabolic requirements per unit mass.

folivirus Any of a family (Filoviridae) of single-stranded chiefly filamentous RNA viruses infecting vertebrates that comprise a single genus (Filovirus) characterized by a helical nucleocapsid and glycoprotein envelope and that include the Marburg virus and the Ebola viruses.

follicle-stimulating hormone (FSH) A hormone from an anterior lobe of the pituitary gland that stimulates the growth of the ovum-containing follicles in the ovary and that activates sperm-forming cells – called also *follitropin*.

follicle 1. a small lymph node; 2. a vesicle in the mammalian ovary that contains a developing egg surrounded by a covering of cells: Ovarian follicle; especially: Graafian follicle.

frugivore Eats fruit and leaves.

fusiform Tapering toward each end.

FWHM (Full width at half maximum): The minimum distance by which two points of radioactivity must be separated to be perceived independently (in a reconstructed image).

gamma ray A photon emitted spontaneously by a radioactive substance; *also* a high-energy photon – usually used in plural.

gastrogavage The introduction of nutriment into the stomach by means of a tube passed through the esophagus.

genetic markers Gene which indicate a particular trait or possibility of disease or mutation.

genetic polymorphisms Short segments of DNA that contain a series of tandem repeats.

genotype All or part of the genetic constitution of an individual or group.

gliosis Excessive development of neuroglia especially interstitially.

gonadotrophin A gonadotropic hormone (as follicle-stimulating hormone).

gradient echo A signal, detected in a nuclear magnetic resonance spectrometer, that is analogous to a spin echo but is produced by varying the external magnetic field following application of a single radiofrequency pulse rather than by application of a series of radiofrequency pulses. Can be used to extract more information from a test sample or subject than is possible in conventional nuclear magnetic resonance spectroscopy.

granulomas A mass or nodule of chronically inflamed tissue, with granulations, that is usually associated with an infective process.

half-life The time required for half of something to undergo a process, e.g. the time required for half of the atoms of a radioactive substance to become disintegrated.

haplotype A group of alleles of different genes on a single chromosome that are closely enough linked to be inherited, usually as a unit.

haustra One of the pouches or sacculations into which the large intestine is divided.

hemangioma A usually benign tumor made up of blood vessels. Typically occurs as a purplish or reddish slightly elevated area of skin.

hematometra An accumulation of blood or menstrual fluid in the uterus.

hemicellulose Any of various plant polysaccharides less complex than cellulose and easily hydrolyzable to simple sugars and other products.

hemithorax A lateral half of the thorax.

hemoperitoneum	Blood in the peritoneal cavity.
hemosiderin	A yellowish brown granular pigment formed by breakdown of hemoglobin, found in phagocytes and in tissues especially in disturbances of iron metabolism.
heterotopic	Occurring in an abnormal place.
homozygous	Having the two genes at corresponding loci on homologous chromosomes identical for one or more loci.
humoral immune response	1. pertaining to elements dissolved in the blood or body fluids, e.g., humoral immunity from antibodies in the blood as opposed to cellular immunity; 2. pertaining to one of the humors of the body or to humoralism.
hydrocele	An accumulation of serous fluid in a sacculated cavity (as the scrotum).
hydrops	1. Oedema; 2. distension of a hollow organ with fluid, e.g. hydrops of the gallbladder.
hydrosalpinx	Abnormal distension, with fluid, of one or both fallopian tubes with fluid usually due to inflammation.
hydroureter	Abnormal distension of the ureter with urine.
hyperglycaemia	An excess of sugar in the blood.
hyperinsulinaemia	The presence of excess insulin in the blood.
hypernatremia	The presence of an abnormally high concentration of sodium in the blood.
hyperphosphatemia	An excessive amount of phosphates in the blood; it is usually asymptomatic.
hypertension	Abnormally high arterial blood pressure.
hypertoma	The quality or state of being hypertonic.
hypertrophic	Pertaining to or marked by the enlargement or overgrowth of an organ or part due to an increase in size of its constituent cells.
hypoplasias	A condition of arrested development in which an organ or part remains below the normal size or in an immature state.
hytadid cyst	The larval cyst of a tapeworm of the genus Echinococcus. Usually occurs as a fluid-filled sac containing daughter cysts in which scolices develop but that occasionally forms a proliferating spongy mass which actively metastasises in the host's tissues.
immunocompetence	The capacity for a normal immune response.
immunohistochemical staining	A laboratory process of detecting an organism or structure, in tissues, with antibodies. These antibodies are labeled with a compound that is seen as a colored deposit when viewed microscopically.
imprinting	Infant recognition of face, social gestures.
infarct	An area of coagulation necrosis in a tissue, due to local ischemia resulting from obstruction of circulation to the area, most commonly by a thrombus or embolus.
interferon	Any of a group of heat-stable soluble basic antiviral glycoproteins, of low molecular weight, that are usually produced by cells exposed to the action of a virus, sometimes to the action of another intracellular parasite (e.g. a bacterium), or experimentally to the action of some chemicals. Also include some used medically as antiviral or antineoplastic agents.
intubation	The introduction of a tube into a hollow organ (as the trachea or intestine) to keep it open or restore its patency if obstructed.
ipsilateral damage	Situated, appearing on or affecting the same side of the body.
ischial callosities	Calluses on rear, so the animals can sit and sleep, e.g. baboons.
isotonic	Denoting a condition of equal tone, tension or activity, e.g. a solution in which body cells can be bathed without a net flow of water across the semipermiable membrane; a solution having the same tonicity as another with which it is compared.
kinesics	How an animal moves, how an animal holds its body.

kyphosis	Exaggerated outward curvature of the thoracic region of the spinal column.
laparotomy	Surgical section of the abdominal wall.
Lapmoscopy	Above procedure in the abdomen.
leiomyoma	A benign tumor (e.g. a fibroid) consisting of smooth muscle fibers.
leiomyosarcoma	A sarcoma composed, in part, of smooth muscle cells.
leptomeninges	The pia mater and the arachnoid considered together as investing the brain and spinal cord.
Lesch-Hyhan syndrome	A rare X-linked disorder of purine metabolism due to deficient hypoxanthine phosphoribosyltransferase, characterized by physical and mental retardation, compulsive self-mutilation of the fingers and lips by biting, choreoathetosis, spastic cerebral palsy, impaired renal function; and by excessive purine synthesis and consequent hyperuricemia and uricaciduria.
lesion	Any pathological or traumatic discontinuity of tissue or loss of function of a part.
leukemogenesis	Induction or production of leukemia.
leukoencephalopothy	Any of various diseases affecting the brain's white matter.
Levodopa	(L-DOPA) the levorotatory form of dopa that is converted to dopamine in the brain, and is used in treating Parkinson's disease.
Lewy body	An eosinophilic inclusion body found in the cytoplasm of neurons of the cortex and brain stem in Parkinson's disease and some forms of dementia.
ligands	A group, ion or molecule coordinated to a central atom or molecule in a complex.
lipofuscin	A dark brown lipochrome found especially in the tissue of the aged.
lissencephaly	The condition of having a smooth cerebrum without convolutions.
lordosis	Exaggerated forward curvature of the lumbar and cervical regions of the spinal column.
luteinizing hormone (LH)	A hormone of protein-carbohydrate composition that is obtained from the adenohypophysis of the pituitary gland. In the female it stimulates the development of the corpora lutea and, together with follicle-stimulating hormone, the secretion of progesterone. In the male it stimulates the development of interstitial tissue in the testis and the secretion of testosterone.
lymph node	A small organ in the body which produces white blood cells needed to fight infection.
lymphadenopathy	Abnormal enlargement of the lymph nodes.
lymphedema	Edema due to faulty lymphatic drainage.
lymphopenia	Reduction in the number of lymphocytes circulating in the blood of humans or animals.
lymphoplasmacytic	Of, relating to, or consisting of lymphocytes and plasma cells.
macrophages	A phagocytic tissue cell of the mononuclear phagocyte system that may be fixed or freely motile, is derived from a monocyte, and functions in the protection of the body against infection and noxious substances.
mammillary bodies	Either of two small rounded eminences on the underside of the brain behind the tuber cinereum forming the terminals of the anterior pillars of the fornix.
matrix	The intercellular substance in which tissue cells are embedded.
meconium	A dark greenish mass of desquamated cells, mucus, and bile that accumulates in the bowel of a fetus and is typically discharged shortly after birth.
menarche	The beginning of the menstrual function.
menorrhagia	Abnormally profuse menstrual flow.
mesangial	Of or relating to the thin membrane which helps to support the capillary loops in a renal glomerulus.
metalloproteinases	A protein that has one or more tightly bound metal ions forming part of its structure.

metastasis	Change of position, state, or form: as in transfer of a disease-producing agent from an original site of disease to another part of the body with development of a similar lesion in the new location.
MHC alleles	Major histocompatibility complex of any alternate form of a gene that may occur at a given locus.
microcephaly	A condition of abnormal smallness of the head, usually associated with mental retardation.
monocytes	A large white blood cell with finely granulated chromatin dispersed throughout the nucleus; formed in the bone marrow, enters the blood, and migrates into the connective tissue where it differentiates into a macrophage.
mononucleosis	An abnormal increase of mononuclear white blood cells in the blood.
mucoid vasculopathy	Pertaining or relating to, or resembling mucus disorder of blood vessels.
Mutagen	A substance that causes mutations. A mutation is a change in the genetic material in a body cell. Mutations can lead to birth defects, miscarriages, or cancer.
myoclonic seizures/epilepsy	An inherited form of epilepsy characterized by myoclonic seizures, progressive mental deterioration and the presence of Lafora bodies in parts of the central nervous system.
myorelaxation	Relaxation of muscle.
nacardiosis	Actinomycosis caused by actinomycetes of the genus *Nocardia* and characterized by production of spreading granulomatous lesions.
neutropenia	Leukopenia in which the decrease in white blood cells is chiefly in neutrophils.
nigrostriatal system	Of, relating to, or joining the corpus striatum and the substantia nigra.
nongravid	Not pregnant.
nystagmus	A rapid involuntary oscillation of the eyeballs occurring normally with dizziness, during and after bodily rotation, or abnormally after injuries.
odontoma	A tumor originating from a tooth and containing dental tissue.
oedema	An abnormal excess accumulation of serous fluid in connective tissue or in a serous cavity — called also dropsy.
olecranon	The large process of the ulna that projects behind the elbow joint, forms the bony prominence of the elbow, and receives the insertion of the triceps muscle.
oligodendrocyte	A neuroglial cell resembling an astrocyte but smaller with few and slender processes having few branches.
oligohydramnios	Deficiency of amniotic fluid sometimes resulting in an embryonic defect through adherence between embryo and amnion.
omentum	A fold of peritoneum connecting or supporting abdominal structures.
opisthotonus	A condition of spasm of the muscles of the back, causing the head and lower limbs to bend backward and the trunk to arch forward.
optokinetic	Of, relating to, or involving movements of the eyes.
oropharynx	The part of the pharynx that is below the soft palate and above the epiglottis and is continuous with the mouth.
osteomalacia	A disease of adults that is characterized by softening of the bones and is analogous to rickets in the immature.
osteopenia	Reduction in bone volume to below normal levels especially due to inadequate replacement of bone lost to normal lysis.
oviductal fimbria	Oviductal: a bordering fringe at the entrance of the fallopian tubes.
oxalosis	An autosomal recessive trait characterized by excretion of excessive amounts of oxalate and the formation of calcium oxalate deposits in tissues throughout the body.
palpebral (reflex)	Of, relating to, or located on or near the eyelids.
pandemic	Worldwide.

papilledema Swelling and protrusion of the blind spot of the eye caused by edema.

parakeratosis An abnormality of the horny layer of the skin resulting in a disturbance in the process of keratinization.

paramesonephric ducts (also Muellerian duct) Either of a pair of ducts parallel to the Wolffian ducts in vertebrate animals and giving rise, in the female, to the oviducts.

paraparesis Partial paralysis affecting the lower limbs.

paratracheal Adjacent to the trachea.

parenteral Situated or occurring outside the intestine.

Parkinson's disease A chronic progressive nervous disease, chiefly of later life, that is linked to decreased dopamine production in the substantia nigra and is marked by tremor and weakness of resting muscles and by a shuffling gait.

pathogens A specific causative agent of disease.

pedal (reflex) Of, or relating to the foot.

pedunculated Having, growing on, or being attached by a stemlike connecting part; a general term for collections of nerve fibers coursing between different areas in the central nervous system.

peracute Very acute and violent.

periaquaductal Relating to, the gray matter which surrounds the aqueduct of Sylvius.

periovular Surrounding an ovum.

perivascular The tissues surrounding a blood vessel.

perivitelline space The fluid-filled space between the fertilization membrane and the ovum after the entry of a sperm into the egg.

petechial Characterized by or of the nature of a pinpoint, nonraised, perfectly round, purplish red spot caused by intradermal or submucous hemorrhage.

phagocyte A cell which destroys microbes and other foreign matter by ingesting it. Part of the immune system.

phagocytosis The engulfing and usually the destruction of particulate matter by phagocytes. Serves as an important bodily defense mechanism against infection by microorganisms and against occlusion of mucous surfaces or tissues by foreign particles and tissue debris.

pharmacokinetic The action of a drug in the body including bodily absorption, distribution, metabolism and excretion.

phenotype The visible properties of an organism that are produced by the interaction of the genotype and the environment.

phosphataemia The occurrence of phosphate in the blood, especially in excessive amounts.

photon A unit of intensity of light at the retina equal to the illumination received per square millimeter of a pupillary area from a surface having a brightness of one candle per square meter.

phylogenetics Based on natural evolutionary relationships acquired in the course of phylogenetic development.

piloerection Hair stands up on end, makes animal look bigger.

plasma The fluid portion of the blood in which the particulate components are suspended.

pleiotropic Producing many effects in the phenotype.

pleocytosis An abnormal increase in the number of cells (e.g. as lymphocytes) in the cerebrospinal fluid.

polyandry One female has several mates – males help with offspring – marmosets and tamarins.

polydipsia Excessive or abnormal thirst.

polyhydramnios Excessive accumulation of the amniotic fluid.

polymorphism The quality or character of occurring in several different forms.

polyphagia Excessive appetite or eating.

positron	A positively charged particle having the same mass and magnitude of charge as the electron and constituting the antiparticle of the electron – called also *positive electron.*
precocial	Born well developed – pr:mates are physically precocial (opposite of altricial).
primiparous	Bearing or having borne only one offspring.
proceptive behaviour	Female goes to male and attempts to instigate mating.
progesterone	The principal progestational hormone of the body, liberated by the corpus luteum, placenta, and in minute amounts by the adrenal cortex; it prepares the uterus for the reception and development of the fertilized ovum by transforming the endometrium from the proliferative to the secretory stage and maintains an optimal intrauterine environment for sustaining pregnancy.
proxemics	An animal's use of space.
radionuclide	A radioactive species of atom characterized by the atomic number, mass number and quantum state of its nucleus, and capable of existing for a measurable lifetime (generally greater than 10–10 sec).
radiotracer	A radioactive dissecting instrument for isolating vessels and nerves.
raphe	The seamlike union of the two lateral halves of a part or organ (as of the tongue, perineum, or scrotum) having externally a ridge or furrow and internally usually a fibrous connective tissue septum.
recrudescent	Breaking out again: renewing disease after abatement, suppression, or cessation.
retrovirus replication	Intergration of DNA "provirus" into the last genome.
reverse transcriptase	Enzyme that allows retograde transfer of genetic information from RNA to DNA.
rigid endoscopy	Minimally unvasive surgery using small instruments to penetrate body cavities.
saccular	Resembling a sac.
sagittal	The median plane of the body or any plane parallel to it.
salpingo-oophorectomy	Surgical excision of a fallopian tube and an ovary.
secundipara	A woman who has had two pregnancies which resulted in viable offspring.
seronegative	Having a negative serum reaction especially in a test for the presence of an antibody.
seroprevalence	The frequency of individuals in a population that have a particular element (e.g. antibodies to HIV) in their blood serum.
serositis	Inflammation of one or more serous membranes.
social contraception	Negative effect of behavior on reproduction. Occurs in some nonhuman primates.
somatosensory	Sensory activity having its origin elsewhere than in the special sense organs (as eyes and ears) and conveying information about the state of the body proper and its immediate environment.
spermiogram	A diagram of the various cells formed during the development of the sperm.
SPF	1. Specific-pathogen free, a term applied to animals reared for use in laboratory experiments, and known to be free of specific pathogenic microorganisms; 2. About source animals for xenotransplantation, maintained through a combination of limited access and strict attention to personal hygiene and sanitation.
spin echo	A signal that is detected in a nuclear magnetic resonance spectrometer, that is an echo-like replication of a free induction decay produced by a planned series of radiofrequency pulses, and that can be used to extract more information from a test sample or subject than conventional nuclear magnetic resonance spectroscopy – usually used attributively (*spin-echo* magnetic resonance imaging of the cervical spine).

splenomegaly Abnormal enlargement of the spleen.

spondyloarthropathy Any of several diseases (as ankylosing spondylitis) affecting the joints of the spine.

steatitis Inflammation of fatty tissue.

stem cells Stem cells are primitive, undifferentiated cells which have the ability to reproduce and become any cell, tissue or organ and thus have the ability to replace damaged or diseased body parts without fear of rejection by the body.

stereotactic Precise direction of the tip of a delicate instrument or beam of radiation in three planes using coordinates provided by medical imaging (as computed tomography) in order to reach a specific locus in the body.

strongyloidiasis Infection of humans or domestic animals with *Strongyloides stercoralis.* The female worm and her larvae inhabit the mucosa and submucosa of the small intestine (see *intestinal s.*), and the larvae expelled in the feces develop in the soil and can penetrate skin on contact. They later are carried in the bloodstream to the lungs (see *pulmonary s.*), and from there they travel to the intestine via the trachea and esophagus.

subchondral bone Situated beneath cartilage.

sustentacular Serving to support or sustain.

syncytiotrophoblast The outer syncytial layer of the trophoblast that actively invades the uterine wall forming the outermost fetal component of the placenta.

telangiectasis An abnormal dilatation of capillary vessels and arterioles that often forms an angioma.

telomere (length) The natural end of a chromosome.

teratogen An agent or factor that causes the production of physical defects in the developing embryo.

teratoma A tumor neoplasm made up of a heterogeneous mixture of tissues, none of which is native to the area in which it occurs.

theca interna The inner layer of the theca folliculi that is highly vascular and that contributes theca lutein cells to the formation of the corpus luteum.

thermoplastic Capable of softening or fusing when heated and of hardening again when cooled.

thoracoscopy The direct examination of the pleutal cavity by means of an endoscope.

thrombocytopenia Persistent decrease in the number of blood platelets that is often associated with hemorrhagic conditions.

thromboxane Substances produced, especially by platelets and which cause constriction of vascular and bronchial smooth muscle, and promote blood clotting.

tomograph An X-ray machine used to show detailed images of structures in a predetermined plane of tissue.

toxoplasmosis Infection with, or disease caused by, a sporozoan of the genus *Toxoplasma* (*T. gondii*) that invades the tissues and may seriously damage the central nervous system especially of infants.

tracer A mechanical or radioactive means of identifying structures or substances.

tracheobronchial Pertaining to the trachea and bronchi.

transaxial Directed at right angles to the long axis of the body or a part.

transcriptional suppression 1. The act of holding back or checking; 2. Sudden stoppage or inhibition, e.g. of a secretion, excretion, normal discharge, or other function; 3. In psychiatry, conscious inhibition of an unacceptable impulse or idea as contrasted with repression, which is unconscious; 4. In genetics, masking of the phenotypic expression of a mutation by the occurrence of a second (suppressor) mutation at a different site from the first; the organism appears to be reverted but is in fact doubly mutant.

transporter genes	A segment of a DNA molecule that contains all the information required for synthesis of a product (polypeptide chain or RNA molecule), including both coding and noncoding sequences. It is the biologic unit of heredity, self-reproducing, and transmitted from parent to progeny.
Trendelenburg positioning	A position of the body for medical examination or operation in which the patient is placed head down on a table inclined at about 45 degrees from the floor with the knees uppermost and the legs hanging over the end of the table.
trichotillomania	Abnormal desire to pull out one's hair – called also *hairpulling*.
trichroma	A person who has normal colour vision.
trocar	A sharp-pointed surgical instrument fitted with a cannula and used especially to insert the cannula into a body cavity as a drainage outlet.
typhilitis	Inflammation of the cecum.
ultrasound	Vibrations of the same physical nature as sound but with frequencies above the range of human hearing; the diagnostic or therapeutic use of ultrasound and especially as a noninvasive technique involving the formation of a two-dimensional image used for the examination and measurement of internal body structures and the detection of bodily abnormalities.
uterovaginal primorduim	The rudiment or commencement of the uterus and the vagina.
valvular endocardiosis	Pertaining to chronic fibrosis of atrioventricular valves in usually the mitral valve; it may lead to congestive heart failure.
virions	A complete virus particle that consists of an RNA or DNA core with a protein coat sometimes with external envelopes and that is the extracellular infective form of a virus.
visuospatial memory	Relating to, thought processes that involve visual and spatial awareness.
voxel	Each defined volume unit of an element being scanned in computerized axial tomography.
western blot	A blot consisting of a sheet of cellulose nitrate or nylon that contains spots of protein for identification by a suitable molecular probe and is used especially for the detection of antibodies.
xenodiagnosis	The detection of a parasite by feeding test material (as blood) from a suspected host (a human) to a suitable intermediate host (an insect) and later examining the intermediate host for the parasite.
xenografts	A graft of tissue transplanted between animals of different species.
xenotransplantation	Transplantation of an organ, tissue, or cells between two different species (as a human and a domestic swine).
xerophthalmia	A dry thickened lusterless condition of the eyeball resulting especially from a severe systemic deficiency of vitamin A.
zona pellucida	Mucopolysaccharide layer surrounding the ovum; the transparent elastic outer layer or envelope of a mammalian ovum often traversed by numerous radiating striae.
zoonoses	A disease of animals that may be transmitted to humans under natural conditions.

Index